AF654433

Complete Practice of NCERT for

**NEET · JEE · 11 & 12**

# MASTER THE NCERT

# PHYSICS-II

Thousands of MCQs Based on
Each & Every Topic of Latest NCERT

Class–XII

Digvijay Singh
Atique Hassan

*Co-authored by*

Mansi Garg
Manish Dangwal

arihant

ARIHANT PRAKASHAN (Series), MEERUT

arihant

Arihant Prakashan (Series), Meerut





卐 **Administrative & Production Offices**

**Regd. Office**

'Ramchhaya' 4577/15, Agarwal Road, Darya Ganj, New Delhi -110002
Tele: 011- 47630600, 43518550

**Head Office**

Kalindi, TP Nagar, Meerut (UP) - 250002
Tel: 0121-7156203, 7156204

卐 **Sales & Support Offices**

Agra, Ahmedabad, Bengaluru, Bareilly, Chennai, Delhi, Guwahati, Hyderabad, Jaipur, Jhansi, Kolkata, Lucknow, Nagpur & Pune.

卐 **ISBN** 978-93-26192-82-8

卐 **PRICE** ₹375.00

**PO No :** TXT-XX-XXXXXXX-X-XX

Published by Arihant Publications (India) Ltd.

For further information about the books published by Arihant, log on to www.arihantbooks.com or e-mail at info@arihantbooks.com

Follow us on

# PREFACE

In the present dynamic scenario of competitions, NCERT Books are gaining importance day by day. On an average about 30% questions asked in various competitions are from NCERT. Similarly with the incorporation of Board weightage in competitions the necessity of dual synchronous preparation i.e., to study NCERT for School Exams and for Competitive Exams, is increasing day by day. **Arihant's Master the NCERT Series** is framed with the philosophy of such synchronous preparation in mind.

The book consists of 14 chapters, in each of which questions are framed on each & every line of NCERT text. Some special features of Master the NCERT books which make these books stand apart from other NCERT based books are

- Each chapter has a Key Notes for quick revision of whole NCERT content.
- Each chapter has topically divided objective questions based on NCERT content to cover all the topics of NCERT text.
- Separate section in each chapter having special types questions for NEET, JEE and Other exams.
- Complete coverage of NCERT & NCERT Exemplar objective questions in each chapter.
- Detailed explanations for selected questions.

The variety in types of questions framed will be helpful in analysis of self-performance and exposures to face tough problems of competitions. **Previous Years' Medical, Engineering & Other Entrances Questions** have also been incorporated at appropriate places so that the students get the exposure of type of questions asked in various competitions on the same topic.

Huge efforts have been made from our side to keep this book error free, but inspite of that if any error or whatsoever is skipped in the book then that is purely incidental, apology for the same, please write to us about that so that it can be corrected in the further edition of the book. Suggestions for further improvement of the book will also be welcomed & incorporated in further editions.

**Publisher**

# CONTENTS

MASTER THE NCERT

PHYSICS-II

MASTER THE NCERT
PHYSICS-II



**NEET**
**Solved Paper 2021**

*Scan the QR Code*

CHAPTER > 01

# Electric Charge and Fields

## KEY NOTES

- Electrostatics deals with the study of forces, fields and potentials arising from static charges.

## Electric Charge

- The physical property of a matter that causes it to experience a force when placed in an electromagnetic field is called **charge**.
- Like charges repel each other while unlike charges attract each other.
- The property which differentiates the two kinds of charges is called the **polarity of charge.**
- If an object possesses an electric charge, it is said to be **electrified** or **charged**. When it has no charge, it is said to be **electrically neutral.**
- A body can be charged positively by losing some of its electrons and similarly, can be charged negatively by gaining electrons.

## Conductors and Insulators

- Those substances which allow electricity to pass through them easily are called **conductors**.
- They have electric charges (electrons) that are comparatively free to move inside the material.
- Metals, human & animal bodies and earth are conductors.
- Most of the non-metals like glass, porcelain, plastic, nylon, wood offer high resistance to the passage of electricity through them, they are called **insulator**.
- When we bring a charged body in contact with the earth, all the excess charge on the body disappears by causing a momentary current to pass to the ground through the connecting conductor (such as our body).
- The process of sharing the charges with the earth is called **grounding** or **earthing**. It provides a safety measure for electrical circuits and appliances.

### Charging by Induction

- When a charged body is brought near a conductor, then temporary electrification of conductor is called **electrostatic induction.**

  A body can be charged positively or negatively by the method of induction.

## Basic Properties of Electric Charge

- **Additivity of Charges** The net electric charge on a body is equal to the algebraic sum of all the electric charges distributed on the different parts of the body.

  If a system contains $n$-charges $q_1, q_2, q_3 \ldots q_n$, then

$$q_{\text{total}} = q_1 + q_2 + q_3 + \ldots + q_n$$

- **Charge is conserved** in an isolated system consisting of many charged bodies. Charges may get redistributed due to interactions among the bodies but it is found that the total charge of the isolated system is always conserved.
- It is not possible to create or destroy net charge carried by any isolated system although the charge carrying particles may be created or destroyed in a process.
- **Quantisation of Charge** The charge on a body is always an integral multiple of $e$ (basic unit of charge, i.e. charge on an electron). This is called quantisation of charge.

  i.e. $q = ne$

  The charge on an electron is taken to be negative.

  The value of basic unit of charge is given as

$$e = 1.602192 \times 10^{-19} \text{ C}$$

## Coulomb's Law

- Coulomb measured the force between two point charges and found that it varied inversely as the square of the distance between the charges and directly proportional to the product of the magnitude of the two charges and acted along the line joining the two charges.
  The magnitude of the force $F$ between them is given by
  $$F = k\frac{|q_1 q_2|}{r^2}$$
  where, $k = \frac{1}{4\pi\varepsilon_0} = 9\times10^9 \text{ N-m}^2/\text{C}^2$
  and $\varepsilon_0 =$ permittivity of free space $= 8.854\times10^{-12}\text{C}^2\text{N}^{-1}\text{m}^{-2}$.
- The force between two charges $q_1$ and $q_2$ located at a distance $r$ in a medium is expressed as
  $$F_{\text{medium}} = \frac{1}{4\pi\varepsilon}\cdot\frac{q_1 q_2}{r^2}$$
  where, $\varepsilon =$ absolute permittivity of the medium.
  $$\frac{F_{\text{vacuum}}}{F_{\text{medium}}} = \frac{1}{4\pi\varepsilon_0}\cdot\frac{q_1 q_2}{r^2} \bigg/ \frac{1}{4\pi\varepsilon}\cdot\frac{q_1 q_2}{r^2}$$
  $$= \frac{\varepsilon}{\varepsilon_0} = \varepsilon_r = K$$
  where, $\varepsilon_r =$ relative permittivity of the medium,
  i.e. the ratio of permittivity of the medium to the permittivity of free space is called **relative permittivity.**
- **Coulomb's Law in Vector Form** Let the position vectors of charges $q_1$ and $q_2$ be $\mathbf{r}_1$ and $\mathbf{r}_2$ respectively as (see given figure)

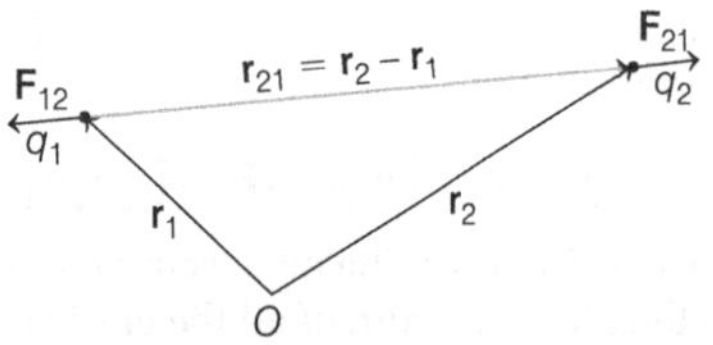


- The force $F_{21}$ on charge $q_2$ due to charge $q_1$ is given as
  $\mathbf{F}_{21} = \frac{1}{4\pi\varepsilon_0}\cdot\frac{q_1 q_2}{r_{21}^2}\hat{\mathbf{r}}_{21}$, where $\hat{\mathbf{r}}_{21}$ is unit vector.
- The force $F_{12}$ on charge $q_1$ due to charge $q_2$ is given as
  $$\mathbf{F}_{12} = \frac{1}{4\pi\varepsilon_0}\cdot\frac{q_1 q_2}{r_{12}^2}\hat{\mathbf{r}}_{12}$$
- Coulomb's law agrees with the Newton's third law
  i.e. $\mathbf{F}_{12} = -\mathbf{F}_{21}$

**Forces Between Multiple Charges : Superposition's Principle** According to this principle, "total force on a given charge due to number of charges is the vector sum of individual forces acting on that charge due to the presence of other charges".

KEY NOTES

If the charges $q_1, q_2, q_3 .....q_n$ are present in a system, then total force $F_1$ on charge $q_1$ due to all other charges is given by

$$\mathbf{F}_1 = \mathbf{F}_{12} + \mathbf{F}_{13} + \mathbf{F}_{14} + ..... + \mathbf{F}_{1n}$$

i.e. $$\mathbf{F}_1 = \frac{1}{4\pi\varepsilon_0}\left(\frac{q_1 q_2}{r_{12}^2}\hat{\mathbf{r}}_{12} + \frac{q_1 q_2}{r_{13}^2}\hat{\mathbf{r}}_{13} + ..... + \frac{q_1 q_2}{r_{1n}^2}\hat{\mathbf{r}}_{1n}\right)$$
$$= \frac{q}{4\pi\varepsilon_0}\sum_{i=1}^{n}\frac{q_i}{r_{1i}^2}\hat{\mathbf{r}}_{1i}$$

## Electric Field

- It is the region surrounding an electric charge or a group of charges, in which another charge experiences a force.
  Its unit is N/C and it is a vector quantity.
- The electric field produced by the charge $Q$ at a point $r$ is given as $\mathbf{E}(\mathbf{r}) = \frac{1}{4\pi\varepsilon_0}\cdot\frac{Q}{r^2}\hat{\mathbf{r}}$
  where, $\hat{\mathbf{r}} = \frac{\mathbf{r}}{r}$ is a unit vector from the origin to the point $\mathbf{r}$.
- If the position of charge $q$ is denoted by vector $\mathbf{r}$, then it experiences a force $\mathbf{F}$ equal to the charge $q$ multiplied by the electric field $\mathbf{E}$ at the location of $q$.
  Thus, $\mathbf{F}(\mathbf{r}) = q\mathbf{E}(\mathbf{r})$.
- The charge $Q$ which is producing the electric field is called a **source charge** and the charge $q$, which tests the effect of a source charge is called a **test charge.**
- **Electric field vector E** (also known as electric field intensity) at any point is given by
  $$\mathbf{E} = \lim_{q_0\to0}\frac{\mathbf{F}}{q_0}$$
  where, $q_0$ is a small positive test charge which experiences a force $\mathbf{F}$ at given point.
- Direction of electric field intensity is taken from positive charge to negative charge.
- **Electric Field due to a System of Charges** If a system of charges $q_1, q_2, q_3....q_n$ with position vector $\mathbf{r}_1, \mathbf{r}_2, \mathbf{r}_3 ....\mathbf{r}_n$ relative to origin $O$, then Coulomb's law and the superposition principle can be used to determine the net field at a point $P$ denoted by the position vector $\mathbf{r}$.

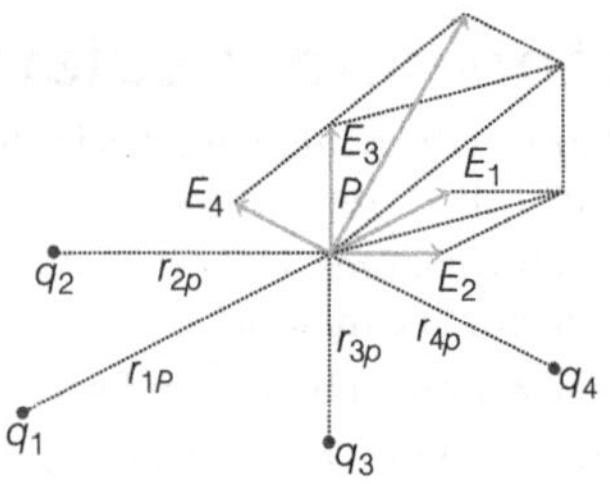

By the superposition principle, the electric field **E** at **r** due to system of charges is (as shown in above figure)

$$\mathbf{E}(\mathbf{r}) = \mathbf{E}_1(\mathbf{r}) + \mathbf{E}_2(\mathbf{r}) ...... + \mathbf{E}_n(\mathbf{r})$$

$$= \frac{1}{4\pi\varepsilon_0} \frac{q_1}{r_{1p}^2} \hat{\mathbf{r}}_{1p} + \frac{1}{4\pi\varepsilon_0} \frac{q_2}{r_{2p}^2} \hat{\mathbf{r}}_{2p} .....+ \frac{1}{4\pi\varepsilon_0} \frac{q_n}{r_{np}^2} \hat{\mathbf{r}}_{np}$$

$$\Rightarrow \quad \mathbf{E}(\mathbf{r}) = \frac{1}{4\pi\varepsilon_0} \sum_{i=1}^{n} \frac{q_i}{r_{ip}^2} \hat{\mathbf{r}}_{ip}$$

## Electric Field Lines

- These are a way of pictorially mapping the electric field around a configuration of charge(s). The tangent on these lines at any point gives direction of field at that point.
- The field lines follow some important general properties
  - (i) Field lines start from positive charges and ends at negative charges. If there is a single charge, they may start or end at infinity.
  - (ii) In a charge free region, electric field lines can be taken to be continuous curves without any breaks.
  - (iii) Two field lines can never cross each other (if they did, the field at the point of intersection will not have a unique direction, which is absurd).
  - (iv) Electrostatic field lines do not form any closed loops. This follows from the conservative nature of electric field.

# Electric Flux and Electric Dipole

- **Electric flux** is a measure of flow of electric field through a surface. Mathematically, electric flux is the product of an area element $\Delta S$ and normal component of **E** integrated over a surface,

  i.e. $$\phi_E = \int \mathbf{E} \cdot \Delta \mathbf{S} = \int \mathbf{E} \cdot \hat{\mathbf{n}} \, \Delta S = \int E \, \Delta S \cos\theta$$

  where, $\hat{\mathbf{n}}$ is the unit vector normal to area element $\Delta S$.

  The SI unit of electric flux is $Nm^2C^{-1}$ or V-m.
- If the area element $dS$ is normal to $E$ at a point, then $\theta = 0°$, hence number of field lines crossing it is proportional to $E \, \Delta S$.
- If the area element $dS$ is parallel to $E$ at a point, then $\theta = 90°$, hence number of field lines crossing it is zero.
- A pair of two equal and opposite point charges at a fixed distance is called an **electric dipole**.
- The total charge of the electric dipole is obviously zero, this does not mean that the field of the electric dipole is zero ($p$).
- The product of the magnitude of either charge $q$ and the distance between the charges $2a$ is called **electric dipole moment (p)**.

  i.e. $$\mathbf{p} = q \times 2\,\mathbf{a}$$

  It is a vector quantity and its SI unit is C-m.

  The direction of electric dipole moment is from $-q$ to $q$.
- **Electric field at a point distance $r$ from the centre of dipole along its axial line** is given by

$$\mathbf{E} = \frac{1}{4\pi\varepsilon_0} \cdot \frac{2pr}{(r^2 - a^2)^2}$$

(direction of **E** is same as of **p**)

$$\mathbf{E} = \frac{1}{4\pi\varepsilon_0} \cdot \frac{2\mathbf{p}}{r^3} \qquad [\text{when } r >> a]$$

- **Electric field at a point distance $r$ from centre of dipole along its equatorial line** is given by

$$\mathbf{E} = -\frac{1}{4\pi\varepsilon_0} \cdot \frac{\mathbf{p}}{(r^2 + a^2)^{3/2}}$$

(direction of **E** is opposite to that of **p**)

$$\mathbf{E} = \frac{\mathbf{p}}{4\pi\varepsilon_0 r^3} \qquad [\text{when } r >> a]$$

## Physical Significance of Dipoles

- In most molecules, the centres of positive charges and of negative charges lie at the same place. Therefore, their dipole moment is zero, e.g. $CO_2$ and $CH_4$.
- But in some molecules, the centres of negative charges and of positive charges do not coincide. Therefore, they have a permanent electric dipole moment, even in the absence of an electric field. Such molecules are called **polar molecules,** e.g. $H_2O$.

## Dipole in a Uniform External Field

- When a dipole is placed at an angle $\theta$ from the direction of a uniform electric field $E$, it experiences a torque given by $\tau = pE \sin\theta$

  or $\quad \tau = \mathbf{p} \times \mathbf{E}$
- Direction of torque is perpendicular to the plane of paper and inwards.
- When dipole is parallel or anti-parallel to $E$, then torque is zero at $\theta = 0°$ or $180°$ and maximum at $\theta = 90°$.

# Continuous Charge Distribution

There are three types of continuous charge distribution and according to charge distribution, there are three types of charge density as given below

(i) **Surface Charge Density** If $\Delta Q$ charge is distributed uniformly on area element $\Delta S$, then surface charge density ($\sigma$) is given as $\sigma = \frac{\Delta Q}{\Delta S}$ C/m$^2$.

(ii) **Linear Charge Density** If $\Delta Q$ charge is distributed uniformly on line element $\Delta l$ of the wire, then linear charge density ($\lambda$) is given as $\lambda = \frac{\Delta Q}{\Delta l}$ C/m.

(iii) **Volume Charge Density** If $\Delta Q$ charge is distributed uniformly on volume element $\Delta V$, then volume charge density ($\rho$) is given as $\rho = \frac{\Delta Q}{\Delta V}$ C/m$^3$.

- Electric field due to charge $\rho \Delta V$ is given by

$$\Delta E = \frac{1}{4\pi\varepsilon_0} \cdot \frac{\rho \Delta V}{r'^2} \hat{\mathbf{r}}'$$

where, $r'$ is the distance between the charge element and $P$, and $\hat{\mathbf{r}}'$ is a unit vector in the direction from charge element to $P$.

- The total electric fields due to different volume elements is given as

$$\mathbf{E} \cong \frac{1}{4\pi\varepsilon_0} \sum_{\text{all}\Delta r} \frac{\rho \Delta V}{r'^2} \hat{\mathbf{r}}'$$

## Gauss's Law and its Applications

- It states that the surface integral of the electric field intensity (i.e. total electric flux) over any closed surface (called Gaussian surface) in free space is equal to $\frac{1}{\varepsilon_0}$ times. Hence, the net charge enclosed within the surface.

$$\phi_E = \oint \mathbf{E} \cdot d\mathbf{S} = \frac{1}{\varepsilon_0} \sum_{i=1}^{n} q_i = \frac{q}{\varepsilon_0}$$

where, $q = \sum_{i=1}^{n} q_i$ is the algebraic sum of all the charges inside the closed surface.

- Some important points regarding to Gauss's law
  (i) It is true for any closed surface, no matter what is shape or size.
  (ii) The term $q$ on the right side of Gauss's law, includes the sum of all charges enclosed by the surface.
  (iii) The surface which is chosen for the application of Gauss's law is called **Gaussian surface**.
  The Gaussian surface can pass through a continuous charge distribution.
  (iv) Gauss's law is often useful towards a much easier calculation of the electrostatic field when the system has some symmetry.
  (v) It is based on the inverse square dependence on distance contained in the Coulomb's law. Any violation of Gauss's law will indicate departure from the inverse square law.

**Electric field due to various charged symmetrical bodies can be obtained by applying Gauss's law as given in the following table**

| | Uniformly Charged Symmetrical Bodies | Electric Field |
|---|---|---|
| (i) | Infinitely long wire | $E = \frac{\lambda}{2\pi\varepsilon_0 r}$ where, $\lambda$ = linear charge density. |
| (ii) | Infinite plane sheet | $E = \frac{\sigma}{2\varepsilon_0}$ where, $\sigma$ = surface charge density. |
| (iii) | Conducting sphere or shell of radius $R$ and charge $Q$ | $E = 0$ for $r < R$, $E = \frac{Q}{4\pi\varepsilon_0 R^2}$ for $r = R$ and $E = \frac{Q}{4\pi\varepsilon_0 r^2}$ for $r > R$ |

# Mastering NCERT

## MULTIPLE CHOICE QUESTIONS

## TOPIC 1 ~ *Electric Charge and its Basic Properties*

**1** An electric shock is experienced on opening the door of a car due to
(a) heating of car engine
(b) discharge of electric charge
(c) motion of car
(d) None of the above

**2** If a plastic rod rubbed with fur is made to touch two small pith balls suspended nearby, then which figure shows their final configuration?

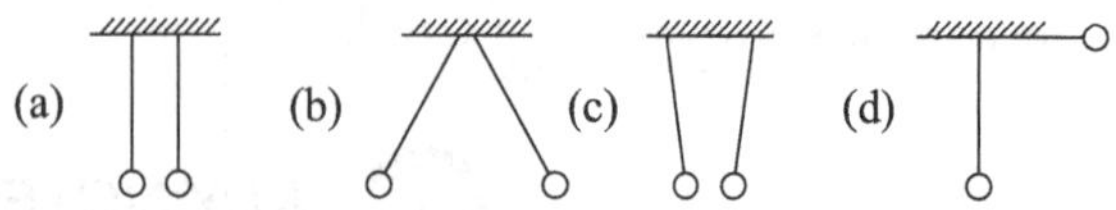

**3** A soap bubble is given a negative charge, then its radius
(a) decreases
(b) increases
(c) remains unchanged
(d) nothing can be predicted as information is insufficient

**4** One metallic sphere $A$ is given positive charge, whereas another identical metallic sphere $B$ of exactly same mass as that of $A$ is given equal amount of negative charge. Then,
(a) mass of $A$ and mass of $B$ still remain equal
(b) mass of $A$ increases
(c) mass of $B$ decreases
(d) mass of $B$ increases

**5** If two bodies are rubbed and one of them acquires $q_1$ charge and another acquires $q_2$ charge, then ratio of $q_1 : q_2$ is
(a) $1 : 2$ (b) $2 : 1$
(c) $-1 : 1$ (d) $1 : 4$

**6** In general, metallic ropes are suspended from the carriers to the ground which take inflammable material. The reason is
(a) their speed is controlled
(b) to keep the gravity of the carrier nearer to the earth
(c) to keep the body of the carrier in contact with the earth
(d) nothing should be placed under the carrier

**7** In charging by induction,
(a) body to be charged must be an insulator
(b) body to be charged must be a semiconductor
(c) body to be charged must be a conductor
(d) any type of body can be charged by induction

**8** When a glass rod is rubbed with a silk cloth, charge appears on both. This observation is consistent with law of conservation of charge as
(a) charge on both causes attraction
(b) charge on both causes repulsion
(c) charges appearing on both bodies are equal and opposite
(d) charge on first body is more than that of second body

**9** Suppose charge on a metallic sphere is $q$. If the sphere is put in contact with an identical uncharged sphere, the charge will spread over the two spheres. By symmetry, charge on each sphere will be
(a) $q$ (b) $\frac{q}{2}$
(c) $2q$ (d) zero

**10** If a body gives out $10^9$ electrons per second, how much time is required to get a total charge of 1 C from it?
(a) Around 198 min
(b) Around 198 h
(c) Around 198 days
(d) Around 198 yr

**11** The number of electrons that must be removed from an electrically neutral silver dollar to give it a charge of $+2.4$ C is
(a) $2.5 \times 10^{19}$ (b) $1.5 \times 10^{19}$
(c) $1.5 \times 10^{-19}$ (d) $2.5 \times 10^{-19}$

**12** An object of mass 1 kg contains $4 \times 10^{20}$ atoms. If one electron is removed from every atom of the solid, the charge gained by the solid of 1 g is
(a) 2.8 C (b) $6.4 \times 10^{-2}$ C
(c) $3.6 \times 10^{-3}$ C (d) $9.2 \times 10^{-4}$ C

## TOPIC 2 ~ *Coulomb's Law*

**13** The Coulomb's law is based on the assumption that charges are treated as
(a) positive charges
(b) negative charges
(c) neutral charges
(d) point charges

**14** Two electrical charged particles of different magnitudes, when placed at a distance $d$ from each other, experience a force of attraction $F$.

These two particles are then put in contact and again placed at the same distance from each other.

What is the nature of new force between them?
(a) Attractive
(b) Repulsive
(c) Attractive or repulsive depending upon magnitude of charges present on them
(d) Cannot predicted

**15** Force between two charges varies with distance between them as

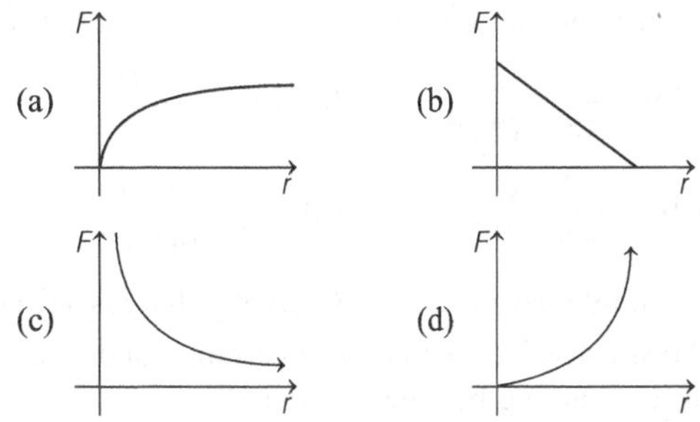


**16** If point charges $Q_1 = 2 \times 10^{-7}$ C and $Q_2 = 3 \times 10^{-7}$ C are at 30 cm separation, then find electrostatic force between them. ***JIPMER 2018***
(a) $2 \times 10^{-3}$ N (b) $6 \times 10^{-3}$ N
(c) $5 \times 10^{-3}$ N (d) $1 \times 10^{-3}$ N

**17** For two charges $q_1$ and $q_2$, if force between them for some separation in air is $F$, then force between them in a medium of permittivity $\varepsilon$ will be

(a) $\frac{\varepsilon_0}{\varepsilon}F$ (b) $\frac{\varepsilon}{\varepsilon_0}F$ (c) $\varepsilon\varepsilon_0 F$ (d) $\frac{F}{\varepsilon_0\varepsilon}$

**18** The ratio of the forces between two small spheres with constant charge (i) in air (ii) in a medium of dielectric constant $K$ is

(a) $1:K$ (b) $K:1$ (c) $1:K^2$ (d) $K^2:1$

**19** Two point charges placed at a certain distance $r$ in air exert a force $F$ on each other. Then, the distance $r'$ at which these charges will exert the same force in a medium of dielectric constant $K$ is given by

(a) $r$ (b) $r/K$ (c) $r/\sqrt{K}$ (d) $r\sqrt{K}$

**20** $F_G$ and $F_e$ represents gravitational and electrostatic force respectively, between electron and proton at a distance of 10 cm. The ratio of $F_G/F_e$ is of the order of

(a) $10^{42}$ (b) $10^{-39}$ (c) 1 (d) $10^{-43}$

**21** Suppose the charge of a proton and an electron differ slightly. One of them is $-e$ and the other is $(e+\Delta e)$. If the net of electrostatic force and gravitational force between two hydrogen atoms placed at a distance $d$ (much greater than atomic size) apart is zero, then $\Delta e$ is of the order [given, mass of hydrogen, $m_h = 1.67\times 10^{-27}$ kg] **NEET 2017**

(a) $10^{-20}$ C (b) $10^{-23}$ C (c) $10^{-37}$ C (d) $10^{-47}$ C

**22** Two small spheres each having the charge $+Q$ are suspended by insulating threads of length $L$ from a hook. This arrangement is taken in space, where there is no gravitational effect, then the angle between the two suspensions and the tension in each thread will be

(a) $180°, \frac{1}{4\pi\varepsilon_0}\frac{Q^2}{(2L)^2}$ (b) $90°, \frac{1}{4\pi\varepsilon_0}\frac{Q^2}{L^2}$

(c) $180°, \frac{1}{4\pi\varepsilon_0}\frac{Q^2}{2L^2}$ (d) $180°, \frac{1}{4\pi\varepsilon_0}\frac{Q^2}{L^2}$

**23** Two point charges $A$ and $B$, having charges $+Q$ and $-Q$ respectively, are placed at certain distance apart and force acting between them is $F$.
If 25% charge of $A$ is transferred to $B$, then force between the charges becomes **NEET (National) 2019**

(a) $\frac{9F}{16}$ (b) $\frac{16F}{9}$ (c) $\frac{4F}{3}$ (d) $F$

**24** Suppose the spheres $A$ and $B$ with charge $6.5\times 10^{-7}$C have identical sizes and distance between centres is 50 cm. A third sphere of the same size but uncharged is brought in contact with the first, then brought in contact with the second and finally removed from both. What is the new force of repulsion between $A$ and $B$?

(a) $5.7\times 10^{-3}$ N (b) $1.5\times 10^{-2}$ N
(c) 0.24 N (d) $0.24\times 10^{-2}$ N

**25** Charge $q_2$ of mass $m$ revolves around a stationary charge $q_1$ in a circular orbit of radius $r$. The orbital periodic time of $q_2$ would be

(a) $\left(\frac{4\pi^3 mr^2}{kq_1q_2}\right)^{1/2}$ (b) $\left(\frac{kq_1q_2}{4\pi^2 mr^2}\right)^{1/2}$

(c) $\left(\frac{4\pi^2 mr^4}{kq_1q_2}\right)^{1/2}$ (d) $\left(\frac{4\pi^2 mr^3}{kq_1q_2}\right)^{1/2}$

**26** Two charges, each equal to $q$ are kept at $x=-a$ and $x=a$ on the $X$-axis. A particle of mass $m$ and charge $q_0 = -\frac{q}{2}$ is placed at the origin. If charge $q_0$ is given a small displacement ($y << a$) along the $Y$-axis, the net force acting on the particle is proportional to

(a) $y$ (b) $y^2$ (c) $\frac{1}{y}$ (d) $\frac{1}{y^2}$

**27** Two identical charged spheres suspended from a common point by two massless strings of lengths $l$, are initially at a distance $x$ ($x << l$) apart because of their mutual repulsion. The charges begin to leak from both the spheres at a constant rate. As a result, the spheres approach each other with a velocity $v$. Then, $v$ varies as a function of the distance $x$ between the sphere, as **NEET 2016**

(a) $v \propto x$ (b) $v \propto x^{-1/2}$
(c) $v \propto x^{-1}$ (d) $v \propto x^2$

**28** Three charges $q_1, q_2$ and $q_3$ each of 1 C are at the vertices of an equilateral triangle of side $l$. Force on a charge $q = 2\mu$C placed at the centroid of the triangle is

(a) 3 N (b) 3 μN
(c) zero (d) $3\times 10^{-2}$ N

**29** If charges $q$, $q$ and $-q$ are placed at vertices of an equilateral triangle of side $l$. If $\mathbf{F}_1$, $\mathbf{F}_2$ and $\mathbf{F}_3$ are the forces on the charges respectively, then **JEE Main 2013**

(a) $|\mathbf{F}_1+\mathbf{F}_2+\mathbf{F}_3| = \sqrt{3}\frac{kq^2}{l^2}$

(b) $|\mathbf{F}_1+\mathbf{F}_2+\mathbf{F}_3| = 0$

(c) $|\mathbf{F}_1+\mathbf{F}_2+\mathbf{F}_3| = 3\sqrt{2}\frac{kq^2}{l^2}$

(d) $|\mathbf{F}_1+\mathbf{F}_2+\mathbf{F}_3| = \sqrt{2}\frac{kq^2}{l^2}$

**30** Three charges $+Q$, $q$, $+Q$ are placed respectively at distance 0, $\frac{d}{2}$ and $d$ from the origin on the $X$-axis. If the net force experienced by $+Q$ placed at $x=0$ is zero, then value of $q$ is **JEE Main 2019**

(a) $\frac{+Q}{2}$ (b) $\frac{+Q}{4}$ (c) $\frac{-Q}{2}$ (d) $\frac{-Q}{4}$

**31** Four charges equal to $-Q$ are placed at the four corners of a square and a charge $q$ is at its centre. If the system is in equilibrium, the value of $q$ is

(a) $-\frac{Q}{4}(1+2\sqrt{2})$ (b) $\frac{Q}{4}(1+2\sqrt{2})$

(c) $-\frac{Q}{2}(1+2\sqrt{2})$ (d) $\frac{Q}{2}(1+2\sqrt{2})$

# TOPIC 3 ~ Electric Field and Field Lines

**32** The electric field at a point is

(a) always continuous

(b) continuous, if there is no charge at that point

(c) continuous, if there is a charge at that point

(d) None of the above

**33** When the charge of a body becomes half, the electric field becomes

(a) half (b) twice

(c) thrice (d) No change

**34** The electric field in a certain region is acting radially outward and is given by $E = Aa$. A charge contained in a sphere of radius $a$ centred at the origin of the field will be **CBSE AIPMT 2015**

(a) $4\pi\varepsilon_0 Aa^2$ (b) $A\varepsilon_0 a^2$

(c) $4\pi\varepsilon_0 Aa^3$ (d) $\varepsilon_0 Aa^3$

**35** The bob of a simple pendulum has mass 2g and a charge of 5.0 $\mu$C. It is at rest in a uniform horizontal electric field of intensity 2000 V/m. At equilibrium, the angle that the pendulum makes with the vertical is (take $g = 10\,\text{m/s}^2$) **JEE Main 2019**

(a) $\tan^{-1}(2.0)$ (b) $\tan^{-1}(0.2)$

(c) $\tan^{-1}(5.0)$ (d) $\tan^{-1}(0.5)$

**36** An electron falls from rest through a vertical distance $h$ in a uniform and vertically upward directed electric field $E$. The direction of electric field is now reversed, keeping its magnitude the same. A proton is allowed to fall from rest in it through the same vertical distance $h$. The time of fall of the electron, in comparison to the time of fall of the proton is **NEET 2018**

(a) 10 times greater (b) 5 times greater

(c) smaller (d) equal

**37** An electron of mass $m_e$ initially at rest moves through a certain distance in a uniform electric field in time $t_1$. A proton of mass $m_p$ also initially at rest takes time $t_2$ to move through an equal distance in this uniform electric field. Neglecting the effect of gravity, the ratio of $t_2/t_1$ is nearly equal to

(a) 1836 (b) $(m_e/m_p)^{1/2}$

(c) $(m_p/m_e)^{1/2}$ (d) 1

**38** Two point charges $q_1$ and $q_2$ of $+10^{-8}$C and $-10^{-8}$C, respectively are placed 0.1 m apart.

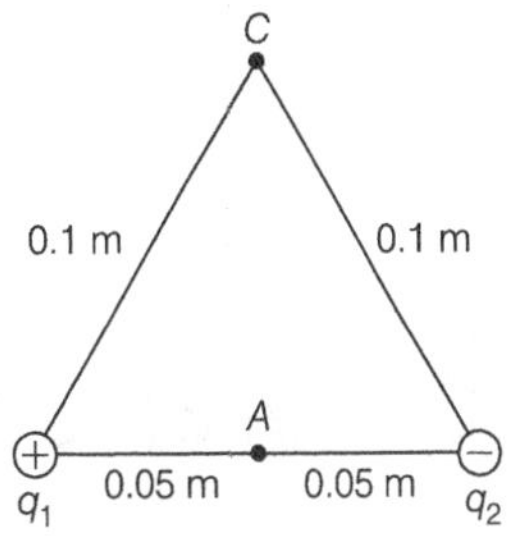


Then, ratio of magnitudes of electric fields at $A$ and $C$ is

(a) 4 : 1 (b) 1 : 4

(c) 8 : 1 (d) 1 : 8

**39** Three identical positive point charges below are placed at the vertices of an isosceles right angled triangle as shown. Which of the vectors numbered coincides in direction with the electric field at the mid-point $M$ of the hypotenuse due to the given system of charges?

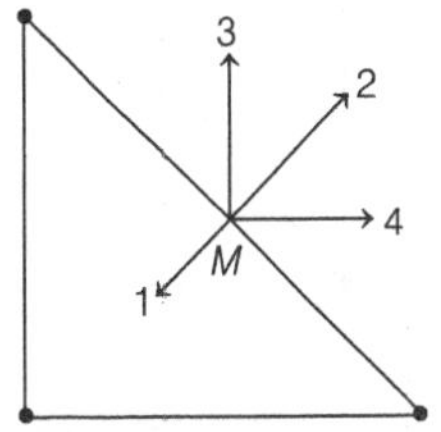


(a) 1 (b) 2

(c) 3 (d) 4

**40** Charges $q, 2q, 3q$ and $4q$ are placed at the corners $A, B, C$ and $D$ of a square as shown in the figure. The direction of electric field at the centre of the square is along

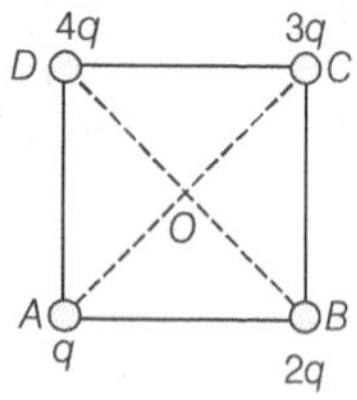


(a) $AB$ (b) $CB$ (c) $BD$ (d) $AC$

**41** Find the correct diagram of electric lines of forces for negative charge. ***JIPMER 2019***

(a) 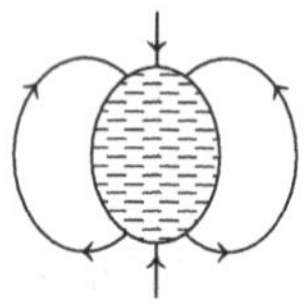

(b) 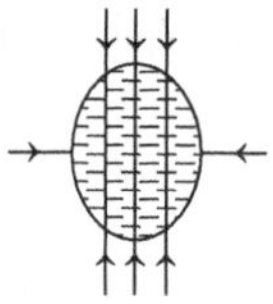

(c) 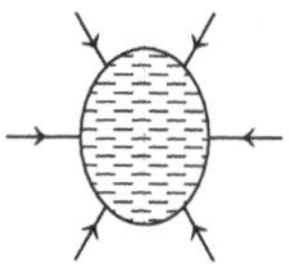

(d) 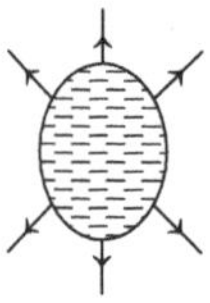

**42** A charged particle is free to move in an electric field. It will travel

(a) always along a line of force
(b) along a line of force, if its initial velocity is zero
(c) along a line of force, if it has some initial velocity in the direction of an acute angle with the line of force
(d) None of the above

**43** Two field lines can never cross each other because

(a) field lines are closed curves
(b) field lines repels each other
(c) field lines crowded only near the charge
(d) field has a unique direction at each point

**44** In the diagram shown below,

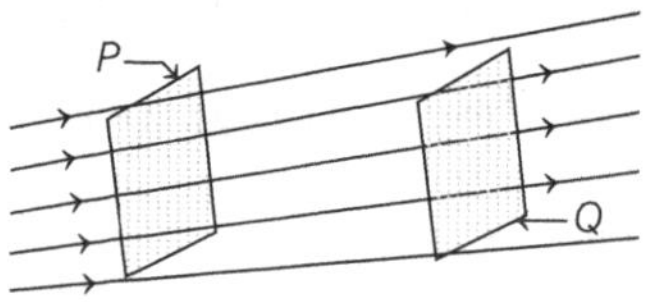


(a) field strength at $P$ is less than field strength at $Q$
(b) field strength at $P$ and $Q$ are equal
(c) field is more strong at $P$ and less strong at $Q$
(d) Cannot be determined from the figure

## TOPIC 4 ~ *Electric Flux and Electric Dipole*

**45** If an area is tilted in the direction of field such that normal to plane of area makes $\theta$ angle with the direction of field, then number of field lines passing through the area is

(a) equal to $E\Delta S$ (b) proportional to $E\Delta S\cos\theta$
(c) equal to $E\Delta S\cos\theta$ (d) proportional to $E\Delta S\cos\theta/r^2$

**46** Determine the electric dipole moment of the system of three charges, placed on the vertices of an equilateral triangle as shown in the figure. ***JEE Main 2019***

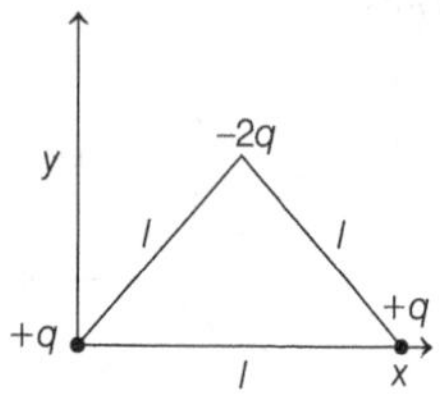


(a) $\sqrt{3}\,ql\,\frac{\hat{j}-\hat{i}}{\sqrt{2}}$ (b) $2ql\,\hat{j}$
(c) $-\sqrt{3}\,ql\,\hat{j}$ (d) $(ql)\frac{\hat{i}+\hat{j}}{\sqrt{2}}$

**47** Two electric dipoles of moment $p$ and $64p$ are placed in opposite direction on a line at a distance of 25 cm. The electric field will be zero at point between the dipoles whose distance from the dipole of moment $p$ is

(a) 5 cm (b) $\frac{25}{9}$ cm (c) 10 cm (d) $\frac{4}{13}$ cm

**48** What is the angle between the electric dipole moment and the electric field strength due to it on the equatorial line?

(a) 0° (b) 90°
(c) 180° (d) None of these

**49** Consider a region inside which there are various types of charges but the total charge is zero. At points outside the region,

(a) the electric field is necessarily zero
(b) the electric field is due to the dipole moment of the charge distribution only
(c) the dominant electric field is $\propto \frac{1}{r_3}$, for large $r$, where $r$ is the distance from a origin in this region
(d) None of the above

**50** The ratio of electric fields on the axis and at equator of an electric dipole will be

(a) 1 : 1 (b) 2 : 1
(c) 4 : 1 (d) None of these

**51** In given figures, $OP = OQ = 15$ cm, $OA = OB = 2.5$ mm

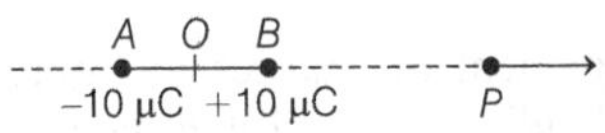


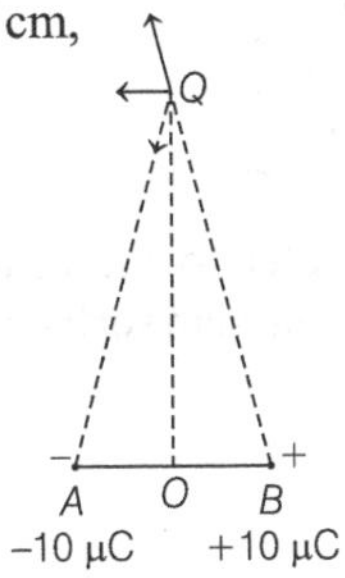


Magnitudes of electric field at $P$ and $Q$ are respectively

(a) $2.6 \times 10^5$ NC$^{-1}$, $2.6 \times 10^5$ NC$^{-1}$
(b) $1.3 \times 10^5$ NC$^{-1}$, $1.3 \times 10^5$ NC$^{-1}$
(c) $2.6 \times 10^5$ NC$^{-1}$, $1.3 \times 10^5$ NC$^{-1}$
(d) $1.3 \times 10^5$ NC$^{-1}$, 2.6 NC$^{-1}$

**52** Four point charges $-q$, $+q$, $+q$ and $-q$ are placed on $Y$-axis at $y = -2d$, $y = -d$, $y = +d$ and $y = +2d$, respectively. The magnitude of the electric field $E$ at a point on the $X$-axis at $x = D$, with $D >> d$, will behave as ***JEE Main 2019***

(a) $E \propto \frac{1}{D}$ (b) $E \propto \frac{1}{D^3}$
(c) $E \propto \frac{1}{D^2}$ (d) $E \propto \frac{1}{D^4}$

**53** When a dipole is placed in an uniform external field, then

(a) net force on dipole is zero but torque is non-zero
(b) torque is zero but net force is non-zero
(c) torque and net force both are non-zero
(d) both torque and net force are zero

**54** When an electric dipole **p** is placed in a uniform electric field **E**, then at what angle between **p** and **E** the value of torque will be maximum?

(a) 90° (b) 0°
(c) 180° (d) 45°

**55** Two small identical electrical dipoles $AB$ and $CD$, each of dipole moment $p$ are kept at an angle of 120° as shown in the figure. If this system is subjected to electric field (**E**) directed along direction which makes $\theta°$ with $xy$-plane. Then, the magnitude and direction of the torque acting on this is

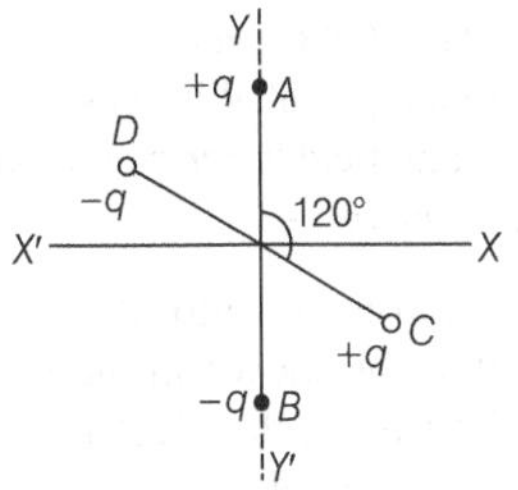


(a) $pE \sin\theta$, along $Z$-axis
(b) $2pE \sin\theta$, along $Z$-axis
(c) $\frac{\sqrt{3}}{2} pE \sin\theta$, along $Z$-axis
(d) $\frac{1}{2} pE \sin\theta$, along $Z$-axis

**56** An electric dipole is situated in an electric field of uniform intensity $E$ whose dipole moment is $p$ and moment of inertia is $I$. If the dipole is displaced slightly from the equilibrium position, then the angular frequency of its oscillations is

(a) $\left(\frac{pE}{I}\right)^{1/2}$ (b) $\left(\frac{pE}{I}\right)^{3/2}$
(c) $\left(\frac{I}{pE}\right)^{1/2}$ (d) $\left(\frac{p}{IE}\right)^{1/2}$

# TOPIC 5 ~ *Gauss's Law and its Applications*

**57** The surface charge density $\sigma$ of an area element ignores

(a) the quantisation of charge
(b) discontinuity in charge distribution at macroscopic level
(c) discontinuity in charge distribution at microscopic level
(d) Both (a) and (c)

**58** A uniformly charged conducting sphere of 2.4 m diameter has a surface charge density of 80 μCm$^{-2}$. What is the charge on the sphere?

(a) $0.7 \times 10^{-1}$ C (b) $1.4 \times 10^{-2}$ C
(c) $1.4 \times 10^{-3}$ C (d) $1.7 \times 10^{4}$ C

**59** Two non-conducting solid spheres of radii $R_1$ & $R_2$ and carrying uniform volume charge densities $+\rho$ & $-\rho$ respectively, are placed such that they partially overlap, as shown in the figure.
At all points in the overlapping region,

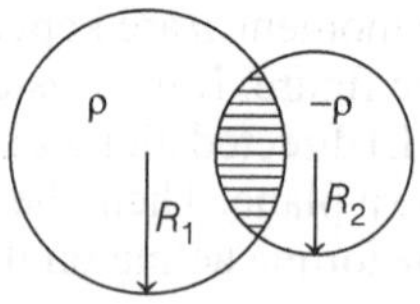


(a) the electrostatic field is zero
(b) the electrostatic potential is constant
(c) the electrostatic field is constant in magnitude
(d) None of the above

**60** Consider the charge configuration and spherical Gaussian surface as shown in the figure. When calculating the flux of the electric field over the spherical surface, the electric field will be due to

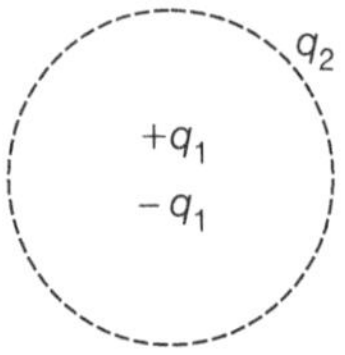


(a) $q_2$
(b) only the positive charges
(c) all the charges
(d) $+q_1$ and $-q_1$

**61** If $\int_s \mathbf{E} \cdot d\mathbf{S} = 0$ over a surface, then
(a) the electric field inside the surface and on it is zero
(b) the electric field inside the surface is necessarily uniform
(c) the number of flux lines entering the surface must be equal to the number of flux lines leaving it
(d) None of the above

**62** Total electric flux coming out of a unit positive charge put in air is
(a) $\varepsilon_0$ (b) $\varepsilon_0^{-1}$
(c) $(4p\varepsilon_0)^{-1}$ (d) $4\pi\varepsilon_0$

**63** If same charge $q$ is placed inside a sphere and cube having radius 1m and side 2m, respectively. What will be the ratio of flux passing through them?
(a) 1 : 1 (b) 1 : 8 (c) 8 : 1 (d) 1 : 2

**64** A sphere encloses an electric dipole with charge $\pm 3 \times 10^{-6}$ C. What is the total electric flux across the sphere? ***NEET (Odisha) 2019***
(a) $-3 \times 10^{-6}$ N-m$^2$/C (b) Zero
(c) $3 \times 10^6$ N-m$^2$/C (d) $6 \times 10^{-6}$ N-m$^2$/C

**65** Eight dipoles of charges of magnitude $e$ each are placed inside a cube. The total electric flux coming out of the cube will be
(a) $\frac{8e}{\varepsilon_0}$ (b) $\frac{16e}{\varepsilon_0}$
(c) $\frac{e}{\varepsilon_0}$ (d) zero

**66** Refer to the arrangement of charges in figure and a Gaussian surface of radius $R$ with $Q$ at the centre, then

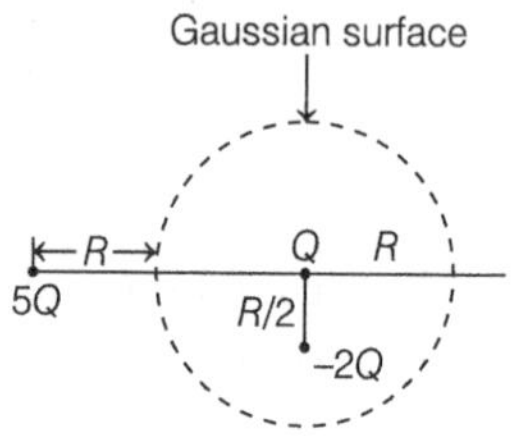


(a) total flux through the surface of the sphere is $\frac{-Q}{\varepsilon_0}$
(b) field on the surface of the sphere is $\frac{-Q}{4\pi\varepsilon_0 R^2}$
(c) flux through the surface of sphere due to $5Q$ is $-\frac{5Q}{\varepsilon_0}$
(d) field on the surface of sphere due to $-2Q$ is same everywhere

**67** Shown in the figure is a shell made of a conductor. It has inner radius $a$ and outer radius $b$ and carries charge $Q$. At its centre is a dipole $\mathbf{p}$ as shown.
In this case, ***JEE Main 2019***

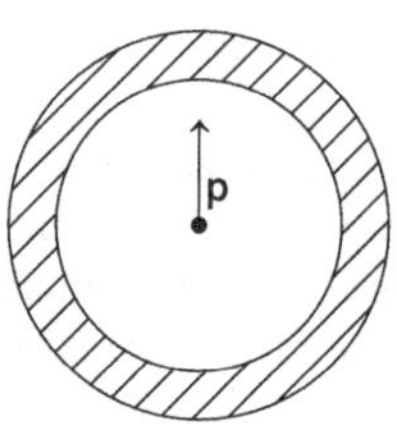


(a) surface charge density on the inner surface is uniform and equal to $\frac{\left(\frac{Q}{2}\right)}{4\pi a^2}$
(b) electric field outside the shell is the same as that of a point charge at the centre of the shell
(c) surface charge density on the outer surface depends on $|\mathbf{p}|$
(d) surface charge density on the inner surface of the shell is zero everywhere

**68** A cube is placed in a uniform magnetic field. If net flux through the cube is 1.05 Nm$^2$C$^{-1}$, the charge within the cube is
(a) $9.27 \times 10^{-12}$ C (b) $9.27 \times 10^{12}$ C
(c) $6.97 \times 10^{-12}$ C (d) $6.97 \times 10^{12}$ C

**69** Two parallel infinite line charges with linear charge densities $+\lambda$ (C/m) and $-\lambda$ (C/m) are placed at a distance of $2R$ in free space.
What is the electric field mid-way between the two line charges? ***NEET (National) 2019***

(a) $\frac{2\lambda}{\pi\varepsilon_0 R}$ (N/C)
(b) $\frac{\lambda}{\pi\varepsilon_0 R}$ (N/C)
(c) $\frac{\lambda}{2\pi\varepsilon_0 R}$ (N/C)
(d) Zero

**70** Two infinite plane parallel sheets separated by a distance $d$ have equal and opposite uniform charge densities $\sigma$. Electric field at a point between the sheets, is

(a) zero
(b) $\frac{\sigma}{\varepsilon_0}$
(c) $\frac{\sigma}{2\varepsilon_0}$
(d) depends upon the location of the point

**71** Two large, thin metal plates are parallel and close to each other. On their inner faces, the plates have surface charge densities of opposite signs and of magnitude $17.0 \times 10^{-22}\,\text{Cm}^{-2}$. What is **E** in the outer region of the first plate? ***JEE Main 2016***

(a) $17 \times 10^{-22}$ N/C
(b) $1.5 \times 10^{-15}$ N/C
(c) $1.9 \times 10^{-10}$ N/C
(d) Zero

**72** A hollow metal sphere of radius $R$ is uniformly charged. The electric field due to the sphere at a distance $r$ from the centre ***NEET (National) 2019***

(a) zero as $r$ increases for $r < R$, decreases as $r$ increases for $r > R$
(b) zero as $r$ increases for $r < R$, increases as $r$ increases for $r > R$
(c) decreases as $r$ increases for $r < R$ and for $r > R$
(d) increases as $r$ increases for $r < R$ and for $r > R$

# SPECIAL TYPES QUESTIONS

## I. Assertion and Reason

**Direction** (Q. Nos. 73-79) *In the following questions, a statement of Assertion is followed by a corresponding statement of Reason. Of the following statements, choose the correct one.*

(a) Both Assertion and Reason are correct and Reason is the correct explanation of Assertion.
(b) Both Assertion and Reason are correct but Reason is not the correct explanation of Assertion.
(c) Assertion is correct but Reason is incorrect.
(d) Assertion is incorrect but Reason is correct.

**73 Assertion** When we produce charge $q_1$ on a body by rubbing it against another body which gets a charge $q_2$ in the process, then $q_1 + q_2 = 0$.
**Reason** Charge on an isolated system remains constant.

**74 Assertion** At macroscopic level, quantisation of charge has no practical consequence and can be ignored.
**Reason** 1 $\mu$C charge contains $10^{13}$ times electronic charge $e$ approximately.

**75 Assertion** When a neutral body acquires positive charge, its mass decreases.
**Reason** A body acquires positive charge when it loses electrons.

**76 Assertion** Mass of a body decreases slightly when it is negatively charged. ***AIIMS 2018***
**Reason** Charging is due to transfer of electrons.

**77 Assertion** A positive point charge initially at rest in a uniform electric field starts moving along electric lines of force. (Neglect all other forces except electric forces).
**Reason** A point charge released from rest in an electric field always moves along the line of force.

**78 Assertion** **E** in outside vicinity of a conductor depends only on the local charge density $\sigma$ and it is independent of the other charges present anywhere on the conductor.
**Reason** **E** in outside vicinity of a conductor is given by $\frac{\sigma}{\varepsilon_0}$.

**79 Assertion** Upon displacement of charges within a closed surface, **E** at any point on the surface does not change.
**Reason** The flux crossing through a closed surface is independent of the location of charge within the surface.

## II. Statement Based Questions

**80** I. When we rub a glass rod with silk, some of electrons from the rod are transferred to the silk cloth.

II. Rod gets positively charged and silk gets negatively charged.

III. Only less tightly bound electrons in rod are transferred to silk by rubbing.

Which of the following statement(s) is/are correct?

(a) Both I and II (b) Both II and III
(c) Both I and III (d) I, II and III

**81** I. If a system contains two point charges $q_1$ and $q_2$, the total charge of the system is obtained simply by adding algebraically $q_1$ and $q_2$, e.g. (– 2units charge) + (6 units charge) = 4 units charge.

II. Total charge of a system having charges to value –2units and + 6units is 8 units charge.

III. Charge has magnitude but no direction, similar to mass.

IV. Mass of a body is always positive, whereas a charge can be either positive or negative.

Which of the following statement(s) is/are incorrect?

(a) Only I (b) Only II
(c) Only III (d) Only IV

**82** I. Charge is conserved.

II. When bodies are charged by rubbing, there is a transfer of electrons from one body to other, no new charges are created nor destroyed.

III. Within an isolated system consisting of many charged bodies, due to interaction among the bodies charges may get redistributed but it is found that the total charge of the isolated system is always conserved.

IV. When a neutron turns into a proton and an electron. The proton and electron thus created have equal and opposite charges and the total charge is zero, before and after the creation.

Which of the following statement(s) is/are correct?

(a) Both I and IV
(b) Both I and II
(c) I, II and IV
(d) I, II, III and IV

**83** I. The charge $q$ on a body is always given by $q = ne$, where $n$ is any integer, positive or negative.

II. By convention, the charge on an electron is taken to be negative.

III. The fact that electric charge is always an integral multiple of $e$ is termed as quantisation of charge.

IV. The quantisation of charge was experimentally demonstrated by Newton in 1912.

Which of the following statement(s) is/are incorrect?

(a) Only I (b) Only II
(c) Only IV (d) Only III

**84** I. Magnitude of electric field at a point decreases inversely as square of distance of that point from the charge.

II. **E** is strong near the charge.

III. Away from the charge field gets weaker and density of field lines is less, resulting in well separated lines.

IV. Only a finite number of lines can be drawn from a charge.

Which of the following statement(s) is/are incorrect?

(a) Only I (b) Only II (c) Only III (d) Only IV

**85** Choose the incorrect statement with respect to properties of charge.

(a) Charges are additive in nature.
(b) Charges are conservative in nature.
(c) Charges are quantised in nature.
(d) Charges can be transformed from one type to another.

**86** 10 C of charge is given to a conducting spherical shell and – 3 C point charge is placed inside the shell. For this arrangement, mark out the correct statement(s).

(a) The charge on the inner surface of the shell will be + 3 C and it can be distributed uniformly or non-uniformly.
(b) The charge on the inner surface of the shell will be + 3 C and its distribution would be uniform.
(c) The net charge on outer surface of the shell will be + 4 C and its distribution can be uniform or non-uniform.
(d) The net charge on outer surface of the shell will be + 4 C and its distribution would be uniform.

**87** Consider Gauss's law, $\oint \mathbf{E} \cdot d\mathbf{S} = \dfrac{q}{\varepsilon_0}$

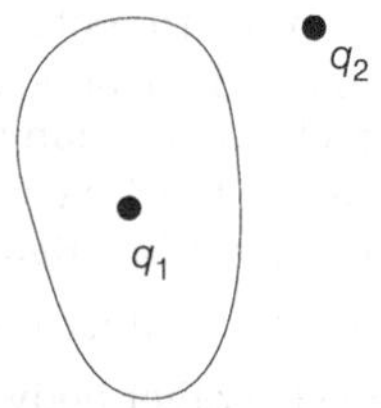


Then for the situation shown above at the Gaussian surface, which of the following statement is correct?

(a) **E** due to $q_2$ would be zero.
(b) **E** due to both $q_1$ and $q_2$ would be non-zero.
(c) $\phi$ due to both $q_1$ and $q_2$ would be non-zero.
(d) $\phi$ due to $q_2$ would be non-zero.

**88** Which of the following statement(s) is/are correct regarding electric dipoles?

(a) Electric dipoles are natural.
(b) Electric dipoles are induced.
(c) Both (a) and (b)
(d) Electric dipole exists only in the presence of a strong electric field.

**89** Figure shows electric field lines in which an electric dipole $p$ is placed as shown. Which of the following statements is correct?

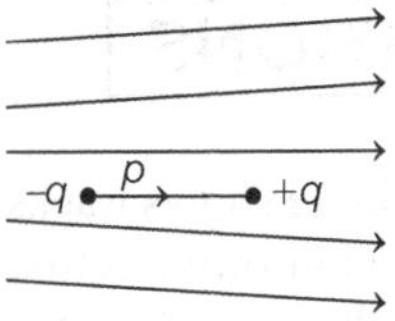


(a) The dipole will not experience any force.
(b) The dipole will experience a force towards right.
(c) The dipole will experience a force towards left.
(d) The dipole will experience a force upwards.

**90** For a thin spherical shell of radius $R$ having a surface charge density $\sigma$, which of these statement is correct?

(a) Field at a distance $r$ from the centre $r > R$ is given by $E = \frac{\sigma}{\varepsilon_0}$.
(b) Field at a distance $r$ from the centre $r < R$ given by $E = \frac{kq}{r^2}$, where $q = \sigma\,(4\pi R^2)$.
(c) Field at a distance $r$ from the centre $r > R$ is given by $E = \frac{1}{4\pi\varepsilon_0} \cdot \frac{q}{r^2}$, where $q = \sigma \cdot 4\pi R^2$.
(d) Field at surface of shell is given by $E = \frac{\sigma}{2\varepsilon_0}$.

## III. Matching Type

**91** Match the Column I (charge) with Column II (property) and select the correct answer from the codes given below.

| | Column I | | Column II |
|---|---|---|---|
| A. | Like charges | 1. | Of two types |
| B. | Unlike charges | 2. | Repel each other |
| C. | Charge can be | 3. | Attract each other |
| D. | Charges are | 4. | Neutralised, if they are equal and opposite |

| | A | B | C | D |
|---|---|---|---|---|
| (a) | 1 | 2 | 3 | 4 |
| (b) | 2 | 1 | 3 | 4 |
| (c) | 4 | 1 | 3 | 2 |
| (d) | 2 | 3 | 4 | 1 |

**92** Match the Column I (type of body) with Column II (electric field) and select the correct answer from the codes given below.

| | Column I | | Column II |
|---|---|---|---|
| A. | Infinite plane sheet of charge | 1. | 0 |
| B. | Infinite plane sheet of uniform thickness | 2. | $\frac{\sigma}{2\varepsilon_0}$ |
| C. | Non-conducting charged solid sphere at its surface | 3. | $\frac{R\rho}{3\varepsilon_0}$ |
| D. | Conducting charged solid sphere at its centre | 4. | $\frac{\sigma}{\varepsilon_0}$ |

| | A | B | C | D |
|---|---|---|---|---|
| (a) | 2 | 4 | 3 | 1 |
| (b) | 3 | 2 | 1 | 4 |
| (c) | 1 | 4 | 2 | 3 |
| (d) | 2 | 1 | 4 | 3 |

**93** Match the Column I (electric lines of force) with Column II (types of charge) and select the correct answer from the codes given below.

| | Column I | | Column II |
|---|---|---|---|
| A. | 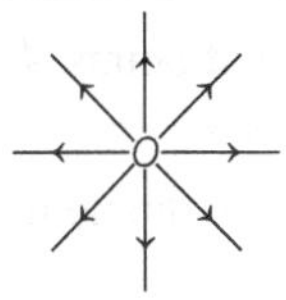 | 1. | A pair of equal and opposite charges |
| B. | 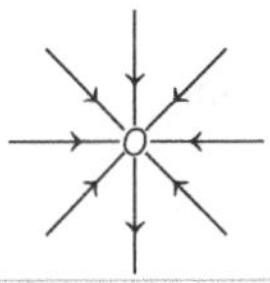 | 2. | A pair of positive charges |
| C. | 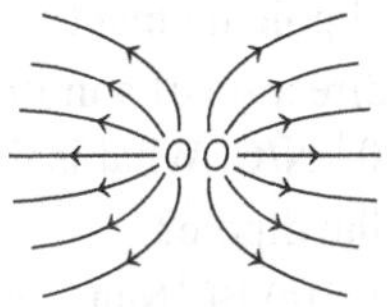 | 3. | A single positive charge |
| D. | 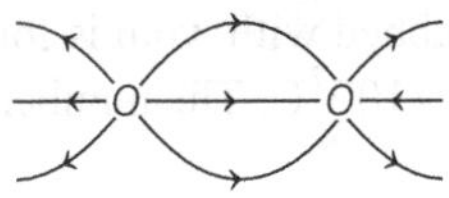 | 4. | A single negative charge |

| | A | B | C | D |
|---|---|---|---|---|
| (a) | 1 | 2 | 3 | 4 |
| (b) | 3 | 2 | 1 | 4 |
| (c) | 3 | 4 | 1 | 2 |
| (d) | 3 | 4 | 2 | 1 |

# NCERT & NCERT Exemplar

## MULTIPLE CHOICE QUESTIONS

### NCERT

**94** What is the force between two small charged spheres having charges of $2\times10^{-7}$ C and $3\times10^{-7}$ C placed 30 cm apart in air?
(a) $5\times10^{-2}$ N (b) $6\times10^{-3}$ N (c) $7\times10^{-4}$ N (d) $8\times10^{-4}$ N

**95** The electrostatic force on a small sphere of charge 0.4 μC due to another small sphere of charge −0.8 μC in air is 0.2 N. What is the distance between the two spheres?
(a) 5 cm (b) 10 cm (c) 12 cm (d) 15 cm

**96** Two point charges $q_A = 3\,\mu C$ and $q_B = -3\,\mu C$ are located 20 cm apart in vacuum. What is the electric field at the mid-point *O* of the line *AB* joining the two charges?
(a) 0
(b) $2.7\times10^6$ (N/C)
(c) $5.4\times10^6$ (N/C)
(d) $10.2\times10^6$ (N/C)

**97** A system has two charges $q_A = 2.5\times10^{-7}$ C and $q_B = -2.5\times10^{-7}$ C located at points *A* (0, 0, −15 cm) and *B* (0, 0, +15 cm), respectively.
The total charge and electric dipole moment of the system are
(a) $5.0\times10^{-7}$ C, $7.5\times10^{-8}$ C-m
(b) $2.5\times10^{-7}$ C, $7.5\times10^{-8}$ C-m
(c) 0,0
(d) 0, $7.5\times10^{-8}$ C-m

**98** An electric dipole with dipole moment $4\times10^{-9}$ C-m is aligned at 30° with the direction of a uniform electric field of magnitude $5\times10^4$ N/C. What is the magnitude of the torque acting on the dipole?
(a) $10^{-2}$ N-m (b) $10^{-3}$ N-m (c) $10^{-4}$ N-m (d) $10^{-5}$ N-m

**99** A polythene piece rubbed with wool is found to have a negative charge of $3\times10^{-7}$ C. The number of electrons transferred is
(a) $1.6\times10^9$
(b) $1.8\times10^{10}$
(c) $1.6\times10^{11}$
(d) $1.8\times10^{12}$

**100** Two insulated charged copper spheres *A* and *B* have their centres separated by a distance of 50 cm. The charge on each sphere is $6.5\times10^{-7}$ C.
What is the force of repulsion, if each sphere is charged double the above amount and the distance between them is halved?
(a) 0 (b) 0.12 N (c) 0.24 N (d) 0.48 N

**101** The given figure shows tracks of three charged particles in a uniform electrostatic field. Which particle has the highest charge to mass ratio?

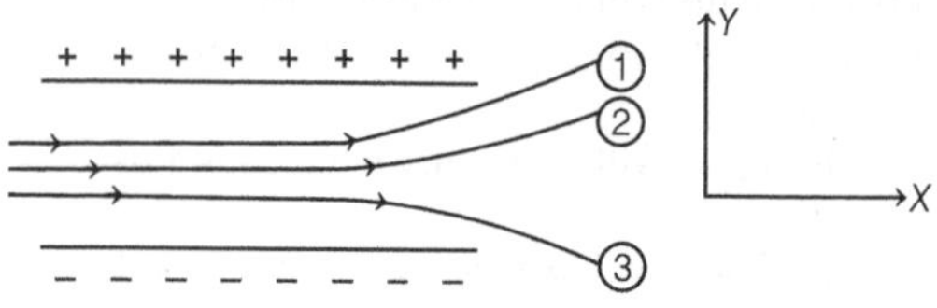


(a) 1
(b) 2
(c) 3
(d) All are equal

**102** Consider a uniform electric field $\mathbf{E} = 3\times10^3\,\hat{\mathbf{i}}$ NC$^{-1}$.
What is the flux through a square of side 10 cm, if the normal to its plane makes a 60° angle with the *X*-axis?
(a) 0
(b) 5 Nm$^2$C$^{-1}$
(c) 10 Nm$^2$C$^{-1}$
(d) 15 Nm$^2$C$^{-1}$

**103** A point charge $+10\,\mu C$ is at a distance 5 cm directly above the centre of a square of side 10 cm, as shown in figure. What is the magnitude of the electric flux through the square?

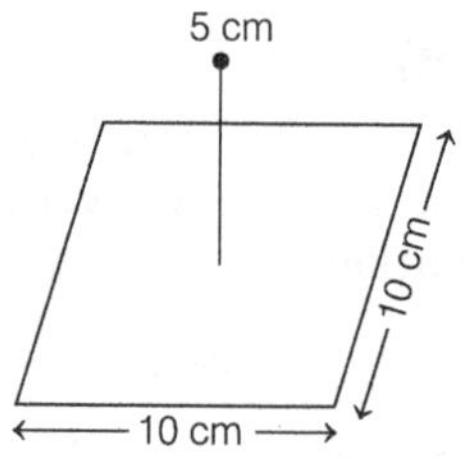


(a) 0
(b) $1.8\times10^2$ Nm$^2$C$^{-1}$
(c) $1.8\times10^4$ Nm$^2$C$^{-1}$
(d) $1.8\times10^5$ Nm$^2$C$^{-1}$

**104** A point charge of 2.0 μC is at the centre of a cubic Gaussian surface 9.0 cm on edge. What is the net electric flux through the surface?
(a) 0
(b) $2\times10^2$ Nm$^2$/C
(c) $2\times10^4$ Nm$^2$/C
(d) $2\times10^5$ Nm$^2$/C

**105** A conducting sphere of radius 10 cm has an unknown charge. If the electric field 20 cm from the centre of the sphere is $1.5\times10^3$ N/C and points radially inwards, what is the net charge on the sphere?
(a) 0 (b) $5\times10^{-5}$ C (c) $6\times10^{-9}$ C (d) $8\times10^{-10}$ C

**106** An infinite line charge produces a field of $9 \times 10^4$ NC$^{-1}$ at a distance of 2 cm. Calculate the linear charge density.

(a) $10^{-5}$ Cm$^{-1}$ (b) $10^{-6}$ Cm$^{-1}$

(c) $10^{-7}$ Cm$^{-1}$ (d) $10^{-8}$ Cm$^{-1}$

## NCERT Exemplar

**107** In figure two positive charges $q_2$ and $q_3$ fixed along the $Y$-axis, exert a net electric force in the positive $x$-direction on a charge $q_1$ fixed along the $X$-axis. If a positive charge $Q$ is added at $(x, 0)$, the force on $q_1$

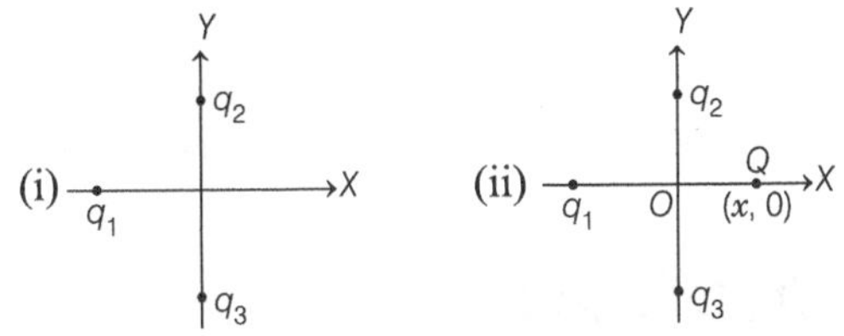


(a) shall increase along the positive $X$-axis

(b) shall decrease along the positive $X$-axis

(c) shall point along the negative $X$-axis

(d) shall increase but the direction changes because of the intersection of $Q$ with $q_2$ and $q_3$

**108** A positive point charge is brought near an isolated conducting sphere (figure). The best representation of electric field is given by

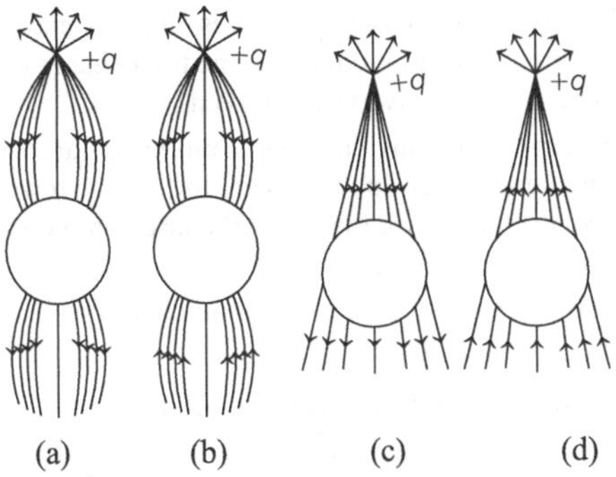


**109** The electric flux through the surface

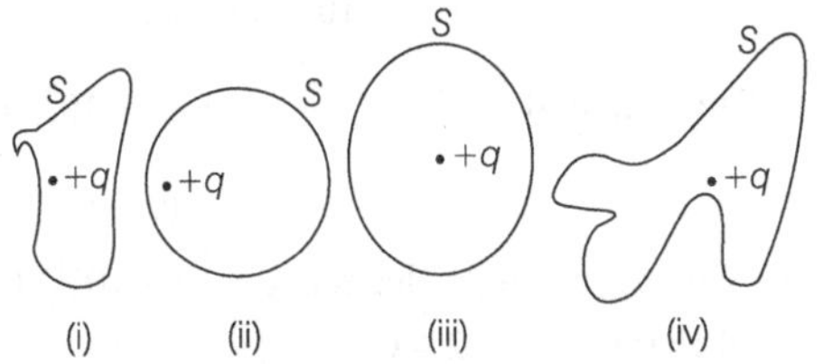


(a) in Fig. (iv) is the largest

(b) in Fig. (iii) is the least

(c) in Fig. (ii) is same as Fig. (iii) but is smaller than Fig. (iv)

(d) is the same for all the figures

**110** A metallic spherical shell has an inner radius $R_1$ and outer radius $R_2$. A charge $Q$ is placed at the centre of the spherical cavity. What will be surface charge density on the outer surface?

(a) $\frac{-Q}{4\pi R_1^2}$ (b) $\frac{Q}{4\pi R_1^2}$ (c) $\frac{Q}{4\pi R_2^2}$ (d) $\frac{-Q}{4\pi R_2^2}$

**111** What will be the total flux through the faces of the cube as given in the figure with side of length $a$, if a charge $q$ is placed at $B$, mid-point of an edge of the cube?

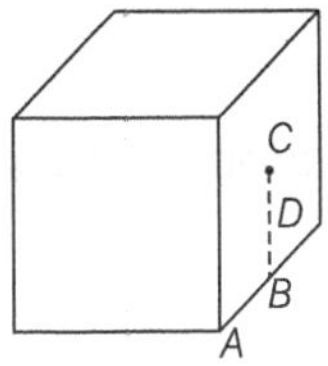


(a) $\frac{q}{8\varepsilon_0}$ (b) $\frac{q}{3\varepsilon_0}$

(c) $\frac{q}{4\varepsilon_0}$ (d) $\frac{q}{2\varepsilon_0}$

**112** Two charges $q$ and $-3q$ are placed fixed on $X$-axis separated by distance $d$. Where should a third charge $2q$ be placed such that it will not experience any force?

(a) $\frac{d}{3}(1+\sqrt{3})$ to the right of $q$

(b) $\frac{d}{3}(1+\sqrt{3})$ to the left of $q$

(c) $\frac{d}{2}(1+\sqrt{3})$ to the left of $q$

(d) $\frac{d}{2}(1+\sqrt{3})$ to the right of $q$

**113** A point charge $+q$ is placed at a distance $d$ from an isolated conducting plane. The field at a point $P$ on the other side of the plane is

(a) directed perpendicular to the plane and away from the plane

(b) directed perpendicular to the plane but towards the plane

(c) directed radially away from the point charge

(d) directed radially towards the point charge

**114** A hemisphere is charged uniformly positively. The electric field at a point on a diameter away from the centre is directed

(a) perpendicular to the diameter

(b) parallel to the diameter

(c) at an angle tilted towards the diameter

(d) at an angle tilted away from the diameter

# Answers

## > Mastering NCERT with MCQs

| | | | | | | | | | |
|---|---|---|---|---|---|---|---|---|---|
| 1 (b) | 2 (b) | 3 (b) | 4 (d) | 5 (c) | 6 (c) | 7 (c) | 8 (c) | 9 (b) | 10 (d) |
| 11 (b) | 12 (b) | 13 (d) | 14 (b) | 15 (c) | 16 (b) | 17 (a) | 18 (b) | 19 (c) | 20 (b) |
| 21 (c) | 22 (a) | 23 (a) | 24 (a) | 25 (d) | 26 (a) | 27 (b) | 28 (c) | 29 (b) | 30 (d) |
| 31 (b) | 32 (b) | 33 (a) | 34 (c) | 35 (d) | 36 (c) | 37 (c) | 38 (c) | 39 (b) | 40 (b) |
| 41 (c) | 42 (b) | 43 (d) | 44 (c) | 45 (b) | 46 (c) | 47 (a) | 48 (c) | 49 (c) | 50 (b) |
| 51 (c) | 52 (d) | 53 (a) | 54 (a) | 55 (a) | 56 (a) | 57 (d) | 58 (c) | 59 (c) | 60 (c) |
| 61 (c) | 62 (b) | 63 (a) | 64 (b) | 65 (d) | 66 (a) | 67 (b) | 68 (a) | 69 (b) | 70 (b) |
| 71 (d) | 72 (a) | | | | | | | | |

## > Special Types Questions

| | | | | | | | | | |
|---|---|---|---|---|---|---|---|---|---|
| 73 (a) | 74 (a) | 75 (a) | 76 (d) | 77 (c) | 78 (d) | 79 (d) | 80 (d) | 81 (b) | 82 (d) |
| 83 (c) | 84 (d) | 85 (d) | 86 (b) | 87 (b) | 88 (c) | 89 (c) | 90 (c) | 91 (d) | 92 (a) |
| 93 (d) | | | | | | | | | |

## > NCERT & NCERT Exemplar MCQs

| | | | | | | | | | |
|---|---|---|---|---|---|---|---|---|---|
| 94 (b) | 95 (c) | 96 (c) | 97 (d) | 98 (c) | 99 (d) | 100 (c) | 101 (c) | 102 (d) | 103 (d) |

# Hints & Explanations

**2** (*b*) When a plastic rod is rubbed with fur, an electric charge is developed in it. On touching this rod to pith balls, these charges are transferred from the rod to the pith ball in contact. As the charge on both pith balls is same, so they repel each other with same force.

Thus, figure given in option (b) shows final configuration.

**3** (*b*) When a soap bubble is given a negative charge, then the charges at diametrically opposite ends like at the end *AB* or *CD* as shown in figure alongside repel. Hence, its radius increases.

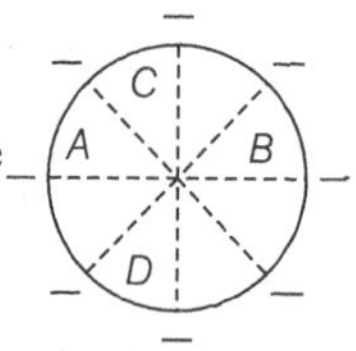


**4** (*d*) A body can be charged positively by losing some of its electrons. Similarly, a body can be charged negatively by gaining electrons. As, electrons possess mass, so mass of sphere *A* (being positively charged) decreases, while that of sphere *B* (being negatively charged) increases.

**5** (*c*) Due to friction, one body loses some electrons and other gains them. As, charges appearing on both will be equal and opposite, so ratio of $q_1 : q_2 = -1 : 1$.

**6** (*c*) During the motion of the carriers, its body gets charged because of rubbing with dry air and dust particle of the atmosphere. Due to this excess charge, the possibility of generation of spark increases and the inflammable material may catch fire.

In order to prevent this, metallic ropes are suspended from the carrier to keep its body in the contact with earth. So, that excess charge may flow from the carrier body to the ground (for earthing).

**7** (*c*) Induction requires shifting of free charge carrier which are present only in conductors.

Hence, in charging by induction, body to be charged must be a conductor

**8** (*c*) Initially, both the glass rod and silk cloth are electrically neutral, so net charge is zero.

Finally, after rubbing, the positive charge on glass rod is exactly equal to the negative charge on the silk cloth. So, net charge is again zero. This is consistent with law of conservation of charge.

**9** (*b*) According to the charge conservation, within the isolated system consisting of many charged bodies, due to the interaction among the bodies, charges may get redistributed but it is found that the total charge of the isolated system is always conserved.

So, if a sphere having charge *q* is put in contact with an identical uncharged sphere. Then, the charge *q* would get redistributed such that overall charge on the system remains conserved.

Also, by symmetry we can say that, charge on each sphere, then would be $q/2$.

**10** (*d*) 1 coulomb of charge is made of $n = \frac{q}{e}$

$$= \frac{1}{1.6 \times 10^{-19}} \qquad [\because e = 1.6 \times 10^{-19}\ \text{C}]$$

$$= 6.25 \times 10^{18} \text{ electrons}$$

$$\therefore \text{Time required} = \frac{6.25 \times 10^{18}}{10^9}\ \text{s} = 6.25 \times 10^9\ \text{s} \approx 198 \text{ yr}$$

$$(\because 1\text{yr} = 3.17 \times 10^7\ \text{s})$$

**11** (*b*) Charge on single electron, $e = 1.6 \times 10^{-19}$ C

Total charge, $q = +\ 2.4$ C

Then, by quantisation of charge, $q = ne$

$\therefore$Number of electrons, to be removed from silver dollar, $n = \frac{q}{e} = \frac{2.4\ \text{C}}{1.6 \times 10^{-19}\ \text{C}} = 1.5 \times 10^{19}$

**12** (*b*) Here, number of electrons removed

= number of atoms in 1 g

or $n = \frac{4 \times 10^{20}}{10^3} = 4 \times 10^{17}$

$\therefore$ Charge, $q = ne = 4 \times 10^{17} \times 1.6 \times 10^{-19}$
$= 6.4 \times 10^{-2}$ C

**14** (*b*) When two charges (says $q_1$ and $q_2$) of different magnitudes are placed at a distance $d$, they experience a force of attraction $F$. If they are placed in contact with each other and then separated, charges on them will be redistributed and net charge on them will be the difference of two, i.e. $q_{net} = |q_1 - q_2|$.
This net charge have same polarity, so after placing them at a distance $d$, they experience a force of repulsion.

**15** (*c*) According to Coulomb's law, force between two point charges, varies with distance as $F \propto \frac{1}{r^2}$.
Therefore, the graph between $F$ and $r$ will be as shown in Fig.(c).

**16** (*b*) Given, point charges, $Q_1 = 2 \times 10^{-7}$ C,
$Q_2 = 3 \times 10^{-7}$ C and $r = 30\text{ cm} = 30 \times 10^{-2}$ m

$\because$ Electrostatic force, $F = k\frac{Q_1Q_2}{r^2}$

where, $k = 9 \times 10^9\ \text{Nm}^2\text{C}^{-2}$.

$$\Rightarrow \quad F = 9 \times 10^9 \times \frac{2 \times 10^{-7} \times 3 \times 10^{-7}}{(30 \times 10^{-2})^2} = 6 \times 10^{-3}\ \text{N}$$

**17** (*a*) Force in air, $F = \frac{1}{4\pi\varepsilon_0} \cdot \frac{q_1 q_2}{r^2}$

where, $\varepsilon_0$ is the permittivity of free space and force in medium, $F_m = \frac{1}{4\pi\varepsilon_0\varepsilon_r} \cdot \frac{q_1 q_2}{r^2}$

where, $\varepsilon_r$ is relative permittivity.

$$F_{\text{medium}} = \frac{\varepsilon_0}{\varepsilon} \cdot \frac{1}{4\pi\varepsilon_0} \cdot \frac{q_1 q_2}{r^2} \qquad \left(\because \frac{\varepsilon}{\varepsilon_0} = \varepsilon_r\right)$$

$$\Rightarrow \quad F_{\text{medium}} = \frac{\varepsilon_0}{\varepsilon} F$$

**18** (*b*) Relative permittivity $\varepsilon_r$ is also called dielectric constant $K$.
The force between charges in air is

$$F_{\text{air}} = \frac{1}{4\pi\varepsilon_0} \cdot \frac{q_1 q_2}{r^2} \qquad \ldots\text{(i)}$$

The force between charges in a medium of dielectric constant $K$,

$$F_{\text{medium}} = \frac{1}{4\pi\varepsilon_0 K} \cdot \frac{q_1 q_2}{r^2} \qquad \ldots\text{(ii)}$$

On dividing Eq. (i) by Eq. (ii), we get

$$\Rightarrow \quad \frac{F_{\text{air}}}{F_{\text{medium}}} = \frac{\frac{1}{4\pi\varepsilon_0} \cdot \frac{q_1 q_2}{r^2}}{\frac{1}{4\pi\varepsilon_0 K} \cdot \frac{q_1 q_2}{r^2}} = \frac{K}{1}$$

So, the ratio of the forces between two small spheres is $K : 1$.

**19** (*c*) Force exerted between two point charges in air

$$F = \frac{q_1 q_2}{4\pi\varepsilon_0 r^2}$$

So, force exerted between two point charges in a medium of dielectric constant $K$ is $F' = \frac{q_1 q_2}{4\pi\varepsilon_0 K (r')^2}$

where, $r'$ is the distance.
Given, $F = F'$

$$\Rightarrow \quad \frac{q_1 q_2}{4\pi\varepsilon_0 r^2} = \frac{q_1 q_2}{4\pi\varepsilon_0 K (r')^2} \Rightarrow \frac{1}{r^2} = \frac{1}{K(r')^2}$$

$$\Rightarrow \quad r^2 = K(r')^2 \Rightarrow r = \sqrt{K}\, r'$$

$$\therefore \quad r' = \frac{r}{\sqrt{K}}$$

**20** (*b*) The electric force between an electron and proton at a distance $r$ is

$$F_e = \frac{-ke^2}{r^2} \qquad \ldots\text{(i)}$$

where, constant, $k = 9 \times 10^9\ \text{Nm}^2\ \text{C}^{-2}$
and charge of electron, $e = 1.6 \times 10^{-19}$ C.
Here, the negative sign indicates that the force is attractive.
The corresponding gravitational force between electron and proton is

$$F_G = -G\frac{m_p m_e}{r^2} \qquad \ldots\text{(ii)}$$

where, $m_p$ & $m_e$ are the masses of the proton & electron and $G$ is gravitational constant.
From Eqs. (i) and (ii), we get

$$\left|\frac{F_e}{F_G}\right| = \frac{k e^2}{G m_p m_e} = \frac{(9 \times 10^9)(1.6 \times 10^{-19})^2}{(6.67 \times 10^{-11})(9.1 \times 10^{-31})(1.67 \times 10^{-27})}$$
$$= 2.3 \times 10^{39}$$

$$\Rightarrow \quad F_G / F_e \approx 10^{-39}$$

So, the ratio of $F_G / F_e$ is of the order of $10^{-39}$.

**21** (*c*) Net charge on one H-atom,

$$q = -e + (e + \Delta e) = \Delta e$$

Net electrostatic repulsive force between two H-atoms,

$$F_r = \frac{kq^2}{d^2} = \frac{k(\Delta e)^2}{d^2}$$

where, $q$ = charge on H-atom = $\Delta e$ and $d$ = distance between H-atom.
Similarly, net gravitational attractive force between two H-atoms,

$$F_G = \frac{Gm_h^2}{d^2}$$

where, $m_h$ = mass of H-atom and $G$ = gravitational constant.
It is given that, $F_r - F_G = 0$

$$\Rightarrow \quad \frac{k(\Delta e)^2}{d^2} - \frac{Gm_h^2}{d^2} = 0$$

$$\Rightarrow \quad (\Delta e)^2 = \frac{Gm_h^2}{k}$$

$$(\Delta e)^2 = \frac{(6.67\times 10^{-11})\,(1.67\times 10^{-27})^2}{9\times 10^9}$$

$[\because G = 6.67\times 10^{-11}\,\text{m}^3\text{kg}^{-1}\text{s}^{-2},\ m_h = 1.67\times 10^{-27}\ \text{kg}$
and $k = 9\times 10^9\ \text{Nm}^2\text{C}^{-2}.]$

$\Rightarrow \qquad \Delta e = 1.437\times 10^{-37}$ C

$\therefore$ So, the order of $\Delta e$ is $10^{-37}$C.

**22** (*a*) Two small spheres of charge $+Q$ are suspended by threads of length $L$ as shown below

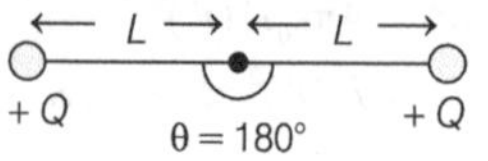


When this arrangement is in space, i.e. there is no gravitational force, so angle between two suspensions will be 180° due to repulsive force.

$\therefore$ Tension in each thread,

i.e. $F = \dfrac{kQQ}{(2L)^2} = \dfrac{1}{4\pi\varepsilon_0}\dfrac{Q^2}{(2L)^2}$ $\qquad \left[\because k = \dfrac{1}{4\pi\varepsilon_0}\right]$

**23** (*a*) The force between two point charges $A$ and $B$ having charge $+Q$ and $-Q$ respectively, placed at a certain distance $r$ as shown in figure is given by

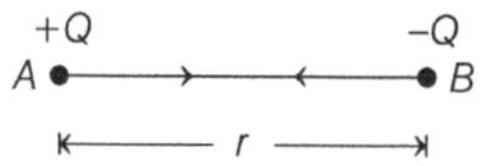


$$F = \frac{kQ_AQ_B}{r^2} = \frac{kQ(-Q)}{r^2} = -\frac{kQ^2}{r^2} \qquad \ldots\text{(i)}$$

where, $k = \text{constant} = \dfrac{1}{4\pi\varepsilon_0}$

When 25% charge of $A$ is transferred to $B$, then new amount of charge on $A$ and $B$ respectively becomes

$$Q'_A = \frac{75}{100}(Q_A) = \frac{3}{4}Q$$

$$Q'_B = \left(\frac{25}{100}Q_A + Q_B\right) = \left(\frac{1}{4}Q - Q\right) = \frac{-3}{4}Q$$

So, the force between the two charges now becomes,

$$F' = \frac{kQ'_AQ'_B}{r^2} = \frac{k\left(\frac{3}{4}Q\right)\left(-\frac{3}{4}Q\right)}{r^2}$$

$$= \frac{-9kQ^2}{16r^2} = \frac{9}{16}F \qquad \text{[from Eq. (i)]}$$

Thus, the new force between the charges is (9/16) times the initial force between the charges.

**24** (*a*) When $A$ and $C$ are touched, charge is distributed equally on them as shown below

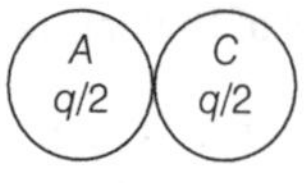


i.e. $\qquad q_A = q_C = \dfrac{q}{2}$

When $C$ is touched with $B$, total charge of $B$ and $C$

$$= q + q/2 = \frac{3q}{2}$$

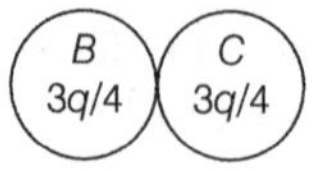


When $B$ and $C$ are separated, then charge is equally shared between them, i.e. $q_B = q_C = \dfrac{3q}{4}$

So, finally we have following situation

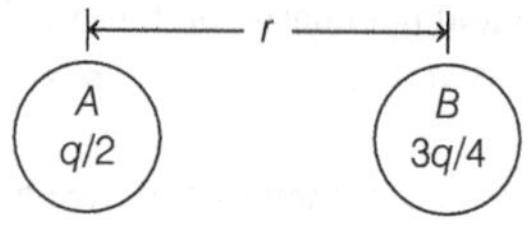


Given, $q_A = q_B = 6.5\times 10^{-7}$C and $r = 50\times 10^{-2}$ m

$\therefore$ New force of repulsion between $A$ and $B$ is

$$F = \frac{9\times 10^9 \times q_Aq_B}{r^2}$$

$$= \frac{9\times 10^9 \times \frac{1}{2}\times\frac{3}{4}\times (6.5\times 10^{-7})^2}{(0.5)^2} = 5.7\times 10^{-3}\ \text{N}$$

**25** (*d*) As, coulomb's attraction = centripetal force

$\therefore \quad \dfrac{1}{4\pi\varepsilon_0}\cdot\dfrac{q_1q_2}{r^2} = mr\omega^2 = \dfrac{4\pi^2mr}{T^2}$ $\qquad \left[\because \omega = \dfrac{2\pi}{T}\right]$

$\Rightarrow \quad T^2 = \dfrac{4\pi^2mr^3}{kq_1q_2}$ $\qquad \left[\because k = \dfrac{1}{4\pi\varepsilon_0}\right]$

$\Rightarrow \quad T = \left(\dfrac{4\pi^2mr^3}{kq_1q_2}\right)^{1/2}$

**26** (*a*) The given situation is as shown below

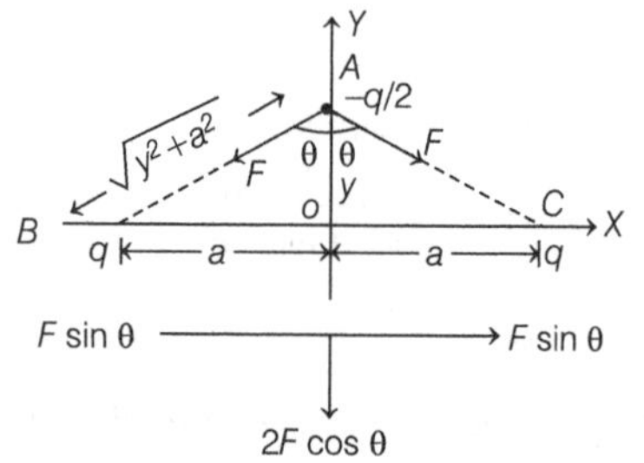


Net force acting on $\dfrac{-q}{2}$ due to two charges $q$ will be in negative $y$-direction, $F_{\text{net}} = 2F\cos\theta$. As forces acting along the $X$-axis are equal in magnitude and opposite in direction. Thus, they cancel each other.

So, $\qquad F_{\text{net}} = \dfrac{2kq_1q_2\cos\theta}{r^2}$

Here, $\qquad q_1 = q,\ q_2 = \left(\dfrac{-q}{2}\right),\ r = \sqrt{y^2 + a^2}$

From $\Delta AOB$, $\cos\theta = \dfrac{AO}{AB} = \dfrac{y}{\sqrt{y^2 + a^2}}$

So, $\qquad F_{\text{net}} = \dfrac{2kq\left(\dfrac{-q}{2}\right)}{(\sqrt{y^2 + a^2})^2}\cdot\dfrac{y}{\sqrt{y^2 + a^2}}$

$$F_{\text{net}} = \frac{-2kq\left(\frac{q}{2}\right)y}{(y^2+a^2)^{3/2}}$$

As $(y << a)$, so $y^2$ can be neglected, w.r.t. $a^2$.

$$\Rightarrow \quad F_{\text{net}} = -\frac{kq^2 y}{a^3} \Rightarrow F_{\text{net}} \propto y \qquad (\because y << a)$$

where, negative sign implies that the force acts in negative $y$-direction.

**27** (*b*) According to question, two identical charged spheres suspended from a common point by two massless strings of length $l$ as shown below

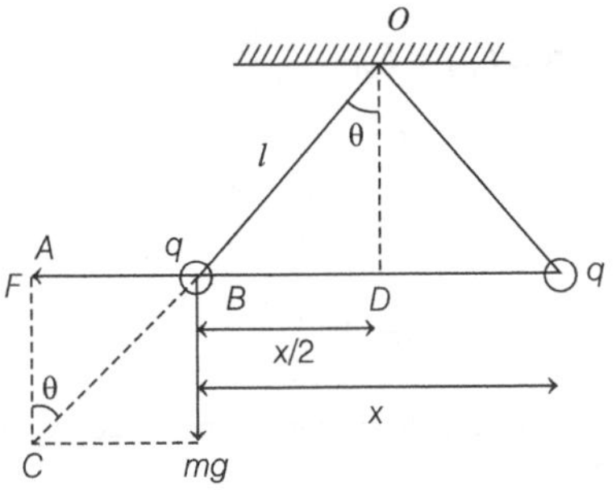


In $\Delta ABC$, $\quad \tan\theta = \dfrac{AB}{AC} = \dfrac{F}{mg}$ ...(i)

In $\Delta BOD$, $\quad \tan\theta = \dfrac{BD}{OD} = \dfrac{x/2}{\sqrt{l^2 - \frac{x^2}{4}}}$ ... (ii)

From Eqs. (i) and (ii), we get

$$\frac{F}{mg} = \frac{x/2}{\sqrt{l^2 - \frac{x^2}{4}}}$$

or $\quad \dfrac{kq^2}{x^2 mg} = \dfrac{x/2}{\sqrt{l^2 - \frac{x^2}{4}}}$ $\quad \left[\because \text{here, force } F = \dfrac{kq^2}{x^2}\right]$

As $(x << l)$, so $\dfrac{x^2}{4}$ can be neglected in $\left(l^2 - \dfrac{x^2}{4}\right)$,

So, $\quad \dfrac{kq^2}{x^2 mg} = \dfrac{x}{2l} \Rightarrow q^2 \propto x^3$

or $\quad q \propto x^{3/2}$ ... (iii)

Since, the charges begin to leak from both the spheres at a constant rate, so they approaches each other with a velocity $v$, then from Eq. (iii), we get

$$\Rightarrow \quad \frac{dq}{dt} \propto \frac{d(x^{3/2})}{dx} \cdot \frac{dx}{dt}$$

$$\Rightarrow \quad \frac{dq}{dt} \propto x^{1/2} v$$

$$\Rightarrow \quad v \propto \frac{1}{x^{1/2}} \quad \left(\because \frac{dq}{dt} = \text{constant}\right)$$

or $\quad v \propto x^{-1/2}$

**28** (*c*) Three charges $q_1$, $q_2$ and $q_3$ each of 1 C are at the vertices of an equilateral triangle of side $l$ as shown below

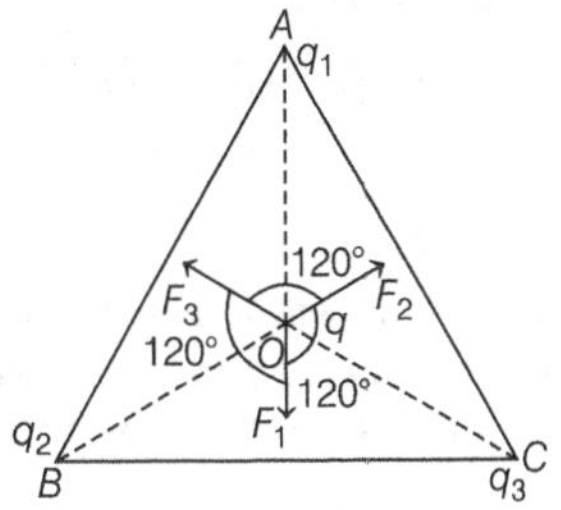


According to figure, force on charge $q$ which is placed at the centroid of the triangle due to charge $q_1$,

$$F_1 = \frac{1}{4\pi\varepsilon_0} \cdot \frac{q_1 q}{AO^2}$$

Similarly, $F_2 = \dfrac{1}{4\pi\varepsilon_0} \cdot \dfrac{q_2 q}{BO^2}$ and $F_3 = \dfrac{1}{4\pi\varepsilon_0} \cdot \dfrac{q_3 q}{CO^2}$

Let, $\quad AO = BO = CO = r$ (say)

and $\quad q_1 = q_2 = q_3 = 1\,\text{C}$ (given)

$\therefore \quad F_1 = F_2 = F_3$

So, $\quad$ force $= \dfrac{1}{4\pi\varepsilon_0} \cdot \dfrac{q}{r^2} = F$ (say)

Since, the angle between any two forces is 120°.

$\therefore$ Taking resultant force of any of the two forces says $F_1$ and $F_2$, we get

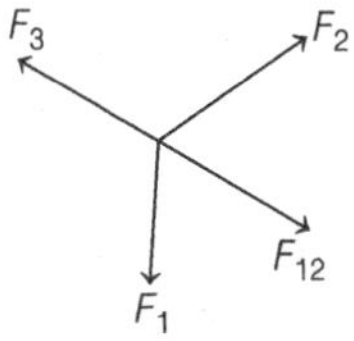


$$F_{12} = \sqrt{F_1^2 + F_2^2 + 2F_1F_2\cos 120^\circ}$$

$$= \sqrt{F^2 + F^2 + 2F^2\left(-\frac{1}{2}\right)}$$

$$F_{12} = F$$

Since, forces $F_{12}$ and $F_3$ are equal in magnitude but opposite in direction, hence they cancel each other. Therefore, net force on $q$ will be zero.

**29** (*b*) Given situation is shown in following figure. $ABC$ is an equilateral triangle of side $l$, i.e.

$$AB = BC = CA = l$$

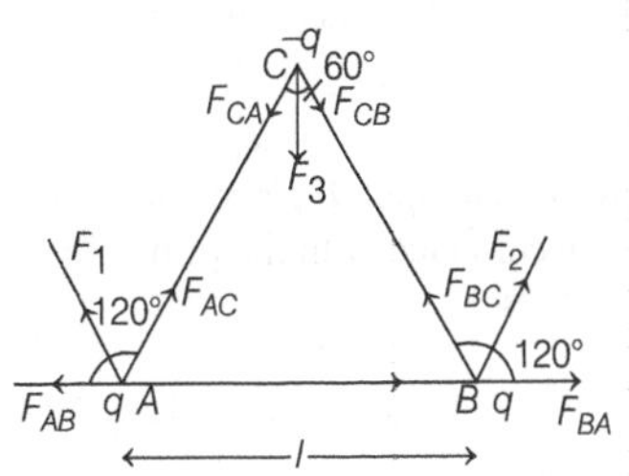


The magnitude of forces $F_{AB}$, $F_{BC}$ and $F_{AC}$ are equal,

i.e. $F_{AB} = F_{BC} = F_{AC} = k \cdot \dfrac{q^2}{l^2} = F$ (say)

$\therefore$ Force on charge $q$ at $A$,

$$F_1 = \sqrt{F_{AB}^2 + F_{AC}^2 + 2F_{AB}F_{AC}\cos 120^\circ}$$

$$= \sqrt{F^2 + F^2 + 2F^2(-1/2)}$$

$$F_1 = F \text{ (directed along } BC)$$

Similarly, force on charge $q$ at $B$,

$$F_2 = F \text{ (directed along } AC)$$

Force on charge $-q$ at $C$,

$$F_3 = \sqrt{F_{CA}^2 + F_{CB}^2 + 2F_{CA}\,F_{CB}\cdot\cos 60^\circ}$$

$$= \sqrt{F^2 + F^2 + 2F^2(1/2)}$$

$$F_3 = F\sqrt{3}$$

(directed vertically downward as shown in the figure)
Resolving the forces $F_1$, $F_2$ and $F_3$ in two perpendicular directions ($X$ and $Y$-axes) as shown in the following figure

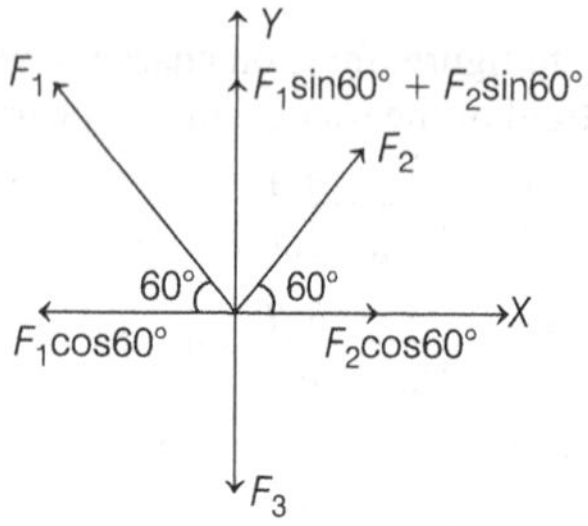


Since, $F_1$ and $F_2$ are equal in magnitude, hence $F_1 \cos 60^\circ$ and $F_2 \cos 60^\circ$ both cancel each other.

$$\therefore\ |\mathbf{F}_1 + \mathbf{F}_2 + \mathbf{F}_3| = F_3 - (F_1 \sin 60^\circ + F_2 \sin 60^\circ)$$

$$= F_3 - [F \sin 60^\circ + F \sin 60^\circ] \qquad [\because F_1 = F_2 = F]$$

$$= F_3 - 2F \sin 60^\circ$$

$$= F_3 - 2F\cdot\frac{\sqrt{3}}{2} = F_3 - F\sqrt{3}$$

$$= F\sqrt{3} - F\sqrt{3} \qquad [\because F_3 = F\sqrt{3}]$$

$$= 0$$

**30** (*d*) The given condition is shown in the figure given below,

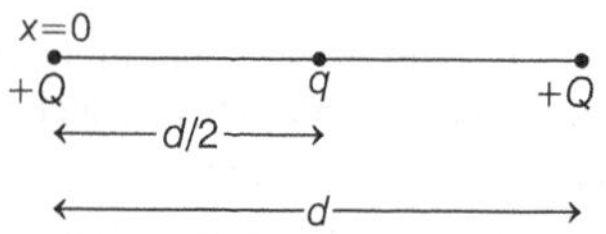


Then, according to the Coulomb's law, the electrostatic force between two charges $q_1$ and $q_2$ such that the distance between them is $(r)$ given as,

$$F = \frac{1\cdot q_1 q_2}{4\pi\varepsilon_0\cdot r^2}$$

$\therefore$Net force on charge '$Q$' placed at origin i.e. at $x = 0$ in accordance with the principle of superposition can be given as

$$F_{\text{net}} = \frac{1}{4\pi\varepsilon_0}\cdot\frac{Q\times q}{\left(\frac{d}{2}\right)^2} + \frac{1}{4\pi\varepsilon_0}\cdot\frac{Q\times Q}{(d)^2}$$

Since, it has been given that, $F_{\text{net}} = 0$.

$$\Rightarrow \frac{1}{4\pi\varepsilon_0}\cdot\frac{Q\times q}{\left(\frac{d}{2}\right)^2} + \frac{1}{4\pi\varepsilon_0}\cdot\frac{Q\times Q}{(d)^2} = 0$$

$$\Rightarrow \frac{1}{4\pi\varepsilon_0}\cdot\frac{Q\times q}{\left(\frac{d}{2}\right)^2} = -\frac{1}{4\pi\varepsilon_0}\cdot\frac{Q\times Q}{(d)^2} \text{ or } q = -\frac{Q}{4}$$

**31** (*b*) The given condition can be seen in the figure below

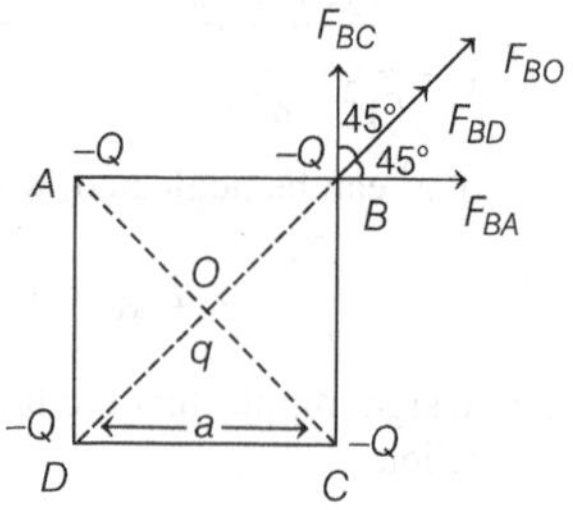


Let $AB = BC = CD = DA = a$

So, from the figure, $BD = \sqrt{2}a$ and $OB = \dfrac{a}{\sqrt{2}}$

Now, force, $F_{BA} = F_{BC} = \dfrac{Q^2}{4\pi\varepsilon_0 a^2}$ ...(i)

Force, $F_{BD} = \dfrac{Q^2}{4\pi\varepsilon_0(\sqrt{2}a)^2} = \dfrac{Q^2}{4\pi\varepsilon_0(2a^2)}$ ...(ii)

and force, $F_{BO} = \dfrac{-qQ}{4\pi\varepsilon_0\left(\frac{a}{\sqrt{2}}\right)^2} = \dfrac{-2qQ}{4\pi\varepsilon_0 a^2}$ ...(iii)

As system is in equilibrium, so no force experienced by any charge.
Therefore, the net force acting on $-Q$ at point $B$ will be zero

i.e. $F_{BA}\cos 45^\circ + F_{BC}\cos 45^\circ + F_{BD} + F_{BO} = 0$

$$\frac{Q^2}{4\pi\varepsilon_0 a^2}\times\frac{1}{\sqrt{2}} + \frac{Q^2}{4\pi\varepsilon_0 a^2}\times\frac{1}{\sqrt{2}} + \frac{Q^2}{4\pi\varepsilon_0(2a^2)} - \frac{2qQ}{4\pi\varepsilon_0 a^2} = 0$$

On solving, we get, $q = \dfrac{Q}{4}(1 + 2\sqrt{2})$

**32** (*b*) The electric field due to a charge $Q$ at a point in space may be defined as the force that a unit positive charge would experience if placed at that point.
Thus, electric field due to the charge $Q$ will be continuous, if there is no charge at that point.
It will be discontinuous if there is a charge at that point.

**34** (*c*) Given, electric field, $E = Aa$ ...(i)

A charge $q$ contained in a sphere of radius $a$ centred at the origin as shown below

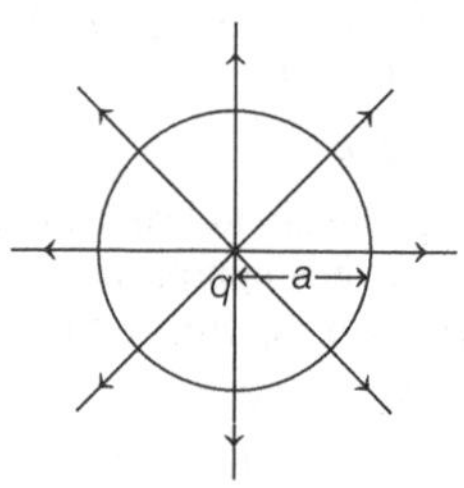


Here, radius , $r = a$

$\therefore$Electric field, $E = \dfrac{1}{4\pi\varepsilon_0}\cdot\dfrac{q}{a^2}$ ...(ii)

From Eqs. (i) and (ii), we get

$$\frac{1}{4\pi\varepsilon_0}\cdot\frac{q}{a^2} = Aa \Rightarrow q = 4\pi\varepsilon_0 A a^3$$

**35** (*d*) Forces on the bob are as shown

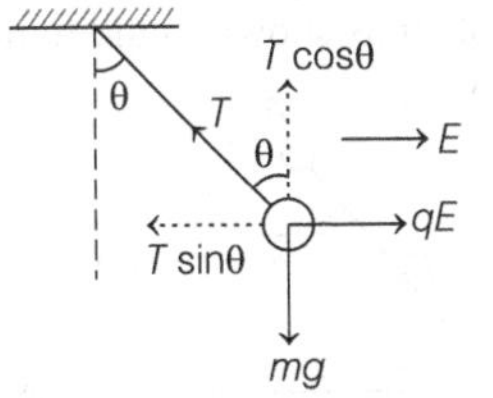


For equilibrium, $T\cos\theta = mg$ ...(i)

and $T\sin\theta = qE$ ...(ii)

Dividing Eq. (ii) by Eq. (i), we get

$$\tan\theta = \frac{qE}{mg}$$

Here, $q = 5\mu C = 5\times 10^{-6}$ C, $E = 2000$ V / m,

$$m = 2g = 2\times 10^{-3} \text{ kg}, g = 10 \text{ ms}^{-2}$$

$$\therefore \quad \tan\theta = \frac{5\times 10^{-6}\times 2000}{2\times 10^{-3}\times 10} = \frac{1}{2} = 0.5$$

So, the angle made by the string of the pendulum with the vertical is

$$\theta = \tan^{-1}(0.5)$$

**36** (*c*) Force on a charged particle in the presence of an electric field is given as

$$F = qE \quad ...(i)$$

where, $q$ is the charge on the particle and $E$ is the electric field.

From Newton's second law of motion, force on a particle with mass $m$ is given as

$$F = ma \quad ...(ii)$$

where, $a$ is the acceleration.

From Eqs. (i) and (ii), we get

$$F = ma = qE \Rightarrow a = \frac{qE}{m} \quad ...(iii)$$

Now, consider that a particle falls from rest through a vertical distance $h$. Therefore, $u = 0$ and the second equation of motion becomes

$$s = ut + \frac{1}{2}at^2 \text{ or } h = 0\times t + \frac{1}{2}at^2$$

$$\Rightarrow \quad h = \frac{1}{2}\times\frac{qE}{m}t^2 \quad \text{[from Eq. (iii)]}$$

$$\Rightarrow \quad t^2 = \frac{2hm}{qE} \text{ or } t = \sqrt{\frac{2hm}{qE}}$$

Since, the particles given in the question is electron and proton. The quantity $\sqrt{\frac{2h}{qE}}$ (here, $q_p = q_e = e$) for both of them is constant. Thus, we can write

$$t = k\sqrt{m}$$

where, $k = \sqrt{\frac{2h}{qE}}$ or $t \propto \sqrt{m}$

As, mass of proton ($m_p$) >> mass of electron ($m_e$), so the time of fall of an electron would be smaller than the time of fall of a proton.

**37** (*c*) For electron, acceleration $= \frac{eE}{m_e}$

$$\left[\because a = \frac{qE}{m}, \text{ here } q = e \text{ and } m = m_e\right]$$

We know that, $s = ut + \frac{1}{2}at^2$

$$s = 0(t_1) + \frac{1}{2}\frac{eEt_1^2}{m_e} \quad [\because u = 0]$$

$$s = \frac{eE}{2m_e}\times t_1^2 \quad ...(i)$$

Similarly for proton, $s = 0(t_2) + \frac{1}{2}\frac{eEt_2^2}{m_p}$

$$s = \frac{eE}{2m_p}\times t_2^2 \quad ...(ii)$$

Dividing Eq. (ii) by Eq. (i), we get

$$\therefore \quad \frac{t_2^2}{t_1^2} = \frac{m_p}{m_e} \Rightarrow \frac{t_2}{t_1} = \sqrt{\frac{m_p}{m_e}} = \left(\frac{m_p}{m_e}\right)^{1/2}$$

**38** (*c*) Given, point charges, $q_1 = +10^{-8}$ C,

$$q_2 = -10^{-8} \text{ C}$$

and distance, $r = 0.1$ m

The electric field vector $\mathbf{E}_1$ at $A$ due to the positive charge $q_1$ points towards the right and has a magnitude

$$E_1 = \frac{kq_1}{r^2} = \frac{(9\times 10^9 \text{ Nm}^2\text{ C}^{-2})\times(10^{-8}\text{C})}{(0.05 \text{ m})^2}$$

$$= 3.6\times 10^4 \text{NC}^{-1}$$

The electric field vector $\mathbf{E}_2$ at $A$ due to the negative charge $q_2$ points towards the right and has a magnitude

$$E_2 = \frac{kq_2}{r^2} = \frac{(9\times 10^9 \text{ Nm}^{-2}\text{C}^{-2})\times(10^{-8}\text{C})}{(0.05 \text{ m})^2}$$

$$= 3.6\times 10^4 \text{NC}^{-1}$$

The magnitude of the total electric field at $A$ is

$$E_A = E_1 + E_2 = 7.2\times 10^4 \text{ NC}^{-1} \quad ...(ii)$$

This $\mathbf{E}_A$ is directed towards the right.

Now, consider the following given figure

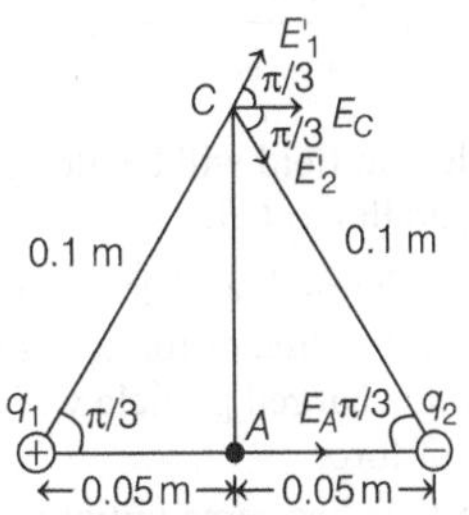


The magnitude of each electric field vector at point $C$, due to charge $q_1$ and $q_2$ is

$$E_1' = E_2' = \frac{(9\times 10^9 \text{ Nm}^2\text{ C}^{-2})\times(10^{-8}\text{C})}{(0.1 \text{ m})^2}$$

$$= 9\times 10^3 \text{ NC}^{-1}$$

The resultant of these two vectors is

$$E_C = E_1' \cos\frac{\pi}{3} + E_2' \cos\frac{\pi}{3}$$

[$\because$ the given figure is equilateral triangle, so angle will be $\frac{\pi}{3}$]

$$= 9\times 10^3 \text{ NC}^{-1} \qquad \ldots\text{(ii)}$$

This $\mathbf{E}_C$ points towards the right

Dividing Eq (i) by Eq. (ii), we get

$$\therefore \qquad \frac{E_A}{E_C} = \frac{7.2\times 10^4}{9\times 10^3} = \frac{8}{1}$$

**39** (*b*) The given condition with the respective directions of electric field at point $M$ due to the given charges are as shown below

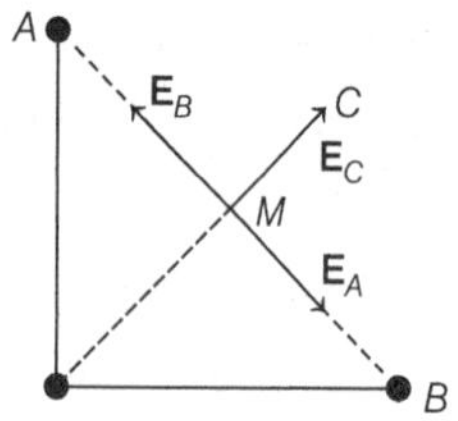


where,

$E_A$ = electric field at $M$ due to charge placed at $A$,

$E_B$ = electric field at $M$ due to charge placed at $B$

and $E_C$ = electric field at $M$ due to charge placed at $C$.

As seen from figure $|\mathbf{E}_B| = |\mathbf{E}_A|$, so net electric field at $M, E_{\text{net}} = |\mathbf{E}_C|$, in the direction of vector 2.

**40** (*b*) Charges at $A$ and $C$ produce net electric field along $OA$ (as $q_C > q_A$). Similarly, charges at $B$ and $D$ produce net electric field along $OB$ (as $q_D > q_B$).

The direction of the fields is as shown in the figure below

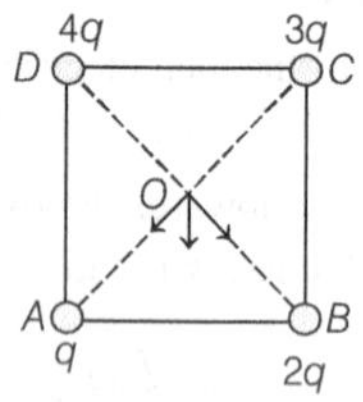


So, the net electric field will be along bisector of angle $AOB$ which parallel to $CB$.

**42** (*b*) Force on a charged particle (positive or negative) acts along a electric line of force, if its initial velocity is zero. Therefore, charged particle will travel along electric lines of force.

If charged particle has some initial velocity at any angle from the direction of electric field, then charged particle moves on parabolic path. Hence, charged particle will not move along electric line of force.

**44** (*c*) Areas of $P$ and $Q$ are equal but more lines pass through area at $P$. So, field is stronger at $P$ as compared to $Q$.

**46** (*c*) Given system is equivalent to two dipoles inclined at 60° to each other as shown in the figure below

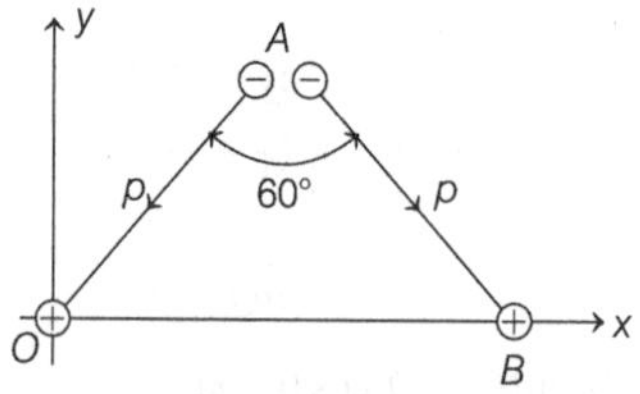


Now, magnitude of resultant of these dipole moments is

$$p_{\text{net}} = \sqrt{p^2 + p^2 + 2p\cdot p\cos 60^\circ} = \sqrt{3}p = \sqrt{3}ql$$

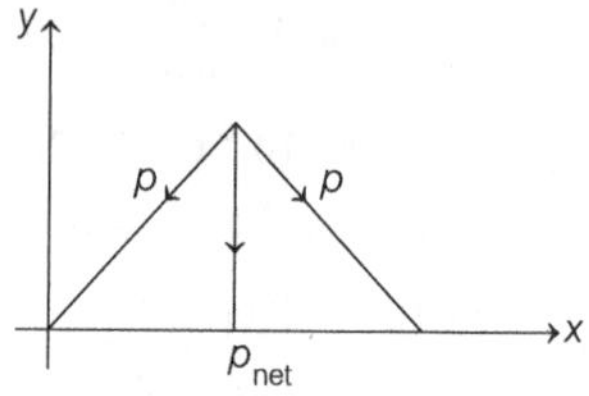


As, resultant is directed along negative $y$-direction

$$p_{\text{net}} = -\sqrt{3}p\hat{\mathbf{j}} = -\sqrt{3}ql\hat{\mathbf{j}}$$

**47** (*a*) Suppose neutral point $N$ lies at a distance $x$ from dipole of moment $\mathbf{p}$ or at a distance $(25-x)$ from dipole of moment 64 $\mathbf{p}$.

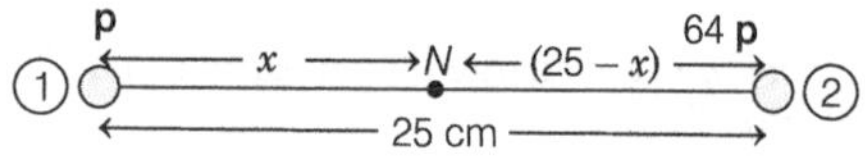


$\because$ Electric field due to dipole is $E = \dfrac{2p}{4\pi\varepsilon_0 r^3}$.

where, $p$ is dipole moment and $r$ is distance.

$\therefore$ At $N$, | Electric field due to dipole (1) |

= |Electric field due to dipole (2)|

$$\Rightarrow \quad \frac{1}{4\pi\varepsilon_0}\cdot\frac{2p}{x^3} = \frac{1}{4\pi\varepsilon_0}\cdot\frac{2\,(64\,p)}{(25-x)^3}$$

$$\Rightarrow \quad \frac{1}{x^3} = \frac{64}{(25-x)^3} \Rightarrow x = 5\text{ cm}$$

**48** (*c*) The direction of electric dipole moment and electric field strength on equatorial lines are shown as

From figure, $\mathbf{E}_{\text{net}}$ and $\mathbf{p}$ are in opposite directions, so angle between them is 180°.

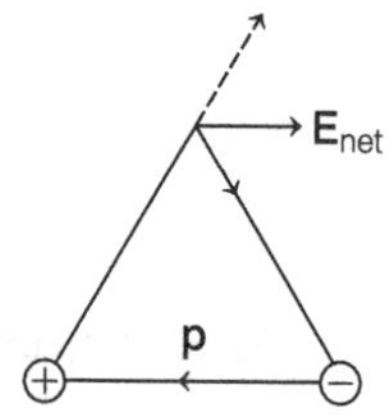


**49** (*c*) When there are various types of charges in a region, but the total charge is zero, the region can be supposed to contain a number of electric dipoles. Therefore, at points outside the region (may be anywhere w.r.t. electric dipoles), the dominant electric field $\propto \frac{1}{r^3}$

(as electric field due to dipoles is, $E = \frac{k2p}{r^3}$) for large $r$.

**50** (*b*) As we know,

$$\frac{E_{\text{axis}}}{E_{\text{equator}}} = \frac{2p/4\pi\varepsilon_0 r^3}{p/4\pi\varepsilon_0 r^3} = \frac{2}{1}$$

The ratio of electric fields on the axis and at equator is 2 : 1.

**51** (*c*) Given, $a = 2.5\,\text{mm} = 2.5\times10^{-3}$ m, $r = 15$ cm $= 15\times10^{-2}$ m and $q = 10\mu\text{C} = 10\times10^{-6}$ C

As, $r >> a$

∴ Electric field at a point $P$ on dipole axis is

$$\therefore \quad E_P = \frac{2p}{4\pi\varepsilon_0 r^3} = \frac{2(2a\times q)}{4\pi\varepsilon_0 r^3} \quad [\because p = q\times 2a]$$

$$= \frac{2\,(5\times10^{-3}\times10\times10^{-6})\times(9\times10^{9})}{(15\times10^{-2})^3}$$

$$= 2.6\times10^{5}\ \text{NC}^{-1}$$

Now, electric field at a point $Q$ on equatorial plane is

$$E_Q = \frac{p}{4\pi\varepsilon_0 r^3} = \frac{1}{2}E_P = \frac{1}{2}\times 2.6\times10^{5}$$

$$= 1.3\times10^{5}\ \text{NC}^{-1}$$

**52** (*d*) The given distribution of charges can be shown as the figure below

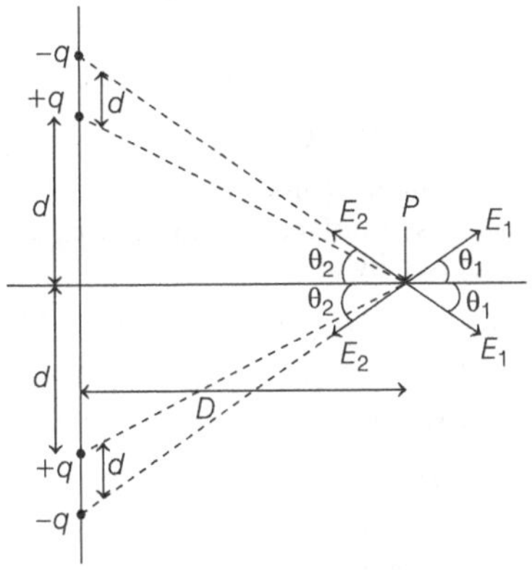


Electric field at point $P$,

$$E = E_1\cos\theta_1 + E_1\cos\theta_1 - E_2\cos\theta_2 - E_2\cos\theta_2$$

$$= 2E_1\cos\theta_1 - 2E_2\cos\theta_2$$

$$= \frac{2kq}{(d^2 + D^2)}\cos\theta_1 - \frac{2kq}{(2d)^2 + D^2}\cos\theta_2$$

As, $\cos\theta_1 = \dfrac{D}{(d^2 + D^2)^{1/2}}$

Similarly,

$$\cos\theta_2 = \frac{D}{[(2d)^2 + D^2]^{1/2}}$$

$$= 2kqD[(d^2 + D^2)^{-3/2} - (4d^2 + D^2)^{-3/2}]$$

$$= \frac{2kqD}{D^3}\left[\left(1 + \frac{d^2}{D^2}\right)^{-3/2} - \left(1 + \frac{4d^2}{D^2}\right)^{-3/2}\right]$$

As $D >> d$, then by applying binomial approximation, we get

$$= \frac{2kq}{D^2}\left[1 - \frac{3}{2}\frac{d^2}{D^2} - \left(1 - \frac{3}{2}\left(\frac{4d^2}{D^2}\right)\right)\right]$$

$$= \frac{2kq}{D^2}\left[\frac{9d^2}{2D^2}\right] = \frac{9kqd^2}{D^4}$$

$$\Rightarrow \quad E \propto \frac{1}{D^4}$$

**54** (*a*) If an electric dipole $p$ is placed in a unifrom electric field $E$, then the torque will be

$$\tau = \mathbf{p}\times\mathbf{E} = pE\sin\theta$$

The value of torque will be maximum, if $\theta = 90°$

i.e. $\tau_{\max} = pE$

**55** (*a*) Given, dipole moment of dipoles $AB$ & $CD = p$ and angle, $\theta = 120°$.

If $p_1$ and $p_2$ are the dipole moments of $AB$ and $CD$, then net dipole moment,

$$p_{\text{net}} = \sqrt{p_1^2 + p_2^2 + 2p_1 p_2\cos\theta}$$

$$p_{\text{net}} = \sqrt{p^2 + p^2 + 2p^2\cos 120°} \quad [\because p_1 = p_2 = p]$$

$$p_{\text{net}} = \sqrt{p^2 + p^2 - p^2} = p \quad \left[\because \cos 120° = -\frac{1}{2}\right]$$

When this system is subjected to electric field $E$ at an angle $\theta$,

$$\tau = p_{\text{net}}E\sin\theta$$

$\Rightarrow \tau = pE\sin\theta = \mathbf{p}\times\mathbf{E}$ and it is along positive $Z$-axis, because $\tau$ acts perpendicular to plane containing $\mathbf{p}$ and $\mathbf{E}$, i.e. perpendicular to $xy$-plane.

**56** (*a*) An electric dipole of dipole moment $p$ and moment of inertia $I$ is situated in an electric field of intensity $E$. When dipole is given a small angular displacement $\theta$ about its equilibrium position, then restoring torque will be $\tau = -pE\sin\theta = I\alpha$

where, $\alpha$ is angular acceleration,

For small $\theta$, $I\alpha = -pE\theta$ $\quad [\because \sin\theta = \theta]$

$$I\frac{d^2\theta}{dt^2} = -pE\theta \quad \left[\because \alpha = \frac{d^2\theta}{dt^2}\right]$$

or

$$\frac{d^2\theta}{dt^2} = \frac{-pE}{I}\theta$$

Also,

$$\frac{d^2\theta}{dt^2} = -\omega^2\theta \Rightarrow \omega^2 = \frac{pE}{I}$$

∴ Angular frequency, $\omega = \sqrt{\dfrac{pE}{I}}$

**58** (*c*) Given, radius of sphere, $r = \dfrac{d}{2} = \dfrac{2.4}{2} = 1.2$ m

Surface charge density, $\sigma = 80\ \mu\text{Cm}^{-2} = 80\times10^{-6}\ \text{Cm}^{-2}$

$$\because \text{Surface charge density} = \frac{\text{Charge}}{\text{Surface area}}$$

$$\sigma = \frac{q}{4\pi r^2}$$

$\therefore$ Charge, $q = \sigma \times 4\pi r^2 = 80 \times 10^{-6} \times 4 \times 3.14 \times 1.2 \times 1.2$

$$q = 1.4 \times 10^{-3}\,\text{C}$$

**59** (*c*) Two non-conducting solid spheres of radii $R_1$ & $R_2$ and carrying uniform volume charge densities $+\rho$ & $-\rho$, respectively are placed such that they partially overlap as shown in the figure

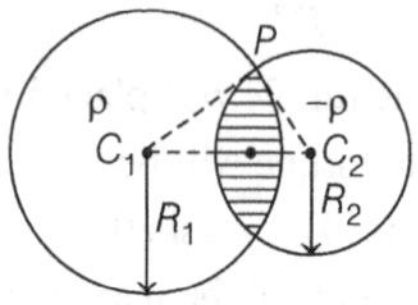


According to the question,
The electric field at $P$ due to charge $\rho\Delta V$ is

$$\mathbf{E}_1 = \frac{1}{4\pi\varepsilon_0}\frac{\rho\Delta V}{(C_1P)^2} \qquad \left[\because \rho = \frac{\Delta Q}{\Delta V}\right]$$

$$= \frac{1}{4\pi\varepsilon_0}\frac{\rho\frac{4}{3}\pi(C_1P)^3}{(C_1P)^2} = \frac{\rho}{3\varepsilon_0}C_1P, \text{ along } C_1P$$

Similarly, electric field at $P$ due to charge $-\rho dV$ is

$$\mathbf{E}_2 = \frac{1}{4\pi\varepsilon_0}\frac{-\rho\Delta V}{(C_2P)^2}$$

$$= -\frac{1}{4\pi\varepsilon_0}\cdot\frac{\rho(4/3)\pi(C_2P)^3}{(C_2P)^2}$$

$$= \frac{(-\rho)}{3\varepsilon_0}C_2P, \text{ along } PC_2$$

So, net electric field, $|\mathbf{E}_P| = |\mathbf{E}_1| + |\mathbf{E}_2|$

$$= \left|\frac{\rho}{3\varepsilon_0}C_1P\right| + \left|\frac{(-\rho)}{3\varepsilon_0}C_2P\right| = \frac{\rho}{3\varepsilon_0}(C_1P + C_2P)$$

Since, these two spheres partially overlap.
So the distance, $C_1P + C_2P \approx C_1C_2$

$$\therefore \qquad \mathbf{E}_P = \frac{\rho}{3\varepsilon_0}C_1C_2, \text{ along } C_1C_2$$

Hence, electrostatic field is constant in magnitude.

**60** (*c*) As flux $= \frac{q_{\text{enclosed}}}{\varepsilon_0}$, so flux is due to only charges $+q_1$ and $-q_1$ that makes a sum zero. But $q_2$ produces its own flux and net flux linked with sphere is zero.

Hence, electric field will be due to all the charges.

**61** (*c*) $\oint_S \mathbf{E}\cdot d\mathbf{S} = \phi_E$ represents electric flux over the closed surface.

In general, $\oint_S \mathbf{E}\cdot d\mathbf{S}$ means the algebraic sum of number of flux lines entering the surface and number of flux lines leaving the surface.

When $\oint_S \mathbf{E}\cdot d\mathbf{S} = 0$, it means that the number of flux lines entering the surface must be equal to the number of flux lines leaving it.

**62** (*b*) By Gauss's law,
electric flux through closed surface area, $\phi = \frac{q_{\text{enclosed}}}{\varepsilon_0}$.

If $q_{\text{enclosed}} = 1$ unit, then

$$\phi = \frac{1}{\varepsilon_0} = \varepsilon_0^{-1}$$

**63** (*a*) According to Gauss's law, electric flux passing through a closed surface is equal to $\frac{1}{\varepsilon_0}$ times the charge $q$ enclosed by the surface,

i.e. $\qquad \phi = \frac{q}{\varepsilon_0}$

Since, cube and sphere both enclose same charge, therefore flux passing through cube and sphere is same,

i.e. $\qquad \phi_{\text{sphere}} = \phi_{\text{cube}}$

$\therefore \qquad \phi_{\text{sphere}} : \phi_{\text{cube}} = 1:1$

**64** (*b*) When a sphere encloses an electric dipole, the situation is as shown

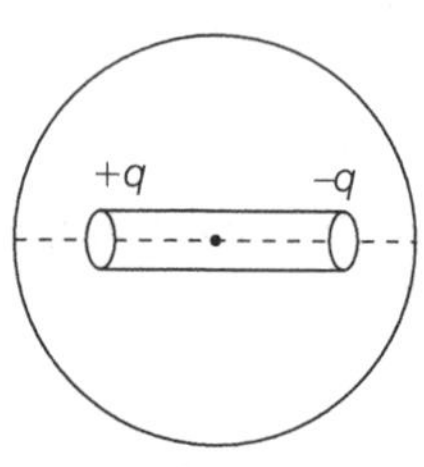


Given, $q = \pm 3 \times 10^{-6}$ C

Thus, according to Gauss's law, the net electric flux across the closed surface is equal to $1/\varepsilon_0$ times the net charge enclosed by it, i.e.

$$\phi_E = \frac{q_{\text{enclosed}}}{\varepsilon_0} = \frac{q_+ + q_-}{\varepsilon_0} = \frac{+3\times 10^{-6} - 3\times 10^{-6}}{\varepsilon_0} = 0$$

Hence, electric flux across the sphere is zero.

**65** (*d*) Net charges of one dipole $= -e + e = 0$
Net charge of 8 dipoles $= 8 \times 0 = 0$
$\therefore$ Net charge inside cube, $q = 0$
By Gauss's law,
Total flux emerging from surface, $\phi = \frac{q}{\varepsilon_0} = \frac{0}{\varepsilon_0} = 0$

**66** (*a*) Total charge inside the given surface is

$$= Q - 2Q = -Q$$

$\therefore$ Total flux through the surface of the sphere, $\phi = \frac{-Q}{\varepsilon_0}$

As, charge $5Q$ lies outside the surface, thus it makes no contribution to electric flux through the given surface.

The field on the surface of sphere is $E = \frac{Q}{4\pi\varepsilon_0 R^2}$

As $E \propto \frac{1}{R^2}$, so the field due to $-2Q$ is not same everywhere on the sphere.

**67** (*b*) Using Gauss's law at outer surface, let charge on dipole is $q$,

$$\phi = \frac{\Sigma q}{\varepsilon_0} = E \cdot A$$

or $$E = \frac{1}{A\varepsilon_0} \Sigma q$$

$$= \frac{(+Q + q - q)}{A\varepsilon_0} = \frac{Q}{A\varepsilon_0} = \frac{\sigma}{\varepsilon_0} = \text{constant}$$

**68** (*a*) Given, net flux, $\phi = 1.05 \, \text{Nm}^2\text{C}^{-1}$

$\therefore$ By Gauss's law, $\phi = \dfrac{q}{\varepsilon_0}$

We know that, $\varepsilon_0 = 8.854 \times 10^{-12}$ C

Charge, $q = \phi\varepsilon_0 = 1.05 \times 8.854 \times 10^{-12}$ C

$= 9.27 \times 10^{-12}$ C

**69** (*b*) Consider two infinite line charges with linear charge densities $+\lambda$ (C/m) and $-\lambda$ (C/m) respectively, which are lying in $y$-direction as shown in the figure below.

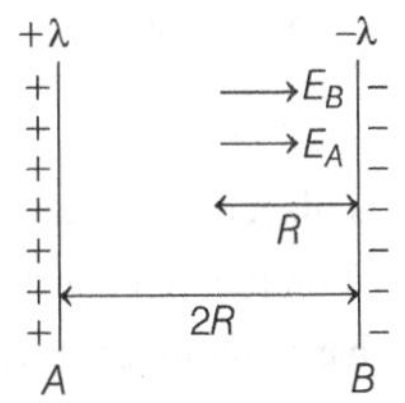


Then, the electric field due to line $A$ at the mid-way between the two line charges, i.e. at distance $R$ is

$$|\mathbf{E}_A| = \frac{\lambda}{2\pi\varepsilon_0 R} \text{ N/C} \quad \ldots \text{(i)}$$

which lies along +ve $X$-axis (outward), i.e. from $A$ to $B$.

Similarly, the electric field due to line $B$ at the mid-way between the two line charges, i.e. at distance $R$ is

$$|\mathbf{E}_B| = \frac{\lambda}{2\pi\varepsilon_0 R} \text{ (N/C)} \quad \ldots \text{(ii)}$$

Due to negative charge on $B$, $\mathbf{E}_B$ also lies along +ve $X$-axis (inward), i.e. from $A$ to $B$.

So, the resultant electric field at distance $R$ is given as

$$|\mathbf{E}_R| = |\mathbf{E}_A| + |\mathbf{E}_B|$$

Substituting the values from Eqs. (i) and (ii), we get

$$|\mathbf{E}_R| = \frac{\lambda}{2\pi\varepsilon_0 R} + \frac{\lambda}{2\pi\varepsilon_0 R} = \frac{\lambda}{\pi\varepsilon_0 R} \text{ (N/C)}$$

which also lies along the + ve $X$-axis, i.e. from $A$ to $B$.

**70** (*b*) The electric field between two parallel plates is as shown below

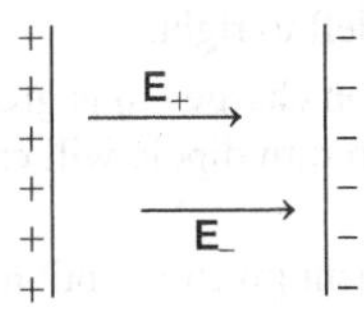


Here, $$E_+ = E_- = \frac{\sigma}{2\varepsilon_0}$$

$$\Rightarrow \quad E_{\text{net}} = 2\left(\frac{\sigma}{2\varepsilon_0}\right) = \left(\frac{\sigma}{\varepsilon_0}\right)$$

**71** (*d*) There are two plates $A$ and $B$ having surface charge densities, $\sigma_A = 17.0 \times 10^{-22}$ (Cm$^{-2}$) on $A$ and $\sigma_B = -17.0 \times 10^{-22}$ (Cm$^{-2}$) on $B$, respectively as shown below

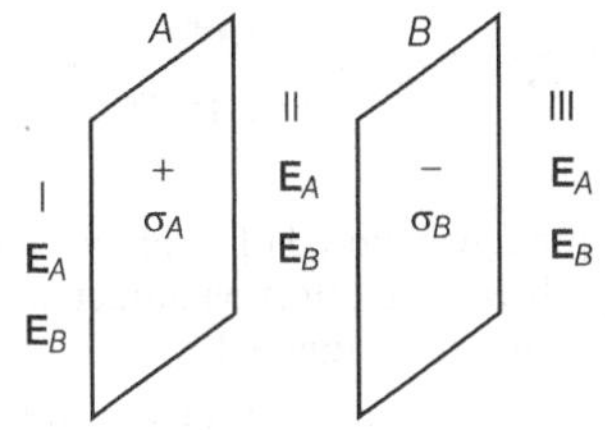


According to Gauss's theorem, if the plates have same surface charge density but having opposite signs, then the electric field in the outer region of the first plate (region I) is zero.

$$\mathbf{E}_{\text{I}} = \mathbf{E}_A + \mathbf{E}_B = \frac{\sigma}{2\varepsilon_0} + \left(-\frac{\sigma}{2\varepsilon_0}\right) = 0$$

**72** (*a*) As the hollow sphere is uniformly charged, so the net charge will appear on the surface of the sphere.

(i) The electric field at a point outside the hollow sphere $(r > R)$ is

$$\phi = \oint_S \mathbf{E} \cdot d\mathbf{S} = \frac{Q_{\text{enclosed}}}{\varepsilon_0} \quad \text{[from Gauss's law]}$$

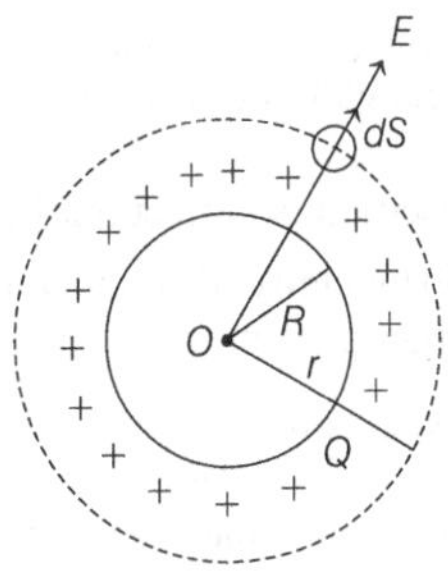


$$\Rightarrow \quad E\,(4\pi r^2) = \frac{Q}{\varepsilon_0}$$

$$\Rightarrow \quad E = \frac{Q}{4\pi\varepsilon_0 r^2}$$

$$\Rightarrow \quad E \propto \frac{1}{r^2}$$

(ii) The electric field inside hollow sphere $(r < R)$ is

$$E = \frac{Q_{\text{inside}}}{4\pi\varepsilon_0 r^2} = 0 \quad [\because Q_{\text{inside}} = 0]$$

Thus, the electric field due to the sphere at distance $r$ is zero as $r$ increases for $r < R$ and decreases as $r$ increases for $r > R$.

**73** (*a*) When we produce charge $q_1$ on a body by rubbing it against another body which gets a charge $q_2$ in the process, then $q_1$ and $q_2$ both are equal in magnitude but opposite in polarities. Hence, $q_1 + q_2 = 0$.
This is because charge on an isolated system remains constant.
Therefore, Assertion and Reason are correct and Reason is the correct explanation of Assertion.

**74** (*a*) At macroscopic level, we deal with charges of order $1\mu C = 10^{-6}$ C, which has $\frac{10^{-6}}{1.6 \times 10^{-19}} = 0.625 \times 10^{13} = 10^{13}$ charges.
Addition of a few hundred of *e* charges do not make any physically observable effect on attraction or repulsion, so quantisation can be ignored.
Therefore, Assertion and Reason are correct and Reason is the correct explanation of Assertion.

**75** (*a*) A neutral body acquires positive charge, when it loses some electrons, hence its mass decreases.
Therefore, Assertion and Reason are correct and Reason is the correct explanation of Assertion.

**76** (*d*) A body acquire charge by the transfer of electrons.
When a body acquire negative charge by gaining more electrons, then its mass increases slightly from its neutral state.
Therefore, Assertion is incorrect but Reason is correct.

**77** (*c*) When a positive charge initially at rest is placed in a uniform electric field, then it moves along the electric lines of force.
But if a point charge is released from rest in a direction making an angle with the field, then it follows a parabolic path.
Therefore, Assertion is correct but Reason is incorrect.

**78** (*d*) **E** in outside vicinity of conductor surface depends on all the charges present in the space
and its expression is $\mathbf{E} = \frac{\sigma}{\varepsilon_0}$.
Therefore, Assertion is incorrect but Reason is correct.

**79** (*d*) Due to displacement of charge within a closed surface, **E** at any point may change. But net flux crossing the surface will not change.
The flux crossing through a closed surface is independent of the location of charge within the surface, as it is same everywhere on its surface.
Therefore, Assertion is incorrect but Reason is correct.

**81** (*b*) The statement II is incorrect and it can be corrected as,
$-2$ units charge $+ 6$ unit charge $= 4$ units charge.
Rest statements are correct.

**83** (*c*) The statement IV is incorrect and it can be corrected as,
The quantisation of charge was first suggested by the experimental laws of electrolysis discovered by English expermentalist Faraday. It was experimentally demonstrated by Millikan in 1912.
Rest statements are correct.

**84** (*d*) The statement IV is incorrect and it can be corrected as,
There are infinite number of electric lines of forces can be drawn from a charge.
Rest statements are correct.

**85** (*d*) The statement given in option (d) is incorrect and it can be corrected as,
Charges can be transferred from one body to another but they cannot be transformed into another type.
Rest statements are correct.

**86** (*b*) Due to induction, charge on various surfaces are as shown in given figure below

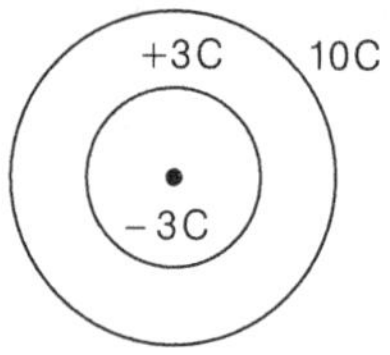


As $-3$ C charge is placed inside the shell,
so charge induced on the inner surface of shell $= +3$ C
Distribution of charge on inner surface would be uniform.
Thus, the statement given in option (b) is correct, rest are incorrect.

**87** (*b*) In the Gauss's law, **E** is due to all point charges present in space, i.e. $q_1$ and $q_2$ is non-zero.
$\phi$ depends only on the enclosed charges, i.e. on $q_1$ and is non-zero.
Thus, the statement given in option (b) is correct, rest are incorrect.

**88** (*c*) Natural occurring molecular dipoles have atoms with different electronegativity and possess permanent electric dipole moment, e.g. HCl, $H_2O$ etc.
Induced dipoles are formed when a neutral atom is placed in a region of strong electric field.
Thus, the statement given in options (a) and (b) are correct.

**89** (*c*) The space between the given electric field lines is increasing from left to right. The strength of electric field decreases with the increase in the space between electric field lines. As a result, force on charges also decreases from left to right.
Thus, the force on charge $-q$ is greater than the force on charge $+q$, in turn dipole will experience a force towards left.
Thus, the statement given in option (c) is correct, rest are incorrect.

**90** (*c*)

(i) The electric field due to a spherical shell at a distance $r$ from its centre at $r > R$ is given by

$$E = \frac{q}{4\pi\varepsilon_0 r^2}, \text{ where } q = \sigma \cdot 4\pi R^2$$

(ii) The electric field at a distance $r$ from the centre at $r < R$ is given by

$$E = \frac{q}{4\pi\varepsilon_0 r^2} = 0 \qquad [\because q_{\text{inside}} = 0]$$

(iii) The electric field at the surface of shell ($r = R$) is given by

$$E = \frac{q}{4\pi\varepsilon_0 R^2}, \text{ where } q = 4\pi R^2 \cdot \sigma$$

$$\Rightarrow \quad E = \frac{\sigma \cdot 4\pi R^2}{4\pi\varepsilon_0 R^2} = \frac{\sigma}{\varepsilon_0}$$

Thus, the statement given in option (c) is correct, rest are incorrect.

**93** (*d*)

A. Electric lines of forces emerge from a single positive charge and goes to infinity.

B. Electric lines of forces enter into a single negative charge coming from infinity.

C. Electric lines of forces are repelled away, when they are produced by the pair of positive charge.

D. The electric line of forces produced by a pair of equal and opposite charges.

Hence, A $\rightarrow$3, B$\rightarrow$4, C$\rightarrow$2 and D$\rightarrow$1.

**94** (*b*) Given, charges, $q_1 = 2\times 10^{-7}$ C

$q_2 = 3\times 10^{-7}$ C

and distance, $d = 30$ cm $= 0.3$ m

Using Coulomb's law, the force between two charges is given by

$$F = \frac{1}{4\pi\varepsilon_0} \cdot \frac{q_1 q_2}{d^2}$$

$$F = \frac{9\times 10^9 \times 2\times 10^{-7} \times 3\times 10^{-7}}{0.3\times 0.3}$$

$$= \frac{9\times 2\times 3\times 10^{-5}}{3\times 3\times 10^{-2}} = 6\times 10^{-3} \text{ N (repulsion)}$$

**95** (*c*) Given, force, $F = 0.2$N, charges, $q_1 = 0.4\times 10^{-6}$ C and $q_2 = -0.8\times 10^{-6}$C.

Using Coulomb's law, the magnitude of force between two charges is

$$F = \frac{1}{4\pi\varepsilon_0} \cdot \frac{q_1 q_2}{r^2}$$

$$\therefore \quad 0.2 = 9\times 10^9 \times \frac{0.4\times 10^{-6} \times |-0.8| \times 10^{-6}}{r^2}$$

$$\left[\because \frac{1}{4\pi\varepsilon_0} = 9\times 10^9\right]$$

$$r^2 = 16\times 9\times 10^{-4}\text{m} \Rightarrow r = 12 \text{ cm}$$

**96** (*c*) Given, $AB = 20$ cm, $AO = OB = 10$ cm $= 0.1$ m

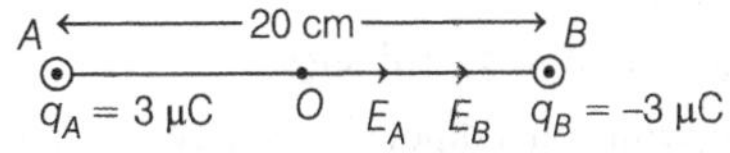


$$q_A = 3\,\mu\text{C} = 3\times 10^{-6} \text{ C}$$

$$q_B = -3\,\mu\text{C} = -3\times 10^{-6} \text{ C}$$

The electric field at a point due to a charge $q$ is

$$E = \frac{1}{4\pi\varepsilon_0} \cdot \frac{q}{r^2}$$

where, $r$ is the distance between charge and the point.

Electric field due to $q_A$ at $O$ is

$$E_A = \frac{1}{4\pi\varepsilon_0} \cdot \frac{q_A}{(AO)^2}$$

$$E_A = \frac{9\times 10^9 \times 3\times 10^{-6}}{(0.1)^2} = \frac{27\times 10^3}{0.1\times 0.1}$$

$$= 2.7\times 10^{-6} \text{ (N/C)}$$

The direction of $E_A$ is $A$ to $O$ or towards $OB$ as the electric field is always directed away from positive charge.

Electric field due to $q_B$ at $O$ is

$$E_B = \frac{1}{4\pi\varepsilon_0} \cdot \frac{q_B}{(OB)^2}$$

$$E_B = \frac{9\times 10^9 \times 3\times 10^{-6}}{(0.1)^2} = \frac{27\times 10^3}{0.1\times 0.1}$$

$$= 2.7\times 10^6 \text{ (N/C)}$$

The direction of $E_B$ is $O$ to $B$ or towards $OB$ as the electric field is always directed towards the negative charge.

Now, we see that both $E_A$ and $E_B$ are in same direction. So, the resultant electric field at $O$ is $E$.

Hence, $\quad E = E_A + E_B = 2.7\times 10^6 + 2.7\times 10^6$

$$= 5.4\times 10^6 \text{ (N/C)}$$

The direction of $E$ (resultant electric field) will be from $O$ to $B$ or towards $B$.

**97** (*d*) A system of two charges, $q_A = 2.5\times 10^{-7}$ C and $q_B = -2.5\times 10^{-7}$ C is as shown below

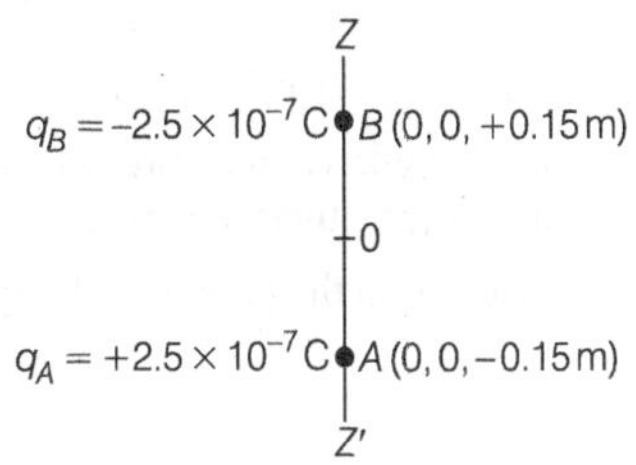


Total charge $= 2.5\times 10^{-7} - 2.5\times 10^{-7} = 0$

Electric dipole moment,

$p$ = either charge × separation between charges

$= 2.5\times 10^{-7} \times (0.15 + 0.15)$ C-m

$= 7.5\times 10^{-8}$ C-m

The direction of dipole moment is from $B$ to $A$, i.e. along negative $Z$-axis.

**98** (*c*) Given, dipole moment, $p = 4\times 10^{-9}$C-m
Angle, $\theta = 30°$
Electric field, $E = 5\times 10^4$ N/C
Torque acting on a dipole in the electric field,

$$\tau = \mathbf{p}\times \mathbf{E} = pE \sin\theta$$

or $$\tau = 4\times 10^{-9}\times 5\times 10^4 \times \sin 30°$$

$$= \frac{20\times 10^{-5}}{2} = 10^{-4} \text{ N-m}$$

The direction of torque is perpendicular to both electric field and dipole moment.

**99** (*d*) The charge on an object is given by $q = \pm\, ne$
The number of electrons transferred,

$$n = \frac{\text{Total charge } q}{\text{Charge of electron } e}$$

$$n = \frac{-3\times 10^{-7}}{-1.6\times 10^{-19}} = 1.875\times 10^{12}$$

$$\left[\because q = 3\times 10^{-7}\text{C},\ e = 1.6\times 10^{-19}\text{C}\right]$$

$$n = 1.8\times 10^{12}$$

**100** (*c*) If $q_A$ and $q_B$ are charges of copper spheres $A$ and $B$, respectively them.
According to the question, if the charge is doubled

$$q'_A = 2q_A \quad\text{and}\quad q'_B = 2q_B$$

and distance between them is halved, i.e. $r' = \frac{r}{2}$

Now, the force between the two spheres is

$$F' = \frac{1}{4\pi\varepsilon_0}\cdot\frac{q'_A q'_B}{r'} = \frac{1}{4\pi\varepsilon_0}\frac{(2q_A)(2q_B)}{(r/2)^2}$$

$$= \frac{1}{4\pi\varepsilon_0}\cdot\frac{4q_A q_B}{r^2/4} = 16\frac{1}{4\pi\varepsilon_0}\frac{q_A q_B}{r^2}$$

$$= \frac{16\times 9\times 10^9\times (6.5\times 10^{-7})^2}{(50\times 10^{-2})^2}$$

$$\left[\because q_1 = q_2 = 6.5\times 10^{-7}\text{C and } r = 50\text{cm}\right]$$

$$= 16\times 1.521\times 10^{-2} = 0.24 \text{ N}$$

This force is also repulsive in nature because both the charges are similar (positive) in nature.

**101** (*c*) As the deflection in the path of a charged particle is directly proportional to the charge/mass ratio,

i.e. $y \propto \frac{q}{m}$

Here, the deflection in particle 3 is maximum, so the charge to mass ratio of particle 3 is maximum.

**102** (*d*) Given, the area vector makes an angle of 60° with $X$-axis.
Electric field, $\mathbf{E} = 3\times 10^3\,\hat{\mathbf{i}}\ \text{NC}^{-1}$, $\theta = 60°$ and side of square, $a = 10$ cm
Then, area of square, $S = (10)^2 = 100\,\text{cm}^2 = 10^{-2}\text{m}^2$

Using the formula of electric flux, $\phi = \mathbf{E}\cdot\mathbf{S}$

$$\phi = ES\cos\theta$$
$$= 3\times 10^3\times 10^{-2}\cos 60°$$
$$= 3\times 10\times\frac{1}{2}$$
$$= 15\,\text{Nm}^2\text{C}^{-1}$$

**103** (*d*) A point charge + 10 μC is at a distance 5 cm above the centre of a square of side 10 cm is shown below

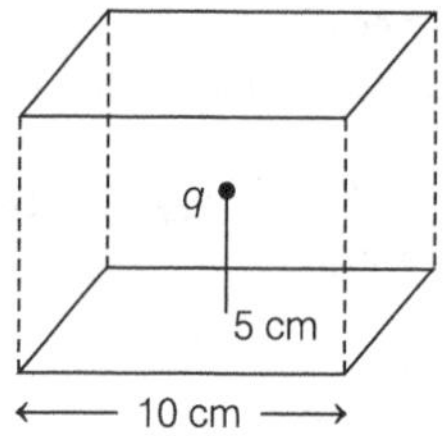


Given, $q = 10\,\mu\text{C} = 10\times 10^{-6}$ C
According to Gauss's theorem, the total flux enclosed is

$$\phi = \frac{q}{\varepsilon_0} \qquad \text{...(i)}$$

The flux enclosed by one face of square is (1/6) of total flux (because the cube has six square shaped faces).
The flux linked with each face,

$$\phi' = \frac{\phi}{6} = \frac{1}{6}\cdot\frac{q}{\varepsilon_0} \qquad \text{[from Eq. (i)]}$$

$$\phi' = \frac{1}{6}\times\frac{10\times 10^{-6}}{8.854\times 10^{-12}}$$
$$= 1.88\times 10^5\ \text{Nm}^2\text{C}^{-1}$$

Thus, the flux linked with the square is $1.88\times 10^5\ \text{Nm}^2\text{C}^{-1}$.

**104** (*d*) Given, charge, $q = 2.0\mu\text{C} = 2.0\times 10^{-6}$C
According to Gauss' theorem, the net electric flux ($\phi$) through the surface is

$$\phi = \frac{q}{\varepsilon_0} = \frac{2\times 10^{-6}}{8.854\times 10^{-12}} = 2.26\times 10^5\ \text{Nm}^2/\text{C}$$

$$\simeq 2\times 10^5\ \text{Nm}^2/\text{C}$$

Thus, the net electric flux through the surface is $2\times 10^5\ \text{Nm}^2/\text{C}$.

**105** (*c*) Electric field, $E = \frac{1}{4\pi\varepsilon_0}\cdot\frac{q}{r^2}$

$$1.5\times 10^3 = \frac{9\times 10^9\times q}{(20\times 10^{-2})^2}$$

$$\left[\text{given, electric field, } E = 1.5\times 10^3\,\text{N/C},\ \text{distance, } r = 20\,\text{cm} = 20\times 10^{-2}\,\text{m}\right]$$

$$q = \frac{1.5\times 10^3\times 20\times 20\times 10^{-4}}{9\times 10^9}$$

$$= 6.67\times 10^{-9}\text{ C} \simeq 6\times 10^{-9}\text{ C}$$

**106** (*c*) Given, electric field, $E = 9\times 10^4$ NC$^{-1}$ and distance, $r = 2$ cm.

Electric field due to infinite line charge,

$$E = \frac{\lambda}{2\pi\varepsilon_0 r}$$

$$E = \frac{2}{2}\times\frac{\lambda}{2\pi\varepsilon_0 r} = \frac{2\lambda}{4\pi\varepsilon_0 r}$$

Putting the values, we get

$$9\times 10^4 = \frac{2\times 9\times 10^9 \times \lambda}{2\times 10^{-2}}$$

$$\lambda = \frac{9\times 10^4 \times 2\times 10^{-2}}{2\times 9\times 10^9} = 10^{-7}\ \text{Cm}^{-1}$$

**107** (*a*) The net force on $q_1$ by $q_2$ and $q_3$ is along the $+x$-direction, hence $q_1$ is a negative charge. Thus, nature of force between $q_1$ and newly introduced charge $Q$ (positive) is attractive and net force on $q_1$ by ($q_2, q_3$ and $Q$) are along the same direction as given in the diagram below, i.e. force shall increase along the positive $X$-axis.

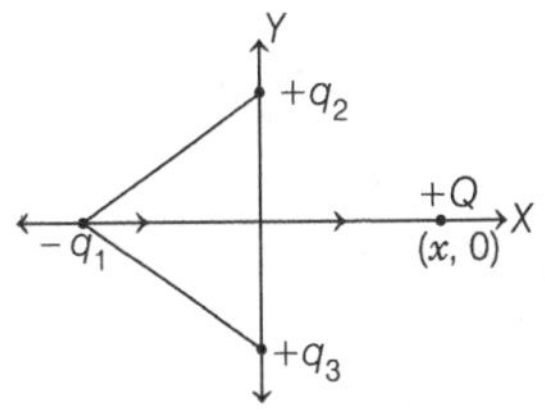


**108** (*a*) When a positive point charge is brought near an isolated conducting sphere without touching the sphere, then the free electrons in the sphere are attracted towards the positive charge. This leaves an excess of positive charge on the far (right) surface of sphere and an equivalent negative charge in the near (left) side as given in the figure below

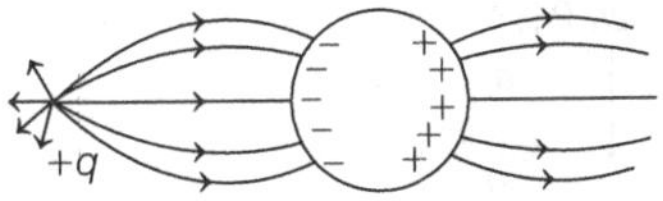


An electric field line starts from positive charge and ends at negative charge (in this case from positive point charge to negative charge created inside the sphere). Also, electric field line emerges from a positive charge, (in case of single charge) and ends at infinity.
Here, all these conditions are fulfilled in Fig. (a).

**109** (*d*) Electric flux through a surface does not depend on the shape, size or area of a surface but it depends on charges enclosed by the surface.
As there is same charge $+q$ is present in all the given figures, so electric flux is same for all the figures.

**110** (*c*) The given condition is as shown below

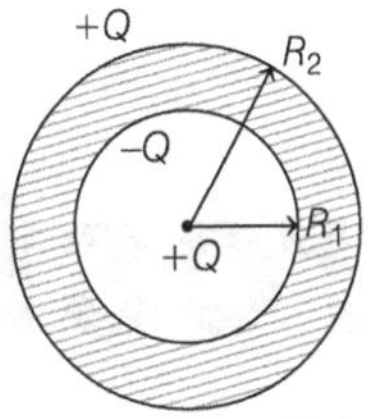


Now, surface charge density on the inner surface of spherical shell is $\sigma = \frac{q}{S} = \frac{-Q}{4\pi R_1^2}$

An equal and opposite charge would appear on outer surface,

$\therefore$ Surface charge density on the outer surface $= \frac{+Q}{4\pi R_2^2}$

**111** (*c*) When the charge $q$ is placed at $B$, mid-point of an edge of the cube, it is being shared equally by 4 faces. Therefore, total flux through the faces of the given cube $= q/4\varepsilon_0$.

**112** (*c*) Here, let us keep the charge $2q$ at a distance $x$ from $A$ as shown below

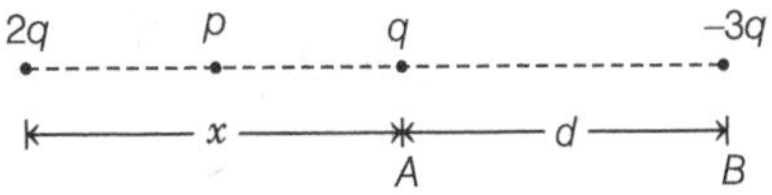


The charge $2q$ will not experience any force, when force of repulsion on it due to $q$ at $A$, is balanced by force of attraction on it due to $-3q$, at $B$.
Thus, force of attraction by $-3q$ = force of repulsion by $q$

$$\Rightarrow \frac{2q\times 3q}{4\pi\varepsilon_0 (x+d)^2} = \frac{2q\times q}{4\pi\varepsilon_0 x^2}$$

$$\Rightarrow (x+d)^2 = 3x^2$$

$$\Rightarrow x^2 + d^2 + 2xd = 3x^2$$

$$\Rightarrow 2x^2 - 2dx - d^2 = 0$$

$$\Rightarrow x = \frac{d}{2} \pm \frac{\sqrt{3}d}{2}$$

(neglecting negative sign)

$$\therefore x = \frac{d}{2} + \frac{\sqrt{3}d}{2} = \frac{d}{2}(1+\sqrt{3}) \text{ to the left of } q.$$

**113** (*a*) When a positive point charge brought near an isolated conducting plane, some negative charge develops on the surface of the plane towards the charge and an equal positive charge develops on opposite side of the plane. Thus, the electric field lines are away from positive charge and perpendicular to the surface.
Hence, the field due to it at a point $P$ on other side is directed perpendicular and away from the plane.

CHAPTER > 02

# Electrostatic Potential and Capacitance

## KEY NOTES

- The work done by external force in moving a charged particle $q$ from $R$ to $P$ against electrostatic repulsive force (as shown in figure) is given as

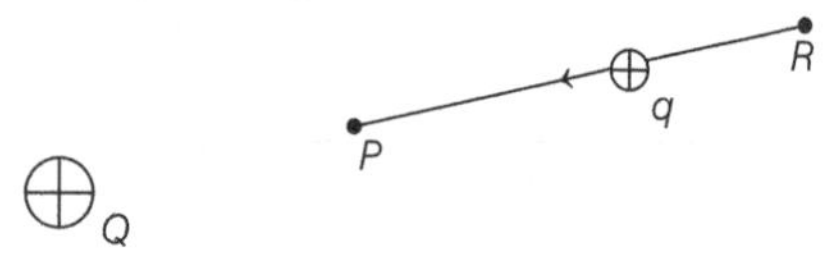


$$W_{RP} = \int_R^P \mathbf{F}_{\text{ext}} \cdot d\mathbf{r} = -\int_R^P \mathbf{F}_E \cdot d\mathbf{r}$$

This work done gets stored as potential energy.

- Therefore, electric potential energy difference between two points is defined as the work required to be done by an external force in moving (without accelerating) charge $q$ from one point to another for electric field of any arbitrary charge configuration.

i.e. $\Delta U = U_P - U_R = W_{RP}$

- The work done by an electrostatic field in moving a charge from one point to another depends only on the initial and the final points and is independent of the path taken to go from one point to the other.

This is the fundamental characteristics of a conservative force.

## Electrostatic Potential

- The work done by external force in bringing a unit positive charge from one point $R$ to another point $P$ is called **potential difference** between those points ($P$ and $R$).

i.e. $V_{RP} = V_P - V_R = \dfrac{U_P - U_R}{q} = \dfrac{W_{RP}}{q}$

- Work done by an external force in bringing a unit positive charge from infinity to a point is equal to **electrostatic potential** $V$ at that point.

i.e. $V = \dfrac{W}{q_0}$ volt

- **Electrostatic potential due to a point charge** $q$ at any point $P$ lying at a distance $r$ as shown in the figure below.

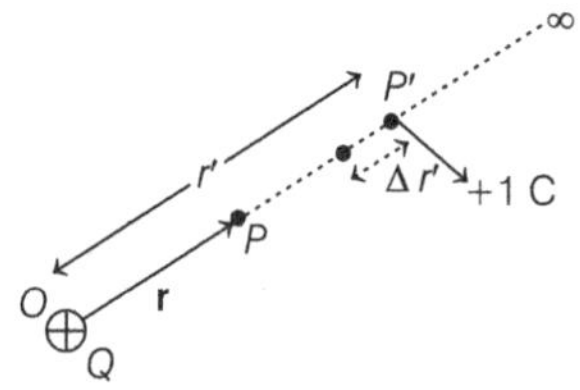


It is given by $V = \dfrac{1}{4\pi\varepsilon_0} \cdot \dfrac{q}{r}$.

- **Electrostatic potential due to an electric dipole** at any point $P$ whose position vector is $\mathbf{r}$ w.r.t. mid-point of dipole is given by

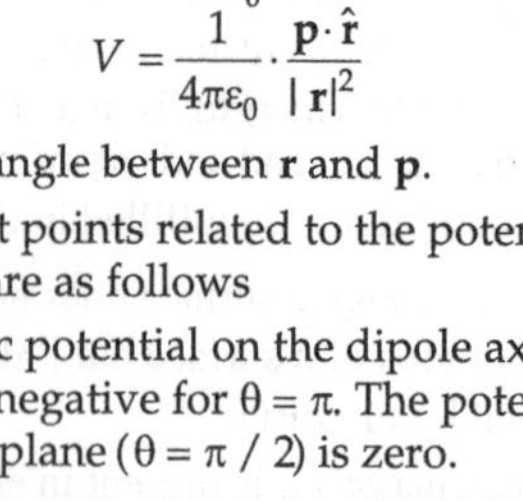


$$V = \frac{1}{4\pi\varepsilon_0} \cdot \frac{p\cos\theta}{r^2}$$

or

$$V = \frac{1}{4\pi\varepsilon_0} \cdot \frac{\mathbf{p}\cdot\hat{\mathbf{r}}}{|\mathbf{r}|^2}$$

where, $\theta$ is the angle between $\mathbf{r}$ and $\mathbf{p}$.

- Some important points related to the potential due to an electric dipole are as follows
  (i) The electric potential on the dipole axis is positive for $\theta = 0°$ and negative for $\theta = \pi$. The potential in the equatorial plane ($\theta = \pi / 2$) is zero.
  (ii) The electric potential due to dipole falls off at large distance, as $\dfrac{1}{r^2}$ not as $\dfrac{1}{r}$.

- **Electric potential at any point $P$ due to system of charges**
  It is equal to the algebraic sum of potentials due to individual charges.

i.e. $$V = \frac{1}{4\pi\varepsilon_0}\cdot\left(\frac{q_1}{r_{1P}} + \frac{q_2}{r_{2P}} + \ldots\frac{q_n}{r_{nP}}\right)$$

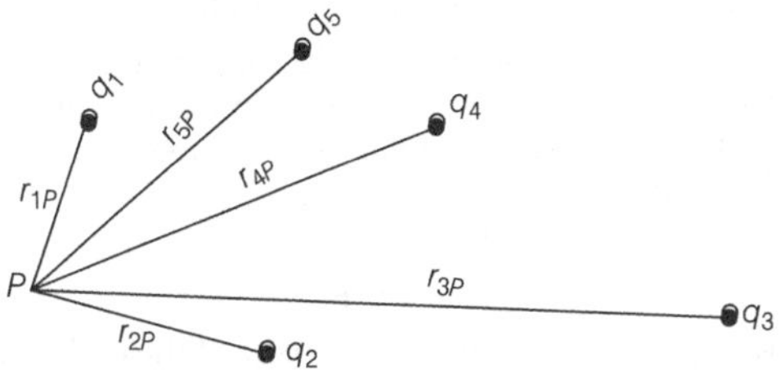


- **Electrostatic potential due to a thin charged spherical shell** carrying charge $q$ and radius $R$ at any point $P$ lying
  (i) inside the shell, $V = \frac{1}{4\pi\varepsilon_0}\cdot\frac{q}{R}$, for $r < R$
  (ii) on the surface of shell, $V = \frac{1}{4\pi\varepsilon_0}\cdot\frac{q}{R}$, for $r = R$
  (iii) outside the shell, $V = \frac{1}{4\pi\varepsilon_0}\cdot\frac{q}{r}$, for $r > R$
  where, $r$ is the distance of point $P$ from the centre of the shell.

## Equipotential Surface

- An equipotential surface is a surface with a constant value of potential at all points on the surface.
- Equipotential surfaces of a single point charge are concentric spherical surface centred at the charge.
- The electric field at every point is normal to the equipotential surface passing through that point.
- For any charge configuration, equipotential surface through a point is normal to the electric field at that point.
- There is no potential difference between any two points on the equipotential surface and no work is required to move a test charge on the equipotential surface.
- **Relation between electric field and electric potential** is given as
  $$|\mathbf{E}| = \frac{-\delta V}{\delta l} = \frac{+|\delta V|}{\delta l}$$
  where, $\delta l$ is the perpendicular distance between two equipotential surfaces and $\delta V$ is the change in $V$ in the direction of electric field $E$.
  There are two important conclusions concerning the relation between electric field and electric potential :
  (i) Electric field is in the direction in which the potential decreases steepest.
  (ii) Its magnitude is given by the change in the magnitude of potential per unit displacement normal to the equipotential surface at the point.

## Potential Energy of a System of Charges

- Electrostatic potential energy of a system of two point charges is given by
  $$U = \frac{1}{4\pi\varepsilon_0}\cdot\frac{q_1q_2}{r}$$
  where, $q_1$ and $q_2$ are charges and $r$ is the separation between the point charges.
- The total work done in assembling the charges at the given locations, as shown in the figure below is given as
  $$U = \frac{1}{4\pi\varepsilon_0}\left(\frac{q_1q_2}{r_{12}} + \frac{q_1q_3}{r_{13}} + \frac{q_2q_3}{r_{23}}\right)$$

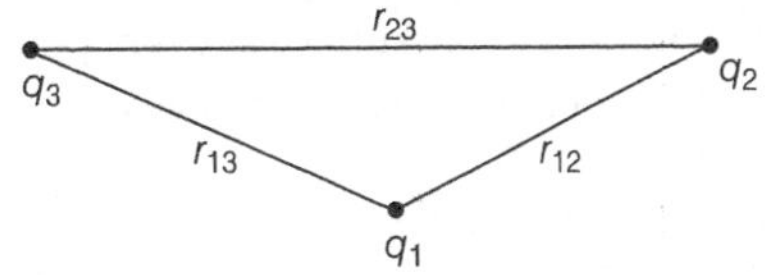


- The potential energy is a characteristics of the present state of configuration and not the way the state is achieved.

## Potential Energy in an External Field

- **Potential energy of a single charge** $q$ at a point with position vector $\mathbf{r}$, in an external field is $qV(\mathbf{r})$, where $V(\mathbf{r})$ is the potential at the point due to external electric field $E$.
- **Potential energy of a system of two charges in an external field** can be given as
  $$U = q_1V(\mathbf{r}_1) + q_2V(\mathbf{r}_2) + \frac{q_1q_2}{4\pi\varepsilon_0 r_{12}}$$
  where, $q_1$ and $q_2$ = two point charges at position vectors $\mathbf{r}_1$ and $\mathbf{r}_2$ respectively,
  $V(\mathbf{r}_1)$ = potential at $\mathbf{r}_1$ due to the external field
  and $V(\mathbf{r}_2)$ = potential at $\mathbf{r}_2$ due to the external field.
- **Potential Energy of a Dipole in an External Field**
  If an external torque $\tau_{ext}$ is applied in such a manner that it just neutralises this torque and rotates it in the plane of paper from angle $\theta_0$ to angle $\theta_1$ at an infinitesimal angular speed and without angular acceleration.
  The amount of work done by the external torque will be given by,
  $$W = pE(\cos\theta_0 - \cos\theta_1)$$
- Potential energy of a dipole in a uniform electric field,
  $$U(\theta) = -\mathbf{p}\cdot\mathbf{E} = -pE\cos\theta$$
- The potential energy $U'(\theta)$ with an inclination $\theta$ of the dipole in an external field is given by
  $$U'(\theta) = q[V(\mathbf{r}_1) - V(\mathbf{r}_2)] - \frac{q^2}{4\pi\varepsilon_0 \times 2a}$$
  where, $q$ is magnitude of either charge of electric dipole and $\mathbf{r}_1$ & $\mathbf{r}_2$ denote the position vectors of $+q$ & $-q$.

or $U'(\theta) = -\mathbf{p}\cdot\mathbf{E} - \dfrac{q^2}{4\pi\varepsilon_0 \times 2a}$

- When $\theta_0 = \frac{\pi}{2}$, work done against the external field **E** in bringing $+q$ & $-q$ are equal, opposite and cancel out.
  i.e. $q[V(\mathbf{r}_1) - V(\mathbf{r}_2)] = 0$

## Electrostatics of Conductors

- **Conductors** are the materials through which electric charge can flow easily.
- In electrostatics, the conductors have following properties
  (i) Inside a conductor, electrostatic field is zero.
  (ii) At the surface of a charged conductor, electrostatic field must be normal to the surface at every point.
  (iii) The interior of a conductor can have no excess charge in the static situation.
  (iv) Electrostatic potential is constant throughout the volume of the conductor and has the same value (as inside) on its surface.
  (v) Electric field at the surface of a charged conductor is given as

$$E = \frac{\sigma}{\varepsilon_0}\hat{\mathbf{n}}$$

  where, $\sigma$ is the surface charge density and $\hat{\mathbf{n}}$ is a unit vector normal to the surface in the outward direction.
  For $\sigma > 0$, electric field is normal to the surface outward and for $\sigma < 0$, the electric field is normal to the surface inward.
  (vi) For any charge and field configuration outside, any cavity in a conductor remains shielded from outside electric influence. The field inside the cavity is always zero, this is known as **electrostatic shielding.**

## Dielectrics and Polarisation

- **Dielectrics** are non-conducting substances. In contrast to conductors, they have no (or negligible number of) charge carriers.
- In a dielectric, the free movement of charges is not possible. It turns out that the external field induces dipole moment by stretching or re-orienting molecules of the dielectric.
- The collective effect of all the molecular dipole moments is net charges on the surface of the dielectric which produces a field that opposes the external field.
- On the basis of their behaviour in external field, dielectric are of two types
  (i) In a **non-polar molecule**, the centres of positive and negative charges coincide. The molecule has no permanent dipole moment, e.g. $O_2$, $H_2$, etc.
  (ii) The **polar molecule** is one in which the centres of positive and negative charges are separated even when there is no external field. Such molecules have a permanent dipole moment. e.g. HCl, $H_2O$, etc.
- The dielectric in which the induced dipole moment is in the direction of the field and is proportional to the field strength is called **linear isotropic dielectrics.**
- When a polar or non-polar dielectric develops a net dipole moment in the presence of an external electric field, the dipole moment per unit volume is called **polarisation** and is denoted by **p**.
- For linear isotropic dielectrics,

$$\mathbf{p} = \chi_e \mathbf{E}$$

  where, $\chi_e$ is a constant characteristic of the dielectric and is known as the **electric susceptibility** of the dielectric medium.
- When dielectric slab is placed inside the electric field, then due to polarisation, every volume element $\Delta V$ of slab has a dipole moment $\mathbf{p}\Delta V$ in the direction of the field.
- The polarised dielectric is equivalent to two charged surfaces with induced surface charge densities, says $\sigma_p$ and $-\sigma_p$.
  The electric field produced by these surface charges opposes the external field. The total electric field in the dielectric is thereby reduced from the case when no dielectric is present.
- Surface charge density $\pm\sigma_p$ arises from bound (not free) charges in the dielectric.

## Capacitors and Capacitance

- A capacitor is a system of two conductors separated by an insulator. Its capacitance is given by $C = \frac{Q}{V}$ (Farad)
  where, $Q$ is charge given to one of the conductor of capacitor and $V$ is the potential difference between the conductors.
- The capacitance $C$ depends only on the geometrical configuration (shape, size and separation) of the system of two conductors.
  It also depends on the nature of the insulator (dielectric) separating the two conductors.
- The maximum electric field that a dielectric medium can withstand without breakdown (of its insulating property) is called its **dielectric strength.**
- For a capacitor to store a large amount of charge without leaking, its capacitance should be high enough so that the potential difference and hence the electric field do not exceed the breakdown limits.

## Parallel Plate Capacitor

- It consists of two large plane parallel conducting plates separated by a small distance as shown in the figure below

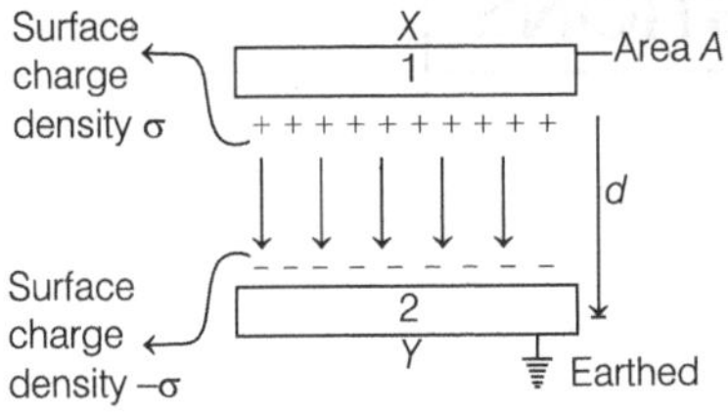


- Intensity of the electric field between the plates of a capacitor,
$$E = \frac{\sigma}{\varepsilon_0} = \frac{q}{A\varepsilon_0}$$
- Potential difference between the plates,
$$V = Ed = \frac{qd}{A\varepsilon_0}$$
where, $d$ is the distance between the conductor plates.
- The capacitance $C$ of the parallel plate capacitor is given by
$$C = \frac{q}{V} = \frac{\varepsilon_0 A}{d}$$

## Effect of Dielectric on Capacitance

- Electric potential between the plates of capacitor, when dielectric material of dielectric constant $K$ is introduced, is given by
$$V = \frac{\sigma d}{\varepsilon_0 K} = \frac{Qd}{A\varepsilon_0 K}$$
- The capacitance $C$ with dielectric between the plates, is given as
$$C = \frac{Q}{V} = \frac{\varepsilon_0 KA}{d}$$
and $$K = \frac{\varepsilon}{\varepsilon_0} = \frac{C}{C_0}$$
where, $\varepsilon$ is permittivity of dielectric medium.
- When a dielectric fully occupies the intervening region between the plates of a capacitor, the dielectric is polarised by the field, the effect is equivalent to two charged sheets having surface charge densities $+\sigma_p$ and $-\sigma_p$.
The electric field in the dielectric is
$$E = \frac{\sigma - \sigma_p}{\varepsilon_0}.$$
- The quantity $\varepsilon_0 E + P$ is called the **electric displacement** and is denoted by $D$.
Thus, $$D = \varepsilon_0 E + P$$
- The ratio of the magnitudes of $D$ and $E$ is given as
$$\frac{D}{E} = \varepsilon_0 K \Rightarrow D = \varepsilon_0 KE$$
and $$P = D - \varepsilon_0 E = \varepsilon_0 (K-1)E$$
Also, for electric susceptibility, $\chi_e = \varepsilon_0 (K-1)$.

## Combination of Capacitors

- In the **series combination of capacitors**, equivalent capacitance $C_s$ of $n$ capacitors is given by

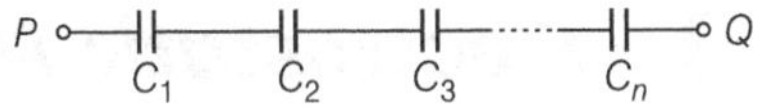


$$\frac{1}{C_s} = \frac{1}{C_1} + \frac{1}{C_2} + \frac{1}{C_3} + \dots + \frac{1}{C_n}$$

- In the **parallel combination of capacitors**, the equivalent capacitance $C_p$ of $n$ capacitors is given by

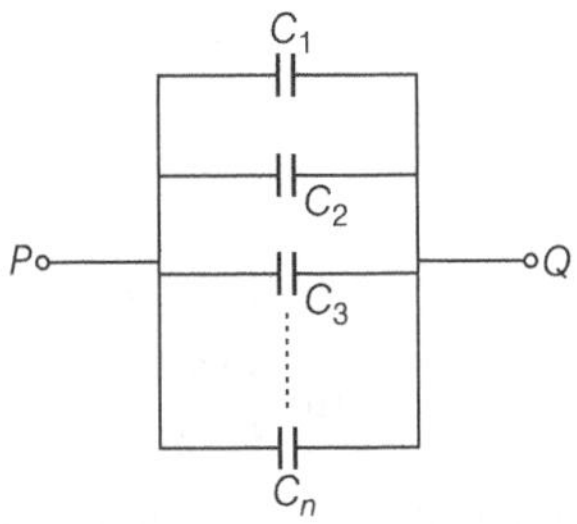


$$C_p = C_1 + C_2 + C_3 + \dots C_n$$

## Energy Stored in a Capacitor

- The work done in the transfer of charge $q$ between two plates of capacitor is
$$W = \frac{1}{2} qV = \frac{1}{2} CV^2 = \frac{q^2}{2C}$$
This work done is stored in the form of potential energy of the system.
- Total energy stored in the capacitor is given by
$$U = \frac{1}{2}\varepsilon_0 E^2 \times Ad = \frac{1}{2}\varepsilon_0 E^2 V$$
where, $V$ is volume of capacitor.
- The electrostatic energy density (energy per unit volume) in a region with electric field $E$ is $U = \frac{1}{2}\varepsilon_0 E^2$.

## van de graaff Generator

It is a device which is used for generating high electric potential of the order of 10 million volts, for accelerating charged particles to a very high speed.

- Its working is based on following two points
  (i) The action of sharp points (corona discharge).
  (ii) Total charge given to a spherical shell resides on its outer surface.

# Mastering NCERT

## MULTIPLE CHOICE QUESTIONS

## TOPIC 1 ~ Electrostatic Potential and Equipotential Surfaces

**1** Work done by external forces in moving a charge $q$ from $R$ to $P$ as shown in figure is

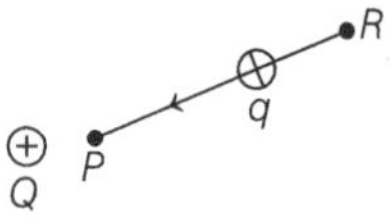


(a) $-\int_R^P \mathbf{F}_E \cdot d\mathbf{r}$ (b) $\int_R^P \mathbf{F}_E \cdot d\mathbf{r}$

(c) $\int_R^P 2\mathbf{F}_E \cdot d\mathbf{r}$ (d) $\int_R^P \frac{\mathbf{F}_E}{2} \cdot d\mathbf{r}$

**2** While bringing the unit positive charge as shown in the figure below, potential energy difference between points $R$ and $P$ is equal to

(a) $\Delta U = -W_{RP}$

(b) $\Delta U = W_{RP}$

(c) $\Delta U = 2W_{RP}$

(d) $\Delta U = 4\,W_{RP}$

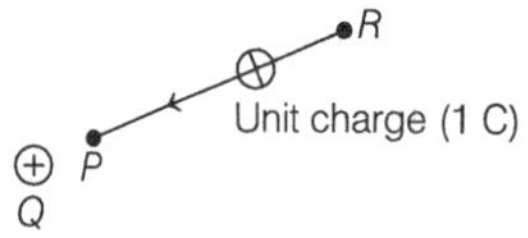


**3** A point charge $(+q)$ is taken along two different paths say $AMB$ (path 1) and $ACB$ (path 2) such that $A$, $M$ and $B$ lie on the circle and $+q$ (another charge) is placed at the centre (as shown in the figure), so the work done along path $ACB$ and path $AMB$ are related as

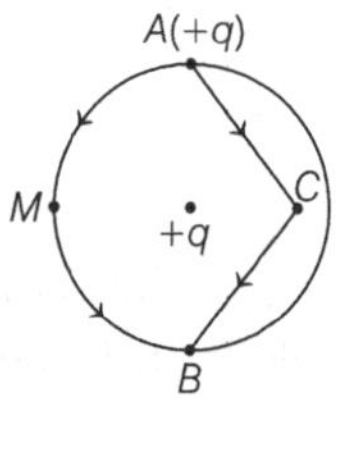


(a) $W_{AMB} > W_{ACB}$ (b) $W_{ACB} > W_{AMB}$

(c) $W_{ACB} = W_{AMB}$ (d) None of these

**4** A uniform electric field $E$ exists between two charged plates. What would be the work done in moving a charge $q$ along the closed path $ABCDA$?

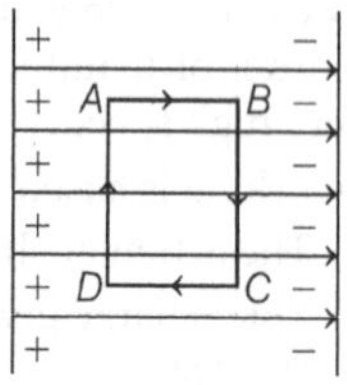


(a) $\oint \mathbf{E} \cdot d\mathbf{l}$ (b) $-\oint \mathbf{E} \cdot d\mathbf{l}$

(c) Zero (d) None of these

**5** If 100 J of work has to be done in moving an electric charge of 4 C from a place, where potential is $-10$ V to another place where potential is $V$ volt, find the value of $V$.

(a) 5 V (b) 10 V (c) 25 V (d) 15 V

**6** If $\delta W$ is the work done in bringing an infinitesimal small test charge $\delta q$ from infinity to a point $P$, then potential at point $P$ is

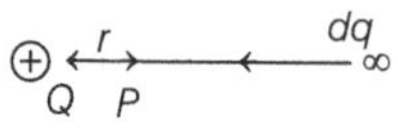


(a) $\frac{\delta W}{\delta q}$ (b) $-\frac{\delta W}{\delta q}$

(c) $\delta W \cdot \delta q$ (d) None of these

**7** To find the value of potential at a point, the external force at every point of the path is to be equal and opposite to the

(a) work done

(b) electrostatic force on the test charge at that point

(c) potential energy

(d) None of the above

**8** In the given figure, total work done $W$ by the external force from $r' = \infty$ to $r' = r$ is

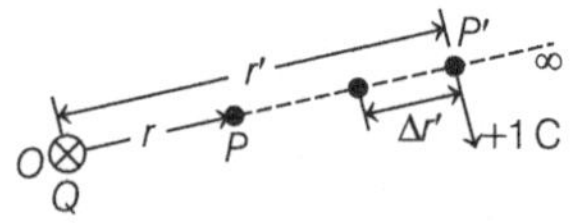


(a) $-\frac{Q}{4\pi\varepsilon_0 r}$ (b) $\frac{Q}{4\pi\varepsilon_0 r}$

(c) $\frac{2Q}{4\pi\varepsilon_0 r}$ (d) $-\frac{2Q}{4\pi\varepsilon_0 r}$

**9** As per this diagram, a point charge $+q$ is placed at the origin $O$. Work done in taking another point charge $-Q$ from the point $A$ [coordinates $(0, a)$] to another point $B$ [coordinate $(a,0)$] along the straight path $AB$ is

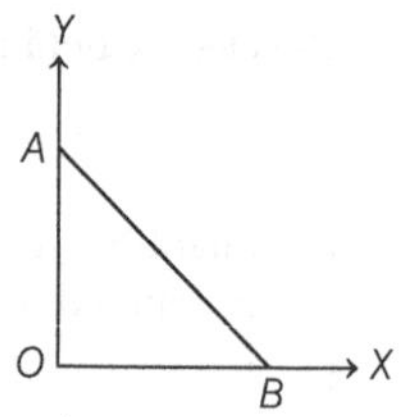


(a) zero (b) $\left(\frac{-qQ}{4\pi\varepsilon_0 a^2}\right)\sqrt{2}a$

(c) $\left(\frac{qQ}{4\pi\varepsilon_0 a^2}\right)\cdot\frac{a}{\sqrt{2}}$ (d) $\left(\frac{qQ}{4\pi\varepsilon_0}\frac{1}{a^2}\right)\sqrt{2}a$

**10** The figure given below shows the variation of potential $V$ or electric field $E$ with $r$ for a point charge $Q$.

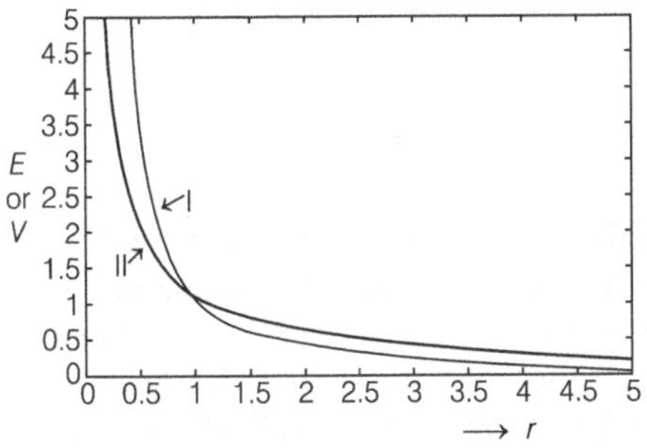


Which of the given curves represent $E$ *versus* $r$ or $V$ *versus* $r$?

(a) I→ $E$ *versus* $r$, II→ $V$ *versus* $r$
(b) I→ $V$ *versus* $r$, II→ $V$ *versus* $r$
(c) Both represent $E$ *versus* $r$
(d) None of the above

**11** The variation of electric potential of an electric dipole with distance $r$ is

(a) $E \propto \frac{1}{r^3}$ (b) $E \propto \frac{1}{r^2}$
(c) $E \propto \frac{1}{r}$ (d) $E \propto r$

**12** The potential in the equatorial plane of a dipole of a dipole moment $p$ is

(a) infinite (b) zero
(c) maximum (d) equal to $p$

**13** Two charges $3 \times 10^{-8}$ C and $-2 \times 10^{-8}$ C located 15 cm apart. At what point on the line joining the two charges is the electric potential zero?

(a) 9 cm (b) 42 cm
(c) 18 cm (d) 20 cm

**14** In the figure given below, $O$ is the centre of line joining $AB$. Two charges of opposite nature and same magnitude are placed at points $A$ and $B$, respectively.

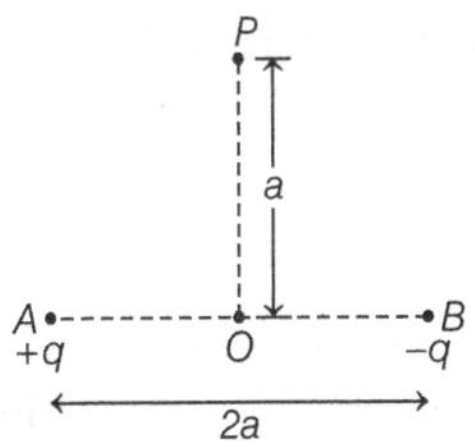


The potential at point $P$ is

(a) $\frac{2kq}{a\sqrt{2}}$ (b) zero
(c) $\frac{kq}{a\sqrt{2}}$ (d) cannot be determined

**15** A charge $2Q$ is placed at each corner of a cube of side $a$. The potential at the centre of the cube is

(a) $\frac{8Q}{\pi\varepsilon_0 a}$ (b) $\frac{4Q}{4\pi\varepsilon_0 a}$ (c) $\frac{8Q}{\sqrt{3}\,\pi\varepsilon_0 a}$ (d) $\frac{2Q}{\pi\varepsilon_0 a}$

**16** The electrostatic potential of a uniformly charged thin spherical shell of charge $Q$ and radius $R$ at a distance $r$ from the centre is

(a) $\frac{Q}{4\pi\varepsilon_0 r}$ for outside points and $\frac{Q}{4\pi\varepsilon_0 R}$ for inside points of the shell
(b) $\frac{Q}{4\pi\varepsilon_0 r}$ for both points inside and outside the shell
(c) zero for points outside and $\frac{Q}{4\pi\varepsilon_0 r}$ for points inside the shell
(d) zero for both points inside and outside the shell

**17** A hollow metal sphere of radius 10 cm is charged such that the potential on its surface becomes 80 V. The potential at the centre of the sphere is

(a) 80 V (b) 800 V (c) 8 V (d) zero

**18** The potential of a large liquid drop when eight liquid drops are combined is 20 V , then the potential of each single drop was

(a) 10 V (b) 7.5 V (c) 5 V (d) 2.5 V

**19** Three concentric metal shells $A$, $B$ and $C$ of respective radii $a$, $b$ and $c$ $(a < b < c)$ have surface charge densities $+\sigma$, $-\sigma$ and $+\sigma$, respectively. The potential of shell $B$ is ***JEE Main 2018***

(a) $\frac{\sigma}{\varepsilon_0}\left[\frac{a^2-b^2}{a}+c\right]$ (b) $\frac{\sigma}{\varepsilon_0}\left[\frac{a^2-b^2}{b}+c\right]$
(c) $\frac{\sigma}{\varepsilon_0}\left[\frac{b^2-c^2}{b}+a\right]$ (d) $\frac{\sigma}{\varepsilon_0}\left[\frac{b^2-c^2}{c}+a\right]$

**20** A solid conducting sphere having a charge $Q$ is surrounded by an uncharged concentric hollow spherical shell. Let the potential difference between the surface of the solid sphere and that of the outer surface of hollow shell be $V$. What will be the new potential difference between the same two surfaces, if the shell is given a charge $-3Q$?

(a) $V$ (b) $-3V$ (c) $2V$ (d) $4V$

**21** A small sphere of radius $r_1$ having charge $q_1$ is enclosed by a spherical shell of radius $r_2$ having charge $q_2$. Which charge will necessarily flow from the sphere to the shell, when connected by a wire?

(a) $q_1$ (b) $q_2$
(c) Both (a) and (b) (d) May be $q_2$

**22** Equipotential surface between two equal and opposite charges passing through the middle point is

(a) a plane (b) curved surface
(c) Both (a) and (b) (d) None of these

**23** For a uniform electric field **E**, along the $X$-axis, the equipotential surfaces are plane
(a) perpendicular to $Z$-axis
(b) parallel to the $yz$-plane
(c) perpendicular to $yz$-plane
(d) perpendicular to $Y$-axis

**24** Three equipotential surfaces are shown in the figure

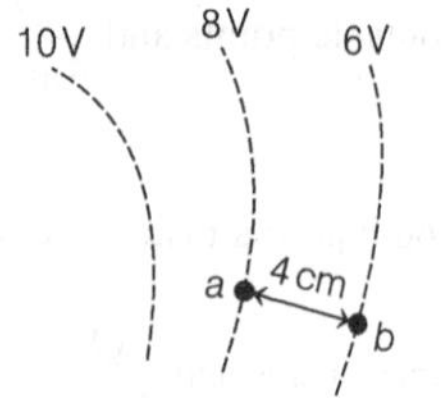


Which of the following option correctly represents the corresponding field lines?

(a)
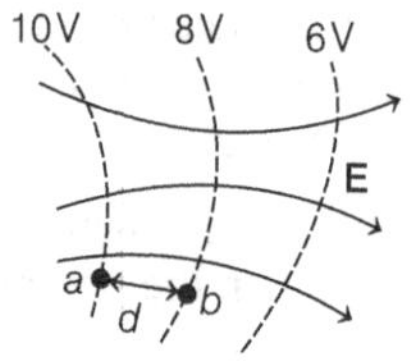


(b)
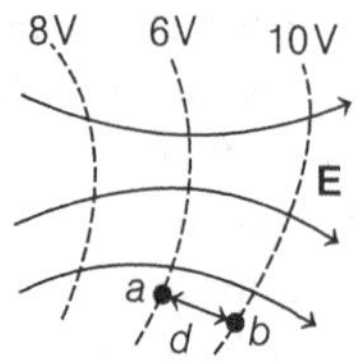


(c)
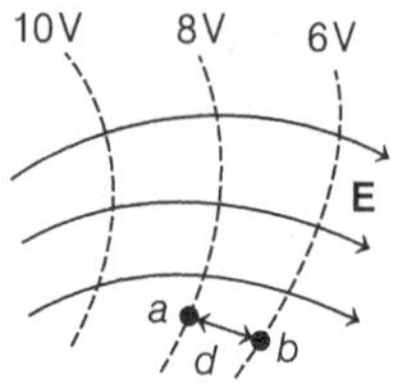


(d) None of these

**25** The figure given below shows various equipotential surfaces.

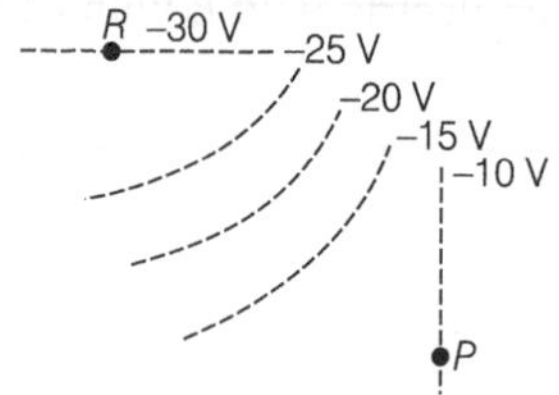


What is the direction of electric field **E** at $P$ and $R$?
(a) At $P$, **E** is to the left and at $R$, **E** is upward
(b) At $P$, **E** is to the right and at $R$, **E** is downward
(c) At $P$, **E** is to downward and at $R$, **E** is to the left
(d) At $P$, **E** is to upward and at $R$, **E** is to the right

**26** The diagrams below show regions of equipotentials.

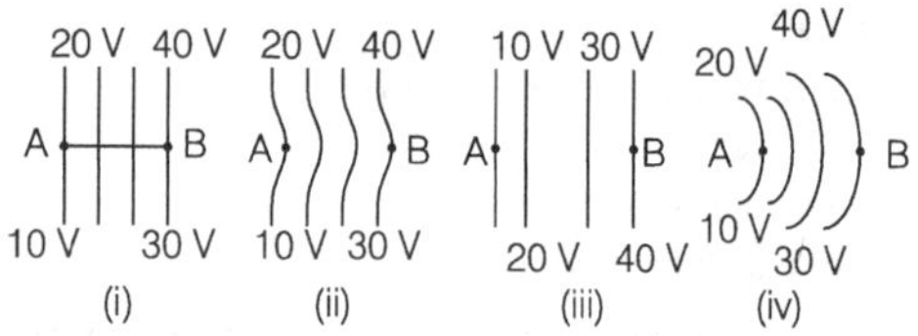


A positive charge is moved from $A$ to $B$ in each diagram. Then, ***NEET 2017***
(a) maximum work is required to move $q$ in Fig. (iii)
(b) in all the four cases, the work done is the same
(c) minimum work is required to move $q$ in Fig. (i)
(d) maximum work is required to move $q$ in Fig. (ii)

**27** The electric potential at a certain distance from a point charge is 600 V and the electric field is 200 $NC^{-1}$. The distance of the point charge (in metres) is
(a) 2 (b) 3 (c) 1 (d) 0

# TOPIC 2 ~ *Electrostatic Potential Energy*

**28** What is the work done in moving a 2 $\mu C$ point charge from corner $A$ to corner $B$ of a square $ABCD$ in figure, when a 10 $\mu C$ charge exists at the centre of the square?

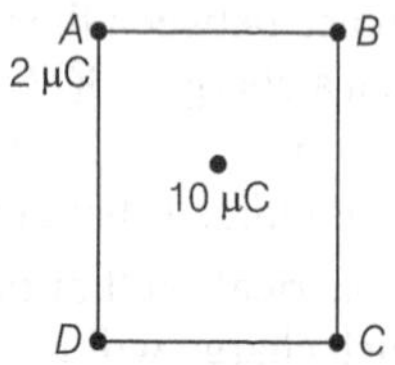


(a) Zero (b) 5
(c) 2 (d) 20

**29** In the figure, proton moves a distance $d$ in a uniform electric field $E$ as shown in the figure. Does the electric field do a positive or negative work on the proton? Does the electric potential energy of the proton increase or decrease?

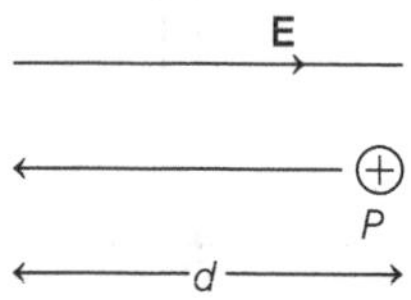


(a) Negative, increase (b) Positive, decrease
(c) Negative, decrease (d) Positive, increase

**30** The total potential energy of a system of three charges $q_1, q_2$ and $q_3$ located at $r_1, r_2$ and $r_3$ respectively is

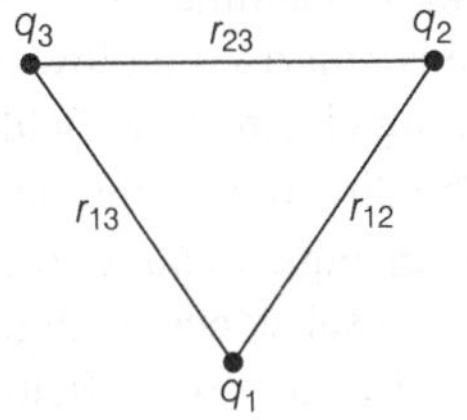


(a) $\frac{1}{4\pi\varepsilon_0}\left(\frac{q_1q_2}{r_{12}}+\frac{q_1q_3}{r_{13}}+\frac{q_2q_3}{r_{23}}\right)$

(b) $\frac{1}{4\pi\varepsilon_0}\left(\frac{q_1q_2}{r_{12}}-\frac{q_1q_3}{r_{13}}+\frac{q_2q_3}{r_{23}}\right)$

(c) $\frac{1}{4\pi\varepsilon_0}\left(\frac{q_1q_2}{r_{12}}+\frac{q_1q_3}{r_{13}}-\frac{q_2q_3}{r_{23}}\right)$

(d) $\frac{1}{4\pi\varepsilon_0}\left(\frac{q_1q_2}{r_{12}}-\frac{q_1q_3}{r_{13}}-\frac{q_2q_3}{r_{23}}\right)$

**31** Three charges $Q$, $+q$ and $+q$ are placed at the vertices of a right angle isosceles triangle as shown below. The net electrostatic energy of the configuration is zero, if the value of $Q$ is ***JEE Main 2019***

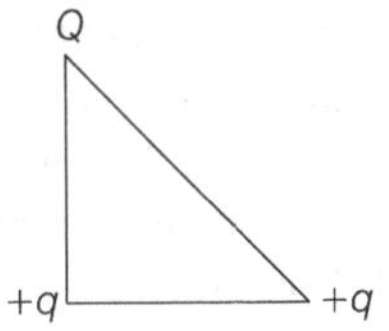


(a) $-2q$ (b) $\frac{-q}{1+\sqrt{2}}$

(c) $+q$ (d) $\frac{-\sqrt{2}q}{\sqrt{2}+1}$

**32** Two charges of equal magnitude $q$ are placed in air at a distance $2a$ apart and third charge $-2q$ is placed at mid-point. The potential energy of the system is ($\varepsilon_0$ = permittivity of free space)

(a) $-\frac{q^2}{8\pi\varepsilon_0 a}$ (b) $-\frac{3q^2}{8\pi\varepsilon_0 a}$

(c) $-\frac{5q^2}{8\pi\varepsilon_0 a}$ (d) $-\frac{7q^2}{8\pi\varepsilon_0 a}$

**33** Three point charges $+q_1, -2q$ and $-2q$ are placed at the vertices of an equilateral triangle of side $a$. The work done by some external force to increase their separation to $2a$ will be

(a) $\frac{1}{4\pi\varepsilon_0}\frac{2q^2}{a}$ (b) $\frac{1}{4\pi\varepsilon_0}\frac{q^2}{2a}$

(c) $\frac{1}{4\pi\varepsilon_0}\frac{\delta q}{a^2}$ (d) zero

**34** Two charges $q_1$ and $q_2$ are placed 30 cm apart, as shown in the figure. A third charge $q_3$ is moved along the arc of a circle of radius 40 cm from $C$ to $D$. The change in the potential energy of the system is $\frac{q_3}{4\pi\varepsilon_0}K$, where $K$ is

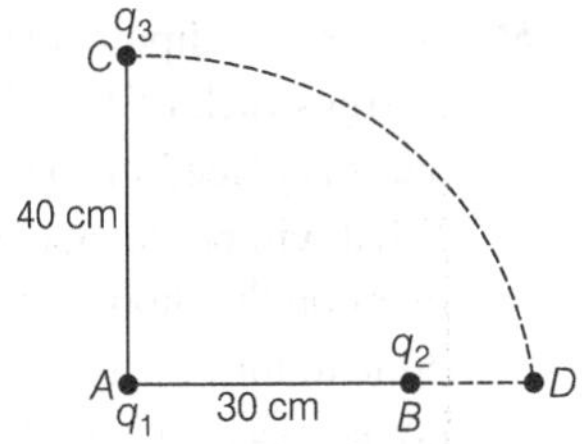


(a) $8q_2$ (b) $8q_1$ (c) $6q_2$ (d) $6q_1$

**35** Four charges are arranged at the corners of a square $ABCD$ of side $d$, as shown in the figure. The work required to put together this arrangement is

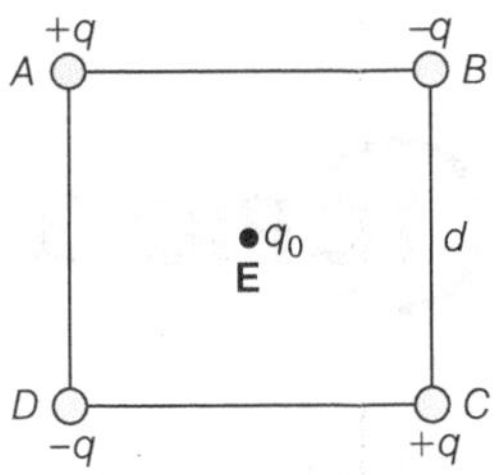


(a) $\frac{-q^2}{4\pi\varepsilon_0 d}(4-\sqrt{2})$

(b) $\frac{-q^3}{4\pi\varepsilon_0 d}(4-\sqrt{2})$

(c) $\frac{-q^4}{4\pi\varepsilon_0 d}(4-\sqrt{2})$ (d) $\frac{q^2}{4\pi\varepsilon_0 d}(4-\sqrt{2})$

**36** The potential difference between a cloud and the earth is $10^7$ V. Calculate the amount of energy dissipated when the charge of 100 C is transferred from the cloud to the ground due to lightning bolt.

(a) zero (b) $10^9$ J (c) 60 J (d) $10^7$ J

**37** Suppose a system of two charges as shown below placed in an external electric field

$$E = A\,(1/r^2),\text{ where } A = 9\times10^5\ \text{C/m}^2.$$

What would be the electrostatic energy of the configuration?

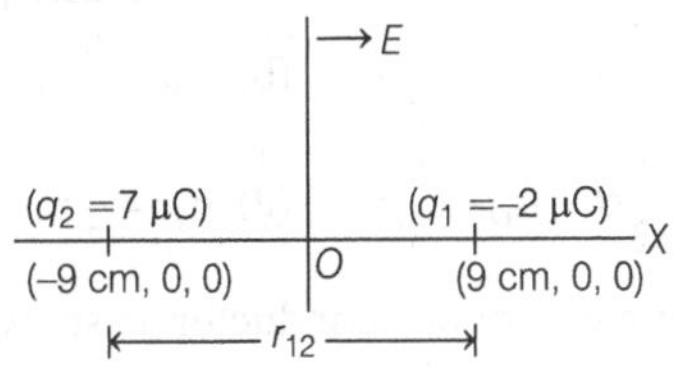


(a) 49 J (b) 49.3 J (c) − 49.3 J (d) 45 J

**38** An electric field of 1000 V/m is applied to an electric dipole at angle of 45°. The value of electric dipole moment is $10^{-29}$ C-m. What is the potential energy of the electric dipole? ***JEE Main 2019***

(a) $-9\times10^{-20}$ J (b) $-10\times10^{-29}$ J

(c) $-20\times10^{-18}$ J (d) $-7\times10^{-27}$ J

**39** An electric dipole, made up of positive and negative charges each of 3.5 μC and placed at a distance 4.2 cm apart is placed in an electric field of $5.87 \times 10^5$ NC$^{-1}$. What will be the maximum torque which the field can exert on the dipole? Also, obtain the work that must be done to turn the dipole from 0° to 180°.

(a) $86.289 \times 10^{-3}$ Nm$^{-1}$, $1.70 \times 10^{-1}$ J

(b) $75.26 \times 10^{-3}$ Nm$^{-1}$, $1.70 \times 10^{-1}$ J

(c) $86.289 \times 10^{-3}$ Nm$^{-1}$, $3.70 \times 10^{-3}$ J

(d) $55.26 \times 10^{-3}$ Nm$^{-1}$, $1.70 \times 10^{-3}$ J

**40** A molecule of a substance has a permanent electric dipole moment of magnitude $10^{-29}$ C-m. A mole of this substance is polarised (at low temperature) by applying a strong electrostatic field of magnitude $10^6$ Vm$^{-1}$. The direction of the field is suddenly changed from an angle of 60° to 0°. Estimate the heat released from the substance in aligning its dipoles along the new direction of the field.

(a) − 3 J (b) 3 J
(c) − 6 J (d) 6 J

# TOPIC 3 ~ Electrostatics of Conductors, Dielectrics and Polarisation

**41** What happens when a conductor is placed in an external electric field?

(a) The electric field due to induced charge opposes the external field within the conductor

(b) In the static situation, the two fields cancel each other and the electrostatic field in the conductor is zero

(c) Both (a) and (b)

(d) Neither (a) nor (b)

**42** If a conductor is given an excess of charge, then it

(a) resides on its surface

(b) move to its centre

(c) escape from its surface

(d) distributes equally inside it

**43** Two metal spheres, one of radius $R$ and the other of radius $2R$ respectively have the same surface charge density $\sigma$. They are brought in contact and separated. What will be the new surface charge densities on them? *NEET (Odisha) 2019*

(a) $\sigma_P = \frac{5}{6}\sigma, \ \sigma_Q = \frac{5}{2}\sigma$ (b) $\sigma_P = \frac{5}{2}\sigma, \ \sigma_Q = \frac{5}{6}\sigma$

(c) $\sigma_P = \frac{5}{2}\sigma, \ \sigma_Q = \frac{5}{3}\sigma$ (d) $\sigma_P = \frac{5}{3}\sigma, \ \sigma_Q = \frac{5}{6}\sigma$

**44** Consider a cavity in a conductor as shown in the figure below, then the electric field and charge density in the cavity are

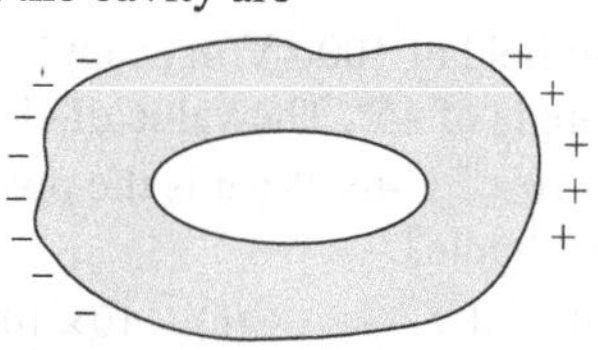

(a) $E > 0, \sigma > 0$ (b) $E < 0, \sigma < 0$
(c) $E = 0, \sigma = 0$ (d) $E > 0, \sigma < 0$

**45** Which of the following is/are the example of non-polar molecules?

(a) Oxygen (b) Hydrogen (c) Nitrogen (d) All of these

**46** The total dipole moment of dielectric with polar molecules in absence of an external electric field is

(a) zero (b) negative (c) infinite (d) None of these

**47** If dielectric constant and dielectric strength be denoted by $K$ and $X$ respectively, then a material suitable for using it as a dielectric in a capacitor must have

(a) high $K$ and high $X$ (b) high $K$ and low $X$
(c) low $K$ and high $X$ (d) low $K$ and low $X$

**48** The extent of polarisation depends on

(a) the dipole potential energy in the external field tending to align the dipoles with the field

(b) thermal energy tending to disrupt the alignment

(c) Both (a) and (b)

(d) Neither (a) nor (b)

**49** When a dielectric is placed in an electric field, the electric field inside the dielectric

(a) increases (b) decreases (c) constant (d) zero

**50** When **P** is normal to the surface, the polarisation is numerically equal to the

(a) surface density of the induced charge

(b) dipole moment

(c) Both (a) and (b)

(d) Neither (a) nor (b)

**51** The dielectric constant of helium at 0°C and 1 atmospheric pressure is 1.00074. Find the dipole moment induced in each helium atom when the gas is in an electric field of intensity 10 Vm$^{-1}$.

(a) $2.4 \times 10^{40}$ C-m (b) $2.4 \times 10^{-39}$ C-m
(c) $2.6 \times 10^{-39}$ C-m (d) $1.5 \times 10^{40}$ C-m

## TOPIC 4 ~ Capacitance, Combination of Capacitors and Energy Stored in Capacitors

**52** The graph shows the variation of voltage $V$ across the plates of two capacitors $A$ and $B$ *versus* increase of charge $Q$ stored in them. Which of the capacitors has higher capacitance?

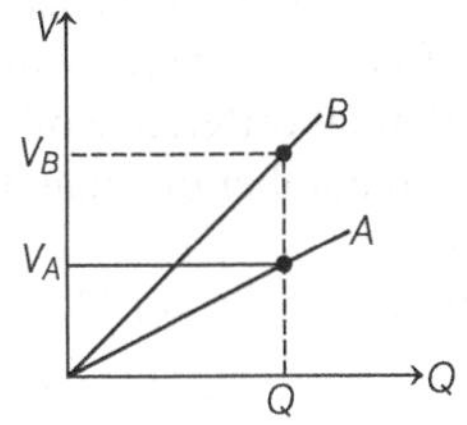


(a) Capacitor $A$
(b) Capacitor $B$
(c) Both have same capacitance
(d) None of the above

**53** The maximum electric field that a dielectric medium of a capacitor can withstand without breakdown (of its insulating property) is called its

(a) polarisation (b) capacitance
(c) dielectric strength (d) None of these

**54** Consider two parallel plates of area $A$ each having surface charge densities $\sigma$ and $-\sigma$ separated by distance $d$ as shown in figure below. The electric field in outer region I (region above the plate 1) is

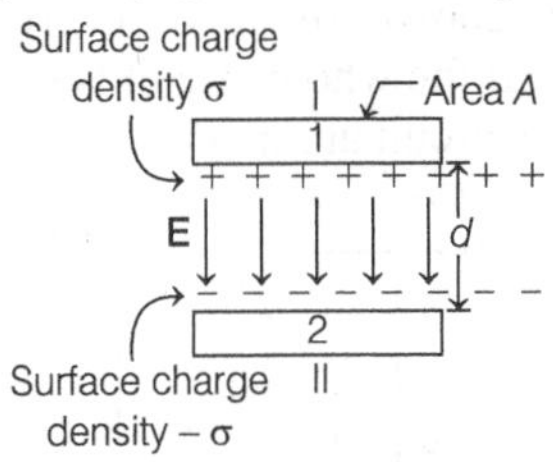


(a) $\frac{\sigma}{2\varepsilon_0}$ (b) $-\frac{\sigma}{2\varepsilon_0}$
(c) zero (d) $\frac{2\sigma}{4\varepsilon_0}$

**55** A parallel plate capacitor with plates of area 1 $m^2$ each, are at a separation of 0.1 m. If the electric field between the plates is 100 N/C, the magnitude of charge on each plate is

$$\left(\text{Take}, \varepsilon_0 = 8.85 \times 10^{-12}\ \frac{C^2}{\text{N-m}^2}\right)$$

**JEE Main 2019**

(a) $9.85 \times 10^{-10}$ C (b) $8.85 \times 10^{-10}$ C
(c) $7.85 \times 10^{-10}$ C (d) $6.85 \times 10^{-10}$ C

**56** A parallel plate capacitor has 1μF capacitance. One of its two plates is given $+2\,\mu$C charge and the other plate $+4\,\mu$C charge. The potential difference developed across the capacitor is **JEE Main 2019**

(a) 1 V (b) 5 V
(c) 2 V (d) 3 V

**57** A parallel plate capacitor is made of two circular plates separated by a distance of 5 mm and with a dielectric of dielectric constant 2.2 between them. When the electric field in the dielectric is $3 \times 10^4$ V/m, the charge density of the positive plate will be closed to

(a) $6 \times 10^{-7}$ $Cm^{-2}$ (b) $3 \times 10^{-7}$ $Cm^{-2}$
(c) $3 \times 10^4$ $Cm^{-2}$ (d) $6 \times 10^4$ $Cm^{-2}$

**58** A parallel plate air capacitor has a capacitance 18 μF. If the distance between the plates is tripled and a dielectric medium is introduced, the capacitance becomes 72 μF. The dielectric constant of the medium is

(a) 4 (b) 9 (c) 12 (d) 2

**59** In a parallel plate capacitor, the separation between the plates is 3 mm with air between them. Now, a dielectric of dielectric constant 2 is introduced between the plates due to which the capacity increases. In order to bring its capacity of the original value, the separation between the plates must be made

(a) 1.5 mm (b) 2.5 mm (c) 4 mm (d) 6 mm

**60** The capacitance of a parallel plate capacitor with air in between the plates is 3 μF. With the introduction of a dielectric medium between the plates, the capacitance becomes 15 μF. The permittivity of the medium is

(a) 5 $C^2 N^{-1} m^{-2}$ (b) 15 $C^2 N^{-1} m^{-2}$
(c) $0.44 \times 10^{-10}$ $C^2 N^{-1} m^{-2}$ (d) $8.845 \times 10^{-11}$ $C^2 N^{-1} m^{-2}$

**61** A capacitor of capacitance 15nF having dielectric slab of $\varepsilon_r = 2.5$, dielectric strength = 30 MV/m and potential difference = 30 V. Calculate the area of the plate.

(a) $6.7 \times 10^{-4}$ $m^2$ (b) $4.2 \times 10^{-4}$ $m^2$ **AIIMS 2019**
(c) $8.0 \times 10^{-4}$ $m^2$ (d) $9.85 \times 10^{-4}$ $m^2$

**62** A slab of material of dielectric constant $K$ has the same area as the plates of a parallel plate capacitor but has a thickness $(3/4)d$, where $d$ is the separation of the plates. How is the capacitance changed, when the slab is inserted between the plates?

(a) $\frac{4K}{K+3}C_0$ (b) $\frac{K+3}{4K}C_0$ (c) $\frac{K-3}{4K}C_0$ (d) $\frac{4K}{K-3}C_0$

**63** Two capacitors $C_1$ and $C_2$ are charged to 120 V and 200 V, respectively. It is found that by connecting them together, the potential on each one can be made zero. Then,

(a) $5C_1 = 3C_2$ (b) $3C_1 = 5C_2$
(c) $3C_1 + 5C_2 = 0$ (d) $9C_1 = 4C_2$

**64** The parallel combination of two air filled parallel plate capacitors of capacitance $C$ and $nC$ is connected to a battery of voltage, $V$. When the capacitors are fully charged, the battery is removed and after that a dielectric material of dielectric constant $K$ is placed between the two plates of the first capacitor. The new potential difference of the combined system is ***JEE Main 2019***

(a) $\frac{(n+1)V}{(K+n)}$ (b) $\frac{nV}{K+n}$ (c) $V$ (d) $\frac{V}{K+n}$

**65** Find the capacitance of the parallel plate capacitor as shown in the figure below. ***JIPMER 2018***

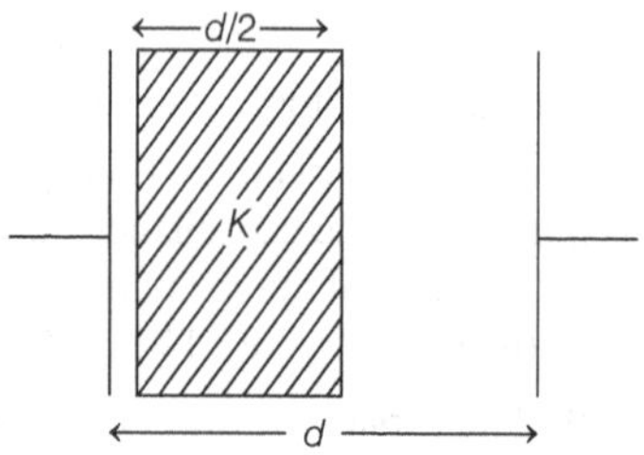


(a) $\frac{2KA\varepsilon_0}{(K+1)d}$ (b) $\frac{2KA\varepsilon_0}{d}$
(c) $\frac{(K+1)A\varepsilon_0}{2d}$ (d) $\frac{2KA\varepsilon_0}{(K^2+1)d}$

**66** Across each of two capacitors of capacitance $1\,\mu F$ and a potential difference of 10 V is applied. Then, positive plate of one is connected to the negative plate of the other and negative plate of one is connected to the positive plate of the other. After contact,

(a) charge on each is zero
(b) charge on each is same but non-zero
(c) charge on each is different but non-zero
(d) None of the above

**67** A gang capacitor is formed by interlocking nine plates with each other. The distance between the consecutive plates is 0.885 cm and the overlapping area of the plate is 5 cm$^2$. The capacity of the unit is

(a) 1.06 pF (b) 4 pF
(c) 6.36 pF (d) 12.72 pF

**68** A parallel plate capacitor has the space between its plates filled by the two slabs of thickness $\frac{d}{2}$ each, having dielectric constants $K_1$ & $K_2$ and $d$ is the plate separation of the capacitor. The capacity of the capacitor is ***JEE Main 2016***

(a) $\frac{2\varepsilon_0 d}{A}\left(\frac{K_1+K_2}{K_1K_2}\right)$ (b) $\frac{2\varepsilon_0 A}{d}\left(\frac{K_1K_2}{K_1+K_2}\right)$

(c) $\frac{2\varepsilon_0 A}{A}(K_1+K_2)$ (d) $\frac{2\varepsilon_0 A}{d}\left(\frac{K_1+K_2}{K_1K_2}\right)$

**69** A capacitance of $2\,\mu F$ is required in an electrical circuit across a potential difference of 1kV. A large number of $1\,\mu F$ capacitors are available which can withstand a potential difference of not more than 300 V. The minimum number of capacitors required to achieve this is ***JEE Main 2017***

(a) 16 (b) 24 (c) 32 (d) 2

**70** An infinite number of identical capacitors each of capacitance $1\,\mu F$ are connected as shown in the figure. Then, the equivalent capacitance between $A$ and $B$ is ***AIIMS 2018***

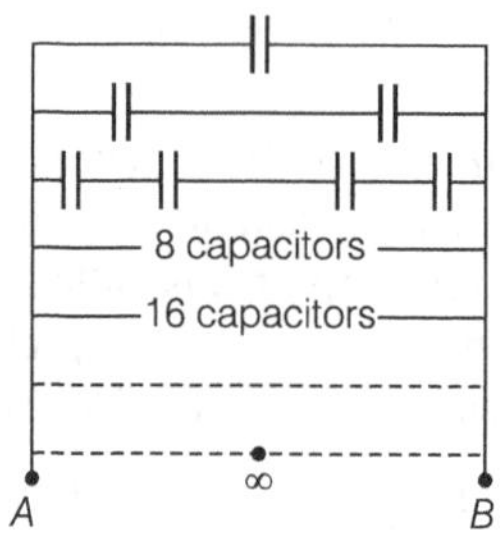


(a) $1\,\mu F$ (b) $2\mu F$ (c) $\frac{1}{2}\,\mu F$ (d) $\infty$

**71** In the circuit shown, find $C$ if the effective capacitance of the whole circuit is to be $0.5\,\mu F$. All values in the circuit are in $\mu F$. ***JEE Main 2019***

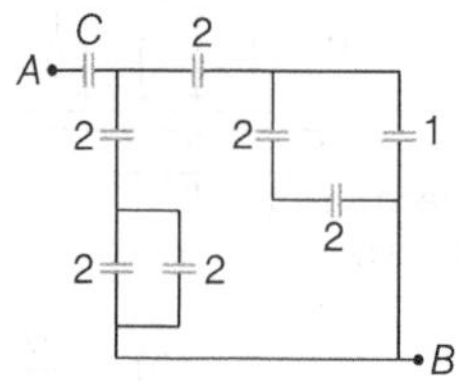


(a) $\frac{6}{5}\,\mu F$ (b) $4\,\mu F$ (c) $\frac{7}{10}\,\mu F$ (d) $\frac{7}{11}\,\mu F$

**72** A number of condensers, each of the capacitance $1\,\mu F$ and each one of which gets punctured, if a potential difference just exceeding 500 V is provided. An arrangement suitable for giving capacitance of $2\,\mu F$ across which 3000 V may be applied requires at least

(a) 6 component capacitors
(b) 12 component capacitors
(c) 72 component capacitors
(d) 2 component capacitors

**73** In the circuit shown in the figure, the potential difference across the 4.5 μF capacitor is

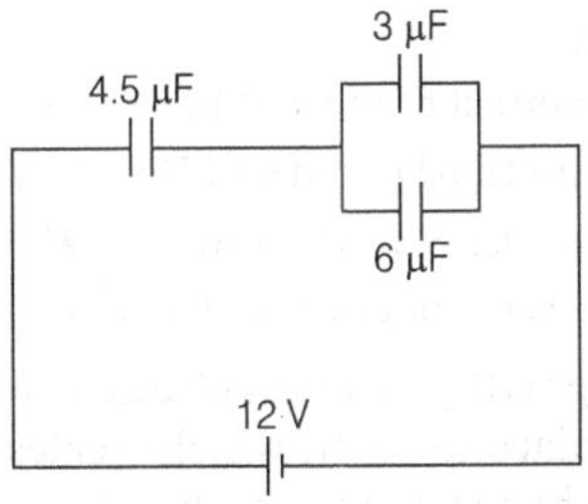


(a) $\frac{8}{3}$ V (b) 4 V (c) 6 V (d) 8 V

**74** The electrostatic force between the metal plates of an isolated parallel plate capacitor $C$ having a charge $Q$ and area $A$, is **NEET 2018**

(a) proportional to the square root of the distance between the plates
(b) linearly proportional to the distance between the plates
(c) independent of the distance between the plates
(d) inversely proportional to the distance between the plates

**75** A 900 pF capacitor is charged by 100 V battery in the figure. How much electrostatic energy is stored by the capacitor?

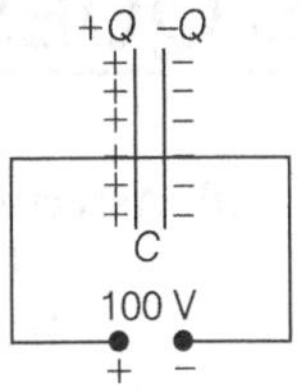


(a) $45 \times 10^{-6}$ J
(b) $4.5 \times 10^{6}$ J
(c) $4.5 \times 10^{-6}$ J
(d) $0.45 \times 10^{5}$ J

**76** A parallel plate capacitor having capacitance 12 pF is charged by a battery to a potential difference of 10 V between its plates. The charging battery is now disconnected and a porcelain slab of dielectric constant 6.5 is slipped between the plates. The work done by the capacitor on the slab is **JEE Main 2019**

(a) 560 pJ (b) 508 pJ
(c) 692 pJ (d) 600 pJ

**77** A capacitor with capacitance 5 μF is charged to 5 μC. If the plates are pulled apart to reduce the capacitance to 2 μF, how much work is done? **JEE Main 2019**

(a) $6.25 \times 10^{-6}$ J (b) $2.16 \times 10^{-6}$ J
(c) $2.55 \times 10^{-6}$ J (d) $3.75 \times 10^{-6}$ J

**78** A capacitor is charged by a battery. The battery is removed and another identical uncharged capacitor is connected in parallel. The total electrostatic energy of resulting system **NEET 2017**

(a) increases by a factor of 4
(b) decreases by a factor of 2
(c) remains the same
(d) increases by a factor of 2

**79** In the given circuit, find the heat produced between $A$ and $B$ (given $C = 1\,\mu$F). **JIPMER 2019**

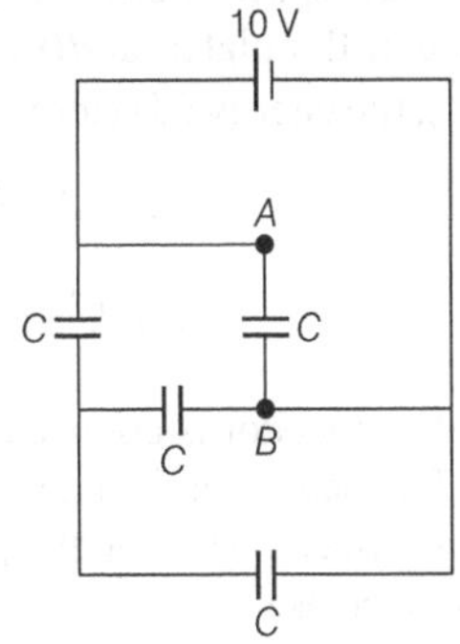


(a) 50 μJ (b) 60 μJ (c) 70 μJ (d) 100 μJ

**80** Two identical capacitors $C_1$ and $C_2$ of equal capacitance are connected as shown in the circuit. Terminals $a$ and $b$ of the key $K$ are connected to charge capacitor $C_1$ using battery of emf $V$ volt. Now, disconnecting $a$ and $b$, the terminals $b$ and $c$ are connected. Due to this, what will be the percentage loss of energy? **NEET (Odisha) 2019**

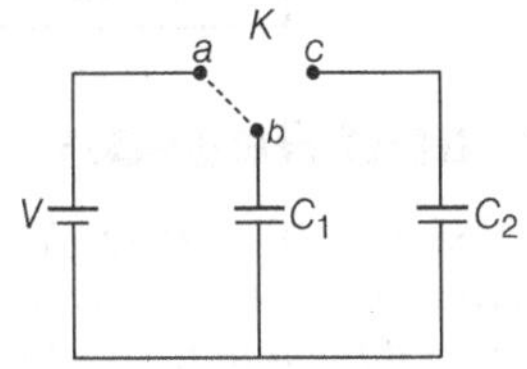


(a) 75% (b) 0% (c) 50% (d) 25%

**81** A capacitor of 2μF is charged as shown in the figure. When the switch $S$ is turned to position 2, the percentage of its stored energy dissipated is **NEET 2016**

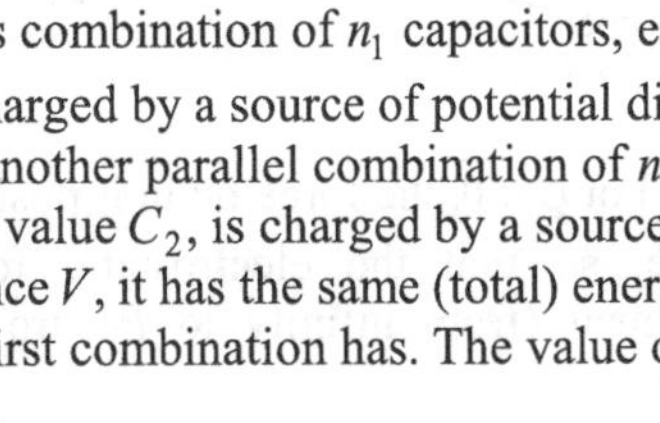


(a) 20% (b) 75% (c) 40% (d) 0%

**82** A series combination of $n_1$ capacitors, each of value $C_1$, is charged by a source of potential difference 4V. When another parallel combination of $n_2$ capacitors, each of value $C_2$, is charged by a source of potential difference $V$, it has the same (total) energy stored in it as the first combination has. The value of $C_2$ in terms of $C_1$ is

(a) $\frac{2C_1}{n_1 n_2}$ (b) $16\frac{n_1}{n_2}$
(c) $2\frac{n_2}{n_1}C_1$ (d) $\frac{16C_1}{n_1 n_2}$

**83** A parallel plate capacitor has a uniform electric field $E\ (\text{Vm}^{-1})$ in the space between the plates. If the distance between the plates is $d(\text{m})$ and area of each plate is $A(\text{m}^2)$, the energy (J) stored in the capacitor is

(a) $\frac{1}{2}\varepsilon_0 E^2$ (b) $\varepsilon_0 EAd$
(c) $\frac{1}{2}\varepsilon_0 E^2 Ad$ (d) $E^2 Ad/\varepsilon_0$

**84** A parallel plate capacitor is connected to a 5V battery and charged. The battery is then disconnected and a glass slab is introduced between the plates, then the quantities that decrease are

(a) charge and potential difference
(b) charge and capacitance
(c) capacitance and potential difference
(d) energy stored and potential difference

**85** van de graaff generator is based on the phenomenon of corona discharge, which is used for developing

(a) electric current of order of $10\times 10^6$ A
(b) electric potential of order of $10\times 10^6$ V
(c) electric power of order of $10\times 10^6$ W
(d) electric charge of order of $10\times 10^6$ C

**86** In van de graaff generator, charge is moved from one place to other, which creates a potential difference between two points. The potential produces

(a) accelerated charged particles
(b) heating in generator
(c) decelerated charged particles
(d) None of the above

# SPECIAL TYPES QUESTIONS

## I. Assertion and Reason

**Direction** (Q. Nos. 87-96) *In the following questions, a statement of Assertion is followed by a corresponding statement of Reason. Of the following statements, choose the correct one.*

(a) Both Assertion and Reason are correct and Reason is the correct explanation of Assertion.
(b) Both Assertion and Reason are correct but Reason is not the correct explanation of Assertion.
(c) Assertion is correct but Reason is incorrect.
(d) Assertion is incorrect but Reason is correct.

**87 Assertion** Work done by the electrostatic force in bringing the unit positive charge from infinity to the point $P$ is positive.

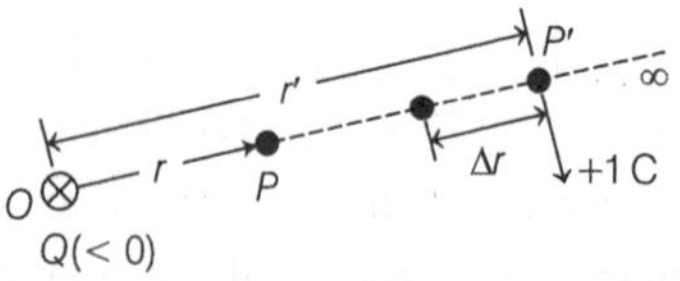


**Reason** For $Q < 0$, the force on unit positive charge is attractive, so that the electrostatic force and the displacement (from infinity to $P$) are in the same direction.

**88 Assertion** An electron moves from a region of lower potential to a region of higher potential.
**Reason** An electron has a negative charge.

**89 Assertion** The expression of potential energy $U = \frac{1}{4\pi\varepsilon_0}\frac{q_1 q_2}{r_{12}}$, is unaltered whatever way the charges are brought to the specified locations.

**Reason** Path-independence of work for electrostatic force.

**90 Assertion** In the absence of an external electric field, the dipole moment per unit volume of a polar dielectric is zero.

**Reason** The dipoles of a polar dielectric are randomly oriented.

**91 Assertion** Polar molecules have permanent dipole moment.

**Reason** In polar molecules, the centre of positive and negative charges coincides even when there is no external field.

**92 Assertion** A capacitor can be given only a limited quantity of charge.

**Reason** Charge stored by a capacitor depends on the shape and size of the plates of capacitor and the surrounding medium.

**93 Assertion** Capacity of a parallel plate capacitor increases when distance between the plates is decreased.

**Reason** Capacitance of a capacitor is directly proportional to distance between them.

**94 Assertion** A parallel plate capacitor is connected across a battery through a key. A dielectric slab of dielectric constant $K$ is introduced between the plates. The electric field inside the capacitor becomes $K$ times.

**Reason** The surface density of charge on the plate remains constant or unchanged.

**95 Assertion** Charge on all the condensers connected in series is the same.

**Reason** Capacitance of capacitor is directly proportional to charge on it.

**96 Assertion** If three capacitors of capacitances $C_1 < C_2 < C_3$ are connected in parallel and in series, then their equivalent capacitances.

$$C_p > C_s$$

where, $C_p$ and $C_s$ are the equivalent capacitances in parallel and series, respectively.

**Reason** $\frac{1}{C_p} = \frac{1}{C_1} + \frac{1}{C_2} + \frac{1}{C_3}$ and $C_s = C_1 + C_2 + C_3$

where, $C_p$ and $C_s$ is the equivalent capacitance in parallel and series, respectively.

## II. Statement Based Questions

**97** I. In a metal, the outer (valence) electrons part away from their atoms and are free to move.

II. The valence electrons are within the metal but not free to leave the metal.

III. The free electrons form a kind of gas; they collide with each other and with the ions and move randomly in different directions.

Which of the following statement(s) is/are correct?

(a) Both I and II (b) Both I and III
(c) I, II and III (d) None of these

**98** I. In metals, the positive ions are made up of the nuclei and the bound electrons remain held in their fixed positions.

II. In electrolytic conductors, the charge carriers are both positive and negative ions.

Which of the following statement(s) is/are correct?

(a) Only I
(b) Only II
(c) Both I and II
(d) None of the above

**99** I. In an external electric field, the positive and negative charges of a non-polar molecule are displaced in opposite directions.

II. In non-polar molecules, displacement stops when the external force on the constituent charges of the molecule is balanced by the restoring force.

III. The non-polar molecule develops an induced dipole moment.

Which of the following statement(s) is/are correct?

(a) Only I (b) Only II
(c) Both I and II (d) I, II and III

**100** The electric field due to a plane sheet of charge having charge $q$ and area $A$ is given by $E = \frac{q}{2\varepsilon_0 A}$.

Further, with usual notation, in case of a parallel plate capacitor,

$$E = \frac{q}{\varepsilon_0 A}, V = \frac{qd}{\varepsilon_0 A} \text{ and } C = \frac{\varepsilon_0 A}{d}$$

Let us now consider that the charges given to the two plates $A$ and $B$ of a parallel plate capacitor are different, i.e. $q_1$ and $q_2$ as shown in figure.

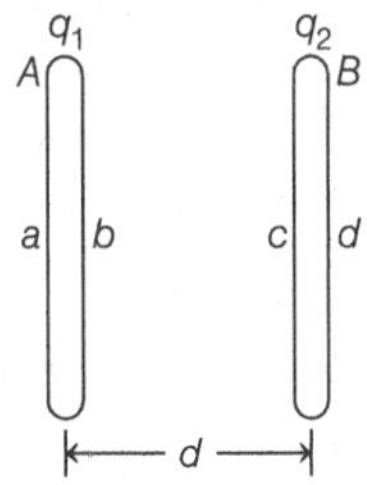


With the help of the paragraph given above, which of the following statement is correct?

I. The charge on the surfaces $a$ and $b$ of the plate $A$ are $\frac{q_1 + q_2}{2}$ and $\frac{q_1 - q_2}{2}$.

II. The charges on the surfaces $c$ and $d$ of the plate $B$ are $\frac{q_1 - q_2}{2}$ and $\frac{q_1 + q_2}{2}$.

III. The electric field between the two plates $A$ and $B$ is $\frac{q_1 - q_2}{2\varepsilon_0 A}$.

IV. The capacitance of the capacitor is given by $\frac{\varepsilon_0 A}{d}$.

(a) Both I and II (b) Both II and III
(c) I, III and IV (d) I, II, III and IV

**101** I. Electric field, electric displacement and electric polarisation are related by the relation as $D = \varepsilon_0 E - P$.

II. Electric field vector (**E**), electric displacement vector (**D**) and electric polarisation vector (**P**) are mutually perpendicular to each other.

III. Electric field vector (**E**), electric displacement vector (**D**) and electric polarisation vector (**P**) are mutually parallel to each other.

Which of the following statement(s) is/are correct?

(a) Only I (b) Both II and III
(c) Both I and III (d) Only III

**102** The electrostatic potential on the surface of a charged conducting sphere is 100 V. Two statements are made in this regard,

$S_1$: At any point inside the sphere, electric intensity is zero.

$S_2$: At any point inside the sphere, the electrostatic potential is 100 V.

Which of the following statement (s) is/are correct?

(a) $S_1$ is true but $S_2$ is false.
(b) Both $S_1$ and $S_2$ are false.
(c) $S_1$ is true, $S_2$ is also true and $S_1$ is the cause of $S_2$.
(d) $S_1$ is true, $S_2$ is also true but the statements are independent.

**103** Choose the incorrect statement. Equipotential surfaces

(a) are closer in regions of large electric fields compared to regions of lower electric fields.
(b) will be more crowded near sharp edges of a conductor.
(c) will be more crowded near regions of large charge densities.
(d) will always be equally spaced.

**104** Read the following statements regarding to electrostatics of conductors. Which of the following is correct?

(a) Inside the conductor, electrostatic field is zero.
(b) At the surface of a charged conductor, electrostatic field must be parallel to the surface at every point.
(c) Electrostatic potential continuously varies throughout the volume of the conductor.
(d) Electrostatic field at the surface of a charged conductor is zero.

**105** The two plates of a parallel plate capacitor having charges $Q_1$ and $Q_2$ are given. The capacitance of the capacitor is $C$. When the switch is closed, mark the incorrect statement(s). Assuming both $Q_1, Q_2$ to be positive.

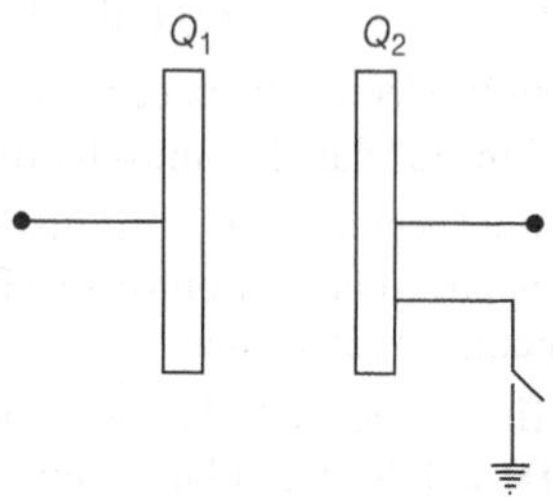


(a) The charge flown through switch is zero.
(b) The charge flown through switch is $Q_1 + Q_2$.
(c) Potential difference across the capacitor plates is $\frac{Q_1}{C}$.
(d) The charge of the capacitor is $Q_1$.

**106** A parallel plate capacitor is charged and the charging battery is then disconnected. If the plates of the capacitor are moved further apart by means of insulating handles, then which of the following statement is correct?

(a) The charge on the capacitor increases.
(b) The voltage across the plates increases.
(c) The capacitance increases.
(d) The electrostatic energy stored in the capacitor decreases.

**107** A parallel plate capacitor is connected to a battery as shown in figure. Read the statements I and II and choose the correct option(s).

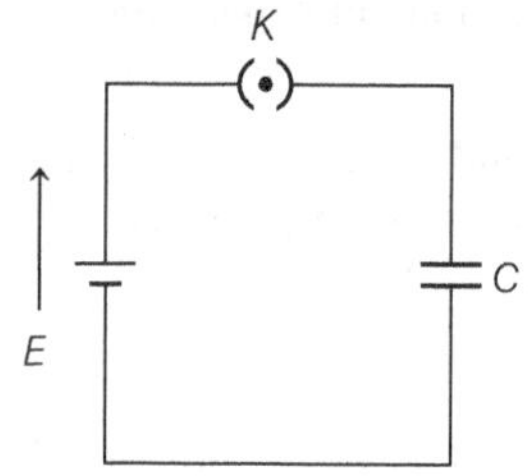


I. Key $K$ is kept closed and plates of capacitors are moved apart using insulating handle.
II. Key $K$ is opened and plates of capacitors are moved apart using insulating handle.

(a) In I $Q$ remains same but $C$ changes.
(b) In II $V$ remains same but $C$ changes.
(c) In I $V$ remains same and but $Q$ changes.
(d) In II $Q$ and $V$ both changes.

**108** A 900 pF capacitor is charged by 100 V battery, then which of the following statement is correct?

(a) The charge on the capacitor is $4.5 \times 10^{-8}$ C.
(b) The energy stored by the capacitor is $9 \times 10^{-6}$ J.
(c) If capacitor is disconnected from the battery and connected to another 900 pF capacitor, then energy stored by the system is $2.25 \times 10^{-6}$ J.
(d) None of the above

**109** A parallel plate air capacitor of capacitance $C_0$ is connected to a cell of emf $V$ and then disconnected from it. A dielectric slab of dielectric constant $K$, which can just fill the air gap of the capacitor, is now inserted in it.

Which of the following statement is incorrect?

(a) The potential difference between the plates decreases $K$ times.
(b) The energy stored in the capacitor decreases $K$ times.
(c) The change in energy is $\frac{1}{2}C_0V^2(K-1)$.
(d) The change in energy is $\frac{1}{2}C_0V^2\left(\frac{1}{K}-1\right)$.

## III. Matching Type

**110** Match the Column I (energy in eV) with Column II (energy in J) and select the correct answer from the codes given below.

| | Column I | | Column II |
|---|---|---|---|
| A. | 1 keV | 1. | $1.6\times10^{-7}$ J |
| B. | 1 MeV | 2. | $1.6\times10^{-10}$ J |
| C. | 1 GeV | 3. | $1.6\times10^{-16}$ J |
| D. | 1 TeV | 4. | $1.6\times10^{-13}$ J |

| | A | B | C | D |
|---|---|---|---|---|
| (a) | 4 | 2 | 3 | 1 |
| (b) | 3 | 4 | 2 | 1 |
| (c) | 2 | 3 | 4 | 1 |
| (d) | 1 | 4 | 3 | 2 |

**111** Match the Column I (potential) with Column II (value) for electric potential at different points $P_1, P_2, P_3$ and $P_4$ as shown in the figure with respect to an electric dipole.

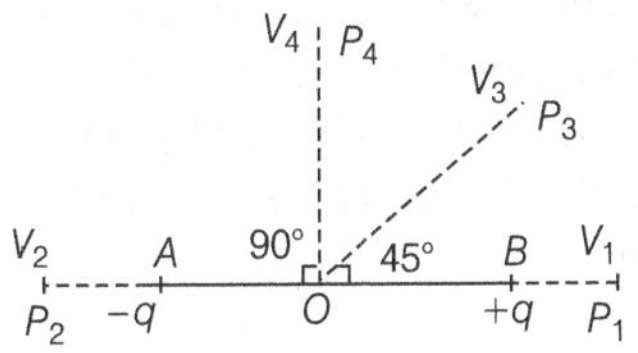


Each points $P_1, P_2, P_3$ and $P_4$ are equal distance $r$ from mid-point $O$ of dipole.

| | Column I | | Column II |
|---|---|---|---|
| A | $V_1$ | 1. | 0 |
| B | $V_2$ | 2. | $\frac{p}{4\sqrt{2}\pi\varepsilon_0 r^2}$ |
| C | $V_3$ | 3. | $\frac{-p}{4\pi\varepsilon_0 r^2}$ |
| D | $V_4$ | 4. | $\frac{p}{4\pi\varepsilon_0 r^2}$ |

| | A | B | C | D |
|---|---|---|---|---|
| (a) | 3 | 4 | 2 | 1 |
| (b) | 1 | 2 | 3 | 4 |
| (c) | 4 | 3 | 2 | 1 |
| (d) | 4 | 3 | 1 | 2 |

**112** Match the Column I (parameter of capacitor) with Column II (value) and select the correct answer from the codes given below. Consider a dielectric inserted between the plates fully occupying the intervening region. The surface charge densities on the charged plates are $\pm\sigma$.

| | Column I | | Column II |
|---|---|---|---|
| A. | Electric field $E$ | 1. | $Qd/A\varepsilon_0 K$ |
| B. | Potential difference across the plates $V$ | 2. | $(\sigma-\sigma_p)/\varepsilon_0$ |
| C. | The capacitance $C$, with dielectric between the plates | 3. | $C/C_0$ |
| D. | Dielectric constant of the substance ($K$) | 4. | $\varepsilon_0 KA/d$ |

| | A | B | C | D | | A | B | C | D |
|---|---|---|---|---|---|---|---|---|---|
| (a) | 1 | 2 | 3 | 4 | (b) | 2 | 1 | 4 | 3 |
| (c) | 3 | 4 | 1 | 2 | (d) | 4 | 2 | 3 | 1 |

**113** Match the Column I (energy of capacitor) with Column II (value) and select the correct answer from the codes given below.
A capacitor $C_1$ of capacitance $C$ is charged to a potential difference $V_0$. The terminals of the charged capacitor are then connected to an uncharged capacitor $C_2$ of capacitance $C/2$.

| | Column I | | Column II |
|---|---|---|---|
| A. | Final energy of capacitor $C_1$ | 1. | $-(1/6)\,CV_0^2$ |
| B. | Final energy of capacitor $C_2$ | 2. | $(1/6)\,CV_0^2$ |
| C. | Final energy of the system | 3. | $(1/3)\,CV_0^2$ |
| D. | Change in energy on joining the capacitors | 4. | $(2/9)\,CV_0^2$ |
| | | 5. | $(1/9)\,CV_0^2$ |

| | A | B | C | D | | A | B | C | D |
|---|---|---|---|---|---|---|---|---|---|
| (a) | 1 | 2 | 3 | 5 | (b) | 4 | 5 | 3 | 1 |
| (c) | 5 | 4 | 3 | 1 | (d) | 1 | 2 | 3 | 4 |

**114** Match the Column I (charge on capacitor) with Column II (value) and select the correct answer from the codes given below.

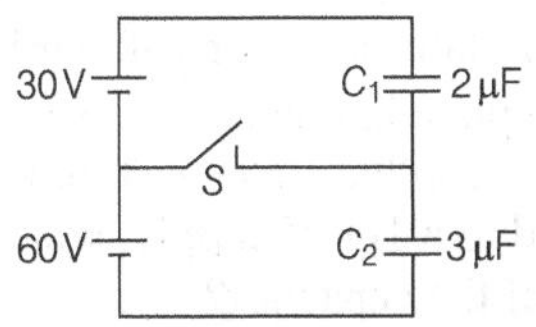


| | Column I | | Column II |
|---|---|---|---|
| A. | Charge on both the capacitors ($S$ open) | 1. | 108 μC |
| B. | Charge on $C_2$ ($S$ closed) | 2. | 240 μC |
| C. | Charge on $C_1$ ($S$ closed) | 3. | 60 μC |
| D. | Charge on $C_1$ and $C_2$ ($S$ closed) | 4. | 180 μC |

| | A | B | C | D | | A | B | C | D |
|---|---|---|---|---|---|---|---|---|---|
| (a) | 1 | 2 | 3 | 4 | (b) | 4 | 3 | 2 | 1 |
| (c) | 1 | 4 | 3 | 2 | (d) | 1 | 3 | 2 | 4 |

# NCERT & NCERT Exemplar

## MULTIPLE CHOICE QUESTIONS

### NCERT

**115** Two charges $5 \times 10^{-8}$ C and $-3 \times 10^{-8}$ C are located 16 cm apart. At what point from positive charge on the line joining the two charges is the electric potential zero? (Take, the potential at infinity to be zero).

(a) 6 cm (b) 8 cm
(c) 10 cm (d) 30 cm

**116** A regular hexagon of side 10 cm has a charge 5 μC at each of its vertices. Calculate the potential at the centre of the hexagon.

(a) $3.7 \times 10^{-6}$ V (b) $2.7 \times 10^{6}$ V
(c) $2.7 \times 10^{-4}$ V (d) $3.7 \times 10^{4}$ V

**117** A parallel plate capacitor with air between the plates has a capacitance of 8 pF (1 pF $= 10^{-12}$ F). What will be the capacitance, if the distance between the plates is reduced by half and the space between them is filled with a substance of dielectric constant 6?

(a) 24 pF (b) 96 pF (c) 8 pF (d) 12 pF

**118** Three capacitors each of capacitance 9 pF are connected in series. What is the potential difference across each capacitor, if the combination is connected to a 120 V supply?

(a) 40 V (b) 60 V
(c) 80 V (d) 50 V

**119** In a parallel plate capacitor with air between the plates, each plate has an area of $6 \times 10^{-3}$ m$^2$ and the distance between the plates is 3 mm. If this capacitor is connected to a 100 V supply, what is the charge on each plate of the capacitor?

(a) $1.77 \times 10^{-9}$ C (b) $2.77 \times 10^{-9}$ C
(c) $1 \times 10^{-8}$ C (d) $4.7 \times 10^{-6}$ C

**120** A 600 pF capacitor is charged by a 200 V supply, then it is disconnected from the supply and is connected to another uncharged 600 pF capacitor. How much electrostatic energy is lost in the process?

(a) $12 \times 10^{-6}$ J (b) $8 \times 10^{-6}$ J
(c) $6 \times 10^{-6}$ J (d) $4 \times 10^{-6}$ J

**121** A charge of 8 mC is located at the origin. Calculate the work done in taking a small charge of $-2 \times 10^{-9}$ C from a point $P(0, 0, 3)$ (in cm) to a point $Q(0, 4, 0)$ (in cm), *via* a point $R(0, 6, 9)$ (in cm).

(a) 2.4 J (b) 1.2 J
(c) 2 J (d) 3.6 J

**122** A cube of side $b$ has a charge $q$ at each of its vertices. Determine the potential due to this charge array at the centre of the cube.

(a) $\frac{4\sqrt{3}q}{\pi\varepsilon_0 b}$ (b) $\frac{3q}{\pi\varepsilon_0 b}$
(c) $\frac{4q}{\sqrt{3}\pi\varepsilon_0 b}$ (d) zero

**123** Two tiny spheres carrying charges 1.5 μC and 2.5 μC are located 30 cm apart. Find the potential at a point 10 cm from this mid-point in a plane normal to the line and passing through the mid-point.

(a) $10^5$ V (b) $2 \times 10^5$ V
(c) $2 \times 10^8$ V (d) Zero

**124** In a hydrogen atom, the electron and proton are bound at a distance of about 0.53 Å.
What is the potential energy of the system in eV, taking the zero of the potential energy at infinite separation of the electron from proton?

(a) + 27.16 eV (b) − 27.16 eV
(c) 13.58 eV (d) − 13.58 eV

### NCERT Exemplar

**125** A positively charged particle is released from rest in an uniform electric field. The electric potential energy of the charge

(a) remains constant because the electric field is uniform
(b) increases because the charge moves along the electric field
(c) decreases because the charge moves along the electric field
(d) decreases because the charge moves opposite to the electric field

**126** Figure shows some equipotential lines distributed in space. A charged object is moved from point $A$ to point $B$.

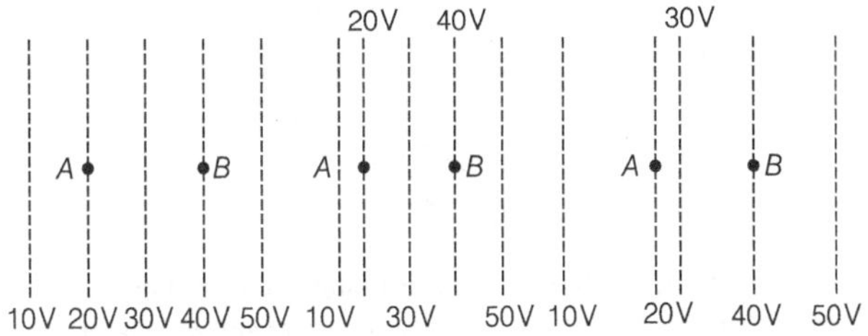

Fig. (i) Fig. (ii) Fig. (iii)

(a) The work done in Fig. (i) is the greatest
(b) The work done in Fig. (ii) is least
(c) The work done is the same in Figs. (i), (ii) and (iii)
(d) The work done in Fig. (iii) is greater than Fig. (ii) but equal to that in

**127** Equipotentials at a great distance from a collection of charges whose total sum is not zero are approximately
(a) spheres (b) planes (c) paraboloids (d) ellipsoids

**128** A parallel plate capacitor is made of two dielectric blocks in series. One of the blocks has thickness $d_1$ and dielectric constant $K_1$ and the other has thickness $d_2$ and dielectric constant $K_2$ as shown in figure. This arrangement can be thought as a dielectric slab of thickness $d\ (= d_1 + d_2)$ and effective dielectric constant $K$, then $K$ is` **JEE Main 2016**

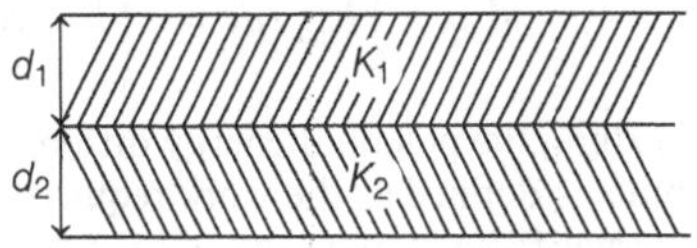


(a) $\dfrac{K_1 d_1 + K_2 d_2}{d_1 + d_2}$ (b) $\dfrac{K_1 d_1 + K_2 d_2}{K_1 + K_2}$
(c) $\dfrac{K_1 K_2 (d_1 + d_2)}{(K_1 d_2 + K_2 d_1)}$ (d) $\dfrac{2K_1 K_2}{K_1 + K_2}$

# Answers

## > Mastering NCERT with MCQs

| | | | | | | | | | |
|---|---|---|---|---|---|---|---|---|---|
| 1 (a) | 2 (b) | 3 (c) | 4 (c) | 5 (d) | 6 (a) | 7 (b) | 8 (b) | 9 (a) | 10 (a) |
| 11 (b) | 12 (b) | 13 (a) | 14 (b) | 15 (c) | 16 (a) | 17 (a) | 18 (c) | 19 (b) | 20 (a) |
| 21 (a) | 22 (a) | 23 (b) | 24 (c) | 25 (a) | 26 (b) | 27 (b) | 28 (a) | 29 (a) | 30 (a) |
| 31 (d) | 32 (d) | 33 (d) | 34 (a) | 35 (a) | 36 (b) | 37 (b) | 38 (d) | 39 (a) | 40 (b) |
| 41 (c) | 42 (a) | 43 (d) | 44 (c) | 45 (d) | 46 (a) | 47 (a) | 48 (c) | 49 (b) | 50 (a) |
| 51 (b) | 52 (a) | 53 (c) | 54 (c) | 55 (b) | 56 (a) | 57 (a) | 58 (c) | 59 (d) | 60 (c) |
| 61 (a) | 62 (a) | 63 (b) | 64 (a) | 65 (a) | 66 (a) | 67 (b) | 68 (b) | 69 (c) | 70 (b) |
| 71 (d) | 72 (c) | 73 (d) | 74 (c) | 75 (c) | 76 (b) | 77 (d) | 78 (d) | 79 (a) | 80 (c) |
| 81 (c) | 82 (d) | 83 (c) | 84 (d) | 85 (b) | 86 (a) | | | | |

## > Special Types Questions

| | | | | | | | | | |
|---|---|---|---|---|---|---|---|---|---|
| 87 (a) | 88 (a) | 89 (a) | 90 (a) | 91 (c) | 92 (a) | 93 (c) | 94 (c) | 95 (c) | 96 (c) |
| 97 (c) | 98 (c) | 99 (d) | 100 (c) | 101 (d) | 102 (c) | 103 (d) | 104 (a) | 105 (a) | 106 (b) |
| 107 (c) | 108 (c) | 109 (c) | 110 (b) | 111 (c) | 112 (b) | 113 (b) | 114 (c) | | |

## > NCERT & NCERT Exemplar MCQs

| | | | | | | | | | |
|---|---|---|---|---|---|---|---|---|---|
| 115 (c) | 116 (b) | 117 (b) | 118 (a) | 119 (a) | 120 (c) | 121 (b) | 122 (c) | 123 (b) | 124 (b) |
| 125 (c) | 126 (c) | 127 (b) | 128 (c) | | | | | | |

# Hints & Explanations

**1** (*a*) According to the question, a charge $q$ is brought from a point $R$ to a point $P$ against the repulsive force on it due to charge $Q$. In bringing the charge $q$ from $R$ to $P$, an external force $\mathbf{F}_{ext}$ is applied just to counter the repulsive electric force $\mathbf{F}_E$ (i.e. $\mathbf{F}_{ext} = -\mathbf{F}_E$). In this situation, work done by the external force is negative of the work done by the electric force. Thus, the work done by external force in moving the charge $q$ from $R$ to $P$ is

$$W_{RP} = \int_R^P \mathbf{F}_{ext} \cdot d\mathbf{r} = -\int_R^P \mathbf{F}_E \cdot d\mathbf{r}$$

**2** (*b*) Here, the potential energy difference between two points is defined as the work required to be done by an external force in moving (without accelerated) charge of $+1\,C$ from point $R$ to $P$ for electric field of any arbitrary charge configuration.

i.e. $\Delta U = U_{final} - U_{initial} = U_P - U_R = W_{RP}$

Hence, option (b) is correct.

**3** (*c*) As per the question, a point charge $+q$ is taken along paths $AMB$ and $ACB$, so that $A$, $M$ and $B$ lie on the circle and another charge, i.e. $+q$ is placed at the centre. It is known that the electrostatic force is a conservative force depending only on the initial and the final points and is independent of the path taken to go from one point to the other.

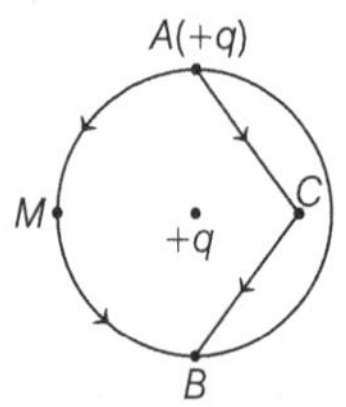


Therefore, work done along the paths are equal.

i.e. $\quad W_{ACB} = W_{AMB}$

**4** (*c*) Since, work done by an electrostatic field in moving charge is dependent on the initial and final points only. So, in the given situation in the question, it is clearly seen that initial point and the final point is same, as the charge $q$ is moving along a closed path $ABCDA$ for figure given in question.

Thus, net work done in doing so is zero.

**5** (*d*) Given, $W_{AB} = 100\text{ J}, q_0 = 4\text{C}$

$$V_A = -10\text{ V and } V_B = V$$

Since, $V_B - V_A = \dfrac{W_{AB}}{q_0}$, by external force

$$V - (-10\text{V}) = \frac{100\text{ J}}{4\text{C}} = 25\text{ V}$$

or $\quad V = 25\text{ V} - 10\text{ V} = 15\text{ V}$

**8** (*b*) Electrostatic force between any two positive charges $q_1$ and $q_2$ separated by a distance $r$ is given by

$$\mathbf{F}_{21} = \frac{q_1 q_2}{4\pi\varepsilon_0 r^2}\hat{\mathbf{r}} \qquad \text{...(i)}$$

Here, $\mathbf{F}_{21}$ is the electrostatic force on charge $q_2$ due to charge $q_1$ which is along the vector $\mathbf{r}$.

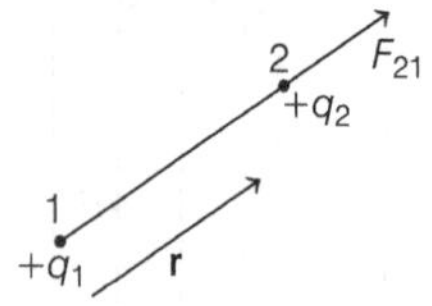


Also, $\hat{\mathbf{r}}$ is the unit vector along $\mathbf{r}$ and $r = |\mathbf{r}|$.

From Eq. (i), in this case $q_1 = +Q$ and $q_2 = +1\,C$

$$\Rightarrow \quad \mathbf{F} = \frac{Q \times 1}{4\pi\varepsilon_0 (r')^2}\hat{\mathbf{r}}' \qquad \text{...(ii)}$$

Total work done $W$ by the external force is obtained by integrating Eq. (ii) from $r' = \infty$ to $r' = r$

$$\therefore \quad W = -\int_\infty^r \frac{Q}{4\pi\varepsilon_0 (r')^2} dr'$$

$$= \frac{Q}{4\pi\varepsilon_0 r'}\Bigg|_\infty^r = \frac{Q}{4\pi\varepsilon_0 r}$$

**9** (*a*) Potential at point $A$ due to a point charge $q$ at origin $O$, $V_A = \dfrac{1}{4\pi\varepsilon_0} \cdot \dfrac{q}{a}$

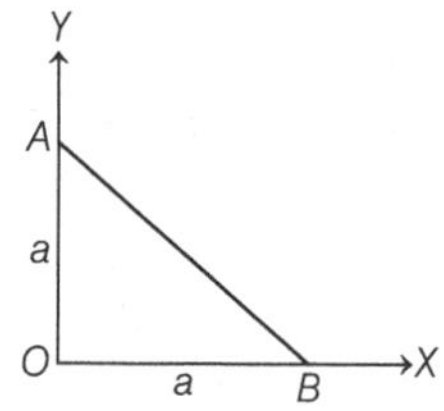


Similarly,

Potential at point $B$, $V_B = \dfrac{1}{4\pi\varepsilon_0} \cdot \dfrac{q}{a}$

So, work done in taking a test charge $-Q$ from $A$ to $B$,

$$W = V_A - V_B / -Q = 0$$

**10** (*a*) The electric field due to a point charge $Q$ at a distance $r$ is given as

$$E = \frac{KQ}{r^2}$$

$$\Rightarrow \quad E \propto \frac{1}{r^2}$$

and potential, $\quad V = \dfrac{KQ}{r}$

$$\Rightarrow \quad V \propto \frac{1}{r}$$

So, for the values of $r$ such that $0 < r < 1$, the value of $E$ would be more than that of $V$.

And for $r > 1$, the value of $E$ would be less as compared to that of $V$.

$\therefore$ From the given graph, it can be concluded that curve I represents $E$ *versus* $r$ and curve II represents $V$ *versus* $r$.

**12** (*b*) The potential due to an electric dipole is

$$V = \frac{1}{4\pi\varepsilon_0}\frac{\mathbf{p}\cdot\mathbf{r}}{r^2} = \frac{1}{4\pi\varepsilon_0}\cdot\frac{pr\cos\theta}{r^2}$$

$$= \frac{1}{4\pi\varepsilon_0}\cdot\frac{pr\cos\left(\frac{\pi}{2}\right)}{r^2} \qquad \left[\because \theta = \frac{\pi}{2}\right]$$

$$= 0$$

Thus, the potential in the equatorial plane of a dipole of dipole moment $p$ is zero.

**13** (*a*) Let $P$ be the point on line joining two charges at a distance $x$ from positive charge

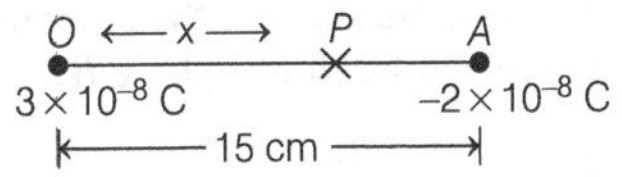


According to question,

$$\frac{1}{4\pi\varepsilon_0}\left[\frac{3\times 10^{-8}}{x\times 10^{-2}} - \frac{2\times 10^{-8}}{(15-x)\times 10^{-2}}\right] = 0$$

$$\Rightarrow \qquad \frac{3}{x} - \frac{2}{15-x} = 0$$

which gives $x = 9$ cm

If $x$ lies on the extended line $OA$, the required condition is

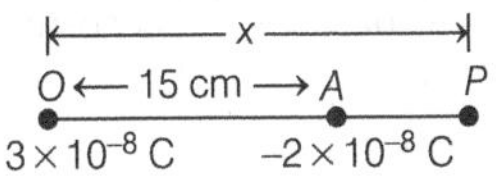


$$\frac{3}{x} - \frac{2}{x-15} = 0$$

which gives $x = 45$ cm

Thus, electric potential is zero at 9 cm and 45 cm away from the positive charge on the side of the negative charge.

**14** (*b*) As $O$ is the centre of line $AB$, so $AO = a$ and the distance of point $P$ from $A$ and $B$ is

$$AP = BP = \sqrt{a^2 + a^2} = \sqrt{2}a$$

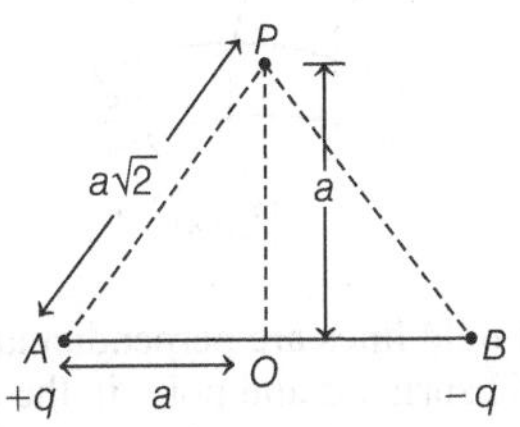


The potential at $P$ due to charge $+q$ at $A$ and $-q$ at $B$ is

$$V_P = V_A + V_B = \frac{Kq}{AP} + \frac{K(-q)}{BP}$$

$$= \left(\frac{Kq}{a\sqrt{2}}\right) + \left(\frac{-Kq}{a\sqrt{2}}\right) = \text{zero}$$

**15** (*c*) Let $V_1$ be the potential at the centre of the cube due to one charge

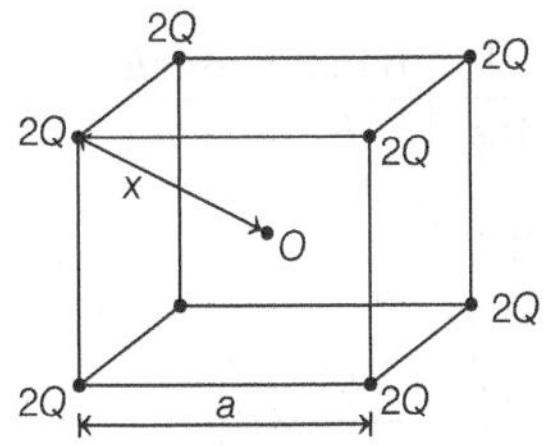


$$\Rightarrow \qquad V_1 = \frac{1}{4\pi\varepsilon_0}\cdot\frac{2Q}{x}$$

Distance of centre of cube from its corner, $x = \frac{a\sqrt{3}}{2}$.

Potential at centre of cube due to all eight corners of the cube,

$$V = 8V_1 = 8\left[\frac{1}{4\pi\varepsilon_0}\frac{2Q\times 2}{\sqrt{3}a}\right]$$

$$= \frac{32Q}{4\pi\varepsilon_0\sqrt{3}\times a} = \frac{8Q}{\sqrt{3}\pi\varepsilon_0 a}$$

**17** (*a*) Since, for a hollow metallic sphere, the potential is constant inside the sphere and equals to its value at the surface.

So, if potential on the surface of the sphere is equals to 80V, then the potential at the centre of the sphere is also 80V.

**18** (*c*) Volume of eight drops = Volume of a big drop

$$\therefore \qquad \left(\frac{4}{3}\pi r^3\right)\times 8 = \frac{4}{3}\pi R^3$$

$$\Rightarrow \qquad 2r = R \qquad \ldots\text{(i)}$$

According to charge conservation, $8q = Q$ ...(ii)

Potential of one small drop, $V' = \frac{q}{4\pi\varepsilon_0 r}$

Similarly, potential of big drop,

$$V = \frac{Q}{4\pi\varepsilon_0 R}$$

Now, $\frac{V'}{V} = \frac{q}{Q}\times\frac{R}{r}$

$$\Rightarrow \qquad \frac{V'}{20} = \frac{q}{8q}\times\frac{2r}{r} \qquad \text{[from Eqs. (i) and (ii)]}$$

$$\therefore \qquad V' = 5\text{ V}$$

So, the potential of each single drop was 5V.

**19** (*b*) Potential of $B$ = Potential due to charge on $A$ + Potential due to charge on $B$ + Potential due to charge on $C$.

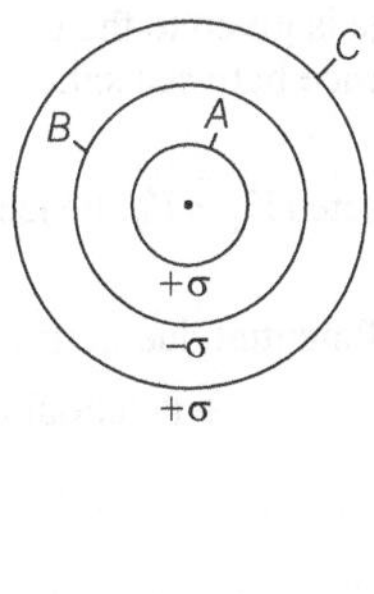

$$\therefore \quad V_B = \frac{k(Q_A + Q_B)}{b} + \frac{kQ_C}{c}$$

$$= \frac{1}{4\pi\varepsilon_0}\left[\frac{\sigma 4\pi a^2}{b} - \frac{\sigma 4\pi b^2}{b} + \frac{\sigma 4\pi c^2}{c}\right]$$

$$= \frac{\sigma\varepsilon}{\varepsilon_0}\left(\frac{a^2 - b^2}{b} + \frac{c^2}{c}\right) = \frac{\sigma}{\varepsilon_0}\left(\frac{a^2 - b^2}{b} + c\right)$$

$$V_B = \frac{\sigma}{\varepsilon_0}\left(\frac{a^2 - b^2}{b} + c\right)$$

**20** (*a*) Initially, the potential on the solid sphere having charge $Q$ is

$$V_{\text{sphere}} = \frac{1}{4\pi\varepsilon_0} \cdot \frac{Q}{r_1}$$

As the spherical shell surrounds the solid sphere, so an equal charge will be developed on it. Hence, potential on spherical shell is

$$V_{\text{shell}} = \frac{1}{4\pi\varepsilon_0} \cdot \frac{Q}{r_2}$$

$\therefore$ Potential difference between them,

$$V_{\text{sphere}} - V_{\text{shell}} = \frac{Q}{4\pi\varepsilon_0}\left[\frac{1}{r_1} - \frac{1}{r_2}\right] = V \quad \ldots\text{(i)}$$

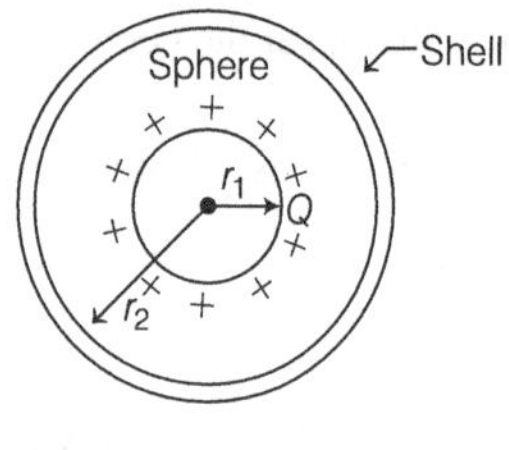


The extra potential on shell due to the charge $-3Q$,

$$V_0 = \frac{1}{4\pi\varepsilon_0}\left(-\frac{3Q}{r_2}\right)$$

This potential is equally distributed between sphere and shell.

So, $$V'_{\text{sphere}} = \frac{1}{4\pi\varepsilon_0}\left(\frac{Q}{r_1}\right) + V_0$$

and $$V'_{\text{shell}} = \frac{1}{4\pi\varepsilon_0}\left(\frac{Q}{r_2}\right) + V_0$$

Hence, new potential difference,

$$V'_{\text{sphere}} - V'_{\text{shell}} = \frac{Q}{4\pi\varepsilon_0}\left(\frac{1}{r_1} - \frac{1}{r_2}\right) = V \quad \text{\{from Eq. (i)\}}$$

i.e. if any charge is given to the external shell, the potential difference between sphere and shell will not change.

**21** (*a*) When connected $V_A = V_B$, hence $q_1$ will flow to outer surface of $B$.

$V_B$ = Potential due to its own charge $q_2$ + Potential due to charge $q_1$ on $A$

$$= \frac{1}{4\pi\varepsilon_0}\left(\frac{q_2}{r_2} + \frac{q_1}{r_2}\right)$$

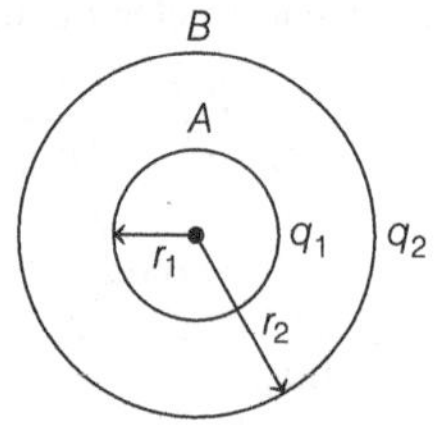


Potential on the inner sphere $A$ due to its own charge,

$$V_1 = \frac{1}{4\pi\varepsilon_0} \cdot \frac{q_1}{r_1}$$

As the potential at every point inside charged sphere is the same as that on its surface, the potential on the inner sphere $A$ due to charge $q_2$ on sphere $B$,

$$V_2 = \frac{1}{4\pi\varepsilon_0} \cdot \frac{q_2}{r_2}$$

Total potential on the inner sphere $A$,

$$V_A = V_1 + V_2 = \frac{1}{4\pi\varepsilon_0}\left(\frac{q_1}{r_1} + \frac{q_2}{r_2}\right)$$

Thus, $V_A - V_B = \dfrac{q_1}{4\pi\varepsilon_0}\left(\dfrac{1}{r_1} - \dfrac{1}{r_2}\right)$

As $r_1 < r_2, \dfrac{1}{r_1} > \dfrac{1}{r_2}$ or $\left(\dfrac{1}{r_1} - \dfrac{1}{r_2}\right)$ is positive, further since $q_1$ is positive, $V_A > V_B$.

Thus, when the two spheres are connected by a wire, charge $q_1$ on $A$ will flow entirely to $B$, irrespective of the charge $q_2$ already present on $B$.

**23** (*b*) For a uniform electric field **E** say, along the $X$-axis, the equipotential surfaces are planes perpendicular to the $X$-axis, i.e. planes parallel to the $yz$-plane.

**24** (*c*) The field lines are perpendicular to the equipotential surfaces. So, in the given figure, this condition, is correctly depicted in option (c) only.

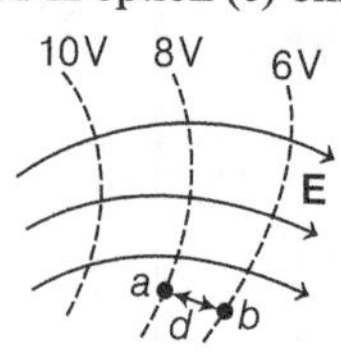


**25** (*a*) Electric field lines are perpendicular to the equipotential surfaces and point in the direction of decreasing potential. So, at $P$, electric field **E** is to the left and at $R$, **E** is upward.

**26** (*b*) We know that, work done, $W = q\,\Delta V$

As, $\Delta V$ is same in all the cases, so work done will be same in all four cases.

**27** (*b*) As electric field, $E = \frac{V}{r}$

$\therefore$ The distance of point charge,

$$r = \frac{V}{E} = \frac{600\text{ V}}{200\text{ NC}^{-1}} = 3\text{ m}$$

**28** (*a*) As charges at $A$ and $B$ are at equal distance from $10\,\mu$C at the centre. So, potential at $A$ and $B$ due to $10\,\mu$C charge are equal.

$\therefore$ Potential difference,

$$(V_A - V_B) = 0$$

Thus, work done, $W_{AB} = q\,(V_B - V_A) = 0.$

**29** (*a*) Since, the proton is moving against the direction of electric field, so work is done on the proton against electric field. It implies that electric field does negative work on the proton.

Again, proton is moving in electric field from low potential region to high potential region, hence its potential energy increases.

**30** (*a*) To bring charge $q_1$ first from infinity to $\mathbf{r}_1$, no work is required. So, while bringing charge $q_2$ from infinity to $\mathbf{r}_2$, $q_1$ produces a potential $V_1$, so work is done in this step which can be given as

$$W_1 = q_2 V_1(\mathbf{r}_2) = \frac{1}{4\pi\varepsilon_0}\frac{q_1 q_2}{r_{12}} \quad \text{...(i)}$$

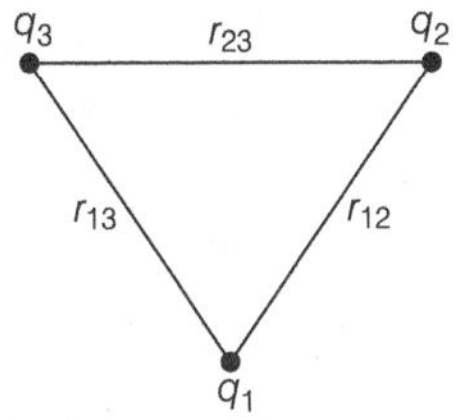


The charges $q_1$ and $q_2$ produce a potential, which at any point $P$ is given by

$$V_{12} = \frac{1}{4\pi\varepsilon_0}\left(\frac{q_1}{r_{1p}} + \frac{q_2}{r_{2p}}\right) \quad \text{...(ii)}$$

Now, work done in bringing $q_3$ from infinity to the point $\mathbf{r}_3$ is

$$W_2 = q_3 V_{12}(\mathbf{r}_3) = \frac{1}{4\pi\varepsilon_0}\left(\frac{q_1 q_3}{r_{13}} + \frac{q_2 q_3}{r_{23}}\right) \quad \text{...(iii)}$$

The total work done in assembling the charges at the given locations is obtained by combining Eqs. (i) and (iii).

This work done is stored as potential energy and given by

$$U = W_1 + W_2 = \frac{1}{4\pi\varepsilon_0}\left(\frac{q_1 q_2}{r_{12}} + \frac{q_1 q_3}{r_{13}} + \frac{q_2 q_3}{r_{23}}\right)$$

**31** (*d*) Electrostatic energy between two charges $q_1$ and $q_2$ such that the distance between them is $r$ is given as

$$U = \frac{K\,q_1 q_2}{r}$$

In accordance to the principle of superposition, total energy of the charge system as shown in the figure below is

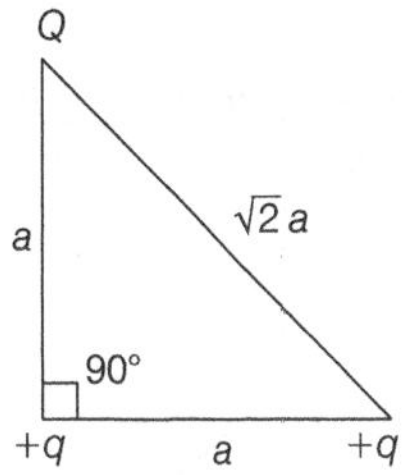


$$U = \frac{Kq^2}{a} + \frac{KQq}{a} + \frac{KQq}{\sqrt{2}a}$$

It is given that, $U = 0$

$$\therefore \quad \frac{Kq}{a}\left[q + Q + \frac{Q}{\sqrt{2}}\right] = 0$$

$$\Rightarrow \quad Q = \frac{-\sqrt{2} \times q}{(\sqrt{2}+1)}$$

**32** (*d*) Here, potential energy of the system,

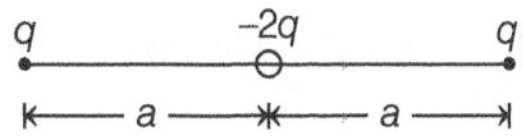


$$U = \frac{1}{4\pi\varepsilon_0}\left[\frac{q_1 q_2}{r_{12}} + \frac{q_1 q_3}{r_{13}} + \frac{q_2 q_3}{r_{23}}\right]$$

$$= \frac{1}{4\pi\varepsilon_0}\left[\frac{qq}{2a} + \frac{q(-2q)}{a} + \frac{q(-2q)}{a}\right]$$

$$= \frac{1}{4\pi\varepsilon_0}\left[\frac{q^2}{2a} + \frac{-2q^2}{a} - \frac{2q^2}{a}\right]$$

$$= \frac{1}{4\pi\varepsilon_0}\left[\frac{-4q^2}{a} + \frac{q^2}{2a}\right] = \frac{-7q^2}{8\pi\varepsilon_0 a}$$

**33** (*d*) Work done in increasing the separation from $a$ to $2a$ is

$$W = U_f - U_i \quad \text{...(i)}$$

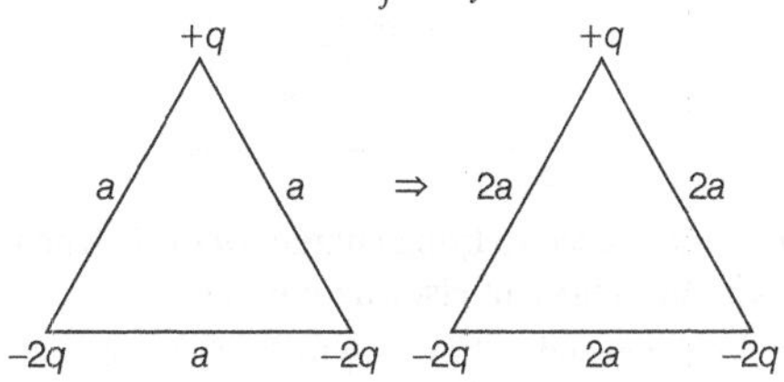


Here,

$$U_i = \frac{1}{4\pi\varepsilon_0}\left[\frac{q(-2q)}{a} + \frac{q(-2q)}{a} + \frac{(-2q)(-2q)}{a}\right]$$

$$= \frac{1}{4\pi\varepsilon_0 a}[-2q^2 - 2q^2 + 4q^2] = 0$$

Similarly, $U_f = \frac{1}{4\pi\varepsilon_0(2a)}(-2q^2 - 2q^2 + 4q^2) = 0$

Hence, from Eq. (i)

$$W = U_f - U_i = 0$$

**34** (*a*) When charge $q_3$ is at $C$, then potential energy of the system is

$$U_C = \frac{1}{4\pi\varepsilon_0}\left[\left(\frac{q_1q_3}{AC}+\frac{q_2q_3}{BC}\right)+\frac{q_1q_2}{AB}\right]$$
$$= \frac{1}{4\pi\varepsilon_0}\left\{\left(\frac{q_1q_3}{0.4}+\frac{q_2q_3}{0.5}\right)+\frac{q_1q_2}{0.3}\right\}$$

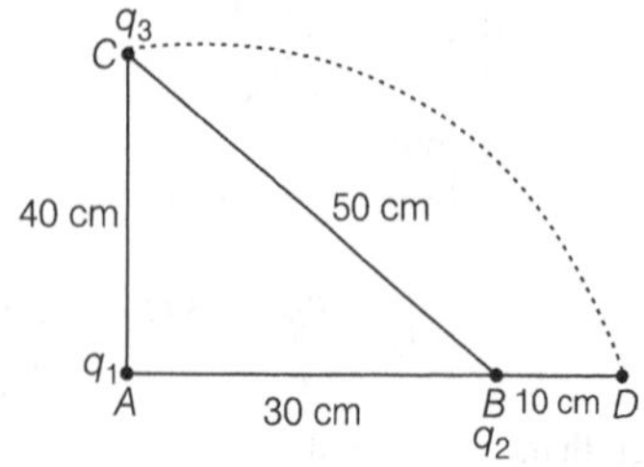


Similarly, when charge $q_3$ is at $D$, then potential energy is

$$U_D = \frac{1}{4\pi\varepsilon_0}\left[\left(\frac{q_1q_3}{AC}+\frac{q_2q_3}{BD}\right)+\frac{q_1q_2}{AB}\right]$$
$$= \frac{1}{4\pi\varepsilon_0}\left\{\left(\frac{q_1q_3}{0.4}+\frac{q_2q_3}{0.1}\right)+\frac{q_1q_2}{0.3}\right\}$$

Here, change in potential energy,

$$\Delta U = U_D - U_C$$
$$\frac{q_3}{4\pi\varepsilon_0}K = \frac{1}{4\pi\varepsilon_0}\left(\frac{q_2q_3}{0.1}-\frac{q_2q_3}{0.5}\right)\left[\because \Delta U = \frac{q_3K}{4\pi\varepsilon_0}\right]$$
$$\Rightarrow \quad K = \left(\frac{q_2}{0.1}-\frac{q_2}{0.5}\right) = 10q_2 - \frac{10q_2}{5}$$
$$= 10q_2 - 2q_2$$
$$\Rightarrow \quad K = 8q_2$$

**35** (*a*) Suppose, first the charge $+q$ is brought to $A$ and then the charges $-q$, $+q$ and $-q$ are brought to $B$, $C$ and $D$, respectively. The total work needed can be calculated in steps

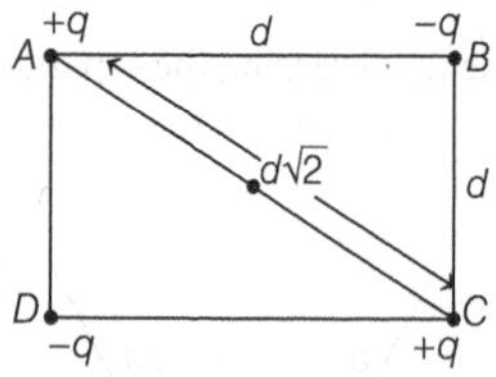


(i) Work needed to bring charge $+q$ to $A$, when no charge is present elsewhere is zero.

(ii) Work needed to bring $-q$ to $B$ when $+q$ is at $A$. This is given by (charge at $B$) × (electrostatic potential at $B$ due to charge $+q$ at $A$)

$$= -q \times \left(\frac{q}{4\pi\varepsilon_0 d}\right) = -\frac{q^2}{4\pi\varepsilon_0 d}$$

(iii) Work needed to bring charge $+q$ to $C$ when $+q$ is at $A$ and $-q$ is at $B$. This is given by (charge at $C$) × (potential at $C$ due to charges at $A$ and $B$)

$$= +q\left(\frac{+q}{4\pi\varepsilon_0 d\sqrt{2}}+\frac{-q}{4\pi\varepsilon_0 d}\right) = \frac{-q^2}{4\pi\varepsilon_0 d}\left(1-\frac{1}{\sqrt{2}}\right)$$

(iv) Work needed to bring $-q$ to $D$ when $+q$ at $A$, $-q$ at $B$ and $+q$ at $C$. This is given by (charge at $D$) × (potential at $D$ due to charges at $A$, $B$ and $C$)

$$= -q\left(\frac{+q}{4\pi\varepsilon_0 d}+\frac{-q}{4\pi\varepsilon_0 d\sqrt{2}}+\frac{q}{4\pi\varepsilon_0 d}\right)$$
$$= \frac{-q^2}{4\pi\varepsilon_0 d}\left(2-\frac{1}{\sqrt{2}}\right)$$

Adding the work done in steps (i), (ii), (iii) and (iv), the total work required is

$$= \frac{-q^2}{4\pi\varepsilon_0 d}\left\{(0)+(1)+\left(1-\frac{1}{\sqrt{2}}\right)+\left(2-\frac{1}{\sqrt{2}}\right)\right\}$$
$$= \frac{-q^2}{4\pi\varepsilon_0 d}(4-\sqrt{2})$$

**36** (*b*) Given, $q = 100\,\text{C}$

and potential difference between the cloud and the earth,

$$V = 10^7\ \text{V}$$

Thus, energy dissipated, $W = qV = 100\text{C} \times 10^7\text{V} = 10^9\,\text{J}$

**37** (*b*) In the presence of external electric field, we first find the potentials at position1 and 2, respectively using

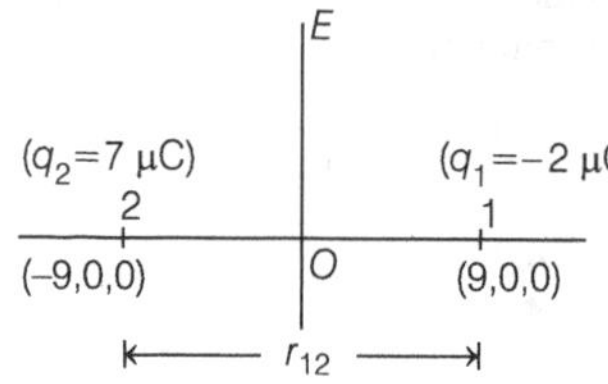


$$E = -\frac{dV}{dr}$$
$$\Rightarrow \quad \int_\infty^{r_1}\frac{A}{r^2}dr = -\int_0^{V_1}dV \quad \Rightarrow \quad \frac{A}{r_1} = V_1$$

or $\quad V_1 = A/r_1$ and $V_2 = A/r_2$

Total electrostatic energy of the system

$$= q_1V_1 + q_2V_2 + \frac{kq_1q_2}{r_{12}} \qquad \text{...(i)}$$

Given, $q_1 = -2\,\mu\text{C} = -2\times10^{-6}$ C,

$q_2 = 7\,\mu\text{C} = 7\times10^{-6}$ C, $r_1 = 9\,\text{cm} = 0.09$ m

$r_2 = 9\,\text{cm} = 0.09$ m

$A = 9\times10^5\ \text{C/m}^2$, $r_{12} = (r_1 - r_2) = (9+9)\,\text{cm} = 18\,\text{cm}$ $= 0.18$ m

Substituting these values in Eq. (i), we get

$$= A\left(\frac{-2\times10^{-6}}{0.09}\right)+A\left(\frac{7\times10^{-6}}{0.09}\right)+\left(\frac{-9\times10^9\times7\times2\times10^{-12}}{0.18}\right)$$
$$= \frac{9\times10^5(5\times10^{-6})}{0.09} - 0.7\ \text{J}$$
$$= 50\ \text{J} - 0.7\ \text{J}$$
$$= 49.3\ \text{J}$$

Thus, electrostatic energy is 49.3 J.

**38** (*d*) Given, $\mathbf{E} = 1000\ \text{V/m}, \theta = 45°$ and $\mathbf{p} = 10^{-29}\text{C-m}$

We know that, electric potential energy stored in an electric dipole kept in uniform electric field is given by the relation

$$U = -\mathbf{p}\cdot\mathbf{E} = -PE\cos\theta$$
$$= -10^{-29} \times 1000 \times \cos 45°$$
$$\Rightarrow \quad U \approx -7\times 10^{-27}\ \text{J}$$

**39** (*a*) Given, $q = 3.5\mu\text{C}, 2l = 4.2\ \text{cm}$

and $E = 5.87\times 10^5\text{NC}^{-1}$

When a dipole is placed in an electric field **E**, torque exerted, $\tau = pE\sin\theta$, where $p$ = dipole moment $(= q\times 2l)$ and $\theta$ is the angle which dipole makes with the field.

For maximum torque, $\theta = 90°$

$$\tau_{\text{max}} = pE\sin 90°$$

i.e. $$\tau_{\text{max}} = pE$$

$$\Rightarrow \quad \tau_{\text{max}} = q\times 2l\times E = (3.5\times 10^{-6})\times 4.2 \times 10^{-2}\times 5.87\times 10^5$$
$$= 86.289\times 10^{-3}\ \text{Nm}^{-1}$$

Work done in rotating the dipole from an angle $\theta_0$ to $\theta$,

$$W = \int_{\theta_0}^{\theta} pE\sin\theta\, d\theta = pE\,[\cos\theta_0 - \cos\theta]$$

Here, $\theta_0 = 0°$ and $\theta = 180°$

$$W = pE\,(\cos 0° - \cos 180°)$$
$$\Rightarrow \quad W = 2pE$$
$$= 2\times q\times 2l\times E$$
$$= 2\times(3.5\times 10^{-6})\times(4.2\times 10^{-2})\times 5.8\times 10^5$$
$$= 1.705\times 10^{-1}\ \text{J} \simeq 1.70\times 10^{-1}\ \text{J}$$

Hence, the maximum torque is $86.289\times 10^{-3}\ \text{Nm}^{-1}$ and work done is $1.70\times 10^{-1}$ J.

**40** (*b*) Given, dipole moment of each molecule $= 10^{-29}\text{C-m}$

and number of molecules in 1 mole of the substance $= 6\times 10^{23}$.

Therefore, total dipole moment of all molecules,

$$p = 6\times 10^{23}\times 10^{-29}\ \text{C-m} = 6\times 10^{-6}\ \text{C-m}$$

Initial potential energy,

$$U_i = -pE\cos\theta$$
$$= -6\times 10^{-6}\times 10^{6}\times\cos 60°$$
$$= -6\times\frac{1}{2} = -3\ \text{J}$$

Final potential energy (when $\theta = 0°$),

[finally $p$ and $E$ are aligned in the same direction]

$$U_f = -6\times 10^{-6}\times 10^{6}\times\cos 0° = -6\ \text{J}$$

Change in potential energy $= -6\text{J} - (-3\text{J}) = -3$ J

So, there is a loss in potential energy. This must be the energy released by the substances in the form of heat in aligning its dipole.

$\therefore$ Heat energy $= -\text{PE} = -(-3\ \text{J}) = 3$ J

**43** (*d*) The surface charge density of a closed surface area having charge $Q$ is given by

$$\sigma = \frac{\text{Charge}}{\text{Area}} = \frac{Q}{A} \text{ or } Q = \sigma A$$

Thus, the charges on spheres $P$ and $Q$ having same charge density as shown in the figure below is given by

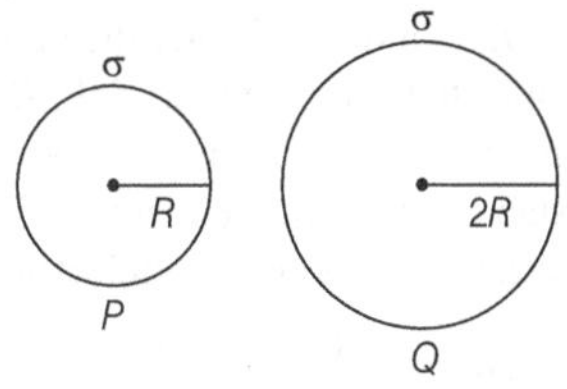


$$Q_P = \sigma\times 4\pi R^2 = 4\pi\sigma R^2 \quad \text{...(i)}$$

and $$Q_Q = \sigma\times 4\pi(2R)^2 = 16\pi\sigma R^2 \quad \text{...(ii)}$$

When they are brought in contact with each other, the total charge will be

$$Q_t = Q_P + Q_Q$$
$$= 4\pi\sigma R^2 + 16\pi\sigma R^2 \quad \text{[from Eqs. (i) and (ii)]}$$
$$= 20\pi\sigma R^2 \quad \text{...(iii)}$$

In connection of two charged conducting bodies, the potential will become same on both, i.e.

$$\frac{Q_P}{4\pi\varepsilon_0 R} = \frac{Q_Q}{4\pi\varepsilon_0 2R}$$

$$\Rightarrow \quad \frac{Q_P}{R} = \frac{Q_Q}{2R} \Rightarrow \frac{Q_P}{Q_Q} = \frac{1}{2}$$

So, the charges on the sphere $P$ and $Q$ after separation will be distributed as

$$\Rightarrow \quad Q'_P = \frac{Q_P}{Q_P + Q_Q}\times Q_t = \frac{1}{3}Q_t$$

and $$Q'_Q = \frac{Q_Q}{Q_P + Q_Q}\times Q_t = \frac{2}{3}Q_t$$

After separation, the new surface charge densities on $P$ and $Q$ will be

$$\sigma_P = \frac{Q'_P}{\text{Area}} = \frac{1}{3}\frac{Q_t}{\text{Area}} = \frac{1}{3}\frac{20\pi\sigma R^2}{4\pi R^2} = \frac{5}{3}\sigma$$

and $$\sigma_Q = \frac{Q'_Q}{\text{Area}} = \frac{2}{3}\frac{Q_t}{\text{Area}} = \frac{2}{3}\times\frac{20\pi\sigma R^2}{4\pi(2R)^2}$$
$$= \frac{2}{3}\times\frac{5}{4}\sigma = \frac{5}{6}\sigma$$

**44** (*c*) Consider a cavity in a conductor as shown in the following figure

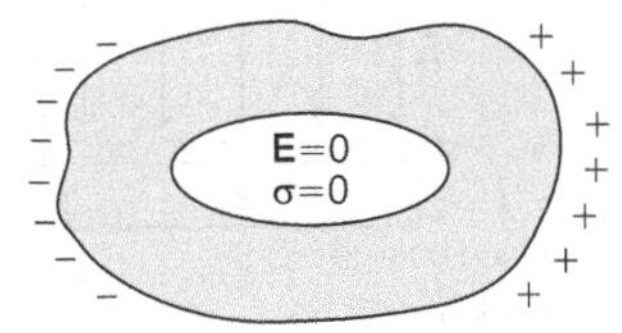

We know that, whatever be the charge and the field configuration outside a conductor, any cavity inside it remains shielded from outside electric influence. This means, inside the cavity, conductivity ($\sigma$) and electric field ($E$) are zero.

**49** ($b$) When a dielectric is placed in an electric field, an electric field $E_i$ is induced inside the dielectric in a direction opposite to the external field $E_0$ due to polarisation. On account of this, the net electric field $E$ inside the slab is less than the external field as $E = E_0 - E_i$, i.e., the electric field inside the dielectric decreases.

**51** ($b$) As, $P = \varepsilon_0 \chi_e E$ and $\varepsilon_r = 1 + \chi_e$

$\Rightarrow \quad P = \varepsilon_0 (\varepsilon_r - 1) E$

If $n$ is the number of helium atoms/$m^3$ and $P$ is the dipole moment per atom, then

$$P = np$$

or $$p = \frac{P}{n} = \frac{\varepsilon_0 (\varepsilon_r - 1) E}{n} \quad \ldots(\text{i})$$

Since, at NTP, 1 mole ($6.02 \times 10^{23}$ atoms) of a gas occupies a volume of 22.4 L (i.e. $22.4 \times 10^{-3}$ $m^3$).

$$n = \frac{6.02 \times 10^{23}}{22.4 \times 10^{-3} \text{m}^3}$$

$$= 2.69 \times 10^{25} / \text{m}^3 \quad \ldots(\text{ii})$$

From Eqs. (i) and (ii), we get

$$p = \frac{(8.85 \times 10^{-12} \text{C}^2/\text{Nm}^2)(1.00074 - 1)(10\text{V/m})}{2.69 \times 10^{25}/\text{m}^3}$$

$$= 2.4 \times 10^{-39} \text{ C-m}$$

Thus, the dipole moment is $2.4 \times 10^{-39}$ C-m.

**52** ($a$) From the given graphs, the voltages $V_A$ and $V_B$ of capacitors $A$ and $B$ corresponding to charge $Q$ on each of the capacitors are $V_A = \frac{Q}{C_A}$ and $V_B = \frac{Q}{C_B}$

or $$\frac{V_B}{V_A} = \frac{Q/C_B}{Q/C_A} = \frac{C_A}{C_B}$$

Since, $V_B > V_A \Rightarrow C_A > C_B$, i.e. the capacitor $A$ has the higher capacitance than $B$.

**54** ($c$) It is given that two parallel plates of area $A$ each having surface charge densities $\sigma$ and $-\sigma$ separated by distance $d$. This can be shown as

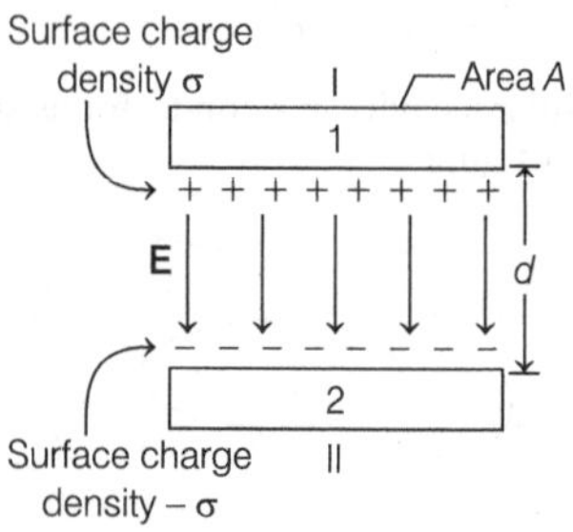


The electric field in outer region I (region above the plate 1) is

$$E = E_\sigma + E_{-\sigma} = \frac{\sigma}{2\varepsilon_0} - \frac{\sigma}{2\varepsilon_0} = 0$$

**55** ($b$) If $Q$ = charge on each plate, then

$$Q = CV = \frac{\varepsilon_0 A}{d} \cdot Ed = \varepsilon_0 AE$$

Here, $A = 1 \text{ m}^2$, $E = 100$ N/C

and $$\varepsilon_0 = 8.85 \times 10^{-12} \frac{\text{C}^2}{\text{N-m}^2}$$

So, by substituting given values, we get

$Q = 8.85 \times 10^{-12} \times 1 \times 100 = 8.85 \times 10^{-10}$ C

**56** ($a$) Net value of charge on plates of capacitor after steady state is reached is

$$q_{\text{net}} = \frac{q_2 - q_1}{2}$$

where, $q_2$ and $q_1$ are the charges given to plates. (Note that this formula is valid for any polarity of charge.)

Here, $q_2 = 4\,\mu\text{C}$, $q_1 = 2\,\mu\text{C}$

$\therefore$ Charge of capacitor is

$$q = \Delta q_{\text{net}} = \frac{4-2}{2} = 1\,\mu\text{C}$$

Potential difference between capacitor plates is

$$V = \frac{Q}{C} = \frac{1\,\mu\text{C}}{1\,\mu\text{F}} = 1\text{V}$$

**57** ($a$) When free space is in between parallel plates of capacitor, $E = \frac{\sigma}{\varepsilon_0}$

When dielectric is introduced between parallel plates of capacitor, $E' = \frac{\sigma}{K\varepsilon_0}$

Electric field inside dielectric $= \frac{\sigma}{K\varepsilon_0} = 3 \times 10^4 \quad \ldots(\text{i})$

where, $K$ = dielectric constant of medium = 2.2

and $\varepsilon_0$ = permittivity of free space $= 8.85 \times 10^{-12}$ $\text{C}^2/\text{Nm}^2$

Thus, charge density,

$$\sigma = K\varepsilon_0 \times 3 \times 10^4 = 2.2 \times 8.85 \times 10^{-12} \times 3 \times 10^4$$

$$= 6.6 \times 8.85 \times 10^{-8} = 5.841 \times 10^{-7} \approx 6 \times 10^{-7} \text{ Cm}^{-2}$$

**58** ($c$) Initially, $$C_0 = \frac{\varepsilon_0 A}{d} = 18\mu\text{F} \quad \ldots(\text{i})$$

When $d' = 3d$ and a dielectric medium is introduced,

$$C = \frac{K\varepsilon_0 A}{3d} = 72\mu\text{F} \quad \ldots(\text{ii})$$

On dividing Eq. (ii) by Eq. (i), we get

$$\frac{K}{3} = \frac{72}{18} = 4$$

$\therefore$ Dielectric constant, $K = 12$

**59** (*d*) Given, initial separation, $d = 3$ mm

Let the new separation between the plates be $d'$.

$$\therefore \quad \frac{C'd'}{K} = Cd \qquad (\because \varepsilon_0 A = \text{constant})$$

Given, $\quad C' = C \Rightarrow d' = Kd$

$\Rightarrow \quad d' = 2 \times 3 = 6$ mm

**60** (*c*) Capacitance of a parallel plate capacitor with air is $C = \frac{\varepsilon_0 A}{d}$ and capacitance of a same parallel plate capacitor with the introduction of a dielectric medium is

$$C' = \frac{K\varepsilon_0 A}{d}$$

where, $K$ is the dielectric constant of the medium.

Given, $C = 3\,\mu\text{F}$ and $C' = 15\,\mu\text{F}$

$$\therefore \quad K = \frac{C'}{C} = \frac{15}{3} = 5$$

Also, $\quad K = \frac{\varepsilon}{\varepsilon_0}$

or $\quad \varepsilon = K\varepsilon_0 = 5 \times 8.854 \times 10^{-12}$

$$= 0.4427 \times 10^{-10}\ \text{C}^2\ \text{N}^{-1}\text{m}^{-2}$$

**61** (*a*) Given, capacitance of a capacitor, $C = 15$ nF $= 15 \times 10^{-9}$ F

$\varepsilon_r = 2.5$, electric field, $E = 30 \times 10^6$ V/m

and potential difference, $V = 30$ V

If $d$ be the distance between the plates, then

$$d = \frac{V}{E} = \frac{30}{30 \times 10^6} = 10^{-6}\ \text{m}$$

Capacitance of capacitor, $C = \frac{\varepsilon_0\, \varepsilon_r A}{d}$

$$15 \times 10^{-9} = \frac{8.85 \times 10^{-12} \times 2.5 \times A}{10^{-6}}$$

$\Rightarrow \quad A = 6.7 \times 10^{-4}\ \text{m}^2$

Thus, the area is $6.7 \times 10^{-4}\ \text{m}^2$.

**62** (*a*) Let, $E_0 = V_0/d$ be the electric field between the plates, when there is no dielectric and the potential difference is $V_0$. If the dielectric is inserted, the electric field in the dielectric will be $E = E_0/K$ and the potential difference will be

$$V = \left[\begin{array}{l}\text{Potential due to}\\ \text{the empty space left } \left(d - \frac{3}{4}\right)d\\ \text{in between the plates}\end{array}\right] + \left(\begin{array}{l}\text{Potential due to}\\ \text{the slab between the plates}\end{array}\right)$$

$$= E_0\left(\frac{1}{4}d\right) + \frac{E_0}{K}\left(\frac{3}{4}d\right) \qquad (\because V = E \cdot ds)$$

$$= E_0 d\left(\frac{1}{4} + \frac{3}{4K}\right) = V_0 \cdot \frac{K+3}{4K}$$

The potential difference decreases by the factor $\frac{K+3}{4K}$ while the free charge $Q_0$ on the plates remains unchanged.

$\therefore$ The capacitance become,

$$C = \frac{Q_0}{V} = \frac{4K}{K+3}\frac{Q_0}{V_0} = \frac{4K}{K+3}C_0 \qquad \left[\because C_0 = \frac{Q_0}{V_0}\right]$$

**63** (*b*) Two capacitors are given as below

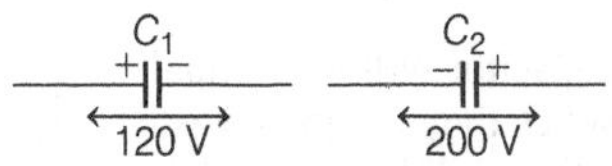


For potential to be made zero after connection, the charge on both capacitors should be equal.

$$\therefore \quad q_1 = q_2$$

$$C_1V_1 = C_2V_2 \text{ or } 120\,C_1 = 200C_2$$

$\Rightarrow \quad 3C_1 = 5C_2$

**64** (*a*) When a dielectric slab is inserted to fill the space between the plates, then the charge on the capacitor plates remains same. However, the capacitance increase, i.e. $C = KC_0$, where $C_0$ is the capacitance of the capacitor without slab and $K$ is the dielectric constant of slab.

When parallel combination is fully charged, charge on the combination is

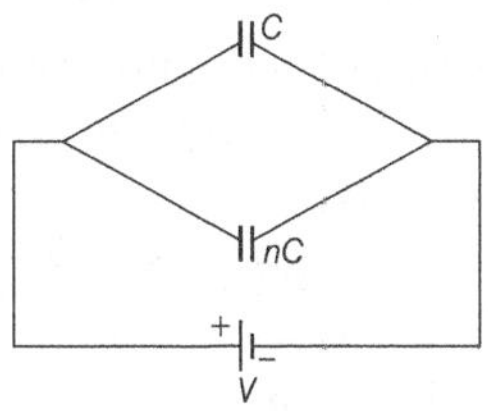


$$Q = C_{eq}V = C(1+n)V$$

When battery is removed and a dielectric slab is placed between two plates of first capacitor, then charge on the system remains same. Now, equivalent capacitance after insertion of dielectric is

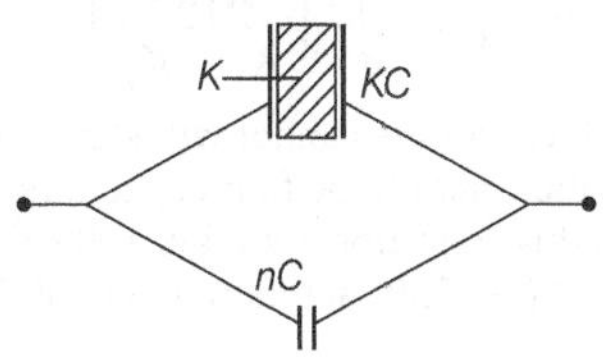


$$C_{eq} = KC + nC = (n+K)C$$

If potential value after insertion of dielectric is $V'$, then charge on system is

$$Q' = C_{eq}V' = (n+K)CV'$$

As $Q = Q'$, we have

$$C(1+n)V = (n+K)CV'$$

$$\therefore \quad V' = \frac{(1+n)V}{(n+K)}$$

**65** (*a*) This combination is same as the two capacitors are connected in series and distance between plates of each capacitor is *d*/2.

$$\text{So, } C_1 = \frac{K\varepsilon_0 A}{d/2} \text{ and } C_2 = \frac{\varepsilon_0 A}{d/2}$$

$$\text{Hence, } C_{\text{net}} = \frac{C_1 C_2}{C_1 + C_2} = \frac{\left(\frac{2K\varepsilon_0 A}{d}\right)\left(\frac{2\varepsilon_0 A}{d}\right)}{\left(\frac{2K\varepsilon_0 A}{d} + \frac{2\varepsilon_0 A}{d}\right)} = \frac{2KA\varepsilon_0}{(K+1)d}$$

**66** (*a*) Each capacitor will have equal charges. When they are joined as described, they will be in parallel combination. The positive charge will cancel out negative charge of each other.

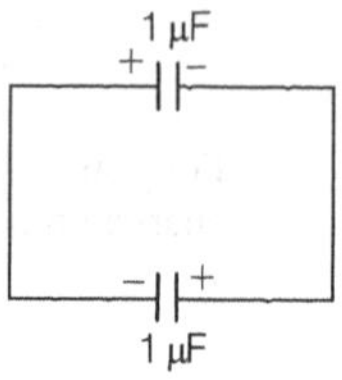


So, net charge in each capacitor will be zero.

**67** (*b*) According to the question, the given arrangement of nine plates is equivalent to the parallel combination of 8 capacitors.

The capacity of each capacitor,

$$C = \frac{\varepsilon_0 A}{d} = \frac{8.854 \times 10^{-12} \times 5 \times 10^{-4}}{0.885 \times 10^{-2}} = 0.5 \text{ pF}$$

Hence, the capacity of the unit of 8 capacitors

$$= 8C = 8 \times 0.5 = 4 \text{pF}$$

**68** (*b*) Here, due the insertion of two slabs between the capacitor plates, two new capacitors are formed whose new capacitances are

$$\text{Capacitance, } C_1 = \frac{K_1 \varepsilon_0 A}{d/2} = \frac{2K_1 \varepsilon_0 A}{d} \text{ and } C_2 = \frac{2K_2 \varepsilon_0 A}{d}$$

$$\therefore \frac{1}{C_s} = \frac{1}{C_1} + \frac{1}{C_2} = \frac{d}{2K_1 \varepsilon_0 A} + \frac{d}{2K_2 \varepsilon_0 A}$$

$$= \frac{d}{2\varepsilon_0 A}\left(\frac{K_1 + K_2}{K_1 K_2}\right)$$

$$\therefore \text{ Capacity, } C_s = \frac{2\varepsilon_0 A}{d}\left(\frac{K_1 K_2}{K_1 + K_2}\right)$$

**69** (*c*) As each capacitors cannot withstand more than 300 V, so there should be four capacitors in each row became in this condition 1 kV i.e. 1000 V will be divided by 4 (i.e. 250 not more than 300 V).

Now, equivalent capacitance of one row

$$= \frac{1}{4} \times 1\mu\text{F} = 0.25\mu\text{F}$$

[∵ in series combination, $C_{eq} = c/n$]

Now, we need equivalent of 2μF, so let we need *n* such rows

$$\therefore \quad n \times 0.25 = 2\mu\text{F}$$

[∵ in parallel combination $C_{eq} = nc$]

$$n = \frac{2}{0.25} = 8$$

∴Total number of capacitors = number of rows × number of capacitors in each row = 8 × 4 = 32

**70** (*b*) Consider the figure given in the question.

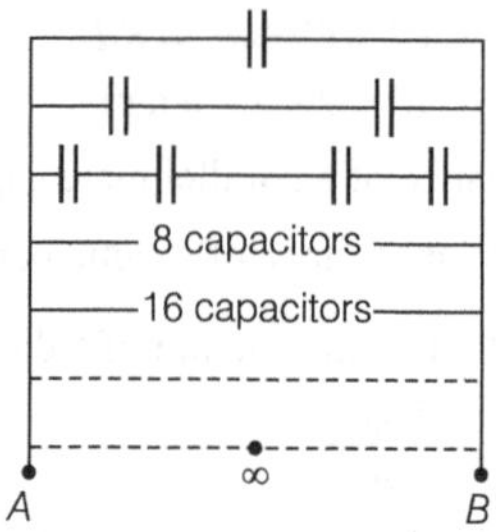


This combination forms a GP,

$$s = 1 + \frac{1}{2} + \frac{1}{4} + \frac{1}{8} + \ldots$$

Sum of infinite GP,

$$s = \frac{a}{1-r} \Rightarrow s = \frac{1}{1-\frac{1}{2}} \Rightarrow s = \frac{1}{\frac{1}{2}}$$

$$\therefore \quad s = 2$$

Hence, capacitance of the combination,

$$C_\infty = 2 \times 1\mu\text{F} = 2\mu\text{F}$$

**71** (*d*) Two capacitors $C_1$ and $C_2$, if connected in series, then their equivalent capacitance is

$$C_{\text{eq}} = \frac{C_1 \times C_2}{C_1 + C_2}$$

If they are connected in parallel, then their equivalent capacitance is, $C_{\text{eq}} = C_1 + C_2$.

We simplify given circuit as

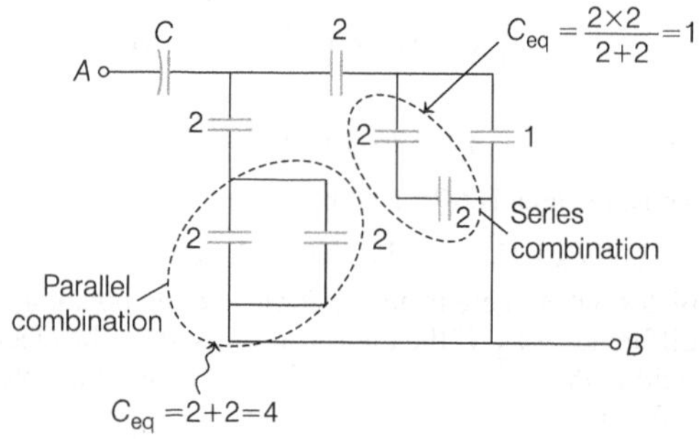


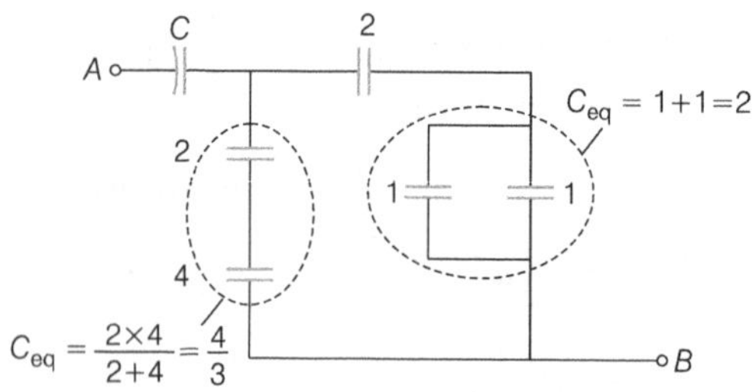

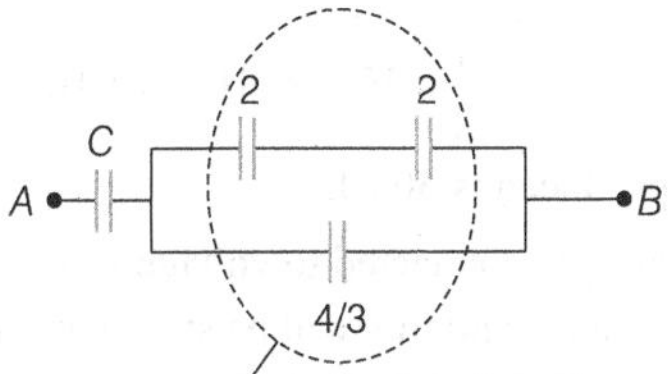


So, $C_{AB} = \frac{C \times 7/3}{C + 7/3} = \frac{1}{2}$ (given)

$\Rightarrow \frac{7C}{3} = \frac{C}{2} + \frac{7}{6}$

$\Rightarrow \left(\frac{14-3}{6}\right)C = \frac{7}{6} \Rightarrow C = \frac{7}{11}\mu F$

**72** (*c*) Given, $V = 500\,V, V_{max} = 3000\,V, C = 1\mu F$

and $C_{net} = 2\mu F$

Minimum number of conductors in each row

$$= \frac{V_{max}}{V} = \frac{3000}{500} = 6$$

If $C_s$ is capacity of 6 conductors in a row, then

$$\frac{1}{C_s} = \frac{1}{1} + \frac{1}{1} + \frac{1}{1} + \frac{1}{1} + \frac{1}{1} + \frac{1}{1} = 6$$

$\Rightarrow C_s = \frac{1}{6}\mu F$

Let there be $m$ such rows in parallel.

Total capacity, $C_{net} = m \times C_s$

$$2 = m \times \frac{1}{6}$$

$\therefore m = 12$

Total number of capacitors $= 6 \times 12 = 72$

**73** (*d*) In the given circuit, net capacitance of 3 μF and 6 μF capacitors being connected in parallel is

$$C' = 3\,\mu F + 6\,\mu F = 9\,\mu F$$

Now, 4.5 μF capacitor and capacitance 9 μF are in series. So, the total capacity of capacitor,

$$C = \frac{9 \times 4.5}{9 + 4.5} = \frac{9 \times 4.5}{13.5} = 3\,\mu F$$

Charge, $q = CV = 3 \times 12 = 36\,\mu C$ [given, $V = 12\,V$]

$\therefore$Potential difference across 4.5 μF,

$$V' = \frac{q}{4.5} = \frac{36}{4.5} = 8\,V$$

**74** (*c*) As we know that, the total work done in transferring a charge to a parallel plate capacitor is given as

$$W = \frac{Q^2}{2C} \quad \text{...(i)}$$

where, $C$ is the capacitance of the capacitor.

We can also write a relation for work done as

$$W = F \cdot d \quad \text{...(ii)}$$

where, $F$ is the electrostatic force between the plates of capacitor and $d$ is the distance between the plates.

From Eqs. (i) and (ii), we get

$$W = \frac{Q^2}{2C} = Fd$$

$\Rightarrow F = \frac{Q^2}{2Cd}$ ...(iii)

As, the capacitance of a parallel plate is given as

$$C = \frac{\varepsilon_0 A}{d}$$

Substituting the value of $C$ in Eq. (iii), we get

$$F = \frac{Q^2 d}{2\varepsilon_0 A d} = \frac{Q^2}{2\varepsilon_0 A}$$

This means, electrostatic force is independent of the distance between the plates.

**75** (*c*) Given, $C = 900\,pF = 900 \times 10^{-12}\,F$ and $V = 100\,V$

The charge on the capacitor is

$$Q = CV = 900 \times 10^{-12}\,F \times 100\,V = 9 \times 10^{-8}\,C$$

$\therefore$ The energy stored by the capacitor

$$= (1/2)\,CV^2 = (1/2)\,QV$$
$$= (1/2) \times 9 \times 10^{-8}\,C \times 100\,V$$
$$= 4.5 \times 10^{-6}\,J$$

**76** (*b*) Energy stored in a charged capacitor is given by

$$U = \frac{1}{2}CV^2 = \frac{1}{2}\cdot\frac{Q^2}{C} \quad \text{... (i)}$$

Here, $C = 12 \times 10^{-12}\,F$ and $V = 10\,V$.

$\Rightarrow U = \frac{1}{2} \times 12 \times 10^{-12} \times 100$

$U = 6 \times 10^{-10}\,J$ ... (ii)

After insertion of slab, capacitance will be

$C' = KC$ and final energy,

$$U' = \frac{1}{2}\cdot\frac{Q^2}{C'} = \frac{1}{2}\frac{Q^2}{KC}$$

$\Rightarrow U' = \frac{1}{K}U = \frac{1}{6.5} \times 6 \times 10^{-10}\,J$ ... (iv)

($\because$ given, $K = 6.5$)

So, energy dissipated in the process will be equal to work done on the slab, i.e.

$$\Delta U = U - U' = \left(1 - \frac{1}{6.5}\right) \times 6 \times 10^{-10}\,J$$

$\Rightarrow \Delta U = \frac{5.5}{6.5} \times 6 \times 10^{-10}\,J$

$\cong 5.08 \times 10^{-10}\,J$ or 508 pJ

**77** (*d*) Potential energy stored in a capacitor is

$$U = \frac{1}{2}QV = \frac{1}{2}\frac{Q^2}{C}$$

So, initial energy of the capacitor, $U_i = \frac{1}{2}Q^2/C_1$

Final energy of the capacitor, $U_f = \frac{1}{2}Q^2/C_2$

As we know, work done, $W = \Delta U = U_f - U_i$

$$= \frac{1}{2}Q^2\left[\frac{1}{C_2} - \frac{1}{C_1}\right]$$

Here, $Q = 5\,\mu C = 5 \times 10^{-6}$ C,

$C_1 = 5\,\mu F = 5 \times 10^{-6}$F,

$C_2 = 2\mu F = 2 \times 10^{-6}$ F

$$\Rightarrow \quad \Delta U = \frac{1}{2} \times (5 \times 10^{-6})^2 \left[\frac{1}{2\times 10^{-6}} - \frac{1}{5\times 10^{-6}}\right]$$

$$= \frac{1}{2} \times \frac{5 \times 5 \times 10^{-12}}{10^{-6}} \times \frac{3}{10}$$

$$= \frac{25 \times 3}{20} \times 10^{-6}\text{ J}$$

$$\Rightarrow \quad \Delta U = 3.75 \times 10^{-6}\text{ J}$$

$\therefore$ Work done in reducing the capacitance from 5 μF to 2 μF by pulling plates of capacitor apart is $3.75 \times 10^{-6}$ J.

**78** (*d*) Energy stored in capacitors

$$= \frac{1}{2}CV^2$$

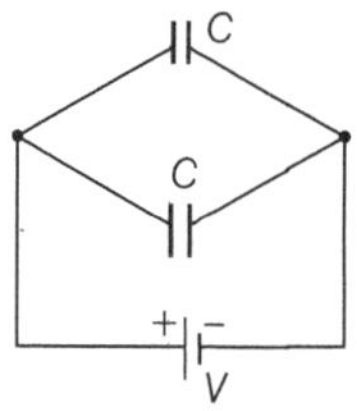


Also, potential drop remains same across both capacitors in parallel.

So, initially stored energy,

$$U_1 = \frac{1}{2}CV^2$$

Finally, potential drop across each capacitor will be still $V$.

So, finally stored energy

$$U_2 = \frac{1}{2}CV^2 + \frac{1}{2}CV^2 = \frac{1}{2}(2C)V^2$$

$$= 2\left(\frac{1}{2}CV^2\right) = 2U_1$$

Thus, final stored energy is increased by a factor of 2.

**79** (*a*) As, potential drop is zero along an ideal conductor, because it has zero resistance.

So, potential at point, $A = 10$ V

and potential at point, $B = 0$ V

$\therefore$Potential difference across $AB = (10 - 0)$ V $= 10$ V

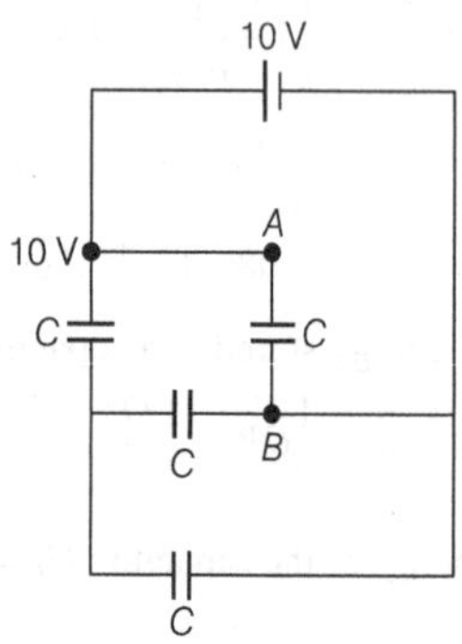


$\because$ Maximum heat produced = Energy stored $= \frac{1}{2}CV^2$

$$= \frac{1}{2} \times 10^{-6} \times (10)^2 = 50\,\mu\text{J}$$

$\therefore$ Heat produced is 50 μJ.

**80** (*c*) When $C_1$ is connected to voltage source, it is charged to a potential $V$ and this will be stored as a potential energy in the capacitor given by

$$U = \frac{1}{2}CV^2$$

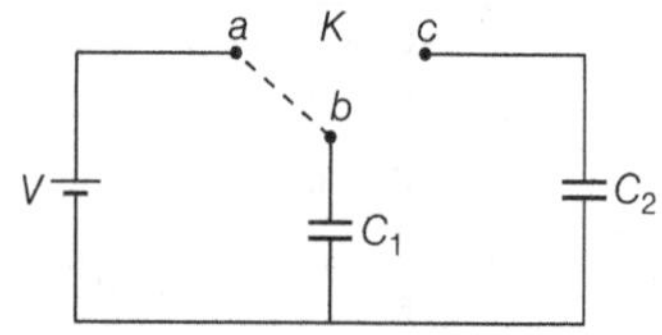


When key is disconnected from battery *a* & *b* and *b* & *c* are connected, the charge will be transferred from the capacitor $C_1$ to capacitor $C_2$, then

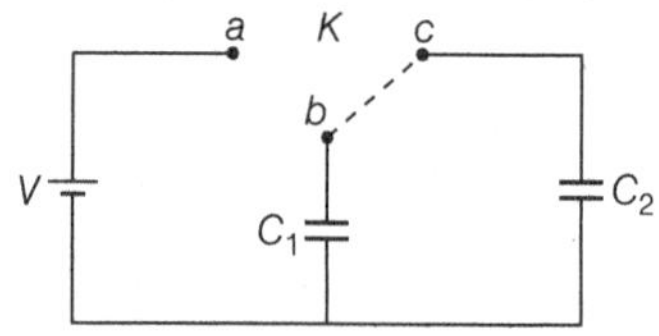


The loss of energy due to redistribution of charge is given by

$$\Delta U = \frac{C_1C_2}{2(C_1 + C_2)}(V_1 - V_2)^2$$

According to the question, $C_1 = C_2$

$$= \frac{C \times C}{2(C + C)}(V - 0)^2 = \frac{1}{4}CV^2$$

$$\therefore \text{Percentage loss} = \frac{\Delta U}{U} \times 100 = \frac{\frac{1}{4}CV^2}{\frac{1}{2}CV^2} \times 100 = 50\%$$

**81** (*c*) Stored energy in the capacitor can be given as

$$= \frac{1}{2}CV^2$$

Initially, the energy dissipated, when 2 μF is charged, is

$$E_i = \frac{1}{2} \times 2 \times V^2$$

When the switch $S$ is connected to position 2, energy dissipated across 8 μF will be

$$E_f = \frac{1}{2}C_{\text{net}}V^2 = \frac{1}{2}\left(\frac{C_1C_2}{C_1 + C_2}\right)V^2$$

$$= \frac{1}{2}\left(\frac{2 \times 8}{10}\right)V^2 = \frac{1}{2} \times \frac{16}{10} \times V^2$$

Therefore, % loss of energy

$$= \frac{E_i}{E_f} \times 100\% = \left(\frac{\frac{1}{2} \times \frac{16}{20}V^2}{\frac{1}{2} \times 2 \times V^2}\right) \times 100\% = 40\%$$

**82** (*d*) Given, potential difference across capacitor system,

$$\Delta V = 4\text{V}$$

Energy stored in a capacitor system,

$$U = \frac{1}{2} C_{\text{eq}} \Delta V^2$$

**Case I** When the capacitors are joined in series,

$$U_{\text{series}} = \frac{1}{2} \frac{C_1}{n_1} (4V)^2$$

**Case II** When the capacitors are joined in parallel,

$$U_{\text{parallel}} = \frac{1}{2} (n_2 C_2) V^2$$

Given, $U_{\text{series}} = U_{\text{parallel}}$

or $$\frac{1}{2} \frac{C_1}{n_1} (4V)^2 = \frac{1}{2} (n_2 C_2) V^2 \Rightarrow C_2 = \frac{16 C_1}{n_1 n_2}$$

**83** (*c*) According to the question,

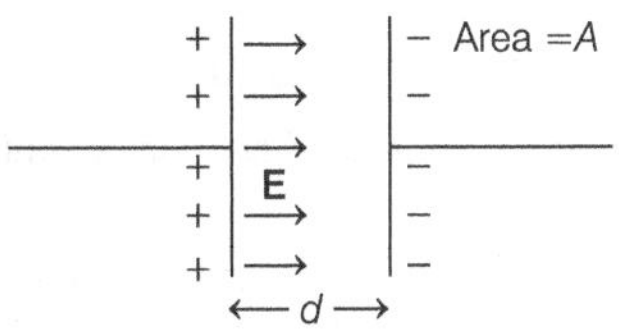


where, $E$ = electric field between the parallel plates.

The energy stored in the capacitor,

$$U = \frac{1}{2} CV^2 = \frac{1}{2} \left( \frac{A\varepsilon_0}{d} \right) (Ed)^2$$

$$\left( \because C = \frac{A\varepsilon_0}{d} \text{ and } V = Ed \right)$$

$$= \frac{1}{2} \varepsilon_0 E^2 Ad$$

**84** (*d*) The quantities, energy stored and potential difference decreases, because $U = \frac{1}{2} \frac{q^2}{KC}$ and $V = \frac{q}{KC}$ decreases. Because on inserting a dielectric, the capacitance will increase $(KC_0)$, where $C_0$ is the capacitance when no glass slab is present and $K$ is dielectric constant, $q$ is constant here.

**86** (*a*) A van de graaff generator consists of a large spherical conducting shell. By mean of a moving belt and suitable brushes, charge is continuously transferred to the shell and potential difference of the order of several million volt is built up, which can be used for accelerating charged particles.

**87** (*a*) Among the given figure under consideration, it can be said that for $Q < 0$, the force on unit positive charge is attractive, so that the electrostatic force and displacement are in same direction.

So, due to this reason, work done by the electrostatic force in bringing the unit positive charge from infinity to point $P$ is positive.

Therefore, Assertion and Reason are correct and Reason is the correct explanation of Assertion.

**88** (*a*) As electric field is set up from higher potential to lower potential because it is in the direction in which the potential decreases steeply.

So, if an electron having negative charge is placed in electric field, then it will move opposite to the direction of electric field,

i.e. from lower potential to higher potential.

Therefore, Assertion and Reason are correct and Reason is the correct explanation of Assertion.

**89** (*a*) The potential energy $U = \frac{1}{4\pi\varepsilon_0} \frac{q_1 q_2}{r_{12}}$ is unaltered whatever way the charges are brought to the specified locations, because of path-independence of work for electrostatic force.

Therefore, Assertion and Reason are correct and Reason is the correct explanation of Assertion.

**90** (*a*) Polar dielectric molecules are the one in which centres of positive and negative charges are separated even when there is no external field. It means, they have permanent dipoles.

So, in the absence of any external electric field, the different permanent dipoles are oriented randomly due to thermal agition. So, the total dipole moment is zero.

Therefore, Assertion and Reason are correct and Reason is the correct explanation of Assertion.

**91** (*c*) The molecules of a substance may be polar or non-polar. In a non-polar molecule, the centre of positive and negative charges coincides.

On the other hand, a polar molecule is one in which the centres of positive and negative charges are separated, even when there is no external field.

Such molecules have a permanent dipole moment.

Therefore, Assertion is correct but Reason is incorrect.

**92** (*a*) The maximum amount of charge a capacitor can have depends on the shape and size of capacitor and also on the surrounding medium.

Thus, a capacitor can be given only a limited quantity of charge.

Therefore, Assertion and Reason are correct and Reason is the correct explanation of Assertion.

**93** (*c*) Capacitance of a capacitor is inversely proportional to distance between the plates.

So, capacity of a parallel plate capacitor increases when distance between the plates is decreased.

Therefore, Assertion is correct but Reason is incorrect.

**94** (*c*) Surface density of parallel plate capacitor without a dielectric slab is given as

$$\sigma = \frac{q}{A} = \frac{CV}{A} \quad \text{...(i)}$$

Surface density on a parallel plate capacitor in the presence of dielectric slab is,

$$\sigma' = \frac{q'}{A} = \frac{C'V'}{A} = \frac{(KC)V}{A}$$

$$(\text{as } C' = KC, V' = V)$$

So from Eq. (i), the above equation can be written as

$$\sigma' = \frac{Kq}{A} = K\sigma$$

The electric field on introduction of dielectric between plates of a capacitor becomes,

$$E' = \frac{K\sigma}{\varepsilon_0} = KE \qquad \left[\because E = \frac{\sigma}{\varepsilon_0}\right]$$

Thus, the electric field and surface density both become $K$ times on introducing the dielectric inside the plates.

Therefore, Assertion is correct but Reason is incorrect.

**95** (*c*) In a series connection of capacitors charge on each capacitor remains same and equals to the main charge supplied by the battery, but potential difference across each capacitors is different.

Also, the capacitance of a capacitor is dependent only on its geometrical configuration and is independent of the charge on it.

Therefore, Assertion is correct but Reason is incorrect.

**96** (*c*) Net capacitance of the capacitors when connected in series is given as

$$\frac{1}{C_s} = \frac{1}{C_1} + \frac{1}{C_2} + \frac{1}{C_3} \qquad \ldots \text{(i)}$$

and when they are connected in parallel, net capacitance is given as

$$C_p = C_1 + C_2 + C_3 \qquad \ldots \text{(ii)}$$

Since, $C_1 < C_2 < C_3$, then comparing Eqs. (i) and (ii), we can say that, $C_p < C_s$.

Therefore, Assertion is correct but Reason is incorrect.

**100** (*c*) At the point $P$, the electric field due to charge $q_a$ is towards right, whereas it will be towards left due to charges $q_b$, $q_c$ and $q_d$.

Since, the resultant electric field at a point inside a conductor is zero,

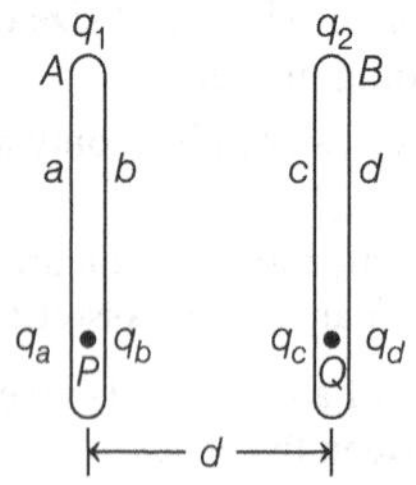


$$\frac{q_a}{2\varepsilon_0 A} - \frac{q_b}{2\varepsilon_0 A} - \frac{q_c}{2\varepsilon_0 A} - \frac{q_d}{2\varepsilon_0 A} = 0$$

or
$$q_a = q_b + q_c + q_d$$
$$= (q_1 - q_a) + (q_2 - q_d) + q_d$$
$$= q_1 - q_a + q_2$$

where, $q_a = \dfrac{q_1 + q_2}{2}$

and $q_b = q_1 - q_a = q_1 - \dfrac{q_1 + q_2}{2} = \dfrac{q_1 - q_2}{2}$

Similarly, the electric field at $Q$ is zero.

$$\frac{q_a}{2\varepsilon_0 A} + \frac{q_b}{2\varepsilon_0 A} + \frac{q_c}{2\varepsilon_0 A} - \frac{q_d}{2\varepsilon_0 A} = 0$$

$\Rightarrow$
$$q_d = q_a + q_b + q_c$$
$$= (q_1 - q_b) + q_b + (q_2 - q_d) = q_1 + q_2 - q_d$$

where, $q_d = \dfrac{q_1 + q_2}{2}$

and $q_c = q_2 - q_d$

$\therefore$ $q_c = \dfrac{q_2 - q_1}{2}$

The charges on the inner surfaces $(q_b, q_c)$ which are equal and opposite are responsible for creating the electric field between the plates of the capacitor.

Thus, $E = \dfrac{\sigma}{\varepsilon_0} = \dfrac{q_1 - q_2}{2\varepsilon_0 A}$

$$\text{Capacitance, } C = \frac{q_b}{V} = \frac{\left(\frac{q_1 - q_2}{2}\right)}{\frac{(q_1 - q_2)d}{2\varepsilon_0 A}} = \frac{\varepsilon_0 A}{d}$$

So, statements I, III and IV are correct but II is incorrect.

**101** (*d*) The statement III is correct but rest are incorrect and these can be corrected as,

Electric field vector (**E**), electric displacement vector (**D**) and electric polarisation vector (**P**) are related as

$$\mathbf{D} = \varepsilon_0 \mathbf{E} + \mathbf{P} \qquad \ldots \text{(i)}$$

where, $\varepsilon_0$ is permittivity of free space.

From Eq. (i), it is clear that **D**, **E** and **P** are in same direction, i.e. they are mutually parallel to each other.

**102** (*c*) Electric field intensity inside a conducting sphere is always zero, so electrostatic potential inside the sphere is same as on the surface, i.e. 100 V.

Thus, statement given in option (c) is correct but rest are incorrect.

**103** (*d*) The statement given in option (d) is incorrect and it can be corrected as,

Because the electric field intensity is large near sharp edges of charged conductor and near regions of large charge densities.

Therefore, equipotential surfaces are not always equally spaced.

Rest statements are correct.

**104** (*a*) The statement given in option (a) is correct but rest are incorrect and these can be corrected as,

At the surface of a charged conductor, electrostatic field must be normal to the surface at every point.

Electrostatic potential is constant throughout the volume of the conductor.

Electrostatic field at the surface of a charged conductor is non-zero and equal to $\left(\dfrac{\sigma}{\varepsilon_0}\right)$.

**105** (*a*) When we earth plate $B$, whole of its charge will go to earth. Hence, the charge flown through switch is not zero.

Now, charge of plate $A$ will induce charge $-Q_1$, and $+Q_1$ on inner side and outer side of plate $B$. Again charge $+Q_1$ of plate $B$ will go to earth. Charge $-Q_1$ will not go to earth. It will be bounded to $+Q_1$, charge of plate $A$ (see figure)

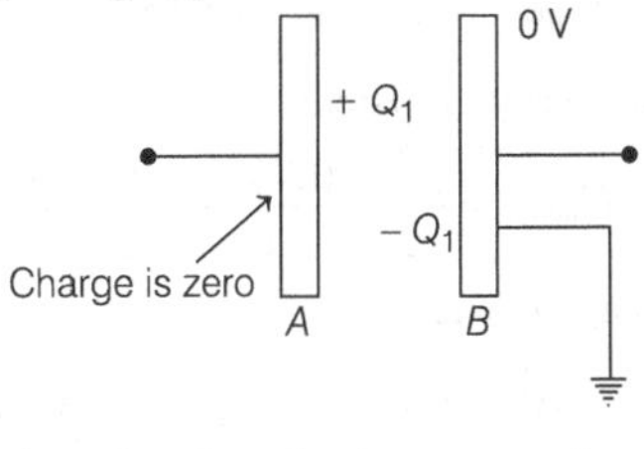


So, it is clear that $Q_1 + Q_2$ charge goes from 2nd plate to earth.

Charge of capacitor is $Q_1$ and hence its potential is $\frac{Q_1}{C}$.

Thus, the statement given in option (a) is incorrect, rest are correct.

**106** (*b*) The statement given in option (b) is correct but rest are incorrect and these can be corrected as,
When battery is disconnected, charge remains conserved.

The plates are pulled apart, so capacity will decrease, so there is an increase of voltage.

Now, capacity decreases,
hence energy will increase as charge is constant
$\left(E = \frac{Q^2}{2C}\right)$.

**107** (*c*) According to the question, a parallel plate capacitor is connected to a battery

I. When key is closed, $C = \frac{\varepsilon_0 A}{d}$, changes as separation changes, hence $q = CV$ changes. The voltage $V$ remains same

II. When key is opened and source is disconnected charge remains same, hence

$V = \frac{Q}{C}$ changes as $C = \frac{\varepsilon_0 A}{d}$ changes.

Thus, the statement given in option (c) is correct but rest are incorrect.

**108** (*c*) Given, $C = 900\,\text{pF}$ and $V = 100\,\text{V}$.

The charge on the capacitor is

$$Q = CV = 900 \times 10^{-12}\ \text{F} \times 100\ \text{V}$$
$$= 9 \times 10^{-8}\,\text{C}$$

The energy stored by the capacitor is

$$= \frac{1}{2}CV^2 = \frac{1}{2}QV$$
$$= (1/2) \times 9 \times 10^{-8}\ \text{C} \times 100\ \text{V}$$
$$= 4.5 \times 10^{-6}\,\text{J}$$

In the steady situation, the two capacitors have their positive plates at the same potential and their negative plates at the same potential. Let the common potential difference be $V'$. The charge on each capacitor is $Q' = CV'$. By charge conservation law, $Q' = Q$, this implies $V' = V/2$. The total energy of the system is

$$= 2\frac{1}{2}Q'V' = \frac{1}{4}QV$$
$$= 2.25 \times 10^{-6}\,\text{J}$$

Thus, the statement given in option (c) is correct, rest are incorrect.

**109** (*c*) Initial energy, $U_i = \frac{1}{2}C_0V^2$

When dielectric slab is introduced the capacitance becomes $K$ times and potential difference decreases $K$ times.

$\therefore$ Final energy, $U_f = \frac{1}{2}(KC_0)(V/K)^2$

or $$U_f = \frac{1}{K}\left(\frac{1}{2}C_0V^2\right)$$

So, energy stored in the capacitor decreases $K$ times.

$\therefore$ Change in energy $= U_f - U_i = \frac{1}{2}C_0V^2\,(1/K - 1)$

Thus, the statement given in option (c) is incorrect but rest are correct.

**111** (*c*)

A. Electric potential at any point due to electric dipole is given as

$$V = \frac{p\cos\theta}{4\pi\varepsilon_0 r^2}$$

At point $P_1$, $\theta = 0°$

$$\Rightarrow \quad V_1 = \frac{1}{4\pi\varepsilon_0}\cdot\frac{p\cos 0°}{r^2} = \frac{p}{4\pi\varepsilon_0 r^2}$$

B. At point $P_2$, $\theta = 180°$

$$\Rightarrow \quad V_2 = \frac{1}{4\pi\varepsilon_0}\cdot\frac{p\cos 180°}{r^2} = \frac{-p}{4\pi\varepsilon_0 r^2}$$

C. At point $P_3$, $\theta = 45°$

$$\Rightarrow \quad V_3 = \frac{1}{4\pi\varepsilon_0}\cdot\frac{p\cos 45°}{r^2} = \frac{p}{4\sqrt{2}\pi\varepsilon_0 r^2}$$

D. At point $P_4$, $\theta = 90°$

$$\Rightarrow \quad V_4 = \frac{1}{4\pi\varepsilon_0}\cdot\frac{p\cos 90°}{r^2} = 0$$

Hence, A $\rightarrow$ 4, B $\rightarrow$ 3, C $\rightarrow$ 2 and D $\rightarrow$ 1.

**112** (*b*)

A. The electric field in the dielectric corresponds to the case when the net surface charge density on the plates is $\pm(\sigma - \sigma_p)$, where $\pm\sigma_p$ = surface charge density of dielectric

i.e. $$E = \frac{\sigma - \sigma_p}{\varepsilon_0} \quad \text{...(i)}$$

B. So the potential difference across the plates is

$$V = Ed = \frac{\sigma - \sigma_p}{\varepsilon_0} d \qquad \ldots\text{(ii)}$$

For linear dielectric, we except $\sigma_p$ to be proportional to $E_0$, i.e., to $\sigma$. Thus, $(\sigma - \sigma_p)$ is proportional to $\sigma$ and we can write

$$\sigma - \sigma_p = \frac{\sigma}{K} \qquad \ldots\text{(iii)}$$

where, $K$ is a constant characteristic of the dielectric. Clearly, $K > 1$, then we have

$$V = \frac{\sigma d}{\varepsilon_0 K} = \frac{Qd}{A\varepsilon_0 K} \qquad \ldots\text{(iv)}$$

C. The capacitance $C$ with dielectric between the plates is

$$C = \frac{Q}{V} = \frac{\varepsilon_0 KA}{d} \qquad \ldots\text{(v)}$$

D. The product $\varepsilon_0 K$ is called the permittivity of the medium and is denoted by $\varepsilon$.

$$\varepsilon = \varepsilon_0 K \qquad \ldots\text{(vi)}$$

For vacuum, $K = 1$ and $\varepsilon = \varepsilon_0$.

The dimensionless ratio, $K = \dfrac{\varepsilon}{\varepsilon_0}$ ...(vii)

is called the dielectric constant of the substance. As remarked before, from Eq. (iii), it is clear that $K$ is greater than 1.

From equations, $(C_0 = \varepsilon_0 \dfrac{A}{d})$

and $C = \dfrac{\varepsilon_0 KA}{d}$,

$$K = \frac{C}{C_0} \qquad \ldots\text{(viii)}$$

Hence, A $\rightarrow$ 2, B $\rightarrow$ 1, C $\rightarrow$ 4 and D $\rightarrow$ 3.

**113** (*b*)

A. As, $\dfrac{Q_1}{C_1} = \dfrac{Q_2}{C_2}, Q_2 = \left(\dfrac{C_2}{C_1}\right) Q_1 = \dfrac{(C/2)}{C} Q_1 = \dfrac{Q_1}{2}$

Further as, $Q = Q_1 + Q_2 = Q_1 + \dfrac{Q_1}{2} = \dfrac{3}{2} Q_1$

$$Q_1 = \frac{2}{3} Q \text{ and } Q_2 = \frac{1}{3} Q$$

Final energy of capacitor $C_1$,

$$U_1 = \frac{Q_1^2}{2C_1} = \frac{\left(\frac{2}{3}Q\right)^2}{2C}$$

$$= \frac{2}{9}\left(\frac{Q^2}{C}\right) = \frac{2}{9}\left(\frac{C^2 V_0^2}{C}\right)$$

$$= \frac{2}{9} C V_0^2 \qquad (\text{as } Q = CV_0)$$

B. Final energy of capacitor $C_2$,

$$U_2 = \frac{Q_2^2}{2C_2} = \frac{\left(\frac{1}{3}Q\right)^2}{2(C/2)} = \frac{1}{9}\left(\frac{Q^2}{C}\right) = \frac{1}{9} C V_0^2$$

C. Final energy of system,

$$U_{\text{final}} = U_1 + U_2 = \frac{1}{3} C V_0^2$$

D. Inital energy of system,

$$U_{\text{initial}} = \frac{Q^2}{2C} = \frac{(CV_0)^2}{2C} = \frac{1}{2} C V_0^2$$

$\therefore$ Change in energy,

$$\Delta U = U_{\text{final}} - U_{\text{initial}} = -\frac{1}{6} C V_0^2$$

Hence, A $\rightarrow$ 4, B $\rightarrow$ 5, C $\rightarrow$ 3 and D $\rightarrow$ 1.

**114** (*c*)

A. When $S$ is open, then

$$C_{\text{eq}} = \frac{C_1 C_2}{C_1 + C_2} = \frac{(2\mu\text{F})(3\mu\text{F})}{(2\mu\text{F} + 3\mu\text{F})} = \frac{6}{5}\mu\text{F}$$

Charge $q$ on both the capacitors will be the same (as these are in series), i.e.

$$q = C_{\text{eq}} V = \left(\frac{6}{5}\mu\text{F}\right)(30 + 60)\text{ V} = 108\,\mu\text{C}$$

B. Charge on $C_2$, i.e. $q_2 = C_2 V_2 = (3\,\mu\text{F})(60\text{ V}) = 180\,\mu\text{C}$

C. When $S$ is closed, charge on $C_1$, i.e.

$$q_1 = C_1 V_1 = (2\mu\text{ F})(30 VC) = 60\,\mu\text{C}$$

D. Charge on $C_1$ and $C_2 = 60\,\mu\text{C} + 180\,\mu\text{C} = 240\,\mu\text{C}$

Hence, A $\rightarrow$ 1, B $\rightarrow$ 4, C $\rightarrow$ 3 and D $\rightarrow$ 2.

**115** (*c*) Let $C$ be the point at which potential is zero.

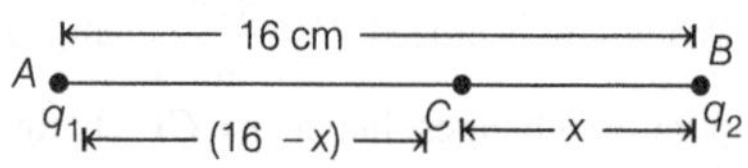


Given, $q_1 = 5 \times 10^{-8}$ C, $q_2 = -3 \times 10^{-8}$ C and distance, $AB = 16$ cm

The potential at point $C$ due to charge $q_1$,

$$V_A = \frac{1}{4\pi\varepsilon_0} \cdot \frac{q_1}{AC} = \frac{9 \times 10^9 \times 5 \times 10^{-8}}{(16 - x) \times 10^{-2}} \qquad \ldots\text{(i)}$$

The potential at point $C$ due to charge $q_2$,

$$V_B = \frac{1}{4\pi\varepsilon_0} \cdot \frac{q_2}{BC} = \frac{-9 \times 10^9 \times 3 \times 10^{-8}}{x \times 10^{-2}} \qquad \ldots\text{(ii)}$$

As, $V_A + V_B = 0$

$$\Rightarrow \frac{9 \times 10^9 \times 5 \times 10^{-8}}{(16 - x) \times 10^{-2}} + \left(\frac{-9 \times 10^9 \times 3 \times 10^{-8}}{x \times 10^{-2}}\right) = 0$$

$$\Rightarrow \frac{5}{16 - x} - \frac{3}{x} = 0 \Rightarrow x = 6\text{ cm}$$

**116** (*b*) Given, $q = 5\mu C = 5\times 10^{-6}$ C

and $\quad r = 10\,\text{cm} = 10\times 10^{-2}$ m

The regular hexagon is shown as follows

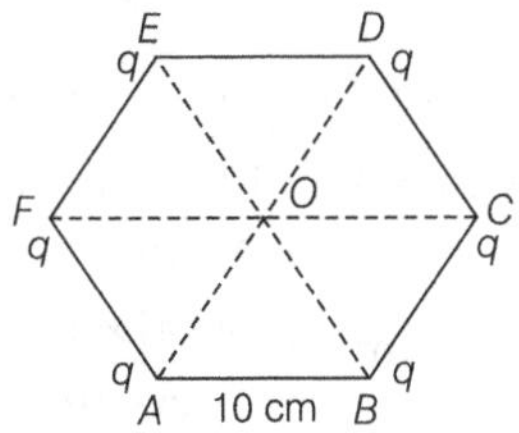


i.e. $OA = OB = OC = OD = OE = OF = 10$ cm

Potential at point $O$ = Sum of potential at centre $O$ due to individual point charges

$$\therefore \quad V_O = V_A + V_B + V_C + V_D + V_E + V_F$$

$$\left(\because V = \frac{1}{4\pi\varepsilon_0}\cdot\frac{q}{r}\right)$$

As charge is same, so

$$V_O = V_A \times 6 = \frac{9\times 10^9 \times 5\times 10^{-6}\times 6}{10\times 10^{-2}}$$

$$= 2.7\times 10^6 \text{ V}$$

**117** (*b*) It is given that, capacitor with air between the plates has a capacitance of 8 pF, i.e. initially,

$$C_0 = \frac{\varepsilon_0 A}{d} = 8\,\text{pF}$$

When dielectric is inserted and the distance between the plates is reduced by half ($d = d/2$),then

$$C = \frac{K\varepsilon_0 A}{d'} = \frac{6\varepsilon_0 A\times 2}{d} = 6\left(\frac{\varepsilon_0 A}{d}\right)\times 2$$

$$= 6(8)\times 2 = 96\,\text{pF}$$

**118** (*a*) In series, $\dfrac{1}{C_s} = \dfrac{1}{C_1} + \dfrac{1}{C_2} + \dfrac{1}{C_3}$

$$= \frac{1}{9} + \frac{1}{9} + \frac{1}{9}$$

$$\frac{1}{C_s} = \frac{3}{9} \Rightarrow C_s = 3\,\text{pF}$$

Charge, $q = C_s V = 3\times 120 = 360$ pC

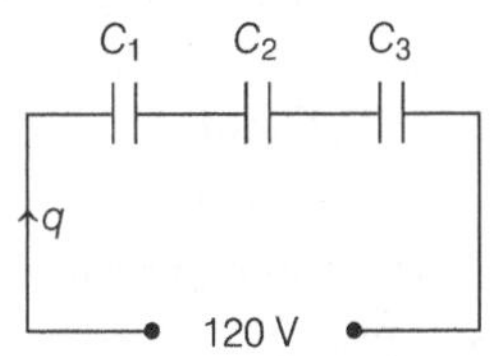


Potential difference across $C_1$,

$$V_1 = \frac{q}{C_1} = \frac{360}{9}$$

$$= 40\,\text{V} = V_2 = V_3$$

**119** (*a*) Given, $A = 6\times 10^{-3}\,\text{m}^2$, $d = 3$ mm and $V = 100$ V

$$C = \frac{\varepsilon_0 A}{d} = \frac{8.854\times 10^{-12}\times 6\times 10^{-3}}{3\times 10^{-3}}$$

$$C = 1.77\times 10^{-11}\,\text{F}$$

When the capacitor is connected to a 100 V supply, charge on each plate of the capacitor

$$q = CV = 1.77\times 10^{-11}\times 100$$

Thus, $\quad q = 1.77\times 10^{-9}$ C

**120** (*c*) Given, capacitance of capacitor, $C_1 = 600$ pF $= 600\times 10^{-12}$ F, supply voltage, $V_1 = 200$ V,

$C_2 = 600\,\text{pF} = 600\times 10^{-12}$ F and $V_2 = 0$

Loss in energy, $E = \dfrac{C_1C_2(V_1 - V_2)^2}{2(C_1 + C_2)}$

$$\Rightarrow \quad E = \frac{600\times 10^{-12}\times 600\times 10^{-12}(200-0)^2}{2(600+600)\times 10^{-12}}$$

$$= 6\times 10^{-6}\,\text{J}$$

Thus, the $6\times 10^{-6}$ J amount of electrostatic energy is lost in the sharing of charges.

**121** (*b*) According to question, charge at origin $O$ is

$$q_O = 8\,\text{mC} = 8\times 10^{-3}\,\text{C}$$

Charge $q_P$ at point $P = -2\times 10^{-9}$ C

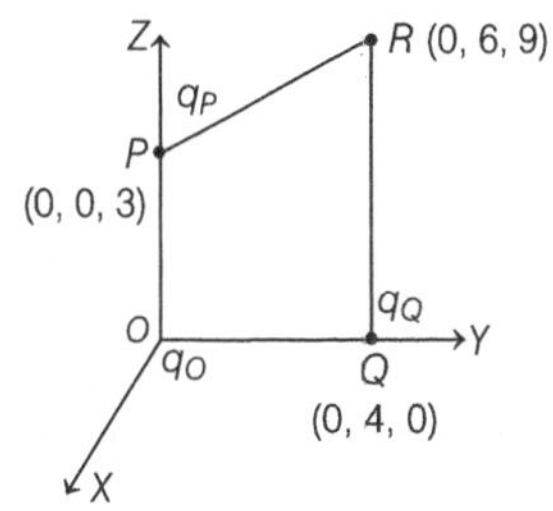


Distance $OP = r_P = 3\,\text{cm} = 0.03$ m

Distance $OQ = r_Q = 4\,\text{cm} = 0.04$ m

Work done in bringing the charge $q_P$ from $P$ to $Q$

$= q_P \times$ Potential difference between $P$ and $Q$

$$W_{PQ} = q_P(V_Q - V_P)$$

$$= -2\times 10^{-9}\left(\frac{1}{4\pi\varepsilon_0}\cdot\frac{q_O}{OQ} - \frac{1}{4\pi\varepsilon_0}\cdot\frac{q_O}{OP}\right)$$

$$W_{PQ} = -2\times 10^{-9}\left(\frac{9\times 10^9\times 8\times 10^{-3}}{0.04} - \frac{9\times 10^9\times 8\times 10^{-3}}{0.03}\right)$$

$$= 1.2\,\text{J}$$

Thus, the work done is 1.2 J.

**122** (*c*) Length of the main diagonal of the cube

$$= \sqrt{b^2 + b^2 + b^2} = \sqrt{3}b$$

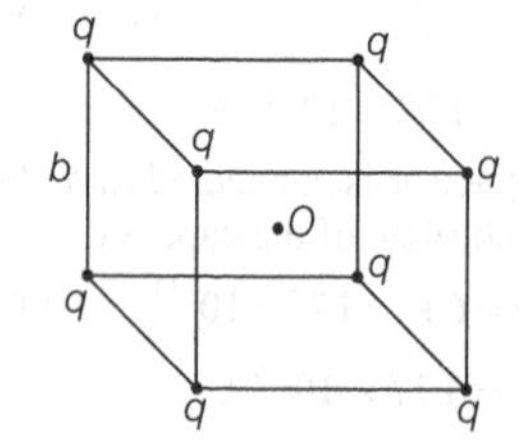


Distance of centre $O$ from each of the vertices is $r$

$$r = \frac{b\sqrt{3}}{2}$$

Potential at point $O$ due to one charge,

$$V = \frac{1}{4\pi\varepsilon_0} \cdot \frac{q}{r}$$

Therefore, potential at point $O$ due to all charges placed at the vertices of the cube,

$$V' = 8V = \frac{8 \times 1 \times q}{4\pi\varepsilon_0 \cdot r}$$

$$= \frac{8q \times 2}{4\pi\varepsilon_0 \cdot b\sqrt{3}} = \frac{4q}{\sqrt{3}\pi\varepsilon_0 b}$$

**123** (*b*) The distance from point $P$ to point $A$ is equal to the distance from point $P$ to point $B$.

$(BC = AC = 15 \text{ cm} = 0.15 \text{ m}, PC = 10 \text{ cm} = 0.1 \text{ m})$

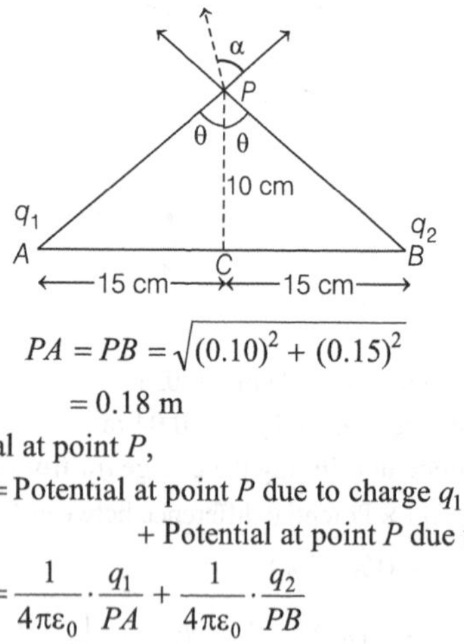


$$\therefore \quad PA = PB = \sqrt{(0.10)^2 + (0.15)^2}$$

$$= 0.18 \text{ m}$$

Potential at point $P$,

$V_P$ = Potential at point $P$ due to charge $q_1$ + Potential at point $P$ due to charge $q_2$

$$V_P = \frac{1}{4\pi\varepsilon_0} \cdot \frac{q_1}{PA} + \frac{1}{4\pi\varepsilon_0} \cdot \frac{q_2}{PB}$$

$$= 9 \times 10^9 \left( \frac{1.5 \times 10^{-6}}{0.18} + \frac{2.5 \times 10^{-6}}{0.18} \right)$$

$$= \frac{9 \times 10^9}{0.18} \times 10^{-6} \times 4$$

$$= 2 \times 10^5 \text{ V}$$

Thus, the potential, $V_p = 2 \times 10^5$ V.

**124** (*b*) The potential energy is $U = \frac{1}{4\pi\varepsilon_0} \cdot \frac{q_e q_p}{r}$

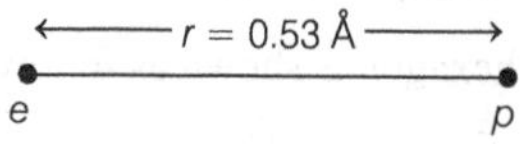


We know, $e = -1.6 \times 10^{-19}$ C and $p = +1.6 \times 10^{-19}$ C

$$U = \frac{9 \times 10^9 \times 1.6 \times 10^{-19} \times (-1.6) \times 10^{-19}}{0.53 \times 10^{-10}}$$

$$= 43.47 \times 10^{-19} \text{ J} = \frac{43.47 \times 10^{-19}}{1.6 \times 10^{-19}} \text{ eV}$$

$(\because 1 \text{ eV} = 1.6 \times 10^{-19} \text{ J})$

$$= -27.16 \text{ eV}$$

Thus, potential energy of system is −27.16 eV.

**125** (*c*) Electric field always acts along decreasing potential. Thus, when a positively charged particle is released from rest in a uniform electric field, the work done or electric potential energy of the charge decreases. Because it moves along the electric field, i.e. from high potential to low potential.

**126** (*c*) The work done by a electrostatic force is given by $W_{12} = q(V_2 - V_1)$. Here, initial and final potentials are same in all three cases and same charge is moved, so work done is same in all three cases.

**128** (*c*) The capacitance of parallel plate capacitor filled with dielectric block has thickness $d_1$ and dielectric constant $K_1$ is given by $\quad C_1 = \frac{K_1 \varepsilon_0 A}{d_1}$

Similarly, capacitance of parallel plate capacitor filled with dielectric block has thickness $d_2$ and dielectric constant $K_2$ is given by

$$C_2 = \frac{K_2 \varepsilon_0 A}{d_2}$$

Since, the two capacitors are in series combination, the equivalent capacitance is given by

$$\frac{1}{C} = \frac{1}{C_1} + \frac{1}{C_2}$$

$$C = \frac{C_1 C_2}{C_1 + C_2} = \frac{\frac{K_1 \varepsilon_0 A}{d_1} \cdot \frac{K_2 \varepsilon_0 A}{d_2}}{\frac{K_1 \varepsilon_0 A}{d_1} + \frac{K_2 \varepsilon_0 A}{d_2}}$$

$$= \frac{K_1 K_2 \varepsilon_0 A}{K_1 d_2 + K_2 d_1} \quad \text{...(i)}$$

But the equivalent capacitances is given by

$$C = \frac{K \varepsilon_0 A}{d_1 + d_2} \quad \text{... (ii)}$$

On comparing Eqs. (i) and (ii), we have

$$K = \frac{K_1 K_2 (d_1 + d_2)}{K_1 d_2 + K_2 d_1}$$

CHAPTER > 03

# Current Electricity

KEY NOTES

## Electric Current

- The rate of flow of electric charge through any cross-sectional area of a conductor is known as electric current.

$$I = \frac{q}{t} = \frac{ne}{t} \qquad (\because q = ne)$$

where, $n$ = number of charged particles constituting the current

and $e$ = electronic charge.

Its SI unit is ampere.

## Electric Current in Conductors

- The electric current in a conductor can be explained by electron theory. **Free electrons** or **conduction electrons**, carry the charge in the substance from one place to the other.
- Therefore, the electrical conductivity of a solid substance depends upon the number of free electrons in it.
- Electrolytic solutions are the type of conductors in which electric current flows due to the movement of both positive and negative charges.
- When no electric field is present in conductors, the average number of electrons travelling in any direction will be equal to the number of electrons travelling in the opposite direction. So, there will be no net electric current.

## Ohm's Law

- It states that, the current flowing through a conductor is directly proportional to the potential difference across its ends.

i.e. $I \propto V$ or $V \propto I$

$\Rightarrow$ $V = IR$

where, $R$ = resistance of conductor, which is measured in ohm ($\Omega$).

- The resistance $R$ not only depends on the material of the conductor but also on the dimensions of the conductor.
- Resistance of a conductor can also be written as $R = \rho \frac{l}{A}$

where, $l$ = length of the conductor, $A$ = area of cross-section

and $\rho$ = constant, known as **resistivity** of the material.

It depends upon the nature of the material and temperature of the conductor but not on the dimensions of conductor.

## Limitations of Ohm's law

There are some materials and devices used in electric circuits, where the proportionality of $V$ and $I$ does not hold. The deviations broadly are one or more of the following types

(i) $V$ ceases to be proportional to $I$.

(ii) If $I$ is the current for a certain $V$, then reversing the direction of $V$ keeping its magnitude fixed, does not produce a current of the same magnitude as $I$ in the opposite direction. This happens in a diode.

(iii) The relation between $V$ and $I$ is not unique, i.e. there is more than one value of $V$ for the same current $I$.

A material exhibiting such behaviour is GaAs.

## Current Density

- It is the current per unit area ($I/A$) and given by

$$J = \frac{I}{A}$$

Its SI unit is $A/m^2$.

- The **relation between electric field and current density** is given as

$$\mathbf{E} = \mathbf{J}\rho$$

It can be written in another form as $\mathbf{J} = \frac{1}{\rho}\mathbf{E} = \sigma\mathbf{E}$

where, $\sigma = \frac{1}{\rho}$ is called the **conductivity**.

## Drift of Electrons and Origin of Resistivity

- **Drift Velocity** It is defined as the average velocity with which the free electrons in a conductor get drifted towards the positive end of a conductor under the influence of an external electric field applied. It is given by

$$\Rightarrow \quad \mathbf{v}_d = -\frac{e\mathbf{E}}{m}\tau$$

where, $\tau$ = relaxation time, $\mathbf{E}$ = electric field,

$m$ = mass and $e$ = charge on electron.

- **Relaxation Time** It is a time interval between two successive collisions of electrons in a conductor, when current flows in it. It is given by

$$\tau = \frac{\text{mean free path}}{\text{rms velocity of electron}} = \frac{\lambda}{v_{rms}}$$

With rise in temperature, $v_{rms}$ increases, consequently $\tau$ decreases.

- **Electric current in terms of drift velocity,** $I = neA\,v_d$.

where, $n$ = number density of free electrons

and $A$ = cross-sectional area.

- **Current density at any point of conductor,** $J = \frac{I}{A} = nev_d$

- The vector $\mathbf{J}$ is parallel to $\mathbf{E}$ and hence $\mathbf{J} = \frac{ne^2}{m}\tau\mathbf{E}$.

- The **relation between conductivity and relaxation time** is given as $\sigma = \frac{ne^2}{m}\tau$.

## Mobility

- The magnitude of the drift velocity per unit electric field is called **mobility** of charge carriers.

i.e. $$\mu = \frac{|v_d|}{E} = \frac{e\tau}{m}$$

Its SI unit is $m^2s^{-1}V^{-1}$.

## Resistivity of Various Materials

- Materials are classified as conductors, semiconductors are insulators depending on their resistivities.
- Metals have low resistivities in the range of $10^{-8}$ Ω-m to $10^{-6}$ Ω-m.
- Insulators like ceramic, rubber and plastic have resistivities $10^{18}$ times greater than metals or more.
- The resistivities of semiconductors lie between the resistivities of metals and insulators.

### Temperature Dependence of Resistivity

- Resistivity of a metallic conductor depends on the temperature and approximately is given by

$$\rho_T = \rho_0\,[1 + \alpha(T - T_0)]$$

where, $\rho_T$ is the resistivity at a temperature $T$,

$\rho_0$ is resistivity at a reference temperature $T_0$

and $\alpha$ is called the **temperature coefficient of resistivity.**

**Note** The value of $\alpha$ is positive for metallic conductor and in terms of resistance, it is given as $\alpha = \frac{R_2 - R_1}{R_1(t_2 - t_1)}$.

- Some materials like nichrome, manganin and constantan exhibit a very weak dependence of resistivity with temperature.
- From the relation $\rho = \frac{1}{\sigma} = \frac{m}{ne^2\tau}$, $\rho$ depends inversely both on the number $n$ of free electrons per unit volume and on the average time $\tau$ between collisions.
- In a metal, $n$ is not dependent on temperature to any appreciable extent and thus, the decrease in the value of $\tau$ with rise in temperature causes $\rho$ to increase.
- For insulators and semiconductors, $n$ increases with temperature. This increase in $n$ compensates any decrease in $\tau$, so that for such materials, $\rho$ decreases with temperature.

## Colour Coding of Carbon Resistors

The resistance of a carbon resistor can be calculated by the code given on it in the form of coloured strips as shown below

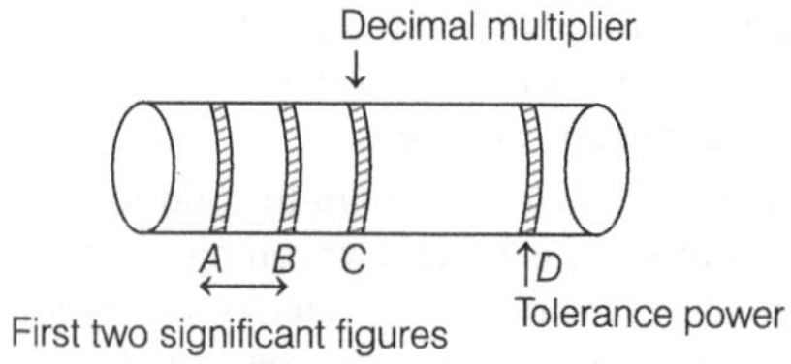


**KEY NOTES**

**Carbon Resistor Colour Codes**

| Colour | Number | Multiplier |
|---|---|---|
| Black | 0 | 1 |
| Brown | 1 | $10^1$ |
| Red | 2 | $10^2$ |
| Orange | 3 | $10^3$ |
| Yellow | 4 | $10^4$ |
| Green | 5 | $10^5$ |
| Blue | 6 | $10^6$ |
| Violet | 7 | $10^7$ |
| Grey | 8 | $10^8$ |
| White | 9 | $10^9$ |

**Tolerance**

| **Colour** | Gold | Silver | No colour |
|---|---|---|---|
| **Tolerance** | 5% | 10% | 20% |

This colour coding can be easily learned in the sequence "BB ROY Great Britain has Very Good Wife".

From colour codes resistance of resistor, $R = AB \times C \pm D\%$.

# Combination of Resistors

## 1. In Series

(*i*) Equivalent resistance, $R = R_1 + R_2 + R_3$.

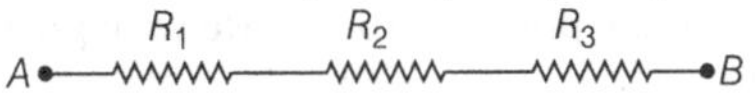


(*ii*) Current through each resistor is same.

(*iii*) Sum of potential differences across individual resistors is equal to the potential difference applied by the source.

## 2. In Parallel

(*i*) Equivalent resistance, $\dfrac{1}{R} = \dfrac{1}{R_1} + \dfrac{1}{R_2} + \dfrac{1}{R_3}$.

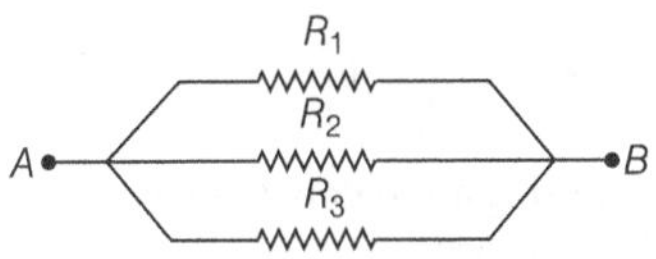


(*ii*) Potential difference across each resistor is same.

(*iii*) Sum of electric currents flowing through individual resistors is equal to the electric current drawn from the source.

# Electrical Energy and Power

- **Electrical Energy** It is the total work done by the source of emf in maintaining the electric current in the given circuit for a specified time.
- It is given by $E = W = VIt = V^2t/R$.
- Its SI unit is joule (J) or watt-second.
- To express electrical energy consumed commercially, a special unit **kilowatt-hour** is used in place of joule.

  It is also called **1 unit of electrical energy**.
- **1 kilowatt-hour** or **1 unit** of electrical energy is the amount of energy dissipated in 1 hour in a circuit, when the electric power in the circuit is 1 kilowatt.

  $$1 \text{ kilowatt-hour (kWh)} = 3.6 \times 10^6 \text{ joule (J)}$$
- **Electric Power** It is the rate at which electrical energy is supplied per unit time to maintain the flow of current in a conductor and given by $P = VI = I^2R = V^2/R$.

  Its SI unit is watt (W) and its commercial unit is horse power (HP), where 1 HP = 746 W.

# Cells, Emf and Internal Resistance

It is a device which converts chemical energy into electrical energy. The combination of cell produces a battery.

- **Emf of a Cell** It is the amount of work done in moving a unit positive charge through the whole circuit and given by

  $$\varepsilon = \frac{W}{q}$$

  Its SI unit is volt or joule/coulomb.
- **Terminal Potential Difference of a Cell** In a closed circuit, it is the potential difference between the two terminals of the cell. It is given by $V = IR$.

  Its value is always less than the emf of the cell.
- **Internal Resistance of a Cell** It is the resistance offered by the electrolyte of the cell to the flow of current through it.

  In terms of $\varepsilon$ and $V$, it is represented as $r = \left(\dfrac{\varepsilon}{V} - 1\right) R$.
- If cell is in charging state, then $\varepsilon = V - Ir$.
- Electric current flowing through the cell across finite resistance $R$ is given as $I = \dfrac{\varepsilon}{R + r}$.

  In a cell, current is maximum, for $R = 0$, i.e. $I_{max} = \dfrac{\varepsilon}{r}$.

# Cells in Series and in Parallel

## 1. Series Combination

- The series combination of two cells is shown below

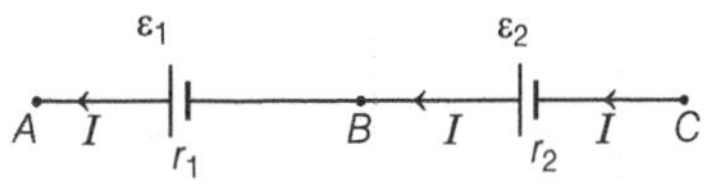


- The equivalent emf of a series combination of the above shown cells is just the sum of their individual emf's, i.e. $\varepsilon_{eq} = \varepsilon_1 + \varepsilon_2$ and the equivalent internal resistance of this combination is just the sum of their internal resistances, i.e. $r_{eq} = r_1 + r_2$.
- If polarity of one of the batteries is reversed, then equivalent emf is $|\varepsilon_1 - \varepsilon_2|$ and net internal resistance continue to be same.

KEY NOTES

## 2. Parallel Combination

- The equivalent emf of parallel combination of two cells as shown below and is given by $\varepsilon_{eq} = \frac{\varepsilon_1 r_2 + \varepsilon_2 r_1}{r_1 + r_2}$

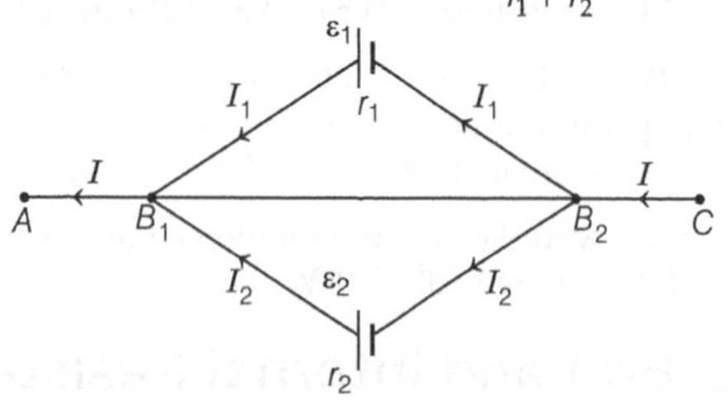


and internal resistance of combination, $r_{eq} = \frac{r_1 r_2}{r_1 + r_2}$.

# Kirchhoff's Rules

- There are two rules given by Kirchhoff as given below
  - (i) **Junction Rule** At any junction, the sum of the currents entering the junction is equal to the sum of currents leaving the junction, i.e. $\Sigma I = 0$.

    This rule is based on the law of **conservation of charges.**
  - (ii) **Loop Rule** The algebraic sum of changes in potential around any closed loop involving resistors and cells in the loop is zero, i.e. $\Sigma IR + \Sigma \varepsilon = 0$.

## Wheatstone Bridge

- It is an arrangement of four resistors used to measure the resistance of one of them in terms of other three.
- The Wheatstone bridge circuit is shown in figure below

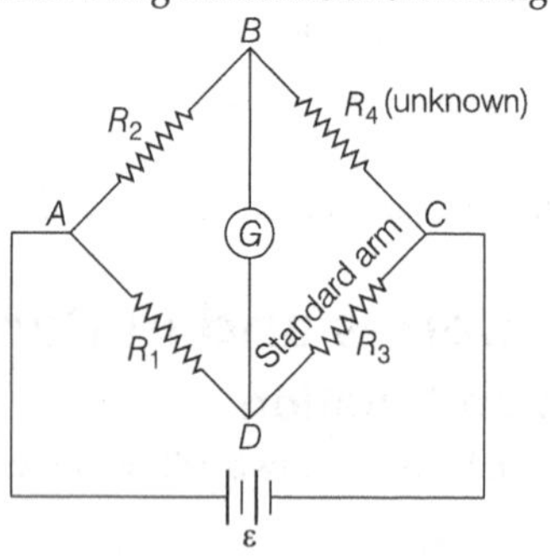


- Across one pair of diagonally opposite points (*A* and *C*) a source is connected, so as shown in figure, AC is called the **battery arm**.
- Between the other two vertices *B* and *D*, a galvanometer *G* is connected. This line shown as *BD* in the figure, is called the **galvanometer arm**.
- In the case of balanced bridge, the resistors are such that galvanometer current, $I_g = 0$.

  In this case, $$\frac{R_2}{R_1} = \frac{R_4}{R_3}$$

  This equation is called the **balance condition** for the galvanometer to give zero or null deflection.

## Meter Bridge

- It is a device which is used to measure resistance of a unknown wire.
- Its working principle is based on Wheatstone bridge principle, i.e. $R = S\frac{l_1}{100 - l_1}$

  where, $R$ = resistance of unknown wire, $S$ = known resistance and $l_1$ = distance of null point from first end.

## Potentiometer

- It is an electrical device which can be used to
  - (i) measure the potential difference with greater accuracy.
  - (ii) compare the emfs of two cells.

    i.e. $$\frac{E_1}{E_2} = \frac{l_1}{l_2}$$

    where, $l_1$ and $l_2$ are the balancing lengths corresponding to cells of emf $E_1$ and $E_2$, respectively.
  - (iii) determine the internal resistance of a primary cell.

    i.e. $$r = R\left(\frac{l_1}{l_2} - 1\right)$$

    where, $l_1$ and $l_2$ are balancing lengths and $R$ is the applied resistance.
- The potentiometer has the advantage that it draws no current from the voltage source being measured. As such, it is unaffected by the internal resistance of the source.

KEY NOTES

# Mastering NCERT

## MULTIPLE CHOICE QUESTIONS

## TOPIC 1 ~ Electric Current, Ohm's Law and Drift of Electrons

**1** An electric current in a conductor is due to the motion of
(a) charges (b) atoms
(c) molecules (d) conductor itself

**2** In a hydrogen atom, an electron moves in an orbit of radius $5.0\times 10^{-11}$ m with a speed of $2.2\times 10^{6}$ $ms^{-1}$. The equivalent current is
(a) $11.2\times 10^{-3}$ A (b) $1.9\times 10^{-3}$ A
(c) $1.12\times 10^{-3}$ A (d) $11.2\times 10^{4}$ A

**3** Two boys $A$ and $B$ are sitting at two points in a field near an assemblage of charged balls each carrying charge $+3e$. $A$ throws 100 balls per second towards $B$ while $B$ throws 50 balls per second towards $A$ simultaneously. Find the current at the mid-point of $A$ and $B$.
(a) $i=3.4\times 10^{-17}$ A from $A$ to $B$
(b) $i=2.4\times 10^{-17}$ A from $A$ to $B$
(c) $i=3.4\times 10^{-17}$ A from $B$ to $A$
(d) None of the above

**4** The current in a wire varies with time according to the equation $i=4+2t$, where $i$ is in ampere and $t$ is in second. The quantity of charge which passes through a cross-section of the wire during the time $t=2$ s to $t=6$ s is
(a) 40 C (b) 48 C (c) 38 C (d) 43 C

**5** Twenty million electrons reaches from point $X$ to point $Y$ in two microsecond as shown in the figure. Direction and magnitude of the current is

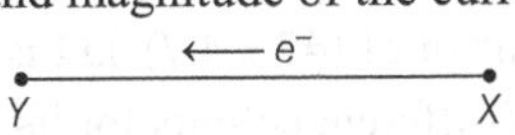


(a) $1.5\times 10^{-10}$ A from $X$ to $Y$
(b) $1.6\times 10^{-6}$ A from $Y$ to $X$
(c) $1.5\times 10^{-13}$ A from $Y$ to $X$
(d) $1.6\times 10^{-4}$ A from $X$ to $Y$

**6** In a region $10^{19}$ $\alpha$-particles and $10^{19}$ protons move to the left, while $10^{19}$ electrons moves to the right per second. The current is
(a) 3.2 A towards left (b) 3.2 A towards right
(c) 6.4 A towards left (d) 6.4 A towards right

**7** In which conductors, both positive and negative charges can move?
(a) Non-electrolytic solution (b) Electrolytic solution
(c) Both (a) and (b) (d) Neither (a) nor (b)

**8** If we consider a mechanism, where the ends of the cylinder are supplied with fresh charges to make up for any charges neutralised by electrons moving inside the conductor. In that case,
(a) there will be a steady electric field in the body of the conductor
(b) there will be in a continuous current rather than a current for a short period of time
(c) Both (a) and (b)
(d) Neither (a) nor (b)

**9** For a wire, $\frac{R}{l}=\frac{1}{2}$ and length of wire, $l=5$ cm. If potential difference of 1 V is applied across it, then current through wire will be ($R$ = resistance) ***AIIMS 2019***
(a) 40 A (b) 4 A (c) 25 A (d) 2.5 A

**10** The $V$-$I$ graph for a good conductor makes angle 40° with $V$-axis. Here, $V$ denotes voltage and $I$ denotes current. The resistance of the conductor will be
(a) sin 40° (b) cos 40° (c) tan 40° (d) cot 40°

**11** A wire of resistance $4\Omega$ is stretched to twice its original length. The resistance of stretched wire would be ***NEET 2013***
(a) $2\Omega$ (b) $4\Omega$ (c) $8\Omega$ (d) $16\Omega$

**12** A wire is stretched so as to change its diameter by 0.25%. The percentage change in resistance is
(a) 4.0% (b) 2.0% (c) 1.0% (d) 0.5%

**13** Two plates $R$ and $S$ are in the form of a square and have the same thickness. A side of $S$ is twice the side of $R$. The direction of current is shown by an arrow head in figure. Then, we can say that

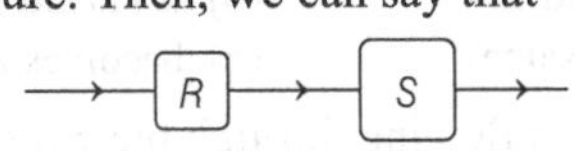


(a) the resistance of $R$ is twice that of $S$
(b) both have the same resistance
(c) the resistance of $S$ is four times that of $R$
(d) the resistance of $R$ is half that of $S$

**14** All the edges of a block are unequal. Its longest edge is twice its shortest edge, then ratio of the maximum and minimum resistance between parallel faces is
(a) 8 (b) 4
(c) 2 (d) None of these

**15** A conductor with rectangular cross-section has dimensions $(a \times 2a \times 4a)$ as shown in the figure. Resistance across $AB$ is $R_1$, across $CD$ is $R_2$ and across $EF$ is $R_3$. Then,

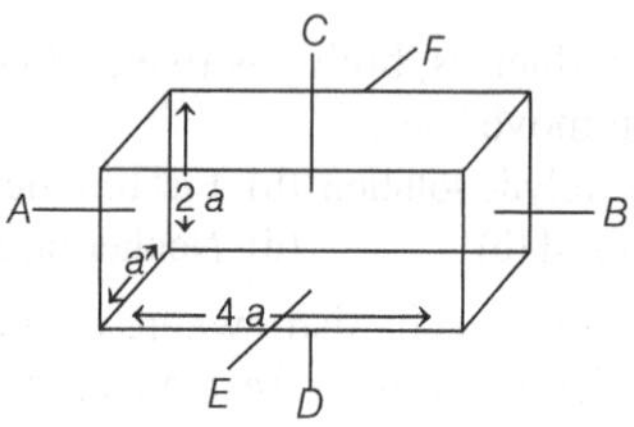


(a) $R_1 = R_2 = R_3$ (b) $R_1 > R_2 > R_3$
(c) $R_2 > R_3 > R_1$ (d) $R_1 > R_3 > R_2$

**16** A copper wire of cross-sectional area 2.0 mm$^2$, resistivity $= 1.7 \times 10^{-8}$ Ω-m, carries a current of 1 A. The electric field in the copper wire is
(a) $8.5 \times 10^{-5}$ Vm$^{-1}$ (b) $8.5 \times 10^{-4}$ Vm$^{-1}$
(c) $8.5 \times 10^{-3}$ Vm$^{-1}$ (d) $8.5 \times 10^{-2}$ Vm$^{-1}$

**17** In cosmic rays, 0.15 protons cm$^{-2}$ s$^{-1}$ are entering the earth's atmosphere. If the radius of the earth is 6400 km, the current received by the earth in the form of cosmic rays is nearly
(a) 0.12 A (b) 1.2 A
(c) 12 A (d) 120 A

**18** If resistivity of copper conductor is $1.7 \times 10^{-8}$ Ω-m and electric field is 100 Vm$^{-1}$, then current density will be
(a) $6 \times 10^9$ Am$^{-2}$ (b) $1.7 \times 10^{-6}$ Am$^{-2}$
(c) $1.7 \times 10^{-10}$ Am$^{-2}$ (d) $6 \times 10^7$ Am$^{-2}$

**19** A charge of 4C is flowing through a copper wire of cross-section 2 mm$^2$ in 2s. If the conductivity of copper wire be $6.25 \times 10^7$ Sm$^{-1}$, then electric field in copper wire is
(a) $0.8 \times 10^{-2}$ N/C (b) $1.6 \times 10^{-3}$ N/C
(c) $1.6 \times 10^{-2}$ N/C (d) $3.2 \times 10^{-3}$ N/C

**20** A potential difference $V$ is applied to a copper wire of length $l$ and diameter $d$. If $V$ is doubled, then the drift velocity
(a) is doubled (b) is halved
(c) remains same (d) becomes zero

**21** A current $i$ is flowing through the wire of diameter $d$ having drift velocity of electrons $v_d$ in it. What will be new drift velocity when diameter of wire is made $d/4$? **JIPMER 2018**
(a) $4v_d$ (b) $\frac{v_d}{4}$ (c) $16v_d$ (d) $\frac{v_d}{16}$

**22** When a current $I$ flows through a wire, the drift velocity of the electrons is $v$. When current $2I$ flows through another wire of the same material having double the length and double the area of cross-section, the drift velocity of the electrons will be
(a) $v/8$ (b) $v/4$
(c) $v/2$ (d) $v$

**23** A potential difference of 100 V is applied to the ends of a copper wire of 1 m length. What is the average drift velocity of electrons? **JEE Main 2016**
(Take, $\sigma = 5.81 \times 10^7 \Omega^{-1}$ or $n_{Cu} = 8.5 \times 10^{28}$ m$^{-3}$)
(a) 0.43 ms$^{-1}$ (b) 0.83 ms$^{-1}$
(c) 0.52 ms$^{-1}$ (d) 0.95 ms$^{-1}$

**24** A current of 10 A is passing through a metallic wire of cross-sectional area $4 \times 10^{-6}$ m$^2$. If the density of the aluminium conductor is 2.7 g/cc. Considering aluminium gives 1 electron per atom for conduction, then find the drift velocity of the electrons, if molecular weight of aluminium is 27 g. **AIIMS 2019**
(a) $1.6 \times 10^{-4}$ m/s (b) $3.6 \times 10^{-4}$ m/s
(c) $2.6 \times 10^{-4}$ m/s (d) $1.5 \times 10^{-4}$ m/s

**25** If temperature of a metallic wire increases keeping constant electric field, then current density through wire will be
(a) increases (b) decreases
(c) remains same (d) None of these

**26** If resistivity of copper is $1.72 \times 10^{-8}$ Ω-m and number of free electrons in copper is $8.5 \times 10^{28}$ / m$^3$. Find the mobility. **JIPMER 2019**
(a) $4.27 \times 10^{-3}$ m$^2$ / C-Ω (b) $6.8 \times 10^{-3}$ m$^2$ / C-Ω
(c) $8.5 \times 10^{-3}$ m$^2$ / C-Ω (d) $3.4 \times 10^{-3}$ m$^2$ / C-Ω

**27** A 200 Ω resistor has a certain colour code. If one replaces the red colour by green in the code, the new resistance will be **JEE Main 2019**
(a) 100 Ω (b) 400 Ω
(c) 300 Ω (d) 500 Ω

**28** A carbon resistor of $(47 \pm 4.7)$ kΩ is to be marked with rings of different colours for its identification. The colour code sequence will be **NEET 2018**
(a) Yellow - Green - Violet - Gold
(b) Yellow - Violet - Orange - Silver
(c) Violet - Yellow - Orange - Silver
(d) Green - Orange - Violet - Gold

**29** A voltage of 30V is applied across a colour coded carbon resistor with first, second and third rings of blue, black and yellow colours. What is the current flowing through the resistor?
(a) $5 \times 10^{-4}$ A (b) $5 \times 10^4$ A
(c) $0.5 \times 10^{-4}$ A (d) $0.5 \times 10^4$ A

**30** Resistance of a wire at 20°C is $20\Omega$ and at 500°C is $60\Omega$. At what temperature, its resistance is $25\Omega$?
(a) 160°C (b) 250°C (c) 100°C (d) 80°C

**31** The current $i$ and voltage $V$ graph for a given metallic wire at two different temperatures $T_1$ and $T_2$ are shown in the figure. It is concluded that

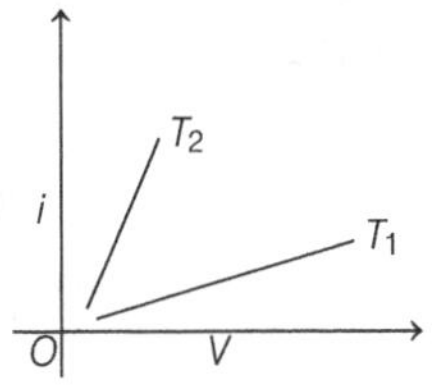


(a) $T_1 > T_2$ (b) $T_1 < T_2$
(c) $T_1 = T_2$ (d) $T_1 = 2T_2$

**32** The temperature dependence of resistances of Cu and undoped Si in the temperature range 300-400 K, is best described by ***JEE Main 2016***
(a) linear increase for Cu, linear increase for Si
(b) linear increase for Cu, exponential increase for Si
(c) linear increase for Cu, exponential decrease for Si
(d) linear decrease for Cu, linear decrease for Si

**33** The relaxation time of electrons in copper at room temperature is (take, resistivity of copper $= 1.7 \times 10^{-8}\ \Omega$-m, density of electron in copper $= 8.5 \times 10^{28}\ m^{-3}$, mass of electrons $= 9.1 \times 10^{-31}$ kg and drift velocity of electrons $= 1.6 \times 10^{-4}\ ms^{-1}$)
(a) $3.2 \times 10^{-12}$ s (b) $2.5 \times 10^{-14}$ s
(c) $1.5 \times 10^{-10}$ s (d) $6.2 \times 10^{-11}$ s

# TOPIC 2 ~ *Electric Energy and Power*

**34** If charges move without collisions through the conductor, their kinetic energy would also change, so that the total energy is
(a) changed (b) unchanged
(c) doubled (d) halved

**35** A current of 2 mA was passed through an unknown resistor which dissipated a power of 4.4 W. Dissipated power when an ideal power supply of 11 V is connected across it is ***JEE Main 2019***
(a) $11 \times 10^{-4}$ W (b) $11 \times 10^{-5}$ W
(c) $11 \times 10^{5}$ W (d) $11 \times 10^{-3}$ W

**36** Two cities are 150 km apart. Electric power is sent from one city to another city through copper wires. The fall of potential per km is 8 V and the average resistance per km is $0.5\ \Omega$. The power loss in the wire is ***CBSE AIPMT 2014***
(a) 19.2 W (b) 19.2 kW
(c) 19.2 J (d) 12.2 kW

**37** An electric bulb is marked 100 W, 230 V. If the supply voltage drops to 115 V, what is the heat and light energy produced by the bulb in 20 min?
(a) 20000 J (b) 25000 J
(c) 30000 J (d) 35000 J

**38** The number of electrons moving per second through the filament of a lamp of 100 W, operating at 200 V is
(a) $3.1 \times 10^{18}$ (b) $3.1 \times 10^{19}$
(c) $4.8 \times 10^{18}$ (d) $4.8 \times 10^{19}$

**39** An electric motor operating on a 50 V DC supply draws a current of 12 A. If the efficiency of the motor is 30%, then estimate the resistance of the windings of the motor.
(a) $1.5\Omega$ (b) $2\Omega$ (c) $2.9\Omega$ (d) $1\Omega$

**40** Which of the following acts as a circuit protection device? ***NEET (National) 2019***
(a) Inductor (b) Switch
(c) Fuse (d) Conductor

# TOPIC 3 ~ *Combination of Resistors : Series and Parallel*

**41** When two or more resistors are connected in series, then
(a) same current flows through each resistor
(b) same voltage is applied across each resistor
(c) different current flows through each resistor
(d) both current and voltage are different in each resistor

**42** Two wires of same metal have the same length but areas of their cross-sections are in the ratio 3 : 1. They are joined in series. The resistance of the thicker wire is $10\Omega$. The total resistance of the combination is
(a) $(5/2)\Omega$ (b) $(40/2)\Omega$ (c) $40\Omega$ (d) $100\Omega$

**43** A circuit contain two resistances $R_1$ and $R_2$ are in series. Find the ratio of input voltage to voltage of $R_2$.

***JIPMER 2019***

(a) $\frac{R_2}{R_1 + R_2}$ (b) $\frac{R_1 + R_2}{R_2}$
(c) $\frac{R_1 + R_2}{R_1}$ (d) $\frac{R_1}{R_1 + R_2}$

**44** Two resistances 400 Ω and 800 Ω are connected in series with 6 V battery. The potential difference measured by voltmeter of 10 kΩ across 400 Ω resistor is
(a) 2 V (b) 1.95 V
(c) 3.8 V (d) 4 V

**45** A uniform metallic wire has a resistance of 18 Ω and is bent into an equilateral triangle. Then, the resistance between any two vertices of the triangle is

***JEE Main 2019***

(a) 12 Ω (b) 8 Ω (c) 2 Ω (d) 4 Ω

**46** Two resistances are joined in parallel whose equivalent resistance is (3/5) Ω. One of the resistance wire is broken and the effective resistance becomes 3 Ω. The resistance (in ohm) of the wire that got broken was
(a) 4/3 (b) 2
(c) 6/5 (d) 3/4

**47** A wire of uniform cross-section and length $l$ has a resistance of 16 Ω. It is cut into four equal parts. Each part is stretched uniformly to length $l$ and all the stretched parts are connected in parallel. Calculate the total resistance of the combination so formed. (Assume that the stretching of the wire does not cause any change in the density of the material.)
(a) 10 Ω (b) 12 Ω
(c) 16 Ω (d) 18 Ω

**48** Find $i$ in shown figure. ***JIPMER 2018***

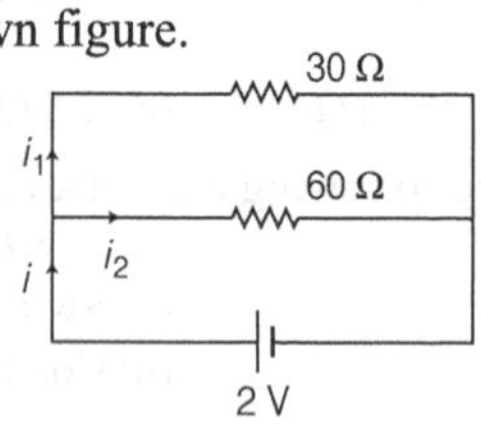


(a) 0.2 A (b) 0.1 A
(c) 0.3 A (d) 0.4 A

**49** Two resistances $R$ and $2R$ are connected in parallel in an electric circuit. The thermal energy developed in $R$ and $2R$ in the ratio
(a) 1:2 (b) 1:4
(c) 4:1 (d) 2:1

**50** The equivalent resistance of $n$-resistors each of same resistance when connected in series is $R$. If the same resistances are connected in parallel, the equivalent resistances will be
(a) $R/n^2$ (b) $R/n$
(c) $n^2R$ (d) $nR$

**51** Five resistors are connected as shown in figure. Find the equivalent resistance between the points $B$ and $C$.

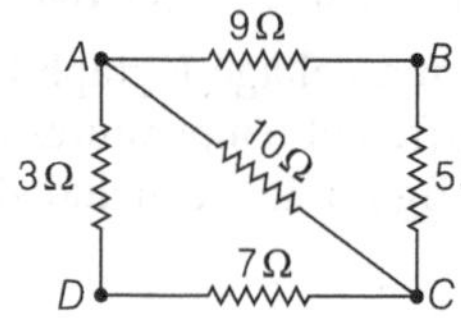


(a) $\frac{70}{19}\,\Omega$ (b) $\frac{19}{70}\,\Omega$
(c) $\frac{16}{5}\,\Omega$ (d) $\frac{15}{8}\,\Omega$

**52** Six equal resistances are connected between points $P$, $Q$ and $R$ as shown in the figure. Then, the net resistance will be maximum between

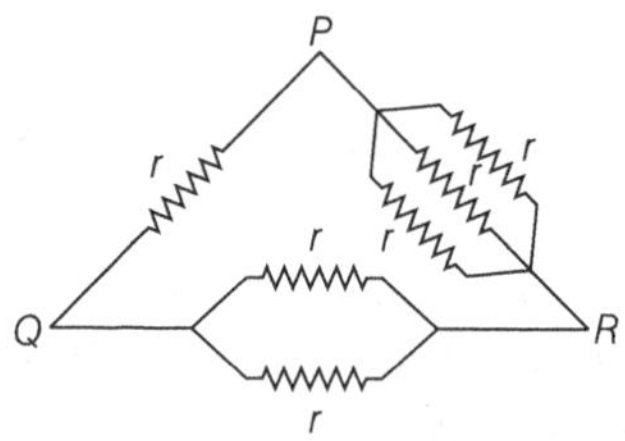


(a) $P$ and $Q$ (b) $Q$ and $R$
(c) $P$ and $R$ (d) $R$ and $S$

**53** The resistance across $P$ and $Q$ in the figure is

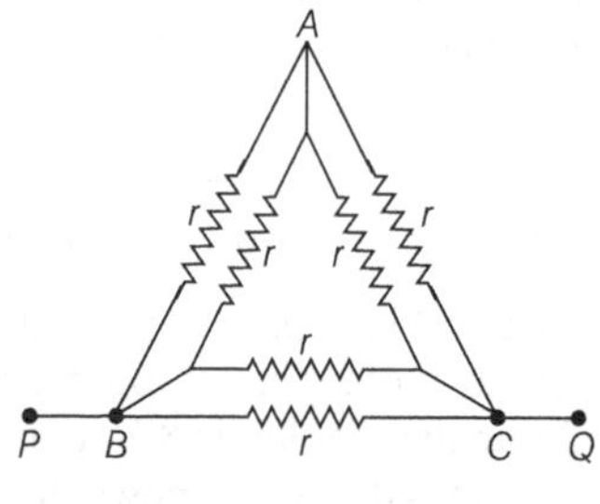


(a) $r/3$ (b) $r/2$ (c) $2r$ (d) $6r$

**54** A ring is made of a wire having a resistance $R_0 = 12\,\Omega$. Find the points $A$ and $B$ as shown in the figure, at which a current carrying conductor should be connected, so that the resistance $R$ of the sub-circuit between these points is equal to (8/3) Ω. ***NEET 2012***

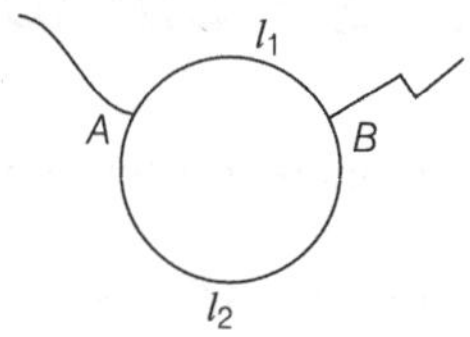


(a) $\frac{l_1}{l_2} = \frac{5}{8}$ (b) $\frac{l_1}{l_2} = \frac{1}{3}$
(c) $\frac{l_1}{l_2} = \frac{3}{8}$ (d) $\frac{l_1}{l_2} = \frac{1}{2}$

**55** What is the equivalent resistance across the points $A$ and $B$ in the circuit given below?

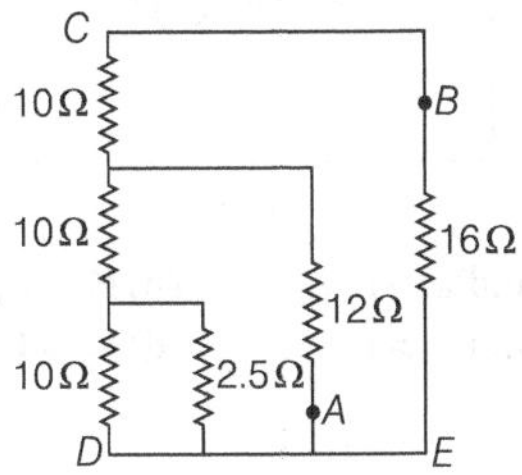


(a) 8Ω (b) 12Ω
(c) 16Ω (d) 32Ω

**56** Find $R_{net}$ between $A$ and $B$. ***JIPMER 2018***

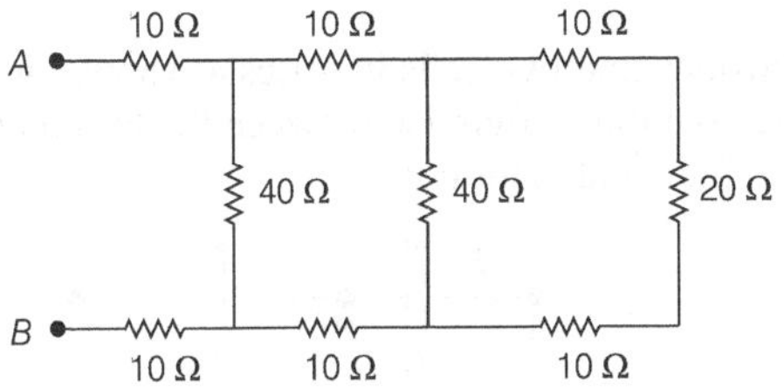


(a) 60 Ω (b) 40 Ω (c) 70 Ω (d) 20 Ω

**57** The actual value of resistance $R$, shown in the figure is 30 Ω. This is measured in an experiment as shown using the standard formula $R = \frac{V}{I}$, where $V$ and $I$ are the readings of the voltmeter and ammeter, respectively. If the measured value of $R$ is 5% less, then the internal resistance of the voltmeter is ***JEE Main 2019***

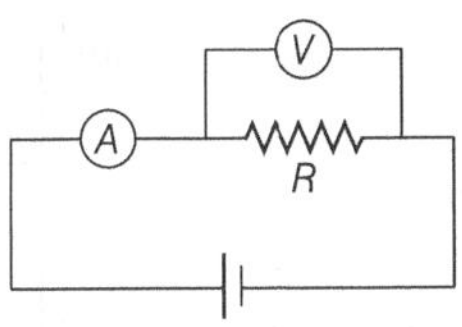


(a) 600 Ω (b) 570 Ω (c) 350Ω (d) 35Ω

**58** The reading of an ideal voltmeter in the circuit shown is ***NEET (Odisha) 2019***

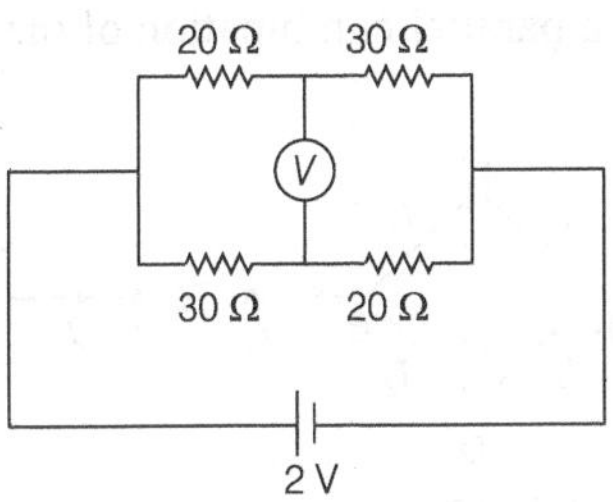


(a) 0.6 V (b) 0 V (c) 0.5 V (d) 0.4 V

**59** Find the current in the 8Ω resistance in the given circuit. ***JIPMER 2019***

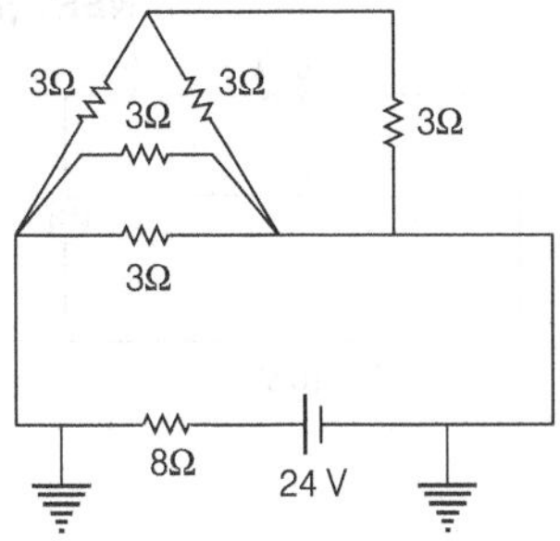


(a) 2 A (b) 3A
(c) 4 A (d) 5 A

**60** In the given circuit, find voltage across 12 Ω resistance. ***JIPMER 2019***

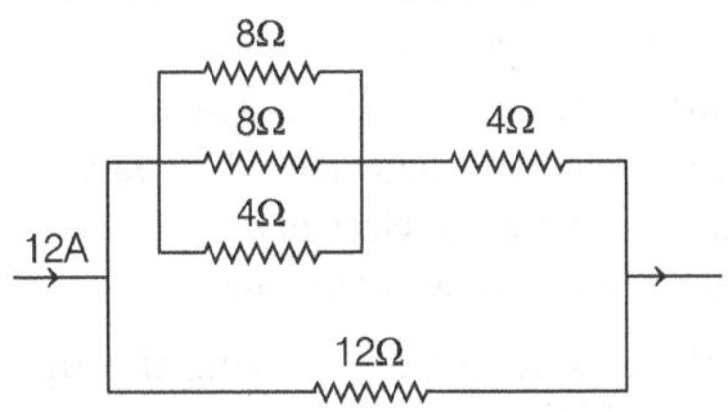


(a) 12 V (b) 36 V
(c) 72 V (d) 48 V

**61** Find $V_P - V_Q$ in the circuit shown in figure. ***JIPMER 2018***

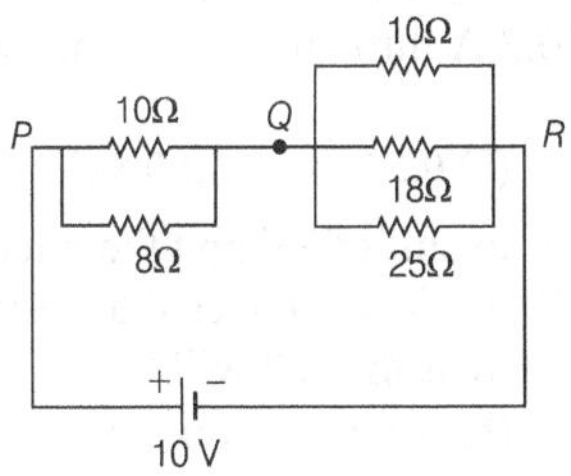


(a) 6.68 V (b) 8 V
(c) 4.65 V (d) 7 V

**62** The resistive network shown below is connected to a DC source of 16 V. The power consumed by the network is 4 W. The value of $R$ is ***JEE Main 2019***

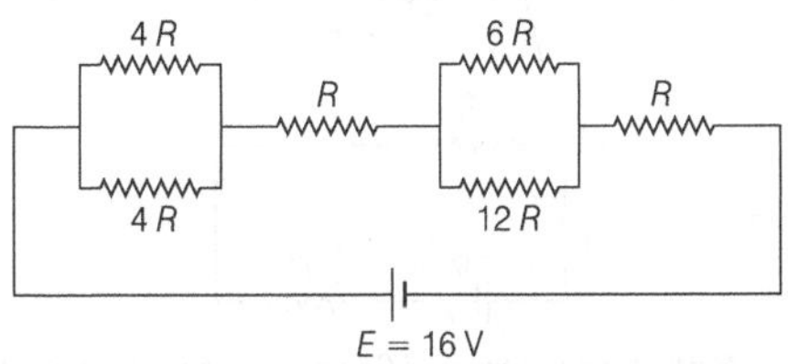


(a) 6Ω (b) 8Ω
(c) 1Ω (d) 16Ω

**63** In the circuits shown below, the readings of voltmeters and the ammeters will be

**NEET (National) 2019**

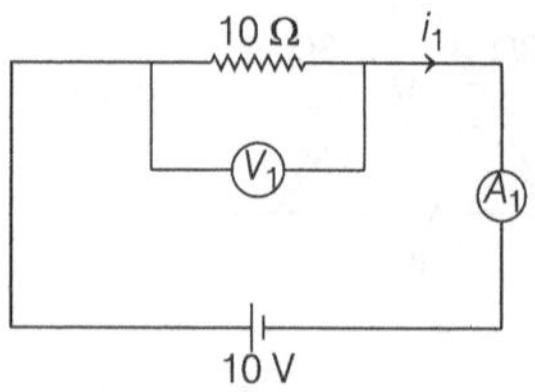

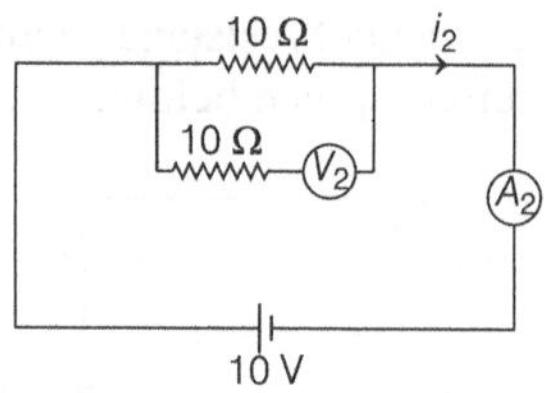

(a) $V_1 = V_2$ and $i_1 > i_2$ (b) $V_1 = V_2$ and $i_i = i_2$
(c) $V_2 > V_1$ and $i_1 > i_2$ (d) $V_2 > V_1$ and $i_1 = i_2$

# TOPIC 4 ~ Cells EMF and Internal Resistance

**64** When two electrodes (positive and negative) of a cell are immersed in an electrolytic solution, the charges are exchanged between

(a) positive electrode and electrolyte only
(b) negative electrode and electrolyte only
(c) Both electrodes and electrolyte
(d) directly between two electrodes

**65** When there is no current flowing through the cell, the potential differences between the positive and negative electrodes of cell is

(a) equal to emf $\varepsilon$ (b) equal to $Ir$
(c) equal to $\varepsilon - Ir$ (d) greater than $\varepsilon$

**66** The internal resistance of a 2.1 V cell which gives a current of 0.2 A through a resistance of 10 $\Omega$ is

**NEET 2013**

(a) 0.2 $\Omega$ (b) 0.5 $\Omega$ (c) 0.8 $\Omega$ (d) 1.0 $\Omega$

**67** The cell has an emf of 2V and the internal resistance of this cell is $0.1\,\Omega$, it is connected to resistance of $3.9\,\Omega$, the voltage across the cell will be

(a) 1.95 V (b) 1.5 V (c) 2 V (d) 1.8 V

**68** Consider a simple circuit shown in figure carrying a variable resistance $R'$, which can vary from $R$ to infinity. If $r$ is the internal resistance of the battery ($r << R' << R$), then

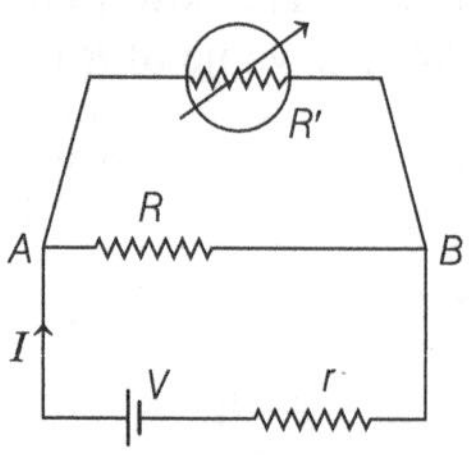

(a) potential drop across $AB$ is nearly constant as $R'$ is varied
(b) current through $R'$ is nearly a constant as $R'$ is varied
(c) current $I$ depends sensitively on $R'$
(d) $I \leq \dfrac{V}{r+R}$ always

**69** Consider first two cells in series as shown in figure. The potential difference between the terminals $A$ and $C$ of the combination is

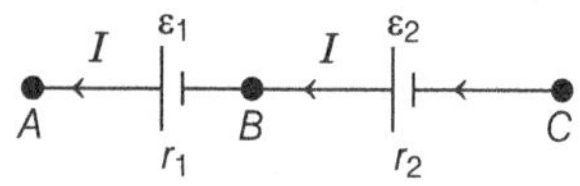

(a) $V_{AC} = \varepsilon_1 - Ir_1$ (b) $V_{AC} = \varepsilon_2 - Ir_2$
(c) $V_{AC} = \varepsilon_{eq} - Ir_{eq}$ (d) $V_{AC} = 2\varepsilon_{eq} - Ir_{eq}$

**70** A battery consists of a variable number $n$ of identical cells (having internal resistance $r$ each) which are connected in series. The terminals of the battery are short-circuited and the current $I$ is measured. Which of the graphs shows the correct relationship between $I$ and $n$?

**NEET 2018**

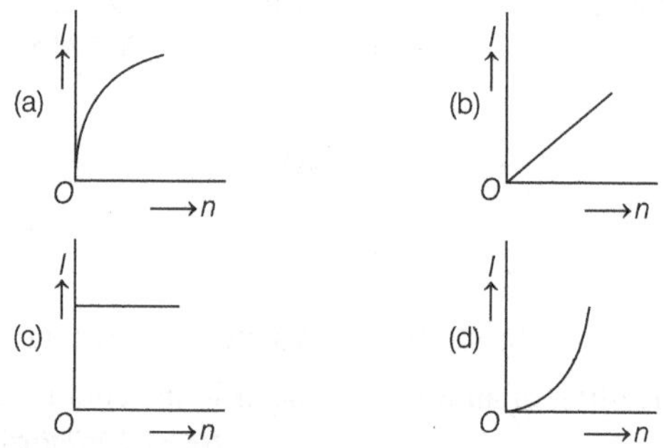

**71** Consider a parallel combination of the cells in the figure.

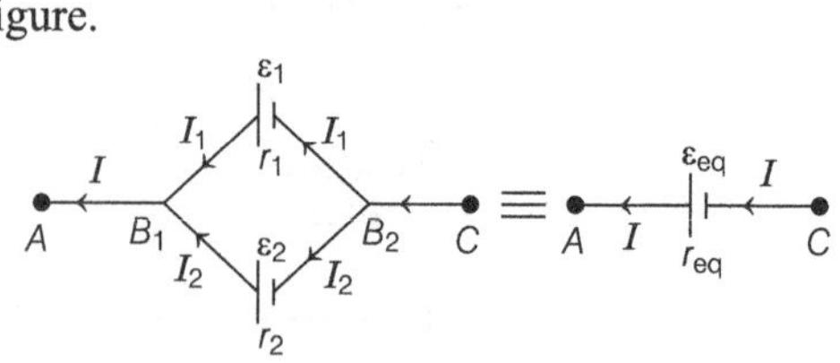

The potential difference across its terminals $B_1$ and $B_2$ is

(a) $V = \varepsilon_{eq} - Ir_{eq}$ (b) $V = \varepsilon_2 - Ir_2$
(c) $V = 2\varepsilon_{eq} - Ir_{eq}$ (d) $V = \varepsilon_1 - 2Ir_1$

**72** Two cells having the internal resistance 0.2 Ω and 0.4Ω are connected in parallel. The voltage across the battery terminal is 1.5 V. The emf of first cell is 1.2 V. The emf of second cell is

(a) 2.7 V (b) 2.1 V (c) 3 V (d) 4.2 V

**73** Is it possible that any battery has some constant non-zero value of emf but the potential difference between the plates is zero?

(a) Not possible
(b) Yes, if another identical battery is joined in series
(c) Yes, if another identical battery is joined in opposition
(d) Yes, possible, if another similar battery is joined in parallel

**74** In the given circuit diagram, the emf and internal resistance of a single battery which is equivalent to a combination of three batteries respectively, are

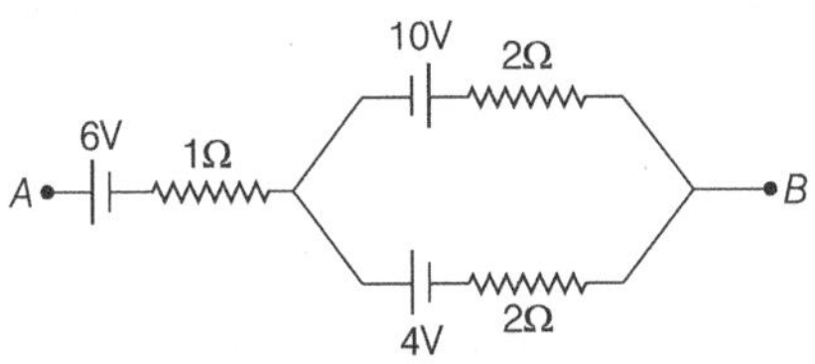


(a) 10V, 4 Ω (b) 12 V, 2Ω (c) 3V, 6Ω (d) 3V, 2Ω

**75** A set of $n$-equal resistors, of value $R$ each, are connected in series to a battery of emf $E$ and internal resistance $R$. The current drawn is $I$. Now, the $n$-resistors are connected in parallel to the same battery, then the current drawn from battery becomes $10I$.

The value of $n$ is **NEET 2018**

(a) 20 (b) 11 (c) 10 (d) 9

**76** To draw maximum current from a combination of cells, how should the cells be grouped?

(a) Parallel
(b) Series
(c) Mixed grouping
(d) Depends upon the relative values of internal and external resistances

**77** The maximum power dissipated in an external resistance $R$ when connected to a cell of emf $E$ and internal resistance $r$ will be

(a) $\frac{E^2}{r}$ (b) $\frac{E^2}{2r}$ (c) $\frac{E^2}{3r}$ (d) $\frac{E^2}{4r}$

**78** To get a maximum current through a resistance of 2.5 Ω, one can use $m$ rows of cells each row having $n$ cells. The internal resistance of each cell is 0.5 Ω. What are the values of $m$ and $n$, if the total number of cells are 20?

(a) $m = 2, n = 10$ (b) $m = 4, n = 5$
(c) $m = 5, n = 4$ (d) $n = 2, m = 10$

# TOPIC 5 ~ Kirchhoff's Laws, Wheatstone Bridge and Meter Bridge

**79** Kirchhoff's junction rule is a reflection of

(a) conservation of current density vector
(b) conservation of charge
(c) the fact that the momentum with which a charged particle approaches a junction is unchanged (as a vector) as the charged particle leaves the junction
(d) the fact that there is a net accumulation of charges at a junction

**80** Five conductors are meeting at a point $x$ as shown in the figure. What is the value of current in fifth conductor?

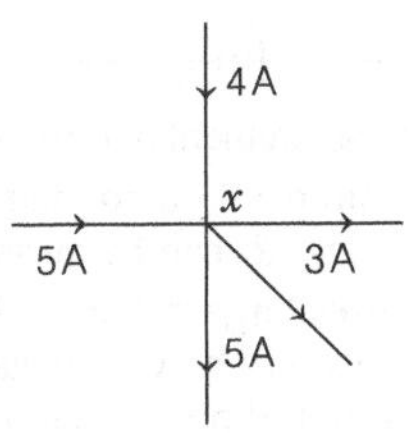


(a) 3A away from $x$ (b) 1A away from $x$
(c) 4A away from $x$ (d) 1A towards $x$

**81** The algebraic sum of changes in potential around any closed loop involving resistor and cells in the loop is

(a) more than zero
(b) less than zero
(c) zero
(d) constant

**82** In the given circuit diagram, the currents $I_1 = -0.3$ A, $I_4 = 0.8$ A and $I_5 = 0.4$ A, are flowing as shown. The currents $I_2$, $I_3$ and $I_6$ respectively, are **JEE Main 2019**

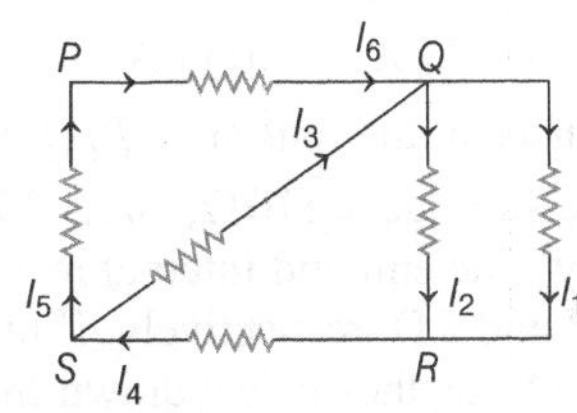


(a) 1.1 A, 0.4 A, 0.4 A (b) 1.1 A, – 0.4 A, 0.4 A
(c) 0.4 A, 1.1 A, 0.4 A (d) – 0.4 A, 0.4 A, 1.1 A

**83** In the circuit given here, the points $A$, $B$ and $C$ are at potentials 70 V, zero, 10 V, respectively. Then,

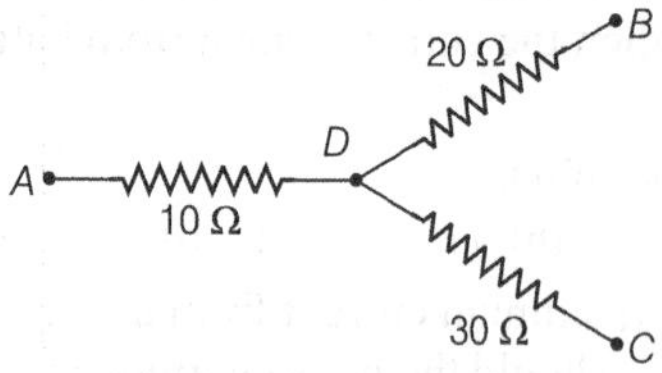


(a) the point $D$ will be at a potential of 60 V
(b) the point $D$ will be at a potential of 20 V
(c) currents in the paths $AD$, $DB$ and $DC$ are in the ratio of 1 : 2 : 3
(d) currents in the paths $AD$, $DB$ and $DC$ are in the ratio of 3 : 2 :1

**84** The emf of the battery shown in figure is **JEE Main 2015**

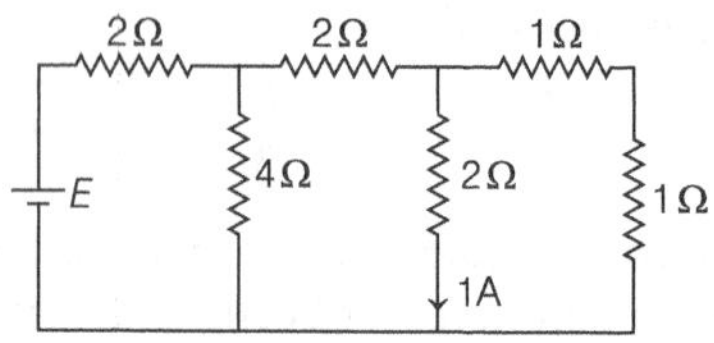


(a) 12 V (b) 13 V (c) 16 V (d) 18 V

**85** A battery of emf ε and negligible internal resistance is connected across the diagonally opposite corners of a cubical network consisting of 12 resistors each of resistance $R$. The equivalence resistance of the network is

(a) $\frac{5}{6}R$ (b) $\frac{6}{5}R$ (c) $3R$ (d) $12\,R$

**86** The Wheatstone bridge and its balance condition provide a practical method for determination of an

(a) known resistance (b) unknown resistance
(c) Both (a) and (b) (d) None of these

**87** Find current $i$ in circuit shown in the figure. **JIPMER 2018**

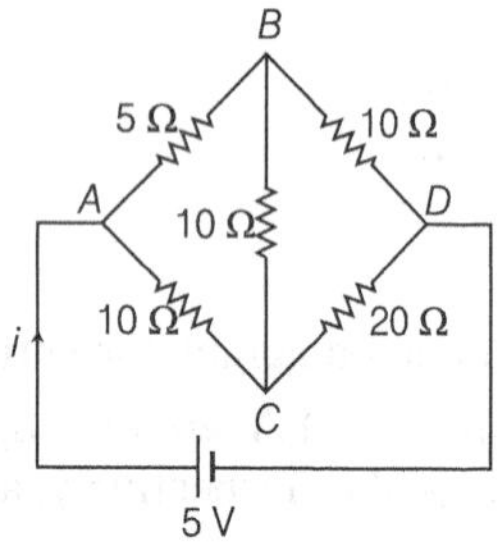


(a) 0.5 A (b) 0.2 A (c) 1 A (d) 2 A

**88** The resistances of the four arms $P$, $Q$, $R$ and $S$ in a Wheatstone's bridge are 10 Ω, 30 Ω, 30 Ω and 90 Ω, respectively. The emf and internal resistance of the cell are 7 V and 5 Ω, respectively. If the galvanometer resistance is 50 Ω, the current drawn from the cell will be **NEET 2013**

(a) 1.0 A (b) 0.2 A (c) 0.1 A (d) 2.0 A

**89** If each of the resistance in the network in figure is $R$, the equivalent resistance between terminals $A$ and $B$ (in ohm) is

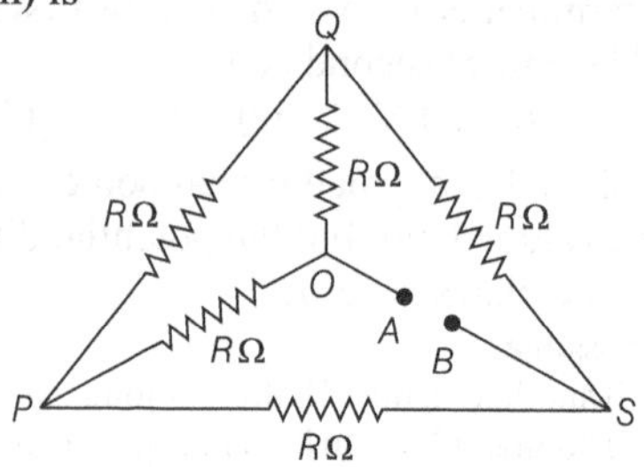


(a) $5R$ (b) $2R$ (c) $4R$ (d) $R$

**90** In a Wheatstone bridge (see figure), resistances $P$ and $Q$ are approximately equal. When $R = 400\Omega$, the bridge is balanced. On interchanging $P$ and $Q$, the value of $R$ for balance is $405\Omega$. The value of $X$ is close to **JEE Main 2019**

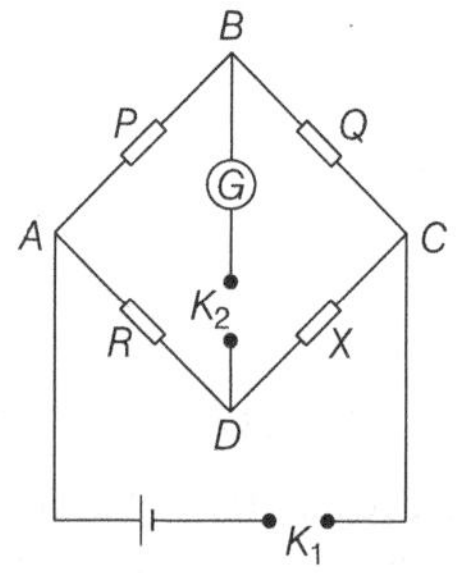


(a) 404.5 Ω (b) 401.5 Ω (c) 402.5 Ω (d) 403.5 Ω

**91** The measurement of an unknown resistance $R_4$ is to be carried out using Wheatstone bridge as given in the figure. Two students perform an experiment in two ways. The first students takes $R_2 = 10\Omega$ and $R_1 = 5\Omega$ and the second student takes $R_2 = 1000\Omega$ and $R_1 = 500\Omega$. In the standard arm, both take $R_3 = 5\Omega$. Both find $R_4 = \frac{R_2}{R_1} \times R_3 = 10\Omega$ within errors.

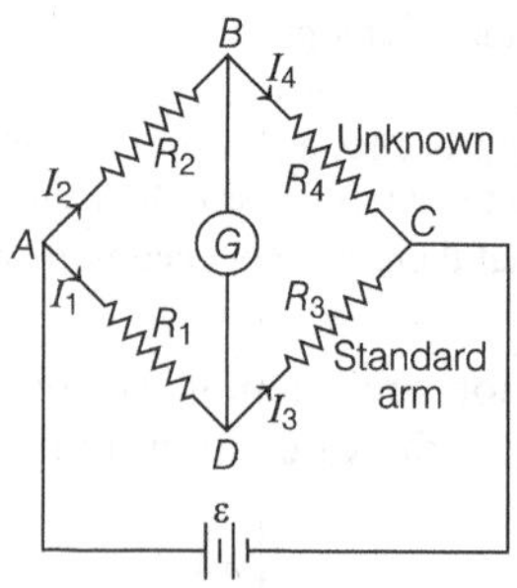


(a) The errors of measurement of the two students are the same
(b) Errors of measurement do not depend on the accuracy with which $R_2$ and $R_1$ can be measured
(c) If the student uses large values of $R_2$ and $R_1$ the currents through the arms will be very weak. This will make determination of null point accurately more difficult
(d) Wheatstone bridge is a very accurate instrument and has no errors of measurement

**92** The four arms of a Wheatstone bridge in figure have following resistances $AB = 100\,\Omega$, $BC = 10\,\Omega$, $CD = 5\,\Omega$ and $DA = 60\,\Omega$.

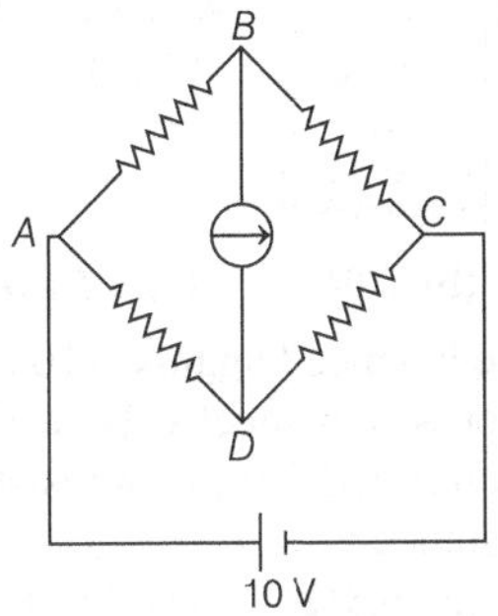


A galvanometer of $15\,\Omega$ resistance is connected across $BD$. Calculate the current through the galvanometer when a potential difference of 10 V is maintained across $AC$.

(a) 4.87 A (b) 4.87 mA
(c) 4.0 A (d) 4.0 mA

**93** The meter bridge shown in the balance position with $\frac{P}{Q} = \frac{l_1}{l_2}$. If we now interchange the positions of galvanometer and cell, will the bridge work? If yes, that will be the balanced condition? **NEET (Odisha) 2019**

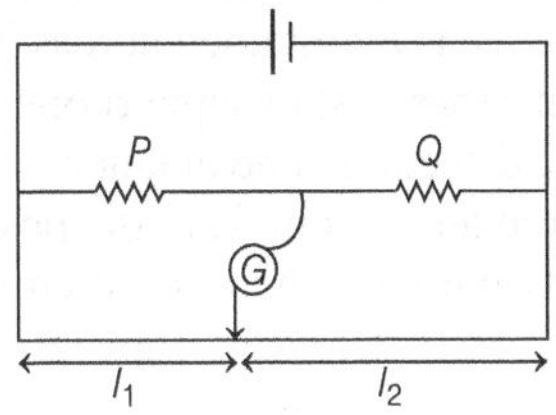


(a) yes, $\frac{P}{Q} = \frac{l_2 - l_1}{l_2 + l_1}$ (b) no, no null point
(c) yes, $\frac{P}{Q} = \frac{l_2}{l_1}$ (d) yes, $\frac{P}{Q} = \frac{l_1}{l_2}$

**94** The resistances in the two arms of the meter bridge are $5\,\Omega$ and $R\,\Omega$, respectively. When the resistance $R$ is shunted with an equal resistance, the new balance point is at $1.6\,l_1$. The resistance $R$ is **CBSE AIPMT 2014**

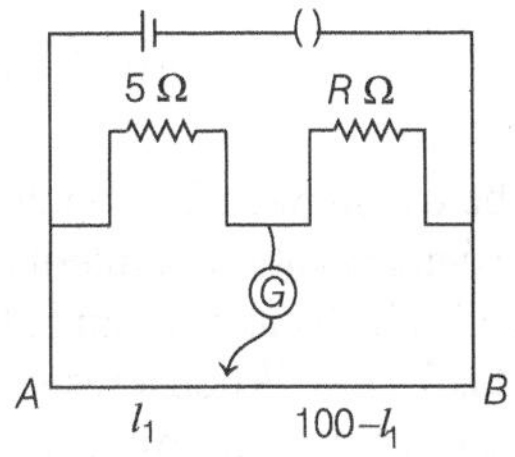


(a) $10\,\Omega$ (b) $15\,\Omega$ (c) $20\,\Omega$ (d) $25\,\Omega$

# TOPIC 6 ~ *Potentiometer*

**95** The potentiometer works on the condition of
(a) no deflection in galvanometer
(b) small deflection in galvanometer
(c) large deflection in galvanometer
(d) None of the above

**96** In a potentiometer, the null point is received at 7th wire. If now we have to change the null point at the 9th wire, what should we do?
(a) Attach resistance in series with battery
(b) Increase resistance in main circuit
(c) Decrease resistance in main circuit
(d) Decrease applied emf

**97** A potentiometer is an accurate and versatile device to make electrical measurement of emf because the method involves **NEET 2017**
(a) cells
(b) potential gradients
(c) a condition of no current flow through the galvanometer
(d) a combination of cells, galvanometer and resistances

**98** Potentiometer measures the potential difference more accurately than a voltmeter because
(a) it has a wire of high resistance
(b) it has a wire of low resistance
(c) it does not draw current from external circuit
(d) it draws a heavy current from external circuit

**99** The circuit shown here is used to compare the emf of two cells $E_1$ and $E_2$ $(E_1 > E_2)$. The null point is at $C$ when the galvanometer is connected to $E_1$. When the galvanometer is connected to $E_2$, the null point will be

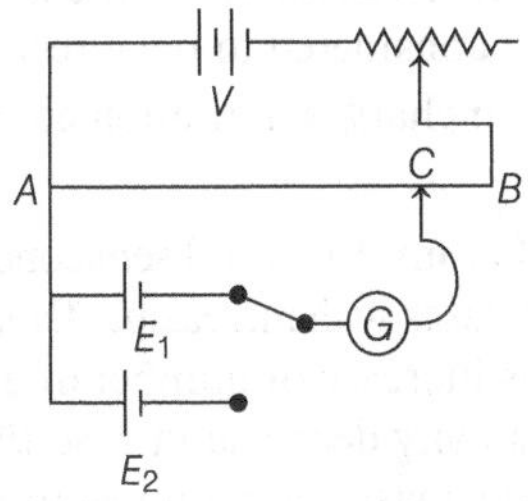


(a) to the left of $C$ (b) to the right of $C$
(c) at $C$ itself (d) no change in null point

**100** A potentiometer wire is 100 cm long and a constant potential difference is maintained across it. Two cells are connected in series first to support one another and then in opposite direction. The balance points are obtained at 50 cm and 10 cm from the positive end of the wire in the two cases. The ratio of emf is **NEET 2016**

(a) 5 : 4 (b) 3 : 4
(c) 3 : 2 (d) 5 : 1

**101** A potentiometer wire of length $L$ and a resistance $r$ are connected in series with a battery of emf $E_0$ and a resistance $r_1$. An unknown emf is balanced at a length $l$ of the potentiometer wire. The emf $E$ will be given by **CBSE AIPMT 2015**

(a) $\frac{LE_0 r}{l r_1}$ (b) $\frac{E_0}{(r+r_1)} \cdot \frac{rl}{L}$
(c) $\frac{E_0 l}{L}$ (d) $\frac{LE_0 r}{(r+r_1) l}$

**102** The figure below shows a 2.0 V potentiometer used for the determination of internal resistance of a 2.5 V cell. The balance point of the cell in the open circuit is 75 cm. When a resistor of 10 Ω is used in the external circuit of the cell, the balance point shifts to 65 cm length of potentiometer wire. The internal resistance of the cell is **JEE Main 2018**

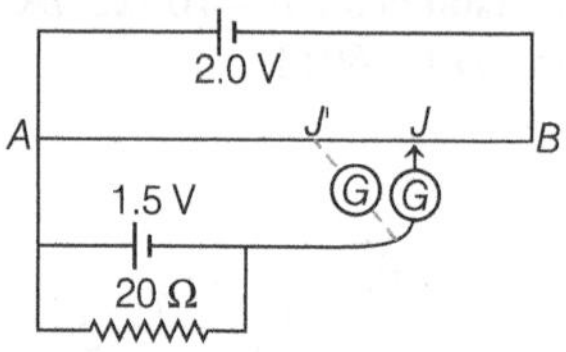


(a) 2.5 Ω (b) 2.0 Ω (c) 1.54 Ω (d) 1.0 Ω

**103** A cell can be balanced against 110 cm and 100 cm of potentiometer wire respectively, with and without being short circuited through a resistance of 10 Ω. Its internal resistance is
(a) 1.0 Ω (b) 0.5 Ω (c) 2.0 Ω (d) zero

**104** A resistance of $R$ Ω draws current from a potentiometer. Potentiometer has a total resistance $R_0$ Ω as shown in figure.

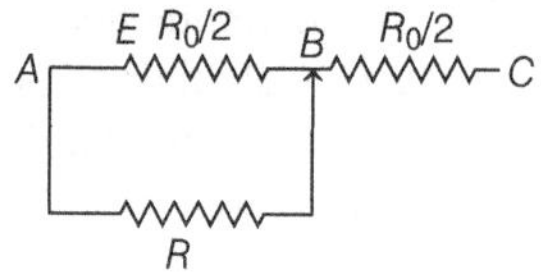


A voltage $V$ is supplied to the potentiometer, then for the voltage across $R$ when the sliding contact is in the middle of the potentiometer is

(a) $\frac{2VR}{R_0+4R}$ (b) $\frac{2VR}{R_0}$ (c) $\frac{4VR}{R_0+2R}$ (d) $\frac{VR}{R_0+2R}$

# SPECIAL TYPES QUESTIONS

## I. Assertion and Reason

**Direction** (Q. Nos. 105-112) *In the following questions, a statement of Assertion is followed by a corresponding statement of Reason. Of the following statements, choose the correct one.*

(a) Both Assertion and Reason are correct and Reason is the correct explanation of Assertion.
(b) Both Assertion and Reason are correct but Reason is not the correct explanation of Assertion.
(c) Assertion is correct but Reason is incorrect.
(d) Assertion is incorrect but Reason is correct.

**105** **Assertion** In electrostatics, all charges whether free or bound are considered to be at rest.
**Reason** The charges in motion constitute an electric current.

**106** **Assertion** For insulators and semiconductors, number of electrons increases with increasing temperature.
**Reason** This increase of number of electrons is more effective than any decrease in $\tau$, so that for such materials $\rho$ decreases with temperature.

**107** **Assertion** Conductivity arises from mobile charge carriers.
**Reason** In metals, these mobile charge carriers are electrons with fixed positive ions in background; these can be both positive and negative ions in an electrolyte.

**108** **Assertion** The average time of collisions $\tau$ decreases with increasing temperature.
**Reason** At increased temperature, average speed of the electrons, which act as the carriers of current increases, resulting in more frequent collisions.

**109** **Assertion** Charge carriers do not move with acceleration but with a steady drift velocity.
**Reason** Charge carriers undergo collisions with ions and atoms during transit.

**110** **Assertion** If we bend an insulated conducting wire, the resistance of the wire increases.
**Reason** The drift velocity of electron in bended wire remains same.

**111** **Assertion** The drift velocity of electrons in a metallic wire decreases, when temperature of the wire increases.
**Reason** On increasing temperature, conductivity of metallic wire decreases.

**112** **Assertion** Manganin and constantan are widely used in standard resistors.

**Reason** For manganin and constantan resistance, values would change very little with temperature.

## II. Statement Based Questions

**113** When no electric field is applied across a conductor, then consider the following statements.

I. The electrons will be moving due to thermal motion during which they collide with the fixed ions.

II. The direction of thermal velocity after the collision is completely random.

III. On average, the number of electrons travelling in any direction will be equal to the number of electrons travelling in the opposite direction. So, there will be no net electric current.

Which of the following statement (s) is/are correct?

(a) Both I and II (b) Both II and III
(c) Both I and III (d) I, II and III

**114** I. The resistance $R$ of a conductor depends on its length $l$ and cross-sectional area $A$ through the relation $R = \frac{\rho A}{l}$.

II. Metals have low resistivity in the range of $10^{-8}$ Ω-m to $10^{-6}$ Ω-m.

III. In most substances, the carriers of current are electrons, while in some cases, e.g. in ionic crystals and electrolytic liquids, positive and negative ions.

Which of the following statement(s) is/are correct?

(a) Only I (b) Both II and III
(c) Both I and II (d) Both I and III

**115** I. In the ionosphere and magnetosphere, strong electric current is generated from the solar terrestrial interaction.

II. In the lower atmosphere, the current is weaker and is maintained by thunderstorm.

III. The smaller ice particles in the clouds acquire negative charge and the larger ones positive charge.

Which of the following statement(s) is/are correct?

(a) Only I (b) Both II and III
(c) Both I and II (d) I, II and III

**116** Choose the correct statement for which Ohm's law is valid.

(a) For metallic conductors at low temperature.
(b) For metallic conductors at high temperature.
(c) For electrolytes when current passes through them.
(d) All of the above

**117** Which of the following statements is false?

***JEE Main 2017***

(a) In a balanced Wheatstone bridge, if the cell and the galvanometer are exchanged, the null point is disturbed.
(b) A rheostat can be used as a potential divider.
(c) Kirchhoff's second law represents energy conservation.
(d) Wheatstone bridge is the most sensitive when all the four resistances are of the same order of magnitude.

**118** Two bulbs consume same energy operating at 200 V and 300 V respectively, when these are connected in series across a DC source of 500 V, then which of the following statement is correct?

(a) Ratio of potential difference across them is 3/2.
(b) Ratio of potential difference across them is 4/9.
(c) Ratio of power produced in them is 2/3.
(d) Ratio of power produced in them is 1/3.

**119** For the circuit shown in figure, which of the following statement is correct?

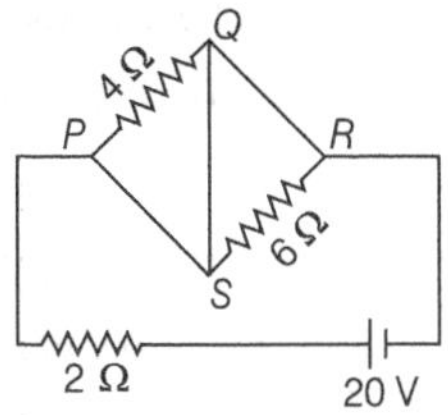


(a) Power supplied by the battery is 100 W.
(b) Current flowing in the circuit is 5A.
(c) Potential difference across the 4 Ω resistance is equal to the potential difference across the 6 Ω resistance.
(d) Current in wire $QS$ is zero.

## III. Matching Type

**120** Match the Column I (relation) with Column II (graph) and select the correct answer from the codes given below.

| Column I | | Column II |
|---|---|---|
| A. $V$ ceases to be proportional to $I$. | 1. | 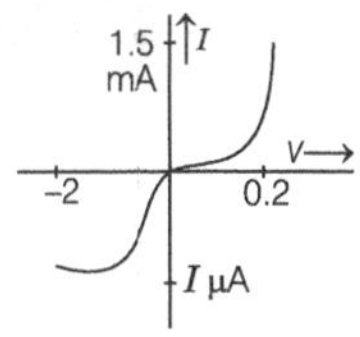  |
| B. The relation between $V$ and $I$ depends on the sign of $V$. | 2. | 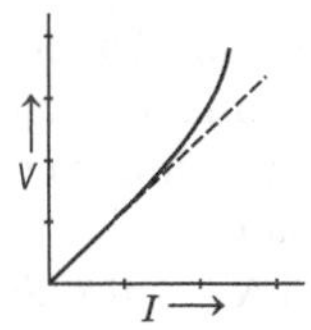  |
| C. The relation between $V$ and $I$ is not unique. | 3. | 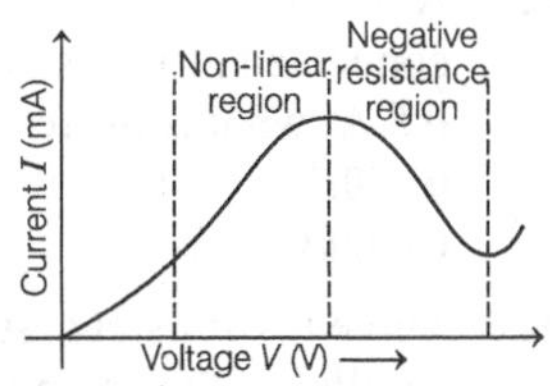  |

| | A | B | C | | A | B | C |
|---|---|---|---|---|---|---|---|
| (a) | 2 | 3 | 1 | (b) | 2 | 1 | 3 |
| (c) | 1 | 2 | 3 | (d) | 1 | 3 | 2 |

**121** Consider the figure.

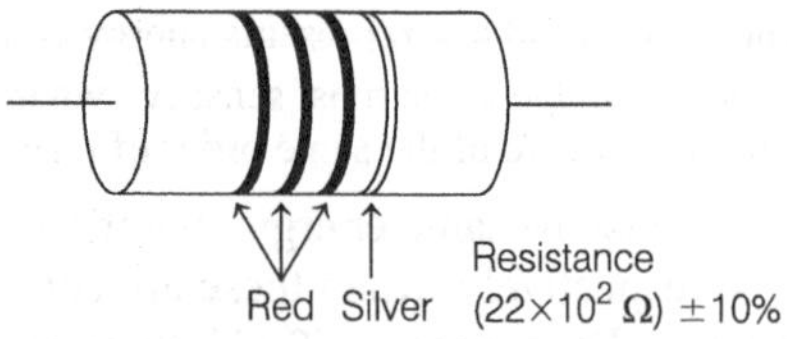


Match the Column I (band) with Column II (significance) and select the correct answer from the codes given below.

| Column I | Column II |
|---|---|
| A. The first two bands from the end | 1. A tolerance of 20% |
| B. The third band | 2. Tolerance or possible variation in percentage about the indicated values |
| C. The last band | 3. The decimal multiplier |
| D. Absence of last band | 4. First two significant figures of the resistance (in ohm) |

| | A | B | C | D |
|---|---|---|---|---|
| (a) | 4 | 3 | 1 | 2 |
| (b) | 3 | 1 | 2 | 4 |
| (c) | 2 | 4 | 1 | 3 |
| (d) | 4 | 3 | 2 | 1 |

**122** Match the Column I (graph) with Column II (relation) and select the correct answer from the codes given below.

| Column I | Column II |
|---|---|
| A. 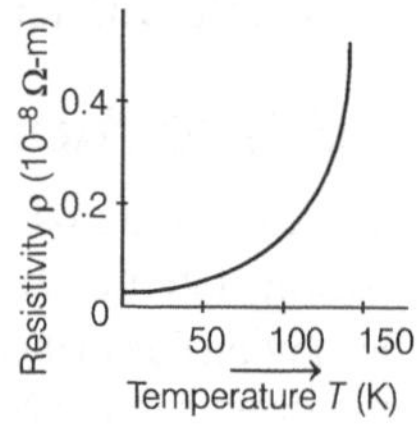 | 1. Temperature dependence of resistivity for a typical semiconductor |
| B. 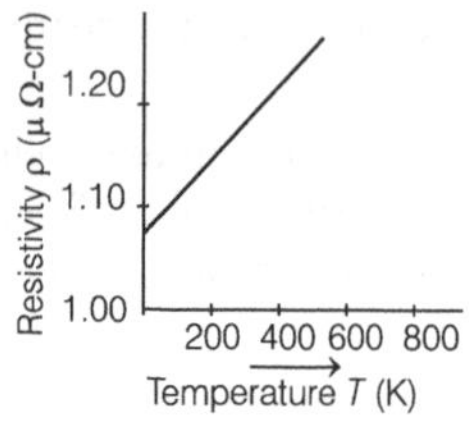 | 2. Resistivity ρ of copper as a function of temperature $T$ |
| C. 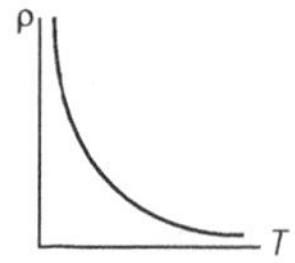 | 3. Resistivity ρ of nichrome as a function of absolute temperature $T$. |

| | A | B | C |
|---|---|---|---|
| (a) | 3 | 2 | 1 |
| (b) | 2 | 1 | 3 |
| (c) | 2 | 3 | 1 |
| (d) | 3 | 1 | 2 |

# NCERT & NCERT Exemplar

## MULTIPLE CHOICE QUESTIONS

### NCERT

**123** The storage battery of a car has an emf of 12 V. If the internal resistance of the battery is 0.4 Ω, what is the maximum current that can be drawn from the battery?

(a) 10 A (b) 20 A (c) 30 A (d) 40 A

**124** A battery of emf 10 V and internal resistance 3 Ω is connected to a resistor. If the current in the circuit is 0.5 A, what is the resistance of the resistor?

(a) 10 Ω (b) 17 Ω
(c) 15 Ω (d) None of these

**125** Three resistors 2 Ω, 4 Ω and 5 Ω are combined in parallel. If the combination is connected to a battery of emf 20 V and negligible internal resistance, determine the total current drawn from the battery.

(a) 19 A (b) 29 A (c) 13 A (d) 23 A

**126** At room temperature (27°C), the resistance of a heating element is 100 Ω. What is the temperature of the element, if the resistance is found to be 117 Ω? Given that, the temperature coefficient of the material of the resistor is $1.70 \times 10^{-4}$/°C.

(a) 1000°C (b) 1027°C (c) 2027°C (d) 3027°C

**127** A negligible small current is passed through a wire of length 15 m and uniform cross-section $6.0 \times 10^{-7}$ m$^2$ and its resistance is measured to be 5.0 Ω. What is the resistivity of the material at the temperature of the experiment?

(a) $2 \times 10^{-9}$ Ω-m (b) $3 \times 10^{-9}$ Ω-m
(c) $2 \times 10^{-7}$ Ω-m (d) $4 \times 10^{-7}$ Ω-m

**128** A silver wire has a resistance of 2.1 Ω at 27.5 °C and a resistance of 2.7 Ω at 100 °C. Determine the temperature coefficient of resistance of silver.

(a) 0.0039/°C (b) 0.02/°C (c) 0.05/°C (d) 0.06/°C

**129** A heating element using nichrome connected to a 230 V supply draws an initial current of 3.2 A which settles after a few seconds to a steady value of 2.8 A. What is the steady temperature of the heating element, if the room temperature is 27.0°C? Temperature coefficient of resistance of nichrome averaged over the temperature range involved is $1.70 \times 10^{-4}$ °C$^{-1}$.

(a) 676°C (b) 876°C
(c) 867°C (d) 500°C

**130** Determine the current drawn from source for the network shown below.

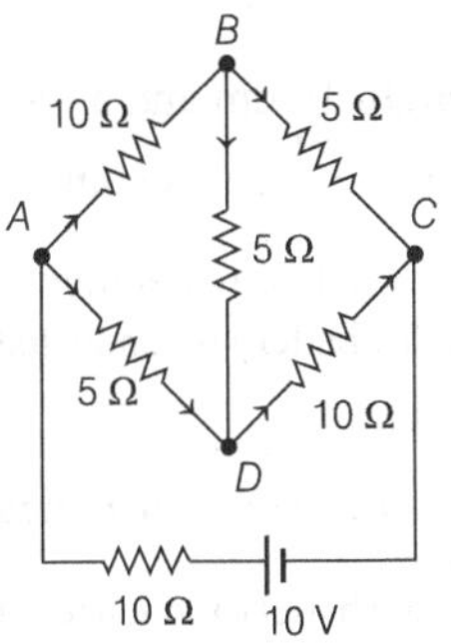


(a) $\frac{4}{17}$ A (b) $\frac{10}{17}$ A
(c) $\frac{6}{17}$ A (d) $\frac{5}{17}$ A

**131** In a meter bridge, the balance point is found to be at 39.5 cm from the end $A$, when the resistor $Y$ is of 12.5 Ω. Determine the balance point of the bridge above, if $X$ and $Y$ are interchanged.

(a) 39.5 cm (b) 12.5 cm
(c) 60.5 cm (d) None of these

**132** A storage battery of emf 8 V and internal resistance 0.5 Ω is being charged by a 120 V DC supply using a series resistor of 15.5 Ω. What is the terminal voltage of the battery during charging?

(a) 10 V (b) 11.5 V
(c) 7 V (d) 16 V

**133** In a potentiometer arrangement, a cell of emf 1.25 V gives a balance point at 35.0 cm length of the wire. If the cell is replaced by another cell and the balance point shifts to 63.0 cm, what is the emf of the second cell?

(a) 2.25 V (b) 3.25 V (c) 4.5 V (d) 6 V

**134** The number density of free electrons in a copper conductor estimated is $8.5 \times 10^{28}$ m$^{-3}$. How long does an electron take to drift from one end of a wire 3.0 m long to its other end? The area of cross-section of the wire is $2.0 \times 10^{-6}$ m$^2$ and it is carrying a current of 3.0 A.

(a) 6 h 33 min (b) 7 h 33 min
(c) 6 h (d) 5 h 20 min

**135** Determine the current drawn from a 12 V supply with internal resistance 0.5 Ω by the infinite network shown in given figure. Each resistor has 1 Ω resistance.

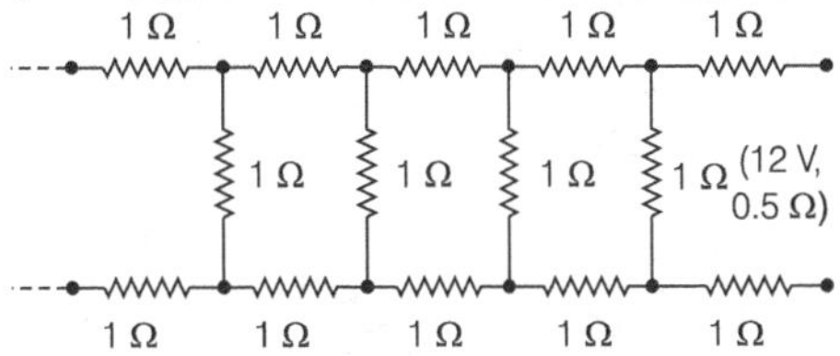


(a) 3A (b) 2.7A (c) 3.7A (d) 4.2A

**136** Figure shows a potentiometer circuit for comparison of two resistances. The balance point with a standard resistor $R = 10.0\Omega$ is found to be 58.3 cm, while that with the unknown resistance $X$ is 68.5 cm. Determine the value of $X$.

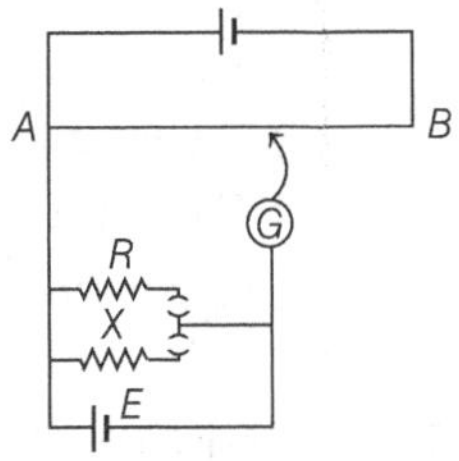


(a) 9.5 Ω (b) 10.5 Ω (c) 11.75 Ω (d) 13 Ω

**137** Figure shows a 2.0 V potentiometer used for the determination of internal resistance of a 1.5 V cell. The balance point of the cell in open-circuit is 76.3 cm. When a resistor of 9.5 Ω is used in the external circuit of the cell, the balance point shifts to 64.8 cm length of the potentiometer wire. Determine the internal resistance of the cell.

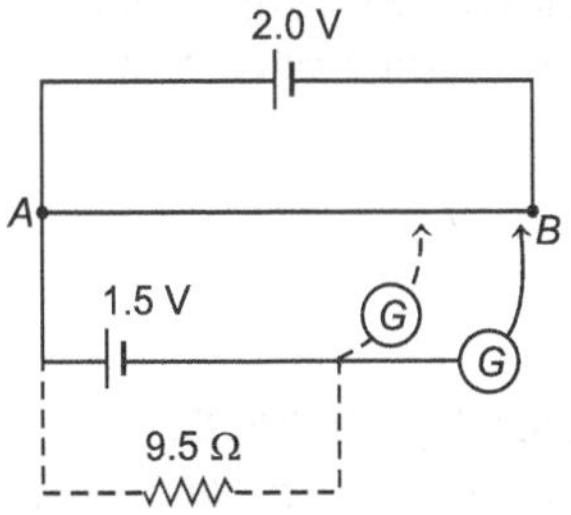


(a) 3 Ω (b) 1.68 Ω (c) 4.5 Ω (d) 3.6 Ω

## NCERT Exemplar

**138** Consider a current carrying wire (current $I$) in the shape of a circle. Note that as the current progresses along the wire, the direction of $J$ (current density) changes in an exact manner, while the current $I$ remains unaffected. The agent that is essentially responsible for

(a) source of emf
(b) electric field produced by charges accumulated on the surface of wire
(c) the charges just behind a given segment of wire which push them just the right way by repulsion
(d) the charges ahead

**139** Two batteries of emfs $\varepsilon_1$ and $\varepsilon_2 (\varepsilon_2 > \varepsilon_1)$ and internal resistances $r_1$ and $r_2$, respectively are connected in parallel as shown in figure.

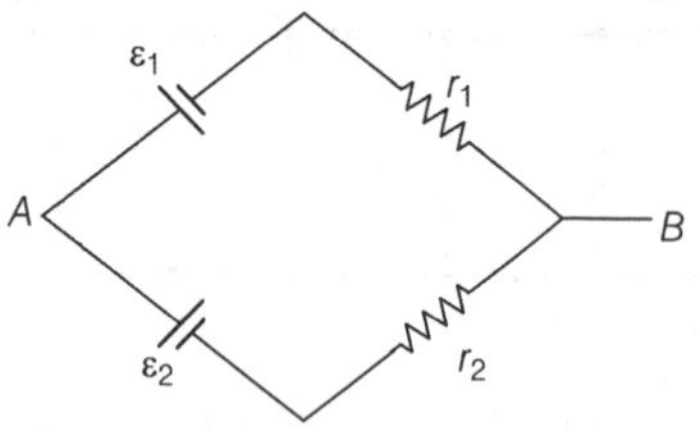


(a) The equivalent emf $\varepsilon_{eq}$ of the two cells is between $\varepsilon_1$ and $\varepsilon_2$, i.e. $\varepsilon_1 < \varepsilon_{eq} < \varepsilon_2$
(b) The equivalent emf $\varepsilon_{eq}$ is smaller than $\varepsilon_1$
(c) The $\varepsilon_{eq}$ is given by $\varepsilon_{eq} = \varepsilon_1 + \varepsilon_2$ always
(d) $\varepsilon_{eq}$ is independent of internal resistances $r_1$ and $r_2$

**140** A resistance $R$ is to be measured using a meter bridge. Student chooses the standard resistance $S$ to be $100\,\Omega$. He finds the null point at $l_1 = 2.9$ cm. He is told to attempt to improve the accuracy. Which of the following is a useful way?
(a) He should measure $I_1$ more accurately
(b) He should change $S$ to $1000\,\Omega$ and repeat the experiment
(c) He should change $S$ to $3\,\Omega$ and repeat the experiment
(d) He should give up hope of a more accurate measurement with a meter bridge

**141** Two cells of emfs approximately 5 V and 10 V are to be accurately compared using a potentiometer of length 400 cm. Which of the following statement is correct in support to the given situation?
(a) The battery that runs the potentiometer should have voltage of 8 V.
(b) The battery of potentiometer can have a voltage of 15 V and $R$ adjusted, so that the potential drop across the wire slightly exceeds 10 V.
(c) The first portion of 50 cm of wire itself should have a potential drop of 10 V.
(d) Potentiometer is usually used for comparing resistances and not voltages.

**142** A metal rod of length 10 cm and a rectangular cross-section of $1\text{cm} \times \frac{1}{2}$ cm is connected to a battery across opposite faces. The resistance will be
(a) maximum when the battery is connected across $1\text{ cm} \times \frac{1}{2}\text{ cm}$ faces
(b) maximum when the battery is connected across $10\text{ cm} \times 1\text{ cm}$ faces
(c) maximum when the battery is connected across $10\text{ cm} \times \frac{1}{2}$/cm faces
(d) same irrespective of the three faces

**143** Which of the following characteristics of electrons determines the current in a conductor?
(a) Drift velocity
(b) Thermal velocity
(c) Both drift velocity and thermal velocity
(d) Neither drift velocity nor thermal velocity

# Answers

## > Mastering NCERT with MCQs

| | | | | | | | | | |
|---|---|---|---|---|---|---|---|---|---|
| 1 (a) | 2 (c) | 3 (b) | 4 (b) | 5 (b) | 6 (c) | 7 (b) | 8 (c) | 9 (a) | 10 (d) |
| 11 (d) | 12 (c) | 13 (b) | 14 (b) | 15 (b) | 16 (c) | 17 (a) | 18 (a) | 19 (c) | 20 (a) |
| 21 (c) | 22 (d) | 23 (a) | 24 (c) | 25 (b) | 26 (a) | 27 (d) | 28 (b) | 29 (c) | 30 (d) |
| 31 (a) | 32 (c) | 33 (b) | 34 (b) | 35 (b) | 36 (b) | 37 (c) | 38 (a) | 39 (c) | 40 (c) |
| 41 (a) | 42 (c) | 43 (b) | 44 (b) | 45 (d) | 46 (d) | 47 (c) | 48 (b) | 49 (d) | 50 (a) |
| 51 (a) | 52 (a) | 53 (a) | 54 (d) | 55 (a) | 56 (b) | 57 (b) | 58 (d) | 59 (b) | 60 (d) |
| 61 (c) | 62 (b) | 63 (b) | 64 (c) | 65 (a) | 66 (b) | 67 (a) | 68 (a) | 69 (c) | 70 (c) |
| 71 (a) | 72 (b) | 73 (c) | 74 (d) | 75 (c) | 76 (d) | 77 (d) | 78 (a) | 79 (b) | 80 (b) |
| 81 (c) | 82 (a) | 83 (d) | 84 (b) | 85 (a) | 86 (b) | 87 (a) | 88 (b) | 89 (d) | 90 (c) |
| 91 (c) | 92 (b) | 93 (d) | 94 (b) | 95 (a) | 96 (b) | 97 (c) | 98 (c) | 99 (a) | 100 (c) |
| 101 (b) | 102 (c) | 103 (a) | 104 (a) | | | | | | |

## > Special Types Questions

| | | | | | | | | | |
|---|---|---|---|---|---|---|---|---|---|
| 105 (b) | 106 (b) | 107 (b) | 108 (a) | 109 (a) | 110 (d) | 111 (b) | 112 (a) | 113 (d) | 114 (b) |
| 115 (c) | 116 (a) | 117 (a) | 118 (b) | 119 (c) | 120 (b) | 121 (d) | 122 (c) | | |

## > NCERT & NCERT Exemplar MCQs

| | | | | | | | | | |
|---|---|---|---|---|---|---|---|---|---|
| 123 (c) | 124 (b) | 125 (a) | 126 (b) | 127 (c) | 128 (a) | 129 (c) | 130 (b) | 131 (c) | 132 (b) |
| 133 (a) | 134 (b) | 135 (c) | 136 (c) | 137 (b) | 138 (b) | 139 (a) | 140 (c) | 141 (b) | 142 (a) |
| 143 (a) | | | | | | | | | |

# Hints & Explanations

**2** (*c*) Given, $r = 5.0\times 10^{-11}$ m

and $v = 2.2\times 10^{6}$ ms$^{-1}$

$\because$ Time period of revolution,

$$T = \frac{2\pi r}{v} = \frac{2\times 3.14\times (5.0\times 10^{-11})}{2.2\times 10^{6}}$$

$$= 14.27\times 10^{-17}\text{s}$$

$\therefore$ Equivalent current $= \frac{\text{Charge}}{\text{Time}} = \frac{e}{T}$

$$= \frac{1.6\times 10^{-19}}{14.27\times 10^{-17}}$$

$$= 0.112\times 10^{-2}\text{A}$$

$$= 1.12\times 10^{-3}\text{A}$$

**3** (*b*) Let mid-point be $C$ as shown in the figure

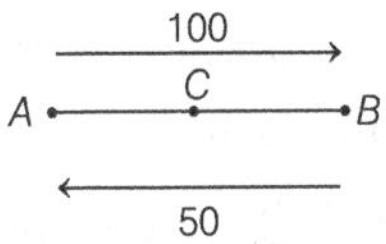


Current from $A$ to $B$, $i_1 = 100\times 3e = 300e$ $\left(\because i = \frac{ne}{t}\right)$

Current from $B$ to $A$, $i_2 = 50\times 3e = 150\,e$

Net current, $i = 300\,e - 150e = 150e$, towards right

$i = 150\times 1.6\times 10^{-19} = 2.4\times 10^{-17}$A, from $A$ to $B$

**4** (*b*) The instantaneous charge, $dq = i\,dt = (4 + 2t)\,dt$

$$\Rightarrow \quad q = \int_2^6 (4 + 2t)\,dt = [4t + t^2]_2^6$$

$$= (4\times 6 + 6^2) - (4\times 2 + 2^2)$$

$$= 60 - 12 = 48\text{C}$$

**5** (*b*) Given, number of electrons, $n = 20000000 = 2\times 10^{7}$

Total charge on twenty million electrons is

$$q = ne$$

$$= 2\times 10^{7}\times 1.6\times 10^{-19}\text{C}$$

$$(\because e = 1.6\times 10^{-19}\text{C})$$

$$= 3.2\times 10^{-12}\text{C}$$

Now, time taken by twenty million electrons to pass from point $X$ to point $Y$ is

$$t = 2\,\mu\text{s} = 2\times 10^{-6}\text{s}$$

$$\Rightarrow \quad i = \frac{q}{t} = \frac{3.2\times 10^{-12}}{2\times 10^{-6}} = 1.6\times 10^{-6}\text{ A}$$

Since, the direction of the current is always opposite to the direction of flow of electrons. Therefore, due to flow of electrons from point $X$ to point $Y$, the current will flow from point $Y$ to point $X$.

**6** (*c*) An $\alpha$-particle has a charge equal to that of 2 protons. The motion of $\alpha$-particle and proton towards left and the motion of electrons towards right, all will produce conventional current towards left.

The total current will be

$$i = n_\alpha q_\alpha + n_p q_p + n_e q_e \quad \ldots\text{(i)}$$

Here, $n_\alpha = n_p = n_e = 10^{19}, q_p = q_e = 1.6\times 10^{-19}$ C

and $q_\alpha = 2q_p = 2\times 1.6\times 10^{-19}$ C

Substituting these values in Eq. (i), we get

$$i = 10^{19}\times (2\times 1.6\times 10^{-19}) + 10^{19}\times (1.6\times 10^{-19}) + 10^{19}\times (1.6\times 10^{-19})$$

$= 6.4$ A, towards left

**9** (*a*) Given, for a wire, $\frac{R}{l} = \frac{1}{2}$

Length of wire, $l = 5\text{ cm} = 5\times 10^{-2}$ m

$$\therefore \quad R = \frac{l}{2} = 2.5\times 10^{-2}\,\Omega$$

Potential difference, $V = 1$V

or $\quad IR = 1 \quad [\because V = IR]$

$$\Rightarrow \quad I = \frac{1}{R} = \frac{1}{2.5\times 10^{-2}} = \frac{100}{2.5} = 40\text{ A}$$

**10** (*d*) The V-I graph for a conductor is as shown in the figure

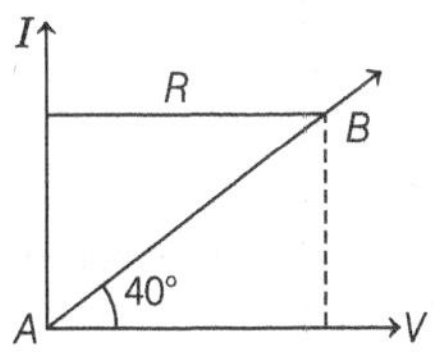


According to Ohm's law, we get

$$\text{Current, } I = \frac{V}{R}.$$

Also, slope $AB$ = resistance $R = \frac{V}{I}$

$$\Rightarrow \quad \frac{1}{\tan 40°} = R = \frac{V}{I}$$

or $\quad R = \cot 40°$

**11** (*d*) Initially, $R = \frac{\rho l}{A} = 4\Omega \quad \ldots\text{(i)}$

As the length of wire becomes twice, its area becomes half at its previous value. So, the new resistance,

$$R = \frac{\rho(2l)}{(A/2)} = 4\left(\frac{\rho l}{A}\right) = 4\times R \quad \text{[from Eq. (i)]}$$

$$= 4\times 4 = 16\Omega$$

**12** (c) On stretching, diameter of wire decreases but volume $V$ remains constant.

So, $V = Al$ or $l = V/A$

Now, $R = \frac{\rho l}{A} = \frac{\rho V}{A^2} = \frac{\rho V}{\pi^2 D^4/16} = \frac{16\rho V}{\pi^2 D^4}$

Taking logarithm of both sides and differentiating it, we get

$$\frac{\Delta R}{R} = -4\frac{\Delta D}{D}$$

or $\frac{\Delta R}{R} = -4 \times (-0.25\%) = 1.0\%$

$\therefore$ Resistance will increase and percentage change in it is 1%.

**13** (b) Both plates have same thickness, let it is $d$.

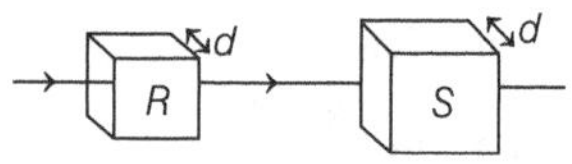


Then, $R_R = \frac{\rho l}{ld}$ and $R_S = \frac{\rho 2l}{2ld} = \frac{\rho l}{ld}$

$\therefore$ $\frac{R_R}{R_S} = 1$

Therefore, both plates have the same resistance.

**14** (b) Let the length of various edges of the block shown below in increasing order be $l, x, 2l$, respectively.

The resistance is given by $R = \rho l/A$

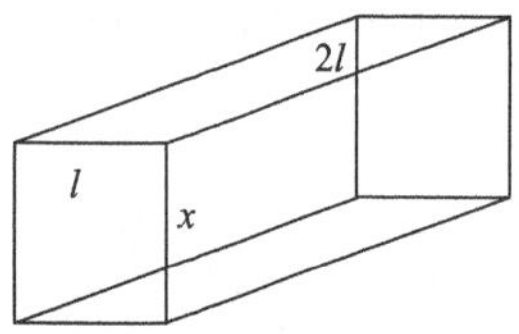


So, from figure the resistance between parallel faces having area of $(l \times x)$,

$$R_1 = \frac{\rho \times 2l}{l \times x} = \frac{2\rho}{x} \quad \text{...(i)}$$

Resistance between parallel faces having area of $l \times 2l$,

$$R_2 = \frac{\rho \times x}{l \times 2l} = \frac{\rho x}{2l^2} \quad \text{...(ii)}$$

and resistance between parallel faces having area of $2l \times x$,

$$R_3 = \frac{\rho \times l}{2l \times x} = \frac{\rho}{2x} \quad \text{...(iii)}$$

From Eqs. (i), (ii) and (iii), we get

$$R_{max} = R_1 = \frac{2\rho}{x}$$

and $R_{min} = R_3 = \frac{\rho}{2x}$

$\therefore$ $\frac{R_{max}}{R_{min}} = \frac{2\rho/x}{\rho/2x} = \frac{2}{1/2} = 4$

The ratio of the maximum and minimum resistance between parallel faces is 4.

**15** (b) As, resistance, $R = \frac{\rho l}{A}$

So, resistance across $AB$,

$$R_1 = \frac{\rho(4a)}{(2a)(a)} = \frac{2\rho}{a}$$

$\{\because$ for $R_1$, $l = 4a$ and $A = (2a)(a)\}$

Resistance across $EF$, $\Rightarrow R_3 = \frac{\rho(a)}{(4a)(2a)} = \frac{\rho}{8a}$

$\{\because$ for $R_3$, $l = a$ and $A = (4a)(2a)\}$

Resistance across $CD$,

$$R_2 = \frac{\rho(2a)}{(4a)a} = \frac{\rho}{2a}$$

$\{$for $R_2$, $l = 2a$ and $A = (4a)(a)\}$

$\therefore$ $R_1 > R_2 > R_3$

**16** (c) Electric field, $E = \frac{V}{l} = \frac{IR}{l}$ $[\because V = IR]$

$= \frac{I\rho l}{Al} = \frac{I\rho}{A}$ $[\because R = \rho l/A]$

Given, $I = 1\text{A}$, $A = 2.0\ \text{mm}^2 = 2 \times 10^{-6}\ \text{m}^2$

$\rho = 1.7 \times 10^{-8}\ \Omega\text{-m}$

$\Rightarrow$ $E = \frac{1 \times 1.7 \times 10^{-8}}{2 \times 10^{-6}} = 8.5 \times 10^{-3}\ \text{Vm}^{-1}$

**17** (a) Surface area of earth, $A = 4\pi r^2$

where, $r = 6400\ \text{km} = 6.4 \times 10^6$ m.

As charge is entering the earth per second per unit area, so current density is given by

$J = \frac{ne}{tA} = 0.15 \times 1.6 \times 10^{-19}\ \text{cm}^{-2}\ \text{s}^{-1}$ $\left\{\because \frac{n}{tA} = 0.15\right\}$

$= 0.15 \times 1.6 \times 10^{-19} \times 10^4\ \text{m}^{-2}\ \text{s}^{-1}$

$\therefore$ Current, $i = JA = J \times 4\pi r^2$

$= 0.15 \times 1.6 \times 10^{-19} \times 10^4 \times 4 \times 3.14 \times (6.4 \times 10^6)^2$

$= 0.12$ A

**18** (a) Current density, $J = \sigma E$

where, $\sigma = \text{conductivity} = \frac{1}{\text{resistivity}}$

$= \frac{1}{\rho} = \frac{1}{1.7 \times 10^{-8}\Omega\text{-m}}$

$\Rightarrow$ $J = \frac{1}{1.7 \times 10^{-8}} \times 100 = \frac{100}{17} \times 10^9 \simeq 6 \times 10^9\ \text{Am}^{-2}$

**19** (c) Given, $A = 2\ \text{mm}^2 = 2 \times 10^{-6}\ \text{m}^2$, $q = 4$ C, $t = 2$ s,

$\sigma = 6.25 \times 10^7\ \text{Sm}^{-1}$

$\therefore$ $i = \frac{q}{t} = \frac{4}{2} = 2$ A

$\therefore$ Current density, $J = \frac{i}{A} = \sigma E$

$\Rightarrow$ $E = \frac{i}{A\sigma} = \frac{2}{2 \times 10^{-6} \times 6.25 \times 10^7}$

$= 1.6 \times 10^{-2}$ N/C

**20** (*a*) Drift velocity, $v_d = \frac{eE}{m}\tau = \frac{e}{m}\tau\left(\frac{V}{l}\right)$ $[\because E = V/l]$

or $v_d \propto V$

Above relation states that, if $V$ is doubled, so drift velocity $v_d$ will also be doubled.

**21** (*c*) Current flowing in the wire, $i = nAev_d$

or $v_d \propto \frac{1}{A}$

If diameter of wire is $\frac{d}{4}$, then area will be $\frac{A}{16}$, so new drift velocity $= 16\, v_d$.

**22** (*d*) Current, $I = neAv_d$

$\Rightarrow$ Drift velocty, $v_d = \frac{I}{neA}$

$\Rightarrow$ $v_d \propto \frac{I}{A}$ $(\because n$ and $e$ are constants)

According to question, when a current $I$ flows through a wire, the drift velocity of electrons is $v$ and when the current $2I$ flows through another wire of double the area of cross section, the drift velocity of electrons will be $v_d'$, then

$$\frac{v'_d}{v_d} = \frac{I'/A'}{I/A} = \frac{2I/2A}{I/A} = 1$$

$\Rightarrow$ $v_d' = v_d = v$

**23** (*a*) Given, $\Delta V = 100\text{ V}$, $l = 1\text{ m}$, $\sigma = 5.81 \times 10^7 \Omega^{-1}$ and $n_{Cu} = 8.5 \times 10^{28} \text{m}^{-3}$

$\therefore$Electric field $= \frac{\Delta V}{l} = \frac{100}{1} = 100\text{ Vm}^{-1}$

$\therefore$Drift velocity, $v_d = \frac{i}{n_{Cu}eA}$

and current density, $J = \sigma E = \frac{i}{A}$

$$\Rightarrow v_d = \frac{\sigma}{en_{Cu}}E = \frac{5.81 \times 10^7 \times 100}{1.6 \times 10^{-19} \times 8.5 \times 10^{28}} = 0.43\text{ ms}^{-1}$$

**24** (*c*) Given, current, $i = 10\text{ A}$

Area of cross-section, $A = 4 \times 10^{-6}\text{ m}^2$

Density of conductors, $\rho = 2.7\text{ g/cc}$
$= 2.7 \times 10^3\text{ kg/m}^3$

Molecular weight of aluminium,

$M_w = 27\text{g} = 27 \times 10^{-3}\text{ kg}$

If $n$ be the total number of electrons in the conductor per unit volume, then

$$n = \frac{\text{Total number of electrons}}{\text{Volume of conductor}}$$

$$= \frac{\text{Number of atoms per mole} \times \text{Number of moles}}{V}$$

$$= \frac{\text{Avogadro's number}}{V} \times \left(\frac{M}{M_w}\right)$$

$$= 6 \times 10^{23} \times \frac{\rho}{M_w} \qquad \left[\because \rho = \frac{M}{V}\right]$$

$$= 6 \times 10^{23} \times \frac{2.7 \times 10^3}{27 \times 10^{-3}}$$

$\therefore$ $n = 6 \times 10^{28}$

We know that, drift velocity,

$$v_d = \frac{i}{neA} = \frac{10}{6 \times 10^{28} \times 1.6 \times 10^{-19} \times 4 \times 10^{-6}}$$

$= 2.6 \times 10^{-4}\text{ m/s}$

**25** (*b*) We know that, current density, $J = nev_d$ ...(i)

But drift velocity, $v_d = \frac{eE\tau}{m}$ ...(ii)

From Eqs. (i) and (ii), we get

$$J = ne \cdot \frac{eE\tau}{m}$$

$$J = \frac{ne^2E\tau}{m} \Rightarrow J \propto \tau$$

Since, on increasing temperature, relaxation time ($\tau$) decreases, therefore current density through wire decreases.

**26** (*a*) Given, resistivity of copper,

$\rho = 1.72 \times 10^{-8} \Omega\text{-m}$

Electrons density, $n = 8.5 \times 10^{28}/\text{m}^3$

$$\therefore \text{Mobility, } \mu = \frac{v_d}{E} = \frac{i}{neAE} = \frac{J}{neE} = \frac{1}{\rho\, ne} \qquad \left[\because J = \frac{i}{A} = \frac{E}{\rho}\right]$$

$$= \frac{1}{1.72 \times 10^{-8} \times 8.5 \times 10^{28} \times 1.6 \times 10^{-19}}$$

$= 4.27 \times 10^{-3}\text{ m}^2/\text{C-}\Omega$

**27** (*d*) Given, resistance is $200\,\Omega = 20 \times 10^1 \Omega$

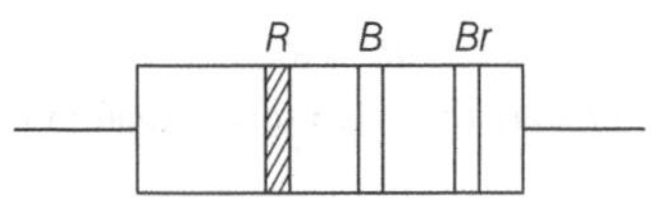


So, colour scheme will be red, black and brown. Significant figure of red band is 2 and for green is 5. When red (2) is replaced with green (5), new resistance will be

200 ohm $\longrightarrow$ 500 ohm.

**28** (*b*) Given, $R = (47 \pm 4.7)\text{ k}\Omega = 47 \times 10^3 \Omega \pm 10\%\,\Omega$

As per the colour code for carbon resistors, the colour assigned to numbers are as

$4 \rightarrow$ Yellow
$7 \rightarrow$ Violet
$10^3 \rightarrow$ Orange
$10\% \rightarrow$ Silver

Hence, the bands of colour on carbon resistor in sequence are; yellow violet, orange and silver.

**29** (*c*) First significant figure (blue) = 6, second significant figure (black) = 0, number of zeros to be attached (yellow) = 4

Thus, $R = 60 \times 10^4 \Omega$

As, $V = 30$ V

$$\Rightarrow \quad I = \frac{V}{R} = \frac{30}{60 \times 10^4} = 0.5 \times 10^{-4} \text{ A}$$

**30** (*d*) Given, $R_{20} = 20\Omega$, $R_{500} = 60\Omega$, $R_t = 25\Omega$

$$\therefore \quad R_t = R_0 (1 + \alpha t)$$

where, $\alpha$ is the temperature coefficient of resistance.

$$\Rightarrow \quad 20 = R_0 (1 + \alpha \times 20) \quad \text{...(i)}$$

$$\Rightarrow \quad 60 = R_0 (1 + \alpha \times 500) \quad \text{...(ii)}$$

On dividing Eq. (ii) by Eq. (i), we get

$$\frac{1 + 500\alpha}{1 + 20\alpha} = 3$$

$$\Rightarrow \quad \alpha = \frac{2}{440} = \frac{1}{220} \, {}^\circ\text{C}^{-1}$$

From Eq (i), we get

$$20 = R_0 \left(1 + \frac{1}{220} \times 20\right) \quad \text{...(iii)}$$

At $R = 25\Omega$, $\quad 25 = R_0 \left(1 + \frac{1}{220} \times t\right) \quad$ ...(iv)

On dividing Eq. (iv) by Eq. (iii), we get

$$\frac{25}{20} = \frac{\left(1 + \frac{1}{220} \times t\right)}{\left(1 + \frac{1}{220} \times 20\right)} \Rightarrow \frac{5}{4} = \frac{220 + t}{240}$$

$$\therefore \quad t = 80^\circ\text{C}$$

**31** (*a*) From *i*-*V* graph, we find that

$$(\text{Slope})_2 > (\text{Slope})_1$$

$$\Rightarrow \quad \frac{1}{R_2} > \frac{1}{R_1} \quad \left[\because \frac{i}{V} = \frac{1}{R}\right]$$

$$\Rightarrow \quad R_1 > R_2$$

As $R_1$ and $R_2$ are directly proportional to the $T_1$ and $T_2$, respectively.

$$\Rightarrow \quad T_1 > T_2$$

**32** (*c*) As we know, Cu is conductor, so with increase in temperature, resistance will increase linearly. While, Si is a semiconductor, so with increase in temperature, resistance will decreases exponentially.

So, the temperature dependence of resistances of Cu and undoped Si in temperature range 300-400 K is best described by linear increase for Cu and exponential decrease for Si.

**33** (*b*) Given, $\rho = 1.7 \times 10^{-8} \Omega$-m, $n = 8.5 \times 10^{28}$ m$^{-3}$,

$e = 1.6 \times 10^{-19}$ C, $m_e = 9.1 \times 10^{-31}$ kg

and $v_d = 1.6 \times 10^{-4}$ ms$^{-1}$

We know that, $\rho = \dfrac{m_e}{ne^2\tau}$

$$\Rightarrow \quad \tau = \frac{m_e}{e^2 n \rho}$$

$$= \frac{9.1 \times 10^{-31}}{(1.6 \times 10^{-19})^2 \times 8.5 \times 10^{28} \times 1.7 \times 10^{-8}}$$

$$= 2.5 \times 10^{-14} \text{s}$$

**35** (*b*) Power dissipated by any resistor *R*,

when *I* current flows through it is,

$$P = I^2 R \quad \text{... (i)}$$

Given $I = 2$ mA $= 2 \times 10^{-3}$ A and $P = 4.4$ W

Using Eq. (i), we get

$$4.4 = (2 \times 10^{-3})^2 \times R$$

$$\text{or} \quad R = \frac{4.4}{4 \times 10^{-6}} \cdot \Omega \quad \text{...(ii)}$$

When this resistance *R* is connected with 11 V supply then power dissipated is

$$P = \frac{V^2}{R}$$

$$\text{or} \quad P = \frac{(11)^2}{4.4} \times 4 \times 10^{-6} \quad [\because \text{Using Eq. (ii)}]$$

$$\Rightarrow \quad P = \frac{11 \times 11 \times 4 \times 10^{-6}}{44 \times 10^{-1}} \text{ W}$$

$$\text{or} \quad P = 11 \times 10^{-5} \text{ W}$$

**36** (*b*) Distance between two cities, $d = 150$ km

Fall of potential per km, $V = 8$ V

Average resistance per km, $R = 0.5\Omega$

Power loss per km, $P_1 = \dfrac{V^2}{R} = \dfrac{8^2}{0.5} = 128$ W

$\therefore$ Total power loss in wire $= P_1 \times d = 128 \times 150$

$= 19200$ W

$= 19.2$ kW

**37** (*c*) Resistance of bulb, $R = \dfrac{V^2}{P} = \dfrac{(230)^2}{100} = 529\Omega$

When the voltage drops to 115 V, heat and light energy produced by the bulb in 20 min is given by

$$W = \frac{V^2}{R} t = \frac{(115)^2}{529} \times 20 \times 60 = 30000 \text{ J}$$

**38** (*a*) Given, $P = 100$ W and $V = 200$ V

As, power, $P = Vi$

$$\Rightarrow \quad i = \frac{P}{V} = \frac{100}{200} = 0.5 \text{ A}$$

Charge passing through the lamp in 1s,

$$q = i \times t = 0.5 \times 1 = 0.5 \text{ C} \quad (\because t = 1\text{ s})$$

Therefore, number of electrons moving through the filament per second,

$$n = q/e = \frac{0.5}{1.6 \times 10^{-19}} = 3.1 \times 10^{18}$$

**39** (*c*) Given, $\varepsilon = 50$ V and $i = 12$ A

$\therefore \quad P = \varepsilon i = 50 \times 12 = 600$ W

Since, the efficiency of the motor is 30%.

Power dissipated as heat in the motor windings

$= (100\% - 30\%) = 70\%$

Total power dissipated as heat in the windings

$$= \left(\frac{70}{100}\right) \times 600 = 420 \text{ W}$$

If $R$ is the resistance of the windings of the motor, total power dissipated as heat $= i^2 R$

$\Rightarrow \quad i^2 R = 420$ W

$$\Rightarrow \quad R = \frac{420 \text{ W}}{(12 \text{ A})^2} = 2.9\,\Omega$$

**40** (*c*) Among given devices, fuse is used in electric circuit as a protection device.

It helps in preventing excessive amount of current to flow in the circuit or from short circuiting.

It has low melting point and low resistivity, so when excess amount of current flow in the circuit, it melts and breaks the circuit.

**42** (*c*) For the same length and same material,

$$\frac{R_2}{R_1} = \frac{A_1}{A_2} = \frac{3}{1}$$

$\Rightarrow \quad R_2 = 3R_1$

Let first wire is thick and second wire in thin, then the resistance of thick wire,

$$R_1 = 10\,\Omega$$

The resistance of thin wire,

$$R_2 = 3R_1 = 3 \times 10 = 30\,\Omega$$

Total resistance for series

$$= R_1 + R_2 = 10 + 30 = 40\Omega$$

**43** (*b*) The situation is shown in the circuit diagram below

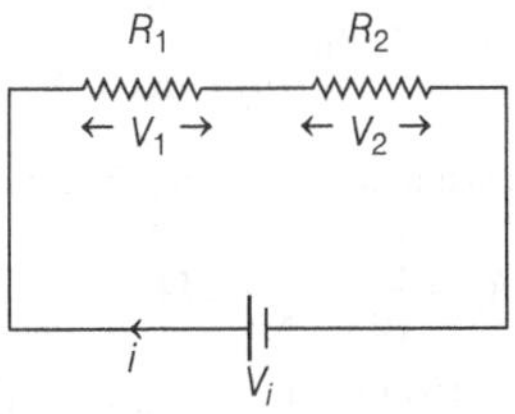


Current flowing through the circuit,

$$i = \frac{V_i}{R_1 + R_2}$$

where, $V_i$ = input voltage.

Voltage across $R_2$, $V_2 = iR_2$

$$\Rightarrow \quad V_2 = \frac{V_i R_2}{R_1 + R_2}$$

$$\therefore \quad \frac{V_i}{V_2} = \frac{R_1 + R_2}{R_2}$$

**44** (*b*) Given, the resistance of 400 Ω and (10 kΩ) 10000 Ω are in parallel, their effective resistance $R_p$ will be

$$R_p = \frac{400 \times 10000}{400 + 10000} = \frac{5000}{13}\,\Omega$$

Total resistance of circuit, $R = \frac{5000}{13} + 800 = \frac{15400}{13}\,\Omega$

Current in the circuit, $i = \frac{V}{R} = \frac{6}{15400/13} = \frac{39}{7700}$ A $\quad [\because V = 6\text{V}]$

Potential difference across voltmeter,

$$V = iR_p = \frac{39}{7700} \times \frac{5000}{13} = 1.95 \text{ V}$$

**45** (*d*) Resistance of each arm of equilateral triangle will be

$$R = \frac{18}{3} = 6\,\Omega$$

So we have following combination will be

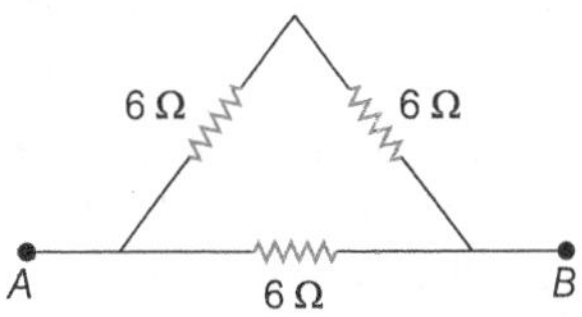


Equivalent resistance is

$$\therefore \quad R_{AB} = \frac{12 \times 6}{12 + 6} = \frac{12 \times 6}{18} = 4\,\Omega$$

**46** (*d*) The equivalent resistance of parallel combination,

$$R_p = \frac{R_1 R_2}{R_1 + R_2} = \frac{3}{5} \text{ and } R_1 = 3\,\Omega, \text{ then}$$

$$\frac{3 \times R_2}{3 + R_2} = \frac{3}{5} \Rightarrow 15R_2 = 9 + 3R_2 \Rightarrow 12R_2 = 9$$

$$\therefore \quad R_2 = \frac{3}{4}\,\Omega$$

**47** (*c*) Resistance of each part, $R' = \frac{R}{4} = \frac{16}{4} = 4\Omega$

and length of each part, $l' = l/4$.

In stretching $l'$ to $l$, it is stretched 4 times, so resistance become $R'' = \frac{\rho(4l)}{(A/4)} = 4^2 \frac{\rho l}{A} = 4^2 R'$

$\Rightarrow \quad R'' = 4^2 \times 4 = 64\Omega$

For parallel combination,

$$\frac{1}{R_{eq}} = \frac{1}{64} + \frac{1}{64} + \frac{1}{64} + \frac{1}{64} = \frac{4}{64}$$

$$\Rightarrow \quad R_{eq} = \frac{64}{4} = 16\,\Omega$$

**48** (*b*) $\because$ 60 Ω and 30 Ω resistor are connected in parallel, so their net resistance,

$$R_{net} = \frac{R_1 R_2}{R_1 + R_2} = \frac{30 \times 60}{30 + 60} = \frac{1800}{90} = 20\,\Omega$$

$$\therefore \quad i = \frac{V}{R_{net}} = \frac{2}{20} = 0.1 \text{ A} \quad [\because V = 2\text{ V}]$$

**49** (*d*) The resistances are connected in parallel, hence voltage will remain constant.

Thermal energy developed, $E = \dfrac{V^2 t}{R} \Rightarrow E \propto \dfrac{1}{R}$

$\therefore \qquad \dfrac{E_1}{E_2} = \dfrac{R_2}{R_1} = \dfrac{2R}{R} = \dfrac{2}{1}$

Thus, the thermal energy developed in $R$ is $2R$ in the ratio is 2 : 1.

**50** (*a*) Effective resistance of $n$ resistors each of the resistance $r$ in series, $R_s = r \times n = R$, so $r = R/n$.

When these resistances are connected in parallel, the effective resistance, $R_p = r/n = \dfrac{R/n}{n} = R/n^2$

**51** (*a*) Resistance in the branch $ADC$,

$$R_1 = 3\,\Omega + 7\,\Omega = 10\,\Omega$$

Since, arms $ADC$ $(10\,\Omega)$ and $AC$ $(10\,\Omega)$ are in parallel their equivalent resistance, $R_2 = \dfrac{10 \times 10}{10 + 10}\,\Omega = 5\,\Omega$

Since, $R_2$ is in series with $9\Omega$ in arm $AB$, equivalent resistance, $R_3 = 5\,\Omega + 9\,\Omega = 14\,\Omega$.

As $R_3$ is in parallel with $5\,\Omega$ in arm $BC$, equivalent resistance between $B$ and $C$,

i.e. $\qquad R_{BC} = \dfrac{14 \times 5}{14 + 5} = \dfrac{70}{19}\,\Omega$

**52** (*a*)

(i) Equivalent circuit when the equivalent resistance is calculated between points $P$ and $Q$ is as shown below

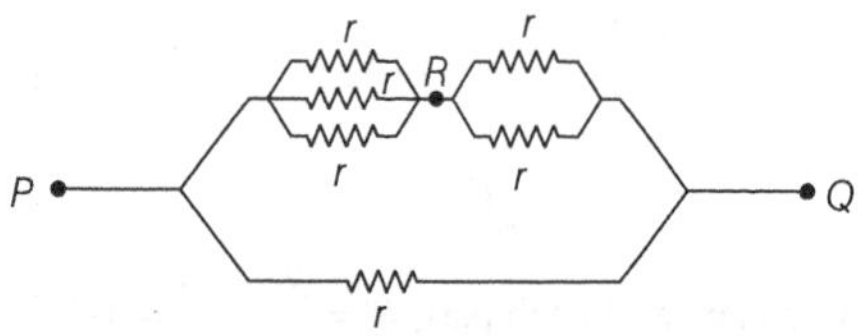


$$R_{PQ} = \frac{r \times \left(\frac{r}{3} + \frac{r}{2}\right)}{r + \frac{r}{3} + \frac{r}{2}} = \frac{5}{11} r$$

(ii) Equivalent circuit when the equivalent resistance is calculated between points $Q$ and $R$ is as shown below

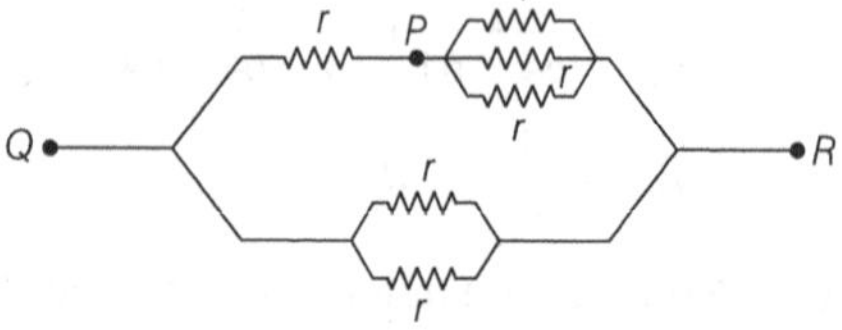


$$R_{QR} = \frac{\frac{r}{2} \times \left(r + \frac{r}{3}\right)}{\frac{r}{2} + r + \frac{r}{3}} = \frac{4}{11} r$$

(iii) Equivalent circuit when the equivalent resistance is calculated between points $P$ and $R$ is as shown below

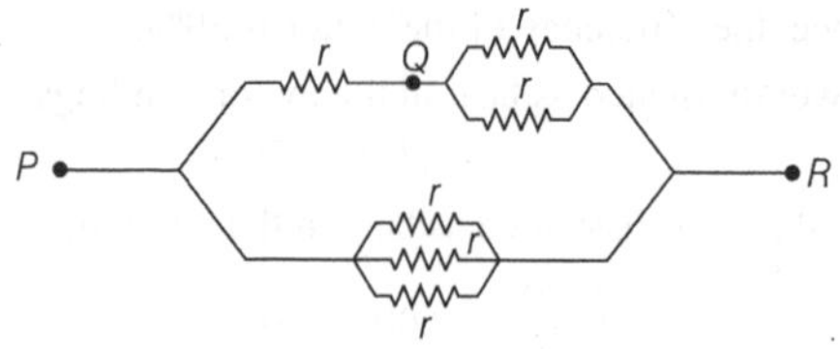


$$R_{PR} = \frac{\frac{r}{3} \times \left(r + \frac{r}{2}\right)}{\frac{r}{3} + r + \frac{r}{2}} = \frac{3}{11} r$$

Hence, net resistance will be maximum between $P$ and $Q$.

**53** (*a*) Two resistances of each side of triangle are connected in parallel, therefore the effective resistance of each arm of the triangle would be $\dfrac{r \times r}{r + r} = r/2$.

The two arms $AB$ and $AC$ are in series,

$$R' = (r/2) + (r/2) = r$$

$\therefore$ Total resistance, $\dfrac{1}{R} = \dfrac{1}{r} + \dfrac{2}{r} = \dfrac{3}{r}$

$\Rightarrow \qquad R = r/3$

**54** (*d*) Given, resistance of wire,

$$R_0 = R_1 + R_2 = 12\,\Omega$$

Resistance between $A$ and $B$,

$$R = \frac{R_1 \times R_2}{R_1 + R_2} = \frac{8}{3}\Omega$$

$\Rightarrow \qquad R_1 R_2 = \dfrac{8}{3}(R_1 + R_2)$

$\qquad = \dfrac{8}{3} \times 12 = 32\,\Omega$

We get, $\qquad R_1 = 4\Omega$ and $R_2 = 8\Omega$

As $\qquad R \propto l$, then $\dfrac{l_1}{l_2} = \dfrac{R_1}{R_2} = \dfrac{4}{8} = \dfrac{1}{2}$

**55** (*a*) Given, resistance $10\,\Omega$ and $2.5\,\Omega$ are in parallel,

$$R_1 = \frac{10 \times 2.5}{10 + 2.5} = 2\,\Omega$$

Now, $2\,\Omega$ and $10\,\Omega$ are in series, $R_2 = 10 + 2 = 12\,\Omega$

$R_2$ and $12\,\Omega$ are in parallel, $\dfrac{1}{R_3} = \dfrac{1}{12} + \dfrac{1}{12}$

$\Rightarrow \qquad R_3 = 6\,\Omega$

Now, $R_3$ and $10\,\Omega$ are in series, $R_4 = 10 + 6 = 16\,\Omega$

Now, $R_4$ and $16\,\Omega$ are in parallel,

$$\frac{1}{R} = \frac{1}{16} + \frac{1}{16} \Rightarrow R = \frac{16}{2} = 8\,\Omega$$

$\therefore$ Net equivalent resistance across the points $A$ and $B$ is $8\,\Omega$.

**56** (*b*) The circuit is as shown below

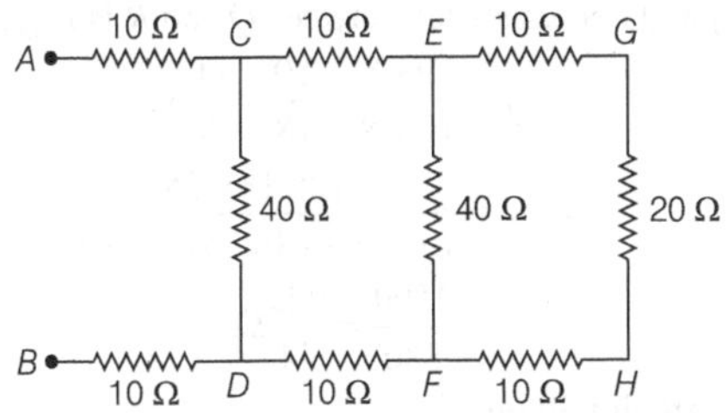


In the circuit, the branch *EGHF* have three resistances of 10 Ω, 20 Ω and 10 Ω respectively, which are connected in series combination. So, their equivalent resistance is given by

$$R_1 = 10 + 20 + 10 = 40\Omega$$

This $R_1$ resistance is parallel with 40 Ω resistance in the branch *EF*, so their equivalent resistance,

$$R_2 = \frac{40 \times 40}{40 + 40} = 20\Omega$$

Now, circuit becomes

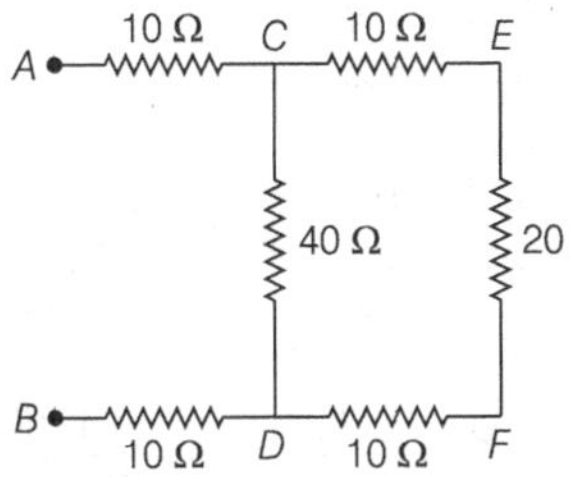


Now, in the branch *CEFD*, 10 Ω, 20 Ω and 10 Ω resistances are connected in series combination, so their equivalent resistance is given by

$$R_3 = 10\Omega + 20\Omega + 10\Omega = 40\Omega$$

This $R_3$ is in parallel with 40 Ω resistance in branch *CD*. Their equivalent resistance,

$$R_4 = \frac{40 \times 40}{40 + 40} = 20\Omega$$

Now, circuit becomes

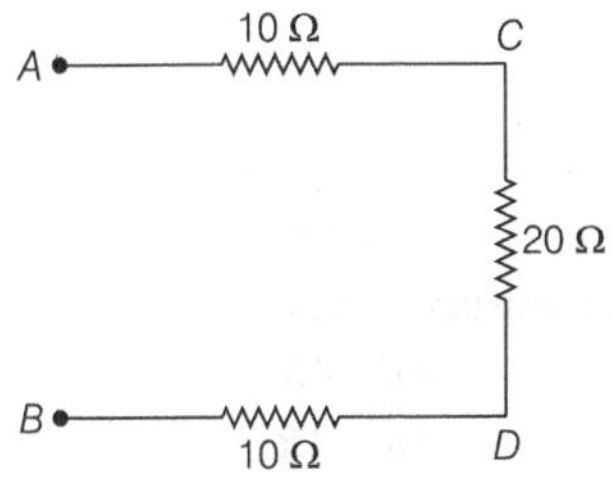


The net resistance between *A* and *B*,

$$R_{net} = 10 + 20 + 10 = 40\Omega$$

**57** (*b*) Measured value of $R$ = 5% less than actual value of $R$.

Actual values of $R = 30\Omega$

So, measured value of $R$ is

$$R' = 30 - (5\% \text{ of } 30) = 30 - \frac{5}{100} \times 30$$

$$\Rightarrow \quad R' = 28.5\Omega \qquad \dots \text{(i)}$$

Now, let us assume that internal resistance of voltmeter $R_V$. Replacing voltmeter with its internal resistance, we get following circuit.

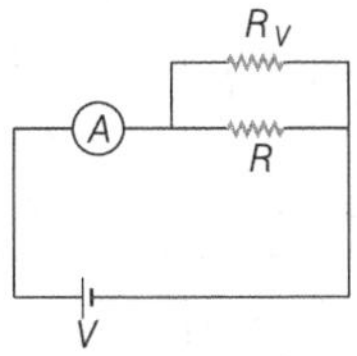


It is clear that the measured value, $R'$ should be equal to parallel combination of $R$ and $R_V$. Mathematically,

$$R' = \frac{RR_V}{R + R_V} = 28.5\Omega$$

Given, $\quad R = 30\Omega$

$$\Rightarrow \quad \frac{30R_V}{30 + R_V} = 28.5$$

$$\Rightarrow \quad 30R_V = (28.5 \times 30) + 28.5\,R_V$$

$$\Rightarrow \quad 1.5R_V = 28.5 \times 30$$

$$\Rightarrow \quad R_V = \frac{28.5 \times 30}{1.5} = 19 \times 30$$

or $\quad R_V = 570\Omega$

**58** (*d*) The given circuit diagram can be drawn as shown below

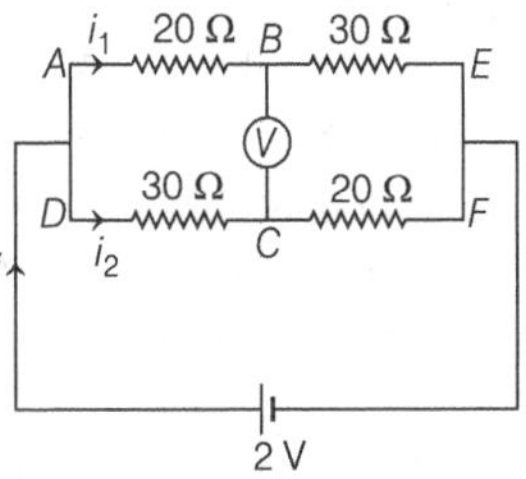


The equivalent resistance of circuit is given by

$$\frac{1}{R_{eq}} = \frac{1}{R_{AE}} + \frac{1}{R_{DF}}$$

$$= \frac{1}{(20 + 30)} + \frac{1}{(30 + 20)}$$

$$= \frac{1}{50} + \frac{1}{50} = \frac{2}{50}$$

$$\Rightarrow \quad R_{eq} = 25\Omega$$

The current in circuit, $i = \frac{V}{R} = \frac{2}{25}$ A

As the resistance of two branches is same, i.e. 50 Ω.

So, the current, $i_1 = i_2$

$$\Rightarrow \quad i = i_1 + i_2$$

$$\frac{2}{25} = 2i_1 \Rightarrow i_1 = \frac{1}{25}\text{A}$$

$$\Rightarrow \quad i_2 = i_1 = \frac{1}{25}\text{A}$$

∴ The voltage across *AB*,

$$V_1 = i_1 R_1 = \frac{1}{25} \times 20$$

and voltage across $CD$,

$$V_2 = i_2 R_2 = \frac{1}{25} \times 30$$

$$\therefore \text{Voltmeter reading} = V_2 - V_1 = \frac{30}{25} - \frac{20}{25} = \frac{10}{25} = 0.4 \text{ V}$$

**59** (*b*) The circuit can be redrawn by breaking into two parts,

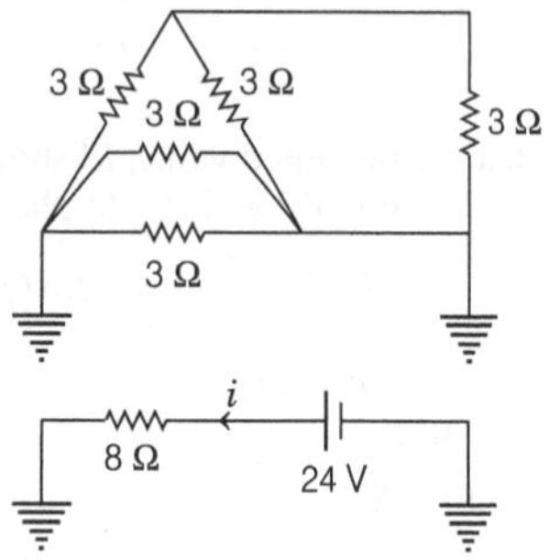


As the other end of 8Ω resistor is grounded, so it is at zero potential. The potential difference across 8 Ω resistor = 24 − 0 = 24 V

$$\therefore \text{Current in } 8\,\Omega \text{ resistor, } i = \frac{V}{R} = \frac{24-0}{8} = 3\text{A}$$

**60** (*d*) If $R_1$ be the equivalent resistance of parallel resistors 8Ω, 8Ω and 4Ω, then

$$\frac{1}{R_1} = \frac{1}{8} + \frac{1}{8} + \frac{1}{4} = \frac{4}{8} \Rightarrow R_1 = 2\Omega$$

∴Total resistance of upper branch,

$$R_2 = R_1 + 4 = 2 + 4 = 6\Omega$$

Now, circuit can be redrawn as

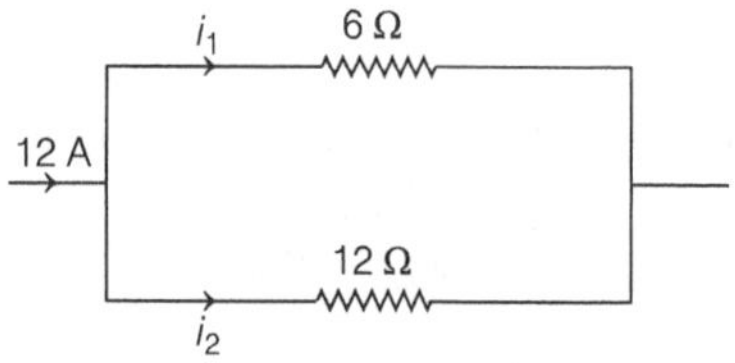


Here, $i_1 + i_2 = 12$

As voltage across parallel resistors is same, so

$$\frac{V}{6} + \frac{V}{12} = 12 \qquad \left[\because I = \frac{V}{R}\right]$$

$$V = 48 \text{ V}$$

or voltage across 12 Ω resistor = 48 V

**61** (*c*) The circuit is as shown below

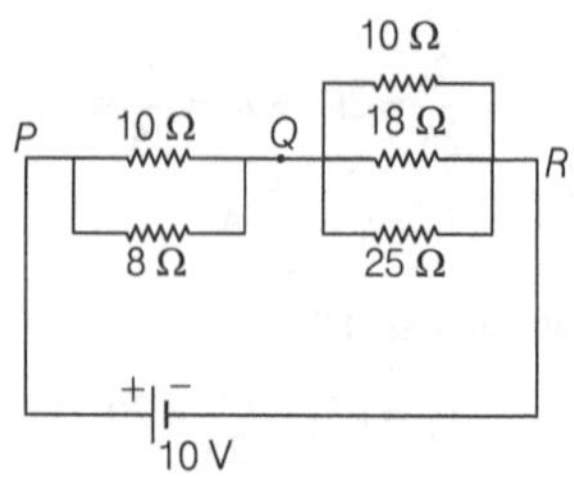


In the given circuit, net resistance between $P$ and $Q$ is $R_{PQ}$ and net resistance between $Q$ and $R$ is $R_{QR}$, then

$$R_{PQ} = \frac{10 \times 8}{10+8} = \frac{80}{18} = \frac{40}{9}\Omega$$

and
$$R_{QR} = \frac{10 \times 18 \times 25}{10 \times 18 + 18 \times 25 + 25 \times 10} = \frac{4500}{880} = \frac{225}{44}\Omega$$

So, circuit will be

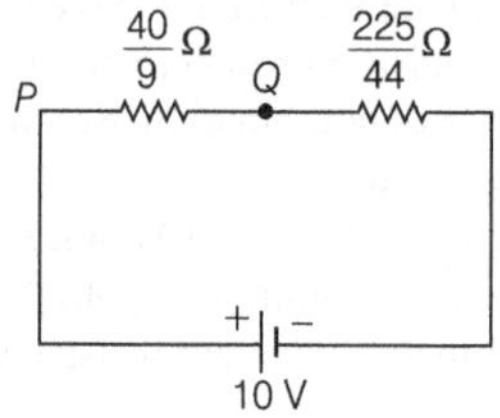


In the series circuit, voltages will be divide in the ratio of their resistances.

$$\therefore \quad \frac{R_{PQ}}{R_{QR}} = \frac{V_{PQ}}{V_{QR}}$$

$$\therefore \quad \frac{\left(\frac{40}{9}\right)}{\left(\frac{225}{44}\right)} = \frac{V_{PQ}}{V_{QR}} \text{ or } \frac{1760}{2025} = \frac{V_{PQ}}{V_{QR}}$$

Now,
$$\frac{V_{PQ}}{V_{PR}} = \frac{V_{PQ}}{V_{PQ} + V_{QR}} = \frac{1760}{1760 + 2025}$$

$$\Rightarrow \quad \frac{V_{PQ}}{10} = \frac{1760}{3785} \Rightarrow V_{PQ} = 4.65 \text{ V}$$

or $\quad V_P - V_Q = 4.65$ V

**62** (*b*) Given circuit is

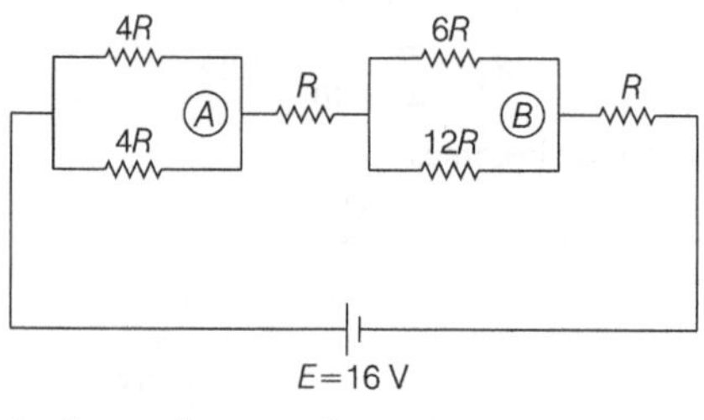


Equivalent resistance of part $A$,

$$R_A = \frac{4R \times 4R}{4R + 4R} = 2R$$

Equivalent resistance of part $B$,

$$R_B = \frac{6R \times 12R}{6R + 12R} = \frac{72}{18}R = 4R$$

∴Equivalent circuit is

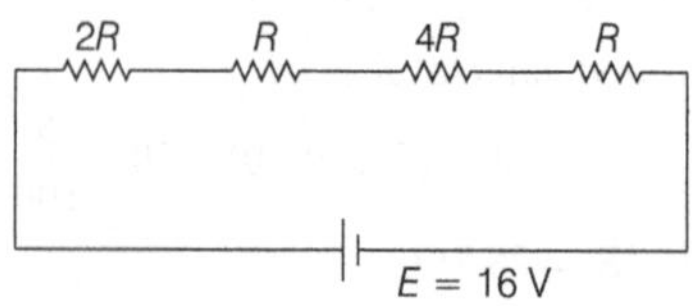

$\therefore$ Total resistance of the given network is

$$R_s = 2R + R + 4R + R = 8R$$

As we know, power of the circuit,

$$P = \frac{E^2}{R_s} = \frac{(16)^2}{8R} = \frac{16 \times 16}{8R} \quad \ldots\text{(i)}$$

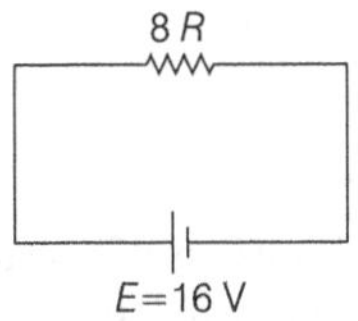


According to question, power consumed by the network, $P = 4$ W

From Eq. (i), we get

$$\therefore \quad \frac{16 \times 16}{8R} = 4 \Rightarrow R = \frac{16 \times 16}{8 \times 4} = 8\Omega$$

**63** (*b*) For an ideal voltmeter, the resistance is infinite and for an ideal ammeter, the resistance is zero.

For circuit 1,

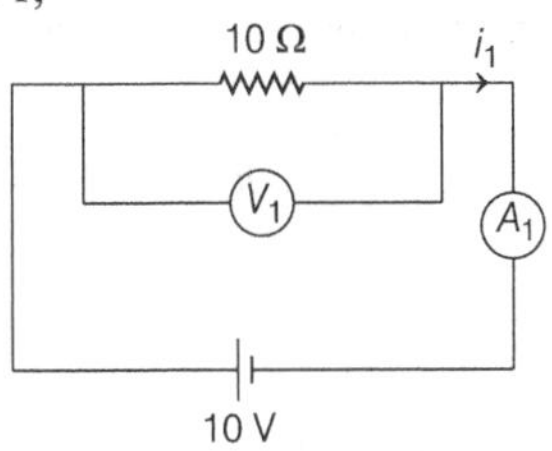


$$R \times i_1 = V \quad \text{or} \quad 10\, i_1 = 10$$

$$\Rightarrow \quad i_1 = \frac{10}{10} = 1\,\text{A}$$

$$\therefore \quad V_1 = i_1 \times R = 1 \times 10 = 10\,\text{V}$$

Similarly, for circuit 2, the addition of 10 Ω to voltmeter does not affect the current and hence

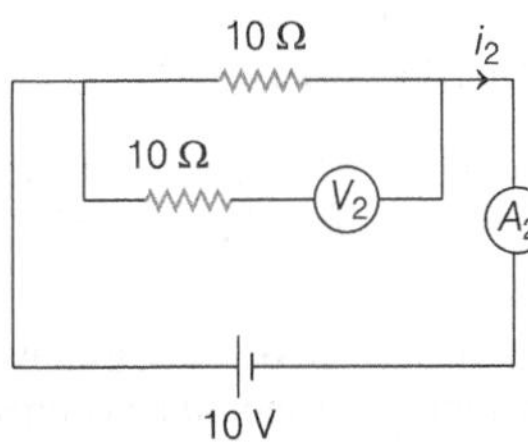


$$R \times i_2 = V \text{ or } 10 i_2 = 10$$

$$\Rightarrow \quad i_2 = \frac{10}{10} = 1\,\text{A}$$

$$\Rightarrow \quad V_2 = i_2 R = 1 \times 10 = 10\,\text{V}$$

$$\therefore \quad V_1 = V_2 \text{ and } i_1 = i_2$$

**66** (*b*) As, $i = \frac{E}{R + r}$ or $E = i(R + r)$

Given, $i = 0.2$ A, $E = 2.1$ V and $R = 10\Omega$

$$\Rightarrow \quad 2.1 = 0.2(10 + r) \Rightarrow 10 + r = \frac{2.1}{0.2}$$

$$\therefore \quad r = 10.5 - 10 = 0.5\,\Omega$$

**67** (*a*) The situation can be drawn as

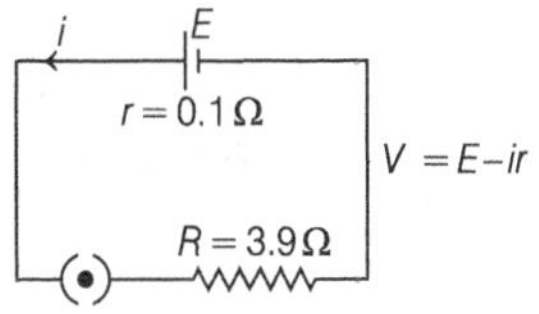


$$\therefore \quad V = E - ir$$

where, $r$ is the internal resistance of cell.

Also, current, $i = \frac{E}{R + r} \Rightarrow V = E - \left(\frac{E}{R + r}\right) r$

Putting numerical values, we have

$$E = 2\text{ V}, \ r = 0.1\Omega \text{ and } R = 3.9\,\Omega$$

$$\Rightarrow \quad V = 2 - \left(\frac{2}{3.9 + 0.1}\right) \times 0.1 \Rightarrow V = 1.95\text{ V}$$

**68** (*a*) Here, the potential drop is taking place across $AB$. As $R$ & $R'$ are parallel, so potential drop $R$ & $R'$ will be equal whatever may be the value of $R'$. Since, potential drop across $R$ is $V$ which is constant (the value of $r$ is very small), so potential drop across $R'$ will also be constant, when $R'$ is varied Since, the equivalent resistance of parallel combination of $R$ and $R'$ is always less than $R$, therefore $I \geq \frac{V}{r + R}$, always. Thus, $I$ is independent of $R'$. Also, as $R'$ varies the value of current through $R'$ changes linearly.

**69** (*c*) The given circuits are as shown below

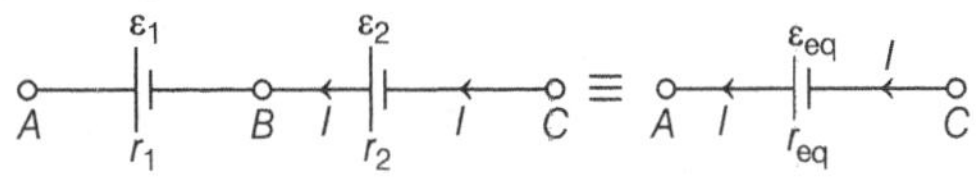


So, potential difference between the terminals $A$ and $C$ is,

$$V_{AC} = V_A - V_C = (V_A - V_B) + (V_B - V_C)$$
$$= \varepsilon_1 - Ir_1 + \varepsilon_2 - Ir_2$$
$$= (\varepsilon_1 + \varepsilon_2) - I\,(r_1 + r_2) = \varepsilon_{eq} - Ir_{eq}$$

**70** (*c*) If $n$ identical cells are connected in series, then equivalent emf of the combination, $E_{eq} = nE$

Equivalent internal resistance, $r_{eq} = nr$

$$\therefore \text{ Current, } I = \frac{E_{eq}}{r_{eq}} = \frac{nE}{nr} \text{ or } I = \frac{E}{r} = \text{constant}$$

Thus, current $I$ is independent of the number of cells $n$ present in the circuit.

Therefore, the graph showing the relationship between $I$ and $n$ would be as shown below

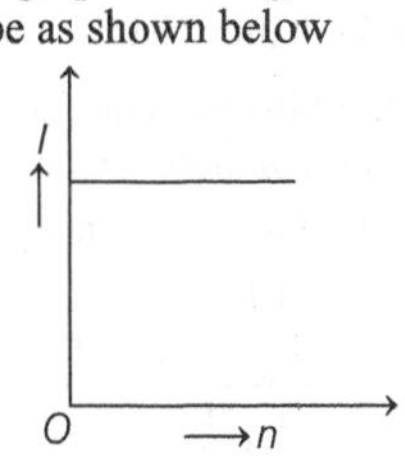


This is correctly shown in option (c).

**71** (*a*) The given circuit is as shown below

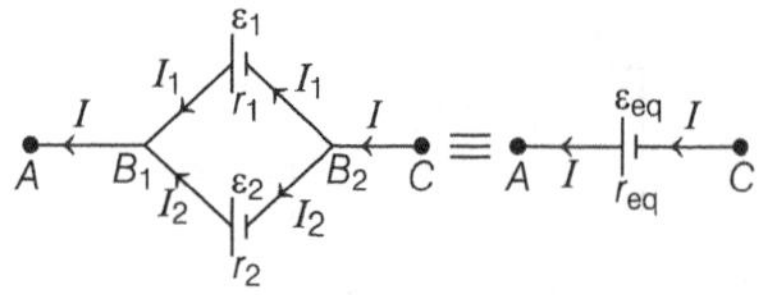


$$I = I_1 + I_2$$

If $V$ be the potential difference across $B_1$ and $B_2$, then

$$I_1 = \frac{\varepsilon_1 - V}{r_1} \Rightarrow I_2 = \frac{\varepsilon_2 - V}{r_2}$$

Combining these three equations, we get

$$I = I_1 + I_2 = \frac{\varepsilon_1 - V}{r_1} + \frac{\varepsilon_2 - V}{r_2}$$

$$= \left(\frac{\varepsilon_1}{r_1} + \frac{\varepsilon_2}{r_2}\right) - V\left(\frac{1}{r_1} + \frac{1}{r_2}\right)$$

Hence, $V = \dfrac{\varepsilon_1 r_2 + \varepsilon_2 r_1}{r_1 + r_2} - I\,\dfrac{r_1 r_2}{r_1 + r_2}$

If we want to replace the combination by a single cell, between $B_1$ and $B_2$ of emf $\varepsilon_{eq}$ and internal resistance $r_{eq}$, we would have, $V = \varepsilon_{eq} - I\, r_{eq}$.

**72** (*b*) Given that, the resultant voltage across the battery terminal = 1.5 V

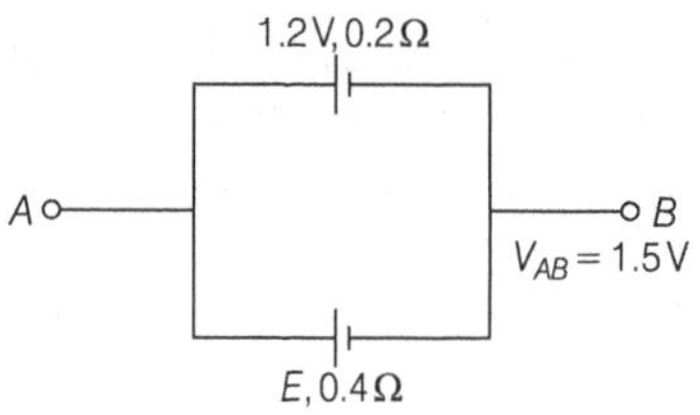


For circuit having two cells of emf $E_1$ and $E_2$ with internal resistances of $r_1$ and $r_2$ in parallel, the equivalent emf of battery is given by

$$E_{eq} = \frac{E_1/r_1 + E_2/r_2}{1/r_1 + 1/r_2} = \frac{\frac{1.2}{0.2} + \frac{E}{0.4}}{\frac{1}{0.2} + \frac{1}{0.4}} = 1.5$$

$\Rightarrow \quad 2.4 + E = 1.5 \times 3$

Solving this, we get $E = 2.1$ V.

**73** (*c*) Yes, when an identical battery is connected in opposition, i.e. in opposite order to a battery as shown below, then the potential difference between its two ends $A$ and $B$ is zero ($V_{net} = E - E = 0$)

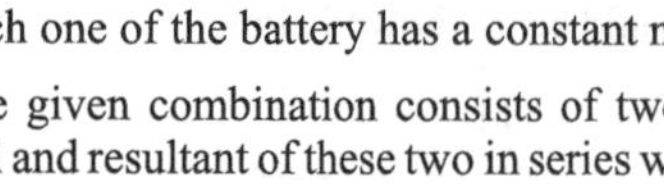


But each one of the battery has a constant non-zero emf.

**74** (*d*) The given combination consists of two batteries in parallel and resultant of these two in series with third one.

For parallel combination, we can apply

$$E_{eq} = \frac{\frac{E_1}{r_1} + \frac{E_2}{r_2}}{\frac{1}{r_1} + \frac{1}{r_2}} = \frac{-\frac{10}{2} + \frac{4}{2}}{\frac{1}{2} + \frac{1}{2}} = -\frac{6}{2} = -3 \text{ V}$$

Further, $\quad \dfrac{1}{r'_{eq}} = \dfrac{1}{r_1} + \dfrac{1}{r_2} = \dfrac{1}{2} + \dfrac{1}{2} = 1$

$\Rightarrow \quad r'_{eq} = 1\,\Omega$

Now, this is in series with third one

i.e.,

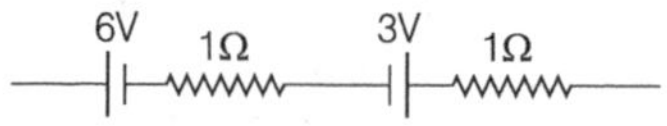


$\Rightarrow r_{eq} = 1 + 1 = 2\Omega$

and $\quad E_{eq} = 6 - 3 = 3$ V

**75** (*c*) When $n$-equal resistors of resistance $R$ are connected in series, then the current drawn is given as

$$I = \frac{E}{nR + r}$$

where, $nR$ = equivalent resistance of $n$ resistors in series

and $\quad r$ = internal resistance of battery.

Given, $\quad r = R$

$$\Rightarrow \quad I = \frac{E}{nR + R} = \frac{E}{R(n+1)} \quad \ldots\text{(i)}$$

Similarly, when $n$ equal resistors are connected in parallel, then the current drawn is given as

$$I' = \frac{E}{\frac{R}{n} + R}$$

where, $\dfrac{R}{n}$ = equivalent resistance of $n$-resistors in parallel.

Given, $\quad I' = 10I$

$$\Rightarrow \quad 10I = \frac{E}{\frac{R}{n} + R} = \frac{nE}{(n+1)R} \quad \ldots\text{(ii)}$$

Substituting the value of $I$ from Eq. (i) in Eq. (ii), we get

$$10\left\{\frac{E}{R(n+1)}\right\} = \frac{nE}{R(n+1)} \Rightarrow n = 10$$

**76** (*d*) When internal resistance of cells are high, then to draw maximum current, cells should be combined in parallel.

When internal resistance of cells are low, then to draw maximum current, cells should be connected in series.

To draw maximum current in mixed combination (grouping) of cells, external resistance (load) of the circuit should be equal to the total internal resistance of cells.

Hence, to draw maximum current, combination of cells depends upon the relative values of internal and external resistances.

**77** (*d*) Current drawn from cell, $i = \dfrac{E}{r + R}$

As, $\quad P = i^2 R$

$$\Rightarrow \quad P = \frac{E^2 R}{(r + R)^2}$$

Power is maximum, when $r = R$

$$\therefore \quad P_{max} = \frac{E^2}{4R} \text{ or } P_{max} = \frac{E^2}{4r}$$

**78** (*a*) Total number of cells $= mn = 20$ ...(i)

For maximum current, $R = nr/m$

$$\Rightarrow \quad 2.5 = n \times 0.5/m \text{ or } n = 5m$$

From Eq. (i), we get

$$m \times 5m = 20 \text{ or } m^2 = 4 \Rightarrow m = 2$$

Therefore, $n = 5 \times 2 = 10$

**80** (*b*) According to Kirchhoff's first law,

$$(5\text{A}) + (4\text{A}) + (-3\text{A}) + (-5\text{A}) + I = 0 \Rightarrow I = -1\text{ A}$$

So, current in fifth conductor is 1A flowing away from *x*.

**82** (*a*) Given circuit with currents as shown in the figure below, [In the question $I_1 = 0.3$ A is given, due to it we change the direction of $I_1$, in this figure]

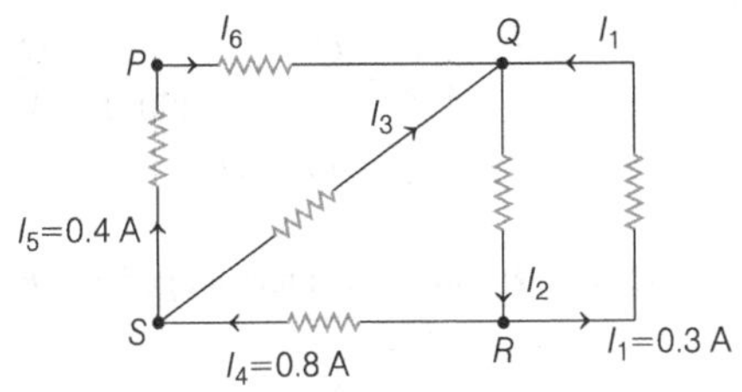


From Kirchoff's junction rule, $\Sigma I = 0$

At junction $S$, $I_4 = I_5 + I_3$

$$\Rightarrow \quad 0.8 = 0.4 + I_3$$

$$\Rightarrow \quad I_3 = 0.4\text{ A}$$

At junction $P$, $I_5 = I_6$

$$\Rightarrow \quad I_6 = 0.4\text{ A}$$

At junction $Q$, $I_2 = I_1 + I_3 + I_6 = 0.3 + 0.4 + 0.4 = 1.1\text{A}$

**83** (*d*) Applying Kirchhoff's law at point $D$ as shown, we get

$$i_1 = i_2 + i_3$$

$$\frac{V_A - V_D}{10} = \frac{V_D - V_B}{20} + \frac{V_D - V_C}{30}$$

$$\Rightarrow \quad 70 - V_D = \frac{V_D}{2} + \frac{V_D - 10}{3}$$

$$\Rightarrow \quad V_D = 40\text{ V}$$

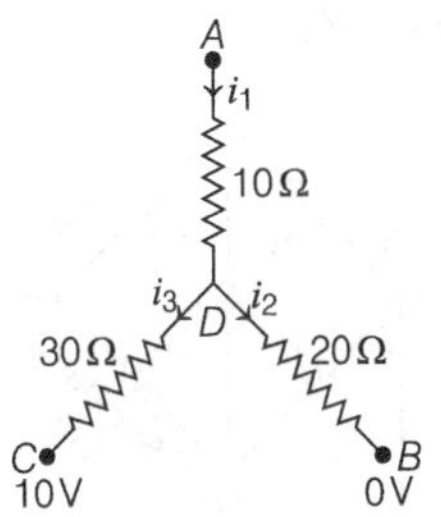


$$i_1 = \frac{V_A - V_D}{10} = \frac{70 - 40}{10} = 3\text{A},$$

$$i_2 = \frac{V_D - V_B}{20} = \frac{40 - 0}{20} = 2\text{A}$$

and

$$i_3 = \frac{V_D - V_C}{30} = \frac{40 - 10}{30} = 1\text{A}$$

Hence, the currents in the paths *AD*, *DB* and *DC* are in the ratio of 3 : 2 : 1.

**84** (*b*) The distribution of current is shown in figure

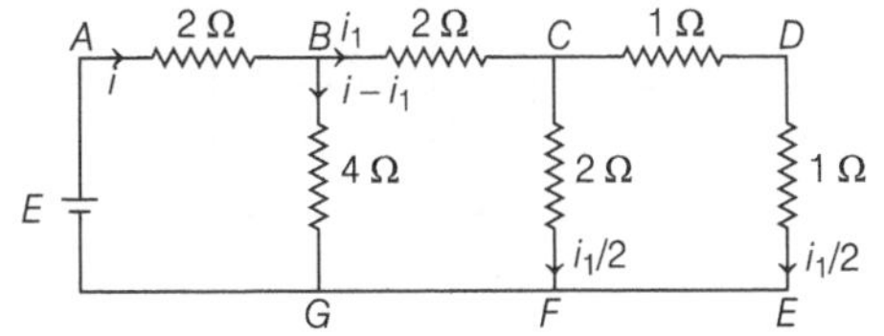


As per question, $\frac{i_1}{2} = 1\text{A}$ (given)

or $\quad i_1 = 2\text{A}$

In a closed circuit *BCFGB*,

$$2i_1 + 2 \times \frac{i_1}{2} - 4(i - i_1) = 0$$

$$\Rightarrow \quad 4 + 2 = 4(i - 2)$$

$$\Rightarrow \quad \frac{3}{2} + 2 = i$$

$$\Rightarrow \quad i = \frac{7}{2}\text{ A}$$

Again, in the closed circuit *ABGA*,

$$2i + 4(i - i_1) = E$$

$$\Rightarrow \quad E = 2 \times \frac{7}{2} + 4\left(\frac{7}{2} - 2\right) = 13\text{ V}$$

**85** (*a*) In a closed loop, say $ABCC'EA$, apply Kirchhoff's second rule,

$$-IR - \left(\frac{I}{2}\right)R - IR + \varepsilon = 0 \quad ...(i)$$

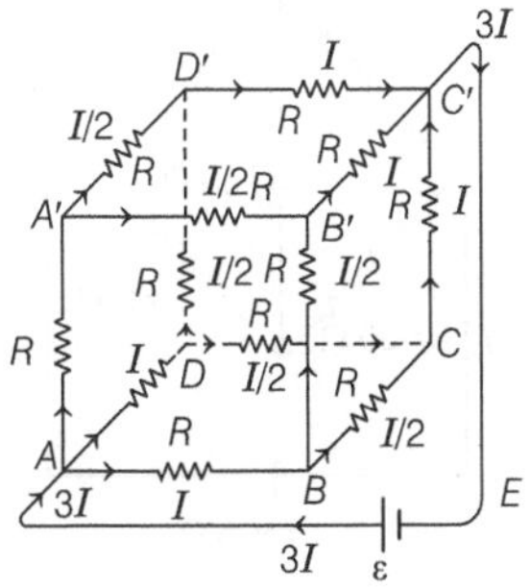


where, $R$ is the resistance of each edge and $\varepsilon$ the emf of battery.

Thus, $\quad \varepsilon = \frac{5}{2}IR$ [from Eq. (i)]

The equivalent resistance $R_{eq}$ of the network is

$$R_{eq} = \frac{\varepsilon}{3I} = \frac{(5/2)IR}{3I} = \frac{5}{6}R$$

The effective resistance between two diagonally opposite ends $= 5R/6$.

**87** (*a*) The given circuit satisfies the Wheatstone bridge condition, so no current is flow in the branch $BC$ and it behaves like an open circuit.

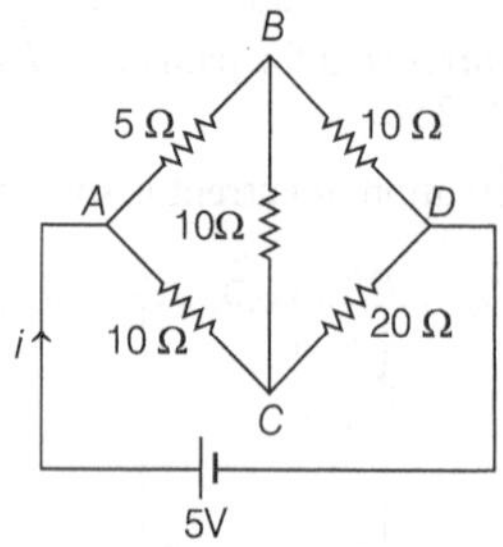


So, circuit becomes as shown below

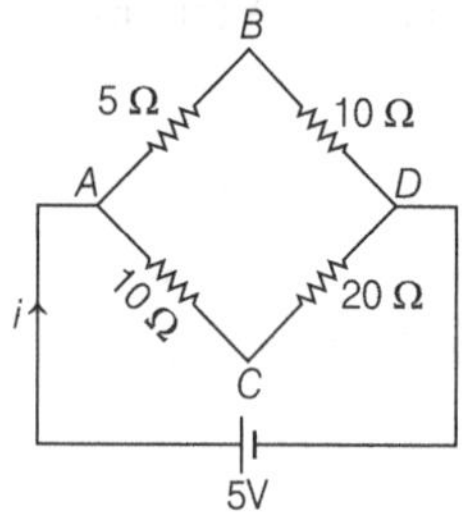


Now, in the above circuit,

$$R_{ABD} = 5 + 10 = 15\,\Omega$$
$$R_{ACD} = 10 + 20 = 30\,\Omega$$

Resistances $R_{ABD}$ and $R_{ACD}$ are in parallel, so

$$R_{\text{net}} = \frac{15 \times 30}{15 + 30} = \frac{450}{45} = 10\,\Omega$$

$\therefore$ Current, $i = \dfrac{E}{R_{\text{net}}} = \dfrac{5}{10} = 0.5\text{ A}$

**88** (*b*) Effective resistance of the network shown below

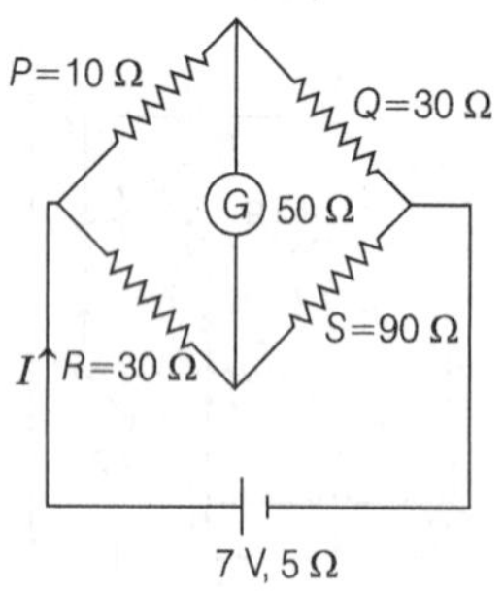


$$\therefore \quad R_{\text{eff}} = \frac{(P+Q)(R+S)}{(P+Q)+(R+S)}$$
$$= \frac{40 \times 120}{120 + 40} = \frac{4800}{160} = 30\,\Omega$$

$\therefore$ Current, $I = \dfrac{7}{30+5}$ $\quad \left(\because I = \dfrac{E}{R+r}\right)$

$$= \frac{7}{35} = 0.2\text{ A}$$

**89** (*d*) The given network can be drawn as a balanced Wheatstone bridge as shown below

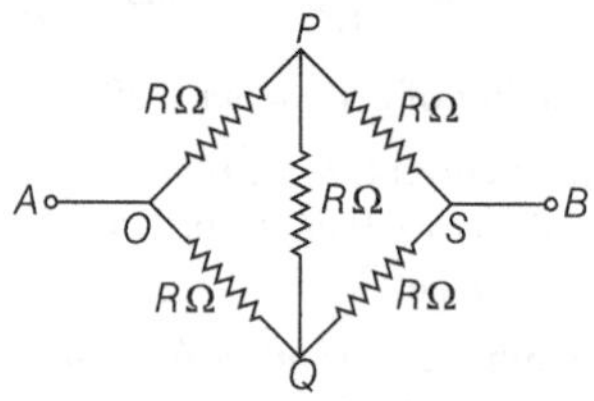


Since, the resistance between $P$ and $Q$ is ineffective at balanced condition.

$\therefore$ Equivalent resistance $= \dfrac{2R \times 2R}{2R + 2R} = R\Omega$

**90** (*c*) For a balanced Wheatstone bridge, $\dfrac{P}{R} = \dfrac{Q}{X}$

In first case when $R = 400\,\Omega$, the balancing equation will be

$$\frac{P}{R} = \frac{Q}{X} \Rightarrow \frac{P}{400\,\Omega} = \frac{Q}{X}$$

$$\Rightarrow \qquad P = \frac{400 \times Q}{X} \qquad \text{...(i)}$$

In second case, $P$ and $Q$ are interchanged and $R = 405\,\Omega$

$$\therefore \qquad \frac{Q}{R} = \frac{P}{X}$$
$$\Rightarrow \qquad \frac{Q}{405} = \frac{P}{X} \qquad \text{...(ii)}$$

Substituting the value of $P$ from Eq. (i) in Eq. (ii), we get

$$\frac{Q}{405} = \frac{Q \times 400}{X^2}$$
$$\Rightarrow \qquad X^2 = 400 \times 405$$
$$\Rightarrow \qquad X = \sqrt{400 \times 405} = 402.5$$

The value of $X$ is close to $402.5\,\Omega$.

**91** (*c*) Given, for first student,

$$R_2 = 10\,\Omega,\ R_1 = 5\,\Omega \text{ and } R_3 = 5\,\Omega$$

For second student, $R_1 = 500\,\Omega$, $R_3 = 5\,\Omega$ and $R_2 = 1000\Omega$

Now, according to Wheatstone bridge rule,

$$\frac{R_2}{R_4} = \frac{R_1}{R_3} \Rightarrow R_4 = R_3 \times \frac{R_2}{R_1} \qquad \text{...(i)}$$

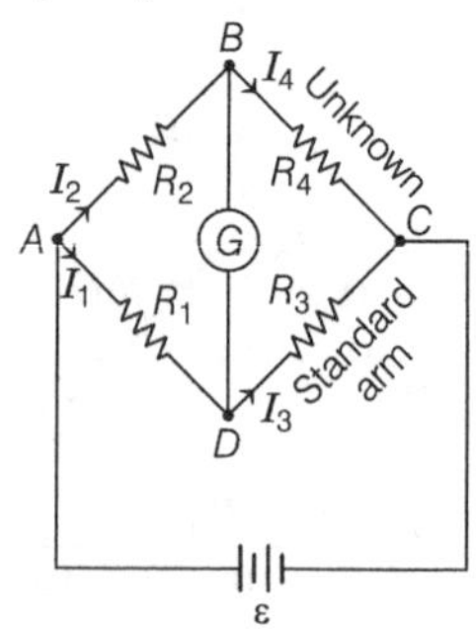

Now, putting all the values in Eq. (i), we get $R_4 = 10\,\Omega$, for both students. Thus, we can analyse that the Wheatstone bridge is most sensitive and accurate, if resistances are of same value.

Thus, the errors of measurement of the two students depend on the accuracy and sensitivity of the bridge, which in turn depends on the accuracy with which $R_2$ and $R_1$ can be measured.

When $R_2$ and $R_1$ are larger, the currents through the arms of bridge is very weak. This can make the determination of null point accurately more difficult.

**92** (*b*) Considering the mesh *BADB*, we have

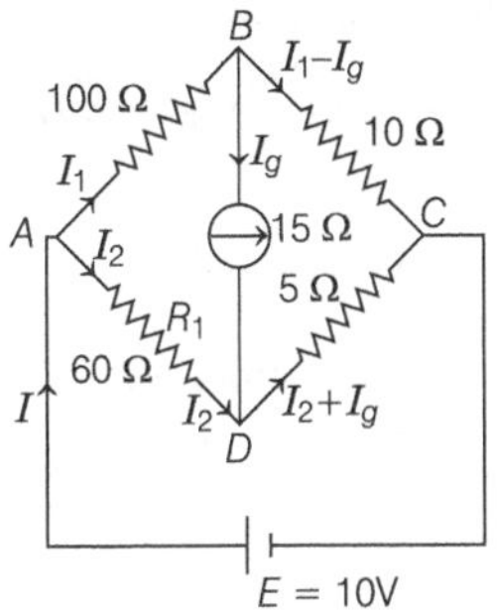


$$100\,I_1 + 15\,I_g - 60\,I_2 = 0$$

$$\Rightarrow \quad 20\,I_1 + 3I_g - 12\,I_2 = 0 \quad \text{...(i)}$$

Considering the mesh *BCDB*, we have

$$10\,(I_1 - I_g) - 15I_g - 5\,(I_2 + I_g) = 0$$

$$\Rightarrow \quad 10\,I_1 - 30\,I_g - 5I_2 = 0$$

$$\Rightarrow \quad 2I_1 - 6I_g - I_2 = 0 \quad \text{...(ii)}$$

Considering the mesh *ADCEA*, we have

$$60I_2 + 5\,(I_2 + I_g) = 10$$

$$\Rightarrow \quad 65\,I_2 + 5I_g = 10$$

$$\Rightarrow \quad 13I_2 + I_g = 2 \quad \text{...(iii)}$$

On multiplying Eq. (ii) by 10, we get

$$20\,I_1 - 60\,I_g - 10I_2 = 0 \quad \text{...(iv)}$$

Subtracting Eq. (iv) from Eq. (i), we get

$$63I_g - 2I_2 = 0$$

$$\Rightarrow \quad I_2 = 31.5\,I_g$$

On substituting the value of $I_2$ into Eq. (iii), we get

$$13\,(31.5\,I_g) + I_g = 2$$

$$\Rightarrow \quad 410.5\,I_g = 2$$

$$\Rightarrow \quad I_g = 4.87 \text{ mA}$$

**93** (*d*) For balanced position in a meter bridge.

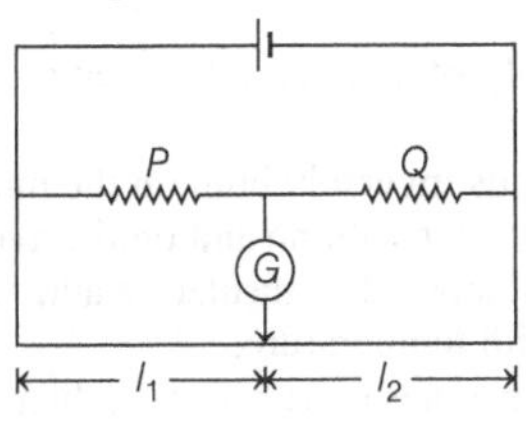


$$\frac{P}{Q} = \frac{l_1}{l_2}$$

Now, if position of *G* and cell is interchanged,

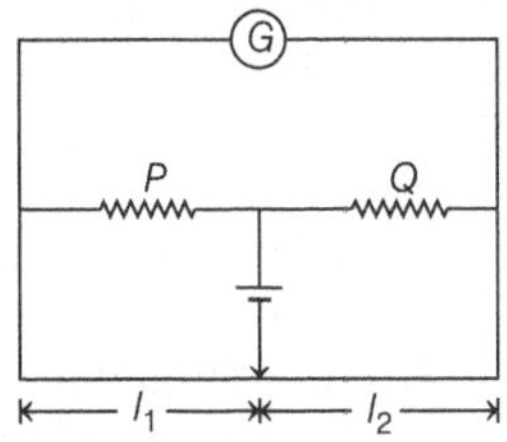


The balance condition still remains the same, if the jockey points at the same point as given in the initial condition, for which there is no deflection in the galvanometer or no current will be drawn from the cell. Thus, the bridge will work as usual and balance condition remains same, i.e. $P/Q = l_1/l_2$.

**94** (*b*) For first case, $\dfrac{5}{l_1} = \dfrac{R}{(100 - l_1)}$ ...(i)

Now, by shunting resistance *R* with an equal resistance *R*, new resistance in that arm become $R/2$, so

$$\frac{5}{1.6l_1} = \frac{R/2}{(100 - 1.6l_1)} \quad \text{...(ii)}$$

From Eqs. (i) and (ii), we get

$$\frac{1.6}{1} = \frac{(100 - 1.6l_1)}{100 - l_1} \times 2$$

$$\Rightarrow \quad 160 - 1.6l_1 = 200 - 3.2l_1$$

$$\Rightarrow \quad 1.6\,l_1 = 40$$

$$l_1 = \frac{400}{16} = 25 \text{ m}$$

From Eq. (i), $\dfrac{5}{25} = \dfrac{R}{75} \Rightarrow R = 15\Omega$

**96** (*b*) The working of potentiometer is based on the fact that the fall of potential across any portion of the wire is directly proportional to the length of that portion, provided the wire is of uniform area of cross-section and a constant current is flowing through it.

To shift the balance point on higher length, the potential gradient of the wire is to be decreased. The same can be obtained by decreasing the current of the main circuit, which is possible by increasing the resistance in series with potentiometer wire, i.e. in main circuit.

**97** (*c*) When a cell is balanced against potential drop across a certain length of potentiometer wire, no current flows through the cell.

$\therefore$ emf of cell = potential drop across balance length of potentiometer wire.

So, potentiometer is a more accurate device for measuring emf of a cell as no current flows through the galvanometer during measurement of emf.

**99** (*a*) In case of potentiometer, $E_1/E_2 = l_1/l_2$. As given that $E_1 > E_2$, therefore $l_1 > l_2$. Therefore, the null point for the cell of emf $E_2$ must be at shortest length than that of cell $E_1$.

Thus, the null point on potentiometer wire should shift towards left of *C*.

**100** (*c*) According to question, emf of the cell is directly proportional to the balancing length.

i.e. $E \propto l$ ...(i)

Now, in the first case, cells are connected in series to support one another, i.e.

$$\text{Net emf} = E_1 + E_2$$

From Eq. (i), we get

$$E_1 + E_2 = 50 \text{ (given)} \quad \text{...(ii)}$$

Again, cells are connected in series in opposite direction, i.e.

$$\text{Net emf} = E_1 - E_2$$

From Eq. (i), $E_1 - E_2 = 10$ ...(iii)

From Eqs. (ii) and (iii), we get

$$\frac{E_1 + E_2}{E_1 - E_2} = \frac{50}{10}$$

$$\Rightarrow \quad \frac{E_1}{E_2} = \frac{5+1}{5-1} = \frac{6}{4} = \frac{3}{2}$$

Thus, the ratio of emf is 3 : 2.

**101** (*b*) Consider a potentiometer wire of length $L$ and a resistance $r$ are connected in series with a battery of emf $E_0$ and a resistance $r_1$ as shown in figure.

Current in wire $AB$, $I = \dfrac{E_0}{r_1 + r}$

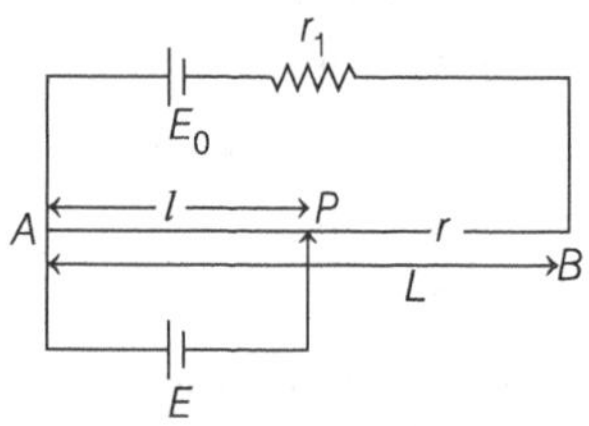


Potential gradient, $x = \dfrac{Ir}{L} = \left[\dfrac{E_0}{r_1 + r}\right]\dfrac{r}{L}$

and emf $E$ will be given by

$$E = x \cdot l = \left[\frac{E_0 r}{r_1 + r}\right]\frac{l}{L}$$

**102** (*c*) For a potentiometer, the internal resistance $r$ is given by $r = R\left(\dfrac{l_1}{l_2} - 1\right)$

Given, $R = 10\,\Omega$, $l_1 = 75$ cm, $l_2 = 65$ cm

$$\therefore \quad r = 10\left(\frac{75}{65} - 1\right) = 10 \times 0.154 = 1.54\,\Omega$$

**103** (*a*) In potentiometer experiment, we find internal resistance of a cell.

Let $E$ be the emf of the cell and $V$ the terminal potential difference, then

$$\frac{E}{V} = \frac{l_1}{l_2}$$

where, $l_1$ and $l_2$ are lengths of potentiometer wire with and without short circuited through a resistance.

Given, $l_1 = 110$ cm and $l_2 = 100$ cm, $R = 10\Omega$

So, $\dfrac{E}{V} = \dfrac{R + r}{R}$ $[\because E = I(R + r) \text{ and } V = IR]$

$$\Rightarrow \quad \frac{R + r}{R} = \frac{l_1}{l_2}$$

or $$1 + \frac{r}{R} = \frac{110}{100}$$

or $$\frac{r}{R} = \frac{10}{100} = \frac{1}{10}$$

or $$r = \frac{R}{10} = \frac{1}{10} \times 10 = 1.0\,\Omega$$

**104** (*a*) The circuit can be drawn as

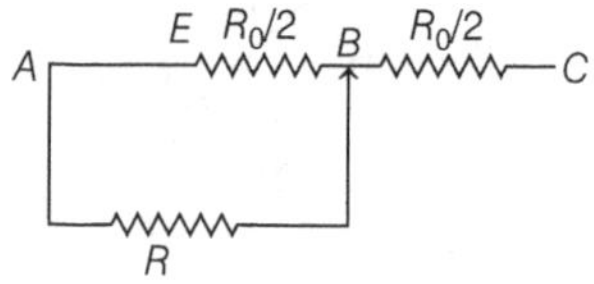


Equivalent resistance across $AB$,

$$\frac{1}{R_1} = \frac{1}{R} + \frac{1}{(R_0/2)}$$

$$\Rightarrow \quad R_1 = \frac{R_0 R}{R_0 + 2R} \quad \text{...(i)}$$

The total resistance between $A$ & $C$ will be the sum of resistances between $A$ & $B$ and $B$ & $C$, i.e. $R_1 + R_0/2$

So, the current flowing through the potentiometer will be

$$I = \frac{V}{R_1 + R_0/2} = \frac{2V}{2R_1 + R_0}$$

The voltage $V_1$ taken from the potentiometer will be the product of current $I$ and resistance $R_1$,

$$V_1 = IR_1 = \left(\frac{2V}{2R_1 + R_0}\right) \times R_1$$

$$= \frac{2V}{2\left(\dfrac{R_0 R}{R_0 + 2R}\right) + R_0} \times \frac{R_0 R}{R_0 + 2R}$$

$$\Rightarrow \quad V_1 = \frac{2VR}{R_0 + 4R} \quad \text{[using Eq. (i)]}$$

**105** (*b*) In electrostatics, all charges whether free or bound are considered to be at rest.

But net charges in motion constitute an electric current.

Therefore, Assertion and Reason are correct but Reason is not the correct explanation of Assertion.

**106** (*b*) Resistivity of a material is given by $\rho = \dfrac{1}{\sigma} = \dfrac{m}{ne^2\tau}$

$\rho$ thus depends inversely both on the number $n$ of free electrons per unit volume and on the average time $\tau$ between collisions. For insulators and semiconductors, $n$ increases with temperature.

This increase of $n$ is more effective than any decrease in $\tau$, so that for such materials $\rho$ decreases with temperature.

Therefore, Assertion and Reason are correct but Reason is not the correct explanation of Assertion.

**108** (*a*) With increase in temperature, average speed of the electrons, which acts as the carriers of current increases, resulting in more frequent collisions.

Thus, the average time of collisions $\tau$ decreases with increasing temperature.

Therefore, Assertion and Reason are correct and Reason is the correct explanation of Assertion.

**109** (*a*) Charge carriers do not move with acceleration but with a steady drift velocity.

This is because of the collisions with ions and atoms during transit.

Therefore, Assertion and Reason are correct and Reason is the correct explanation of Assertion.

**110** (*d*) Bending will not increase the resistance of the conducting wire.

Also, drift velocity of electron is independent of bending of conductor, i.e. it remains same.

Therefore, Assertion is incorrect but Reason is correct.

**111** (*b*) On increasing the temperature of a conductor, the kinetic energy of free electrons increases.

On account of this, they collide more frequently with each other (and with the ions of the conductor) and consequently their drift velocity decreases.

So, on increasing temperature, conductivity of metallic wire decreases.

Therefore, Assertion and Reason are correct but Reason is not the correct explanation of Assertion.

**112** (*a*) Manganin and constantan have very low temperature coefficient of resistance. So, their resistance values change very little with temperature.

This is the reason, they are widely used in standard resistors.

Therefore, Assertion and Reason are correct and Reason is the correct explanation of Assertion.

**114** (*b*) The statements II and III are correct but I is incorrect and it can be corrected as,

The resistance $R$ of a conductor is directly proportional to length $l$ and inversely proportional to cross-sectional area $A$ and given by relation,

$$R = \rho \cdot \frac{l}{A}$$

**115** (*c*) The statements I and II are correct but III is incorrect and it can be corrected as,

The smaller ice particles in the clouds acquires positive charge and the larger ones negative charge.

**116** (*a*) The statement given in option (a) is correct but rest are incorrect and these can be corrected as,

Ohm's law is valid for metallic conductors at low temperature but not valid at high temperature.

When current passes through the electrolyte, then it violates Ohm's law.

**117** (*a*) In a balanced Wheatstone bridge, there is no effect on position of null point, if we exchange the battery and galvanometer. So, option (a) is incorrect.

**118** (*b*) As, power of a bulb, $P = \dfrac{V^2}{R}$

$$\Rightarrow \quad R = \frac{V^2}{P} \text{ or } R \propto V^2$$

i.e. $$\frac{R_1}{R_2} = \left(\frac{200}{300}\right)^2 = \frac{4}{9}$$

When connected in series, potential drop is in the ratio of their resistance.

So, $$\frac{V_1}{V_2} = \frac{R_1}{R_2} = \frac{4}{9}$$

Now, $$P = I^2 R$$

or $$P \propto R \quad \text{(in series } I \text{ is same)}$$

$$\Rightarrow \quad \frac{P_1}{P_2} = \frac{R_1}{R_2} = \frac{4}{9}$$

$\therefore$ The statement given in option (b) is correct.

**119** (*c*)

Since, current passes through the least resistive path. So, in the given circuit we can say that current can follow the path *PBAR*.

Therefore, no current will flow through $4\Omega$ and $6\Omega$ resistances.

So, potential difference across the $4\ \Omega$ resistance = potential difference across the $6\Omega$ resistance.

Current passing through the battery is

$$I = (20/2) = 10\text{ A}$$

But there is a current passing in wire $QS$ from $S$ to $Q$.

$\therefore$ Power supplied by the battery,

$$P = EI = (20)(10) = 200\text{ W}$$

$\therefore$ The statement given in option (c) is correct.

**122** (*c*)

A. The relation of resistivity of a metal (like copper) with temperature is $\rho_T = \rho_0[1 + \alpha(T - T_0)]$

It implies that a graph of $\rho_T$ plotted against $T$ would be a straight line. At temperature much lower than 0° C, the graph, however deviates considerably from a straight line as shown.

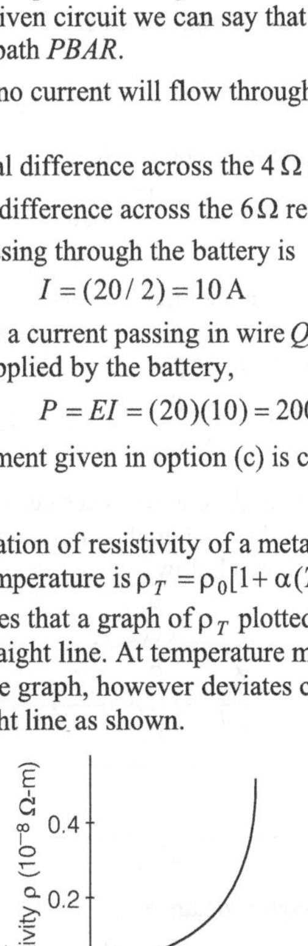

B. Some materials like nichrome, exhibit a very weak dependence of resistivity with temperature as shown

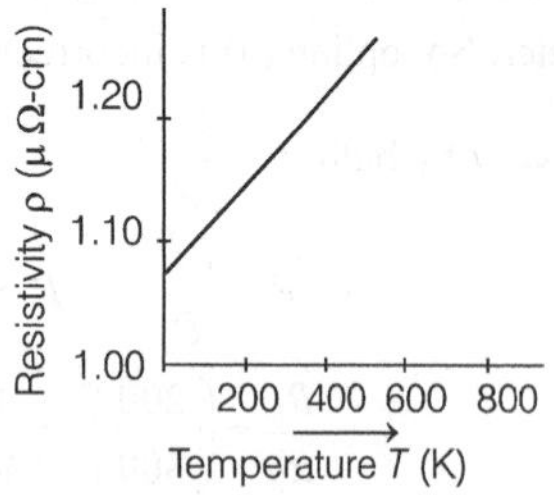


C. Unlike metals, the resistivites of semiconductors decreases with increasing temperatures.Its dependence is as shown

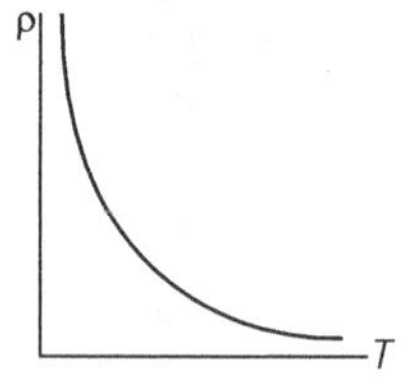


Hence, A → 2, B → 3 and C → 1.

**123** (*c*) Given $E = 12\text{ V}$ and $r = 0.4\,\Omega$

Current drawn from the battery, $I = \dfrac{E}{R + r}$

In case of maximum current, $R = 0$

$$\therefore \quad I_{\max} = \frac{E}{r} = \frac{12}{0.4} = 30\text{ A}$$

**124** (*b*) The current in the circuit, $I = \dfrac{E}{R + r}$

Given, $E = 10\text{ V}$, $I = 0.5\text{ A}$ and $r = 3\Omega$

$$\therefore \quad 0.5 = \frac{10}{R + 3} \Rightarrow R = 17\Omega$$

**125** *(a)* The potential drop across each resistor is same as they are in parallel and applied potential difference, $V = 20$ V as shown below

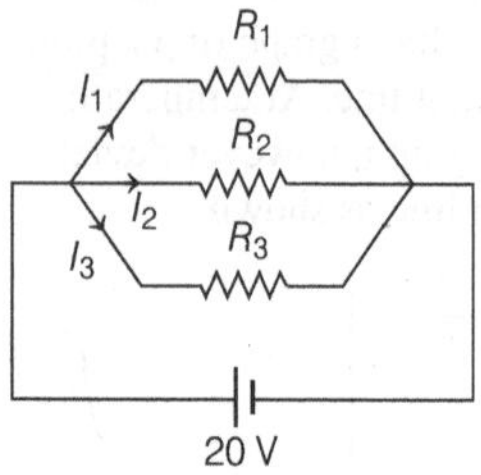


Current through resistance $R_1$,

$$I_1 = \frac{V}{R_1} = \frac{20}{2} = 10\text{ A} \qquad (\because R_1 = 2\Omega)$$

Current through resistance $R_2$,

$$I_2 = \frac{V}{R_2} = \frac{20}{4} = 5\text{ A} \qquad (\because R_2 = 4\Omega)$$

Current through resistance $R_3$,

$$I_3 = \frac{V}{R_3} = \frac{20}{5} = 4\text{ A} \qquad (\because R_3 = 5\Omega)$$

Total current drawn, $I = I_1 + I_2 + I_3$

$$= 10 + 5 + 4 = 19\text{ A}$$

**126** (*b*) The temperature coefficient of resistor,

$$\alpha = \frac{R_t - R_{27}}{R_{27}(t - 27)}$$

Given, $R_{27} = 100\Omega$, $R_t = 117\Omega$ and $\alpha = 1.7 \times 10^{-4}$ /°C

$$1.70 \times 10^{-4} = \frac{117 - 100}{100\,(t - 27)}$$

or $$t - 27 = \frac{17}{100 \times 1.70 \times 10^{-4}}$$

or $$t = 1000 + 27 = 1027\,°\text{C}$$

**127** (*c*) Resistance of wire, $R = \rho \dfrac{l}{A}$

Given, $l = 15$ m, $A = 6.0 \times 10^{-7}\text{ m}^2$

and $R = 5.0\,\Omega$

$$\Rightarrow \text{Resistivity, } \rho = \frac{RA}{l} = \frac{5.0 \times 6.0 \times 10^{-7}}{15}$$

$$= 2 \times 10^{-7}\,\Omega\text{-m}$$

**128** (*a*) Let the temperature coefficient of silver be α, then

$$\alpha = \frac{R_{t_2} - R_{t_1}}{R_{t_1}(t_2 - t_1)}$$

$$\Rightarrow \quad \alpha = \frac{R_{100} - R_{27.5}}{R_{27.5}\,(100 - 27.5)}$$

$$= \frac{2.7 - 2.1}{2.1 \times 72.5}$$

$$\Rightarrow \quad \alpha = 0.0039/\,°\text{C}$$

**129** (*c*) Resistance at 27 °C,

$$R_{27\,°\text{C}} = \frac{V}{I_{27°\text{C}}} = \frac{230}{3.2} = \frac{2300}{32}\,\Omega$$

Resistance at *t*°C, $R_{t\,°\text{C}} = \dfrac{V}{I_{t\,°\text{C}}} = \dfrac{230}{2.8} = \dfrac{2300}{28}\,\Omega$

Temperature coefficient of resistance,

$$\alpha = \frac{R_t - R_{27}}{R_{27}(t - 27)}$$

$$\Rightarrow \quad 1.7 \times 10^{-4} = \frac{\frac{2300}{28} - \frac{2300}{32}}{\frac{2300}{32}(t - 32)}$$

or $$t - 27 = \frac{82.143 - 71.875}{71.875 \times 1.7 \times 10^{-4}}$$

$$= 840.347$$

or $$t = 840.3 + 27$$

$$= 867.3\,°\text{C}$$

$$\approx 867°\text{ C}$$

**130** (*b*) In the given circuit, distribution of current is as shown

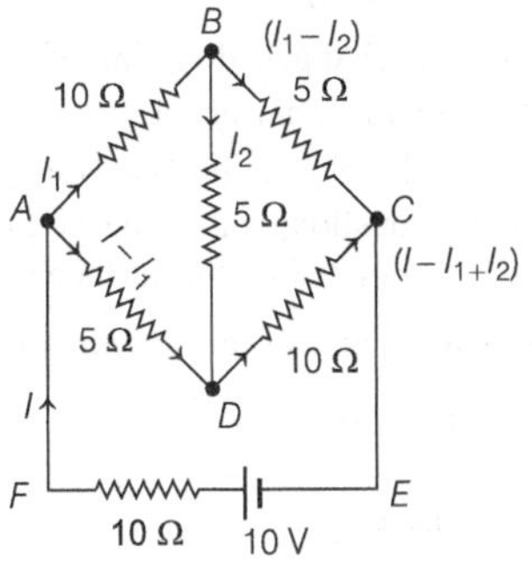


In loop *ABDA*, applying KVL,

$$10I_1 + 5I_2 - 5(I - I_1) = 0$$

$$\Rightarrow \quad 2I_1 + I_2 - I + I_1 = 0$$

$$\Rightarrow \quad 3I_1 + I_2 = I \quad \text{...(i)}$$

In loop *BCDB*, applying KVL,

$$5(I_1 - I_2) - 10(I - I_1 + I_2) - 5I_2 = 0$$

$$\Rightarrow \quad I_1 - I_2 - 2I + 2I_1 - 2I_2 - I_2 = 0$$

$$\Rightarrow \quad 3I_1 - 4I_2 = 2I \quad \text{...(ii)}$$

By solving the Eqs. (i) and (ii), we get

$$I_1 = \frac{2I}{5} \text{ and } I_2 = -\frac{I}{5} \quad \text{...(iii)}$$

In loop *ABCEFA*, applying KVL,

$$10 = 10I + 10I_1 + 5(I_1 - I_2)$$

$$\Rightarrow \quad 2 = 2I + 3I_1 - I_2 \quad \text{...(iv)}$$

Putting the values of $I_1$ and $I_2$ from Eq. (iii) in Eq. (iv), we get

$$2 = 2I + 3\left(\frac{2I}{5}\right) - \left(-\frac{I}{5}\right)$$

or $$2 = \frac{17}{5}I \text{ or } I = \frac{10}{17}\text{ A}$$

$\therefore$ Current drawn from source, $I = \frac{10}{17}$ A

**131** (*c*) In a meter bridge, the balance point is found to be at 39.5 cm from end *A* when the resistor *Y* is of 12.5Ω. Since, we know that in balancing condition,

$$\frac{X}{Y} = \frac{l}{100 - l}$$

$$\Rightarrow \quad X = \frac{l}{100 - l}Y$$

$$\Rightarrow \quad X = \frac{39.5 \times 12.5}{100 - 39.5}$$

$$= 8.16\,\Omega$$

The resistance of resistor *X* is 8.16 Ω.

If *X* and *Y* are interchanged, then the balance length will also interchanged. Thus, the balance length becomes $100 - 39.5 = 60.5$ cm.

**132** (*b*) Effective emf in the circuit,

$$E = 120 - 8 = 112\text{ V}$$

Current in circuit, $I = \frac{\text{Effective emf}}{\text{Total resistance}} = \frac{E}{r + R}$

Given, $r = 0.5\Omega$ and $R = 15.5\Omega$

$$\Rightarrow \quad I = \frac{112}{0.5 + 15.5} = 7\text{ A}$$

The battery of 8 V is being charged by 120 V, so the terminal potential across of 8 V will be greater than its emf or terminal potential difference, i.e. $V = E + Ir$ $= 8 + 7(0.5) = 11.5$ V

**133** (*a*) As we know that in case of potentiometer, the potential gradient remains constant.

i.e. $$E \propto l$$

$$\therefore \quad \frac{E_1}{E_2} = \frac{l_1}{l_2}$$

Given, $E_1 = 1.25$ V, $l_1 = 35$ cm and $l_2 = 63$ cm

$$\Rightarrow \quad \frac{1.25}{E} = \frac{35}{63}$$

$$\Rightarrow \quad E = \frac{1.25 \times 63}{35} = 2.25\text{ V}$$

**134** (*b*) Time taken by electron to drift from one end to another of the wire,

$$t = \frac{\text{Length of the wire}}{\text{Drift velocity}} = \frac{l}{v_d} \quad \text{...(i)}$$

Current in wire, $I = ne\,A\,v_d$

$$\Rightarrow \quad v_d = \frac{I}{ne\,A} \quad \text{...(ii)}$$

$$\Rightarrow \quad t = \frac{l\,ne\,A}{I}$$

Given, $n = 8.5 \times 10^{28}\text{ m}^{-3}$, $l = 3$ m, $A = 2 \times 10^{-6}\text{ m}^2$

and $I = 3$ A

$$= \frac{3.0 \times 8.5 \times 10^{28} \times 1.6 \times 10^{-19} \times 2.0 \times 10^{-6}}{3}$$

or $$t = 2.72 \times 10^4\text{ s}$$

$$= 7\text{ h } 33\text{ min}$$

**135** (*c*) Let the effective resistance of the network is *x*. If one part of the network has resistance $(1\Omega, 1\Omega, 1\Omega)$ is separated as shown, the effective resistance remains *x* (as it is infinite network). Here, *x* and 1Ω are in parallel.

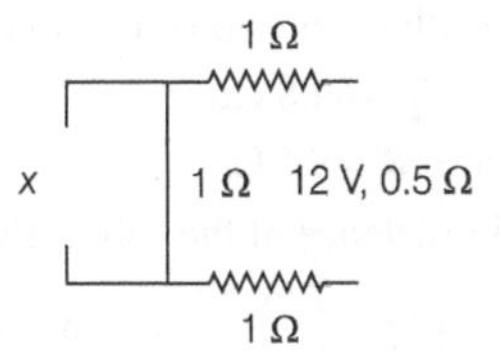


$$\therefore \quad \frac{1}{R_p} = \frac{1}{x} + \frac{1}{1} = \frac{1 + x}{x} \Rightarrow R_p = \frac{x}{1 + x}$$

Now, resistances $R_p$, $1\Omega$ and $1\Omega$ are in series. So, the resultant resistance,

$$R = R_p + 1 + 1 = \frac{x}{1+x} + 1 + 1$$

$$= \frac{x}{1+x} + 2 \quad \ldots\text{(i)}$$

In case of infinite resistances, the value of $R$ remains $x$.

$$\therefore \quad x = \frac{x}{1+x} + 2$$

$$\Rightarrow \quad x(x+1) = x + 2 + 2x$$

$$\Rightarrow \quad x^2 - 2x - 2 = 0$$

$$x = \frac{-(-2) \pm \sqrt{4+8}}{2}$$

$$= \frac{2 \pm \sqrt{12}}{2} = 1 \pm \sqrt{3}$$

The value of resistance cannot be negative, so the resistance of network,

$$x = 1 + \sqrt{3} = 1 + 1.732$$

$$x = 2.732\,\Omega$$

Total resistance of the circuit $= 2.732 + 0.5$

$$= 3.232\,\Omega$$

Current drawn from the supply,

$$I = \frac{V}{3.232} = \frac{12}{3.232} = 3.72\text{ A}$$

**136** (*c*) Given, $l_1 = 58.3$ cm, $l_2 = 68.5$ cm, $R = 10.0\,\Omega$, $X = ?$

$E_1$ and $E_2$ be the potential drops across $R$ and $X$, respectively.

Then, $$\frac{E_2}{E_1} = \frac{iX}{iR} = \frac{X}{R}$$

or $$X = \frac{E_2}{E_1} \cdot R \quad \ldots\text{(i)}$$

According to the principle of potentiometer,

$$\frac{E_2}{E_1} = \frac{l_2}{l_1} \quad \ldots\text{(ii)}$$

From Eqs. (i) and (ii), we get

$$X = \frac{l_2}{l_1} \cdot R = \frac{68.5}{58.3} \times 10 = 11.75\,\Omega$$

**137** (*b*) Given, balancing length when cell is in open-circuit,

$$l_1 = 76.3\text{ cm}$$

Balancing length when cell is in closed-circuit,

$$l_2 = 64.8\text{ cm}$$

and resistance, $R = 9.5\,\Omega$

The internal resistance of the cell is given by

$$r = \left(\frac{l_1}{l_2} - 1\right) R = \left(\frac{76.3}{64.8} - 1\right) \times 9.5 = 1.68\,\Omega$$

**138** (*b*) The current density is directed along $E$ and the relationship is given by

$$\mathbf{J} = \sigma \mathbf{E}$$

As direction of $E$ changes, direction of $J$ also changes but current $I$ remains unaffected.

Thus, the agent that is responsible for the given situation in the question, is the electric field produced by charges accumulated on the surface of wire.

**139** (*a*) The equivalent emf of this combination is given by

$$\varepsilon_{eq} = \frac{\varepsilon_2 r_1 + \varepsilon_1 r_2}{r_1 + r_2}$$

Also, $\varepsilon_2 > \varepsilon_1$

This suggest that the equivalent emf $\varepsilon_{eq}$ of the two cells is

$$\varepsilon_1 < \varepsilon_{eq} < \varepsilon_2$$

$\because$ $\varepsilon_{eq}$ is weighted mean of $\varepsilon_1$ and $\varepsilon_2$, so it will lie between $\varepsilon_1$ and $\varepsilon_2$.

**140** (*c*) The percentage error in $R$ can be minimised by adjusting the balance point near the middle of the bridge, i.e. when $l_1$ is close to 50 cm. This requires a suitable choice of $S$.

Since, $$\frac{R}{S} = \frac{l_1}{(100 - l_1)}$$

Since, here $R : S :: 2.9 : 97.1$, imply that the $S$ is nearly 33 times to that of $R$. In order to make this ratio 1:1, it is necessary to reduce the value of $S$ nearly (1/33) times, i.e. change the value of $S$ to 3 $\Omega$.

**141** (*b*) In a potentiometer experiment, the emf of a cell can be measured, if the potential drop along the potentiometer wire is more than the emf of the cell to be determined. Here, values of emfs of two cells are given as 5V and 10V, therefore the potential drop along the potentiometer wire must be more than 10V.

This is done by using a battery of voltage of 15V and adjusting $R$.

**142** (*a*) The resistance of wire is given by $R = \rho \dfrac{l}{A}$

For greater value of $R$, $l$ must be higher and $A$ should be lower and it is possible only when the battery is connected across 1 cm $\times \left(\frac{1}{2}\right)$ cm (area of cross-section $A$).

**143** (*a*) The relationship between current and drift speed is given by $I = neAv_d$

where, $I$ is the current and $v_d$ is the drift velocity.

So, $$I \propto v_d$$

Thus, only drift velocity determines the current in a conductor, as current is independent of thermal velocity.

CHAPTER > 04

# Moving Charges and Magnetism

## KEY NOTES

- Danish physicist Hans Oersted concluded that moving charges or currents produce a magnetic field in the surrounding space.
- The magnetic field of several sources is the vector addition of magnetic field of each individual source.

**Note** A current or a field (electric or magnetic) emerging out of the plane of the paper is depicted by a dot ⊙ and a current or a field going into the paper is depicted by a cross ⊗.

## Magnetic Force

- The magnetic force on a moving charge $q$ in a uniform magnetic field $\mathbf{B}$ is $\mathbf{F} = q\,[\mathbf{v} \times \mathbf{B}] = qvB\,\sin\theta\hat{\mathbf{n}}$, where $\theta$ is the angle between $\mathbf{v}$ and $\mathbf{B}$.
- Unit of magnetic field $\mathbf{B}$ is newton-second/coulomb-metre. This unit is called **tesla** (T).
- A smaller unit (non-SI) called gauss ($1\text{ G} = 10^{-4}\text{ T}$) is also often used.
- Magnetic force has following features
  - (i) It depends on $q$, $\mathbf{v}$ and $\mathbf{B}$ (charge of the particle, the velocity and the magnetic field). Force on a negative charge is opposite to that on a positive charge.
  - (ii) The magnetic force $q[\mathbf{v} \times \mathbf{B}]$ includes a vector product of velocity and magnetic field. The force acts in a (side ways) direction perpendicular to both the velocity and the magnetic field. Its direction is given by the screw rule or right-hand rule for vector product.
  - (iii) The magentic force is zero, if charge is not moving (as then $|\mathbf{v}| = 0$). Only a moving charge feels the magnetic force.

### Lorentz Force

- It is the force on an electric charge $q$ (moving with a velocity $v$ and located at $r$ at a given time $t$) due to both, electric and magnetic fields and can be written as

$$\mathbf{F} = q[\mathbf{E}(\mathbf{r}) + \mathbf{v} \times \mathbf{B}(\mathbf{r})]$$
$$= \mathbf{F}_{\text{electric}} + \mathbf{F}_{\text{magnetic}}$$

### Magnetic Force on a Current Carrying Conductor

- When the total number of mobile charge carriers in a current carrying conductor is $nlA$, then in the presence of an external magnetic field $\mathbf{B}$, the force on these carriers is

$$\mathbf{F} = I\,(\mathbf{l} \times B) \qquad [\because I = (nqv_d)A]$$

where, $\mathbf{l}$ is a vector of magnitude $l$, the length of the rod and with a direction identical to the current $I$.

- If the wire has an arbitrary shape, we can calculate the Lorentz force on it by considering it as a collection of linear strips $d\mathbf{l}_j$ as $\mathbf{F} = \sum_j I d\mathbf{l}_j \times \mathbf{B}$

## Motion in a Magnetic Field

- A moving charge particle $q$ in the presence of external magnetic field $\mathbf{B}$ experiences a perpendicular force $q[\mathbf{v} \times \mathbf{B}]$, which acts as a centripetal force and produces a circular motion perpendicular to the magnetic field.

- The particle will describe a circle, if **v** and **B** are perpendicular to each other. This motion has following features
  - (i) The radius of the circle described by the charged particle is
  $$r = \frac{mv}{qB} = \frac{p}{qB} = \frac{\sqrt{2mK}}{qB}$$
  where, $p$ = momentum of charged particle
  and $K$ = kinetic energy of charged particle.
  The larger the momentum, the larger is the radius and bigger the circle described.
  - (ii) Angular frequency $\omega$ of the particle is,
  $$\omega = \frac{qB}{m}$$
  which is independent of the velocity or energy.
  - (iii) The time taken by the particle to complete one revolution is $T = \frac{2\pi m}{qB}$.
- If velocity of the charged particle has a component along **B**, then it will make the particle to move along the field, thereby producing a helical motion. However, this component of velocity remains unchanged. As, the motion along the magnetic field will not be affected by the magnetic field.
- The distance moved along the magnetic field in one rotation is called **pitch** $P$.
$$P = v_{\|}T = \frac{2\pi m v_{\|}}{qB}$$
where, $v_{\|}$ = component of the velocity parallel to the magnetic field.
The radius of the circular component of motion is also called the **radius of the helix.**

## Motion in Combined Electric and Magnetic Fields

### Velocity Selector

- When values of **E** and **B** are equal, total force on the charge is zero and the charge will move in the fields undeflected, i.e.
$$qE = qvB$$
or
$$v = \frac{E}{B}$$
This condition can be used to select charged particles of a particular velocity out of a beam.
Thus, the crossed $E$ and $B$ fields, therefore serve as a velocity selector.

### Cyclotron

- It is a machine used to accelerate charged particle or ions to high energies.
- Important points related to cyclotron are as given below
  - (i) It uses both electric and magnetic fields in combination, to increase the energy of charged particles. As the fields are perpendicular to each other, they are called **crossed fields.**
  - (ii) It further uses the fact that the frequency of revolution of the charged particle in a magnetic field is independent of its energy.
  - (iii) A schematic view of the cyclotron is as shown below

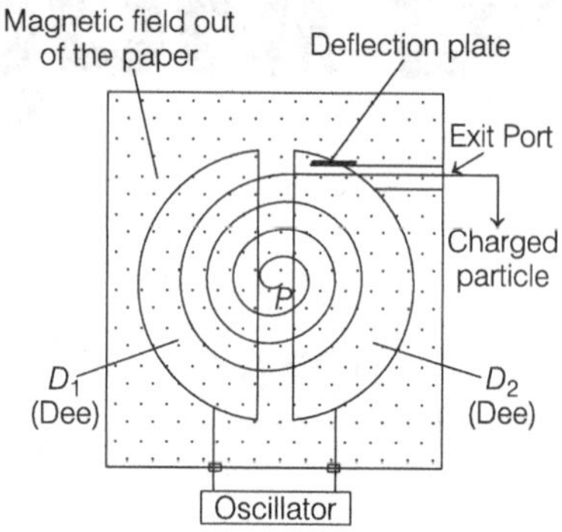


  - (iv) The particle is always accelerated by the electric field. Each time the acceleration increases, the energy of the particle increases and the radius of the circular path increases. So, the path is a spiral one.
  - (v) The period of revolution is given by
  $$T = \frac{1}{\nu_c} = \frac{2\pi m}{qB} \quad \text{or} \quad \nu_c = \frac{qB}{2\pi m}$$
  This frequency is called the **cyclotron frequency.**
  - (vi) The frequency $\nu_a$ of the applied voltage is adjusted, so that the polarity of the dees is reversed in the same time that it takes the ions to complete one-half of the revolution. The requirement $\nu_a = \nu_c$ is called the **resonance condition.**
  - (vii) The velocity of the charged particle or ions when it leave the system *via* an exit slit is
  $$v = \frac{qBR}{m}$$
  where, $R$ is the radius of the trajectory at exit and equals the radius of a dee.
  - (viii) The kinetic energy of the ions is
  $$E = \frac{1}{2}mv^2 = \frac{q^2B^2R^2}{2m}$$

KEY NOTES

(ix) The operation of the cyclotron is based on the fact that the time for one revolution of an ion is independent of its speed or radius of its orbit.

- The cyclotron has following uses
  - (i) It is used to bombard nuclei with energetic particles, so accelerated by it and study the resulting nuclear reactions.
  - (ii) It is used to implant ions into solids and modify their properties or even synthesise new materials.
  - (iii) It is used in hospitals to produce radioactive substances which can be used in diagnosis and treatment.

## Magnetic Field Due to a Current Element : Biot-Savart's Law

- According to this law, the magnitude of the magnetic field $d\mathbf{B}$ is proportional to the current $I$, the element length $|d\mathbf{l}|$ and inversely proportional to the square of the distance $r$.

  Its direction is perpendicular to the plane containing $d\mathbf{l}$ and $\mathbf{r}$.

  Thus, $$d\mathbf{B} = \frac{\mu_0}{4\pi} \frac{I d\mathbf{l} \times \mathbf{r}}{r^3}$$

  $$|d\mathbf{B}| = \frac{\mu_0}{4\pi} \frac{I dl \sin\theta}{r^2}$$

  where, $\frac{\mu_0}{4\pi} = 10^{-7}$ T-m/A and $\mu_0$ is the permeability of free space or vacuum.

### Magnetic Field on the Axis of a Circular Current Loop

- Magnetic field at any point $P$ along the **axis of circular current carrying loop** as shown in the figure below is

$$\mathbf{B} = \frac{\mu_0 I R^2}{2(x^2 + R^2)^{3/2}} \hat{\mathbf{i}}$$

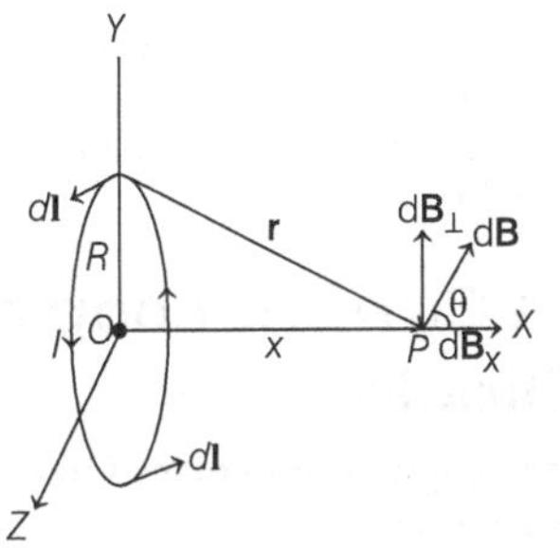


- Magnetic field at the **centre of a circular current carrying conductor/ coil**

$$\mathbf{B} = \frac{\mu_0 I}{2R} \hat{\mathbf{i}}$$

where, $R$ is the radius of a circular loop.

- The magnetic field lines due to a circular wire, form closed loops and the direction of the magnetic field is given by right-hand thumb rule.
- According to right-hand thumb rule, curl the palm of your right-hand around the circular wire with the fingers pointing in the direction of the current.

  The right-hand thumb gives the direction of the magnetic field.

### Magnetic Field due to a Long Straight Wire

- Magnetic field at point $P$ due to long straight wire having current $I$ is given by

$$B = \frac{\mu_0}{4\pi} \cdot \frac{I}{r} [\sin \phi_1 + \sin \phi_2]$$

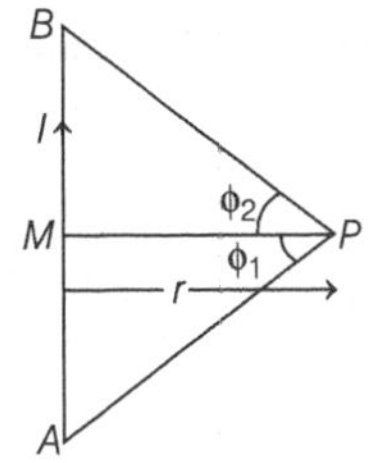


## Ampere's Circuital Law

- The line integral of the magnetic field $\mathbf{B}$ around any closed loop is equal to $\mu_0$ times the total current flowing through the loop.

  i.e. $$\oint \mathbf{B} \cdot d\mathbf{l} = \mu_0 I$$

- An Amperian loop is possible such that at each point of the loop, either
  - (i) $\mathbf{B}$ is tangential to the loop and is a non-zero constant $B$, or
  - (ii) $\mathbf{B}$ is normal to the loop, or
  - (iii) $\mathbf{B}$ vanishes.
- The magnetic field at a distance $r$ outside the wire is tangential and given by

$$B = \frac{\mu_0 I}{2\pi r}$$

- From the result, some important points can be derived,
  - (i) The magnetic field at every point on a circle of radius $r$ is same in magnitude. The magnetic field around a wire possesses cylindrical symmetry.
  - (ii) The field direction at any point on this circle is tangential to it. The lines of constant magnitude of magnetic field forms concentric circles. The circular lines are called **magnetic field lines.**
  - (iii) Even though the wire is of infinite length, the field due to it at a non-zero distance is not infinite.

(iv) The right-hand rule is used to determine the direction of magnetic field due to a long wire and is given as

"Grasp the wire in your right hand with your extended thumb pointing in the direction of the current. Your fingers will curl around in the direction of the magnetic field.

- The Ampere's circuital law holds for steady currents, which do not fluctuate with time.

## Solenoid

- It consists of a long wire wound in the form of a helix, where the neighbouring turns are closely spaced. The net magnetic field is the vector sum of the fields due to all the turns.
- **Magnetic field due to a straight solenoid** can be given as

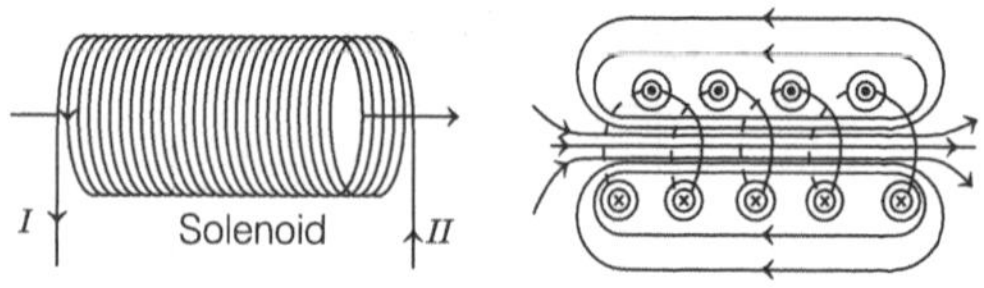


(i) At any point inside the solenoid,

$$B = \mu_0 nI$$

where, $n$ = number of turns per unit length.

(ii) At the ends of the solenoid,

$$B = \frac{1}{2}\mu_0 nI$$

(iii) The direction of the field is given by the right-hand rule. The solenoid is commonly used to obtain a uniform magnetic field.

## Toroid

- It is a hollow circular ring on which a large number of turns of a wire are closely wound.
- It can be viewed as a solenoid which has been bent into a circular shape to close on itself as shown below.

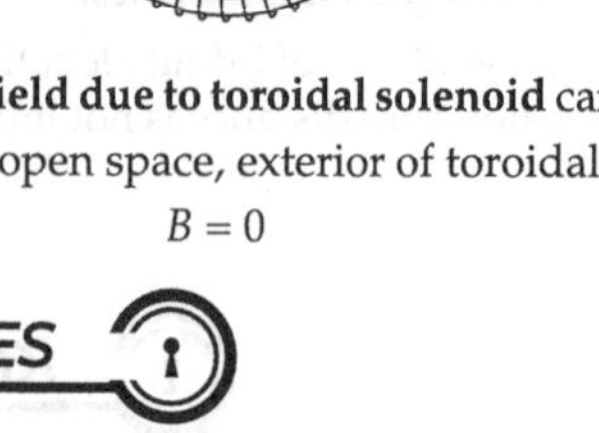


- **Magnetic field due to toroidal solenoid** can be given as

(i) In the open space, exterior of toroidal solenoid,

$$B = 0$$

(ii) Inside the toroidal solenoid,

$$B = \frac{\mu_0 NI}{2\pi r}$$

where, $N$ = total number of turns = $2\pi rn$.

Then, $B = \mu_0 nI$

## Force between Two Parallel Current Carrying Conductors

- If two infinitely long parallel conductors carrying the currents $I_a$ and $I_b$ respectively and are separated by a distance $d$, then magnetic force experienced by length $L$ of any one conductor due to the other current carrying conductor is $\mathbf{F}_{ba} = \frac{\mu_0}{2\pi} \cdot \frac{I_a I_b}{d} \cdot L$

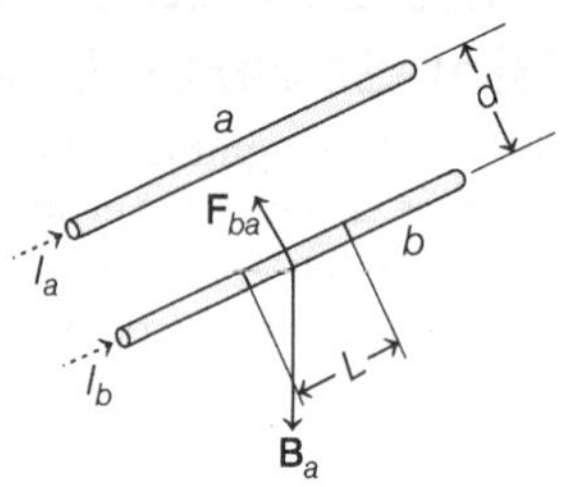


- The force $\mathbf{F}_{ba}$ per unit length, $f_{ba} = \frac{\mu_0 I_a I_b}{2\pi d}$
- The direction of force depends on the direction of current in them as

(i) attractive, if current flow in same direction.

(ii) repulsive, if current flow in opposite direction.

### Ampere

- One ampere is the current which flows through each of the two parallel uniform long linear conductors, which are placed in free space at a distance of 1 m from each other and attract or repel each other with a force of $2 \times 10^{-7}$ N/m.
- The SI unit of charge, the Coulomb can be defined in terms of ampere, when a steady current of 1 A is setup in conductor, the quantity of charge that flows through its cross-section in 1 s is one coulomb (1 C).

## Torque on a Current Loop and Magnetic Dipole

- The **torque on a rectangular loop placed in a uniform magnetic field** is given as

$$\tau = IAB \sin\theta$$

where, $\theta$ = angle between plane of coil & magnetic field lines and $A = ab$ is the area of the rectangular coil.

- The magnetic moment of the current loop is $\mathbf{m} = I\mathbf{A}$, where the direction of area vector $\mathbf{A}$ is given by the right-hand thumb rule. Its unit is A-m$^2$.

- The torque can be expressed as vector product of the magnetic moment of the coil and magnetic field.

$$\tau = \mathbf{m} \times \mathbf{B}$$

- Torque $\tau$ vanishes when $m$ is either parallel or anti-parallel to the magnetic field **B**. This indicates a state of equilibrium as there is no torque on the coil.
- If the loop has $N$-closely wound turns, then $\mathbf{m} = NI\mathbf{A}$.

## Circular Current Loop as a Magnetic Dipole

- The magnetic field due to a current carrying circular coil on the axial line is given by

$$\mathbf{B} = \frac{\mu_0}{4\pi} \cdot \frac{2m}{x^3}$$

and in the plane of the loop of at a distance $x$ is $\mathbf{B} = \frac{\mu_0}{4\pi} \cdot \frac{m}{x^3}$.

- A planar current loop is equivalent to a magnetic dipole of dipole moment $\mathbf{m} = I\mathbf{A}$.

## Magnetic Dipole Moment of a Revolving Electron

- Elementary particles such as an electron or a proton also carry an intrinsic magnetic moment.
- The electron of charge $e$ performs uniform circular motion around a stationary heavy nucleus, this constitutes a current.

i.e. $$I = \frac{e}{T}$$

where, $T$ = time period of revolution $= \frac{2\pi r}{v}$.

- The magnetic moment associated with this circulating current is

$$\mu_l = \frac{evr}{2}$$

or $$\mu_l = -\frac{e}{2m_e}\mathbf{l}$$

where, $l$ = magnitude of angular momentum of electron. The negative sign indicates that the angular momentum of the electron is opposite in the direction to the magnetic moment.

**Note** The Bohr magneton value, $(\mu_l)_{min} = 9.27 \times 10^{-24}$ A-m$^2$.

- The ratio $\frac{\mu_l}{l} = \frac{e}{2m_e}$ is called the **gyromagnetic ratio** and is constant. Its value is $8.8 \times 10^{10}$ C / kg for an electron.
- The magnetic moment $\mu_l$ is labelled as the orbital magnetic moment. The electron also has an intrinsic magnetic moment called the **spin magnetic moment.**

# Moving Coil Galvanometer

- This device is used to measure current upto nanoampere.
- When a current flows through the coil of a galvanometer, a torque acts on it. This torque is given by

$$\tau = NIAB$$

- Due to this torque, a deflection $\phi$ is indicated on the scale by a pointer attached to the spring as

$$\phi = \left(\frac{NAB}{k}\right)I$$

where, $k$ = torsional constant of the spring or the restoring torque per unit twist.

- The galvanometer cannot as such be used as an ammeter to measure the value of the currents in a given circuit. This is for two reasons
  - (i) Galvanometer is a very sensitive device, it gives a full-scale deflection for a current of the order of $\mu$A.
  - (ii) For measuring currents, the galvanometer has to be connected in series and as it has a large resistance, this will change the value of current in the circuit.
- To use a galvanometer as an ammeter, a small resistance $r_s$ is attached (called **shunt resistance**) in parallel with galvanometer coil. The resistance of this arrangement is

$$\frac{R_G r_s}{R_G + r_s} \simeq r_s \qquad (\text{if } R_G \geq r_s)$$

- The **current sensitivity** of the galvanometer is the deflection per unit current and given by

$$\frac{\phi}{I} = \frac{NAB}{k}$$

The sensitivity increases by increasing the number of turns $N$.

- The galvanometer can be used as a voltmeter to measure the voltage across a given section of the circuit. For this, it must be connected in parallel with that section of the circuit.
- A large resistance $R$ is connected in series with the galvanometer to keep the disturbance below one per cent.

So, resistance of the voltmeter is now

$$R_G + R \simeq R$$

- **Voltage sensitivity** as the deflection per unit voltage is

$$\frac{\phi}{V} = \left(\frac{NAB}{k}\right)\frac{1}{R}$$

# Mastering NCERT

## MULTIPLE CHOICE QUESTIONS

## TOPIC 1 ~ Magnetic Force

**1** When a magnetic needle is placed near the current carrying conductor, then deflection in magnetic needle
(a) increases on increasing the current in the conductor
(b) increases on bringing the needle closer to the current carrying conductor
(c) increases on bringing the needle away from the current carrying conductor
(d) Both (a) and (b)

**2** When an electron is kept in a strong magnetic field, then
(a) electron moves in opposite direction of applied magnetic field
(b) electron moves in the direction of applied magnetic field
(c) electron moves in the perpendicular direction of magnetic field
(d) electron remains stationary at same position

**3** Magnetic field in a region is produced by
(a) only static charges
(b) only moving charges
(c) Both static and moving charges
(d) Neither static charges nor moving charges

**4** The Lorentz force on a charged particle of charge $q$ moving with velocity $\mathbf{v}$ in presence of both the electric field $\mathbf{E}(\mathbf{r})$ and the magnetic field $\mathbf{B}(\mathbf{r})$ is given by
(a) $q\left[\mathbf{E}(\mathbf{r})+\frac{1}{2}\mathbf{v}\times\mathbf{B}(\mathbf{r})\right]$ (b) $q[\mathbf{E}(\mathbf{r})+\mathbf{v}\times\mathbf{B}(\mathbf{r})]$
(c) $q[\mathbf{E}(\mathbf{r})-\mathbf{v}\times\mathbf{B}(\mathbf{r})]$ (d) $q[\mathbf{v}\times\mathbf{B}(\mathbf{r})-\mathbf{E}(\mathbf{r})]$

**5** A proton is projected with velocity $\mathbf{v}=2\hat{\mathbf{i}}$ in a region, where magnetic field $\mathbf{B}=(\hat{\mathbf{i}}+3\hat{\mathbf{j}}+4\hat{\mathbf{k}})\,\mu\text{T}$ and electric field $\mathbf{E}=10\hat{\mathbf{i}}\,\mu\text{V/m}$, then find out the net force on proton. *AIIMS 2019*
(a) $22.6\times10^{-25}$ N (b) $14.14\times10^{-25}$ N
(c) $1.6\times10^{-25}$ N (d) $10.5\times10^{-25}$ N

**6** A proton enters a magnetic field of flux density 1.5 Wb $\text{m}^{-2}$ with a velocity of $2\times10^7\ \text{ms}^{-1}$ at an angle of $30°$ with the field. The force on the proton will be
(a) $2.4\times10^{-12}$ N (b) $0.24\times10^{-12}$ N
(c) $24\times10^{-12}$ N (d) $0.024\times10^{-12}$ N

**7** Which of the following diagram is correct for the direction of magnetic force on a charge moving in a uniform magnetic field?

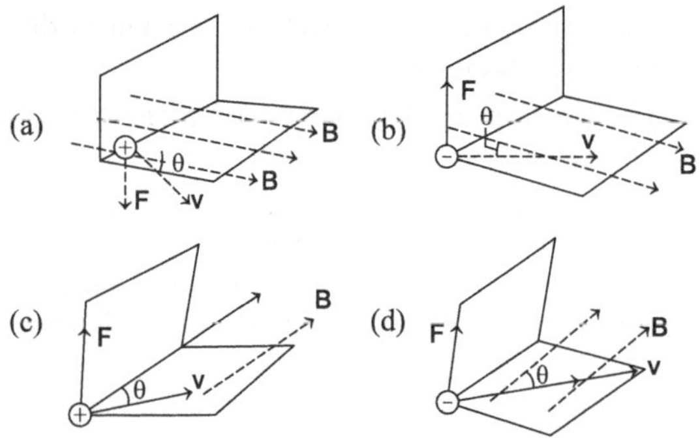


**8** If a particle of charge $10^{-12}$ C moving along the $x$-direction with a velocity $10^5\ \text{ms}^{-1}$ experiences a force of $10^{-10}$ N in $y$-direction due to magnetic field, then the minimum magnetic field is
(a) $6.25\times10^3$ T in $z$-direction
(b) $10^{-15}$ T in $z$-direction
(c) $6.25\times10^{-3}$ T in $z$-direction
(d) $10^{-3}$ T in $z$-direction

**9** An electron is travelling horizontally towards east. A magnetic field in vertically downward direction exerts a force on the electron along
(a) east (b) west
(c) north (d) south

**10** A straight wire of mass 200 g and length 1.5 m carries a current of 2 A. It is held in mid-air by a uniform horizontal magnetic field. What is the magnitude of magnetic field?
(a) 1 T (b) 0.65 T
(c) 0.35 T (d) 0.85 T

**11** When a charged particle moves in the region of magnetic field, then
(a) magnitude of its velocity keeps on changing
(b) velocity of particle remains constant
(c) direction of motion keeps on changing
(d) kinetic energy of particle keeps on changing

# TOPIC 2 ~ Motion of Charged Particle in Magnetic Field and in Combined Electric and Magnetic Fields

**12** An electron moves in a circular orbit with a uniform speed $v$. It produces a magnetic field $B$ at the centre of the circle. The radius of the circle is proportional to

(a) $\frac{B}{v}$  (b) $\frac{v}{B}$

(c) $\sqrt{\frac{v}{B}}$  (d) $\sqrt{\frac{B}{v}}$

**13** In an experiment, electrons are accelerated, from rest by applying a voltage of 500 V. Calculate the radius of the path, if a magnetic field 100 mT is then applied. (Take, charge of the electron $= 1.6 \times 10^{-19}$ C and mass of the electron $= 9.1 \times 10^{-31}$ kg) **JEE Main 2019**

(a) $7.5 \times 10^{-2}$ m  (b) $7.5 \times 10^{-4}$ m

(c) $7.5 \times 10^{-3}$ m  (d) $7.5 \times 10^{-2}$ m

**14** Two particles of masses $m_a$ and $m_b$ with same charge are projected in a perpendicular magnetic field. They travel along circular path of radii $r_a$ and $r_b$ such that $r_a > r_b$. Then, which of the following option is correct?

(a) $m_a v_a > m_b v_b$

(b) $m_a > m_b$ and $v_a > v_b$

(c) $m_a = m_b$ and $v_a > v_b$

(d) $m_b v_b > m_a v_a$

**15** A deuteron of kinetic energy 50 keV is describing a circular orbit of radius 0.5 m in a plane perpendicular to magnetic field $B$. The kinetic energy of the proton that describes a circular orbit of radius 0.5 m in the same plane with the same magnetic field $B$ is

(a) 25 keV  (b) 50 keV

(c) 200 keV  (d) 100 keV

**16** A proton, a deuteron and an $\alpha$-particle enter a region of perpendicular magnetic field (to their velocities) with same kinetic energy. If $r_p$, $r_d$ and $r_\alpha$ are the radii of circular paths of these particles, then **JEE Main 2018**

(a) $r_\alpha = r_d < r_p$  (b) $r_\alpha = r_d = r_p$

(c) $r_\alpha < r_d < r_p$  (d) $r_\alpha > r_d > r_p$

**17** An electron enters a region of magnetic field perpendicularly with a speed of $3 \times 10^7$ ms$^{-1}$. Strength of magnetic field is $6 \times 10^{-4}$ T. Frequency is

(a) 2 MHz  (b) 17 MHz

(c) 20 MHz  (d) 2 MHz

**18** A particle of mass $m$ and charge $q$ moves with a constant velocity $v$ along the positive $x$-direction. It enters a region containing a uniform magnetic field $B$ directed along the negative $z$-direction, extending from $x = a$ to $x = b$.

The minimum value of $v$ required, so that the particle can just enter the region $x > b$ is

(a) $qb\,B/m$

(b) $q\,(b - a)\,B/m$

(c) $q\,aB/m$

(d) $q\,(b + a)\,B/2m$

**19** A proton of mass $1.67 \times 10^{-27}$ kg and charge $1.6 \times 10^{-19}$C is projected with a speed of $2 \times 10^6$ ms$^{-1}$ at an angle of $60°$ to the $X$-axis. If a magnetic field of 0.104 T is applied along $Y$-axis, then path of proton is

(a) a circle of radius 0.2 m and time period of $\pi \times 10^{-7}$ s

(b) a circle of radius 0.1 m and time period $2\pi \times 10^{-7}$ s

(c) a helix of radius 0.2 m and time period of $2\pi \times 10^{-7}$ s

(d) a helix of radius 0.2 m and time period of $4\pi \times 10^{-7}$ s

**20** A proton of energy 1 MeV describes a circular path in plane at right angles to a uniform magnetic field of $6.28 \times 10^{-4}$ T. The mass of proton is $1.67 \times 10^{-27}$ kg. The frequency of proton is nearly equal to

(a) $10^7$ Hz  (b) $10^5$ Hz

(c) $10^6$ Hz  (d) $10^4$ Hz

**21** Which one is a correct figure to represent path of a charge moving at an angle (other than $0°$ or $90°$) in a magnetic field?

(a)
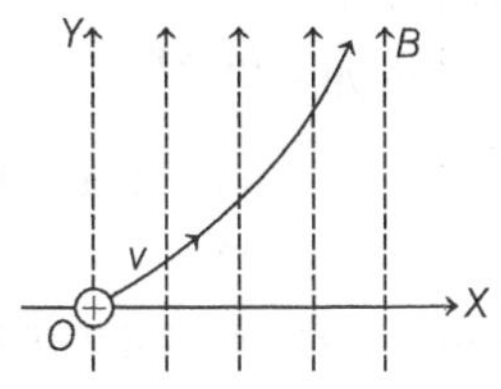


(b)
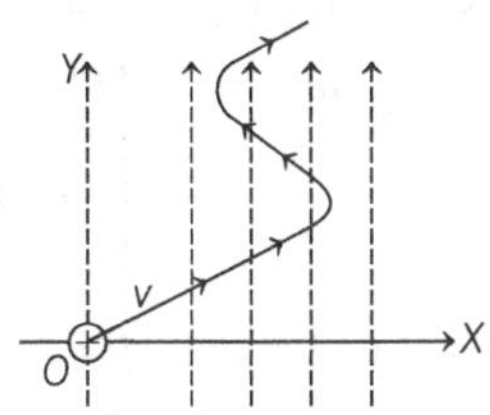


(c)
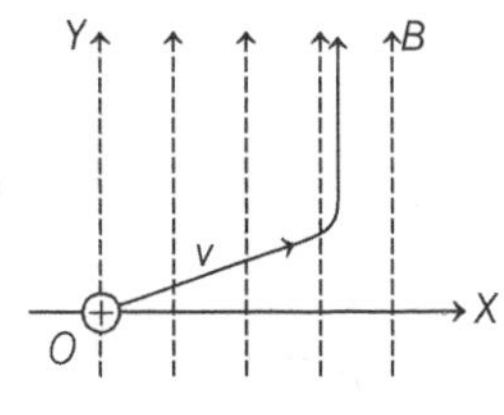


(d)
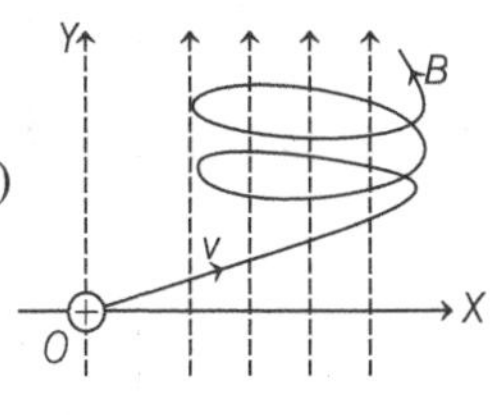

**22** In the given figure, $E$ and $B$ are magnitudes of electric and magnetic fields which are acting on a moving electron as shown. The direction of velocity of electron is along $X$-axis.

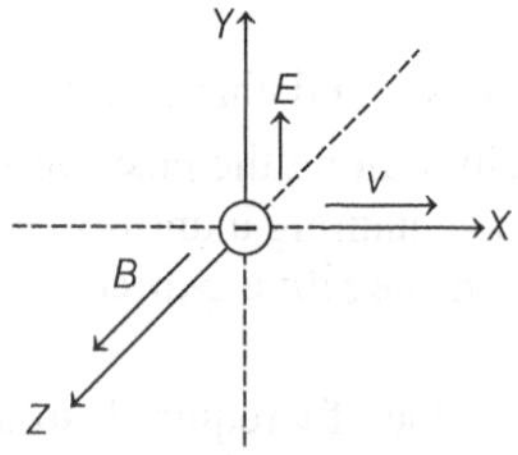


Then, force on electron is given by

(a) $\mathbf{F} = q(E + vB)\,\hat{\mathbf{j}}$ (b) $\mathbf{F} = -q(E + vB)\,\hat{\mathbf{j}}$

(c) $\mathbf{F} = q(E + vB)\,\hat{\mathbf{k}}$ (d) $\mathbf{F} = q(E - vB)\,\hat{\mathbf{k}}$

**23** An ionised gas contains both positive and negative ions. If it is subjected simultaneously to an electric field along the $+x$-direction and a magnetic field along the $+z$-direction, then

(a) positive ions deflect towards $+y$-direction and negative ions towards $-y$-direction

(b) all ions deflect towards $+y$-direction

(c) all ions deflect towards $-y$-direction

(d) positive ions deflect towards $-y$-direction and negative ions towards $+y$-direction

**24** An electron moves straight inside a charged parallel plate capacitor. Plates of capacitor have charge density $\sigma$. The space between plates is filled with constant magnetic field of induction **B**.

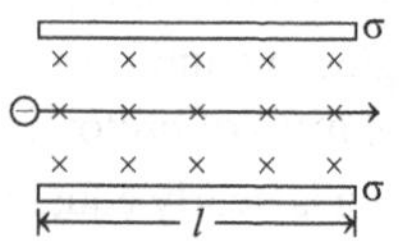


Time taken by electron to pass through the region is

(a) $\dfrac{e\sigma}{\varepsilon_0 lB}$ (b) $\dfrac{\varepsilon_0 lB}{\sigma}$ (c) $\dfrac{\sigma e}{\varepsilon_0 B}$ (d) $\dfrac{\varepsilon_0 B}{e\sigma}$

**25** A cyclotron is used to accelerated charged particles or ions to high energies. It uses

(a) only electric field

(b) only magnetic field

(c) Both electric and magnetic fields

(d) None of the above

**26** Cyclotron frequency of an electron circulating in a magnetic field of 1T is

(a) 28 MHz (b) 280 MHz (c) 2.8 GHz (d) 28 GHz

**27** An alternating electric field of frequency $\nu$, is applied across the dees (radius $= R$) of a cyclotron that is being used to accelerate protons (mass $= m$). The operating magnetic field $B$ used in the cyclotron and the kinetic energy $K$ of the proton beam produced by it, are given by ***CBSE AIPMT 2012***

(a) $B = \dfrac{m\nu}{e}$ and $K = 2m\pi^2\nu^2R^2$

(b) $B = \dfrac{2\pi m\nu}{e}$ and $K = m^2\pi\nu R^2$

(c) $B = \dfrac{2\pi m\nu}{e}$ and $K = 2m\pi^2\nu^2R^2$

(d) $B = \dfrac{m\nu}{e}$ and $K = m^2\pi\nu R^2$

**28** A proton and an $\alpha$-particle both enter a region of uniform magnetic field $B$, moving at right angles to the field $B$. If the radius of circular orbits for both the particles is equal and the kinetic energy acquired by proton is 1 MeV, the energy acquired by the $\alpha$-particle will be ***CBSE AIPMT 2015***

(a) 4 MeV (b) 0.5 MeV (c) 1.5 MeV (d) 1 MeV

## TOPIC 3 ~ *Biot-Savart's Law and Ampere's Circuital Law*

**29** Vector form of Biot-Savart's law is

(a) $d\mathbf{B} = \dfrac{\mu_0}{4\pi} \cdot \dfrac{I \times d\mathbf{l}}{r^2}$ (b) $d\mathbf{B} = \dfrac{Id\mathbf{l} \times \mathbf{r}}{r^3}$

(c) $d\mathbf{B} = \dfrac{\mu_0}{4\pi} \cdot \dfrac{Id\mathbf{l} \times \mathbf{r}}{r^3}$ (d) $d\mathbf{B} = \dfrac{\mu_0}{4\pi} \cdot \dfrac{Id\mathbf{l} \times \mathbf{r}}{r^2}$

**30** An element $\Delta l = \Delta x\,\hat{\mathbf{i}}$ is placed at the origin and carries a current $I = 10$ A.

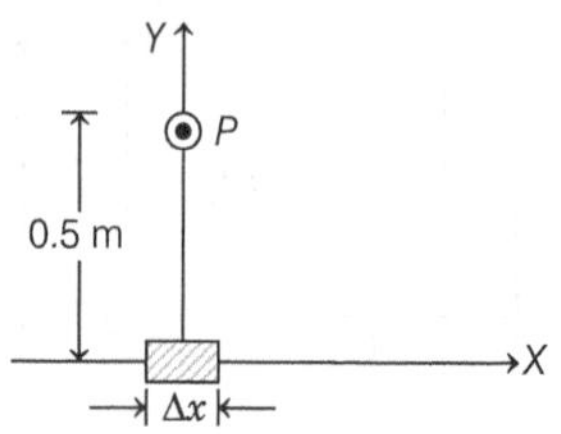


If $\Delta x = 1$ cm, magnetic field at point $P$ is ***AIIMS 2018***

(a) $4 \times 10^{-8}\,\hat{\mathbf{k}}$ T (b) $4 \times 10^{-8}\,\hat{\mathbf{i}}$ T

(c) $4 \times 10^{-8}\,\hat{\mathbf{j}}$ T (d) $-4 \times 10^{-8}\,\hat{\mathbf{j}}$ T

**31** A long, straight wire of radius $a$ carries a current distributed uniformly over its cross-section. The ratio of the magnetic fields due to the wire at distance $\dfrac{a}{3}$ and $2a$ respectively, from the axis of the wire is ***JEE Main 2020***

(a) $\dfrac{3}{2}$ (b) 2

(c) $\dfrac{2}{3}$ (d) $\dfrac{1}{2}$

**32** Two circular loops having same radius ($R = 10$ cm) and same current ($I = 7/2$ A) are placed along same axis as shown. If distance between their centres is 10 cm, then find net magnetic field at point $P$. ***AIIMS 2019***

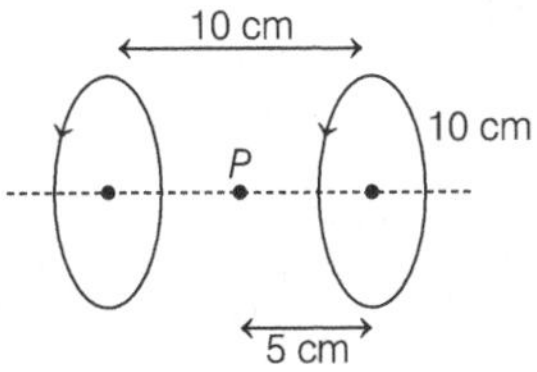


(a) $\frac{50\mu_0}{\sqrt{5}}$ T (b) $\frac{28\mu_0}{\sqrt{5}}$ T (c) $\frac{56\mu_0}{\sqrt{5}}$ T (d) $\frac{56\mu_0}{\sqrt{3}}$ T

**33** A tightly wound coil of 100 turns with radius 10 cm, magnitude of carrying-current in coil is 1 A, what will be the magnitude of magnetic field at centre of coil?

(a) $6.28\times 10^{-6}$ T (b) $6.28\times 10^{-7}$ T
(c) $6.28\times 10^{-5}$ T (d) $6.28\times 10^{-4}$ T

**34** A straight wire carrying a current of 12 A is bent into a semi-circular loop of radius 2.0 cm. What will be magnetic field at the centre of the semi-circular loop?

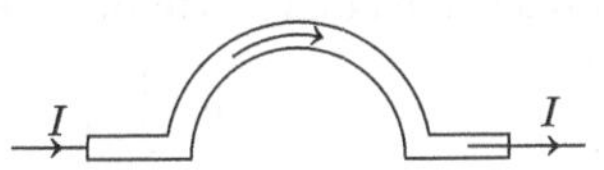


(a) $\mathbf{B} = 1.9\times 10^{-4}$ T $\odot$ (b) $\mathbf{B} = 1.9\times 10^{4}$ T $\odot$
(c) $\mathbf{B} = 1.9\times 10^{-4}$ T, $\otimes$ (d) $\mathbf{B} = 1.9\times 10^{4}$ T, $\otimes$

**35** The magnetic field at the centre of a circular current carrying-conductor of radius $r$ is $B_c$. The magnetic field on its axis at a distance $r$ from the centre is $B_a$. The value of $B_c : B_a$ will be

(a) $1:\sqrt{2}$ (b) $1:2\sqrt{2}$
(c) $2\sqrt{2}:1$ (d) $\sqrt{2}:1$

**36** Two circular coils of radii 5 cm and 10 cm carry equal currents of 2 A. The coils have 50 and 100 turns respectively and are placed in such a way that their planes as well as their centres coincide. Magnitude of magnetic field at the common centre of coils is

(a) $8\pi\times 10^{-4}$ T, if currents in the coils are in same sense
(b) $4\pi\times 10^{-4}$ T, if currents in the coils are in opposite sense
(c) zero, if currents in the coils are in same sense
(d) $8\pi\times 10^{-4}$ T, if currents in the coils are in opposite sense

**37** Two concentric circular coils $X$ and $Y$ of radii 16 cm and 10 cm lie in same vertical plane containing north to south direction. Coil $X$ has 20 turns, coil $Y$ has 25 turns and current in coil $X$ is 16 A whereas in coil $Y$ is 18 A. Current in $X$ is anti-clockwise and in $Y$ is clockwise. For an observer facing west and looking at coils, magnetic field at the centre of assembly of coils is

(a) $1.57\times 10^{-3}$ T (towards east)
(b) $1.57\times 10^{-3}$ T (towards west)
(c) $1.57\times 10^{-3}$ T (towards north)
(d) $15.7\times 10^{-3}$ T (towards south)

**38** Find the magnetic field at point $P$ due to a straight line segment $AB$ of length 6 cm carrying a current of 5 A (See figure). (Take, $\mu_0 = 4\pi\times 10^{-7}$ N-A$^{-2}$) ***JEE Main 2019***

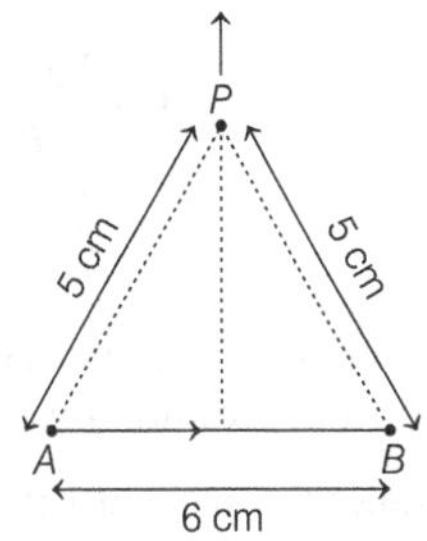


(a) $2.0\times 10^{-5}$ T (b) $1.5\times 10^{-5}$ T
(c) $3.0\times 10^{-5}$ T (d) $2.5\times 10^{-5}$ T

**39** Two identical wires $A$ and $B$, each of length $L$, carry the same current $I$. Wire $A$ is bent into a circle of radius $R$ and wire $B$ is bent to form a square of side $a$. If $B_A$ and $B_B$ are the values of magnetic field at the centres of the circle and square respectively, then the ratio $B_A/B_B$ is ***JEE Main 2016***

(a) $\frac{\pi^2}{8}$ (b) $\frac{\pi^2}{16\sqrt{2}}$ (c) $\frac{\pi^2}{16}$ (d) $\frac{\pi^2}{8\sqrt{2}}$

**40** The arc's $BC$ (radius = $b$) and $AD$ (radius = $a$) of loop are joined by the straight wires $AB$ and $CD$. A steady current $I$ is flowing in the loop. Angle made by $AB$ and $CD$ at the origin is 30°.Another straight thin wire with steady current $I$, flowing out of plane of the paper is kept at the origin.

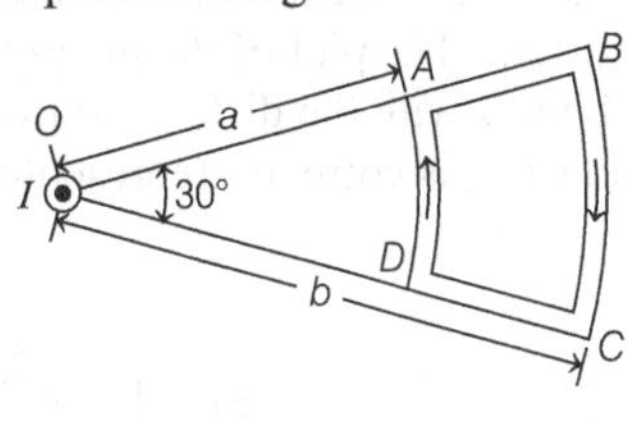


The magnitude of the magnetic field $B$ due to loop $ABCD$ at the origin $O$ is

(a) zero (b) $\frac{\mu_0 I (a-b)}{24\, ab}$
(c) $\frac{\mu_0 I}{4\pi}\left(\frac{a-b}{ab}\right)$ (d) $\frac{\mu_0 I}{4\pi}\left[2(a-b)+\frac{\pi}{3}(a+b)\right]$

**41** A straight conductor carrying current $i$ splits into two parts as shown in the figure. The radius of the circular loop is $R$. The total magnetic field at the centre $P$ of the loop is **NEET (Odisha) 2019**

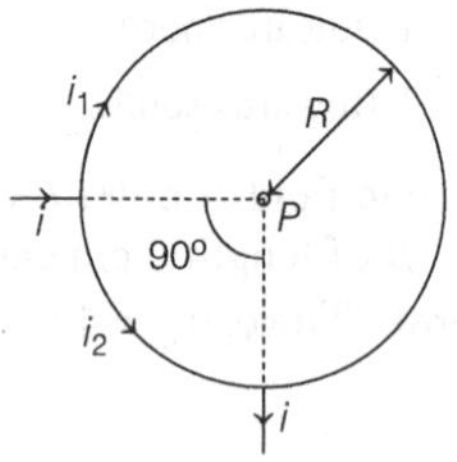


(a) zero
(b) $\frac{3\mu_0 i}{32R}$, outward
(c) $\frac{3\mu_0 i}{32R}$, inward
(d) $\frac{\mu_0 i}{2R}$, inward

**42** According to Ampere's circuital law, the line integral of the magnetic field around any closed loop is

(a) equal to $\frac{1}{\mu_0}$ times the total current flowing through the loop
(b) equal to $\mu_0$ times the total current flowing through the loop
(c) equal $\frac{\mu_0}{\varepsilon_0}$ times the total current flowing through the loop
(d) equal to $\frac{\varepsilon_0}{\mu_0}$ times the total current flowing through the loop

**43** A long straight wire of radius $a$ carries a steady current $I$. The current is uniformly distributed over its cross-section. The ratio of the magnetic fields **B** and **B′** at radial distances $a/2$ and $2a$ respectively, from the axis of the wire is **NEET 2016**

(a) $\frac{1}{2}$ (b) 1 (c) 4 (d) $\frac{1}{4}$

**44** A cylindrical conductor of radius $R$ is carrying a constant current. The plot of the magnitude of the magnetic field $B$ with the distance $d$ from the centre of the conductor, is correctly represented by the figure **NEET 2019**

(a)
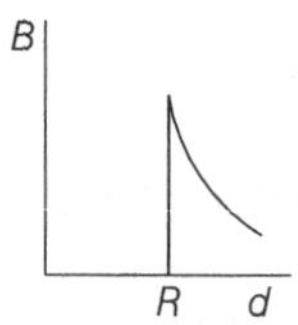


(b)
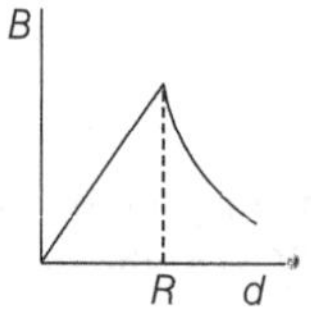


(c)
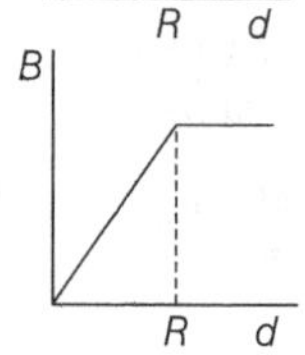


(d)
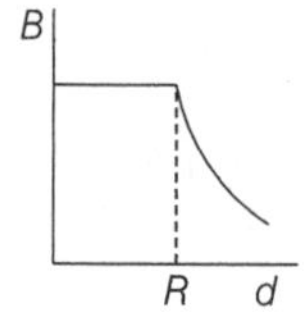


**45** An infinitely long conductor $PQR$ is bent to form a right angle as shown. A current $I$ flows through $PQR$. The magnetic field due to current in $PQ$ at the point $M$ is $B_1$. The ratio of net field at $M$ due to $PQR$ and $B_1$ is given by

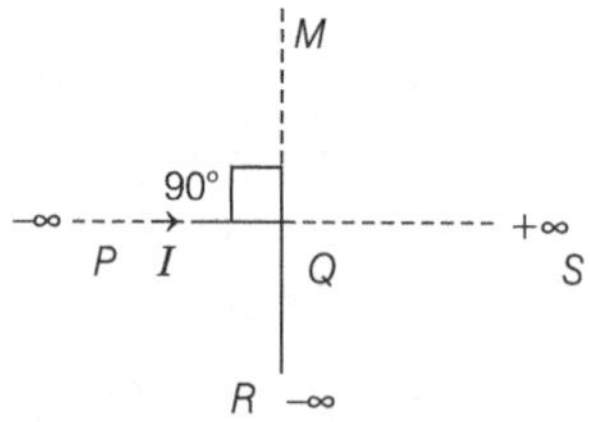


(a) $\frac{1}{2}$ (b) 1
(c) $\frac{2}{3}$ (d) 2

**46** Two identical long conducting wires $AOB$ and $COD$ are placed at right angle to each other, with one above other such that $O$ is their common point for the two. The wires carry $I_1$ and $I_2$ currents, respectively. Point $P$ is lying at distance $d$ from $O$ along a direction perpendicular to the plane containing the wires. The magnetic field at the point $P$ will be **CBSE AIPMT 2014**

(a) $\frac{\mu_0}{2\pi d}\left(\frac{I_1}{I_2}\right)$ (b) $\frac{\mu_0}{2\pi d}(I_1 + I_2)$
(c) $\frac{\mu_0}{2\pi d}(I_1^2 - I_2^2)$ (d) $\frac{\mu_0}{2\pi d}(I_1^2 + I_2^2)^{1/2}$

**47** A long solenoid has 20 turns $cm^{-1}$. The current necessary to produce a magnetic field of 20 mT inside the solenoid is approximately

(a) 1A (b) 2 A
(c) 4 A (d) 8 A

**48** A solenoid of length 0.5 m has a radius of 1 cm and is made up of 500 turns. It carries a current of 5A. What is the magnetic field inside the solenoid?

(a) $6.28 \times 10^{-3}$ T (b) $6.28 \times 10^{3}$ T
(c) $6.28 \times 10^{-5}$ T (d) $6.28 \times 10^{5}$ T

**49** A toroid of core of inner radius 0.25 m and outer radius 0.26 m around which 3500 turns of a wire are wound. If the current in the wire is 11 A, then magnetic field inside the core of the toroid is

(a) $3 \times 10^{2}$ T (b) $3 \times 10^{-2}$ T
(c) $3 \times 10^{-7}$ T (d) $3 \times 10^{7}$ T

**50** Two toroids 1 and 2 have total number of turns 200 and 100 respectively with average radii 40 cm and 20 cm, respectively. If they carry same current $i$, the ratio of the magnetic fields along the two loops is **NEET (Odisha) 2019**

(a) 1 : 1 (b) 4 : 1
(c) 2 : 1 (d) 1 : 2

## TOPIC 4 ~ Force between two Parallel Currents, Torque on Current Loop and Magnetic Dipole

**51** If two parallel current carrying conductors placed 1 m apart in vacuum are placed such that each carries 1 A current, then there is a force of

(a) $2\times10^{-7}\ \text{Nm}^{-1}$

(b) $2\times10^{7}\ \text{Nm}^{-1}$

(c) $9\times10^{9}\ \text{Nm}^{-1}$

(d) $9\times10^{-9}\ \text{Nm}^{-1}$

**52** Two long parallel straight wires $A$ and $B$ carrying currents of 4.0 A and 5.0 A in same direction separated by a distance of $4\times10^{-2}$ m.
The force on a 0.20 m section of wire $A$ is

(a) $2\times10^{-5}$ N (towards $B$)

(b) $2\times10^{-5}$ N (away from $B$)

(c) $2\times10^{-5}$ N (perpendicular to $B$)

(d) $2\times10^{-5}$ N (parallel to $B$)

**53** Two very long, straight and parallel wires carry same steady current $I$ flowing in same direction. The distance between the wires is $d$. At a certain instant of time, a point charge $q$ is at a point equidistant from the two wires in the plane of the wires. Its instantaneous velocity $v$ is perpendicular to this plane. The magnitude of the force due to the magnetic field acting on the charge at this instant is

(a) $\dfrac{\mu_0 Iqv}{2\pi d}$ (b) $\dfrac{\mu_0 Iqv}{\pi d}$

(c) $\dfrac{2\mu_0 Iqv}{\pi d}$ (d) 0

**54** In figure, two parallel infinitely long current carrying wires are shown. If resultant magnetic field at point $A$ is zero, then determine the value of current $I$.

***AIIMS 2019***

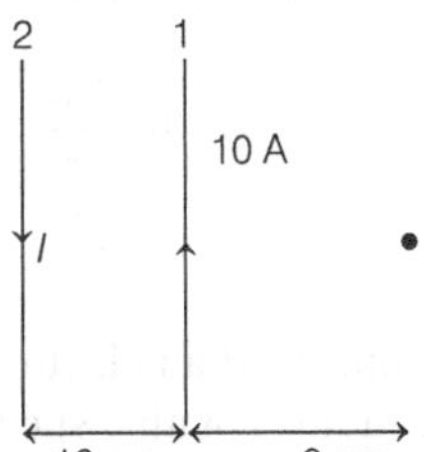


(a) 50 A (b) 15 A

(c) 30 A (d) 25 A

**55** A rigid square loop of side $a$ and carrying current $I_2$ is lying on a horizontal surface near a long current $I_1$ carrying wire in the same plane as shown in figure. The net force on the loop due to the wire will be

***JEE Main 2019***

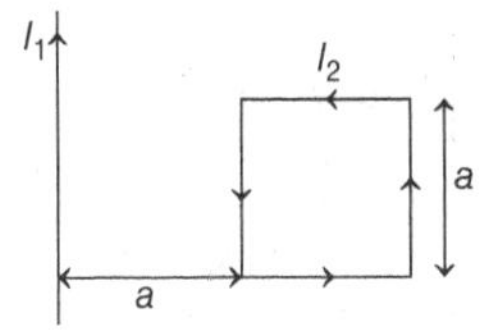


(a) repulsive and equal to $\dfrac{\mu_0 I_1 I_2}{2\pi}$

(b) attractive and equal to $\dfrac{\mu_0 I_1 I_2}{3\pi}$

(c) zero

(d) repulsive and equal to $\dfrac{\mu_0 I_1 I_2}{4\pi}$

**56** A current carrying closed loop in the form of a right angled isosceles $\Delta\, ABC$ is placed in a uniform magnetic field acting along $AB$. If the magnetic force on the arm $BC$ is $F$, then force on the arm $AC$ is

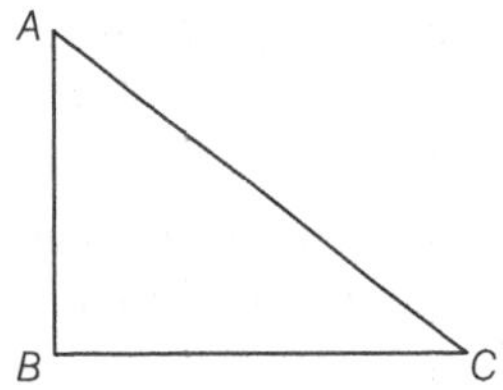


(a) $-F$ (b) $F$ (c) $\sqrt{2}F$ (d) $-\sqrt{2}F$

**57** Three long, straight and parallel wires carrying currents are arranged as shown in the figure.

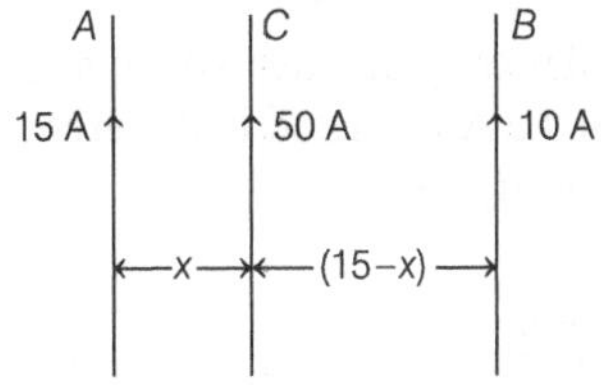


The wire $C$ which carries a current of 50 A is so replaced that it experiences no force. The distance of wire $C$ from wire $A$ is

(a) 9 cm (b) 7 cm

(c) 5 cm (d) 3 cm

**58** An arrangement of three parallel straight wires placed perpendicular to plane of paper carrying same current $I$ along the same direction is shown in figure. Magnitude of force per unit length on the middle wire $B$ is given by **NEET 2017**

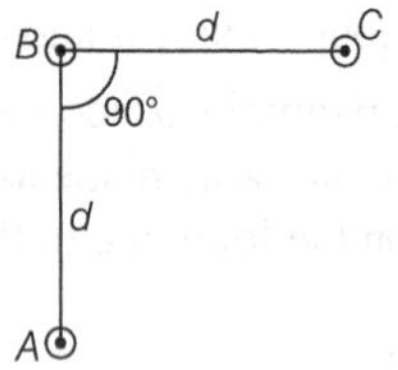


(a) $\frac{\mu_0 I^2}{2\pi d}$ (b) $\frac{2\mu_0 I^2}{\pi d}$ (c) $\frac{\sqrt{2}\mu_0 I^2}{\pi d}$ (d) $\frac{\mu_0 I^2}{\sqrt{2}\,\pi d}$

**59** A current carrying square loop is suspended in a uniform magnetic field acting in the plane of the loop. If the force on one arm of the loop is **F**, then the net force on the remaining three arms of the loop is

(a) 3 **F** (b) −**F** (c) −3 **F** (d) **F**

**60** Area enclosed by a planar loop is $2\times10^{-3}\,m^2$. If loop carries a current of 2 A, then magnitude of magnetic dipole moment of the loop is

(a) $4\times10^{3}\,JT^{-1}$ (b) $8\times10^{-3}\,JT^{-1}$
(c) $4\times10^{-3}\,JT^{-1}$ (d) $2\times10^{-3}\,JT^{-1}$

**61** A circular coil of 25 turns and radius 12 cm is placed in a uniform magnetic field of 0.5 T normal to the plane of the coil. If the current in the coil is 6 A, then total torque acting on the coil is

(a) zero (b) 3.4 N-m (c) 3.8 N-m (d) 4.4 N-m

**62** A current loop in a magnetic field **NEET 2013**

(a) experiences a torque whether the field is uniform or non-uniform in all orientations
(b) can be in equilibrium in one orientations
(c) can be equilibrium in two orientations, both the equilibrium states are unstable
(d) can be in equilibrium in two orientations, one stable while the other is unstable

**63** A rectangular loop of sides 10 cm and 5 cm carrying a current $I$ of 12 A is placed in different orientations as shown in the figures.

(i)
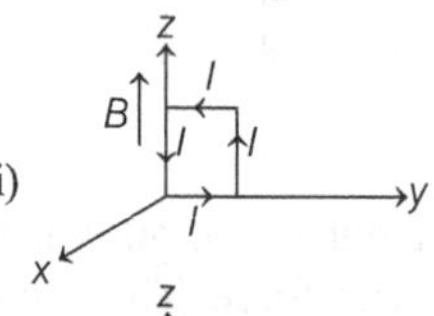

(ii)
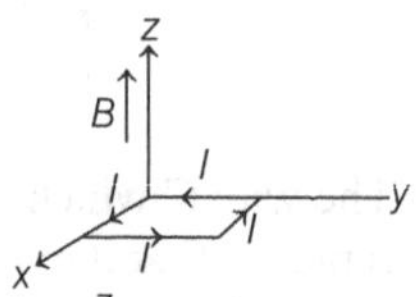

(iii)
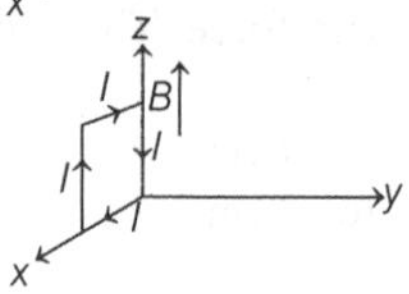

(iv)
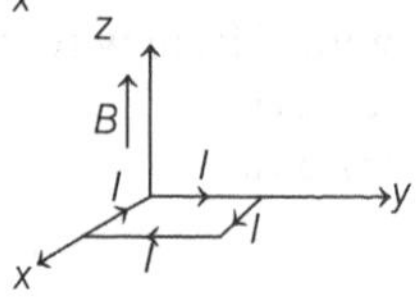

If there is a uniform magnetic field of 0.3 T in the positive $z$-direction, in which orientations the loop would be in (i) stable equilibrium and (ii) unstable equilibrium?

(a) (i) and (ii) respectively
(b) (i) and (iii) respectively
(c) (ii) and (iv) respectively
(d) (ii) and (iii) respectively

**64** A steady current $i$ flows in a small square loop of wire of side $L$ in a horizontal plane. The loop is now folded about its middle such that half of it lies in a vertical plane. Let $\mu_1$ and $\mu_2$ denotes the magnetic moments due to the current loop before and after folding, respectively. Then,

(a) $\mu_2 = 0$
(b) $\mu_1$ and $\mu_2$ are in the same direction
(c) $\frac{|\mu_1|}{|\mu_2|} = \sqrt{2}$
(d) $\frac{|\mu_1|}{|\mu_2|} = \left(\frac{1}{\sqrt{2}}\right)$

**65** An electron is revolving around the nucleus of hydrogen atom of radius 1 Å with a speed of $10^6$ m/s. Magnetic dipole moment associated with the electron is

(a) $8\times10^{-24}\,A\text{-}m^2$ (b) $2\times10^{-24}\,A\text{-}m^2$
(c) $4\times10^{-24}\,A\text{-}m^2$ (d) $3.25\times10^{-24}\,A\text{-}m^2$

**66** Magnetic field at the centre of a circular loop of area $A$ is $B$. The magnetic moment of the loop will be

(a) $\frac{BA^2}{\mu_0\pi}$ (b) $\frac{BA^{3/2}}{\mu_0\pi}$
(c) $\frac{BA^{3/2}}{\mu_0\pi^{1/2}}$ (d) $\frac{2BA^{3/2}}{\mu_0\pi^{1/2}}$

**67** A square loop is carrying a steady current $I$ and the magnitude of its magnetic dipole moment is $m$. If this square loop is changed to a circular loop and it carries the same current, the magnitude of the magnetic dipole moment of circular loop (in A-m) will be **JEE Main 2019**

(a) $\frac{4m}{\pi}$ (b) $\frac{3m}{\pi}$
(c) $\frac{2m}{\pi}$ (d) $\frac{m}{\pi}$

**68** The magnetic moment of an electron orbiting in a circular orbit of radius $r$ with a speed $v$ is equal to

(a) $evr/2$ (b) $evr$ **JIPMER 2018**
(c) $er/2v$ (d) None of these

**69** The value of Bohr magneton is

(a) $9.27\times10^{-24}$ A-m$^2$ (b) $9.27\times10^{-27}$ A-m$^2$
(c) $8.8\times10^{-10}$ A-m$^2$ (d) $6.25\times10^{-12}$ A-m$^2$

# TOPIC 5 ~ Moving Coil Galvanometer

**70** A moving coil galvanometer is an instrument which
(a) is used to measure emf of cell
(b) is used to measure potential difference
(c) is used to measure resistance
(d) is a deflection type instrument that gives a deflection when a current flows through its coil

**71** The moving coil galvanometer consists of a coil with many turns, free to rotate about a fixed axis in a radial magnetic field. There is a cylindrical soft iron core which not only makes the field radial but also
(a) decreases the strength of the magnetic field
(b) increases the strength of the magnetic field
(c) provide constant magnetic field
(d) None of the above

**72** A galvanometer of resistance $70\,\Omega$ is converted into an ammeter by a shunt resistance $r_s = 0.03\,\Omega$. The value of its resistance will become
(a) $0.025\,\Omega$ (b) $0.022\,\Omega$
(c) $0.035\,\Omega$ (d) $0.030\,\Omega$

**73** The resistance of a galvanometer is 50 ohm and the maximum current which can be passed through it is 0.002 A. What resistance must be connected to it in order to convert it into an ammeter of range 0-0.5 A? ***JEE Main 2019***
(a) 0.2 ohm (b) 0.5 ohm
(c) 0.002 ohm (d) 0.02 ohm

**74** In the circuit shown, the galvanometer $G$ of resistance $60\,\Omega$ is shunted by a resistance $r = 0.02\,\Omega$. The current through $R$ is nearly 1 A. The value of resistance $R$ (in ohm) is nearly ***AIIMS 2018***

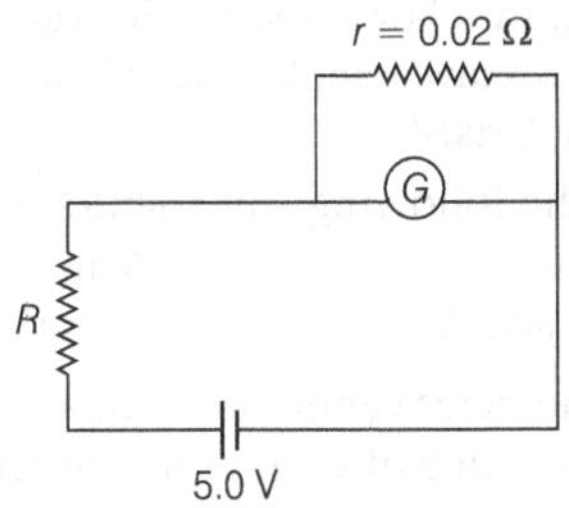


(a) $1.00\,\Omega$ (b) $5.00\,\Omega$
(c) $11.0\,\Omega$ (d) $6.0\,\Omega$

**75** A galvanometer of $50\Omega$ resistance has 25 divisions. A current of $4\times10^{-4}$ A gives a deflection of one division. To convert this galvanometer into a voltmeter having a range of 25V, it should be connected with a resistance of ***JIPMER 2019; JEE Main 2017***
(a) $2500\,\Omega$ as a shunt (b) $245\,\Omega$ as a shunt
(c) $2550\,\Omega$ in series (d) $2450\,\Omega$ in series

**76** A galvanometer with a coil of resistance $12\,\Omega$ shows full scale deflection for a current of 2.5 mA. The ratio of net resistance of an ammeter of range 0 to 7.5 A and a voltmeter of range 0 to 10 V is
(a) $10^{-12}$ (b) $10^{-7}$ (c) $10^{-6}$ (d) $10^{-8}$

**77** In two galvanometers $A$ and $B$, to produce a deflection of 10 divisions, currents of 3 mA and 5 mA are required respectively, then
(a) $A$ is more sensitive than $B$
(b) $B$ is more sensitive than $A$
(c) Both are equally sensitive
(d) sensitivity of $B$ is $5/3$ that $A$

**78** The coil of a galvanometer consists of 100 turns and effective area of 1 $cm^2$. The restoring couple is $10^{-8}$ N-m $rad^{-1}$. The magnetic field between poles is of 5 T. Current sensitivity of this galvanometer is
(a) $5\times10^4$ rad/μ A (b) $5\times10^6$ $A^{-1}$
(c) $2\times10^{-7}$ $A^{-1}$ (d) 5 rad/μ A

**79** Relation between voltage sensitivity $S_V$ and current sensitivity $S_i$ for a galvanometer of resistance $G$ ohms is
(a) $\frac{S_i}{G} = S_V$ (b) $\frac{S_V}{G} = S_i$ (c) $S_i S_V = G$ (d) $\sqrt{S_i S_V} = G$

**80** To increase current sensitivity of a moving coil galvanometer by 50%, its resistance is increased, so that its new resistance is twice of its initial resistance. Its voltage sensitivity changes by a factor of
(a) increases by a factor of 2
(b) decreases by a factor of (1/2)
(c) increases by a factor of 4
(d) decreases by a factor of (1/4)

**81** Current sensitivity of a moving coil galvanometer is 5 div/mA and its voltage sensitivity (angular deflection per unit voltage applied) is 20 div/V. The resistance of the galvanometer is ***NEET 2018***
(a) $250\,\Omega$ (b) $25\,\Omega$ (c) $40\,\Omega$ (d) $500\,\Omega$

# SPECIAL TYPES QUESTIONS

## I. Assertion and Reason

**Direction** (Q. Nos. 82-89) *In the following questions, a statement of Assertion is followed by a corresponding statement of Reason. Of the following statements, choose the correct one.*

(a) Both Assertion and Reason are correct and Reason is the correct explanation of Assertion.
(b) Both Assertion and Reason are correct but Reason is not the correct explanation of Assertion.
(c) Assertion is correct but Reason is incorrect.
(d) Assertion is incorrect but Reason is correct.

**82** **Assertion** A magnetic field interacts with a moving charge and not with a stationary charge.

**Reason** A moving charge produce a magnetic field. ***AIIMS 2018***

**83** **Assertion** Electron moving perpendicular to **B** will perform circular motion.

**Reason** Force exerted by magnetic field on the electron is perpendicular to its velocity. ***AIIMS 2019***

**84** **Assertion** A charge particle is released from rest in magnetic field, then it will move in a circular path.

**Reason** Work done by magnetic field is zero. ***AIIMS 2019***

**85** **Assertion** When a charged particle moves in a region of magnetic field such that its velocity is at some acute angle with the direction of field, its trajectory is a helix.

**Reason** Perpendicular component of velocity causes a rotating centripetal force and parallel component of velocity does not produces any force.

**86** **Assertion** Cyclotron is a device which is used to accelerate the positive ions.

**Reason** Cyclotron frequency depends upon the velocity.

**87** **Assertion** The magnetic field produced by a current-carrying solenoid is independent of its length and cross-sectional area.

**Reason** The magnetic field inside the solenoid is uniform.

**88** **Assertion** Magnetic properties of iron like materials cannot be linked to intrinsic magnetic moment of electrons of its atoms.

**Reason** The electron is an elementary particle, it does not spin around an axis and it does not possess any intrinsic magnetic moment.

**89** **Assertion** If we increase the current sensitivity of a galvanometer by increasing number of turns, its voltage sensitivity also increases.

**Reason** Resistance of a wire also increases with $N$.

## II. Statement Based Questions

**90** Consider a moving charged particle in a region of magnetic field.

I. If **v** is parallel to **B**, then path of particle is spiral.
II. If **v** is perpendicular to **B**, then path of particle is a circle.
III. If **v** has a component along **B**, then path of particle is helical.
IV. If **v** is along **B**, then path of particle is a circle.

Which of the following statement(s) is/are correct?

(a) Both I and II (b) Both II and III
(c) Both III and IV (d) Both IV and I

**91** For a velocity selector; (a region of perpendicular electric and magnetic fields).

I. Allows charged particles to pass straight when $v = E/B$.
II. Deflects particles in a direction perpendicular to both **v** and **B**, when $v > E/B$.
III. Deflects particles in the direction of electric field when $v < E/B$.
IV. Deflects all particlcs in a direction perpendicular to both $E$ and $B$.

Which of the following statement(s) is/are correct?

(a) I, III and IV (b) II, III and IV
(c) I, II and III (d) I, II and IV

**92** Difference between Coulomb's force and Biot-Savart's force are

I. Electrostatic force is along the displacement vector joining the source and the field but magnetic force is perpendicular to the plane containing the displacement vector and source.
II. There is an angle dependence in the Biot-Savart's law which is not present in Coulomb's law.
III. Biot-Savart's law can be expressed in vector form but Coulomb's law cannot be expressed in vector form.
IV. Biot-Savart's law is applicable in all medium but Coulomb's law is applicable only in medium which can be polarised.

Which of the following statement(s) is/are correct?

(a) Both I and II (b) Both II and III
(c) Both III and IV (d) Both IV and I

**93** The galvanometer cannot be used as an ammeter to measure the value of current in a given circuit. The following reasons are

I. galvanometer gives full scale deflection for a small current.
II. galvanometer has a large resistance.
III. a linear scale cannot be designed so that $I \propto \alpha$.
IV. a galvanometer can give inaccurate values.

Which of the following statement(s) is/are correct?

(a) Both I and IV
(b) Both II and III
(c) Both I and II
(d) Both III and IV

**94** Which of the following statement (s) is/are correct ?

(a) Moving charges or currents produced a magnetic field in the surrounding space.
(b) The electric field can convey energy and momentum and is established instantaneously.
(c) The magnetic field of several sources is the arithmetic addition of magnetic field of each individual source.
(d) The Lorentz force depends on $q$, **v** and **B**. Force on a negative charge is same as that on a positive charge.

**95** Which of the following statement (s) is/are incorrect ?

(a) The vector product makes the force due to magnetic field vanish (become zero), if velocity and magnetic field are parallel or anti-parallel.
(b) The magnetic force is zero, if charge is not moving (as then $|v| = 0$).
(c) The magnitude of magnetic field $B$ is 1 SI unit, when the force acting on a unit charge (1 C), moving parallel to **B** with a speed of 1 m/s is one newton.
(d) In the presence of an external magnetic field **B**, the force on mobile charge carriers is $F = (nlA)\, qv_d \times B$.

**96** If a proton is projected in a direction perpendicular to a uniform magnetic field with velocity $v$ and an electron is projected along the lines of force, then which of the following statement(s) is/are correct regarding their motion?

(a) The electron will travel along a circle with constant speed and the proton will move along a straight line.
(b) Proton will move in a circle with constant speed and there will be no effect on the motion of electron.
(c) There will not be any effect on the motion of electron and proton.
(d) The electron and proton both will follow the path of a parabola.

**97** Which of the following statement(s) is/are correct ?

(a) The larger the momentum the lesser is the radius and smaller the circle described.
(b) The distance moved along the magnetic field in one rotation is called pitch $P$.
(c) In a cyclotron, both the fields (electric or magnetic) are parallel to each other.
(d) The frequency $\nu_a$ of the applied voltage is adjusted so that the polarity of the dees remains same.

**98** Which of the following statement(s) is/are incorrect?

(a) If a moving charged particle enters into a region of magnetic field from outside, it does not complete a circular path.
(b) If a moving charged particle traces a helical path in a uniform magnetic field, the axis of the helix is parallel to the magnetic field.
(c) The power associated with the force exerted by a magnetic field on a moving charged particle is always equal to unity.
(d) If in a region a uniform magnetic field and a uniform electric field both exist, a charged particle moving in this region cannot trace a circular path.

**99** Which of the following statement(s) is/are incorrect ?

(a) The operation of the cyclotron is based on the fact that the time for one revolution of an ion is dependent on its speed or radius of its orbit.
(b) The cyclotron is used to bombard nuclei with energetic particles, so accelerated by it and study the resulting nuclear reactions.
(c) The cyclotron is used to implantions into solids and modify their properties or even synthesis new materials.
(d) The cyclotron is used in hospitals to produce radioactive substances which can be used in diagnosis and treatment.

**100** Which of the following statement(s) is/are correct?

(a) The magnitude of the magnetic field $d\mathbf{B}$ is proportional to the current $I$, the element length $|d\mathbf{l}|$ and to the square of the distance $r$.
(b) The electric and magnetic fields are of short range.
(c) The principle of superposition is not applicable to magnetic field.
(d) In the right-hand thumb rule, the thumb gives the direction of the magnetic field.

**101** Which of the following statement(s) is/are incorrect ?

(a) The magnetic field possesses what is called a cylindrical symmetry.
(b) The lines of constant magnitude of magnetic field form concentric circles called magnetic field lines, originating from negative charges.
(c) Even though the wire is infinite, the field due to it at a non-zero distance is not infinite.
(d) The direction of magnetic field is given by right-hand rule in which fingers will curl around in the direction of the magnetic field.

**102** Which of the following statement(s) is/are incorrect ?

(a) For Ampere's law relation, a sign convention is given by the right-hand rule, in which direction of the thumb gives the sense in which the current $I$ is regarded as positive.
(b) An amperian loop is that at each point of which either **B** is tangential to the loop and is a non-zero constant $B$ or **B** is normal to the loop or **B** vanishes.
(c) When there is a system with a symmetry such as for a straight infinite current carrying wire, the Ampere's law enables an easy evaluation of the magnetic fields.
(d) Ampere's circuital law holds for currents that are accelerated with time.

**103** Which of the following statement(s) is/are incorrect ?

(a) Ampere studied the nature of magnetic force between two conductors and its dependence on the magnitude of the current, on shape and size of the conductors as well as the distances between them.

(b) An instrument called the current balance is used to measure the mechanical force between two current carrying conductors.

(c) When a steady current of 1 A is set up in conductor, the quantity of charge that flows through its cross-section in 1 s is one coulomb (1 C).

(d) The torques can be expressed as vector sum of the magnetic moment of the coil and the magnetic field.

**104** An electron is projected with uniform velocity along the axis of a current carrying long solenoid. Which of the following statement(s) is/are correct?

(a) The electron will be accelerated along the axis.

(b) The electron path will be circular about the axis.

(c) The electron will experience a force at 45° to the axis and hence execute a helical path.

(d) The electron will continue to move with uniform velocity along the axis of the solenoid.

**105** Which of the following statement(s) is/are incorrect ?

(a) In a solenoid, the net magnetic field is the vector sum of the fields due to all the turns.

(b) Enamelled wires are used for winding, so that turns are insulated from each other.

(c) The field inside the solenoid approaches zero and everywhere outside becomes parallel to the axis.

(d) The solenoid is commonly used to obtain a uniform magnetic field.

**106** Which of the following statement(s) is/are correct ?

(a) A toroid is a solid circular ring on which a large number of turns of a wire are closely wound.

(b) The field **B** inside the toroid is not constant in magnitude for the ideal toroid of closely wound turns.

(c) The magnetic field at any point in the open space inside the toroid is maximum.

(d) The number of turns in a toroid is equal to the product of average area of the toroid and number of turns per unit length.

**107** Which of the following statement(s) is/are correct?

(a) The magnetic field due to a current loop on axial line is $\mathbf{B} = \frac{\mu_0}{4\pi} \cdot \frac{2\mathbf{m}}{x^3}$.

(b) The magnetic field for a point in the plane of the loop at a distance $x$ from the centre is $\mathbf{B} \simeq \frac{\mu_0}{4\pi} \cdot \frac{\mathbf{m}}{x^2}$.

(c) A planar current loop is different from a magnetic dipole of dipole moment $\mathbf{m} = I \cdot \mathbf{A}$.

(d) None of the above

**108** Which of the following statement(s) is/are incorrect ?

(a) The deflection $\phi$ indicated on the scale of a moving coil galvanometer by a pointer attached to the spring is given by $\phi = \left(\frac{NAB}{k}\right) I$.

(b) The galvanometer can be used as a detector to check, if a current is flowing in the circuit.

(c) Current sensitivity of the galvanometer is the deflection per unit current while voltage sensitivity is the deflection per unit voltage.

(d) In voltage measurement, to keep the disturbance due to the measuring device below one per cent a small resistance $R$ is connected in parallel with the galvanometer.

## III. Matching Type

**109** A charged particle with some initial velocity is projected in a region where non-zero electric and/or magnetic fields are present. In Column I, information about the existence of electric and/or magnetic field and direction of initial velocity of charged particle are given, while in Column II the probable path of the charged particle is mentioned. Match the entries of Column I with the entries of Column II and select the correct answer from the codes given below.

| Column I | Column II |
|---|---|
| A. $\mathbf{E} = 0$, $\mathbf{B} \neq 0$, $\mathbf{E} \neq \mathbf{B}$ and initial velocity is at any angle with $\mathbf{B}$ ($\theta \neq 90°$) | 1. Straight line |
| B. $\mathbf{E} = 0, \mathbf{B} \neq 0$ and initial velocity is perpendicular to $\mathbf{B}$ | 2. Helical path with uniform pitch |
| C. $\mathbf{E} \neq 0, \mathbf{B} \neq 0$, $\mathbf{E} \parallel \mathbf{B}$ and initial velocity is $\perp$ to both | 3. Circular |
| D. $\mathbf{E} \neq 0, \mathbf{B} \neq 0, \mathbf{E}$ perpendicular to $\mathbf{B}$ and $\mathbf{v}$ perpendicular to both $\mathbf{E}$ & $\mathbf{B}$ | 4. Helical path with non-uniform pitch |

| | A | B | C | D | | A | B | C | D |
|---|---|---|---|---|---|---|---|---|---|
| (a) | 2 | 3 | 4 | 1 | (b) | 1 | 3 | 4 | 2 |
| (c) | 2 | 3 | 1 | 4 | (d) | 4 | 1 | 2 | 3 |

**110** For a current carrying wire loop of $N$-turns, placed in a region of a uniform magnetic field **B**. Match the Column I (torque and magnetic moment of loop) with Column II (values) and select the correct answer from the codes given below.

| Column I | Column II |
|---|---|
| A. Torque on loop | 1. $\mathbf{mB}$ |
| B. Torque on loop when $\mathbf{m}$ is either parallel or anti-parallel to $\mathbf{B}$ | 2. $NI\mathbf{A}$ |
| C. Magnetic moment of loop | 3. zero |
| D. Torque on loop when $\mathbf{m}$ is perpendicular to $\mathbf{B}$ | 4. $\mathbf{m} \times \mathbf{B}$ |

| | A | B | C | D | | A | B | C | D |
|---|---|---|---|---|---|---|---|---|---|
| (a) | 1 | 2 | 3 | 4 | (b) | 2 | 1 | 4 | 3 |
| (c) | 4 | 3 | 2 | 1 | (d) | 1 | 3 | 4 | 2 |

**111** Figure shows the construction details of a moving coil galvanometer.

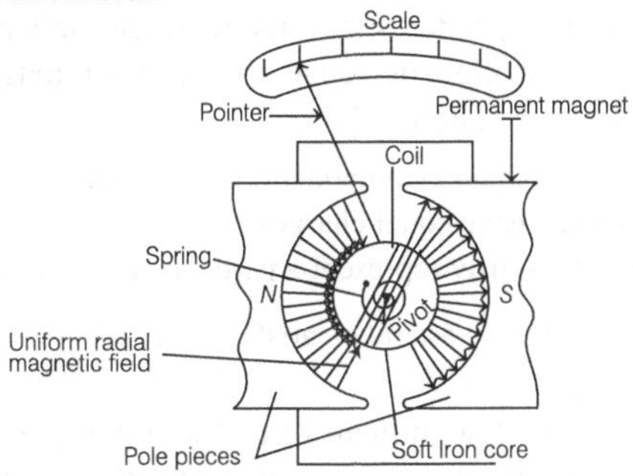


Match the Column I (parts) with Column II (effect) and select the correct answer from the codes given below.

| Column I | Column II |
|---|---|
| A. Soft iron core | 1. produces deflecting torque |
| B. Pole pieces | 2. produces restoring torque |
| C. Spring | 3. produces radial field |
| D. Coil | 4. increases field strength |

| | A | B | C | D | | A | B | C | D |
|---|---|---|---|---|---|---|---|---|---|
| (a) | 1 | 2 | 3 | 4 | (b) | 1 | 2 | 4 | 3 |
| (c) | 2 | 1 | 3 | 4 | (d) | 4 | 3 | 2 | 1 |

# NCERT & NCERT Exemplar

## MULTIPLE CHOICE QUESTIONS

### NCERT

**112** The horizontal component of the earth's magnetic field at a certain place is $3.0 \times 10^{-5}$ T and the direction of the field is from the geographic south to the north pole. A very long straight conductor is carrying a steady current of 1 A.
What is the force per unit length on it when it is placed on a horizontal table and the direction of current is east to west?
(a) $3 \times 10^{-5}$ $Nm^{-1}$ (b) $5 \times 10^{-2}$ $Nm^{-1}$
(c) $6 \times 10^{-1}$ $Nm^{-1}$ (d) $8 \times 10^{-3}$ $Nm^{-1}$

**113** A circular coil of wire consisting of 100 turns, each of radius 8.0 cm carries a current of 0.40 A. What is the magnitude of the magnetic field $B$ at the centre of the coil?
(a) 4 T (b) $3.1 \times 10^{-4}$ T (c) $2 \times 10^{-3}$ T (d) $10^{-4}$ T

**114** A long straight wire in the horizontal plane carries a current of 50 A in north to south direction. Give the magnitude and direction of $B$ at a point 2.5 m east of the wire.
(a) $2 \times 10^{-6}$ T, south (b) $3 \times 10^{-6}$ T, downward
(c) $4 \times 10^{-6}$ T, upward (d) 6 T, north

**115** A horizontal overhead power line carries a current of 90 A in east to west direction. What are the magnitude and direction of the magnetic field due to the current 1.5 m below the line?
(a) $1.2 \times 10^{-5}$ T, vertically down
(b) $1.2 \times 10^{-5}$ T, vertically up
(c) 4 T, vertically up
(d) 4 T, vertically down

**116** What is the magnitude of magnetic force per unit length on a wire carrying a current of 8 A and making an angle of 30° with the direction of a uniform magnetic field of 0.15 T?
(a) 0.4 $Nm^{-1}$ (b) 0.6 $Nm^{-1}$ (c) 4 $Nm^{-1}$ (d) 6 $Nm^{-1}$

**117** A 3.0 cm wire carrying a current of 10 A is placed inside a solenoid perpendicular to its axis. The magnetic field inside the solenoid is given to be 0.27 T. What is the magnetic force on the wire?
(a) $7 \times 10^{-2}$ N (b) $8.1 \times 10^{-2}$ N
(c) $6.4 \times 10^{-2}$ N (d) $4 \times 10^{2}$ N

**118** Two long and parallel straight wires $A$ and $B$ carrying currents of 8.0 A and 5.0 A in the same direction are separated by a distance of 4.0 cm. Estimate the force on a 10 cm section of wire $A$.
(a) $5 \times 10^{-5}$ N, repulsive (b) $5 \times 10^{-5}$ N, attractive
(c) $2 \times 10^{-5}$ N, repulsive (d) $2 \times 10^{-5}$ N, attractive

**119** A closely wound solenoid 80 cm long has 5 layers of windings of 400 turns each. The diameter of the solenoid is 1.8 cm. If the current carried is 8.0 A, estimate the magnitude of $B$ inside the solenoid near its centre.
(a) $2.5 \times 10^{-2}$ T (b) $3.5 \times 10^{-2}$ T
(c) $4.5 \times 10^{-2}$ T (d) $5 \times 10^{-2}$ T

**120** A square coil of side 10 cm consists of 20 turns and carries a current of 12 A. The coil is suspended vertically and the normal to the plane of the coil makes an angle of 30° with the direction of a uniform horizontal magnetic field of magnitude 0.80 T. What is the magnitude of torque experienced by the coil?
(a) 1.96 N-m (b) 0.96 N-m (c) 2.0 N-m (d) 4 N-m

**121** Two moving coil meters $M_1$ and $M_2$ having the following particulars :
$R_1 = 10\,\Omega$, $N_1 = 30$, $A_1 = 3.6 \times 10^{-3}\ m^2$, $B_1 = 0.25$ T
$R_2 = 14\,\Omega$, $N_2 = 42$, $A_2 = 1.8 \times 10^{-3} m^2$, $B_2 = 0.50$ T
(The spring constants are identical for the two meters).
Determine the ratio of voltage sensitivity of $M_2$ and $M_1$.
(a) 4 (b) 5
(c) 6 (d) 1

**122** In a chamber, a uniform magnetic field of 6.5 G ($1G = 10^{-4}$ T) is maintained. An electron is shot into the field with a speed of $4.8 \times 10^6\ ms^{-1}$ normal to the field. Determine the radius of the circular orbit. (Take, $e = 1.6 \times 10^{-19}$ C and $m_e = 9.1 \times 10^{-31}$ kg)
(a) 2 cm (b) 8 cm
(c) 6 cm (d) 4.2 cm

**123** A circular coil of 30 turns and radius 8.0 cm carrying a current of 6.0 A is suspended vertically in a uniform horizontal magnetic field of magnitude 1.0 T. The field lines make an angle of $60^\circ$ with the normal of the coil. Calculate the magnitude of the counter torque that must be applied to prevent the coil from turning.
(a) 4 N-m (b) 6 N-m
(c) 3.133 N-m (d) 2.8 N-m

## NCERT Exemplar

**124** Two charged particles traverse identical helical paths in a completely opposite sense, in a uniform magnetic field $\mathbf{B} = B_0\hat{\mathbf{k}}$. Then,
(a) they have equal $z$-components of momenta.
(b) they must have equal charges.
(c) they necessarily represent a particle, anti-particle pair.
(d) the charge to mass ratio satisfy $\left(\frac{e}{m}\right)_1 + \left(\frac{e}{m}\right)_2 = 0.$

**125** Biot-Savart's law indicates that the moving electrons (velocity $v$) produce a magnetic field **B** such that
(a) **B** is perpendicular to $v$
(b) **B** is parallel to $v$
(c) it obeys inverse cube law
(d) it is along the line joining the electron and point of observation

**126** In a cyclotron, a charged particle
(a) undergoes acceleration all the time
(b) speeds up between the dees because of the magnetic field
(c) speeds up within a dee
(d) slows down within a dee and speeds up between dees

**127** A circular current loop of magnetic moment $M$ is in an arbitrary orientation in an external magnetic field **B.** The work done to rotate the loop by 30° about an axis perpendicular to its plane is
(a) $MB$ (b) $\sqrt{3}\,\frac{MB}{2}$ (c) $\frac{MB}{2}$ (d) zero

# Answers

### > Mastering NCERT with MCQs

| | | | | | | | | | |
|---|---|---|---|---|---|---|---|---|---|
| 1 (d) | 2 (d) | 3 (b) | 4 (b) | 5 (a) | 6 (a) | 7 (c) | 8 (d) | 9 (c) | 10 (b) |
| 11 (c) | 12 (b) | 13 (b) | 14 (a) | 15 (d) | 16 (a) | 17 (b) | 18 (b) | 19 (c) | 20 (d) |
| 21 (d) | 22 (b) | 23 (c) | 24 (b) | 25 (c) | 26 (d) | 27 (c) | 28 (d) | 29 (c) | 30 (a) |
| 31 (c) | 32 (c) | 33 (d) | 34 (c) | 35 (c) | 36 (a) | 37 (b) | 38 (b) | 39 (d) | 40 (b) |
| 41 (a) | 42 (b) | 43 (b) | 44 (b) | 45 (b) | 46 (d) | 47 (d) | 48 (a) | 49 (b) | 50 (a) |
| 51 (a) | 52 (a) | 53 (d) | 54 (c) | 55 (d) | 56 (b) | 57 (a) | 58 (d) | 59 (b) | 60 (c) |
| 61 (a) | 62 (d) | 63 (c) | 64 (c) | 65 (a) | 66 (d) | 67 (a) | 68 (a) | 69 (a) | 70 (d) |
| 71 (b) | 72 (d) | 73 (a) | 74 (b) | 75 (d) | 76 (c) | 77 (a) | 78 (b) | 79 (a) | 80 (d) |
| 81 (a) | | | | | | | | | |

### > Special Types Questions

| | | | | | | | | | |
|---|---|---|---|---|---|---|---|---|---|
| 82 (a) | 83 (a) | 84 (d) | 85 (a) | 86 (c) | 87 (b) | 88 (d) | 89 (d) | 90 (b) | 91 (c) |
| 92 (a) | 93 (c) | 94 (a) | 95 (c) | 96 (b) | 97 (b) | 98 (c) | 99 (a) | 100 (d) | 101 (b) |
| 102 (d) | 103 (d) | 104 (d) | 105 (c) | 106 (b) | 107 (a) | 108 (d) | 109 (a) | 110 (c) | 111 (d) |

### > NCERT & NCERT Exemplar MCQs

| | | | | | | | | | |
|---|---|---|---|---|---|---|---|---|---|
| 112 (a) | 113 (b) | 114 (c) | 115 (b) | 116 (b) | 117 (b) | 118 (d) | 119 (a) | 120 (b) | 121 (d) |
| 122 (d) | 123 (c) | 124 (d) | 125 (a) | 126 (a) | 127 (d) | | | | |

# Hints & Explanations

**2** (*d*) When a charged particle is kept stationary ($v = 0$) in the strong magnetic field, then no force acts on the charged particle.

i.e. $$F = Bqv = Bq \times 0$$
$$F = 0$$

Hence, when an electron is kept in a strong magnetic field, then electron remains stationary at same position.

**3** (*b*) Static charges produce only electric field, whereas moving charges produce electric and magnetic fields both.

**4** (*b*) When a point charge $q$ is moving with velocity $v$ and located at $r$ at a given time $t$ in presence of both the electric field $\mathbf{E}(\mathbf{r})$ and the magnetic field $\mathbf{B}(\mathbf{r})$, then it experiences a force called Lorentz force and given by

$$\mathbf{F} = \mathbf{F}_{\text{electric}} + \mathbf{F}_{\text{magnetic}}$$
$$\mathbf{F} = q[\mathbf{E}(\mathbf{r}) + \mathbf{v} \times \mathbf{B}(\mathbf{r})]$$

**5** (*a*) Given, velocity of proton, $\mathbf{v} = 2\hat{\mathbf{i}}$

Magnetic field, $\mathbf{B} = (\hat{\mathbf{i}} + 3\hat{\mathbf{j}} + 4\hat{\mathbf{k}})\,\mu\text{T}$

Electric field, $\mathbf{E} = 10\hat{\mathbf{i}}\mu$ V/m

Charge on proton, $q = 1.6 \times 10^{-19}$ C

Applied Lorentz force on the proton,

$$\mathbf{F} = q\mathbf{E} + q(\mathbf{v} \times \mathbf{B}) = q[\mathbf{E} + (\mathbf{v} \times \mathbf{B})]$$
$$= 1.6 \times 10^{-19}[10 \times 10^{-6}\hat{\mathbf{i}} + 2\hat{\mathbf{i}} \times (\hat{\mathbf{i}} + 3\hat{\mathbf{j}} + 4\hat{\mathbf{k}}) \times 10^{-6}]$$
$$= 1.6 \times 10^{-19} \times 10^{-6}[10\hat{\mathbf{i}} + 6\hat{\mathbf{k}} - 8\hat{\mathbf{j}}]$$
$$\mathbf{F} = 1.6 \times 10^{-25}[10\hat{\mathbf{i}} - 8\hat{\mathbf{j}} + 6\hat{\mathbf{k}}]\,\text{N}$$
$$|F| = 1.6 \times 10^{-25} \times \sqrt{(10)^2 + (-8)^2 + (6)^2}$$
$$= 1.6 \times 10^{-25} \times 14.14 = 22.6 \times 10^{-25}\ \text{N}$$

**6** (*a*) Given, magnetic field, $B = 1.5\ \text{Wbm}^{-2}$,

velocity, $v = 2 \times 10^7\ \text{ms}^{-1}$ and angle, $\theta = 30°$

Net force on the proton,

$$F = Bqv\sin\theta$$
$$= 1.5 \times 1.6 \times 10^{-19} \times 2 \times 10^7 \times \sin 30°$$
$$= 2.4 \times 10^{-12}\ \text{N}$$

**7** (*c*) Direction of force on a moving charge in magnetic field can be found by right hand-rule.

As we know, the direction of **F** is the direction of cross product of velocity **v** and magnetic field **B**, which is perpendicular to the plane containing **v** and **B**.

So for positive charge it acts upward and for negative charge it acts downward, hence diagram (c) is correct.

**8** (*d*) Given, charge, $q = 10^{-12}$ C, velocity, $v = 10^5\ \text{ms}^{-1}$

Force, $F = 10^{-10}$ N

As we know, force, $F = qvB$

$$\therefore \text{Magnetic field, } B = \frac{F}{qv} = \frac{10^{-10}}{10^{-12} \times 10^5}$$
$$= 10^{-3}\ \text{Wb m}^{-2} \text{ or T}$$

So, according to right hand rule, as particle is moving in $x$-direction and experiences a force in $y$-direction, then magnetic field is acting along $z$-direction.

**9** (*c*) If an electron is travelling horizontally towards east and magnetic field in vertically downward as shown below, then according to Fleming's left hand rule, force on electron is along north.

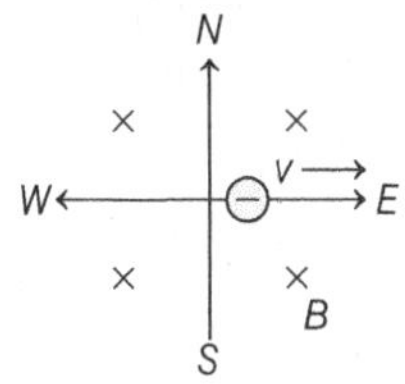


**10** (*b*) A straight wire of mass 200 g and length 1.5 m carries a current of 2A. It is held in a mid-air by a uniform horizontal magnetic field.

There is an upward force **F** of magnitude $IlB$. For mid-air suspension, this must be balanced by the force due to gravity as shown in figure.

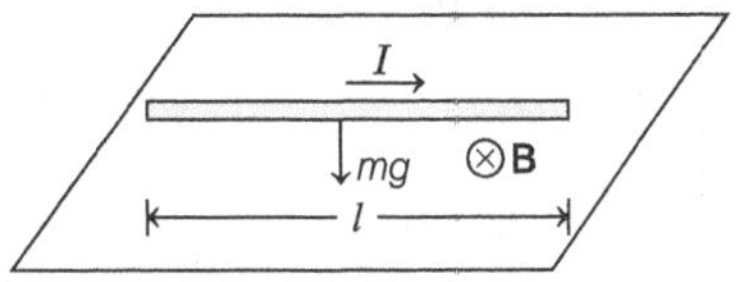


$$mg = IlB \qquad [\because m = 200\ \text{g} = 0.2\ \text{kg}]$$
$$\Rightarrow \quad B = \frac{mg}{Il} = \frac{0.2 \times 9.8}{2 \times 1.5} = 0.65\ \text{T}$$

**11** (*c*) When a charged particle moves in the region of magnetic field, then force is perpendicular to the velocity and it produces a change of direction.

Therefore, the direction of motion keeps on changing.

**12** (*b*) $\because$ The radius of circular path,

$$r = \frac{mv}{qB} \Rightarrow r \propto \frac{v}{B}$$

The radius of the circle is directly proportional to $\frac{v}{B}$.

**13** (*b*) During the circular motion of accelerated electron in the presence of magnetic field, its radius is given by

$$r = \frac{mv}{Be} = \frac{\sqrt{2meV}}{eB}$$

where, $v$ is velocity and $V$ is voltage.

After substituting the given values, we get

$$= \frac{\sqrt{2 \times 9.1 \times 10^{-31} \times 1.6 \times 10^{-19} \times 500}}{1.6 \times 10^{-19} \times 100 \times 10^{-3}}$$
$$= 10\left[\frac{2 \times 9.1 \times 500}{1.6} \times 10^{-12}\right]^{1/2}$$
$$r = 7.5 \times 10^{-4}\ \text{m}$$

**14** (a) Radius of path, $r = \frac{mv}{qB}$

Given, $r_a > r_b$

$\Rightarrow \quad \frac{m_a v_a}{qB} > \frac{m_b v_b}{qB} \quad [\because q_a = q_b \text{ (given)}]$

$\Rightarrow \quad m_a v_a > m_b v_b$

**15** (d) Radius of the circular orbit that a charged particle traverses is given as

$\because \quad r = \frac{mv}{Bq} = \frac{\sqrt{2mK}}{Bq}$

where, $K$ is kinetic energy.

When $B$ is same,we can write $r \propto \sqrt{\frac{2mK}{q}}$

$\Rightarrow \quad \frac{r_{\text{proton}}}{r_{\text{deuteron}}} = \sqrt{\frac{m_p K_p}{m_d K_d}}$

$\Rightarrow \quad \frac{0.5}{0.5} = \sqrt{\frac{m_p K_p}{2m_p K_p}} \quad [\because r_p = r_d = 0.5 \text{ m}]$

[∵ mass of deuteron = 2 × mass of proton]

$1 = \frac{m_p K_p}{2m_p K_d}$

$K_p = 2K_d \quad [\because \text{Given, } K_d = 50 \text{ keV}]$

$\Rightarrow \quad K_p = 2 \times 50$

$= 100 \text{ keV}$

**16** (a) As we know that, radius of circle,

$$r = \frac{mv}{Bq} = 2 \cdot \frac{\frac{1}{2}mv^2}{Bqv}$$

$\Rightarrow \quad r = \frac{2E}{Bqv} \quad \left[\because E = \frac{1}{2}mv^2\right]$

As, energy $E$ is same, then $r \propto 1/q$.
Since, charge of deuteron, $q_d = 2q_p$,
and charge of α-particle, $q_\alpha = 2q_p$
where, $q_p$ is charge of proton.

$\Rightarrow \quad r_p > r_d = r_\alpha$

or $\quad r_\alpha = r_d < r_p$

**17** (b) Given, speed, $v = 3 \times 10^7 \text{ ms}^{-1}$

and magnetic field, $B = 6 \times 10^{-4}$ T.

Radius, $r = \frac{mv}{qB} = \frac{9 \times 10^{-31} \text{kg} \times 3 \times 10^7 \text{ ms}^{-1}}{1.6 \times 10^{-19} \text{C} \times 6 \times 10^{-4} \text{T}}$

$= 28 \times 10^{-2} \text{m}$

Frequency of an electron,

$$\nu = \frac{v}{2\pi r} = \frac{3 \times 10^7}{2 \times \frac{22}{7} \times 28 \times 10^{-2}}$$

$= \frac{300 \times 10^7}{176} = 1.7 \times 10^7$

$= 17$ MHz

**18** (b) The particle moves in a circular path of radius $r$ in the magnetic field. It can just enter the region $x > b$ for $r \geq (b - a)$.

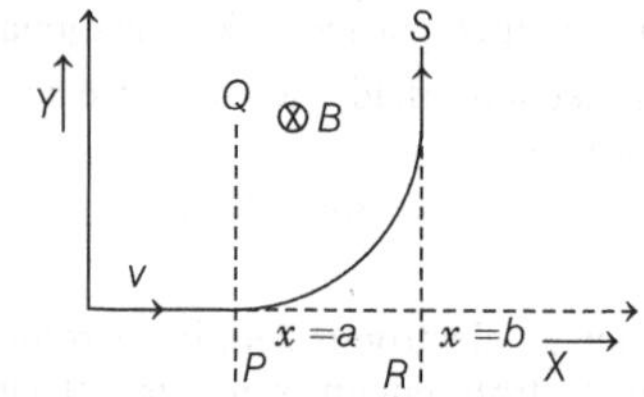


Now, $\quad r = \frac{mv}{qB} \geq (b - a)$ or $v \geq \frac{q(b-a)B}{m}$

$\Rightarrow \quad v_{\min} = \frac{q(b-a)B}{m}$

**19** (c) Given, mass of proton, $m = 1.67 \times 10^{-27}$ kg

charge of proton, $q = -1.6 \times 10^{-19}$ C,

speed, $v = 2 \times 10^6 \text{ ms}^{-1}$

magnetic field, $B = 0.104$ T and angle, $\theta = 60°$.

As, component of velocity $\mathbf{v}$ is along magnetic field $\mathbf{B}$. So, path is helical.

Radius of a helix, $r = \frac{mv_\perp}{Bq} = \frac{1.67 \times 10^{-27} \times 2 \times 10^6}{0.104 \times 1.6 \times 10^{-19}}$

$\Rightarrow \quad r = 0.2$ m

Time period, $T = \frac{2\pi m}{Bq} = \frac{2\pi \times 1.67 \times 10^{-27}}{0.104 \times 1.6 \times 10^{-19}}$

$= 2\pi \times 10^{-7}$ s

**20** (d) Given, magnetic field, $B = 6.28 \times 10^{-4}$ T

and mass of proton, $m_p = 1.67 \times 10^{-27}$ kg

As we know that, frequency of proton,

$$f = \frac{1}{T} = \frac{qB}{2\pi m} = \frac{1.6 \times 10^{-19} \times 6.28 \times 10^{-4}}{2 \times \frac{22}{7} \times 1.67 \times 10^{-27}} \approx 10^4 \text{ Hz}$$

**21** (d) When a charge moving at an angle (other than 0° or 90°) in a magnetic field, then it will move in a helical path as shown in figure.

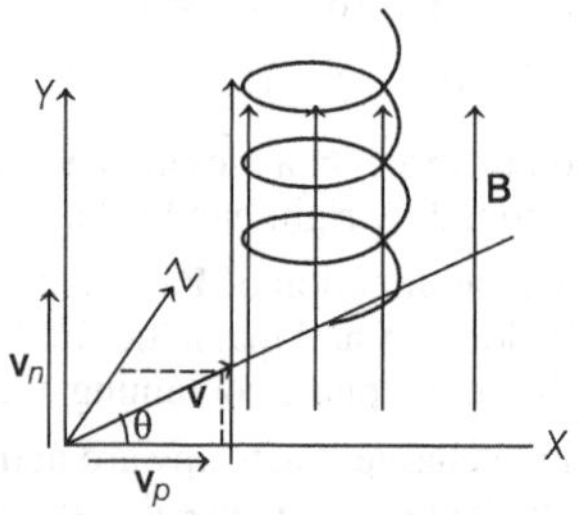


Hence, option (d) is correct.

**22** (b) From the figure given in the question, we get

Force due to electric field, $\mathbf{F}_E = q\mathbf{E} = -qE\,\hat{\mathbf{j}}$,

Force due to magnetic field, $\mathbf{F}_B = q(\mathbf{v} \times \mathbf{B})$

$= q(v\hat{\mathbf{i}} \times B\hat{\mathbf{k}}) = qvB\,(\hat{\mathbf{i}} \times \hat{\mathbf{k}})$

$= -qvB\hat{\mathbf{j}} \quad (\because \hat{\mathbf{i}} \times \hat{\mathbf{k}} = -\hat{\mathbf{j}})$

Therefore, force on electron is given by

$$\mathbf{F} = \mathbf{F}_E + \mathbf{F}_B = -qE\hat{\mathbf{j}} - qvB\hat{\mathbf{j}}$$
$$= -q(E + vB)\hat{\mathbf{j}}$$

**23** (*c*) For an ionised gas, as the electric field is switched ON, positive ion will start to move along positive $x$-direction and negative ion along negative $x$-direction. Current associated with motion of both types of ions is along positive $x$-direction.

According to Fleming's left hand rule, force on both types of ions will be along negative $y$-direction ($-y$ direction).

**24** (*b*) If electric and magnetic fields are $E$ and $B$ respectively, then

Velocity of a particle, $v = \dfrac{E}{B} = \dfrac{\sigma}{\varepsilon_0 B}$

$$\left[\because \text{ for parallel plate capacitor, } E = \frac{\sigma}{\varepsilon_0}\right]$$

$\therefore$ Time taken by electron to pass through the region,

$$t = \frac{l}{v} = \frac{\varepsilon_0 lB}{\sigma}$$

**25** (*c*) The cyclotron is a device used to accelerate charged particles or ions to high energies. The cyclotron use both electric and magnetic fields in combination to increase the energy of charged particles.

**26** (*d*) Given, magnetic field, $B = 1\,\text{T}$

Frequency of an electron, $\nu_e = \dfrac{Bq}{2\pi m}$

$$= \frac{1 \times 1.6 \times 10^{-19}}{2 \times 3.14 \times 9.1 \times 10^{-31}}$$
$$= 28\,\text{GHz}$$

**27** (*c*) Frequency of cyclotron, $\nu = \dfrac{eB}{2\pi m}$

$\text{KE} = \dfrac{1}{2}mv^2$ and radius, $R = \dfrac{mv}{eB}$

Here, as proton complete only half rotation, so velocity,

$$v = \frac{\pi R}{T/2} = \frac{2\pi R}{T} = 2\pi R\nu$$

$\therefore$ Radius, $R = \dfrac{mv}{eB} = \dfrac{m(2\pi R\nu)}{eB}$

$\Rightarrow$ Magnetic field, $B = \dfrac{2\pi m\nu}{e}$

Kinetic energy, $K = \dfrac{1}{2}m(v)^2 = \dfrac{1}{2}m(2\pi R\nu)^2$

$$= 2m\pi^2\nu^2R^2$$

**28** (*d*) Radius in magnetic fields of circular orbit,

$$R = \frac{mv}{qB} = \frac{\sqrt{2mE}}{qB} \quad \left[\begin{array}{l}\because \text{ Kinetic energy}, E = \frac{1}{2}mv^2 \\ \therefore mv = \sqrt{2mE}\end{array}\right]$$

and total energy of a moving particle in a circular orbit,

$$E = \frac{q^2B^2R^2}{2m}$$

For a proton entering in a region of magnetic field,

$$E_1 = \frac{e^2 \times B^2 \times R^2}{2 \times m_p} \quad \text{...(i)}$$

where, $m_p$ is the mass of proton.
Similarly, for an $\alpha$-particle moving in an uniform magnetic field,

$$E_2 = \frac{(2e)^2 \times B^2 \times R^2}{2 \times (4m_p)} \quad (\because m_\alpha = 4m_p) \text{...(ii)}$$

Dividing Eq. (ii) by Eq. (i), we get

$$\frac{E_2}{E_1} = \frac{(2e)^2 \times B^2 \times R^2}{2 \times (4m_p)} \times \frac{2 \times m_p}{e^2 \times B^2 \times R^2}$$

$$\frac{E_2}{E_1} = 1 \Rightarrow E_2 = E_1 = 1\,\text{MeV}$$

**30** (*a*) The magnitude of magnetic field,

$$|d\mathbf{B}| = \frac{\mu_0}{4\pi} \cdot \frac{I\,dl\sin\theta}{r^2}$$

Given, $dl = \Delta x = 10^{-2}$ m, $I = 10$ A, $r = 0.5$ m,

$$\frac{\mu_0}{4\pi} = 10^{-7}\ \text{Tm/A},\ \theta = 90° \Rightarrow \sin 90° = 1$$

$$\Rightarrow |d\mathbf{B}| = \frac{10^{-7} \times 10 \times 10^{-2}}{25 \times 10^{-2}} = 4 \times 10^{-8}\ \text{T}$$

Since, $dl \times \mathbf{r} = \Delta x\hat{\mathbf{i}} \times y\hat{\mathbf{j}} = y\,\Delta x\,(\hat{\mathbf{i}} \times \hat{\mathbf{j}}) = y\,\Delta x\hat{\mathbf{k}}$

So, the direction of the field is in the + $z$-direction.
Therefore, $d\mathbf{B} = 4 \times 10^{-8}\ \hat{\mathbf{k}}\text{T}$

**31** (*c*) By Ampere's circuital rule, magnetic field inside and outside a wire carrying a uniformly distributed current is,

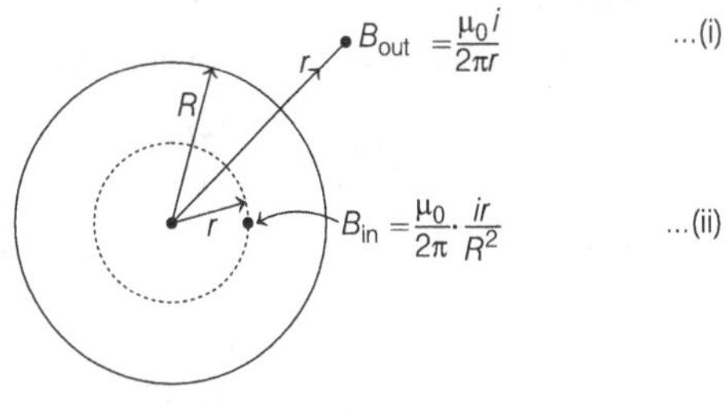


In given case, radius of wire is $a$. So, magnetic field due to wire at a distance $\dfrac{a}{3}$ from centre,

$$B_1 = \frac{\mu_0}{2\pi}\frac{i\left(\frac{a}{3}\right)}{a^2} = \frac{\mu_0 i}{6\pi a} \quad \text{[from Eq. (ii)]}$$

and magnetic field at a distance $2a$ from centre,

$$B_2 = \frac{\mu_0 i}{2\pi(2a)} = \frac{\mu_0 i}{4\pi a} \quad \text{[from Eq. (i)]}$$

Ratio of fields will be

$$\frac{B_1}{B_2} = \frac{\left(\frac{\mu_0 i}{6\pi a}\right)}{\left(\frac{\mu_0 i}{4\pi a}\right)} = \frac{2}{3}$$

**32** (*c*) Given, radius of identical circular loops,

$$R = 10\,\text{cm} = 10^{-1}\,\text{m}$$

Current, $I = \frac{7}{2}$ A

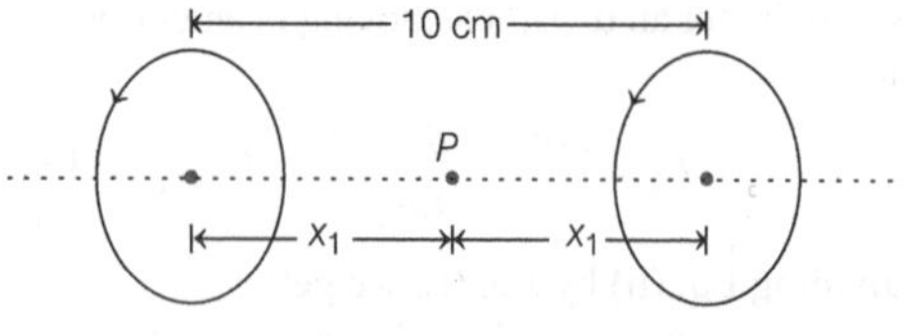


Distance from both circular loops at point $P$,

$$x_1 = x_2 = x = 5\,\text{cm} = 5 \times 10^{-2}\,\text{m}$$

From figure, according to Maxwell's right-hand thumb rule, it is clear that magnetic field will be in same direction by both the coils.

i.e. $B = B_1 + B_2 = \dfrac{\mu_0 I R^2}{2(R^2 + x^2)^{3/2}} + \dfrac{\mu_0 I R^2}{2(R^2 + x^2)^{3/2}}$

$$= \frac{\mu_0 I R^2}{(R^2 + x^2)^{3/2}} = \frac{\mu_0 \times (7/2) \times (10^{-1})^2}{[(0.1)^2 + (0.05)^2]^{3/2}} = \frac{56\mu_0}{\sqrt{5}}\,\text{T}$$

**33** (*d*) Since, the coil is tightly wound, we may take each circular element to have the same radius $R = 10$ cm $= 10^{-1}$ m, number of turns, $N = 100$ and current, $I = 1$ A.

The magnitude of the magnetic field at centre of coil is

$$B = \frac{\mu_0 N I}{2R} = \frac{4\pi \times 10^{-7} \times 10^2 \times 1}{2 \times 10^{-1}}$$

$$= 2\pi \times 10^{-4} = 6.28 \times 10^{-4}\,\text{T}$$

**34** (*c*) Given, current, $I = 12$ A,

radius, $R = 2$ cm $= 2 \times 10^{-2}$ m

Magnetic field at the centre of current carrying circular loop,

$$B_c = \frac{\mu_0 I}{2R} = \frac{\mu_0}{2\pi} \cdot \frac{\pi I}{R}$$

Magnetic field at the centre of semi-circular loop,

$$B = \frac{B_c}{2} = \frac{\mu_0}{4\pi}\frac{\pi I}{R}$$

$$= 10^{-7} \times \frac{3.14 \times 12}{2 \times 10^{-2}}$$

$$= 1.9 \times 10^{-4}\,\text{T}$$

Thus, **B** is $1.9 \times 10^{-4}$ T normal to the plane of the paper going into it, i.e. ⊗.

**35** (*c*) Magnetic field at the centre of a current-carrying conductor,

$$B_c = \frac{\mu_0 I}{2r} \quad \ldots\text{(i)}$$

Magnetic field at axial point due to a current-carrying conductor at distance of $r$,

$$B_a = \frac{\mu_0 I r^2}{2(r^2 + d^2)^{3/2}}$$

$$\Rightarrow \quad B_a = \frac{\mu_0 I r^2}{2(2r^2)^{3/2}} \quad (\because d = r) \ldots\text{(ii)}$$

Now, $\dfrac{B_c}{B_a} = \dfrac{\mu_0 I}{2r} \times \dfrac{2(2r^2)^{3/2}}{\mu_0 I r^2} = 2\sqrt{2}$

$$B_c : B_a = 2\sqrt{2} : 1$$

**36** (*a*) Magnetic field due to the current carrying coil at the centre is given as $B = \dfrac{\mu_0 N I}{2R}$

For coil with $R = 5$ cm, $N = 50$, $I = 2$ A,

$$|\mathbf{B}_1| = \frac{4\pi \times 10^{-7} \times 50 \times 2}{2 \times 5 \times 10^{-2}} = 4\pi \times 10^{-4}\,\text{T}$$

Similarly, for coil with $R = 10$ cm, $I = 2$A, $N = 100$,

$$|\mathbf{B}_2| = \frac{4\pi \times 10^{-7} \times 100 \times 2}{2 \times 10 \times 10^{-2}} = 4\pi \times 10^{-4}\,\text{T}$$

$$\Rightarrow \quad |\mathbf{B}_1| = |\mathbf{B}_2| = 4\pi \times 10^{-4}\,\text{T}$$

If currents are in same sense, then

$$B_{\text{net}} = |\mathbf{B}_1| + |\mathbf{B}_2| = 8\pi \times 10^{-4}\,\text{T}.$$

If currents are in opposite sense, $B_{\text{net}} = |\mathbf{B}_1| - |\mathbf{B}_2| = 0$.

**37** (*b*) The magnitude of magnetic field in current-carrying circular loop,

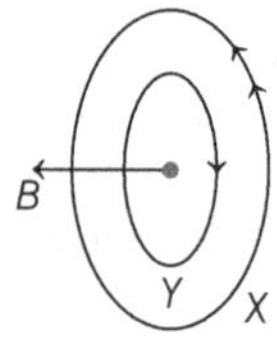


$$B = \frac{\mu_0 N I}{2R}$$

For coil $X$, $N = 20$, $I = 16$ A and $R = 16$ cm $= 16 \times 10^{-2}$ m

$$B_X = \frac{4\pi \times 10^{-7} \times 20 \times 16}{2 \times 0.16}$$

$$= 4\pi \times 10^{-4}\,\text{T} \quad \text{(towards east)}$$

Similarly, for coil $Y$, $N = 25$, $I = 18$ A and $R = 10$ cm $= 10 \times 10^{-2}$ m

$$B_Y = \frac{4\pi \times 10^{-7} \times 25 \times 18}{2 \times 0.1} = 9\pi \times 10^{-4}\,\text{T}$$

$$B_{\text{net}} = B_Y - B_X = 9\pi \times 10^{-4}\,\text{T} - 4\pi \times 10^{-4}\,\text{T}$$

$$= 5\pi \times 10^{-4}\,\text{T}$$

$$= 1.57 \times 10^{-3}\,\text{T} \quad \text{(towards west)}$$

**38** (*b*) The given figure can be drawn as shown below

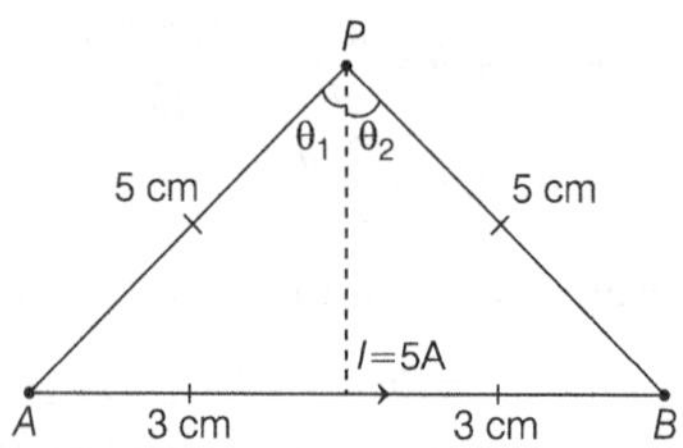


By Biot-Savart's law, magnetic field due to a wire segment at point $P$ is

$$B = \frac{\mu_0 I}{4\pi d}(\sin\theta_1 + \sin\theta_2)$$

Here, $\theta_1 = \theta_2 = \theta$ (say)

Then, $B = \dfrac{\mu_0 I}{4\pi d} \times 2\sin\theta \quad \ldots\text{(i)}$

From given data,

$$I = 5\text{A},$$
$$\mu = 4\pi \times 10^{-7}\ \text{NA}^{-2}$$
$$d = \sqrt{5^2 - 3^2} = \sqrt{16} = 4\ \text{cm}$$
$$\sin\theta = \frac{3}{5}$$

On substituting these values in Eq. (i), we get

$$B = \frac{\mu_0 I}{4\pi d} \times 2\sin\theta$$
$$= \frac{4\pi \times 10^{-7} \times 5}{4\pi \times 4 \times 10^{-2}} \times 2 \times \left(\frac{3}{5}\right)$$
$$= \frac{5 \times 2 \times 3}{4 \times 5} \times 10^{-5} = 1.5 \times 10^{-5}\ \text{T}$$

**39** (*d*) Two identical wires *A* and *B*, each of length *L*, carry the same current *I*. Wire *A* is bent into a circle and wire *B* is bent to form a square as shown below

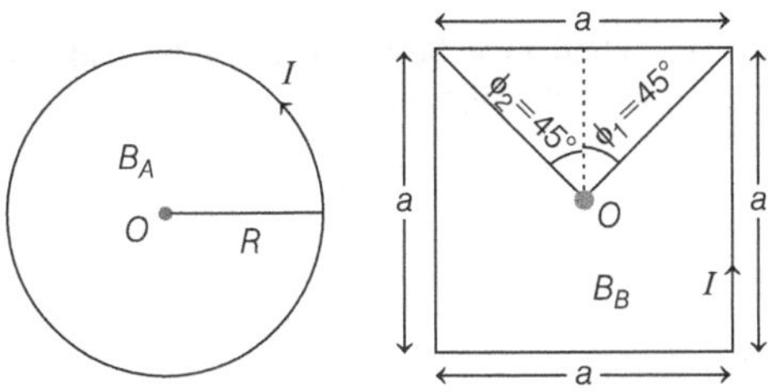


$$B \text{ at centre of a circle} = \frac{\mu_0 I}{2R}$$
$$B \text{ at centre of a square} = \frac{\mu_0 I}{4\pi r}(\sin\phi_1 + \sin\phi_2)$$
$$= 4 \times \frac{\mu I}{4\pi \cdot \frac{a}{2}}(\sin 45° + \sin 45°)$$
$$= 4\sqrt{2}\,\frac{\mu_0 I}{2\pi a}$$

Now, for circle, radius, $R = \frac{L}{2\pi}$

and for square, side, $a = \frac{L}{4}$ (as $L = 2\pi R = 4a$)

where, $L$ = length of wire.

$\therefore$ Magnetic field for wire *A*,

$$B_A = \frac{\mu_0 I}{2 \cdot \frac{L}{2\pi}} = \frac{\pi\mu_0 I}{L} = \pi\left(\frac{\mu_0 I}{L}\right) \quad \ldots\text{(i)}$$

Magnetic field for wire *B*,

$$B_B = 4\sqrt{2}\,\frac{\mu_0 I}{2\pi\left(\frac{L}{4}\right)} = \frac{8\sqrt{2}\mu_0 I}{\pi L} = \frac{8\sqrt{2}}{\pi}\left(\frac{\mu_0 I}{L}\right) \quad \ldots\text{(ii)}$$

From Eqs. (i) and (ii), we get

$$\therefore\ B_A : B_B = \pi^2 : 8\sqrt{2}$$

**40** (*b*) Consider the figure given below

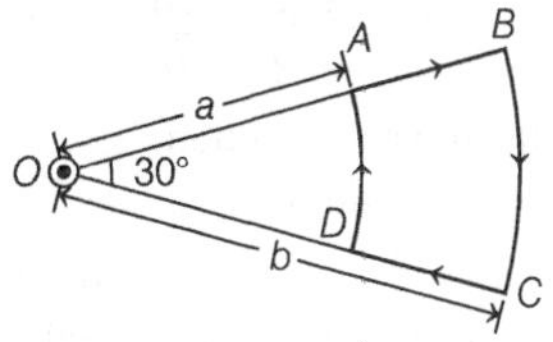


The magnetic field at the centre of an arc subtended at an angle θ is given by $B = \frac{\mu_0 I}{2R} \times \frac{\theta}{2\pi}$

where, *I* is current and *R* is radius.

Now, magnetic field due to wire *AB* and *CD* is

$$B_{AB} = B_{CD} = 0$$
$$[\because\ \theta = 0° \text{ for } AB \text{ and } CD]$$

and due to arc *BC*,

$$B_{BC} = \frac{\mu_0 I}{2b} \times \left(\frac{\pi}{6}\right) \times \frac{1}{2\pi}, \text{ downward}$$
$$\left[\because\ \theta = \frac{\pi}{6}, R = b\right]$$

Similarly due to arc *DA*,

$$B_{DA} = \frac{\mu_0 I}{2a} \times \left(\frac{\pi}{6}\right) \times \frac{1}{2\pi}, \text{ upward}$$
$$\left[\because\ \theta = \frac{\pi}{6}, R = a\right]$$

$\therefore$ Net magnetic field at *O*,

$$B = B_{AB} + B_{BC} + B_{CD} + B_{DA}$$
$$\Rightarrow \quad B = 0 + \frac{\mu_0 I}{4b} \times \left(\frac{\pi}{6\pi}\right) + 0 - \frac{\mu_0 I}{4a} \times \left(\frac{\pi}{6\pi}\right)$$
$$= \frac{\mu_0 I}{24b} - \frac{\mu_0 I}{24a} = \frac{\mu_0 I}{24\,ab}(a - b)$$

**41** (*a*) The magnetic field at the centre of an arc subtended at an angle θ is given by

$$B = \frac{\mu_0 i}{2R} \times \frac{\theta}{2\pi}$$

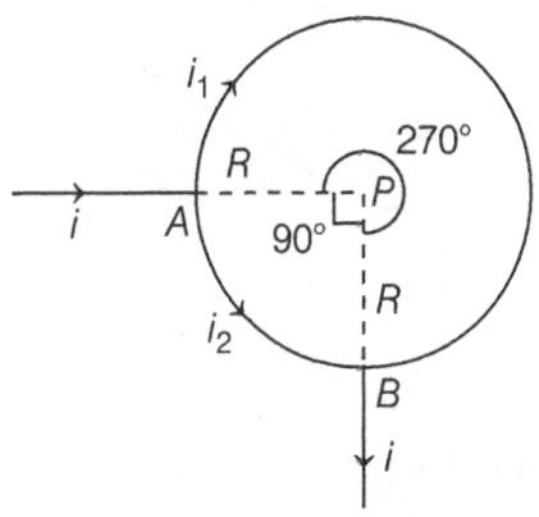


Then, the magnetic field due to larger arc *AB* is

$$B_1 = \frac{\mu_0 i_1}{2R} \times \frac{270}{2\pi} \quad \ldots\text{(i)}$$

which acts in inward direction according to right-hand thumb rule.

Magnetic field due to smaller arc $AB$ is

$$B_2 = \frac{\mu_0 i_2}{2R} \times \frac{90}{2\pi} \quad \ldots\text{(ii)}$$

which acts in outward direction.
The resultant magnetic field,

$$B_R = B_1 + B_2 \quad \text{[from Eqs. (i) and (ii)]}$$

$$= -\frac{\mu_0 i_1 \times 270}{4\pi R} + \frac{\mu_0 i_2 \times 90}{4\pi R} \quad \ldots\text{(iii)}$$

which acts in inward direction as $B_1 > B_2$.
Two arcs can also be seen as the two resistances in parallel combination.
So, the potential across them will be same, i.e.

$$V_1 = V_2$$

$$i_1 R_1 = i_2 R_2 \quad \ldots\text{(iv)}$$

where, $R_1$ and $R_2$ = resistance of respective segments.
The wire is uniform, so

$$\frac{L_1}{L_2} = \frac{R \times 270}{R \times 90}$$

[$\because$ length of arc = radius × angle]

From Eq. (iv), we get

$$\Rightarrow \quad \frac{i_1}{i_2} = \frac{R_2}{R_1} = \frac{90}{270} = \frac{1}{3} \text{ or } 3i_1 = i_2 \quad \ldots\text{(v)}$$

From Eqs. (iii) and (v), we get

$$B_R = \frac{\mu_0}{4\pi R}(-270 i_1 + 90 i_2)$$

$$= \frac{\mu_0}{4\pi R}[-270 i_1 + 90(3i_1)]$$

$$= \frac{\mu_0}{4\pi R}(-270 i_1 + 270 i_1) = 0$$

**43** (*b*) Consider two circular loops of radius $a/2$ and $2a$ as shown in the diagram. Applying Ampere's circuital law for these loops, we get

$$\oint \mathbf{B} \cdot d\mathbf{l} = \mu_0 I_{\text{enclosed}}$$

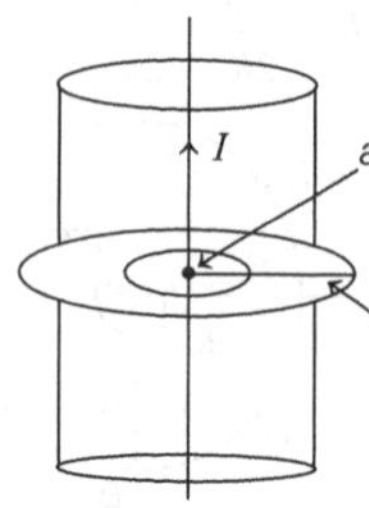


For the smaller loop,

$$B \oint dl = \mu_0 I_{\text{enclosed}} \quad \ldots\text{(i)}$$

$$\because \quad \oint dl = 2\pi\left(\frac{a}{2}\right)$$

and $I_{\text{enclosed}}$ = current per unit mass of sectional area of cylinder × cross section of loop

$$= \frac{I}{\pi a^2} \times \pi\left(\frac{a}{2}\right)^2$$

$$\therefore \mathbf{B} \times 2\pi\left(\frac{a}{2}\right) = \mu_0 \times \frac{I}{\pi a^2} \times \pi\left(\frac{a}{2}\right)^2 \quad \text{[from Eq. (i)]}$$

$$= \mu_0 I \times \frac{1}{4} = \frac{\mu_0 I}{4}$$

$$\Rightarrow \mathbf{B} = \frac{\mu_0 I}{4\pi a}, \text{ at distance } \frac{a}{2} \text{ from the axis of the wire.}$$

Similarly, for bigger loop

$$\mathbf{B}' \times 2\pi(2a) = \mu_0 I$$

(where, $I$ is total current enclosed by Amperian loop is 2)

$$\Rightarrow \quad \mathbf{B}' = \frac{\mu_0 I}{4\pi a}$$

at distance $2a$ from the axis of the wire.

So, ratio of $\dfrac{\mathbf{B}}{\mathbf{B}'} = \dfrac{\mu_0 I}{4\pi a} \times \dfrac{4\pi a}{\mu_0 I} = 1$

**44** (*b*) The cylinder can be considered to be made from concentric circles of radius $R$.

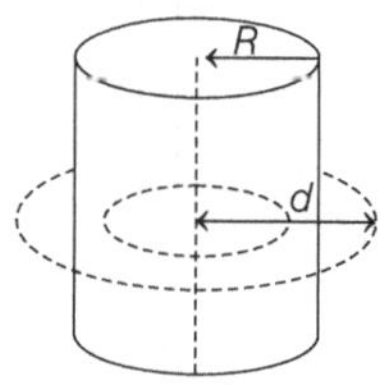


(i) The magnetic field at point outside cylinder, i.e. $d > R$.

From Ampere's circuital law, $\oint B \cdot dl = \mu_0 I$

$$\Rightarrow \quad B\int dl = \mu_0 I$$

$$B(2\pi d) = \mu_0 I$$

$$\Rightarrow \quad B = \frac{\mu_0 I}{2\pi d}$$

where, $\mu_0$ = permeability of free space.

$$\Rightarrow \quad B \propto \frac{1}{d}$$

(ii) The magnetic field at surface, i.e. $d = R$

$$B = \frac{\mu_0 I'}{2\pi R}$$

(iii) The magnetic field at inside point. The current for a point inside the cylinder is given by

$I'$ = current per unit cross-sectional area of cylinder × cross-section of loop

$$= \frac{I}{\pi R^2}\pi d^2 = \frac{Id^2}{R^2}$$

$$\therefore \quad B = \frac{\mu_0 I'}{2\pi d} = \frac{\mu_0 I d^2}{2\pi R^2 d} = \frac{\mu_0 I}{2\pi R^2} d$$

$$\Rightarrow \quad B \propto d$$

So, the variation of magnetic field can be plotted as

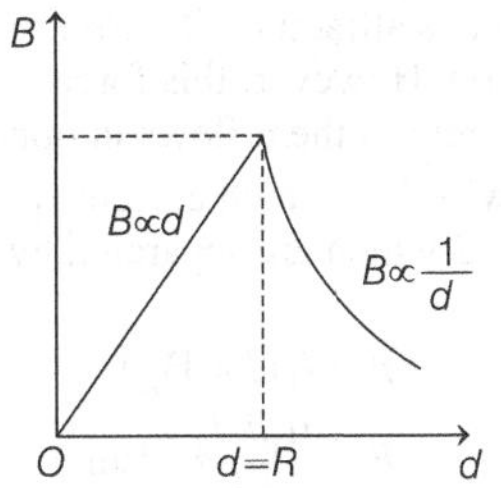


**45** (*b*) The magnetic field due to a infinite straight wire is given by

$$|dB| = \frac{\mu_0}{4\pi} \cdot \frac{Idl \sin \theta}{r^2}$$

where, $I$ = current in wire and $r$ = distance from wire.
Initially, the magnetic field at $M$ due to $PQ$,

$$|dB|_{PQ} = \frac{\mu_0}{4\pi} \cdot \frac{Idl \sin 90°}{r^2}$$

$$= \frac{\mu_0}{4\pi} \cdot \frac{Idl}{r^2} = B_1 \quad \text{(given)}$$

and due to $QR$,

$$|dB|_{QR} = \frac{\mu_0}{4\pi} \cdot \frac{Idl \sin 0°}{r^2} = \frac{\mu_0}{4\pi}(0) = 0$$

∴ Net magnetic field at $M$, due to $PQR$,

$$B_{net} = |dB|_{PQ} + |dB|_{QR} = |dB|_{PQ} + 0 = B_1$$

$$\therefore \quad \frac{B_{net}}{B_1} = 1$$

The ratio of net field at $M$ due to $PQR$ and $B_1$ is given by 1.

**46** (*d*) Two given identical long conducting wires are as shown below

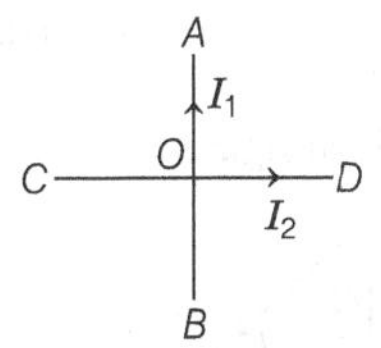


Since, the point $P$ lies at a distance $d$ such that it is perpendicular to the plane containing the wire. This means, if the wires lies in $xy$-plane then this point lies along $Z$-axis.

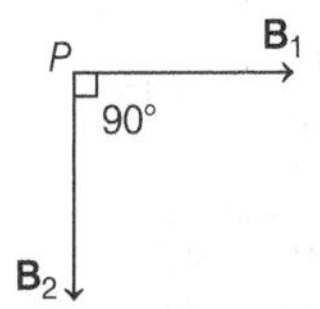


∴ Magnetic field, $|\mathbf{B}_1| = \frac{\mu_0}{2\pi} \frac{I_1}{d}$ and $|\mathbf{B}_2| = \frac{\mu_0}{2\pi} \frac{I_2}{d}$

$$B_{net} = \sqrt{B_1^2 + B_2^2}$$

$$\Rightarrow \quad B_{net} = \frac{\mu_0}{2\pi d} (I_1^2 + I_2^2)^{1/2}$$

**47** (*d*) Given, $n = 20$ turns cm$^{-1}$ $= 20 \times 100$ turns m$^{-1}$

$$= 2000 \text{ turns m}^{-1}$$

and magnetic field, $B = 20 \text{ mT} = 20 \times 10^{-3}$ T

By the formula, magnetic field, $B = \mu_0 nI$

or $\quad 20 \times 10^{-3} = 4\pi \times 10^{-7} \times 2000 \times I$

or $$I = \frac{20 \times 10^{-3}}{4\pi \times 10^{-7} \times 2000} \Rightarrow I \simeq 8 \text{ A}$$

**48** (*a*) Given, length, $l = 0.5$ m, radius, $r = 1$ cm $= 0.01$ m and current, $I = 5$ A.

The number of turns per unit length is

$$n = \frac{500}{0.5} = 1000 \text{ turns m}^{-1}$$

$\because \quad l/r = 50$, i.e. $l >> r$.

Hence, we can use the long solenoid formula,

Magnetic field, $B = \mu_0 nI$

$$= 4\pi \times 10^{-7} \times 10^3 \times 5$$

$$= 6.28 \times 10^{-3} \text{ T}$$

**49** (*b*) Given, inner radius of toroid = 0.25 m,
Outer radius of toroid = 0.26 m
Number of turns = 3500
and current = 11 A.

$$\therefore \text{ Mean radius, } r_m = \frac{0.25 + 0.26}{2} = 25.5 \times 10^{-2} \text{ m}$$

Number of turns in unit length,

$$n = \frac{N}{2\pi r_m} = \frac{3500}{2\pi \times 25.5 \times 10^{-2}}$$

Magnetic field of a toroid,

$$B = \mu_0 nI = 4\pi \times 10^{-7} \times \frac{3500}{2\pi \times 25.5 \times 10^{-2}} \times 11$$

$$= 3 \times 10^{-2} \text{ T}$$

**50** (*a*) The magnetic field within the turns of toroid is

$$B = \frac{\mu_0 NI}{2\pi r}$$

where, $N$ = number of turns, $I$ = current in loops
and $\quad r$ = radius of each turn.

Given, $N_1 = 200$, $N_2 = 100$, $r_1 = 40$ cm, $r_2 = 20$ cm and current $I$ is same, then

$$\frac{B_1}{B_2} = \frac{\mu_0 N_1 I}{2\pi r_1} \times \frac{2\pi r_2}{\mu_0 N_2 I}$$

Substituting the given values in the above relation, we get

$$\frac{B_1}{B_2} = \left(\frac{N_1}{N_2}\right)\left(\frac{r_2}{r_1}\right)$$

$$= \left(\frac{200}{100}\right)\left(\frac{20}{40}\right) = 2 \times \frac{1}{2} = 1$$

$$\therefore \quad B_1 : B_2 = 1 : 1$$

The ratio of their magnetic fields along the two loops is 1 : 1.

**51** (*a*) The force acting per unit length of a current carrying conductor,

$$\frac{F}{l} = \frac{\mu_0}{4\pi} \frac{2I_1 I_2}{d}$$

When $d = 1\text{m}$, $l = 1\text{m}$ and $I_1 = I_2 = 1$ A

$$F = 10^{-7} \times \frac{2 \times 1 \times 1}{1^2}$$

Then, $\quad F = 2 \times 10^{-7}\ \text{Nm}^{-1}$

**52** (*a*) Given, $I_1 = 4.0\text{A}, I_2 = 5.0\text{ A}, d = 4 \times 10^{-2}\text{m}$

and $l = 0.20$ m

Force on a current carrying wire,

$$F = \frac{\mu_0 I_1 I_2 l}{2\pi d} = \frac{4\pi \times 10^{-7} \times 4.0 \times 5.0 \times 0.20}{2 \times 3.14 \times 4 \times 10^{-2}}$$

$$= 2 \times 10^{-5}\ \text{N (towards } B)$$

**53** (*d*) The given situation can be drawn as below

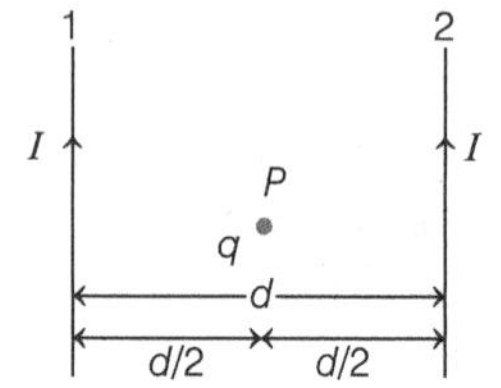


The magnetic field $\mathbf{B}_1$ at mid-point $P$ due to first wire is in downward direction, according to right-hand rule. Similarly, magnetic field $\mathbf{B}_2$ due to second wire is in upward direction.

Let the charge $q$ is moving with instantaneous velocity $\mathbf{v}$ in upward direction, then magnitude of force on $q$, due to first wire is

$F_1 = q(\mathbf{v} \times \mathbf{B}_1) = qvB_1 \sin\theta = qvB_1 \sin 180°$

Similarly, magnitude of force on $q$, due to second wire is

$$F_2 = q(\mathbf{v} \times \mathbf{B}_2) = qvB_2 \sin\theta$$
$$= qvB_2 \sin 0° = 0$$

Thus, net magnitude of force on $q$,

$$F_{\text{net}} = F_1 + F_2 = 0$$

**54** (*c*) Resultant magnetic field due to current carrying parallel wire 1 (10 A) and wire 2 ($I$A) at point $A$ is zero.

i.e. $\quad B_1 = B_2$

$$\frac{\mu_0 I_1}{2r_1} = \frac{\mu_0 I_2}{2r_2}$$

$$\Rightarrow \quad \frac{I_1}{r_1} = \frac{I_2}{r_2} \quad \text{...(i)}$$

As, given, $\quad r_1 = 9\text{ cm} = 9 \times 10^{-2}$ m

$r_2 = 18 + 9 = 27\text{ cm} = 27 \times 10^{-2}$ m

$\therefore$From Eq. (i),

$$\frac{10}{9 \times 10^{-2}} = \frac{I}{27 \times 10^{-2}} \Rightarrow I = 30\text{ A}$$

**55** (*d*) Net force experienced by two wires separated by same distance is attractive, if current flow in them in same direction. However, this force is repulsive in nature, if current in them flows in opposite direction.

Force on a wire 1 in which current $I_1$ is flowing due to another wire 2 which are separated by a distance $r$ is given as

$$\mathbf{F} = I_1 (l \times \mathbf{B}_2)$$

or $\quad F = \frac{\mu_0 I_1 I_2}{2\pi r} \cdot l \sin\theta \quad \left[\because \mathbf{B}_2 = \frac{\mu_0 I_2}{2\pi r}\right]$

Thus, the given square loop can be drawn as shown below

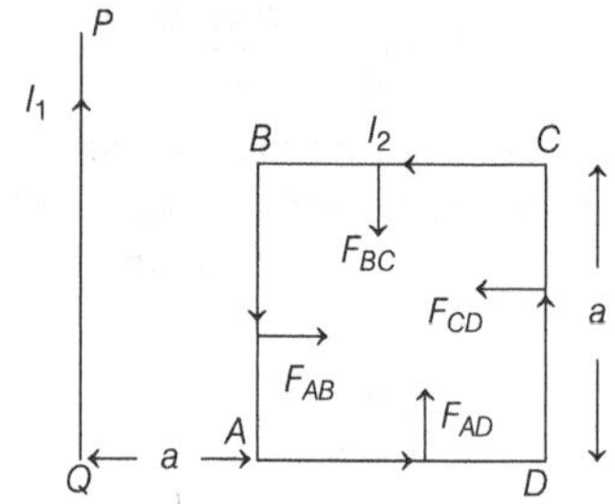


$$F_{AB} = \frac{\mu_0 I_1}{2\pi a} \cdot I_2 a \text{(away from wire } PQ)$$

$$F_{BC} = F_{AD} = 0 \quad [\because \theta = 0°]$$

$$F_{CD} = \frac{\mu_0 I_1}{2\pi(2a)} \cdot I_2 a$$

$$= \frac{\mu_0 I_1}{4\pi a} \cdot I_2 a \text{(towards the wire } PQ)$$

$\therefore \quad F_{\text{net}} = F_{AB} - F_{CD}$

$$= \frac{\mu_0 I_1 I_2}{2\pi} - \frac{\mu_0 I_1 I_2}{4\pi} \quad \text{(away from wire)}$$

$$= \frac{\mu_0 I_1 I_2}{4\pi} \quad \text{(repulsive in nature)}$$

**56** (*b*) Consider the given figure

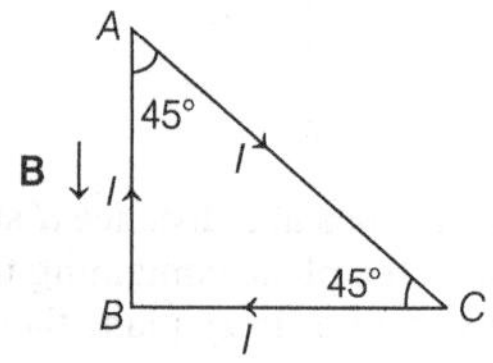


Take, $BC = l$

From figure, $\quad \frac{BC}{AC} = \sin 45°$

$$\Rightarrow \quad \frac{l}{AC} = \frac{1}{\sqrt{2}}$$

$$\Rightarrow \quad AC = \sqrt{2}l \quad \text{...(i)}$$

Since, magnetic field is along $AB$, therefore arm $BC$ is perpendicular to the direction of magnetic field $B$, i.e. $\theta = 90°$.

$\therefore \quad F_{BC} = IB \times BC \times \sin 90° = IBl \sin 90°$

$F_{BC} = IBl \quad \text{...(ii)}$

Force on wire $AC$,

$$F_{AC} = IB \times AC \times \sin 45°$$
$$= IB \times \sqrt{2}l \times \frac{1}{\sqrt{2}} \quad \text{[From Eq. (i)]}$$
$$= IBl = F_{BC}$$
$$\therefore \quad F_{AC} = F_{BC} = F \quad [\text{given, } F_{BC} = F]$$

**57** (*a*) Consider the given figure

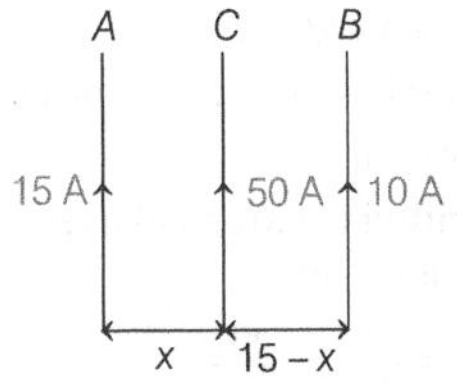


Since, wire $C$ is so replaced that, it experiences no net force, so $\frac{F_{AC}}{l} = \frac{F_{BC}}{l}$

$$\frac{\mu_0 2I_A I_C}{4\pi d} = \frac{\mu_0 2I_B I_C}{4\pi d'}$$
$$\frac{\mu_0}{4\pi} \cdot \frac{2 \times 15 \times 50}{x} = \frac{\mu_0}{4\pi} \times \frac{2 \times 50 \times 10}{15 - x}$$
$$\frac{15}{x} = \frac{10}{(15 - x)}$$
$$225 - 15x = 10x$$
$$\Rightarrow \quad 25x = 225 \Rightarrow x = 9\,\text{cm}$$

**58** (*d*) Consider the given figure.

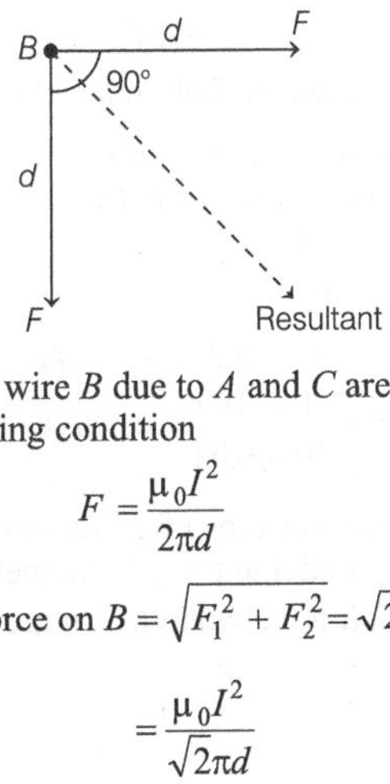


As force on wire $B$ due to $A$ and $C$ are attractive, so we have following condition

$$F = \frac{\mu_0 I^2}{2\pi d}$$

Resultant force on $B = \sqrt{F_1^2 + F_2^2} = \sqrt{2}\,F = \sqrt{2} \times \frac{\mu_0 I^2}{2\pi d}$

$$= \frac{\mu_0 I^2}{\sqrt{2}\pi d}$$

**59** (*b*) When a current carrying closed loop is suspended in a uniform magnetic field, $\mathbf{F}_{\text{net}} = 0$, only torque acts.

So, if force on one arm is **F**, then force of remaining three arms must be –**F**.

**60** (c) Given, area of the loop, $A = 2 \times 10^{-3}\,\text{m}^2$

Current, $I = 2\text{A}$

$\therefore$ Magnitude of magnetic dipole moment associated with the loop,

$$m = IA = 2 \times 2 \times 10^{-3} = 4 \times 10^{-3}\,\text{JT}^{-1}$$

**61** (*a*) The torque acting on the coil

$$|\tau| = |\mathbf{m} \times \mathbf{B}| = mB \sin\theta$$

Here, the circular coil is placed normal to the direction of magnetic field, then the angle between the direction of magnetic moment **m** and magnetic field **B** is zero, so

$$\tau = mB\sin\theta = mB\sin 0° = 0$$
$$\Rightarrow \quad \tau = 0$$

**62** (*d*) A current loop in a magnetic field can be in equilibrium in two orientations. For parallel $m$ and $B$, current loop is stable and for anti-parallel it is unstable.

**63** (*c*) Since, **B** is uniform, then only torque acts on current carrying conductor.

As, we know that torque, $\tau = \mathbf{m} \times \mathbf{B}$

$$\Rightarrow \quad |\tau| = |\mathbf{m}||\mathbf{B}|\sin\theta$$

where, $m$ = magnetic moment of coil.

For given orientation (anti-clockwise) shown in Fig. (ii), $\theta = 0° \Rightarrow \tau = 0$

So, for this orientation the loop would be in stable equilibrium.

and for Fig. (iv), $\theta = \pi \Rightarrow \tau = 0$

Therefore, for this orientation (clockwise) the loop would be in unstable equilibiurm.

**64** (*c*) Consider the following given figure

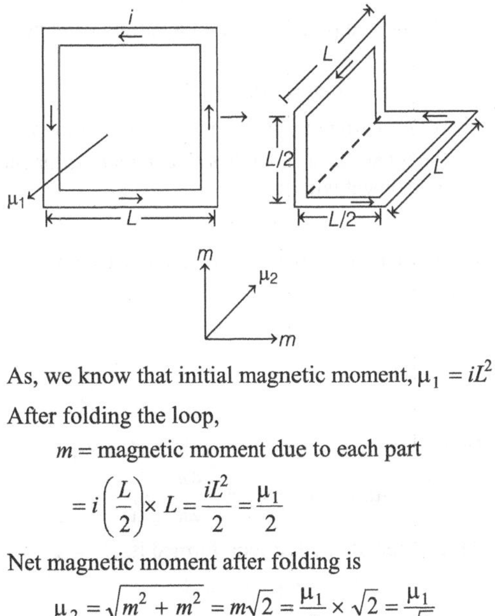


As, we know that initial magnetic moment, $\mu_1 = iL^2$

After folding the loop,

$m$ = magnetic moment due to each part

$$= i\left(\frac{L}{2}\right) \times L = \frac{iL^2}{2} = \frac{\mu_1}{2}$$

Net magnetic moment after folding is

$$\mu_2 = \sqrt{m^2 + m^2} = m\sqrt{2} = \frac{\mu_1}{2} \times \sqrt{2} = \frac{\mu_1}{\sqrt{2}}$$
$$\therefore \quad \frac{|\mu_1|}{|\mu_2|} = \sqrt{2}$$

**65** (*a*) Given, radius, $r = 1\text{Å} = 10^{-10}$ m

and speed, $v = 10^6$ m/s

Current associated with moving electron

$$I = \frac{e}{T} = \frac{e}{\left(\frac{2\pi r}{v}\right)} = \frac{ev}{2\pi r} \quad \ldots\text{(i)}$$

$\therefore$ Magnetic dipole moment associated with revolving electron,

$$m = IA$$
$$= \frac{ev}{2\pi r} \times \pi r^2 \quad \text{[from Eq. (i)]}$$
$$= \frac{evr}{2}$$
$$= \frac{1.6 \times 10^{-19} \times 10^6 \times 10^{-10}}{2}$$
$$= 0.8 \times 10^{-23} \text{A-m}^2 = 8 \times 10^{-24} \text{A-m}^2$$

**66** (*d*) Magnetic field at the centre of circular loop,

$$B = \frac{\mu_0}{4\pi} \frac{2\pi I}{r} = \frac{\mu_0 I}{2r}$$

[where, $I$ is current and $r$ is radius]

$$\Rightarrow \quad I = \frac{2Br}{\mu_0}$$

Also, $A = \pi r^2$ or $r = (A/\pi)^{1/2}$

Magnetic moment, $m = IA = \frac{2Br}{\mu_0} A$

On putting the value of $r$, we get

$$m = \frac{2BA}{\mu_0} \times \left(\frac{A}{\pi}\right)^{1/2} = \frac{2BA^{3/2}}{\mu_0 \pi^{1/2}}$$

**67** (*a*) Magnetic dipole moment of a current carrying loop is

$$m = IA (\text{A - m}^2)$$

where, $I$ = current in loop and $A$ = area of loop.

Let the given square loop has side $a$, then its magnetic dipole moment will be

$$m = Ia^2$$

When square is converted into a circular loop of radius $r$,

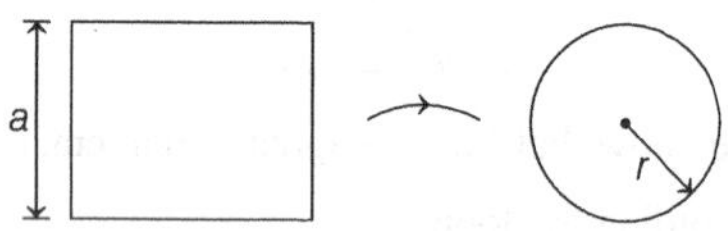


Then, wire length will be same in both area,

$$\Rightarrow \quad 4a = 2\pi r \Rightarrow r = \frac{4a}{2\pi} = \frac{2a}{\pi}$$

Hence, area of circular loop formed is, $A' = \pi r^2$

$$= \pi \left(\frac{2a}{\pi}\right)^2 = \frac{4a^2}{\pi}$$

Magnitude of magnetic dipole moment of circular loop will be

$$m' = IA' = I\frac{4a^2}{\pi}$$

Ratio of magnetic dipole moments of both shapes is,

$$\frac{m'}{m} = \frac{I \cdot \frac{4a^2}{\pi}}{Ia^2} = \frac{4}{\pi} \Rightarrow m' = \frac{4m}{\pi} (\text{A-m})$$

**68** (*a*) An electron orbiting in a circular orbit of radius $r$ with speed $v$ as shown below.

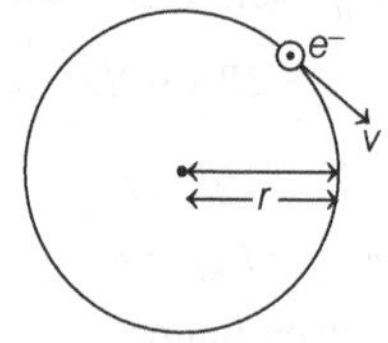


Magnetic moment, $M = NiA$

where, $N$ = number of turns of the current loop

and $i$ = current.

Since, the orbiting electron behaves as a current loop of current $i$, we can write

$$i = \frac{e}{T} = \frac{e}{\frac{2\pi r}{v}} = \frac{ev}{2\pi r}$$

where, $A$ = area of the loop = $\pi r^2$.

$$\Rightarrow \quad M = NiA = (1)\left(\frac{ev}{2\pi r}\right)(\pi r^2) = \frac{evr}{2}$$

**69** (*a*) For $n = 1$, in an $H_2$ atom,

$$\text{Bohr magneton, } \mu_B = \frac{e}{4\pi m_e} h$$
$$= \frac{1.60 \times 10^{-19} \times 6.63 \times 10^{-34}}{4 \times 3.14 \times 9.1 \times 10^{-31}}$$

$$\left[\because h = 6.63 \times 10^{-34} \text{ J-s}, \; m_e = 9.1 \times 10^{-31} \text{ kg}\right]$$

$$= 9.27 \times 10^{-24} \text{ A-m}^2$$

This value is called the Bohr magneton.

**72** (*d*) To convert a galvanometer into an ammeter resistance is connected in parallel,

$$\text{Resistance, } R = \frac{R_G r_s}{R_G + r_s}$$

Given, $R_G = 70\,\Omega$ and $r_s = 0.03\,\Omega$

$$\therefore \quad R = \frac{70 \times 0.03}{70 + 0.03} = 0.02998 = 0.030\,\Omega$$

**73** (*a*) An ammeter is a type of galvanometer with a shunt connected in parallel to the galvanometer.

Ammeter circuit is shown in the figure below

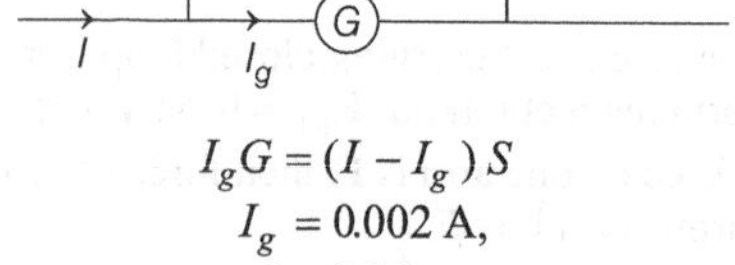


So, $I_g G = (I - I_g) S$

Here, $I_g = 0.002$ A,

$I = 0.5$ A,

$G = 50\,\Omega$

So, shunt resistance required is

$$S = \frac{I_g G}{I - I_g} = \frac{0.002 \times 50}{(0.5 - 0.002)} \approx 0.2\,\Omega$$

**74** (*b*) Here, resistance of the galvanometer,

$$R_g = 60\,\Omega$$

When the galvanometer is shunted by a resistance *r*, its effective resistance,

$$R_P = \frac{R_g r}{R_g + r} = \frac{60 \times 0.02}{60 + 0.02} \approx 0.02\,\Omega \quad [\because r = 0.02\,\Omega]$$

Total resistance of the circuit,

$$R_{\text{net}} = R + R_P = R + 0.02$$

$$\text{Current, } I = \frac{V}{R_{\text{net}}} = \frac{5}{R + 0.02}$$

$$\Rightarrow \quad 1 = \frac{5}{R + 0.02} \quad [\because I = 1\,\text{A}]$$

$$R + 0.02 = 5 \Rightarrow R = 5 - 0.02 = 4.98 \approx 5\,\Omega$$

**75** (*d*) Given, resistance of galvanometer, $R_g = 50\Omega$ and $I = 4 \times 10^{-4}$ A

Number of division = 25, $I_g = 25 \times I$

$$= 25 \times 4 \times 10^{-4} = 10^{-2}\text{ A}$$

Consider the given figure

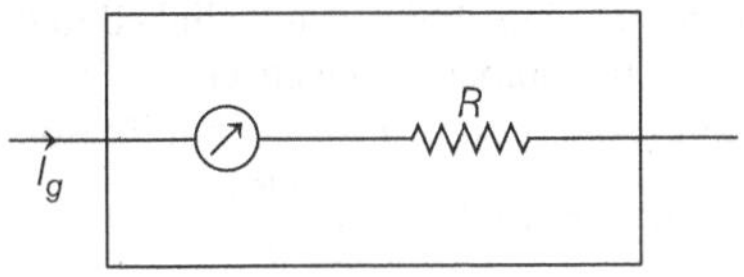


*R* is the resistance connected in series of galvanometer to convert into voltmeter.

$$\therefore \quad V = I_g (R_g + R)$$

$$25 = 10^{-2} (50 + R)$$

$$\Rightarrow \quad R = \frac{25}{10^{-2}} - 50 = 2500 - 50 = 2450\Omega \text{ in series}$$

**76** (*c*) Given, $G = 12\,\Omega$, $I_g = 2.5 \times 10^{-3}$ A and $V = 10$ V

$$\text{For ammeter, } S = \frac{GI_g}{I - I_g} = \frac{12 \times 2.5 \times 10^{-3}}{7.5 - (2.5 \times 10^{-3})}$$

$$\Rightarrow \quad S = 4 \times 10^{-3}\,\Omega$$

$$\text{Now,} \quad \frac{1}{R_a} = \frac{1}{G} + \frac{1}{S}$$

$$\text{Resistance of ammeter, } R_a = \frac{SG}{S + G} = \frac{4 \times 10^{-3} \times 12}{4 \times 10^{-3} + 12}$$

$$= 4.0 \times 10^{-3}\,\Omega$$

For a galvanometer as voltmeter, resistance,

$$R = \frac{V}{I_g} - G = \frac{10}{2.5 \times 10^{-3}} - 12$$

$$= 3988\,\Omega$$

Voltmeter's resistance,

$$R_V = R + G = 4000\,\Omega = 4 \times 10^3\,\Omega$$

$$\text{Ratio,} \quad \frac{R_a}{R_V} = \frac{4 \times 10^{-3}}{4 \times 10^3} = 10^{-6}$$

Hence, the ratio of net resistance of an ammeter of range 0 to 7.5 A and a voltmeter of range 0 to 10 V is $10^{-6}$.

**77** (*a*) Two galvanometers *A* and *B* produce a deflection of 10 divisions, then minimum current is measured by $A \rightarrow 0.3$ mA/div and by $B \rightarrow 0.5$ mA/div, so *A* is more sensitive than *B*.

**78** (*b*) Given, spring constant, $k = 10^{-8}$ N-m rad$^{-1}$,

Number of turns, $N = 100$,

Area, $A = 1\text{ cm}^2 = (1 \times 10^{-2})^2\text{ m}^2 = 1 \times 10^{-4}\text{ m}^2$

and magnetic field, $B = 5$ T.

$$\therefore \quad S_i = \frac{NBA}{k} = \frac{100 \times 5 \times 1 \times 10^{-4}}{10^{-8}} = 5 \times 10^6\text{ A}^{-1}$$

**79** (*a*) As we know, current sensitivity, $S_i = \dfrac{NBA}{k}$ ...(i)

and voltage sensitivity, $S_V = \dfrac{NBA}{kR}$ ...(ii)

From Eqs. (i) and (ii), we get

$$\frac{S_i}{S_V} = R \Rightarrow \frac{S_i}{G} = S_V \quad (\because R = G)$$

**80** (*d*) Voltage sensitivity,

$$S_V = \frac{\text{Current sensitivity } S_i}{\text{Galvanometer resistance } R}$$

Given, $R_{\text{initial}} = R$, $R_{\text{final}} = 2R_{\text{initial}} = 2R$

Since, $S_i$ has been increased to 50% of initial value, so

$$(S_i)_{\text{final}} = S_i + 50\% \text{ of } S_i$$

$$= 1.5\,S_i$$

$$\Rightarrow \text{New voltage sensitivity, } S_V' = \frac{1.5S_i}{2R} = \frac{3}{4}S_V$$

$$\left[\because S_V = \frac{S_i}{R}\right]$$

$$\text{Decrease in voltage sensitivity} = \left[1 - \left(\frac{3}{4}\right)\right] S_V = \left(\frac{1}{4}\right) S_V$$

**81** (*a*) Current sensitivity of a moving coil galvanometer is the deflection θ per unit current *I* flowing through it, i.e.

$$I_S = \frac{\theta}{I} = \frac{NAB}{k} \quad \text{...(i)}$$

where, $N$ = number of turns in the coil,

$A$ = area of each turn of coil, $B$ = magnetic field

and $k$ = restoring torque per unit twist of the fibre strip.

Similarly, voltage sensitivity is the deflection per unit voltage,

$$\text{i.e.} \quad V_S = \frac{\theta}{V} = \left(\frac{NAB}{k}\right)\frac{I}{V} = \frac{NAB}{kR_G} \quad \text{...(ii)}$$

where, $R_G$ is the resistance of the galvanometer.

From Eqs. (i) and (ii), we get

$$R_G = \frac{I_S}{V_S} \quad \text{...(iii)}$$

Given, $I_S = 5$ div/mA

$$= 5 \times 10^{-3}\text{ div/A}$$

and $\quad V_S = 20$ div/V

Substituting the given values in Eq. (iii), we get

$$R_G = \frac{5 \times 10^3}{20} = 250$$

$\therefore$ The resistance of the galvanometer is $250\,\Omega$.

**82** (*a*) A stationary charge produces electric field only. However, a moving charge which is equivalent to a current is produce a magnetic field in the surrounding space and can interacts with external magnetic field.

Therefore, Assertion and Reason are correct and Reason is the correct explanation of Assertion.

**83** (*a*) When an electron enters perpendicular to a uniform magnetic field, then a magnetic force acts on the electron in a direction perpendicular to both, the directions of magnetic field and the direction of velocity of electron.

We know that, when direction of force on a particle is perpendicular to the direction of velocity, then particle moves in uniform circular motion. Therefore, electron moving perpendicular to magnetic field **B** will perform circular motion.

Therefore, Assertion and Reason are correct and Reason is the correct explanation of Assertion.

**84** (*d*) When a charge particle of charge $q$ is released from rest $(u = 0)$ in the uniform magnetic field $B$, then magnetic force on the charged particle $F = Bqu\sin\theta = 0$, because $u = 0$. Therefore, the charge particle will not move on circular path.

In the magnetic field, force on charge particle always acts in perpendicular direction to the direction of velocity of charge particle, therefore work done by magnetic field on charge particle is zero.

Therefore, Assertion is incorrect but Reason is correct.

**85** (*a*) When a charged particle moves in a region of magnetic field, such that its velocity is at some acute angle with the direction of field, then this velocity has a component parallel to the magnetic field and a component perpendicular to the field.

There is no force on a charged particle in the direction of the field. Hence, the particle continues to travel with the given velocity parallel to the field.

The perpendicular component of velocity results in a Lorentz force $(\mathbf{v} \times \mathbf{B})$, which is also perpendicular to the plane containing **v** and **B**. This provides the necessary centripetal force and thus, the particle has a tendency to perform circular motion.

Hence, the perpendicular component of velocity causes a rotating centripetal force and parallel component of velocity does not produced any force and the resulting trajectory is a helix as shown

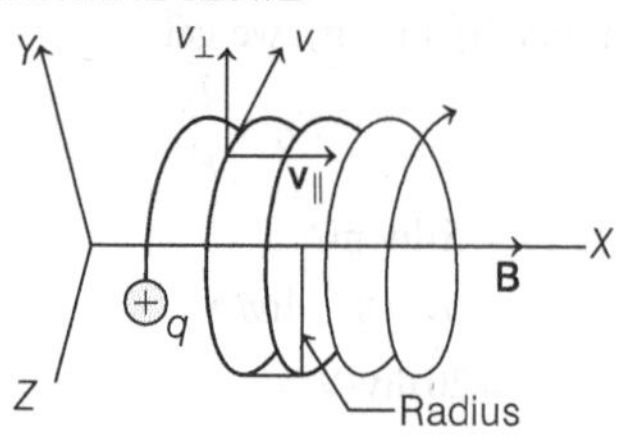


Therefore, Assertion and Reason are correct and Reason is the correct explanation of Assertion.

**86** (*c*) Cyclotron is utilised to accelerate the positive ions.

Cyclotron frequency is given by $\nu = \frac{Be}{2\pi m}$.

It means cyclotron frequency does not depend upon velocity.

Therefore, Assertion is correct but Reason is incorrect.

**87** (*b*) The magnetic field due to solenoid having $n$-number of turns/metre and carrying current $I$ is $B = \mu_0 nI$.

It is obvious that magnetic field is independent of length and cross-sectional area.

Also, magnetic field is uniform inside the solenoid.

Therefore, Assertion and Reason are correct but Reason is not the correct explanation of Assertion.

**88** (*d*) The electron is an elementary particle and it does not have an axis to spin like a top of our earth.

Also, it does not possess any intrinsic magnetic moment. The microscopic roots of magnetism in iron and other materials which can be linked to this intrinsic spin magnetic moment of electron.

Therefore, Assertion is incorrect but Reason is correct.

**89** (*d*) Current sensitivity, $S_i = \frac{NBA}{k}$

$$S_i \propto N$$

and voltage sensitivity, $S_V = \frac{NBA}{kR}$

$$S_V \propto \frac{N}{R}$$

So, when $S_i$ is increased by increasing number of turns $N$, length of wire used also increases and so resistance $R$ increases.

Hence, $S_V$ may remain same or decrease whereas $S_i$ increases.

Therefore, Assertion is incorrect but Reason is correct.

**90** (*b*) Statements II and III are correct but I and IV are incorrect and these can be corrected as,

When **v** is parallel to **B** or **v** is along **B**, then path of the charged particle is straight line.

**91** (*c*) In a velocity selector, where $F_e$ and $F_B$ are electric and magnetic forces.

**Case I** $F_e = F_B \Rightarrow v = E/B$

i.e. particles pass straight undeflected through the region.

**Case II** $F_e < F_B \Rightarrow v > E/B$

i.e. particles deflects in a direction perpendicular to both **v** and **B**.

**Case III** $F_e > F_B \Rightarrow v < E/B$

i.e. particles deflects in the direction of electric field.

So, statements I, II and III are correct but IV is incorrect.

**92** (*a*) Statements I and II are correct but III and IV are incorrect and these can be corrected as,

Biot-Savart's law and Coulomb's law both can be expressed in vector form.

Biot-Savart's law and Coulomb's law, both are applicable in all type of medium.

**93** (*c*) Statements I and II are correct but III and IV are incorrect and these can be corrected as,

As the deflection ($\alpha$) of the coil is proportional to the current passed through it, so a linear scale can be used to measure the deflection.

i.e. $I \propto \alpha$

A galvanometer can give accurates values.

**94** (*a*) Statement given in option (a) is correct while rest are incorrect, and these can be corrected as,

The electric field can convey energy and momentum and is not established instantaneously but takes finite time to propagate.

The magnetic field of several sources is the vector addition of magnetic field of each individual source.

The Lorentz force depends on $q$, **v** and **B**. Force on a negative charge is opposite to that on a positive charge.

**95** (*c*) Statement given in option (c) is incorrect and it can be corrected as

The magnitude of magnetic field $B$ is 1 SI unit, when the force acting on a unit charge (1 C), moving perpendicular to **B** with a speed of 1 m/s is one newton.

Rest statements are correct.

**96** (*b*) For proton, $\mathbf{v} \perp \mathbf{B}$ and for electron $\mathbf{v} \parallel \mathbf{B}$. So, force on proton $= q\,(\mathbf{v} \times \mathbf{B}) = Bqv$, whereas force on electron = 0.

So, the proton will move in a circle with constant speed and there is no effect on the motion of electron.

Thus, the statement given in option (b) is correct, rest are incorrect.

**97** (*b*) The radius of the circle described by a charged particle in a magnetic field is given by

$$r = \frac{mv}{qB}$$

The larger the momentum, the larger is the radius and bigger the circle described.

The distance moved along the magnetic field in one rotation is called pitch $P$ and given by

$$P = v_{\parallel} T = 2\pi m v_{\parallel} / qB.$$

In cyclotron as the fields are perpendicular to each other, they are called crossed fields.

The frequency $\nu_a$ of the applied voltage is adjusted, so that the polarity of the dees is reversed in the same time that it takes the ions to complete one half of the revolution.

Thus, the statement given in option (b) is correct, rest are incorrect.

**98** (*c*) Statement given in option (c) is incorrect and it can be corrected as,

As in case of moving charged particle in magnetic field $\mathbf{F}_{\text{mag}} \perp \mathbf{v}$, hence power associated will be zero.

Rest statements are correct.

**99** (*a*) Statement given in option (a) is incorrect and it can be corrected as,

The operation of the cyclotron is based on the fact that the time for one revolution of an ion is independent of its speed or radius of its orbit.

Rest statements are correct.

**100** (*d*) Statement given in option (d) is correct while rest are incorrect and these can be corrected as,

The magnitude of the magnetic field $d\mathbf{B}$ is proportional to the current $I$, the element length $|d\mathbf{l}|$ and inversely proportional to the square of the distance $r$. Its direction is perpendicular to the plane containing $d\mathbf{l}$ and **r**.

Both electric and magnetic fields are long range, since both depend inversely on the square of distance from the source to the point of interest.

The principle of superposition applies to both fields.

**101** (*b*) Statement given in option (b) is incorrect and it can be corrected as,

The lines of constant magnitude of magnetic field form concentric circles called magnetic field lines. It originates from positive charges and ends at negative charges.

Rest statements are correct.

**102** (*d*) Statement given in option (d) is incorrect and it can be corrected as,

Ampere's circuital law holds for steady currents which are not accelerated with time.

Rest statements are correct.

**103** (*d*) Statement given in option (d) is incorrect and it can be corrected as,

The torque can be expressed as vector product of magnetic moment $m$ of the coil and the magnetic field $B$, i.e. $\tau = \mathbf{m} \times \mathbf{B}$, where $\mathbf{m} = I \cdot \mathbf{A}$.

Rest statements are correct.

**104** (*d*) When an electron is projected with uniform velocity $v$ along the axis of solenoid force on it,

$$F = qvB\sin\theta = qvB\sin 0° = 0, \text{ as } \theta = 0 \text{ or } \pi$$

So, the electron will continue to move with uniform velocity along the axis of the solenoid.

Thus, the statement given in option (d) is correct, rest are incorrect.

**105** (*c*) Statement given in option (c) is incorrect and it can be corrected as,

The field outside the solenoid approaches zero and everywhere inside becomes parallel to the axis.

Rest statements are correct.

**106** (*b*) Statement given in option (b) is correct but rest are incorrect and these can be corrected as,

A toroid is a hollow circular ring on which a large number of turns of a wire are closely wound.

The magnetic field at any point in the open space inside the toroid is zero.

The field **B** inside the toroid is constant in magnitude for the ideal toroid of closely wound turns.

The number of turns in a toroid is given by $N = 2\pi rn$ = (average) perimeter of the toroid × number of turns per unit length.

**107** (*a*) Statement given in option (a) is correct, rest are incorrect and these can be corrected as,

The magnetic field for a point in the plane of the loop at a distance $x$ from its centre is $\mathbf{B} \approx \frac{\mu_0}{4\pi}\frac{\mathbf{m}}{x^3}$ $(x >> R)$.

A planar current loop is equivalent to a magnetic dipole of dipole moment $\mathbf{m} = I \cdot \mathbf{A}$.

**108** (*d*) Statement given in option (d) is incorrect and it can be corrected as,

When a galvanometer is used as a voltmeter, it must draw a very small current, otherwise the voltage measurement will disturb the original set up by an amount which is very large. So, to keep the disturbance due to the measurement device below one per cent, large resistance $R$ is connected in series with the galvanometer.

Rest statements are correct.

**109** (*a*)

A. Since, $\mathbf{E} = 0$ and $\mathbf{B} \neq 0$ and $\mathbf{E} \neq \mathbf{B}$, the magnetic force act on it and the path will be helical (with uniform pitch) if $\mathbf{v}$ is at any angle to $\mathbf{B}$ $(\theta \neq 90°)$.

B. Since, $\mathbf{E} = 0$, $\mathbf{B} \neq 0$ and $v \perp \mathbf{B}$, then a force called Lorentz force, $F = q(\mathbf{v} \times \mathbf{B})$, acts on charged particle, which provide centripetal force to it and hence the path will be circular.

C. Since, $\mathbf{E} \neq 0$, $\mathbf{B} \neq 0$, $\mathbf{E} \parallel \mathbf{B}$, so path will be helical with non-uniform pitch as both electric and magnetic forces act on it.

D. Since, $\mathbf{E} \neq 0$, $\mathbf{B} \neq 0$, $\mathbf{E} \perp \mathbf{B}$ and $\mathbf{v}$ is perpendicular to both $\mathbf{E}$ and $\mathbf{B}$, then the particle will move in a straight line due to the resultant force of magnetic and electric fields.

Hence, A → 2, B → 3, C → 4 and D → 1.

**112** (*a*) Given, $B = 3.0 \times 10^{-5}$ T and $I = 1$ A

Force acting on a horizontal table, $F = IlB \sin\theta$

So, the force per unit length is, $f = \frac{F}{l} = IB\sin\theta$

When the current is flowing from east to west, then $\theta = 90°$,

$\Rightarrow \quad f = IB\sin 90° = 1 \times 3.0 \times 10^{-5} \times 1$

$= 3 \times 10^{-5}$ Nm$^{-1}$

**113** (*b*) Given, $n = 100$, $r = 8\text{ cm} = 8 \times 10^{-2}$ m and $I = 0.40$ A

The magnetic field $B$ at the centre,

$$B = \frac{\mu_0}{4\pi}\cdot\frac{2\pi In}{r} = \frac{10^{-7} \times 2 \times 3.14 \times 0.40 \times 100}{8 \times 10^{-2}}$$

$= 3.1 \times 10^{-4}$ T

**114** (*c*) Given, $I = 50$ A and $r = 2.5$ m

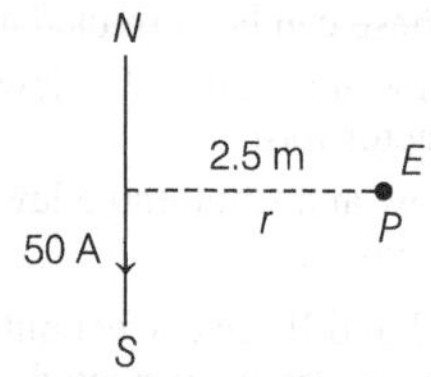


The magnitude of magnetic field,

$$B = \frac{\mu_0}{4\pi}\cdot\frac{2I}{r}$$

$$= 10^{-7}\cdot\frac{2 \times 50}{2.5}$$

$= 4 \times 10^{-6}$ T, in upward direction

**115** (*b*) Given, $I = 90$ A and $r = 1.5$ m

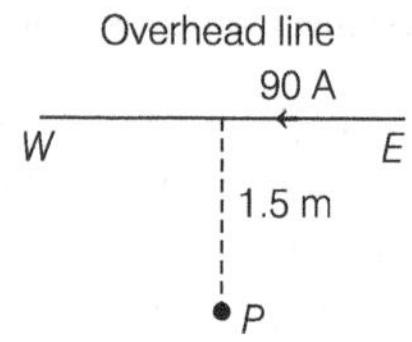


The magnitude of magnetic field,

$$B = \frac{\mu_0}{4\pi}\cdot\frac{2I}{r} = \frac{10^{-7} \times 2 \times 90}{1.5} = 1.2 \times 10^{-5}\text{ T}$$

The direction of $B$ will be vertically upward.

**116** (*b*) Given, $I = 8$ A, $\theta = 30°$, and $B = 0.15$ T

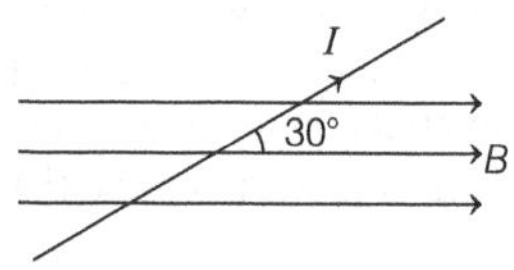


The magnitude of magnetic force,

$$|\mathbf{F}| = I(\mathbf{l} \times \mathbf{B}) = I\,l\,B\sin\theta$$

$\Rightarrow \quad F = 8 \times l \times 0.15 \times \sin 30°$

The magnitude force per unit length,

$$f = \frac{F}{l} = \frac{8 \times 0.15}{2} = 0.6\text{ Nm}^{-1}$$

**117** (*b*) Given, $l = 3.0\text{ cm} = 3.0 \times 10^{-2}$m, $I = 10$ A and $B = 0.27$ T.

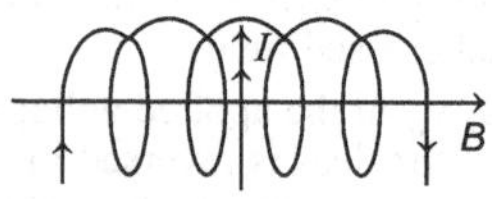


From the figure, the angle between the magnetic field and the direction of flow of current is 90°.

The magnitude of magnetic force on the wire,

$$F = IlB\sin 90°$$

$= 10 \times 3.0 \times 10^{-2} \times 0.27 \times \sin 90°$

$= 8.1 \times 10^{-2}$ N

**118** (*d*) Given, $I_1 = 8.0$ A, $I_2 = 5.0$ A

and $r = 4.0$ cm $= 0.04$ m

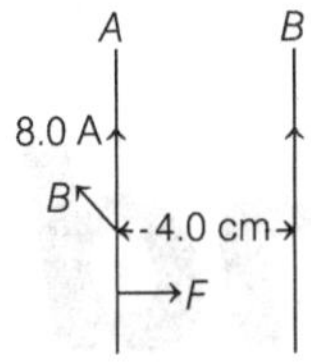


Force per unit length on two parallel wire carrying current,

$$F = \frac{\mu_0}{4\pi} \cdot \frac{2I_1 \cdot I_2}{r}$$

$$= \frac{10^{-7} \times 2 \times 8.0 \times 5.0}{0.04}$$

$$= 2 \times 10^{-4} \text{ N}$$

The force on a 10 cm section of wire $A$ is

$$F' = F \times 0.1 \quad (\because 10 \text{ cm} = 0.1 \text{ m})$$

$$\Rightarrow F' = 2 \times 10^{-4} \times 0.1$$

$$= 2 \times 10^{-5} \text{ N}$$

This force is attractive as currents are in same direction.

**119** (*a*) Given, $l = 80$ cm $= 0.8$ m and $I = 8.0$ A.

The total number of turns, $N = 400 \times 5 = 2000$

and number of turns/length, $n = \dfrac{2000}{0.8} = 2500$

The magnitude of magnetic field inside the solenoid,

$$B = \mu_0 nI = 4\pi \times 10^{-7} \times nI$$

$$= 4 \times 3.14 \times 10^{-7} \times 2500 \times 8$$

$$= 2.5 \times 10^{-2} \text{ T}$$

The direction of magnetic field is along the axis of solenoid.

**120** (*b*) Given, $a = 10$ cm, $A = a^2 = (10 \times 10^{-2})^2$ m$^2$,

$N = 20, I = 12$ A, $\theta = 30°$ and $B = 0.80$ T

The magnitude of torque experienced by the coil,

$\tau = NI\,AB \sin\theta$

$$= 20 \times 12 \times (10 \times 10^{-2})^2 \times 0.80 \times \sin 30°$$

$$\tau = 2.4 \times 0.80 \times \sin 30° = \frac{2.4 \times 0.80}{2} = 0.96 \text{ N-m}$$

**121** (*d*) Using the formula of voltage sensitivity, $V = \dfrac{NAB}{kR}$

$$\therefore \quad \frac{V_{s_2}}{V_{s_1}} = \frac{N_2 B_2 A_2 k_1 R_1}{k_2 . R_2 N_1 B_1 A_1}$$

Given, $R_1 = 10\,\Omega, N_1 = 30, A_1 = 3.6 \times 10^{-3}$ m$^2$,

$B_1 = 0.25$ T, $R_2 = 14\,\Omega, N_2 = 42, A_2 = 1.8 \times 10^{-3}$ m$^2$,

$B_2 = 0.50$ T and $k_2 = k_1 = k$

$$\Rightarrow \frac{V_{s_2}}{V_{s_1}} = \frac{42 \times 0.50 \times 1.8 \times 10^{-3} \times k \times 10}{k \times 14 \times 30 \times 0.25 \times 3.6 \times 10^{-3}} = 1$$

**122** (*d*) Here, $\mathbf{F} = q\,(\mathbf{v} \times \mathbf{B}) = e\,(\mathbf{v} \times \mathbf{B}) = evB \sin 90°$

As this force provides the centripetal force, i.e.

$$evB \sin 90° = \frac{mv^2}{r}$$

$$\Rightarrow r = \frac{mv}{eB \times 1}$$

Given, $B = 6.5$ G $= 6.5 \times 10^{-4}$ T, $v = 4.8 \times 10^6$ ms$^{-1}$

$e = 1.6 \times 10^{-19}$ C and $m_e = 9.1 \times 10^{-31}$ kg

$$\Rightarrow r = \frac{9.1 \times 10^{-31} \times 4.8 \times 10^6}{1.6 \times 10^{-19} \times 6.5\ 10^{-4}}$$

$$= 4.2 \times 10^{-2} \text{ m} = 4.2 \text{ cm}$$

**123** (*c*) Given, $N = 30, r = 8$ cm $= 0.08$ m, $I = 6$ A,

$B = 1.0$ T and $\theta = 60°$.

Magnitude of torque acting on the current carrying coil due to the magnetic field,

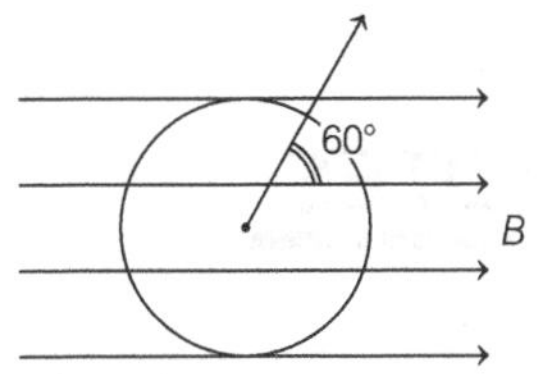


$$\tau = NIAB \sin\theta$$

$$= 30 \times 6 \times \pi (0.08)^2 \times 1 \times \sin 60°$$

$$= 30 \times 6 \times 3.14 \times 0.08 \times 0.08 \times \frac{\sqrt{3}}{2}$$

$$= 3.133 \text{ N-m}$$

**124** (*d*) For given pitch $d$ of a helical path corresponding to motion of a charged particle in a uniform magnetic field, we have,

$$\frac{q}{m} = \frac{2\pi v \cos\theta}{qB} = \text{constant}$$

$$\Rightarrow \left(\frac{e}{m}\right)_1 + \left(\frac{e}{m}\right)_2 = 0$$

**125** (*a*) Magnetic field associated with moving charge,

$$\mathbf{B} = \frac{\mu}{4\pi} q \frac{\mathbf{v} \times \mathbf{r}}{r^2}$$

According to thumb's rule, **B** is perpendicular to **v**.

**126** (*a*) The charged particle in a cyclotron undergoes acceleration as

(i) It speeds up between the dees because of the oscillating electric field and

(ii) Its speed remains the same inside the dees, because of the magnetic field but direction undergoes change continuously.

**127** (*d*) The rotation of the loop by 30° about an axis perpendicular to its plane make no change in the angle made by magnetic moment with the direction of magnetic field, therefore the work done to rotate the loop is zero.

CHAPTER > 05

# Magnetism and Matter

## KEY NOTES

- A naturally occurring ore of iron, magnetite attracts small pieces of iron towards it. The phenomenon of attraction of small bits of iron, steel, cobalt, nickel, etc. towards the ore, is called **magnetism**.

### Bar Magnet

- A magnet is a material or an object that produces a magnetic field. A bar magnet consists of two equal and non-separable magnetic poles.
- When a bar magnet is freely suspended, it points in the north-south direction.
- The tip which points to the geographic north is called the **north pole** and the tip which points to the geographic south is called the **south pole** of the magnet.
- There is a repulsive force when north pole (or south pole) of two magnets are brought close together. Conversely, there is an attractive force between the north pole of one magnet and the south pole of the other.
- We cannot isolate the north or south pole of a magnet. If a bar magnet is broken into two halves, we get two similar bar magnets with somewhat weaker properties. Unlike electric charges, isolated magnetic north and south pole known as **magnetic monopoles** do not exist.

### Magnetic Field Lines

- The magnetic field lines of a magnet are the imaginary lines which continuously represent the direction of that magnetic field.
- Important properties of magnetic field lines are given below
  (i) The magnetic field lines of a magnet (or a solenoid) form continuous closed loops.
  (ii) The tangent to the field line at a given point represents the direction of the net magnetic field **B** at that point.
  (iii) The larger the number of field lines crossing per unit area, the stronger is the magnitude of the magnetic field **B**.
  (iv) The magnetic field lines do not intersect, for if they do, the direction of the magnetic field would not be unique at the point of intersection.

### Magnetic Dipole

- It is an arrangement which consists of two magnetic poles of equal and opposite strengths separated at a small distance. A bar magnet, a compass needle, etc. are the examples of magnetic dipoles.
- **Pole Strength** ($q_m$) It can be defined as the strength of a magnetic pole to attract magnetic materials towards itself. It depends on the nature of material and area of cross-section of the magnet.

  The force between two magnetic poles separated by some distance is given by

$$F = \frac{\mu_0}{4\pi} \frac{q_{m_1} \cdot q_{m_2}}{r^2}$$

  It is called **Coulomb's law of magnetic force.**
- **Magnetic Dipole Moment** It is the product of the strength of either of the pole strength ($\pm q_m$) and the magnetic length of the magnet. It is represented by **m**.

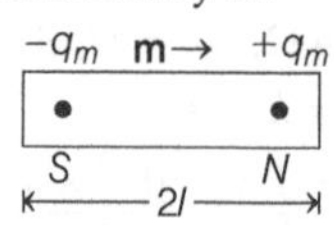


The direction of magnetic dipole moment is same as that of 2*l*. Therefore, $\mathbf{m} = q_m(2\mathbf{l})$.

## Bar Magnet as an Equivalent Solenoid

- The resemblance of magnetic field lines of a bar magnet and a solenoid suggest that a bar magnet may be thought of as a large number of circulating currents in analogy with a solenoid.
- Cutting a bar magnet in half is like cutting a solenoid, we get two smaller solenoids with weaker magnetic properties.
- The field lines remain continuous, emerging from one face of the solenoid and entering into the other face.
- The magnitude of the magnetic moment of the solenoid is $m$ = (total number of turns × current × cross sectional area) $= n(2l)\ I(\pi a^2)$.
- A bar magnet and a solenoid produce similar magnetic fields. The magnetic moment of a bar magnet is thus equal to the magnetic moment of an equivalent solenoid that produces the same magnetic field.

## Dipole in Uniform Magnetic Field

- A magnetic dipole when placed in a uniform magnetic field does not experience any net force. However, it experiences a torque given by

$$\tau = \mathbf{m} \times \mathbf{B} = mB\sin\theta$$

where, $\theta$ is the angle from magnetic field along which the dipole has been placed.

- The equation of simple harmonic motion is represented as

$$\frac{d^2\theta}{dt^2} = -\frac{mB}{I}\theta$$

where negative sign shows that the restoring torque is in opposition to deflecting torque.

- The time period of the oscillation is

$$T = 2\pi\sqrt{\frac{I}{mB}}$$

and frequency is, $f = \frac{1}{2\pi}\sqrt{\frac{mB}{I}}$

- Due to the torque, dipole align itself in the direction of field and so, some work has to be done to rotate it. This work done is stored as potential energy of the dipole and given by

$$U_m = \int \tau(\theta)d\theta = -mB\cos\theta$$

or $$U_m = -\mathbf{m}\cdot\mathbf{B}$$

(i) $U_m = 0$, at $\theta = 90°$

(ii) $U_{min} = -mB$, at $\theta = 0°$ (most stable position)

(iii) $U_{max} = +mB$, at $\theta = 180°$ (most unstable position)

- **Work done** in rotating the dipole in a uniform magnetic field from $\theta_1$ to $\theta_2$ is given by

$$W = mB(\cos\theta_1 - \cos\theta_2)$$

## Electrostatic Analogy

The analogy between electric and magnetic dipoles is as shown below

| Physical quantities | Electrostatics | Magnetism |
|---|---|---|
| | $1/\varepsilon_0$ | $\mu_0$ |
| Dipole moment | $\mathbf{p}$ | $\mathbf{m}$ |
| Equatorial field for a short dipole | $-\mathbf{p}/4\pi\varepsilon_0 r^3$ | $-\mu_0\mathbf{m}/4\pi r^3$ |
| Axial field for a short dipole | $2\mathbf{p}/4\pi\varepsilon_0 r^3$ | $-\mu_0 2\mathbf{m}/4\pi r^3$ |
| External field : Torque | $\mathbf{p}\times\mathbf{E}$ | $\mathbf{m}\times\mathbf{B}$ |
| External field : Energy | $-\mathbf{p}\cdot\mathbf{E}$ | $-\mathbf{m}\cdot\mathbf{B}$ |

## Magnetism and Gauss's Law

The net magnetic flux $\phi_B$ through any closed surface is always zero, i.e. the number of magnetic field lines leaving the surface is balanced by the number of lines entering it, i.e.

$$\phi_B = \Sigma\mathbf{B}\cdot\Delta\mathbf{S} = 0$$

where, $B$ = magnetic field and $\Delta S$ = area element.

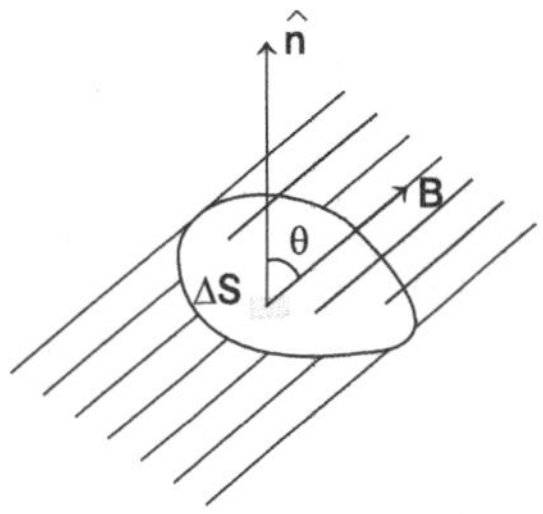


## Earth's Magnetism

- Our earth behaves as a huge powerful magnet and its strength varies from place to place on the earth's surface.
- Some definitions related to earth's magnetism are as follows
  - (*i*) **Geographic Axis** The straight line passing through the geographical north and south poles of the earth is called its geographic axis. It is the axis of rotation of the earth.
  - (*ii*) **Magnetic Axis** The straight line passing through the magnetic north and south poles of the earth is called the magnetic axis.
  - (*iii*) **Magnetic Equator** It is the huge circle on the earth perpendicular to the magnetic axis.
  - (*iv*) **Magnetic Meridian** The vertical plane in the direction of $B$ is called magnetic meridian.
  - (*v*) **Geographic Meridian** The vertical plane passing through the line joining the geographical north and south poles is called the geographic meridian.

**Note** The magnetic field of earth arises due to electrical currents produced by convective motion of metallic fluid (consisting mostly of molten iron and nickel) in the outer core of the earth. This is known as the **dynamo effect**.

## Magnetic Elements of Earth

- The earth's magnetic field at a place can be completely described by three parameters which are called elements of earth's magnetic field.
- There are three magnetic elements of the earth as given below
  - (i) **Magnetic Declination** Magnetic declination or simply declination at a place is the angle between the geographic meridian and magnetic meridian. It is denoted by $\theta$.
  - (ii) **Magnetic Inclination or Dip** Dip at a place is defined as the angle made by the direction of the earth's total magnetic field with the horizontal direction. It is denoted by $\delta$.
  - (iii) Horizontal and Vertical Component of the Earth's Magnetic Field These are the components of the earth's magnetic field along the horizontal and vertical directions, respectively. These are denoted by $B_H$ and $B_V$.

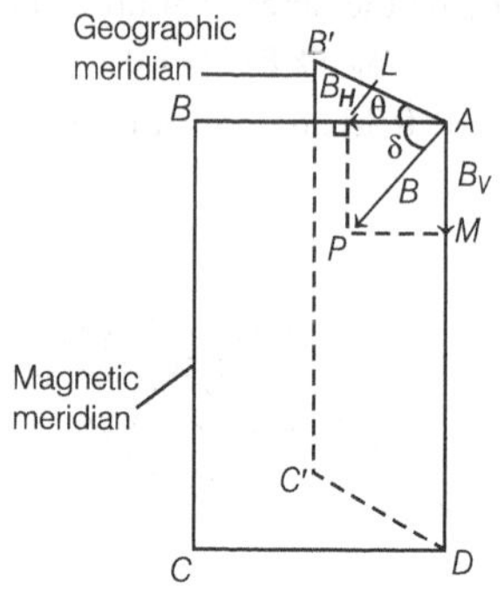


In the figure, $B_H = B\cos\delta$, $B_V = B\sin\delta$ and $\tan\delta = \frac{B_V}{B_H}$.

## Terms Used in Magnetism

The various terms are used in magnetism as given below

(i) **Magnetic induction (*B*)** is the number of magnetic field lines passing normally through a cross-sectional area of material. Its SI unit is tesla (T) or weber/metre$^2$ (Wm$^{-2}$) and CGS unit is gauss (G).
Also, $1\text{T} = 10^4$ G.

(ii) **Intensity of magnetisation (*I*)** is the magnetic moment per unit volume of the magnetic substance.

i.e. $$I = \frac{m_{\text{net}}}{V}$$

Its SI unit is Am$^{-1}$.

(iii) **Magnetic permeability** ($\mu$) is the ratio of the magnetic induction to the magnetic intensity.

i.e. $$\mu = \frac{B}{H}$$

Its SI unit is Wb/A-m or TmA$^{-1}$.

(iv) **Relative magnetic permeability** ($\mu_r$) is the ratio of magnetic permeability of the substance to the permeability of free space and given as $\mu_r = \frac{\mu}{\mu_0} = \frac{B}{B_0}$.

(v) **Magnetic intensity (*H*)** is a measure of the capability of external magnetic field to magnetise the given substance and given by $H = \frac{B_0}{\mu_0} = \frac{B}{\mu_0} - I$.

Its SI unit is Am$^{-1}$ and CGS unit is oersted.

(vi) **Magnetic susceptibility ($\chi_m$)** is the ratio of intensity of magnetisation in a substance to the magnetic intensity.

i.e. $$\chi_m = \frac{I}{H}$$

$\mu_r$ and $\chi_m$ are related as $\mu_r = \chi_m + 1$.

## Magnetic Properties of Materials

In terms of magnetic behaviour, all substances are categorised in three types

### Diamagnetic

- These substances have tendency to move from stronger to the weaker part of the external magnetic field.
- Their features are given below
  - (i) The magnetic field inside them reduces, when placed in an external magnetic field.
  - (ii) Diamagnetic substances are the ones in which resultant magnetic moment in an atom is zero.
  - (iii) Some diamagnetic materials are bismuth, copper, lead, silicon, nitrogen (at STP), water and sodium chloride.
  - (iv) Superconductors are metals, cooled to very low temperature which exhibits both perfect conductivity and perfect diamagnetism.
  - (v) The phenomenon of perfect diamagnetism in superconductors is called the **Meissner effect**.
  - (vi) Superconducting magnets can be gainfully exploited in variety of situation, e.g. for running magnetically levitated superfast trains.

### Paramagnetic

- These substances get weakly magnetised when placed in an external magnetic field. They have a tendency to move from a region of weak magnetic field to strong magnetic field.
- Their features are given below
  - (i) The individual atoms (or ions or molecules) of a paramagnetic materials possess a permanent magnetic dipole moment of their own.
  - (ii) The field lines gets concentrated inside the paramagnetic material.
  - (iii) Some paramagnetic materials are aluminium, sodium, calcium, oxygen (at STP) and copper chloride.
  - (iv) The magnetisation of a paramagnetic material is inversely proportional to the absolute temperature *T*.

$$I = C\frac{B_0}{T} \text{ and } \chi = \frac{C\mu_0}{T}$$

KEY NOTES

This is known as **Curie's law** and the constant $C$ is called **Curie's constant.**

(v) For a paramagnetic material, both $\chi$ and $\mu$ depend not only on the materials but also on the sample temperature.

## Ferromagnetic

These substances get strongly magnetised when placed in an external magnetic field. They have strong tendency to move from a region of weak magnetic field to strong magnetic field.

Their features are given below

(i) The individual atoms (or ions or molecules) in a ferromagnetic materials possess a dipole moment which align themselves in a common direction over a macroscopic volume called **domain**.

(ii) In a ferromagnetic material, the field lines are highly concentrated.

(iii) In some ferromagnetic materials, the magnetisation persists, such materials are called **hard magnetic materials** or **hard ferromagnets**. Alnico, an alloy of iron, aluminium, nickel cobalt and copper, are such materials, the naturally occurring lodestone is another.

(iv) On the other hand, there is a class of ferromagnetic materials in which the magnetisation disappears on removal of the external field. Soft iron is one such material, such materials are called **soft ferromagnetic materials**. Iron, cobalt, nickel, gadolinium, etc are such materials.

(v) The relative magnetic permeability ($\mu_r$) is > 1000.

(vi) The ferromagnetic property depends on temperature. At high enough temperature, a ferromagnet becomes a paramagnet. The temperature of transition from ferromagnetism to paramagnetism is called the **Curie temperature $T_C$.**

(vii) The susceptibility above the Curie temperature, i.e. in the paramagnetic phase is described by

$$\chi = \frac{C}{T - T_C} \qquad (T > T_C)$$

- The curve between **B** and **H** for ferromagnetic material is depicted below

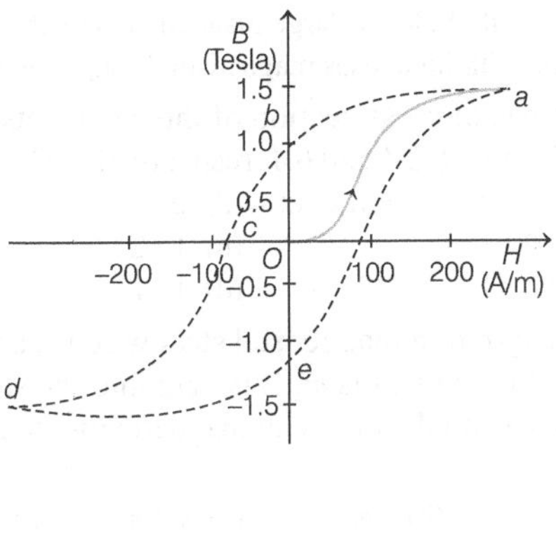


Some features of this graph are given below

(i) The value of $B$ at $H = 0$ is called **retentivity** or **remanence.**

(ii) The value of $H$ at $B = 0$ is called **coercivity**.

(iii) For a given value of $H$, $B$ is not unique but depends on previous history of the sample. This phenomenon is called **hysteresis.** The word hysteresis means lagging behind.

**Note** In terms of the susceptibility $\chi$, a material is diamagnetic if $\chi$ is negative, para if $\chi$ is positive & small and ferro if $\chi$ is large & positive.

## Permanent Magnet and Electromagnet

- Permanent magnets are those substances which at room temperature retain their ferromagnetic property for a long period of time.
- The different ways of making a permanent magnets are
  (i) by holding an iron rod in the north-south direction and hammer it repeatedly.
  (ii) by striking a steel rod with one end of a bar magnet a large number of times, always in the same sense.
  (iii) by placing a ferromagnetic rod in a solenoid and pass a current. The magnetic field of the solenoid magnetises the rod.
- A suitable material for making permanent magnet should have
  (i) high retentivity, so that the magnet is strong.
  (ii) high coercivity, so that the magnetisation is not erased by stray magnetic fields, temperature fluctuations or minor mechanical damage.
  (iii) high permeability.
- Steel is favoured for making permanent magnet. It has a slightly smaller retentivity than soft iron but this is outweighted by the much smaller coercivity of soft iron. Other suitable materials for permanent magnets are alnico, cobalt steel and ticonal.
- Core of electromagnets are made of ferromagnetic materials which have high permeability and low retentivity. Then, materials must have a high resistivity to reduce eddy current losses. Soft iron is suitable material for electromagnets.
- The hysteresis curve of electromagnets must be norrow, so that the energy dissipated and the heating will consequently be small.
- Electromagnets are used in electric bells, loudspeakers and telephone diaphragms. Giant electromagnets are used in cranes to lift machinery and bulk quantities of iron & steel.

# Mastering NCERT

## MULTIPLE CHOICE QUESTIONS

## TOPIC 1 ~ Magnet and Magnetisation

**1** A natural bar magnet
(a) is always suspended in north-south direction
(b) always has two poles
(c) is made of iron and its alloys
(d) All of the above

**2** The patterns of iron fillings permits us
(a) to plot the north-south poles
(b) to plot the geographic north-south poles
(c) to plot the magnetic field lines
(d) to plot the electric dipole

**3** Which of the following point lies in the region of highest magnetic field strength ?

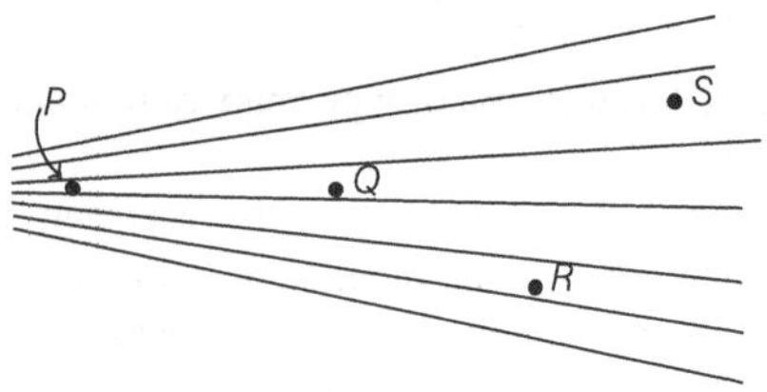


(a) $P$ (b) $Q$
(c) $R$ (d) $S$

**4** Many of the diagrams given in figure below, show magnetic field lines (thick lines in the figure). Which of them correctly show this?

(i)
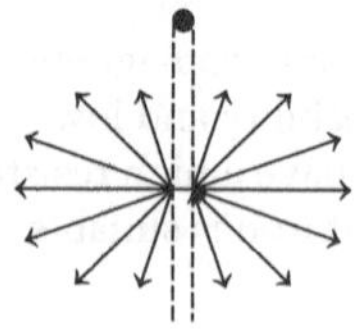

(ii)
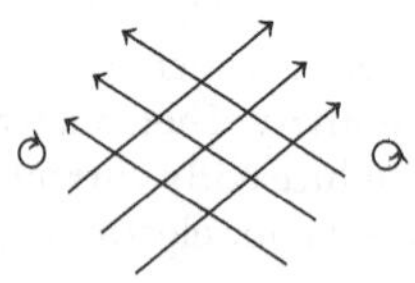

(iii)
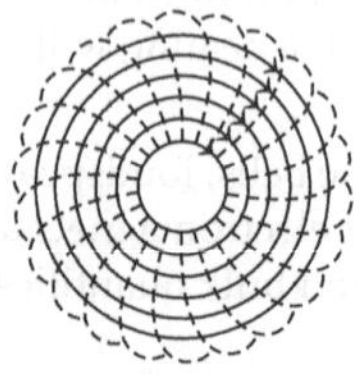

(iv)
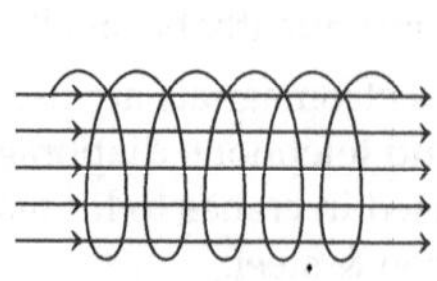

(a) Both (i) and (iv) (b) Both (ii) and (iv)
(c) Both (iii) and (iv) (d) Only (iii)

**5** Two identical bar magnets are fixed with their centres at a distance $x$ apart. A stationary charge $+q$ is placed at a point $A$ as shown in the figure between two magnets. The force on the charge is

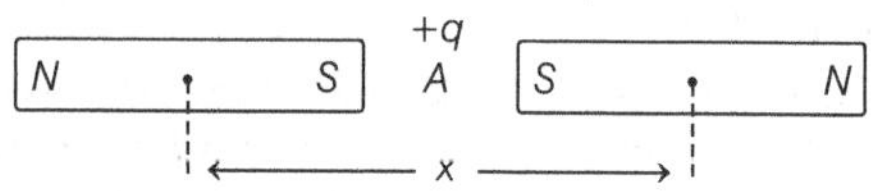


(a) towards left
(b) towards right
(c) perpendicular to the common axis of both magnets
(d) zero

**6** If the bar magnet of magnetic moment 0.40 A-m$^2$ is replaced by a solenoid of cross-sectional area $2 \times 10^{-4}$ m$^2$ and 1000 turns but of the same magnetic moment, the current flowing through the solenoid is
(a) $-2$ A (b) 2 A
(c) 20 A (d) 1 A

**7** A large magnet is broken into two pieces, so that their lengths are in the ratio 2 : 1. The pole strengths of the two pieces will have ratio
(a) 2 : 1 (b) 1 : 2 (c) 4 : 1 (d) 1 : 1

**8** Two magnets have the same length and the same pole strength. But one of the magnets has a small hole at its centre, then
(a) both have equal magnetic moment
(b) one with hole has small magnetic moment
(c) one with hole has large magnetic moment
(d) one with hole loses magnetism through the hole

**9** The areas of cross-section of three magnets of same lengths are $A$, $2A$ and $6A$, respectively. The ratio of their magnetic moments will be
(a) 6 : 2 : 1 (b) 1 : 2 : 6
(c) 2 : 6 : 1 (d) 1 : 1 : 1

**10** The length of a magnetised steel wire is $l$ and the magnetic moment is $m$. It is bent into the shape of $L$ with two equal sides. The magnetic moment now will be
(a) $m/2$ (b) $2m$ (c) $\sqrt{2}\,m$ (d) $m/\sqrt{2}$

**11** A bar magnet of length $l$ and magnetic dipole moment $M$ is bent in the form of an arc as shown in figure. The new magnetic dipole moment will be **NEET 2013**

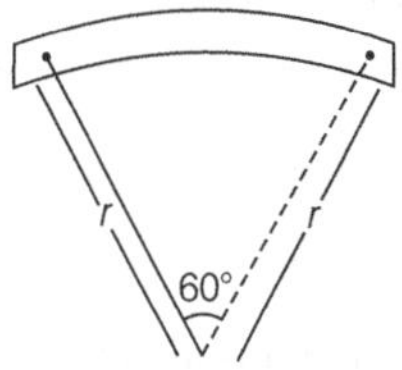


(a) $3\pi$ (b) $\frac{3M}{\pi}$ (c) $\frac{\pi}{3M}$ (d) $3M$

**12** A magnet with moment $m$ is given. If it is bent into a semi-circular form, its new magnetic moment will be

(a) $\frac{m}{\pi}$ (b) $\frac{m}{2}$ (c) $m$ (d) $\frac{2m}{\pi}$

**13** Following figures show the arrangement of bar magnets in different configurations. Each magnet has magnetic dipole moment $m$. Which configuration has highest net magnetic dipole moment? **CBSE AIPMT 2014**

(a)

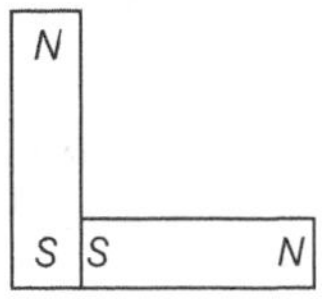


(b)

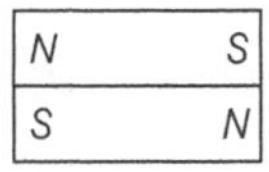


(c)

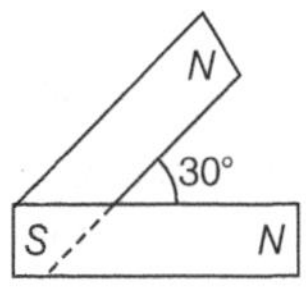


(d)

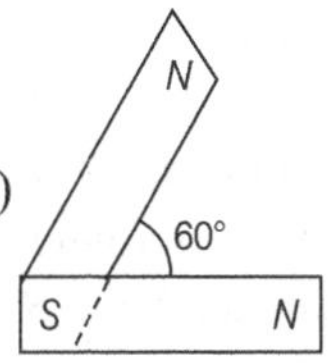


# TOPIC 2 ~ The Dipole in a Uniform Magnetic Field, Magnetism and Gauss's Law

**14** When a magnetic dipole is placed in a uniform magnetic field in unstable condition, then it experiences a/an

(a) impulse (b) moment of pole strength
(c) torque (d) None of these

**15** A small circular loop of conducting wire has radius $a$ and carries current $i$. It is placed in a uniform magnetic field $B$ perpendicular to its plane such that when rotated slightly about its diameter and released, it starts performing simple harmonic motion of time period $T$. If the mass of the loop is $m$, then **JEE Main 2020**

(a) $T = \sqrt{\frac{\pi m}{iB}}$ (b) $T = \sqrt{\frac{2m}{iB}}$
(c) $T = \sqrt{\frac{2\pi m}{iB}}$ (d) $T = \frac{\pi m}{2iB}$

**16** A magnetic needle of magnetic moment $6.7 \times 10^{-2}$ Am$^2$ and moment of inertia $7.5 \times 10^{-6}$ kg m$^2$ is performing simple harmonic oscillations in a magnetic field of 0.01 T. Time taken for 10 complete oscillations is **JEE Main 2017**

(a) 8.89 s (b) 6.98 s (c) 8.76 s (d) 6.65 s

**17** A magnetic instrument placed in magnetic meridian has a small bar magnet. The magnet executes oscillations with a time period of 2 s in the earth's horizontal magnetic field of 24 $\mu$T. When a horizontal field of $18\,\mu$T is produced opposite to the earth's field by placing a current carrying wire, the new time period of magnet will be

(a) 1 s (b) 2s (c) 3s (d) 4 s

**18** A bar magnet is oscillating in the earth's magnetic field with a period $T$. What happens to its period of motion, if its mass is quadrupled?

(a) Motion remains simple harmonic with new period $= \frac{T}{2}$
(b) Motion remains simple harmonic with new period $= 2T$
(c) Motion remains simple harmonic with new period $= 4T$
(d) Motion remains simple harmonic and the period stays nearly constant

**19** Rate of change of torque $\tau$ with deflection $\theta$ is maximum for a magnet suspended freely in a uniform magnetic field of induction $B$, when

(a) $\theta = 0°$ (b) $\theta = 45°$ (c) $\theta = 60°$ (d) $\theta = 90°$

**20** A short bar magnet of magnetic moment $m = 0.32$ JT$^{-1}$ is placed in a uniform magnetic field of 0.15 T. If the bar is free to rotate in the plane of the field, then the potential energy which would corresponds to its unstable equilibrium is

(a) $4.8 \times 10^{-2}$ J (b) $-4.8 \times 10^{-2}$ J
(c) $7.5 \times 10^{-2}$ J (d) $3.6 \times 10^{5}$ J

**21** A dipole or current carrying loop is placed in a uniform magnetic field in four different orientations as shown in figure. The potential energy increases in order of

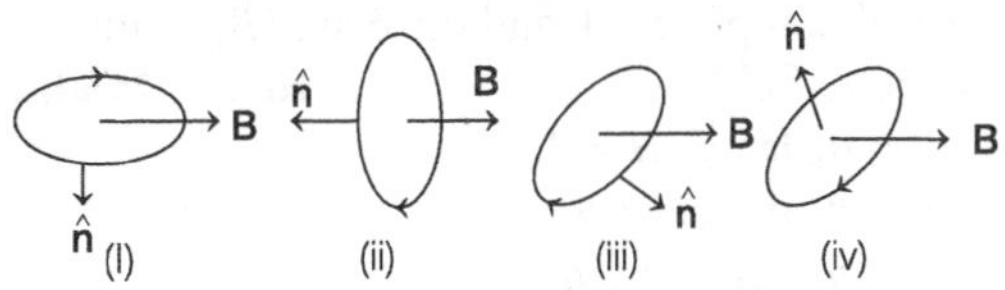


(a) (iii), (i), (iv) and (ii) (b) (i), (ii), (iii) and (iv)
(c) (ii), (iii), (i) and (iv) (d) (iv), (iii), (i) and (ii)

**22** A magnet of dipole moment $m$ is aligned in equilibrium position in a magnetic field of intensity $B$. The work done to rotate it through an angle $\theta$ in the magnetic field is

(a) $mB \sin\theta$ (b) $mB \cos\theta$ (c) $mB(1-\cos\theta)$ (d) $mB(1-\sin\theta)$

**23** A thin bar magnet is placed in a uniform magnetic field and is aligned with the field. The needle is now rotated by an angle of 45° and the work done is $W$. The torque on the magnetic needle at this position is

(a) zero (b) $\frac{W}{\sqrt{2}-1}$ (c) $\frac{W}{\sqrt{2}}$ (d) $\sqrt{2}W$

**24** A bar magnet having a magnetic moment of $2\times10^4$ $JT^{-1}$ is free to rotate in a horizontal plane. A horizontal magnetic field $B=6\times10^{-4}$ T exists in the space. The work done in taking the magnet slowly from a direction parallel to the field to a direction 60° from the field is

(a) 0.6 J (b) 12 J (c) 6 J (d) 2 J

**25** The work done in turning a magnet of magnetic moment $M$ by an angle of 90° from the meridian, is $n$ times the corresponding work done to turn it through an angle of 60°. The value of $n$ is given by

(a) 2 (b) 1 (c) 0.5 (d) 0.25

**26** A bar magnet of length 3 cm has points $A$ and $B$ along axis at a distance of 24 cm and 48 cm on the opposite ends. Ratio of magnetic fields at these points will be

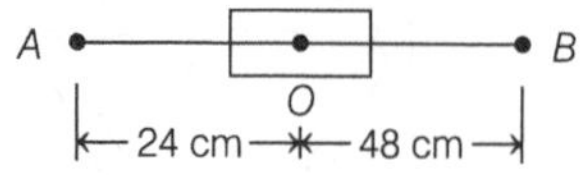


(a) 8 (b) 3 (c) 4 (d) $1/2\sqrt{2}$

**27** What is the magnitude of the equatorial and axial fields due to a bar magnet of length 5.0 cm at a distance of 50 cm from its mid-point ? The magnetic moment of the bar magnet is 0.40 A-$m^2$.

(a) $3.2\times10^{-6}$ T, $6.4\times10^{-7}$ T
(b) $4.2\times10^{5}$ T, $6\times10^{-7}$ T
(c) $5\times10^{-6}$ T, $3.2\times10^{-7}$ T
(d) $3.2\times10^{-7}$ T, $6.4\times10^{-7}$ T

**28** The net magnetic flux passing through a hypothetical closed surface enclosing a bar magnet is (in magnitude only)

(a) 1 (b) less than 1 (c) greater than 1 (d) zero

## TOPIC 3 ~ *The Earth's Magnetism*

**29** The strength of the earth's magnetic field varies from place to place on the earth's surface, its value being of the order of

(a) $10^5$ T (b) $10^{-6}$ T (c) $10^{-5}$ T (d) $10^8$ T

**30** The magnetic field is now arise due to electrical currents produced by convective motion of metallic fluids (consisting mostly of molten iron and nickel) in the core of the earth. This is known as the

(a) Tyndall effect (b) dynamo effect (c) Compton effect (d) Raman effect

**31** In the magnetic meridian of a certain place, the horizontal component of the earth's magnetic field is 0.26 G and the dip angle is 60°. The magnetic field of the earth at this location is

(a) 5.2 G (b) 5.00 G (c) 0.52 G (d) 0.5 G

**32** The relations amongst the three elements of earth's magnetic field, namely horizontal component $H$, vertical component $V$ and dip $\delta$ are ($B_E$ = total magnetic field) ***NEET (Odisha) 2019***

(a) $V=B_E\tan\delta$, $H=B_E$
(b) $V=B_E\sin\delta$, $H=B_E\cos\delta$
(c) $V=B_E\cos\delta$, $H=B_E\sin\delta$
(d) $V=B_E$, $H=B_E\tan\delta$

**33** At a certain place, horizontal component is $\sqrt{3}$ times the vertical component. The angle of dip at this place is

(a) zero (b) $\pi/3$ (c) $\pi/6$ (d) None of these

**34** The horizontal component of the earth's magnetic field at any place is $0.36\times10^{-4}$ Wb/$m^2$. If the angle of dip at that place is 60°, then the value of vertical component of the earth's magnetic field will be (in Wb/$m^2$) ***AIIMS 2018***

(a) $0.12\times10^{-4}$ (b) $0.24\times10^{-4}$ (c) $0.40\times10^{-4}$ (d) $0.622\times10^{-4}$

**35** The angle of dip at a certain place where the horizontal and vertical components of the earth's magnetic field are equal is

(a) 30° (b) 90° (c) 60° (d) 45°

**36** The value of angle of dip is zero at the magnetic equator because on it

(a) $B_E$ and $H_E$ are equal
(b) the values of $B_E$ and $H_E$ are zero
(c) the value of $B_E$ is zero
(d) the value of $H_E$ is zero

**37** If a magnet is suspended at an angle of 30° to the magnetic meridian, the dip needle makes an angle of 45° with the horizontal. The real dip is
(a) $\tan^{-1}(\sqrt{3/2})$ (b) $\tan^{-1}(\sqrt{3})$
(c) $\tan^{-1}(\sqrt{3}/2)$ (d) $\tan^{-1}(2/\sqrt{3})$

**38** At a point $A$ on the earth's surface, the angle of dip, $\delta = +25°$. At a point $B$ on the earth's surface, the angle of dip, $\delta = -25°$.
We can interpret that **NEET 2019**
(a) $A$ is located in the southern hemisphere and $B$ is located in the northern hemisphere
(b) $A$ is located in the northern hemisphere and $B$ is located in the southern hemisphere
(c) $A$ and $B$ are both located in the southern hemisphere
(d) $A$ and $B$ are both located in the northern hemisphere

**39** If $\theta_1$ and $\theta_2$ be the apparent angles of dip observed in two vertical planes at right angles to each other, then the true angle of dip $\theta$ is given by **NEET 2017**
(a) $\cot^2\theta = \cot^2\theta_1 + \cot^2\theta_2$
(b) $\tan^2\theta = \tan^2\theta_1 + \tan^2\theta_2$
(c) $\cot^2\theta = \cot^2\theta_1 - \cot^2\theta_2$
(d) $\tan^2\theta = \tan^2\theta_1 - \tan^2\theta_2$

**40** The time of vibration of a magnetic needle vibrating in the vertical plane is 3 s. When magnetic needle is made to vibrate in the horizontal plane, the time of vibration is $3\sqrt{2}$ s. Then, the angle of dip is
(a) 30° (b) 45° (c) 60° (d) 90°

# TOPIC 4 ~ *Magnetisation and Magnetic Intensity*

**41** Property of a bulk material called intensity of magnetisation is
(a) magnetic moment per unit area
(b) magnetic moment per unit volume
(c) a vector quantity and having unit of $Am^{-1}$
(d) Both (b) and (c)

**42** Obtain the earth's magnetisation assuming that the earth's field can be approximated by a giant bar magnet of magnetic moment $8.0 \times 10^{22}$ A-m$^2$. The earth's radius is 6400 km.
(a) $7.9\times 10^5$ $Am^{-1}$ (b) 73 $Am^{-1}$
(c) $1.8\times 10^{-5}$ $Am^{-1}$ (d) 1.0 $Am^{-1}$

**43** The horizontal component of flux density of earth's magnetic field is $1.7\times10^{-6}$ T. The value of horizontal component of intensity of earth's magnetic field will be
(a) 2.45 $Am^{-1}$ (b) 1.35 $Am^{-1}$
(c) 1.53 $Am^{-1}$ (d) 0.35 $Am^{-1}$

**44** A paramagnetic substance in the form of a cube with sides 1 cm has a magnetic dipole moment of $20\times10^{-6}$ J / T when a magnetic intensity of $60\times10^3$ A / m is applied. Its magnetic susceptibility is **JEE Main 2019**
(a) $3.3\times 10^{-4}$ (b) $3.3\times 10^{-2}$
(c) $4.3\times 10^{-2}$ (d) $2.3\times 10^{-2}$

**45** A rod of ferromagnetic material has dimension 10 cm× 0.5 cm× 0.2 cm. It is placed in a magnetic field of strength $0.5\times 10^4$ $Am^{-1}$. The magnetic moment produced in the rod is 5 A-m$^2$. The value of magnetic induction will be
(a) 0.54 T (b) 6.28 T (c) 0.358 T (d) 2.591 T

**46** A solenoid has core of a material with relative permeability 400. The windings of the solenoid are insulated from the core and carry a current of 2A. If the number of turns is 1000 per metre, calculate $H$ and $M$.
(a) $2\times 10^3$ $Am^{-1}$, $8\times 10^5$ $Am^{-1}$
(b) 1.0 $Am^{-1}$, $1.5\times 10^5$ $Am^{-1}$
(c) $6.8\times 10^{-5}$ $Am^{-1}$, $1.2\times 10^{-5}$ $Am^{-1}$
(d) $2.1\times 10^{-4}$ $Am^{-1}$, $-2.6\times 10^5$ $Am^{-1}$

**47** The core of a toroid having 3000 turns has inner and outer radii 11 cm and 12 cm, respectively. The magnetic field in the core for a current of 0.70 A is 2.5 T. The relative permeability of the core is
(a) 685 (b) 880
(c) 448 (d) 790

**48** An iron rod of 0.2 cm$^2$ cross-sectional area is subjected to a magnetising field of 1200 $Am^{-1}$. The susceptibility of iron is 599. The permeability will be
(a) $7.9\times 10^5$ $TmA^{-1}$ (b) $8.0\times 10^{22}$ $TmA^{-1}$
(c) $7.5\times 10^{-4}$ $TmA^{-1}$ (d) $1.8\times 10^{-5}$ $TmA^{-1}$

# TOPIC 5 ~ Magnetic Properties of Materials

**49** Which of the following substance have tendency to move from stronger to the weaker part of the external magnetic field?
(a) Ferromagnetic substance
(b) Paramagnetic substance
(c) Diamagnetic substance
(d) All of the above

**50** If a diamagnetic substance is brought near the north or the south pole of a bar magnet, then it is
(a) attracted by the both poles
(b) repelled by both the poles
(c) repelled by the north pole and attracted by the south pole
(d) attracted by the north pole and repelled by the south pole

**51** The variation of magnetic susceptibility $\chi$ with temperature for a diamagnetic substance is best represented by figure

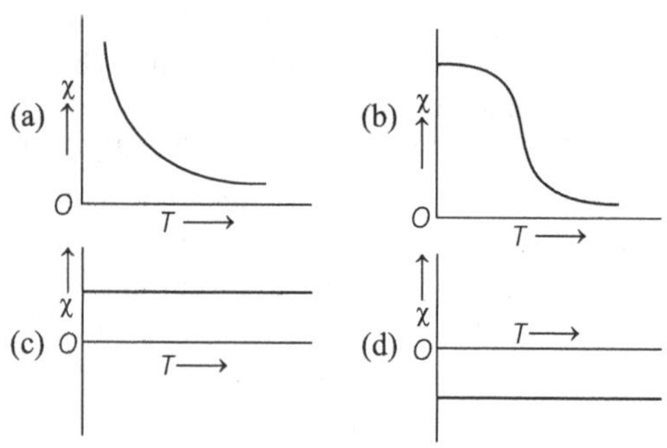


**52** If a diamagnetic solution is poured into a U-tube and one arm of this U-tube is placed between poles of a strong magnet, with the meniscus in line with the field, then the level of solution will
(a) rise
(b) fall
(c) oscillate slowly
(d) remain as such

**53** A thin diamagnetic rod is placed vertically between the poles of an electromagnet. When the current in the electromagnet is switched ON, then the diamagnetic rod is pushed up, out of the horizontal magnetic field. Hence, the rod gains gravitational potential energy. The work required to do this comes from **NEET 2018**
(a) the lattice structure of the material of the rod
(b) the magnetic field
(c) the current source
(d) the induced electric field due to the changing magnetic field

**54** The relative permeability of a substance $X$ is slightly less than unity and that of substance $Y$ is slightly more than unity, then
(a) $X$ is paramagnetic and $Y$ is ferromagnetic
(b) $X$ is diamagnetic and $Y$ is ferromagnetic
(c) $X$ and $Y$ both are paramagnetic
(d) $X$ is diamagnetic and $Y$ is paramagnetic

**55** The $M$-$H$ curve for a paramagnetic material is represented by figure.

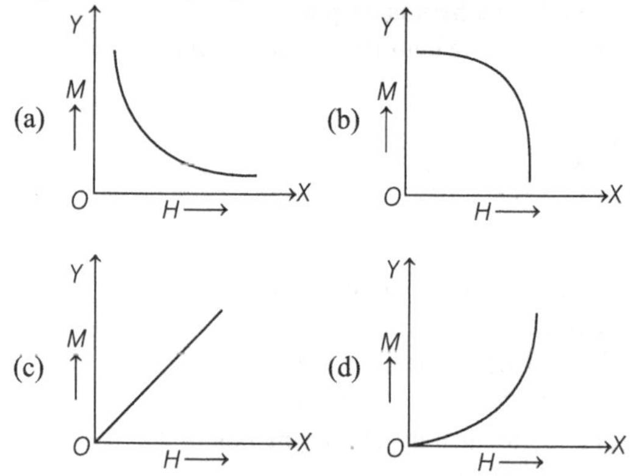


**56** A domain in ferromagnetic iron is in the form of a cube of side length 1 μm. Estimate the number of iron atoms in the domain. The molecular mass of iron is 55 g/mol and its density is 7.9 g cm$^{-3}$.
(a) $8.65 \times 10^{-10}$ atoms
(b) $8.0 \times 10^{-13}$ atoms
(c) $8.0 \times 10^{5}$ atoms
(d) $8.65 \times 10^{10}$ atoms

**57** The susceptibility of magnesium at 300 K is $1.2 \times 10^{-5}$. At what temperature will the susceptibility increase to $1.8 \times 10^{-5}$?
(a) 100 K (b) 200 K (c) 300 K (d) 400 K

**58** A bar magnet is demagnetised by inserting it inside a solenoid of length 0.2 m, 100 turns and carrying a current of 5.2 A. The coercivity of the bar magnet is
(a) 1200 A/m
(b) 285 A/m **JEE Main 2019**
(c) 2600A/m
(d) 520A/m

**59** A sample of paramagnetic salt contains $2.0 \times 10^{24}$ atomic dipoles, each of dipole moment $1.5 \times 10^{-23}$ JT$^{-1}$. The sample is placed under a homogeneous magnetic field of 0.84 T and cooled to a temperature of 4.2 K. The degree of magnetic saturation achieved is equal to 15%. The total dipole moment of the sample for a magnetic field of 0.98 T and a temperature of 2.8 K is
(a) $3 \times 10^{3}$ JT$^{-1}$
(b) $4\pi \times 10^{-6}$ JT$^{-1}$
(c) 7.9 JT$^{-1}$
(d) $7 \times 10^{4}$ JT$^{-1}$

**60** The variation of magnetic susceptibility $\chi$ with absolute temperature $T$ for a ferromagnetic material is given in figure by

(a) 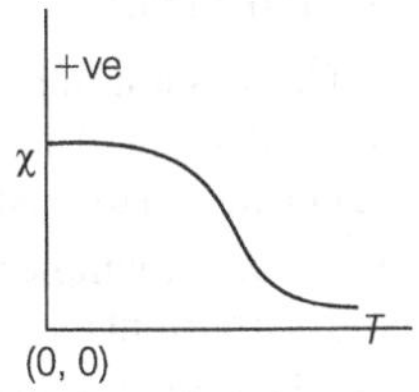


(b) 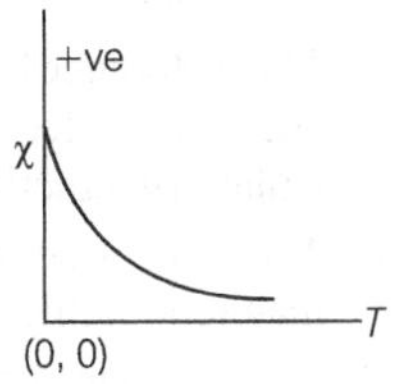


(c) 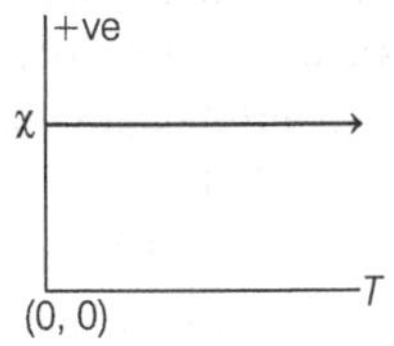


(d) 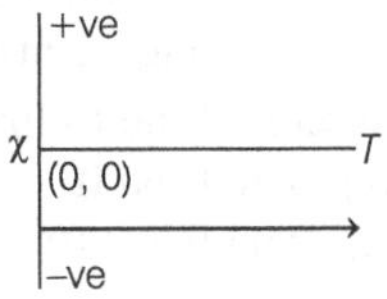


**61** Hysteresis loops for two magnetic materials $A$ and $B$ are as given below

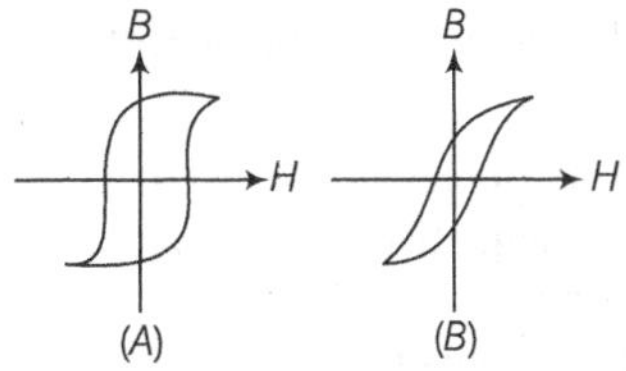


These materials are used to make magnets for electric generators, transformer core and electromagnet core. Then, it is proper to use **JEE Main 2016**

(a) $A$ for electric generators and transformers
(b) $A$ for electromagnets and $B$ for electric generators
(c) $A$ for transformers and $B$ for electric generators
(d) $B$ for electromagnets and transformers

# SPECIAL TYPES QUESTIONS

## I. Assertion and Reason

**Direction** (Q. Nos. 62-74) *In the following questions, a statement of Assertion is followed by a corresponding statement of Reason. Of the following statements, choose the correct one.*

(a) Both Assertion and Reason are correct and Reason is the correct explanation of Assertion.
(b) Both Assertion and Reason are correct but Reason is not the correct explanation of Assertion.
(c) Assertion is correct but Reason is incorrect.
(d) Assertion is incorrect but Reason is correct.

**62** **Assertion** The poles of magnet cannot be separated by breaking into two pieces.

**Reason** The magnetic moment will be reduced to half when a magnet is broken into two equal pieces.

**63** **Assertion** According to Gauss's law for magnetism, the net magnetic flux through any closed surface is zero.

**Reason** The number of magnetic field lines leaving the surface is balanced by the number of lines entering it.

**64** **Assertion** The true geographic north direction cannot be found by using a compass needle.

**Reason** The magnetic meridian of the earth is along the axis of rotation of the earth.

**65** Consider the figure,

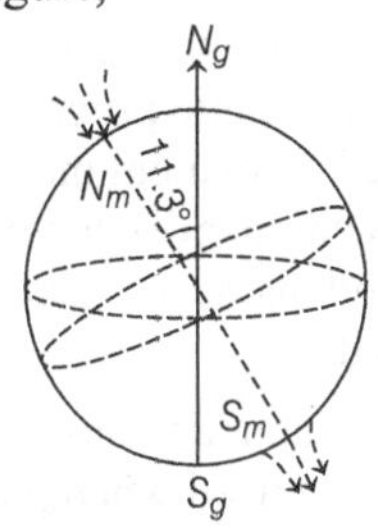


**Assertion** Unlike in the case of bar magnet, the field lines in above figure go into the earth at the north magnetic pole $N_m$ and come out from the south magnetic pole $S_m$.

**Reason** The magnetic north was the direction in which the north pole of a magnetic needle pointed and similarly the north pole of a magnet was so named as it was the North seeking pole.

**66** **Assertion** A magnetic needle which is free to swing horizontally, would lie in the magnetic meridian and the north pole of the needle would point towards the magnetic north pole.

**Reason** The line joining the magnetic poles is tilted with respect to the geographic axis of the earth. The magnetic meridian at a point makes some angle with the geographic meridian.

**67** **Assertion** Susceptibility is defined as the ratio of intensity of magnetisation $M$ to magnetic intensity $H$.

**Reason** Greater the value of susceptibility smaller is the value of intensity of magnetisation $M$.

**68** **Assertion** A paramagnetic sample displays greater magnetisation (for the same magnetising field) when cooled.
**Reason** The magnetisation does not depends on temperature.

**69** **Assertion** Paramagnetic substances are those which get weakly magnetised when placed in an external magnetic field.
**Reason** Paramagnetic substances have tendency to move from a region of weak magnetic field to strong magnetic field, i.e. they get weakly attracted to a magnet.

**70** **Assertion** Paramagnetic substances get poorly attracted in magnetic field.
**Reason** Because magnetic dipoles are weakly aligned along external magnetic field. ***AIIMS 2019***

**71** **Assertion** Ferromagnetic substances are those which gets strongly magnetised when placed in an external magnetic field.
**Reason** The individual atoms (or ions or molecules) in a ferromagnetic material possess a dipole moment as in a paramagnetic material.

**72** **Assertion** Substances which at room temperature retain their ferromagnetic property for a long period of time are called permanent magnets.
**Reason** Permanent magnet can be made by placing a ferromagnetic rod inside a solenoid and passing current through it.

**73** **Assertion** Ferromagnetic substances are those which gets strongly magnetised when placed in an external magnetic field.
**Reason** Ferromagnetic substances have strong tendency to move from a region of weak magnetic field to strong magnetic field, i.e. they get strongly attracted to a magnet.

**74** **Assertion** The ferromagnetic substances do not obey Curie law.
**Reason** At Curie point, a ferromagnetic substance start behaving as a paramagnetic substance.

## II. Statement Based Questions

**75** I. The magnetic field lines of a magnet (or a solenoid) form continuous closed loops. This is unlike the electric dipole, where these field lines begin from a positive charge and end on the negative charge or escape to infinity.
II. The tangent to the field line at a given point represents the direction of the net magnetic field $B$ at that point.
III. The larger the number of field lines crossing per unit area, the stronger the magnitude of the magnetic field $B$.
IV. The magnetic field lines intersect each other.
Which of the following statement(s) is/are correct?
(a) I, II and IV (b) I, III and IV
(c) II, III and IV (d) I, II and III

**76** I. The difference between the Gauss's law of magnetism and that for electrostatics is a reflection of the fact that isolated magnetic poles are not known to exist.
II. There are no sources or sinks of $B$, the simplest magnetic element is a dipole or a current loop.
III. All magnetic phenomena can be explained in terms of an arrangement of dipoles and/or current loops.
Which of the following statement(s) is/are correct?
(a) Both I and II (b) Only II
(c) Both II and III (d) I, II and III

**77** I. The earth behaves as a magnet with the magnetic field pointing approximately from the geographic south to the north.
II. When a bar magnet is freely suspended, it points in the north-south direction. The tip which points to the geographic north is called the north pole and the tip which points to geographic south is called the south pole of magnet.
III. There is a repulsive force when north poles (or south poles) of two magnets are brought close together. Conversely there is an attractive force between the north pole of one magnet and the south pole of other.
IV. We can isolate the north or south pole of a magnet.
Which of the following statement(s) is/are correct?
(a) I, II and IV (b) Both I and II
(c) Both III and IV (d) I, II and III

**78** I. The magnetic field lines of the earth resemble that of a (hypothetical) electrical dipole located at the centre of the earth.
II. The location of the north magnetic pole is at a latitude of $79.74'' N$ and a longitude of $71.8'' W$, a place somewhere in north Canada.
III. The magnetic south pole is at $79.74'' S$, $108.22'' E$ in the Antarctica.
Which of the following statement(s) is/are correct?
(a) I, II and III (b) Both II and III
(c) Only III (d) Both I and II

**79** Consider the statements given with respect to the figure showing a bar of diamagnetic material placed in an external magnetic field.

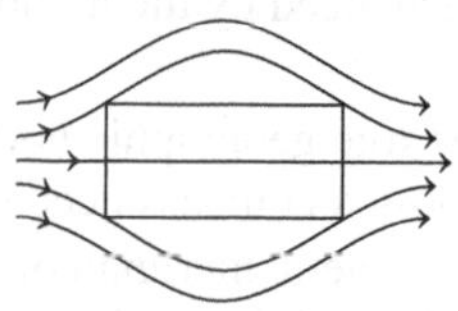

I. The field lines are attracted and result in the field inside the material will get increased.

II. When placed in a non-uniform magnetic field, the bar will tend to move from high to low field.

III. Reduction in the field inside the material is slight, being one part in $10^5$.

Which of the following statement(s) is/are correct?

(a) Both I and II (b) Both I and III
(c) Both II and III (d) I, II and III

**80** I. When an iron rod in the north-south direction is hammered repeatedly, it becomes an electromagnet.

II. A steel rod is striked with one end of a bar magnet a large number of times, always in the same sense to make a permanent magnet.

III. The hysteresis curve allows us to select suitable materials for permanent magnet.

Which of the following statement(s) is/are correct?

(a) Both I and II (b) Both II and III
(c) I, II and III (d) None of these

**81** Which of the following statement(s) is/are correct ?

(a) Magnetic phenomena are not universal in nature.
(b) The name lodestone (or loadstone) given to a naturally occurring ore of iron magnetite means leading stone.
(c) A bar magnet is not analogous to a solenoid.
(d) The magnetic field lines are discontinuous, emerging from one face of the solenoid to the infinity.

**82** A horizontal circular loop carries a current that loops anti-clockwise when viewed from above. It is replaced by an equivalent magnetic dipole *N-S*. Which of the following statement(s) is/are correct?

(a) The line *N-S* should be along a diameter of the loop.
(b) The line *N-S* should be parallel to the plane of the loop.
(c) South pole should be below the loop.
(d) North pole should be below the loop.

**83** A magnetic compass needle of known magnetic moment **m** and moment of inertia $I$ is allowed to oscillate in a magnetic field **B**. Which of the following statement(s) is/are incorrect ?

(a) The equation of simple harmonic motion is represented by $\frac{d^2\theta}{dt^2} = \frac{-mB}{I}\theta$.
(b) The negative sign in the equation implies that restoring torque is in opposite to deflecting torque.
(c) The square of the angular frequency is $\omega^2 = mBI$.
(d) The torque on it in case of magnetism is $\mathbf{m} \times \mathbf{B}$ while in case of electrostatics, it is $\mathbf{p} \times \mathbf{E}$ (where **p** is electric dipole moment).

**84** A circular coil of 16 turns and radius 10 cm carrying a current of 0.75 A rests with its plane normal to an external field of magnitude $5.0 \times 10^{-2}$ T. Which of the following statement(s) is/are correct?

(a) Magnetic moment of the coil is 3.77 $JT^{-1}$.
(b) If coil oscillates about its stable equilibrium with a frequency of 2/s, then moment of inertia of the coil will be $1.2 \times 10^{-4}$ kg-m$^2$.
(c) If coil oscillates about its stable equilibrium with a frequency of 3/s, then moment of inertia of the coil will be $3.2 \times 10^{-4}$ kg-m$^2$.
(d) If the moment of inertia is $4 \times 10^{-2}$, then coil oscillates with frequency of 3/s.

**85** Which of the following statement(s) is/are incorrect ?

(a) The axial field for a short dipole is $2\mathbf{p}/4\pi\varepsilon_0 r^3$ in electrostatics while in magnetism its value is $\mu_0 2\mathbf{m}/4\pi r^3$.
(b) The equatorial field for a short dipole is $\frac{1}{4\pi\varepsilon_0}\cdot\frac{\mathbf{p}}{r^2}$ in electrostatics, while in magnetism its value is $\frac{\mu_0}{4\pi}\cdot\frac{\mathbf{m}}{r^2}$.
(c) The strength of the earth's magnetic field varies from place to place on the earth's surface, its value being of the order of $10^{-5}$ T.
(d) The magnetic field arise due to electrical currents produced by convective motion of metallic fluids.

**86** Which of the following statement(s) is/are correct ?

(a) The horizontal plane containing the longitude circle and the axis of rotation of the earth is called magnetic meridian.
(b) The angle between the true geographic north and the south shown by a compass needle is called the magnetic declination or simple declination.
(c) The declination is smaller at higher latitudes and greater near the equator.
(d) Dip is the angle that the total magnetic field $\mathbf{B}_E$ of the earth makes with the surface of the earth.

**87** A long solenoid has 1000 turns per metre and carries a current of 1 A. It has a soft iron core of $\mu_r = 1000$. The core is heated beyond the Curie temperature, $T_C$. Which of the following statement(s) is/are correct?

(a) The **H** field in the solenoid is (nearly) unchanged but the **B** field increases drastically.
(b) The **H** and **B** fields in the solenoid are nearly unchanged.
(c) The magnetisation in the core reverses direction.
(d) The susceptibility in the core diminishes by a factor of about $10^8$.

**88** *S* is the surface of a lump of magnetic material. Which of the following statement(s) is/are correct?

(a) Lines of **B** are necessarily continuous across *S*.
(b) Some lines of **B** must be discontinuous across *S*.
(c) Lines of **H** are necessarily continuous across *S*.
(d) Magnetic intensity **H** remains same for inside and outside jump.

**89** Which of the following statement(s) is/are incorrect ?

(a) Diamagnetic substances are those which have tendency to move from stronger to the weaker part of the external magnetic field.

(b) Superconductors are metals, cooled to very low temperature which exhibits both perfect conductivity and perfect diamagnetism.

(c) Superconducting magnets can be used for running magnetically levitated superfast trains.

(d) For a paramagnetic material, both $\chi$ and $\mu_r$ depends only on the material.

**90** Which of the following statement(s) is/are incorrect associated with a ferromagnetic material?

(a) In a ferromagnetic material, the field lines are highly concentrated.

(b) Hard ferromagnets are those which persists magnetisation.

(c) Soft ferromagnetic materials are those in which the magnetisation disappears on removal of the external field.

(d) The relative magnetic permeability of ferromagnets is less than 1.

**91** Which of the following statement(s) is/are incorrect for thc *B*-*H* curve shown below? ***JEE Main 2013***

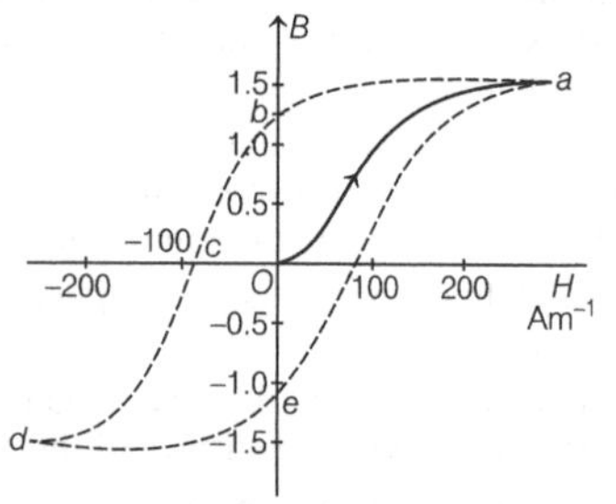


(a) From the figure, the retentivity or remanence is value of $B$ at $H = -100$.

(b) The coercivity of the material is value of $H$ at c.

(c) For a given value of $H$, $B$ is not unique but depends on the magnetic history of the sample. This phenomenon is called hysteresis.

(d) None of the above

**92** Which of the following statement(s) is/are incorrect ?

(a) The susceptibility above the Curie temperature, i.e. in the paramagnetic phase is $\chi = \dfrac{C}{T - T_C}$ $(T > T_C)$.

(b) For permanent magnets, the materials should have high retentivity, high coercivity and high permeability.

(c) For electromagnets, the material should have high permeability, low retentivity and wide hysteresis curve.

(d) Electromagnets are used in electric bells, loudspeakers, telephone diaphragms and in cranes, etc.

## III. Matching Type

**93** Match the Column I (magnetic/electric field pattern) with Column II (name of object producing magnetic/electrical field) and select the correct answer from the codes given below.

| Column I | Column II |
|---|---|
| A. 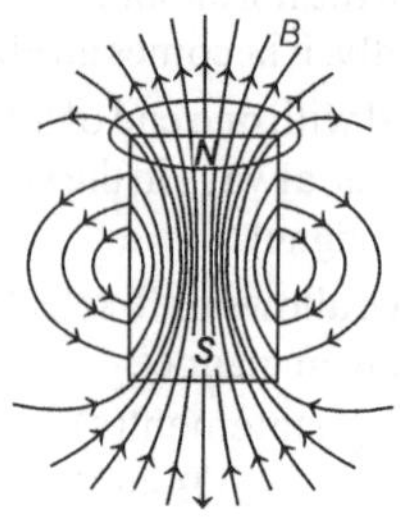  | 1. The field lines of a current carrying finite solenoid |
| B. 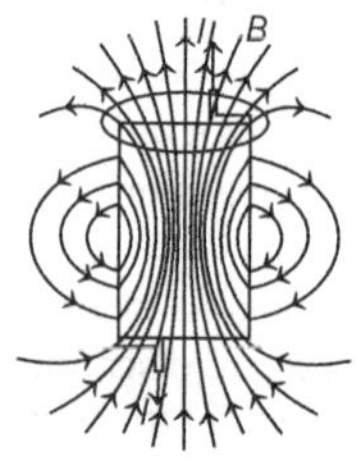  | 2. A field lines of electric dipole |
| C. 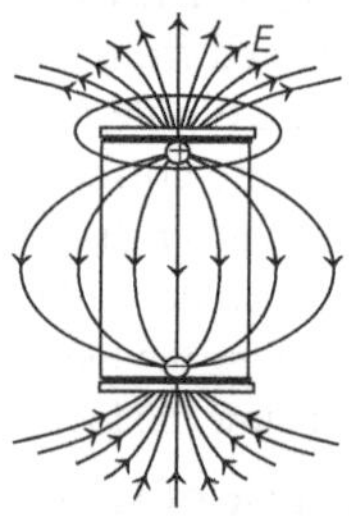  | 3. The field lines of a bar magnet |

|  | A | B | C |  | A | B | C |
|---|---|---|---|---|---|---|---|
| (a) | 2 | 1 | 3 | (b) | 3 | 1 | 2 |
| (c) | 1 | 2 | 3 | (d) | 2 | 3 | 1 |

**94** Match the Column I (potential energy of magnetic dipole at different orientation) with Column II (value of potential energy in magnetic field) and select the correct answer from the codes given below.

| Column I | Column II |
|---|---|
| A. Potential energy at $\theta = 90°$ | 1. Minimum |
| B. Potential energy at $\theta = 0°$ | 2. Maximum |
| C. Potential energy at $\theta = 180°$ | 3. Zero |

|  | A | B | C |
|---|---|---|---|
| (a) | 3 | 1 | 2 |
| (b) | 1 | 2 | 3 |
| (c) | 2 | 1 | 3 |
| (d) | 1 | 3 | 2 |

**95** Match the Column I (magnetic quantities) with Column II (their formula) and select the correct answer from the codes given below.

| | Column I | | Column II |
|---|---|---|---|
| A. | Equatorial field for a short dipole | 1. | $-\mathbf{m}\cdot\mathbf{B}$ |
| B. | Axial field for a short dipole | 2. | $\mathbf{m}\times\mathbf{B}$ |
| C. | External field : torque | 3. | $-\mu_0\mathbf{m}/4\pi r^3$ |
| D. | External field : energy | 4. | $\mu_0 2\mathbf{m}/4\pi r^3$ |

| | A | B | C | D |
|---|---|---|---|---|
| (a) | 3 | 4 | 2 | 1 |
| (b) | 3 | 4 | 1 | 2 |
| (c) | 3 | 1 | 2 | 4 |
| (d) | 4 | 2 | 3 | 1 |

**96** Match the Column I (types of magnetic materials) with Column II (their magnetic properties) and select the correct answer from the codes given below.

| | Column I | | Column II |
|---|---|---|---|
| A. | Diamagnetic | 1. | $\mu >> \mu_0, \mu_r >> 1$ and $\chi >> 1$ |
| B. | Paramagnetic | 2. | $-1 \le \chi < 0, 0 \le \mu_r < 1$ and $\mu < \mu_0$ |
| C. | Ferromagnetic | 3. | $0 < \chi < \varepsilon, 1 < \mu_r < 1 + \varepsilon$ and $\mu > \mu_0$ |

| | A | B | C |
|---|---|---|---|
| (a) | 1 | 2 | 3 |
| (b) | 2 | 3 | 1 |
| (c) | 2 | 1 | 3 |
| (d) | 3 | 2 | 1 |

# NCERT & NCERT Exemplar

## MULTIPLE CHOICE QUESTIONS

## NCERT

**97** The earth's magnetic field at the equator is approximately 0.4 G, the earth's dipole moment is

(a) $1\times 10^{23}$ A-m$^2$
(b) $1.05\times 10^{23}$ A-m$^2$
(c) $8\times 10^{22}$ A-m$^2$
(d) $4\times 10^{2}$ A-m$^2$

**98** A short bar magnet placed with its axis at 30° with a uniform external magnetic field of 0.25 T experiences a torque of magnitude equal to $4.5\times 10^{-2}$ J. What is the magnitude of magnetic moment of the magnet?

(a) 0.36 JT$^{-1}$
(b) 0.63 JT$^{-1}$
(c) 0.64 JT$^{-1}$
(d) 0.60 JT$^{-1}$

**99** A short bar magnet of magnetic moment, $m = 0.32$ JT$^{-1}$ is placed in a uniform magnetic field of 0.15 T. If the bar is free to rotate in the plane of the field, then the potential energy of the magnet in (i) stable and (ii) unstable equilibrium are

(a) $8.4\times 10^{-2}$ J, $4.8\times 10^{-2}$ J
(b) $-4.8\times 10^{-2}$ J, $4.8\times 10^{-2}$ J
(c) $0.32\times 10^{-2}$ J, $-0.32\times 10^{-2}$ J
(d) $-23\times 10^{-2}$ J, $0.23\times 10^{-2}$ J

**100** A bar magnet of magnetic moment 1.5 JT$^{-1}$ lies aligned with the direction of a uniform magnetic field of 0.22 T. What is the amount of work required by an external torque to turn the magnet, so as to align its magnetic moment to the field direction?

(a) 0.66 J
(b) 0.44 J
(c) 0.33 J
(d) None of these

**101** A magnetic needle free to rotate in a vertical plane parallel to the magnetic meridian has its north tip pointing down at 22° with the horizontal. The horizontal component of the earth's magnetic field at the place is known to be 0.35 G. The magnitude of the earth's magnetic field at the place will be

(a) 0.68 G
(b) 0.38 G
(c) 0.83 G
(d) 0.86 G

**102** At a certain location in Africa, a compass points 12° *W* of the geographic north. The north tip of the magnetic needle of a dip circle placed in the plane of magnetic meridian points 60° above the horizontal. The horizontal component of the earth's field is measured to be 0.16 G. The magnitude of the earth's field at the location will be

(a) $0.23\times 10^{-4}$ T
(b) $0.18\times 10^{-4}$ T
(c) $0.32\times 10^{-4}$ T
(d) $0.81\times 10^{-4}$ T

**103** A short bar magnet of magnetic moment $5.25\times 10^{-2}$ JT$^{-1}$ is placed with its axis perpendicular to the earth's field direction. Magnitude of the earth's field at the place is given to be 0.42 G. Ignore the length of the magnet in comparison to the distance involved.

At what distance from the centre of the magnet, the resultant field is inclined at 45° with the earth's field on normal bisector?

(a) 5 cm
(b) 6 cm
(c) 7 cm
(d) 8 cm

**104** A compass needle free to turn in a horizontal plane is placed at the centre of circular coil of 30 turns and radius 12 cm. The coil is in a vertical plane making an angle of 45° with the magnetic meridian. When the current in the coil is 0.35 A, the needle points west to east. The horizontal component of the earth's magnetic field at the location will be

(a) $3.9 \times 10^{-5}$ T
(b) $9.3 \times 10^{-5}$ T
(c) $4.9 \times 10^{-5}$ T
(d) $9.5 \times 10^{-5}$ T

## NCERT Exemplar

**105** A toroid of $n$ turns, mean radius $R$ and cross-sectional radius $a$ carries current $I$. It is placed on a horizontal table taken as $xy$-plane, its magnetic moment **m**

(a) is non-zero and points in the $z$-direction by symmetry
(b) points along the axis of the toroid ($\mathbf{m} = m\,\phi$)
(c) is zero, otherwise there would be a field falling as $\frac{1}{r^3}$ at large distances outside the toroid
(d) is pointing radially outwards

**106** The magnetic field of the earth can be modelled by that of a point dipole placed at the centre of the earth. The dipole axis makes an angle of 11.3° with the axis of the earth. At Mumbai, declination is nearly zero, then

(a) the declination varies between 11.3° $W$ to 11.3° $E$
(b) the least declination is 0°
(c) the plane defined by dipole axis and the earth axis passes through Greenwich
(d) declination averaged over the earth must be always negative

**107** In a permanent magnet at room temperature,

(a) magnetic moment of each molecule is zero
(b) the individual molecules have non-zero magnetic moment which are all perfectly aligned
(c) domains are partially aligned
(d) domains are all perfectly aligned

**108** A paramagnetic sample shows a net magnetisation of 8 $Am^{-1}$, when placed in an external magnetic field of 0.6 T at a temperature of 4 K. When the same sample is placed in an external magnetic field of 0.2 T at a temperature of 16 K, the magnetisation will be

(a) (32/3) $Am^{-1}$
(b) (2/3) $Am^{-1}$
(c) 6 $Am^{-1}$
(d) 2.4 $Am^{-1}$

# Answers

### > Mastering NCERT with MCQs

| | | | | | | | | | |
|---|---|---|---|---|---|---|---|---|---|
| 1 (d) | 2 (c) | 3 (a) | 4 (d) | 5 (d) | 6 (b) | 7 (d) | 8 (b) | 9 (b) | 10 (d) |
| 11 (b) | 12 (d) | 13 (c) | 14 (c) | 15 (c) | 16 (d) | 17 (d) | 18 (b) | 19 (a) | 20 (a) |
| 21 (a) | 22 (c) | 23 (b) | 24 (c) | 25 (a) | 26 (a) | 27 (d) | 28 (d) | 29 (c) | 30 (b) |
| 31 (c) | 32 (b) | 33 (c) | 34 (d) | 35 (d) | 36 (a) | 37 (c) | 38 (b) | 39 (a) | 40 (c) |
| 41 (d) | 42 (b) | 43 (b) | 44 (a) | 45 (b) | 46 (a) | 47 (a) | 48 (c) | 49 (c) | 50 (b) |
| 51 (d) | 52 (b) | 53 (c) | 54 (d) | 55 (c) | 56 (d) | 57 (b) | 58 (c) | 59 (c) | 60 (a) |
| 61 (d) | | | | | | | | | |

### > Special Types Questions

| | | | | | | | | | |
|---|---|---|---|---|---|---|---|---|---|
| 62 (b) | 63 (a) | 64 (c) | 65 (a) | 66 (b) | 67 (c) | 68 (c) | 69 (a) | 70 (a) | 71 (b) |
| 72 (b) | 73 (b) | 74 (d) | 75 (d) | 76 (d) | 77 (d) | 78 (b) | 79 (c) | 80 (b) | 81 (b) |
| 82 (c) | 83 (c) | 84 (b) | 85 (b) | 86 (d) | 87 (d) | 88 (a) | 89 (d) | 90 (d) | 91 (a) |
| 92 (c) | 93 (b) | 94 (a) | 95 (a) | 96 (b) | | | | | |

### > NCERT & NCERT Exemplar MCQs

| | | | | | | | | | |
|---|---|---|---|---|---|---|---|---|---|
| 97 (b) | 98 (a) | 99 (b) | 100 (c) | 101 (b) | 102 (c) | 103 (a) | 104 (a) | 105 (c) | 106 (a) |
| 107 (d) | 108 (b) | | | | | | | | |

# Hints & Explanations

**3** (*a*) The larger the number of field lines crossing per unit area, the stronger is the magnitude of the magnetic field strength. Since, here point $P$ among the given point lies in the region which has maximum number of field lines crossing per unit area. Hence, point $P$ lies in the region of highest magnetic field strength.

**4** (*d*)

(i) **Wrong** Magnetic field lines can never emanate from a point, as shown in figure (i). Over any closed surface, the net flux must always be zero, i.e. pictorially as many field lines should seem to enter the surface, while as the number of lines leaving it. The correct magnetic field lines should be encircling the straight conductor.

(ii) **Wrong** Magnetic field line (like electric field lines) can never cross each other, otherwise the direction of field at the point of intersection is ambiguous.

(iii) **Right** Magnetic lines are completely confined within a toroid. Nothing wrong here in field lines forming closed loops, since each loop encloses a region across which a current passes. The entire region enclosed by the windings of a toroid contains magnetic field lines.

(iv) **Wrong** Field lines due to a solenoid at its ends and outside cannot be so straight and confined; such a pattern violates Ampere's law. The lines should curve out at both ends and meet eventually to form closed loops.

**5** (*d*) A moving charge in a uniform magnetic field $B$ experiences a force. It is given by

$$F = Bqv\sin\theta \qquad \ldots(\text{i})$$

When charge particle is stationary, then $v = 0$.

$\therefore$ From Eq. (i), we get $F = 0$

Hence, no force will acts on the charge.

**6** (*b*) Given, magnetic moment of bar magnet,

$$m = 0.40 \text{ A-m}^2$$

Number of turns, $N = 1000$

and area, $A = 2\times 10^{-4}\text{ m}^2$

Magnetic moment of solenoid, $m = NIA$

$\Rightarrow \quad 0.40 = 1000\times I\times 2\times 10^{-4}$

$\Rightarrow \quad I = 0.40\times 10^4/(1000\times 2) = 2\text{A}$

Hence, the current flowing, $I = 2\text{A}$.

**7** (*d*) Pole strength does not depend on length it depends upon cross-section area of magnet.
So, strength of the two pieces will remain same.

**8** (*b*) As, we know, magnetic dipole moment $\mathbf{m} = q_m(2\mathbf{l})$, so hole reduces the effective length of the magnet, hence magnetic moment of magnet with hole, is small.

**9** (*b*) As, magnetic moment, $m = q_m\cdot 2l$

where, $q_m =$ pole strength and $2l =$ length of bar magnet

$$\therefore \quad m \propto q_m \qquad \ldots(\text{i})$$

Since, pole strength of magnet is directly proportional to the area of cross-section of a bar magnet ($A$)

i.e. $$q_m \propto A \qquad \ldots(\text{ii})$$

From Eqs. (i) and (ii), we get

$$m \propto A$$

Given, $A_1 = A$, $A_2 = 2A$ and $A_3 = 6A$

Hence, the ratio of magnetic moments

$$m_1 : m_2 : m_3 = 1:2:6$$

**10** (*d*) If $q_m$ is strength of each pole and $l$ is the length of steel wire, then $m = q_m \times l$.

When the wire is bent into $L$-shape, effective distance between the poles

$$= \sqrt{(l/2)^2 + (l/2)^2} = l/\sqrt{2}$$

$$\therefore\ m' = q_m \times \frac{l}{\sqrt{2}} = \frac{m}{\sqrt{2}} \ (m \text{ will remain unchanged})$$

**11** (*b*) The magnetic moment, $M = q_m l \qquad \ldots(\text{i})$

When the bar magnet is bent in the form of an arc of radius $r$, then from the formula,

$$\text{Angle} = \frac{\text{Arc}}{\text{Radius}}$$

$$\Rightarrow \quad \frac{\pi}{3} = \frac{l}{r} \Rightarrow r = \frac{3l}{\pi}$$

Also, new distance of the poles of bar magnet is given as $AB = OA = OB = r$ (i.e. equilateral triangle)

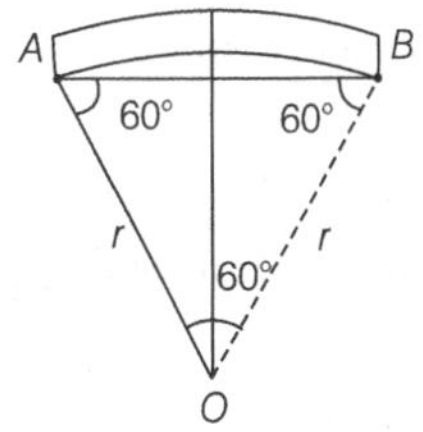


Hence, new magnetic moment,

$$M' = q_m \cdot AB = q_m \cdot r = \frac{q_m\cdot 3l}{\pi} = \frac{3M}{\pi} \qquad [\text{from Eq. (i)}]$$

**12** (*d*) As magnetic moment, $m = q_m \times 2l \Rightarrow q_m = \dfrac{m}{2l}$

Further, as length of semi-circular magnet, $\pi r = 2l$

or $$r = 2l/\pi$$

Distance between the two poles, $2l' = 2r = \dfrac{4l}{\pi}$

Magnetic moment, $m' = q_m \times 2l' = \left(\dfrac{m}{2l}\right)\left(\dfrac{4l}{\pi}\right) = 2m/\pi$

**13** (*c*) Magnetic moment is from $S$ to $N$,

$$M_{\text{net}} = \sqrt{m^2 + m^2 + 2m^2\cos\theta}$$

Therefore, $M_{\text{net}}$ will be maximum, if $\cos\theta$ is maximum, and $\cos\theta$ will be maximum when $\theta$ will be minimum.
So, at $\theta = 30°$ $M_{\text{net}}$ will be maximum.

**15** (c) Time period of oscillation of a magnetic dipole in a uniform magnetic field,

$$T = 2\pi\sqrt{\frac{I}{\mu B}}$$

where, $\mu$ = magnetic dipole.

Moment of dipole = Current × Area = $\pi a^2 i$

$$I = \text{moment of inertia of a coil} = \frac{ma^2}{2}$$

So, $$T = 2\pi\sqrt{\frac{\frac{ma^2}{2}}{\pi a^2 iB}}$$

$$= 2\pi\sqrt{\frac{m}{2\pi iB}} = \sqrt{\frac{2\pi m}{iB}}$$

**16** (d) Time period of oscillation is

$$T = 2\pi\sqrt{\frac{I}{MB}}$$

$$\Rightarrow \quad T = 2\pi\sqrt{\frac{7.5\times 10^{-6}}{6.7\times 10^{-2}\times 0.01}} = 0.665 \text{ s}$$

Hence, time for 10 oscillations is $t = 6.65$ s.

**17** (d) Time period in vibration magnetometer is given by

$$T = 2\pi\sqrt{\frac{I}{m\times B_H}}$$

$$\Rightarrow \quad T \propto \frac{1}{\sqrt{B_H}} \quad \dots (i)$$

So, for two different cases

$$\Rightarrow \quad \frac{T_1}{T_2} = \sqrt{\frac{(B_H)_2}{(B_H)_1}} \quad \text{[From Eq. (i)]}$$

Given, $T_1 = 2\text{s}$, $B_{H_1} = 24\ \mu\text{T}$ and $B_{H_2} = 24 - 18 = 6\ \mu\text{T}$

$$\Rightarrow \quad \frac{2}{T_2} = \sqrt{\frac{6}{24}} = \frac{1}{2}$$

$$\therefore \quad T_2 = 4 \text{ s}$$

**18** (b) The time period of a bar magnet in a magnetic field is given by

$$T = 2\pi\sqrt{\left(\frac{I}{mB}\right)}$$

where, $I$ is moment of inertia of bar magnet, $m$ is magnetic moment and $B$ is magnetic induction.

When mass is made 4 times, moment of inertia $I$ becomes 4 times (as $I = Mr^2$, then $I \propto M$). From the above equation of time period, $T \propto \sqrt{I}$, so $T$ becomes twice, as mass $M$ is quadrupled.

**19** (a) As we know, torque, $\tau = mB \sin\theta$

Hence, rate of change of torque with deflection,

$\frac{d\tau}{d\theta} = mB \cos\theta$. It will be maximum when $\cos\theta = 1$, i.e. $\theta = 0°$.

**20** (a) For unstable equilibrium, **m** is to be anti-parallel to **B**. In this case, $\theta = 180°$, thus

$$U = -mB\cos 180° = -mB(-1) = mB$$

Given, $m = 0.32\text{JT}^{-1}$ and $B = 0.15\text{T}$

Hence, $U = (0.32)(0.15) = 4.8\times 10^{-2}$ J

**21** (a) The potential energy of a dipole or a current carrying coil in a magnetic field is

$U = -\mathbf{m}\cdot\mathbf{B} = -mB\cos\theta$

where, $\theta$ is the angle between magnetic field **B** and magnetic moment **m** along normal vector $\hat{\mathbf{n}}$.

In given cases,

(i) $U = -mB\cos 90° = 0$ $[\because \cos 90° = 0]$

(ii) $U = -mB\cos 180° = mB$ $[\because \cos 180° = -1]$

(iii) $U = -mB\cos(360° - \theta) = -mB\cos\theta$ $[\because \cos(360° - \theta) = +\cos\theta]$

(iv) $U = -mB\cos(90° + \theta) = mB\sin\theta$ $[\because \cos(90° + \theta) = -\sin\theta]$

So, correct order of potential energy in increasing order is (iii) > (i) > (iv) > (ii).

Hence, option (a) matches the order.

**22** (c) Work done to rotate a magnet of magnetic dipole moment $m$ in uniform magnetic field $B$ from equilibrium position to angle $\theta$,

$$W = \int dW = \int_0^\theta \tau\, d\theta = \int_0^\theta mB\sin\theta\, d\theta = mB(1 - \cos\theta)$$

**23** (b) Given, work done = $W$ and $\theta = 45°$

We know that, $W = mB(\cos 0° - \cos\theta)$

$$= mB(1 - \cos 45°)$$

$$= mB\left(1 - \frac{1}{\sqrt{2}}\right) = \left(\frac{\sqrt{2}-1}{\sqrt{2}}\right)mB$$

$$\Rightarrow \quad mB = \frac{\sqrt{2}W}{\sqrt{2}-1} \quad \dots (i)$$

Hence, torque, $\tau = mB\sin\theta$

$$\tau = mB\sin 45° = \frac{W\times\sqrt{2}}{\sqrt{2}-1}\times\frac{1}{\sqrt{2}} \quad \text{[from Eq. (i)]}$$

$$= \frac{W}{\sqrt{2}-1}$$

**24** (c) When magnetic dipole is rotated from initial position $\theta_1$ to final position $\theta_2$, then work done

$$= MB(\cos\theta_1 - \cos\theta_2)$$

Given, $\theta_1 = 0°$ and $\theta_2 = 60°$

magnetic moment, $m = 2\times 10^4\ \text{JT}^{-1}$

and magnetic field, $B = 6\times 10^{-4}$ T

$$\therefore \quad W = mB(\cos 0° - \cos 60°)$$

$$\Rightarrow \quad W = mB\left(1 - \frac{1}{2}\right) \quad \left(\because \cos 0° = 1 \text{ and } \cos 60° = 1/2\right)$$

$$= \frac{2\times 10^4\times 6\times 10^{-4}}{2} = 6 \text{ J}$$

∴ Work done to rotate bar magnet is 6J.

**25** (*a*) Work done in rotating the dipole from $\theta_1$ to $\theta_2$ is

$$W = -MB(\cos\theta_2 - \cos\theta_1)$$

**Case I** $W_1 = -MB(\cos 90° - \cos 0°) = MB$ ...(i)

**Case II** $W_2 = -MB(\cos 60° - \cos 0°)$

$$= -MB\left(\frac{1}{2} - 1\right) = \frac{1}{2}MB \quad ...(ii)$$

From Eqs. (i) and (ii), we get

$$W_2 = \frac{1}{2}W_1 \quad ...(iii)$$

Given, $W_1 = nW_2$ ... (iv)

By comparing Eqs. (iii) and (iv), we get

$\therefore$ $n = 2$

**26** (*a*) Given, $2l = 3$ cm, $r_1 = 24$ cm and $r_2 = 48$ cm

The axial magnetic field, $B = \frac{\mu_0}{4\pi} \cdot \frac{2m}{r^3}$

$$\Rightarrow \quad B \propto \frac{1}{r^3}$$

Hence, $\frac{B_1}{B_2} = \frac{r_2^3}{r_1^3} = \left(\frac{48}{24}\right)^3 = 8$

Thus, the ratio of magnetic fields is 8.

**27** (*d*) Given, $m = 0.40\,\text{A-m}^2$ and $r = 5.0\text{ cm} = 0.5$ m

Magnetic field at equatorial line,

$$B_e = \frac{\mu_0 m}{4\pi r^3}$$

$$= \frac{10^{-7} \times 0.40}{(0.5)^3} = \frac{10^{-7} \times 0.40}{0.125}$$

$$= 3.2 \times 10^{-7}\,\text{T}$$

Magnetic field at axial line,

$$B_a = \frac{\mu_0 2m}{4\pi r^3} = 6.4 \times 10^{-7}\,\text{T}$$

Hence, the magnetic field at equatorial line and axial line are $3.2 \times 10^{-7}$ T and $6.4 \times 10^{-7}$ T, respectively.

**28** (*d*) Net magnetic flux passing through a hypothetical closed surface enclosing a bar magnet (dipole) is zero because bar magnet has two poles (north and south poles), hence number of magnetic field lines leaving the surface is balanced by the number of magnetic field lines entering it.

**31** (*c*) Given, dip angle, $\delta = 60°$ and horizontal component of earth's magnetic field, $H_E = 0.26$ G

We know that,

$$\cos\delta = \frac{H_E}{B_E}$$

$\therefore$ Magnetic field, $B_E = \frac{H_E}{\cos 60°} = \frac{0.26}{(1/2)} = 0.52$ G

**32** (*b*) Let $B_E$ be the net magnetic field at some point and $H$ & $V$ be the horizontal & vertical components of $B_E$. Let $\delta$ be the angle of dip, which is the angle between direction of earth's magnetic field $B_E$ and horizontal line in the magnetic meridian.

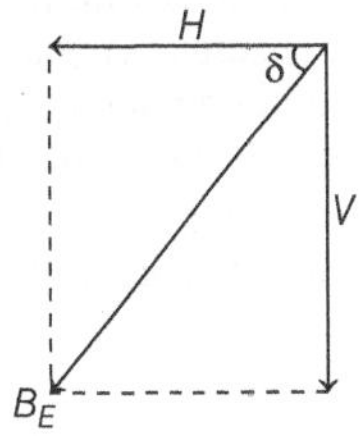


Thus, from figure, we can see that

$H = B_E \cos\delta$ and $V = B_E \sin\delta$

**33** (*c*) Given that, $H_E = \sqrt{3}V_E$, where $H_E$ = horizontal component of earth's magnetic field and $V_E$ = vertical component of earth's magnetic field.

Angle of dip, $\tan\delta = \frac{V_E}{H_E}$

$$\tan\delta = \frac{V_E}{H_E} = \frac{V_E}{\sqrt{3}V_E} = \frac{1}{\sqrt{3}}$$

$\therefore$ Angle of dip, $\delta = 30° = (\pi/6)$ rad

**34** (*d*) Given, $H_E = 0.36 \times 10^{-4}\,\text{Wb/m}^2$ and $\delta = 60°$

Vertical component of earth's magnetic field,

$$V_E = H_E \tan\delta = 0.36 \times 10^{-4} \times \tan 60°$$

$$= 0.36 \times 10^{-4} \times \sqrt{3}$$

$\therefore$ $V_E = 0.622 \times 10^{-4}\,\text{Wb/m}^2$

**35** (*d*) Given that, $V_E = H_E$

Angle of dip, $\tan\delta = \frac{V_E}{H_E} = \frac{H_E}{H_E} = 1 = \tan 45°$

Angle of dip, $\delta = 45°$

**36** (*a*) If $V_E$ and $H_E$ are vertical and horizontal components of earth's magnetic field.

At magnetic equator, $V_E = 0$

$\therefore$ $\tan\delta = \frac{V_E}{H_E} = 0 \Rightarrow \delta = 0°$

Since, value of angle of dip is zero at equator.

So, $H_E = B_E \cos\delta = B_E \cos 0° = B_H$

Hence, $H_E = B_E$

**37** (*c*) Given, angle of suspension to the magnetic maridian $\alpha = 30°$ and dip angle $\delta' = 45°$.

Hence, $\tan\delta' = \frac{V_E}{H'_E} = \frac{V_E}{H_E \cos\alpha}$

$$= \frac{\tan\delta}{\cos\alpha} = \frac{\tan\delta}{\cos\alpha} \quad \left[\because \tan\delta = \frac{V_E}{H_E}\right]$$

$$\Rightarrow \quad \tan\delta = \cos\alpha \tan\delta'$$

$$= \cos 30° \tan 45° = \frac{\sqrt{3}}{2}$$

$$\delta = \tan^{-1}\frac{\sqrt{3}}{2}$$

$\therefore$ Angle of the real dip is $\tan^{-1}\frac{\sqrt{3}}{2}$.

**38** (*b*) The angle of dip $\delta$ is the angle between the horizontal component of earth's magnetic field and the total magnetic field of the earth. Its value is different at different places. It is zero at equator, i.e. the dip neddle becomes parallel to horizontal component. It varies from $-90^\circ$ in south pole to $+90^\circ$ in the north pole.

This means the values $\delta$ is positive in northern hemisphere and is negative in southern hemisphere.

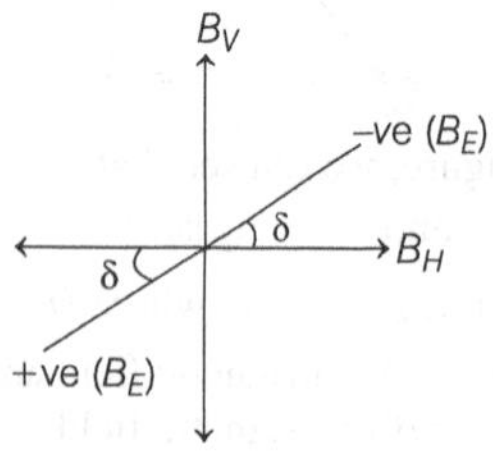


$\therefore$ For point $A$, $\delta = +25^\circ$, so $A$ lies in the northern hemisphere.

Similarly, for $B$, $\delta = -25^\circ$, so $B$ lies in the southern hemisphere.

**39** (*a*) Let the $H_E$ and $V_E$ be the horizontal and vertical components of earth's magnetic field $B$, then

$$\tan\theta = \frac{V_E}{H_E} \Rightarrow \cot\theta = \frac{H_E}{V_E} \quad \text{...(i)}$$

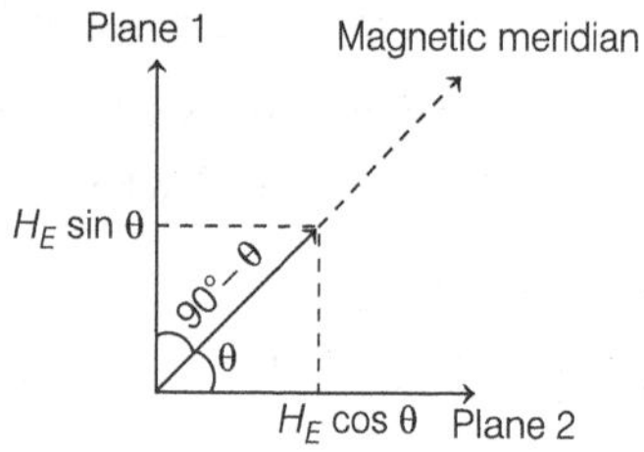


Let planes 1 and 2 are mutually perpendicular planes making angles $\theta$ and $(90^\circ - \theta)$ with magnetic meridian. The vertical component of earth's magnetic field remain same in two planes but effective horizontal components in the two planes is given by

$$B_1 = H_E \cos\theta \quad \text{...(ii)}$$

and

$$B_2 = H_E \sin\theta \quad \text{...(iii)}$$

Then, $\tan\theta_1 = \frac{V_E}{B_1} = \frac{V_E}{H_E\cos\theta}$

$$\cot\theta_1 = \frac{H_E\cos\theta}{V_E} \quad \text{...(iv)}$$

Similarly, $\tan\theta_2 = \frac{V_E}{B_2} = \frac{V_E}{H_E\sin\theta}$

$$\Rightarrow \quad \cot\theta_2 = \frac{H_E\sin\theta}{V_E} \quad \text{...(v)}$$

From Eqs. (iv) and (v), we get

$$\Rightarrow \cot^2\theta_1 + \cot^2\theta_2 = \frac{H_E^2\cos^2\theta}{V_E^2} + \frac{H_E^2\sin^2\theta}{V_E^2}$$

$$\Rightarrow \cot^2\theta_1 + \cot^2\theta_2 = \frac{H_E^2}{V_E^2}(\cos^2\theta + \sin^2\theta)$$

$$\Rightarrow \cot^2\theta_1 + \cot^2\theta_2 = \cot^2\theta \quad \text{[from Eq. (i)]}$$

**40** (*c*) Given, times of vibration in different planes, $T_1 = 3\text{s}$ and $T_2 = 3\sqrt{2}\text{s}$

As, we know time of vibration, $T = 2\pi\sqrt{\frac{I}{MB}}$

Time of vibration in verticle plane, $T_1 = 2\pi\sqrt{\frac{I}{MR}} = 3$ ... (i)

where, $R$ is resultant intensity of earth's field and $T$ is time of vibration of a magnetic needle.

Time of vibration in horizontal plane,

$$T_2 = 2\pi\sqrt{\frac{I}{MB_H}} = 3\sqrt{2} \quad \text{... (ii)}$$

where, $B_H$ = horizontal component of earth's magnetic field.

On dividing Eq. (i) by Eq. (ii), we get

$$\frac{1}{\sqrt{2}} = \sqrt{\frac{B_H}{R}} = \sqrt{\frac{R\cos\delta}{R}} = \sqrt{\cos\delta} \quad [\because B_H = R\cos\delta]$$

$$\Rightarrow \quad \cos\delta = \frac{1}{2} \Rightarrow \delta = 60^\circ$$

Hence, the angle of dip is 60°.

**41** (*d*) Intensity of magnetisation of a bulk material is given by magnetic dipole moment developed per unit volume when it is placed in magnetic field, i.e.

$$I = \frac{m_{\text{net}}}{V}$$

where, $m_{\text{net}}$ = magnetic moment and $V$ = volume.

Unit of $I = \text{Am}^{-1}$ and it is a vector quantity.

**42** (*b*) Given, $m = 8.0 \times 10^{22}\ \text{A-m}^2$ and $R = 6400$ km.

$$\text{Magnetisation, } M = \frac{m_{\text{net}}}{V} = \frac{8.0\times 10^{22}}{(4/3)\pi R^3} \quad \left[\because V = \frac{4}{3}\pi R^3\right]$$

$$= \frac{8.0\times 10^{22}}{(4/3)(3.14)(6.4\times 10^6)^3} = 73\ \text{Am}^{-1}$$

Hence, the earth's magnetisation is 73 $\text{Am}^{-1}$.

**43** (*b*) Given, $B = 1.7\times 10^{-6}$ T

Horizontal component of intensity of earth is magnetic field,

$$H = \frac{B}{\mu_0} = \frac{1.7\times 10^{-6}}{4\pi\times 10^{-7}} = 1.35\ \text{Am}^{-1}$$

**44** (*a*) Given, side of cube = 1cm = $10^{-2}$ m

$\therefore$ Volume, $V = 10^{-6}\text{m}^3$

Dipole moment, $\text{M} = 20\times 10^{-6}$ J/T

Applied magnetic intensity, $\text{H} = 60\times 10^3$ A/m

Intensity of magnetisation

$$I = \frac{\text{M}}{V} = \frac{20\times 10^{-6}}{10^{-6}} = 20\ \text{A/m}$$

Now, magnetic susceptibility $\chi$ is

$$\chi = \frac{\text{Intensity of magnetisation}}{\text{Applied magnetic intensity}} = \frac{I}{H}$$

$$= \frac{20}{60 \times 10^3}$$

$$\Rightarrow \quad \chi = \frac{1}{3} \times 10^{-3} = 3.33 \times 10^{-4}$$

**45** (*b*) Here, $V = (10 \times 0.5 \times 0.2)\ \text{cm}^3$

$$= 1\ \text{cm}^3 = 10^{-6}\text{m}^3$$

$$H = 0.5 \times 10^4\ \text{Am}^{-1} \text{ and } m = 5\ \text{A-m}^2$$

Intensity of magnetisation, $M = \frac{m}{V} = \frac{5}{10^{-6}}$

$$= 5 \times 10^6\ \text{Am}^{-1}$$

Magnetic induction, $B = \mu_0 (M + H)$

$$B = 4\pi \times 10^{-7}\ (5 \times 10^6 + 0.5 \times 10^4) = 6.28\ \text{T}$$

Hence, the magnetic induction will be 6.28 T.

**46** (*a*) Given, $\mu_r = 400$, $I = 2\text{A}$ and $n = 1000$

(i) The field $H$ is depend on the material core and it is

$$H = nI = 1000 \times 2 = 2 \times 10^3\ \text{Am}^{-1}$$

(ii) Magnetisation, $M = (B - \mu_0 H)/\mu_0$

$$\Rightarrow \quad M = (\mu_r \mu_0 H - \mu_0 H)/\mu_0 \quad [\because B = \mu_r \mu_0 H]$$

$$= (\mu_r - 1)H = (400 - 1)H = 399 \times H$$

$$= 399 \times 2 \times 10^3 \approx 8 \times 10^5\ \text{Am}^{-1}$$

Hence, the value of $H$ and $M$ are $2 \times 10^3 \text{Am}^{-1}$ and $8 \times 10^5 \text{Am}^{-1}$, respectively.

**47** (*a*) Given, $R_1 = 11\ \text{cm}$, $R_2 = 12\ \text{cm}$,

$$I = 0.70\ \text{A}, B = 2.5\ \text{T} \text{ and } N = 3000$$

Mean radius of the toroid,

$$R = \frac{R_1 + R_2}{2} = \frac{(11 + 12)\ \text{cm}}{2}$$

$$= 11.5 \times 10^{-2}\ \text{m}$$

As, we know $B = \mu H, H = nI$ and $\mu_r = \mu/\mu_0$

So, $\quad B = \mu_r \mu_0 nI$

$$\Rightarrow \quad \mu_r = \frac{B}{\mu_0 nI} = \frac{B}{\mu_0 (N/2\pi R) I} \quad \left[\because n = \frac{N}{2\pi R}\right]$$

$$= \frac{2.5 \times 2\pi\ (11.5 \times 10^{-2})}{(4\pi \times 10^{-7})\ (3000) \times (0.70)} = 685$$

Hence, $\mu_r = 685$.

**48** (*c*) Permeability of core,

$$\mu = \mu_0 \mu_r = \mu_0 (1 + \chi_m) \quad (\because \mu_r = 1 + \chi_m)$$

Given, $\chi_m = 599$

$$= (4\pi \times 10^{-7})\ (1 + 599)$$

$$= 7.5 \times 10^{-4} \text{TmA}^{-1}$$

**50** (*b*) When a diamagnetic substance is brought near the north or south pole of a bar magnet, it is feebly repelled by both the poles of magnet such as antimony, copper, gold, silver, etc.

**51** (*d*) For diamagnetic substances, the magnetic susceptibility is negative and it is independent of temperature. Therefore, choice (d) is correct in figure.

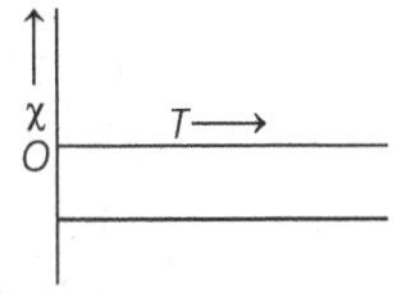


**52** (*b*) A diamagnetic liquid moves from stronger part of magnetic field to weaker part. Therefore, the meniscus of the level of solution will fall.

**53** (*c*) When the source of current is switched ON, a magnetic field sets up in between the poles of the electromagnet.

As we know that, a diamagnetic substance when placed in a magnetic field acquires a feeble magnetism opposite to the direction of magnetic field. Also, in the presence of the field (non-uniform), these substances are attracted towards the weaker field, i.e. they move from stronger to weaker magnetic field.

Due to these reasons, the rod is repelled or pushed up by the field produced due to the current source. Hence, the work required to do this comes from current source.

**54** (*d*) As, $\mu_r < 1$ for substance $X$, then it must be diamagnetic. Similarly $\mu_r > 1$ for substance $Y$, then it must be paramagnetic.

**55** (*c*) In a paramagnetic material, $M = C\frac{B}{T} = C\frac{\mu H}{T}$

$$[\because B = \mu H]$$

Hence, $M \propto H$. Therefore, the graph between $H$ and $M$ is a straight line represented by in figure

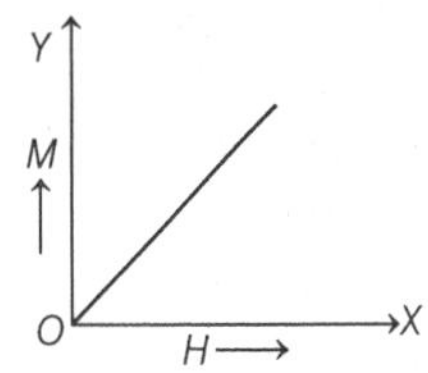


**56** (*d*) The volume of cubic domain is

$$V = (10^{-6})^3\ \text{m}^3 = 10^{-18}\text{m}^3 = 10^{-12}\text{cm}^3$$

$$[\because \text{side} = 1\ \mu\text{m}]$$

Mass = Density × Volume $\quad (\because \text{Density} = 7.9\ \text{gcm}^{-3})$

$$= 7.9\ \text{gcm}^{-3} \times 10^{-12}\ \text{cm}^3$$

$$= 7.9 \times 10^{-12}\ \text{g}$$

It is given that Avogadro's number i.e. $(6.023 \times 10^{23})$ of iron atoms have a mass of 55 g. Hence, the number of atoms in the domain is

$$N = \frac{\text{Mass}}{\text{Molecular mass}}$$

$$= \frac{7.9 \times 10^{-12} \times 6.023 \times 10^{23}}{55}$$

$$= 8.65 \times 10^{10} \text{atoms}$$

**57** (*b*) Given, $\chi = 1.2\times 10^5$, $\chi' = 1.8\times 10^{-5}$ and $T = 300\,\text{K}$

As magnetic susceptibility, $\chi = \frac{C}{T}$

$$\Rightarrow \quad \frac{\chi}{\chi'} = \frac{T'}{T}$$

$$\Rightarrow \quad T' = \left(\frac{\chi}{\chi'}\right)T = \left(\frac{1.2\times 10^{-5}}{1.8\times 10^{-5}}\right)300 = 200\ \text{K}$$

So, temperature of magnesium is 200 K.

**58** (*c*) Coercivity of a bar magnet is the value of magnetic field intensity ($H$) that is needed to reduce magnetisation to zero.

Since, for a solenoid magnetic induction is given as,

$$B = \mu_0 nI \quad \text{...(i)}$$

where, $n$ is the number of turns ($N$) per unit, length ($l$) and $I$ is the current.

Also, $$B = \mu_0 H \quad \text{...(ii)}$$

∴From Eqs. (i) and (ii), we get

$$\mu_0 nI = \mu_0 H \text{ or } H = nI = \frac{N}{l}I$$

Substituting the given values, we get

$$H = \frac{100}{0.2}\times 5.2 = 2600\ \text{A/m}$$

Thus, the value of coercivity of the bar magnet is 2600 A/m.

**59** (*c*) According to Curie's law, $\chi = \frac{C}{T}$ ... (i)

and magnetic susceptibility, $\chi = \frac{I}{H}$ ... (ii)

From Eqs. (i) and (ii), we get

$$\frac{C}{T} = \frac{I}{H} = \frac{m/V}{B/\mu} \quad \left[\because I = \frac{m}{V} \text{ and } H = \frac{B}{\mu}\right]$$

$$\Rightarrow \quad m = \frac{CV}{\mu}\left(\frac{B}{T}\right)$$

For a given sample, $CV/\mu$ = constant.

Thus, $m \propto B/T$

or $$\frac{m_1}{m_2} = \frac{B_1/T_1}{B_2/T_2}$$

Given, $B_1 = 0.84$ T, $B_2 = 0.98$ T

$T_1 = 4.2$ K and $T_2 = 2.8$ K

Thus, $$\frac{m_1}{m_2} = \frac{0.84/4.2}{0.98/2.8} = 4/7$$

or $m_2 = (7/4)\, m_1$

Initial total magnetic moment of the sample, i.e.

$m_1 = 15\%$ of $nm$, where $n$ is the number of atomic dipoles and $m$ is the dipole moment of each dipole.

Given, $n = 2\times 10^{24}$ and $m = 1.5\times 10^{-23}$ JT$^{-1}$

∴ $m_1 = 15\%$ of $(2.0 \times 10^{24})(1.5\times 10^{-23}) = 4.5$ JT$^{-1}$

Thus, $m_2 = (7/4)(4.5) = 7.9$ JT$^{-1}$

**60** (*a*) As temperature of ferromagnetic material is raised, its susceptibility $\chi$ remains constant upto curie temperature and then decreases as shown in figure.

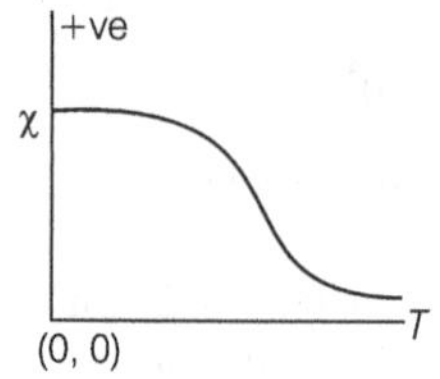


Hence, the correct variation is shown by figure in option (a).

**61** (*d*) Area of hysteresis loop is proportional to the net energy absorbed per unit volume by the material, as it is taken over a complete cycle of magnetisation.

For electromagnets and transformers, energy loss should be low.

i.e. thin hysteresis curves.

Also, $|B| \to 0$ when $H = 0$ and $|H|$ should be small when $B \to 0$.

**62** (*b*) The poles of magnet cannot be separated by breaking into two pieces.

If a bar magnet is broken into two halves, we get two similar bar magnets with somewhat weaker properties. As magnetic moment will be reduced to half, because magnetic dipole moment is directly proportional to the length between the poles.

Therefore, Assertion and Reason are correct but Reason is not the correct explanation of Assertion.

**63** (*a*) The net magnetic flux passes through a closed surface enclosing a bar magnet is zero because a bar magnet has two poles (one north and other south) and the magnetic flux leaving the north pole is equal to the magnetic flux entering the south pole of bar magnet.

Thus, according to Gauss's law for magnetism, the net magnetic flux througth any closed surface is zero i.e. the net flux $\phi_B$ is $\phi_B = \sum_{\text{all}} \mathbf{B}\cdot\Delta\mathbf{S} = 0$.

Therefore, Assertion and Reason are correct and Reason is the correct explanation of Assertion.

**64** (*c*) There is always a angle of declination between magnetic meridian and the geographic meridian. A compass is simply a needle shaped magnet, which always points in magnetic north-south poles (i.e. in magnetic meridian). Thus, true geographic north direction cannot be found by using a compass needle. The magnetic meridian of the earth is defined by some angle from the axis of rotation of the earth.

Therefore, Assertion is correct but Reason is incorrect.

**65** (*a*) The pole convention of earth arose, so because the magnetic north was the direction in which the north pole of a magnetic needle pointed and similarly the north pole of a magnet was so named as it was the north seeking pole.

In reality, the north magnetic pole of earth behave like the south pole of a bar magnet inside the earth and *vice-versa.* That is why, if one looks at the magnetic field lines of the earth one sees that unlike in the case of a bar magnet, the field lines go into the earth at the north magnetic pole $N_m$ and come out from the south magnetic pole $S_m$.

Therefore, Assertion and Reason are correct and Reason is the correct explanation of Assertion.

**66** (*b*) A magnetic needle, which is free to swing horizontally, would lie in the magnetic meridian and the north pole of the needle would point towards the magnetic north pole.

The line joining the magnetic poles is tilted with respect to the geographic axis of the earth. Hence, the magnetic meridian at a point makes some angle with the geographic meridian.

Therefore, Assertion and Reason are correct but Reason is not the correct explanation of Assertion.

**67** (*c*) From the relation, susceptibility of the material is

$$\chi = \frac{M}{H} \Rightarrow M = \chi H$$

Thus, it is obvious that greater the value of susceptibility of a material greater will be the value of intensity of magnetisation, i.e. more easily it can be magnetised.

Therefore, Assertion is correct but Reason is incorrect.

**68** (*c*) The magnetisation of a paramagnetic material is inversely proportional to the absolute temperature $T$.

$$M = C\frac{B_0}{T} \qquad \text{[Curie's law]}$$

where, $C =$ Curie constant.

Thus a paramagnetic sample displays greater magnetisation (for the same magnetic field) when cooled.

Therefore, Assertion is correct but Reason is incorrect.

**69** (*a*) Paramagnetic substances are those which get weakly magnetised when placed in an external magnetic field, because these substances have tendency to move from a region of weak magnetic field to strong magnetic field, i.e. they get weakly attracted to a magnet.

Therefore, Assertion and Reason are correct and Reason is the correct explanation of Assertion.

**70** (*a*) When paramagnetic substances are kept in an external magnetic field, then it develops feeble magnetisation in the direction of externally applied magnetic field, i.e. magnetic dipoles are weakly aligned along external magnetic field.

Therefore, Assertion and Reason are correct and Reason is the correct explanation of Assertion.

**71** (*b*) Ferromagnetic substances are those which gets strongly magnetised when placed in an external magnetic field. They have strong tendency to move from a region of weak magnetic field to strong magnetic field.

The individual atoms (or ions or molecules) in a ferromagnetic material possess a dipole moment as in a paramagnetic material. However, they interact with one another in such a way that they spontaneously align themselves in a common direction over a microscopic volume called domain.

Therefore, Assertion and Reason are correct but Reason is not the correct explanation of Assertion.

**72** (*b*) Substances which at room temperature retain their ferromagnetic property for a long period of time are called permanent magnets.

Permanent magnets can be made in a variety of ways. An efficient way to make a permanent magnet is to place a ferromagnetic rod in a solenoid and pass a current. The magnetic field of the solenoid magnetises the rod.

Therefore, Assertion and Reason are correct but Reason is not the correct explanation of Assertion.

**73** (*b*) Ferromagnetic substances have property that they gets strongly magnetised when placed in an external magnetic field.

They have strong tendency to move from a region of weak magnetic field to strong magnetic field, i.e. they get strongly attracted to a magnet.

Therefore, Assertion and Reason are correct but Reason is not the correct explanation of Assertion.

**74** (*d*) The temperature of transition from ferromagnetism to paramagnetism is called the Curie temperature $T_C$, i.e. at Curie point a ferromagnetic substance start behaving as a paramagnetic substance.

Thus, ferromagnetic substances obey Curie law.

Therefore, Assertion is incorrect but Reason is correct.

**75** (*d*) The statements I, II and III are correct but IV is incorrect and it can be corrected as,

The magnetic field lines never intersect, for if they did, the direction of the magnetic field would not be unique at the point of intersection.

**77** (*d*) The statements I, II and III are correct but IV is incorrect and it can be corrected as,

We cannot isolate the north or south pole of a magnet. If a bar magnet is broken into two halves, we get two similar bar magnets with somewhat weaker properties. Unlike electric charges, isolated magnetic north and south poles known as magnetic monopoles do not exist.

**78** (*b*) The statements II and III are correct but I is incorrect and it can be corrected as,

The magnetic field lines of the earth resemble that of a (hypothetical) magnetic dipole located at the centre of the earth.

**79** (*c*) The statements II and III are correct but I is incorrect and it can be corrected as,

Figure shows a bar of diamagnetic material placed in an external magnetic field. The field lines are repelled or expelled by diamagnetic material, results in the field inside the material is get reduced.

**80** (*b*) The statements II and III are correct but I is incorrect and it can be corrected as,

When an iron rod in the north-south direction is hammered repeatedly, it becomes a permanent magnet.

**81** (*b*) The statement given in option (b) is correct but rest are incorrect and these can be corrected as,

Magnetic phenomena are universal in nature.

The resemblance of magnetic field lines of a bar magnet and a solenoid suggest that a bar magnet may be thought of as a large number of circulating currents in analogy with a solenoid.

The magnetic field lines remain continuous, emerging from one face of the solenoid and entering into the other face.

**82** (*c*) The statement given in option (c) is correct but rest are incorrect and these can be corrected as,

Since, a current carrying circular loop behaves as a magnetic dipole whose magnetic moment's direction is perpendicular to the plane of the loop.

Therefore, when dipole is replaced by equivalent bar magnet, then the line joining *N-S* will be perpendicular to the plane of loop. As current is anti-clockwise, so north pole lies above the loop and south pole lies below the loop.

**83** (*c*) The time period of a compass needle in uniform magnetic field, $T = 2\pi\sqrt{\frac{I}{mB}}$

$$\Rightarrow \quad \frac{T}{2\pi} = \sqrt{\frac{T}{mB}}$$

$$\Rightarrow \quad \frac{2\pi}{T} = \sqrt{\frac{mB}{I}}$$

$$\Rightarrow \quad \omega^2 = \frac{mB}{I} \qquad \left(\because \omega = \frac{2\pi}{T}\right)$$

Hence, the square of angular frequency is $\frac{mB}{I}$.

Thus, the statement given in option (c) is incorrect, rest are correct.

**84** (*b*) Given, $N = 16, R = 10$ cm, $i = 0.75$A and $B = 5\times 10^{-2}$ T

(a) Magnetic moment of the coil

$$m = nIA = 16\times 0.75\times \pi(0.1)^2$$
$$= 16\times 0.75\times 3.14\times 0.1\times 0.1$$
$$= 0.377 \text{ JT}^{-1}$$

(b) Frequency of oscillation of the coil,

$$f = \frac{1}{2\pi}\sqrt{\frac{m\times B}{I}}$$

Squaring on both the sides, we get

$$f^2 = \frac{1}{4\pi^2}\cdot\frac{mB}{I}$$

$$\Rightarrow \quad I = \frac{mB}{4\pi^2 f^2} = \frac{0.377\times 5\times 10^{-2}}{4\times 3.14\times 3.14\times 2\times 2}$$

$$\Rightarrow \quad I = 1.2\times 10^{-4} \text{ kg-m}^2$$

(c) Moment of inertia,

$$I = \frac{mB}{4\pi^2 f^2} = \frac{0.377\times 5\times 10^{-2}}{4\times 3.14\times 3.14\times 3\times 3}$$
$$= 5.3\times 10^{-5} \text{ kg-m}^2$$

(d) Frequency of oscillation of the coil

$$f = \frac{1}{2\pi}\sqrt{\frac{m\times B}{I}} = \frac{1}{2\times 3.14}\sqrt{\frac{0.377\times 5\times 10^{-2}}{4\times 10^{-2}}}$$

$$\Rightarrow \quad f = 0.109/\text{s}$$

Thus, the statement given in option (b) is correct, rest are incorrect.

**85** (*b*) Magnetic field at large distances due to a bar magnet of magnetic moment **m** can be obtained from the equation for electric field due to an electric dipole of dipole moment **p**, by making the following replacements

$$\mathbf{E}\to\mathbf{B}, \mathbf{p}\to\mathbf{m}, \frac{1}{4\pi\varepsilon_0}\to\frac{\mu_0}{4\pi}$$

So, the equatorial field for a short dipole is $\frac{1}{4\pi\varepsilon_0}\cdot\frac{\mathbf{p}}{r^3}$ in electrostatics, while in magnetism its value is $\frac{\mu_0}{4\pi}\cdot\frac{\mathbf{m}}{r^3}$.

Thus, the statement given in option (b) is incorrect, rest are correct.

**86** (*d*) The statement given in option (d) is correct but rest are incorrect and these can be corrected as,

The vertical plane containing the longitude circle and the axis of rotation of the earth is called the geographic meridian. While, magnetic meridian of a place is the vertical plane which passes through the imaginary line joining the magnetic north and the south poles.

The angle between the true geographic north and the north shown by a compass needle is called the magnetic declination or simply declination.

The declination is greater at higher latitudes and smaller near the equator.

**87** (*d*) Here, for solenoid $\mathbf{H} = ni$

Given, $n = 1000\,\text{m}^{-1}, \mu_r = 1000$ and $i = 1$A

$$\Rightarrow \quad \mathbf{H} = 1000\times 1 = 1000 \text{ Am}^{-1}$$

Thus, when the core is heated, **H** remains constant, because **H** does not depend on $\mu_r$.

But, as we know, $B = \mu_0\mu_r ni = \mu_0 ni\mu_r = k\mu_r$

where, $k = \mu_0 ni$ = constant.

Thus, from above equation, we find that **B** varies with the variation in $\mu_r$.

Now, for magnetisation in the core, when temperature of the iron core of solenoid is raised beyond Curie temperature, then it will behave as paramagnetic material, i.e. magnetisation does not reverses.

As, we know $(\chi)_{\text{ferro}} \approx 10^3$ and $(\chi)_{\text{Para}} \approx 10^{-5}$

$(\chi)_{\text{ferro}} \rightarrow$ susceptibility of ferromagnetic material.

$(\chi)_{\text{para}} \rightarrow$ susceptibility of paramagnetic material.

$$\Rightarrow \quad \frac{(\chi)_{\text{ferro}}}{(\chi)_{\text{para}}} = \frac{10^3}{10^{-5}} = 10^8 \Rightarrow (\chi)_{\text{para}} = \frac{\chi_{\text{ferro}}}{10^8}$$

Thus, the statement given in option (d) is correct, rest are incorrect.

**88** (*a*) The statement given in option (a) is correct but rest are incorrect and these can be corrected as,

Magnetic field lines for magnetic material form continuous lines, so lines of **B** are necessarily continuous across *S*.

Magnetic intensity (**H**) varies about inside and outside the lump.

So, lines of **H** cannot all be continuous across *S*.

**89** (*d*) The susceptibility of paramagnetic material is

$$\chi = C\frac{\mu_0}{T}$$

This is known as Curie's law.

Thus, for a paramagnetic material both $\chi$ and $\mu_0$ depend not only on the type of material but also on the sample's temperature.

Thus, the statement given in option (d) is incorrect, rest are correct.

**90** (*d*) The statement given in option (d) is incorrect and it can be corrected as,

There are a number of elements, which are ferromagnetic like iron, cobalt, nickel, gadolinium, etc. The relative magnetic permeability is > 1000.

Rest statements are correct.

**91** (*a*) The statement given in option (a) is incorrect and it can be corrected as,

The value of magnetic field left behind in the magnetic material after the magnetising field has been removed is called retentivity or remanence. Hence, retentivity is the value of $B$ at $H = 0$

At $H = 0, B \neq 0.$

This is represented by the curve *Ob*. In curve, $B_R \sim 1.2$ T, where the subscript *R* denotes retentivity.

Rest statements are correct.

**92** (*c*) The statement given in option (c) is incorrect and it can be corrected as,

Core of electromagnets are made of ferromagnetic materials which have high permeability and low retentivity. Soft iron is a suitable material for electromagnets. The hysteresis curve of such materials must be narrow. The energy dissipated and the heating will consequently be small.

Rest statements are correct.

**93** (*b*)

A. Field lines shown in this field pattern are closed lines. The strongest external magnetic fields are near the poles. These field lines are less concentrated as compared to magnetic field pattern of [*B*] in Column I. Hence, it is the magnetic field pattern of a bar magnet.

B. Field lines shown in this field pattern are highly concentrated and form continuous close pattern. The strongest external magnetic field, is near the poles. Hence, this pattern of magnetic field resembles with field pattern of a solenoid where the ends of a solenoid behave as north and south poles of a magnet.

C. In this field pattern, the field lines are curved and extend from the positive to negative charge. In general, electric field lines always begin on a positive charge and end on a negative charge and do not start or stop in mid-space.

Hence, this is pattern of electric field of an electric dipole.

Hence, A $\rightarrow$ 3, B $\rightarrow$ 1 and C $\rightarrow$ 2.

**94** (*a*) Since, the magnetic potential energy $U_m$ is given by

$$U_m = \mathbf{m} \cdot \mathbf{B} = -mB\cos\theta$$

A. For $\theta = 90°$, $U_m = -mB\cos 90° = 0$

B. For $\theta = 0°$, $U_m = -mB\cos 0° = -mB$

The potential energy is minimum and hence the needle is in most stable position.

C. For $\theta = 180°$, $U_m = -mB\cos 180° = +mB$

The potential energy is maximum and hence the needle is in unstable position.

Hence, A $\rightarrow$ 3, B $\rightarrow$ 1 and C $\rightarrow$ 2.

**97** (*b*) The equatorial field is $B_E = \dfrac{\mu_0 m}{4\pi r^3}$

Given, $B_E = 0.4\text{ G} = 4 \times 10^{-5}$ T $\quad (\because 1\text{G} = 10^{-4}\text{ T})$

and $r = 6.4 \times 10^6$ m

Dipole moment, $m = \dfrac{4\pi r^3 B_E}{\mu_0} = \dfrac{B_E . r^3}{(\mu_0 / 4\pi)}$

$$m = \frac{4 \times 10^{-5} \times (6.4 \times 10^6)^3}{10^{-7}} \quad (\because \mu_0/4\pi = 10^{-7})$$

$$= 4 \times 10^2 \times (6.4 \times 10^6)^3$$

$$= 1.05 \times 10^{23}\text{ A-m}^2$$

Hence, the dipole moment of earth is $1.05 \times 10^{23}$ A-m$^2$.

**98** (*a*) Given, uniform magnetic field, $B = 0.25$ T,

Magnitude of torque, $\tau = 4.5 \times 10^{-2}$ J

and angle between magnetic moment and magnetic field, $\theta = 30°$.

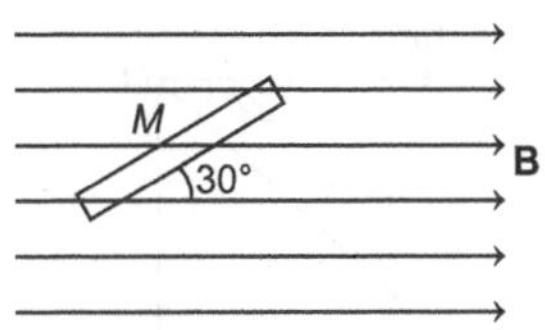

Torque experienced on a magnet placed in external magnetic field,

$$\tau = \mathbf{m} \times \mathbf{B}$$

$$\tau = mB \sin \theta \quad (\because \mathbf{A} \times \mathbf{B} = AB \sin \theta)$$

$$4.5 \times 10^{-2} = m \times 0.25 \times \sin 30^\circ$$

Magnitude of magnetic moment of the magnet,

$$m = \frac{4.5 \times 10^{-2}}{0.25 \times \sin 30^\circ} = \frac{4.5 \times 10^{-2} \times 2}{0.25 \times 1} \quad \left(\because \sin 30^\circ = \frac{1}{2}\right)$$

$$= 0.36 \text{ JT}^{-1}$$

Thus, the magnitude of magnetic moment of the magnet is $0.36 \text{ JT}^{-1}$.

**99** (*b*) Given, magnetic moment of magnet, $m = 0.32 \text{ JT}^{-1}$ and magnitude of magnetic field, $B = 0.15$ T

(i) For stable equilibrium, the angle between magnetic moment **m** and magnetic field **B** is $\theta = 0^\circ$

$\therefore$ In this position, it will be in a direction parallel to magnetic field thus no torque will act on it.

$\therefore$ The potential energy of the magnet,

$$U = -\mathbf{m} \cdot \mathbf{B}$$

$$= -mB \cos \theta \quad (\because \mathbf{A} \cdot \mathbf{B} = AB \cos \theta)$$

$$= -0.32 \times 0.15 \cos 0^\circ = -4.8 \times 10^{-2} \text{ J}$$

Thus, for the stable equilibrium the potential energy is $-4.8 \times 10^{-2}$ J.

(ii) For the unstable equilibrium, the angle between the magnetic moment and magnetic field is 180°.

$\therefore$ In this position, it will be in a direction perpendicular to magnetic field thus maximum torque will act on it.

$\therefore$ Potential energy of the magnet,

$$U = -mB \cos 180^\circ = -0.32 \times 0.15 (-1)$$

$$= 4.8 \times 10^{-2} \text{ J}$$

Thus, for the unstable equilibrium the potential energy is $4.8 \times 10^{-2}$ J.

**100** (*c*) Given, magnetic moment of magnet, $m = 1.5 \text{ JT}^{-1}$,
Uniform magnetic field, $B = 0.22$ J
and angles, $\theta_1 = 0^\circ$ and $\theta_2 = 90^\circ$

Work done in rotating the magnet from angle $\theta_1$ to angle $\theta_2$,

$$W = -mB (\cos \theta_2 - \cos \theta_1)$$

$$= -1.5 \times 0.22 (\cos 90^\circ - \cos 0^\circ) = 0.33 \text{ J}$$

Thus, the work done is 0.33 J.

**101** (*b*) Given, angle of dip, $\delta = 22^\circ$ and horizontal component of the earth's magnetic field, $H_E = 0.35$ G

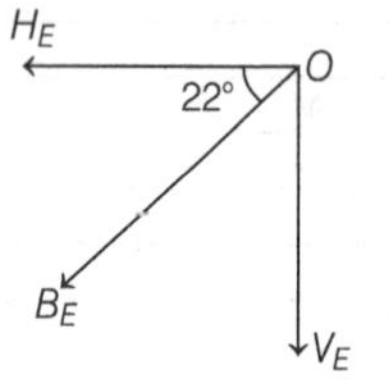


Let the magnitude of the earth's magnetic field at the place is $B_E$.

Using the formula, $H_E = B_E \cos \delta$

or
$$B_E = \frac{H_E}{\cos \delta} = \frac{0.35}{\cos 22^\circ}$$

$$= \frac{0.35}{0.9272} = 0.38 \text{ G}$$

Thus, the value of the earth's magnetic field at that place is 0.38 G.

**102** (*c*) Given, angle of declination, $\theta = 12^\circ W$,

Angle of dip, $\delta = 60^\circ$

and horizontal component of earth's magnetic field,

$$H_E = 0.16 \text{ G}$$

Let the magnitude of earth's magnetic field at that place is $B_E$.

Using the formula, $H_E = B_E \cos \delta$

$$B_E = \frac{H_E}{\cos \delta} = \frac{0.16}{\cos 60^\circ} = \frac{0.16 \times 2}{1}$$

$$= 0.32 \text{ G} = 0.32 \times 10^{-4} \text{ T}$$

$$[\because 1\text{G} = 10^{-4} \text{ T}]$$

Thus, earth's magnetic field is $0.32 \times 10^{-4}$ T.

**103** (*a*) **At normal bisector**

Let $r$ is the distance between axial line and point $P$. The magnetic field at point $P$, due to a short magnet

$$B = \frac{\mu_0}{4\pi} \cdot \frac{m}{r^3} \quad \ldots\text{(i)}$$

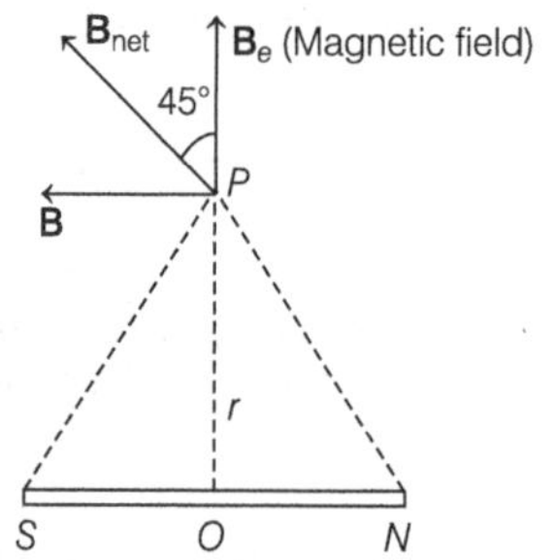


The direction of $B$ is along $PB$, i.e. along $N$-pole to $S$-pole.

According to the vector analysis,

$$\tan 45^\circ = \frac{B \sin 90^\circ}{B \cos 90^\circ + B_E}$$

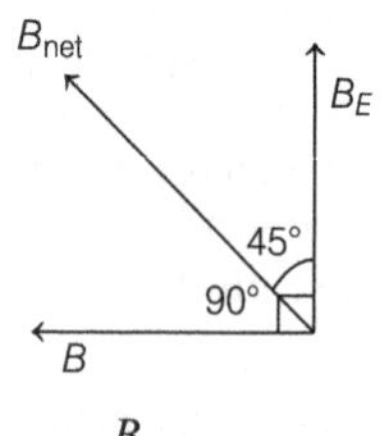


$$\Rightarrow \quad 1 = \frac{B}{B_E} \quad \text{or} \quad B = B_E$$

$$B_E = \frac{\mu_0}{4\pi} \cdot \frac{m}{r^3} \quad \text{[from Eq. (i)]}$$

Given, $B_E = 0.42\,G = 0.42 \times 10^{-4}\,T$

and $m = 5.25 \times 10^{-2}\,JT^{-1}$

$$0.42 \times 10^{-4} = \frac{10^{-7} \times 5.25 \times 10^{-2}}{r^3}$$

$$r^3 = \frac{5.25 \times 10^{-9}}{0.42 \times 10^{-4}} = 12.5 \times 10^{-5}$$

$$r = 0.05\,m = 5\,cm$$

Thus, the distance from the centre of magent is 5 cm.

**104** (*a*) Number of turns in the coil, $N = 30$

Current in the coil, $I = 0.35$ A

Radius of circular coil, $r = 12\,cm = 0.12\,m$

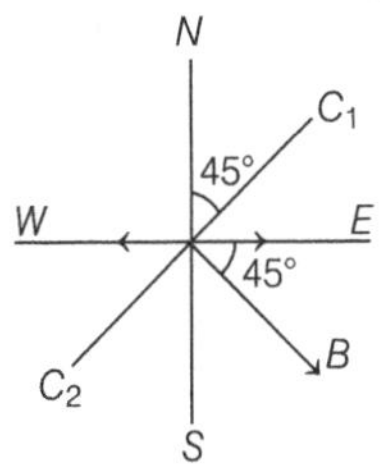


Let *N-S* be the line of magnetic meridian and the coil is placed at an angle of 45° with the magnetic meridian. $C_1$ and $C_2$ be the plane of coil. The needle points west to east. The magnetic field produced due to the coil is *B*. The direction of *B* is along the axis of coil, i.e. it makes an angle of 45° with east. The needle points west to east only, if the direction of magnetic field *B* is at 45° from east.

Using the formula, the magnetic field produced by a current carrying coil,

$$B = \frac{\mu_0}{4\pi} \cdot \frac{2\pi IN}{r}$$

$$= \frac{10^{-7} \times 2 \times 3.14 \times 0.35 \times 30}{0.12}$$

$$= 1.83 \times 10^{-6} \times 30$$

$$= 5.49 \times 10^{-5}\,T$$

Restoring torque due to $H_E = mH \sin 90° = mH_E$

where, $H_E$ = horizontal component of earth's magnetic field.

Deflecting torque due to *B*

$$= mB \sin 45°$$

So, $mH_E = mB \sin 45°$

The horizontal component of magnetic field

$$H_E = B \sin 45°$$

$$= 5.49 \times 10^{-5} \times \frac{1}{\sqrt{2}}$$

$$= 3.9 \times 10^{-5}\,T$$

Thus, $H_E = 3.9 \times 10^{-5}\,T$

**105** (*c*) In case of toroid, the magnetic field is only confined inside the body of toroid in the form of concentric magnetic lines of force and there is no magnetic field outside the body of toroid. This is because the loop encloses no current. Thus, the magnetic moment of toroid is zero.

In general, if we take *r* as a large distance outside the toroid, then $m \propto \frac{1}{r^3}$. But this case is not possible here.

Thus, only statement given in option (c) is correct.

**106** (*a*) For the earth's magnetism, the magnetic field lines of the earth resemble that of a hypothetical magnetic dipole located at the centre of the earth. Here, at Mumbai the axis of the dipole does not coincide with the axis of rotation of the earth but is presently tilted by approximately 11.3° with respect to the axis. This results into two situations as given in the figure below.

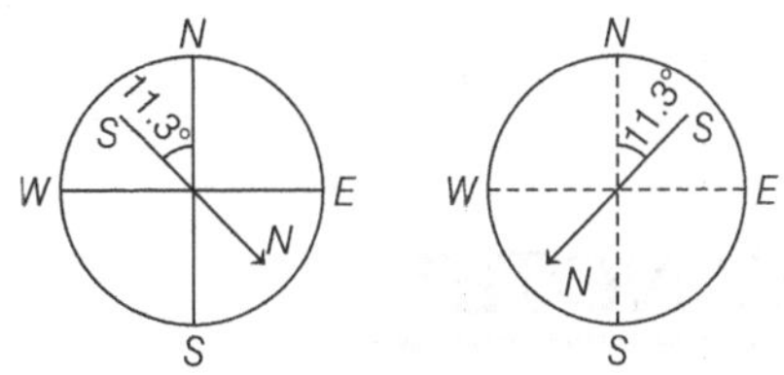


Hence, the declination varies between 11.3° *W* to 11.3° *E*.

**107** (*d*) As we know, a permanent magnet is a substance which at room temperature retain ferromagnetic property for a long period of time.

The individual atoms or molecule in a ferromagnetic material possess a dipole moment as in a paramagnetic material.

However, they interact with one another in such a way that they spontaneously align themselves in a common direction over a macroscopic volume called domain. Thus, we can say that in a permanent magnet at room temperature, domains are all perfectly aligned.

Thus, statements given in option (a), (b) and (c) are incorrect, but statement given in option (d) is correct.

**108** (*b*) Using Curie's law, we can deduce a formula for the relation between magnetic field induction, temperature and magnetisation.

i.e. $M \text{ (magnetisation)} \propto \frac{B \text{ (magnetic field induction)}}{T \text{ (temperature in kelvin)}}$

$$\Rightarrow \quad \frac{M_2}{M_1} = \frac{B_2}{B_1} \times \frac{T_1}{T_2}$$

Given, $M_1 = 8\,Am^{-1}$, $B_1 = 0.6\,T$ and $T_1 = 4\,K$, $B_2 = 0.2T$ and $T_2 = 16\,K$

$$\Rightarrow \quad \frac{0.2}{0.6} \times \frac{4}{16} = \frac{M_2}{8}$$

$$\Rightarrow \quad M_2 = 8 \times \frac{1}{12} = \frac{2}{3}\,Am^{-1}.$$

Thus, magnetisation is $(2/3)Am^{-1}$.

CHAPTER > 06

# Electromagnetic Induction

## KEY NOTES

- The phenomenon of generation of current or emf by changing the magnetic fields is known as **Electromagnetic Induction** (EMI). The emf developed in the conductor by the process of EMI is known as **induced emf** and if the conductor is in the form of a closed loop, then the current flowing in the conductor is known as **induced current**.

### Magnetic Flux

- The total number of magnetic field lines of force crossing through any surface normally, when it is placed in a magnetic field is known as the magnetic flux of that surface.
- Magnetic flux through a plane of area **A** placed in a uniform magnetic field **B** can be written as $\phi_B = \mathbf{B} \cdot \mathbf{A} = BA\cos\theta$ where, $\theta$ is angle between **B** and **A**.

  If the magnetic field has different magnitudes and directions at various parts of a surface, then

$$\phi_B = \mathbf{B}_1 \cdot d\mathbf{A}_1 + \mathbf{B}_2 \cdot d\mathbf{A}_2 + \ldots\ldots = \sum_{\text{all}} \mathbf{B}_i \cdot d\mathbf{A}_i$$

  The SI unit of magnetic flux is weber (Wb) or tesla metre square ($Tm^2$). It is a scalar quantity.

### Faraday's Law of Induction

- Whenever the amount of magnetic flux linked with a circuit changes, an emf is induced in the circuit.
- The magnitude of the induced emf in a circuit is equal to the time rate of change of magnetic flux through the circuit.

  Mathematically, the induced emf is given by

$$\varepsilon = \frac{-d\phi_B}{dt}$$

  The negative sign indicates the direction of $\varepsilon$ and hence the direction of current in a closed loop.
- In case of a closely wound coil of $N$ turns, change of flux associated with each turn is the same. Therefore, the expression for the total induced emf is given by

$$\varepsilon = \frac{-Nd\phi_B}{dt}$$

- The induced emf can be increased by increasing the number of turns $N$ of a closed coil.

### Lenz's Law and Conservation of Energy

- Lenz's law states that, the direction of any magnetic induction effect is such as to oppose the cause of effect.
- This law is based upon law of conservation of energy.

### Direction of Induced Current with the help of Lenz's Law

- Direction of induced current can be determined by checking whether the flux through a conducting loop or circuit is increasing or decreasing.
- If flux is decreasing, the magnetic field due to induced current will be along the existing magnetic field.
- If flux is increasing, the magnetic field due to induced current will be opposite to existing magnetic field.

## Motional emf

- Due to motion of conductor in magnetic field, a potential difference is maintained between the ends of the conductor as long as the conductor continues to move through the uniform magnetic field. This potential difference is called motional electromagnetic force, which is denoted by $\varepsilon$.
- When a conducting rod of length $l$ is moving with a uniform magnetic field $B$ (as shown below), then an induced emf is set up. The magnitude of induced emf will be $|\varepsilon| = Blv$.

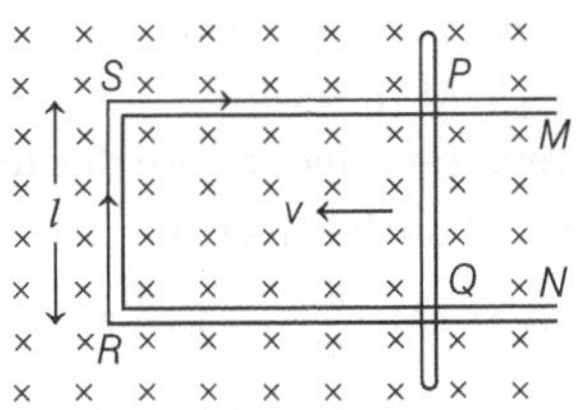


- In the case of a stationary conductor, the force on its charges is given by

$$F = q(\mathbf{E} + \mathbf{v} \times \mathbf{B}) = q\mathbf{E}$$

Since, $\mathbf{v} = 0$. Thus any force on the charge must arise from the electric field term $\mathbf{E}$ alone.

**Some important points related to emf are given below**

- If the rod is moving, making an angle $\theta$ with the direction of magnetic field, then $|\varepsilon| = Blv \sin\theta$

  $|\varepsilon| = 0$, for motion parallel to $\mathbf{B}$, i.e. $\mathbf{v} \,||\, \mathbf{B}$.
- If a conducting rod moves on two parallel conducting rails, then again an emf is induced whose magnitude is given by

$$|\varepsilon| = Blv \qquad [\text{since } \mathbf{v} \perp \mathbf{B}]$$

- The direction of induced current is given by Fleming's right hand rule.
- The induced current in the loop is given by

$$I = \frac{|\varepsilon|}{R} = \frac{Blv}{R}$$

where, $R$ = resistance of loop.
- Due to presence of current and magnetic field , arm $PQ$ (as shown in the above figure) experiences a magnetic force given by

$$F = BIl = \frac{B^2 l^2 v}{R}$$

- The force given above is directed opposite to the velocity of the rod and hence the power required to maintain constant speed $v$ is given as

$$P = Fv = \frac{B^2 l^2 v^2}{R}$$

- An emf is induced between the ends of rod, when a conducting rod of length $l$ rotates about an axis passing through one of its ends (that end may be fixed) with an angular velocity $\omega$ in a plane perpendicular to the magnetic field $B$ and is given by

$$|\varepsilon| = \frac{1}{2} Bl^2\omega$$

## Eddy Current

- The currents induced in bulk pieces of conductors when the magnetic flux linked with the conductor changes are known as eddy currents.
- Eddy currents are always produced in a plane which is perpendicular to the direction of magnetic field. Eddy current shows both heating and magnetic effects.
- Eddy currents can be reduced by
  (i) Reducing the area of metallic core of transformers, electric motor and other such devices.
  (ii) Using laminations of metal to make a metal core. The plane of the laminations must be arranged parallel to the magnetic field.
- Various applications of eddy currents are as follows
  (i) **Magnetic Breaking in Trains** Strong electromagnets are situated above the rails in some electrically powered trains. When electromagnets situated above rails are activated, the eddy currents induced in the rails opposes the motion of train.
  (ii) **Electromagnetic Damping** Certain galvanometer have a fixed core made of non-magnetic metallic material. When the coil oscillates, the eddy currents generated in the core opposes the motion and bring the coil to rest quickly.
  (iii) **Induction Furnace** It can be used to produce high temperatures and can be utilised to prepare alloys, by melting the constituent metals.
  (iv) **Electric Power Meters** Electric currents are induced in the shiny metal disc by magnetic fields produced by sinusoidally varying currents in a coil.

## Inductance

- It is a scalar quantity which plays same role in an electrical circuit as played by inertia in mechanics. It depends only on the geometry of the coil and intrinsic material properties.
- It is the ratio of flux linkage to the current.

$$L = \frac{N\phi}{I} = \frac{NBA}{I}$$

The SI unit of $L$ is henry (H).

## Self-Inductance

- It is the phenomenon of induction of emf in a single isolated coil due to change of flux through the coil by means of varying the current through the same coil.

  In this case, flux linkage through the coil of $N$ turns is proportional to the current through the coil and is expressed as

$$N\,\phi_B \propto I \text{ or } N\phi_B = LI$$

  where constant of proportionality $L$ is called the coefficient of self-induction of the coil.

- Induced emf due to self-induction is given by

$$\varepsilon = -L\frac{dI}{dt}$$

- For a solenoid coil having length $l$, total number of turns $N$ and cross-sectional area $A$, the value of $L$ is

$$L = \frac{\mu_0 N^2 A}{l} = \mu_0 n^2 Al$$

  where, $n = N/l$

- If we fill the inside of solenoid with a material of relative permeability $\mu_r$, then $L = \mu_r \mu_0 n^2 Al$.

  The self-inductance of the coil depends on its geometry and on the permeability of the medium.

- The self-induced emf is also called the **back emf**, as it opposes any change in the current in a circuit. The self-inductance plays the role of inertia.
- Work needs to be done against back emf $\varepsilon$ in establishing the current and given by

$$\frac{dW}{dt} = |\varepsilon|\, I$$

$$W = \int_0^I LI\, dI = \frac{1}{2}LI^2$$

  This work done is stored as **magnetic potential energy.**

## Mutual Inductance

- The phenomenon according to which an opposing emf is produced in a coil as a result of change in current or magnetic flux linked with a neighbouring coil is called mutual induction.
- For two long co-axial solenoids each having length $l$, the mutual inductance or coefficient of mutual induction is

$$M_{12} = M_{21} = M = \mu_0 n_1 n_2 \pi r_1^2 l$$

  where, $n_1$ & $n_2$ are number of turns per unit length of inner & outer coils of solenoid respectively,

  $r_1$ = radius of inner coil

  and $l$ = length of solenoid.

- If a medium of relative permeability $\mu_r$ is present, the mutual inductance would be $M = \mu_r \mu_0 n_1 n_2 \pi r_1^2 l$.
- The induced emf in the inner coil is given by

$$\varepsilon = -M\frac{dI_2}{dt}$$

- This shows that varying current in a coil can induce emf in a neighbouring coil. The magnitude of the induced emf depends upon the rate of change of current and mutual inductance of the two coils.
- In case of current flowing simultaneously in two nearby coils, the flux linked with one coil will be the sum of two fluxes which exists independently.

$$N_1\phi_1 = M_{11}I_1 + M_{12}I_2$$

  where, $M_{11}$ represents inductance due to the same coil.

- The emf induced can be given by

$$\varepsilon_1 = -M_{11}\frac{dI_1}{dt} - M_{12}\frac{dI_2}{dt}$$

  $M_{11}$ is the **self-inductance** and is written as $L_1$

$$\therefore \quad \varepsilon_1 = -L_1\frac{dI_1}{dt} - M_{12}\frac{dI_2}{dt}$$

## AC Generator or Dynamo

- It is an electrical machine which is used to convert mechanical energy into electrical energy.
- It is based on the principle of electromagnetic induction and the emf induced is given by

$$\varepsilon = -N\frac{d\phi}{dt} = -NBA\omega \sin \omega t$$

$$= \varepsilon_0 \sin \omega t$$

  where, $\varepsilon_0 = NBA\omega$.

- The current induced in the circuit is given by

$$i' = \frac{\varepsilon}{R} = \frac{\varepsilon_0}{R}\sin \omega t$$

$$= i_0 \sin \omega t$$

- There are various types of generators depending on the source of mechanical energy
  - (i) In **hydroelectric generators**, the mechanical energy required for rotation of the armature is provided by water falling from a height, e.g. dams.
  - (ii) In **thermal generators,** water is heated to produce steam using coal or other sources. The steam at high pressure produces the rotation of the armature.
  - (iii) In **nuclear power generators,** instead of coal, nuclear fuel is used as a source.

KEY NOTES

# Mastering NCERT

## MULTIPLE CHOICE QUESTIONS

## TOPIC 1 ~ The Experiments of Faraday and Henry : Magnetic Flux

**1** The phenomenon in which electric current is generated by varying magnetic field is appropriately known as electromagnetic

(a) conduction (b) induction
(c) pressure (d) radiation

**2** The direction of deflection of pointer of galvanometer when the bar magnet is pulled away from the coil as shown in figure, is

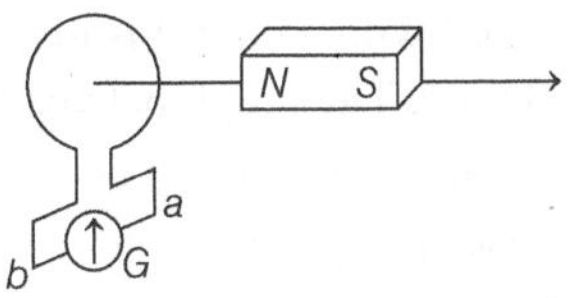


(a) towards $a$
(b) no deflection
(c) towards $b$
(d) indicator oscillates between $a$ and $b$

**3** Current in the coil is larger, when the magnet is

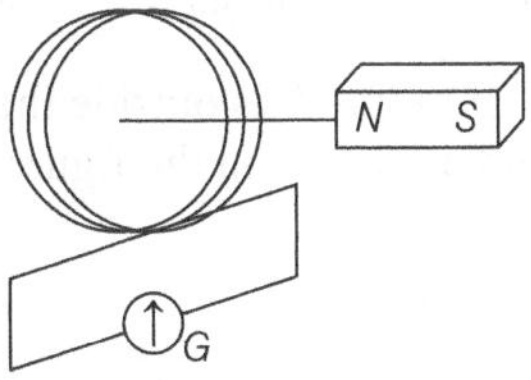


(a) pushed towards the coil faster
(b) pulled away from the coil faster
(c) Both (a) and (b)
(d) Neither (a) nor (b)

**4** A circular loop of radius $R$ carrying current $I$ lies in $xy$-plane with its centre at origin. The total magnetic flux through $xy$-plane is

(a) directly proportional to $R$
(b) directly proportional to $I$
(c) inversely proportional to $I$
(d) zero

**5** Consider a circular coil of wire carrying constant current $I$, forming a magnetic dipole. The magnetic flux through an infinite plane that contains the circular coil and excluding the circular coil area is given by $\phi_i$. The magnetic flux through the area of the circular coil area is given by $\phi_o$. Which of the following option is correct ? **JEE Main 2020**

(a) $\phi_i > \phi_o$ (b) $\phi_i < \phi_o$ (c) $\phi_i = \phi_o$ (d) $\phi_i = -\phi_o$

**6** A circular disc of radius 0.2 m is placed in a uniform magnetic field of induction $(1/\pi)$ $Wbm^{-2}$ in such a way that its axis makes an angle of 60° with $\mathbf{B}$. The magnetic flux linked with the disc is

(a) 0.02 Wb (b) 0.06 Wb (c) 0.08 Wb (d) 0.01 Wb

**7** At time $t = 0$ magnetic field of 1000 G is passing perpendicularly through the area defined by the closed loop shown in the figure. If the magnetic field reduces linearly to 500 gauss, in the next 5 s, then induced emf in the loop is **JEE Main 2020**

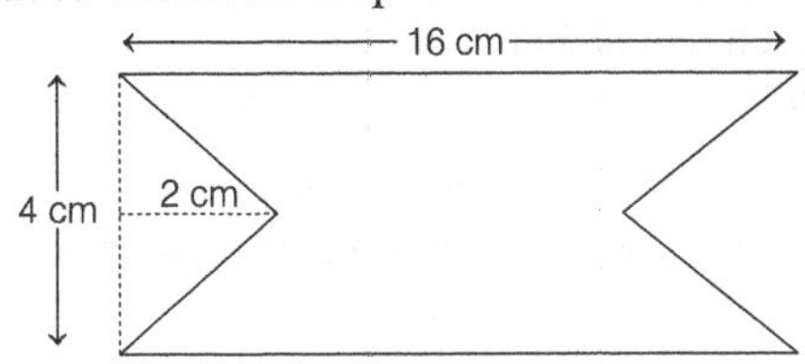


(a) 48 $\mu$V (b) 28 $\mu$V
(c) 56 $\mu$V (d) 36 $\mu$V

**8** A square of side $a$ metre lies in the $yz$-plane in a region, where the magnetic field is given $\mathbf{B} = B_0\,(3\hat{\mathbf{i}} + 3\hat{\mathbf{j}} - 4\hat{\mathbf{k}})$ T, where $B_0$ is constant. The magnitude of flux passing through the square in Wb is

(a) $2B_0a^2$ (b) $5B_0a^2$ (c) $3B_0a^2$ (d) $4\,B_0a^2$

**9** Consider the following figure, a uniform magnetic field of 0.2 T is directed along the positive $X$-axis. The magnetic flux through top surface of the figure is **AIIMS 2018**

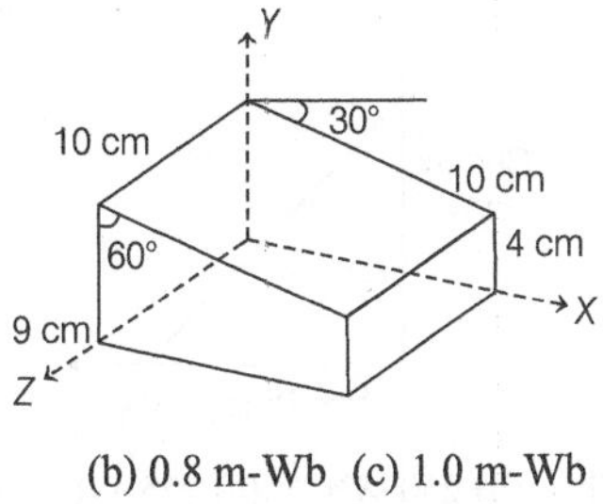


(a) zero (b) 0.8 m-Wb (c) 1.0 m-Wb (d) –1.8 m-Wb

# TOPIC 2~ *Faraday's Law of Induction*

**10** According to Faraday's law of electromagnetic induction, the magnitude of the induced emf in a circuit is equal to the time rate of change of
(a) magnetic field induction
(b) magnetising field intensity
(c) magnetic flux
(d) electric flux

**11** As shown in the figure, it is observed that the galvanometer shows a momentary deflection when the tapping key $K$ is pressed. When the key is released, the current in coil $C_2$ and the resulting magnetic field is

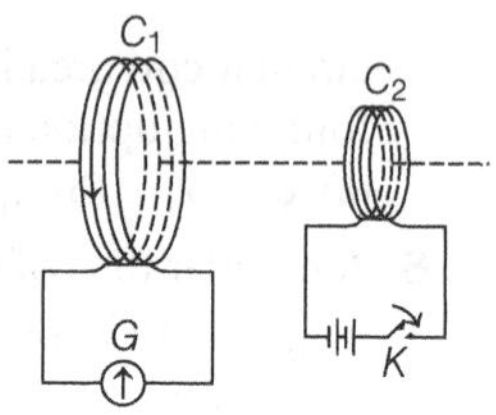


(a) increase from zero to maximum value
(b) first increase, then decrease
(c) remains same
(d) decreases from the maximum value to zero

**12** The magnetic flux linked with the coil varies with time as $\phi_B = 3t^2 + 4t + 9$. The magnitude of the induced emf at 2s is
(a) 9 V (b) 16 V (c) 3 V (d) 4 V

**13** A varying magnetic flux linking a coil is given by $\phi_B = xt^2$. If at time $t = 3$ s, the emf induced is 9 V, then the magnitude of $x$ is
(a) 0.66 $\text{Wbs}^{-2}$ (b) 1.5 $\text{Wbs}^{-2}$
(c) $-0.66$ $\text{Wbs}^{-2}$ (d) None of these

**14** A copper disc of radius 0.1 m is rotated about its centre with 20 rev/ s in a uniform magnetic field of 0.1 T with its plane perpendicular to the field. The emf induced across the radius of the disc is
(a) $\frac{\pi}{20}$ V (b) $\frac{\pi}{10}$ V
(c) $20\pi$ mV (d) None of these

**15** In a coil of resistance 100 Ω, a current is induced by changing the magnetic flux through it as shown in the figure. The magnitude of change in flux through the coil is **JEE Main 2017**

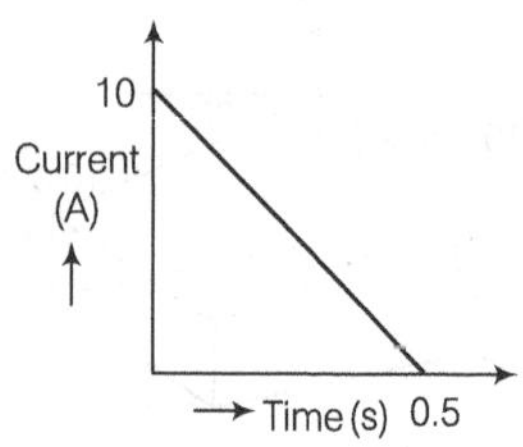


(a) 225 Wb (b) 250 Wb (c) 275 Wb (d) 200 Wb

**16** A coil of 800 turns effective area 0.05 $m^2$ is kept perpendicular to a magnetic field $5 \times 10^{-5}$ T. When the plane of the coil is rotated by 90° around any of its co-planar axis in 0.1 s, the emf induced in the coil will be **NEET 2019**
(a) 0.2 V (b) $2 \times 10^{-3}$ V (c) 0.02 V (d) 2 V

**17** A circular coil of radius 10 cm, 500 turns and resistance 2 Ω is placed with its plane perpendicular to the horizontal component of the earth's magnetic field. It is rotated about its vertical diameter through 180° in 0.25 s. Find the magnitude of the emf induced in the coil. Horizontal component of the earth's magnetic field at the place is $3.0 \times 10^{-5}$ T.
(a) $3 \times 10^3$ V (b) $3.8 \times 10^{-3}$ mV
(c) $3.8 \times 10^3$ mV (d) $3.8 \times 10^{-3}$ V

**18** The flux linked with a circuit is given by $\phi_B = t^3 + 3t - 7$. The graph between time ($X$-axis) and induced emf ($Y$-axis) will be a
(a) straight line through the origin
(b) straight line with positive intercept
(c) straight line with negative intercept
(d) parabola not through the origin

**19** A uniform magnetic field is restricted within a region of radius $r$. The magnetic field changes with time at a rate of $\frac{dB}{dt}$. Loop 1 of radius $R > r$ encloses the region $r$ and loop 2 of radius $R$ is outside the region of magnetic field as shown in the figure. Then the emf generated is **NEET 2016**

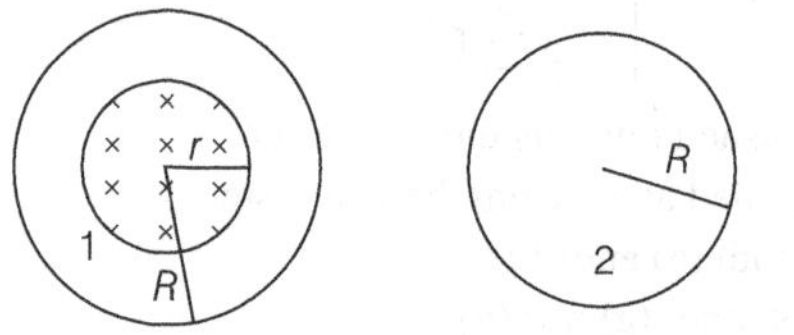


(a) zero in loop 1 and zero in loop 2
(b) $\frac{-dB}{dt}\pi r^2$ in loop 1 and $\frac{-dB}{dt}\pi r^2$ in loop 2
(c) $\frac{-dB}{dt}\pi R^2$ in loop 1 and zero in loop 2
(d) $\frac{-dB}{dt}\pi r^2$ in loop 1 and zero in loop 2

**20** A coil of resistance 400 Ω is placed in a magnetic field. If the magnetic flux $\phi_B$ (Wb) linked with the coil varies with time $t$(s) as $\phi_B = 50t^2 + 4$. Current at 2 s is **CBSE AIPMT 2012**
(a) 0.5 A (b) 0.1 A (c) 2 A (d) 1 A

**21** A square loop of wire of side length 10 cm is placed at an angle of 45° with a magnetic field that changes uniformly from 0.1 T to zero in 0.7 s. The induced current in the loop (its resistance is $1\Omega$) is

(a) 1.0 mA (b) 2.5 mA (c) 3.5 mA (d) 4.0 mA

**22** A circular coil of radius $R = 10$ cm having 500 turns and total resistance $2\Omega$ is placed initially perpendicular to the earth's magnetic field of $3\times 10^{-5}$ T.

The coil is rotated about its vertical diameter by an angle $\pi$ in 0.5 s. Then,

(a) the initial flux through the coil is $5\pi\times 10^{-7}$ Wb

(b) the final flux through the coil is $-7\pi\times 10^{-7}$ Wb

(c) the emf induced in the coil is $2\times 10^{-3}$ V

(d) induced current in the coil is 2.3 mA

**23** A square loop of side 10 cm and resistance 0.5 $\Omega$ is placed vertically in the east-west plane. A uniform magnetic field of 0.10 T is set-up across the plane in the north-east direction. The magnetic field is decreased to zero in 0.70 s at a steady rate. The magnitudes of induced emf and current during this time interval are

(a) 0.1 mV, 2A (b) 1 mV, 2 mA

(c) 1 V, 2 A (d) 1 mV, 0.2 A

**24** If a coil of 40 turns and area 4.0 $cm^2$ is suddenly removed from a magnetic field, it is observed that a charge of $2.0\times 10^{-4}$ C flows into the coil. If the resistance of the coil is 80 $\Omega$, then magnetic flux density (in $Wbm^{-2}$) is

(a) 0.5 (b) 1.0 (c) 1.5 (d) 2.0

## TOPIC 3 ~ *Lenz's Law and Conservation of Energy*

**25** The polarity of induced emf is such that it tends to produce a current which opposes the change in magnetic flux that produce it, is the statement of

(a) Faraday's law (b) Lenz's law

(c) Biot-Savart's law (d) Fleming's left hand rule

**26** An infinitely long cylinder is kept parallel to an uniform magnetic field $B$ directed along positive $Z$-axis. This direction of induced current as seen from the $Z$-axis will be

(a) clockwise of the positive $Z$-axis

(b) anti-clockwise of the positive $Z$-axis

(c) zero, no current is induced

(d) along the magnetic field

**27** Near a circular loop of conducting wire as shown in the figure an electron moves along a straight line. The direction of the induced current, if any in the loop is

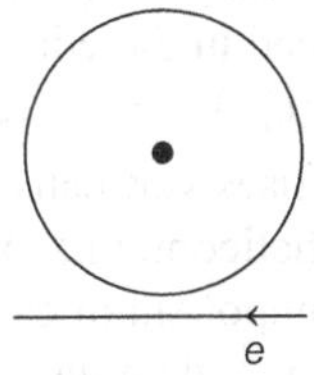


(a) variable (b) clockwise

(c) anti-clockwise (d) zero

**28** A closed loop moves normal to the constant electric field between the plates of a large capacitor. Is the current induced in the loop when it is wholly inside the region between the capacitor plates?

(a) Yes (b) No

(c) May be possible (d) May not be possible

**29** A rectangular loop and a circular loop are moving out of a uniform magnetic field region in the given figure, to a field free region with a constant velocity $v$. In which loop do you expect the induced emf to be constant during the passage out of the field region?

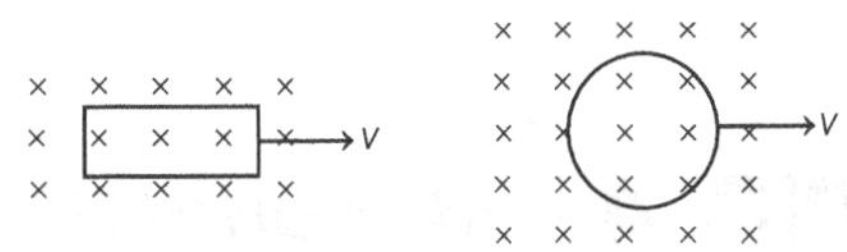


(a) Rectangular loop (b) Circular loop

(c) Both (a) and (b) (d) Neither (a) nor (b)

**30** The north pole of a bar magnet is moved towards a coil along the axis passing through the centre of the coil and perpendicular to the plane of the coil. The direction of the induced current in the coil when viewed in the direction of the motion of the magnet is

(a) clockwise

(b) anti-clockwise

(c) no current in the coil

(d) Either clockwise or anti-clockwise

**31** The north pole of a magnet is falling on a metallic ring as shown in the figure. The direction of induced current, if looked from upside, in the ring will be

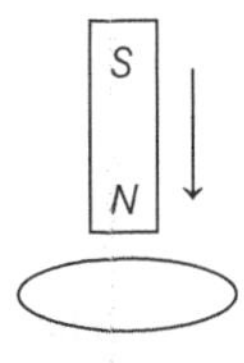


(a) clockwise or anti-clockwise depending on metal of the ring

(b) no induced current

(c) anti-clockwise

(d) clockwise

**32** There is a uniform magnetic field directed perpendicular and into the plane of the paper. An irregular shaped conducting loop is slowly changing its shape into a circular loop in the plane of the paper. Then,
(a) current is induced in the loop in the anti-clockwise direction
(b) current is induced in the loop in the clockwise direction
(c) AC is induced in the loop
(d) no current is induced in the loop

**33** When the north-pole of a bar magnet is being pushed towards the closed coil, current is induced in the coil in counter-clockwise direction.

What will happen, if an open circuit is used in place of the closed loop (in figure below $CD$ is an open circuit)?

(a) Induced current flows form $C$ to $D$
(b) Induced current flows form $D$ to $C$
(c) No induced current flow
(d) None of the above

**34** The magnet in figure rotates as shown on a pivot through its centre. At the instant shown, what are the directions of the induced current?

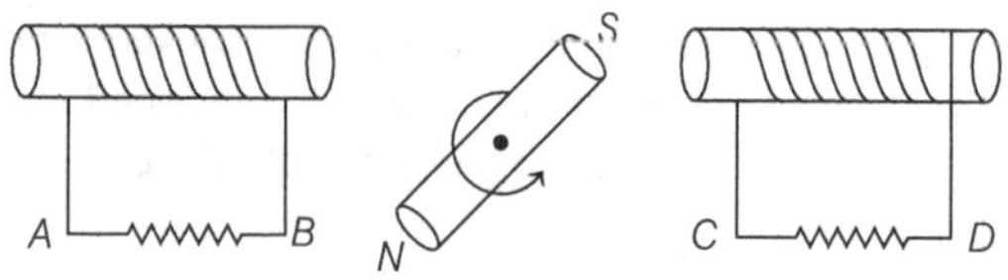


(a) $A$ to $B$ and $C$ to $D$
(b) $B$ to $A$ and $C$ to $D$
(c) $A$ to $B$ and $D$ to $C$
(d) $B$ to $A$ and $D$ to $C$

**35** Predict the polarity of the capacitor in the situation described in the given figure.

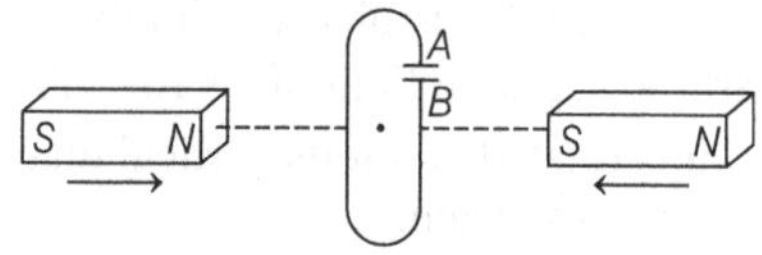


(a) Plate $A$ will be positive
(b) Plate $B$ will be positive
(c) Plate $A$ will be negative
(d) No charge is developed

**36** A magnet $N$-$S$ is suspended from a spring and when it oscillates, the magnet moves in and out of the coil $C$. Then, as magnet oscillates

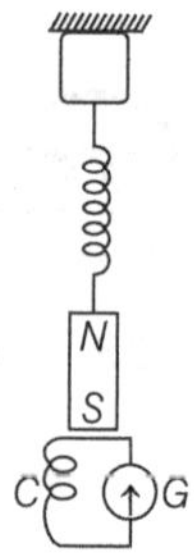


(a) $G$ shows no deflection but the amplitude steadily decreases
(b) $G$ shows deflection to the left and right with decreasing amplitude
(c) $G$ shows deflection to the left and right with constant amplitude
(d) $G$ shows deflection on one side

# TOPIC 4 ~ Motional Electromotive Force

**37** A conducting rod of length $l$ is moving in a transverse magnetic field of strength $B$ with velocity $v$. The induced voltage across the conducting rod is

(a) $\sqrt{\frac{Bvl}{2}}$ (b) $\sqrt{2Bvl}$ (c) $Bvl$ (d) $\frac{Bvl}{2}$

**38** A rectangular conductor $PQRS$ in which the conductor $PQ$ is free to move. The rod $PQ$ is moved towards the left with a constant velocity $v$ as shown in figure. Assume that, there is no loss of energy due to friction.

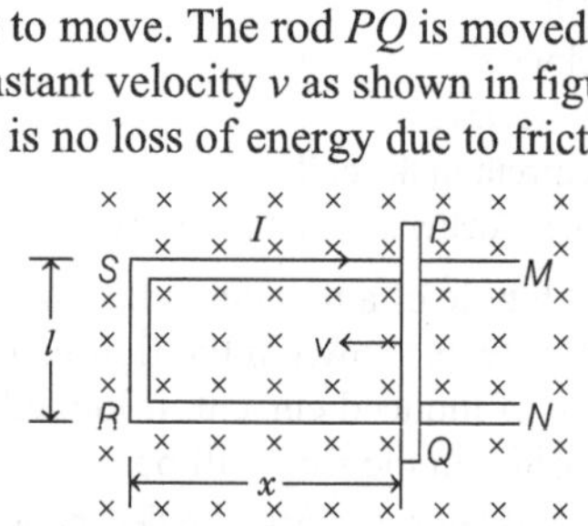


The rate of change of flux $\phi_B$ will induce an emf given by

(a) $\varepsilon = -\frac{d\phi_B}{dt}$ (b) $\varepsilon = -\frac{d}{dt}(Blx)$

(c) $\varepsilon = -Bl\frac{dx}{dt}$ (d) All of these

**39** A horizontal straight wire 20 m long extending from east to west is falling with a speed of 5.0 $ms^{-1}$ at right angles to the horizontal component of the earth's magnetic field $0.30 \times 10^{-4}$ $Wbm^{-2}$. The instantaneous value of the emf induced in the wire will be

(a) 6.0 mV (b) 3 mV (c) 4.5 mV (d) 1.5 mV

**40** A helicopter rises vertically with a speed of 100 $ms^{-1}$. If helicopter has length 10 m and horizontal component of earth's magnetic field is $5 \times 10^{-3}$ $Wbm^{-2}$, then the induced emf between the tip of nose and tail of helicopter is

(a) 50 V (b) 0.5 V (c) 5 V (d) 25 V

**41** Two parallel rails of a railway track insulated from each other and with the ground are connected to a millivoltmeter. The distance between the rails is 1 m. A train is travelling with a velocity of 72 $kmh^{-1}$ along the track. The reading of the millivoltmeter (in mV) is (vertical component of the earth's magnetic induction is $2 \times 10^{-5}$ T)

(a) 1.44 (b) 0.72 (c) 0.4 (d) 0.2

**42** A horizontal straight wire 10 m long extending from east to west is falling with a speed of 4.0 $ms^{-1}$, at right angles to the horizontal component of the earth's magnetic field of strength $0.30\times 10^{-2}$ $Wbm^{-2}$. The instantaneous value of the induced potential gradient in the wire from west to east is

(a) $+ 1.5\times 10^{-3}$ $Vm^{-1}$ (b) $- 1.5\times 10^{-3}$ $Vm^{-1}$
(c) $+ 1.2\times 10^{-2}$ $Vm^{-1}$ (d) $- 1.5\times 10^{-4}$ $Vm^{-1}$

**43** The magnitude of the earth's magnetic field at a place is $B_0$ and the angle of dip is $\delta$. A horizontal conductor of length $L$ lying in magnetic north-south moves eastwards with a velocity $v$. The emf induced across the conductor is

(a) zero (b) $B_0 Lv \sin \delta$
(c) $B_0 Lv$ (d) None of these

**44** A rectangular conductor $SPQR$ is placed in a uniform magnetic field perpendicular to the plane and extending from $x = 0$ to $x = b$ as shown below. The arm $PQ$ of length $l$ is moving from $x = 0$ to $x = 2b$ with constant speed $v$. Find the flux and emf induced in the conductor.

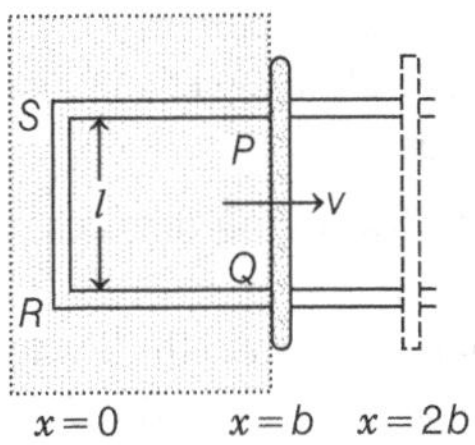


(a) $Blx, - Blv$; when $0 \le x < b$ (b) $2Bl, - 2Bl$
(c) $Blb, 0, b \le x < 2b$ (d) Both (a) and (c)

**45** A circular coil expands radially in a region of magnetic field and no electromotive force is produced in the coil. This can be because

(a) the magnetic field is constant
(b) the magnetic field is not in the same plane as the circular coil and it may or may not vary
(c) the magnetic field has a perpendicular (to the plane of the coil) component whose magnitude is decreasing suitably
(d) there is a constant magnetic field in the perpendicular (to the plane of the coil) direction

**46** A thin semi-circular conducting ring $PQR$ of radius $r$ is falling with its plane vertical in a horizontal magnetic field $B$, as shown in figure. The potential difference developed across the ring when its speed is $v$, is **CBSE AIPMT 2014**

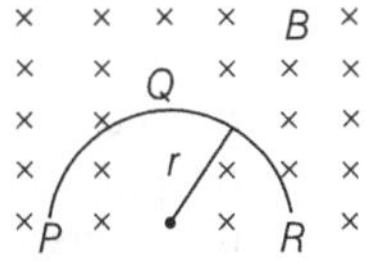


(a) zero
(b) $Bv\pi r^2/2$ and $P$ is at higher potential
(c) $\pi rBv$ and $R$ is at higher potential
(d) $2rBv$ and $R$ is at higher potential

**47** An $\angle AOB$ made of a conducting wire moves along its bisector through a magnetic field $B$ as suggested by figure. Find the emf induced between $A$ and $B$, if the magnetic field is perpendicular to the plane of the paper. (where, $AO = BO = l$)

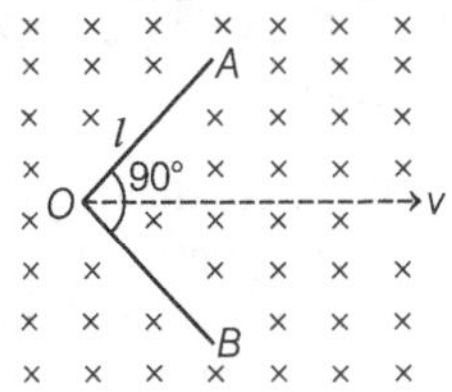


(a) $\frac{Bvl}{\sqrt{2}}$ (b) $Bv$ (c) $\sqrt{2}Blv$ (d) $Blv$

**48** A simple pendulum with bob of mass $m$ and conducting wire of length $L$ swings under gravity through an angle $\theta$. The earth's magnetic field component in the direction perpendicular to swing is $B$. Maximum emf induced across the pendulum is

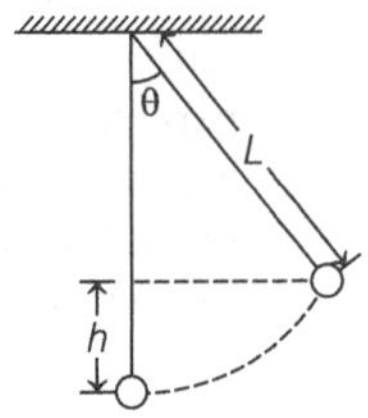


(a) $2BL \sin\left(\frac{\theta}{2}\right)(gL)^{1/2}$ (b) $BL \sin\left(\frac{\theta}{2}\right)(gL)$
(c) $BL \sin\left(\frac{\theta}{2}\right)(gL)^{3/2}$ (d) $BL \sin\left(\frac{\theta}{2}\right)(gL)^2$

**49** A metallic rod of 1 m length is rotated with a frequency of 50 $revs^{-1}$, with one end hinged at the centre and other end at the circumference of a circular metallic ring of radius 1 m. It is rotating about an axis passing through the centre and perpendicular to the plane of the ring as shown in figure. A constant and uniform magnetic field of 1 T parallel to the axis is present everywhere. What is the emf between the centre and the metallic ring?

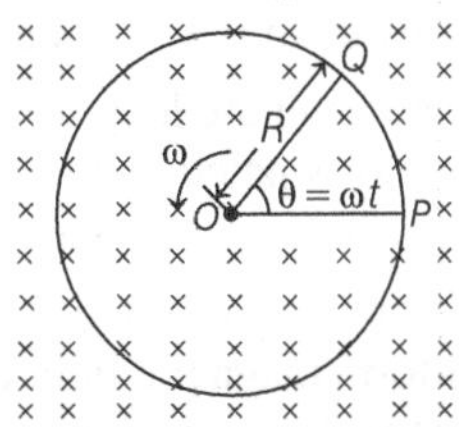


(a) 147 V (b) 157 V (c) 140 V (d) 164 V

**50** A cycle wheel of radius 0.5 m is rotated with constant angular velocity of 10 $rads^{-1}$ in a region of magnetic field of 0.1 T which is perpendicular to the plane of the wheel. The emf generated between its centre and the rim is **NEET (Odisha) 2019**

(a) 0.25 V (b) 0.125 V (c) 0.5 V (d) zero

**51** A wheel with 10 metallic spokes each 0.5 m long is rotated with a speed of 120 rev/min in a plane normal to the horizontal component of earth's magnetic field $H_E$ at a place. If $H_E = 0.4$ G at the place, what is the induced emf between the axle and the rim of the wheel? (Take, $1\text{G} = 10^{-4}$ T)

(a) $6.28\times 10^{-5}$ V (b) $62.8\times 10^{-5}$ V
(c) $0.628\times 10^{-5}$ V (d) $62.8\times 10^{-5}$ mV

## TOPIC 5 ~ *Energy Consideration and Eddy Current*

**52** A conducting rod of length $L$ is moving in a transverse magnetic field of strength $B$ with velocity $v$. The resistance of the rod is $R$. The current in the rod is

(a) $\frac{BLv}{R}$ (b) $BLv$ (c) zero (d) $\frac{B^2v^2L^2}{R}$

**53** Figure shows a conducting rod $PQ$ in contact with metal rails $RP$ and $SQ$, which are 25 cm apart in a uniform field of flux density 0.4 T acting perpendicular to the plane of the paper. Ends $R$ and $S$ are connected through a $5\,\Omega$ resistance. What is the magnitude and the direction of the current through the $5\,\Omega$ resistor when the rod moves to the right with a velocity of $5\ \text{ms}^{-1}$?

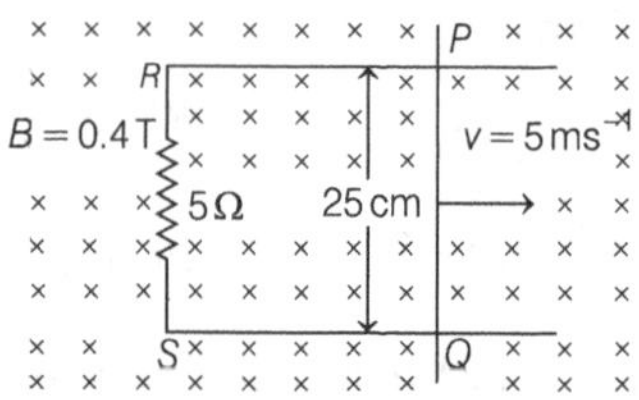


(a) 0. 1 A from $R$ to $S$ (b) 0.1 A from $S$ to $R$
(c) 0.5 A from $R$ to $S$ (d) 0.5 A from $S$ to $R$

**54** A rectangular loop of sides 10 cm and 5 cm with a cut is stationary between the pole pieces of an electromagnet. The magnetic field of the magnet is normal to the loop. The current feeding the electromagnet is reduced so that the field is decreased from its initial value of 0.3 T at the rate of $0.02\ \text{Ts}^{-1}$.

If the cut is joined and the loop has a resistance of $2.0\,\Omega$, the power dissipated by the loop as heat is

(a) 5 nW (b) 4 nW (c) 3 nW (d) 2 nW

**55** A rectangular wire loop of resistance $1.5\,\Omega$ placed in a region of uniform magnetic field of magnitude 0.3 T directed normal to the loop. If the area of loop decreases at a rate of $0.05\ \text{m}^2/\text{s}$ in this magnetic field, the heat produced in it is

(a) $1.5\times 10^{-6}$ W (b) $15\times 10^{-6}$ W
(c) $2\times 10^{-6}$ W (d) $150\times 10^{-6}$ W

**56** A copper plate is allowed to swing like a simple pendulum between the pole pieces of a strong magnet (as shown in figure).

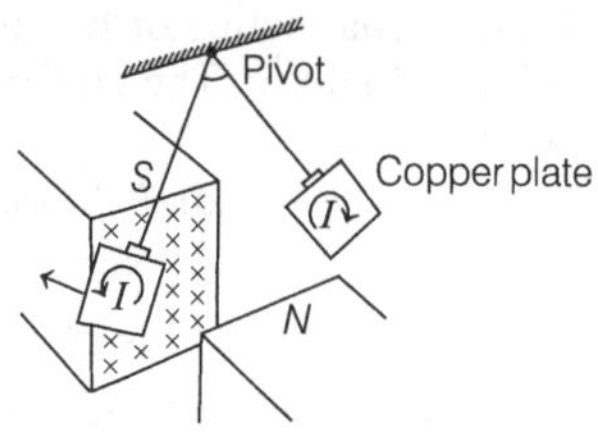


The motion of the plate and the phenomenon behind this are

(a) damped, electromagnetic induction
(b) accelerated, induced current
(c) accelerated, electromagnetic induction
(d) damped, Gauss' law

**57** In which of the following devices, the eddy current effect is not used? ***NEET 2019***

(a) Magnetic braking in train (b) Electromagnet
(c) Electric heater (d) Induction furnace

**58** In induction furnace, heat is produced due to ***JIPMER 2019***

(a) eddy current (b) resistance
(c) capacitor (d) None of these

**59** The shiny metal disc in the electric power meter rotates due to

(a) continuously decreasing current
(b) motor fitted inside
(c) continuously increasing current
(d) eddy current

## TOPIC 6 ~ *Inductance and AC Generator*

**60** Which of the following is equivalent to self-induction?

(a) Inertia of mass (b) Inertia of energy
(c) Inertia of moment (d) Inertia of current

**61** An electric current can be induced in a coil by flux change produced by another coil in its vicinity or flux change produced by the same coil. In both cases, the flux through a coil is

(a) proportional to the square of current
(b) inversely proportional to the square of current
(c) proportional to the current
(d) inversely proportional to the current

**62** The mutual inductance $M_{12}$ of coil 1 with respect to coil 2

(a) increases when they are brought nearer
(b) depends on the current passing through the coils
(c) increases when one of them is rotated about an axis
(d) is different from $M_{21}$ of coil 2 with respect to coil 1

**63** There are two identical coils $A$ and $B$ placed parallel along common axis, separated by some distance. A current of 2A flows through coil $A$ and a magnetic flux of $10^{-2}$ Wb passes through $B$ (no current through $B$). If no current passes through $A$ and a current of 1 A passes through $B$, what is the flux through $A$?

(a) $5\times 10^{-3}$ Wb (b) $0.5\times 10^{-3}$ Wb
(c) $50\times 10^{-3}$ Wb (d) $00.5\times 10^{-3}$ Wb

**64** $O$ is the centre of two coplanar concentric circular conductors $A$ and $B$, of radii $r$ and $R$ respectively as shown in the figure (here $r <<< R$). The mutual inductance of the system of the conductor can be given by

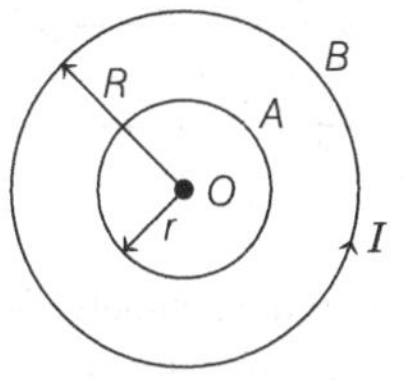


(a) $\frac{\mu_0 \pi r^2}{2R}$ (b) $\frac{\mu_0 \pi R^2}{2r}$ (c) $\frac{\pi R^2}{\mu_0 r}$ (d) $\frac{\mu_0 \pi r}{2R}$

**65** Two coils $P$ and $Q$ are separated by some distance. When a current of 3 A flows through coil $P$, a magnetic flux of $10^{-3}$ Wb passes through $Q$. No current is passed through $Q$. When no current passes through $P$ and a current of 2 A passes through $Q$, the flux through $P$ is **JEE Main 2019**

(a) $6.67\times 10^{-3}$ Wb (b) $6.67\times 10^{-4}$ Wb
(c) $3.67\times 10^{-3}$ Wb (d) $3.67\times 10^{-4}$ Wb

**66** A long solenoid of diameter 0.1 m has $2\times 10^4$ turns per metre. At the centre of the solenoid, a coil of 100 turns and radius 0.01 m is placed with its axis coinciding with the solenoid axis. The current in the solenoid reduces at a constant rate to 0 A from 4 A in 0.05 s. If the resistance of the coil is $10\pi^2\ \Omega$, the total charge flowing through the coil during this time is **NEET 2017**

(a) $32\pi\ \mu C$ (b) $16\ \mu C$ (c) $32\ \mu C$ (d) $16\pi\ \mu C$

**67** According to the phenomenon of mutual induction,

(a) the mutual inductance does not dependent on geometry of the two coils involved
(b) the mutual inductance depends on the intrinsic magnetic property, like relative permeability of the material
(c) the mutual inductance is independent of the magnetic property of the material
(d) ratio of magnetic flux produced by the coil 1 at the place of the coil 2 and the current in the coil 2 will be different from that of the ratio defined by interchanging the coils

**68** There are two long coaxial solenoids of same length $l$. The inner and outer coils have radii $r_1$ and $r_2$ and number of turns per unit length $n_1$ and $n_2$, respectively. The ratio of mutual inductance to the self-inductance of the inner coil is **JEE Main 2019**

(a) $\frac{n_2}{n_1}\cdot\frac{r_1}{r_2}$ (b) $\frac{n_2}{n_1}\cdot\frac{r_2^2}{r_1^2}$ (c) $\frac{n_2}{n_1}$ (d) $\frac{n_1}{n_2}$

**69** If coil is open, then $L$ and $R$ respectively become

(a) $\infty, 0$ (b) $0, \infty$ (c) $\infty, \infty$ (d) $0, 0$

**70** A current passing through a coil of self-inductance of 2 mH changes at the rate of $20\ \text{mAs}^{-1}$. The emf induced in the coil is

(a) $10\ \mu V$ (b) $40\ \mu V$ (c) 10 mV (d) 40 mV

**71** The inductance of a coil is $L = 10$ H and resistance $R = 5\ \Omega$. If applied voltage of battery is 10 V and if circuit switches off in 1 ms, find induced emf in inductor.

(a) $2\times 10^4$ V (b) $1.2\times 10^4$ V
(c) $2\times 10^{-4}$ V (d) None of these

**72** In 0.2 s, the current in a coil increases from 1.5 A to 2.5 A. If inductance of this coil is 60 mH, then induced current in an external resistance of $3\ \Omega$ will be

(a) 1 A (b) 0.5 A (c) 0.2 A (d) 0.1 A

**73** A copper wire is wound on a wooden frame, whose shape is that of an equilateral triangle. If the linear dimension of each side of the frame is increased by a factor of 3, keeping the number of turns of the coil per unit length of the frame the same, then the self-inductance of the coil **JEE Main 2019**

(a) increases by a factor of 3
(b) decreases by a factor of $9\sqrt{3}$
(c) increases by a factor of 27
(d) decreases by a factor of 9

**74** A long solenoid has 1000 turns. When a current of 4A flows through it, the magnetic flux linked with each turn of the solenoid is $4\times 10^{-3}$ Wb. The self-inductance of the solenoid is **NEET 2016**

(a) 3 H (b) 2 H (c) 1 H (d) 4 H

**75** A solenoid 60 cm long has 50 turns on it and is wound on an iron rod of 7.5 mm radius. Find the flux through the solenoid when the current in it is 3A and the relative permeability of iron is 600.

(a) 1.66 Wb (b) 2.66 Wb
(c) 1.66 mWb (d) $1.66\ \mu$ Wb

**76** Two identical induction coils each of inductance $L$ joined in series are placed very close to each other such that the winding direction of one is exactly opposite to that of other, what is the net inductance?

(a) $L^2$ (b) $2L$ (c) $L/2$ (d) Zero

**77** Three solenoid coils of same dimension, same numbers of turns and same numbers of layers of winding are taken. Coil 1 of inductance $L_1$ was wound using a wire of resistance $11\Omega m^{-1}$, coil 2 of inductance $L_2$ was wound using the similar wire but the direction of winding was reversed in each layer and coil 3 of inductance $L_3$ was wound using a superconducting wire. The self- inductance of the coils $L_1$, $L_2$ and $L_3$ are such that

(a) $L_1 = L_2 = L_3$ (b) $L_1 = L_2, L_3 = 0$
(c) $L_1 = L_3, L_2 = 0$ (d) $L_1 > L_2 > L_3$

**78** Which method is used to induce an emf or current in a loop in AC generator?

(a) A change in the loop's orientation
(b) A change in its effective area
(c) Both (a) and (b)
(d) Neither (a) nor (b)

**79** When the coil is rotated with a constant angular speed ω, the angle θ between the magnetic field vector **B** and the area vector **A** of the coil at any instant *t*, is

(a) $\theta = AB$ (b) $\theta = At$ (c) $\theta = \omega t$ (d) $\theta = Bt$

**80** Wire loop is rotated in a magnetic field.
The frequency of change of direction of the induced emf is ***NEET 2013***

(a) once per revolution (b) twice per revolution
(c) four times per revolution (d) six times per revolution

**81** A square coil of side 25 cm having 1000 turns is rotated with a uniform speed in a magnetic field about an axis perpendicular to the direction of the field. At an instant *t*, the emf induced in the coil is $\varepsilon = 200 \sin 100\,\pi t$. The magnetic induction is

(a) 0.50 T (b) 0.2 T (c) $10^{-3}$ T (d) 0.01 T

**82** Kamla paddles a stationary bicycle, the paddles of the bicycle are attached to a 100 turns coil of area $0.10\ m^2$. The coil rotates at half a revolution per second and it is placed in a uniform magnetic field of 0.01 T perpendicular to the axis of rotation of the coil. What is the maximum voltage generated in the coil?

(a) 3.14 V (b) 31.4 V (c) 0.314 V (d) 314 V

**83** When a circular coil of radius 1 m and 100 turns is rotated in a horizontal uniform magnetic field, the peak value of emf induced is 100 V. The coil is unwound and then rewound into a circular coil of radius 2 m. If it is rotated now with the same speed, under similar conditions, the new peak value of emf developed is

(a) 50 V (b) 25 V (c) 100 V (d) 200 V

**84** In a region of uniform magnetic induction $B = 10^{-2}$ T, a circular coil of radius 30 cm and resistance $\pi^2\Omega$ is rotated about an axis which is perpendicular to the direction of $B$ and which forms a diameter of the coil. If the coil rotates at 200 rpm, the amplitude of the alternating current induced in the coil is

(a) $4\pi^2$ mA (b) 30 mA (c) 6 mA (d) 200 mA

**85** A rectangular coil *ABCD* which is rotated at constant angular velocity about an horizontal axis as shown in the figure. The axis of rotation of the coil as well as the magnetic field *B* are horizontal. Maximum current will flow in the circuit when the plane of the coil is

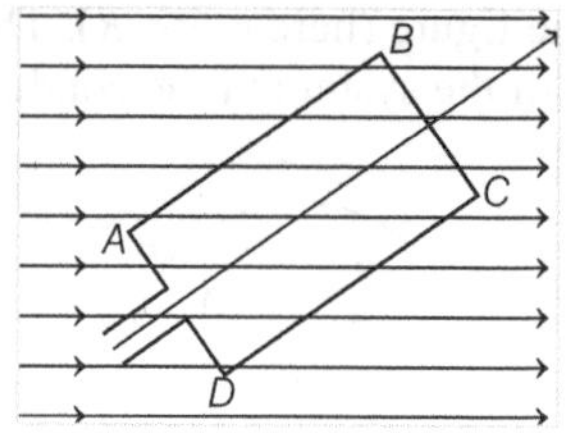


(a) inclined at 30° to the magnetic field
(b) perpendicular to the magnetic field
(c) inclined at 45° to the magnetic field
(d) parallel to the magnetic field

**86** The variation of emf with time for four types of generators are shown in the figures. Which amongst them can be called AC?

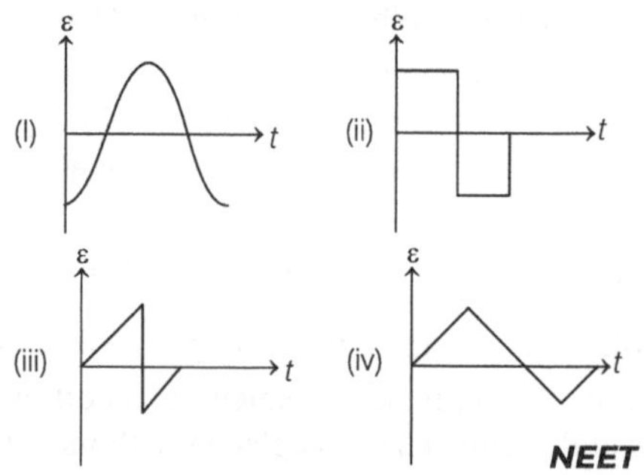


***NEET (Odisha) 2019***

(a) Both (i) and (iv) (b) (i), (ii), (iii) and (iv)
(c) Both (i) and (ii) (d) Only (i)

**87** The self-induced emf of a coil is 25 V. When the current in it is changed at uniform rate from 10 A to 25 A in 1s, the change in the energy of the inductance is ***JEE Main 2019***

(a) 437.5 J (b) 740 J (c) 637.5 J (d) 540 J

**88** The expression for the magnetic energy stored in a solenoid in terms of magnetic field *B*, area *A* and length *l* of the solenoid, is

(a) $\frac{1}{2\mu_0}BAl$ (b) $\frac{1}{2\mu_0}B^2Al$ (c) $\frac{1}{\mu_0}B^2Al$ (d) $\frac{1}{\mu_0}BA^2l$

**89** The magnetic potential energy stored in a certain inductor is 25 mJ, when the current in the inductor is 60 mA. This inductor is of inductance ***NEET 2018***

(a) 1.389 H (b) 138.88 H (c) 0.138 H (d) 13.89 H

# SPECIAL TYPES QUESTIONS

## I. Assertion and Reason

**Direction** (Q. Nos. 90-98) *In the following questions, a statement of Assertion is followed by a corresponding statement of Reason. Of the following statements, choose the correct one.*

(a) Both Assertion and Reason are correct and Reason is the correct explanation of Assertion.
(b) Both Assertion and Reason are correct but Reason is not the correct explanation of Assertion.
(c) Assertion is correct but Reason is incorrect.
(d) Assertion is incorrect but Reason is correct.

**90 Assertion** Magnetic flux of a magnetic field **B** passing through an area **A** is given by the relation

$$\phi = \mathbf{B} \cdot \mathbf{A}$$

**Reason** Quantity obtained by product of two vectors is always scalar.

**91 Assertion** Faraday's laws are consequence of conservation of energy.

**Reason** Earth's magnetic field can induce emf.

**92 Assertion** Lenz's law violates the principle of conservation of energy.

**Reason** Induced emf always opposes the change in magnetic flux responsible for its production.

**93 Assertion** Eddy currents are undesirable.

**Reason** Eddy currents heat up the core and dissipate electrical energy in the form of heat.

**94 Assertion** When a metallic conductor is moved in and out in a magnetic field, then an induced emf is produced across it.

**Reason** Eddy current will be produced in a metallic surface moving in and out of magnetic field.

*AIIMS 2019*

**95 Assertion** When coil in galvanometer with metallic core oscillates, then electromagnetic damping occurs.

**Reason** Eddy currents generated in the core oppose the motion and bring the coil to rest quickly.

**96 Assertion** If the inner solenoid is much shorter than (and placed well inside) the outer solenoid, then the flux linkage $N_1\phi_1$ can still be calculated.

**Reason** The inner solenoid is effectively immersed in a uniform magnetic field due to the outer solenoid.

**97 Assertion** In a system of co-axial solenoids, it is extremely difficult to calculate the flux linkage with the outer solenoid when current flows in inner solenoid.

**Reason** The magnetic field due to the inner solenoid would vary across the length as well as cross-section of the outer solenoid (when inner solenoid is smaller in length and radius).

**98 Assertion** The self-induced emf is also called the back emf.

**Reason** The self-induced emf opposes any change in the current in a circuit.

## II. Statement Based Questions

**99** Consider a coil and magnet as shown below

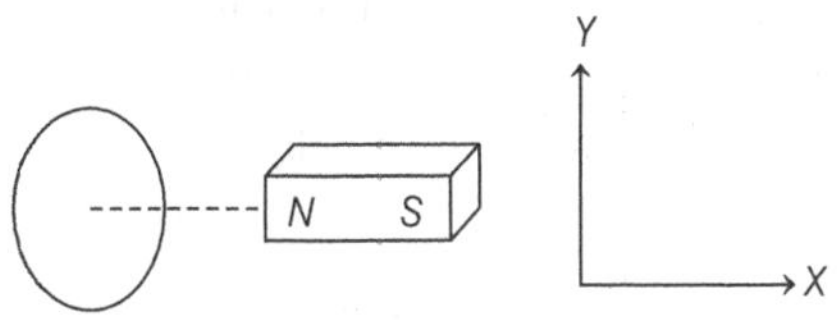


Current is induced in coil, when

I. coil and magnet both are at rest.
II. coil is at rest and magnet moves along $x$.
III. magnet is at rest and coil moves along $x$.
IV. Both coil and magnet move along y with same speed.

Which of the following statement(s) is/are correct?

(a) Both I and IV
(b) Both I and II
(c) Both III and IV
(d) Both II and III

**100** Consider the following statements.

I. Inductance is a scalar quantity.
II. Inductance has the dimensions of $[ML^2T^2A^2]$.
III. The SI unit of inductance is henry and is denoted by H.
IV. Inductance depends only on the geometry of the coil and intrinsic material properties.

Which of these statement(s) is/are Correct?

(a) Only I (b) Only II
(c) Only III (d) Only IV

**101** Consider the situation shown below.

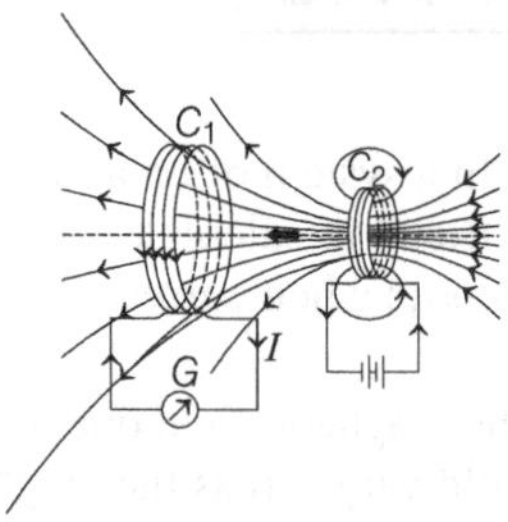


I. Because of change in current in $C_2$, emf is induced in $C_1$.

II. The steady current in the coil $C_2$ produces a steady magnetic field.

III. If coil $C_2$ is moved towards the coil $C_1$, the galvanometer shows a deflection. This indicates that electric current is induced in coil $C_1$.

Which of the above statements are correct for the figure shown?

(a) Only I (b) Both II and III

(c) Both I and III (d) I, II and III

**102** An AC generator is shown below

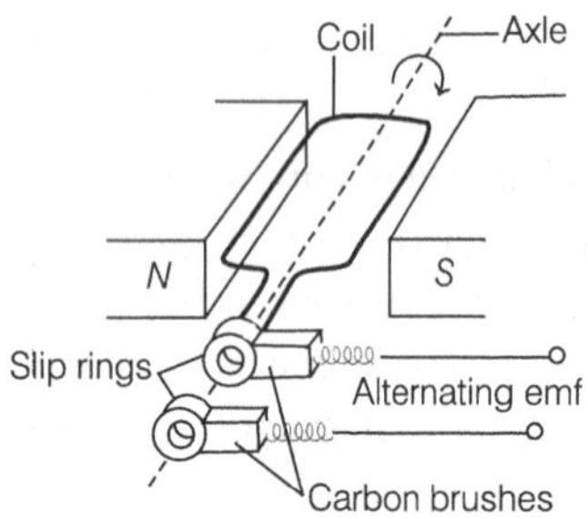


I. AC generator consists of a coil mounted on a rotor shaft.

II. The axis of rotation of the coil is perpendicular to the direction of the magnetic field.

III. The coil (called armature) is mechanically rotated in the uniform magnetic field by some external means.

IV. Magnetic flux associated with armature coil changes continuously due to its rotation.

Which of the following statement(s) is/are incorrect?

(a) I, II and III (b) I, II and IV

(c) II, III and I (d) None of these

**103** I. Modern day generators produce electric power as high as 500 MW.

II. In most generators, the coils are held stationary and it is the electromagnets which are rotated.

III. The frequency of rotation of armature in India is 50MHz.

Which of thc following statement(s) is/are incorrect?

(a) Only I (b) Only III

(c) Both I and II (d) Only II

**104** Consider the motion of a magnet towards or away from coil $C_1$ in Fig. (i) and moving a current carrying coil $C_2$ towards or away from coil $C_1$ in Fig. (ii).

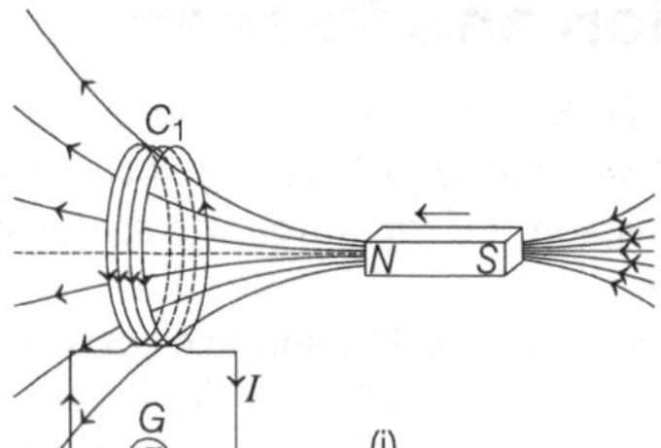


(i)

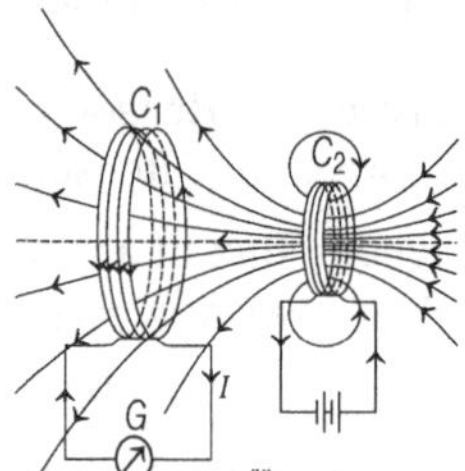


(ii)

For magnetic flux associated with coil $C_1$, which of the following statement is correct?

(a) Changes in both situations (i) and (ii).

(b) Changes in situation (i) but do not change in situation (ii).

(c) Does not change in situation (i) but changes in situation (ii).

(d) Does not change in both situations (i) and (ii).

**105** Which of the following statement(s) is/are incorrect ?

(a) The phenomenon in which electric current is generated by varying magnetic field is appropriately called electromagnetic induction.

(b) If the magnetic field has different magnitudes and directions at various parts of a surface, then the magnetic flux through the surface is, $\phi_B = \sum_{au} \mathbf{B}_i d\mathbf{A}_i$

(c) An emf is induced in a coil when magnetic flux through the coil changes with time, whose magnitude is equal to the time rate of change of magnetic flux through the circuit.

(d) The induced emf can be increased by decreasing the number of turns $N$ of a closed coil.

**106** Which of the following statement(s) is/are incorrect ?

(a) Flux can be varied by changing any one or more of the terms $B$, $A$ and $\theta$.

(b) Flux can also be altered by changing the shape of a coil (i.e. by shrinking it or stretching it) in a magnetic field.

(c) A gentle push on the magnet in magnetic field initiates the motion in magnet and its kinetic energy continuously increases without expanding any energy.

(d) When a bar magnet experiences a repulsive force due to the induced current, a person has to do work in moving the magnet. The energy spent by the person is dissipated by joule heating produced by the induced current.

**107** Which of the following statement(s) is/are correct ?

(a) The induced currents produced in a conductor due to changing magnetic flux is called eddy current.

(b) If rectangular slots are made in a copper plate, the electromagnetic damping increases.

(c) The plane of the lamination of metal core must be arranged perpendicular to the magnetic field.

(d) The magnetic breaking effect in trains is not smooth.

**108** Which of the following statement(s) is/are incorrect?

(a) A varying element in a coil can induce emf in a neighbouring coil, the magnitude of which depends upon the rate of change of current and mutual inductance of the two coils.

(b) It is also possible that emf is induced in a single isolated coil due to change of flux through the coil by means of varying the current through the same coil. This phenomenon is called self-inductance.

(c) For the current $I$ at an instant in a circuit, the rate of work done is, $\frac{dW}{dt} = |\varepsilon| I$ and total work done is,

$W = \int dW = \int_0^I LI \, dI$

(d) The flux linked with one coil will be the difference of two fluxes which exists independently (when current is flowing simultaneously in two nearby coils).

**109** Which of the following statement(s) is/are correct?

(a) For a closely wound coil of $N$-turns, the magnetic flux linked with all turns is different.

(b) When the flux $\phi_B$ through the coil changes, only few turn of them contributes to the induced emf.

(c) If the geometry of the coil vary with time, then $\frac{dB}{dt} \propto \frac{dI}{dt}$.

(d) If a medium of relative permeability $\mu_r$ had been present instead of air, the mutual inductance remains same.

**110** Which of the following statement(s) is/are correct?

(a) The emf has its minimum value, when $\theta = 90°$ or $\theta = 270°$, as the change of flux is lowest at these points.

(b) In commercial generator, the mechanical energy required for rotation of the armature is provided by water falling from a height, e.g. from dams. These are called thermal generators.

(c) Water is heated to produce steam using coal or other sources. The steam at high pressure produces the rotation of the armature. These are called nuclear power generators.

(d) A magnet dropped in an aluminium pipe takes much longer time to come out of the pipe, due to electromagnetic damping.

## III. Matching Type

**111** Match the Column I (static and dynamic positions of coils and relative motion of magnet with respect to coil) with Column II (cause of induced current in coil $C_1$) and select the correct answer from the codes given below.

| | Column I | | Column II |
|---|---|---|---|
| A. | 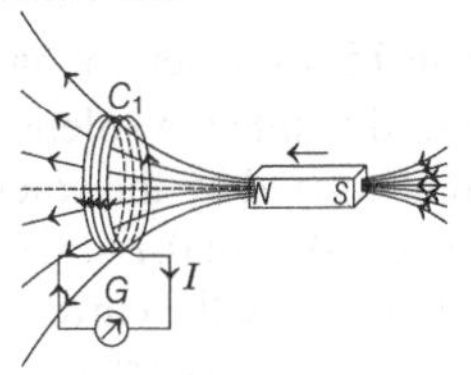 | 1. | Current is induced in coil $C_1$ due to motion of the current carrying coil $C_2$ |
| B. | 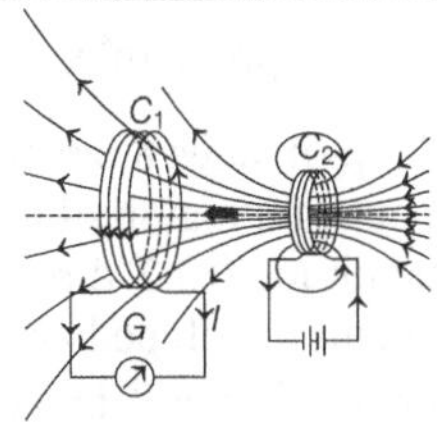 | 2. | Coils $C_1$ and $C_2$ held stationary, current in $C_1$ is induced by changing current in $C_2$ |
| C. | 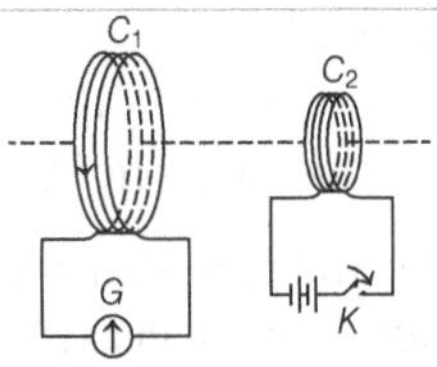 | 3. | When the bar magnet is pushed towards the coil, the pointer in galvanometer $G$ deflects |

| | A | B | C | | A | B | C |
|---|---|---|---|---|---|---|---|
| (a) | 2 | 1 | 3 | (b) | 1 | 2 | 3 |
| (c) | 3 | 2 | 1 | (d) | 3 | 1 | 2 |

**112** Match the Column I (planar loops of different shapes) with Column II (direction of induced current) and select the correct answer from the codes given below.

| | Column I | | Column II |
|---|---|---|---|
| A. | 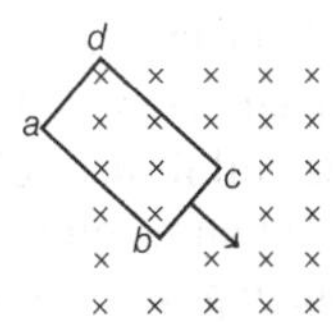 | 1. | $b\ a\ c\ b$ |
| B. | 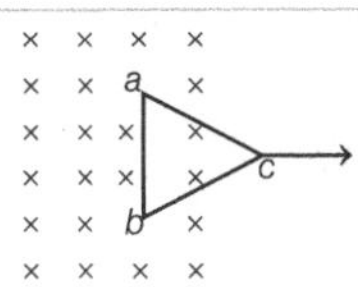 | 2. | $c\ d\ a\ b\ c$ |
| C. | 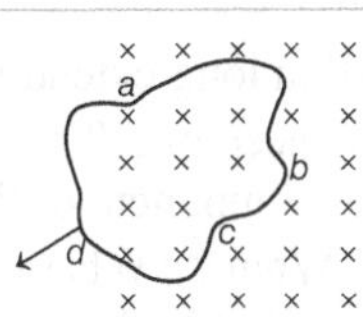 | 3. | $b\ c\ d\ a\ b$ |

| | A | B | C | | A | B | C |
|---|---|---|---|---|---|---|---|
| (a) | 3 | 1 | 2 | (b) | 2 | 1 | 3 |
| (c) | 1 | 2 | 3 | (d) | 1 | 3 | 2 |

# NCERT & NCERT Exemplar

## MULTIPLE CHOICE QUESTIONS

### NCERT

**113** A long solenoid with 15 turns per cm has a small loop of area 2.0 $cm^2$ placed inside the solenoid normal to its axis. If the current carried by the solenoid changes steadily from 2.0 A to 4.0 A in 0.1 s, what is the induced emf in the loop while the current is changing?

(a) $7.5\times 10^{-6}$ V (b) $5\times 10^{6}$ V
(c) $7\times 10^{5}$ V (d) $5\times 10^{5}$ V

**114** A rectangular wire loop of sides 8 cm and 2 cm with a small cut is moving out of a region of uniform magnetic field of magnitude 0.3 T directed normal to the loop. What is the emf developed across the cut, if the velocity of the loop is 1 $cms^{-1}$ in a direction normal to the longer side?

(a) $3\times 10^{-4}$ V (b) $2.4\times 10^{-4}$ V
(c) $5\times 10^{-4}$ V (d) $2\times 10^{-4}$ V

**115** A 1.0 m long metallic rod is rotated with an angular frequency of 400 $rads^{-1}$ about an axis normal to the rod passing through its one end. The other end of the rod is in contact with a circular metallic ring. A constant and uniform magnetic field of 0.5 T parallel to the axis exists everywhere. The emf developed between the centre and the ring is

(a) 200 V (b) 50 V
(c) 100 V (d) None of these

**116** A circular coil of radius 8.0 cm and 20 turns is rotated about its vertical diameter with an angular speed of 50 $rads^{-1}$ in a uniform horizontal magnetic field of magnitude $3.0\times 10^{-2}$ T. If the coil forms a closed loop of resistance 10 $\Omega$, the maximum emf induced, average emf induced and average power loss due to heating are respectively

(a) 0.603 V, 0 V and 0.018 W
(b) 0.603 V, 0.02 V and 0.015 W
(c) 0 V, 0.0603 V and 0 W
(d) 0.05 V, 0.02 V and 0.018 W

**117** A horizontal straight wire 10 m long extending from east to west is falling with a speed of 5.0 $ms^{-1}$, at right angles to the horizontal component of the earth's magnetic field, $0.30\times 10^{-4}$ $Wbm^{-2}$. What is the instantaneous value of the emf induced in the wire?

(a) $1.9\times 10^{-4}$ V (b) $1.5\times 10^{-3}$ V
(c) $2.0\times 10^{-4}$ V (d) $3.0\times 10^{-4}$ V

**118** Current in a circuit falls from 5.0 A to 0.0 A in 0.1 s. If an average emf of 200 V induced, then an estimate of the self-inductance of the circuit is

(a) 5 H (b) 4 H (c) 3 H (d) 2 H

**119** A pair of adjacent coils has a mutual inductance of 1.5 H. If the current in one coil changes from 0 to 20 A in 0.5 s. The change of flux linkage with the other coil is

(a) 60 Wb (b) 65 Wb (c) 30 Wb (d) 50 Wb

**120** An air-cored solenoid with length 30 cm, area of cross-section 25 $cm^2$ and number of turns 500, carries a current of 2.5 A. The current is suddenly switched OFF in a brief time of $10^{-3}$ s. How much is the average back emf induced across the ends of the open switch in the circuit? Ignore the variation in magnetic field near the ends of the solenoid.

(a) 6.5 V (b) 7.5 V (c) 6.0 V (d) 8.0 V

### NCERT Exemplar

**121** A square of side $L$ metres lies in the $xy$-plane in a region, where the magnetic field is given by $\mathbf{B} = B_0(2\hat{\mathbf{i}} + 3\hat{\mathbf{j}} + 4\hat{\mathbf{k}})$ T, where $B_0$ is constant. The magnitude of flux passing through the square is

(a) $2B_0L^2$ Wb (b) $3B_0L^2$ Wb
(c) $4B_0L^2$ Wb (d) $\sqrt{29}B_0L^2$ Wb

**122** A loop, made of straight edges has six corners at $A\,(0, 0, 0)$, $B\,(L, 0, 0)$, $C\,(L, L, 0)$, $D\,(0, L, 0)$, $E\,(0, L, L)$ and $F\,(0, 0, L)$. A magnetic field $\mathbf{B} = B_0(\hat{\mathbf{i}} + \hat{\mathbf{k}})$ T is present in the region. The flux passing through the loop $ABCDEFA$ (in that order) is

(a) $B_0L^2$ Wb (b) $2B_0L^2$ Wb
(c) $\sqrt{2}B_0L^2$ Wb (d) $4B_0L^2$ Wb

**123** A cylindrical bar magnet is rotated about its axis. A wire is connected from the axis and is made to touch the cylindrical surface through a contact. Then,

(a) a direct current flows in the ammeter $A$
(b) no current flows through the ammeter $A$
(c) an alternating sinusoidal current flows through the ammeter $A$ with a time period, $T = \frac{2\pi}{\omega}$
(d) a time varying non-sinusoidal current flows through the ammeter $A$

**124** There are two coils $A$ and $B$ as shown in figure. A current starts flowing in $B$ as shown, when $A$ is moved towards $B$ and stops, when $A$ stops moving. The current in $B$ is counter-clockwise. $B$ is kept stationary when $A$ moves.

We can infer that

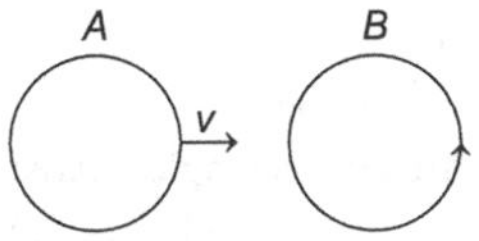


(a) there is a constant current in the clockwise direction in $A$
(b) there is a varying current in $A$
(c) there is no current in $A$
(d) there is a constant current in the counter-clockwise direction in $A$

**125** Coil $A$ is made to rotate about a vertical axis (shown in figure). No current flows in $B$, if $A$ is at rest. The current in coil $A$,when the current in $B$ (at $t = 0$) is counter-clockwise and the coil $A$ is as shown at this instant, $t = 0$, is

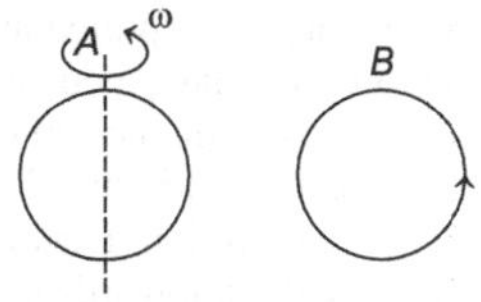


(a) constant current clockwise
(b) varying current clockwise
(c) varying current counter-clockwise
(d) constant current counter-clockwise

**126** The self-inductance $L$ of a solenoid of length $l$ and area of cross-section $A$, with a fixed number of turns $N$, increases as
(a) $l$ and $A$ increase
(b) $l$ decreases and $A$ increases
(c) $l$ increases and $A$ decreases
(d) Both $l$ and $A$ decrease

# Answers

## > Mastering NCERT with MCQs

| | | | | | | | | | |
|---|---|---|---|---|---|---|---|---|---|
| 1 (b) | 2 (c) | 3 (c) | 4 (d) | 5 (d) | 6 (a) | 7 (c) | 8 (c) | 9 (c) | 10 (c) |
| 11 (d) | 12 (b) | 13 (b) | 14 (c) | 15 (b) | 16 (c) | 17 (d) | 18 (d) | 19 (d) | 20 (a) |
| 21 (a) | 22 (c) | 23 (b) | 24 (b) | 25 (b) | 26 (c) | 27 (a) | 28 (b) | 29 (a) | 30 (b) |
| 31 (c) | 32 (a) | 33 (b) | 34 (a) | 35 (a) | 36 (b) | 37 (c) | 38 (d) | 39 (b) | 40 (c) |
| 41 (c) | 42 (c) | 43 (b) | 44 (a) | 45 (c) | 46 (c) | 47 (c) | 48 (a) | 49 (b) | 50 (b) |
| 51 (a) | 52 (c) | 53 (a) | 54 (a) | 55 (d) | 56 (a) | 57 (c) | 58 (a) | 59 (d) | 60 (a) |
| 61 (c) | 62 (a) | 63 (a) | 64 (a) | 65 (b) | 66 (c) | 67 (b) | 68 (c) | 69 (b) | 70 (b) |
| 71 (a) | 72 (d) | 73 (c) | 74 (c) | 75 (c) | 76 (d) | 77 (b) | 78 (c) | 79 (c) | 80 (b) |
| 81 (d) | 82 (c) | 83 (d) | 84 (c) | 85 (d) | 86 (b) | 87 (a) | 88 (b) | 89 (d) | |

## > Special Types Questions

| | | | | | | | | | |
|---|---|---|---|---|---|---|---|---|---|
| 90 (c) | 91 (c) | 92 (d) | 93 (a) | 94 (b) | 95 (a) | 96 (a) | 97 (a) | 98 (a) | 99 (d) |
| 100 (b) | 101 (d) | 102 (d) | 103 (b) | 104 (a) | 105 (d) | 106 (c) | 107 (a) | 108 (d) | 109 (c) |
| 110 (d) | 111 (d) | 112 (a) | | | | | | | |

## > NCERT & NCERT Exemplar MCQs

| | | | | | | | | | |
|---|---|---|---|---|---|---|---|---|---|
| 113 (a) | 114 (b) | 115 (c) | 116 (a) | 117 (b) | 118 (b) | 119 (c) | 120 (a) | 121 (c) | 122 (b) |
| 123 (b) | 124 (d) | 125 (a) | 126 (b) | | | | | | |

# Hints & Explanations

**2** (*c*) When the magnet is pulled away from the coil, then according to Lenz's law, a current in induced in the coil such that it opposes the motion of the magnet.
Therefore, direction of induced current in the coil will be in clockwise direction. Hence, deflection in the galvanometer will be towards *b*.

**3** (*c*) Current will be larger, when the magnet is pushed towards or pulled away from the coil faster, because rate of change of magnetic flux increases and hence induced emf across the coil increases.

**4** (*d*) As the circular loop lies in *xy*-plane, so the magnetic field or flux is perpendicular to its plane, i.e. in *z*-direction.
Therefore, the total magnetic flux through *xy*-plane is

$$\phi = BA\cos\theta = BA\cos 90^\circ \quad [\because B \perp A]$$
$$= 0$$

**5** (*d*) We are given with following situation,

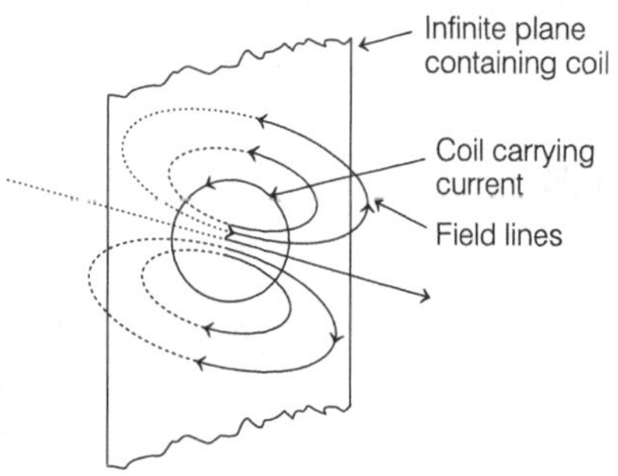


From the diagram of field lines, we can observe that whatever be the number of field lines emitted from coil, all of them goes back into the infinite plane only.
So, magnetic flux emanating from coil is equal and opposite to the flux linked with infinite plane.
So, $\phi_i = -\phi_o$

**6** (*a*) The magnetic flux $\phi_B$ passing through a plane surface of area *A* placed in a uniform magnetic field *B* is given by

$$\phi_B = BA\cos\theta$$

where, θ is the angle between the direction of *B* and the normal (axis) to the plane.

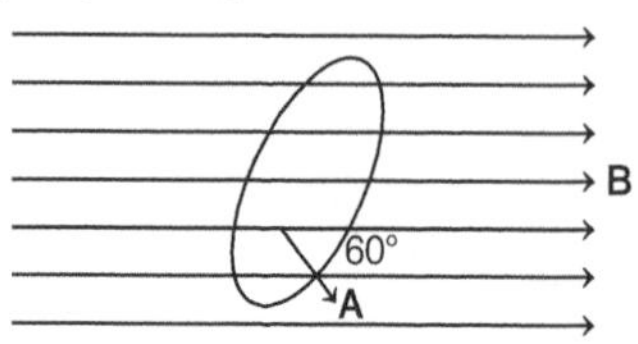


Given, $\theta = 60^\circ, B = (1/\pi)\ \text{Wbm}^{-2}, A = \pi(0.2)^2$

$$\therefore \quad \phi_B = \frac{1}{\pi}\times \pi(0.2)^2 \times \cos 60^\circ = (0.2)^2 \times \frac{1}{2} = 0.02\ \text{Wb}$$

**7** (*c*) Induced emf in the loop,

$$E = -\frac{\Delta\phi}{\Delta t} = -\left(\frac{\Delta B}{\Delta t}\right)(A) \qquad [\because \phi = BA]$$

$$= -\left(\frac{B_2 - B_1}{\Delta t}\right)A \qquad \ldots\text{(i)}$$

Here, $B_2 = 500\,\text{G} = 500 \times 10^{-4}\ \text{T}$,
$B_1 = 1000\,\text{G} = 1000 \times 10^{-4}\ \text{T}$,
$\Delta t = 5\text{s}$

$\therefore$ $A$ = area of loop
= Area of rectangle − Area of two triangles

$$= \left(16\times 4 - 2\times\frac{1}{2}\times 4\times 2\right)\text{cm}^2$$
$$= 56\times 10^{-4}\ \text{m}^2$$

Using Eq. (i), we get

$$E = \frac{(1000-500)\times 10^{-4}\times 56\times 10^{-4}}{5}$$
$$= 56\times 10^{-6}\ \text{V}$$
$$= 56\,\mu\text{V}$$

**8** (*c*) Given, $\mathbf{B} = B_0\,(3\hat{\mathbf{i}} + 3\hat{\mathbf{j}} - 4\hat{\mathbf{k}})\ \text{T}$

$$\mathbf{B}_x = (3B_0\hat{\mathbf{i}})\text{T}$$

or $B_x = (3B_0)\,\text{T}$ = component of **B** along area **A**
Area of the square, $A = a^2(\text{m}^2)$

Magnetic flux, $\phi_B = BA\cos\theta$

$\Rightarrow$ $\phi_B = (B\cos\theta)A$
= (component of **B** along **A**) *A*

$\phi_B$ (magnitude of magnetic flux)

$$= B_x\,A = 3B_0a^2\ (\text{Tm}^2) = 3\,B_0a^2\ \text{Wb}$$

**9** (*c*) Magnetic flux, $\phi = BA\cos\theta$
Given, magnetic field, $B = 0.2\ \text{T}$
Area of top surface, $A = (10\times 10\times 10^{-4})\,\text{m}^2$
For the top surface, the angle between normal to the surface and the *X*-axis is $\theta = 60^\circ$ as shown below.

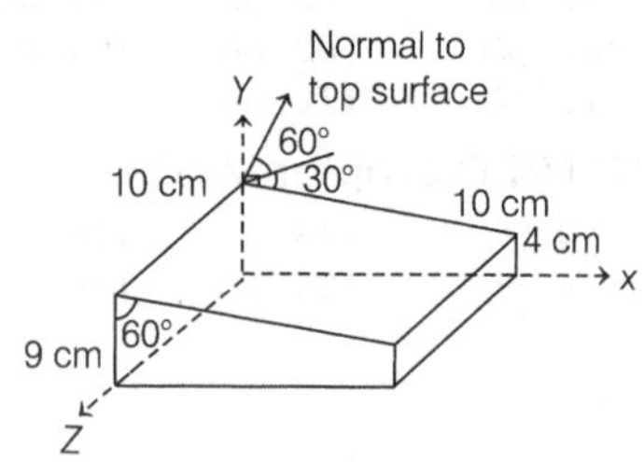


$$\therefore \quad \phi = 0.2\times(10\times 10\times 10^{-4})\times\cos 60^\circ$$
$$= 10^{-3}\ \text{Wb} = 1.0\ \text{m-Wb}$$

**11** (*d*) When the tapping key *K* is pressed, the current in the coil $C_2$ and the resulting magnetic field rises from zero to a maximum value in a short time. So, the galvanometer shows a momentary deflection due to the change in flux along $C_1$. If the key is held pressed continuously, there is no deflection in the galvanometer as the current is constant. So, no change in flux occur across $C_1$.

When the key is released, a momentary deflection is observed again, but in opposite direction due to change in the direction of field or flux across $C_1$.

It is due to the fact that the current in coil $C_2$ and the resulting magnetic field decreases from the maximum value to zero in a short time.

**12** (*b*) Magnitude of induced emf,

$$|\varepsilon|=\frac{d\phi_B}{dt}=\frac{d}{dt}(3t^2+4t+9)=6t+4+0$$

At $t=2\text{s}$, $|\varepsilon|=6\times2+4=16\text{ V}$

**13** (*b*) From Faraday's law, induced emf,

$$\varepsilon=-\frac{d\phi_B}{dt}$$

Given, $\phi_B=xt^2$

$$\therefore \quad \varepsilon=\frac{-d(xt^2)}{dt}=-2tx$$

Given, $t=3\text{s}$ and $\varepsilon=9\text{ V}$

$\therefore$ Magnitude of $x=\dfrac{|\varepsilon|}{2t}=\dfrac{9}{2\times3}=1.5\text{ Wbs}^{-2}$

**14** (*c*) From Faraday's law of electromagnetic induction,

$$\varepsilon=-\frac{d\phi_B}{dt}=-BAN \quad (\because dt=1\text{s and }\phi_B=BA)$$

Given, $B=0.1\text{ T}, N=20, A=\pi r^2=\pi(0.1)^2$

$$\therefore \quad \varepsilon=0.1\times20\times\pi(0.1)^2=20\pi\text{ mV}$$

**15** (*b*) Induced current, $I=\dfrac{e}{R}$

Here, $e=$ induced emf $=\dfrac{d\phi}{dt}$

$$I=\frac{l}{R}=\left(\frac{d\phi}{dt}\right)\cdot\frac{1}{R}$$

$$d\phi=IRdt$$

$$\phi=\int IRdt$$

$\therefore$ Here, $R$ is constant.

$$\therefore \quad \phi=R\int Idt$$

$$\int I\cdot dt=\text{Area under } I\text{-}t \text{ graph}$$

$$=\frac{1}{2}\times10\times0.5=2.5$$

$$\therefore \quad \phi=R\times2.5=100\times2.5=250\text{ Wb}$$

**16** (*c*) Given, area of coil, $A=0.05\text{ m}^2$

magnetic field, $B=5\times10^{-5}\text{ T}$

and number of turns, $N=800$.

The magnetic flux linked with the coil is

$$\phi=N(\mathbf{B}\cdot\mathbf{A})=NBA\cos\theta \quad \ldots\text{(i)}$$

where, $\theta$ is the angle between $\mathbf{B}$ and $\mathbf{A}$.

The emf induced when coil is rotated from $\theta_1=0°$ to $\theta_2=90°$ is

$$\varepsilon=-\frac{\Delta\phi}{\Delta t}=-\frac{\Delta}{\Delta t}(NBA\cos\theta) \quad \text{[using Eq. (i)]}$$

$$=-\frac{NBA}{\Delta t}(\cos\theta_2-\cos\theta_1) \quad \ldots\text{(ii)}$$

Here, $\Delta t=0.1\text{ s}$

Thus, substituting the given values in Eq. (ii), we get

$$\varepsilon=-\frac{800\times5\times10^{-5}\times0.05\times[\cos90°-\cos0°]}{0.1}$$

$$=2000\times10^{-5}$$

$$=0.02\text{ V}$$

**17** (*d*) Given, $r=10\text{ cm}=10^{-1}\text{m}, B=3.0\times10^{-5}\text{ T}$

$N=500, \theta_1=0°, \theta_2=180°$ and $\Delta t=0.25\text{ s}$

Initial flux through the coil,

$$\phi_{B\text{ (initial)}}=BA\cos\theta_1$$

$$=3.0\times10^{-5}\times(\pi\times10^{-2})\times\cos0°$$

$$=3\pi\times10^{-7}\text{ Wb}$$

Final flux after the rotation, about its vertical diameter through $180°$ is given as

$$\phi_{B\text{ (final)}}=BA\cos\theta_2$$

$$=3.0\times10^{-5}\times(\pi\times10^{-2})\times\cos180°$$

$$=-3\pi\times10^{-7}\text{ Wb}$$

$$\Rightarrow \quad \Delta\phi_B=\phi_{B\text{(final)}}-\phi_{B\text{ (initial)}}$$

$$=(-3\pi-3\pi)\times10^{-7}\text{ Wb}$$

$$=-6\pi\times10^{-7}\text{ Wb}$$

Therefore, estimated value of the induced emf is

$$\varepsilon=-N\frac{\Delta\phi_B}{\Delta t}=-500\times(-6\pi\times10^{-7})/0.25$$

$$=3.8\times10^{-3}\text{ V}$$

**18** (*d*) Given, magnetic flux, $\phi_B=t^3+3t-7$

$\therefore$ Induced emf, $\varepsilon=-\dfrac{d\phi_B}{dt}=-(3t^2+3)=-3t^2-3$

At $t=0, \varepsilon=-3\text{ V}$

The above equation is in the form of $y=ax^2+b$, which represents the equation of parabola not passing through origin.

**19** (*d*) Induced emf in the region is given by $|\varepsilon|=\dfrac{d\phi_B}{dt}$

where $\quad \phi_B=BA=\pi r^2B$

$$\Rightarrow \quad \frac{d\phi_B}{dt}=\pi r^2\frac{dB}{dt}$$

Rate of change of magnetic flux associates with loop 1,

$$\varepsilon_1=-\frac{d\phi_1}{dt}=-\pi r^2\frac{dB}{dt}$$

Similarly, $\varepsilon_2=$ emf associated with loop 2

$$=\frac{d\phi_2}{dt}=0 \quad (\because \phi_2=0)$$

**20** (*a*) Magnitude of induced emf of coil, $|\varepsilon|=\left|-\dfrac{d\phi_B}{dt}\right|_t$

Given, $\phi_B=50t^2+4$ and $R=400\Omega$

$$\Rightarrow \quad |\varepsilon|=\left|-\frac{d\phi_B}{dt}\right|_{t=2}=\left|-\frac{d}{dt}(50t^2+4)\right|_{t=2}$$

$$=100t|_{t=2}=200\text{V}$$

Current in the coil, $I=\dfrac{|\varepsilon|}{R}=\dfrac{200}{400}=\dfrac{1}{2}=0.5\text{ A}$

**21** (*a*) Given, side of square, $l = 10\text{ cm} = 0.1\text{ m}$

$\therefore$ Area, $A = l^2 = 10^{-2}\text{ m}^2$

$\theta = 45°$, $dB = 0.1\text{ T}$, $dt = 0.7\text{ s}$ and $R = 1\Omega$

From, Faraday's law, induced emf

$$\varepsilon = -N\frac{d\phi_B}{dt}$$

$$\Rightarrow \quad \varepsilon = -1\cdot\frac{d}{dt}(BA\cos 45°) = -\frac{A}{\sqrt{2}}\frac{dB}{dt}$$

$$= \frac{10^{-2}}{-\sqrt{2}} \times \frac{0.1}{0.7} = -1\text{ mV}$$

Magnitude of current $= \frac{|\varepsilon|}{R} = \frac{1\text{mV}}{1\Omega} = 1.0\text{ mA}$

**22** (*c*) Given, radius, $R = 10\text{ cm} = 10\times 10^{-2}\text{ m}$

Number of turns, $N = 500$

Magnetic field, $B = 3\times 10^{-5}\text{ T}$

Resistance, $r = 2\Omega$, $\theta_1 = 0°$, $\theta_2 = \pi = 180°$ and $\Delta t = 0.5\text{ s}$

Initial flux through the coil $(\phi_B)_i = BA\cos\theta_1$

$$= 3\times 10^{-5}\times\pi\times(10\times 10^{-2})^2\cos 0°$$

$$= 3\pi\times 10^{-7}\text{ Wb}$$

Final flux after rotation,

$$(\phi_B)_f = BA\cos\theta_2 = 3\times 10^{-5}\times\pi\times(10\times 10^{-2})\times\cos 180°$$

$$= -3\pi\times 10^{-7}\text{ Wb}$$

Emf induced in the coil

$$\varepsilon = -N\frac{\Delta\phi_B}{\Delta t} = -N\left\{\frac{(\phi_B)_f - (\phi_B)_i}{\Delta t}\right\}$$

$$= -\frac{500\times(-3\pi\times 10^{-7} - 3\pi\times 10^{-7})}{0.5}$$

$$= \frac{500\times 6\pi\times 10^{-7}}{0.5} \simeq 2\times 10^{-3}\text{ V}$$

Induced current in the coil,

$$I = \frac{\varepsilon}{r} = \frac{2\times 10^{-3}}{2}$$

$$= 10^{-3}\text{ A} = 1\text{ mA}$$

**23** (*b*) The angle $\theta$ made by the area vector of the coil with the magnetic field is 45° as shown below. The initial magnetic flux is

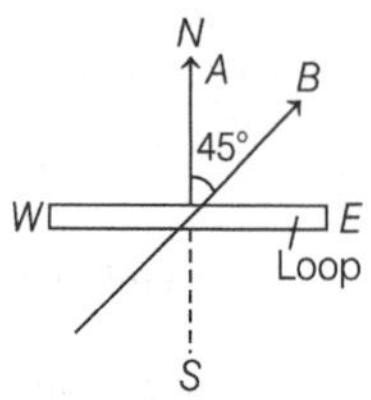


$$\phi_B = BA\cos\theta$$

Given, $B = 0.1\text{ T}$

$A = (10\times 10\times 10^{-4})\text{ m}^2 = 10^{-2}\text{ m}^2$ and $\theta = 45°$

$$\phi_B = 0.1\times 10^{-2}\times\cos 45°$$

$$= \frac{0.1\times 10^{-2}}{\sqrt{2}}\text{ Wb} = \frac{10^{-3}}{\sqrt{2}}\text{ Wb}$$

Final flux, $\phi_{min} = 0$

The change in flux is brought about in 0.70 s. The magnitude of the induced emf is given by

$$\varepsilon = \frac{|\Delta\phi_B|}{\Delta t} = \frac{|(\phi_B - 0)|}{\Delta t} = \frac{10^{-3}}{\sqrt{2}\times 0.70} = 1.0\text{ mV}$$

and the magnitude of the current is

$$I = \frac{\varepsilon}{R} = \frac{1.0\text{mV}}{0.5\,\Omega} = 2\text{ mA} \qquad [\text{given, } R = 0.5\Omega]$$

**24** (*b*) Given, $N = 40$, $A = 4.0\text{ cm}^2 = 4.0\times 10^{-4}\text{ m}^2$

$q = 2.0\times 10^{-4}\text{ C}$ and $R = 80\Omega$

According to Faraday's law of electromagnetic induction, induced emf,

$$|\varepsilon| = \frac{d\phi_B}{dt} \qquad \ldots\text{(i)}$$

We know that,

$$i = \frac{dq}{dt} \Rightarrow q = \int i\,dt = \int\frac{\varepsilon}{R}dt$$

$$\Rightarrow \quad q = \frac{1}{R}\int\frac{d\phi_B}{dt}\cdot dt \qquad [\text{from Eq. (i)}]$$

$$= \frac{1}{R}\int d\phi_B = \frac{\phi_B}{R}$$

For $N$-turns, $q = \frac{N\phi_B}{R} = \frac{N}{R}(BA) \qquad (\because \phi_B = BA)$

$$\Rightarrow \quad qR = NBA$$

$$\Rightarrow \quad B = \frac{qR}{NA} = \frac{2.0\times 10^{-4}\times 80}{40\times 4.0\times 10^{-4}}$$

$$= 1.0\text{ Wb m}^{-2}$$

**26** (*c*) In uniform magnetic field, change in magnetic flux is zero. Therefore, induced current will be zero.

**27** (*a*) Since, electron is moving from right to left, the flux linked with loop will first increase and then decrease as the electron passes. Therefore, induced current $I$ in the loop will be first clockwise and then will move in anti-clockwise direction. Hence, direction of induced current is variable.

**28** (*b*) According to question, there can be two cases as follows

**Case I** Closed coil is moving inside constant electric field.

**Case II** Closed coil is wholly inside constant electric field.

However, no current is induced in either case because current cannot be induced by changing the electric flux.

**29** (*a*) The induced emf is expected to be constant only in the case of the rectangular loop.

In the case of circular loop, the rate of change of area of the loop during its passage out of the field region is not constant, hence induced emf will vary accordingly.

**30** (*b*) When the north pole of the magnet is brought towards one end of the coil, the induced current flows in the coil in such a direction, so as to oppose the motion of the north pole of the magnet towards the coil. Thus, direction of induced current is anti-clockwise as shown below.

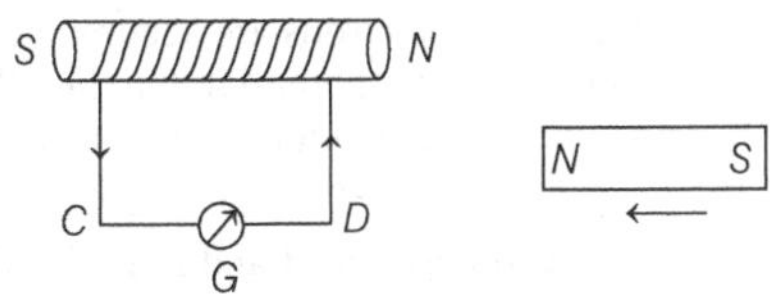


**31** (*c*) By Lenz's law, the direction of induced current in the ring is such as to oppose the falling of *N*-pole of the magnet.

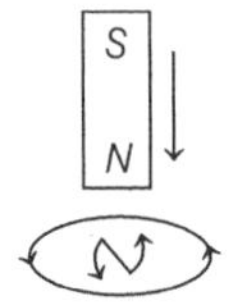


So, the direction of induced current in the loop will be anti-clockwise, because the induced current makes the ring a magnetic dipole, with its *N*-pole upward which opposes (repel) the *N*-pole of falling magnet.

**32** (*a*) As the shape of the loop is changing and hence the flux linked with the loop changes. There will be an induced emf and hence induced current in the coil.

As circle has maximum area, for given perimeter. So, magnetic flux through the coil is increasing. Induced emf opposes it by giving induced current in anti-clockwise direction as field is in downward direction.

**33** (*b*) When the bar magnet is pushed towards the open circuit *CD*, then an emf is induced across the open ends of the circuit. The direction of the induced emf can be found using Lenz's law. For conductor *CD*,

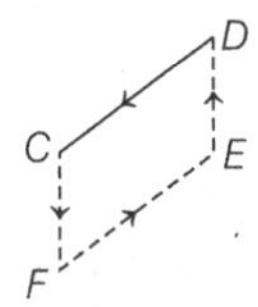


when north-pole of the bar magnet approaches to it, then current is induced in conductor *CD* in such a direction that it opposes the increase in flux. This is possible only, if the current in the coil is in a counter-clockwise direction, i.e. from *D* to *C*.

**34** (*a*) In the rotation of magnet, *N*-pole moves closer to coil *CD* and *S*-pole moves closer to coil *AB*.

As per Lenz's law, *N*-pole should develop at the end corresponding to *C*. So, induced current flows from *C* to *D*.

Again *S*-pole should develop at the end corresponding to *B*. Therefore, induced current in the coil flows from *A* to *B*.

**35** (*a*) The polarity of plate *A* will be positive with respect to plate *B* in the capacitor. Both magnets, by motion, increase rightward magnetic flux through coil, while induced emf with produce leftward flux. Hence,positive charge will come at *A*.

**36** (*b*) When the *S*-pole of a magnet is moved towards the coil, then by Lenz's law, the face of coil towards magnet becomes *S*-pole. Thus, the current flows in clockwise direction to oppose the change in the magnetic flux. So, the galvanometer shows deflection to the left as shown in figure. Now, when the *S*-pole is moved away, current flows in anti-clockwise direction to make the face of the coil towards magnet, a *N*-pole.

Thus, the coil will try to attract the magnet. So, the galvanometer shows the deflection to the right as shown below. Since, flux varies, so the amplitude of deflection will not be constant and the deflection will decrease.

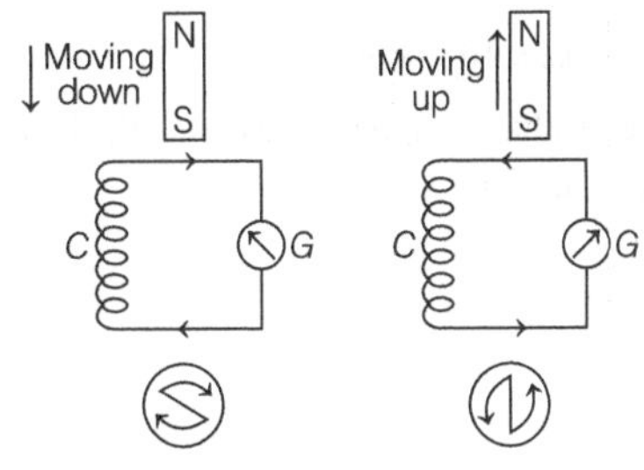


**38** (*d*) A rectangular conductor *PQRS* in which the conductor *PQ* is free to move is shown below.

*PQRS* forms a closed circuit enclosing an area that changes as *PQ* moves.

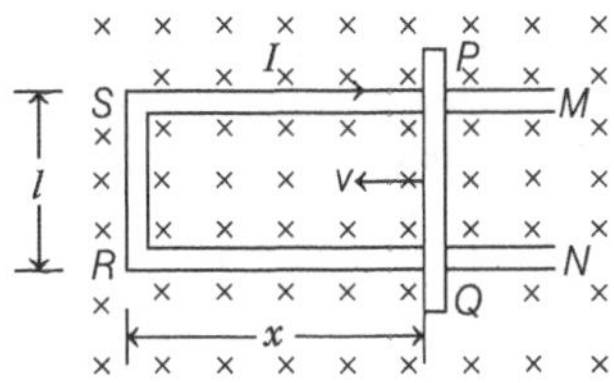


It is placed in a uniform magnetic field **B** which is perpendicular to the plane of this system. If the length $RQ = x$ and $RS = l$, the magnetic flux $\phi_B$ enclosed by the loop *PQRS* will be $\phi_B = Blx$.

Since, $x$ is changing with time, the rate of change of flux $\phi_B$ will induce an emf given by

$$\varepsilon = \frac{-d\phi_B}{dt} = -\frac{d}{dt}(Blx) = -Bl\frac{dx}{dt} = Blv$$

The induced emf $Blv$ is called motional emf.

**39** (*b*) Given, $B_H = 0.30 \times 10^{-4}\ \text{Wbm}^{-2}$,

$$l = 20\ \text{m} \text{ and } v = 5.0\ \text{ms}^{-1}$$

Induced emf across the ends of wire,

$$\varepsilon = B_H lv = 0.30 \times 10^{-4} \times 20 \times 5.0 = 3\ \text{mV}$$

**40** (*c*) Given, $B_H = 5 \times 10^{-3}\ \text{Wbm}^{-2}$

$$l = 10\ \text{m and } v = 100\ \text{ms}^{-1}$$

The induced emf between tip of nose and tail of helicopter is given by

$$\varepsilon = B_H vl = 5 \times 10^{-3} \times 100 \times 10 = 5\ \text{V}$$

**41** (*c*) Given, $v$ = velocity of train $= 72 \times \frac{5}{18} = 20\ \text{ms}^{-1}$,

$l$ = distance between the rails = 1 m

and $B$ = vertical component of earth's magnetic field $= 2 \times 10^{-5}$ T

$\therefore$ Induced emf, $\varepsilon = Bvl = 2 \times 10^{-5} \times 20 \times 1$

$$= 40 \times 10^{-5}\ \text{V}$$

$$= 40 \times 10^{-2}\ \text{mV} = 0.4\ \text{mV}$$

**42** (*c*) Given, $B = 0.30 \times 10^{-2}\ \text{Wb m}^{-2}$, $v = 4.0\ \text{ms}^{-1}$

and $\quad l = 10\ \text{m}$

As induced emf or potential, $\varepsilon = Bvl$

So, induced potential gradient $= \dfrac{\varepsilon}{l} = Bv$

$$= 0.30 \times 10^{-2} \times 4.0 = 0.012\ \text{Vm}^{-1}$$

$$= 1.2 \times 10^{-2}\ \text{Vm}^{-1}$$

**43** (*b*) The resultant magnetic field at a place is $B_0$ as shown in the figure.

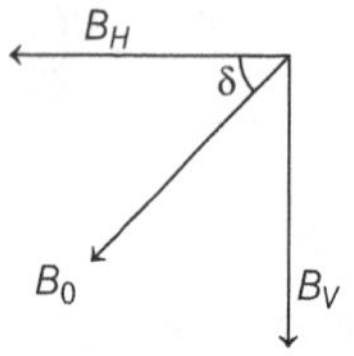


where, $\delta$ = angle of dip,

$B_H$ = horizontal component of earth's magnetic field and $B_V$ = vertical component of earth magnetic field.

When horizontal conductor of length $L$ moves eastwards with velocity $v$, then it cuts vertical component of earth's magnetic field.

$\therefore \quad B_V = B_0 \sin \delta \quad$ ...(i)

$\therefore$ Induced emf across the conductor,

$$\varepsilon = B_V vL = B_0 \sin \delta \cdot vL \quad \text{[from Eq. (i)]}$$

$$= B_0 vL \sin \delta$$

**44** (*a*) Let us consider the forward motion from $x = 0$ to $x = 2b$. The flux $\phi_B$ linked with the circuit *SPQR* is

$$\phi_B = Blx, \quad 0 \le x < b$$

The induced emf is $\varepsilon = -\dfrac{d\phi_B}{dt} = -\dfrac{Blx}{t}$

$$= -Blv, \quad 0 \le x < b$$

This is because, both $\phi_B$ and $\varepsilon$ are zero for $b < x \le 2b$, as $B$ is zero for $x > b$.

**45** (*c*) A circular coil expands radially in a region of magnetic field only, if (i) the magnetic field is in the same plane as the circular coil or (ii) the magnetic field has a perpendicular (to the plane of the coil) component whose magnitude is decreasing suitably in such a way that the cross-product of magnetic field and surface area of plane of coil remain constant at every instant.

Then, no electro-motive force is produced in the coil.

**46** (*c*) According to figure, length of semi-circular conducting ring *PQR*, $l = \pi r$.

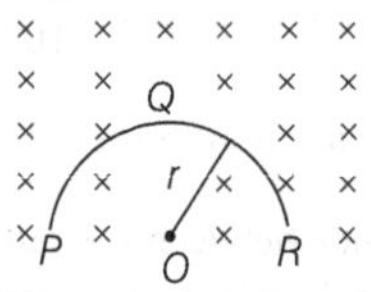


When conducting ring *PQR* is falling with its plane vertical in a horizontal magnetic field $B$ with speed $v$. Then, potential difference developed across the ring as

$$V = Bvl = Bv(\pi r), \quad V = \pi r Bv \text{ volt}$$

According to Fleming right hand rule, direction of induced current is from $P$ to $R$, hence electrons are moving from point $R$ to $P$ in the conducting wire during its motion. Therefore, electrons are accumulated at end $P$, hence this end is at low potential. On the other hand, end $R$ has positively charged ions due to lack of electrons. Therefore, this end $R$ is at higher potential.

**47** (*c*) The emf induced between $A$ and $B$ as shown below is given by

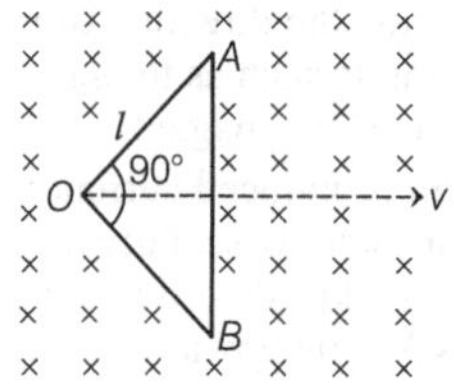


$$\varepsilon = Bv(AB)$$

In $\Delta AOB$, $AB = \sqrt{AO^2 + BO^2} = \sqrt{l^2 + l^2} = \sqrt{2}l$

$$\therefore \quad \varepsilon = Bv(\sqrt{2}l) = \sqrt{2}Bvl$$

**48** (*a*) When bob reaches at point $B$ from point $A$ as shown below, then vertical height travelled by the bob is

$$h = L - L\cos\theta \Rightarrow h = L(1 - \cos\theta)$$

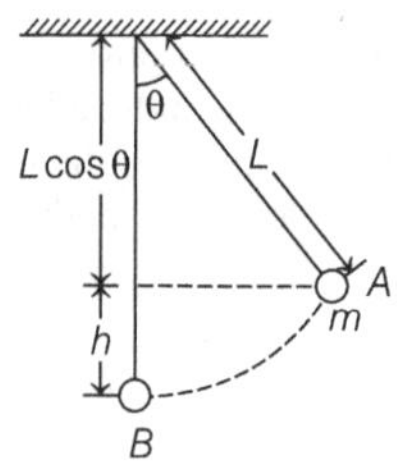


Velocity of bob at point $B$ is $v$ and at $A$ is $u$, which is zero at extreme point, then from equation,

$$v^2 = u^2 + 2gh$$

$$\therefore \quad v^2 = 2gh = 2gL(1 - \cos\theta)$$

$$= 2gL(2\sin^2\theta/2)$$

$$\Rightarrow \quad v = 2\sqrt{gL}\sin\theta/2$$

Thus, maximum emf induced

$$\varepsilon_{\text{max}} = BvL = B \times 2\sqrt{gL}(\sin\theta/2)L$$

$$= 2BL(\sin\theta/2)(gL)^{1/2}$$

**49** (*b*) As the rod is rotated, free electrons in the rod move towards the outer end due to Lorentz force and get distributed over the ring. Thus, the resulting separation of charges produces an emf across the ends of the rod. At a certain value of emf, there is no more flow of electrons and a steady state is reached. The magnitude of the emf generated across a length $dr$ of the rod as it moves at right angles to the magnetic field is given by

$$d\varepsilon = Bvdr$$

Hence,

$$\varepsilon = \int d\varepsilon = \int_0^R Bvdr = \int_0^R B\omega rdr = B\omega\left[\frac{r^2}{2}\right]_0^R = \frac{B\omega R^2}{2}$$

Note that, we have used $v = \omega r$.

Given, $B = 1\ \text{T}$, $v = 50\ \text{revs}^{-1}$

$\Rightarrow \quad \omega = 2\pi\nu = 2\pi \times 50 \text{ rad s}^{-1}$ and $R = 1\text{ m}$

$$\varepsilon = \frac{1}{2} \times 1 \times 2\pi \times 50 \times (1)^2 = 157\text{ V}$$

**50** (*b*) When a conducting disc or wheel of radius $r$ rotates with constant angular velocity of $\omega$ about its axis in a uniform magnetic field perpendicular to its plane and parallel to its axis of rotation, then

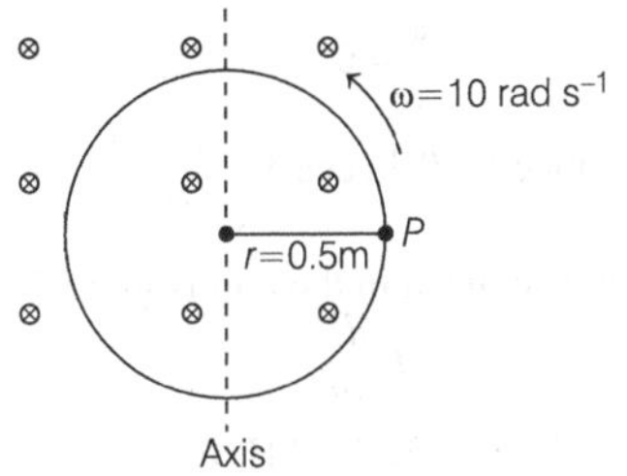


Induced emf is given by $\varepsilon = \frac{1}{2} B\omega r^2$ ...(i)

Given, $B = 0.1\text{ T}$, $\omega = 10 \text{ rads}^{-1}$, $r = 0.5\text{ m}$

Substituting these values in Eq. (i), we get

$$\therefore \quad \varepsilon = \frac{1}{2} \times 0.1 \times 10 \times (0.5)^2$$

$$= \frac{1}{8}\text{ V} = 0.125\text{ V}$$

**51** (*a*) Given, $H_E = 0.4\text{G} = 0.4 \times 10^{-4}\text{T}$ $[\because 1\text{ G} = 1 \times 10^{-4}\text{ T}]$

$\nu = 120$ rev/min and $R = 0.5\text{ m}$

Induced emf, $\varepsilon = (1/2)\,\omega BR^2 = \frac{1}{2}\omega H_E R^2$

$$= (1/2) \times 4\pi \times 0.4 \times 10^{-4} \times (0.5)^2 = 6.28 \times 10^{-5}\text{ V}$$

$$\left(\because \omega = 2\pi\nu = \frac{2\pi \times 120}{60} = 4\pi \text{rads}^{-1}\right)$$

**52** (*c*) Emf across rod is induced and charges get accumulated across its ends. But current flowing in it is zero, because circuit is open.

**53** (*a*) Given, $B = 0.4$ T, $v = 5\text{ ms}^{-1}$, $l = 25$ cm $= 0.25$ m

and $R = 5\Omega$.

Induced emf produced is given by

$$\varepsilon = Blv = 0.4 \times 0.25 \times 5 = 0.5\text{ V}$$

Current through the $5\Omega$ resistance,

$$i = \frac{\varepsilon}{R} = \frac{0.5}{5} = 0.1\text{ A}$$

According to Lenz's law when the rod $PQ$ moves towards right, the induced current should flow in a direction, so that the rod $PQ$ experiences force towards left. According to Fleming's left hand rule, then the current through the rod $PQ$ will flow from end $Q$ to $P$, i.e. from end $R$ to $S$ through the resistance of $5\Omega$.

**54** (*a*) Given, $A = 10 \times 5 = 50\text{ cm}^2 = 50 \times 10^{-4}\text{m}^2$

$\frac{dB}{dt} = 0.02\text{ T s}^{-1}$, $R = 2.0\Omega$

$$|\varepsilon| = \frac{d\phi_B}{dt} = A \cdot \frac{dB}{dt} \quad [\because \phi_B = BA \text{ and } A \text{ is a constant}]$$

$$= 50 \times 10^{-4} \times 0.02 = 10^{-4}\text{ V}$$

Power dissipated in the form of heat $= \frac{\varepsilon^2}{R}$

$$= \frac{10^{-4} \times 10^{-4}}{2} = 0.5 \times 10^{-8}\text{ W}$$

$$= 5 \times 10^{-9}\text{ W} = 5\text{nW}$$

**55** (*d*) Given, $R = 1.5\,\Omega$, $B = 0.3\text{ T}$ and $\frac{dA}{dt} = 0.05\text{ m}^2/\text{s}$.

Let $\varepsilon$ be the magnitude of emf induced in the loop, then

$$\varepsilon = \frac{d\phi_B}{dt} = \frac{d\,(B \cdot A)}{dt} \quad [\because \phi_B = B \cdot A]$$

$$\Rightarrow \quad \varepsilon = B \cdot \frac{dA}{dt} \quad (\because B \text{ is constant})$$

$$= 0.3 \times 0.05 = 0.015\text{ V}$$

or $\quad \varepsilon = 1.5 \times 10^{-2}\text{ V}$

$\therefore$ Heat produced in the coil

$$H = P = \frac{\varepsilon^2}{R} = \frac{(1.5 \times 10^{-2})^2}{1.5}$$

$$= \frac{1.5 \times 1.5 \times 10^{-4}}{1.5}$$

$$= 1.5 \times 10^{-4}\text{ W or } 150 \times 10^{-6}\text{ W}$$

**56** (*a*) A copper plate is allowed to swing like a simple pendulum between the pole pieces of a strong magnet. It is found that the motion is damped and in a little while the plate comes to a halt in the magnetic field. We can explain this phenomenon on the basis of electromagnetic induction.

**57** (*c*) Electric heaters are not based on the eddy current effect. Rather their working is based on Joule's heating effect of current. According to this effect, the passage of an electric current through a resistor produces heat.

**63** (*a*) Given, $I_A = 2\text{ A}$

and $\phi_B$ [flux linked with $B$ due to current (in A)] $= 10^{-2}$ Wb

As, $\quad \phi_B = M_{BA} I_A$

$$M_{BA} = \frac{\phi_B}{I_A} = \frac{10^{-2}}{2} = 5 \times 10^{-3}\text{ H}$$

When $I_B = 1\text{ A}$, $\phi_A = M_{AB} I_B = (5 \times 10^{-3}\text{ H})\,(1\text{ A})$

$$= 5 \times 10^{-3}\text{ Wb}$$

**64** (*a*) Magnetic flux linked with coil $A$, $\phi_A = BA = B \cdot \pi r^2$

$$\Rightarrow \quad \phi_A = \frac{\mu_0 I}{2R} \cdot \pi r^2$$

$\left(\because B = \frac{\mu_0 I}{2R}\right.$, magnetic field at the centre of $B$ and $\left. N_1 = N_2 = 1\right)$

Now, mutual inductance for two concentric coplanar circular coils or the coefficient $M$ is given by

$$\phi_A = MI$$

$$\Rightarrow \quad M = \frac{\phi_A}{I} = \frac{\frac{\mu_0 I \cdot \pi r^2}{2R}}{I} = \frac{\mu_0 \pi r^2}{2R}$$

**65** (*b*) As, coefficient of mutual induction is same for both coils

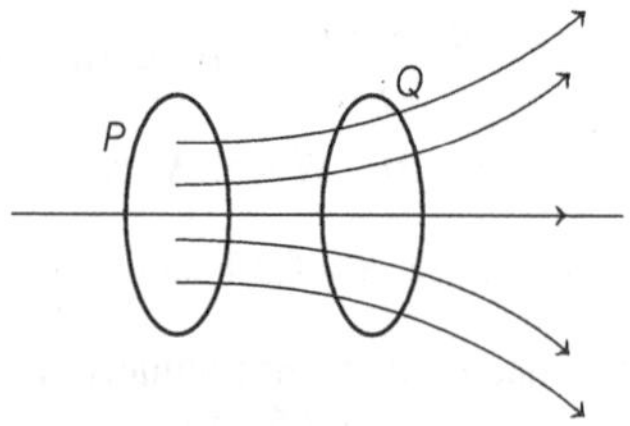


$\therefore \quad M_{PQ} = M_{QP}$

$\Rightarrow \quad \dfrac{N_P \phi_{PQ}}{I_Q} = \dfrac{N_Q \phi_{QP}}{I_P}$ ...(i)

Here, $N_P = N_Q = 1$, $\phi_{PQ} = ?$, $\phi_{QP} = 10^{-3}$ Wb

$I_Q = 2\text{A}, I_P = 3\text{A}$

Substituting values in Eq (i), we get

$$\phi_{PQ} = \frac{I_Q \cdot \phi_{QP}}{I_P} = \frac{2}{3} \times 10^{-3}$$

$$= 0.667 \times 10^{-3} = 6.67 \times 10^{-4} \text{ Wb}$$

**66** (*c*) Current induced in the coil is given by

$$i = \frac{1}{R}\left(\frac{\Delta\phi}{\Delta t}\right) \Rightarrow \frac{\Delta q}{\Delta t} = \frac{1}{R}\left(\frac{\Delta\phi}{\Delta t}\right)$$

$\Rightarrow \quad \Delta q = \dfrac{\Delta\phi}{R}$ ...(i)

Since, $\Delta\phi = \mu_0 N_1 N_2 \Delta i A / l$

Given, resistance of the solenoid, $R = 10\pi^2\ \Omega$

Radius of second coil, $r = 0.01\text{m} = 10^{-2}$ m

$\Delta t = 0.05\text{ s}, \Delta i = 4 - 0 = 4\text{ A}$

Charge flowing through the coil is given by

$$\Delta q = \frac{\Delta\phi}{R} = \frac{\mu_0 N_1 N_2 A \Delta i}{lR} = \frac{\mu_0 N_1 N_2 \pi r^2}{lR}\left(\frac{\Delta i}{\Delta t}\right)\Delta t$$

$$= \mu_0 N_1 N_2 \pi r^2 \left(\frac{\Delta i}{\Delta t}\right)\frac{1}{R}\Delta t \quad [\because l = 1\text{m}]$$

$$= 4\pi \times 10^{-7} \times 2 \times 10^4 \times 100 \times \pi$$

$$\times (10^{-2})^2 \times \left(\frac{4}{0.05}\right) \times \frac{1}{10\pi^2} \times 0.05$$

$$= 32 \times 10^{-6}\text{ C} = 32\,\mu\text{C}$$

**67** (*b*) According to the phenomenon of mutual induction, the mutual inductance depends upon the magnetic property like relative permeability of medium between the coils ($\mu_r$) or nature of material on which two coils are wound. Also, mutual inductance of the pair of coil depends on distance between two coils and geometry of two coils.

**68** (*c*) Mutual inductance for a coaxial solenoid of radius $r_1$ and $r_2$ and number of turns $n_1$ and $n_2$, respectively is given as,

$M = \mu_0 n_1 n_2\, \pi r_1^2\, l$ (for internal coil of radius $r_1$)

Self inductance for the internal coil, $L = \mu_0 n_1^2 \pi r_1^2 l$

$\therefore \quad \dfrac{M}{L} = \dfrac{n_1 n_2}{n_1^2} = \dfrac{n_2}{n_1}$

**69** (*b*) When coil is open, there is no current in it, hence no flux associated with it, i.e. $\phi = 0$.

Also, we know that flux linked with the coil is directly proportional to the current in the coil,

i.e. $\quad \phi \propto i \Rightarrow \phi = Li$

where, $L$ is proportionality constant known as self-inductance.

$\therefore \quad L = \dfrac{\phi}{i} = 0$

Again, since $i = 0$, hence $R = \dfrac{V}{i} = \dfrac{V}{0} = \infty \quad \left(\because R = \dfrac{V}{I}\right)$

**70** (*b*) The emf induced in the coil is given by

$$\varepsilon = -L\frac{dI}{dt}$$

Given, $L = 2\text{ mH} = 2 \times 10^{-3}$ H

and $\dfrac{dI}{dt} = 20\text{ mAs}^{-1} = 20 \times 10^{-3}\text{ As}^{-1}$

$\therefore \quad |\varepsilon| = 2 \times 10^{-3} \times 20 \times 10^{-3}\text{ V} = 40\,\mu\text{V}$

**71** (*a*) Given, $V = 10$ V, $R = 5\Omega$ and $L = 10$ H

Current, $\quad I = \dfrac{V}{R} = \dfrac{10}{5} = 2\text{ A}$

Circuit switches off in 1 ms, i.e.

$dt = 1 \times 10^{-3}$ s and $dI = 2$ A

$\therefore$ Induced emf in inductor is

$$|\varepsilon| = L\frac{dI}{dt} = 10 \times \frac{2}{1 \times 10^{-3}} = 2 \times 10^4\text{ V}$$

**72** (*d*) The emf induced in a coil is given by, $\varepsilon = -L\dfrac{\Delta i}{\Delta t}$

Given, $\quad L = 60\text{ mH} = 60 \times 10^{-3}$ H,

$\Delta i = (2.5 - 1.5)\text{ A} = 1\text{ A}$ and $\Delta t = 0.2$ s

$\therefore \quad \varepsilon = -\dfrac{60 \times 10^{-3} \times 1}{0.2} = -0.3\text{ V}$

Induced current, $i = \dfrac{|\varepsilon|}{R} = \dfrac{0.3}{3} = 0.1\text{ A}$

**73** (*c*) Self-inductance of a coil is given by the relation

$$L = \mu_0 n^2 A \cdot l$$

where, $n$ is number of turns per unit length. Shape of the wooden frame is equilateral triangle.

$\therefore$ Area of equilateral triangle,

$$A = \frac{\sqrt{3}}{4}a^2$$

(where, $a$ is side of equilateral triangle)

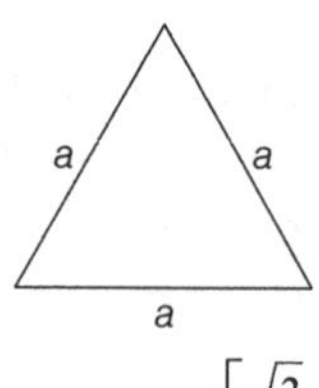


$\therefore$ Self-inductance, $L = \mu_0 n^2 \left[\dfrac{\sqrt{3}}{4}a^2\right].l$

Here, $l = 3a \times N$ (where, $N$ is total turns)

$\therefore \quad L = \mu_0 n^2 \left[\frac{\sqrt{3}}{4} a^2\right] \times 3aN$ or $L \propto a^3$

When each side of frame is increased by a factor 3 keeping the number of turns per unit length of the frame constant.

Then, $a' = 3a$

$\therefore \quad L' \propto (a')^3$

or $L' \propto (3a)^3$

or $L' \propto 27a^3$

or $L' = 27L$

So, self-inductance will becomes 27 times.

**74** (*c*) Given, number of turns of solenoid, $N = 1000$

Current, $I = 4$ A

and magnetic flux, $\phi_B = 4 \times 10^{-3}$ Wb

Self-inductance of solenoid is given by

$$L = \frac{\phi_B \cdot N}{I} \quad \ldots\text{(i)}$$

Substitute the given values in Eq. (i), we get

$$L = \frac{4 \times 10^{-3} \times 1000}{4} = 1\,\text{H}$$

**75** (*c*) Flux through the solenoid is

$$\phi = \mu_r \mu_0 \frac{N^2}{l} AI = \frac{\mu_r \mu_0 N^2 \pi r^2 I}{l} \quad \ldots\text{(i)}$$

Given, $l = 60$ cm $= 6 \times 10^{-1}$ m, $N = 50$,

$r = 7.5$ mm $= 7.5 \times 10^{-3}$ m, $I = 3$A and $\mu_r = 600$

Substitute these values in Eq. (i), we get

$$\phi = 600 \times 4\pi \times 10^{-7} \times 50 \times 50 \times \pi \times \frac{(7.5 \times 10^{-3})^2 \times 3}{6 \times 10^{-1}}$$

$$= 1.66 \times 10^{-3}\ \text{Wb} = 1.66\ \text{m Wb}$$

**76** (*d*) When the two coils are joined in series, such that the winding of one is opposite to the other, then the emf produced in first coil is 180° out of phase of the emf produced in second coil.

Thus, emf produced in first coil is negative and the emf produced in second coil is positive. The net inductance is

$$L = -L_1 + L_2 \quad \ldots\text{(i)}$$

From Faraday's law of electromagnetic induction $\phi = LI$, where $\phi$ is flux and $I$ is the current.

From Eq. (i), $L = -\frac{\phi}{I} + \frac{\phi}{I} \Rightarrow L = 0$

**77** (*b*) Solenoids 1 and 2 were wound using the similar wire and same number of turns $\Rightarrow L_1 = L_2$

Solenoid 3 was wound using superconducting wire (means $R = 0$), so infinite current will flow and resistance to the change of current will be zero, hence $L_3 = 0$.

So, $L_1 = L_2$ and $L_3 = 0$

**81** (*d*) Given, $a = 25$ cm $= 0.25$ m

$\Rightarrow A = a^2 = (0.25)^2$ m$^2$ and $N = 1000$

Emf induced in the coil, $\varepsilon = 200 \sin 100\pi t$

Comparing with standard equation $\varepsilon = \varepsilon_0 \sin \omega t$, we get

$\varepsilon_0 = 200, \omega = 100\pi$

$\because BAN\omega = \varepsilon_0$

$$\therefore \quad B = \frac{\varepsilon_0}{AN\omega} = \frac{200}{(0.25 \times 0.25) \times 1000 \times 100\pi} = 0.01\ \text{T}$$

**82** (*c*) Since, coil rotates half revolution per second, hence frequency of rotation of coil, $\nu = \frac{1/2}{1} = \frac{1}{2} = 0.5$ Hz

Given, $N = 100$, $A = 0.10\text{m}^2$ and $B = 0.01$ T

Emf induced, $\varepsilon = \varepsilon_0 \sin \omega t = NBA\omega \sin \omega t$

$$\varepsilon_{\max} = \varepsilon_0 = NBA\omega \qquad (\because \sin \omega t = 1)$$
$$= NBA\,(2\pi\nu) \qquad [\because \omega = 2\pi\nu]$$
$$= 100 \times 0.01 \times 0.10 \times 2 \times 3.14 \times 0.5$$
$$= 0.314\ \text{V}$$

The maximum voltage is 0.314 V.

**83** (*d*) Induced emf, $\varepsilon = \varepsilon_0 \sin \omega t = NBA\omega \sin \omega t$

Initially,

Peak value, $\varepsilon_0 = N_1 BA_1 \omega = 100$ V

Here, $2\pi r_1 \times 100 = 2\pi r_2 \times N_2$ (given)

$$N_2 = \frac{r_1 \times 100}{r_2} = \frac{1 \times 100}{2} = 50 = \frac{N_1}{2}$$

Finally, $\varepsilon_0 = N_2 BA_2 \omega$

$$\Rightarrow \quad \varepsilon_0 = \frac{N_1}{2} \times B \times 4A_1 \times \omega = 2 \times \text{(initial emf)}$$

$$= 2 \times 100 = 200\ \text{V} \quad (\because r_2 = 2r_1 \Rightarrow A_2 = 4A_1)$$

**84** (*c*) When a coil of $N$ number of turns and area $A$ is rotated in external magnetic field **B**, magnetic flux linked with the coil changes and hence an emf is induced in the coil. At any instant $t$, if $\varepsilon$ is the emf induced in the coil, then alternating emf induced is

$$\varepsilon = \varepsilon_0 \sin \omega t$$

Maximum current, $i_0 = \frac{\varepsilon_0}{R} = \frac{NBA\omega}{R}$

Given, $N = 1, B = 10^{-2}$ T, $r = 30$ cm $= 0.3$ m

$$A = \pi r^2 = \pi (0.3)^2\ \text{m}^2, R = \pi^2\ \Omega$$

$$\nu = 200\text{rpm} = \frac{200}{60}\ \text{s}^{-1} \text{ and } \omega = 2\pi\left(\frac{200}{60}\right)$$

$$\therefore \quad i_0 = \frac{1 \times 10^{-2} \times \pi (0.3)^2 \times 2\pi \times 200}{60 \times \pi^2}$$

$$= 6 \times 10^{-3}\ \text{A} = 6\ \text{mA}$$

**85** (*d*) As the coil is rotated, angle $\theta$ (angle which normal to the coil makes with $B$ at any instant $t$) changes, therefore magnetic flux $\phi$ linked with the coil changes and hence an emf is induced in the coil. At this instant $t$, if $\varepsilon$ is the emf induced in the coil, then

$$\varepsilon = -\frac{d\phi}{dt} = -\frac{d}{dt}(NAB \cos \omega t)$$

where, $N$ is number of turns in the coil.

$$\Rightarrow \quad \varepsilon = -NAB \frac{d}{dt}(\cos \omega t) = -NAB\,(-\sin \omega t)\,\omega$$

$$\Rightarrow \quad \varepsilon = NAB\,\omega \sin \omega t$$

The induced emf will be maximum, when $\sin \omega t$ = maximum = 1

$\therefore \quad \varepsilon_{max} = \varepsilon_0 = NAB\omega \times 1$

Therefore, $\varepsilon$ would be maximum, hence current is maximum (as $i_0 = \varepsilon_0 / R$) when $\theta = 90°$, i.e. normal to plane of coil is perpendicular to the field or plane of coil is parallel to magnetic field.

**86** (*b*) The emf generated due to a rotating conductor in a generator is given by, $\varepsilon = -\frac{d\phi_B}{dt}$

where, $\phi_B$ = magnetic flux linked with conductor.

From the above equation, we can conclude that the emf generated is a varying function of time with opposite polarity. So, all the graphs are correct for the given variation.

**87** (*a*) Energy stored in an inductor of inductance $L$ and current $I$ is given by

$$E = \frac{1}{2}LI^2$$

When current is being changed from $I_1$ to $I_2$, change in energy will be

$$\Delta E = E_2 - E_1 = \frac{1}{2}LI_2^2 - \frac{1}{2}LI_1^2 \quad \ldots(i)$$

As $I_1$ and $I_2$ are given, we need to find value of $L$.

Now, induced emf in a coil is

$$\varepsilon = L\frac{dI}{dt}$$

Here, $\varepsilon = 25$ V,

$$dI = I_2 - I_1 = (25 - 10) = 15 \text{ A and } dt = 1 \text{ s}$$

$$\Rightarrow \quad 25 = L \times \frac{15}{1} \text{ or } L = \frac{25}{15} = 5/3 \text{ H}$$

Putting values of $L$, $I_1$ and $I_2$ in Eq. (i), we get

$$\Delta E = \frac{1}{2} \times \frac{5}{3} \times [25^2 - 10^2] = \frac{1}{2} \times \frac{5}{3} \times 525$$

$$\Delta E = 437.5 \text{ J}$$

**88** (*b*) The magnetic energy is $U_B = \frac{1}{2}LI^2$

Since, for a solenoid, $B = \mu_0 nI$ and $L = \mu_0 n^2 Al$

$$\Rightarrow U_B = \frac{1}{2}(\mu_0 n^2 Al)\left(\frac{B}{\mu_0 n}\right)^2 = \frac{1}{2\mu_0} B^2 Al$$

**89** (*d*) Given, magnetic potential energy stored in an inductor, $U = 25 \text{ mJ} = 25 \times 10^{-3}$ J

Current in an inductor, $I_0 = 60 \text{ mA} = 60 \times 10^{-3}$ A

As, the expression for energy stored in an inductor is given as

$$U = \frac{1}{2}LI_0^2$$

where, $L$ is the inductance of the inductor.

Substituting the given values in above equation., we get

$$25 \times 10^{-3} = \frac{1}{2} \times L \times (60 \times 10^{-3})^2$$

$$\Rightarrow \quad L = \frac{2 \times 25 \times 10^{-3}}{3600 \times 10^{-6}} = \frac{500}{36} \text{ or } L = 13.89 \text{ H}$$

**90** (*c*) Quantity obtained by product of two vectors may be either vector or scalar. Scalar product (dot product) of two vectors gives scalar quantity whereas cross product of two vectors gives vector quantity.

Since, magnetic flux of a magnetic field **B** through an area **A** is equal to scalar product $\mathbf{B} \cdot \mathbf{A}$, therefore magnetic flux is a scalar quantity.

Therefore, Assertion is correct but Reason is incorrect.

**91** (*c*) According to Faraday's law, the conservation of mechanical energy into electrical energy is in accordance with the law of conservation of energy.

The earth's magnetic field also produces a flux through the loop. But it is a steady field and hence does not induce any emf.

Therefore, Assertion is correct but Reason is incorrect.

**92** (*d*) Lenz's law is based on conservation of energy and induced emf always opposes the cause of its production, i.e. change in magnetic flux.

Therefore, Assertion is incorrect but reason is correct.

**93** (*a*) Eddy currents are undesirable,

since they heat up the core and dissipate electrical energy in the form of heat.

Therefore, Assertion and Reason are correct and Reason is the correct explanation of Assertion.

**94** (*b*) When a metallic surface is moved in and out in magnetic field, then magnetic flux linked with metallic surface is changed continuously, hence according to Faraday's law of electromagnetic induction, an induced emf is produced.

Since, metallic plate behaves as closed path, therefore an induced current (eddy current) starts flowing in the metallic surface.

Therefore, Assertion and Reason are correct but Reason is not the correct explanation of Assertion.

**95** (*a*) Certain galvanometers have a fixed core made of non-magnetic metallic material.

When the coil oscillates, the eddy currents generated in the core oppose the motion and bring the coil to rest quickly. This is called electromagnetic damping.

Therefore, Assertion and Reason are correct and Reason is the correct explanation of Assertion.

**96** (*a*) If the inner solenoid is much shorter than (and placed well inside) the outer solenoid, then the flux linkage $N_1\phi_1$ can still be calculated.

It is because the inner solenoid is effectively immersed in a uniform magnetic field due to the outer solenoid.

Therefore, Assertion and Reason are correct and Reason is the correct explanation of Assertion.

**97** (*a*) In a system of co-axial solenoids, it is extremely difficult to calculate the flux linkage with the outer solenoid when current flows in inner solenoid.

It is because the magnetic field due to the inner solenoid would vary across the length as well as cross-section of the outer solenoid (when inner solenoid is smaller in length and radius).

Therefore, Assertion and Reason are correct and Reason is the correct explanation of Assertion.

**98** (*a*) The self-induced emf is also called the back emf as it opposes any change in the current in a circuit. Physically, the self-inductance plays the role of inertia. It is the electromagnetic analogue of mass in mechanics.

Therefore, Assertion and Reason are correct and Reason is the correct explanation of Assertion.

**99** (*d*) According to figure,

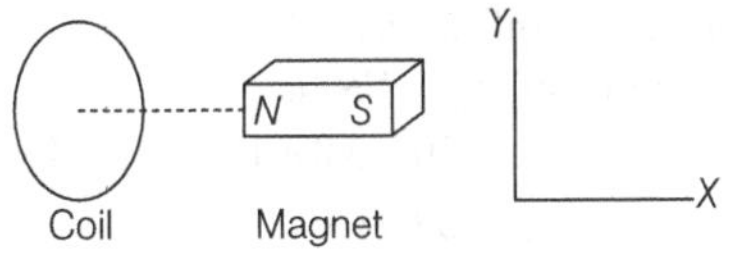


Current is induced in the coil only when there is a relative motion between the magnet and coil i.e. either the coil or the magnet moves along $x$, while the other is at rest.

When both coil and magnet move along $y$ with same speed, then no relative motion occurs between coil and magnet, so no current is induced in the coil.

So, statements II and III are correct but I and IV are incorrect.

**100** (*b*) Statements I, III and IV are correct but II is incorrect and it can be corrected as,

Inductance has dimension $[ML^2T^{-2}A^{-2}]$.

**101** (*d*) When current in coil $C_2$ changes, then magnetic flux linked with coil $C_1$ also changes. Hence an induced emf is produced in the coil $C_1$ such that it opposes any change in current in the coil $C_2$.

The steady current in the coil $C_2$ produces a steady magnetic field.

As coil $C_2$ is moved towards the coil $C_1$, the galvanometer shows a deflection. This indicates that electric current is induced in coil $C_1$.

So, all statements are correct.

**103** (*b*) Statements I and II are correct but statements III is incorrect and it can be corrected as,

The frequency of rotation of armature in India is 50 Hz.

**104** (*a*) The magnetic flux associated with coil $C_1$ changes in both the situations as

(i) when a magnet is moved towards or away from coil $C_1$ or

(ii) when a current carrying coil $C_2$ is moved towards or away from coil $C_1$.

Thus, the statement given in option (a) is correct, rest are incorrect.

**105** (*d*) The induced emf can be given by $\varepsilon = -N\dfrac{d\phi_B}{dt}$

Therefore, it can be increased by increasing the number of turns $N$ of a closed coil.

Thus, the statement given in option (d) is incorrect, rest are correct.

**106** (*c*) The statement ''a gentle push on the magnet in the magnetic field initiates the motion in magnet and its kinetic energy continuously increases without expanding any energy'' violates the law of conservation of energy.

Thus, the statement given in option (c) is incorrect, rest are correct.

**107** (*a*) The statement given in option (a) is correct but rest are incorrect and these can be corrected as,

If rectangular slots are made in the copper plates, then area available to the flow of eddy current is less. Thus, the electromagnetic damping decreases.

Eddy currents are minimised by using laminations of metal to make a metal core. The plane of the laminations must be arranged parallel to the magnetic field.

In magnetic breaking of trains, strong electromagnets are situated above the rails. When the electromagnets are activated, the eddy currents induced in the rails oppose the motion of the train. As there are no mechanical linkages, the breaking effect is smooth.

**108** (*d*) The flux linked with one coil will be the sum of two fluxes which exist independently. (when current is flowing simultaneously in two nearby coils)

$\therefore N_1\phi_1 = M_{11}I_1 + M_{12}I_2$, where $M_{11}$ represents inductance due to the same coil.

Thus, the statement given in option (d) is incorrect, rest are correct.

**109** (*c*) The statement given in option (c) is correct and rest are incorrect and these can be corrected as,

For a closely wound coil of $N$-turns, the magnetic flux linked with all turns is same.

When the flux $\phi_B$ through the coil changes, each turn contributes to the induced emf.

The mutual inductance between two solenoids (in air) of same length $l$, radius $r_1$ and $r_2$ $(r_2 > r_1)$ is given by $M = \mu_0 n_1 n_2 \pi r_1^2 l$, where $n_1$ and $n_2$ are their number of turns per unit length respectively.

If a medium of relative permeability $\mu_r$ had been present instead of air, the mutual inductance would become, $M = \mu_r \mu_0 n_1 n_2 \pi r_1^2 l$.

**110** (*d*) The statement given in option (d) is correct but rest are incorrect and these can be corrected as,

The emf has its extreme value, when $\theta = 90°$ or $\theta = 270°$, as the change of flux is greatest at these points.

In commercial generators, the mechanical energy required for rotation of the armature is provided by water falling from a height, e.g., from dams. These are called hydroelectric generators.

Water is heated to produce steam using coal or other sources. The steam at high pressure produces the rotation of the armature. These are called thermal generators. Instead of coal, if a nuclear fuel is used, we get nuclear power generator.

Thus, the statement given in option (d) is correct, rest are incorrect.

**112** (*a*)

A. The magnetic flux through the rectangular loop *abcd* increases, due to the motion of the loop into the region of magnetic field. Thus, the induced current must flow along the path *bcdab*, so that it opposes the increasing flux.

B. Due to the outward motion, magnetic flux through the triangular loop *abc* decreases. Thus, the induced current must flow along *bacb*, so as to oppose the change in flux.

C. As the magnetic flux decreases due to the motion of the irregular shaped loop *abcd* out of the region of magnetic field, the induced current flows along *cdabc*, so as to oppose the change in flux.

Hence, A $\rightarrow$ 3, B $\rightarrow$ 1 and C $\rightarrow$ 2.

**113** (*a*) Given, number of turns, $n = 15$ turns per cm

$= 1500$ turns per metre

Area of small loop, $A = 2.0\ \text{cm}^2 = 2 \times 10^{-4}\ \text{m}^2$

Change in current, $\frac{dI}{dt} = \frac{4-2}{0.1} = \frac{2}{0.1} = 20\ \text{A/s}$

Let $\varepsilon$ be the magnitude of induced emf, according to Faraday's law,

$$\varepsilon = \frac{d\phi}{dt} = \frac{d}{dt}(BA) \qquad (\because \phi = BA)$$

or $$\varepsilon = A\frac{dB}{dt} = A\frac{d}{dt}(\mu_0 n I)$$

[$\because A$ is a constant]

($\because$ magnetic field inside the solenoid, $B = \mu_0 nI$)

or $\varepsilon = A\,\mu_0 n \frac{dI}{dt}$

$\Rightarrow \varepsilon = 2 \times 10^{-4} \times 4 \times 3.14 \times 10^{-7} \times 1500 \times 20$

$(\because \mu_0 = 4\pi \times 10^{-7})$

$\Rightarrow \varepsilon = 7.5 \times 10^{-6}\ \text{V}$

Thus, the induced emf in the loop is $7.5 \times 10^{-6}$ V.

**114** (*b*) Given, $l = 8\ \text{cm} = 8 \times 10^{-2}\ \text{m}$

Width of the loop, $b = 2\ \text{cm}$

$= 2 \times 10^{-2}\ \text{m}$

Velocity of the loop, $v = 1\ \text{cms}^{-1} = 0.01\ \text{ms}^{-1}$

Magnitude of magnetic field, $B = 0.3$ T

In this case, motional emf,

$\varepsilon = Blv = 0.3 \times 8 \times 10^{-2} \times 0.01 = 2.4 \times 10^{-4}$ V

**115** (*c*) Length of rod, $l = 1.0$ m

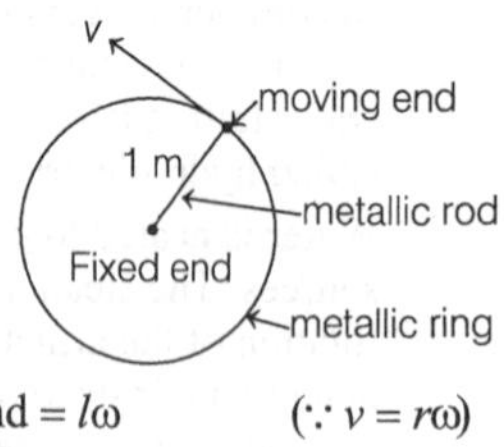


Angular frequency of rod,

$\omega = 400\ \text{rads}^{-1}$

Magnetic field, $B = 0.5$ T

The linear velocity of fixed end = 0

The linear velocity of other end = $l\omega$ $(\because v = r\omega)$

Average linear velocity,

$$v = \frac{0 + l\omega}{2} = \frac{l\omega}{2} \qquad \ldots(\text{i})$$

By using the formula of motional emf,

$$\varepsilon = Bvl = \frac{Bl\omega}{2} \cdot l \qquad \text{[from Eq. (i)]}$$

$$\Rightarrow \varepsilon = \frac{0.5 \times 1.0 \times 400 \times 1.0}{2} = 100\ \text{V}$$

Thus, the emf developed between the centre and ring is 100 V.

**116** (*a*) Given, radius of coil, $r = 8.0\ \text{cm} = 0.08$ m

Number of turns, $N = 20$

Resistance of closed loop, $R = 10\,\Omega$

Angular speed, $\omega = 50\ \text{rads}^{-1}$

Magnitude of magnetic field, $B = 3.0 \times 10^{-2}$ T

Induced emf produced in the coil, $\varepsilon = NBA\omega \sin \omega t$

For maximum emf, $\sin \omega t = 1$

$\therefore$ Maximum emf, $\varepsilon_0 = NBA\omega$

$= 20 \times 3.0 \times 10^{-2} \times 3.14 \times (0.08)^2 \times 50$

$\Rightarrow \varepsilon_0 = 0.603$ V

Maximum current in the coil,

$$I_0 = \frac{\varepsilon_0}{R} = \frac{0.603}{10} = 0.0603\ \text{A}$$

Average induced emf,

$$\varepsilon_{av} = \frac{1}{T}\int_0^{2\pi} \varepsilon\, dt = \frac{1}{T}\int_0^{2\pi} NBA\omega \sin \omega t\, dt$$

$$\Rightarrow \varepsilon_{av} = \frac{1}{T} \cdot NAB\omega \left[\frac{\cos \omega t}{\omega}\right]_0^{2\pi}$$

$$= \frac{NBA}{T}[\cos 2\pi\omega - \cos 0°]$$

$$= \frac{NBA}{T}[\cos 100\pi - \cos 0°]$$

$$\Rightarrow \varepsilon_{av} = \frac{NBA}{T}[1-1] = 0\ \text{V}$$

For full cycle average emf, $\varepsilon_{av} = 0$

Average power loss due to heating

$$= \frac{\varepsilon_0 I_0}{2} = \frac{0.603 \times 0.0603}{2} = 0.018\ \text{W}$$

**117** (*b*) Given, velocity of straight wire, $v = 5\ \text{ms}^{-1}$

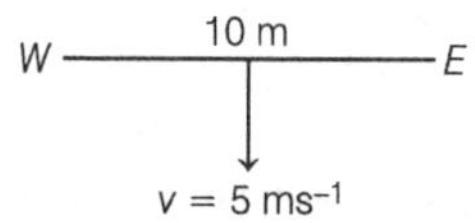


Magnetic field of straight wire, $B = 0.30 \times 10^{-4}\ \text{Wbm}^{-2}$

Length of wire, $l = 10$ m

Emf induced in the wire, $\varepsilon = Blv \sin \theta$

Here, $\theta = 90°$

($\because$ Wire is falling at right angle to earth's horizontal magnetic field component.)

$\therefore \sin \theta = 1$

$\Rightarrow \varepsilon = 0.30 \times 10^{-4} \times 10 \times 5 = 1.5 \times 10^{-3}$ V

**118** (*b*) Change in current, $dI = 5.0 - 0.0 = 5$ A

Time taken in current change, $dt = 0.1$ s

Induced average emf, $\varepsilon_{av} = 200$ V

Induced emf in the circuit, $\varepsilon = L\frac{dI}{dt}$

$$\Rightarrow 200 = L\left(\frac{5}{0.1}\right) \text{ or } L = \frac{200}{50} = 4\ \text{H}$$

**119** (*c*) Given, mutual inductance of coil, $M = 1.5$ H

Current change in coil, $dI = 20 - 0 = 20$ A

Time taken in change, $dt = 0.5$ s

Induced emf in the coil, $\varepsilon = M\frac{dI}{dt} = \frac{d\phi}{dt}$

or $\quad d\phi = M \cdot dI = 1.5 \times 20$

$\Rightarrow \quad d\phi = 30\ \text{Wb}$

Thus, the change of flux linkage is 30 Wb.

**120** (*a*) Given, length of solenoid,

$$l = 30\ \text{cm} = 30 \times 10^{-2}\ \text{m}$$

Area of cross-section,

$$A = 25\ \text{cm}^2 = 25 \times 10^{-4}\ \text{m}^2$$

Number of turns, $N = 500$,

Current, $I_1 = 2.5\ \text{A}, I_2 = 0$

and time, $dt = 10^{-3}\ \text{s}$

Induced emf in the solenoid,

$$\varepsilon = \frac{d\phi}{dt} = \frac{d}{dt}BA \qquad (\because \phi = BA)$$

Magnetic field induction $B$ at a point well inside the long solenoid carrying current $I$ is

$$B = \mu_0 nI$$

$$\left(\text{where, } n = \text{number of turns per unit length} = \frac{N}{l}\right)$$

$$\therefore \quad \varepsilon = NA\frac{dB}{dt} = NA\frac{d}{dt}\left(\mu_0 \frac{N}{l} I\right)$$

$$= NA\frac{\mu_0 N}{l} \cdot \frac{dI}{dt}$$

$$\Rightarrow \varepsilon = 500 \times 25 \times 10^{-4} \times 4 \times 3.14 \times 10^{-7} \times \frac{500}{30 \times 10^{-2}} \times \frac{2.5}{10^{-3}}$$

$\Rightarrow \varepsilon = 6.5\ \text{V}$

The average back emf induced across the ends of the open switch is 6.5 V.

**121** (*c*) Since, square lies in *xy*-plane, hence area vector is directed along *Z*-axis (i.e. along $\hat{\mathbf{k}}$)

$$\therefore \quad A = L^2\hat{\mathbf{k}}$$

Given, $\mathbf{B} = B_0(2\hat{\mathbf{i}} + 3\hat{\mathbf{j}} + 4\hat{\mathbf{k}})\ \text{T}$

$$\phi = \mathbf{B} \cdot \mathbf{A} = B_0(2\hat{\mathbf{i}} + 3\hat{\mathbf{j}} + 4\hat{\mathbf{k}}) \cdot L^2\hat{\mathbf{k}}$$

$$= 4B_0L^2\ \text{Wb}$$

**122** (*b*) The magnetic flux linked with uniform surface of area $A$ in uniform magnetic field is given by

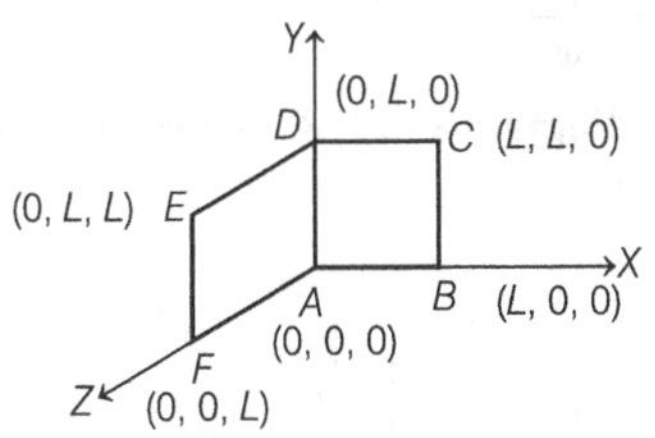


$$\phi = \mathbf{B} \cdot \mathbf{A}$$

Since, loop *ABCD* lies in *xy*-plane, hence area vector of the face *ABCD* is given as

$$\mathbf{A}_1 = L^2\hat{\mathbf{k}}$$

Again, loop *ADEF* lies on *yz*-plane, hence area vector of the face *ADEF* is given as

$$\mathbf{A}_2 = L^2\hat{\mathbf{i}}$$

$\therefore$ Total area, $\mathbf{A} = \mathbf{A}_1 + \mathbf{A}_2$

$$= L^2\hat{\mathbf{k}} + L^2\hat{\mathbf{i}}$$

and $\quad \mathbf{B} = B_0(\hat{\mathbf{i}} + \hat{\mathbf{k}})\ \text{T}$

Now, $\quad \phi = \mathbf{B} \cdot \mathbf{A}$

$$= B_0(\hat{\mathbf{i}} + \hat{\mathbf{k}}) \cdot (L^2\hat{\mathbf{k}} + L^2\hat{\mathbf{i}})$$

$$= 2B_0L^2\ \text{Wb}$$

**123** (*b*) When cylindrical bar magnet is rotated about its axis as shown below, no change in flux linked with the circuit takes place, consequently, no emf induces and hence no current flows through the ammeter *A*.

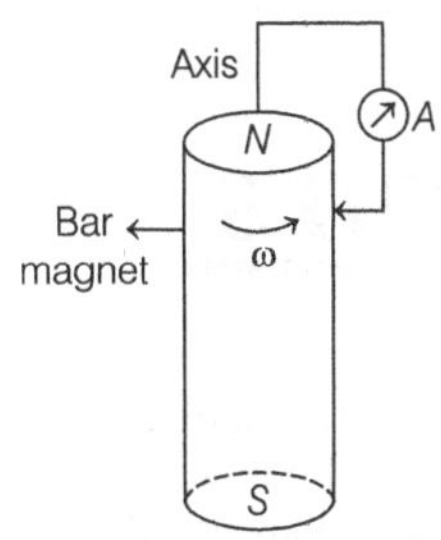


**124** (*d*) When the *A* stops moving, the current in *B* become zero, it possible only if the current in *A* is constant in the counter-clockwise direction.

If the current in *A* would be variable, there must be an induced emf (current) in *B* even, if the *A* stops moving.

**125** (*a*) When the current in *B* (at $t = 0$) is counter-clockwise, then *B* is equivalent to north pole of magnet and magnetic field lines are emanating towards coil *A*. When coil *A* is rotating at $t = 0$, a constant currrent in clockwise direction (by Lenz's law) flow in *A* due to constant change in flux.

**126** (*b*) The self-inductance of a long solenoid of cross-sectional area $A$ and length $l$, having $n$ turns per unit length, filled the inside of the solenoid with a material of relative permeability (e.g. soft iron, which has a high value of relative permeability) is given by

$$L = \mu_r \mu_0 n^2 Al$$

where, $\quad n = N / l$

$$\Rightarrow \quad L = \frac{\mu_r \mu_0 N^2 A}{l}$$

So, the self-inductance $L$ increases as $l$ decreases and $A$ increases.

CHAPTER > 07

# Alternating Current

## KEY NOTES

- The currents whose magnitude and direction changes periodically through a load is known as **alternating currents** and corresponding voltage as **alternating voltage or emf,** which are represented respectively as

$$I = I_m \sin \omega t$$
$$V = V_m \sin \omega t$$

- where, $I_m$ and $V_m$ = peak value of current and voltages, respectively and $\omega$ = angular frequency.

## AC Voltage Applied to a Resistor

When an AC voltage source is connected across a resistor in a circuit, then

(i) Current in the circuit is $i = i_m \sin \omega t$

where, $i_m = \frac{V_m}{R}$.

(ii) The voltage and current are in phase with each other.

(iii) **Phasor diagram** and ***V* and *I versus*** $\omega t$ graph are as shown below

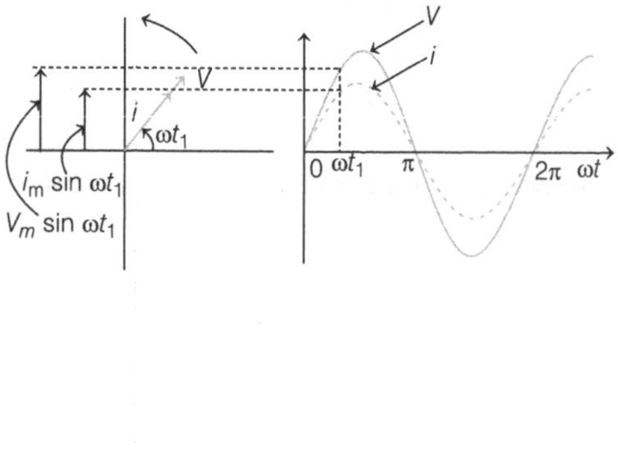


## AC Voltage Applied to an Inductor

- When an AC voltage source is connected to a purely inductive circuit, then

(i) Current in the circuit is $i = i_m \sin\left(\omega t - \frac{\pi}{2}\right)$

where, $i_m = \frac{V_m}{\omega L}$.

(ii) The quantity $\omega L$ is analogous to the resistance and is called **inductive reactance**. It is denoted by $X_L$, i.e.

$$X_L = \omega L = (2\pi\nu) L$$

(iii) The dimension of inductive reactance is same as that of resistance and its SI unit is ohm ($\Omega$).

(iv) The current in this type of circuit lags the voltage in phase by $\frac{\pi}{2}$ or one-quarter $\left(\frac{1}{4}\right)$ cycle. It means the current reaches its maximum value later than the voltage by one-fourth of a period $\left[\frac{T}{4} = \frac{\pi/2}{\omega}\right]$.

(v) The **phasor diagram** and ***V* & *I versus*** $\omega t$ graph are as shown below

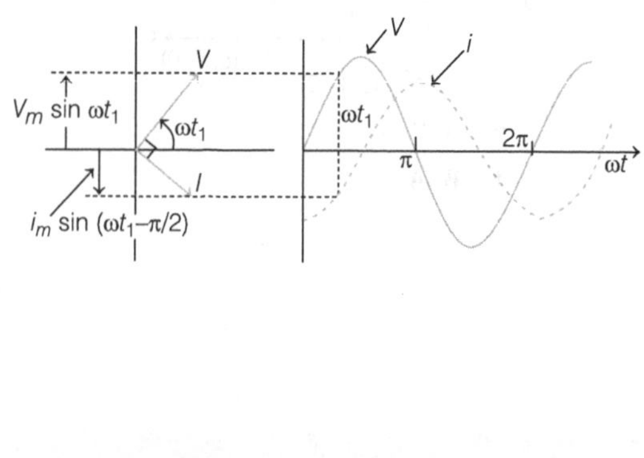

## AC Voltage Applied to a Capacitor

- When a capacitor is connected to an AC voltage source in a circuit, then
  - (i) The current in the circuit is $i = i_m \sin\left(\omega t + \frac{\pi}{2}\right)$

    where, $i_m = \frac{V_m}{1/\omega C}$.
  - (ii) The term $\left(\frac{1}{\omega C}\right)$ is also analogous to resistance and it is called **capacitive reactance**. It is denoted by $X_C$, i.e.
    $$X_C = \frac{1}{\omega C} = \frac{1}{2\pi \nu C}$$
  - (iii) The dimension of capacitive reactance is the same as that of resistance and its SI unit is ohm (Ω).
  - (iv) The current in this circuit is ahead of voltage in phase by $\frac{\pi}{2}$ or one quarter cycle. It means, the current reaches its maximum value earlier than the voltage by one-fourth of a period.
  - (v) The **phasor diagram** and ***V* & *I versus*** $\omega t$ graph are as shown below

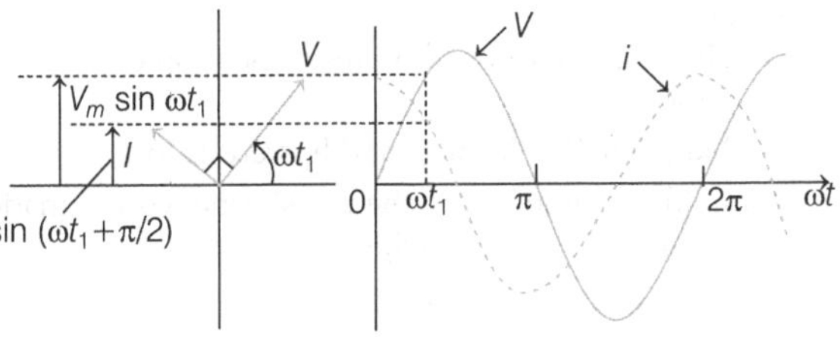


## AC Voltage Applied to a Series *L-C-R* Circuit

When a resistor, capacitor and an inductor are connected in series with an alternating voltage source, then

- (i) voltage across the circuit is $V = \sqrt{V_R^2 + (V_L - V_C)^2}$
- (ii) The current in the circuit is
  $$i_m = \frac{V_m}{\sqrt{R^2 + (X_L - X_C)^2}}$$
- (iii) The **impedance Z** which is the opposition offered by AC circuit to the flow of alternating current through it, is given as $Z = \sqrt{R^2 + (X_L - X_C)^2}$.
- (iv) The phase angle $\phi$ is the angle between $V_R$ & $V$ and can be determined as
  $$\tan \phi = \frac{V_{Cm} - V_{Lm}}{V_{Rm}}$$
  or
  $$\tan \phi = \frac{X_C - X_L}{R}$$
- (v) If $X_C > X_L$, $\phi$ is positive and the circuit is predominantly capacitive, so the current in the circuit leads the source voltage.
- (vi) If $X_C < X_L$, $\phi$ is negative and the circuit is predominantly inductive, so the current in the circuit lags the source voltage.
- (vii) **Phasor diagram** and ***V* & *I versus*** $\omega t$ graph are as shown below

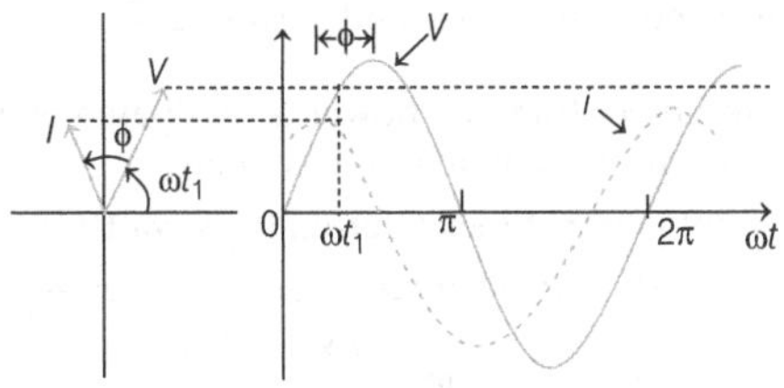


## Resonance

- It is a condition in a series *L-C-R* circuit, when at a particular frequency $\omega_0$, $X_C = X_L$ and the impedance is minimum ($Z = R$).
- This frequency is called **resonant frequency** and is given by $\omega_0 = \frac{1}{\sqrt{LC}}$.
- At resonant frequency, the current amplitude is maximum, i.e. $i_m = \frac{V_m}{R}$.
- **Quality factor** (Q-factor) is the measure of sharpness at the resonance of an series *L-C-R* circuit.
- $Q$ - factor can also be expressed as
  $$Q = \frac{\omega_0 L}{R} \text{ or } Q = \frac{1}{\omega_0 CR} = \frac{1}{R}\sqrt{\frac{L}{C}}$$
- The quantity ($Q = \omega_0 / 2\Delta\omega$) is also the measure of sharpness of resonance, where, $\omega_1 - \omega_2 = 2\Delta\omega$ is called the **bandwidth of the circuit**, such that $\omega_1 = \omega_0 + \Delta\omega$ and $\omega_2 = \omega_0 - \Delta\omega$.

## Power in AC Circuit

- It is the product of voltage and the component of current, which is in phase with voltage and given by $P = VI \cos \phi$, where $V$ and $I$ are rms values of voltage and currents.
- **Instantaneous power** of AC is given by
  $P = VI = V_m I_m \sin \omega t \sin(\omega t + \phi)$
- **Average power** of AC is given by
  $P_{av} = V_{rms} I_{rms} \cos \phi = \frac{1}{2} V_m I_m \cos \phi$
- **Power factor** is the cosine of angle of lag or lead or the ratio of resistance and impedance.

  It is given by $\cos \phi = \frac{\text{True power}}{\text{Apparent power}} = \frac{P_{av}}{V_{rms} I_{rms}}$

- There are following cases, for power dissipation in a circuit

  *Case* **I Resistive Circuit** If the circuit contains only $R$, then in that case, $\phi = 0 \Rightarrow \cos\phi = 1$.

  Thus, there is maximum power dissipation.

  *Case* **II Purely Inductive** or **Capacitive Circuit** If the circuit contains only an inductor or a capacitor, phase difference between voltage and current is $\frac{\pi}{2}$.

  Therefore, $\cos\phi = 0$ and no power is dissipated even though a current is flowing in the circuit.

  This current is sometimes referred to as **wattless current**.

  *Case* **III *L-C-R* Series Circuit** In an *L-C-R* series circuit,

  $$\phi = \tan^{-1}\left(\frac{X_C - X_L}{R}\right)$$

  So, $\phi$ may be non-zero in a *R-L* or *R-C* or *R-C-L* circuits. In this, power is dissipated only in the resistor.

  *Case* **IV Power Dissipated at Resonance in *L-C-R* Circuit**

  At resonance, $X_C - X_L = 0$ and $\phi = 0$, so $\cos\phi = 1$ and $P = I^2Z = I^2R$, i.e. maximum power is dissipated in a circuit (through $R$) at resonance.

## *L-C* Oscillations

- These are the electrical oscillations of constant amplitude and frequency which are produced, when a charged capacitor is allowed to discharge through a non-resistive inductor..
- The charge on the capacitor oscillates with a natural frequency, $\omega_0 = \frac{1}{\sqrt{LC}}$

## Transformer

- It is a device which transforms a strong AC at low voltage to a weak AC at high voltage or *vice-versa* based on the principle of electromagnetic induction.
- For an ideal transformer, $\frac{\varepsilon_s}{\varepsilon_p} = \frac{V_s}{V_p} = \frac{N_s}{N_p} = \frac{I_p}{I_s}$

  (i) If $N_s > N_p$ the voltage is stepped up ($V_s > V_p$). This type of arrangement is called a **step-up transformer**. However, there is less current in the secondary than the primary, i.e. $I_s < I_p$.

  (ii) If $N_s < N_p$, we have a **step-down transformer**. In this case, $V_s < V_p$ and $I_s > I_p$, i.e. voltage is stepped down or reduced and current is increased.
- The efficiency of a transformer is given by

  $$\eta = \frac{\text{output power}}{\text{input power}} = \frac{V_s\, I_s}{V_p\, I_p}$$
- In a transformer, some power is lost due to following reasons

  (*i*) Heating due to winding resistance and eddy current losses.

  (*ii*) Magnetic flux leakage and hysteresis loss.
- To minimise these losses, the transformer core is made up of a laminated soft iron strips.

KEY NOTES

# Mastering NCERT

## MULTIPLE CHOICE QUESTIONS

## TOPIC 1 ~ *AC Voltage Applied to a Resistor*

**1** Which of the following current changes continuously in magnitude and periodically in direction?

(a) Direct current (b) Induced current
(c) Alternating current (d) Displacement current

**2** The frequency of an alternating voltage is 50 cycles/s and its amplitude is 120 V. Then, the rms value of voltage is

(a) 101.3 V (b) 84.8 V
(c) 70.7 V (d) 56.5 V

**3** A generator produces a voltage that is given by $V = 240 \sin 120\, t$, where $t$ is in seconds. The frequency and rms voltage are

(a) 60 Hz and 240 V (b) 19 Hz and 120 V
(c) 19 Hz and 170 V (d) 754 Hz and 70 V

**4** The electric mains in the house is marked 220 V, 50 Hz. Write down the equation for instantaneous voltage.

(a) $3.1 \sin (100\pi)\, t$ (b) $31.1 \cos (100\pi)\, t$
(c) $311.1 \sin (100\pi)\, t$ (d) $311.1 \cos (100\pi)\, t$

**5** In an AC circuit, $I = 100 \sin 200\, \pi t$. The time required for the current to achieve its peak value will be

(a) $\frac{1}{100}$ s (b) $\frac{1}{200}$ s (c) $\frac{1}{300}$ s (d) $\frac{1}{400}$ s

**6** A resistance of 20 $\Omega$ is connected to a source of an alternating potential, $V = 220 \sin (100\, \pi\, t)$. The time taken by current to change from its peak value to rms value is

(a) 0.2 s (b) 0.25 s
(c) $25 \times 10^{-3}$ s (d) $2.5 \times 10^{-3}$ s

**7** An alternating current is given by the equation $i = i_1 \cos \omega t + i_2 \sin \omega t$. The rms current is given by

(a) $\frac{1}{\sqrt{2}} (i_1 + i_2)$ (b) $\frac{1}{\sqrt{2}} (i_1 + i_2)^2$

(c) $\frac{1}{\sqrt{2}} (i_1^2 + i_2^2)^{1/2}$ (d) $\frac{1}{2} (i_1^2 + i_2^2)^{1/2}$

**8** If an AC main supply is given to be 220 V. What would be the average emf during a positive half-cycle?

(a) 198 V (b) 386 V
(c) 256 V (d) None of these

**9** The average value of AC voltage given by,

$$V = V_m \sin \omega t$$

over time interval $t = 0$ to $t = \frac{\pi}{\omega}$ is

(a) 0 (b) $\frac{2V_m}{\pi}$ (c) $\frac{V_m}{\pi}$ (d) $V_m$

**10** The rms value of the alternating current shown in figure is

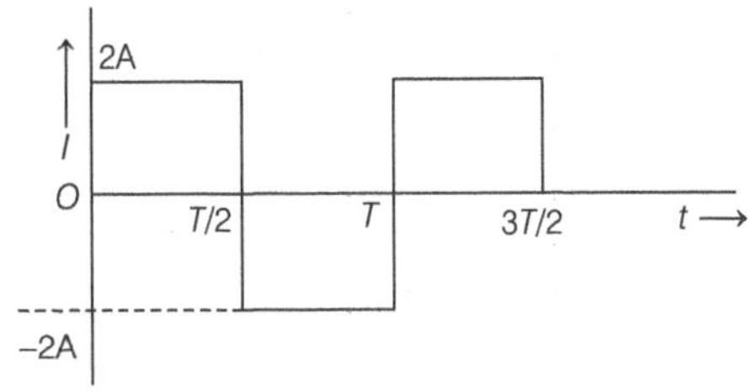


(a) 2A (b) – 2A
(c) 4A (d) 1A

**11** Which of the following is the graphical representation of alternating current and voltage for a purely resistive circuit?

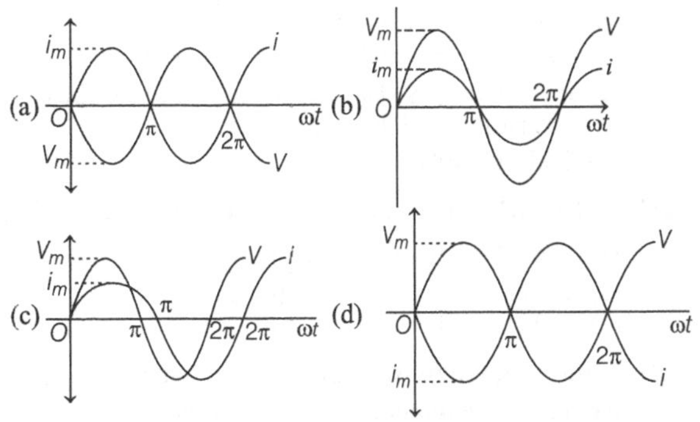


**12** An alternating voltage $V(t) = 220 \sin 100\pi t$ volt is applied to a purely resistive load of 50 $\Omega$. The time taken for the current to rise from half of the peak value to the peak value is **JEE Main 2019**

(a) 5 ms (b) 2.2 ms
(c) 7.2 ms (d) 3.3 ms

# TOPIC 2 ~ AC Voltage Applied to an Inductor

**13** When an alternating voltage is applied to an inductor as shown in the figure, then

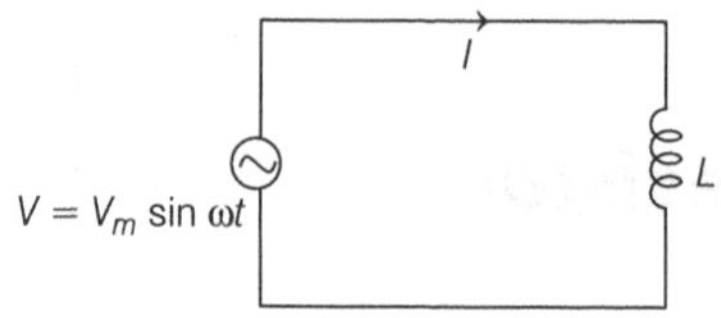


(a) $v + L\frac{dI}{dt} = 0$ (b) $V - L\frac{dI}{dt} = 0$

(c) $L + v\frac{dI}{dt} = 0$ (d) $L - v\frac{dI}{dt} = 0$

**14** Which of the following graphs represents the correct variation of inductive reactance $X_L$ with angular frequency ω?

(a)
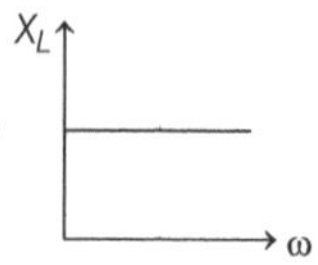


(b)
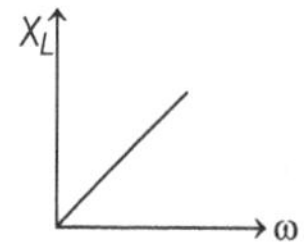


(c)
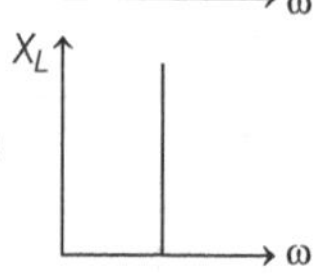


(d)
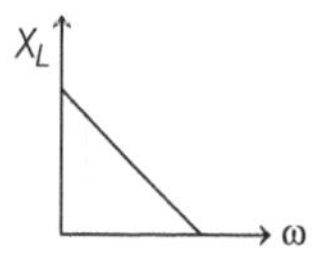


**15** A pure inductor of inductance 25.0 mH is connected to a source of 220 V. Find the inductive reactance, if the frequency of the source is 50 Hz.

(a) 785 Ω (b) 6.50 Ω

(c) 7.85 Ω (d) 8.75 Ω

**16** The rms current in the circuit containing a pure inductor of 40 mH, connected to a source 200 V, 50 Hz is

(a) 25 A (b) 16 A

(c) 11 A (d) 28 A

**17** An inductance of negligible resistance, whose reactance is 120 Ω at 200 Hz is connected to a 240 V, 60 Hz, power line. The current in the inductor is

(a) 6.66 A (b) 6.60 A

(c) 5.45 A (d) 54.5 A

**18** An arc lamp requires a direct current of 10 A at 80 V to function. If it is connected to a 220 V (rms), 50 Hz AC supply, the series inductor needed for it to work is close to ***JEE Main 2016***

(a) 80 H (b) 0.08 H

(c) 0.044 H (d) 0.065 H

# TOPIC 3 ~ AC Voltage Applied to a Capacitor

**19** When a capacitor is connected to an AC source, then

(a) it limits or regulates the current by complete prevention of flow of charge

(b) it limits or regulates the current, but does not completely prevent the flow of charge

(c) It is alternately charged and discharged as the current reverses in each half cycle

(d) Both (b) and (c)

**20** A 15.0 μF capacitor is connected to a 220 V, 50 Hz source. The capacitive reactance is

(a) 220 Ω (b) 215 Ω

(c) 212 Ω (d) 204 Ω

**21** If the frequency is doubled, what happen to the capacitive reactance and the current?

(a) Capacitive reactance is halved, the current is doubled

(b) Capacitive reactance is doubled, the current is halved

(c) Capacitive reactance and the current are halved

(d) Capacitive reactance and the current are doubled

**22** When a capacitor of 36μF is connected to a 240 V, 50 Hz supply the currents (rms and peak) in the circuit are

(a) 1.47 A, 2.04 A (b) 1.95 A, 2.73 A

(c) 2.73 A, 3.85 A (d) 2.4 A, 1.08 A

**23** An alternating voltage, $V = 200\sqrt{2} \sin (100\,t)$ is connected to a 1μF, capacitor through an AC ammeter. The reading of the ammeter shall be

(a) 10 mA (b) 20 mA (c) 40 mA (d) 80 mA

**24** When an alternating voltage source of $V = 200 \sin\left(100\pi t - \frac{\pi}{3}\right)$ is applied to a pure capacitor of capacitance 2μF, then the instantaneous value of current through the capacitor is

(a) $200 \sin\left(100\pi t + \frac{\pi}{6}\right)$ (b) $0.04\pi \sin\left(100\pi t + \frac{\pi}{6}\right)$

(c) $200 \sin\left(100\pi t - \frac{\pi}{6}\right)$ (d) $0.04\pi \sin\left(100\pi t - \frac{\pi}{6}\right)$

**25** A 15μF capacitor is connected to a 220 V, 50 Hz source. The rms value of the current flowing the capacitor is
(a) 2.56 A
(b) 1.53 A
(c) 1.04 A
(d) 1.96 A

**26** Same current is flowing in two alternating circuits. The first circuit contains only inductance and the second contains only a capacitance. If the frequency of the emf of AC is increased, the value of current will
(a) increase in the first circuit and decrease in the second circuit
(b) increase in both the circuits
(c) decrease in both the circuits
(d) decrease in the first circuit and increase in the second circuit

# TOPIC 4 ~ AC Voltage Applied to a Series L-C-R Circuit

**27** For series *L-C-R* circuit,
(a) applied emf and current across resistance are in same phase
(b) applied emf and current across inductor coil have phase difference of $\pi$.
(c) potential difference at capacitor and inductor have phase difference of $\pi/2$
(d) potential difference across resistance and capacitor have phase difference of $\pi$.

**28** In a series *L-C-R* circuit, the potential difference between the terminals of the inductor is 60 V, between the terminals of the capacitor is 30 V and that across the resistor is 40 V. Then, supply voltage will be equal to
(a) 50 V (b) 70 V (c) 130 V (d) 10 V

**29** In a series *L-C-R* circuit, source voltage is 120 V. The voltage across inductor is 50 V and across resistor is 40 V, then determine voltage across the capacitor.
**AIIMS 2019**
(a) $V_C = 10(5 - 8\sqrt{2})$ (b) $V_C = 10(5 + 8\sqrt{2})$
(c) $V_C = 20(5 + 8\sqrt{2})$ (d) $V_C = 10(5 + 7\sqrt{2})$

**30** In a series *L-C-R* AC circuit, the voltage across each of the components *L*, *C* and *R* is 50 V. The voltage across the *C-R* circuit will be
(a) 50 V (b) $50\sqrt{2}$ V
(c) 100 V (d) zero

**31** In a series *L-C-R* AC circuit, the voltage across each of the components, *L*, *C* and *R* is 20 V. The voltage across the *L-C* combination will be
(a) 50 V (b) $50\sqrt{2}$ V (c) 100 V (d) zero

**32** In a series *L-C-R* circuit, the frequency of 10 V AC voltage source is adjusted in such a fashion that the reactance of the inductor measures 15 Ω and that of the capacitor 11 Ω. If $R = 3\,\Omega$, the potential difference across the series combination of *L* and *C* will be
(a) 8 V (b) 10 V (c) 22 V (d) 52 V

**33** If the reading of voltmeter $V_1$ is 30 V, what is the reading of voltmeter $V_2$?

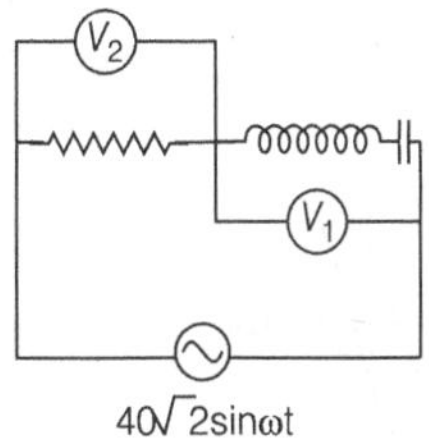


(a) 20.50 V (b) 26.45 V
(c) 32.65 V (d) 15.24 V

**34** In the given circuit, the readings of voltmeters $V_1$ and $V_2$ are 300 V each. The readings of the voltmeter $V_3$ and ammeter *A* are respectively

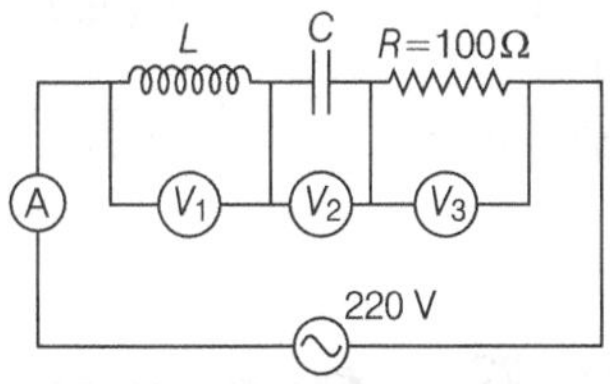


(a) 100 V, 2.0 A (b) 150 V, 2.2 A
(c) 220 V, 2.2 A (d) 220 V, 2.0 A

**35** A resistor and a capacitor are connected in series with an AC source. If the potential drop across the capacitor is 5V and that across the resistor is 12 V, then applied voltage is
(a) 13 V (b) 17 V
(c) 5 V (d) 12 V

**36** A sinusoidal voltage of peak value 300 V and an angular frequency $\omega = 400\,\text{rads}^{-1}$ is applied to a series *L-C-R* circuit, in which $R = 3\,\Omega$, $L = 20$ mH and $C = 625\,\mu$F. The peak current in the circuit is
(a) $30\sqrt{2}$ A (b) 60 A
(c) 100 A (d) $60\sqrt{2}$ A

**37** A resistor of $200\,\Omega$ and a capacitor of $15\,\mu F$ are connected in series to a 220 V, 50 Hz AC source. The current in the circuit is
(a) 755 A (b) 7.55 mA (c) 0.755A (d) 0.755 mA

**38** A circuit when connected to an AC source of 12 V gives a current of 0.2 A. The same circuit when connected to a DC source of 12 V, gives a current of 0.4 A. The circuit is ***NEET (Odisha) 2019***
(a) series *L-R* (b) series *R-C*
(c) series *L-C* (d) series *L-C-R*

**39** Which of the following graph, is correct for a series *L-C-R* circuit, where $X_C > X_L$?

(a)
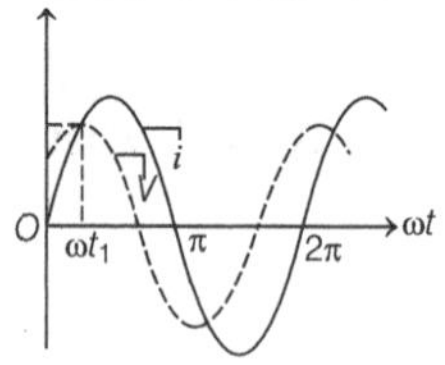


(b)
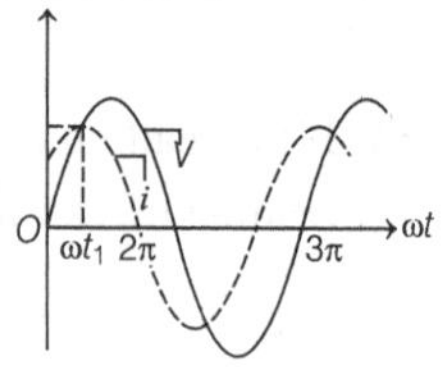


(c)
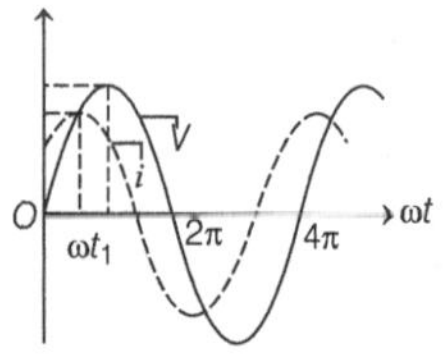


(d)
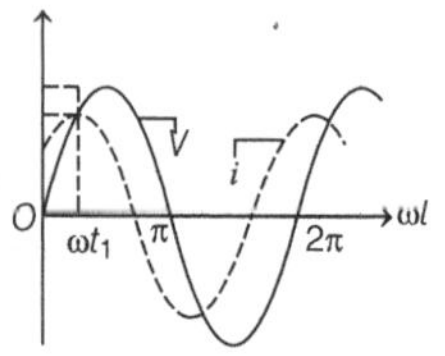


**40** In a circuit *L*, *C* and *R* are connected in series with an alternating voltage source of frequency $f$. The current leads the voltage by 45°. The value of *C* is ***AIIMS 2018***

(a) $\dfrac{1}{2\pi f(2\pi fL + R)}$ (b) $\dfrac{1}{\pi f(2\pi fL + R)}$
(c) $\dfrac{1}{2\pi f(2\pi fL - R)}$ (d) $\dfrac{1}{\pi f(2\pi fL - R)}$

**41** A resistance of $300\,\Omega$ and an inductance of $(1/\pi)$ Henry are connected in a series to an AC voltage of 20 V and 200 Hz frequency. The phase angle between the voltage and current is
(a) $\tan^{-1}(4/3)$ (b) $\tan^{-1}(3/4)$
(c) $\tan^{-1}(3/2)$ (d) $\tan^{-1}(2/5)$

**42** In a series *R-C* circuit shown in figure, the applied voltage is 10 V and the voltage across capacitor is found to be 8V. Then, the voltage across *R* and the phase difference between current and the applied voltage will respectively be

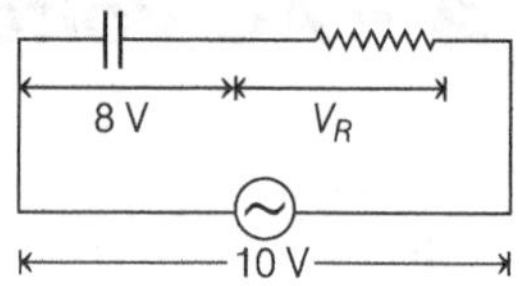


***AIIMS 2018***

(a) 6V, $\tan^{-1}\left(\dfrac{4}{3}\right)$ (b) 3V, $\tan^{-1}\left(\dfrac{3}{4}\right)$
(c) 6V, $\tan^{-1}\left(\dfrac{5}{3}\right)$ (d) None of these

**43** An AC source of angular frequency ω is fed across a circuit having resistor *R* and capacitor *C* in series. The current flowing is *I*. If now the frequency of source is changed to ω/3 (but maintaining the same voltage), the current in the circuit is found to be halved. Calculate the ratio of reactance to resistance at the original frequency ω.
(a) $\sqrt{\dfrac{3}{5}}$ (b) $\sqrt{\dfrac{2}{5}}$
(c) $\sqrt{\dfrac{1}{5}}$ (d) $\sqrt{\dfrac{4}{5}}$

# TOPIC 5 ~ *Resonance*

**44** At resonant frequency, the amplitude of current in *R-L-C* circuit is
(a) maximum (b) minimum
(c) zero (d) infinity

**45** In series *R-L-C* circuit, $L = 1.00$ mH, $C = 1.00$ nF and $R = 200\,\Omega$. For the source applied with $V_m = 100$ V, the resonant frequency is
(a) $1\times 10^3$ rad/s
(b) $1\times 10^6$ rad/s
(c) $1.56\times 10^6$ rad/s
(d) $1.75\times 10^3$ rad/s

**46** Figure shows a series *L-C-R* circuit, connected to a variable frequency source of 200 V with $C = 80\,\mu F$, $L = 5$ H and $R = 40\,\Omega$. The source frequency which drives the circuit at resonance is

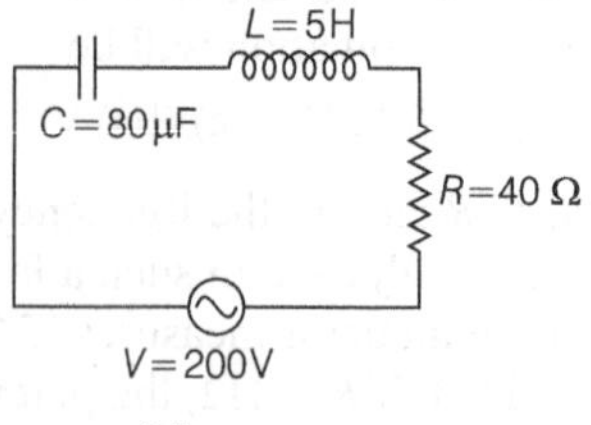


(a) 25 Hz (b) $\dfrac{25}{\pi}$ Hz (c) 50 Hz (d) $\dfrac{50}{\pi}$ Hz

**47** In an *L-C-R* circuit, capacitance is changed from $C$ to $2C$. For the resonant frequency to remain unchanged, the inductance should be changed from $L$ to

(a) $4L$ (b) $2L$ (c) $L/2$ (d) $L/4$

**48** In an *L-C* circuit, angular frequency at resonance is $\omega$. What will be the new angular frequency when inductor's inductance is made two times and capacitor's capacitance is made four times? ***JIPMER 2018***

(a) $\frac{\omega}{2\sqrt{2}}$ (b) $\frac{\omega}{\sqrt{2}}$ (c) $2\omega$ (d) $\frac{2\omega}{\sqrt{2}}$

**49** For the series *L-C-R* circuit shown in the figure, what is the angular resonant frequency and amplitude of the current at the resonating frequency?

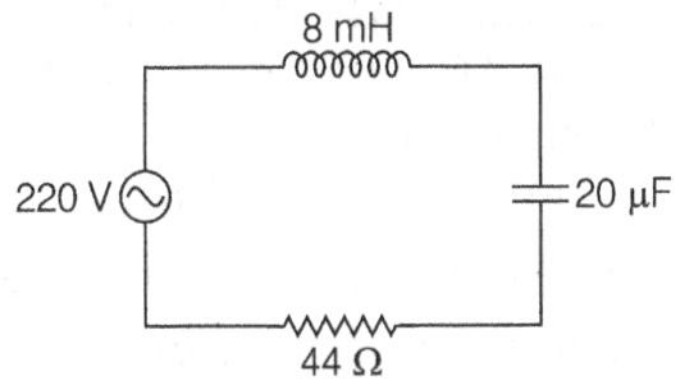


(a) 2500 rads$^{-1}$ and $5\sqrt{2}$ A (b) 2500 rads$^{-1}$ and 5 A

(c) 2500 rads$^{-1}$ and $\frac{5}{\sqrt{2}}$ A (d) 25 rads$^{-1}$ and $5\sqrt{2}$ A

**50** In a series *L-C-R* circuit, the voltage across $R$ is 100 V, $R = 1\,k\Omega$ and $C = 2\,\mu F$. The resonant frequency $\omega$ is 200 rad/s. At resonance, the voltage across $L$ is

(a) 40 V (b) 250 V

(c) $4 \times 10^{-3}$ V (d) $2.5 \times 10^{-2}$ V

**51** An *L-C-R* series circuit consists of a resistance of $10\,\Omega$, a capacitor of reactance $60\,\Omega$ and an inductor coil. The circuit is found to resonate when put across a 300 V, 100 Hz supply. The inductance of coil is (take, $\pi = 3$)

(a) 0.1 H (b) 0.01 H (c) 0.2 H (d) 0.02 H

**52** Find resonant frequency in the given circuit.

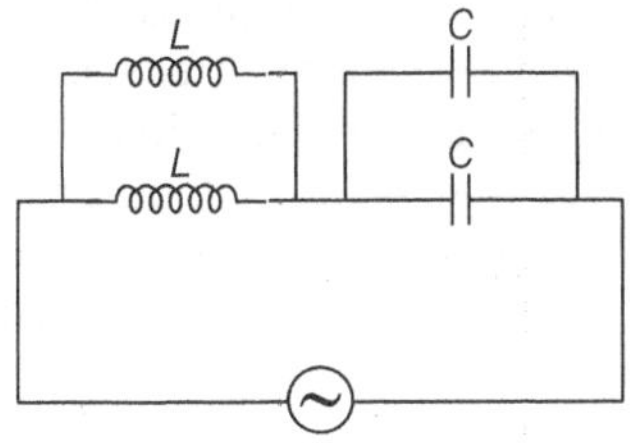


***JIPMER 2019***

(a) $\frac{1}{\sqrt{LC}}$ (b) $\frac{2}{\sqrt{LC}}$

(c) $\frac{1}{2\sqrt{LC}}$ (d) $\frac{4}{\sqrt{LC}}$

**53** The graphs shown below, depict the variation of current $I_m$ *versus* angular frequency $\omega$ for four different *L-C-R* circuits. The graph which has maximum quality factor is

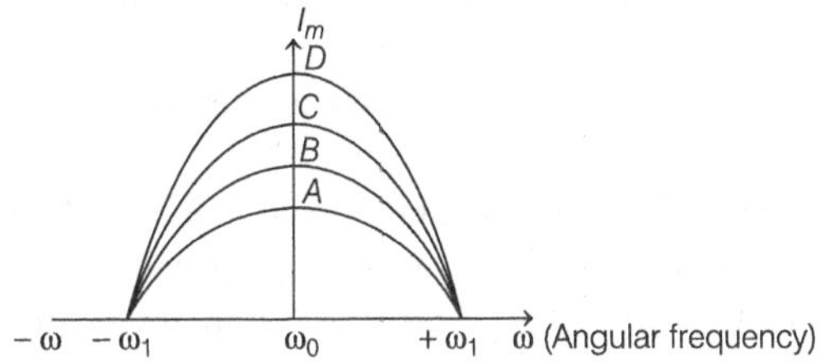


(a) $A$ (b) $B$

(c) $C$ (d) $D$

**54** For an *R-L-C* circuit driven with voltage of amplitude $v_m$ and frequency $\omega_0 = \frac{1}{\sqrt{LC}}$, the current exhibits resonance. The quality factor, $Q$ is given by ***JEE Main 2018***

(a) $\frac{\omega_0 L}{R}$ (b) $\frac{\omega_0 R}{L}$

(c) $\frac{R}{\omega_0 C}$ (d) $\frac{CR}{\omega_0}$

# TOPIC 6 ~ *Power in AC Circuit*

**55** In an AC circuit, the instantaneous values of emf and current are $\varepsilon = 200 \sin 314t$ V and $I = \sin(314t + \pi/3)$ A. The average power consumed is

(a) 200 W (b) 100 W

(c) 50 W (d) 25 W

**56** The potential difference $V$ and the current $i$ flowing through an instruments in an AC circuit of frequency $f$ are given by $V = 5 \cos \omega t$ volts and $i = 2 \sin \omega t$ amperes (where, $\omega = 2\pi f$). The power dissipated in the instrument is

(a) zero (b) 10 W (c) 5 W (d) 2.5 W

**57** The line that draws power supply to your house from street has

(a) zero average current

(b) 220 V average voltage

(c) voltage and current out of phase by 90°

(d) voltage and current possibly differing in phase $\phi$ such that $\phi > \pi/2$

**58** In an electrical circuit $R$, $L$, $C$ and an AC voltage source are all connected in series. When $L$ is removed from the circuit, the phase difference between the voltage and the current in the circuit is $\pi/3$. If instead, $C$ is removed from the circuit, the phase difference is again $\pi/3$. The power factor of the circuit is **CBSE AIPMT 2012**

(a) 1/2 (b) $1/\sqrt{2}$ (c) 1 (d) $\sqrt{3}/2$

**59** A lamp consumes only 50% of peak power in an AC circuit. What is the phase difference between the applied voltage and the circuit current?

(a) $\frac{\pi}{6}$ (b) $\frac{\pi}{3}$
(c) $\frac{\pi}{4}$ (d) $\frac{\pi}{2}$

**60** For a value of $\omega$, the current amplitude becomes $1/\sqrt{2}$ times its maximum value. At this value, the power dissipated by the circuit becomes

(a) double (b) one-fourth
(c) one-third (d) half

**61** A series AC circuit containing an inductor (20 mH), a capacitor (120 $\mu$F) and a resistor (60 $\Omega$) is driven by an AC source of 24 V/50 Hz. The energy dissipated in the circuit in 60 s is **JEE Main 2019**

(a) $3.39\times 10^3$ J
(b) $5.65\times 10^2$ J
(c) $2.26\times 10^3$ J
(d) $5.17\times 10^2$ J

**62** A group of electric lamps having total power rating of 600 W, 220V is supplied by an AC voltage $V = 169\sin(314t + 60°)$. The rms value of the current is

(a) 10 A (b) 9.04 A (c) 1.48 A (d) 8 mA

**63** A coil of self-inductance $L$ is connected in series with a bulb $B$ and an AC source. Brightness of the bulb decreases when **NEET 2013**

(a) frequency of the AC source is decreased
(b) number of turns in the coil is reduced
(c) a capacitance of reactance $X_C - X_L$ is included in the same circuit
(d) an iron rod is inserted in the coil

**64** A coil of inductive reactance 31 $\Omega$ has a resistance of 8 $\Omega$. It is placed in series with a condenser of capacitive reactance 25 $\Omega$. The combination is connected to an AC source of 110 V. The power factor of the circuit is

(a) 0.56 (b) 0.64 (c) 0.80 (d) 0.33

**65** An inductor 20 mH, a capacitor 100 $\mu$F and a resistor 50$\Omega$ are connected in series across a source of emf, $V = 10\sin 314t$. The power loss in the circuit is **NEET 2018**

(a) 2.74 W (b) 0.43 W
(c) 0.79 W (d) 1.13 W

**66** A voltage of peak value 283 V and varying frequency is applied to a series $L$-$C$-$R$ combination in which $R = 3\,\Omega$, $L = 25$ mH and $C = 400\,\mu$F. The frequency (in Hz) of the source at which maximum power is dissipated in the above circuit is

(a) 51.5 Hz (b) 50.7 Hz (c) 51.1 Hz (d) 50.3 Hz

# TOPIC 7 ~ L-C Oscillations and Transformers

**67** A charged 40 $\mu$F capacitor is connected to a 16 mH inductor. What is the angular frequency of free oscillations of the circuit?

(a) 1.1 s (b) $1.25\times 10^3\ s^{-1}$
(c) $2\times 10^3\ s^{-1}$ (d) $2.5\times 10^3\ s^{-1}$

**68** In $L$-$C$ circuit, the inductance $L = 40$ mH and capacitance $C = 100\,\mu$F. If a voltage $V(t) = 10\sin(314t)$ is applied to the circuit, the current in the circuit is given as **JEE Main 2020**

(a) $0.52\cos 314t$ (b) $0.52\sin 314t$
(c) $10\cos 314t$ (d) $5.2\cos 314t$

**69** A 1 $\mu$F capacitor is charged to 25 V of potential. The battery is then disconnected and a pure 100 mH coil is connected across the capacitor, so that $L$-$C$ oscillations are set up. The maximum current in the coil is

(a) 0.25 A (b) 0.01 A (c) 2.5 A (d) 1.6 A

**70** If maximum energy is stored in a capacitor at $t = 0$, then find the time after which, current in the circuit will be maximum? **AIIMS 2019**

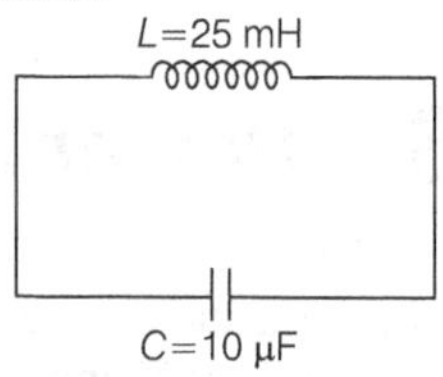


(a) $\frac{\pi}{2}$ m-s (b) $\frac{\pi}{4}$ m-s (c) $\pi$ m-s (d) 2 m-s

**71** A 1.5 mH inductor in an $L$-$C$ circuit stores a maximum energy of 30$\mu$J. The rms value of current in the circuit is

(a) $\frac{1}{\sqrt{2}}\times 10^{-1}$ A (b) $\sqrt{2}\times 10^{-2}$ A
(c) $\frac{1}{\sqrt{2}}\times 10^{-2}$ A (d) $\sqrt{2}\times 10^{-1}$ A

**72** A battery of 10 V is connected to the primary of a transformer of ratio 20. The output across secondary coil is
(a) 20 V (b) 5 V (c) 10 V (d) zero

**73** A power transmission line feeds input power at 2300 V to a step-down transformer with its primary windings having 4000 turns. What should be the number of turns in the secondary in order to get output power at 230 V?
(a) 600 (b) 550 (c) 400 (d) 375

**74** How much current is drawn by the primary coil of a transformer which steps down 220 V to 22 V to operate a device with an impedance of $220\,\Omega$ ?
(a) 1 A (b) 0.25 A (c) 0.01 A (d) 1.50 A

**75** A 110 V AC is connected to a transformer of ratio 10. If resistance of secondary coil is $550\,\Omega$, current through secondary coil will be
(a) 10 A (b) 2 A (c) zero (d) 55 A

**76** The ratio of secondary to primary turns is 4 : 5, for an ideal transformer, then the ratio of power output to power input is
(a) 4 : 9 (b) 9 : 4 (c) 5 : 4 (d) 1 : 1

**77** A transformer consisting of 300 turns in the primary and 150 turns in the secondary gives output power of 2.2 kW. If the current in the secondary coil is 10 A, then the input voltage and current in the primary coil are **JEE Main 2019**
(a) 440 V and 5 A (b) 220 V and 20 A
(c) 220 V and 10 A (d) 440 V and 20 A

**78** A transformer with turns ratio $\frac{N_1}{N_2} = \frac{50}{1}$ is connected to a 120 V AC supply. If primary and secondary circuit resistances are $1.5\,k\Omega$ and $1\,\Omega$ respectively, then find out power of output. **AIIMS 2019**
(a) 5.76 W (b) 11.4 W
(c) 2.89 W (d) 7.56 W

**79** A transformer is used to light 140 W, 24 V lamp from 240 V AC mains. The current in the mains is 0.7 A. The efficiency of transformer is nearer to
(a) 90% (b) 80%
(c) 70% (d) 60%

**80** A power transmission line feeds input power at 2300 V to a step-down transformer with its primary windings having 4000 turns. The output power is delivered at 230 V by the transformer. If the current in the primary of the transformer is 5A and its efficiency is 90%, the output current would be **JEE Main 2019**
(a) 45 A (b) 50 A
(c) 25 A (d) 35 A

**81** A transformer having efficiency of 90% is working on 200 V and 3 kW power supply. If the current in the secondary coil is 6 A, the voltage across the secondary coil and the current in the primary coil respectively, are **CBSE AIPMT 2014**
(a) 300 V, 15 A
(b) 450 V, 15 A
(c) 450 V, 13.5 A
(d) 600 V, 15 A

# SPECIAL TYPES QUESTIONS

## I. Assertion and Reason

**Direction** (Q. Nos. 82-87) *In the following questions, a statement of Assertion is followed by a corresponding statement of Reason. Of the following statements, choose the correct one.*

(a) Both Assertion and Reason are correct and Reason is the correct explanation of Assertion.
(b) Both Assertion and Reason are correct but Reason is not the correct explanation of Assertion.
(c) Assertion is correct but Reason is incorrect.
(d) Assertion is incorrect but Reason is correct.

**82.** **Assertion** The alternating current lags behind the emf by a phase angle of $\pi/2$, when AC flows through an inductor.

**Reason** The inductive reactance increases as the frequency of AC source decreases.

**83** **Assertion** When the capacitor is connected to an AC source, it limits or regulates the current, but does not completely prevent the flow of charge.

**Reason** The capacitor is alternately charged and discharged as the current reverses each half-cycle.

**84** **Assertion** Capacitor serves as a barrier for DC and offers an easy path to AC.

**Reason** Capacitive reactance is inversely proportional to frequency.

**85** **Assertion** If $X_C > X_L$, $\phi$ is positive and the circuit is predominantly capacitive. The current in the circuit leads the source voltage.

**Reason** If $X_C < X_L$, $\phi$ is negative and the circuit is predominantly inductive. The current in the circuit lags the source voltage.

**86** **Assertion** In a series $R$-$L$-$C$ circuit, the voltages across resistor, inductor and capacitor are 8V, 16V and 10V, respectively. The resultant emf in the circuit is 10 V.

**Reason** Resultant emf of the circuit is given by the relation.

$$E = \sqrt{V_R^2 + (V_L - V_C)^2}$$

**87** **Assertion** Resonance phenomenon is exhibited by a circuit only, if both $L$ and $C$ are present in the circuit.

**Reason** Voltage across $L$ & $C$ cancel each other and the current amplitude is $V_m / R$, the total source voltage appearing across $R$ causes resonance.

## II. Statement Based Questions

**88** I. The electric main supply in our home and offices is a voltage that varies like a sine function with time.

II. AC voltages can be easily and efficiently converted from one voltage to the other by means of transformers.

III. Electrical energy can also be transmitted economically over long distances.

Which of the following statement(s) is/are correct?

(a) Only I
(b) Both I and III
(c) Both I and II
(d) I, II and III

**89** Consider the figure and study the statements given below.

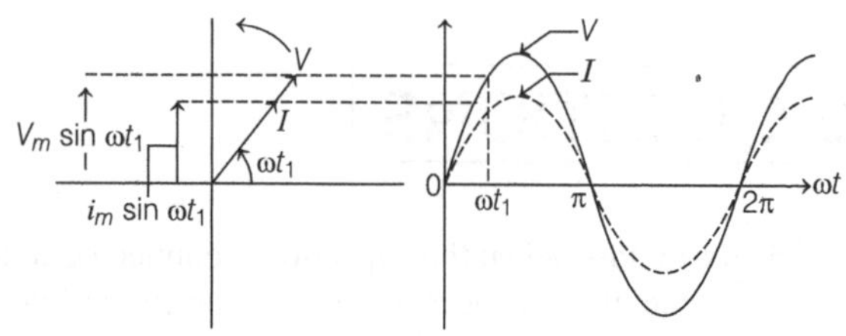


(a) (b)

I. The vertical components of phasors **V** and **I** represent the sinusoidally varying quantities $V$ and $i$.

II. The magnitudes of phasors **V** and **I** represent the amplitudes or the peak values $V_m$ and $i_m$ of these oscillating quantities.

III. The projection of voltage and current phasors on vertical axis, i.e. $V_m \sin \omega t$ and $i_m \sin \omega t$, respectively represent the value of voltage and current at that instant.

Which of the above statement(s) is/are correct?

(a) Both I and II (b) Both I and III
(c) Both II and III (d) I, II and III

**90** I. When a capacitor is connected to a voltage source in a DC circuit, current will flow for the short time, required to charge the capacitor.

II. As charge accumulates on the capacitor plates, the voltage across them increases, opposing the current.

III. A capacitor in a DC circuit will limit or oppose the current as it charges.

IV. When the capacitor is fully charged, the current in the circuit falls to zero.

Which of the above statements are incorrect? Choose the correct option.

(a) I, II and III (b) II, III and IV
(c) Both I and IV (d) None of these

**91** Which of the following statement(s) is/are correct ?

(a) In case of resistor, both $V$ and $i$ reach zero, minimum and maximum values at the different time.
(b) The sum of the instantaneous current values over one complete cycle is unity and the average current is one.
(c) The average power dissipated in a resistance is
$$P_{avg} = \frac{1}{2} I_m^2 R.$$
(d) To show phase relationship between voltage and current in an AC circuit, we use the notion of scalars.

**92** Which of the following statement(s) is/are incorrect?

(a) Phasor diagram say nothing about the initial condition.
(b) At any arbitrary value of $t$, if we draw different phasors which show the relative angle between different phasors. The solution, so obtained is called the steady-state solution.
(c) We do have a transient solution which exists even for $V = 0$. The general solution is the sum of transient solution and the steady-state solution.
(d) After a sufficient long time, the behaviour of the circuit is described by transient solution only.

**93** Which of the following statement(s) is/are correct?

(a) In an inductive circuit, using Kirchhoff's loop rule, we get $$V - L\frac{di}{dt} = 0$$ where, the second term is the mutual induced emf in the inductor.
(b) The quantity ω is analogous to the conductance.
(c) The current phasor **I** is $\frac{\pi}{2}$ behind the voltage phasor
(d) The average power supplied to an inductor over one complete cycle is zero.

**94** Which of the following statement(s) is/are correct about the capacitive reactance?

(a) The reactance of the capacitor is directly proportional to the frequency of the current.
(b) The reactance of the capacitor is inversely proportional to the frequency of the current.
(c) Capacitive reactance is measured in farad.
(d) The reactance of a capacitor in an AC circuit is similar to the resistance of a capacitor in a DC circuit.

**95** Which of the following statement(s) is/are incorrect about capacitor?

(a) The current in a capacitive circuit is
$$i = i_m \sin\left(\omega t - \frac{\pi}{2}\right).$$
where the amplitude of the oscillating current is
$$i_m = \frac{V_m}{\omega C}$$
(b) For a purely capacitive circuit $\left(\frac{1}{\omega C}\right)$ plays the role of resistance. It is called capacitive reactance and denoted by $X_C = \frac{1}{\omega C}$
(c) The current reaches its maximum value earlier than the voltage by one-fourth of a period.
(d) In case of a capacitor, the average power,
$$P_c = \left\langle \frac{i_m V_m}{2} \sin(2\omega t) \right\rangle = 0$$
Since, $< \sin(2\omega t) > = 0$ over a complete cycle.

**96** Which of the following statement(s) is/are correct ?

(a) In a series *L-C-R* circuit, the AC current in each elements is different at any time, having the different amplitude and phase.
(b) By analogy to the resistance in a circuit, we introduce the impedance $Z$ in an *RLC* series circuit given by
$$Z = \sqrt{R^2 + (X_C - X_L)^2}$$
(c) The phasor **I** is always parallel to phasor $\mathbf{V}_R$, the phase angle $\phi$ is the angle between $\mathbf{V}_R$ and **V** and can be determined from
$$\tan\phi = \frac{V_L + V_C}{V_R}, \text{ when } V_L > V_C$$
(d) The impedance triangle is a right triangle with $Z$ as its base.

**97** Which of the following statement(s) is/are incorrect ?

(a) The phenomenon of resonance is common among systems that have a tendency to oscillate at a particular frequency.
(b) If $\omega$ is varied, then at a particular frequency $\omega_0$, $X_C > X_L$ and the impedance become maximum. This frequency is called resonant frequency.
(c) The amplitude of the current in a series *L-C-R* circuit is given by
$$i_m = \frac{V_m}{\sqrt{R^2 + \left(\omega L - \frac{1}{\omega C}\right)^2}}$$
At $\omega = \omega_0 = \frac{1}{\sqrt{LC}}$, $i_m^{\max} = \frac{V_m}{R}$
(d) The difference $\omega_1 - \omega_2 = 2\Delta\omega$ is often called the bandwidth of the circuit. The quantity $\left(\frac{\omega_0}{2\Delta\omega}\right)$ is regarded as a measure of the sharpness of resonance.

**98** Which of the following statement(s) is/are incorrect?

(a) The antenna of a radio accepts signals from many broadcasting stations.
(b) To hear one particular radio station, radio is tuned to a particular frequency.
(c) The signals picked up in the antenna acts as a source in the tuning circuit of the radio, so the circuit can be driven at only one frequency.
(d) In tuning, we vary the capacitance of a capacitor in the tuning circuit such that the resonant frequency of the circuit becomes nearly equal to the frequency of the radio signal received.

**99** Which of the following statement(s) is/are incorrect?

(a) If the resonance is less sharp, the tuning of the circuit will not be good.
(b) Less sharp the resonance, less is the selectivity of the circuit or *vice-versa.*
(c) If quality factor is large, i.e. $R$ is low or $L$ is large, the circuit is more selective.
(d) Larger the value of $Q$, the larger is the value of $2\Delta\omega$ or the bandwidth and sharper is the resonance.

**100** Which of the following statement(s) is/are correct ?

(a) When a capacitor is connected to an inductor, the charge on the capacitor and the current in the circuit exhibit the phenomenon of electrical oscillations similar to oscillation in mechanical system.
(b) The charge on the capacitor, therefore oscillates with a natural frequency, $\omega_0 = \sqrt{LC}$
(c) The charge is given by $q = q_m \sin(\omega_0 t - \phi)$.
The current corresponding to this is $i = i_m \cos(\omega_0 t)$.
(d) In the case of *L-C* circuit, $\omega_0 = \sqrt{LC}$ and for mass on a spring, $\omega_0 = \sqrt{km}$.

**101** An alternating emf of frequency $\nu = \left(\frac{1}{2\pi\sqrt{LC}}\right)$ is applied to a series *L-C-R* circuit. For this frequency of the applied emf, which statement is incorrect?

(a) The circuit is at resonance and its impedance is made up only of a resistive part.

(b) The current in the circuit is in phase with the applied emf and voltage across $R$ equals the applied emf.

(c) The sum of the potential difference across the inductance and capacitance equals the applied emf which is 180° ahead of phase of the current in the circuit.

(d) The quality factor of the circuit is $\omega L/R$ or $1/\omega CR$ and it is the measure of the voltage magnification (Produced by the circuit at reasonance) as well as the sharpness of resonance of the circuit.

**102** $L$-$C$ oscillations is not realistic for the following reasons. Considering this, which of the following statement(s) is/are incorrect?

(a) Every inductor has no resistance.

(b) The effect of resistance is to introduce a damping effect on the charge and current in the circuit and the oscillations finally die away.

(c) Even if the resistance is zero, the total energy of the system would not remain constant. It is radiated away from the system in the form of electromagnetic waves.

(d) Radio and TV transmitters depends on this radiations.

**103** Which of the following statement(s) is/are incorrect ?

(a) A transformer using the principle of mutual induction, change (or transform) an alternating voltage from one to another of greater or smaller value.

(b) If the transformer is assumed to be 100% efficient (no energy losses), the power input is less than the power output.

(c) If $N_s > N_p$ the voltage is stepped up ($V_s \geq V_p$). This type of arrangement is called a step-up transformer. However, $I_s < I_p$.

(d) If $N_s < N_p$, we have a step-down transformer. In this case, $V_s < V_p$ and $I_s > I_p$.

**104** Which of the following statement(s) is/are correct ?

(a) Flux leakage is reduced by winding the primary and secondary coils at different places.

(b) Resistance of windings is minimised by using thin wire.

(c) Eddy currents are reduced by using a laminated core.

(d) Hysteresis is minimised by using a non- magnetic material.

**105** Electrical energy is transmitted over large distances at high alternating voltages. Which of the following statement(s) is/are incorrect?

(a) For a given power level, there is a lower current.

(b) Lower current implies less power loss.

(c) Transmission lines can be made thinner.

(d) It is easy to reduce the voltage at the receiving end, using step-down transformers.

## III. Matching Type

**106** Match the Column I (situation in capacitor) with Column II (circuit diagram) and select the correct answer from the codes given below.

| | Column I | | Column II |
|---|---|---|---|
| A. | The current and voltage are both positive. Energy is absorbed from the source. | 1. | 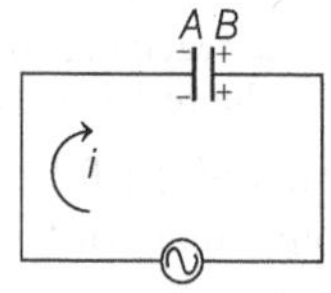 |
| B. | The voltage gets reduced but is still positive. The current is negative. Energy is returned. | 2. | 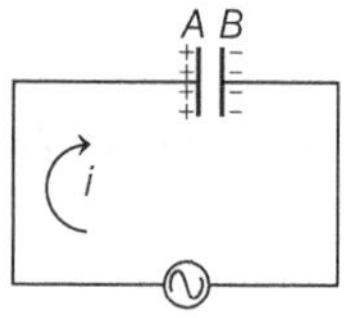 |
| C. | Both the current and voltage are negative. The capacitor absorbs energy. | 3. | 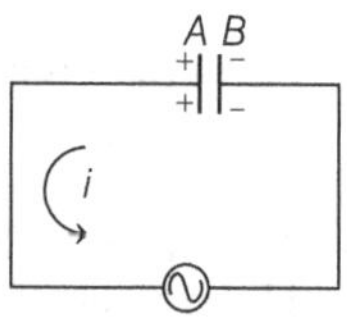 |
| D. | The current is positive but voltage is negative. Energy is returned to the source. | 4. | 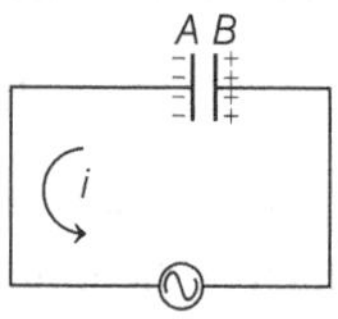 |

| | A | B | C | D |
|---|---|---|---|---|
| (a) | 1 | 3 | 2 | 4 |
| (b) | 2 | 3 | 4 | 1 |
| (c) | 2 | 3 | 1 | 4 |
| (d) | 1 | 2 | 3 | 4 |

**107** Match the Column I (AC circuit condition) with Column II (power dissipation) and select the correct answer from the codes given below.

| | Column I | | Column II |
|---|---|---|---|
| A. | Resistive circuit | 1. | No power dissipation |
| B. | Purely inductive or capacitive circuit | 2. | Maximum power dissipation because of $X_C = X_L$ |
| C. | $L$-$C$-$R$ series circuit | 3. | Power dissipated only in the resistor |
| D. | Power dissipated at resonance in $L$-$C$-$R$ circuit | 4. | Maximum power dissipation |

| | A | B | C | D | | A | B | C | D |
|---|---|---|---|---|---|---|---|---|---|
| (a) | 1 | 2 | 4 | 3 | (b) | 4 | 1 | 3 | 2 |
| (c) | 3 | 1 | 4 | 2 | (d) | 2 | 1 | 3 | 4 |

**108** Match the Column I (electric and magnetic energies) with Column II (circuit diagram) for a $L$-$C$ oscillator and select the correct answer from the codes given below.

| Column I | Column II |
|---|---|
| A. $U_E = \frac{q_m^2}{2C}, U_B = 0$ | 1. 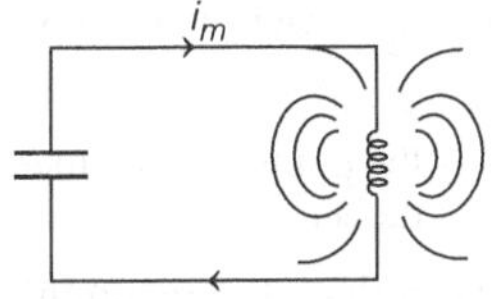 |
| B. $U_E = \frac{q}{2C}$, $U_B = \frac{1}{2}Li^2$ | 2. 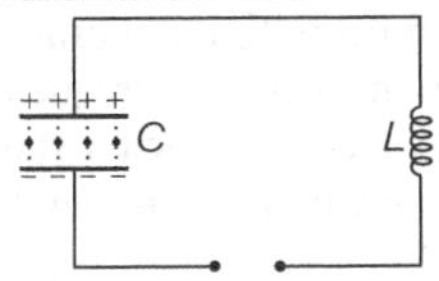 |
| C. $U_E = 0$, $U_B = \frac{1}{2}Li_m^2$ | 3. 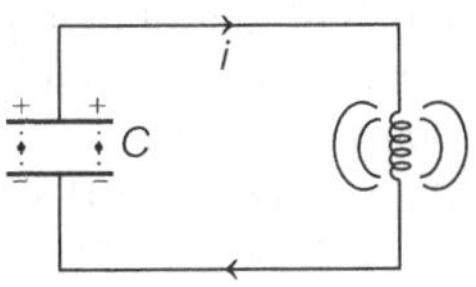 |

| | A | B | C |
|---|---|---|---|
| (a) | 1 | 2 | 3 |
| (b) | 3 | 2 | 1 |
| (c) | 1 | 3 | 2 |
| (d) | 2 | 3 | 1 |

**109** Match the Column I (quantities in spring system) with Column II (quantities in $L$-$C$ circuit) and select the correct answer from the codes given below.

| Column I | Column II |
|---|---|
| A Force constant $k$ | 1. Reciprocal of capacitance i.e., $1/C$ |
| B. Velocity, $v = dx/dt$ | 2. Current, $i = dq/dt$ |
| C. Mechanical energy | 3. Electromagnetic energy |
| D. $E = \frac{1}{2}kx^2 + \frac{1}{2}mv^2$ | 4. $U = \frac{1}{2}q^2/C + \frac{1}{2}Li^2$ |

| | A | B | C | D |
|---|---|---|---|---|
| (a) | 1 | 3 | 2 | 4 |
| (b) | 4 | 3 | 1 | 2 |
| (c) | 1 | 2 | 3 | 4 |
| (d) | 4 | 2 | 3 | 1 |

# NCERT & NCERT Exemplar

## MULTIPLE CHOICE QUESTIONS

### NCERT

**110** A 100 $\Omega$ resistor is connected to a 220 V, 50 Hz AC supply, then the rms value of current in the circuit is

(a) 2.2 A (b) 4.2 A (c) 3.2 A (d) 2.4 A

**111** The peak voltage of an AC supply is 300 V, then the rms voltage will be

(a) 212.1 V (b) 312.1 V (c) 84.2 V (d) 85.2 V

**112** A 44 mH inductor is connected to 220 V, 50 Hz AC supply. Determine the rms value of the current in the circuit.

(a) 20.4 A (b) 15.9 A
(c) 21.4 A (d) 22.4 A

**113** A 60 $\mu$F capacitor is connected to a 110 V, 60 Hz AC supply. The rms value of the current in the circuit will be

(a) 4.49 A (b) 2.29 A
(c) 2.49 A (d) 3.49 A

**114** A series $L$-$C$-$R$ circuit have $L = 2.0$ H, $C = 32\ \mu$F and $R = 10\ \Omega$. What is the $Q$-value of this circuit?

(a) 36 (b) 27 (c) 24 (d) 25

**115** A charged 30 $\mu$F capacitor is connected to a 27 mH inductor. What is the angular frequency of free oscillations of the circuit?

(a) $1.1 \times 10^4$ rads$^{-1}$ (b) $1.1 \times 10^3$ rads$^{-1}$
(c) $1.1 \times 10^2$ rads$^{-1}$ (d) $1.1 \times 10$ rads$^{-1}$

**116** A series $L$-$C$-$R$ circuit with $R = 20\ \Omega$, $L = 1.5$ H and $C = 35\ \mu$F is connected to a variable frequency source of 200 V AC supply. When the frequency of the supply equals the natural frequency of the circuit, what is the average power transferred to the circuit in one complete cycle?

(a) 2 kW (b) 3 kW
(c) 4 kW (d) 5 kW

**117** A radio can tune over the frequency range of a portion of MW broadcast band: (800 kHz to 1200 kHz). If its $L$-$C$ circuit has an effective inductance of 200 $\mu$H, what must be the range of its variable capacitor?

(a) 49 to 79 pF
(b) 88 to 198 pF
(c) 100 to 200 pF
(d) 110 to 200 pF

**118** A coil of inductance 0.50 H and resistance 100 $\Omega$ is connected to a 240 V, 50 Hz AC supply. What is the maximum current in the coil?

(a) 1.824 A (b) 2.824 A
(c) 3.824 A (d) 4.824 A

**119** A 100 $\mu$F capacitor in series with a 40 $\Omega$ resistance is connected to a 110 V, 60 Hz supply. What is the maximum current in the circuit?

(a) 3.00 A (b) 3.24 A (c) 4.24 A (d) 2.24 A

**120** At a hydroelectric power plant, the water pressure head is at a height of 300 m and the water flow available is 100 $m^3$/s. If the turbine generator efficiency is 60%, the electric power available from the plant will be

(a) 184.4 MW (b) 176.4 MW
(c) 190.4 MW (d) 290.4 MW

## NCERT Exemplar

**121** If the rms current in a 50 Hz AC circuit is 5 A, the value of the current 1/300 s after its value becomes zero is

(a) $5\sqrt{2}$ A (b) $5\sqrt{3/2}$ A
(c) $(5/6)$ A (d) $(5/\sqrt{2})$ A

**122** An alternating current generator has an internal resistance $R_g$ and an internal reactance $X_g$. It is used to supply power to a passive load consisting of a resistance $R_g$ and a reactance $X_L$. For maximum power to be delivered from the generator to the load, the value of $X_L$ is equal to

(a) zero (b) $X_g$ (c) $-X_g$ (d) $R_g$

**123** When a voltage measuring device is connected to AC mains, the meter shows the steady input voltage of 220 V. This means

(a) input voltage cannot be AC voltage, but a DC voltage
(b) maximum input voltage is 220 V
(c) the meter reads not $v$ but $<v^2>$ and is calibrated to read $\sqrt{<v^2>}$
(d) the pointer of the meter is stuck by some mechanical defect

**124** To reduce the resonant frequency in an *L-C-R* series circuit with a generator,

(a) the generator frequency should be reduced
(b) another capacitor should be added in parallel to the first
(c) the iron core of the inductor should be removed
(d) dielectric in the capacitor should be removed

**125** Which of the following combinations should be selected for better tuning of an *L-C-R* circuit used for communication?

(a) $R = 20\Omega, L = 1.5\,H, C = 35\mu F$
(b) $R = 25\Omega, L = 2.5\,H, C = 45\mu F$
(c) $R = 15\Omega, L = 3.5\,H, C = 30\mu F$
(d) $R = 25\Omega, L = 1.5\,H, C = 45\mu F$

**126** An inductor of reactance $1\Omega$ and a resistor of $2\Omega$ are connected in series to the terminals of a 6V (rms) AC source. The power dissipated in the circuit is

(a) 8 W (b) 12 W (c) 14.4 W (d) 18 W

**127** The output of a step-down transformer is measured to be 24 V when connected to a 12 W light bulb. The value of the peak current is

(a) $(1/\sqrt{2})$ A (b) $\sqrt{2}$ A (c) 2 A (d) $2\sqrt{2}$ A

# Answers

### > Mastering NCERT with MCQs

| | | | | | | | | | |
|---|---|---|---|---|---|---|---|---|---|
| 1 (c) | 2 (b) | 3 (c) | 4 (c) | 5 (d) | 6 (d) | 7 (c) | 8 (a) | 9 (b) | 10 (a) |
| 11 (b) | 12 (d) | 13 (b) | 14 (b) | 15 (c) | 16 (b) | 17 (a) | 18 (d) | 19 (d) | 20 (c) |
| 21 (a) | 22 (c) | 23 (b) | 24 (b) | 25 (c) | 26 (d) | 27 (a) | 28 (a) | 29 (a) | 30 (b) |
| 31 (d) | 32 (a) | 33 (b) | 34 (c) | 35 (a) | 36 (b) | 37 (c) | 38 (a) | 39 (d) | 40 (c) |
| 41 (a) | 42 (a) | 43 (a) | 44 (a) | 45 (b) | 46 (b) | 47 (c) | 48 (a) | 49 (a) | 50 (b) |
| 51 (a) | 52 (a) | 53 (d) | 54 (a) | 55 (c) | 56 (a) | 57 (a) | 58 (c) | 59 (b) | 60 (d) |
| 61 (d) | 62 (c) | 63 (d) | 64 (c) | 65 (c) | 66 (d) | 67 (b) | 68 (a) | 69 (a) | 70 (b) |
| 71 (d) | 72 (d) | 73 (c) | 74 (c) | 75 (b) | 76 (d) | 77 (a) | 78 (a) | 79 (b) | 80 (a) |
| 81 (b) | | | | | | | | | |

### > Special Types Questions

| | | | | | | | | | |
|---|---|---|---|---|---|---|---|---|---|
| 82 (c) | 83 (a) | 84 (a) | 85 (b) | 86 (a) | 87 (a) | 88 (d) | 89 (d) | 90 (d) | 91 (c) |
| 92 (d) | 93 (d) | 94 (b) | 95 (a) | 96 (b) | 97 (b) | 98 (c) | 99 (d) | 100 (a) | 101 (c) |
| 102 (a) | 103 (b) | 104 (c) | 105 (c) | 106 (c) | 107 (b) | 108 (d) | 109 (c) | | |

### > NCERT & NCERT Exemplar MCQs

| | | | | | | | | | |
|---|---|---|---|---|---|---|---|---|---|
| 110 (a) | 111 (a) | 112 (b) | 113 (c) | 114 (d) | 115 (b) | 116 (a) | 117 (b) | 118 (a) | 119 (b) |
| 120 (b) | 121 (b) | 122 (c) | 123 (c) | 124 (b) | 125 (c) | 126 (c) | 127 (a) | | |

# Hints & Explanations

**2** (*b*) Amplitude of alternating voltage = Peak voltage $(V_m) = 120\text{ V}$, hence rms value of voltage, i.e.

$$V_{\text{rms}} = \frac{V_m}{\sqrt{2}} = \frac{120}{1.414} = 84.8\text{ V}$$

**3** (*c*) The voltage produced by generator, $V = 240 \sin 120t$

Comparing it with standard equation, $V = V_m \sin \omega t$, we get

$$V_m = 240\text{ V and } \omega = 120\text{ rad/s}$$

Frequency of a generator

i.e. $$\nu = \frac{\omega}{2\pi} = \frac{120 \times 7}{2 \times 22} = 19.09 \simeq 19\text{ Hz}$$

$$\Rightarrow \quad V_{\text{rms}} = \frac{V_m}{\sqrt{2}} = \frac{240}{\sqrt{2}} = 120\sqrt{2} \approx 170\text{ V}$$

**4** (*c*) Given, $V_{\text{rms}} = 220\text{ V}, \nu = 50\text{ Hz}$

As, $$V_{\text{rms}} = \frac{V_m}{\sqrt{2}}$$

$\Rightarrow V_m = V_{\text{rms}}\sqrt{2} = (220\text{ V})(1.414) = 311.1\text{ V}$

Further, $\omega = 2\pi\nu = 2\pi \times 50 = 100\,\pi\text{ rads}^{-1}$

Thus, the equation for the instantaneous voltage is given as

$$V = V_m \sin \omega t = 311.1 \sin (100\pi)\,t$$

**5** (*d*) The current takes $\frac{T}{4}$ seconds to reach the peak value, where $T$ is the time period.

Comparing it with standard equation $I = I_0 \sin \omega t$, we get

$$\omega = \frac{2\pi}{T} = 200\,\pi \Rightarrow T = \frac{1}{100}\text{ s}$$

$\therefore$ Time required to reach the peak value $= T/4 = \frac{1}{400}$ s.

**6** (*d*) Since, alternating voltage, $V = 220 \sin (100\,\pi t)$ is connected with $20\,\Omega$ resistor only, hence equation of alternating current is

$$i = i_m \sin(100\pi t)$$

Peak value to rms value means current becomes $1/\sqrt{2}$ times.

If $t$ be the time taken by current to change from its peak value to rms value, then from equation of current,

$$i = i_m \sin (100\,\pi t)$$

$$\frac{i_m}{\sqrt{2}} = i_m \sin(100\pi t)$$

$$\Rightarrow \quad \frac{1}{\sqrt{2}} = \sin(100\pi t)$$

$$\Rightarrow \quad \sin(\pi/4) = \sin(100\pi t)$$

$$\Rightarrow \quad \frac{\pi}{4} = 100\pi t$$

$$\Rightarrow \quad t = \frac{1}{400}\text{ s} = 2.5 \times 10^{-3}\text{ s}$$

**7** (*c*) Equation of alternating current is given as

$$i = i_1 \cos \omega t + i_2 \sin \omega t \quad \ldots\text{(i)}$$

Let, $$i_1 = A \sin \phi \quad \ldots\text{(ii)}$$

and $$i_2 = A \cos \phi \quad \ldots\text{(iii)}$$

$\therefore$ From Eq. (i), we have

$$i = A \sin \phi \cos \omega t + A \cos \phi \sin \omega t$$
$$= A[\cos \omega t \sin \phi + \sin \omega t \cos \phi]$$
$$i = A \sin (\omega t + \phi) \quad \ldots\text{(iv)}$$

Squaring and adding Eqs. (ii) and (iii), we get

i.e. $$i_1^2 + i_2^2 = A^2 \sin^2 \phi + A^2 \cos^2 \phi$$
$$= A^2 (\sin^2 \phi + \cos^2 \phi) = A^2$$

$$\Rightarrow \quad A = (i_1^2 + i_2^2)^{1/2}$$

$\therefore$ From Eq. (iv), we have

$$i = (i_1^2 + i_2^2)^{1/2} \sin (\omega t + \phi) \quad \ldots\text{(v)}$$

$\therefore$ Comparing Eq. (v) with equation, $i = i_m \sin (\omega t + \phi)$, we get

$$i_m = (i_1^2 + i_2^2)^{1/2}$$

$$\therefore \quad i_{\text{rms}} = \frac{i_m}{\sqrt{2}} = \frac{(i_1^2 + i_2^2)^{1/2}}{\sqrt{2}}$$

**8** (*a*) Given, $V_{\text{rms}} = 220\text{ V}$

$$V_{\text{av}} = \frac{2}{\pi} V_m = \frac{2}{\pi} \times (V_{\text{rms}} \times \sqrt{2}) = \frac{2\sqrt{2}}{\pi} V_{\text{rms}}$$
$$= \frac{2\sqrt{2}}{\pi} \times 220 = 198\text{ V}$$

**9** (*b*) Average voltage,

$$V_{\text{av}} = \frac{\int_0^{\pi/\omega} V dt}{\int_0^{\pi/\omega} dt} = \frac{\int_0^{\pi/\omega} V_m \sin \omega t\, dt}{[t]_0^{\pi/\omega}}$$

$$= \frac{V_m \left( \frac{-\cos \omega t}{\omega} \right)_0^{\pi/\omega}}{\pi/\omega}$$

$$= \frac{-V_m}{\pi} (\cos \pi - \cos 0) = \frac{2V_m}{\pi}$$

**10** (*a*) In the given question, there are identical positive and negative half cycles, so the mean value of current is zero for one cycle, but the rms value is not zero. It is calculated as

$$(i^2)_{\text{mean}} = \frac{\int_0^T i^2 dt}{\int_0^T dt} = \frac{1}{T}\left[\int_0^{T/2} (2)^2 dt + \int_{T/2}^T (-2)^2 dt\right]$$

$$= \frac{4}{T}\left(\int_0^{T/2} dt + \int_{T/2}^T dt\right) = \frac{4}{T}([t]_0^{T/2} + [t]_{T/2}^T)$$

$$= \frac{4}{T}\left(\frac{T}{2} + T - \frac{T}{2}\right) = \frac{4T}{T} = 4$$

$$\therefore \quad i_{\text{rms}} = \sqrt{(i^2)_{\text{mean}}} = \sqrt{4} = 2\text{A}$$

**11** (*b*) In a purely resistive circuit, the alternating current and voltage are in phase. This means that the maxima, zero and minima occur at the same time, for both quantities.

This can be graphically represented as,

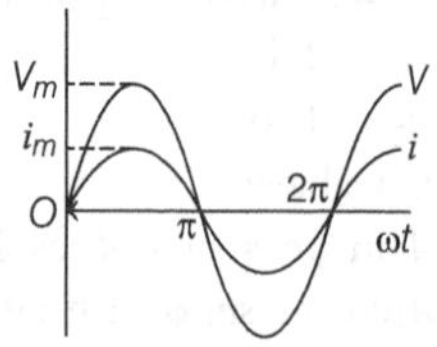


**12** (*d*) In an AC resistive circuit, current and voltage are in phase.

So, $$I = \frac{V}{R}$$

$$\Rightarrow \quad I = \frac{220}{50}\sin(100\pi t) \quad \ldots(i)$$

∴ Time period of one complete cycle of current is

$$T = \frac{2\pi}{\omega} = \frac{2\pi}{100\pi} = \frac{1}{50}\ \text{s}$$

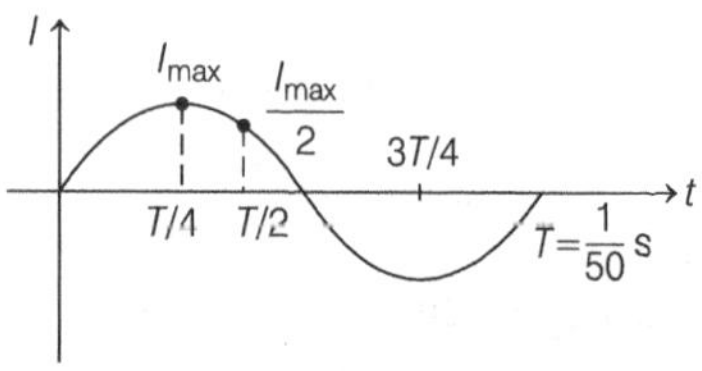


So, current reaches its maximum value at

$$t_1 = \frac{T}{4} = \frac{1}{200}\ \text{s}$$

When current is half of its maximum value, then from Eq. (i), we have

$$I = \frac{I_{max}}{2} = I_{max}\sin(100\pi t_2)$$

$$\Rightarrow \sin(100\pi t_2) = \frac{1}{2} \Rightarrow 100\pi t_2 = \frac{5\pi}{6}$$

So, instantaneous time at which current is half of maximum value is $t_2 = \frac{1}{120}$ s

Hence, time duration in which current reaches half of its maximum value after reaching maximum value is

$$\Delta t = t_2 - t_1 = \frac{1}{120} - \frac{1}{200} = \frac{1}{300}\ \text{s} = 3.3\ \text{ms}$$

**13** (*b*) Using Kirchhoff's rule in given figure,

$$V - L\frac{dI}{dt} = 0$$

where, the second term is the self induced emf in the inductor and $L$ is the self-inductance of coil.

**14** (*b*) Inductive reactance, $X_L = \omega L \Rightarrow X_L \propto \omega$

Hence, inductive reactance increases linearly with angular frequency as shown in graph (b).

**15** (*c*) Given, $L = 25\ \text{mH} = 25 \times 10^{-3}$ H and $\nu = 50$ Hz

The inductive reactance,

$$X_L = 2\pi\nu L = 2 \times 3.14 \times 50 \times 25 \times 10^{-3} = 7.85\ \Omega$$

**16** (*b*) Given, $L = 40\ \text{mH} = 40 \times 10^{-3}$ H, $V = 200$ V and $\nu = 50$ Hz

The rms current in the circuit is

$$i_{rms} = \frac{V}{X_L} = \frac{V}{2\pi\nu L}$$

$$= \frac{200}{2 \times 3.14 \times 50 \times 40 \times 10^{-3}} \approx 16\ \text{A}$$

**17** (*a*) The reactance $X_L$ of the inductance at 200 Hz is $120\ \Omega$.

As, $$X_L = \omega L = 2\pi\nu \times L$$

$$\Rightarrow \quad L = \frac{X_L}{2\pi\nu} = \frac{120\ \Omega}{2\pi \times 200\ \text{s}^{-1}} = \frac{3}{10\pi}\ \text{H}$$

If $X'_L$ denotes the reactance of the same inductance at 60 Hz, then

$$X'_L = \omega' L = 2\pi\nu' L$$

$$\Rightarrow \quad X'_L = (2\pi \times 60)\left(\frac{3}{10\pi}\right) = 36\ \Omega$$

If $i_{rms}$ is the current that flows through the inductance, when connected to 240 V and 60 Hz power line, then

$$i_{rms} = \frac{V_{rms}}{X'_L} = \frac{240\ V}{36\ \Omega} = 6.66\ \text{A}$$

**18** (*d*) Given, $I = 10$ A, $V = 80$ V,

$$R = \frac{V}{I} = \frac{80}{10} = 8\ \Omega \quad \text{and} \quad \omega = 50\ \text{Hz}$$

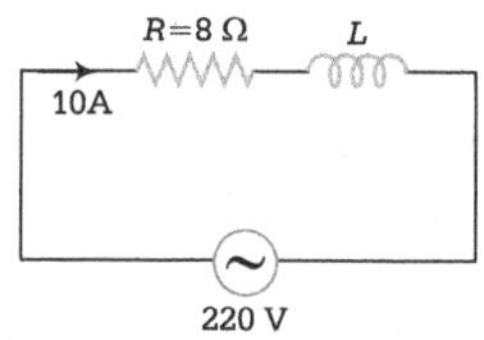


For AC circuit, we have

$$I = \frac{V}{\sqrt{8^2 + X_L^2}} \Rightarrow 10 = \frac{220}{\sqrt{64 + X_L^2}}$$

$$\Rightarrow \quad \sqrt{64 + X_L^2} = 22$$

Squaring on both sides, we get

$$64 + X_L^2 = 484$$

$$\Rightarrow \quad X_L^2 = 484 - 64 = 420$$

$$X_L = \sqrt{420}$$

$$\Rightarrow \quad 2\pi \times \omega L = \sqrt{420}$$

Series inductor on an arc lamp,

$$L = \frac{\sqrt{420}}{(2\pi \times 50)} = 0.065\ \text{H}$$

**20** (*c*) Given, $C = 15\ \mu\text{F} = 15 \times 10^{-6}$ F, $V = 200$ V and $\nu = 50$ Hz

Capacitive reactance,

$$X_C = \frac{1}{2\pi\nu C} = \frac{1}{2\pi\ (50\ \text{Hz})\ (15.0 \times 10^{-6}\ \text{F})} = 212\ \Omega$$

**21** (*a*) Capacitive reactance, $X_C = \frac{1}{\omega C} = \frac{1}{2\pi \nu C}$

$$X_C \propto \frac{1}{\nu} \quad \text{...(i)}$$

Current, $I = \frac{V_{rms}}{X_C} = V_{rms} \cdot \omega C = V_{rms} \cdot 2\pi\nu C$

$$\Rightarrow \quad I \propto \nu \quad \text{...(ii)}$$

From Eqs. (i) and (ii), we conclude that, if the frequency is doubled, the capacitive reactance is halved and the current is doubled.

**22** (*c*) Given, $C = 36\mu F = 36 \times 10^{-6} F, V = 240$ V

and $\nu = 50$ Hz

Capacitive reactance,

$$X_C = \frac{1}{2\pi\nu C} = \frac{1}{2\pi \times 50 \times 36 \times 10^{-6}} = 88\,\Omega$$

The rms current, $i = \frac{V_{rms}}{X_C} = \frac{240 \text{ V}}{88\Omega} = 2.73$ A

The peak current, $i_m = \sqrt{2}\, i = (1.41)\,(2.73 \text{ A}) = 3.85$ A

**23** (*b*) Given, alternating voltage, $V = 200\sqrt{2}\sin(100t)$ and $C = 1\,\mu F = 1 \times 10^{-6}$ F

Comparing the given voltage equation with standard equation $V = V_m \sin \omega t$, we get

$$V_m = 200\sqrt{2} \text{ V and } \omega = 100 \text{ rad/s}$$

Reading of ammeter

$$= i_{rms} = \frac{V_{rms}}{X_C} = \frac{V_m \omega C}{\sqrt{2}} \quad \left(\because X_C = \frac{1}{\omega C}\right)$$

$$= \frac{200\sqrt{2} \times 100 \times (1 \times 10^{-6})}{\sqrt{2}}$$

$$= 2 \times 10^{-2} \text{A} = 20 \text{ mA}$$

**24** (*b*) Alternating voltage source applied to capacitor,

$$V = 200\sin\left(100\pi t - \frac{\pi}{3}\right)$$

$\therefore$ Phase, $\phi_1 = \frac{\pi}{3}, V_m = 200$ V and $\omega = 100\pi$ rad/s

Since, alternating current leads by $\frac{\pi}{2}$ angle from alternating voltage in a purely capacitive circuit, hence phase angle of alternating current is

$$\phi_2 = \frac{\pi}{2} - \phi_1 = \frac{\pi}{2} - \frac{\pi}{3} = \frac{\pi}{6}$$

$\therefore$ Instantaneous value of alternating current through the capacitor is $i = i_m \sin(100\pi t + \phi_2)$

$$= V_m \omega C \sin\left(100\pi t + \frac{\pi}{6}\right) \quad \left[\because i_m = \frac{V_m}{X_C}\right]$$

$$= 200 \times 100\pi \times 2 \times 10^{-6} \sin\left(100\pi t + \frac{\pi}{6}\right)$$

$$[\because C = 2\mu F = 2 \times 10^{-6} F]$$

$$= 0.04\pi \sin\left(100\pi t + \frac{\pi}{6}\right)$$

**25** (*c*) Given, $C = 15\,\mu F = 1.5 \times 10^{-5}$ F, $\nu = 50$ Hz

$$V_{rms} = 220 \text{ V}$$

$\therefore$ Capacitive reactance, $X_C = \frac{1}{\omega C} = \frac{1}{2\pi f C}$

$$= \frac{1}{2 \times 3.14 \times 50 \times 1.5 \times 10^{-5}} = 212.2\,\Omega$$

$$\therefore \quad I_{rms} = \frac{V_{rms}}{X_C} = \frac{220}{212.2} = 1.04 \text{ A}$$

**26** (*d*) For the first circuit, $i = \frac{V}{X_L} = \frac{V}{\omega L} \Rightarrow i \propto \frac{1}{\omega}$

For the second circuit, $i = \frac{V}{X_C} = \frac{V}{1/\omega C} = \omega C V \Rightarrow i \propto \omega$

Hence, if the frequency $\omega$ of the emf of AC is increased, the value of current will decrease in the first circuit and increase in the second circuit.

**27** (*a*) For series *L-C-R* circuit, applied emf and current across resistance are in same phase and across inductor coil have phase difference of $\frac{\pi}{2}$.

The potential difference at capacitor and inductor has phase difference of $\pi$ and across resistance and capacitor is of $\pi/2$.

**28** (*a*) In series *L-C-R* circuit, supply voltage will be

$$V_{rms} = \sqrt{V_R^2 + (V_L - V_C)^2}$$

$$= \sqrt{(40)^2 + (60 - 30)^2}$$

$$= \sqrt{1600 + 900} = \sqrt{2500} = 50 \text{ V}$$

**29** (*a*) Given, source voltage (resultant voltage) $V = 120$ V

Voltage across inductor, $V_L = 50$ V

Voltage across resistor, $V_R = 40$ V

In series *L-C-R* circuit, $V = \sqrt{V_R^2 + (V_L - V_C)^2}$

or $\quad 120 = \sqrt{40^2 + (50 - V_C)^2}$

$$120^2 = 40^2 + (50 - V_C)^2$$

$$\Rightarrow \quad (50 - V_C)^2 = 12800$$

$$\Rightarrow \quad 50 - V_C = 80\sqrt{2}$$

$$\Rightarrow \quad V_C = 10(5 - 8\sqrt{2})$$

**30** (*b*) Given, $V_L = V_C = V_R = 50$ V

Resultant voltage in *C-R* circuit is given as

$$V_{CR} = \sqrt{V_C^2 + V_R^2} = \sqrt{(50)^2 + (50)^2}$$

$$= \sqrt{2500 + 2500} = \sqrt{5000} = 10\sqrt{50} = 50\sqrt{2} \text{ V}$$

**31** (*d*) Given, $V_L = V_C = V_R = 20$ V

Net voltage across *L-C* combination $= V_L - V_C = 0$ V.

**32** (*a*) Given, $R = 3\Omega, X_L = 15\Omega, X_C = 11\Omega$,

and $V_{rms} = 10$ V

$\therefore$ Current through the circuit,

$$i_{rms} = \frac{V_{rms}}{\sqrt{R^2 + (X_L - X_C)^2}}$$

$$= \frac{10}{\sqrt{(3)^2 + (15-11)^2}} = \frac{10}{\sqrt{9+16}} = \frac{10}{5} = 2\text{ A}$$

Since $L, C$ and $R$ are connected in series combination, then potential difference across $R$ is,

$$V_R = i_{rms} R = 2 \times 3 = 6\text{V}$$

Across $L, V_L = i_{rms} X_L = 2 \times 15 = 30$ V

Across $C, V_C = i_{rms} X_C = 2 \times 11 = 22$ V

So, potential difference across series combination of $L$ and $C$,

$$V_{LC} = V_L - V_C = 30 - 22 = 8\text{ V}$$

**33** (*b*) Equation of alternating voltage applied to the circuit,

$$V = 40\sqrt{2} \sin \omega t$$

AC voltmeter reads only rms value of voltage

$$\therefore \quad V_{rms} = \frac{V_m}{\sqrt{2}} = 40\text{ V} \quad [\because V_m = 40\sqrt{2}]$$

Again, according to the question,

$$\sqrt{V_1^2 + V_2^2} = V_{rms} \text{ or } 30^2 + V_2^2 = 40^2$$

$\therefore$ Reading of $V_2 = \sqrt{40^2 - 30^2} = 26.45$ V

**34** (*c*) Given, $V_L = V_C = 300$ V, $R = 100\Omega$ and $V_{rms} = 220$ V

Resultant voltage in $L$-$C$-$R$ circuit,

$$V_{rms} = \sqrt{(V_L - V_C)^2 + V_R^2}$$

$$220 = \sqrt{(300-300)^2 + V_R^2}$$

or $\quad V_R = 220$ V,

Since, $\quad V_L = V_C = 300$ V

$\Rightarrow \quad IX_L = IX_C$

$\Rightarrow X_L = X_C$, hence given circuit is $L$-$C$-$R$ series resonance circuit.

$$\therefore \quad I_{rms} = \frac{V_R}{R} = \frac{220\text{ V}}{100\,\Omega} = 2.2\text{ A}$$

**35** (*a*) Given, $V_R = 12$ V, $V_C = 5$ V

Let the applied voltage be $V$ volt in given $R$-$C$ circuit, then

$$V_{rms} = \sqrt{V_R^2 + V_C^2} = \sqrt{(12)^2 + (5)^2}$$

$$= \sqrt{144 + 25} = \sqrt{169} = 13\text{ V}$$

**36** (*b*) Given, $V_m = 300$ V, $\omega = 400 \text{ rads}^{-1}$

$R = 3\Omega, L = 20\text{ mH} = 20 \times 10^{-3}$ H

and $C = 625\mu\text{F} = 625 \times 10^{-6}$ F.

The impedance of the circuit is,

$$Z = \sqrt{R^2 + (X_L - X_C)^2}$$

Here, $\quad X_L = \omega L = 400 \times 20 \times 10^{-3} = 8\Omega$

and $\quad X_C = \frac{1}{\omega C} = \frac{1}{400 \times 625 \times 10^{-6}} = 4\Omega$

$$\Rightarrow \quad Z = \sqrt{(3)^2 + (8-4)^2} = 5\Omega$$

$\therefore$ Peak current, $i_m = \frac{V_m}{Z} = \frac{300}{5} = 60$ A

**37** (*c*) Given, $R = 200\,\Omega, C = 15\mu\text{F} = 15 \times 10^{-6}$ F

$V_{rms} = 220$ V and $\nu = 50$ Hz

Impedance of the circuit,

$$Z = \sqrt{R^2 + X_C^2} = \sqrt{R^2 + (2\pi\nu C)^{-2}}$$

$$\left[\because X_C = \frac{1}{2\pi\nu C}\right]$$

$$= \sqrt{(200)^2 + (2 \times 3.14 \times 50 \times 15 \times 10^{-6})^{-2}}$$

$$= \sqrt{(200)^2 + (212)^2} = 291.5\Omega$$

Therefore, the current in the circuit is

$$i_{rms} = V_{rms}/Z = \frac{220\text{ V}}{291.5\Omega} = 0.755\text{ A}$$

**38** (*a*) When the circuit is connected to an AC source,

Voltage, $V = 12$ V and current, $I = 0.2$ A .

Impedance, $Z = \frac{V}{I} = \frac{12}{0.2} = 60\,\Omega$

When it is connected to DC source,

Voltage, $V = 12$ V and Current, $I = 0.4$ A.

Resistance, $R = \frac{V}{I} = \frac{12}{0.4} = 30\Omega$

As in case of DC supply, the capacitor act as an open circuit and no current flows through the circuit. So, the given circuit will not have capacitor in series combination. Therefore, the circuit should be a series $L$-$R$ circuit.

**39** (*d*) The AC current flowing in a series $L$-$C$-$R$ circuit is

$$i = i_m \sin(\omega t + \phi)$$

where, $\phi$ = phase difference between the voltage and current in the circuit.

When $X_C > X_L \Rightarrow V_C > V_L$

So, the resultant voltage is $V = \sqrt{V_R^2 + (V_C - V_L)^2}$

and phase angle, $\phi = \tan^{-1}\left(\frac{X_C - X_L}{R}\right)$

As, $X_C > X_L$, $\phi$ is positive and the circuit is predominantly capacitive. Consequently, the current in the circuit leads the source voltage. The phasor diagram and corresponding variation of $V$ and $i$ with $\omega t$ are shown below.

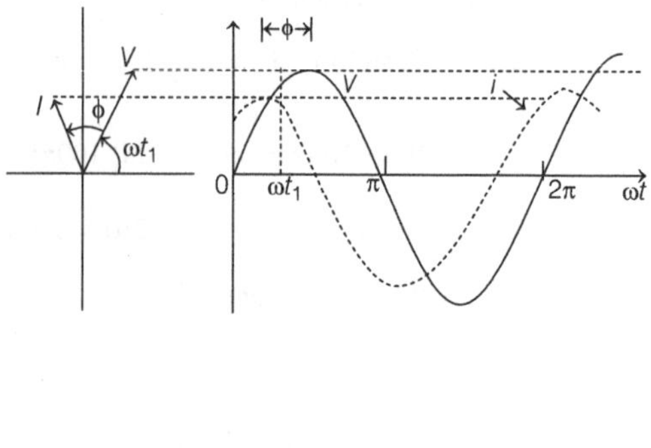

**40** (*c*) $\because \tan\phi = \dfrac{\omega L - \dfrac{1}{\omega C}}{R}$

where, $\phi$ being the angle by which the current leads the voltage.

Given, $\phi = 45°$

$$\therefore \quad \tan 45° = \frac{\omega L - \dfrac{1}{\omega C}}{R} \Rightarrow 1 = \frac{\omega L - \dfrac{1}{\omega C}}{R}$$

$$R = \omega L - \frac{1}{\omega C} \Rightarrow \omega C = \frac{1}{\omega L - R}$$

$$\Rightarrow \quad C = \frac{1}{\omega(\omega L - R)} = \frac{1}{2\pi f(2\pi f L - R)}$$

**41** (*a*) Given, $R = 300\,\Omega$, $L = \left(\dfrac{1}{\pi}\right)$H, $V = 20$ V and $\nu = 200$ Hz

If $\phi$ is the phase difference between $V$ and $I$ for $R$-$L$ circuit, then

$$\tan\phi = \frac{\omega L}{R} = \frac{2\pi\nu L}{R} = \frac{2\pi \times 200}{300} \times \frac{1}{\pi} = \frac{4}{3}$$

$$\Rightarrow \quad \phi = \tan^{-1}(4/3)$$

**42** (*a*) We know that, for series $R$-$C$ circuit

$$V^2 = V_C^2 + V_R^2$$

Given, $V = 10$ V and $V_C = 8$ V

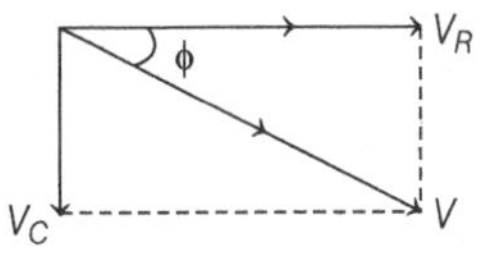


$$100 = 64 + V_R^2$$

$$\Rightarrow \quad V_R^2 = 36 \Rightarrow V_R = \sqrt{36} = 6\text{ V}$$

Also, $\tan\phi = \dfrac{V_C}{V_R} \Rightarrow \tan\phi = \dfrac{8}{6}$

$$\tan\phi = \frac{4}{3}$$

$$\therefore \quad \phi = \tan^{-1}\left(\frac{4}{3}\right)$$

**43** (*a*) At angular frequency $\omega$, the current in $R$-$C$ circuit is given by

$$I_{\text{rms}} = \frac{V_{\text{rms}}}{\sqrt{R^2 + \left(\dfrac{1}{\omega C}\right)^2}} \quad \ldots\text{(i)}$$

When frequency of source is changed to $\dfrac{\omega}{3}$, then

$$\frac{I_{\text{rms}}}{2} = \frac{V_{\text{rms}}}{\sqrt{R^2 + \left[\dfrac{1}{\omega C/3}\right]^2}} = \frac{V_{\text{rms}}}{\sqrt{R^2 + \dfrac{9}{\omega^2 C^2}}} \quad \ldots\text{(ii)}$$

Dividing Eq. (i) by Eq. (ii), we get

$$\frac{I_{\text{rms}}}{\dfrac{I_{\text{rms}}}{2}} = \frac{V_{\text{rms}}}{\sqrt{R^2 + \dfrac{1}{\omega^2 C^2}}} \times \frac{\sqrt{R^2 + 9/\omega^2 C^2}}{V_{\text{rms}}}$$

$$\Rightarrow 4\left(R^2 + \frac{1}{\omega^2 C^2}\right) = R^2 + \frac{9}{\omega^2 C^2}$$

$$3R^2 = \frac{5}{\omega^2 C^2} \Rightarrow \frac{1}{\omega CR} = \sqrt{\frac{3}{5}}$$

$$\Rightarrow \quad \frac{X_C}{R} = \sqrt{\frac{3}{5}} \qquad \left[\because X_C = \frac{1}{\omega C}\right]$$

**45** (*b*) Given, $L = 1\text{ mH} = 1 \times 10^{-3}$ H and $C = 1\text{ nF} = 1 \times 10^{-9}$ F

Resonant frequency of $R$-$L$-$C$ series circuit,

$$\omega_0 = \frac{1}{\sqrt{LC}} = \frac{1}{\sqrt{1 \times 10^{-3} \times 1 \times 10^{-9}}}$$

$$= 1 \times 10^6 \text{ rad/s}$$

**46** (*b*) Given, $C = 80\,\mu\text{F} = 80 \times 10^{-6}$F, $L = 5$ H and $R = 40\,\Omega$

At resonance, source frequency, $\nu = \dfrac{1}{2\pi\sqrt{LC}}$

$$\therefore \quad \nu = \frac{1}{2 \times \pi \times \sqrt{5 \times 80 \times 10^{-6}}}$$

$$= \frac{1}{2 \times \pi\sqrt{(400 \times 10^{-6})}} = \frac{1}{2 \times \pi \times 2 \times 10^{-2}}$$

$$= \frac{100}{\pi \times 4} = \frac{25}{\pi}\text{ Hz}$$

**47** (*c*) At resonance $X_L = X_C \Rightarrow \omega L = \dfrac{1}{\omega C} \Rightarrow \omega = \dfrac{1}{\sqrt{LC}}$

Since, resonant frequency remains unchanged

$$\sqrt{LC} = \text{constant}$$

$$\Rightarrow \quad L_1 C_1 = L_2 C_2 \Rightarrow LC = L_2(2C) \quad \text{(given)}$$

$$\Rightarrow \quad L_2 = L/2.$$

**48** (*a*) $\because$ Angular frequency at resonance, $\omega = \dfrac{1}{\sqrt{LC}}$ ...(i)

According to question, when inductor's inductance is made 2 times and capacitance is 4 times, then

$$\omega' = \frac{1}{\sqrt{2L \times 4C}} = \left(\frac{1}{2\sqrt{2}}\right)\frac{1}{\sqrt{LC}}$$

$$= \frac{\omega}{2\sqrt{2}} \quad \text{[from Eq. (i)]}$$

**49** (*a*) Given, $L = 8\text{ mH} = 8 \times 10^{-3}$ H, $C = 20\,\mu\text{F} = 20 \times 10^{-6}$ F, $R = 44\,\Omega$ and $V_{\text{rms}} = 220$ V

Angular resonant frequency of series $L$-$C$-$R$ circuit,

$$\omega_0 = \frac{1}{\sqrt{LC}} = \frac{1}{\sqrt{8 \times 10^{-3} \times 20 \times 10^{-6}}} = 2500\text{ rads}^{-1}$$

$$\text{Resonant current} = V_m/R = \frac{V_{\text{rms}}\sqrt{2}}{R} = \frac{220\sqrt{2}}{44} = 5\sqrt{2}\text{ A}$$

**50** (*b*) Given, $V_R = 100$ V, $R = 1\text{ k}\Omega = 1000\,\Omega$, $C = 2\mu\text{F} = 2 \times 10^{-6}$F and $\omega = 200$ rad/s

At resonance, $X_L = X_C$ or $\omega L = \dfrac{1}{\omega C}$

or $L=\frac{1}{\omega^2 C}=\frac{1}{(200)^2\times 2\times 10^{-6}}=12.5\text{ H}$

As, $I_0=\frac{V_R}{R}=\frac{100\text{ V}}{1000\,\Omega}=0.1\text{ A}$

$\therefore$ $V_L=I_0X_L=I_0\omega L$
$=(0.1\text{ A})(200\text{ rads}^{-1})(12.5\text{ H})$
$=250\text{ V}$

**51** (*a*) Given, $R=10\,\Omega$, $X_C=60\,\Omega$

$V=300$ V and $\nu=100$ Hz

Angular velocity, $\omega_0=2\pi\nu=2\pi\times 100$

$\omega_0=2\times 3\times 100=600\text{ rads}^{-1}$ $(\because \pi=3)$

Further $\omega_0=\frac{1}{\sqrt{LC}}$ ...(i)

Also, $X_C=\frac{1}{C\omega_0}=60\,\Omega$

$\Rightarrow$ $C=\frac{1}{\omega_0\times 60}=\frac{1}{600\times 60}$

$\Rightarrow$ $C=\frac{1}{36\times 10^3}\text{ F}$

On putting values in Eq. (i), we get

$$600=\frac{1}{\sqrt{L\left(\frac{1}{36\times 10^3}\right)}}$$

$\Rightarrow$ $36\times 10^4=\frac{36\times 10^3}{L}$

$\Rightarrow$ $L=\frac{36\times 10^3}{36\times 10^4}=\frac{1}{10}=0.1\text{ H}$

**52** (*a*) In the given circuit diagram, both inductors of inductance $L$ are parallel to each other, hence

$$L_{eq}=\frac{L_1L_2}{L_1+L_2}=\frac{L\cdot L}{L+L}=\frac{L}{2}$$

Similarly, $C_{eq}=C_1+C_2=C+C=2C$

$\therefore$ Resonant frequency,

$$\omega_r=\frac{1}{\sqrt{L_{eq}\cdot C_{eq}}}=\sqrt{\frac{1}{\frac{L}{2}.2C}}=\sqrt{\frac{1}{LC}}=\frac{1}{\sqrt{LC}}$$

**53** (*d*) As the resonance peak is sharper for circuit $D$, so it has maximum $Q$-factor or quality factor.

**54** (*a*) Sharpness of resonance of a resonant $L$-$C$-$R$ circuit is determined by the ratio of resonant frequency with the selectivity of circuit. This ratio is also called "Quality Factor" or $Q$-factor.

$$Q\text{-factor}=\frac{\omega_0}{2\Delta\omega}=\frac{\omega_0 L}{R}=\frac{1}{\omega_0 CR}$$

**55** (*c*) Given, $\varepsilon=200\sin 314t$ and $I=\sin(314t+\pi/3)$

$\therefore$ $\phi=\frac{\pi}{3}, V_m=200\text{ V}, I_m=1\text{ A}$

$V_{rms}=\frac{V_m}{\sqrt{2}}=\frac{200}{\sqrt{2}}$

$i_{rms}=\frac{I_m}{\sqrt{2}}=\frac{1}{\sqrt{2}}$

$\therefore$ Average power consumed is,

$$P=V_{rms}\,i_{rms}\cos\phi=\frac{200}{\sqrt{2}}\frac{1}{\sqrt{2}}\cos\pi/3$$
$$=\frac{200}{2}\times\frac{1}{2}=50\text{ W}$$

**56** (*a*) Given, $V=5\cos\omega t=5\sin\left(\omega t+\frac{\pi}{2}\right)$ and $i=2\sin\omega t$

Power dissipated in the instrument is,

$P=V_{rms}\times i_{rms}\times\cos\phi=0$

(since, $\phi=\frac{\pi}{2}$, therefore $\cos\phi=\cos\frac{\pi}{2}=0$)

**57** (*a*) For house hold supplies, the line that draws supply power has zero average current over a cycle.

The line is having some resistance, so power factor

$\cos\phi=R/Z\neq 0$

so $\phi\neq\pi/2\Rightarrow\phi<\pi/2$

i.e. phase between voltage and current lies between 0 and $\pi/2$.

**58** (*c*) Given, $\phi=\frac{\pi}{3}$

Here, phase difference, $\tan\phi=\frac{X_L-X_C}{R}$

When $L$ is removed,

$\tan\pi/3=\sqrt{3}=X_C/R\Rightarrow X_C=\sqrt{3}R$

When $C$ is removed ,

$\tan\frac{\pi}{3}=\sqrt{3}=\frac{X_L}{R}\Rightarrow X_L=\sqrt{3}R$

Hence in resonant circuit

$\tan\phi=\frac{\sqrt{3}R-\sqrt{3}R}{R}=0\Rightarrow\phi=0$

$\therefore$ Power factor $=\cos\phi=\cos 0=1$

**59** (*b*) Power consumed by lamp, $P=(1/2)V_m i_m\cos\phi$

$\Rightarrow$ $P=P_{peak}\cdot\cos\phi$

Given, $P=50\%$ of $P_{peak}=\frac{1}{2}P_{peak}$

$\Rightarrow$ $\frac{1}{2}(P_{peak})=P_{peak}\cos\phi$

$\Rightarrow$ $\cos\phi=\frac{1}{2}\Rightarrow\phi=\frac{\pi}{3}$

$\therefore$Phase different between the applied voltage and the circuit current is $\pi/3$.

**60** (*d*) The amplitude of the current in the series $L$-$C$-$R$ circuit is given by

$$i_m=\frac{V_m}{Z}=\frac{V_m}{\sqrt{R^2+\left(\omega L-\frac{1}{\omega C}\right)^2}}\quad\text{...(i)}$$

Since, $i_m=\frac{I_{max}}{\sqrt{2}}\Rightarrow I_{max}=i_m\sqrt{2}$

$$\Rightarrow \quad \frac{V_m}{R} = i_m\sqrt{2}$$

$$\Rightarrow \quad i_m = \frac{V_m}{\sqrt{2}R} \quad \ldots(ii)$$

From Eqs, (i) and (ii), we have

$$\Rightarrow \quad \frac{1}{2}\frac{V_m^2}{R^2} = \frac{V_m^2}{R^2 + (\omega L - 1/\omega C)^2}$$

$$\Rightarrow \quad 2R^2 = R^2 + \left(\omega L - \frac{1}{\omega C}\right)^2$$

$$\Rightarrow \quad R^2 = (\omega L - 1/\omega C)^2$$

$$\Rightarrow \quad R = (\omega L - 1/\omega C) \quad \ldots(iii)$$

If $\phi$ is the phase difference between $V$ and $I$, then

$$\Rightarrow \quad \tan\phi = \frac{\omega L - \frac{1}{\omega C}}{R} = \frac{\omega L - \frac{1}{\omega C}}{\omega L - \frac{1}{\omega C}} = 1 \quad \text{[from Eq. (iii)]}$$

$$\Rightarrow \quad \tan\phi = 1 = \tan 45^\circ$$

$$\phi = 45^\circ$$

Power dissipated, $P = i_m^2 Z\cos\phi$

$$= \frac{I_{max}^2 \times \sqrt{R^2 + \left(\omega L - \frac{1}{\omega C}\right)^2}}{2} \times \cos 45^\circ$$

$$= \frac{I_{max}^2 \times \sqrt{R^2 + R^2}}{2} \times \frac{1}{\sqrt{2}}$$

$$= \frac{I_{max}^2 \times \sqrt{2}R}{2} \times \frac{1}{\sqrt{2}} = \frac{I_{max}^2 R}{2} = \frac{P_{max}}{2}$$

Thus, power dissipated by the circuit becomes half.

**61** (*d*) The given series *R-L-C* circuit is shown in the figure below.

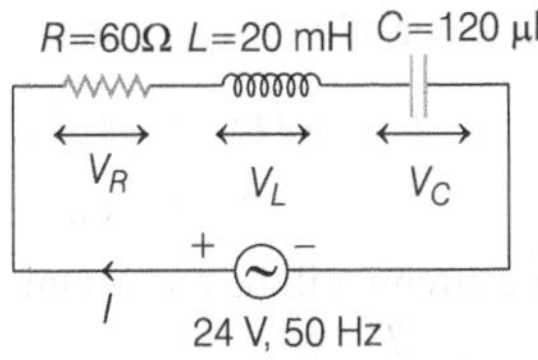


Here, $V_R$ = potential across resistance ($R$)

$V_L$ = potential across inductor ($L$) and

$V_C$ = potential across capacitor ($C$).

Impedance of this series circuit is,

$$Z = \sqrt{R^2 + (X_L - X_C)^2} \quad \ldots(i)$$

$$\because \quad X_L = \omega L = (2\pi f)(L) = 2\pi \times 50 \times 20 \times 10^{-3}\,\Omega$$

$$X_L = 6.28\,\Omega \quad \ldots(ii)$$

and $$X_C = \frac{1}{\omega C} = \frac{1}{2\pi f C}$$

$$= \frac{1}{2\pi \times 50 \times 120 \times 10^{-6}} = \frac{250}{3\pi}\Omega \quad \ldots(iii)$$

and $$X_L - X_C = \left(6.28 - \frac{250}{3\pi}\right) = -20.23\,\Omega \quad \ldots(iv)$$

RMS value of current in circuit is,

$$I_{rms} = \frac{V_{rms}}{Z} = \frac{24}{\sqrt{R^2 + (X_L - X_C)^2}}$$

$$I_{rms} = \frac{24}{\sqrt{60^2 + (-20.23)^2}} = \frac{24}{63.18}$$

$$I_{rms} = 0.379\text{ A}$$

Therefore, energy dissipated is

$$= I_{rms}^2 \times R \times t$$

$$E = (0.379)^2 \times 60 \times 60$$

or $$= 517.10 = 5.17 \times 10^2\text{ J}$$

**62** (*c*) Given, $P = 600\text{ W}, V = 220\text{ V}$

Also, AC voltage, $V = 169\sin(314t + 60^\circ)$

The general equation for the AC voltage is

$$V = V_m \sin(\omega t + \theta)$$

Comparing it with the given equation, we find that

$V_m = 169\text{ V}, \omega = 314\text{ rad s}^{-1}, \theta = 60^\circ$

Let $V_{rms}$ and $i_{rms}$ represent the rms value of AC voltage and current, respectively. Clearly,

$$V_{rms} = \frac{V_m}{\sqrt{2}} = \frac{169}{\sqrt{2}}\text{ V} = 119.5\text{ V}$$

Power of electric lamp, $P = \dfrac{V^2}{R}$

$$\Rightarrow \quad R = \frac{V^2}{P} = \frac{(220)^2}{600}$$

$$i_{rms} = \frac{V_{rms}}{R} = \frac{119.5 \times 600}{(220)^2} \quad (\because V_{rms} = i_{rms}R)$$

$$= 1.48\text{ A}$$

**63** (*d*) As, $Z = \sqrt{R^2 + X_L^2} = \sqrt{R^2 + (2\pi f L)^2}$

Also, $$i = \frac{V}{Z} \text{ and } P = i^2 R$$

When iron rod is inserted, then inductance $L$ of the coil increases which increases impedance $Z$ and consequently, current $i$ and power $P$ of the circuit decreases. So, brightness of bulb decreases.

**64** (*c*) Power factor of AC circuit is given by $\cos\phi = \dfrac{R}{Z}$ ...(i)

where, $R$ is resistance employed and $Z$ the impedance of the circuit.

$$Z = \sqrt{R^2 + (X_L - X_C)^2} \quad \ldots(ii)$$

From Eqs. (i) and (ii), we get

$$\cos\phi = \frac{R}{\sqrt{R^2 + (X_L - X_C)^2}}$$

Given, $R = 8\,\Omega, X_L = 31\,\Omega, X_C = 25\,\Omega$

$$\therefore \quad \cos\phi = \frac{8}{\sqrt{(8)^2 + (31-25)^2}} = \frac{8}{\sqrt{64+36}} = \frac{8}{10}$$

Hence, $\cos\phi = 0.80$

**65** (*c*) Given, inductance, $L = 20\text{ mH} = 20\times 10^{-3}\text{ H}$

Capacitance, $C = 100\mu\text{F} = 100\times 10^{-6}\text{ F} = 10^{-4}\text{F}$

Resistance, $R = 50\,\Omega$

$$\text{emf, } V = 10\sin 314t \quad \ldots\text{(i)}$$

The general equation of emf is given as

$$V = V_m \sin \omega t \quad \ldots\text{(ii)}$$

Comparing Eqs. (i) and (ii), we get

$$V_m = 10\text{V}, \ \omega = 314\text{ rad s}^{-1}$$

The power loss associated with the given AC circuit is given as

$$P = V_{\text{rms}}\, I_{\text{rms}} \cos\phi = V_{\text{rms}}\left(\frac{V_{\text{rms}}}{Z}\right)\left(\frac{R}{Z}\right)$$

$$= \left(\frac{V_{\text{rms}}}{Z}\right)^2 R = \left(\frac{V_m}{\sqrt{2}\cdot Z}\right)^2 R \quad \ldots\text{(iii)}$$

Impedance, $Z = \sqrt{R^2 + (X_L - X_C)^2}$

$$= \sqrt{R^2 + \left(\omega L - \frac{1}{\omega C}\right)^2} \quad \text{(iv)}$$

Substituting the given values in the above Eq. (iv), we get

$$Z = \sqrt{(50)^2 + \left[(314\times 20\times 10^{-3}) - \frac{1}{314\times 10^{-4}}\right]^2}$$

$$= \sqrt{2500 + [6280\times 10^{-3} - 0.00318\times 10^4]^2}$$

$$= \sqrt{2500 + (25.56)^2} = 56.15\,\Omega \simeq 56\,\Omega$$

Now, substituting this values in Eq. (iii), we get

$$P = \left(\frac{10}{\sqrt{2}\times 56}\right)^2 \times 50 = \frac{100}{2\times 3136}\times 50 = 0.79\text{ W}$$

Thus, power loss in the circuit is 0.79 W.

**66** (*d*) Given, $R = 3\,\Omega$, $L = 25\text{ mH} = 25\times 10^{-3}$ H,

$C = 400\,\mu\text{F} = 400\times 10^{-6}$ F and $V = 283$ V

A series resonance circuit admits maximum current, as

$$P = i^2 R$$

So, power dissipated is maximum at resonance.

Therefore, frequency of the source at which maximum power is dissipated in the circuit is

$$\nu = \frac{1}{2\pi\sqrt{LC}} = \frac{1}{2\times 3.14\sqrt{25\times 10^{-3}\times 400\times 10^{-6}}}$$

$$= \frac{1}{2\times 3.14\sqrt{10^{-5}}} = 50.3\text{ Hz}$$

**67** (*b*) Given, $C = 40\,\mu\text{F} = 40\times 10^{-6}\text{F}$,

and $L = 16\text{ mH} = 16\times 10^{-3}$ H

Angular frequency of oscillating circuit,

$$\omega = \frac{1}{\sqrt{LC}} = \frac{1}{\sqrt{(16\times 10^{-3})(40\times 10^{-6})}}$$

$$= \frac{10^4}{8} = 1.25\times 10^3\text{ s}^{-1}$$

**68** (*a*)

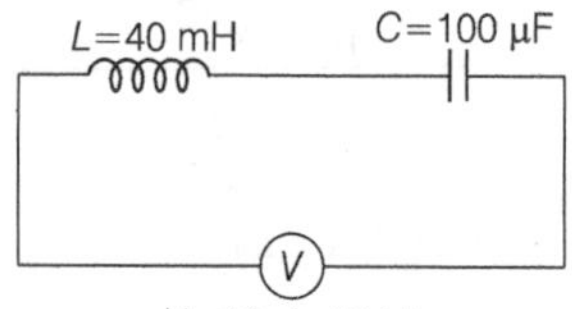


V=10 sin 314 t

Impedance of given ciruit,

$$Z = \sqrt{(X_C - X_L)^2} = X_C - X_L$$

$$= \frac{1}{\omega C} - \omega L = \frac{1}{314\times 100\times 10^{-6}} - 314\times 40\times 10^{-3}$$

$$= 19.28\,\Omega$$

As $X_C > X_L$, circuit is captive, hence current in circuit leads emf by $\frac{\pi}{2}$ radians.

Current in circuit is given by

$$I = I_{\max}\sin\left(\omega t + \frac{\pi}{2}\right) = \frac{V_{\max}}{Z}\cos\omega t = \frac{10}{19.28}\times\cos 314t$$

or $I = 0.52\cos 314t$

**69** (*a*) For *L*-*C* oscillations,

Energy stored in inductor = Energy stored in capacitor

$$\frac{1}{2}Li_m^2 = \frac{1}{2}CV_m^2$$

Given, $V_m = 25$ V, $C = 10\mu\text{F} = 10^{-5}$ F

and $L = 100\text{ mH} = 10^{-1}$ H

or $$i_m = V_m\sqrt{\frac{C}{L}} = 25\sqrt{\frac{10^{-5}}{10^{-1}}}$$

$$= 25\times 10^{-2}\text{ A} = 0.25\text{ A}$$

**70** (*b*) Given, $L = 25\text{ mH} = 25\times 10^{-3}$ H

$$C = 10\,\mu\text{F} = 10^{-5}\text{ F}$$

If *T* be the time period in *L*-*C* oscillation, then

$$T = 2\pi\sqrt{LC} = 2\pi\sqrt{25\times 10^{-3}\times 10^{-5}}$$

$$= \pi\times 10^{-3}\text{ s} = \pi\text{ m-s}$$

Current in the circuit will be maximum, when

$$t = \frac{T}{4} = \frac{\pi}{4}\text{ m-s}$$

**71** (*d*) Given, $L = 1.5\text{ mH} = 1.5\times 10^{-3}$ H

$$E = 30\,\mu\text{J} = 3\times 10^{-5}\text{J}$$

Maximum energy stored in the inductor, $E = \frac{1}{2}LI_m^2$

where, $I_m$ = peak current.

$$\Rightarrow \quad I_m = \sqrt{\frac{2E}{L}} = \sqrt{\frac{2\times 3\times 10^{-5}}{1.5\times 10^{-3}}} = 0.2\text{ A}$$

$$\therefore \quad I_{\text{rms}} = \frac{I_m}{\sqrt{2}} = \frac{0.2}{\sqrt{2}} = \sqrt{2}\times 10^{-1}\text{ A}$$

**72** (*d*) A battery produces direct current (DC). But the transformer does not work on DC supply. Hence, no induction occurs in it and the output across secondary coil is zero.

**73** (*c*) Given, $\varepsilon_p = 2300$ V, $N_p = 4000$ and $\varepsilon_s = 230$ V

Let $N_s$ be the required number of turns in the secondary

As, $\frac{\varepsilon_s}{\varepsilon_p} = \frac{N_s}{N_p}$

$$\Rightarrow N_s = N_p\left(\frac{\varepsilon_s}{\varepsilon_p}\right) = 4000\left(\frac{230\text{ V}}{2300\text{ V}}\right) = 400$$

**74** (*c*) Given, $\varepsilon_p = 220$ V, $\varepsilon_s = 22$ V and $Z = 220\,\Omega$

If $i_s$ is the current through the secondary, then

$$i_s = \frac{\varepsilon_s}{Z} = \frac{22\text{ V}}{220\,\Omega} = 0.1\text{ A}$$

We know that, $\varepsilon_p i_p = \varepsilon_s i_s$

where, $i_p$ is the current drawn by the primary coil.

Thus, $220 \times i_p = 22 \times 0.1$ A

$\Rightarrow \quad i_p = 0.01$ A

**75** (*b*) Given, $V_p = 110$ V, $k = \frac{N_s}{N_p} = 10$

and $R_s = 550\,\Omega$

$$V_s = V_p \times \frac{N_s}{N_p} = 110 \times 10 = 1100\text{ V}$$

Current through secondary coil,

i.e. $\quad i_s = \frac{V_s}{R_s} = \frac{1100}{550} = 2\text{A}$

**76** (*d*) In an ideal transformer, there is no energy loss and flux is completely confined within the magnetic core i.e. perfectly coupled. So, all the input power appears across the output.

i.e., $\quad \frac{P_{\text{out}}}{P_{\text{in}}} = 1$ or $P_{\text{out}} : P_{\text{in}} = 1:1$

**77** (*a*) Given, number of turns in primary, $N_1 = 300$

Number of turns in secondary, $N_2 = 150$

Output power, $P_2 = 2.2$ kW $= 2.2 \times 10^3$ W

Current in secondary coil, $I_2 = 10$ A

Output power, $P_2 = I_2 V_2$

$$\Rightarrow \quad V_2 = \frac{P_2}{I_2} = \frac{2.2 \times 10^3}{10} = 220\text{V} \qquad \ldots\text{(i)}$$

We know that,

$$\frac{N_1}{N_2} = \frac{\text{Input voltage}}{\text{Output voltage}} = \frac{V_1}{V_2} \Rightarrow V_1 = \left(\frac{N_1}{N_2}\right)V_2$$

$$\Rightarrow \quad V_1 = \left(\frac{300}{150}\right) \times (220\text{ V}) \quad \text{[using Eq. (i)]}$$

$$V_1 = 440\text{ V} \qquad \ldots\text{(ii)}$$

Again, $\quad \frac{V_1}{V_2} = \frac{I_2}{I_1}$

$$\Rightarrow \quad I_1 = \left(\frac{V_2}{V_1}\right) I_2 = \frac{220}{440} \times 10$$

[using Eqs. (i) and (ii)]

$\Rightarrow \quad I_1 = 5$A

**78** (*a*) Given, turn ratio of a transformer, $\frac{N_1}{N_2} = \frac{50}{1}$

$\Rightarrow \quad N_1 = 50N_2$

Since, $\quad \frac{N_1}{N_2} = \frac{V_1}{V_2} \Rightarrow 50 = \frac{120}{V_2} \quad [V_1 = 120\text{ V (given)}]$

$\Rightarrow \quad V_2 = \frac{12}{5}$ V

Output power at secondary coil,

$$P_s = \frac{V_2^2}{R_2} = \frac{\left(\frac{12}{5}\right)^2}{1} = \frac{144}{25} = 5.76\text{ W} \quad [R_2 = 1\Omega\text{ (given)}]$$

**79** (*b*) Given, $P_o = 140$ W, $V_0 = 24$ V, $V_i = 240$ V and $i_L = 0.7$ A

Input power of transformer,

$$P_i = V_i \times i_L = 240 \times 0.7 = 168\text{ W}$$

Efficiency of transformer,

i.e. $\quad \eta = \frac{P_o}{P_i} \times 100 = \frac{140}{168} \times 100 \approx 80\%$

**80** (*a*) For a transformer, there are two circuits which have $N_p$ and $N_s$ (number of coil turns), $I_p$ and $I_S$ (currents) respectively as shown below.

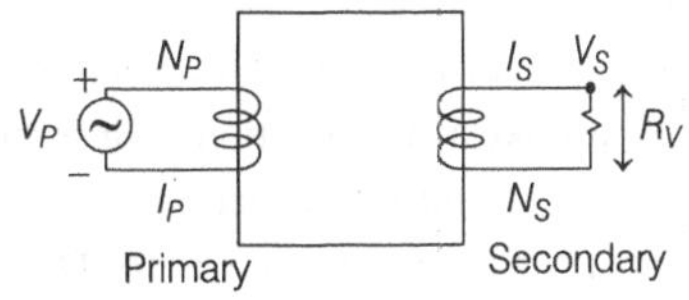


Here, input voltage, $V_p = 2300$ V

Number of turns in primary coil, $N_P = 4000$

Output voltage, $V_S = 230$ volt

Output power, $P_S = V_S \cdot I_S$

Input power, $P_P = V_P I_P$

$\therefore$ The efficiency of the transformer is

$$\eta = \frac{\text{Output (secondary) power}}{\text{Input (primary) power}}$$

$$\Rightarrow \quad \eta = \frac{V_S \cdot I_S}{V_P \cdot I_P} \times 100 \Rightarrow \eta = \frac{(230)(I_S)}{(2300)(5)} \times 100$$

$$90 = \frac{230\, I_S}{(2300) \times 5} \times 100 \Rightarrow I_S = 45\text{ A}$$

**81** (*b*) Given, initial power, $P_1 = 3$ kW $= 3000$ W, $V_1 = 200$ V and $i_2 = 6$ A

As efficiency is 90%, then final power,

$$P_2 = P_1 \times \frac{\eta}{100} = 3000 \times \frac{90}{100} = 2700\text{ W}$$

As, $\quad P_1 = V_1 i_1 = 3000$ W

$$\Rightarrow \quad i_1 = \frac{P_1}{V_1} = \frac{3000}{200} = 15\text{ A}$$

Also, $\quad P_2 = V_2 i_2 = 2700$ W

$$\Rightarrow \quad V_2 = \frac{P_2}{i_2} = \frac{2700}{6} = 450\text{ V}$$

**82** (c) The supply has voltage of $V = V_m \sin \omega t$ and the current through an inductor is

$$i = i_m \sin\left(\omega t - \frac{\pi}{2}\right)$$

Thus, current lags behind the voltage or emf by $\pi/2$, when AC flows through an inductor.

The inductive reactance is directly proportional to the inductance and the frequency of the current as $X_L = \omega L = 2\pi\nu L$. So, it increases as the frequency of AC source increases.

Therefore, Assertion is correct but Reason is incorrect.

**83** (a) When the capacitor is connected to an AC source, it limits or regulates the current, but does not completely prevent the flow of charge.

It is because, the capacitor is alternately charged and discharged as the current reverses each half cycle.

Therefore, Assertion and Reason are correct and Reason is the correct explanation of Assertion.

**84** (a) Capacitive reactance is inversely proportional to frequency as $X_C = \frac{1}{\omega C} = \frac{1}{2\pi f C}$

As DC does not have any frequency, i.e. $f = 0$

hence, $X_C = \infty$.

So, no DC current flows through capacitor.

But AC varies like a sine function with some frequency.

So, it passes easily through capacitor.

Hence, capacitor serves as a barrier for DC and offers an easy path to AC.

Therefore, Assertion and Reason are the correct and Reason is the correct explanation of Assertion.

**85** (b) The phase difference between current and voltage in a series R-L-C circuit is

$$\phi = \tan^{-1}\left(\frac{X_C - X_L}{R}\right)$$

So, if $X_C > X_L$, $\phi$ is positive and the circuit is predominantly capacitive. The current in the circuit leads the source voltage.

If $X_C < X_L$, $\phi$ is negative and the circuit is predominantly inductive. The current in the circuit lags the source voltage.

Therefore, Assertion and Reason are correct but Reason is not the correct explanation of Assertion.

**86** (a) The resultant emf in the L-C-R circuit is given by

$$E = \sqrt{V_R^2 + (V_L - V_C)^2}$$

Given, $V_2 = 8\text{ V}, V_L = 6\text{ V}$ and $V_C = 10\text{ V}$

$\Rightarrow \quad E = \sqrt{(8)^2 + (16-10)^2}$

$\Rightarrow \quad E = \sqrt{64 + 36} \Rightarrow E = 10\text{ V}$

Therefore, Assertion and Reason are correct and Reason is the correct explanation of Assertion.

**87** (a) It is important to note that resonance phenomenon is exhibited by a circuit only, if both L and C are present in the circuit.

Only then do the voltage across L and C cancel each other (both being out of phase) and the current amplitude is $V_m/R$. This means that we cannot have resonance in a R-L or R-C circuit,

$$V = \sqrt{V_R^2 + (V_L - V_C)^2} = V_R \qquad (\because V_L = V_C)$$

Hence, the total source voltage appears across R at resonance.

Therefore, Assertion and Reason are correct and Reason is the correct explanation of Assertion.

**91** (c) Statement given in option (c) is correct but rest are incorrect and these can be corrected as,

In case of a resistor, both V and i reach zero, minimum and maximum values at the same time. Clearly, the voltage and current are in phase with each other.

Like applied voltage, the current varies sinusoidally and has corresponding positive and negative values during each cycle. Thus, the sum of instantaneous current values over one complete cycle is zero and the average current is zero.

To show phase relationship between voltage and current in an AC circuit, we use the notion of phasors. A phasor is a vector which rotates about the origin with angular speed $\omega$.

**92** (d) We can obtaine the amplitude and phase of current for an L-C-R series circuit using the technique of phasors. But this method of analysing AC circuits suffers from certain disadvantages. First, the phasor diagram say nothing about the initial condition.

One can take any arbitrary value of $t$ (say, $t_1$) and draw different phasors which show the relative angle between different phasors. The solution so obtained is called the steady-state solution. This is not a general solution. Additionally, we do have a transient solution which exists even for $V = 0$. The general solution is the sum of the transient solution and the steady-state solution.

After a sufficiently long time, the effects of the transient solution die out and the behaviour of the circuit is described by the steady-state solution.

So, statement given in option (d) is incorrect and rest are correct.

**93** (d) In an inductive circuit, using Kirchhoff's loop rule, we get

$$V - L\frac{di}{dt} = 0$$

where, the second term is the self-inducced emf in the inductor and L is the self-inductance of the inductor.

The quantity $\omega L$ is analogous to the resistance and is called inductive reactance denoted by $X_L (= \omega L)$.

The current phasor **I** is $\frac{\pi}{2}$ behind the voltage phasor **V** i.e. $\phi = \frac{\pi}{2}$. When rotated with frequency $\omega$ counter-clockwise, they generate the voltage and current as given by $V = V_m \sin \omega t$ and $i_m \sin\left(\omega t - \frac{\pi}{2}\right)$ respectively.

Average power supplied to an inductor over one complete cycle is,

$$P_{avg} = V_{rms} \times I_{rms} \times \cos \phi$$
$$= V_{rms} \times I_{rms} \times \cos \frac{\pi}{2} = 0$$

So, statement given in option (d) is correct and rest are incorrect.

**94** (*b*) The capacitive reactance in an AC circuit is given by

$$X_C = \frac{1}{\omega C} = \frac{1}{2\pi f C}$$

where, $f$ = frequency of current.

So, it is inversely proportional to the frequency ($f$).

Here, capacitive reactance plays the role of resistance, so its unit is same as that of resistance, i.e. ohm ($\Omega$).

In a DC circuit, frequency of current is zero, i.e.,

$$f = 0$$
$$\Rightarrow \quad X_C = \frac{1}{2\pi f C} = 0$$

So, its values are different for AC and DC circuits. Hence, statement given in option (b) is correct and rest are incorrect.

**95** (*a*) Statement given in option (a) is incorrect and it can be corrected as

The current in a capacitive circuit is $i = i_m \sin\left(\omega t + \frac{\pi}{2}\right)$ because currents leads $\frac{\pi}{2}$ angle from voltage in capacitive circuit.

where, the amplitude of the oscillating current is

$$i_m = \frac{V_m}{X_C} = \frac{V_m}{1/\omega C} = \omega C V_m$$

Rest statements are correct.

**96** (*b*) In a series *L-C-R* circuit as the resistor, inductor and capacitor are in series, So, the AC current in each element is the same at any time, having the same amplitude and phase.

The phasor **I** is always parallel to phasor $\mathbf{V}_R$, the phase angle $\phi$ is the angle between $\mathbf{V}_R$ and **V** and can be determined from

$$\tan \phi = \frac{V_L - V_C}{V_R}, \quad \text{when } V_L > V_C$$

The impedance diagram is a right triangle with $Z$ as its hypotenuse as shown

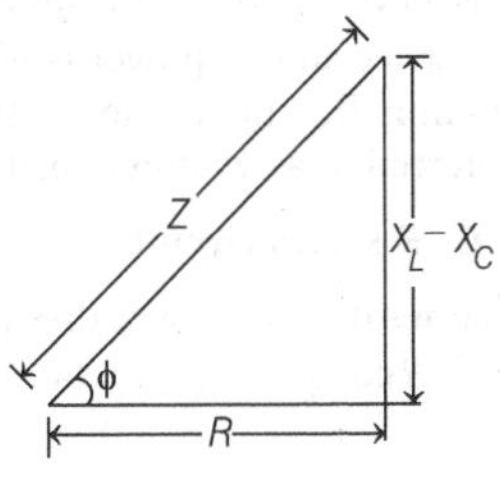


From figure, impedance, $Z = \sqrt{R^2 + (X_L - X_C)^2}$, which is equivalent to the resistance in a circuit.

Hence, statement given in option (b) is correct and rest are incorrect.

**97** (*b*) The statement given in option (b) is incorrect and it can be corrected as,

If $\omega$ is varied, then at a particular frequency $\omega_0$, $X_C = X_L$ and the impedance is minimum ($Z = R$). This frequency is called the resonant frequency and given as

$$\omega_0 = \frac{1}{\sqrt{LC}}$$

Rest statements are correct.

**98** (*c*) The statement given in option (c) is incorrect and it can be corrected as,

The signals picked up in the antenna acts as a source in the tuning circuit of the radio, so the circuit can be driven at many frequencies.

Rest statements are correct.

**99** (*d*) The sharpness of reasonance is given by

$$\frac{\omega_0}{2\Delta\omega} = \frac{\omega_0 L}{R}$$

The ratio $\frac{\omega_0 L}{R}$ is also called the quality factor $Q$ of the circuit i.e., $Q = \frac{\omega_0 L}{R}$

$\Rightarrow 2\Delta\omega = \frac{\omega_0}{Q}$, so larger the value of $Q$, the smaller is the value of $2\Delta\omega$ or the bandwidth and sharper is the resonance.

Thus, statement given in option (d) is incorrect and rest are correct.

**100** (*a*) When a capacitor is connected to an inductor, the charge on the capacitor and the current in the circuit exhibit the phenomenon of electrical oscillations similar to oscillations in mechanical system, whose natural frequency is given by

$$\omega_0 = \frac{1}{\sqrt{LC}}$$

and charge varies sinusoidally with time as

$$q = q_m \cos(\omega_0 t + \phi)$$

where, $q_m$ is the maximum value of $q$ and $\phi$ is the phase constant. Since, $q = q_m$ at $t = 0$, we have $\cos \phi = 1$ or $\phi = 0$. Therefore, in the present case,

$$q = q_m \cos(\omega_0 t)$$

The current $i\left(= -\frac{dq}{dt}\right)$ is given by $i = i_m \sin(\omega_0 t)$

where, $i_m = \omega_0 q_m$

In case of *L-C* circuit, $\omega_0 = \frac{1}{\sqrt{LC}}$ and for mass on a spring, $\omega_0 = \sqrt{\frac{k}{m}}$.

Thus, statement given in option (a) is correct and rest are incorrect.

**101** (c) For the given frequency of applied emf, the circuit is at resonance and its impedance is made up only of a resistive part

$$Z = \sqrt{R^2 + (X_L - X_C)^2} = R \quad [\because X_L = X_C]$$

At resonance condition of *R-L-C* series circuit, the sum of potential difference across the inductor and capacitor is zero, because they are 180° out of phase.

So, the current in the circuit is in phase with the applied emf and voltage across *R* equals the applied emf.

The quality factor of a series *R-L-C* circuit is

$$Q\text{-factor} = \frac{X_L \text{ or } X_C}{R} = \frac{\omega L}{R} \text{ or } \frac{1}{\omega CR}$$

It is a measure of the voltage magnification (produced by the circuit at resonance) as well as the sharpnesss of resonance of the circuit.

Thus, statement given in option (c) is incorrect and rest are correct.

**102** (a) The statement given in option (a) is incorrect and it can be corrected as,

*L-C* oscillation is not realistic because every inductor has some resistance.

Rest statements are correct.

**103** (b) The statement given in option (b) is incorrect and it can be corrected as,

If the transformer is assumed to be 100% efficient (no energy losses), the power input is equal to the power output, i.e.

Power at primary coil = Power at secondary coil

$$I_p V_p = I_s V_s$$

Rest statements are correct.

**104** (c) The statement given in option (c) is correct and rest are incorrect and these can be corrected as,

Flux leakage is reduced by winding the primary and secondary coils one over the other.

Resistance of windings is minimised by using thick wire.

Hysteresis is minimised by using a magnetic material because magnetic flux is passed easily through magnetic material.

**105** (c) We have to transmit energy (power) over large distances at high alternating voltages, so current flowing through the wires will be low, for a given constant power (*P*).

For this, the transmission lines are made thicker.

i.e., $P = E_{rms} i_{rms}$,

$i_{rms}$ is low, when $E_{rms}$ is high.

i.e., Power loss $= i^2_{rms} R =$ low $\quad (\because i_{rms}$ is low)

Now, at the receiving end high voltage is reduced by using step-down transformers.

Hence, statement given in option (c) is incorrect and rest are correct.

**106** (c) **(0-1)** The current *i* flows as shown and from the maximum at 0, reaches a zero value at 1. The plate

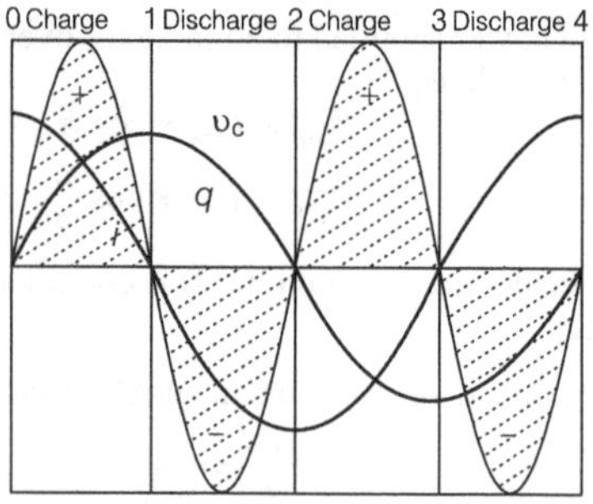


*A* is charged to positive polarity, while negative charge *q* builds up in *B* reaching a maximum at 1 until the current becomes zero. The current and voltage are both positive. So, $P = V_c i$ is positive. Energy is absorbed from the source during this quarter cycle (0-1) as capacitor is charged.

**(1-2)** The current *i* reverses its direction. The accumulated charge is depleted, i.e. the capacitor is discharged during this quarter cycle. The voltage gets reduced but still positive. The current is negative. Their product, the power *P* is negative. The energy absorbed during the $\frac{1}{4}$ cycle in 0-1 is returned during this quarter.

**(2-3)** As *i* continues to flow from *A* to *B* the capacitor is charged to reversed polarity, i.e. the plate *B* acquire positive and *A* acquire negative charge. Both the current and the voltage are negative. Their product *P* is positive. The capacitor absorbs energy during this (1/4) cycle.

**(3-4)** The current *i* reverses its directions at 3 and flows from *B* to *A*. The accumulated charge is depleted and the magnitude of voltage $V_C$ is reduced. $V_C$ becomes zero at 4 when the capacitor is fully discharged. The power is negative. Energy absorbed during 2-3 is returned to the source. Net energy absorbed is zero.

Hence, A → 2, B → 3, C → 1 and D → 4.

**107** (b) **Case I. Resistive circuit**

If the circuit contains only pure *R*, it is called resistive. In that case, $\phi = 0, \cos\phi = 1$. There is maximum power dissipation.

**Case II. Purely inductive or capacitive circuit**

If the circuit contains only an inductor or capacitor, then the phase difference between voltage and current is $\pi/2$. Therefore, $\cos\phi = 0$ and no power is dissipated even though a current is flowing in the circuit. This current is sometimes referred to as wattless current.

**Case III. *L-C-R* series circuit**

In an *L-C-R* series circuit, power dissipated is given by equation $P = I^2 R\cos\phi$, where $\phi = \tan^{-1}(X_C - X_L)/R$.

So, $\phi$ may be non-zero in a $R$-$L$ or $R$-$C$ or $L$-$C$-$R$ circuit. Even in such cases, power is dissipated only in the resistor.

**Case IV. Power dissipated at resonance in *L-C-R* circuit**

At resonance, $X_C - X_L = 0$, and $\phi = 0$. Therefore, $\cos\phi = 1$ and $P = I^2Z = I^2R$. That is maximum power is dissipated in a circuit (through $R$) at resonance.

A $\rightarrow$ 4, B $\rightarrow$ 1, C $\rightarrow$ 3 and D $\rightarrow$ 2.

**108** (*d*) In an $L$-$C$ circuit, initially the electrical energy stored in the charged capacitor is $U_E = \frac{1}{2}\frac{q_m^2}{C}$. Since, there is no current in the circuit, energy in the inductor is zero (i.e. $U_B = 0$). Thus, the total energy of $L$-$C$ circuit is

$$U = U_E = \frac{1}{2}\frac{q_m^2}{C}$$

At $t = 0$, the switch is closed and the capacitor starts to discharge. As the current increases, it sets up a magnetic field in the inductor and thereby, some energy gets stored in the inductor in the form of magnetic energy,

$U_B = \frac{1}{2}Li^2$ and in capacitor, $U_E = \frac{1}{2}\frac{q^2}{C}$.

As the current reaches its maximum value $\left(\text{at } t = \frac{T}{4}\right)$, all the energy is stored in the magnetic field.

i.e., $U_B = \frac{1}{2}Li_m^2$ and $U_E = 0$

Thus, the energy in the system oscillates between the capacitor and the inductor.

Hence, A $\rightarrow$ 2, B $\rightarrow$ 3 and C $\rightarrow$ 1.

**110** (*a*) Given, $R = 100\,\Omega$, $V_{rms} = 220$ V and $\nu = 50$ Hz

Current in the circuit, $i_{rms} = \frac{V_{rms}}{R} = \frac{220}{100} = 2.2$ A

**111** (*a*) Given, $V_0 = 300$ V

The rms value of voltage, $V_{rms} = \frac{V_0}{\sqrt{2}} = \frac{300}{\sqrt{2}} = 212.1$ V

**112** (*b*) Given, $L = 44$ mH $= 44 \times 10^{-3}$ H, $V_{rms} = 220$ V and $f = 50$ Hz

Inductive reactance, $X_L = 2\pi fL$

$$= 2 \times 3.14 \times 50 \times 44 \times 10^{-3} = 13.83\,\Omega$$

The rms value of current in the circuit,

$$i_{rms} = \frac{V_{rms}}{X_L} = \frac{220}{13.83} = 15.9 \text{ A}$$

**113** (*c*) Given, $C = 60\mu\text{F} = 60 \times 10^{-6}$ F

$V_{rms} = 110$ V and $f = 60$ Hz

Capacitive reactance,

$$X_C = \frac{1}{2\pi fC} = \frac{1}{2 \times 3.14 \times 60 \times 60 \times 10^{-6}} = 44.23\,\Omega$$

The rms value of the current in the circuit,

$$i_{rms} = \frac{V_{rms}}{X_C} = \frac{110}{44.23} = 2.49 \text{ A}$$

**114** (*d*) Given, $L = 2$ H, $C = 32\mu\text{F} = 32 \times 10^{-6}$ F and $R = 10\Omega$

$Q$-factor of this circuit,

$$Q = \frac{1}{R}\sqrt{\frac{L}{C}} = \frac{1}{10}\sqrt{\frac{2}{32 \times 10^{-6}}} = \frac{10^3}{40} = 25$$

**115** (*b*) Given, $C = 30\mu\text{F} = 30 \times 10^{-6}$ F

and $L = 27$ mH $= 27 \times 10^{-3}$ H

Resonant angular frequency of oscillation of the circuit,

$$\omega_0 = \frac{1}{\sqrt{LC}} = \frac{1}{\sqrt{27 \times 10^{-3} \times 30 \times 10^{-6}}} = \frac{10^4}{9}$$

$$= 1.1 \times 10^3 \text{ rads}^{-1}$$

**116** (*a*) Given, $R = 20\Omega$, $L = 1.5$ H, $C = 35\mu\text{F} = 35 \times 10^{-6}$ F and $V_{rms} = 200$ V

At the condition of resonance impedance,

$$Z = R = 20\Omega$$

The rms value of current in the circuit,

$$i_{rms} = \frac{V_{rms}}{Z} = \frac{200}{20} = 10 \text{ A}$$

$\Rightarrow$ $\phi = 0°$ (for resonance)

Power transferred to the circuit in one complete cycle,

$$P = i_{rms} \cdot V_{rms} \cos\phi$$

$$= 10 \times 200 \times \cos 0° = 2000 \text{ W}$$

$$= 2 \text{ kW}$$

**117** (*b*) For tuning, the natural frequency is equal to the frequency of oscillations, that means it is the case of resonance.

Frequency of oscillations, $f = \frac{1}{2\pi\sqrt{LC}}$

For capacitance $C_1$, $f_1 = \frac{1}{2\pi\sqrt{LC_1}}$

Here, $f_1 = 800$ kHz $= 8 \times 10^5$ Hz

$L = 200\mu\text{H} = 2 \times 10^{-4}$ H

$$\Rightarrow \quad C_1 = \frac{1}{4\pi^2 f_1^2 L}$$

$$= \frac{1}{4 \times 3.14 \times 3.14 \times (8 \times 10^5)^2 \times 2 \times 10^{-4}}$$

$$= 197.7 \times 10^{-12} \text{ F} = 197.7 \text{ pF}$$

For capacitance $C_2$, $f_2 = \frac{1}{2\pi\sqrt{LC_2}}$

Here, $f_2 = 1200$ kHz $= 12 \times 10^5$ Hz

$$C_2 = \frac{1}{4\pi^2 f_2^2 L} = \frac{1}{4 \times 3.14 \times 3.14 \times (12 \times 10^5)^2 \times 2 \times 10^{-4}}$$

$$= 87.8 \times 10^{-12} \text{ F} = 87.8 \text{ pF}$$

Thus, the range of capacitor is 87.8 pF to 197.7 pF, i.e. 88 to 198 pF.

**118** (*a*) The rms value of voltage, $V_{rms} = 240\,V$, $f = 50\,Hz$, $L = 0.5\,H$ and $R = 100\Omega$

Impedance of circuit, $Z = \sqrt{R^2 + X_L^2} = \sqrt{R^2 + (2\pi f L)^2}$

$$= \sqrt{(100)^2 + (2\times 3.14\times 50\times 0.5)^2}$$

$$= 186.14\,\Omega$$

The rms value of current, $i_{rms} = \frac{V_{rms}}{Z} = \frac{240}{186.14} = 1.29\,A$

The maximum value of current in the circuit,

$$i_0 = \sqrt{2}\, i_{rms} = 1.414\times 1.29 = 1.824\,A$$

**119** (*b*) Given, $C = 100\,\mu F = 100\times 10^{-6}\,F$, $R = 40\Omega$, $V_{rms} = 110\,V$ and $f = 60\,Hz$

Impedance, $Z = \sqrt{R^2 + X_C^2} = \sqrt{R^2 + \left(\frac{1}{2\pi f C}\right)^2}$

$$= \sqrt{(40)^2 + \left(\frac{1}{2\times 3.14\times 60\times 100\times 10^{-6}}\right)^2}$$

$$= \sqrt{1600 + 704.33} = 48\,\Omega$$

The rms value of current, $i_{rms} = \frac{V_{rms}}{Z} = \frac{110}{48}\,A$

The maximum current in the circuit

$$i_0 = \sqrt{2} i_{rms} = 1.414\times \frac{110}{48} = 3.24\,A$$

**120** (*b*) Given, $h = 300\,m$, volume $= 100\,m^3$ and $\eta = 60\%$

$$\text{Power} = \frac{m\times g\times h}{t} = \frac{\text{Volume}\times \text{Density}\times g\times h}{t}$$

$$P_i = 100\times 1000\times 9.8\times 300 = 2.94\times 10^8\,W$$

($\because$ density of water $= 1000\,kg/m^3$)

Suppose, the power output is $P_o$, which is equal to the power available from the plant.

The efficiency of generator,

$$\eta = \frac{P_o}{P_i} \Rightarrow \frac{60}{100} = \frac{P_o}{2.94\times 10^8}$$

$$P_o = \frac{60}{100}\times 2.94\times 10^8$$

$$= 1764\times 10^5\,W$$

$$= 176.4\,MW$$

**121** (*b*) Given, $i_{rms} = 5A$, $f = 50\,Hz$ and $t = \frac{1}{300}\,s$

$$i_0 = \text{Peak value} = \sqrt{2}$$

$$i_{rms} = \sqrt{2}\times 5 = 5\sqrt{2}\,A$$

$\Rightarrow$ $$i = i_0 \sin \omega t = 5\sqrt{2} \sin 2\pi f t$$

$$= 5\sqrt{2} \sin 2\pi\times 50\times \frac{1}{300}$$

$$= 5\sqrt{2}\sin\frac{\pi}{3} = 5\sqrt{2}\times\frac{\sqrt{3}}{2}$$

$$= 5\sqrt{3/2}\,A$$

**122** (*c*) For delivering maximum power from the generator to the load, total internal reactance must be equal to the conjugate of total external reactance.

Hence, $X_{int} = {}^*X_{ext}$

$\Rightarrow$ $X_g = (X_L)^* = -X_L$

or $X_L = -X_g$

**123** (*c*) The voltmeter connected to AC mains reads mean value ($<v^2>$) and is calibrated in such a way that it gives value of $<v^2>$, which is multiplied by form factor to give rms value, i.e. $\sqrt{<v^2>}$.

**124** (*b*) We know that, resonant frequency in an *L-C-R* circuit is given by

$$\nu_0 = \frac{1}{2\pi\sqrt{LC}}$$

Now to reduce $\nu_0$ either we can increase *L* or we can increase *C*.

To increase capacitance, we must connect another capacitor parallel to the first.

**125** (*c*) Quality factor *Q* of an *L-C-R* circuit is given by

$$Q = \frac{1}{R}\sqrt{\frac{L}{C}}$$

where, *R* is resistance, *L* is inductance and *C* is capacitance of the circuit. To make *Q* high, *R* should be low, *L* should be high and *C* should be low.

This is satisfied by values in option (*c*).

**126** (*c*) Given, $X_L = 1\Omega$, $R = 2\Omega$, $V_{rms} = 6\,V$

Average power dissipated in the circuit,

$$P_{av} = V_{rms}\, i_{rms} \cos\phi \quad \text{...(i)}$$

As, $i_{rms} = \frac{i_0}{\sqrt{2}} = \frac{V_{rms}}{Z}$

where, $Z = \sqrt{R^2 + X_L^2} = \sqrt{4+1} = \sqrt{5}$

$\Rightarrow$ $i_{rms} = \frac{6}{\sqrt{5}}\,A$

$$\cos\phi = \frac{R}{Z} = \frac{2}{\sqrt{5}}$$

$$P_{av} = 6\times\frac{6}{\sqrt{5}}\times\frac{2}{\sqrt{5}} \quad \text{[from Eq. (i)]}$$

$$= \frac{72}{\sqrt{5}\sqrt{5}} = \frac{72}{5} = 14.4\,W$$

**127** (*a*) Secondary voltage, $V_s = 24V$

Power associated with secondary, $P_s = 12\,W$

$\Rightarrow$ $i_s = \frac{P_s}{V_s} = \frac{12}{24} = \frac{1}{2}\,A = 0.5\,A$

Peak value of the current in the secondary,

$$i_0 = i_s\sqrt{2} = (0.5)\,(1.414) = 0.707$$

or $i_0 = \frac{1}{\sqrt{2}}\,A$

CHAPTER > 08

# Electromagnetic Waves

## KEY NOTES

### Displacement Current

- It is a current which is produced due to the rate of change of electric flux with respect to time. This is also called **Maxwell's displacement current.** It is given by

$$i_d = \varepsilon_0 \frac{d\phi_E}{dt}$$

- The source of a magnetic field is not just the conduction electric current due to the flowing charges, but also due to the rate of change of electricfield.

  More precisely, the total current $i$ is the sum of the conduction current denoted by $i_c$ and the displacement current denoted by $i_d$.

  So, we have

$$i = i_c + i_d = i_c + \varepsilon_0 \frac{d\phi_E}{dt}.$$

- For a capacitor system outside the capacitor plates, there is only conduction current, i.e. $i_c = i$ and no displacement current, i.e. $i_d = 0$.

  On the other hand, inside the capacitor, there is no conduction current, i.e. $i_c = 0$ and there is only displacement current, i.e. $i_d = i$.

**Note** The displacement current has the same physical effects as the conduction current.

### Maxwell's Equations

- These are the basic laws of electricity and magnetism. These equations give complete description of all electromagnetic interactions.
- There are four Maxwell's equations which are explained as given below

  (i) $\oint \mathbf{E} \cdot d\mathbf{A} = \frac{Q}{\varepsilon_0}$ (Gauss's law for electricity)

  (ii) $\oint \mathbf{B} \cdot d\mathbf{A} = 0$ (Gauss's law for magnetism)

  (iii) $\oint \mathbf{E} \cdot d\mathbf{l} = -\frac{d\phi_B}{dt}$ (Faraday's law)

  (iv) $\oint \mathbf{B} \cdot d\mathbf{l} = \mu_0 i_c + \mu_0 \varepsilon_0 \frac{d\phi}{dt}$ (Ampere-Maxwell law)

## Electromagnetic Waves

- **Sources of EMW** Maxwell found that the accelerated or oscillated charges radiate electromagnetic waves. These charges produce an oscillating electric field in space, which produce an oscillating magnetic field, which in turn, is a source of oscillating electric field and so on.

  The oscillating electric and magnetic fields thus regenerate each other.
- The frequency of the electromagnetic wave naturally equals the frequency of oscillation of the charge and the energy associated with the propagating wave comes at the expense of the energy of the source (the accelerated charge).
- **Nature of EMW** In an electromagnetic wave, electric and magnetic fields are perpendicular to each other and to the direction of propagation.

- A plane electromagnetic wave propagating along the z-direction is shown below

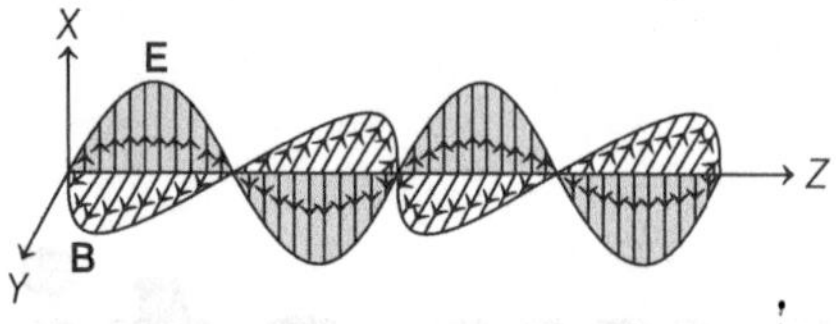


- The electric field $E_x$ is along X-axis and the magnetic field $B_y$ is along the Y-axis are given as

$$E_x = E_0 \sin (kz - \omega t) \text{ and } B_Y = B_0 \sin (kz - \omega t).$$

where, $k$ is **propagation vector** or **wave vector** $= \frac{2\pi}{\lambda}$ and $\omega$ is the angular frequency.

The direction of $k$ describes the direction of propagation of the wave.

- The speed of propagation of the wave is $\omega / k$, which gives $\omega = ck$, where $c = \frac{1}{\sqrt{\mu_0 \varepsilon_0}}$.

In terms of frequency, $\nu = \omega / 2\pi$ and wavelength, $\lambda = \frac{2\pi}{k}$, then we have, $\nu\lambda = c$

- The magnitude of the electric and the magnetic fields in an electromagnetic wave are related as $B_0 = \frac{E_0}{c}$.
- **Some features of electromagnetic waves** are given below
  (i) These waves are self-sustaining oscillations of electric and magnetic fields in free space or vacuum.
  (ii) No material medium is involved in the vibrations of the electric and magnetic fields.
  (iii) In a material, medium of permittivity $\varepsilon$ and permeability $\mu$, then the velocity of light becomes $v = \frac{1}{\sqrt{\mu\varepsilon}}$
  Thus, the velocity of light depends on electric and magnetic properties of the medium.
  (iv) Electromagnetic waves could be diffracted, refracted and polarised. Thus, this established the wave nature of the radiation.
  (v) In a region of free space with the electric field, there is an energy density $(\varepsilon_0 E^2 / 2)$ and with magnetic field, the magnetic energy density $(B^2 / 2\mu_0)$. So, there is a non-zero energy density associated with EM wave.
  (vi) An electromagnetic wave (like other waves) carries energy and momentum.
  The magnitude of total momentum delivered to a surface (for complete absorption) is $p = \frac{U}{c}$
  where, $U$ = total energy transferred to the surface in time $t$ and $c$ = speed of light $(3 \times 10^8 \text{ m/s})$.
  (vii) An electromagnetic wave also exerts pressure, called **radiation pressure**, which is of the order of $7 \times 10^{-6} \text{ N/m}^2$.
  (viii) The energy in electromagnetic wave is divided equally between electric and magnetic fields.

## Electromagnetic Spectrum

- The arrangement of EMW in increasing and decreasing order of wavelength $\lambda$ and frequency $\nu$ is called electromagnetic spectrum. These waves includes visible light waves, X-rays, gamma rays, radio waves, microwaves, ultraviolet and infrared waves.
- Electromagnetic wave spectrum is shown below

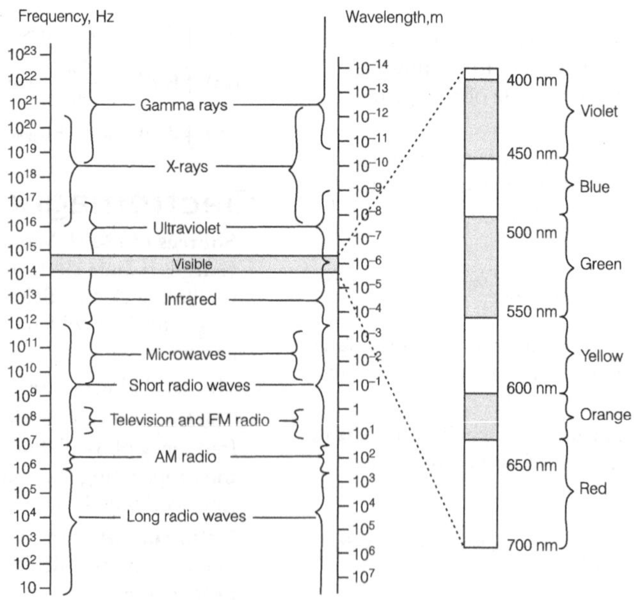


KEY NOTES

**Various Electromagnetic Waves of Electromagnetic Spectrum with their Features and Uses**

| | Name of Wave | Frequency Range | Wavelength Range | Production | Detection | Uses |
|---|---|---|---|---|---|---|
| (*i*) | **Radio waves** | 500 to kHz 1000 MHz | > 0.1 m | Rapid acceleration and decelerations of electrons in aerials. | Receiver's aerials. | • These are used in AM (Amplitude Modulation) from 530 kHz to 1710 kHz and ground wave propagation.<br>• These are used in TV waves ranging from 54 MHz to 890 MHz.<br>• These are used in FM (Frequency Modulation) ranging from 88 MHz to 108 MHz. |
| (*ii*) | **Microwaves** | 1 to GHz 300 GHz | 0.1 to m 1 mm | Klystron valve or magnetron valve. | Point contact diodes. | • These are used in RADAR systems for aircraft navigation.<br>• These are used in microwave oven for cooking purpose.<br>• These are used in study of atomic and molecular structures. |
| (*iii*) | **Infrared waves** (heat waves) | $3 \times 10^{11}$ Hz to $4 \times 10^{14}$ Hz | 1 mm to 700 nm | Vibration of atoms and molecules. | Thermopiles bolometer and infrared photographic film. | • These are used in physical therapy.<br>• These are used in satellites for army purpose.<br>• These are used in weather forecasting. |
| (*iv*) | **Light rays** (visible rays) | $4 \times 10^{14}$ Hz to $7 \times 10^{14}$ Hz | 700 nm to 400 nm | Electrons in atoms emit light, when they move from a higher energy level to a lower energy level. | Eye, photocells and photographic film. | • These are used by the optical organs of humans and animals for three primary purposes given below<br>(i) To see things, avoid bumping into them and escape danger.<br>(ii) To look for food.<br>(iii) To find other living things with which to copulate so as to prolong the species. |
| (*v*) | **Ultraviolet rays** | $10^{14}$ Hz to $10^{16}$ Hz | 400 nm to 1 nm | Inner shell electrons in atoms moving from higher energy level to a lower energy level. | Photocells and photographic film. | • These are used in burglar alarm.<br>• These are used in checking mineral sample.<br>• These are used to study molecular structure.<br>• To kill germs in water purifiers.<br>• Used in LASER eye surgery. |
| (*vi*) | **X-rays** | $3 \times 10^{16}$ Hz to $3 \times 10^{21}$ Hz | 1 nm to $10^{-3}$ nm | X-ray tubes or inner shell electrons, bombarding metals by high energy electrons. | Photographic film, Geiger tubes and ionisation chamber. | • These are used in medicine to detect the fracture, diseased organs, stones in the body etc.<br>• These are used in engineering to detect fault, cracks in bridges and testing of welds.<br>• These are used at metro stations to detect metals or explosive material. |
| (*vii*) | **Gamma (γ) rays** | $3 \times 10^{18}$ Hz to $5 \times 10^{22}$ Hz. | $< 10^{-3}$ nm | Radioactive decay of the nucleus. | Photographic film and ionisation chamber. | • These are used to produce nuclear reactions.<br>• These are used in radio therapy for the treatment of tumour and cancer.<br>• These are used in food industry to kill pathogenic micro-organism. |

# Mastering NCERT

## MULTIPLE CHOICE QUESTIONS

## TOPIC 1 ~ Displacement Current and Maxwell's Equation

**1** Maxwell's work led to the conclusion that light is an electromagnetic wave. His work thus unified
(a) electricity and magnetism
(b) magnetism and light
(c) electricity and light
(d) electricity, magnetism and light

**2** In the given figure, for which configuration the magnetic field (say at point $P$) outside the plates of the capacitor is to be same everywhere?

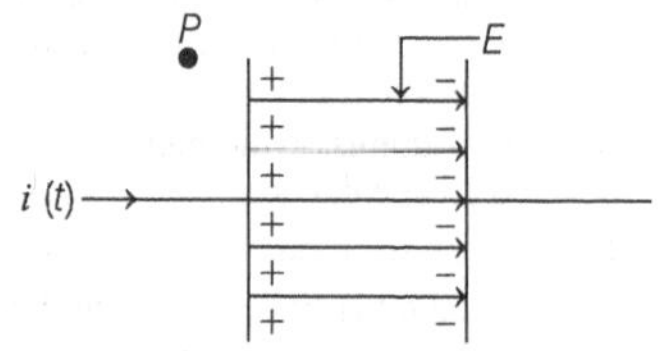


(a) Plane circular loop perpendicular to wire
(b) Plane circular loop parallel to wire
(c) Pot like surface with bottom between the plates
(d) Tiffin box like shape (without the lid)

**3** The charge on a parallel-plate capacitor varies as $q = q_0 \cos 2\pi\nu t$. The plates are very large and close together (area $= A$, separation $= d$). Neglecting the edge effects, find the displacement current through the capacitor.
(a) $-\sin 2\pi\nu t\ (2\pi\nu)$ (b) $q_0 \cos 2\pi\nu t$
(c) $2\pi\, q_0\, \nu \sin\,(2\pi\nu t)$ (d) $-2\pi\, q_0 \nu \sin\,(2\pi\nu t)$

**4** What is an instantaneous rate of change of voltage for displacement current of $10\text{A}$, in the space between the parallel plate capacitor of 1 μF?
(a) $10^5\,\text{Vs}^{-1}$ (b) $10^6\,\text{Vs}^{-1}$
(c) $10^{-6}\,\text{Vs}^{-1}$ (d) $10^7\,\text{Vs}^{-1}$

**5** A parallel-plate capacitor consists of two circular plates with radius $R = 10$ cm separated by distance $d = 0.5$ mm. The capacitor is being charged at a uniform rate by applying a changing potential difference between the two plates. Calculate the displacement current for the capacitor. Assume that, the electric field is due to the displacement current only and rate at which the electric field between the plates changes is $5\times 10^{13}\ \text{Vms}^{-1}$.
(a) 12.8 A (b) 12.6 A
(c) 13.9 A (d) 10.5 A

**6** A sinusoidal voltage is applied directly across an 8 μF capacitor. The frequency of the source is 3.00 kHz and the voltage amplitude is 30.0 V. Find the maximum value of displacement current between the plates of the capacitor.
(a) 42.5 A (b) 4.25 μA
(c) 4.52 A (d) 4.52 μA

**7** A parallel plate capacitor is charged to $100\times 10^{-6}$ C. Due to radiations, falling from a radiating source, the plate loses charge at the rate of $2\times 10^{-7}\ \text{Cs}^{-1}$. The magnitude of displacement current is
(a) $10^{-6}$ A (b) $10^{-4}$ A
(c) $2\times 10^{-7}$ A (d) $2\times 10^{-7}\ \text{mCs}^{-1}$

**8** A parallel plate capacitor of capacitance 20μF is being charged by a voltage source, whose potential is changing at the rate of 3 V/s.
The conduction current through the connecting wires and the displacement current through the plates of the capacitor would be respectively, ***NEET National 2019***
(a) 60 μA, 60 μA (b) 60 μA, zero
(c) zero, zero (d) zero, 60 μA

**9** There may be large regions of space, where there is no conduction current, but there is only
(a) displacement current due to time varying electric fields
(b) induced current due to time varying electric fields
(c) displacement current due to time varying magnetic field
(d) None of the above

**10** "Time-dependent electric and magnetic fields give rise to each other". Which laws give a quantitative expression of this statement?
(a) Faraday's law of electromagnetic induction
(b) Ampere-Maxwell law
(c) Faraday's left hand rule of electromagnetic induction
(d) Both (a) and (b)

# TOPIC 2~ Sources and Nature of EM Waves

**11** Out of the following options, which one can be used to produce a propagating electromagnetic wave? **NEET 2016**
(a) A stationary charge
(b) A chargeless particle
(c) An accelerating charge
(d) A charge moving at constant velocity

**12** A charged particle oscillates about its mean position with a frequency of $10^9$ Hz. The electromagnetic waves produced
(a) will have frequency of $10^9$ Hz
(b) will have frequency of $2 \times 10^9$ Hz
(c) will have wavelength of 0.2 m
(d) will have wavelength of 0.5 m

**13** An EM wave is propagating in a medium with a velocity $\mathbf{v} = v\hat{\mathbf{i}}$. The instantaneous oscillating electric field of this EM wave is along $+Y$-axis. Then, the direction of oscillating magnetic field of EM wave will be along **NEET 2018**
(a) $-y$-direction (b) $+z$-direction
(c) $-z$-direction (d) $-x$-direction

**14** The magnetic field of an electromagnetic wave is given by $B_y = 3 \times 10^{-7} \sin (10^3 x + 6.28 \times 10^{12} t)$, where $t$ and $x$ are in seconds and metres respectively. The wavelength of the electromagnetic wave is
(a) 6.28 cm (b) 3.14 cm (c) 0.63 cm (d) 0.32 cm

**15** The electric field of an electromagnetic wave in free space is given by $\mathbf{E} = 10\cos(10^7 t + kx)\mathbf{j}\ \text{Vm}^{-1}$, where $t$ and $x$ are in seconds and metres, respectively. It can be inferred that,
(a) the wavelength $\lambda$ is 188.4m
(b) the wave number $k$ is 0.33 $\text{radm}^{-1}$
(c) the wave amplitude is 20$\text{Vm}^{-1}$
(d) None of the above

**16** An electromagnetic wave is represented by the electric field $\mathbf{E} = E_0\, \hat{\mathbf{n}} \sin[\omega t + (6y - 8z)]$. Taking unit vectors in $x$, $y$ and $z$- directions to be $\hat{\mathbf{i}}$, $\hat{\mathbf{j}}$, $\hat{\mathbf{k}}$, the direction of propagation $\hat{\mathbf{s}}$, is **JEE Main 2019**
(a) $\hat{\mathbf{s}} = \frac{3\hat{\mathbf{i}} - 4\hat{\mathbf{j}}}{5}$ (b) $\hat{\mathbf{s}} = \frac{-4\hat{\mathbf{k}} + 3\hat{\mathbf{j}}}{5}$
(c) $\hat{\mathbf{s}} = \left(\frac{-3\hat{\mathbf{j}} + 4\hat{\mathbf{k}}}{5}\right)$ (d) $\hat{\mathbf{s}} = \frac{4\hat{\mathbf{j}} - 3\hat{\mathbf{k}}}{5}$

**17** Suppose that the amplitude of electric field of an electromagnetic wave is $E_0 = 120\ \text{NC}^{-1}$ and its frequency is $\nu = 50.0$ MHz. The expressions for $E$ will be (if wave travels along $X$-axis)
(a) $[(120\ \text{NC}^{-1})\sin\{(1.05\ \text{radm}^{-1})x - (3.14 \times 10^8\ \text{rads}^{-1})t\}]\,\hat{\mathbf{i}}$
(b) $[(120\ \text{NC}^{-1})\sin\{(1.05\ \text{radm}^{-1})x - (3.14 \times 10^8\ \text{rads}^{-1})t\}]\,\hat{\mathbf{k}}$
(c) $[(120\ \text{NC}^{-1})\sin\{(1.05\ \text{radm}^{-1})x - (3.14 \times 10^8\ \text{rads}^{-1})t\}]\,\hat{\mathbf{j}}$
(d) $[(120\ \text{NC}^{-1})\cos\{(1.05\ \text{radm}^{-1})x - (3.14 \times 10^8\ \text{rads}^{-1})t\}]\,\hat{\mathbf{j}}$

**18** In an electromagnetic wave in free space, the root mean square value of the electric field is $E_{\text{rms}} = 6$ V/m. The peak value of the magnetic field is **NEET 2017**
(a) $1.41 \times 10^{-8}$ T (b) $2.83 \times 10^{-8}$ T
(c) $0.70 \times 10^{-8}$ T (d) $4.23 \times 10^{-8}$ T

**19** A plane electromagnetic wave of frequency 25 GHz is propagating in vacuum along the $z$-direction. At a particular point in space and time, the magnetic field is given by $\mathbf{B} = 5 \times 10^{-8}\,\hat{\mathbf{j}}$T. The corresponding electric field $\mathbf{E}$ is (speed of light, $c = 3 \times 10^8\ \text{ms}^{-1}$) **JEE Main 2020**
(a) $-1.66 \times 10^{-16}\hat{\mathbf{i}}$ V/m (b) $1.66 \times 10^{-16}\hat{\mathbf{i}}$ V/m
(c) $-15\hat{\mathbf{i}}$ V/m (d) $15\,\hat{\mathbf{i}}$ V/m

**20** In a plane electromagnetic wave, the electric field oscillates sinusoidally at a frequency of $2.0 \times 10^{10}$ Hz and amplitude 48 $\text{Vm}^{-1}$.
What is the amplitude of the oscillating magnetic field?
(a) $0.8 \times 10^{-7}$T (b) $1.6 \times 10^{-7}$T
(c) $3.2 \times 10^{-8}$T (d) $6.4 \times 10^{-8}$T

**21** If the magnetic field in a plane electromagnetic wave is given by $\mathbf{B} = 3 \times 10^{-8} \sin(1.6 \times 10^3 x + 48 \times 10^{10} t)\hat{\mathbf{j}}$T, then what will be expression for electric field? **JEE Main 2020**
(a) $\mathbf{E} = 60\sin(1.6 \times 10^3 x + 48 \times 10^{10} t)\hat{\mathbf{k}}$ V/m
(b) $\mathbf{E} = 3 \times 10^{-8}\sin(1.6 \times 10^3 x + 48 \times 10^{10} t)\hat{\mathbf{i}}$V/m
(c) $\mathbf{E} = 3 \times 10^{-8}\sin(1.6 \times 10^3 x + 48 \times 10^{10} t)\hat{\mathbf{j}}$V/m
(d) $\mathbf{E} = 9\sin(1.6 \times 10^3 x + 48 \times 10^{10} t)\hat{\mathbf{k}}$V/m

**22** Which waves can propagate in a solid, which is rigid and that resists shear?
(a) Electromagnetic waves
(b) Sound waves
(c) Transverse waves of water
(d) Transverse elastic sound waves

**23** An EM wave from air enters a medium. The electric fields are $\mathbf{E}_1 = E_{01}\hat{\mathbf{x}}\cos\left[2\pi\nu\left(\frac{z}{c} - t\right)\right]$ in air and $\mathbf{E}_2 = E_{02}\hat{\mathbf{x}}\cos[k(2z - ct)]$ in medium, where the wave number $k$ and frequency $\nu$ refer to their values in air. The medium is non-magnetic.
If $\varepsilon_{r_1}$ and $\varepsilon_{r_2}$ refer to relative permittivities of air and medium respectively, which of the following options is correct? ***JEE Main 2018***
(a) $\frac{\varepsilon_{r_1}}{\varepsilon_{r_2}} = 4$ (b) $\frac{\varepsilon_{r_1}}{\varepsilon_{r_2}} = 2$
(c) $\frac{\varepsilon_{r_1}}{\varepsilon_{r_2}} = \frac{1}{4}$ (d) $\frac{\varepsilon_{r_1}}{\varepsilon_{r_2}} = \frac{1}{2}$

**24** For a transparent medium relative permeability and permittivity $\mu_r$ and $\varepsilon_r$ are 1.0 and 1.44, respectively. The velocity of light in this medium would be ***NEET (Odisha) 2019***
(a) $2.5 \times 10^8$ ms$^{-1}$ (b) $3 \times 10^8$ ms$^{-1}$
(c) $2.08 \times 10^8$ ms$^{-1}$ (d) $4.32 \times 10^8$ ms$^{-1}$

**25** The wave which had wavelength ten million times that of the light waves, could be diffracted, refracted and polarised. This established that the radiation possesses
(a) particle nature
(b) wave nature
(c) Both particle and wave nature
(d) electromagnetic nature

**26** In free space, the energy of electromagnetic wave in electric field is $U_E$ and in magnetic field is $U_B$. Then ***JEE Main 2019***
(a) $U_E = U_B$ (b) $U_E > U_B$
(c) $U_E < U_B$ (d) $U_E = \frac{U_B}{2}$

**27** Light with an energy flux of 18 Wcm$^{-2}$ falls on a non-reflecting surface at normal incidence. If the surface has an area of 20 cm$^2$, find the average force executed on the surface during a 30 min time span.
(a) $12 \times 10^{-6}$ N (b) $1.2 \times 10^{-7}$ N
(c) $1.2 \times 10^{-6}$ N (d) $12 \times 10^{7}$ N

**28** Light having an energy flux of 40 Wcm$^{-2}$ falls on non-reflecting surface at normal incidence. If the surface has an area of 20 cm$^2$, the total momentum delivered (for complete absorption) during 10 min is
(a) $16 \times 10^{-5}$ kg-ms$^{-1}$ (b) $16 \times 10^{-4}$ kg-ms$^{-1}$
(c) $102 \times 10^{4}$ kg-ms$^{-1}$ (d) $1.03 \times 10^{7}$ kg-ms$^{-1}$

**29** A radiation of energy $E$ falls normally on a perfectly reflecting surface. The momentum transferred to the surface is
(a) $\frac{E}{c}$ (b) $\frac{2E}{c}$ (c) $Ec$ (d) $\frac{E}{c^2}$

**30** A 27 mW laser beam has a cross-sectional area of 10 mm$^2$. The magnitude of the maximum electric field in this electromagnetic wave is given by [Take, permittivity of space, $\varepsilon_0 = 9 \times 10^{-12}$ SI units and speed of light, $c = 3 \times 10^8$ m/s] ***JEE Main 2019***
(a) 1 kV/m (b) 0.7 kV/m
(c) 2 kV/m (d) 1.4 kV/m

**31** The electric field produced by the radiation coming from a 100 W bulb at a distance of 3m is (assume that the efficiency of the bulb is 2.5% and it is a point source.)
(a) 4.5 V/m (b) 2.8 V/m
(c) 5 V/m (d) 4.07 V/m

## TOPIC 3 ~ *Electromagnetic Spectrum*

**32** At the time Maxwell predicted the existence of electromagnetic waves, which were the more familiar electromagnetic waves at that time?
(a) X-rays
(b) $\gamma$-rays
(c) Visible light waves
(d) Radio waves

**33** Which of the following are electromagnetic waves?
(a) Visible light waves and X-rays
(b) Gamma rays and radio waves
(c) Microwaves and ultraviolet rays
(d) All of the above

**34** The condition under which a microwave oven heats up a food item, containing water molecules, most efficiently is ***NEET 2013***
(a) the frequency of the microwave must match the resonant frequency of the water molecules
(b) the frequency of the microwave has no relation with natural frequency of water molecules
(c) microwave are heat waves, so always produce heating
(d) infrared waves produce heating in a microwave oven

**35** Infrared radiation keep the earth's surface warm using
(a) greenhouse effect (b) resonance effect
(c) radioactivity (d) None of these

**36** Visible rays in the spectrum runs from about
(a) $4 \times 10^{14}$Hz to $4 \times 10^{11}$ Hz
(b) $4 \times 10^{14}$ Hz to $7 \times 10^{14}$ Hz
(c) $4 \times 10^{12}$ Hz to $7 \times 10^{14}$ Hz
(d) $4 \times 10^{11}$ Hz to $7 \times 10^{14}$ Hz

**37** UV radiation is absorbed by
(a) ordinary glass (b) prism
(c) black glass (d) Both (b) and (c)

**38** In the electromagnetic spectrum, X-ray region lies
(a) beyond the microwave region
(b) beyond the gamma-rays
(c) beyond the UV region
(d) above the infrared ray region

**39** One common way to generate X-rays is that
(a) bombard a metal target by high energy electrons
(b) bombard a metal target by low energy neutrons
(c) bombard a metal target by low energy protons
(d) bombard a metal target by high energy neutrons

**40** Wavelength of gamma rays are
(a) $10^{-10}$ m to less than $10^{-14}$ m
(b) $10^{-14}$ m to less than $10^{-10}$ m
(c) $10^{-11}$ m to less than $10^{-14}$ m
(d) $10^{-14}$ nm to less than $10^{-10}$ nm

**41** Gamma rays can be detected by
(a) Photocells (b) Geiger tubes
(c) Thermopiles (d) Diodes

**42** The frequencies of X-rays, $\gamma$-rays and UV-rays are respectively $n_1$, $n_2$, $n_3$, then
(a) $n_1 > n_2, n_2 < n_3$ (b) $n_1 < n_2, n_2 < n_3$
(c) $n_1 < n_2, n_2 > n_3$ (d) $n_1 > n_2, n_2 > n_3$

**43** Arrange the following electromagnetic radiations in the order of increasing wavelength. ***JEE Main 2016***
I. Blue light II. Yellow light III. X-ray IV. Radio wave
(a) IV, II, I, III (b) I, II, IV, III
(c) III, I, II, IV (d) II, I, IV, III

# SPECIAL TYPES QUESTIONS

## I. Assertion and Reason

**Direction** (Q. Nos. 44-48) *In the following questions, a statement of Assertion is followed by a corresponding statement of Reason. Of the following statements, choose the correct one.*

(a) Both Assertion and Reason are correct and Reason is the correct explanation of Assertion.
(b) Both Assertion and Reason are correct but Reason is not the correct explanation of Assertion.
(c) Assertion is correct but Reason is incorrect.
(d) Assertion is incorrect but Reason is correct.

**44** **Assertion** The current through surface $P$ is zero in the cases shown in figure below.

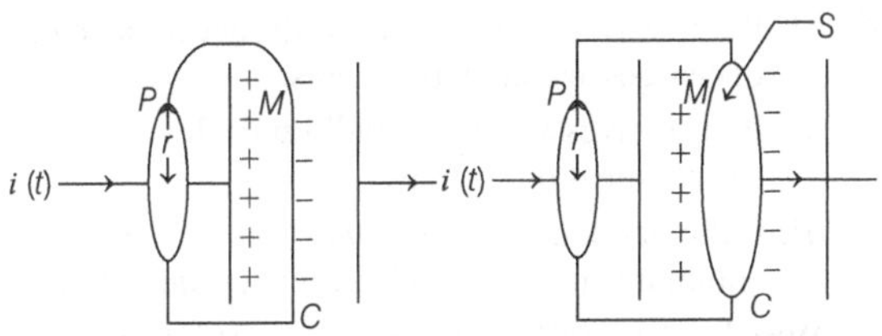


**Reason** While applying Ampere's circuital law to given surfaces with same perimeter, the left hand side of equation $\oint B \cdot dl = \mu_0 i(t)$ has not changed but the right hand side is zero.

**45** **Assertion** An electromagnetic wave is a self-sustaining oscillating wave.

**Reason** An oscillating charge produces an electric field in space, which produces an oscillating magnetic field, which in turn, is a source of electric field and so on.

**46** **Assertion** When the sun shine on our hand, we feel the energy being absorbed from the electromagnetic waves (our hands get warm).

**Reason** Electromagnetic waves transfer momentum to our hand but because $c$ is very large, the amount of momentum transferred is extremely small and we do not feel the pressure.

**47** **Assertion** Infrared waves are sometimes referred to as heat waves.

**Reason** Water molecules present in most materials readily absorb infrared waves. After absorption, their thermal motion increases, i.e. they heat up and heat their surroundings.

**48** **Assertion** UV-radiations are used in LASIK eye surgery.

**Reason** Due to shorter wavelengths, UV-radiations can be focussed into very narrow beams.

## II. Statement Based Questions

**49** Which of the following statement(s) is/are correct?

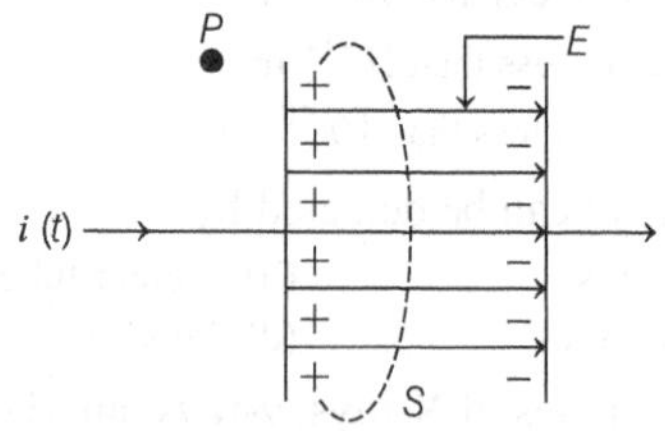


I. The electric field **E** is perpendicular to the surface $S$ in given figure.
II. The magnitude of **E** is same over the area $A$ of the capacitor plates and vanishes outside it.
III. The electric flux $\phi_E$ through the surface $S$ by using the Gauss law, is

$$\phi_E = |\mathbf{E}| A = \frac{1}{\varepsilon_0} \frac{Q}{A} \cdot A = \frac{Q}{\varepsilon_0}$$

Which of the following statement(s) is/are correct?
(a) Both I and II (b) Both II and III
(c) Both I and III (d) I, II and III

**50** I. The total current $i$ is the sum of the conduction current denoted by $i_c$ and the displacement current denoted by $i_d = \varepsilon_0 (d\phi_E / dt)$.
II. Outside the capacitor plates, we have only conduction current $i_c = i$ and displacement current, $i_d \neq 0$.
III. Inside the capacitor, there is no conduction current, i.e. $i_c = 0$ and there is only displacement current, so that $i_d = i$.

Which of the above statement(s) is/are correct?
(a) Both I and III (b) Both II and III
(c) Both I and II (d) I, II and III

**51** I. In all respects, the displacement current has the same physical effects as the conduction current.
II. Due to electric fields in a conducting wire, the displacement current may be zero, since the electric field **E** does not change with time.
III. In a charging capacitor, both conduction and displacement currents may be present in same regions of space.

Which of the above statements is/are correct?
(a) Both I and II
(b) Only III
(c) I, II and III
(d) Only II

**52** I. The source of magnetic field is only the conduction current.
II. An oscillating charge is an example of accelerating charge.
III. The current $i = \varepsilon_0 \frac{d\phi_E}{dt}$ is due to change in electric field, therefore called conduction current.

Which of the following statement(s) is/are incorrect?
(a) Both I and II (b) Both I and III
(c) Both II and III (d) Only III

**53** Consider the figure given below

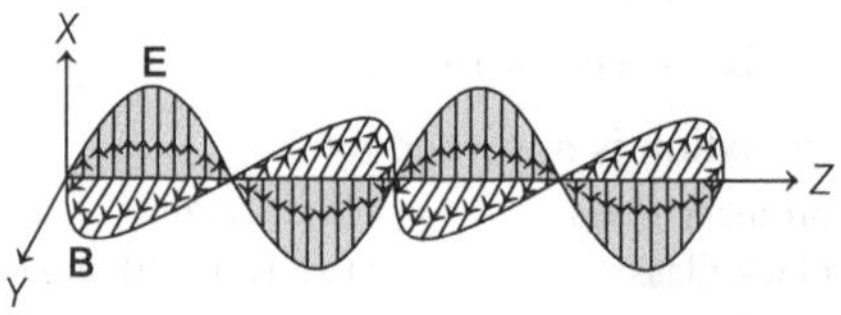


I. A plane electromagnetic wave propagating along the $z$-direction (the fields are shown as a function of the $z$-coordinate, at a given time $t$).
II. The electric field $E$ is along the $Y$-axis and varies sinusoidally with $z$, at a given time.
III. The magnetic field $B$ is along the $Y$-axis and again varies sinusoidally with $z$.

Which of the following statement(s) is/are correct?
(a) Both I and II (b) Both II and III
(c) Both I and III (d) I, II and III

**54** I. Material medium is involved in the vibrations of the electric and magnetic fields.
II. Sound waves in air are longitudinal waves of compression and refraction.

Which of the following statement(s) is/are correct?
(a) Only I (b) Only II
(c) Both I and II (d) None of these

**55** I. The great technological importance of electromagnetic waves stems from their capability to carry energy from one place to another.
II. The radio and TV signals from broadcasting stations carry energy.
III. Light carries energy from the sun to the earth, thus making life possible on the earth.

Which of the following statement(s) is/are correct?
(a) Both I and II (b) Both II and III
(c) Both I and III (d) I, II and III

**56** I. Radio waves are produced by the accelerated motion of charges in conducting wires.
II. Radio waves are used in radio and television communication systems.
III. Cellular phones use radio waves to transmit voice communication in the ultra high frequency band.

Which of the following statement(s) is/are correct?
(a) Both I and III (b) Both II and III
(c) Both I and II (d) I, II and III

**57** Which statement is correct about the symmetrical counterpart and a consequence of the displacement current being a source of a magnetic field?
(a) An electric field changing with time gives rise to a magnetic field.

(b) A magnetic field changing with time gives rise to an electric field.
(c) An emf changing with time gives rise to an electric field.
(d) An displacement current, changing with time gives rise to an electric field.

**58** Which of the following statement is correct in regards to the Maxwell's equations?
(a) The most important prediction to emerge from Maxwell's equations is the existence of magnetic waves.
(b) Maxwell's equations involves only electric and magnetic fields.
(c) The total current has the same value of current $i$ for all surfaces.
(d) We can rephrase Faraday's law of electromagnetic induction that there is an induced emf equal to the rate of change of magnetic flux.

**59** An electromagnetic wave travels in vacuum along $z$-direction $\mathbf{E} = (E_1\hat{\mathbf{i}} - E_2\hat{\mathbf{j}})\cos(kz - \omega t)$. Find the correct statement from the following.
(a) The associated magnetic field is given as $\mathbf{B} = \frac{1}{c}(E_1\hat{\mathbf{i}} - E_2\hat{\mathbf{j}})\sin(kz - \omega t)$
(b) The associated magnetic field is given as $\mathbf{B} = \frac{1}{c}(E_1\hat{\mathbf{i}} + E_2\hat{\mathbf{j}})\cos(kz - \omega t)$
(c) The given electromagnetic field is circularly polarised.
(d) The given electromagnetic wave is plane polarised.

**60** An electromagnetic wave travelling along $Z$-axis is given as $\mathbf{E} = E_0 \cos(kz - \omega t)$. Choose the correct statement from the following.
(a) The associated magnetic field is given as $\mathbf{B} = \frac{1}{c}(\hat{\mathbf{j}} \times \mathbf{E})$.
(b) The electromagnetic field can be written in terms of the associated magnetic field as $\mathbf{E} = c\,(\mathbf{B} \times \hat{\mathbf{j}})$.
(c) $\hat{\mathbf{k}} \cdot \mathbf{E} = 0$, $\hat{\mathbf{k}} \cdot \mathbf{B} = 0$.
(d) $\hat{\mathbf{k}} \times \mathbf{E} = 0$, $\hat{\mathbf{k}} \times \mathbf{B} = 0$.

**61** Which of the following statement is incorrect?
(a) Electromagnetic waves cannot be deflected by any field.
(b) The velocity of light in a medium is $v = \frac{1}{\sqrt{\mu\varepsilon}}$, thus it depends on both the electric and magnetic properties of the medium.
(c) The constancy of the velocity of electromagnetic waves in vacuum is used to define a standard of time.
(d) The direction of propagation vector **k** describes the direction of propagation of the wave.

**62** Which of the following statement is correct for a plane electromagnetic wave?
(a) Average value of electric field is zero only.
(b) Average value of magnetic field is zero only.
(c) Average value of both electric and magnetic fields is zero.
(d) Average value of magnetic energy is zero.

**63** Which of the following statement(s) is/are incorrect about UV-rays?
(a) Ultraviolet rays cover wavelengths ranging from $4 \times 10^{-7}$ m(400 nm) to $6 \times 10^{-10}$ m(0.6 nm).
(b) UV radiation is produced by special lamps and very hot bodies.
(c) The sun is an important source of ultraviolet light.
(d) The ozone layer in the atmosphere cannot absorb UV-rays.

**64** Which of the following statement(s) is/are incorrect?
(a) The wavelength of microwave is greater than that of UV-rays.
(b) The wavelength of infrared rays is lesser than that of UV-rays.
(c) The wavelength of microwave is greater than that of infrared rays.
(d) Gamma rays have shortest wavelengths in the electromagnetic spectrum.

**65** Which of the following statement(s) is/are incorrect?
(a) The classification of electromagnetic waves according to frequency is called electromagnetic spectrum.
(b) The AM band is from 530 kHz to 1710 kHz.
(c) The FM radio band extends from 88 MHz to 108 MHz.
(d) Infrared waves are produced by cold bodies and molecules.

**66** Which of the following statements about EM waves is/are correct?
(a) Gamma rays are produced by in nuclear reactions.
(b) Ultraviolet waves are used in treatment of swollen joints.
(c) Snakes can not detect infrared rays.
(d) Electromagnetic waves do not exhibit polarisation while sound waves.

## III. Matching Type

**67** Match the Column I (equation) with Column II (law) and select the correct answer from the codes given below.

| Column I | | Column II |
|---|---|---|
| A. $\oint \mathbf{E} \cdot d\mathbf{A} = Q/\varepsilon_0$ | 1. | Faraday's law |
| B. $\oint \mathbf{B} \cdot d\mathbf{A} = 0$ | 2. | Ampere-Maxwell law |
| C. $\oint \mathbf{E} \cdot d\mathbf{l} = -\frac{d\phi_B}{dt}$ | 3. | Gauss's law for electricity |
| D. $\oint \mathbf{B} \cdot d\mathbf{l} = \mu_0 i_c + \mu_0\varepsilon_0 \frac{d\phi_E}{dt}$ | 4. | Gauss's law for magnetism |

| | A | B | C | D |
|---|---|---|---|---|
| (a) | 4 | 3 | 2 | 1 |
| (b) | 3 | 2 | 1 | 4 |
| (c) | 3 | 4 | 1 | 2 |
| (d) | 1 | 2 | 3 | 4 |

**68** Match the Column I (EM wave) with Column II (frequency band) and select the correct answer from the codes given below.

| Column I | | Column II | |
|---|---|---|---|
| A. | Radio | 1. | 54 MHz |
| B. | Amplitude modulated | 2. | 88 MHz to 108 MHz |
| C. | Short wave bands | 3. | 530 kHz to 1710 kHz |
| D. | TV wave | 4. | 500 kHz to 1000 MHz |
| E. | Frequency modulated | 5. | 54 MHz to 890 MHz |

| | A | B | C | D | E |
|---|---|---|---|---|---|
| (a) | 3 | 1 | 5 | 2 | 4 |
| (b) | 1 | 5 | 2 | 3 | 4 |
| (c) | 4 | 1 | 5 | 2 | 3 |
| (d) | 4 | 3 | 1 | 5 | 2 |

**69** Match the Column I (EM wave) with Column II (application association) and select the correct answer from the codes given below. **JEE Main 2014**

| Column I | | Column II | |
|---|---|---|---|
| A. | Infrared waves | 1. | To treat muscular strain |
| B. | Radio waves | 2. | For broadcasting |
| C. | X-rays | 3. | To detect fracture of bones |
| D. | Ultraviolet | 4. | Absorbed by the ozone layer of the atmosphere |

| | A | B | C | D |
|---|---|---|---|---|
| (a) | 4 | 3 | 2 | 1 |
| (b) | 1 | 2 | 4 | 3 |
| (c) | 3 | 2 | 1 | 4 |
| (d) | 1 | 2 | 3 | 4 |

# NCERT & NCERT Exemplar

## MULTIPLE CHOICE QUESTIONS

### NCERT

**70** A parallel plate capacitor made of circular plates each of radius $R = 6.0$ cm, has a capacitance $C = 100$ pF. The capacitor is connected to a 230 V AC supply with a frequency of 300 rads$^{-1}$.

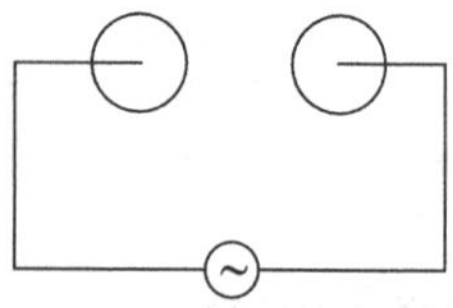

What is the rms value of conduction current?

(a) $6.9\times 10^6$ A (b) 6.9 μA
(c) $5.9\times 10^6$ A (d) 5.9 μA

**71** Suppose that the electric field amplitude of an electromagnetic wave is $E_0 = 120$ N/C and frequency is $\nu = 50$ MHz.

The amplitude of $B_0$ and the value of $k$ are

(a) 200 nT, 0.5radm$^{-1}$ (b) 300 nT, 2 radm$^{-1}$
(c) 400 nT, 1 radm$^{-1}$ (d) 500 nT, 4 radm$^{-1}$

**72** About 5% of the power of a 100W light bulb is converted to visible radiation. What is the average intensity of visible radiation at a distance of 1m from the bulb? Assume that the radiation is emitted isotropically and neglect reflection.

(a) 0.2 Wm$^{-2}$ (b) 0.4 Wm$^{-2}$ (c) 0.8 Wm$^{-2}$ (d) 1.6 Wm$^{-2}$

**73** The amplitude of the magnetic field part of a harmonic electromagnetic wave in vacuum is $B_0 = 510$ nT. What is the amplitude of the electric field part of the wave?

(a) 130 NC$^{-1}$ (b) 153 NC$^{-1}$
(c) 170 NC$^{-1}$ (d) 190 NC$^{-1}$

**74** Suppose that the electric field part of an electromagnetic wave in vacuum is

$$E = \{(3.1\ \text{NC}^{-1})\cos[(1.8\ \text{radm}^{-1})y + (5.4\times 10^6\ \text{rads}^{-1})t]\}$$

What is the value of wavelength $\lambda$ ?

(a) 1.5 m (b) 2.5 m
(c) 3.5 m (d) 4.5 m

### NCERT Exemplar

**75** A linearly polarised electromagnetic wave given as $\mathbf{E} = E_0\hat{\mathbf{i}}\cos(kz - \omega t)$ is incident normally on a perfectly reflecting infinite wall at $z = a$. Assuming that the material of the wall is optically inactive, the reflected wave will be given as

(a) $\mathbf{E}_r = E_0\hat{\mathbf{i}}(kz - \omega t)$
(b) $\mathbf{E}_r = E_0\hat{\mathbf{i}}\cos(kz + \omega t)$
(c) $\mathbf{E}_r = -E_0\hat{\mathbf{i}}\cos(kz + \omega t)$
(d) $\mathbf{E}_r = E_0\hat{\mathbf{i}}\sin(kz - \omega t)$

**76** Light with an energy flux of 20 W $cm^{-2}$ falls on a non-reflecting surface at normal incidence. If the surface has an area of 30 $cm^2$, the total momentum delivered (for complete absorption) during 30 min is

(a) $36\times 10^{-5}$ kg-$ms^{-1}$ (b) $36\times 10^{-4}$ kg-$ms^{-1}$
(c) $108\times 10^{4}$ kg-$ms^{-1}$ (d) $1.08\times 10^{7}$ kg-$ms^{-1}$

**77** The electric field intensity produced by the radiations coming from 100 W bulb at a 3 m distance is $E$. The electric field intensity produced by the radiations coming from 50 W bulb at the same distance is

(a) $\frac{E}{2}$ (b) $2E$ (c) $\frac{E}{\sqrt{2}}$ (d) $\sqrt{2}E$

**78** If **E** and **B** represent electric and magnetic field vectors of the electromagnetic wave, the direction of propagation of electromagnetic wave is along

(a) **E** (b) **B** (c) **B**× **E** (d) **E**× **B**

**79** The ratio of contributions made by the electric and magnetic field components to the intensity of an EM wave is

(a) $c:1$ (b) $c^2:1$ (c) $1:1$ (d) $\sqrt{c}:1$

# Answers

## > Mastering NCERT with MCQs

| | | | | | | | | | |
|---|---|---|---|---|---|---|---|---|---|
| 1 (d) | 2 (a) | 3 (d) | 4 (d) | 5 (c) | 6 (c) | 7 (c) | 8 (a) | 9 (a) | 10 (d) |
| 11 (c) | 12 (a) | 13 (b) | 14 (c) | 15 (a) | 16 (c) | 17 (c) | 18 (b) | 19 (d) | 20 (b) |
| 21 (d) | 22 (d) | 23 (c) | 24 (a) | 25 (b) | 26 (a) | 27 (c) | 28 (b) | 29 (b) | 30 (d) |
| 31 (d) | 32 (c) | 33 (d) | 34 (a) | 35 (a) | 36 (b) | 37 (a) | 38 (c) | 39 (a) | 40 (a) |
| 41 (b) | 42 (c) | 43 (c) | | | | | | | |

## > Special Types Questions

| | | | | | | | | | |
|---|---|---|---|---|---|---|---|---|---|
| 44 (a) | 45 (a) | 46 (b) | 47 (a) | 48 (a) | 49 (d) | 50 (a) | 51 (a) | 52 (b) | 53 (c) |
| 54 (b) | 55 (d) | 56 (d) | 57 (a) | 58 (c) | 59 (d) | 60 (c) | 61 (c) | 62 (c) | 63 (d) |
| 64 (b) | 65 (d) | 66 (a) | 67 (c) | 68 (d) | 69 (d) | | | | |

## > NCERT & NCERT Exempar MCQs

| | | | | | | | | | |
|---|---|---|---|---|---|---|---|---|---|
| 70 (b) | 71 (c) | 72 (b) | 73 (b) | 74 (c) | 75 (b) | 76 (b) | 77 (a) | 78 (d) | 79 (c) |

# Hints & Explanations

**2** (*a*) To find the magnetic field at a point such as $P$, in a region outside the parallel plate capacitor, we consider a plane circular loop of radius $r$, whose plane is perpendicular to the direction of the current carrying wire and which is centred symmetrically with respect to the wire.

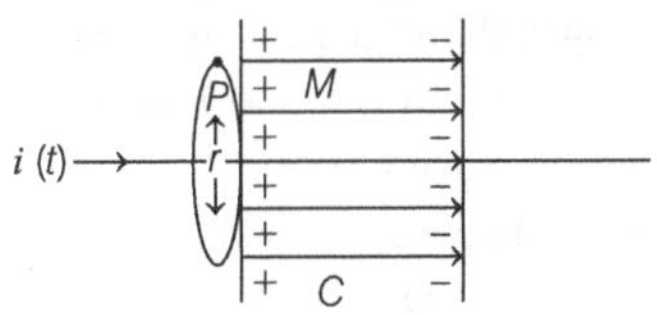


From symmetry, the magnetic field is directing along the circumference of the circular loop and it is same in magnitude at all the points on the loop.

For other shapes, the magnetic field is zero at $P$.

**3** (*d*) Displacement current $i_d$ is equal to charging current ($i_d = dq/dt$).

Given, $q = q_0 \cos 2\pi\nu t$

$$\therefore \quad i_d = \frac{dq}{dt} = \frac{d}{dt}(q) = \frac{d}{dt}(q_0 \cos 2\pi\nu t)$$

$$= q_0(-\sin 2\pi\nu t)(2\pi\nu)$$

$$= -2\pi q_0 \nu \sin(2\pi\nu t)$$

Thus, displacement current is $-2\pi q_0 \nu \sin(2\pi\nu t)$.

**4** (*d*) As, displacement current,

$$\therefore \quad i_d = \varepsilon_0 \frac{d\phi_E}{dt}$$

$$= \varepsilon_0 A \frac{d}{dt}\left(\frac{V}{d}\right) \qquad \left(\because \phi_E = EA = \frac{AV}{d}\right)$$

$$\Rightarrow \quad i_d = \frac{\varepsilon_0 A}{d} \times \frac{dV}{dt} = C\frac{dV}{dt} \qquad \left(\because C = \frac{\varepsilon_0 A}{d}\right)$$

Given, $i_d = 10$ A and $C = 10^{-6}$ F

$$\Rightarrow \quad \frac{dV}{dt} = \frac{i_d}{C} = \frac{10}{10^{-6}} = 10^7 \text{ Vs}^{-1}$$

Thus, an instantaneous rate of voltage is $10^7$ $Vs^{-1}$.

**5** (*c*) Given, radius of plate, $R = 10$ cm $= 0.1$ m, cross-section area of a capacitor,

$$A = \pi R^2 = \pi(0.1)^2$$

$$= 3.14\times 10^{-2} \text{ m}^2$$

Rate of change of electric field,

$$\frac{dE}{dt} = 5\times 10^{13}\ \text{Vms}^{-1}$$

Thus, displacement current,

$$i_d = \varepsilon_0 \frac{d\phi_E}{dt} = \varepsilon_0 A \frac{dE}{dt} \quad [\because \phi_E = EA]$$

$$= (8.85\times 10^{-12})(3.14\times 10^{-2})(5\times 10^{13})$$

$$\therefore \quad i_d = 13.9\ \text{A}$$

**6** (c) Given, $C = 8\,\mu\text{F} = 8\times 10^{-6}$ F, $\nu = 3.00$ kHz

and $V_0 = 30.0$ V

Clearly, $\omega = 2\pi\nu = 2\pi\times(3.00\times 10^3\ \text{s}^{-1})$

$= 6\pi\times 10^3\ \text{s}^{-1}$

Voltage across the capacitor, $V = V_0 \sin \omega t$

$= (30.0)\sin(6\pi\times 10^3\ t)$

Displacement current,

$$i_d = i_c = \frac{dq}{dt} = \frac{d}{dt}(q) = \frac{d}{dt}(CV) = C\frac{dV}{dt}$$

$$= (8\times 10^{-6})\frac{d}{dt}[30.0\sin(6\pi\times 10^3\ t)]$$

$$= (8\times 10^{-6})(30.0)\frac{d}{dt}[\sin(6\pi\times 10^3\ t)]$$

$$= (8\times 10^{-6})(30.0)(6\pi\times 10^3)\cos(6\pi\times 10^3\ t)$$

$$= 4.52\cos(6\pi\times 10^3\ t)\text{A}$$

Hence, the displacement current varies sinusoidally with time and has a maximum value of 4.52 A.

**7** (c) Given, rate of loss of charge is $2\times 10^{-7}\text{Cs}^{-1}$.

Magnitude of displacement current is given by

$$i_d = i_C = \left|\frac{dq}{dt}\right| = 2\times 10^{-7}\ \text{Cs}^{-1} = 2\times 10^{-7}\ \text{A}$$

$\therefore$ Displacement current is $2\times 10^{-7}$ A.

**8** (a) Given, $C = 20\mu\text{F} = 20\times 10^{-6}$ F and $\frac{dV}{dt} = 3$ V/s

The displacement current in a circuit is given by

$$i_d = \varepsilon_0 \frac{d\phi}{dt} \quad \text{[from Maxwell's equation]}$$

$$= \varepsilon_0 \frac{d}{dt}(EA) \quad [\because \phi = EA]$$

$$= \varepsilon_0 A \frac{d}{dt}\left(\frac{V}{d}\right) = \frac{\varepsilon_0 A}{d}\frac{dV}{dt} \quad [\because V = Ed]$$

As the capacitance, $C = \frac{\varepsilon_0 A}{d}$

$$\therefore \quad i_d = C\frac{dV}{dt}$$

Substituting the given values, we get

$$i_d = 20\times 10^{-6}\times 3$$

$$= 60\times 10^{-6}\ \text{A} = 60\,\mu\text{A}$$

As displacement current is in between the plates of capacitor and conduction current is in the connecting wires which are equal to each other.

Hence, $i_c = i_d = 60\,\mu\text{A}$

**12** (a) Given, frequency with which the charged particles oscillates about its mean position $= 10^9$Hz.

So, frequency of electromagnetic waves produced by the charged particle is equals to the frequency of oscillation of the charge $\nu = 10^9$ Hz.

Wavelength of electromagnetic wave,

$$\lambda = \frac{c}{\nu} = \frac{3\times 10^8}{10^9} = 0.3\ \text{m}$$

Hence, option (a) is correct.

**13** (b) Here, given that velocity of EM wave, $\mathbf{v} = v\hat{\mathbf{i}}$

and Instantaneous oscillating electric field, $\mathbf{E} = E\hat{\mathbf{j}}$.

As we already know that, during the propagation of EM waves through a medium oscillating electric and magnetic field vectors are mutually perpendicular to each other and the direction of propagation of the wave $(\mathbf{E}\times\mathbf{B})$.

i.e. $\mathbf{E}\times\mathbf{B} = \mathbf{v}$

$\Rightarrow$ $(E\hat{\mathbf{j}})\times\mathbf{B} = v\hat{\mathbf{i}}$ ...(i)

As we know that, from vector algebra,

$\hat{\mathbf{j}}\times\hat{\mathbf{k}} = \hat{\mathbf{i}}$ ...(ii)

Comparing Eqs. (i) and (ii), we get

$$\mathbf{B} = B\hat{\mathbf{k}}$$

where, $B$ be the magnitude of magnetic field.

Thus, we can say that the direction of oscillating magnetic field of the EM wave will be along $+z$-direction.

**14** (c) Given, $B_y = 3\times 10^{-7}\sin(10^3 x + 6.28\times 10^{12}\ t)$

Comparing with the general equation

$B_y = B_0\sin(kx + \omega t)$, we get

$$k = 10^3 \Rightarrow \frac{2\pi}{\lambda} = 10^3 \quad \left(\because k = \frac{2\pi}{\lambda}\right)$$

$$\Rightarrow \lambda = \frac{2\pi}{10^3} = \frac{2\times 3.14}{10^3} = 6.28\times 10^{-3}\ \text{m} = 0.63\ \text{cm}$$

Thus, wavelength of electromagnetic wave is 0.63 cm

**15** (a) Given, $\mathbf{E} = 10\cos(10^7 t + kx)\hat{\mathbf{j}}\ \text{Vm}^{-1}$

Comparing with the general equation

$E = E_0\cos(\omega t + kx)$, we get

Amplitude of wave $= 10\ \text{Vm}^{-1}$

and angular frequency, $\omega = 10^7$.

$$\because \quad c = \frac{\omega}{k} \Rightarrow 3\times 10^8 = \frac{10^7}{k}$$

$$\Rightarrow \quad k = \frac{1}{30} = 0.033\ \text{rad m}^{-1} \quad \left[\because k = \frac{2\pi}{\lambda}\right]$$

$$\Rightarrow \quad \frac{2\pi}{\lambda} = \frac{1}{30} \Rightarrow \lambda = 2\pi\times 30 = 188.4\ \text{m}$$

Thus, only option (a) is correct.

**16** (c) **Key Idea** Standard expression of electromagnetic wave is given by

$$\mathbf{E} = E_0\hat{\mathbf{n}}\,[\sin(\omega t - \mathbf{k}\cdot\hat{\mathbf{r}})] \quad ...(i)$$

Here, $\mathbf{k}$ is the propagation vector.

Direction of propagation in this case is $\hat{\mathbf{k}}$.

Given expression of electromagnetic wave,

$$\mathbf{E} = E_0\hat{\mathbf{n}} \sin [\omega t + (6y - 8z)]$$

$$\Rightarrow \quad \mathbf{E} = E_0\hat{\mathbf{n}}\sin [\omega t - (8z - 6y)] \quad \text{...(ii)}$$

Comparing Eq. (ii) with Eq. (i), we get

$$\mathbf{k}\cdot\hat{\mathbf{r}} = 8z - 6y \quad \text{...(iii)}$$

Here, $\hat{\mathbf{r}} = x\hat{\mathbf{i}} + y\hat{\mathbf{j}} + z\hat{\mathbf{k}}$ and $\mathbf{k} = k_x\hat{\mathbf{i}} + k_y\hat{\mathbf{j}} + k_z\hat{\mathbf{k}}$

$$\therefore \quad \mathbf{k}\cdot\hat{\mathbf{r}} = xk_x + yk_y + zk_z \quad \text{...(iv)}$$

From Eqs. (iii) and (iv), we get

$$xk_x = \text{zero} \Rightarrow k_x = 0$$
$$yk_y = -6y \Rightarrow k_y = -6$$
$$zk_z = 8z \Rightarrow k_z = 8$$

Hence, $\mathbf{k} = -6\hat{\mathbf{j}} + 8\hat{\mathbf{k}}$

So, direction of propagation,

$$\hat{\mathbf{s}} = \hat{\mathbf{k}} = \frac{\mathbf{k}}{|\mathbf{k}|} = \frac{-6\hat{\mathbf{j}} + 8\hat{\mathbf{k}}}{\sqrt{6^2 + 8^2}} = \frac{-6\hat{\mathbf{j}} + 8\hat{\mathbf{k}}}{10} = \frac{-3\hat{\mathbf{j}} + 4\hat{\mathbf{k}}}{5}$$

**17** (*c*) Given, $E_0 = 120\ \text{NC}^{-1}$, $\nu = 50.0\ \text{MHz} = 50\times 10^6$ Hz

$\therefore$ Angular frequency, $\omega = 2\pi\nu$

$$= (2\times 3.14\ \text{rad})\ (50\times 10^6\ \text{Hz})$$
$$= 3.14\times 10^8\ \text{rads}^{-1}$$

Wave constant, $k = \dfrac{\omega}{c} = \dfrac{3.14\times 10^8\ \text{rads}^{-1}}{3\times 10^8\ \text{ms}^{-1}}$

$$= 1.05\ \text{rad m}^{-1}$$

According to the condition given in the questions, the wave is propagating along *X*-axis, this means **E** should be along *Y*-axis and **B** should be along *Z*-axis.

Clearly, $\mathbf{E} = E_0 \sin\ (kx - \omega t)\hat{\mathbf{j}}$

$$= [120\ \text{N C}^{-1}\ \sin\ \{ (1.05\ \text{rad m}^{-1})x - (3.14\times 10^8\ \text{rads}^{-1})t\}]\ \hat{\mathbf{j}}$$

**18** (*b*) Given, root mean square value of electric field,

$$E_{\text{rms}} = 6\ \text{V/m}$$

We know that, peak value of electric field,

$$E_0 = \sqrt{2}\ E_{\text{rms}} \Rightarrow E_0 = \sqrt{2}\times 6\ \text{V/m}$$

Also, we know that, $c = \dfrac{E_0}{B_0}$

where, $c$ = speed of light in vacuum

and $B_0$ = peak value of magnetic field.

$$\Rightarrow \quad B_0 = \frac{E_0}{c} = \frac{\sqrt{2}\times 6}{3\times 10^8} = \frac{8.48}{3}\times 10^{-8} = 2.83\times 10^{-8}\ \text{T}$$

Thus, peak value of magnetic field $B_0$ is $2.83\times 10^{-8}$ T.

**19** (*d*) Given, $\mathbf{B} = 5\times 10^{-8}\hat{\mathbf{j}}$ and $\mathbf{v} = 3\times 10^8\hat{\mathbf{k}}$

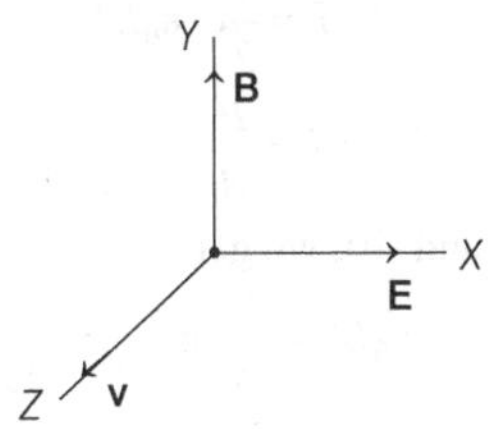


Using $\mathbf{E} = \mathbf{B}\times \mathbf{v}$, we have

$$\mathbf{E} = (5\times 10^{-8}\hat{\mathbf{j}})\times (3\times 10^8\hat{\mathbf{k}}) = 15\hat{\mathbf{i}}\ \text{Vm}^{-1}$$

**20** (*b*) Given, $\nu = 2\times 10^{10}$ Hz and $E_0 = 48\ \text{Vm}^{-1}$.

Using the formula, $c = \dfrac{E_0}{B_0}$

The amplitude of the oscillating magnetic field,

$$B_0 = \frac{E_0}{c} = \frac{48}{3\times 10^8} = 1.6\times 10^{-7}\ \text{T}$$

Thus, amplitude of magnetic field is $1.6\times 10^{-7}$ T.

**21** (*d*) An electromagnetic wave is a travelling wave in which electric vector **E** is perpendicular to magnetic vector **B** and both of these vectors are also perpendicular to the direction of propagation of electromagnetic wave.

So, it is given that

$$\mathbf{B} = 3\times 10^{-8}\sin(1.6\times 10^3 x + 48\times 10^{10} t)\hat{\mathbf{j}}\ \text{T}$$
$$= B_0 \sin(kx + \omega t)\hat{\mathbf{j}}\text{T}$$

Here, $k = 1.6\times 10^3$ units and $\omega = 48\times 10^{10}$ units

Then, it can be implied that electromagnetic wave is travelling along $-X$-axis (this is obtained from phase $kx + \omega t$) and its magnetic vector **B** is directed along $+Y$-axis (this is indicated by unit vector $\hat{\mathbf{j}}$).

Hence, electric field vector is along $+Z$-axis and its magnitude is given by

$$E_0 = c\cdot B_0 = 3\times 10^8 \times 3\times 10^{-8} = 9\ \text{V/m}$$

$\therefore$ Expression for $\mathbf{E} = 9\sin(kx + \omega t)\hat{\mathbf{k}}$

$$= 9\sin(1.6\times 10^3 x + 48\times 10^{10} t)\hat{\mathbf{k}}\ \text{V/m}$$

**23** (*c*) Speed of progressive wave is given by, $v = \dfrac{\omega}{k}$

As electric field in air is,

$$E_1 = E_{01}\mathbf{x}\cos\left(\frac{2\pi\nu z}{c} - 2\pi\nu t\right)$$

$$\therefore \text{ Speed in air} = \frac{2\pi\nu}{\left(\frac{2\pi\nu}{c}\right)} = c$$

Also, $$c = \frac{1}{\sqrt{\mu_0\varepsilon_{r_1}\varepsilon_0}} \quad \text{...(i)}$$

In medium, $E_2 = E_{02}x\cos\ (2kz - kct)$

$$\therefore \text{ Speed in medium} = \frac{kc}{2k} = \frac{c}{2}$$

Also, $$\frac{c}{2} = \frac{1}{\sqrt{\mu_0\varepsilon_{r_2}\varepsilon_0}} \quad \text{...(ii)}$$

As medium is non-magnetic medium,

$$\mu_{\text{medium}} = \mu_{\text{air}}$$

On dividing Eq. (i) by Eq. (ii), we have

$$2 = \sqrt{\frac{\varepsilon_{r_2}}{\varepsilon_{r_1}}} \Rightarrow \frac{\varepsilon_{r_1}}{\varepsilon_{r_2}} = \frac{1}{4}$$

**24** (*a*) The speed of any electromagnetic wave (EM) like light in any medium of permeability $\mu$ and permittivity $\varepsilon$ is given by

$$v = \frac{1}{\sqrt{\mu\varepsilon}}$$

Also, $\mu = \mu_0\mu_r$ and $\varepsilon = \varepsilon_0\varepsilon_r$.

where, $\mu_0$ and $\varepsilon_0$ are permeability and permittivity in free space and $\mu_r$ & $\varepsilon_r$ are relative permeability & relative permittivity of medium, respectively.

Given, $\mu_r = 1.0$ and $\varepsilon_r = 1.44$

$$\therefore \quad v = \frac{1}{\sqrt{\mu_r\varepsilon_r\mu_0\varepsilon_0}} = \frac{c}{\sqrt{\mu_r\varepsilon_r}} \quad \left[\because c = \frac{1}{\sqrt{\mu_0\varepsilon_0}}\right]$$

$$= \frac{3\times 10^8}{\sqrt{1\times 1.44}} = 2.5\times 10^8 \text{ ms}^{-1}$$

Thus, velocity of light in this medium is $2.5\times 10^8 \text{ ms}^{-1}$.

**26** (*a*) Energy density of an electomagnetic wave in electric field,

$$U_E = \frac{1}{2}\varepsilon_0 \cdot E^2 \quad \text{...(i)}$$

Energy density of an electromagnetic wave in magnetic field,

$$U_B = \frac{B^2}{2\mu_0} \quad \text{...(ii)}$$

where, $E$ = electric field, $B$ = magnetic field,

$\varepsilon_0$ = permittivity of medium and

$\mu_0$ = magnetic permeability of medium.

From the theory of electro-magnetic waves, the relation between $\mu_0$ and $\varepsilon_0$ is

$$c = \frac{1}{\sqrt{\mu_0\varepsilon_0}} \quad \text{...(iii)}$$

where, $c$ = velocity of light $= 3\times 10^8$ m/s

and $$\frac{E}{B} = c \quad \text{...(iv)}$$

Dividing Eq. (i) by Eq. (ii), we get

$$\frac{U_E}{U_B} = \frac{\frac{1}{2}\varepsilon_0 E^2}{\frac{1}{2}B^2 \times \frac{1}{\mu_0}} = \frac{\mu_0\varepsilon_0 E^2}{B^2} \quad \text{...(v)}$$

Using Eqs. (iii), (iv) and (v), we get

$$\frac{U_E}{U_B} = \frac{c^2}{c^2} = 1$$

Therefore, $U_E = U_B$

**27** (c) Given, energy flux, $\phi = 18 \text{ Wcm}^{-2}$, area $A = 20\text{cm}^2$

and $t = 30 \text{ min} = (30\times 60)$ s

Therefore, total energy falling on the surface is

$$U = \phi At = 18\times 20\times (30\times 60)$$
$$= 6.48\times 10^5 \text{ J}$$

Therefore, the total momentum delivered (for complete absorption) is

$$p = U/c = \frac{6.48\times 10^5 \text{ J}}{3\times 10^8 \text{ ms}^{-1}} = 2.16\times 10^{-3} \text{ Jms}^{-1}$$

Therefore, average force exerted on the surface is

$$F = \frac{p}{t} = \frac{2.16\times 10^{-3}}{30\times 60} = 1.2\times 10^{-6} \text{ N}$$

**28** (*b*) Given, energy flux, $\phi = 40 \text{ W cm}^{-2}$, surface area, $A = 20 \text{ cm}^2$, time $t = 10$ min $= 10\times 60$ s.

Total energy falling on the surface in time $t$ is

$$U = \phi At = 40\times 20\times (10\times 60) = 4.8\times 10^5 \text{ J}$$

Momentum of the incident light,

$$p = \frac{U}{c} = \frac{4.8\times 10^5}{3\times 10^8} = 16\times 10^{-4} \text{ kg-ms}^{-1}$$

Momentum of the reflected light = 0

(complete absorption)

$\therefore$ Momentum delivered to the surface

$$= 16\times 10^{-4} - 0 = 16\times 10^{-4} \text{ kg-ms}^{-1}$$

**29** (*b*) Initial momentum of surface, $p_i = \frac{E}{c}$

where, $c$ = velocity of light (constant) and $E$ is energy.

Since, the surface is perfectly reflecting, so the same momentum will be reflected completely.

Final momentum, $p_f = \frac{-E}{c}$ (negative value)

$\therefore$ Change in momentum, $\Delta p = p_f - p_i$

$$= -\frac{E}{c} - \frac{E}{c} = -\frac{2E}{c}$$

Thus, momentum transferred to the surface is

$$\Delta p' = |\Delta p| = \frac{2E}{c}$$

**30** (*d*) Given,

Power of laser beam $(P) = 27\text{mW} = 27\times 10^{-3}$ W

Area of cross-section $(A) = 10\text{mm}^2 = 10\times 10^{-6}\text{m}^2$

Permittivity of free space $(\varepsilon_0) = 9\times 10^{-12}$ SI unit

Speed of light $(c) = 3\times 10^8$ m/s

Intensity of electromagnetic wave is given by the relation

$$I = \frac{1}{2}nc\varepsilon_0 E^2$$

where, $n$ is refractive index, for air $n = 1$.

$$\therefore \quad I = \frac{1}{2}c\cdot\varepsilon_0 E^2 \quad \text{...(i)}$$

Also, $$I = \frac{P}{A} \quad \text{...(ii)}$$

From Eq. (i) and (ii), we get

$$\frac{1}{2}c\varepsilon_0 E^2 = \frac{P}{A} \text{ or } E^2 = \frac{2P}{Ac\varepsilon_0}$$

or $$E = \sqrt{\frac{2 \times 27 \times 10^{-3}}{10 \times 10^{-6} \times 3 \times 10^{8} \times 9 \times 10^{-12}}}$$
$$\simeq 1.4 \times 10^3 \text{ V/m} = 1.4 \text{ kV/m}$$

**31** (*d*) Given, $P = 100$ W, $R = 3$ m and efficiency $= 2.5\%$

As, bulb is a point source, so it radiates light uniformly in all directions.

The intensity $I$ at a distance $R$ is
$$I = \frac{\text{power}}{\text{area}} = \frac{100 \times 2.5\%}{4\pi R^2}$$
$$= \frac{100 \times 2.5}{100 \times 4 \times 3.14 \times (3)^2} = 0.022 \text{ W/m}^2$$

As intensity is equally distributed between electric and magnetic fields, so
$$I_E = I_B = \frac{1}{2} I = \frac{1}{2} \times \varepsilon_0 E_{\text{rms}}^2 c$$
$$E_{\text{rms}} = \sqrt{\frac{I}{\varepsilon_0 c}} = \sqrt{\frac{0.022}{(8.85 \times 10^{-12})(3 \times 10^8)}} \text{ V/m}$$
$$= 2.88 \text{ V/m}$$

Therefore, peak electric field,
$$E_0 = \sqrt{2} E_{\text{rms}} = \sqrt{2} \times 2.88 = 4.07 \text{ V/m}$$

**34** (*a*) A microwave oven uses microwaves,which are basically radio waves, to cook food items, containing water molecules. In it, the frequency used by microwave oven, matches the resonant frequency of water molecule, which causes water molecules in the food to vibrate and produce thermal energy in a process known as dielectric heating. Thereby facilitating in the heating up the food items.
Thus, option (a) is correct.

**42** (*c*) As, we know the increasing order of wavelength is
$$\gamma\text{-rays} < \text{X-rays} < \text{UV-rays}$$
So, frequency order with be
$$\gamma\text{-rays} > \text{X-rays} > \text{UV-rays}$$
Given, frequency of X-ray $= n_1$,
frequency of $\gamma$-ray $= n_2$ and frequency of UV-ray $= n_3$.
$$\because \quad n_2 > n_1 > n_3$$
Therefore, $n_1 < n_2$ and $n_2 > n_3$.

**44** (*a*) On applying Ampere's circuital law to such surfaces with the same perimeter, we find that the left hand side of equation $\oint B.dl = \mu_0 i(t)$ has not changed but the right hand side is zero.
Since, no current passes through the surface, i.e. $i = 0$
$\oint B \cdot dl = 0$.

Therefore, Assertion and Reason are correct and Reason is the correct explanation of Assertion.

**45** (*a*) An oscillating charge produces an electric field in space, which produces an oscillating magnetic field, which in turn, is a source of electric field. Thus, EM wave is a self sustaining oscillating wave.
Therefore, Assertion and Reason are correct and Reason is the correct explanation of Assertion.

**46** (*b*) When the sun shines on our hand, we feel the energy being absorbed from the electromagnetic waves (our hands get warm).
Electromagnetic waves also transfer momentum to our hand but because $c$ is very large, the amount of momentum transferred is extremely small and we do not feel the pressure.
Since, momentum, $p = \frac{v}{c} \Rightarrow p \propto \frac{1}{c}$ $\quad (\because v = \text{constant})$

Therefore, Assertion and Reason are correct but Reason is not the correct explanation of Assertion.

**47** (*a*) Infrared waves are sometimes referred to as heat waves. Because water molecules present in most materials readily absorb infrared waves (many other molecules, e.g. $CO_2$, $NH_3$, also absorb infrared waves). After absorption, their thermal motion increases, i.e. they heat up and heat their surroundings.
Therefore, Assertion and Reason are correct and Reason is the correct explanation of Assertion.

**48** (*a*) Due to its shorter wavelengths, UV-radiations can be focused into very narrow beams
for high precision applications such as LASIK eye surgery.
Therefore, Assertion and Reason are correct and Reason is the correct explanation of Assertion.

**50** (*a*) Statements I and III are correct but statement II is incorrect and it can be corrected as,
The total current $i$ is the sum of the conduction current denoted by $i_c$ and the displacement current denoted by $i_d = \varepsilon_0 \, (d\phi_E / dt)$. So, we have
$$i = i_c + i_d = i_c + \varepsilon_0 \frac{d\phi_E}{dt}$$

In explicit terms, this means that outside the capacitor plates, we have only conduction current, i.e. $i_c = i$ and no displacement current, i.e. $i_d = 0$.

**51** (*a*) Statements I and II are correct but statement III is incorrect and it can be corrected as,
In a charging capacitor, both conduction and displacement currents may be present in different regions of space.

**52** (*b*) Statements II is correct but statements I and III are incorrect and these can be corrected as,
The source of a magnetic field is not just the conduction current, due to flowing charges, but also the displacement current.
The current $i = \varepsilon_0 \frac{d\phi_E}{dt}$ is due to change in electric field, therefore called displacement current.

**53** (*c*) Statements I and III are correct but statement II is incorrect and it can be corrected as,
The electric field $E$ is along the $X$-axis and varies sinusoidally with $z$, at a given time.

**54** (*b*) Statement II is correct but statement I is incorrect and it can be corrected as,

Electromagnetic waves are self-sustaining oscillation of electric and magnetic fields in free space or vacuum. They differ from all the other waves, in respect that no material medium is involved in the vibration of the electric and magnetic fields.

**57** (*a*) The fact that an electric field changing with time gives rise to a magnetic field. It is the symmetrical counterpart and consequence of the displacement current, being a source of a magnetic field.

Thus, the statement given in option (a) is correct, rest are incorrect.

**58** (*c*) The most important prediction to emerge from Maxwell's equations is the existence of electromagnetic waves.

The Maxwell's equations involves electric and magnetic fields and their sources, the charge and current densities.

The total current has the same value for all surfaces.

Because if we generalise Ampere's circuital law by adding to the total current carried by conductors through the surface, another term which is $\varepsilon_0$ times the rate of change of electric flux through the same surface.

We can rephrase Faraday's law of electromagnetic induction by saying that a magnetic field, changing with time, give rise to an electric field.

Thus, the statement given in option (c) is correct, rest are incorrect.

**59** (*d*) Here, in electromagnetic wave, the electric field vector is given as $\mathbf{E} = (E_1\hat{\mathbf{i}} - E_2\hat{\mathbf{j}})\cos(kz - \omega t)$

The associated magnetic field vector,

$$\mathbf{B} = \frac{\mathbf{E}}{c} = \frac{E_1\hat{\mathbf{i}} - E_2\hat{\mathbf{j}}}{c}\cos(kz - \omega t)$$

Also, **E** and **B** are perpendicular to each other and the propagation of electromagnetic wave is perpendicular to **E** as well as **B**, so the given electromagnetic wave is plane polarised.

Thus, the statement given in option (d) is correct, rest are incorrect.

**60** (*c*) An electromagnetic wave is travelling along negative *z*-direction. Its electric field is given by

$$\mathbf{E} = E_0\cos(kz - \omega t)$$

which is perpendicular to *Z*-axis. It acts along negative *y*-direction.

The associated magnetic field **B** in electromagnetic wave is along *X*-axis, i.e. along $\hat{\mathbf{k}} \times \mathbf{E}$.

As, $$B_0 = \frac{E_0}{c}$$

$$\Rightarrow \quad \mathbf{B} = \frac{1}{c}(\hat{\mathbf{k}} \times \mathbf{E})$$

The associated electric field can be written in terms of magnetic field as $\mathbf{E} = c\,(\mathbf{B} \times \hat{\mathbf{k}})$.

Angle between $\hat{\mathbf{k}}$ and **E** is 90° and between $\hat{\mathbf{k}}$ and **B** is 90°. Therefore, $\hat{\mathbf{k}} \cdot \mathbf{E} = E\cos 90° = 0$ and $\hat{\mathbf{k}} \cdot \mathbf{B} = E\cos 90° = 0$. $\quad (\because |\hat{\mathbf{k}}| = 1)$

Also, $\hat{\mathbf{k}} \times \mathbf{E} = E\sin 90° = E$ and $\hat{\mathbf{k}} \times \mathbf{B} = B\sin 90° = B$.

Thus, the statement in option (c) is correct, rest are incorrect.

**61** (*c*) The statement in option (c) is incorrect and it can be corrected as,

The constancy of the velocity of electromagnetic waves in vacuum is used to define a standard of length. The meter is defined as the distance travelled by light in vacuum in a time $(1/c)$ second.

**62** (*c*) The average value of both the electric field and magnetic fields is zero as they are sinusoidal waves, whose average comes out to be zero for a complete cycle.

The average value of magnetic energy cannot be zero, since

$$U_{\text{avg}} = \frac{B_0^2}{2\mu_0}, \text{ where } B_0 = \text{peak value.}$$

Thus, the statement given in option (c) is correct, rest are incorrect.

**63** (*d*) The statement given in option (d) is incorrect and it can be corrected as,

Most of ultraviolet rays are absorbed in the ozone layer in the atmosphere at an altitude of about 40-50 km.

Rest statements are correct.

**64** (*b*) The order of wavelength in electromagnetic spectrum is as shown

$$\lambda_{\gamma\text{-rays}} < \lambda_{\text{X-rays}} < \lambda_{\text{UV-rays}} < \lambda_{\text{visible-ray}} < \lambda_{\text{infrared-rays}} < \lambda_{\text{microwaves}} < \lambda_{\text{radio-waves}}$$

From this, we can conclude that wavelength of infrared rays is greater than that of UV-rays.

Thus, the statement given in option (b) is incorrect, rest are correct.

**65** (*d*) The statement given in option (d) is incorrect and it can be corrected as,

Infrared waves are produced by hot bodies and molecules. These waves are emitted by molecules of hot bodies, when they change their rotational-vibrational moments. Hence, they are sometimes referred to as heat waves.

Rest statements are correct.

**66** (*a*) The statement given in option (a) is correct, rest are incorrect and these can be corrected as,

UV-rays are used in the LASIK eye surgery and in water purifiers.

Snakes can detect infrared waves.

Electromagnetic waves exhibit polarisation while sound waves don't.

Rest statements are correct.

**70** (*b*) Given, radius, $R = 6\,\text{cm}$,

capacitance, $C = 100\,\text{pF} = 10^{-10}\,\text{F}$,

Source potential, $V = 230\,\text{V}$ and $\omega = 300\,\text{rad s}^{-1}$

Capacitive reactance, $X_C = \dfrac{1}{\omega C} = \dfrac{1}{300 \times 10^{-10}}$

$$X_C = \frac{10^8}{3}\,\Omega$$

If $i_{rms}$ is the rms value of the conduction current, then

$$i_{rms} = \frac{V_{rms}}{X_C} = \frac{230\text{ V}}{(10^8/3)\Omega} = 6.9\times 10^{-6}\text{ A} = 6.9\mu\text{ A}$$

Thus, conduction current is $6.9\mu$ A.

**71** (*c*) Given, electric field, $E_0 = 120$ N/C

and frequency, $\nu = 50\text{ MHz} = 50\times 10^6$ Hz

Speed of light in vacuum, $c = \frac{E_0}{B_0}$

$$\Rightarrow \quad B_0 = \frac{E_0}{c} = \frac{120}{3\times 10^8} = 40\times 10^{-8}$$

or $\quad B_0 = 400\times 10^{-9}\text{ T} = 400\text{ nT}$

Angular frequency of wave,

$$\omega = 2\pi\nu = 2\times 3.14\times 50\times 10^6$$

$$\Rightarrow \quad \omega = 3.14\times 10^8\text{ rads}^{-1}$$

Therefore, wave number of electromagnetic wave,

$$k = \frac{\omega}{c} = \frac{3.14\times 10^8}{3\times 10^8} = 1.05\text{ rad m}^{-1} \approx 1\text{ rad m}^{-1}$$

**72** (*b*) Given, total power = 100 W

and visible radiation power, $P = 5\%$ of total power

$$\therefore \quad P = \frac{5}{100}\times 100 = 5\text{ W}$$

At a distance of 1m, the energy distributed in the form of sphere.

$\because$ Area of sphere, $A = 4\pi\,(\text{radius})^2$

$\therefore$ Intensity of visible radiation

$$= \frac{P}{A} = \frac{5}{4\times 3.14\times (1)^2} = 0.4\text{ Wm}^{-2}$$

**73** (*b*) Given, magnetic field part of harmonic electromagnetic wave, $B_0 = 510$ nT

Speed of light in vacuum, $c = \frac{E_0}{B_0}$

where, $E_0$ is the electric part of the wave.

$$\Rightarrow \quad 3\times 10^8 = \frac{E_0}{510\times 10^{-9}} \text{ or } E_0 = 153\text{ NC}^{-1}$$

Thus, the amplitude of the electric field part of wave is $153\text{ NC}^{-1}$.

**74** (*c*) Given, electric part of electromagnetic wave in vacuum,

$$E = 3.1\cos(1.8y + 5.4\times 10^6 t)$$

Comparing with standard equation

$E = E_0\cos(ky + \omega t)$, we get,

angular frequency, $\omega = 5.4\times 10^6\text{ rads}^{-1}$,

wave number, $k = 1.8\text{ rad m}^{-1}$

and amplitude of the electric field part of the wave,

$$E_0 = 3.1\text{ NC}^{-1}$$

Wavelength, $\lambda = \frac{2\pi}{k} = \frac{2\pi}{1.8} = 3.491$ m

Thus, wavelength is 3.5 m.

**75** (*b*) Given that, incident electromagnetic wave is

$$\mathbf{E} = E_0\hat{\mathbf{i}}\cos(kz - \omega t)$$

Thus, the reflected electromagnetic wave is given by

$$\mathbf{E}_r = E_0(-\hat{\mathbf{i}})\cos[k(-z) - \omega t + \pi]$$
$$= -E_0\hat{\mathbf{i}}\cos[-(kz + \omega t) + \pi]$$
$$= E_0\hat{\mathbf{i}}\cos[-(kz + \omega t)] = E_0\hat{\mathbf{i}}\cos(kz + \omega t)]$$
$$[\because \cos(\pi - \theta) = -\cos\theta \text{ and } \cos(-\theta) = \cos\theta]$$

**76** (*b*) Given, energy flux, $\phi = 20\text{Wcm}^{-2}$,

surface area, $A = 30\text{cm}^2$

and time, $t = 30\text{ min} = (30\times 60)$ s

Now, total energy falling on the surface in time $t$ is

$$U = \phi At = 20\times 30\times (30\times 60) = 1.08\times 10^6\text{J}$$

Momentum of the incident light,

$$p = \frac{U}{c} = \frac{1.08\times 10^6}{3\times 10^8} = 36\times 10^{-4}\text{ kg-ms}^{-1}$$

Momentum of the reflected light = 0

Thus, momentum delivered to the surface

$$= 36\times 10^{-4} - 0 = 36\times 10^{-4}\text{ kg-ms}^{-1}$$

**77** (*a*) We know that, $E = P/A$

$$\Rightarrow \quad E_0 \propto P_{av}$$

$$\therefore \quad \frac{(E_0)_1}{(E_0)_2} = \frac{(P_{av})_1}{(P_{av})_2} \quad \text{...(i)}$$

Now according to question,

$$(P_{av})_2 = 50\text{ W},\ (P_{av})_1 = 100\text{ W}$$

Putting these values in Eq. (i), we get

$$\frac{E}{E'} = \frac{100}{50} \text{ or } \frac{E'}{E} = \frac{50}{100}$$

$$\Rightarrow \quad \frac{E'}{E} = \frac{1}{2} \Rightarrow E' = \frac{E}{2}$$

Thus, electric field intensity is $\frac{E}{2}$.

**79** (*c*) Intensity of EM wave in terms of electric field,

$$U_{av} = \frac{1}{2}\varepsilon_0 E_0^2$$

Intensity of EM wave in terms of magnetic field,

$$U_{av} = \frac{1}{2}\frac{B_0^2}{\mu_0}$$

As, $(U_{av})_{\text{electric field}} = \frac{1}{2}\varepsilon_0 E_0^2 = \frac{1}{2}\varepsilon_0(cB_0)^2 \quad (\because E_0 = cB_0)$

$$= \frac{1}{2}\varepsilon_0\times c^2 B_0{}^2$$

But, $\quad c = \frac{1}{\sqrt{\mu_0\varepsilon_0}}$

$$\therefore \quad (U_{av})_{\text{electric field}} = \frac{1}{2}\varepsilon_0\times\frac{1}{\mu_0\varepsilon_0}B_0^2 = \frac{1}{2}\frac{B_0^2}{\mu_0}$$
$$= (U_{av})_{\text{magnetic field}}$$

Thus, the energy in electromagnetic wave is divided equally between electric and magnetic vectors.

Therefore, the ratio of contributions by the electric field and magnetic field components to the intensity of an electromagnetic wave is 1 : 1.

CHAPTER > 09

# Ray Optics and Optical Instruments

## KEY NOTES

- Electromagnetic radiation belonging to a region of the spectrum (wavelength of about 400 nm to 750 nm) is called **light**.
- Speed of light in vacuum is $c = 2.99992458 \times 10^8$ m/s or $3 \times 10^8$ m/s. This speed in vacuum is the highest speed attainable in nature.
- The straight line path of a light wave is called **ray of light** and a bundle of such rays constitutes a **beam of light**.

## Reflection of Light

The phenomenon of bouncing back of light rays in the same medium on striking a smooth surface is called reflection of light.

- **Laws of Reflection** There are two laws for reflection of light from a smooth reflecting surface
  - (i) The incident ray, reflected ray and the normal to the reflecting surface at the point of incidence lie in the same plane.
  - (ii) The angle of incidence, $\angle i =$ angle of reflection, $\angle r$.

**Note** These laws are valid at each point on any reflecting surface whether plane or curved.

## Reflection of Light by Spherical Mirrors

- Spherical mirror is a mirror whose reflecting surface is a part of a hollow shpere.

  Spherical mirrors are of two types : concave and convex mirror.
- Some of the terms related to spherical mirrors are as follows
  - (i) The geometric centre of a spherical mirror is called its **pole.**
  - (ii) The line joining the pole and centre of curvature of the spherical mirror is known as the **principal axis** ($P$).
  - (iii) Paraxial rays are those rays that incident at points close to the pole $P$ of the mirror and make small angles with the principal axis.
  - (iv) **Principal focus** of the mirror is a point at which the reflected rays converge for a concave mirror or appears to diverge for a convex mirror.
  - (v) **Focal plane** of the mirror is the plane through which the reflected rays converge (or appears to diverge) when parallel paraxial beam of light were incident making some angle with the principal axis.
  - (vi) The **focal length** (the distance between focus $F$ and pole $P$) of a spherical mirror is half of its radius of curvature, i.e. $f = \frac{R}{2}$.
- These terms are depicted in the spherical (concave and convex) mirrors as shown

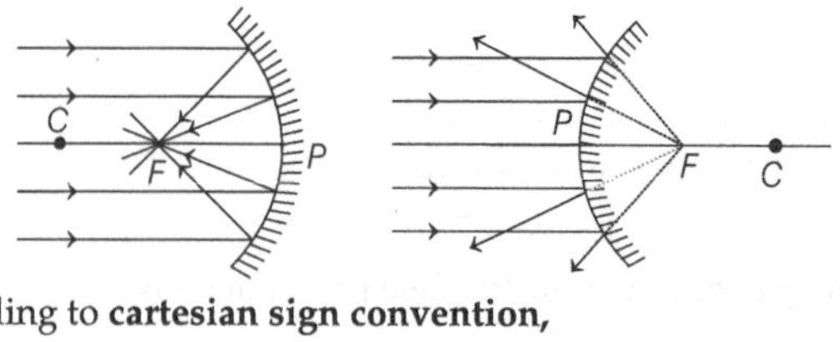


- According to **cartesian sign convention,**
  - (i) All the distances are measured from the pole of the mirror.
  - (ii) The distances measured in the same direction as the incident light are taken as positive and those measured in the direction opposite to the direction of incident light are taken as negative.

(iii) The height measured upwards with respect to X-axis and normal to the principal axis (X-axis) of the mirror/lens are taken as positive. The heights, measured downwards are taken as negative.

- If rays emanating from a point actually meet at another point after reflection and/or refraction, that point is called the **image** of the first point.
  - (i) The image is **real**, if the rays actually converge to the point.
  - (ii) The image is **virtual**, if the rays do not actually meet but appear to diverge from the point when produced backwards.
- **Mirror Equation** If an object is placed at a distance $u$ from the pole of a mirror and its image is formed at a distance $v$ from a pole. Then, according to the mirror equation

$$\frac{1}{v}+\frac{1}{u}=\frac{1}{f}$$

where, $f$ is the focal length of the mirror.

- **Linear magnification** is the ratio of the height of the image $h'$ to the height of the object $h$ and is given by

$$m=\frac{h'}{h}=-\frac{v}{u}$$

## Refraction

It is the phenomena of bending of ray of light when they pass from one transparent medium to another depending on their optical densities.

- **Laws of Refraction** There are two laws of refraction as follows
  - (i) The incident ray, the refracted ray and the normal to the interface at the point of incidence, all lie in the same plane.
  - (ii) The ratio of the sine of angle of incidence $i$ to the sine of angle of refraction $r$ is constant, i.e.

$$\frac{\sin i}{\sin r}=n_{21} \text{ or } {}^1n_2$$

where, $n_{21}$ is a constant, called the **refractive index** of the second medium with respect to the first medium. This is known as **Snell's law of refraction**.

**Note** $n_{21}$ is a characteristic of the pair of media (and also depend on the wavelength of light) but is independent of the angle of incidence.

Following are few important points related to refraction of light

(i) If $n_{21}>1$, $r<i$, then the refracted ray bend towards the normal. In such a case, medium 2 is said to be optically denser than medium 1.

(ii) If $n_{21}<1$, $r>i$, the refracted ray bend away from the normal. This is the case when incident ray in a denser medium refracts into a rarer medium.

**Note** $n_{12}=\frac{1}{n_{21}}$ and $n_{32}=n_{31}\times n_{12}$

- If a light ray passes through a rectangular glass slab, the emergent ray is parallel to the incident ray, then there is no deviation, but it does suffer **lateral displacement/shift** with respect to the incident ray.
- When an object is in denser medium and observer is in rarer medium, then object appears to be at lesser depth (apparent depth) than its actual depth.

  Real and apparent depth are related as follows

$$\frac{\text{real depth }(h)}{\text{apparent depth }(h')}=n_{21}$$

i.e. Real depth > Apparent depth

- The refraction of light through the atmosphere is responsible for many natural phenomena, e.g.
  - (i) Due to refraction, the sun is visible a little before the actual sunrise and until a little after the actual sunset.
  - (ii) The apparent flattening (oval shape) of the sun at sunset and sunrise is also due to the same phenomenon.

## Total Internal Reflection

- The angle of incidence in denser medium corresponding to an angle of refraction 90° in rarer medium is called the **critical angle** $i_C$ for the given pair of media, i.e. for $i=i_c$

$$\sin i_C = n_{21}$$

or
$$n_{12}=\frac{1}{\sin i_C}$$

- A ray of light travelling from denser medium to a rarer medium is incident at the interface of two media at an angle greater than the critical angle for the two media, then the ray is totally reflected back to denser medium and this phenomenon is called **Total Internal Reflection** (TIR).
- Following are some applications of total internal reflection
  - (i) **Mirage** It is the phenomena, in which an inverted image of distant tall objects cause an optical illusion of water. This type of mirage is especially common in hot deserts.
  - (ii) **Diamond** The critical angle for diamond-air interface (≅ 24.4°) is very small, therefore once light enters a diamond, it is very likely to undergo total internal reflection inside it. Due to this, diamond shines brilliantly.
  - (iii) **Prism** Few prisms are designed to bend light by 90° or by 180°, making use of total internal reflection. Such prisms are also used to invert images, without changing their size.

(iv) **Optical Fibres** These fibres are fabricated with high quality composite glass/quartz. Each fibre consists of a core and cladding such that refractive index of the material of the core is higher than that of the cladding. When a signal in the form of light is directed at one end of the fibre at a suitable angle, it undergoes repeated internal reflections along the length of the fibre and finally comes out of the other end.

Thus, these are extensively used for transmitting audio and video signals through long distances.

## Refraction at Spherical Surfaces and by Lenses

- **Refraction at a Spherical Surface** If an object is placed in a medium of refractive index $n$, at a distance $u$ from the pole of a spherical surface of radius of curvature $R$ and after refraction, its image is formed in a medium of refractive index $n_2$ at a distance $v$, then $\frac{n_2}{v}-\frac{n_1}{u}=\frac{n_2-n_1}{R}$

**Note** This equation holds for any curved spherical surface.

- **Lens** A lens is a transparent optical medium bounded by two surfaces, at least one of which should be spherical.

  Lens are of two types ; Convex lens and Concave lens.
- **Lens Maker's formula** is given by
$$\frac{1}{f}=(n_{21}-1)\left(\frac{1}{R_1}-\frac{1}{R_2}\right)$$
- If the lens is thin, then lens Maker's formula is given as
$$\frac{1}{v}-\frac{1}{u}=\frac{1}{f}$$
  This is usually known as **thin lens formula.**
- For lens, the focus on the side of the (original) source of light is called the **first focal point**, whereas the other is called the **second focal point**.
- **Magnification** $m$ produced by lens is defined as the ratio of the size of the image to that of the object.

  Mathematically, $m=\frac{h'}{h}=\frac{v}{u}$

  For erect (and virtual) image, $m$ is positive and for an inverted (and real) image, $m$ is negative.
- **Power of a lens** is a measure of the convergence or divergence, which a lens introduces in the light falling on it.

  It can be expressed mathematically as
$$P=\frac{1}{f(\text{m})}=\frac{100}{f(\text{cm})}\text{ dioptre}$$
  It is also defined as the tangent of the angle by which it converges or diverges a beam of light falling at unit distant from the optical centre, i.e.
$$\tan\delta=h/f$$

KEY NOTES

**Note** It is positive for a converging lens and negative for a diverging lens.

- **Combination of Thin Lenses** If several thin lenses of focal length $f_1, f_2, f_3, \ldots$ are in contact, then the effective focal length and power of their combination is given by
$$\frac{1}{f}=\frac{1}{f_1}+\frac{1}{f_2}+\frac{1}{f_3}, \ldots \text{ and } P=P_1+P_2+P_3, \ldots$$
- Combination of lenses helps to obtain diverging or converging lenses of desired magnifications. It also enhances sharpness of the image. Thus, the total magnification of such a combination is given as
$$m=m_1\,m_2\,m_3\ldots.$$

## Refraction through a Prism

- A **prism** is a portion of a transparent medium bounded by two plane faces inclined to each other at a suitable angle.
- Refraction of light through a prism is shown in the figure given below

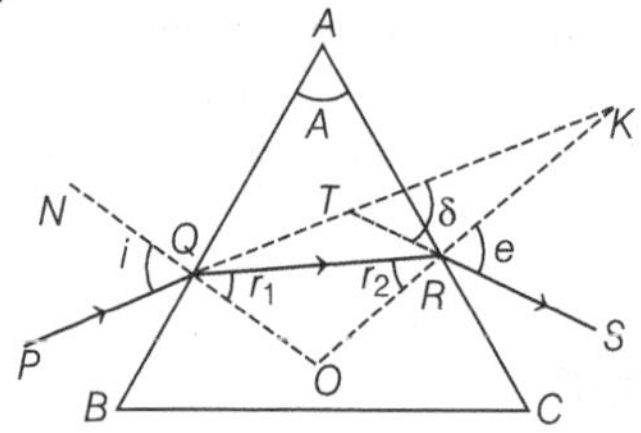


- The angle between emergent ray and incident ray through a prism is called **angle of deviation,** i.e.
$$\delta=i+e-A$$
  where, $i$ is the angle of incidence, $e$ is the angle of emergence and $A$ is the angle of prism.
- The angle of deviation depends on the angle of incidence. At the minimum deviation $D_m$, the refracted ray inside the prism becomes parallel to its base.
- Refractive index of prism material is given as
$$n_{21}=\frac{\sin\left(\frac{A+D_m}{2}\right)}{\sin(A/2)}$$
  For small angle prism, $D_m=(n_{21}-1)A$

# Dispersion of Light

- The phenomenon of splitting of light into different colours is known as dispersion. The pattern of colour components of light thus obtained is called the **spectrum of light.**
- Dispersion takes place because the refractive index of medium for different frequencies (colours) is different. e.g. Bending of red light is least while it is most for the violet. Equivalently, red light travels faster than violet light in a glass prism.
- Thick lenses show **chromatic aberration,** due to dispersion of light. In this red and blue colours, focus at different points after passing through thick lens.

- **Rainbow** is an example of dispersion of sunlight by water drops in atmosphere. It is a phenomenon based on the combined effect of dispersion, refraction and reflection of sunlight by spherical water droplets of rain.

Following are the two types of rainbows

(i) **Primary rainbow** which is a result of three-step process, i.e. refraction, reflection and refraction. In this, red colour is on the top and violet on the bottom.

(ii) **Secondary rainbow** which is formed when light ray undergoes two internal reflections inside a rain drop instead of one as in the primary rainbow. Thus, this type of rainbow is due to four-step process. The order of the colour is reversed in it.

## Scattering of Light

- As sunlight travels through the earth's atmosphere, it gets scattered (changes its direction) by the atmospheric particles.
- Light of shorter wavelength is scattered much more than light of larger wavelengths.
- **Rayleigh Scattering** According to this, the amount of scattering is inversely proportional to the fourth power of the wavelength.

  Some phenomena based on this are as follows

  (i) The bluish colour predominates in a clear sky, since blue has a shorter wavelength than red and is scattered much more strongly.

  In fact, violet gets scattered even more than blue, having a shorter wavelength. But, since our eyes are more sensitive to blue than violet, we see the sky blue.

  (ii) At sunrise or sunset, the sun's rays have to pass through a larger distance in the atmosphere. Due to this, most of the blue and other shorter wavelengths are removed by scattering.

  Thus, the least scattered light, i.e. red coloured light reaches our eyes, therefore the sun appears reddish to us.

## Optical Instruments

Following are the various optical devices and instruments

### Human Eye

- It is an organ which give us sense of sight. It behaves as a convex lens of refractive index 1.437, in which image is formed by this lens is real, inverted and formed at retina.
- The retina contains rod and cones which sense the light intensity and colour respectively and transmit electrical signal *via* the optic nerve to the brain which finally processes this information.
- The shape (curvature) and therefore the focal length of the lens can be modified by ciliary muscles. This property of the eye is called **accommodation.**
- The closest distance for which the lens can focus light on the retina is called the **least distance of distinct vision** or the near point. The standard value for normal vision is 25 cm.
- The eye defects and their remedies are as follows

  (i) The least distance increases with age, because of the decreasing effectiveness of the ciliary muscle and the loss of flexibility of the lens. This condition is called **presbyopia**. It is corrected by using a converging lens for reading.

  (ii) The light from a distant object arriving at the eye-lens may get converged at the point infront of the retina. This type of defect is called **near-sightedness** or **myopia**. To correct this, a concave lens is used.

  (iii) If the eye-lens focusses the incoming light at a point behind the retina, a convergent lens is needed to correct this defect of vision. This defect is called **far-sightedness** or **hypermetropia**.

  (iv) **Astigmatism** occurs when the cornea is not spherical in shape. In this defect, a person cannot focus on horizontal and vertical lines at the same distance at the same time. It is corrected by using a cylindrical lens.

### Microscope

- **A simple magnifier** or **microscope** is a converging lens of small focal length.
- The linear magnification $m$ for the image formed at near point $D$ by a simple microscope can be obtained by using the relation

$$m = \left(1 - \frac{v}{f}\right) = \left(1 + \frac{D}{f}\right)$$

- When the image is at infinity, the angular magnification is $m = \frac{D}{f}$.

  This is one less than the magnification when the image is at the near point.
- A simple microscope has a limited maximum magnification ($\leq 9$) for realistic focal lengths.
- For much larger magnifications, one uses two lenses, one compounding the effect of the other. This is known as a **compound microscope**.
- The lens nearest to the object, called the **objective**, forms a real, inverted, magnified image of the object. This serves as the object for the second lens, the **eyepiece**, which functions essentially like a simple microscope or magnifier, produces the final image, which is enlarged and virtual.

- The magnification due to a compound microscope is

$$m = m_o m_e = \left(\frac{L}{f_o}\right)\left(\frac{D}{f_e}\right)$$

where, $f_o$ = focal length of objective lens,
$f_e$ = focal length of eyepiece
and $L$ = tube length of the compound microscope.

## Telescope

- The telescope is used to provide angular magnification of distant objects. It has objective lens of large focal length and a much larger aperture, than the eyepiece.
- The magnifying power $m$ is the ratio of the angle $\beta$ subtended at the eye by the final image to the angle $\alpha$ which the object subtends at the lens or the eye.

Hence, $m = \frac{\beta}{\alpha} = \frac{f_o}{f_e}$

- The length of the telescope tube is $f_o + f_e$.
- The **resolving power** or **the ability** to observe two objects distinctly, which are in very nearly the same direction, also depends on the diameter of the objective.
  So, the desirable aim in optical telescopes is to make them with objective of large diameter.
- **Terrestrial telescopes** have, in addition, a pair of inverting lenses to make the final image erect.
- **Refracting telescopes** can be used both for terrestrial and astronomical observations.
- Telescopes with mirror objective are called **reflecting telescopes**. There is no chromatic aberration in a mirror.
- A **cassegrain telescope** has the advantage of a large focal length in a short telescope.

# Mastering NCERT

## MULTIPLE CHOICE QUESTIONS

## TOPIC 1 ~ Reflection of Light

**1** Which of the following phenomenon are explained by using the ray picture of light?
(a) Reflection (b) Refraction
(c) Dispersion (d) All of these

**2** The laws of reflection are valid at each point on any reflecting surface, which is
(a) plane (b) curved
(c) Either plane or curved (d) None of these

**3** A ray of light gets reflected from the plane mirror. The incident ray and the reflected ray are perpendicular to each other. The angle of incidence is
(a) 60° (b) 30°
(c) 45° (d) 0°

**4** A ray of light travelling in the direction $\frac{1}{2}(\hat{\mathbf{i}} + \sqrt{3}\,\hat{\mathbf{j}})$ is incident on a plane mirror. After reflection, it travels along the direction $\frac{1}{2}(\hat{\mathbf{i}} - \sqrt{3}\,\hat{\mathbf{j}})$. The angle of incidence is
(a) 30° (b) 45°
(c) 60° (d) 75°

**5** The rays that are incident at points close to the pole of the mirror and make small angles with the principal axis are known as
(a) paraxial rays (b) perpendicular rays
(c) principal rays (d) beam of rays

**6** A convex mirror, if formed from a sphere of radius 16 cm, then the focal length of mirror is
(a) 4 cm (b) 8 cm
(c) 12 cm (d) 16 cm

**7** For a ray passing through the centre of curvature of a concave mirror or appearing to pass through it for a convex mirror, the reflected ray
(a) move parallel to mirror
(b) move parallel to principle axis
(c) move perpendicular to principal axis
(d) retraces its path

**8** If lower half of a concave mirror is blackened, then
(a) image distance increases
(b) image distance decreases
(c) image intensity increases
(d) image intensity decreases

**9** A rod of length 30 cm lies along the principal axis of a concave mirror of focal length 10 cm in such a way that its end closer to the pole is 20 cm away from the mirror. The length of the image is ***CBSE AIPMT 2012***
(a) 10 cm (b) 15 cm (c) 2.5 cm (d) 7.5 cm

**10** An object is placed at a distance of 40 cm from a concave mirror of focal length 15 cm. If the object is displaced through a distance of 20 cm towards the mirror, the displacement of the image will be ***NEET 2018***
(a) 30 cm, towards the mirror
(b) 36 cm, away from the mirror
(c) 30 cm, away from the mirror
(d) 36 cm, towards the mirror

**11** A short linear object of length $b$ lies along the axis of a concave mirror of focal length $f$ at a distance $u$ from the pole of the mirror. The size of the image is approximately equal to
(a) $b\left(\frac{u-f}{f}\right)^{1/2}$ (b) $b\left(\frac{f}{u-f}\right)^{1/2}$
(c) $b\left(\frac{u-f}{f}\right)$ (d) $b\left(\frac{f}{u-f}\right)^{2}$

**12** An object 2 cm high is placed at a distance of 16 cm from a concave mirror, which produces a real image 3 cm high. What is the focal length of the mirror?
(a) – 9.6 cm (b) – 3.6 cm (c) – 6.3 cm (d) – 8.3 cm

**13** An object is gradually moving away from the focal point of a concave mirror along the axis of the mirror. The graphical representation of the magnitude of linear magnification ($m$) *versus* distance of the object from the mirror ($x$) is correctly given by
(graphs are drawn schematically and are not to scale) ***JEE Main 2020***

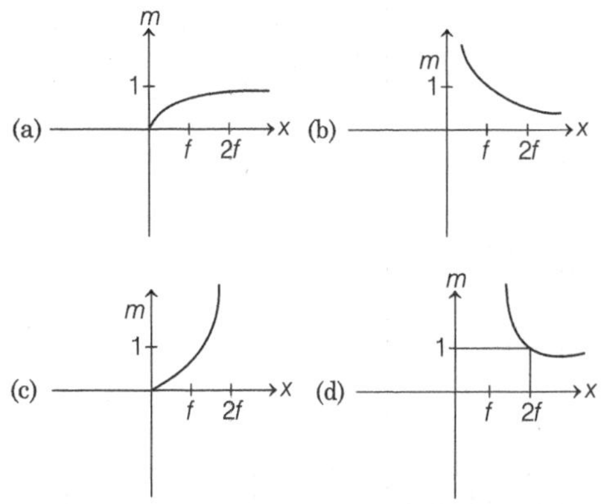


**14** A concave mirror for face viewing has focal length of 0.4 m. The distance at which you hold the mirror from your face in order to see your image upright with a magnification of 5 is ***JEE Main 2019***
(a) 0.16 m (b) 1.60 m (c) 0.32 m (d) 0.24 m

**15** Rear view mirror of a car is of radius of curvature $R = 2$ m. A jogger approaches car (from behind) at a speed of 5 $ms^{-1}$. The speed of image, when jogger is 39 m from the mirror, is
(a) $0.3 \times 10^{-2}$ $ms^{-1}$ (b) $0.1 \times 10^{-2}$ $ms^{-1}$
(c) $0.5 \times 10^{-2}$ $ms^{-1}$ (d) 0.5 $cms^{-1}$

# TOPIC 2 ~ *Refraction*

**16** When a ray of light falls obliquely on a transparent medium, it enters the other medium, then at the interface of two media
(a) its direction changes
(b) its direction does not change
(c) it is completely refracted
(d) it is completely reflected

**17** A ray of light strikes on an air-glass interface at an angle of incidence ($i = 60°$) and gets refracted at an angle of refraction $r$. On increasing the angle of incidence ($i > 60°$), the angle of refraction $r$
(a) decreases (b) remains same
(c) is equal to 60° (d) increases

**18** As you can seen from Fig. (i) and (ii) shows refraction of light from air to glass and from air to water. Find out the value of the angle θ in the case of refraction as shown in Fig. (iii).

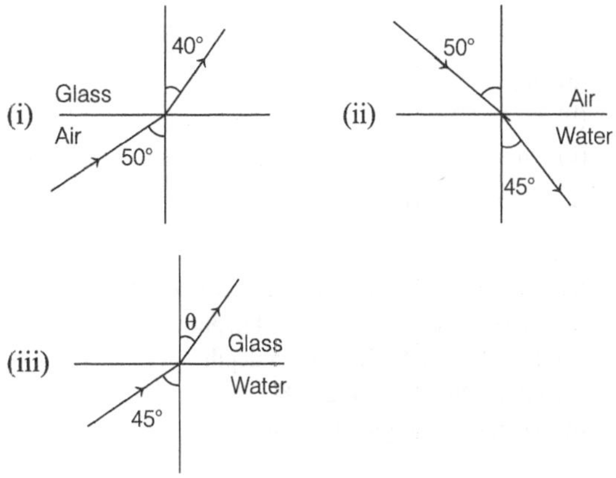


(a) 30° (b) 50°
(c) 40° (d) 45°

**19** A ray of light $AO$ in vacuum is incident on a glass slab at angle 60° and refracted at angle 30° along $OB$ as shown in the figure.

The optical path length of light ray from $A$ to $B$ is **JEE Main 2019**

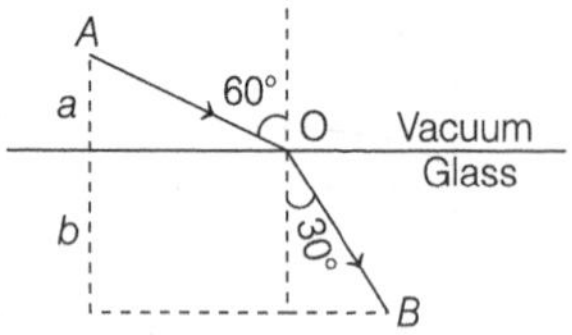


(a) $\frac{2\sqrt{3}}{a} + 2b$ (b) $2a + \frac{2b}{3}$

(c) $2a + 2b$ (d) $2a + \frac{2b}{\sqrt{3}}$

**20** A ray of light strikes a material's slab at an angle of incidence 60°. If the reflected and refracted rays are perpendicular to each other, then which of the given option(s) is/are correct?

(a) The angle of refraction is 45°
(b) The angle of refraction is 30°
(c) The refractive index of the material is $\sqrt{3}$
(d) Both (b) and (c)

**21** A divergent beam of light from a point source $S$ having divergence angle $\alpha$ falls symmetrically on a glass slab as shown in the figure. The angles of incidence of the two extreme rays are equal. If the thickness of the glass slab is $t$ and its refractive index is $n$, then the divergence angle of the emergent beam is

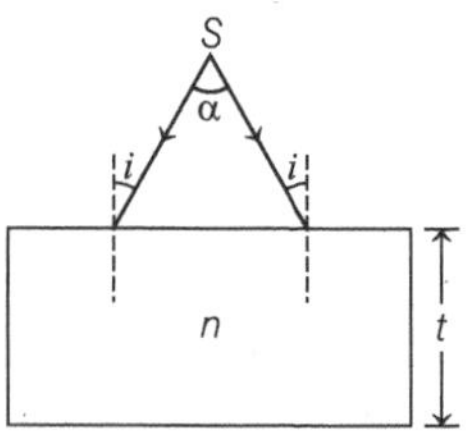


(a) zero
(b) $\alpha$
(c) $\sin^{-1}(1/n)$
(d) $2\sin^{-1}(1/n)$

**22** When an object lying in a denser medium is observed from a rarer medium, then the real depth of the object is

(a) more than that of observed depth
(b) less than that of observed depth
(c) equals to observed depth
(d) depends on angle of vision

**23** A beaker contains water upto height $h_1$ and kerosene of height $h_2$ above water so that the total height (water + kerosene) is $(h_1 + h_2)$. Refractive index of water is $n_1$ and that of kerosene is $n_2$. The apparent shift in position of the bottom of the beaker when it is viewed from above is **JEE Main 2013**

(a) $\left(1-\frac{1}{n_1}\right)h_2 + \left(1-\frac{1}{n_2}\right)h_1$

(b) $\left(1+\frac{1}{n_1}\right)h_1 + \left(1+\frac{1}{n_2}\right)h_2$

(c) $\left(1-\frac{1}{n_1}\right)h_1 + \left(1-\frac{1}{n_2}\right)h_2$

(d) $\left(1+\frac{1}{n_1}\right)h_2 - \left(1+\frac{1}{n_2}\right)h_1$

**24** A container is filled with water ($n = 1.33$) upto a height of 33.25 cm. A concave mirror is placed 15 cm above the water level and the image of an object placed at the bottom is formed 25 cm below the water level. The focal length of the mirror is

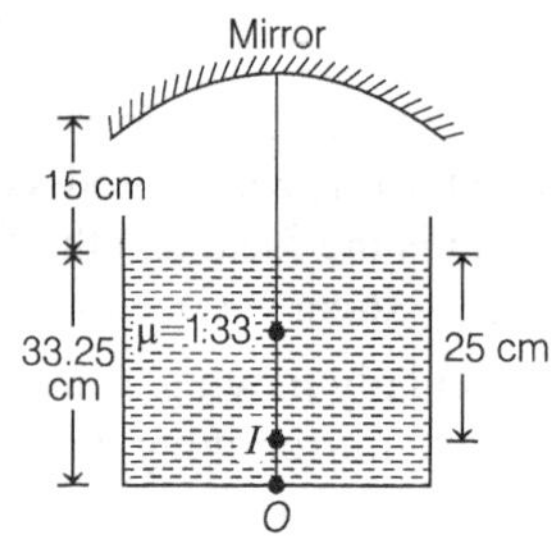


(a) 10 cm (b) 15 cm
(c) −18.3 cm (d) 25 cm

**25** A vessel is filled with water to a height of 13 cm. The apparent depth of a screw lying at the bottom of the vessel is measured by a microscope to be 8.5 cm. If water is replaced by a liquid of refractive index 1.70 upto the same height. Find the distance at which microscope have to be moved to focus on the screw again?

(a) 0.85 cm (b) 0.52 cm
(c) 0.65 cm (d) 1.02 cm

**26** When a light ray enters from oil to glass, then on oil-glass interface, the velocity of light changes by a factor of [given, $n_{oil} = 2$, $n_{glass} = 3/2$] **JIPMER 2019**

(a) $\frac{4}{3}$ (b) $\frac{3}{4}$

(c) 3 (d) 1

# TOPIC 3 ~ *Total Internal Reflection*

**27** In total internal reflection, when the angle of incidence is equal to the critical angle for the pair of media in contact, what will be angle of refraction? ***NEET (Odisha) 2019***

(a) 0°   (b) Equal to angle of incidence
(c) 90°   (d) 180°

**28** Find the value of θ in the given figure ***JIPMER 2019***

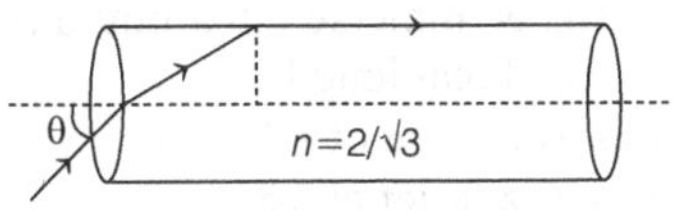


(a) $\sin^{-1}\left(\frac{2}{\sqrt{3}}\right)$   (b) $\sin^{-1}\left(\frac{1}{\sqrt{3}}\right)$
(c) $\sin^{-1}\left(\frac{1}{2}\right)$   (d) $\sin^{-1}\left(\frac{1}{\sqrt{2}}\right)$

**29** If the critical angle for light going from medium $A$ to $B$ is θ. Then find the speed of light in medium $B$, if speed of light is $v$ in medium $A$.

(a) $v(1-\cos\theta)$ (b) $\frac{v}{\cos\theta}$ (c) $\frac{v}{\sin\theta}$ (d) $v(1-\sin\theta)$

**30** Two transparent media $A$ and $B$ are separated by a plane boundary. The speed of light in medium $A$ is $2\times10^8\ \text{ms}^{-1}$ and in medium $B$ is $2.5\times10^8\ \text{ms}^{-1}$. The critical angle for which a ray of light going from $A$ to $B$ is totally internally reflected is

(a) $\sin^{-1}\left(\frac{1}{2}\right)$ (b) $\sin^{-1}\left(\frac{2}{5}\right)$ (c) $\sin^{-1}\left(\frac{4}{5}\right)$ (d) $\sin^{-1}\left(\frac{3}{4}\right)$

**31** A ray of light is incident at the glass-water interface at an angle (as shown below) and it emerges finally parallel to the surface of water, then the value of $n_g$ would be

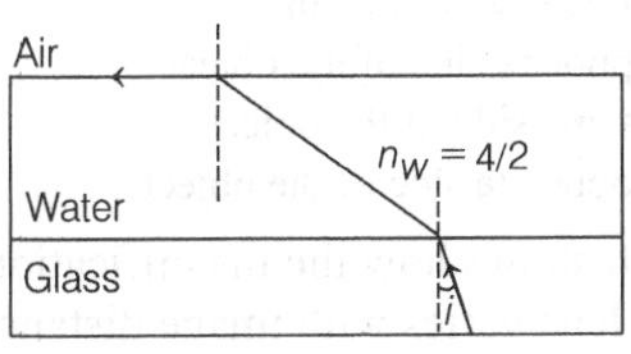


(a) $(4/3)\sin i$ (b) $1/\sin i$ (c) $4/3$ (d) 1

**32** A ray of light from a denser medium strikes a rarer medium at an angle of incidence $i$ (see figure). The reflected and refracted rays make an angle of 90° with each other. The angle of reflection and refraction are $r$ and $r'$. The critical angle is

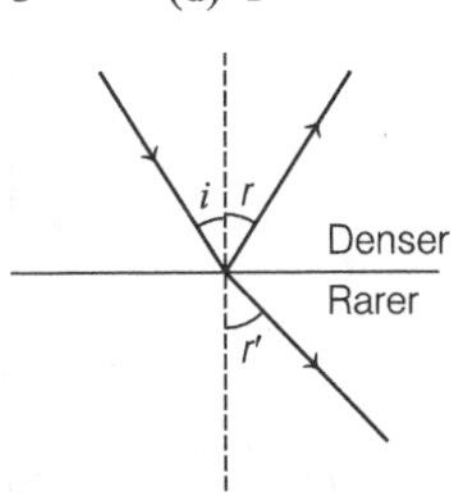


(a) $\sin^{-1}(\tan r)$   (b) $\sin^{-1}(\cot i)$
(c) $\sin^{-1}(\tan r')$   (d) $\tan^{-1}(\sin i)$

**33** A parallel sides slab $ABCD$ of refractive index 2 is sandwiched between two slabs of refractive indices $\sqrt{2}$ and $\sqrt{3}$ as shown in the figure. The minimum value of angle θ such that the ray $PQ$ suffers total internal reflection at both the surfaces $AB$ and $CD$ is

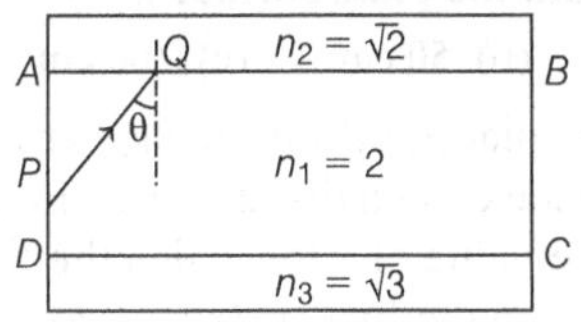


(a) 30°   (b) 45°   (c) 60°   (d) 75°

**34** A light beam is travelling from region I to region IV (refer figure). The refractive index in regions I, II, III and IV are $n_0$, $\frac{n_0}{2}$, $\frac{n_0}{6}$ and $\frac{n_0}{8}$, respectively. The angle of incidence θ for which the beam just misses entering region IV is

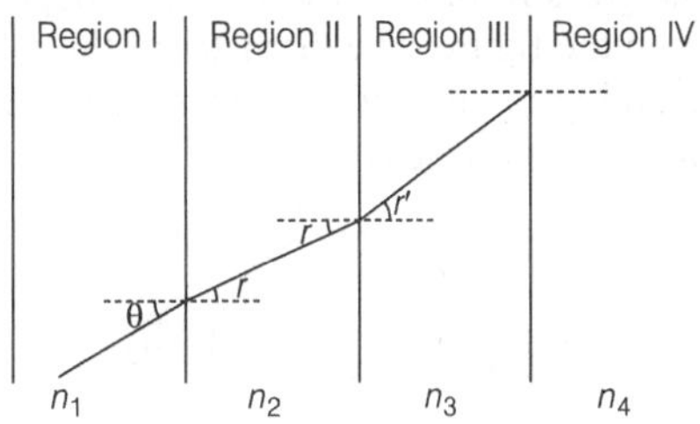


(a) $\sin^{-1}\left(\frac{3}{4}\right)$ (b) $\sin^{-1}\left(\frac{1}{8}\right)$ (c) $\sin^{-1}\left(\frac{1}{4}\right)$ (d) $\sin^{-1}\left(\frac{1}{3}\right)$

**35** While moving in a bus or a car during a hot summer day, a distant patch of road appears to be wet. This is due to

(a) sweating   (b) road material
(c) mirage   (d) sunlight

**36** Using the principle of total internal reflection, the prisms can bend light by

(a) 90°   (b) 180°   (c) 60°   (d) Both 90° or 180°

**37** The optical fibres are used for long distance transmission of signal, the main requirement for this is

(a) optical fibre material should be cheap
(b) it should be made of glass
(c) there should be little absorption of light inside it
(d) it is light in weight

# TOPIC 4 ~ *Refraction at Spherical Surfaces and by Lenses*

**38** Photometry is a measurement of physiological phenomenon. The main physical quantities in photometry is/are

(a) luminous intensity (b) luminous flux
(c) illuminance (d) All of the above

**39** Light from a point source in air falls on a spherical glass surface (whose refractive index $n = 1.5$ and radius of curvature $= 20$ cm). The distance of the light source from the glass surface is 100 cm. Image distance from the glass surface is

(a) 20 cm (b) 50 cm (c) 100 cm (d) 75 cm

**40** An object is placed infront of a sphere of radius $R$ and refractive index $n$ at a distance $x$ from the first surface. The value of $x$ for which the light after refraction from the first surface becomes parallel to the axis is

(a) $\frac{R}{(n-1)}$ (b) $\frac{R}{n+1}$
(c) $\frac{2R}{(1+n)}$ (d) $\frac{Rn}{(n+1)}$

**41** A mark placed on the surface of a sphere is viewed through glass from a position directly opposite as shown in the figure. The diameter of the sphere is 30 cm and refractive index of glass is 1.5. The position of the image is

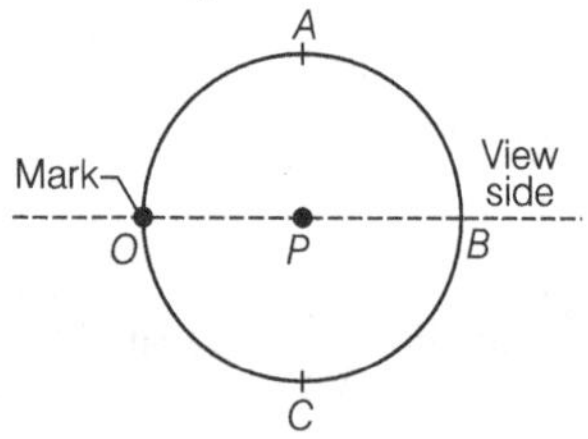


(a) at a distance of 60 cm from surface $ABC$ in the direction of incident light
(b) at a distance of 60 cm from the surface $ABC$ opposite to the direction of incident light
(c) at a distance of 30 cm from the surface $AOC$ opposite to the direction of incident light on $ABC$
(d) Both (b) and (c)

**42** A double convex lens has focal length 25 cm. The radius of curvature of one of the surfaces is double of the other. Find the radii, if the refractive index of the material of the lens is 1.5. ***NEET (Odisha) 2019***

(a) 100 cm, 50 cm
(b) 25 cm, 50 cm
(c) 18.75 cm, 37.5 cm
(d) 50 cm, 100 cm

**43** A thin lens made of glass (refractive index $= 1.5$) of focal length $f = 16$ cm in immersed in a liquid of refractive index 1.42. If its focal length in liquid is $f_1$, then the ratio $f_1 / f$ is closest to the integer ***JEE Main 2020***

(a) 1 (b) 5 (c) 9 (d) 17

**44** A lens is made of flint glass (refractive index = 1.5). When the lens is immersed in a liquid of refractive index 1.25, the focal length

(a) increases by a factor of 1.25
(b) increases by a factor of 2.5
(c) increases by a factor of 1.2
(d) decreases by a factor of 1.2

**45** A concave lens of glass of refractive index 1.5 has both surfaces of same radius of curvature $R$. On immersing it in a medium of refractive index 1.75, it will behave as a

(a) convergent lens of focal length $3.5\,R$
(b) convergent lens of focal length $3.0\,R$
(c) divergent lens of focal length $3.5\,R$
(d) divergent lens of focal length $3.0\,R$

**46** A hollow double concave lens is made of very thin transparent material. It can be filled with air or either of two liquids $L_1$ or $L_2$ having refractive indices $n_{L_1}$ and $n_{L_2}$, respectively (such that $n_{L_1} > n_{L_2} > 1$). The lens will diverge a parallel beam of light, if it is filled with

(a) air and placed in air (b) air and immersed in $L_1$
(c) $L_1$ and immersed in $L_2$ (d) $L_2$ and immersed in $L_1$

**47** A double convex lens whose refractive index is 1.33 has both radii of curvature of magnitude 10 cm. If an object is placed at a distance of 5 cm from this lens, the position of the image formed (in cm) is

(a) 7.46 same side of the object
(b) 7.46 opposite side of the object
(c) 14.45 same side of the object
(d) 14.45 opposite side of the object

**48** The graph shows how the magnification $m$ produced by a thin lens varies with image distance $v$. What is the focal length of the lens used? ***JEE Main 2019***

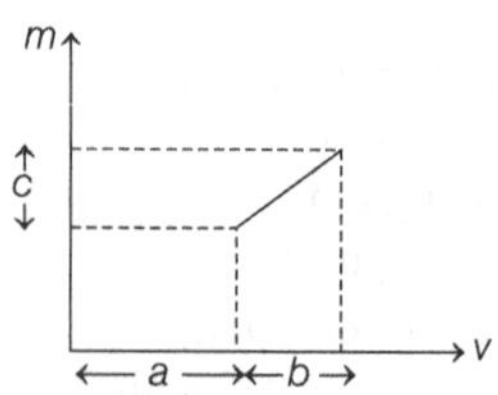


(a) $\frac{b^2c}{a}$ (b) $\frac{b^2}{ac}$ (c) $\frac{a}{c}$ (d) $\frac{b}{c}$

**49** Find the position of final image of the given object which is placed at 40 cm from first lens. Given focal length of each lens is 10 cm. ***JIPMER 2019***

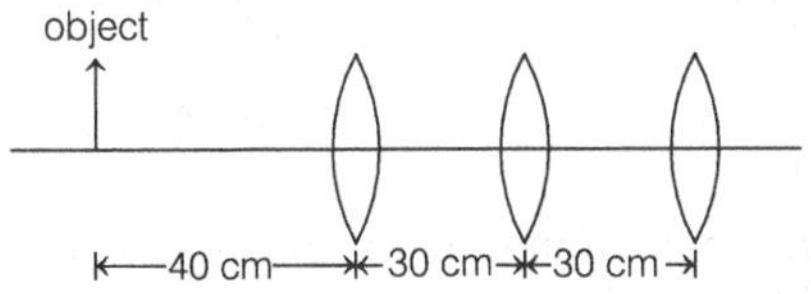


(a) 40 cm (b) 50 cm
(c) 45 cm (d) 55 cm

**50** A diverging lens with magnitude of focal length 25 cm is placed at a distance of 15 cm from a converging lens of magnitude of focal length 20 cm. A beam of parallel light falls on the diverging lens. The final image formed is ***JEE Main 2017***

(a) virtual and at a distance of 40 cm from convergent lens
(b) real and at a distance of 40 cm from the divergent lens
(c) real and at a distance of 6 cm from the convergent lens
(d) real and at a distance of 40 cm from convergent lens

**51** For a given lens, the magnification was found to be twice as large as when the object was 0.15 m distant from it as when the distance was 0.2 m. The focal length of the lens is ***AIIMS 2019***

(a) 1.5 m (b) 0.20 m (c) 0.10 m (d) 0.05 m

**52** The power of a biconvex lens is 20 dioptre and the radius of curvature of each surface is 5 cm. Find the refractive index of the material of lens.

(a) $\frac{4}{3}$ (b) $\frac{3}{2}$ (c) $\frac{9}{8}$ (d) $\frac{3}{5}$

**53** An equi-convex lens has power $P$. It is cut into two symmetrical halves by a plane containing the principal axis. The power of one part will be ***NEET (Odisha) 2019***

(a) 0 (b) $\frac{P}{2}$ (c) $\frac{P}{4}$ (d) $P$

**54** A convex lens of focal length 40 cm, a concave lens of focal length 40 cm and a concave lens of focal length 15 cm are placed in contact. The power of the combination in dioptre is

(a) + 1.5 (b) − 1.5 (c) + 6.67 (d) − 6.67

**55** Two similar thin equi-convex lenses, of focal length $f$ each, are kept co-axially in contact with each other such that the focal length of the combination is $F_1$. When the space between the two lenses is filled with glycerine {which has the same refractive index ($\mu = 1.5$) as that of glass}, then the equivalent focal length is $F_2$. The ratio of $F_1 : F_2$ will be ***NEET 2019***

(a) 1 : 2 (b) 2 : 3 (c) 3 : 4 (d) 2 : 1

**56** A plano-convex lens fits exactly into a plano-concave lens. Their plane surfaces are parallel to each other. If lenses are made of different materials of refractive indices $n_1$ and $n_2$ and $R$ is the radius of curvature of the curved surface of the lenses, then the focal length of the combination is ***CBSE AIPMT 2013***

(a) $\frac{R}{2(n_1 + n_2)}$ (b) $\frac{R}{2(n_1 - n_2)}$
(c) $\frac{R}{n_1 - n_2}$ (d) $\frac{2R}{n_2 - n_1}$

**57** Two identical thin plano-convex glass lenses (refractive index 1.5) each having radius of curvature of 20 cm are placed with their convex surfaces in contact at the centre. The intervening space is filled with oil of refractive index 1.7. The focal length of the combination is ***CBSE AIPMT 2015***

(a) − 20 cm (b) − 25 cm
(c) − 50 cm (d) 50 cm

# TOPIC 5 ~ *Refraction Through a Prism*

**58** A light ray incident normally on one of the face of a triangular prism follow the path as shown below, then the angle of refraction $r_2$ at the second face is

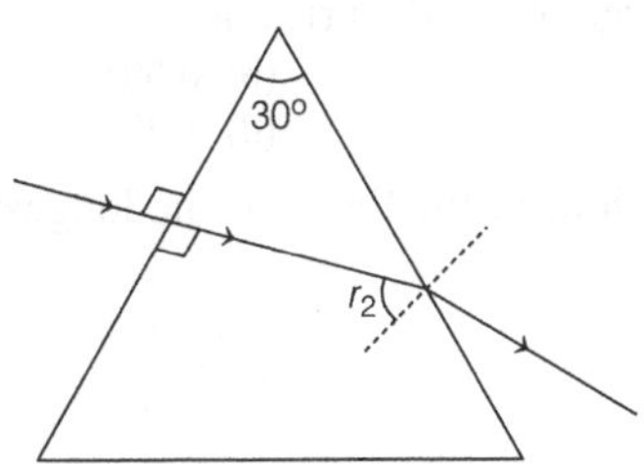


(a) 60° (b) 90°
(c) 30° (d) 45°

**59** The refractive index of the material of a prism is $\sqrt{2}$ and the angle of the prism is 30°. One of the two refracting surfaces of the prism is made a mirror inwards, by silver coating. A beam of monochromatic light entering the prism from the other face will retrace its path (after reflection from the silvered surface), if its angle of incidence on the prism is ***NEET 2018***

(a) 30° (b) 45° (c) 60° (d) zero

**60** A plot of angle of deviation $D$ *versus* angle of incidence for a triangular prism is shown below.

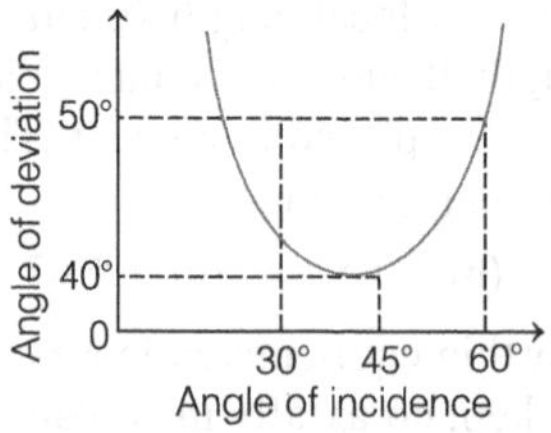


The angle of incidence for which the light ray travels parallel to the base is
(a) 30° (b) 60°
(c) 45° (d) Data insufficient

**61** Which graph correctly represents the relation between angle of incidence and angle of deviation for a prism? **JEE Main 2013**

(a) 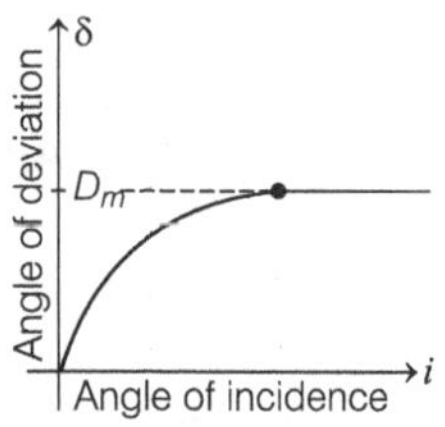


(b) 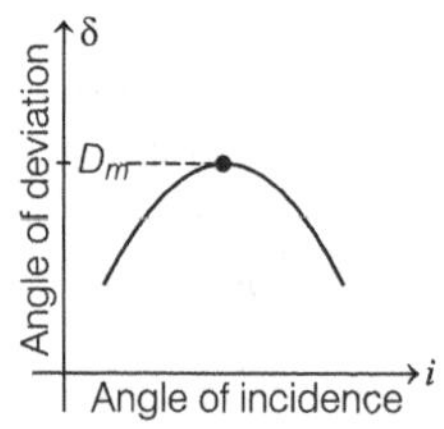


(c) 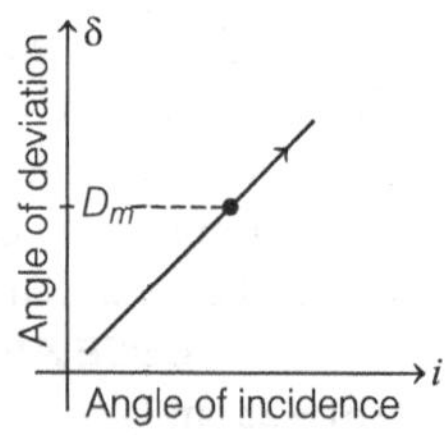


(d) 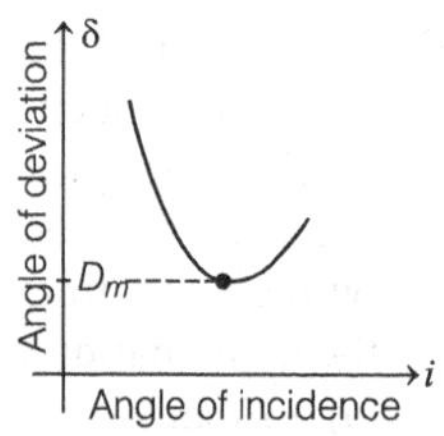


**62** A ray of light passes through an equilateral prism such that, the angle of incidence is equal to the angle of emergence and the latter is equal to 3/4 the angle of prism. The angle of deviation is
(a) 25° (b) 30°
(c) 45° (d) 35°

**63** A monochromatic light is incident at a certain angle on an equilateral triangular prism and suffers minimum deviation. If the refractive index of the material of the prism is $\sqrt{3}$, then the angle of incidence is **JEE Main 2019**
(a) 45° (b) 90° (c) 60° (d) 30°

**64** The refracting angle of a prism is $A$, and refractive index of the material of the prism is $\cot(A/2)$. The angle of minimum deviation is **CBSE AIPMT 2015**
(a) $180° - 3A$ (b) $180° - 2A$
(c) $90° - A$ (d) $180° + 2A$

**65** If a light ray incident on an equilateral prism suffers minimum deviation at an angle 30°, then speed of light in the prism will be **JIPMER 2018**
(a) $\frac{3}{\sqrt{2}} \times 10^8$ m/s (b) $\frac{1}{\sqrt{2}} \times 10^8$ m/s
(c) $\frac{2}{\sqrt{3}} \times 10^8$ m/s (d) $\frac{2}{3} \times 10^8$ m/s

**66** For the angle of minimum deviation of a prism to be equal to its refracting angle, the prism must be made of material whose refractive index
(a) lies between $\sqrt{2}$ and 1 (b) lies between 2 and $\sqrt{2}$
(c) is less than 1 (d) is greater than 2

**67** In an experiment for determination of refractive index of glass of a prism by *i-D* plot, it was found that a ray incident at an angle 35° suffers a deviation of 40° and that it emerges at an angle 79°. In that case, which of the following is closest to the maximum possible value of the refractive index? **JEE Main 2016**
(a) 1.5 (b) 1.6 (c) 1.7 (d) 1.8

**68** The angle of incidence for a ray of light at a refracting surface of a prism is 45°. The angle of prism is 60°. If the ray suffers minimum deviation through the prism, the angle of minimum deviation and refractive index of the material of the prism respectively, are **NEET 2016**
(a) 30°; $\sqrt{2}$ (b) 45°; $\sqrt{2}$
(c) 30°; $\frac{1}{\sqrt{2}}$ (d) 45°; $\frac{1}{\sqrt{2}}$

**69** A prism with refractive index $n = 1.57$ gives a deviation of 5.5°. The prism angle is
(a) 9.65° (b) 7.74°
(c) 5.5° (d) 8.50°

**70** A prism of crown glass with refracting angle of 5° and mean refractive index = 1.51 is combined with a flint glass prism of refractive index = 1.65 to produce no deviation. Find the angle of flint glass. **JIPMER 2018**
(a) 3.92° (b) 4.68°
(c) 5.32° (d) 7.28°

**71** White light is incident on one of the refracting surfaces of a prism of angle 5°. If the refractive indices for red and blue colours are 1.641 and 1.659 respectively, the angular separation between these two colours when they emerge out of the prism is
(a) 0.9° (b) 0.09°
(c) 1.8° (d) 1.2°

**72** Which colour of the light has the longest wavelength? **NEET 2019**
(a) Blue (b) Green
(c) Violet (d) Red

**73** In chromatic aberration, when white light passes through thick lenses
(a) red and blue colours focus at different points
(b) red and blue colour focus at same points
(c) red and yellow colours focus at different points
(d) yellow and green colour focus at same points

**74** The order of colour in primary and secondary rainbows from top to bottom are respectively
(a) violet-red and red-violet (b) red-violet and violet-red
(c) red-violet for both (d) violet-red for both

**75** If the ratio of amount of scattering of two light waves is 1 : 4, then the ratio of their wavelengths is
(a) $\sqrt{2}:1$ (b) 2 : 1
(c) 3 : 1 (d) $2\sqrt{2}:1$

**76** The full moon near the horizon looks reddish due to the reason that red colour light is
(a) most scattered
(b) least scattered
(c) more bright
(d) more intense

# TOPIC 6 ~ Optical Instruments and Defects of Vision

**77** For a normal eye, the cornea of eye provides a converging power of 40 D and the least converging power of the eye lens behind the cornea is 20 D. Using this information, the distance between the retina and the cornea of eye lens can be estimated to be **CBSE AIPMT 2013**
(a) 5 cm (b) 2.5 cm (c) 1.67 cm (d) 1.5 cm

**78** What focal length should the reading spectacles have for a person for whom the least distance of distinct vision is 50 cm?
(a) + 50 cm (b) − 75 cm (c) − 50 cm (d) + 75 cm

**79** The far point of a myopic person is 80 cm infront of the eye. What is the power of the lens required to enable him to see very distant objects clearly?
(a) 2 D (b) 1.50 D (c) 2.75 D (d) − 1.25 D

**80** The nearer point of hypermetropic eye is 20 cm. Find the power of lens which can be used for its correction.
(a) + 2.5 D (b) + 5.0 D (c) − 1.5 D (d) + 1.5 D

**81** The near point of a hypermetropic person is 75 cm from the eye. What is the power of the lens required to enable the person to read clearly a book held at 25 cm from the eye?
(a) + 2.67 D (b) − 3.42 D
(c) 4.62 D (d) 5.42 D

**82** A card sheet divided into square each of size 1 $mm^2$ is being seen through a magnifying glass lens of focal length 10 cm. The card is placed at a distance of 9 cm from the lens. The apparent area of the card through the lens is
(a) 1 $cm^2$ (b) 0.81 $cm^2$
(c) 0.27 $cm^2$ (d) 0.60 $cm^2$

**83** For realistic focal length, a simple microscope has a limited maximum magnification
(a) greater than 9 (b) lesser than 9
(c) equal to 9 (d) Both (b) and (c)

**84** In a microscope, the focal lengths of two lenses are 1.2 cm and 6.25 cm. If an object is placed at 2 cm and the final image is formed at 25 cm from eye lens. The distance between the two lenses is
(a) 6.00 cm (b) 8.00 cm
(c) 7.75 cm (d) 9.25 cm

**85** A compound microscope has an objective with focal length $f_o = 1.0$ cm, an eyepiece with focal length, $f_e = 2.0$ cm and a tube length of 20 cm, then the magnification is
(a) 200 (b) 250
(c) 100 (d) 150

**86** An observer looks at a distant tree of height 10 m with a telescope of magnifying power 20. To observer, the tree appears
(a) 10 times taller (b) 10 times nearer
(c) 20 times taller (d) 20 times nearer

**87** $F_1$ and $F_2$ are focal lengths of objective and eyepiece respectively, of the telescope. The angular magnification of the given telescope is equal to
(a) $\frac{F_1}{F_2}$ (b) $\frac{F_2}{F_1}$ (c) $\frac{F_1F_2}{F_1+F_2}$ (d) $\frac{F_1+F_2}{F_1F_2}$

**88** A telescope consists of two thin lenses of focal lengths 0.3 m and 3 cm, respectively. It is focused on moon which subtends an angle of 0.5° at the objective. Then, the angle subtended at the eye by the final image will be
(a) 5° (b) 0.25°
(c) 0.5° (d) 0.35°

**89** If the focal length of objective lens is increased, then magnifying power of **CBSE AIPMT 2014**
(a) microscope will increase but that of telescope will decrease
(b) microscope and telescope both will increase
(c) microscope and telescope both will decrease
(d) microscope will decrease but that of telescope will increase

**90** A reflecting telescope has an objective lens of focal length 150 cm and an eyepiece of focal length 5.0 cm. What is the separation between the objective and the eyepiece?
(a) 1.25 m (b) 1.55 m
(c) 1.3 m (d) 1.8 m

**91** If focal length of objective and eye lenses are 10 cm and 10 mm respectively, then angular magnification of telescope in normal adjustment is ***AIIMS 2019***
(a) 10 (b) 5
(c) 100 (d) 50

**92** An astronomical telescope has an angular magnification of magnitude 5 for distant objects. The separation between the objective and the eyepiece is 36 cm and the final image is formed at infinity. The focal length $f_o$ of the objective and the focal length $f_e$ of the eyepiece, respectively are
(a) 45 cm and − 9 cm (b) − 7.2 cm and 5 cm
(c) 50 cm and 10 cm (d) 30 cm and 6 cm

**93** In an astronomical telescope in normal adjustment a straight black line of length $L$ is drawn on inside part of objective lens. The eyepiece forms a real image of this line. The length of this image is $I$. The magnification of the telescope is ***CBSE AIPMT 2015***
(a) $\frac{L}{I}+1$ (b) $\frac{L}{I}-1$ (c) $\frac{L+1}{L-1}$ (d) $\frac{L}{I}$

**94** An astronomical telescope has objective and eyepiece of focal lengths 40 cm and 4 cm respectively. To view an object 200 cm away from the objective, the lenses must be separated by a distance ***NEET 2016***
(a) 46.0 cm (b) 50.0 cm (c) 54.0 cm (d) 37.3 cm

**95** To remove the chromatic aberration and distortions, due to lens of large size, the objective of telescope is replaced by
(a) convex mirror (b) plane mirror
(c) concave mirror (d) plano-concave lens

# SPECIAL TYPES QUESTIONS

## I. Assertion and Reason

**Direction** (Q. Nos.96 -109) *In the following questions, a statement of Assertion is followed by a corresponding statement of Reason. Of the following statements, choose the correct one.*
(a) Both Assertion and Reason are correct and Reason is the correct explanation of Assertion.
(b) Both Assertion and Reason are correct but Reason is not the correct explanation of Assertion.
(c) Assertion is correct but Reason is incorrect.
(d) Assertion is incorrect but Reason is correct.

**96 Assertion** A ray of light incident along the normal to the plane mirror retraces its path after reflection from the mirror.

**Reason** A ray of light along the normal has angle of incidence as $\pi/2$.

**97 Assertion** A convex mirror always make a virtual image.

**Reason** The rays always diverge after reflection from the convex mirror.

**98 Assertion** The refractive index of diamond is $\sqrt{6}$ and that of liquid is $\sqrt{3}$. If the light travels from diamond to the liquid, it will totally reflected when the angle of incidence is 30°.

**Reason** $n=\frac{1}{\sin i_c}$, where $n$ is the refractive index of diamond with respect to liquid and $i_c$ is the critical angle.

**99 Assertion** Propagation of light through an optical fibre is due to total internal reflection taking place at the core-clade interface.

**Reason** Refractive index of the material of the core of the optical fibre is greater than that of cladding.

**100 Assertion** The property of convergent lens of converging rays remain same in all media.

**Reason** Property of lens, whether the rays are diverging or converging depends on the surrounding medium.

**101 Assertion** By roughening the surface of a glass sheet, its transparency can be reduced.

**Reason** Glass sheet with rough surface absorbs more light.

**102 Assertion** If the focal length of the mirror is $f$ and distance of the object from the focus is $u$, then magnification of the mirror is $f/u$.

**Reason** Magnification $=\frac{\text{Size of image}}{\text{Size of object}}$

**103 Assertion** Thin prisms do not deviate light much.

**Reason** Thin prism have small angle $A$ and hence, $D_m$ (minimum deviation) is also very small as $D_m=[({}^1n_2-1)A]$, where ${}^1n_2$ is the refractive index of prism w.r.t. medium 1.

**104** **Assertion** Refractive index of glass with respect to air is different for red light and violet light.

**Reason** Refractive index of a pair of media depends on the wavelength of light used.

**105** **Assertion** When white light passes through a prism, bending of red component is most but for the violet it is least.

**Reason** Red colour travels faster than violet in glass.

**106** **Assertion** Sunlight reaches us without dispersion in the form of white light and not as its components.

**Reason** In vacuum, no dispersion takes place.

**107** **Assertion** In case of rainbow, the refracted rays falling on the inner surface of the water droplet get internally reflected.

**Reason** The angle between the refracted ray and normal to the drop surface is greater than the critical angle.

**108.** **Assertion** Secondary rainbow is fainter than primary rainbow.

**Reason** Secondary rainbow is formed by a four step process and hence, the intensity of light is reduced at the second reflection inside the rain drop.

**109** **Assertion** Astigmatism is occurred when the cornea is not spherical in shape.

**Reason** Astigmatism is corrected by using cylindrical lens of desired radius of curvature.

## II. Statement Based Questions

**110** I. The angle of reflection is greater than the angle of incidence.

II. The incident ray, reflected ray and the normal to the reflecting surface at the point of incidence lie in two planes.

III. The line joining the pole and the centre of curvature of the spherical mirror is known as the principal axis.

Which of the following statement is correct with respect to reflection?

(a) Only I (b) Only II (c) Both I and II (d) Only III

**111** I. For concave mirror, the reflected rays appears to diverge from a point $F$ and for convex mirror converge at the point $F$ ($F$ is called principal focus of the mirror).

II. If parallel beam of light were incident, making some angle with the principal axis, the reflected rays would converge (or appear to diverge) from a point in the focal plane through $F$ normal to the principal axis.

III. The distance between the focus $F$ and the pole $P$ of the mirror is called the focal length of the mirror.

Which of the following statement is incorrect?

(a) Only I (b) Only II (c) Both II and III (d) I, II and III

**112** I. The reflected ray goes through the focus of the mirror, when it is incident parallel to the principal axis.

II. The reflected ray is perpendicular to the principal axis when incident ray passes through the focus of the concave mirror or appear to pass through the focus of a convex mirror.

III. For the ray incident at any angle at the pole, the reflected ray follows laws of reflection.

Which of the following statement is incorrect?

(a) Only I (b) Both I and II
(c) Only II (d) Both I and III

**113** Which of the following statement is incorrect regarding optical fibres?

I. These are extensively used for transmitting audio and video signals through long distances.

II. There is appreciable loss in the intensity of light signal.

III. It is used to facilitate visual examination of internal organs like esophagus, stomach and intestines.

(a) Only I (b) Only II
(c) Both I and II (d) Both II and III

**114** I. The phenomenon of splitting of light into different colours is known as dispersion.

II. Dispersion takes place because the refractive index of medium for different frequencies (colours) is different.

III. Thin lenses show chromatic aberration due to dispersion of light.

Which of the following statement is incorrect?

(a) Only I (b) Only II
(c) Only III (d) I, II and III

**115** Which of the following statement is incorrect?

(a) Electromagnetic radiation belonging to the region of the spectrum of wavelength of about 400 nm to 700 nm is called light.
(b) The speed of light in vacuum is least.
(c) The wavelength of light is very small compared to the size of ordinary objects
(d) The straight line path of a light wave is called a ray of light and a bundle of such rays constitutes a beam of light.

**116** Which of the following statement is correct about sign convention used in optics?

(a) All distances are measured from the pole of the mirror or the optical centre of the lens.
(b) The distances measured in the same direction as the incident light are taken as negative and those measured in the direction opposite are taken as positive.
(c) The height measured upward is taken as negative and that measured downward as positive.
(d) All of the above

**117** Which of the following statement is correct?

(a) If ray emanating from a point actually meet at another point after reflection or refraction, then that point is called the image of the first point.
(b) The image is virtual, if the rays actually converge to a point.
(c) The image is real, if the rays do not actually meet but appear to diverge from the point when produced backwards.
(d) All of the above

**118** Which of the following statement is incorrect?

(a) In refraction, the direction of propagation of an obliquely incident ($0° < i < 90°$) ray of light that enters the other medium, changes at the interface of the two media.
(b) The incident ray, the refracted ray and the normal to the interface at the point of incidence, all lie in the same plane.
(c) The ratio of the sine of angle of incidence to the sine of angle of refraction is constant.
(d) The refractive index of a medium depends on the angle of incidence but is independent of the wavelength of light.

**119** Which of the following statement is incorrect?

(a) In case of refraction through a glass slab, when the emergent ray is parallel to the incident ray, there is maximum deviation.
(b) Sun is visible a little before the actual sunrise and until a little after the actual sunset due to refraction of light through the atmosphere.
(c) The apparent shift in the direction of the sun is about half a degree and the corresponding time difference between actual sunset and apparent sunset is about 2 minutes.
(d) The apparent flattening (oval shape) of the sun at sunset and sunrise is due to the refraction of light.

**120** Which of the following statement is true for rays coming from infinity?

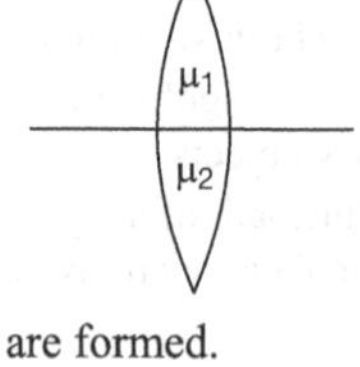


(a) Two images are formed.
(b) Continuous image is formed between focal points of upper and lower lens.
(c) One image is formed.
(d) None of the above

**121** Which of the following statement is incorrect?

(a) The optical density of different layers of air decreases with height.
(b) To a distant observer, the light from a tall object appears to be coming from somewhere below the ground on a hot summer day.
(c) The inverted image of distant tall object cause an optical illusion to the observer on a hot summer day.
(d) The mirage is especially common in hot deserts.

**122** Which of the following statement is correct?

(a) The brilliance in diamond is mainly due to the refraction of light inside them.
(b) The critical angle of diamond-air inerface is very large.
(c) By cutting the diamond suitably, multiple total internal reflection can be made to occur.
(d) Prisms are be used to invert images, by changing there size.

**123** Which of the following statement is incorrect?

(a) In refraction, the measurement of light as perceived by human eye is called photometry.
(b) SI unit of luminous intensity ($I$) is candela (cd).
(c) In photometry, the only parameter, which can be measured directly is luminous flux.
(d) A quantity named luminance ($L$) is used to characterise the brightness of emitting or reflecting flat surfaces.

**124** Which of the following statement is incorrect?

(a) The focus on the side of the (original) source of light is called the first focal point, whereas the other is called the second focal point.
(b) A ray emanating from the object parallel to the principal axis of the lens after refraction passes through the second principal focus (in a convex lens) or appear to diverge (in a concave lens).
(c) A ray of light passing through the optical centre of the lens, emerges without any deviation after refraction.
(d) A ray of light passing through the first principal focus (for a convex lens) or appearing to meet at it (for a concave lens) emerges obliquely to the principal axis after refraction.

**125** Which of the following statement is correct?

(a) Power of a lens is a measure of the convergence or divergence, which a lens introduces in the light falling on it.
(b) Power is defined as the cosine of angle by which it converges or diverges a beam of light falling at unit distant from the optical centre.
(c) Cutting of lenses helps to obtain diverging or converging lenses of desired magnification
(d) Total magnification $m$ of the combination is the sum of magnification ($m_1, m_2, m_3$.....) of individual lenses.

**126** Which of the following statement is correct?

(a) The primary rainbow is a result of four-step process.
(b) The blue colour of sky is due to the fact that the blue gets scattered more than violet.
(c) The white colour of clouds is because they have droplets of water, that absorb all wavelengths.
(d) The sun looks reddish at sunrise or sunset as most of the blue and other shorter wavelengths are removed by scattering.

**127** Pick the incorrect statement in the context with rainbow.
(a) The order of colours is reversed in the secondary rainbow.
(b) An observer can see a rainbow when his front is towards the sun.
(c) Rainbow is a combined effect of dispersion, refraction and reflection of sunlight.
(d) When the light rays undergo two internal reflections in a water drop, a secondary rainbow is formed.

**128** Which of the following statement is incorrect?
(a) Accomodation is the property of eye lens due to which its shape (curvature) and therefore the focal length can be modified.
(b) In presbyopia, the distance of distinct vision of a person decrease with age.
(c) In myopia or nearsightedness, the light from a distant object arriving at the eye-lens get converged at a point in front of retina.
(d) In hypermetropia or farsightedness, the eye-lens focusses the incoming light at a point behind the retina.

**129** Which of the following statement is incorrect?
(a) A simple magnifier or microscope is a converging lens of small focal length.
(b) For microscope the magnification in case of image formed at infinity is one more than the magnification when image is at the near point.
(c) For larger magnifications, one uses two lenses, one compounding the effect of the other. This is known as a compound microscope.
(d) The lens nearest the object is called objective and the lens near eye is called eyepiece.

**130** An astronomical refractive telescope has an objective of focal length 20 m and an eyepiece of focal length 2 cm. Which of the following statement is not correct?
(a) The length of the telescope tube is 20.02 m.
(b) The magnification is 1000.
(c) The image formed is inverted.
(d) An objective of a larger aperture will increase the brightness and reduce chromatic aberration of the image.

**131** Which of the following statement(s) is/are incorrect?
(a) In compound microscope, objective lens forms a real, inverted, magnified image of the object.
(b) Eyepiece lens, produces the final image which is enlarged and virtual.
(c) The final image is inverted with respect to the original object.
(d) None of the above

**132** Which of the following statement is correct?
(a) The telescope is used to provide linear magnification of distant objects.
(b) Terrestrial telescope have in addition, a pair of inverting lenses to make the final image erect.
(c) The resolving power or the ability to observe two objects distinctly, which are in the opposite direction, depends on the distance of the objective.
(d) A cassegrain telescope has the advantage of a small focal length as compare to long telescope.

## III. Matching Type

**133** With reference to the figure given below, match the items in Column I (angle) with Column II (value) and select the correct answer from the codes given below.

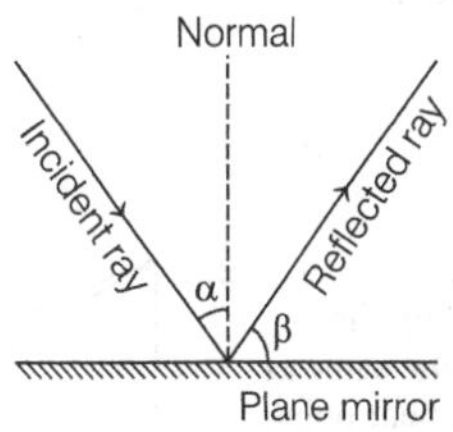


| | Column I | | Column II |
|---|---|---|---|
| A. | Angle of reflection | 1. | $(\pi - 2\alpha)$ |
| B. | Value of angle $\beta$ | 2. | $\alpha$ |
| C. | Angle of deviation of the incident ray | 3. | $\left(\frac{\pi}{2} - \alpha\right)$ |

| | A | B | C | | A | B | C |
|---|---|---|---|---|---|---|---|
| (a) | 1 | 2 | 3 | (b) | 3 | 2 | 1 |
| (c) | 2 | 3 | 1 | (d) | 2 | 1 | 3 |

**134** A layer of ice $(n = 4/3)$ lies on a glass plate $(n = 3/2)$. A ray of light makes an angle of incidence $60°$ on the surface of ice as shown in the figure below.

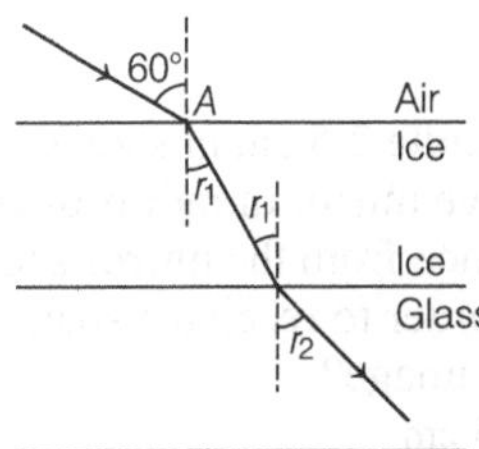


With reference to the above figure, match the Column I (parameter) with Column II (value) and select the correct answer from the codes given below.

| | Column I | | Column II |
|---|---|---|---|
| A. | $\sin r_1$ | 1. | $1/\sqrt{3}$ |
| B. | $\sin r_2$ | 2. | $3\sqrt{3}/8$ |
| C. | Refractive Index of ice w.r.t. glass | 3. | $8/9$ |

| | A | B | C | | A | B | C |
|---|---|---|---|---|---|---|---|
| (a) | 1 | 2 | 3 | (b) | 2 | 1 | 3 |
| (c) | 2 | 3 | 1 | (d) | 1 | 3 | 2 |

**135** Match the Column I (different cases of simple microscope) with Column II (ray diagram) and select the correct answer from the codes given below.

| Column I | Column II |
|---|---|
| A. The magnifying lens is located such that image is at the near point. | 1. 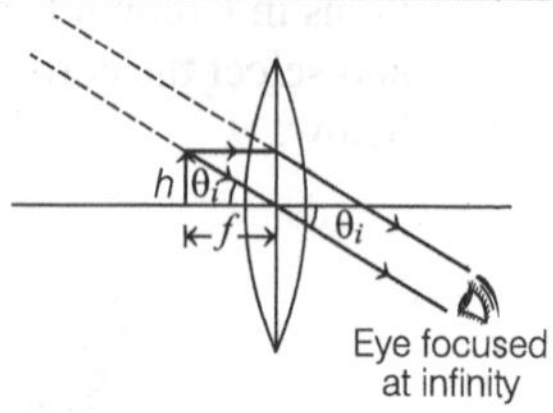  |
| B. The angle subtended by the object, is the same as that at the near point. | 2. 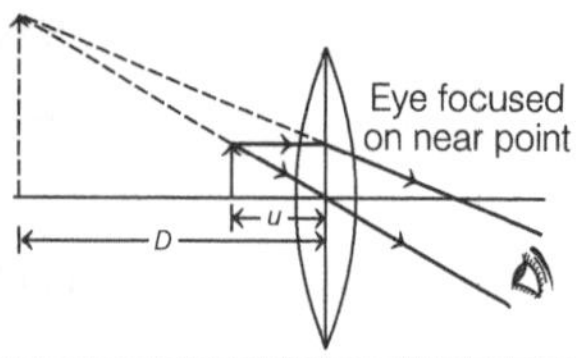  |
| C. The object near the focal point of the lens and the image is far off but closer than infinity. | 3. 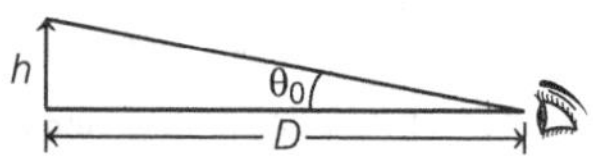  |

| | A | B | C |
|---|---|---|---|
| (a) | 1 | 2 | 3 |
| (b) | 1 | 2 | 3 |
| (c) | 1 | 3 | 2 |
| (d) | 2 | 3 | 1 |

**136** A light ray passes through triangular prism with prism angle $A = 60°$ and $\mu = n = 1.5$. With reference to the above situation, match the Column I (angle) with Column II (value) and select the correct answer from the codes given below.
[Some useful data, $2\sin^{-1}(0.75) = 97°$,
$\sin^{-1}(2/3) = 42°$ and $\sin^{-1}(0.463) = 28°$]

| Column I | Column II |
|---|---|
| A. Minimum deviation ($D_{min}$) | 1. 58° |
| B. Maximum deviation ($D_{max}$) | 2. 37° |
| C. The angle of refraction $r_1$ when maximum deviation occurs | 3. 42° |

| | A | B | C |
|---|---|---|---|
| (a) | 1 | 2 | 3 |
| (b) | 3 | 2 | 1 |
| (c) | 2 | 1 | 3 |
| (d) | 3 | 1 | 2 |

# NCERT & NCERT Exemplar

## MULTIPLE CHOICE QUESTIONS

### NCERT

**137** A small candle 2.5 cm in size is placed 27 cm infront of a concave mirror of radius of curvature 36 cm. At what distance from the mirror should a screen be placed in order to receive a sharp image? What is the size of the image?
(a) 54 cm, 5 cm
(b) – 54 cm, – 5 cm
(c) – 27 cm, 2.5 cm
(d) – 54 cm, – 2.5 cm

**138** A 4.5 cm needle is placed 12 cm away from a convex mirror of focal length 15 cm. Find the magnification of this image.
(a) $\frac{5}{9}$ (b) $\frac{9}{5}$
(c) 5.5 (d) None of these

**139** A tank is filled with water to a height of 12.5 cm. The apparent depth of a needle lying at the bottom of the tank is measured by a microscope to be 9.4 cm. What is the refractive index of water? If water is replaced by a liquid of refractive index 1.63 upto the same height, by what distance would the microscope have to be moved to focus on the needle again?
(a) 1.32, 7.7 cm (b) 1.63, 7.7 cm
(c) 1.32, 1.7 cm (d) 1.63, 1.7 cm

**140** A small bulb is placed at the bottom of a tank containing water to a depth of 80 cm. What is the area of the surface of water through which light from the bulb can emerge out? Refractive index of water is 1.33. (Consider the bulb to be a point source.)
(a) 3.58 $m^2$ (b) 2.58 $m^2$
(c) 6 $m^2$ (d) 4 $m^2$

**141** A prism is made of glass of unknown refractive index. A parallel beam of light is incident on a face of the prism. The angle of minimum deviation is measured to be 40°. What is the refractive index of the material of the prism? The refracting angle of the prism is 60°. If the prism is placed in water (refractive

index 1.33), predict the new angle of minimum deviation of a parallel beam of light.
(a) 1.53, 10°20' (b) 1.2, 35°10'
(c) 1.53, 35°10' (d) 1.2, 10°20'

**142** Double-convex lenses are to be manufactured from a glass of refractive index 1.55, with both faces of the same radius of curvature. What is the radius of curvature required, if the focal length is to be 20 cm?
(a) 20 cm (b) 21 cm (c) 22 cm (d) 23 cm

**143** A beam of light converges at a point $P$. Now a lens is placed in the path of the convergent beam 12 cm from $P$. At what point does the beam converge if the lens is a convex lens of focal length 20 cm?
(a) 7.5 cm (b) 8.5 cm (c) 9.5 cm (d) 6.5 cm

**144** An object of size 3.0 cm is placed 14 cm infront of a concave lens of focal length 21 cm. The image distance is $v$, magnification $m$ and image height $I$. With reference to the above situation, match the Column I with Column II and select the correct answer from the codes given below.

| | Column I | | Column II |
|---|---|---|---|
| A | $v$ (image distance in cm) | 1. | 0.6 |
| B | $I$ (image height in cm) | 2. | 8.4 |
| C | $m$ (magnification) | 3. | 1.8 |

| | A | B | C | | A | B | C |
|---|---|---|---|---|---|---|---|
| (a) | 1 | 2 | 3 | (b) | 3 | 2 | 1 |
| (c) | 2 | 1 | 3 | (d) | 2 | 3 | 1 |

**145** The plano-convex lens of focal length 20 cm and 30 cm are placed together to form a double convex lens. The final focal length will be
(a) 12 cm (b) 60 cm (c) 20 cm (d) 30 cm

**146** A person with a normal near point 25 cm using a compound microscope with an objective of focal length 8.0 mm and eyepiece of focal length 2.5 cm can bring an object placed at 9.0 mm from the objective in sharp focus. What is the separation between the two lenses? How much is the magnifying power of the microscope?
(a) 7.2, 89 (b) 9.4, 88 (c) 8.6, 90 (d) 9.4, 75

**147** A small telescope has an objective lens of focal length 144 cm and an eyepiece of focal length 6.0 cm. What is the magnifying power of the telescope? What is the separation between the objective and the eyepiece?
(a) $-24$, 150 cm (b) $-25$, 150 cm
(c) $-24$, 160 cm (d) $-25$, 160 cm

**148** A giant refracting telescope at an observatory has an objective lens of focal length 15 m. If an eyepiece of focal length 1.0 cm is used, what is angular magnification of the telescope?
(a) 1200 (b) 1000 (c) 1300 (d) 1500

**149** A small pin fixed on a table top is viewed from above, from a distance of 50 cm. By what distance would the pin appear to be raised, if it is viewed from the same point through a 15 cm thick glass slab held parallel to the table? Refractive index of glass $=1.5$.
(a) 15 cm (b) 5 cm (c) 10 cm (d) None of these

**150** A screen is placed 90 cm from an object, the image of the object on the screen is formed by a convex lens at two different locations separated by 20 cm. Determine the focal length of the lens.
(a) 20.5 cm (b) 19.5 cm (c) 23.2 cm (d) 21.4 cm

**151** At what angle should a ray of light be incident on the face of a prism of refracting angle 60°, so that it just suffers total internal reflection at the other face? The refractive index of the material of the prism is 1.524.
(a) 30°75′ (b) 29°75′ (c) 28°75′ (d) 27°75′

**152** For a normal eye, the far point is at infinity and the near point of distinct vision is about 25 cm infront of the eye. The cornea of the eye provides a converging power of about 40 D and the least converging power of the eyelens behind the cornea is about 20 D. From this rough data, estimate the range of accommodation (i.e. range of converging power of its eyelens) of a normal eye.
(a) 20-24 D (b) 18-22 D (c) 64-68 D (d) 35-40 D

**153** A myopic person has been using spectacles of power $-1.0$ D for distant vision. During old age, he also needs to use separate reading glass of power $+2.0$ D. Find the near point of this person.
(a) $+50$ cm (b) $-50$ cm (c) $-100$ cm (d) $+100$ cm

**154** A person looking at a person wearing a shirt with a pattern comprising vertical and horizontal lines is able to see the vertical lines more distinctly than the horizontal ones. What is this defect called? How is such a defect of vision corrected?
(a) Astigmatism, convex lens
(b) Myopia, concave lens
(c) Astigmatism, cylindrical lens
(d) Presbyopia, cylindrical lens

**155** A man with normal near point (25 cm) reads a book with small print using a magnifying glass : a thin convex lens of focal length 5 cm. What is the closest and the farthest distance at which he can read the book when viewing through the magnifying glass?
(a) $-4.2$ cm, $-5$ cm (b) 4.2 cm, 5 cm
(c) $-4.2$ cm, 5 cm (b) 4.2 cm, $-5$ cm

**156** An angular magnification (magnifying power) of 30 $X$ is desired using an objective of focal length 1.25 cm and an eyepiece of focal length 5 cm. So, for this compound microscope,

(a) $u_o = -1.5$ cm, $v_o = 7.5$ cm, $u_e = -4.17$ cm, $L = 11.67$ cm
(b) $u_o = 1.5$ cm, $v_o = -7.5$ cm, $u_e = 4.17$ cm, $L = 2.8$ cm
(c) $u_o = -1.5$ cm, $v_o = -7.5$ cm, $u_e = -4.17$ cm, $L = 10$ cm
(d) $u_o = -1.5$ cm, $v_o = 7.5$ cm, $u_e = 4.17$ cm, $L = 4.82$ cm

**157** A small telescope has an objective lens of focal length 140 cm and an eyepiece of focal length 5.0 cm. What is the magnifying power of the telescope for viewing distant objects when

(i) the telescope is in normal adjustment (i.e. when the final image is at infinity).
(ii) the final image is formed at the least distance of distinct vision (25 cm).

(a) 14, 28 (b) 5, 1.2 (c) 28, 28.7 (d) 28, 33.6

**158** A cassegrain telescope uses two mirrors as shown in figure. Such a telescope is built with the mirrors 20 mm apart. If the radius of curvature of the large mirror is 220 mm and the small mirror is 140 mm, where will the final image of an object at infinity?

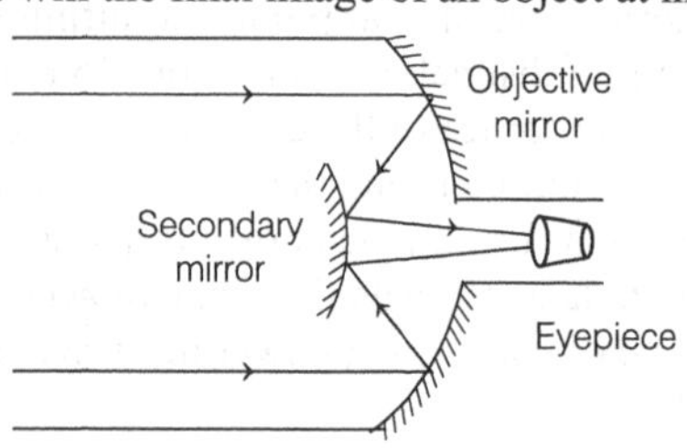


(a) 315 mm (b) 415 mm
(c) 215 mm (d) 115 cm

**159** Figure shows an equi-convex lens (of refractive index 1.50) in contact with a liquid layer on top of a plane mirror. A small needle with its tip on the principal axis is moved along the axis until its inverted image is found at the position of the needle. The distance of the needle from the lens is measured to be 45 cm. The liquid is removed and the experiment is repeated. The new distance is measured to be 30 cm. What is the refractive index of the liquid? **JEE Main 2019**

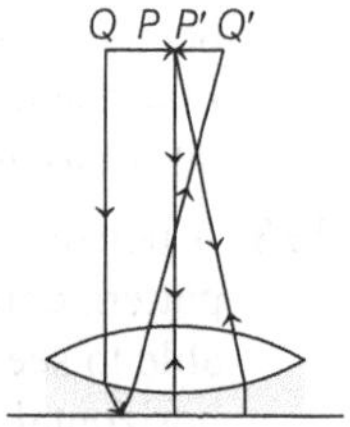


(a) 1.42 (b) 1.50
(c) 1.26 (d) 1.33

## NCERT Exemplar

**160** A ray of light incident at an angle θ on a refracting face of a prism emerges from the other face normally. If the angle of the prism is 5° and the prism is made of a material of refractive index 1.5, the angle of incidence is

(a) 7.5° (b) 5°
(c) 15° (d) 2.5°

**161** A short pulse of white light is incident from air to a glass slab at normal incidence. After travelling through the slab, the first colour to emerge is

(a) blue (b) green (c) violet (d) red

**162** An object approaches a convergent lens from the left of the lens with a uniform speed 5 $ms^{-1}$ and stops at the focus. The image

(a) moves away from the lens with an uniform speed 5 $ms^{-1}$
(b) moves away from the lens with an uniform acceleration
(c) moves away from the lens with a non- uniform acceleration
(d) moves towards the lens with a non-uniform acceleration

**163** The radius of curvature of the curved surface of a plano-convex lens is 20 cm. If the refractive index of the material of the lens be 1.5, it will

(a) act as a convex lens only for the objects that lie on its curved side
(b) act as a concave lens for the objects that lie on its curved side
(c) act as a convex lens irrespective of the side on which the object lies
(d) act as a concave lens irrespective of side on which the object lies

**164** The direction of ray of light incident on a concave mirror is shown by $PQ$ while directions in which the ray would travel after reflection is shown by four rays marked 1,2,3 and 4 as shown in figure. Which of the four rays correctly shows the direction of reflected ray?

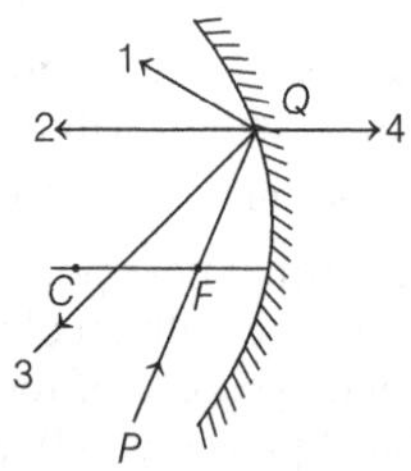


(a) 1 (b) 2 (c) 3 (d) 4

**165** The optical density of turpentine is higher than that of water while its mass density is lower. Figure shows a layer of turpentine floating over water in a container. For which one of the four rays incident on turpentine in figure, the path shown is correct?

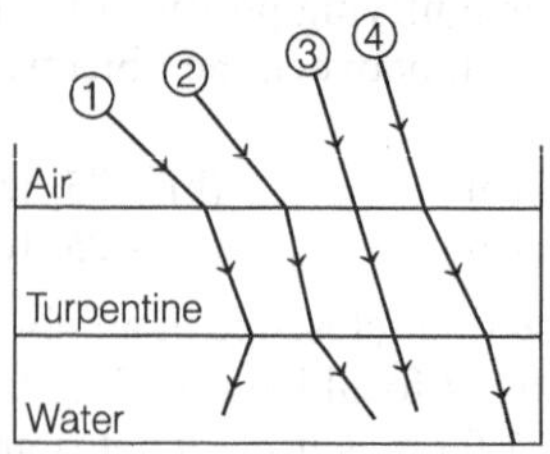


(a) 1 (b) 2 (c) 3 (d) 4

**166** There are certain materials developed in laboratories which have a negative refractive index (figure shown below). A ray incident from air (medium 1) into such a medium (medium 2) shall follow a path given by

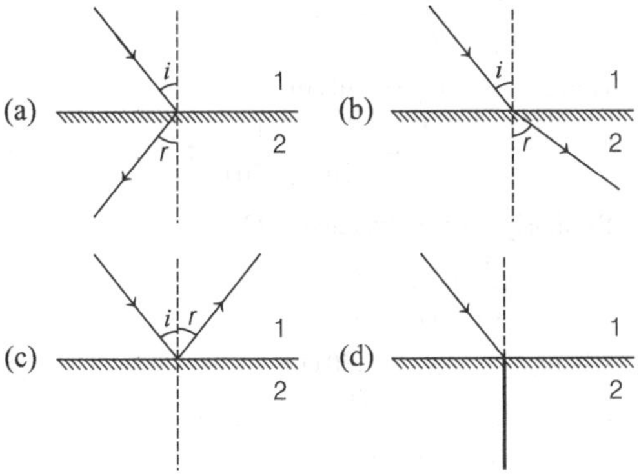

**167** A rectangular block of glass *ABCD* has a refractive index 1.6. A pin is placed midway on the face *AB* figure. When observed from the face *AD*, the pin shall

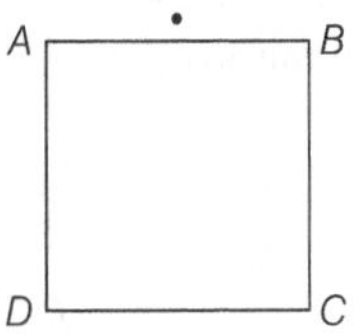


(a) appear to be near *B*
(b) appear to be near *D*
(c) appear to be at the centre of *AD*
(d) not be seen at all

**168** The near vision of an average person is 25 cm. To view an object with an angular magnification of 10, what should be the power of the microscope?
(a) 10 D (b) 20 D
(c) 40 D (d) 30 D

**169** Three immiscible liquids of densities $d_1 > d_2 > d_3$ and refractive indices $n_1 > n_2 > n_3$ are put in a beaker. The height of each liquid column is $\frac{h}{3}$. A dot is made at the bottom of the beaker. For near normal vision, find the apparent depth of the dot?
(a) $\frac{h}{3}\left(\frac{1}{n_1}+\frac{1}{n_2}+\frac{1}{n_3}\right)$ (b) $3h\left(\frac{1}{n_1}+\frac{1}{n_2}-\frac{1}{n_3}\right)$
(c) $\frac{h}{3}\left(\frac{1}{n_1}-\frac{1}{n_2}+\frac{1}{n_3}\right)$ (d) $3h\left(\frac{1}{n_1}-\frac{1}{n_2}-\frac{1}{n_3}\right)$

**170** For a glass prism $(n=\sqrt{3})$, the angle of minimum deviation is equal to the angle of the prism. Find the angle of the prism.
(a) 40° (b) 45°
(c) 60° (d) 75°

**171** A thin convex lens of focal length 25 cm is cut into two pieces 0.5 cm above the principal axis. The top part is placed at (0, 0) and an object placed at $(-50\text{ cm}, 0)$. The coordinates of the image are
(a) (50 cm, −1 cm) (b) (50 cm, +1 cm)
(c) (−50 cm, +1 cm) (d) (−50 cm, −1 cm)

# Answers

## > Mastering NCERT with MCQs

| | | | | | | | | | |
|---|---|---|---|---|---|---|---|---|---|
| 1 (d) | 2 (c) | 3 (c) | 4 (c) | 5 (a) | 6 (b) | 7 (d) | 8 (d) | 9 (d) | 10 (b) |
| 11 (d) | 12 (a) | 13 (d) | 14 (c) | 15 (a) | 16 (a) | 17 (d) | 18 (c) | 19 (c) | 20 (d) |
| 21 (b) | 22 (a) | 23 (c) | 24 (c) | 25 (a) | 26 (a) | 27 (c) | 28 (b) | 29 (c) | 30 (c) |
| 31 (b) | 32 (a) | 33 (c) | 34 (b) | 35 (c) | 36 (d) | 37 (c) | 38 (d) | 39 (c) | 40 (a) |
| 41 (d) | 42 (c) | 43 (c) | 44 (b) | 45 (a) | 46 (d) | 47 (a) | 48 (d) | 49 (b) | 50 (d) |
| 51 (c) | 52 (b) | 53 (d) | 54 (d) | 55 (a) | 56 (c) | 57 (c) | 58 (c) | 59 (b) | 60 (c) |
| 61 (d) | 62 (b) | 63 (c) | 64 (b) | 65 (a) | 66 (b) | 67 (a) | 68 (a) | 69 (a) | 70 (a) |
| 71 (b) | 72 (d) | 73 (a) | 74 (b) | 75 (a) | 76 (b) | 77 (c) | 78 (a) | 79 (d) | 80 (b) |
| 81 (a) | 82 (a) | 83 (d) | 84 (b) | 85 (b) | 86 (d) | 87 (a) | 88 (a) | 89 (d) | 90 (b) |
| 91 (a) | 92 (d) | 93 (d) | 94 (c) | 95 (c) | | | | | |

## > Special Types Questions

| | | | | | | | | | |
|---|---|---|---|---|---|---|---|---|---|
| 96 (c) | 97 (a) | 98 (d) | 99 (b) | 100 (d) | 101 (c) | 102 (d) | 103 (a) | 104 (a) | 105 (d) |
| 106 (a) | 107 (a) | 108 (a) | 109 (b) | 110 (d) | 111 (a) | 112 (c) | 113 (b) | 114 (c) | 115 (b) |
| 116 (a) | 117 (a) | 118 (d) | 119 (a) | 120 (a) | 121 (a) | 122 (c) | 123 (c) | 124 (d) | 125 (a) |
| 126 (d) | 127 (b) | 128 (b) | 129 (b) | 130 (d) | 131 (d) | 132 (b) | 133 (c) | 134 (b) | 135 (d) |
| 136 (c) | | | | | | | | | |

## > NCERT & NCERT Exemplar MCQs

| | | | | | | | | | |
|---|---|---|---|---|---|---|---|---|---|
| 137 (b) | 138 (a) | 139 (c) | 140 (b) | 141 (a) | 142 (c) | 143 (a) | 144 (d) | 145 (a) | 146 (b) |
| 147 (a) | 148 (d) | 149 (b) | 150 (d) | 151 (b) | 152 (a) | 153 (b) | 154 (c) | 155 (a) | 156 (a) |
| 157 (d) | 158 (a) | 159 (d) | 160 (a) | 161 (d) | 162 (c) | 163 (c) | 164 (b) | 165 (b) | 166 (a) |
| 167 (d) | 168 (c) | 169 (a) | 170 (c) | 171 (a) | | | | | |

# Hints & Explanations

**3** (*c*) Let the angle of incidence be $\theta$. From the law of reflection,

Angle of incidence = Angle of reflection = $\theta$

So, the given situation can be depicted as follows

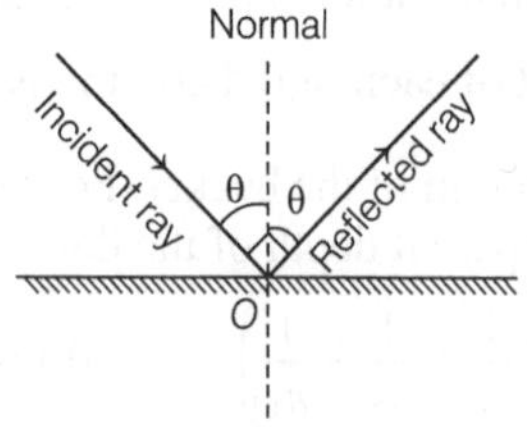


Given, incident ray is perpendicular to reflected ray

$\Rightarrow \theta + \theta = 2\theta = 90°$ (from figure) $\Rightarrow \theta = 45°$

So, angle of incidence = 45°.

**4** (*c*) The given situation can shown as

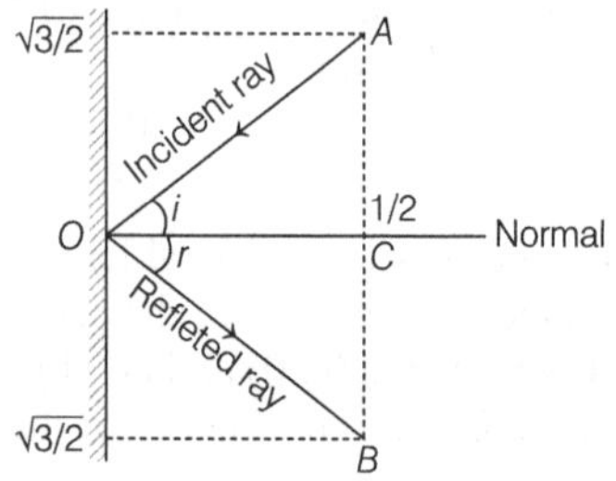


If $i$ be the angle of incidence, then from $\Delta AOC$,

$$\tan i = \frac{AC}{OC} = \left(\frac{\sqrt{3}/2}{1/2}\right) = \sqrt{3} = \tan 60° \Rightarrow i = 60°$$

**6** (*b*) The focal length of a mirror is $f = \frac{R}{2}$

where, $R$ is the radius of curvature of mirror.

Here, $R = 16\,\text{cm}$ $\quad \therefore \quad f = \frac{16}{2} = 8\,\text{cm}$

**8** (*d*) If lower half of a concave mirror is blackened, then taking the laws of reflection to be true for all points of the remaining part of the mirror, the image will be that of the whole object. However, as the area of the reflecting surface has been reduced, so the intensity of the image will be low, i.e. half.

**9** (*d*) The given situation can be depicted as follows

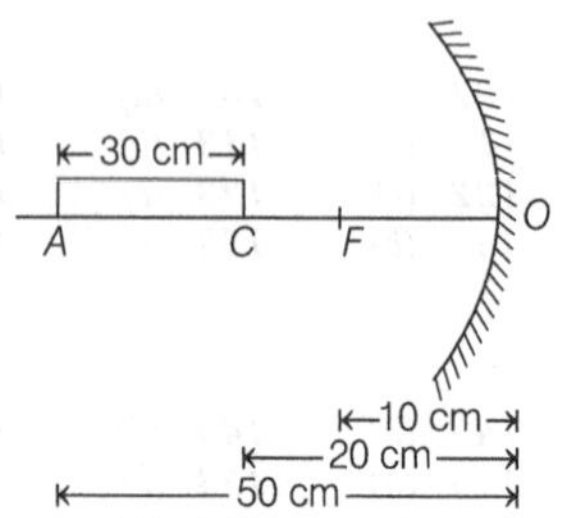


where, focal length of the mirror, $f = -10\,\text{cm}$

By mirror formula, image distance of $A$

$$\frac{1}{v_A} + \frac{1}{u_A} = \frac{1}{f}$$

Here, $u_A = -50\,\text{cm}$

$$\Rightarrow \frac{1}{v_A} = \frac{1}{-10} - \frac{1}{(-50)} \Rightarrow v_A = -12.5\,\text{cm}$$

Similarly, image distance of $C$

$$\frac{1}{v_C} + \frac{1}{u_C} = \frac{1}{f}$$

where, $u_C = -20\,\text{cm}$

$$\Rightarrow \frac{1}{v_C} - \frac{1}{20} = \frac{-1}{10} \Rightarrow v_C = -20\ \text{cm}$$

$\therefore$ The length of image $= |v_A - v_C|$

$$= |-12.5 - (-20)| = 7.5\,\text{cm}$$

**10** (*b*) ***Case* I** When the object distance, $u_1 = -40\,\text{cm}$

Focal length of mirror, $f = -15\,\text{cm}$

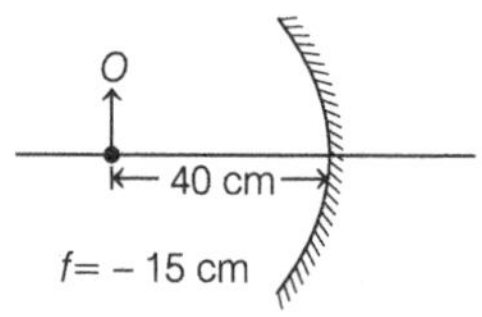


Using the mirror formula, we get

$$\frac{1}{f} = \frac{1}{v_1} + \frac{1}{u_1} - \frac{1}{15} = \frac{1}{v_1} + \left(\frac{-1}{40}\right)$$

$$\Rightarrow \frac{1}{v_1} = \frac{1}{40} - \frac{1}{15} = \frac{3-8}{120} = \frac{-5}{120}$$

$$\Rightarrow v_1 = \frac{-120}{5} = -24\,\text{cm}$$

***Case* II** When the object distance, $u_2 = -20\,\text{cm}$

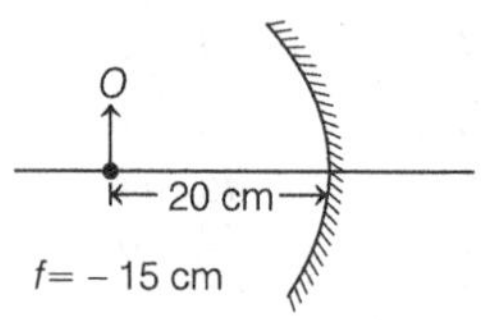


Using the mirror formula again, we get

$$\frac{1}{f} = \frac{1}{v_2} + \frac{1}{u_2} \Rightarrow -\frac{1}{15} = \frac{1}{v_2} + \left(-\frac{1}{20}\right)$$

$$\Rightarrow \frac{1}{v_2} = \frac{1}{20} - \frac{1}{15} = \frac{3-4}{60} = \frac{-1}{60} \Rightarrow v_2 = -60\,\text{cm}$$

As, the net displacement of the images is equal to the difference between the image distance in both the cases.

$\therefore$ The displacement of the image is

$$= v_2 - v_1 = -60 - (-24)$$

$$= -60 + 24 \text{ or } = -36\,\text{cm}$$

= 36 cm, away from the mirror

**11** (*d*) We know that, from the mirror formula,

$$\frac{1}{f}=\frac{1}{v}+\frac{1}{u} \Rightarrow \frac{1}{v}=\frac{1}{f}-\frac{1}{u}$$

$$\frac{1}{v}=\frac{u-f}{uf} \text{ or } v=\left(\frac{uf}{u-f}\right) \quad \ldots(\text{i})$$

On differentiating Eq. (i) w.r.t. $u$, we get

$$\frac{dv}{du}=\frac{\frac{d}{du}(uf)(u-f)-\frac{d}{du}(u-f)(uf)}{(u-f)^2}$$

$$=\frac{(u-f)f-(uf)}{(u-f)^2}$$

$$\Rightarrow \frac{dv}{du}=\frac{uf-f^2-uf}{(u-f)^2} \Rightarrow \frac{dv}{du}=-\frac{f^2}{(u-f)^2}$$

$$\Rightarrow dv=-\left(\frac{f}{u-f}\right)^2 du$$

$$|dv|=\left(\frac{f}{u-f}\right)^2 |du| \quad \ldots(\text{ii})$$

Here, image size $=|dv|$, object size $=|du|=b$

On substituting values in Eq. (ii), we get

$$\text{Image size}=b\left(\frac{f}{u-f}\right)^2$$

**12** (*a*) Given, size (height) of the object, $h_1=2$ cm,

object distance, $u=-16$ cm

and size (height) of the image, $h_2=-3$ cm

(since the image is real and inverted)

As, linear magnification, $m=\frac{-h_2}{h_1}=\frac{v}{u}$

where, $v$ is image distance.

$$\Rightarrow v=\frac{-h_2}{h_1}\times u=-\frac{(-3)}{2}\times(-16)=-24\text{cm}$$

Now, using mirror formula,

$$\frac{1}{f}=\frac{1}{v}+\frac{1}{u}=-\frac{1}{24}-\frac{1}{16}=\frac{-2-3}{48}=\frac{-5}{48}$$

$$\Rightarrow f=\frac{-48}{5}=-9.6\text{ cm}$$

**13** (*d*) In case of a concave mirror, when object is at $f$, image is formed at infinity.

So, magnification when object is at $f$,

$$m=\frac{-v}{u}=\infty$$

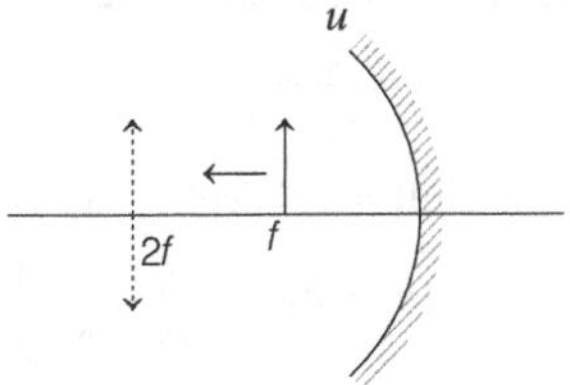


Also, if an object is at of $2f$ from mirror, then the image is also formed at $2f$. So, magnification,

$$m=\frac{-v}{u}=\frac{-(-2f)}{(-2f)}=-1$$

or $|m|=1$

So, these values are correctly depicted in option (d) only.

**14** (*c*) Given, focal length of concave mirror,

$$f=-0.4\text{ m}$$

Magnification = 5

We know that, magnification produced by a mirror,

$$m=-\frac{\text{image distance}}{\text{object distance}}$$

$$\Rightarrow \frac{v}{u}=-5 \text{ or } v=-5u$$

Using mirror formula,

$$\frac{1}{v}+\frac{1}{u}=\frac{1}{f}$$

Substituting the given values in the above equation, we get

$$\frac{1}{-5u}+\frac{1}{u}=-\frac{1}{0.4}$$

$$\Rightarrow \frac{4}{5u}=-\frac{1}{0.4}$$

$$\Rightarrow u=-\frac{1.6}{5}$$

$$=-0.32\text{ m}$$

**15** (*a*) Since, rear view mirror of a car is convex in nature, so the radius of curvature of convex mirror, $R=2$ m (given).

Focal length of the mirror, $f=\frac{R}{2}=\frac{2}{2}=1$ m

Also, image distance, $u=-39$ m

From the mirror equation, we get $v=\frac{fu}{u-f}$ ...(i)

Substituting the given values in Eq. (i), we get

$$v=\frac{(-39)\times 1}{-39-1}=\frac{39}{40}\text{ m}$$

Since, the jogger moves at a constant speed of 5 $ms^{-1}$, after 1s the position of the jogger will be

$$u=-39+(5\times 1)$$

$$=-39+5=-34\text{ m}$$

So, now the image position of the jogger will

$$v=\frac{(-34)\times 1}{-34-1}=\frac{34}{35}\text{m.}$$

The shift in the position of image in 1s is

$$\frac{39}{40}-\frac{34}{35}=\frac{1365-1360}{1400}=\frac{5}{1400}=\frac{1}{280}\text{ m}$$

Therefore, the average speed of the image when the jogger is between 39 m and 34 m from the mirror, is (1/280) $ms^{-1}$ or $0.3\times 10^{-2}$ $ms^{-1}$.

**17** (*d*) From Snell's law of refraction,

$$^a n_g = \frac{\sin i}{\sin r} = \text{constant} \quad \ldots\text{(i)}$$

Since, angle of incidence increases, the angle of refraction has to increase, so that the ratio $\left(\frac{\sin i}{\sin r}\right)$ remains a constant.

**18** (*c*) From Snell's law of refraction refractive index of air-glass interface,

$$^a n_g = \frac{\sin i}{\sin r} = \frac{\sin 50^\circ}{\sin 40^\circ} \quad \ldots\text{(i)}$$

Similarly, refractive index of air-water interface,

$$^a n_w = \frac{\sin 50^\circ}{\sin 45^\circ} \quad \ldots\text{(ii)}$$

Refractive index of water-glass interface,

$$^w n_g = \frac{\sin 45^\circ}{\sin \theta} \quad \ldots\text{(iii)}$$

As, $\quad ^a n_w \times {^w n_g} = {^a n_g}$

$$\Rightarrow \quad \frac{\sin 50^\circ}{\sin 45^\circ} \times \frac{\sin 45^\circ}{\sin \theta} = \frac{\sin 50^\circ}{\sin 40^\circ}$$

[using Eqs. (i), (ii) and (iii)]

$$\sin \theta = \sin 40^\circ \text{ or } \theta = 40^\circ$$

**19** (*c*) From the figure,

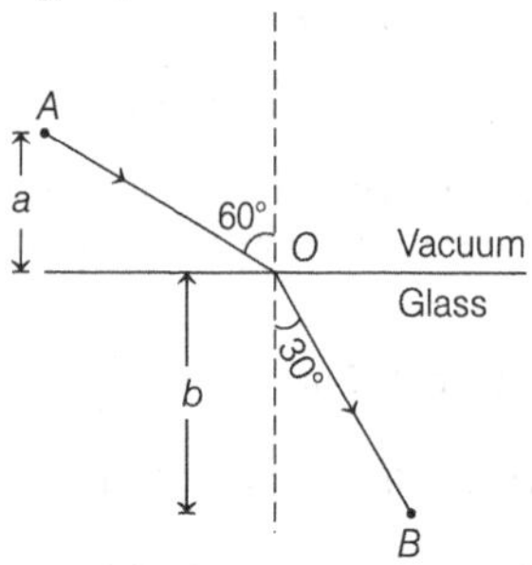


$$\cos 60^\circ = \frac{a}{AO}$$

$$\Rightarrow \quad AO = \frac{a}{\cos 60^\circ} = 2a \quad \ldots\text{(i)}$$

and $\quad \cos 30^\circ = \frac{b}{BO}$

or $\quad BO = \frac{b}{\cos 30^\circ} = \frac{2}{\sqrt{3}} b \quad \ldots\text{(ii)}$

Optical path length of light ray

$$= AO + \mu\,(BO) \quad \ldots\text{(iii)}$$

Here, $\mu$ can be determined using Snell's law, i.e.

$$\mu = \frac{\sin 60^\circ}{\sin 30^\circ} = \frac{\sqrt{3}/2}{1/2} = \sqrt{3} \quad \ldots\text{(iv)}$$

Substituting the values from Eqs. (i), (ii) and (iv) in Eq. (iii), we get

$$\therefore \quad \text{Optical path} = 2a + (\sqrt{3} \times \frac{2}{\sqrt{3}} b)$$

$$= 2a + 2b$$

**20** (*d*) Given, situation is shown in the figure below

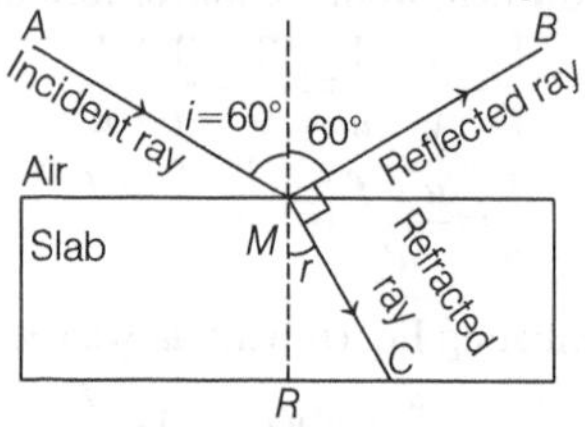


where, angle of incidence, $i = 60^\circ$ and $\angle BMC = 90^\circ$

From the given figure, $\angle CMR$ = angle of refraction,

$$r = 180^\circ - (60^\circ + 90^\circ)$$

$$\Rightarrow \quad r = 30^\circ$$

From Snell's law, $\quad \frac{\sin i}{\sin r} = \frac{n_{\text{slab}}}{n_{\text{air}}}$

$$\Rightarrow \quad \frac{\sin 60^\circ}{\sin 30^\circ} = \frac{n_{\text{slab}}}{1.0} \quad [\because n_{\text{air}} = 1]$$

$$\therefore \quad n_{\text{slab}} = \left(\frac{\sqrt{3}}{2}\right) \times 2 = \sqrt{3}$$

**21** (*b*) When a ray of light passes through the glass slab, the emergent ray is parallel to the incident ray as shown below

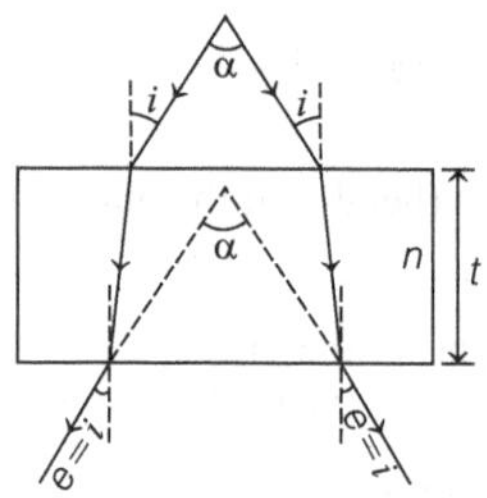


Also, $i = e$.

So, divergent angle of the emergent beam is also same, i.e. $\alpha$.

**23** (*c*) The given beaker is as shown below

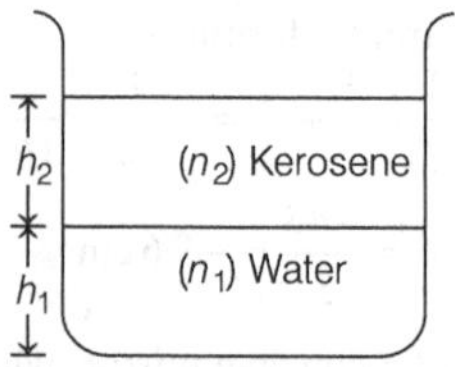


As we know that, apparent depth of an object in a medium is real depth divided by the refractive index of the medium.

Apparent depth of the bottom of the beaker

$$= \frac{h_1}{n_1} + \frac{h_2}{n_2}$$

Thus, apparent shift = real depth − apparent depth

$$= (h_1 + h_2) - \left(\frac{h_1}{n_1} + \frac{h_2}{n_2}\right)$$

$$= h_1\left(1 - \frac{1}{n_1}\right) + h_2\left(1 - \frac{1}{n_2}\right)$$

**24** (c) Apparent depth of the object $= \dfrac{\text{Real depth}}{\text{Refractive index}}$

$$= \frac{33.25}{1.33} \text{ cm}$$

Distance of object from the mirror $= 15 + \dfrac{33.25}{1.33} = 40$ cm

Similarly, distance of image from the mirror $= 15 + \dfrac{25}{1.33}$

$$= 33.8 \text{ cm}$$

For concave mirror, $u = -40$ cm and $v = -33.8$ cm

Applying mirror formula,

$$f = \frac{uv}{u+v} = \frac{(-40)(-33.8)}{(-40-33.8)} = -18.3 \text{ cm}$$

**25** (a) Given, actual depth of the screw in water $h_1 = 13$ cm and apparent depth of screw in water is $h_2 = 8.5$ cm

$$\Rightarrow \quad n_{\text{water}} = \frac{h_1}{h_2} = \frac{13}{8.5} = 1.53$$

When water is replaced by a liquid of refractive index $n' = 1.70$, then the actual depth remains same, but its apparent depth changes.

Let $h_3$ be the new apparent depth of the screw

$$\therefore \quad n' = \frac{h_1}{h_3} \text{ or } h_3 = \frac{h_1}{n'} = \frac{13}{1.70} = 7.65 \text{ cm}$$

Here, $h_3$ is less than $h_2$. Thus to focus the screw, the microscope should be moved up.

∴Distance by which microscope should be moved up

$$= 8.5 - 7.65 = 0.85 \text{ cm}$$

**26** (a) Given, $n_{\text{oil}} = 2$, $n_{\text{glass}} = 3/2$

Velocity of light in oil is given by

$$v_{\text{oil}} = \frac{\text{Velocity of light } c}{n_{\text{oil}}} = \frac{c}{n_{\text{oil}}}$$

Similarly, $v_{\text{glass}} = \dfrac{c}{n_{\text{glass}}}$

$$\therefore \quad \frac{v_{\text{glass}}}{v_{\text{oil}}} = \frac{c/n_{\text{glass}}}{c/n_{\text{oil}}} = \frac{n_{\text{oil}}}{n_{\text{glass}}} = \frac{2}{3/2} = \frac{4}{3}$$

**27** (c) The total internal reflection is the phenomenon of reflection of light in the denser medium when it travels from denser to rarer medium, for the angle of incidence is greater than the critical angle. The critical angle for a pair of given media in contact is the angle of incidence in denser medium for which the angle of refraction in rarer medium is 90°as shown in figure.

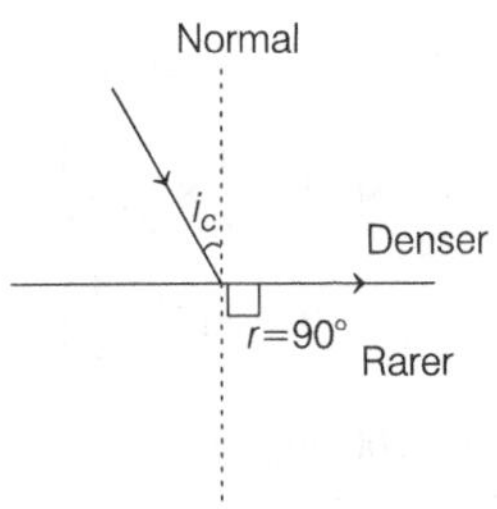


**28** (b) The given figure can be drawn as follows,

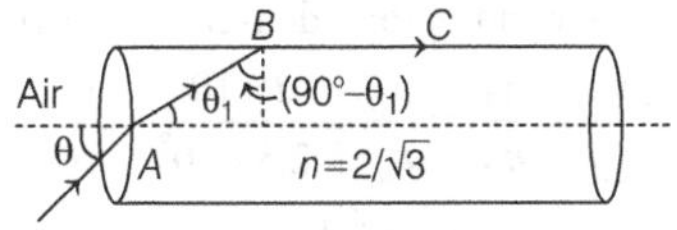


As, the angle of incidence in a denser medium for which the angle of refraction in rarer medium becomes 90° is known as critical angle. Mathematically,

$$i_C = \sin^{-1}\frac{1}{n} \qquad \ldots\text{(i)}$$

where, $n$ = refractive index of denser to rarer medium.

So, ray $BC$ travels along surface, i.e. angle of refraction is 90°.

$$\Rightarrow \quad \text{angle } (90° - \theta_1) = \text{critical angle } i_C$$

$$\sin(90° - \theta_1) = \frac{\sqrt{3}}{2} \qquad \text{(using Eq. (i))}$$

$$\Rightarrow \quad \cos\theta_1 = \frac{\sqrt{3}}{2} \Rightarrow \theta_1 = 30°$$

Now, using Snell's law on 1st interface,

$$n_1 \sin\theta = n_2 \sin\theta_1$$

$$\Rightarrow \quad 1 \sin\theta = \frac{2}{\sqrt{3}} \sin 30° = \frac{2}{\sqrt{3}} \cdot \frac{1}{2}$$

$$\Rightarrow \quad \sin\theta = \frac{1}{\sqrt{3}} \Rightarrow \theta = \sin^{-1}\left(\frac{1}{\sqrt{3}}\right)$$

**29** (c) We know that, $\sin\theta = \dfrac{n_2}{n_1}$

As, refractive index of a medium, $n = \dfrac{c}{v}$, where $c$ and $v$ are the speed of light in vacuum and medium, respectively.

$$\Rightarrow \sin\theta = \frac{v}{v'}$$

Here, $v$ is speed in medium $A$ and $v'$ is speed in medium $B$

$$\Rightarrow \quad v' = \frac{v}{\sin\theta}$$

**30** (c) Given, velocity of light in medium $A$,

$$v_A = 2 \times 10^8 \text{ ms}^{-1}$$

Velocity of light in medium, $B = v_B = 2.5 \times 10^8 \text{ ms}^{-1}$

Since, speed of light in $B$ is more than that in $A$.

⇒ $B$ is rarer and $A$ is denser medium.

So, the totally internally reflected ray is as shown in the figure below

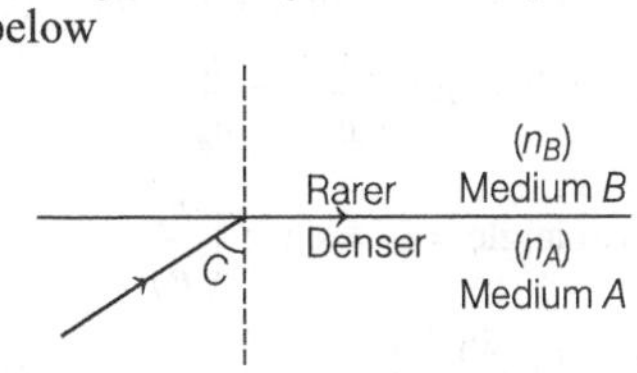


Since, critical angle for a pair of media

$$= i_C = \sin^{-1}\left(\frac{n_r}{n_d}\right)$$

( where $n_r$ = refractive index for rarer medium and $n_d$ = refractive index for a denser medium)

As, $\quad \dfrac{v_A}{v_B} = \dfrac{n_B}{n_A} = \dfrac{n_r}{n_d} = \dfrac{2 \times 10^8}{2.5 \times 10^8} = \dfrac{4}{5}$

$\therefore \quad i_C = \sin^{-1}\left(\dfrac{4}{5}\right)$

**31** (*b*) The given ray of light is as shown below

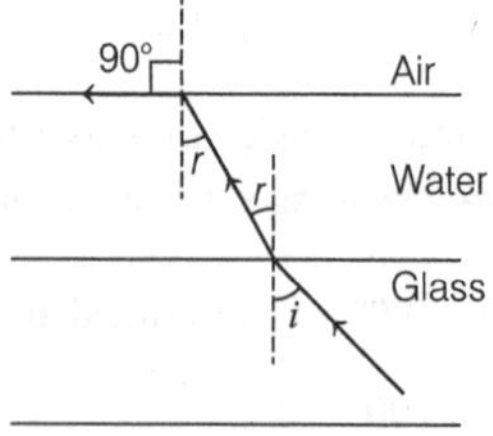


For glass-water interface, applying Snell's law

$$\frac{\sin i}{\sin r} = \frac{n_w}{n_g} \Rightarrow n_g = \frac{n_w \sin r}{\sin i} \quad \text{...(i)}$$

For water-air interface, angle of incidence in water = $r$

Again, $\quad \dfrac{\sin r}{\sin 90°} = \dfrac{n_a}{n_w} = \dfrac{1}{n_w} \Rightarrow \sin r = \dfrac{1}{n_w} \quad \text{...(ii)}$

From Eqs. (i) and (ii), we get

$$n_g = \frac{(n_w) \times \left(\dfrac{1}{n_w}\right)}{\sin i} \Rightarrow n_g = \frac{1}{\sin i}$$

**32** (*a*) The given figure is as shown below

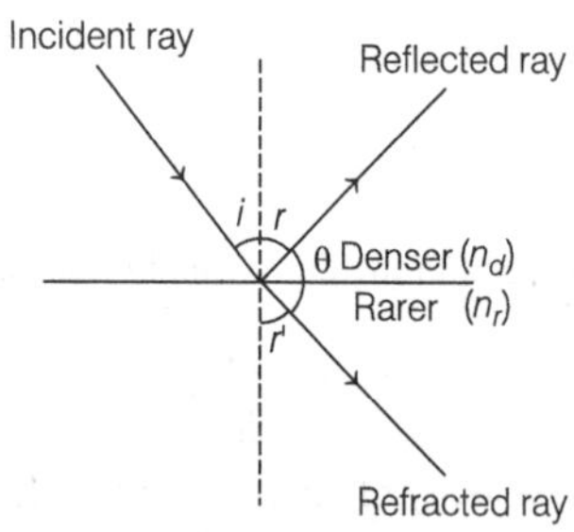


Since, reflected ray and refracted ray make an angle of 90° with each other.

Hence $\theta = 90°$

So, from figure, $r + \theta + r' = 180°$

$$r + r' = 90° \text{ or } i + r' = 90° \quad (\because i = r) \quad \text{...(i)}$$

Applying Snell's law, $\dfrac{\sin i}{\sin r'} = \dfrac{n_r}{n_d} \quad \text{...(ii)}$

Also,critical angle $= i_C = \sin^{-1}\left(\dfrac{n_r}{n_d}\right)$

$\Rightarrow i_C = \sin^{-1}\left(\dfrac{\sin i}{\sin r'}\right) \quad$ [using Eq. (ii)]

As from Eq. (i), $r' = 90° - i$

$\Rightarrow \quad \sin r' = \sin(90° - i) = \cos i$

So, $\quad i_C = \sin^{-1}\left(\dfrac{\sin i}{\cos i}\right) = \sin^{-1}(\tan i)$

or $\quad i_C = \sin^{-1}(\tan r) \quad (\because i = r)$

**33** (*c*) As, critical angle, $i_C = \sin^{-1}\left(\dfrac{n_r}{n_d}\right)$

where, $n_r$ and $n_d$ in the refractive index of rarer and denser medium, respectively

For total internal reflection at surface $AB$, $\theta > i_C$

$$\Rightarrow \quad \theta > \sin^{-1}\left(\frac{\sqrt{2}}{2}\right) \quad \text{...(i)}$$

Also, for total internal reflection at surface $CD$,

$$\theta > \sin^{-1}\left(\frac{\sqrt{3}}{2}\right) \quad \text{...(ii)}$$

For both Eqs. (i) and (ii) to be correct, $\theta > \sin^{-1}\left(\dfrac{\sqrt{3}}{2}\right)$

$\because \quad \sin^{-1}\left(\dfrac{\sqrt{3}}{2}\right) > \sin^{-1}\left(\dfrac{\sqrt{2}}{2}\right)$

$\Rightarrow \quad \theta > \sin^{-1}(\sin 60°) \Rightarrow \theta > 60°$

$\therefore \quad \theta_{\min} = 60°$

**34** (*b*) Since, light beam just misses entering region IV, this mean, the light might have suffered TIR at interface of regions III and IV. For this, it must strike the interface at critical angle ($i_C$).

So, $i_C = \sin^{-1}\left(\dfrac{n_r}{n_d}\right) = \sin^{-1}\left(\dfrac{n_0/8}{n_0/6}\right) = \sin^{-1}\left(\dfrac{3}{4}\right)$

$$\Rightarrow \quad \sin i_C = \frac{3}{4} \quad \text{...(i)}$$

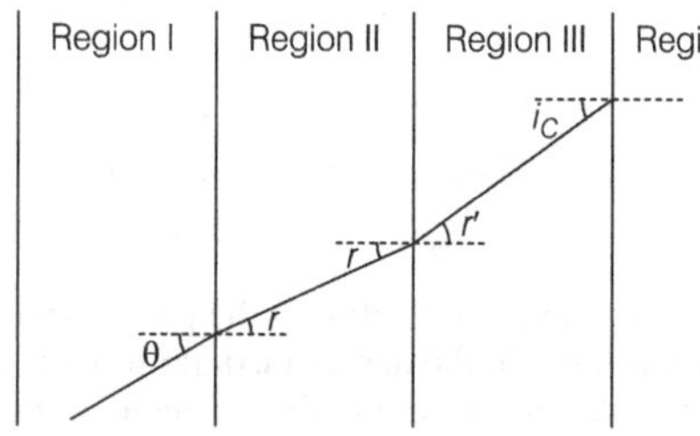


Applying Snell's law at interface,

$$n_1 \sin \theta = n_2 \sin r = n_3 \sin r' = \text{constant}$$

$\Rightarrow \quad n_1 \sin \theta = n_3 \sin r'$

$\Rightarrow \quad n_0 \sin \theta = \dfrac{n_0}{6} \sin i_C \quad [\because r' = i_C]$

$\Rightarrow \quad \sin \theta = \dfrac{1}{6} \times \dfrac{3}{4} = \dfrac{1}{8} \quad$ [from Eq. (i)]

$\Rightarrow \quad \theta = \sin^{-1}\left(\dfrac{1}{8}\right)$

**39** (*c*) Given, $u = -100$ cm, $v = ?$, $R = +20$ cm, $n_1 = 1$ and $n_2 = 1.5$

From refraction formula for curved surface,

$$\frac{n_2}{v}-\frac{n_1}{u}=\frac{n_2-n_1}{R}$$

Substituting the given values in above equation we get

$$\frac{1.5}{v}+\frac{1}{100}=\frac{0.5}{20} \Rightarrow v=+100\,\text{cm}$$

$\therefore$ The image is formed at a distance of 100 cm from the glass surface, in the direction of incident light.

**40** (*a*) The given condition can be shown in the figure below,

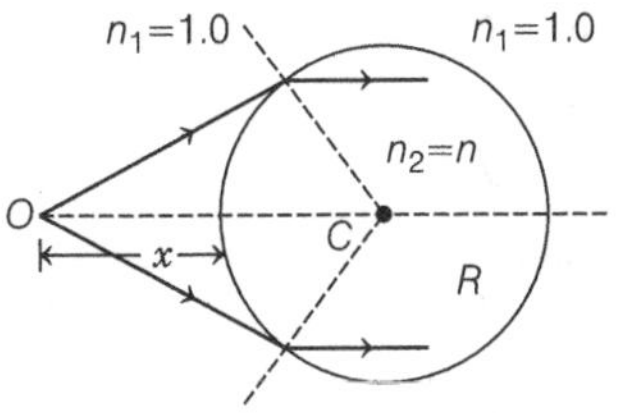


From refraction formula for curved surface, we get

$$\frac{n_2}{v}-\frac{n_1}{u}=\frac{n_2-n_1}{R}$$

$$\Rightarrow \frac{n}{\infty}-\frac{1}{(-x)}=\frac{n-1}{+R} \Rightarrow +\frac{1}{x}=\frac{n-1}{R}$$

or $$x=\frac{R}{(n-1)}$$

**41** (*d*) The given situation can be depicted as

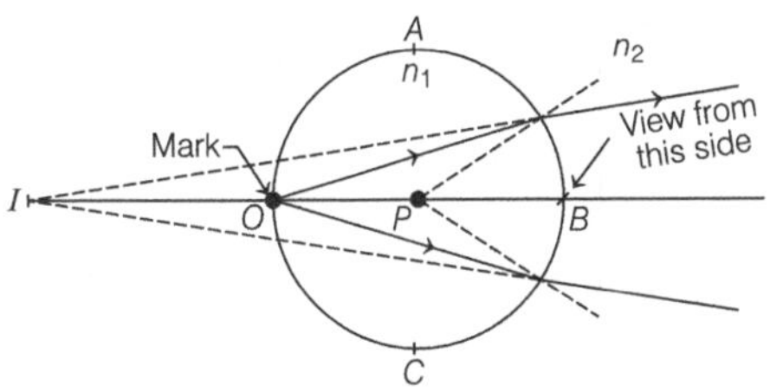


Given, for refraction at spherical surface $ABC$,

$$u=2R=-30\text{ cm}$$

$$n_1=1.5 \text{ and } n_2=1.0\,(\text{for air})$$

$$v=?$$

Applying, refraction formula for curved surface

$$\frac{n_2}{v}-\frac{n_1}{u}=\frac{n_2-n_1}{R}$$

$$\frac{1}{v}-\frac{1.5}{(-30)}=\frac{(1-1.5)}{(-15)}$$

$$\Rightarrow \frac{1}{v}+\frac{1.5}{30}=\frac{-0.5}{-15}$$

$$\Rightarrow \frac{1}{v}=\frac{0.5}{15}-\frac{1.5}{30}$$

$$\Rightarrow v=-60\text{ cm}$$

Thus, the image is formed at a distance of 60 cm from surface $ABC$ and 30 cm from the surface $AOC$ opposite to the direction of incident light.

**42** (*c*) Using the lens Maker's formula,

$$\frac{1}{f}=\left(\frac{n_2-n_1}{n_1}\right)\left(\frac{1}{R_1}-\frac{1}{R_2}\right) \quad \ldots(\text{i})$$

Given, $f=25$ cm

$n_2=1.5$, $n_1=1$ (for air)

Let $R_1=R$ and $R_2=-2R$

Substituting the given values in Eq. (i), we get

$$\Rightarrow \frac{1}{25}=\left(\frac{1.5-1}{1}\right)\left[\frac{1}{R}-\left(-\frac{1}{2R}\right)\right]$$

$$\frac{1}{25\times 0.5}=\frac{1}{R}+\frac{1}{2R}\Rightarrow R=\frac{3}{2}\times 25\times 0.5=18.75\text{ cm}$$

$$\therefore \quad R_1=18.75\text{ cm and } R_2=2\times 18.75=37.5\text{ cm}$$

**43** (*c*) Focal length of lens of refractive index $n_2$ placed in a medium of refractive index $n_1$ is

$$\frac{1}{f}=\left(\frac{n_2}{n_1}-1\right)\left(\frac{1}{R_1}-\frac{1}{R_2}\right)$$

For a glass-lens in air, focal length,

$$\frac{1}{f_{\text{air}}}=\frac{1}{f}=\left(\frac{n_g}{n_a}-1\right)\left(\frac{1}{R_1}-\frac{1}{R_2}\right) \quad \ldots(\text{i})$$

When lens is dipped in a liquid, its focal length,

$$\frac{1}{f_{\text{liquid}}}=\frac{1}{f_1}=\left(\frac{n_g}{n_l}-1\right)\left(\frac{1}{R_1}-\frac{1}{R_2}\right) \quad \ldots(\text{ii})$$

From Eqs. (i) and (ii), we have

$$\frac{f_1}{f}=\frac{\left(\frac{n_g}{n_a}-1\right)}{\left(\frac{n_g}{n_l}-1\right)}$$

As $f=16$ cm, $n_g=1.5$, $n_l=1.42$, we have

$$\frac{f_1}{f}=\frac{\left(\frac{1.5}{1}-1\right)}{\left(\frac{1.5}{1.42}-1\right)}\Rightarrow \frac{f_1}{f}=\frac{\left(\frac{3}{2}-1\right)}{\left(\frac{150}{142}-1\right)}=\frac{\left(\frac{1}{2}\right)}{\left(\frac{4}{71}\right)}=\frac{71}{8}\approx 8.9$$

So, $\frac{f_1}{f}$ is closest to integer 9.

**44** (*b*) When lens is in air,

From lens Maker's formula,

$$\frac{1}{f_1}=({}^a n_g-1)\left(\frac{1}{R_1}-\frac{1}{R_2}\right) \quad \ldots(\text{i})$$

When lens is immersed in liquid,

$$\frac{1}{f_2}=({}^l n_g-1)\left(\frac{1}{R_1}-\frac{1}{R_2}\right) \quad \ldots(\text{ii})$$

On dividing Eq. (i) by Eq. (ii), we get

$$\frac{f_2}{f_1}=\frac{({}^a n_g-1)}{({}^l n_g-1)}$$

$$\Rightarrow \quad \frac{f_2}{f_1} = \frac{(1.5-1)}{\left(\frac{1.5}{1.25}-1\right)} \qquad \left(\because {}^{l}n_g = \frac{n_g}{n_l} \text{ and } n_l = 1.25\right)$$

$$\Rightarrow \quad \frac{f_2}{f_1} = \frac{0.5 \times 1.25}{0.25} = 2.5 \text{ or } f_2 = 2.5\, f_1$$

Hence, focal length increases by a factor of 2.5 when immersed in liquid.

**45** (*a*) Given, for concave lens of glass, $n_g = 1.5$

Refractive index of the medium, $n_m = 1.75$

Applying lens Maker's formula,

$$\frac{1}{f} = ({}^{m}n_g - 1)\left(\frac{1}{R_1} - \frac{1}{R_2}\right) = \left(\frac{n_g}{n_m} - 1\right)\left(\frac{1}{R_1} - \frac{1}{R_2}\right)$$

$$= \left(\frac{1.5}{1.75} - 1\right)\left(\frac{1}{-R} - \frac{1}{+R}\right)$$

$$\Rightarrow \quad \frac{1}{f} = \frac{(1.5 - 1.75)}{1.75} \times \left(-\frac{2}{R}\right)$$

$$\Rightarrow \quad f = \frac{1.75}{0.25 \times 2} R = \frac{7}{2} R = (+\,3.5\,R)$$

Since, $f$ is positive, so it behaves as convergent lens of focal length $(+\,3.5\,R)$.

**46** (*d*) For the given concave lens to behave as a diverging lens its focal length $f$ must be negative. As in the lens Maker's formula given as below

$$\frac{1}{f} = ({}^{1}n_2 - 1)\left(\frac{1}{R_1} - \frac{1}{R_2}\right)$$

where, ${}^{1}n_2$ is the refractive index of the medium 2 (i.e. the medium inside the thin lens) with respect to the refractive index of medium 1 (i.e. the medium outside the lens) the expression $\left(\frac{1}{R_1} - \frac{1}{R_2}\right)$ is negative.

So, $\quad ({}^{1}n_2 - 1) > 0 \Rightarrow \frac{n_2}{n_1} > 1 \text{ or } n_2 > n_1$

Thus, if the lens is filled with $L_2$ having refractive index $n_{L_2}$ and placed in $L_1$ having refractive index $n_{L_1}$, the above condition will be satisfied and it will diverge a parallel beam of light.

**47** (*a*) Using lens Maker's formula,

$$\frac{1}{f} = (n-1)\left(\frac{1}{R_1} - \frac{1}{R_2}\right) \quad \text{...(i)}$$

Given, $R_1 = 10$ cm, $R_2 = -10$ cm, $u = -5$ cm and $n = 1.33$

Substituting the given values in Eq. (i), we get

$$\frac{1}{f} = (1.33 - 1)\left(\frac{1}{10} + \frac{1}{10}\right)$$

$$\frac{1}{f} = 0.33 \times \frac{2}{10} = \frac{0.33}{5} \quad \Rightarrow f = 15.15 \text{ cm}$$

Now, from thin lens formula, $\frac{1}{f} = \frac{1}{v} - \frac{1}{u}$

or image distacne, $v = \frac{uf}{u+f} = \frac{-5 \times 15.15}{-5 + 15.15}$

$$= \frac{-75.75}{10.15} = -7.46 \text{ cm}$$

Since, $v$ is –ve hence image will be formed on the same side of the object.

**48** (*d*) First we find relation of magnification $m$ and focal length $f$. By lens equation,

$$\frac{1}{v} - \frac{1}{u} = \frac{1}{f}$$

$$\Rightarrow \quad 1 - \frac{v}{u} = \frac{v}{f}$$

$$\Rightarrow \quad 1 - m = \frac{v}{f} \qquad \left[\because \frac{v}{u} = m\right]$$

$$\Rightarrow \quad m = 1 - \frac{v}{f} \quad \text{...(i)}$$

Now, from given graph and from Eq. (i),

At $v = a$, magnification is

$$m_1 = 1 - \frac{a}{f} \quad \text{...(ii)}$$

At $v = a + b$, magnification is

$$m_2 = 1 - \frac{a+b}{f} \quad \text{...(iii)}$$

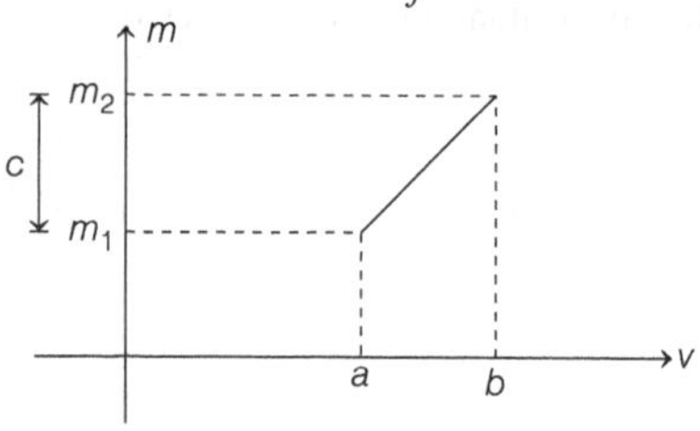


From graph, we can also say that

$$m_2 - m_1 = c \quad \text{...(iv)}$$

So, from Eqs. (ii), (iii) and (iv), we have

$$\left(1 - \frac{(a+b)}{f}\right) - \left(1 - \frac{a}{f}\right) = c$$

$$\Rightarrow \quad \frac{a - (a+b)}{f} = c$$

$$\Rightarrow \quad f = -\frac{b}{c} \text{ or } |f| = \frac{b}{c}$$

**49** (*b*) The combination of the given three lenses says $L_1, L_2$ and $L_3$ is as shown

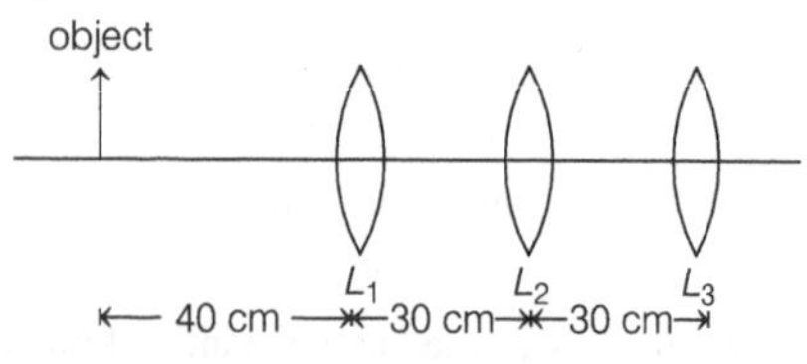


Given, focal length of each lens, $f = 10$ cm

For refraction from 1st lens, $L_1$

Using thin lens formula, $\frac{1}{f}=\frac{1}{v}-\frac{1}{u} \Rightarrow \frac{1}{v}=\frac{1}{f}+\frac{1}{u}$

Given, $u=-40\text{ cm}$

$$\frac{1}{v}=\frac{1}{10}-\frac{1}{40}=\frac{3}{40}\text{ cm} \Rightarrow v=\frac{40}{3}\text{ cm}$$

This image acts as an object for 2nd lens $L_2$.

$\therefore$ Object distance for 2nd lens $L_2$

$$|u|=30-\frac{40}{3}=\frac{50}{3}\text{ cm}$$

Now, for refraction from $L_2$,

Again by using thin lens formula, $\frac{1}{v}-\frac{1}{u}=\frac{1}{f}$

$$\Rightarrow \quad \frac{1}{v}=\frac{1}{u}+\frac{1}{f}=\frac{1}{10}-\frac{3}{50}=\frac{2}{50}$$

$$\Rightarrow \quad v=25\text{ cm}$$

This image acts as an object for 3rd lens $L_3$

$\therefore$ Object distance for 3rd lens $L_3=|u|=30-25=5\text{ cm}$

Similarly, for refraction from lens $L_3$,

$$\frac{1}{v}=\frac{1}{f}+\frac{1}{u}=\frac{1}{10}-\frac{1}{5}$$

or $\quad v=-10\text{ cm}$

Since, $v$ is negative, thus means location of final image $=10\text{ cm}$ to the left of $L_3$

$\therefore$ Distance from $L_1$, i.e. the final image position

$$=30+(30-10)=50\text{ cm}$$

**50** (*d*) Focal length of diverging lens is 25 cm.

As the rays are coming parallel, so the image ($I_1$) will be formed at the focus of diverging lens i.e. at 25 cm towards left of diverging lens.

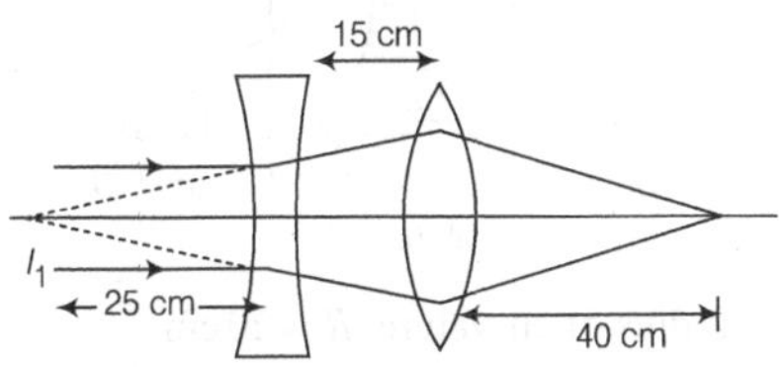


Now, the image ($I_1$) will work as object for converging lens.

For converging lens, distance of object $u$ (i.e. distance of $I_1$) $=-(25+15)=-40\text{ cm}$

$$f=20\text{ cm}$$

$\therefore$ From lens' formula $\frac{1}{f}=\frac{1}{v}-\frac{1}{u}$

$$\frac{1}{20}=\frac{1}{v}-\frac{1}{-40} \Rightarrow \frac{1}{v}=\frac{1}{20}-\frac{1}{40}$$

$$\frac{1}{v}=\frac{1}{90} \Rightarrow v=40\text{ cm}$$

$v$ is positive so image will be real and will form at right side of converging lens at 40 cm.

**51** (*c*) Magnification of a lens, $|m|=\frac{|v|}{|u|}$

**Case I** $u_1=-0.15\text{ m}=-15\text{ cm}$, $v_1=?$ and $m_1=?$

**Case II** $u_2=-0.2\text{ m}=-20\text{ cm}$, $v_2=?$ and $m_2=?$

Given, $\quad |m_1|=2|m_2|$

$$\Rightarrow \quad \frac{|v_1|}{|u_1|}=2\frac{|v_2|}{|u_2|}$$

$$\Rightarrow \quad \frac{|v_1|}{|v_2|}=\frac{2\times 15}{20}=\frac{3}{2}$$

$$\Rightarrow \quad |v_1|=\frac{3}{2}|v_2| \text{ or } v_1=\frac{3}{2}v_2$$

Applying thin lens formula, $\frac{1}{v}-\frac{1}{u}=\frac{1}{f}$

$$\text{or} \quad \frac{1}{f}=\frac{1}{v_1}-\frac{1}{u_1} \quad \ldots\text{(i)}$$

$$\text{and} \quad \frac{1}{f}=\frac{1}{v_2}-\frac{1}{u_2} \quad \ldots\text{(ii)}$$

From Eqs. (i) and (ii), $\frac{1}{v_1}-\frac{1}{u_1}=\frac{1}{v_2}-\frac{1}{u_2}$

$$\Rightarrow \quad \left(\frac{v_2-v_1}{v_1v_2}\right)=\left(\frac{u_2-u_1}{u_1u_2}\right)$$

$$\Rightarrow \quad \frac{v_2-\frac{3}{2}v_2}{\frac{3}{2}v_2\cdot v_2}=\left(\frac{-5}{300}\right)$$

$$\Rightarrow \quad \frac{-v_2}{3v_2^2}=\frac{-5}{300} \text{ or } -v_2=\frac{-300}{15}$$

or $\quad v_2=+20\text{ cm}$

$$\Rightarrow \quad v_1=\frac{3}{2}\times v_2=\frac{3}{2}\times 20=+30\text{ cm}$$

From Eq. (i), $\quad \frac{1}{f}=\frac{1}{+30}-\frac{1}{(-15)}=\frac{3}{30}$

$$\therefore \quad f=10\text{ cm}=0.10\text{ m}$$

**52** (*b*) Given, power of lens, $P=20\text{ D}$

Power of lens (in dioptre) $P=\frac{100}{\text{focal length } f\text{ (in cm)}}$

$$\Rightarrow \quad f=\frac{100}{20}=5\text{ cm}$$

From lens Maker's formula, $\frac{1}{f}=(n-1)\left(\frac{1}{R_1}-\frac{1}{R_2}\right)$

For biconvex lens, $R_1=+R$

and $\quad R_2=-R$

$$\Rightarrow \quad \frac{1}{f}=(n-1)\left(\frac{1}{R}+\frac{1}{R}\right)=(n-1)\left(\frac{2}{R}\right)$$

$$\frac{1}{5}=(n-1)\left(\frac{2}{5}\right) \quad [\because \text{given, } R=5\text{ cm}]$$

$$2n-2=1 \Rightarrow 2n=3 \Rightarrow n=3/2$$

**53** (*d*) When the lens is cut along its principal axis, the focal length of the two halves will remain same because the radius of curvature of both the surfaces are still same. So, the power also remains same as

$$P = \frac{1}{f}$$

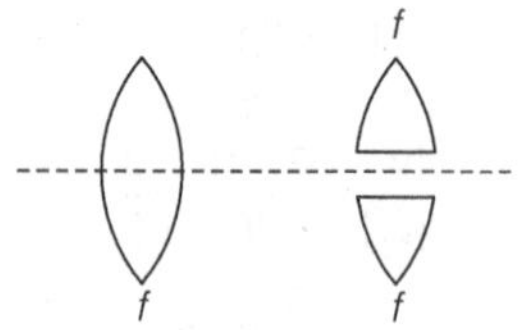


**54** (*d*) For equivalent focal length,

$$\frac{1}{f} = \frac{1}{f_1} + \frac{1}{f_2} + \frac{1}{f_3}$$

Given, $f_1 = 40\text{ cm}$, $f_2 = -40\text{ cm}$ and $f_3 = -15\text{ cm}$

$$\Rightarrow \quad \frac{1}{f} = \frac{1}{(+40)} + \frac{1}{(-40)} + \frac{1}{(-15)}$$

$$\Rightarrow \quad \frac{1}{f} = -\frac{1}{15} \Rightarrow f = -15\text{ cm}$$

Since, power, $P \text{ (in D)} = \frac{1}{f \text{ (in m)}} = -\frac{100}{15} = -\frac{20}{3}$

$$= -6.67\text{ D}$$

**55** (*a*) **Case I** When two equi-convex lens of focal length $f_1$ and $f_2$ respectively, are kept co-axially in contact as shown below, then the equivalent focal length of combination is

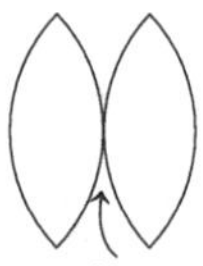


$$\frac{1}{F_1} = \frac{1}{f_1} + \frac{1}{f_2}$$

$$= \frac{1}{f} + \frac{1}{f} \qquad [\because f_1 = f_2 = f]$$

$$= \frac{2}{f}$$

$$\Rightarrow \quad F_1 = \frac{f}{2} \qquad \dots \text{(i)}$$

**Case II** When glycerine of same refractive index as that of the glass is filled in the space between two lenses, then the combination will now comprises of three lenses; first bi-convex, second bi-concave and third is bi-convex. So, the focal length of the combination now is given as shown below

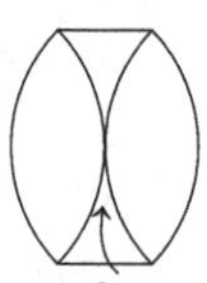


$$\frac{1}{F_2} = \frac{1}{f} + \frac{1}{(-f)} + \frac{1}{f} = \frac{1}{f}$$

$$\Rightarrow \quad F_2 = f \qquad \dots \text{(ii)}$$

From Eqs. (i) and (ii), we get

$$F_1 : F_2 = 1 : 2$$

**56** (*c*) Focal length of the combination of two given lenses,

$$\frac{1}{f} = \frac{1}{f_1} + \frac{1}{f_2} \qquad \dots\text{(i)}$$

where, $f_1$ and $f_2$ are the focal lengths of plano-convex lens and plano-concave lens, respectively

As for plano-convex lens, $R_1 = R$ and $R_1 = \infty$

So, $\frac{1}{f_1} = (n_1 - 1)\left(\frac{1}{R} - \frac{1}{\infty}\right) = \frac{n_1 - 1}{R}$

Similarly, for plano-concave lens,

$R_1 = -R$ and $R_2 = \infty$

So, $\frac{1}{f_2} = (n_2 - 1)\left(\frac{1}{-R} - \frac{1}{\infty}\right) = \frac{-(n_2 - 1)}{R}$

Putting these values of $\frac{1}{f_1}$ and $\frac{1}{f_2}$ in Eq. (i), we get

$$\frac{1}{f} = \frac{(n_1 - 1)}{R} - \frac{(n_2 - 1)}{R}$$

$$= \frac{[n_1 - 1 - n_2 + 1]}{R} = \frac{n_1 - n_2}{R}$$

$$f = \frac{R}{n_1 - n_2}$$

**57** (*c*) Given, situation can be depicted as follows

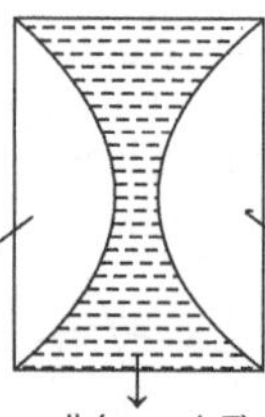


Given, radius of curvature, $R = 20\text{ cm}$

From lens Maker's formula,

$$\frac{1}{f} = (n - 1)\left[\frac{1}{R_1} - \frac{1}{R_2}\right]$$

For the plano-convex lens, $R_1 = R$ and for plane surface, $R_2 = \infty$

$$\therefore \quad \frac{1}{f_{\text{plano-convex lens}}} = (1.5 - 1)\left(\frac{1}{R} - 0\right)$$

$$\Rightarrow \quad \frac{1}{f_{\text{plano-convex lens}}} = \frac{0.5}{R}$$

When the intervening medium is filled with oil, then focal length of the concave lens formed by the oil

$$\frac{1}{f_{\text{concave}}} = (1.7 - 1)\left(-\frac{1}{R} - \frac{1}{R}\right) = -0.7 \times \frac{2}{R} = \frac{-1.4}{R}$$

$$\text{So}, \frac{1}{f_{eq}} = 2 \times \frac{1}{f_{\text{plano-convex lens}}} + \frac{1}{f_{\text{concave}}}$$

$$= 2 \times \frac{0.5}{R} + \left(\frac{-1.4}{R}\right) = \frac{1}{R} - \frac{1.4}{R} = -\frac{0.4}{R}$$

$$\therefore \quad f_{eq} = -\frac{R}{0.4} = -\frac{20}{0.4} = -50 \text{ cm}$$

**58** (c) The given prism can be shown as below

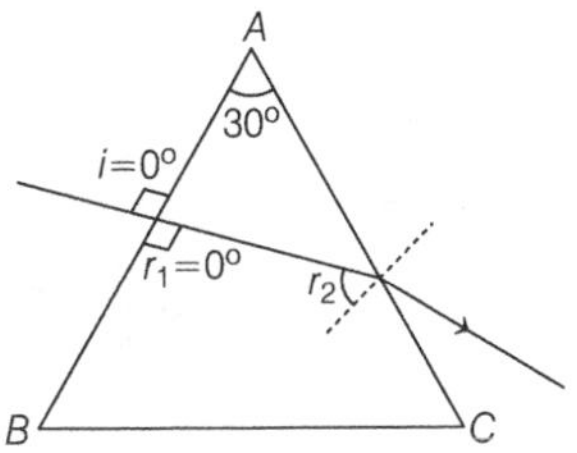


Given, $A = 30°$, $i = 0°$

$\Rightarrow \quad r_1 = 0°$

As, the angle of prism, $A = r_1 + r_2$

$\Rightarrow \quad 30° = 0 + r_2 \Rightarrow r_2 = 30°$

**59** (b) According to the question, the figure of mentioned prism is given as

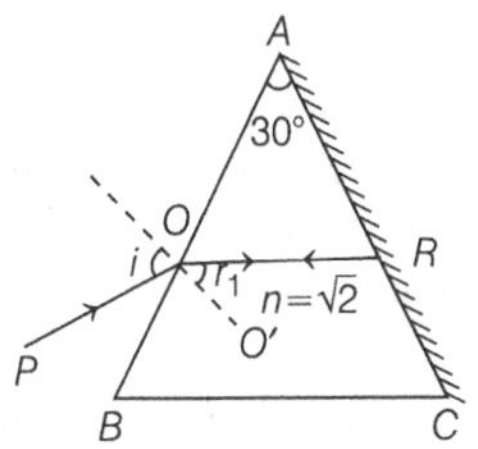


(since, there is no refraction at the face $AC$)

Given, refractive index of the material of prism, $n = \sqrt{2}$

and angle of prism, $A = 30°$

If the ray $OR$ has to retrace its path after reflection (as per the given condition), then the ray has to fall normally on the surface $AC$.

This means, $\quad \angle ARO = \angle ORC = 90°$

In $\Delta AOR$, $\angle AOR + \angle ARO + \angle OAR = 180°$

$\Rightarrow \quad \angle AOR + 90° + 30° = 180°$

$\Rightarrow \quad \angle AOR = 180° - 120° = 60° \quad \text{...(i)}$

From figure, $\quad \angle AOR + \angle r_1 = 90°$

$\Rightarrow \quad \angle r_1 = 90° - 60° = 30° \quad$ [from Eq. (i)]

Applying Snell's law at the face $AB$, we get

$$n = \frac{\sin i}{\sin r_1}$$

Substituting the given values, we get

$$\sqrt{2} = \frac{\sin i}{\sin 30°}$$

$$\Rightarrow \quad \sin i = \sin 30° \times \sqrt{2}$$

$$= \frac{1}{2} \times \sqrt{2} \qquad \left(\because \sin 30° = \frac{1}{2}\right)$$

$$= \frac{1}{\sqrt{2}} \text{ or } i = \sin^{-1}\left(\frac{1}{\sqrt{2}}\right)$$

$$= 45° \qquad \left(\because \sin 45° = \frac{1}{\sqrt{2}}\right)$$

The angle of incidence of the ray on the prism is 45°.

**60** (c) When light ray travels parallel to the base, the light suffers minimum deviations.

So, for minimum deviation, $D_{min} = 40°$

$i = e = 45° \quad$ (from graph)

**61** (d) For a triangular prism with the increase in angle of incidence, angle of deviation first decreases upto angle of minimum deviation $D_m$, when the angle of incidence is equivalent to angle of emergence and then increases with further increase in angle of incidence.

So, the correct graph between angle of incidence and angle of deviation is as shown below,

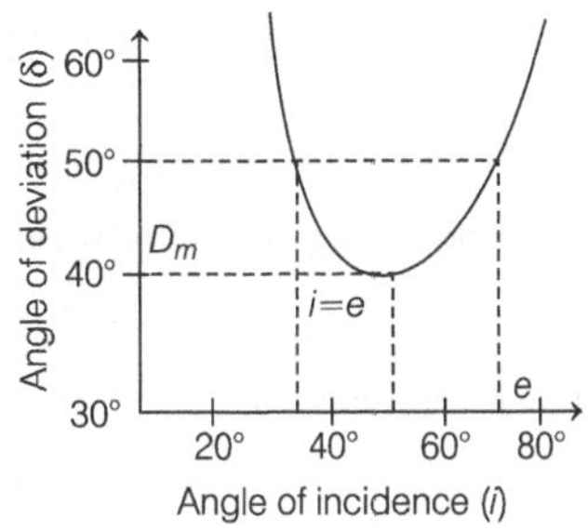


Thus, graph given in option (d) is correct.

**62** (b) For the given equilateral prism, $A = 60°$

$$i = e = \frac{3}{4}A = \frac{3}{4} \times 60° = 45°$$

From relation, $A + D = i + e$

We have, $\quad 60° + D = 45° + 45°$

$\Rightarrow \quad D = 90° - 60° = 30°$

**63** (c) Given, refractive index of material of prism $n = \sqrt{3}$, prism angle $A = 60°$

Using prism formula,

$$n = \frac{\sin\left(\frac{A+\delta}{2}\right)}{\sin\left(\frac{A}{2}\right)}$$

$$\Rightarrow \quad \sqrt{3} = \frac{\sin\left(\frac{60+\delta}{2}\right)}{\sin 30°}$$

$$\Rightarrow \quad \sin\left(\frac{60+\delta}{2}\right) = \frac{\sqrt{3}}{2} \Rightarrow \sin\left(\frac{60+\delta}{2}\right) = \sin 60°$$

$$\text{or} \quad \frac{60+\delta}{2} = 60$$

or angle of minimum deviation $\delta = 60°$

Incident angle, $i = \frac{60+\delta}{2} = 60°$

**64** (*b*) As, we know that refractive index of a prism

$$n = \frac{\sin\left(\frac{A+D_m}{2}\right)}{\sin\frac{A}{2}}$$

Given, $n = \cot\left(\frac{A}{2}\right)$

$$\Rightarrow \cot\frac{A}{2} = \frac{\sin\left(\frac{A+D_m}{2}\right)}{\sin\frac{A}{2}}$$

$$\Rightarrow \frac{\cos\frac{A}{2}}{\sin\frac{A}{2}} = \frac{\sin\left(\frac{A+D_m}{2}\right)}{\sin\frac{A}{2}}$$

$$\Rightarrow \sin\left(\frac{\pi}{2}-\frac{A}{2}\right) = \sin\left(\frac{A+D_m}{2}\right) \Rightarrow \frac{\pi}{2}-\frac{A}{2} = \frac{A}{2}+\frac{D_m}{2}$$

$$\Rightarrow D_m = \pi - 2A \Rightarrow D_m = 180^\circ - 2A$$

**65** (*a*) Given, $A = 60^\circ$ for equilateral prism and $D_m \approx 30^\circ$

According to prism formula, $n = \frac{\sin\left(\frac{A+D_m}{2}\right)}{\sin\left(\frac{A}{2}\right)}$

Substituting the given values in above equation, we get

$$n = \frac{\sin\left(\frac{60^\circ+30^\circ}{2}\right)}{\sin\left(\frac{60^\circ}{2}\right)} = \frac{\sin 45^\circ}{\sin 30^\circ} \text{ or } n = \sqrt{2}$$

$$\because \quad n = \frac{\text{Speed of light in air } (c)}{\text{Speed of light in prism } (v)}$$

$$\Rightarrow v = \frac{c}{n} = \frac{3\times 10^8}{\sqrt{2}} = \frac{3}{\sqrt{2}}\times 10^8 \text{ m/s}$$

**66** (*b*) Let $A$ be the refracting angle of the prism, so angle of minimum deviation, $D_m = A$

As, from prism formula, $n = \frac{\sin\left(\frac{A+D_m}{2}\right)}{\sin\left(\frac{A}{2}\right)}$

$$\Rightarrow n = \frac{\sin\left(\frac{A+A}{2}\right)}{\sin\left(\frac{A}{2}\right)}$$

$$= \frac{\sin A}{\sin\frac{A}{2}} = \frac{2\sin\frac{A}{2}\cos\frac{A}{2}}{\sin\frac{A}{2}}$$

$$n = 2\left\{\cos\left(\frac{A}{2}\right)\right\}$$

In the given situation value of $A$ varies from 0 to 90°.

$$\Rightarrow n = 2\cos\left(\frac{90^\circ}{2}\right) = 2\cos 45^\circ = \sqrt{2}$$

$$n' = 2\cos 0^\circ = 2$$

So, the value of $n$ lies between 2 and $\sqrt{2}$.

**67** (*a*) Given, $D_m = 40^\circ$, $i_1 = 35^\circ$, $i_2 = e = 79^\circ$

As we know, $D_m = (i_1 + i_2) - A$

$$\Rightarrow 40^\circ = (35^\circ + 79^\circ) - A$$

$$\Rightarrow A = 74^\circ$$

Now, from prism formula, $n = \frac{\sin\left(\frac{A+D_m}{2}\right)}{\sin\left(\frac{A}{2}\right)}$

If we take the given deviation as the minimum deviation, then

$$n = \frac{\sin\left(\frac{74^\circ+40^\circ}{2}\right)}{\sin\left(\frac{74^\circ}{2}\right)} \approx 1.5$$

But the given deviation may or may not be the minimum deviation. Rather it will be less than this value. Therefore, $n$ will be less than 1.5.

Hence, maximum possible value of refractive index is 1.5.

**68** (*a*) Consider a ray of light $PQ$ incident on the surface $AB$ and moves along $RS$, after passing through the prism $ABC$ as shown below.

It is given that, the incident ray suffers minimum deviation. Therefore, the ray inside the prism must be parallel to the base $BC$ of the prism.

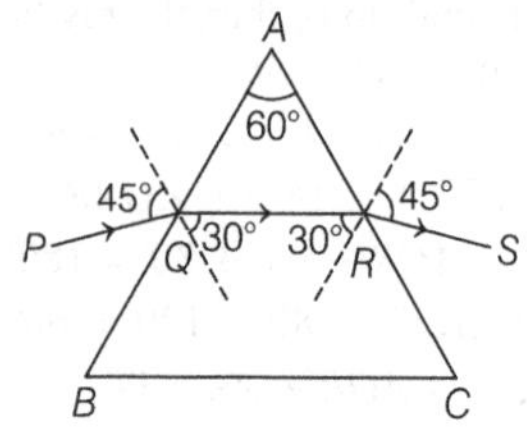


From the geometry of the prism and the ray diagram, it is clear that

angle of incidence, $i = 45^\circ$

angle of refraction $r = r' = 30^\circ$

angle of emergence, $e = 45^\circ$

Therefore, minimum deviation suffered by the ray is

$$D_{\min} = i + e - (r + r') = 90^\circ - 60^\circ = 30^\circ$$

Also we know that, $n = \frac{\sin\left(\frac{A+D_m}{2}\right)}{\sin\frac{A}{2}}$

where, $n$ = refractive index of the material of the prism.

$A$ = angle of prism = 60°

$$\therefore n = \frac{\sin\left(\frac{60° + 30°}{2}\right)}{\sin\left(\frac{60°}{2}\right)} = \frac{\sin 45°}{\sin 30°} = \frac{1/\sqrt{2}}{1/2} = \frac{2}{\sqrt{2}} = \sqrt{2}$$

The angle of minimum deviation and refractive index will be 30° and $\sqrt{2}$, respectively.

**69** (*a*) Given, $n = 1.57, D = 5.5°$

Since, the angle of deviation is very small, so we can use the relation, $D = (n-1)A$

$$\Rightarrow \quad A = \frac{D}{n-1} = \frac{5.5}{1.57-1} = 9.65°$$

**70** (*a*) Let $A'$ be the angle of flint glass prism.

Given, $A = 5°$ and $n = 1.51$ for crown glass prism.

Deviation produced by crown glass,

$$D = (n-1)A = (1.51-1)\times 5 = 2.55°$$

and deviation produced by flint glass,

$$D' = (1.65-1)A' = 0.65\,A'$$

For no deviation, $D' = D$ or $0.65\,A' = 2.55$

$$A' = \frac{2.55}{0.65} = 3.92°$$

**71** (*b*) Given, angle of prism = 5°; $n_r = 1.641, n_b = 1.659$

For this prism, the deviation, $D = (n-1)\,A$

So, for blue colour light, $D_b = (n_b - 1)A$

Similarly, for red colour light, $D_r = (n_r - 1)A$

∴ Angle between the emergent blue and red rays

$$= D_b - D_r = (n_b - n_r)A$$
$$= (1.659 - 1.641)\times 5 = 0.018\times 5° = 0.09°$$

**72** (*d*) Different colours of white light have different wavelengths. The descending order of the wavelength of the component of white light is

$$\lambda_{\text{Red}} > \lambda_{\text{Green}} > \lambda_{\text{Blue}} > \lambda_{\text{Violet}}$$

∴ Red colour light has the longest wavelength.

**75** (*a*) We know that, from Rayleigh's scattering,

Amount of scattering $\propto \frac{1}{\lambda^4}$

$$\Rightarrow \quad \frac{(\text{Amount of scattering})_1}{(\text{Amount of scattering})_2} = \frac{\lambda_2^4}{\lambda_1^4}$$

$$\Rightarrow \quad \frac{1}{4} = \left(\frac{\lambda_2}{\lambda_1}\right)^4 \Rightarrow \frac{\lambda_2}{\lambda_1} = \left(\frac{1}{4}\right)^{1/4} = \frac{1}{\sqrt{2}}$$

$$\therefore \quad \lambda_1 : \lambda_2 = \sqrt{2} : 1$$

**77** (*c*) Given, powers, $P_1 = 40\text{D}$ and $P_2 = 20\text{D}$.

We know that, $P_{\text{eq}} = P_1 + P_2 = 40\text{D} + 20\text{D} = 60\text{ D}$

$$\Rightarrow \quad f_{\text{eq}} = \frac{100}{P_{\text{eq}}} = \frac{100}{60} = 1.67\text{ cm}$$

Thus, the distance between the retina and cornea of eye lens is 1.67 cm.

**78** (*a*) The least distance of distinct vision for a normal person is 25 cm. Now, if the least distance of distinct vision for the person has been changed to 50 cm. So, if a book is at $u = -25$ cm, its image should be formed at $v = -50$ cm. Therefore, the desired focal length is given by

$$\frac{1}{f} = \frac{1}{v} - \frac{1}{u} \Rightarrow \frac{1}{f} = \frac{1}{-50} - \frac{1}{-25} = \frac{1}{50}$$

$$\therefore \quad f = +50\text{cm (convex lens)}$$

**79** (*d*) For myopic eye, the person should use a concave lens of focal length = − (defected far point)

$$= -80\text{ cm} = \frac{-80}{100}\text{ m}$$

i.e. $\quad \text{Power} = \frac{1}{\text{focal length}} = \frac{-100}{80} = -1.25\text{D}$

**80** (*b*) Hypermetropia is corrected by using convex lens.

Focal length of lens used, $f$ = + (defected near point)

$$f = +d = +20\text{ cm}$$

$$\therefore \quad \text{Power of lens} = \frac{100}{f(\text{cm})} = \frac{100}{+20} = 5\text{ D}$$

**81** (*a*) Given, $u = -25$ cm, $v = -75$ cm

$$\Rightarrow \quad \frac{1}{f} = \frac{1}{v} - \frac{1}{u} = \frac{1}{25} - \frac{1}{75} \Rightarrow f = 37.5\text{ cm}$$

$$\therefore \quad P = \frac{1}{f} = \frac{1}{37.5} = 2.67\text{D}$$

So, the corrective lens needs to have a converging power of + 2.67 D.

**82** (*a*) For magnifying an object a simple microscope may be used as magnifying glass having a converging lens.

Given, focal length of the lens, $f = +10$ cm

and $\quad u = -9$ cm

From lens formula, $v = \frac{uf}{u+f} = \frac{-9\times 10}{-9+10} = -90$ cm

Magnification, $m = \frac{v}{u} = \frac{-90}{-9} = 10$

∴ Apparent area of the card through the magnifying lens

$$= (1\text{mm}\times 10)\times(1\text{mm}\times 10) = 100\text{ mm}^2 = 1\text{ cm}^2$$

**84** (*b*) Given, focal length of objective, $f_o = 1.2$ cm,

Focal length of eyepiece, $f_e = 6.25$ cm,

Object distance for objective, $u_o = -2$ cm,

and image distance for eyepiece, $v_e = -25$ cm.

For objective, $\frac{1}{v_o} - \frac{1}{u_o} = \frac{1}{f_o}$

$$\Rightarrow \quad \frac{1}{v_o} = \frac{1}{1.2} - \frac{1}{2}$$

$$\Rightarrow \quad v_o = 3\text{ cm}$$

For eyepiece, $\frac{1}{v_e} - \frac{1}{u_e} = \frac{1}{f_e}$

$$\Rightarrow \quad \frac{1}{-25} - \frac{1}{u_e} = \frac{1}{6.25} \Rightarrow -\frac{1}{u_e} = \frac{1}{6.25} + \frac{1}{25}$$

$$u_e = -5\,\text{cm}$$

Distance between two lenses

$$= |v_o| + |u_e| = 3 + 5 = 8\,\text{cm}$$

**85** (*b*) Given, $f_o = 1.0$ cm, $f_e = 2.0$ cm and tube length, $L = 20$ cm.

The magnification due to objective is

$$m_0 = \frac{L}{f_o} = \frac{20}{1} = 20$$

and due to eyepiece is $m_e = \frac{D}{f_e}$

where, $D$ is the least distance of distinct vision = 25 cm

$$\Rightarrow m_e = \frac{25}{2} = 12.5$$

The magnification of compound microscope is

$$|m| = m_0 m_e = \left(\frac{L}{f_o}\right)\left(\frac{D}{f_e}\right) = 20 \times 12.5 = 250$$

**86** (*d*) Telescope resolves and bring objects closer. Hence, the given telescope with magnifying power 20, will make the tree to appear 20 times nearer to the observer.

**87** (*a*) Given, $f_o = F_1$, $f_e = F_2$

We know that, angular magnification for telescope,

$$|m| = \left|\frac{f_o}{f_e}\right| = \left|\frac{F_1}{F_2}\right| = \frac{F_1}{F_2}$$

**88** (*a*) For a telescope,

Angular magnification, $m = \frac{\beta}{\alpha} = \frac{f_o}{f_e}$

where, $\beta$ is the angle subtended at the eye by the final image and $\alpha$ is the angle subtended by the object at the lens or the eye.

Given, $f_o = 0.3$, $f_e = 3\text{ cm} = 0.03$ m and $\alpha = 0.5°$

$$\therefore \quad \frac{\beta}{0.5°} = \frac{0.3}{0.03} \Rightarrow \beta = 5°$$

**89** (*d*) For microscope, $m = \frac{L}{f_o}\frac{D}{f_e} \Rightarrow m \propto \frac{1}{f_o}$

For telescope, $|m| = \frac{f_o}{f_e} \Rightarrow m \propto f_o$

Hence, with increase in focal length of objective lens, the magnifying power of microscope will decrease but the magnifying power of telescope will increase.

**90** (*b*) The separation between the objective and the eyepiece = length of the telescope tube

$$\Rightarrow \quad L = f_o + f_e$$

Given, $f_o = 150\text{ cm} = 1.5\text{ m}$, $f_e = 5\text{ cm} = 0.05\text{ m}$

$$\Rightarrow \quad L = 1.5 + 0.05 = 1.55\,\text{m}$$

**91** (*a*) Given, focal length of objective lens, $f_o = 10$ cm

Focal length of eye lens, $f_e = 10\text{ mm} = 1\text{ cm}$

As in normal adjustment, the final image is formed at infinity, therefore

Angular magnification, $|m| = \left|\frac{f_o}{f_e}\right| = \frac{10}{1} = 10$

**92** (*d*) Given, angular magnification of telescope, $m = 5$

Separation between the objective and the eyepiece = 36 cm = Length of the telescope tube

$\therefore$ For telescope $|m| = \frac{|f_o|}{|f_e|} = 5$

$$\Rightarrow \quad |f_o| = 5|f_e| \quad \ldots(\text{i})$$

and length of the telescope

$$L = |f_o| + |f_e| = 36 \quad \ldots(\text{ii})$$

From Eqs. (i) and (ii), we get

$$5f_e + f_e = 36 \Rightarrow f_e = 6\,\text{cm} \quad \text{and} \quad f_o = 30\,\text{cm}$$

**93** (*d*) Let $f_o$ and $f_e$ be the focal lengths of objective and eyepiece, respectively. For normal adjustment, distance from objective to eyepiece = $f_o + f_e$.

If a straight black line of $L$ is drawn on inside part of the objective lens, then its real image is formed by the eyepiece length $I$, as shown in the figure below,

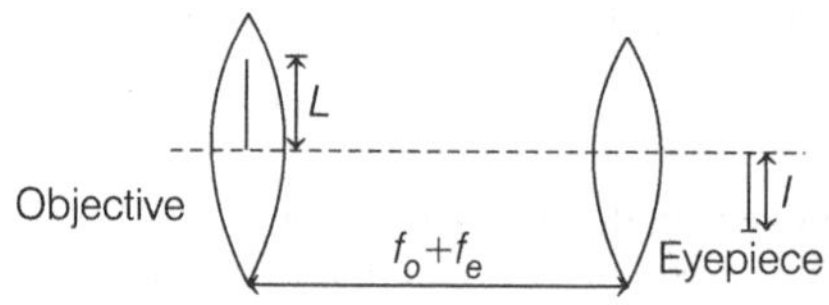


$$\therefore \quad u = (f_o + f_e) \text{ and } f = f_e$$

Using lens formula, $\frac{1}{v} - \frac{1}{u} = \frac{1}{f}$

$$\Rightarrow \quad \frac{1}{v} - \frac{1}{(f_o + f_e)} = \frac{1}{f_e}$$

$$v = \frac{(f_o + f_e)f_e}{f_o}$$

Magnification of lens

$$= \frac{v}{u} = \frac{(f_o + f_e)f_e}{f_o(f_o + f_e)} = \frac{f_e}{f_o} = \frac{I}{L} \quad \ldots(\text{i})$$

In normal adjustment, magnification of telescope,

$$|m| = \left|\frac{f_o}{f_e}\right|$$

$$\Rightarrow \quad m = \frac{L}{I} \quad \text{[using Eq. (i)]}$$

**94** (*c*) According to question,

Focal length of objective lens, $f_o = 40$ cm

Focal length of eyepiece lens, $f_e = 4$ cm

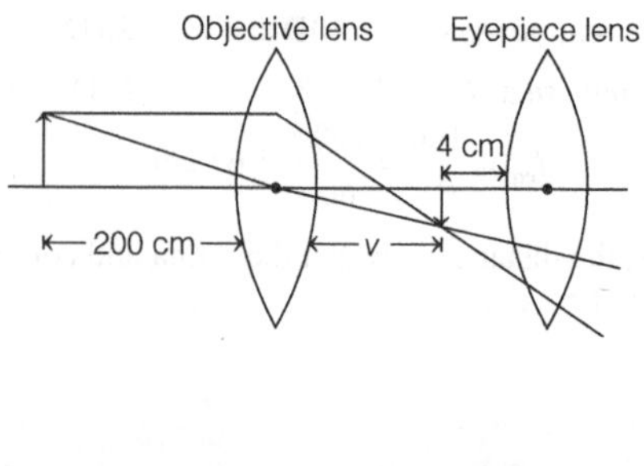

Object distance for objective lens, $u_o = -200$ cm

Applying lens formula for objective lens

$$\frac{1}{v} - \frac{1}{u} = \frac{1}{f} \Rightarrow \frac{1}{v} - \frac{1}{-200} = \frac{1}{40}$$

$$\Rightarrow \quad \frac{1}{v} = \frac{1}{40} - \frac{1}{200} = \frac{5-1}{200} = \frac{4}{200}$$

$$\Rightarrow \quad v = 50\text{ cm}$$

Image will be formed at first focus of eyepiece lens.

So, for normal adjustment distance between objective and eyepiece lens (length of tube) will be

$$v + f_e = 50 + 4 = 54\text{ cm}$$

**96** (*c*) When a ray of light is incident along normal to the plane mirror as shown below, then angle of incidence = angle between incident ray and normal to the mirror = 0°

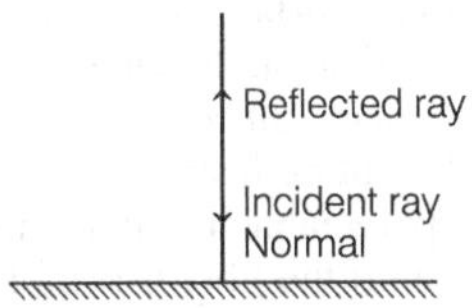


$\Rightarrow$ Angle of reflection = 0° (from laws of reflection)

Hence, the reflected ray retraces its path along the normal at an angle 0° with normal.

Therefore, Assertion is correct but Reason is incorrect.

**97** (*a*) Convex mirror always form virtual image of an object because rays always diverge after reflection from this mirror.

This can be shown in the ray diagram given below

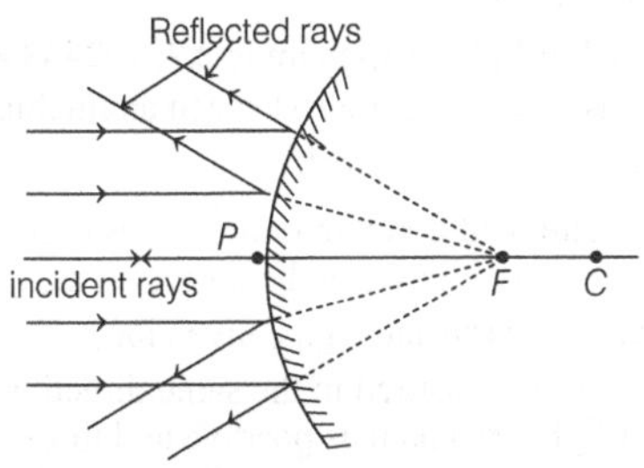


Therefore, Assertion and Reason are correct and Reason is the correct explanation of Assertion.

**98** (*d*) Refractive index of diamond w.r.t. liquid,

$$^{l}n_d = \frac{1}{\sin i_c} = \frac{n_d}{n_l}$$

$$\Rightarrow \quad \frac{\sqrt{6}}{\sqrt{3}} = \frac{1}{\sin i_c} \Rightarrow \sin i_c = \frac{1}{\sqrt{2}} = \sin 45°$$

$$\therefore \quad i_c = 45°$$

This means, that when the ray of light while travelling from diamond to liquid is incident at angle of 45°, then it will be totally reflected due TIR.

Therefore, Assertion is incorrect but Reason is correct.

**99** (*b*) In optical fibre communication propagation of signal through optical fibre takes place, which is based on the phenomenon of total internal reflection at core-clad interface.

The refractive index of the material of the cladding is less than that of core, hence, light striking at core-cladding interface gets totally internally reflected.

Therefore, Assertion and Reason are correct but Reason is not the correct expalnation of Assertion.

**100** (*d*) When a convergent lens is placed inside a transparent medium having refractive index greater than that of material of lens, it behaves as a divergent lens.

Hence, property of a lens, whether the rays are converging or diverging depends on the surrounding medium.

Therefore, Assertion is incorrect but Reason is correct.

**101** (*c*) When glass surface is made rough, then light incident on it is scattered in different directions. Due to which its transparency decreases. However, there is no effect of roughness on absorption of light.

Therefore, Assertion is correct but Reason is incorrect.

**102** (*d*) Magnification produced by mirror,

$$m = \frac{\text{size of image}}{\text{size of object}} = \frac{I}{O} = \frac{f}{f-u} = \frac{f}{x}$$

where, $x$ (say) is distance of the object from focus.

Therefore, Assertion is incorrect but Reason is correct.

**103** (*a*) For thin prisms, angle of prisms $A$ is small

So, for minimum deviation, $$^{1}n_2 = \frac{\sin\left(\frac{A + D_{\min}}{2}\right)}{\sin (A/2)} \quad \ldots\text{(i)}$$

$$\sin\left(\frac{A + D_{\min}}{2}\right) \approx \frac{A + D_{\min}}{2} \quad (\because \sin\theta \approx \theta \text{ for small } \theta)$$

and $$\sin\frac{A}{2} \approx \frac{A}{2}$$

Using above approximations, $^{1}n_2 = \frac{A + D_{\min}}{2}$

$$\Rightarrow \quad D_{\min} = (^{1}n_2 - 1)A$$

Hence, it can be seen that if $A$ is small, $D_{\min}$ is also small. Thus, thin prism do not deviate light much.

Therefore, Assertion and Reason are correct and Reason is the correct explanation of Assertion.

**104** (*a*) Refractive index of any pair of media is inversely proportional to wavelength of light.

As, $\lambda_v < \lambda_r \Rightarrow n_v > n_r$

where, $\lambda_v$ and $\lambda_r$ are the wavelengths of violet and red light, respectively and $n_r$ and $n_v$ are refractive index of violet and red light respectively.

Hence, refractive index of glass with respect to air is different for red light and violet light.

Therefore, Assertion and Reason are correct and Reason is the correct explanation of Assertion.

**105** (*d*) When white light pass through a prism, bending of red component of white light is least while it is most for the violet.

Equivalently, red light travels faster than violet light in a glass prism.

Therefore, Assertion is incorrect but Reason is correct.

**106** (*a*) In vacuum, speed of light is independent of wavelength. Hence, no dispersion takes places in vacuum. Thus, vacuum is a non-dispersive medium in which all colours travel with the same speed.

So, sunlight reaches us without dispersion in the form of white light and not as its component.

Therefore, Assestion and Reason are correct and Reason is the correct explanation of Assertion.

**107** (*a*) The rainbow is formed when the sunlight is first refracted as it enters a water droplet, which causes the different wavelength (colours) of the white light to separate.

Next, these refracted ray falling on the inner surface of the water drop get internally reflected, if the angle between the refracted ray and normal to the drop surface is greater than the critical angle.

Finally, the reflected light is refracted again when it comes out of the droplet.

Therefore, Assertion and Reason are correct and Reason is the correct explanation of Assertion.

**108** (*a*) Secondary rainbow is a result of four-step process. which are as follows

(i) Refraction at the first surface of raindrop.

(ii) Two total internal reflection from the second surface of raindrop.

(iii) Again refraction from the first surface of raindrop from where the light finally emerges out.

Thus, the intensity of light is reduced at the second reflection and

hence the secondary rainbow is fainter than the primary rainbow.

Therefore, Assertion and Reason are correct and Reason is the correct explanation of Assertion.

**109** (*b*) Astigmatism is occurred when the cornea is not spherical in shape.
It can be corrected by using cylindrical lens of desired radius of curvature with an appropriate directed axis.

Therefore, Assertion and Reason are correct but Reason is not the correct explanation of Assertion.

**110** (*d*) Statement III is correct but rest are incorrect and these can be corrected as,

As for the laws of reflection,

The angle of reflection (i.e. the angle between reflected ray and the normal to the reflecting surface or the mirror) equals the angle of incidence (angle between incident ray and the normal).

The incident ray, reflected ray and the normal to the reflecting surface at the point of incidence lie in the same plane.

These laws are valid at each point on any reflecting surface whether plane or curved.

**111** (*a*) The statement I is incorrect and it can be corrected as,

The reflected ray converge at a point $F$ on the principal axis of a concave mirror. For a convex mirror, the reflected rays appear to diverge from a point $F$ on its principal axis. This point $F$ is called the principal focus of the mirror.

Rest statements are correct.

**112** (*c*) The statement II is incorrect and it can be corrected as,

For the ray passing through (or directed towards) the focus of the concave mirror or appering to pass through (or directed towards) the focus of a convex mirror, then the reflected ray is parallel to the principal axis.

Rest statements are correct.

**113** (*b*) Optical fibres are fabricated such that light reflected at one side of inner surface strikes the other at an angle larger than the critical angle. Due to this total internal reflection takes place. Since, light undergoes total internal reflection at each stage in an optical fibre, so there is no appreciable loss in the intensity of the light signal. Even if the fibre is bent, light can easily travel along its length.

So, statement II is incorrect but I and III are correct.

**114** (*c*) The statement III is incorrect and it can be corrected as,

Thick lenses show chromatic aberration due to dispersion of light.

Rest statements are correct.

**115** (*b*) The statement given in option (b) is incorrect and it can be corrected as,

The speed of light in vacuum is $2.99792458 \times 10^8$ m/s or $3 \times 10^8$ ms$^{-1}$. It is the highest speed attainable in nature.

Rest statements are correct.

**116** (*a*) The statement given in option (a) is correct but rest are incorrect and these can be corrected as,

According to cartesian sign conversion,

The distances measured in the same direction as the incident light are taken as positive and those measured in the direction opposite to the direction of incident light are taken as negative.

The height measured upwards with respect to $X$-axis and normal to the principal axis ($X$-axis) of the mirror/lens are taken as positive. The heights measured downward are taken as negative.

**117** (*a*) The statement given in option (a) is correct but rest are incorrect and these can be corrected as,

The image is said to be real, if the rays actually converge to the point.

However, it is said to be virtual, if the rays do not actually meet but appear to diverge from the point when produced backwards.

**118** (*d*) The statement given in option (d) is incorrect and it can be corrected as,

The refractive index is a characteristic of a pair of media and depends on the wavelength of light, but is independent of the angle of incidence.

Rest statements are correct.

**119** (*a*) The statement given in option (a) is incorrect and it can be corrected as,

In case of refraction through a rectangular glass slab, if the emergent ray is parallel to the incident ray, then there is no deviation, but it does suffer lateral displacement/shift with respect to the incident ray.

Rest statements are correct.

**120** (*a*) The statement given in option (a) is correct but rest are incorrect and these can be corrected as,

Since, lens is made up of two layers of different refractive indices, so for a given wavelength of light, it will have two different focal lengths or will have two images at two different points as $\frac{1}{f} \propto (n-1)$ (from lens Maker's formula).

**121** (*a*) The statement given in option (a) is incorrect and it can be corrected as,

The optical density at different layers of air increases with height.

Rest statements are correct.

**122** (*c*) The statement given in option (c) is correct but rest are incorrect and these can be corrected as,

The brilliance of diamond is mainly due to the total internal reflection of light inside them.

The critical angle for diamond-air interface ($\cong 24.4^\circ$) is very small, therefore once light enters a diamond, it is very likely to undergo total internal reflection inside it.

Prisms are designed to bend light by 90° or by 180° by making use of total internal reflection. Such a prism is also used to invert images, without changing their size.

**123** (*c*) The statement given in option (c) is incorrect and it can be corrected as,

In photometry, the only parameter, which can be measured directly is illuminance. It is defined as luminous flux incident per unit area on a surface.

Rest statements are correct.

**124** (*d*) The statement given in option (d) is incorrect and it can be corrected as,

A ray of light passing through the first principal focus (for a convex lens) or appearing to meet at it (for a concave lens) emerges parallel to the principal axis after refraction.

Rest statements are correct.

**125** (*a*) The statement given in option (a) is correct but rest are incorrect and these can be corrected as,

Power is defined as the tangent of angle by which it converges or diverges a beam of light falling at unit distant from the optical centre.

Combination of lenses helps to obtain diverging or converging lenses of desired magnification. It also enhances sharpness of the image.

Also, total magnification $m$ of the combination of lenses is a product of magnification ($m_1, m_2, m_3 \ldots$) of individual lenses i.e. $m = m_1 \, m_2 \, m_3 \ldots$ .

**126** (*d*) The statement given in option (d) is correct but rest are incorrect and these can be corrected as,

The primary rainbow is a result of three step process, i.e. refraction, reflection and refraction.

The violet gets scattered more than blue, as it has shorter wavelength, but since our eyes are more sensitive to blue than violet, we see the sky blue.

Clouds which have droplets of water with dimensions very much greater than the wavelength of the light scatter all wavelength nearly equal, thus they generally appear white.

**127** (*b*) The statement given in option (b) is incorrect and it can be corrected as,

The necessary conditions for the rainbow to take place is

(i) Sun should be shining in part of the sky while it is raining in the opposite part of the sky.

(ii) The observer must stand with his back towards the sun.

Rest statements are correct.

**128** (*b*) The statement given in option (b) is incorrect and it can be corrected as,

The least distance of distinct vision or near point of eye increases with age, because of the decreasing effectiveness of the ciliary muscle and the loss of flexibility of the lens. This condition (defect of the eye) is called presbyopia. It is corrected by using a convergings lens.

Rest statements are correct.

**129** (*b*) The statement given in option (b) is incorrect and it can be corrected as,

For microscope, the magnification in case of image formed at infinity $\left(m = \frac{D}{f}\right)$ is one less than the magnification when the image is at the near point $\left[m = \left(1 + \frac{D}{f}\right)\right]$.

Rest statements are correct.

**130** (*d*) The statement given in option (d) is incorrect and it can be corrected as,

For objective, with large sized lenses the brightness increases and the chromatic aberration also increases.

Given, $f_o = 20$ m and $f_e = 2$ cm $= 0.02$ m.

In normal adjustment, length of telescope,

$$L = f_o + f_e = 20 + 0.02 = 20.02 \text{m}$$

Magnification, $m = \frac{f_o}{f_e} = \frac{20}{0.02} = 1000$

Also, the image formed is inverted with respect to the object.

Rest statements are correct.

**131** (*d*) In compound microscope objective lens forms a real, inverted, magnified image of the object.

This serves as the object for the second lens, the eyepiece, which functions essentially like a simple microscope or magnifier, produces the final image, which is enlarged and virtual.

The first inverted image is thus meet (at or within) the focal plane of the eyepiece, at a distance appropriate for final image formation at infinity, or a little closer for image formation at the near point.

Clearly, the final image is inverted with respect to the original object.

Thus, none of the statement is incorrect.

**132** (*b*) The statement given in option (b) is correct but rest are incorrect and these can be corrected as,

The telescope is used to provide angular magnification of distant objects.

The resolving power or the ability to observe two objects distinctly, which are in very nearly the same direction, depends on the diameter of the objective.

A cassegrain telescope has the advantage of a large focal length in a short telescope.

**133** (*c*) Applying the laws of reflection in the given figure as shown below

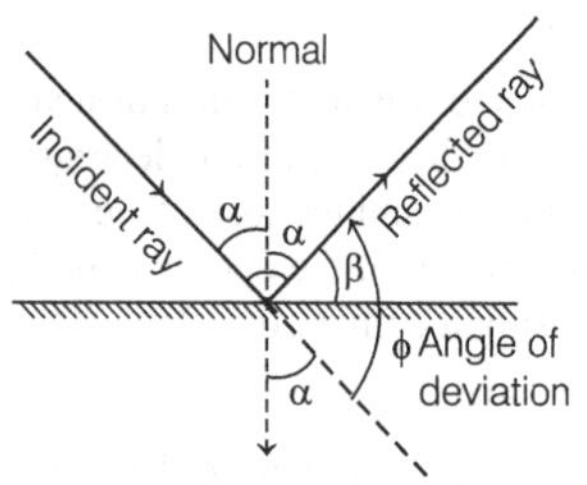


A. Angle of incidence = Angle of reflection = $\alpha$

B. The value of angle, $\beta = \pi/2 - \alpha$

C. Angle of deviation, $\phi = \pi - 2\alpha$ (from figure)

Hence, A $\to$ 2, B $\to$ 3 and C $\to$ 1.

**134** (*b*) In the given case, the light gets refracted twice from air to ice at *A* and from ice to glass at *B* as shown below.

At *A*, Air - ice interface

Applying Snell's law, $\dfrac{\sin i_1}{\sin r_1} = {}^{\text{air}}n_{\text{ice}} = \dfrac{n_{\text{ice}}}{n_{\text{air}}}$

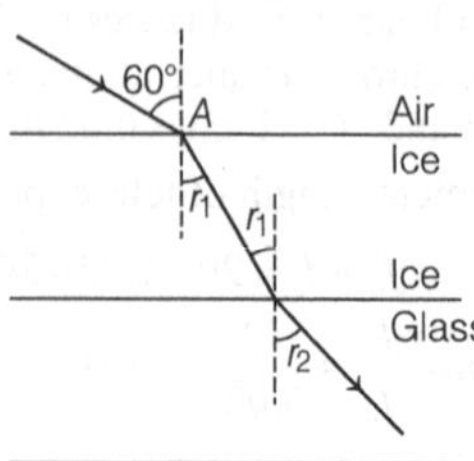


$$\Rightarrow \quad \frac{\sin 60^\circ}{\sin r_1} = \frac{n_{\text{ice}}}{1} \qquad \text{...(i)}$$

$$\Rightarrow \quad \sin r_1 = \frac{\sin 60^\circ}{n_{\text{ice}}} = \frac{\sqrt{3}/2}{4/3} = \frac{3\sqrt{3}}{8}$$

Similarly, at *B*, Ice - glass interface

Applying Snell's law, $\dfrac{\sin r_1}{\sin r_2} = {}^{\text{ice}}n_{\text{glass}} = \dfrac{n_{\text{glass}}}{n_{\text{ice}}}$ ...(ii)

On multiplying Eqs. (i) and (ii), we get

$$\frac{\sin 60^\circ}{\sin r_1} \times \frac{\sin r_1}{\sin r_2} = \frac{n_{\text{ice}}}{1} \times \frac{n_{\text{glass}}}{n_{\text{ice}}} \Rightarrow \frac{\sin 60^\circ}{\sin r_2} = n_{\text{glass}}$$

$$\Rightarrow \quad \sin r_2 = \frac{\sin 60^\circ}{n_{\text{glass}}} = \frac{\sqrt{3}/2}{3/2} = \frac{.1}{\sqrt{3}}$$

Also, refractive index of ice w.r.t. glass

$$\Rightarrow \quad {}^{\text{glass}}n_{\text{ice}} = \frac{{}^{\text{air}}n_{\text{ice}}}{{}^{\text{air}}n_{\text{glass}}} = \frac{4/3}{3/2} = \frac{8}{9}$$

Hence, A $\to$ 2, B $\to$ 1 and C $\to$ 3.

**135** (*d*) When the lens of simple microscope is placed near an object, one focal length away or less and the eye is positioned close to the lens on the other side. Then, an erect, magnified and virtual image of the object can be seen comfortably, i.e. at 25 cm or more.

Also, the angle subtended by the object is the same as that at the near point.

If the object is at a distance *f*, the image is at infinity. However, if the object is at a distance sightly less than the focal length of the lens, the image is virtual and closer than infinity.

Hence, A $\to$ 2, B $\to$ 3 and C $\to$ 1.

**136** (*c*) **Minimum Deviation**

The angle of minimum deviation occurs, when $i = e$ and $r_1 = r_2$ and is given by

$$n = \frac{\sin \dfrac{(A + D_{\min})}{2}}{\sin \dfrac{A}{2}}$$

$$\Rightarrow \quad D_{\min} = 2\sin^{-1}\left[\sin\left(\frac{A}{2}\right) \times n\right] - A$$

Substituting $n = 1.5$ and $A = 60^\circ$

$$D_{\min} = 2\sin^{-1}(\sin 30^\circ \times 1.5) - 60^\circ$$
$$= 2\sin^{-1}(0.75) - 60^\circ$$
$$= (97^\circ - 60^\circ) = 37^\circ$$

Hence, the angle of minimum deviation is 37°.

**Maximum Deviation**

The deviation is maximum, when $i = 90^\circ$ or $e = 90^\circ$, i.e. at grazing incidence or grazing emergence.

Let we take, $i = 90^\circ$

The path of the incident ray will be as shown in the figure below,

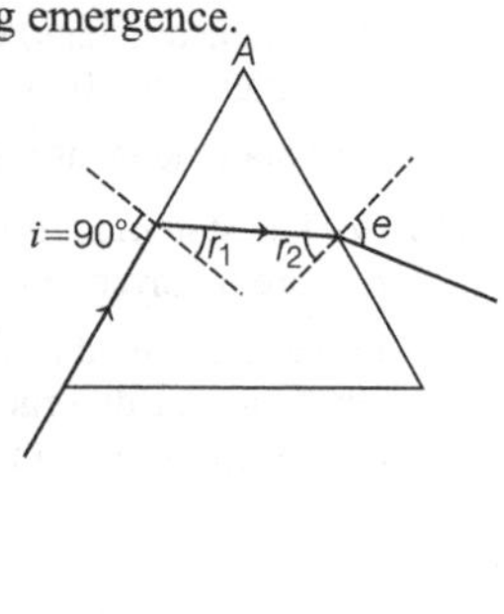


So, $r_1$ must be at critical angle for air-glass interface.

$\Rightarrow \quad r_1 = i_C = \sin^{-1}\left(\frac{n_r}{n_d}\right) = \sin^{-1}\left(\frac{1}{n}\right) = \sin^{-1}\left(\frac{1}{1.5}\right)$

$\Rightarrow \quad r_1 = \sin^{-1}\left(\frac{2}{3}\right) = 42°$

Also, $\quad r_1 + r_2 = A$

$\Rightarrow \quad r_2 = A - r_1 = 60° - 42° = 18°$

Using $\quad \frac{\sin r_2}{\sin e} = \frac{1}{n}$

$\Rightarrow \quad \sin e = n \sin r_2 = 1.5 \sin 18°$

$\Rightarrow \quad \sin e = 0.463 \Rightarrow e = 28°$

Deviation $= D_{max} = i + e - A = 90° + 28° - 60° = 58°$

Hence, the maximum deviation for the prism is 58° and angle of refraction $r_1$ in this case is 42°.

Hence, A $\rightarrow$ 2, B $\rightarrow$ 1 and C $\rightarrow$ 3.

**137** (*b*) Given, $h_1 = 2.5$ cm, $u = -27$ cm, $R = -36$ cm

($\because$ $R$ is –ve for a concave mirror)

As, $\quad f = \frac{R}{2} = -36/2 = -18$ cm

By mirror formula, $\frac{1}{v} = \frac{1}{f} - \frac{1}{u} = \frac{1}{-18} + \frac{1}{27}$

$= \frac{-3+2}{54} = -\frac{1}{54}$ or $v = -54$ cm

Thus, the screen should be placed at 54 cm from the mirror on the same side as the object.

Magnification, $m = \frac{h_2}{h_1} = -\frac{v}{u} = -\frac{-54}{-27} = -2$

Size of image, $\quad h_2 = -2 \times 2.5 = -5$ cm

Negative sign shows that the image is real and inverted.

**138** (*a*) Given, focal length of convex mirror, $f = +15$ cm (focal length of convex mirror is taken as positive)

Distance of object, $u = -12$ cm

Size of object, $O = 4.5$ cm

Using the mirror formula, $\frac{1}{f} = \frac{1}{v} + \frac{1}{u}$

$\Rightarrow \frac{1}{15} = \frac{1}{v} - \frac{1}{12} \Rightarrow \frac{1}{v} = \frac{1}{15} + \frac{1}{12} = \frac{4+5}{60} = \frac{9}{60}$

Distance of image from the mirror, $v = 6.7$ cm

The positive sign shows that the image is formed behind the mirror.

Using the formula of magnification.

$m = -\frac{v}{u} = \frac{I}{O} \quad \Rightarrow \frac{-6.7}{-12} = \frac{I}{4.5}$

Size of images, $I = 2.5$ cm

As $I$ is positive, so image is erect and virtual.

Magnification $m$ is given by

$m = \frac{I}{O} = \frac{2.5}{4.5} = \frac{25}{45} = \frac{5}{9}$

**139** (*c*) When tank is filled with water:

Given, real depth = 12.5 cm and apparent depth = 9.4 cm

Refractive index of water is

$n_w = \frac{\text{Real depth}}{\text{Apparent depth}} = \frac{12.5}{9.4} = 1.32$

When tank is filled with liquid

Given, real depth = 12.5 cm

and refractive index of liquid = 1.63

Refractive index of liquid $= \frac{\text{Real depth}}{\text{Apparent depth}}$

or $\quad 1.63 = \frac{12.5}{\text{Apparent depth}}$

Apparent depth in liquid $= \frac{12.5}{1.63}$ cm $= 7.669$ cm $\simeq$ 7.7 cm

Distance through which the microscope has to be moved

$= 9.4 - 7.7 = 1.7$ cm

**140** (*b*) If the light falls at an angle of incidence equal to critical angle $i_C$, the angle of refraction would be 90° and light will pass through the interface of the two media, forming a circular area of radius $r$.

It is as shown below, where bulb is placed at $O$ and $AB = AC = r$.

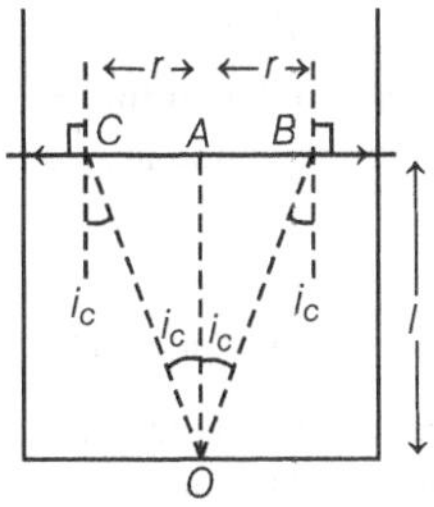


The source of light is 80 cm below the surface of water

i.e. $\quad AO = l = 80$ cm and $n_w = 1.33$

Using the formula for critical angle, $\sin i_C = \frac{1}{n_w}$

$\Rightarrow \sin i_C = \frac{1}{1.33} = 0.75$

$i_C = 48.6°$

In $\Delta OAB$, $\tan i_C = \frac{AB}{AO}$ or $\tan i_C = \frac{r}{l}$

$r = l \tan i_C = 80 \tan 48.6°$

$r = 80 \times 1.1345 = 90.7$ cm

Area of circular surface of water, through which light will emerge, $\quad A = \pi r^2$

$A = 3.14 \times (90.7)^2 = 25831.18 \text{ cm}^2$

$A = 2.58 \text{ m}^2$

**141** (*a*) When the prism is placed in air,

$D_m = 40°$ and $A = 60°$

$\therefore$ Refractive index of the prism material is

$${}^a n_g = \frac{\sin\left(\frac{A + D_m}{2}\right)}{\sin\left(\frac{A}{2}\right)} = \frac{\sin\left(\frac{60°+40°}{2}\right)}{\sin\left(\frac{60°}{2}\right)}$$

$$= \frac{\sin 50°}{\sin 30°} = \frac{0.7660}{0.5000} = 1.532$$

Similarly, when the prism is placed in water,

$$^{w}n_g = \frac{\sin\left(\frac{A + D'_m}{2}\right)}{\sin\frac{A}{2}}$$

or $$\frac{^{a}n_g}{^{a}n_w} = \frac{\sin\left(\frac{60° + D'_m}{2}\right)}{\sin\left(\frac{60°}{2}\right)}$$

or $$\frac{1.532}{1.33} = \frac{\sin\left(\frac{60° + D'_m}{2}\right)}{\sin 30°}$$

or $$\sin\left(\frac{60° + D'_m}{2}\right) = \frac{1.532}{1.33} \times 0.5 = 0.5759$$

$$\therefore \quad \left(30° + \frac{D'_m}{2}\right) = \sin^{-1}(0.5759) = 35°10'$$

or $$D'_m = 10°20'$$

Thus, the refractive index of the material and the new angle of minimum deviation are 1.53 and 10°20′, respectively.

**142** (*c*) Given, the refractive index of glass with respect to air, $^{a}n_g = 1.55$

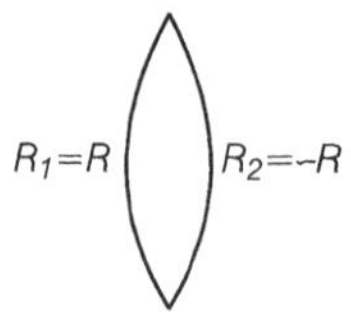


∵ Both faces have same radius of curvature, so for double convex lenses $R_1 = R$ and $R_2 = -R$.

Focal length of lens, $f = +20$ cm

Using the lens Maker's formula

$$\frac{1}{f} = (^{a}\mu_g - 1)\left(\frac{1}{R_1} - \frac{1}{R_2}\right)$$

Substituting the given values in above equation, we get

$$\Rightarrow \quad \frac{1}{20} = (1.55 - 1)\left(\frac{1}{R} + \frac{1}{R}\right)$$

$$\frac{1}{20} = 0.55 \times \frac{2}{R}$$

$$\Rightarrow \quad R = 0.55 \times 2 \times 20 = 22 \text{ cm}$$

Thus, the required radius of curvature is 22 cm.

**143** (*a*) Here, the point $P$ is on the right side of lens acts as virtual object as shown in the figure below.

Given, distance of object from the lens, $u = 12$ cm

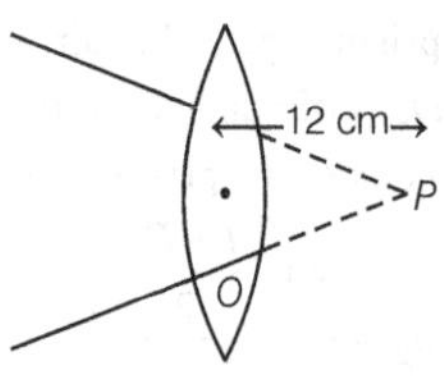


Focal length of convex lens, $f = +20$ cm

Using lens formula, $\frac{1}{v} - \frac{1}{u} = \frac{1}{f} \Rightarrow \frac{1}{v} - \frac{1}{12} = \frac{1}{20}$

$$\frac{1}{v} = \frac{1}{20} + \frac{1}{12} = \frac{3+5}{60} = \frac{8}{60}$$

$$v = 7.5 \text{ cm}$$

Thus, the beam converges on the right side of lens at a distance of 7.5 cm.

**144** (*d*) Given, concave lens, $f = -21$ cm, $u = -14$ cm

Object height, $O = 3$ cm

Applying lens formula,

$$v = \frac{uf}{u+f} = \frac{(-14)\times(-21)}{(-14)+(-21)}$$

$$= \frac{-42}{5} \text{ cm} = -8.4 \text{ cm}$$

Using the formula of magnification of lens,

$$m = \frac{v}{u} = \frac{I}{O}, \text{where } I = \text{height of image.}$$

$$\Rightarrow \quad \frac{-8.4}{-14} = \frac{I}{+3}$$

$$\Rightarrow \quad I = 1.8 \text{ cm}$$

Magnification of a lens, $m = \frac{v}{u} = \frac{-8.4}{-14} = 0.6$

Hence, A → 2, B → 3 and C → 1.

**145** (*a*) Given, $f_1 = +20$ cm, $f_2 = +30$ cm

So, for equivalent focal length $\frac{1}{f} = \frac{1}{f_1} + \frac{1}{f_2}$

$$\Rightarrow \quad \frac{1}{f} = \frac{1}{20} + \frac{1}{30}$$

$$\therefore \quad f = \frac{20 \times 30}{20 + 30} = \frac{600}{50} = 12 \text{ cm}$$

**146** (*b*) Here, for objective

$f_o = 8$ mm $= 0.8$ cm, $u_o = 9$ mm $= -0.9$ cm, $v_0 = ?$

As $$\frac{1}{v_o} - \frac{1}{u_o} = \frac{1}{f_o}$$

$$\therefore \quad \frac{1}{v_o} = \frac{1}{f_o} + \frac{1}{u_o} = \frac{1}{0.8} - \frac{1}{0.9}$$

$$= \frac{0.9 - 0.8}{0.9 \times 0.8} = \frac{0.1}{0.8 \times 0.9}$$

or $$v_o = \frac{0.8 \times 0.9}{0.1} = 7.2 \text{ cm}$$

Now for the eyepiece, we have

$$f_e = 2.5 \text{ cm}, v_e = -D = -25 \text{ cm}, u_e = ?$$

$$\therefore \quad \frac{1}{u_e} = \frac{1}{v_e} - \frac{1}{f_e} = -\frac{1}{25} - \frac{1}{2.5} = \frac{-1-10}{25} = \frac{-11}{25}$$

or $$u_e = -\frac{25}{11} = -2.27 \text{ cm}$$

Hence, the separation between the two lenses

$$= |v_0| + |u_e| = 7.2 + 2.27 = 9.47 \text{ cm}$$

Magnifying power, $m = m_0 \times m_e = \frac{v_0}{|u_0|}\left(1+\frac{D}{f_e}\right)$

$$= \frac{7.2}{0.9}\left(1+\frac{25}{2.5}\right) = 88$$

The separation between the two lenses and magnifying power of the microscope are 9.4 and 88, respectively.

**147** (*a*) Given, focal length of objective lens $f_o = 144$ cm.

Focal length of eye-piece $f_e = 6$ cm

Magnifying power of the telescope in normal adjustment (i.e., when the final image is formed at $\infty$)

$$m = -\frac{f_o}{f_e} = -\frac{144}{6} = -24$$

$\therefore$ Separation between lenses $L = f_o + f_e = 144 + 6$
$= 150$ cm

**148** (*d*) Here, $f_0 = 15$ m, $f_e = 1.0$ cm $= 0.01$ m

Angular magnification, $|m| = \left|\frac{f_0}{f_e}\right| = \frac{15}{0.01} = 1500$

**149** (*b*) Given, thickness of glass slab (real depth) = 15 cm

Refractive index of glass with respect to air ${}^a n_g = 1.5$

Using the formula, ${}^a n_g = \frac{\text{Real depth}}{\text{Apparent depth}}$

$\Rightarrow$ Apparent depth of pin $= \frac{15}{1.5} = 10$ cm

Distance by which the pin appears to be raised
= Real depth – Apparent depth = 15 – 10 = 5 cm

**150** (*d*) As shown in figure, let $O$ and $I$ be the positions of object and image respectively and $L_1$ and $L_2$ be the two conjugate positions of the lens.

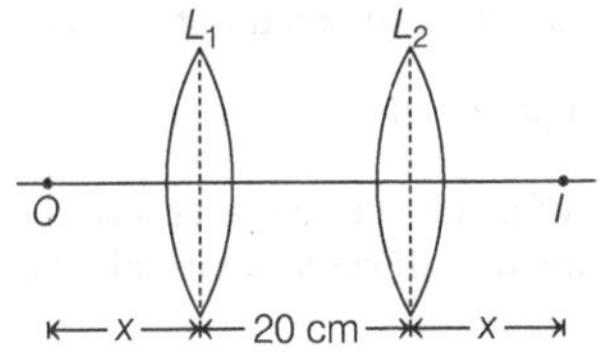


Obviously, $x + 20 + x = 90$ cm

or $x = 35$ cm

When the lens is in position $L_1$, we have

$$u = -x = -35 \text{ cm}$$

$$v = 20 + x = 20 + 35 = 55 \text{ cm}$$

$$\therefore \quad \frac{1}{f} = \frac{1}{v} - \frac{1}{u} = \frac{1}{55} + \frac{1}{35} = \frac{7+11}{385} = \frac{18}{385}$$

or $f = \frac{385}{18} = 21.4$ cm

**151** (*b*) Given, angle of prism, $A = 60°$

and refractive index of prism, $n = 1.524$

Let $i_1$ be the angle of incidence of the ray of light.

Since, it just suffers total internal refraction at the other face of the prism, as shown in the figure below. So, critical angle $i_c$ is given as

$$\sin i_C = \frac{1}{n} = \frac{1}{1.524} = 0.6561 \Rightarrow i_C = 41°$$

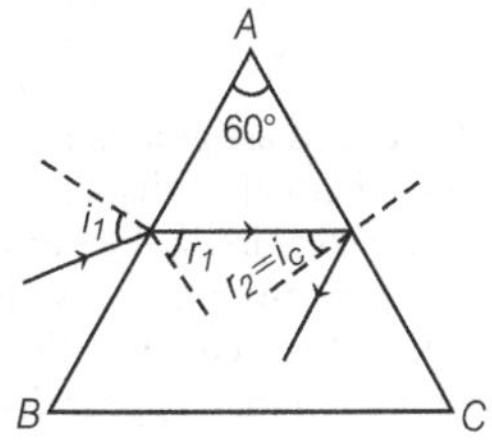


For a prism $r_1 + r_2 = A$, here $r_2 = i_C$

$\therefore \quad r_1 + i_C = A \Rightarrow r_1 + 41° = 60° \Rightarrow r_1 = 19°$

Using the formula, $n = \frac{\sin i_1}{\sin r_1}$

or $\sin i_1 = 1.524 (\sin 19°) = 1.524 \times 0.3256$

or $i_1 = \sin^{-1}(0.4962) \Rightarrow i_1 = 29°75'$

Thus, the angle of incidence should be 29°75′.

**152** (*a*) To see objects at infinity, the eye uses its least converging power = 40D + 20D = 60D

$\therefore$ Approximate distance between the retina and the cornea eyelens

= focal length of the eyelens

$$= \frac{100}{P} = \frac{100}{60} = \frac{5}{3} \text{ cm}$$

To focus an object at the near point on the retina, we have

$$u = -25 \text{ cm}, v = \frac{5}{3} \text{ cm}$$

$\therefore$ Focal length $f$ should be given by

$$\frac{1}{f} = \frac{1}{v} - \frac{1}{u} = \frac{3}{5} + \frac{1}{25} = \frac{15+1}{25} = \frac{16}{25}$$

$$\Rightarrow \quad f = \frac{25}{16} \text{ cm}$$

Corresponding converging power $= \frac{1600}{25} = 64$ D

Power of the eye-lens = 64 – 40 = 24 D

Thus, the range of accommodation of the eyelens is roughly 20 to 24 D.

**153** (*b*) Initially power of spectacles = – 1D, i.e. focal length is $= \frac{100}{(-1)} = -100$ cm. It means that far point of the person is 100 cm and near point is normal, i.e. 25 cm. Due to the old age, the person use + 2 D spectacles, i.e. focal length, $f = \frac{100}{P} = \frac{100}{2} = 50$ cm

So, $u = -25$ cm and $f = 50$ cm

From the lens formula, $\frac{1}{f} = \frac{1}{v} - \frac{1}{u}$

$$\Rightarrow \quad \frac{1}{50}=\frac{1}{v}+\frac{1}{25}$$

$$\Rightarrow \quad \frac{1}{v}=\frac{1}{50}-\frac{1}{25}=\frac{1-2}{50}=-\frac{1}{50}$$

$$\Rightarrow \quad v=-50\text{ cm}$$

The near point is –50 cm.

**154** (*c*) This defect is called astigmatism. This occurs when the cornea is not spherical in shape, e.g. the cornea could have larger curvature in the vertical plane than in horizontal plane or *vice-versa*. It results in lines in one direction being well focused while those in perpendicular direction may appear distorted.

This defect can be corrected by using a cylindrical lens with its axis along the vertical.

**155** (*a*) For the closest distance, $v=-25$ cm, $f=5$ cm, $u=?$

$$\text{As} \quad \frac{1}{v}-\frac{1}{u}=\frac{1}{f}$$

$$\therefore \quad \frac{1}{u}=\frac{1}{v}-\frac{1}{f}=\frac{1}{-25}-\frac{1}{5}=\frac{-1-5}{25}=\frac{-6}{25}$$

$$\text{or} \quad u=-\frac{25}{6}\text{ cm}=-4.2\text{ cm}$$

This is the closest distance at which the man can read the book.

For the farthest image,

$$v=\infty,\ f=5\text{ cm},\ u=?$$

$$\therefore \quad \frac{1}{u}=\frac{1}{v}-\frac{1}{f}=\frac{1}{\infty}-\frac{1}{5}=0-\frac{1}{5}=-\frac{1}{5}$$

$$u=-5\text{ cm}$$

This is the farthest distance at which the man can read the book.

**156** (*a*) We assume the microscope in common usage, i.e. the final image is formed at the least distance of distinct vision,

$$D=25\text{ cm},\ f_e=5\text{ cm}$$

∴ Angular magnification of the eyepiece is

$$m_e=1+\frac{D}{f_e}=1+\frac{25}{5}=6$$

As total magnification, $m=m_e\times m_o$

∴ Angular magnification of the objective,

$$m_o=\frac{m}{m_e}=\frac{30}{6}=5$$

As real image is formed by the objective, therefore,

$$m_o=-\frac{v_o}{u_o}=-5 \text{ or } v_o=-5u_o$$

$$\text{Now,} \quad \frac{1}{v_o}-\frac{1}{u_o}=\frac{1}{f_o}$$

Given, $f_o=1.25$ cm

$$\Rightarrow \quad \frac{1}{-5u_o}-\frac{1}{u_o}=\frac{1}{1.25} \text{ or } \frac{-6}{5u_o}=\frac{1}{1.25}$$

$$\Rightarrow \quad u_o=-\frac{6\times1.25}{5}=-1.5\text{ cm}$$

Thus, the object should be held at 1.5 cm infront of the objective lens.

$$\text{Also,} \quad v_o=-5u_o =-5\times(-1.5)=7.5\text{ cm}$$

$$\text{As} \quad \frac{1}{v_e}-\frac{1}{u_e}=\frac{1}{f_e}$$

$$\therefore \quad \frac{1}{u_e}=\frac{1}{v_e}-\frac{1}{f_e}=\frac{1}{-25}-\frac{1}{5} \quad (\because v_e=-D=-25\text{ cm})$$

$$=\frac{-1-5}{25}=-\frac{6}{25}$$

$$u_e=\frac{-25}{6}=-4.17\text{ cm}$$

∴ Separation between the objective and the eyepiece, $L$

$$=|u_e|+|v_0|=4.17+7.5=11.67\text{ cm}$$

**157** (*d*) Given, $f_o=140$ cm and $f_e=5$ cm

(i) In normal adjustment, magnifying power,

$$|m|=\left|\frac{f_o}{f_e}\right|=\frac{140}{5}=28$$

(ii) When the final image is formed at the least distance of distinct vision (25 cm)

$$|m|=\frac{f_o}{f_e}\left(1+\frac{f_e}{D}\right)=\frac{140}{5}\left(1+\frac{5}{25}\right)=28\times1.2=33.6$$

In normal adjustment, magnifying power and the final image is formed at the least distance of distinct vision are 28 and 33.6, respectively.

**158** (*a*) Given, distance between objective mirror and another mirror, $d=20$ mm

Radius of curvature of objective mirror, $R_1=220$ mm

∴ Focal length of objective mirror, $f_1=\frac{220}{2}=110$ mm

Radius of curvature of small mirror, $R_2=140$ mm

∴ Focal length of small mirror, $f_2=\frac{140}{2}=70$ mm

The image of an object placed at infinity, formed by the objective mirror will act as a virtual object for small mirror.

So, the image of the object formed by the objective will be at its focus $f_1$.

Thus, the object distance for small mirror $u=f_1-d$,

i.e. $u=110-20=90$ mm

Using mirror formula,

$$\frac{1}{v}=\frac{1}{f_2}-\frac{1}{u}=\frac{1}{70}-\frac{1}{90}=\frac{9-7}{630}=\frac{2}{630}$$

$v=315$ mm or $v=31.5$ cm

Thus, the final image is formed at 315 mm away from small mirror.

**159** (*d*) Distance of the needle from the lens in the first case = focal length $F$ of the combination of the convex lens and plano-concave lens formed by the liquid

i.e. $\quad F=45$ cm

Distance measured in second case

$= \text{Focal length of the convex lens}$

i.e. $f_1 = +30\text{ cm}$

The focal length $f_2$ of the plano-concave lens is given by

$$\frac{1}{f_1} + \frac{1}{f_2} = \frac{1}{F}$$

or $$\frac{1}{f_2} = \frac{1}{F} - \frac{1}{f_1} = \frac{1}{45} - \frac{1}{30} = \frac{2-3}{90} = -\frac{1}{90}$$

$\therefore \quad f_2 = -90\text{ cm}$

Now for the equi-convex lens, we have

$$R_1 = R, R_2 = -R, f = 30\text{ cm}, \mu = n = 1.5$$

Using lens Maker's formula,

$$\frac{1}{f} = (n-1)\left[\frac{1}{R_1} - \frac{1}{R_2}\right] \text{ or } \frac{1}{30} = (1.5-1)\left[\frac{1}{R} + \frac{1}{R}\right]$$

$$\frac{1}{30} = 0.5 \times \frac{2}{R}$$

or $R = 0.5 \times 2 \times 30\text{ cm} = 30\text{ cm}$

For plano-convex lens, $f = -90$ cm

For concave surface, $R_1 = -R = -30\text{ cm}$

For plane surface, $R_2 = \infty$

As $$\frac{1}{f} = (n-1)\left[\frac{1}{R_1} - \frac{1}{R_2}\right]$$

$\therefore$ $$\frac{1}{-90} = (n-1)\left[\frac{1}{-30} - \frac{1}{\infty}\right]$$

or $n - 1 = \dfrac{-30}{-90} = +\dfrac{1}{3}$ or $n = 1 + \dfrac{1}{3} = 1.33$

**160** (*a*) The situation is as shown

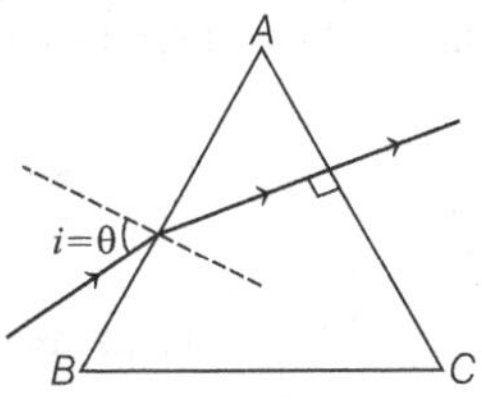


Given, $i = \theta;\ A = 5°;\ n = 1.5$

For ray emerging normally from $AC$, $e = 0$

From relation, $A + D = i + e$, we have

$\Rightarrow \quad 5° + D = \theta + 0$

$\Rightarrow \quad i = \theta = 5° + D$

Also, for small angles,

$$D = (n-1)A = (1.5-1)5° = 2.5°$$

So, $i = \theta = 5° + 2.5° = 7.5°$

**161** (*d*) Since, $v \propto \lambda$, thus the light with maximum wavelength will have maximum speed. Since, red colour has maximum wavelength amongst the constituent colours of white light. Therefore, after travelling through the slab, the red colour will emerge first.

**163** (*c*) Given, $n = 1.5$

If object lies on plane side as shown below

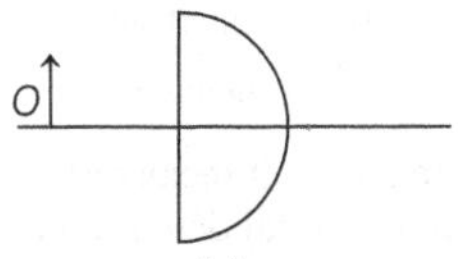


(a)

$R_1 = \infty, R_2 = -20\text{ cm}$

$$\Rightarrow \quad \frac{1}{f} = (n-1)\left(\frac{1}{R_1} - \frac{1}{R_2}\right)$$

$$= (1.5-1)\left(\frac{1}{\infty} + \frac{1}{20}\right) = \frac{1}{40} \Rightarrow f = +40\text{ cm}$$

Since, the focal length is positive, so the lens behaves as convex.

If object lies on its convex side as shown below,

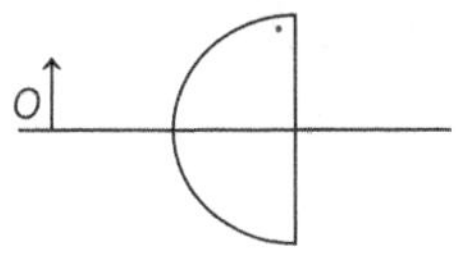


(b)

$R_1 = 20\text{cm}, R_2 = \infty.$

$$\Rightarrow \quad \frac{1}{f'} = (n-1)\left(\frac{1}{R_1} - \frac{1}{R_2}\right) = (1.5-1)\left(\frac{1}{20} - \frac{1}{\infty}\right) = \frac{1}{40}$$

$\Rightarrow \quad f' = 40\text{ cm}$

Then, also the lens behaves as convex.

**164** (*b*) In the given figure, $PQ$ is a ray of light passing through focus and falling on the surface of a concave mirror. On reflection from the mirror, the ray becomes parallel to principal axis of the mirror. Hence, the ray would travel along 2.

**165** (*b*) Ray number 2 follows the correct path according to the laws of refraction.

Since, $\mu_{\text{turpentine}} > \mu_{\text{water}} > \mu_{\text{air}}$

So, as light travels from air to turpentine (i.e. from rarer to denser medium), it bends towards the normal after refraction.

Again when it strikes the turpentine water interface it must bend away from the normal as it travels from denser to rarer medium.

**166** (*a*) For negative refractive index, i.e. $n_2$ is negative

Applying Snell's law,

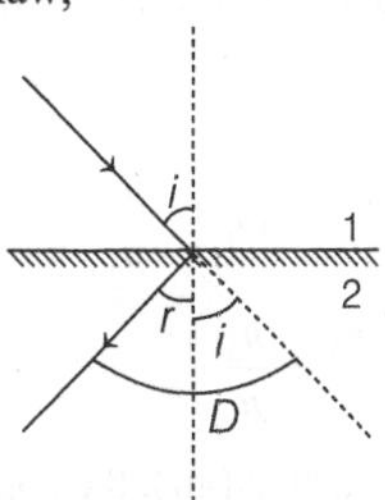


$$\frac{\sin i}{\sin r} = {}^1n_2 \text{ (refractive index of medium 2}$$

$$\text{w.r.t. medium } 1 = {}^1n_2) = \text{negative}$$

$$\Rightarrow \quad \sin r = \frac{\sin i}{{}^1n_2} = \frac{\text{positive}}{\text{negative}} = \text{negative}$$

Since, $\sin r$ is negative, it means $r$ must be on other side of normal such that, total deviation of ray is $D = i + r$ as shown in option(a).

**167** (*d*) Here, in the figure below, a pin had been held at $L$, which is the mid way of face $AB$. When seen from face $AD$ (as long as $i < i_C$) image of $L$ appears to be at $L'$ which is closer to $A$ as shown in the figure below.

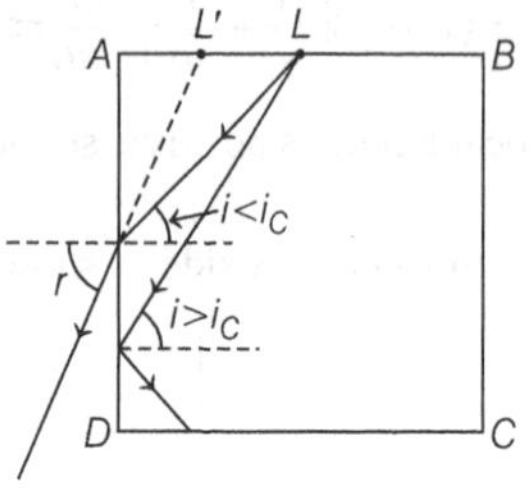


For $\mu = 1.6$, $\sin i_C = \frac{1}{n} = \frac{1}{1.6} = 0.625$

$$\Rightarrow \quad i_C = \sin^{-1}(0.625) = 38.7°$$

So, when angle of incidence becomes greater than $i_C$ $(= 38.7°)$ the rays emanating from $L$ will undergo total internal reflection and the pin shall not be seen at all.

**168** (*c*) The least distance of distinct vision of an average person (i.e. $D$) is 25 cm.

Given, $v = D = 25$, $m = 10$ and $u = f$

But for microscope, the magnification, $m = v/u = D/f$

$$\Rightarrow \quad f = \frac{D}{m} = \frac{25}{10} = 2.5\text{cm} = 0.025\text{ m}$$

As, $$P = \frac{1}{f} = \frac{1}{0.025} = 40\text{ D}$$

This is the required power of lens.

**169** (*a*) Let the apparent depth be $O_1$ for the object seen from medium 2, then $O_1 = \frac{n_2}{n_1}\frac{h}{3}$ ...(i)

$\because$ Apparent depth = Real depth/Refractive index, $n$.

Since, the image formed by medium 1, $O_1$ will act as an object for medium 2.

If seen from medium 3, the apparent depth is $O_1$.

Similarly, the image formed by medium 2, $O_2$ will act as an object for medium 3.

So, real depth = height of liquid column + $O_1$.

$$\Rightarrow \quad O_2 = \frac{n_3}{n_2}\left(\frac{h}{3} + O_1\right) = \frac{n_3}{n_2}\left(\frac{h}{3} + \frac{n_2 h}{n_1 3}\right) \quad \text{[using Eq. (i)]}$$

$$= \frac{h}{3}\left(\frac{n_3}{n_2} + \frac{n_3}{n_1}\right)$$

If seen from outside (air), the apparent height of the dot at the bottom of the beaker is

$$O_3 = \frac{1}{n_3}\left(\frac{h}{3} + O_2\right) = \frac{1}{n_3}\left[\frac{h}{3} + \frac{h}{3}\left(\frac{n_3}{n_2} + \frac{n_3}{n_1}\right)\right]$$

$$= \frac{h}{3}\left(\frac{1}{n_1} + \frac{1}{n_2} + \frac{1}{n_3}\right)$$

This is the required expression for apparent depth.

**170** (*c*) The relationship between refractive index $n$, prism angle $A$ and angle of minimum deviation $D_m$ is given by

$$n = \frac{\sin\left[\frac{(A + D_m)}{2}\right]}{\sin\left(\frac{A}{2}\right)}$$

Here, $D_m = A$

$$\therefore \quad n = \frac{\sin A}{\sin A/2} \Rightarrow = \frac{2\sin\frac{A}{2}\cos\frac{A}{2}}{\sin\frac{A}{2}}$$

[$\because$ using trigonometric identity, $\sin 2\theta = 2\sin\theta\cos\theta$]

$$= 2\cos\frac{A}{2}$$

For the given value of refractive index, $n = \sqrt{3}$

we have,

$$\cos\frac{A}{2} = \frac{\sqrt{3}}{2} \text{ or } \frac{A}{2} = 30°$$

$$\therefore \quad A = 60°$$

This is the required value of prism angle.

**171** (*a*) Given, focal length of the convex lens, $f = +25$ cm.

Since, the lens is cut along $XX'$, 0.5 cm above principal axis $PP'$ as shown below. Let us imagine the lens as if there is no cut. The point object placed at $A(-50, 0)$ can be treated as linear object $AB$, where $u = -50$ cm.

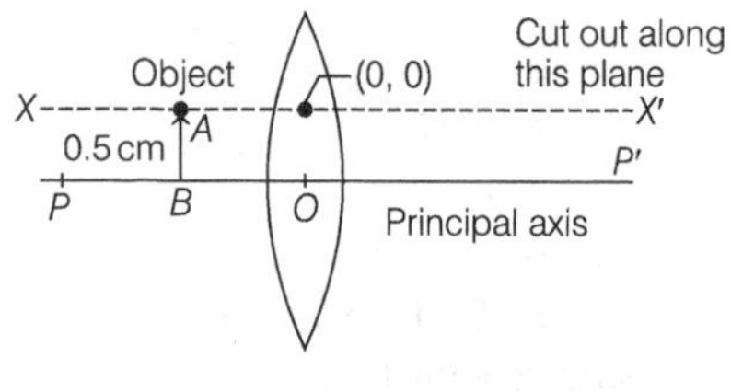


Using thin lens formula, $\frac{1}{v} - \frac{1}{u} = \frac{1}{f}$

$$\Rightarrow \quad v = \frac{uf}{u+f} = \frac{(-50)\times(+25)}{-50+25} = \frac{-1250}{-25} = +50\text{ cm}$$

Also, $$m = \frac{v}{u} = \frac{+50}{-50} = -1$$

Thus, the image of $A$ would have been formed at 50 cm from the pole and 1cm below principal axis.

So, coordinates of image $\equiv$ (+ 50 cm, − 1 cm).

# CHAPTER > 10

# Wave Optics

## KEY NOTES

- Ray optics is based on rectilinear propagation of light and deals with mirrors, lenses, reflection, refraction, etc. while wave optics deals with it that it travels as a wave, can bend around objects, can diffract and interfere, etc.

### Wavefront

- The locus of all those particles which are vibrating in the same phase at any instant is called wavefront. Thus, a wavefront is a surface of constant phase.
- The shape of wavefront depends on the shape of source of light. Wavefront can be of three types
  (i) Spherical wavefront
  (ii) Cylindrical wavefront
  (iii) Plane wavefront

### Huygens' Principle

- It is essentially a geometrical construction, which given the shape of the wavefront at any time allows us to determine the shape of the wavefront at a later time.
- According to Huygens' principle,
  (i) Each point of the wavefront is the source of a secondary disturbance (wavelets) and these wavelets spread out in all directions with the speed of the wave.
  (ii) The envelope of wavefront of secondary wavelets, after a given time, along forward direction gives the new position of wavefront.

### Refraction and Reflection of Plane Waves

- **Refraction of plane waves** is derived using Huygens' principle. This implies that refraction of plane wave follows Snell's law.

  i.e. $n_1 \sin i = n_2 \sin r$

  where, $i$ = angle of incidence of plane wave

  and $r$ = angle of refraction of plane wave.
- Speed and wavelength of plane wave in first medium are $v_1$ and $\lambda_1$ respectively and in second medium are $v_2$ and $\lambda_2$ respectively, then

$$\frac{\lambda_1}{\lambda_2} = \frac{v_1}{v_2}$$

The above equation implies that, when a wave gets refracted into a denser medium ($v_1 > v_2$), the wavelength and the speed of propagation decrease but the frequency $\nu\left(=\frac{v}{\lambda}\right)$ remains the same.

- **Reflection of plane waves** is derived using Huygens' principle. This implies that plane wave follows the law of reflection.
- When angle of incidence $i$ of plane wave (in rarer medium) is greater than critical angle $i_c$, then plane wave will not have any refracted wave and the wave will undergo what is known as **total internal reflection**.

### Doppler Effect

- When the source moves away from the observer, then the frequency as measured by the source will be smaller. This is known as Doppler effect.
- The increase in wavelength due to doppler effect is called **red shift** because wavelength in the middle of the visible region of the spectrum moves towards the red end of the spectrum.
- When waves are received from a source moving towards the observer, there is an apparent decrease in wavelength, this is referred to as **blue shift**.
- The doppler shift can be expressed as $\frac{\Delta\nu}{\nu} = \frac{-v_{\text{radial}}}{c}$

  where, $\frac{\Delta\nu}{\nu}$ is fractional change in frequency and $v_{\text{radial}}$ is the component of the source velocity along the line joining the observer to the source relative to the observer and considered as positive, when the source moves away from the observer.

### Coherent and Incoherent Sources

- **Coherent sources** of light are the sources which emit light waves of same frequency, same wavelength and have a constant initial phase difference.
- Two such sources of light, which do not emit light waves with a constant phase difference are called **incoherent sources**.

## Interference of Light Waves

- The phenomenon of redistribution of energy in the region of superposition of waves is called **interference.**
  The points of maximum intensity in the regions of superposition of waves are said to be in **constructive interference** whereas the points of minimum intensity are said to be in **destructive interference.**
- Two waves of amplitudes $a_1$ and $a_2$ interfere at a point, where phase difference is $\phi$, then resultant amplitude is given by
$$A^2 = a_1^2 + a_2^2 + 2a_1 a_2 \cos \phi$$
  For constructive interference, $A_{max} = (a_1 + a_2)^2$
  For destructive interference, $A_{min} = (a_1 - a_2)^2$
- Resultant intensity, $I = I_1 + I_2 + 2\sqrt{I_1 I_2} \cos \phi$
  When $I_1 = I_2 = I_0$, then resultant intensity,
$$I = I_0 + I_0 + 2I_0 \cos \phi$$
$$= 2I_0 (1 + \cos \phi)$$
$$I = 4I_0 \cos^2 \left(\frac{\phi}{2}\right)$$
  For constructive interference, $\phi = 0, \pm 2\pi, \pm 4\pi \dots$ .
  For destructive interference, $\phi = \pm \pi, \pm 3\pi, \pm 5\pi \dots$ .
- For constructive interference, if initial phase difference is zero, then the interference waves must have
  (i) phase difference $= 2n\pi$,
  where, $n = 0, 1, 2, 3, \dots$ .
  (ii) path difference $= n\lambda$,
  where, $n = 0, 1, 2, 3, \dots$ .
- For destructive interference, assuming initial phase difference $= 0$
  (i) phase difference $= (2n - 1)\pi$
  where, $n = 1, 2, 3, \dots$ .
  (ii) path difference $= \left(n + \frac{1}{2}\right)\frac{\lambda}{2}$
  where, $n = 1, 2, 3, \dots$ .

### Young's Double Slit Experiment

- Young's double slit arrangement for experiment is shown below

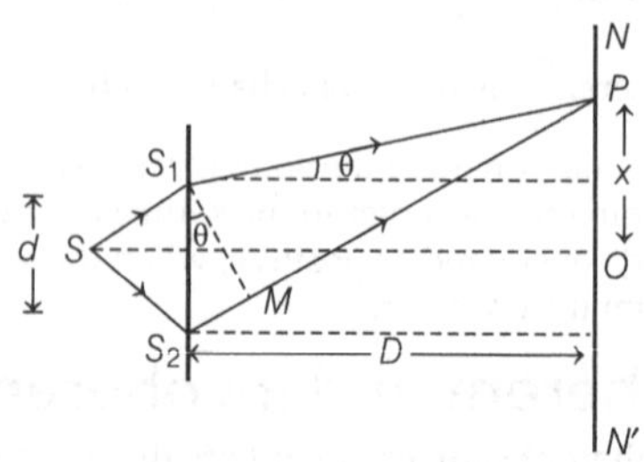


- For light waves reaching at a point $P$, situated at a distance $x$ from central point $O$, the path difference $\Delta y$ is given by
$$\Delta y = S_2P - S_1P = \frac{xd}{D}$$
- If $\frac{xd}{D} = n\lambda$, then we get $n$th **bright fringe**, where $n = 1, 2, 3, 4 \dots$ .
- If $\frac{xd}{D} = (2n - 1)\frac{\lambda}{2}$, then we get $n$th **dark fringe**, where $n = 1, 2, 3, 4 \dots$ .
- **Fringe Width $\beta$** The separation between any two consecutive bright or dark fringes is called fringe width $\beta$ and it is given by
$$\beta = \frac{D\lambda}{d}$$
  where, $\lambda$ = wavelength of wave,
  $D$ = distance between slit and screen
  and $d$ = distance between two slits.
- **Angular Fringe Width $\theta$** This width is expressed as
$$\theta = \frac{\beta}{D} = \frac{\lambda}{d}$$

## Diffraction

- The phenomenon of bending of light around the sharp corners and the spreading of light within the geometrical shadow of the opaque obstacles is called diffraction of light.
  This is shown below

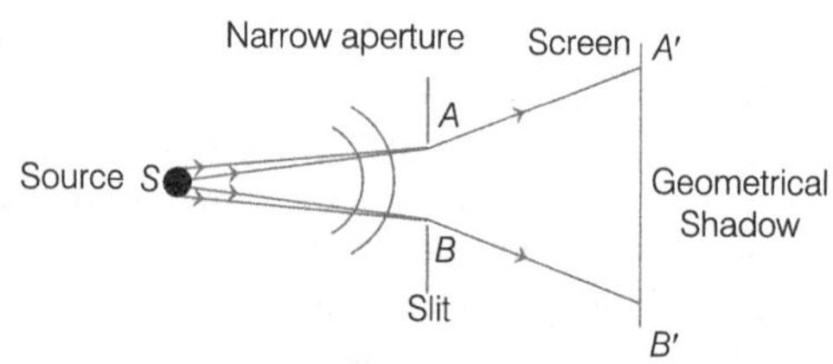


- For most diffraction, dimension of aperture or the obstacle should be comparable to the wavelength of light.
- When the double slit in Young's experiment is replaced by a single narrow slit (illuminated by a monochromatic source), a broad pattern with central bright region is seen. On both sides, there are alternate dark and bright regions of intensity becoming weaker away from the centre.
- **Diffraction due to a Single Slit of Width (*a*)** A parallel beam of light with a plane wavefront $WW'$ is made to fall on a single slit $AB$. Width of the slit is of

the order of wavelength of light, therefore diffraction occurs on passing through the slit, as shown in the diagram given below.

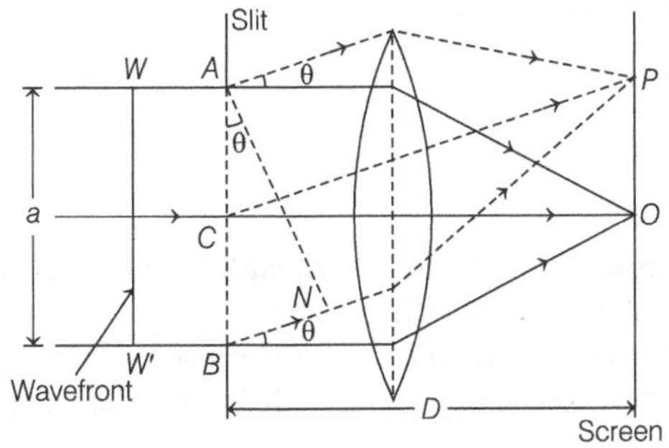


(i) **nth order secondary minima** is obtained when

$a \sin \theta = n\lambda$, where $n = 1, 2, 3, \ldots$.

(ii) **nth order secondary maxima** is obtained when

$a \sin \theta = \left(n + \frac{1}{2}\right) \lambda$, where $n = 0, 1, 2, 3, \ldots$.

(iii) **Angular separation** for $n$th minima,

$\theta_n = \frac{n\lambda}{a}$, where $n = 1, 2, 3, \ldots$.

(iv) **Linear separation** of $n$th secondary minima,

$x_n = \frac{Dn\lambda}{a}$

(v) **Angular position** of $n$th order secondary maxima,

$\theta_n = \left(n + \frac{1}{2}\right) \frac{\lambda}{a}$

(vi) **Intensity of central maxima** is maximum and intensity of secondary maxima decreases with the increase of their order. The diffraction pattern is graphically shown below

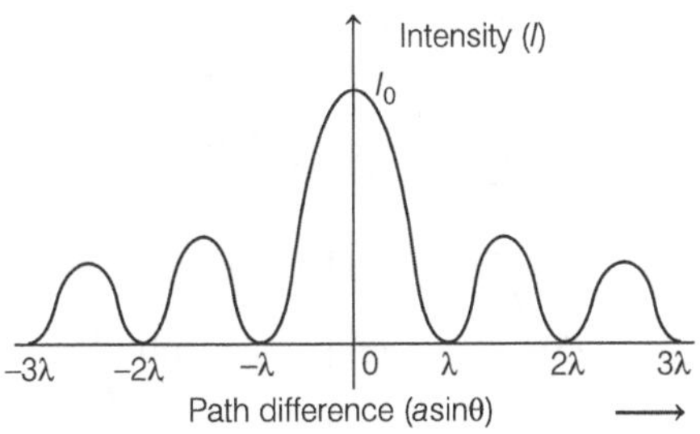


## Resolving Power of Optical Instruments

- It is the ability of the instrument to produce distinctly separate images of two close objects.
- Resolving power of a telescope $= \frac{1}{d\theta} = \frac{D}{1.22\lambda}$

  where, $d\theta$ = angle subtended by the two distinct objects on objective.
- Resolving power of microscope $= \frac{1}{\Delta d} = \frac{2n \sin \beta}{1.22\lambda}$

  where, $n$ = refractive index of objective lens,

  $D$ = diameter of objective lens

  and $\beta$ = half the angle subtended by the diameter of the objective lens at its focus.

The product $n \sin \beta$ is called the **numerical aperture** and is sometimes marked on the objective.

- If $d$ be the width of white stripe (gap) which separates the two regions and $D$ be the distance of screen from eye, then $\frac{d}{D}$ is the limit of resolution of our eye.
- **Fresnel's Distance** The distance at which diffraction spread equal to the size of aperture, $Z_F = \frac{a^2}{\lambda}$

## Polarisation

- The phenomenon of restricting the vibrations of light in a particular direction, perpendicular to the direction of wave motion is called polarisation of light.
- **Malus' Law** According to this law

$$I = I_0 \cos^2 \theta$$

  where, $I_0$ = intensity of plane polarised light,

  $I$ = intensity of transmitted light from the analyser

  and $\theta$ = angle between axis of the polariser and the analyser.
- Thin film of ultramicroscopic crystals used to produce the plane polarised light is known as **polaroid**.
- Polaroids can be used to control the intensity, in sunglasses, windowpanes, photographic cameras and 3D movie cameras, etc.
- The scattered light seen in a direction perpendicular to the direction of incidence is plane polarised. This is called **polarisation by scattering**.
- When a maximum and a mininum of intensity is observed through an analyser without complete darkness, then the wave is called **partially polarised**.

### Polarisation by Reflection

- The angle of incidence at which the reflected light is completely plane polarised is called **polarising angle** or **Brewster's angle** $i_B$.
- **Brewster's Law** According to this law, when unpolarised light is incident at polarising angle $i_B$ on an interface separating air from a medium of refractive index $\mu$, then the reflected light is plane polarised (perpendicular to the plane of incidence), provided $\mu = \tan i_B$.

  At polarising angle, $i_B + r = \frac{\pi}{2}$

  i.e. Reflected plane polarised light is at right angle to the refracted light.

KEY NOTES

# Mastering NCERT

## MULTIPLE CHOICE QUESTIONS

## TOPIC 1 ~ Huygens' Principle

**1** Light propagates rectilinearly, due to
(a) wave nature (b) large wavelength
(c) high velocity (d) high frequency

**2** The direction of wavefront of a wave with the wave motion is
(a) parallel (b) perpendicular
(c) opposite (d) at an angle of $\theta$

**3** If a source is at infinity, then wavefronts reaching to observer are
(a) cylindrical (b) spherical (c) plane (d) conical

**4** According to Huygens' principle, each point of the wavefront is the source of
(a) secondary disturbance (b) primary disturbance
(c) third disturbance (d) fourth disturbance

**5** Huygens' principle of secondary wavelets may be used to
(a) find the velocity of light in vacuum
(b) explain the particle's behaviour of light
(c) find the new position of a wavefront
(d) explain photoelectric effect

## TOPIC 2 ~ Refraction and Reflection of Plane Waves

**6** In given figure, light passes from denser medium 1 to rare medium 2.

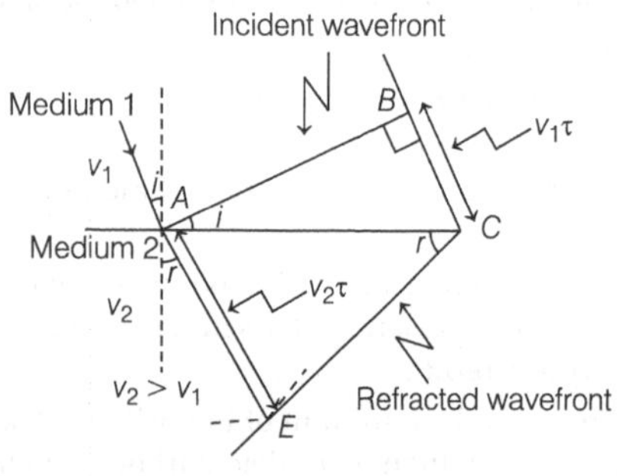


When $i > i_C$ (critical angle of incidence), then wavefront $EC$ is
(a) formed further deep in medium 2
(b) formed closer to surface line $AC$
(c) formed perpendicular to $AC$
(d) formed in medium 1 (on same side of $AB$)

**7** In the given figure, a plane wave $AB$ incident at an angle $i$ on a reflecting surface $MN$.

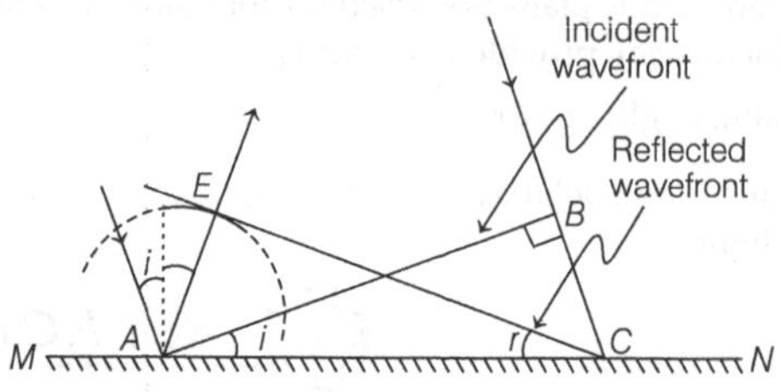


The reflected wavefront is obtained by (take, $v$ be the speed of light and $t$ be the time taken by it from $B$ to $C$)
(a) drawing a sphere of radius $vt$ from point $A$
(b) drawing a sphere of radius $vt$ from point $C$
(c) drawing a perpendicular from point $A$
(d) drawing a perpendicular from point $C$

**8** If $ABC$ is incident wavefront in figure shown below, then refracted wavefront is

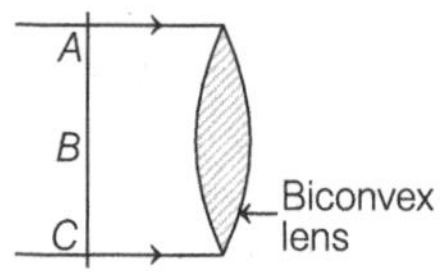


(a) 

(b) 

(c) 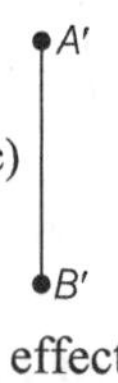

(d) 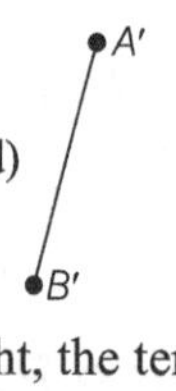


**9** In the context of doppler effect in light, the term red shift signifies
(a) decrease in frequency (b) increase in frequency
(c) decrease in intensity (d) increase in intensity

**10** The source of light is moving towards observer with relative velocity of 3 $kms^{-1}$. The fractional change in frequency of light observed is
(a) $3 \times 10^{-3}$ (b) $3 \times 10^{-5}$
(c) $10^{-5}$ (d) None of these

**11** A planet moves with respect to us, so that light of 475 nm is observed at 475.6 nm. The speed of the planet is
(a) 206 $kms^{-1}$ (b) 378 $kms^{-1}$
(c) 108 $kms^{-1}$ (d) 100 $kms^{-1}$

**12** The earth is moving towards a fixed star with a velocity of 50 $kms^{-1}$. An observer on the earth observes a shift of 0.50Å in the wavelength of light coming from the star. The actual wavelength of light emitted by the star is
(a) 3000 Å (b) 2400 Å
(c) 6000 Å (d) 5800 Å

**13** The wavelength of spectral line coming from a distant star shift from 400 nm to 400.1 nm. The velocity of the star relative to earth is
(a) 75 $kms^{-1}$ (b) 100 $kms^{-1}$ (c) 50 $kms^{-1}$ (d) 200 $kms^{-1}$

## TOPIC 3 ~ *Coherent and Incoherent Addition of Waves*

**14** Two coherent sources of light can be obtained by
(a) two different lamps
(b) two different lamps but of the same power
(c) two different lamps of same power and having the same colours
(d) None of the above

**15** Suppose displacement produced at some point $P$ by a wave is $y_1 = a\cos\omega t$ and by another wave is $y_2 = a\cos\omega t$. Let $I_0$ represents intensity produced by each one of individual wave, then resultant intensity due to overlapping of both waves is
(a) $I_0$ (b) $2I_0$ (c) $\frac{I_0}{2}$ (d) $4I_0$

**16** If the amplitude ratio of two sources producing interference is 3 : 5, then the ratio of intensities produced by each is
(a) 3 : 5 (b) 5 : 3 (c) 9 : 25 (d) 25 : 9

**17** Two waves are emanating from sources $S_1$ and $S_2$ and their displacements are given by
$y_1 = a\cos\omega t$ & $y_2 = a\cos(\omega t - 4\pi)$.
Then, on superimposing they will give rise to
(a) constructive interference
(b) destructive interference
(c) diffraction
(d) polarisation

**18** Two coherent point sources $S_1$ and $S_2$ vibrating in phase emit light of wavelength $\lambda$. The separation between the sources is $2\lambda$. The smallest distance from $S_2$ on a line passing through $S_2$ and perpendicular to $S_1S_2$, where minimum of intensity occurs, is
(a) $\frac{7\lambda}{12}$ (b) $\frac{15\lambda}{4}$ (c) $\frac{\lambda}{2}$ (d) $\frac{3\lambda}{2}$

**19** Two light waves superimposing at the mid-point of the screen are coming from coherent sources of light with phase difference $\pi$ rad. Their amplitudes are 2 cm each. The resultant amplitude at the given point will be
(a) 8 cm (b) 2 cm (c) 4 cm (d) zero

**20** Two identical light waves, propagating in the same direction, have a phase difference $\phi$. Then, after superposing, the intensity of the resultant wave will be proportional to
(a) $\cos\phi$ (b) $\cos\phi/2$ (c) $\cos^2\phi/2$ (d) $\cos^2\phi$

**21** If two incoherent sources each of intensity $I_0$ produce wave, which overlaps at some common points, then resultant intensity obtained is
(a) $4I_0$ (b) $2I_0$
(c) $\frac{I_0}{2}$
(d) dependent on phase difference

## TOPIC 4 ~ *Young's Double Slit Experiment*

**22** In Young's double slit experiment, the phase of two waves were locked by
(a) making two pinholes very close to each other
(b) making two pinholes very far from each other
(c) making one pinhole only
(d) making three pinholes very close to each other

**23** Two sources of light of wavelengths 1500Å and 2500Å are used in Young's double slit experiment simultaneously. Which orders of fringes of two wavelength patterns coincide?
(a) 3rd order of 1st source and 5th order of 2nd source
(b) 7th order of 1st source and 5th order of 2nd source
(c) 5th order of 1st source and 3rd order of 2nd source
(d) 5th order of 1st source and 7th order of 2nd source

**24** In a Young's double slit experiment with slit separation 0.1 mm, one observes a bright fringe at angle $\frac{1}{40}$ rad by using light of wavelength $\lambda_1$. When the light of the wavelength $\lambda_2$ is used a bright fringe is seen at the same angle in the same set up. Given that $\lambda_1$ and $\lambda_2$ are in visible range (380 n-m to 740 n-m), their values are **JEE Main 2019**

(a) 380 n-m, 525 n-m (b) 400 n-m, 500 n-m
(c) 380 n-m, 500 n-m (d) 625 n-m, 500 n-m

**25** Two slits are separated by a distance of 0.5 mm and illuminated with light of $\lambda = 6000$ Å. If the screen is placed 2.5 m from the slits, the distance of the third bright image from the centre will be **JIPMER 2018**

(a) 1.5 mm (b) 3 mm
(c) 6 mm (d) 9 mm

**26** The Young's double slit experiment is performed with blue and green light of wavelengths 4360 Å and 5460 Å, respectively. If $x$ is the distance of 4th maximum from the central one, then

(a) $x(\text{blue}) = x(\text{green})$ (b) $x(\text{blue}) > x(\text{green})$
(c) $x(\text{blue}) < x(\text{green})$ (d) $\frac{x(\text{blue})}{x(\text{green})} = \frac{5460}{4360}$

**27** Young's double slit experiment is performed with sodium (yellow) light of wavelength 589.3 nm and the interference pattern is observed on a screen 100 cm away. The 10th bright fringe has its centre at a distance of 12 mm from the central maximum. The separation between the slits is

(a) 0.49 mm (b) 0.6 mm
(c) 0.7 mm (d) 0.53 mm

**28** In a Young's double slit experiment, Slits are separated by 0.5 mm and the screen is placed 150 cm away. A beam of light consisting of two wavelengths, 650 nm and 520 nm, is used to obtain interference fringes on the screen. The least distance from the common central maximum to the point where the bright fringes due to both the wavelengths coincide, is **JEE Main 2017**

(a) 7.8 mm (b) 9.75 mm
(c) 15.6 mm (d) 1.56 mm

**29** In Young's double slit experiment, the slits are 2 mm apart and are illuminated by photons of two wavelengths $\lambda_1 = 12000$ Å and $\lambda_2 = 10000$ Å. At what minimum distance from the common central bright fringe on the screen, 2 m from the slit, will a bright fringe from one interference pattern coincide with a bright fringe from the other? **NEET 2013**

(a) 8 mm (b) 6 mm
(c) 4 mm (d) 3 mm

**30** The Young's double slit experiment is performed with light of wavelength 6000 Å, wherein 16 fringes occupy a certain region on the screen. If 24 fringes occupy the same region with another light of wavelength $\lambda$, then $\lambda$ is

(a) 6000 Å (b) 4500 Å (c) 5000 Å (d) 4000 Å

**31** In a Young's double slit experiment, the slit separation is 1mm and the screen is 1m from the slit. For a monochromatic light of wavelength 500 nm, the distance of 3rd minima from the central maxima is

(a) 0.50 mm (b) 1.25 mm (c) 1.50 mm (d) 1.75 mm

**32** In Young's double slit experiment, the slit width and the distance of slits from the screen both are doubled. The fringe width will

(a) increases (b) decreases
(c) remain unchanged (d) None of these

**33** The fringe width in a Young's double slit interference pattern is $3.2 \times 10^{-4}$ m, when red light of wavelength 5600Å is used. How much will it change, if blue light of wavelength 4200Å is used?

(a) $8 \times 10^{-4}$ m (b) $0.8 \times 10^{-4}$ m
(c) $4.2 \times 10^{-4}$ m (d) $0.45 \times 10^{-4}$ m

**34** In Young's double slit experiment the fringe width is $1 \times 10^{-4}$ m. If the distance between the slit and screen is doubled and distance between the two slit is reduced to half and wavelength is changed from $6.4 \times 10^{-7}$ m to $4.0 \times 10^{-7}$ m, the value of new fringe width will be

(a) $0.15 \times 10^{-4}$ m (b) $2.0 \times 10^{-4}$ m
(c) $1.25 \times 10^{-4}$ m (d) $2.5 \times 10^{-4}$ m

**35** Distance of 5th dark fringe from centre is 4 mm. If $D = 2$ m and $\lambda = 600$ nm, then distance between slits is

(a) 1.35 mm (b) 2.00 mm **AIIMS 2019**
(c) 3.25 mm (d) 10.35 mm

**36** In Young's double slit experiment, let $S_1$ and $S_2$ be the two slits and $C$ be the centre of the screen. If the $\angle S_1CS_2 = \theta$ and $\lambda$ is the wavelength, then fringe width will be

(a) $\frac{\lambda}{\theta}$ (b) $\lambda\theta$ (c) $\frac{2\lambda}{\theta}$ (d) $\frac{\lambda}{2\theta}$

**37** In Young's double slit experiment, the separation $d$ between the slits is 2 mm, the wavelength $\lambda$ of the light used is 5896 Å and distance $D$ between the screen and slits is 100 cm. It is found that the angular width of the fringes is 0.20°. To increase the fringe angular width to 0.21° (with same $\lambda$ and $D$), the separation between the slits needs to be changed to **NEET 2018**

(a) 2.1 mm (b) 1.9 mm (c) 1.8 mm (d) 1.7 mm

**38** In a two slit experiment with monochromatic light, fringes are obtained on a screen placed at some distance from the slits. If the screen is moved by $5\times10^{-2}$ m towards the slits, the change in fringe width is $3\times10^{-5}$ m. If separation between the slits is $10^{-3}$ m, the wavelength of light used is

(a) 6000Å (b) 5000Å
(c) 3000Å (d) 4500Å

**39** A light of wavelength 500 nm is incident on a Young's double slit. The distance between slit and screen is $D = 1.8$ m and distance between slits is $d = 0.4$ mm. If screen moves with a speed of 4 m/s, then with what speed first maxima will move? ***AIIMS 2019***

(a) 5 mm/s (b) 4 mm/s
(c) 3 mm/s (d) 2 mm/s

**40** In Young's double slit experiment, if source $S$ is shifted by an angle $\phi$ as shown in figure. Then, central bright fringe will be shifted by angle $\phi$ towards

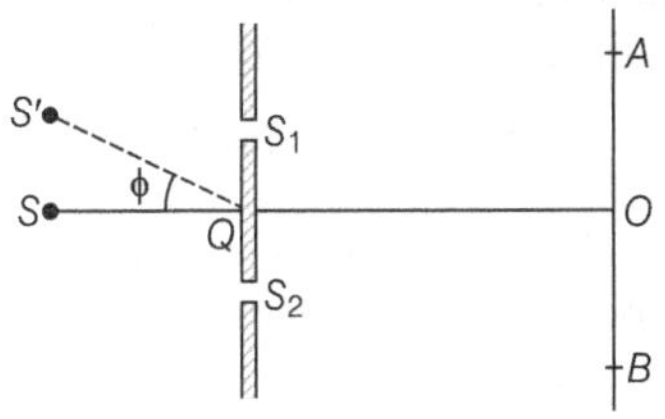


(a) end $A$ of screen
(b) end $B$ of screen
(c) does not shift at all
(d) Either end $A$ and $B$ depending on extra phase difference caused by shifting of source

**41** In Young's double slit experiment two disturbance arriving at a point $P$ have phase difference of $\pi/2$. The intensity of this point expressed as a fraction of maximum intensity $I_0$ is

(a) $\frac{3}{2}I_0$ (b) $\frac{1}{2}I_0$ (c) $\frac{4}{3}I_0$ (d) $\frac{3}{4}I_0$

**42** If the intensities of the two interferring beams in Young's double slit experiment be $I_1$ and $I_2$, then the contrast between the maximum and minimum intensity is good when

(a) $I_1$ is much greater than $I_2$
(b) $I_2$ is smaller than $I_2$
(c) $I_1 = I_2$
(d) Either $I_1 = 0$ or $I_2 = 0$

**43** In Young's double slit experiment, if the point sources are replaced by slits as shown below, then the intensity pattern observed have

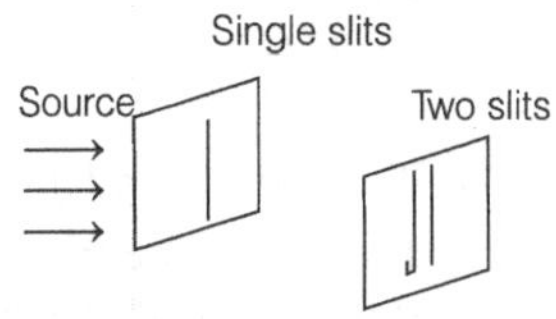


(a) point fringes with decreasing intensity
(b) point fringes with increasing intensity
(c) straight line fringes with decreasing intensities
(d) straight line fringes with increasing intensities

**44** When the double slit in Young's experiment is replaced by a single narrow slit, a broad pattern of alternate dark and bright regions is observed on, either side of the central bright region. As we move from centre on either side, the intensity becomes

(a) larger (b) weaker
(c) remain same (d) zero

# TOPIC 5 ~ *Diffraction*

**45** In a double-slit experiment, green light (5303 Å) falls on a double slit having a separation of 19.44 μ-m and a width of 4.05 μ-m. The number of bright fringes between the first and the second diffraction minima is ***JEE Main 2019***

(a) 5 (b) 10
(c) 9 (d) 4

**46** A parallel beam of light of wavelength 6000 Å gets diffracted by a single slit width 0.3 mm. The angular position of the first minima of diffracted light is

(a) $6\times10^{-3}$ rad (b) $1.8\times10^{-3}$ rad
(c) $3.2\times10^{-3}$ rad (d) $2\times10^{-3}$ rad

**47** In a diffraction pattern due to a single slit of width $a$, the first minimum is observed at an angle 30° when light of wavelength 5000 Å is incident on the slit. The first secondary maximum is observed at an angle of ***NEET 2016***

(a) $\sin^{-1}\left(\frac{2}{3}\right)$ (b) $\sin^{-1}\left(\frac{1}{2}\right)$ (c) $\sin^{-1}\left(\frac{3}{4}\right)$ (d) $\sin^{-1}\left(\frac{1}{4}\right)$

**48** A single slit diffraction pattern is formed with white light. For what wavelength of light, the third secondary maximum in the diffraction pattern coincides with the second secondary maximum in the pattern for red light of wavelength 6500 Å?

(a) 4400 Å (b) 4100Å (c) 4642.8 Å (d) 9100 Å

**49** If we observe the single-slit diffraction with wavelength $\lambda$ and slit width $e$, the width of the central maximum is $2\theta$. On decreasing the slit width for the same $\lambda$, then **CBSE AIPMT 2014**

(a) $\theta$ increases
(b) $\theta$ remains unchanged
(c) $\theta$ decreases
(d) $\theta$ increases or decreases depending on the intensity of light

**50** In a double-slit experiment, at a certain point on the screen the path difference between the two interfering waves is $\frac{1}{8}$ th of a wavelength. The ratio of the intensity of light at that point to that at the centre of a bright fringe is **JEE Main 2020**

(a) 0.568 (b) 0.853
(c) 0.760 (d) 0.672

**51** If a converging lens of focal length $f$ is placed after the slit and the fringes are obtained on screen placed at the focus of lens in single-slit diffraction. Then, the size of screen is

(a) $\lambda / a$ (b) $f\lambda / a$
(c) $a / \lambda$ (d) $f a / \lambda$

**52** In telescope, when a parallel beam of light from distant objects like stars is incident on a convex lens, because of diffraction, it gets focused to a spot. For two stars to be just resolved, the radius of the central bright region ($r_0$) is

(a) $\frac{0.61\lambda f}{a}$ (b) $\frac{0.75\lambda f}{a}$
(c) $\frac{1.94\lambda f}{a}$ (d) $\frac{2.43\lambda f}{a}$

**53** Assume that light of wavelength 6000 Å is coming from a star. What is the limit of resolution of a telescope whose objective has a diameter of 100 inch?

(a) $1.2\times 10^{-5}$ rad (b) $2.9\times 10^{-7}$ rad
(c) $3.5\times 10^{-10}$ rad (d) $2.9\times 10^{-10}$ rad

**54** The diameter of objective lens of a telescope is 6 cm and wavelength of light used is 540 nm. The resolving power of telescope is

(a) $9.1\times 10^{4}$ rad$^{-1}$ (b) $10^{5}$ rad$^{-1}$
(c) $3\times 10^{4}$ rad$^{-1}$ (d) None of these

**55** The aperture of a telescope is 5m. The separation between the moon and the earth is $4\times 10^{5}$ km. With light of wavelength of 5500 Å, the minimum separation between objects on the surface of moon, so that they are just resolved, is close to **JEE Main 2020**

(a) 600 m (b) 20 m (c) 60 m (d) 200 m

**56** An eye can resolve two dark bands separated by white strip of width 0.01 cm. If dark band with white strips are made on the screen at a distance of 2 m from eye, then limit of resolution of eye is

(a) $5\times 10^{-3}$ rad (b) $2\times 10^{2}$ rad
(c) $5\times 10^{-5}$ rad (d) $2\times 10^{-2}$ rad

**57** What will be the ratio of $D/f$ in microscope, where $D$ is the diameter of the aperture and $f$ is the focal length of the objective lens?

(a) $\tan\beta$ (b) $\tan\frac{\beta}{2}$ (c) $2\tan\beta$ (d) $\tan\frac{\beta}{6}$

**58** The ratio of resolving powers of an optical microscope for two wavelengths $\lambda_1 = 4000$ Å and $\lambda_2 = 6000$ Å is **NEET 2017**

(a) 8 : 27 (b) 9 : 4 (c) 3 : 2 (d) 16 : 81

**59** The resolving power of a compound microscope increases, when

(a) refractive index $n$ of the medium is increased keeping wavelength of light $\lambda$ constant (or same)
(b) for the same medium, wavelength of light is decreased
(c) refractive index of the medium and wavelength of light used both are decreased
(d) Both (a) and (b)

**60** For what distance is ray optics a good approximation when the aperture is 3mm wide and wavelength is 500 nm?

(a) 18 m (b) 25 m (c) 30 m (d) 35 m

## TOPIC 6 ~ *Polarisation*

**61** The waves produced by a vibrating string in a plane are

(a) unpolarised wave
(b) plane polarised wave
(c) linearly polarised wave
(d) circularly polarised wave

**62** An unpolarised beam of light of intensity $I_0$ falls on a polaroid. The intensity of the emergent beam is

(a) $\frac{I_0}{2}$ (b) $I_0$
(c) $\frac{I_0}{4}$ (d) zero

**63** A beam of unpolarised light of intensity $I_0$ is passed through a polaroid $A$ and then through another polaroid $B$, which is oriented so that its principal plane makes an angle of 45° relative to that of $A$. The intensity of the emergent light is
(a) $I_0$ (b) $I_0/2$ (c) $I_0/4$ (d) $I_0/8$

**64** At which angle, the intensity of transmitted light is maximum when a polaroid sheet is rotated between two crossed polaroids?
(a) $\pi/4$ (b) $\pi/2$ (c) $\pi/3$ (d) $\pi$

**65** An unpolarised beam intensity $I_0$ is incident on a pair of nicol prism making an angle of 60° with each other. The intensity of light emerging from the pair is
(a) $I_0$ (b) $I_0/2$ (c) $I_0/4$ (d) $I_0/8$

**66** Two polaroids $P_1$ and $P_2$ are placed with their axis perpendicular to each other. Unpolarised light $I_0$ is incident on $P_1$. A third polaroid $P_3$ is kept in between $P_1$ and $P_2$ such that its axis makes an angle 45° with that of $P_1$. The intensity of transmitted light through $P_2$ is **NEET 2017, JEE Main 2018, 13**
(a) $\frac{I_0}{2}$ (b) $\frac{I_0}{4}$ (c) $\frac{I_0}{8}$ (d) $\frac{I_0}{16}$

**67** A system of three polarisers $P_1, P_2, P_3$ is set up such that the pass axis of $P_3$ is crossed with respect to that of $P_1$. The pass axis of $P_2$ is inclined at 60° to the pass axis of $P_3$. When a beam of unpolarised light of intensity$I_0$ is incident on $P_1$, the intensity of light transmitted by the three polarisers is $I$. The ratio $(I_0/I)$ equals (nearly) **JEE Main 2019**
(a) 5.33 (b) 16.00
(c) 10.67 (d) 1.80

**68** Which of the following can be used to control the intensity, in sunglasses, window-panes, etc?
(a) Transverse wave (b) Polaroids
(c) Plane polarised wave (d) Polarised wave

**69** The unpolarised incident sunlight is shown below. The dots stand for polarisation perpendicular to the plane of the figure and the double arrows show polarisation in the plane of the figure. If an observer looks at 90° to the direction of the sun, then the radiations scattered by the air molecules are represented by

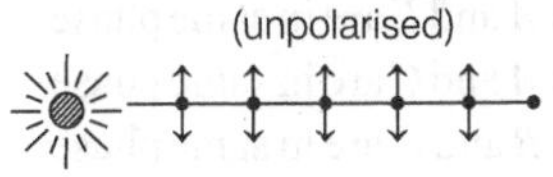


(a) dots only
(b) double arrow parallel to incident radiation
(c) double arrow perpendicular to incident radiation
(d) remains are same

**70** Unpolarised light is incident from air on a plane surface of a material of refractive index $\mu$. At a particular angle of incidence $i$, it is found that the reflected and refracted rays are perpendicular to each other. Which of the following option is correct for this situation? **NEET 2018**
(a) $i = \sin^{-1}\left(\frac{1}{\mu}\right)$
(b) Reflected light is polarised with its electric vector perpendicular to the plane of incidence
(c) Reflected light is polarised with its electric vector parallel to the plane of incidence
(d) $i = \tan^{-1}\left(\frac{1}{\mu}\right)$

**71** When unpolarised light beam is incident from air to glass ($\mu = 1.5$) at the polarising angle, then
(a) reflected beam is 100% polarised
(b) reflected and refracted beams are partially polarised
(c) the reason for (a) is that almost all the light is reflected
(d) All of the above

**72** Unpolarised light is incident on a plane glass surface. What should be the angle of incidence, so that the reflected and refracted rays are perpendicular to each other? [$\mu$ for glass $= 1.5$]
(a) 60° (b) 90° (c) 0° (d) 57°

**73** The refractive index of a medium is 1. If the unpolarised light is incident on it at the polarising angle of the medium, the angle of refraction is
(a) 60° (b) 45° (c) 30° (d) 0°

**74** The critical angle of a certain medium is $\sin^{-1}\left(\frac{4}{5}\right)$. The polarising angle of medium is
(a) $\tan^{-1}\left(\frac{5}{4}\right)$ (b) $\sin^{-1}\left(\frac{4}{5}\right)$
(c) $\sin^{-1}\left(\frac{5}{4}\right)$ (d) $\tan^{-1}\left(\frac{4}{3}\right)$

**75** When the angle of incidence is 45° on the surface of a glass slab, it is found that the reflected ray is completely polarised. The velocity is light in glass is
(a) $\sqrt{3} \times 10^8 \text{ ms}^{-1}$ (b) $3 \times 10^8 \text{ ms}^{-1}$
(c) $2 \times 10^8 \text{ ms}^{-1}$ (d) $\sqrt{2} \times 10^8 \text{ ms}^{-1}$

**76** When both the components of electric field of light waves are present such that one is stronger than the other and such light is viewed through a rotating analyser, one sees a maximum and a minimum of intensity but not complete darkness. This kind of light is called
(a) polarised (b) linearly polarised
(c) partially polarised (d) None of these

# SPECIAL TYPES QUESTIONS

## I. Assertion and Reason

**Direction** (Q. Nos. 77-86) *In the following questions, a statement of Assertion is followed by a corresponding statement of Reason. Of the following statements, choose the correct one.*

(a) Both Assertion and Reason are correct and Reason is the correct explanation of Assertion.
(b) Both Assertion and Reason are correct but Reason is not the correct explanation of Assertion.
(c) Assertion is correct but Reason is incorrect.
(d) Assertion is incorrect but Reason is correct.

**77 Assertion** In the field of geometrical optics, light can be assumed to approximately travel in straight line.

**Reason** The wavelength of visible light is very small in comparison to the dimensions of typical mirrors and lenses.

**78 Assertion** When monochromatic light is incident on a surface separating two media, the reflected and refracted light both have the same frequency as the incident frequency.

**Reason** Speed of light and wavelength of light both changes in refraction and hence the ratio $\nu = c/\lambda$ is a constant.

**79 Assertion** If we have a point source emitting waves uniformly in all directions, the locus of point which have the same amplitude and vibrate in the same phase are spheres.

**Reason** Each point of the wavefront is the source of a secondary disturbance and the wavelets emanating from these points spread out in all directions with the speed of the wave.

**80 Assertion** The emergent plane wavefront is tilted on refraction of a plane wave by a thin prism.

**Reason** The speed of light waves is more in glass and the base of the prism is thicker than the top.

**81 Assertion** Speed of light is independent of its colour only in vacuum.

**Reason** Red colour light travels slower than violet colour light in glass.

**82 Assertion** Increase in the wavelength of light due to doppler's effect is red shift.

**Reason** When the wavelength increases, then wavelength in the middle of the visible region of the spectrum moves towards the red end of the spectrum.

**83 Assertion** No interference pattern is detected when two coherent sources are infinitely close to each other.

**Reason** The fringe width is inversely proportional to the distance between the two slits.

**84 Assertion** The intensity at the bright band on the screen is maximum and equal to $4I_0$, where $I_0$ is the intensity of light from each sources.

**Reason** The intensity at the dark band is always zero irrespective of the intensity of light waves coming from the two sources.

**85 Assertion** In Young's double slit experiment, the fringe width for dark fringes is same as that for white fringes.

**Reason** In Young's double slit experiment, when the fringes are formed with a source of white light, then only dark and bright fringes are observed.

**86 Assertion** Diffraction determines the limitations of the concepts of light rays.

**Reason** A beam of width $a$ starts to spread out due to diffraction after it has travelled a distance $(2a^2/\lambda)$.

## II. Statement Based Questions

**87** I. A point source emits waves uniformly in all directions.
II. In spherical wave, the locus of point which have the same amplitude and vibrate in same phase are spheres.
III. At a small distance from the source, a small portion of sphere can be considered as plane wave.

Which of the following statement(s) is/are correct?

(a) Only I (b) Both I and II (c) Only III (d) I, II and III

**88** In case of reflection of a wavefront from a reflecting surface as shown below.

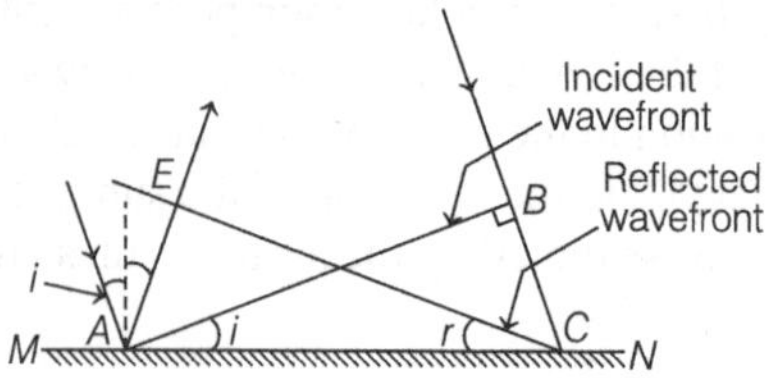


I. Points $A$ and $E$ are in same phase.
II. Points $A$ and $C$ are in same phase.
III. Points $B$ and $A$ are in same phase.
IV. Points $C$ and $E$ are in same phase.

Which of the following statement(s) is/are correct?

(a) Both I and II (b) Both II and III
(c) Both III and IV (d) Both I and IV

**89** I. Two coherent sources emit light waves of same wavelength.
II. Two coherent sources emit light waves of same frequency.
III. Two coherent sources have zero or constant initial phase difference with respect to time.

Which of the following statement(s) is/are correct for coherent sources?

(a) Only I (b) Both I and III
(c) Both II and III (d) I, II and III

**90** I. The interference pattern has a number of equally spaced bright and dark bands, while the diffraction pattern has a central bright maximum, which is twice as wide as the other maxima.
II. The interference pattern is superposition of two waves originating from the two narrow slits, while the diffraction pattern is a superposition of a continuous family of waves originating from each point on a single slit.
III. For a single slit of width $a$, the first null of the interference pattern occurs at an angle of $\lambda / a$ while at the same angle of $\lambda / a$, a maximum is obtained (not a null) for two narrow slits separated by a distance $a$.

Which of the following statement(s) is/are correct?

(a) Both I and II (b) Both I and III
(c) I, II and III (d) None of these

**91** I. For diffraction through a single slit, the wavelength of wave must be comparable to the size of slit.
II. The diffraction is very common in sound waves but not so common in light waves.
III. Diffraction is only observed in electromagnetic waves.

Which of the given statement(s) is/are correct for phenomenon of diffraction?

(a) Only I (b) Both II and III
(c) Both I and II (d) I, II and III

**92** I. Dots and arrows indicates that both polarisations are present in the incident and refracted waves.
II. The reflected light is not linearly polarised.
III. Transmitted intensity will be zero when the axis of the analyser is in the plane of the figure, i.e. the plane of incidence.

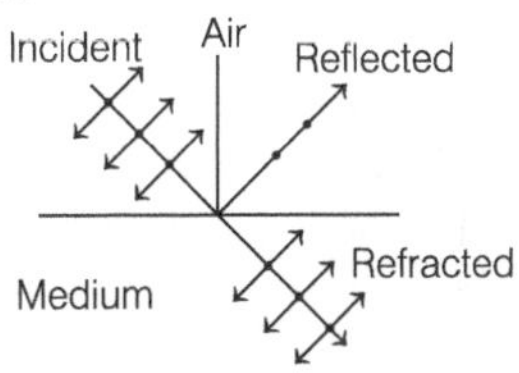


Which of the following statement(s) is/are correct with reference to the figure given?

(a) Only I (b) Only II
(c) Both I and III (d) Both I and II

**93** Let $S_1$ and $S_2$ are two sources and if wave from $S_1$ reaches some common point $P$ be covering seven time of wavelength $\lambda$ and from $S_2$ by covering nine times of wavelength $\lambda$.

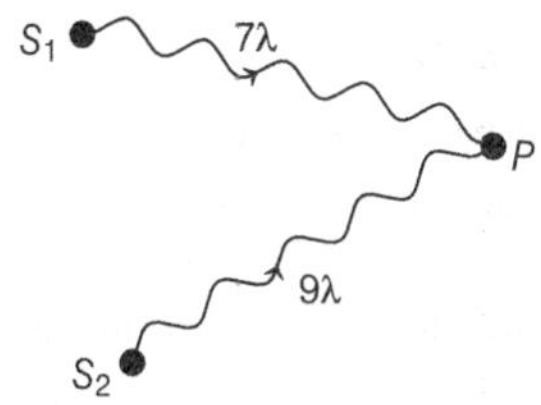


Which of the following statement(s) is/are incorrect?

(a) $S_2P - S_1P = 2\lambda$
(b) Waves from $S_1$ arrives exactly two cycles earlier than waves from $S_2$.
(c) At point $P$, waves from $S_1$ and $S_2$ are in phase.
(d) At point $P$, waves from $S_1$ and $S_2$ are out of phase.

**94** Which of the following statement is incorrect conditions for producing sustained interference?

(a) Phase difference between interferring waves remains constant with time.
(b) Interferring waves have nearly same amplitude levels.
(c) Interferring waves are of same frequency.
(d) Interferring waves are moving only in opposite directions.

**95** Two sources $S_1$ and $S_2$ of intensity $I_1$ and $I_2$ are placed infront of a screen Fig. (a). The pattern of intensity distribution seen in the central portion is given by Fig. (b).

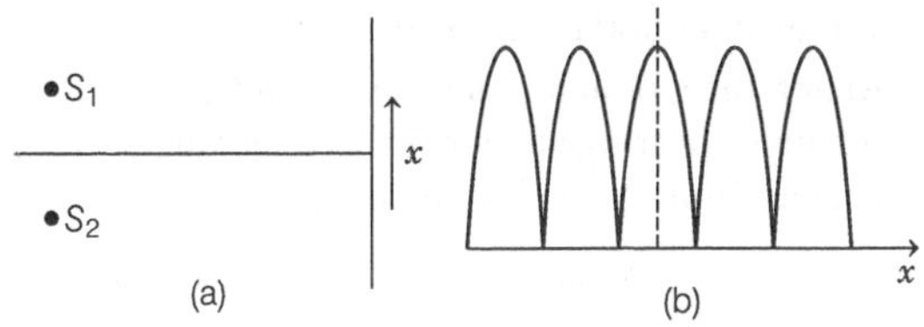


In this case, which of the following statements are incorrect?

(a) $S_1$ and $S_2$ have the same intensities.
(b) $S_1$ and $S_2$ have a constant phase difference.
(c) $S_1$ and $S_2$ have the same phase.
(d) $S_1$ and $S_2$ have the same wavelength.

**96** Sunlight is incident on a pinhole of width $10^3$ Å. Choose the correct statement among the following.

(a) The image of the pinhole seen on a screen shall be a sharp white ring.
(b) The image of the pinhole seen on a screen shall be different from a geometrical image.
(c) The image of the pinhole seen on a screen shall be a diffused central spot with white in colour.
(d) None of the above

## III. Matching Type

**97** Match the Column I (light source) with Column II (shape of wavefront) and select the correct answer from the codes given below.

| | Column I | | Column II |
|---|---|---|---|
| A. | Light from a point source | 1. | Plane |
| B. | Light emerging out of convex lens when point source is of its focus | 2. | Spherical |
| C. | Light emerging from a thin prism | 3. | Plane and tilted |

| | A | B | C | | A | B | C |
|---|---|---|---|---|---|---|---|
| (a) | 1 | 2 | 3. | (b) | 1 | 3 | 2 |
| (c) | 2 | 1 | 3 | (d) | 2 | 3 | 1 |

**98** Match the Column I (interference) with Column II (relation) and select the correct answer from the codes given below.

| | Column I | | Column II |
|---|---|---|---|
| A. | Constructive interference | 1. | $n\lambda$ |
| B. | Destructive interference | 2. | $(2n+1)\lambda/2$ |
| C. | Path difference for constructive interference | 3. | Waves are in phase at point of interference |
| D. | Path difference for destructive interference | 4. | Waves are out of phase at point of interference |

| | A | B | C | D | | A | B | C | D |
|---|---|---|---|---|---|---|---|---|---|
| (a) | 4 | 1 | 3 | 2 | (b) | 4 | 3 | 1 | 2 |
| (c) | 3 | 4 | 1 | 2 | (d) | 3 | 4 | 2 | 1 |

**99** Two slits are made one millimetre apart and the screen is placed 1 m away. A blue-green light of wavelength 500 nm is used.

Match the Column I (activity) with Column II (changes in fringe pattern) and select the correct answer from the codes given below.

| | Column I | | Column II |
|---|---|---|---|
| A. | Screen is moved away from the plane of slits | 1. | Linear width of fringes $\beta$ increases |
| B. | Source is replaced by another source of shorter wavelength | 2. | Angular separation of fringe remains constant |
| C. | Set up of experiment is dipped completely in water | 3. | Fringe separation decreases |
| D. | Distance between slits is reduced | 4. | Fringe width become (3/4)th |

| | A | B | C | D | | A | B | C | D |
|---|---|---|---|---|---|---|---|---|---|
| (a) | 1 | 3 | 4 | 2 | (b) | 3 | 1 | 4 | 2 |
| (c) | 2 | 3 | 4 | 1 | (d) | 2 | 3 | 1 | 4 |

**100** Consider the arrangement shown in the figure. The distance $D$ is large compared to the separation $d$ between the slits. For this arrangement, match the Column I (characteristics) with Column II (value) and select the correct answer from the codes given below.

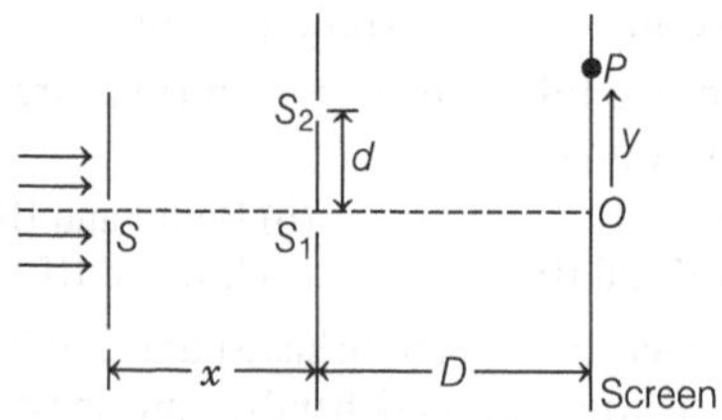


| | Column I | | Column II |
|---|---|---|---|
| A. | The minimum value of $d$, so that there is a dark fringe at $O$ for $x=D$ is | 1. | $\sqrt{\frac{\lambda D}{3}}$ |
| B. | For $x=D$ and $d$ minimum, such that there is dark fringe at $O$, the distance $y$ at which next bright fringe is located is | 2. | $2d$ |
| C. | Fringe width for $x=D$ | 3. | $d$ |
| D. | The minimum value of $d$ so that there is a dark fringe at $O$ for $x = D/2$ is | 4. | $\sqrt{\frac{\lambda D}{2}}$ |

| | A | B | C | D |
|---|---|---|---|---|
| (a) | 4 | 3 | 2 | 1 |
| (b) | 1 | 2 | 3 | 4 |
| (c) | 1 | 3 | 2 | 4 |
| (d) | 4 | 2 | 3 | 1 |

**101** Match the Column I (change in YDSE) with Column II (change in fringes) and select the correct answer from the codes given below.

| | Column I | | Column II |
|---|---|---|---|
| A. | If sodium light is replaced by red light of same intensity | 1. | All fringes are coloured except central fringe |
| B. | Monochromatic light is replaced by white light | 2. | Fringe width will became quadrupled |
| C. | Distance between slits and screen is doubled and the distance betwen slits is halved | 3. | The bright fringe will became less bright |
| D. | If one of the slits is covered by cellophone paper | 4. | The fringe width will increase |

| | A | B | C | D |
|---|---|---|---|---|
| (a) | 4 | 1 | 2 | 3 |
| (b) | 1 | 2 | 3 | 4 |
| (c) | 2 | 3 | 1 | 4 |
| (d) | 3 | 2 | 4 | 1 |

# NCERT & NCERT Exemplar

## MULTIPLE CHOICE QUESTIONS

### NCERT

**102** Monochromatic light of wavelength 589 nm is incident from air on a water surface. What are the wavelength, frequency and speed of

(i) reflected and

(ii) refracted light?(Refractive index of water is 1.33).

(a) Reflected light - $589\times10^{-9}$ m, $5.09\times10^{14}$ Hz, $3\times10^{8}$ ms$^{-1}$
Refracted light - $4.42\times10^{-7}$ m, $5.09\times10^{14}$ Hz, $2.25\times10^{8}$ ms$^{-1}$

(b) Reflected light - $475\times10^{-9}$ m, $509\times10^{-14}$ Hz, $2\times10^{5}$ ms$^{-1}$.
Refracted light - $5\times10^{-5}$ m, $2.09\times10^{14}$, $3\times10^{8}$ ms$^{-1}$

(c) Reflected light - 1m, 1ms$^{-1}$
Refracted light - 1m, 2Hz, $3\times10^{6}$ ms$^{-1}$

(d) None of the above

**103** In a Young's double slit experiment, the slits are separated by 0.28 mm and screen is placed 1.4 m away. The distance between the central bright fringe and the fourth bright fringe is measured to be 1.2 cm.

Determine the wavelength of light used in the experiment.

(a) $5\times10^{-7}$ m (b) $6\times10^{-7}$ m

(c) $0.05\times10^{-7}$ m (d) $0.06\times10^{-7}$ m

**104** A beam of light consisting of two wavelengths 650 nm and 520 nm is used to obtain interference fringes in a Young's double slit experiment with slit width = 2 mm and distance of screen = 1.2 m.

(i) Find the distance of the third bright fringe on the screen from the central maximum for wavelength 650 nm.

(ii) What is the least distance from the central maximum where the bright fringes due to both the wavelength coincide?

(a) $1.17\times10^{-3}$ m, $1.56\times10^{-3}$ m

(b) $2.25\times10^{-2}$ m, 1.25 m

(c) $0.05\times10^{-2}$ m, $1.3\times10^{-4}$ m

(d) None of the above

**105** Light of wavelength 5000Å falls on a plane reflecting surface. What are the wavelength and frequency of the reflected light?

For what angle of incidence is the reflected ray normal to the incident ray?

(a) $\lambda$=3000Å, $\nu=5\times10^{15}$ Hz, $\angle i=45°$

(b) $\lambda$=5000Å, $\nu=6\times10^{14}$ Hz, $\angle i=45°$

(c) $\lambda$=8800Å, $\nu=5\times10^{15}$ Hz, $\angle i=60°$

(d) None of the above

**106** The 6563 Å $H_\alpha$-line emitted by hydrogen in a star is found to be red shifted by 15Å. Estimate the speed with which the star is receding from the earth.

(a) $-5.04\times10^{2}$ ms$^{-1}$ (b) $-6.86\times10^{5}$ ms$^{-1}$

(c) $5.84\times10^{2}$ ms$^{-1}$ (d) $8.8\times10^{3}$ ms$^{-1}$

**107** In Young's double slit experiment using light of wavelength 600 nm, the angular width of a fringe formed on a distant screen is 0.1°. What is the spacing between the two slits?

(a) 1 m (b) $1.5\times10^{-2}$ m

(c) $3.44\times10^{-4}$ m (d) $0.05\times10^{-2}$ m

**108** Two towers on top of two hills are 40 km apart. The line joining them passes 50 m above a hill half-way between the towers. What is the longest wavelength of radiowaves which can be sent between the towers without appreciable diffraction effects?

(a) 0.125 m (b) 2.5 m (c) 0.05 m (d) 0 m

**109** A parallel beam of light of wavelength 500 nm falls on a narrow slit and the resulting diffraction pattern is observed on a screen 1m away. It is observed that the first minimum is at a distance of 2.5 mm from the centre of the screen. Find the width of the slit.

(a) 2 mm (b) 1 mm (c) 0.2 mm (d) 0.1 mm

### NCERT Exemplar

**110** Consider sunlight incident on a slit of width $10^4$ Å. The image seen through the slit shall

(a) be a fine sharp slit white in colour at the centre

(b) a bright slit white at the centre diffusing to zero intensities at the edges.

(c) a bright slit white at the centre diffusing to regions of different colours

(d) only be a diffused slit white in colour

**111** In a Young's double slit experiment, the source is white light. One of the holes is covered by a red filter and another by a blue filter. In this case,

(a) there shall be alternate interference patterns of red and blue

(b) there shall be an interference pattern for red distinct from that for blue

(c) there shall be no distinct interference fringes

(d) there shall be an interference pattern for red mixing with one for blue

**112** Figure shows a standard two slits arrangement with slits $S_1, S_2$. $P_1$ and $P_2$ are the two minima points on either side of $P$ as shown in figure.

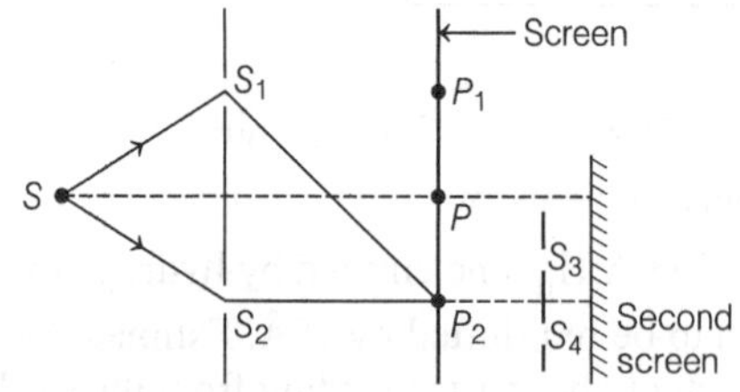


At $P_2$ on the screen, there is a hole and behind $P_2$ is a second 2-slit arrangement with slits $S_3$, $S_4$ and a second screen behind them. What pattern would you find on second screen?

(a) There would be no interference pattern on the second screen but it would be lighted
(b) The second screen would be totally dark
(c) There would be a single bright point on the second screen
(d) There would be a regular two slit pattern on the second screen

**113** The human eye has an approximate angular resolution of $\phi = 5.8 \times 10^{-4}$ rad and a typical photoprinter prints a minimum of 300 dpi (dots per inch, 1 inch = 2.54 cm). At what minimum distance $z$ should a printed page be held so that one does not see the individual dots?

(a) 14.5 cm (b) 12.5 cm (c) 19.8 cm (d) 10.25 cm

# Answers

## > Mastering NCERT with MCQs

| | | | | | | | | | |
|---|---|---|---|---|---|---|---|---|---|
| 1 (a) | 2 (b) | 3 (c) | 4 (a) | 5 (c) | 6 (d) | 7 (a) | 8 (b) | 9 (a) | 10 (c) |
| 11 (b) | 12 (a) | 13 (a) | 14 (d) | 15 (d) | 16 (c) | 17 (a) | 18 (a) | 19 (d) | 20 (c) |
| 21 (b) | 22 (a) | 23 (c) | 24 (d) | 25 (d) | 26 (c) | 27 (a) | 28 (a) | 29 (b) | 30 (d) |
| 31 (b) | 32 (c) | 33 (b) | 34 (d) | 35 (a) | 36 (a) | 37 (b) | 38 (a) | 39 (a) | 40 (b) |
| 41 (b) | 42 (c) | 43 (d) | 44 (b) | 45 (a) | 46 (d) | 47 (c) | 48 (c) | 49 (a) | 50 (b) |
| 51 (b) | 52 (a) | 53 (b) | 54 (a) | 55 (c) | 56 (a) | 57 (c) | 58 (c) | 59 (d) | 60 (a) |
| 61 (b) | 62 (a) | 63 (c) | 64 (a) | 65 (d) | 66 (c) | 67 (c) | 68 (b) | 69 (a) | 70 (b) |
| 71 (a) | 72 (d) | 73 (b) | 74 (a) | 75 (b) | 76 (c) | | | | |

## > Special Types Questions

| | | | | | | | | | |
|---|---|---|---|---|---|---|---|---|---|
| 77 (a) | 78 (b) | 79 (a) | 80 (c) | 81 (c) | 82 (a) | 83 (a) | 84 (c) | 85 (c) | 86 (c) |
| 87 (b) | 88 (c) | 89 (d) | 90 (c) | 91 (c) | 92 (c) | 93 (d) | 94 (d) | 95 (c) | 96 (b) |
| 97 (c) | 98 (c) | 99 (c) | 100 (a) | 101 (a) | | | | | |

## > NCERT & NCERT Exemplar MCQs

| | | | | | | | | | |
|---|---|---|---|---|---|---|---|---|---|
| 102 (a) | 103 (b) | 104 (a) | 105 (b) | 106 (b) | 107 (c) | 108 (a) | 109 (c) | 110 (a) | 111 (c) |
| 112 (d) | 113 (a) | | | | | | | | |

# Hints & Explanations

**3** (*c*) Rays reaching from a source which is at infinity are parallel and when we draw a surface perpendicular to each ray, we get a plane wavefront as shown.

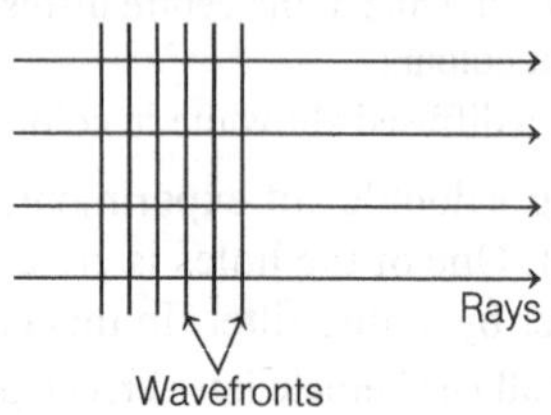


**5** (*c*) Every point on a given wavefront act as a secondary source of light and emits secondary wavelets which travels in all directions with the speed of light in the medium. A surface touching all these secondary wavelets tangentially in the forward direction, gives new wavefront (new position of wavefront) at that instant of time.

**6** (*d*) As we know, for given figure from Snell's law

$$\frac{\sin i}{\sin r} = \frac{n_2}{n_1} \quad \text{...(i)}$$

where, $n_1, n_2$ = refractive indices of medium 1 and medium 2, respectively.

We define critical angle $i_c$ by the formula,

$$\sin i_c = \frac{n_2}{n_1} \quad \text{...(ii)}$$

Thus, if $i = i_c$, then from Eqs. (i) and (ii) $\sin r = 1$ and $r = 90°$. Obviously for $i > i_C$, there cannot be any refracted wave. So, for all angles of incidence greater than the critical angle, we will not have any wavefront in medium 2. Hence, it is formed in medium 1
(on same side of $AB$).

**7** (a) A plane wave $AB$ incident at an angle $i$ on a reflecting surface $MN$, then $t$ be the time taken by the wavefront to advance from point $B$ to $C$. So, from the given figure the distance

$$BC = vt$$

where, $v$ is the speed of light.

For obtaining reflected wavefront, a sphere of radius $vt$ should be drawn from the point $A$ (see figure given in question). The tangent $CE$ drawn on this sphere represents the reflected wavefront of $AB$.

**8** (b) When $ABC$ wavefront passes through glass lens, its velocity is reduced.

$$v_{glass} = \frac{v_{air}}{n_{ga}}$$

where $v_{glass}$ & $v_{air}$ are velocities in glass & air respectively and $n_{ga}$ is refractive index from air to glass.

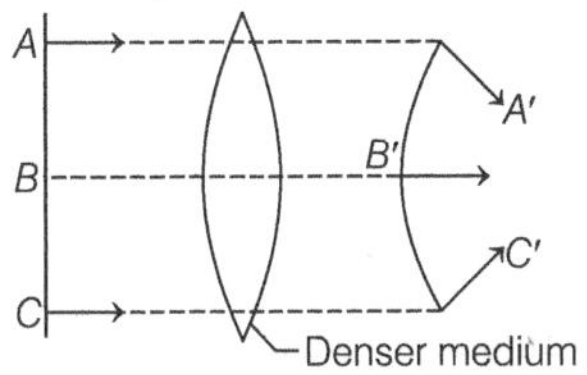


At points $A$ and $C$ waves remain in glass for a short duration, so they travel a larger distance while wave at $B$ covers a small distance as it remains in glass for a longer duration (middle portion of glass is thick) and finally $A'B'C'$ is the position of new wavefront. It is concave in shape as shown in figure.

**9** (a) In doppler effect when source moves away from the observer, frequency observed is smaller than that emitted from the source.

The increase in wavelength (decrease in frequency) due to this effect is known as red shift. Therefore, red shift signifies decrease in frequency.

**10** (c) Given, relative velocity, $\Delta v = 3\text{kms}^{-1} = 3\times 10^3\ \text{ms}^{-1}$

Doppler's shift is given by

$$\frac{\Delta v}{v} = \frac{v_{radial}}{c} \Rightarrow \frac{\Delta v}{v} = \frac{3\times 10^3}{3\times 10^8} = 10^{-5}$$

$$[\because v = 3\times 10^8\ \text{ms}^{-1}]$$

**11** (b) Given, wavelengths of light are 475nm and 475.6nm.

Now, the relation between $v$, $c$ and $\lambda$ is $v\lambda = c$

For small changes in $v$ and $\lambda$,

$$\frac{\Delta v}{v} = \frac{-\Delta\lambda}{\lambda} = \frac{-v_{radial}}{c}$$

$$\left[\because \text{Doppler shift, } \frac{\Delta v}{v} = \frac{-v_{radial}}{c}\right]$$

As, $\Delta\lambda = 475.6 - 475 = +0.6\,\text{nm}$

$$\Rightarrow \quad v_{radial} = c\left(\frac{\Delta\lambda}{\lambda}\right) = c\left(\frac{0.6}{475}\right) = \frac{0.6}{475}\times 3\times 10^8$$

$$= 3.78\times 10^5\ \text{ms}^{-1} = 378\ \text{kms}^{-1}$$

**12** (a) Given, velocity of earth,

$$v = 50\text{kms}^{-1} = 50\times 10^3\ \text{ms}^{-1}$$

Speed of light, $c = 3\times 10^8\ \text{ms}^{-1}$

Wavelength, $\Delta\lambda = 0.50\text{Å}$

Using, $\Delta\lambda = \frac{v}{c}\lambda$

$\therefore$ Wavelength, $\lambda = \frac{c}{v}\Delta\lambda = \frac{3\times 10^8}{50\times 10^3}\times 0.50 = 3000\text{Å}$

**13** (a) Given, wavelengths, $\lambda = 400$ nm and 400.1nm

$$\therefore \quad \Delta\lambda = 400.1\ \text{nm} - 400\ \text{nm} = 0.1\ \text{nm}$$

As, $\frac{v_s}{c} = \frac{\Delta\lambda}{\lambda} \Rightarrow v_s = \frac{\Delta\lambda}{\lambda}c = \frac{0.1}{400}\times 3\times 10^8$

$$= 75\times 10^3\ \text{ms}^{-1} = 75\,\text{kms}^{-1}$$

**14** (d) Two sources are said to be coherent, if they have exactly same frequency and have zero or constant phase difference. Two coherent source cannot be obtained from two different lamps (light source).

**15** (d) The displacement produced by the source $S_1$ at the point $P$ is given by $y_1 = a\cos\omega t$

The displacement produced by the sources $S_2$ (at the point $P$) is given by $y_2 = a\cos\omega t$

Thus, the resultant of displacement at point $P$ would be given by

$$y = y_1 + y_2 = 2a\cos\omega t$$

where, $2a$ is amplitude of wave.

Since, the intensity is proportional to the square of the amplitude,

i.e. intensity, $I \propto (2a)^2$

$$I = 4ka^2$$

where, $I_0 = ka^2$

$$I = 4I_0$$

where, $I_0$ represents the intensity produced by each one of the individual waves.

**16** (c) Given, ratio of amplitudes, $\frac{a_1}{a_2} = \frac{3}{5}$

Since, the intensity is proportional to the square of amplitude.

i.e. $I \propto a^2$

$$\therefore \quad \frac{I_1}{I_2} = \left(\frac{a_1}{a_2}\right)^2 = \left(\frac{3}{5}\right)^2 = \frac{9}{25} \text{ or } 9:25$$

**17** (a) Given, the displacement produced by source $S_1$,

$$y_1 = a\cos\omega t$$

The displacement produced by source $S_2$,

$$y_2 = a\cos(\omega t - 4\pi) = a\cos\omega t$$

As two displacements are in phase, so on superimposing they will give rise to constructive interference.

**18** (a) Given, separation between sources $S_1$ and $S_2 = 2\lambda$.

Two coherent point sources $S_1$ and $S_2$ vibrating in

phase having separation $2\lambda$ are as shown

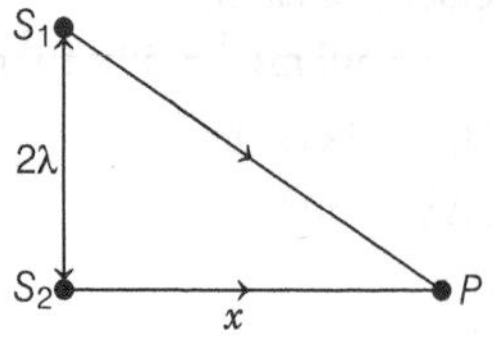


For minimum intensity at $P$, destructive interference must take place at $P$.

So, $S_1P - S_2P = \Delta x$ (path difference)

$= (2n+1)\frac{\lambda}{2}$ (for destructive interference)

From $\Delta S_1S_2P$, for minimum distance,

$$S_1P - S_2P = \frac{3\lambda}{2} \neq \frac{\lambda}{2} \quad \ldots(i)$$

$[\because S_2P = x \text{ and } S_1S_2 = 2\lambda]$

$$\Rightarrow \sqrt{x^2+(2\lambda)^2} - x = \frac{3\lambda}{2}$$

$$\Rightarrow x^2+(2\lambda)^2 = \left[x+\left(\frac{3\lambda}{2}\right)\right]^2$$

$$\Rightarrow x^2+4\lambda^2 = x^2+\frac{9\lambda^2}{4}+2\cdot x\cdot\frac{3\lambda}{2}$$

$$\Rightarrow x\cdot 3\lambda = 4\lambda^2 - \frac{9\lambda^2}{4} = \frac{7\lambda^2}{4}$$

$$\Rightarrow x = \frac{7\lambda}{12}$$

**Note** If we proceed with Eq. (i) taking $S_1P - S_2P = \frac{\lambda}{2}$, $x = \frac{15\lambda}{4}$ which is more than $\frac{7\lambda}{12}$.

**19** (*d*) Given, amplitude, $a = 2\,\text{cm}$

and phase difference, $\phi = \pi\,\text{rad}$

$\because$ Resultant amplitude, $A = 2a\cos\frac{\phi}{2}$

$$A = 2\times 2\times \cos\frac{\pi}{2}$$

$$A = 0 \quad \left[\because \cos\frac{\pi}{2} = 0\right]$$

**20** (*c*) The resultant amplitude of two identical superposing wave is given by $A = 2a\cos\frac{\phi}{2}$ ... (i)

where, $a$ = amplitude of individual wave

and $\phi$ = phase difference between them.

The resultant intensity, $I \propto A^2 \Rightarrow I \propto \cos^2\frac{\phi}{2}$

[from Eq. (i)]

**21** (*b*) When the sources are incoherent there is no interference and resultant intensity is $I = I_1 + I_2$.

For sources of same intensity $I_0$, resultant intensity will be $2I_0$.

**23** (*c*) Let $n$th fringe of 1500 Å coincide with $(n-2)$th fringe of 2500 Å.

$$\therefore \quad 1500\times n = 2500\times(n-2)$$
$$3n = 5(n-2)$$
$$3n = 5n - 10$$
$$2n = 10$$
$$\Rightarrow \quad n = 5$$
$$(n-2) = 3$$

$\therefore$ 5th order of 1st source and 3rd order of 2nd source will coincide.

**24** (*d*) In a YDSE, path difference between 2 rays, reaching at some common point $P$ located at angular position '$\theta$' as shown in the figure below is

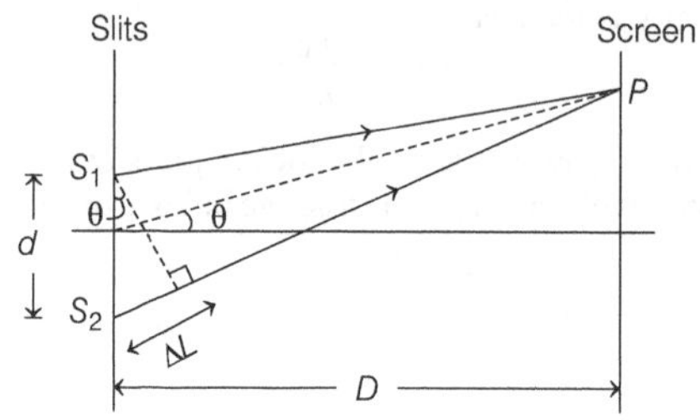


$\Delta L = d\sin\theta$

For small value of $\theta$; $\sin\theta \approx \theta$

So, path difference $= \Delta L = d\theta$

For a bright fringe at same angular position '$\theta$', both of the rays from slits $S_1$ and $S_2$ are in phase.

Hence, path difference is an integral multiple of wavelength of light used.

i.e., $\Delta L = n\lambda$

or $d\theta = n\lambda \Rightarrow \lambda = \frac{d\theta}{n}$

Here, $\theta = \frac{1}{40}$ rad, $d = 0.1$ mm

Hence, $\lambda = \frac{0.1}{40n}\,\text{mm} = \frac{0.1\times 10^{-3}}{40n}\,\text{m}$

$$= \frac{0.1\times 10^{-3}\times 10^9}{40n}\,\text{n-m}$$

$$\Rightarrow \lambda = \frac{2500}{n}\,\text{n-m}$$

So, with light of wavelength $\lambda_1$ we have

$$\lambda_1 = \frac{2500}{n_1}\,(\text{n-m})$$

and with light of wavelength $\lambda_2$, we have

$$\lambda_2 = \frac{2500}{n_2}\,(\text{n-m})$$

Now, choosing different integral values for $n_1$ and $n_2$, (i.e., $n_1, n_2 = 1, 2, 3\ldots$ etc) we find that for

$n_1 = 4$, $\lambda_1 = \frac{2500}{4} = 625$ n-m

and for $n_2 = 5$,

$$\lambda_2 = \frac{2500}{5} = 500\,\text{n-m}$$

These values lie in given interval 500 n-m to 625 n-m.

**25** (*d*) Given, wavelength, $\lambda = 6000$ Å
Distance of screen, $D = 2.5$ m
Distance between two slits, $d = 0.5$ mm
Distance of $n$th bright fringe from the centre, $y_n = \dfrac{nD\lambda}{d}$

$$\therefore \quad y_3 = \frac{3\times 2.5\times 6000\times 10^{-10}}{0.5\times 10^{-3}} \quad [\because n = 3]$$
$$= 9\times 10^{-3} \text{ m} = 9 \text{ mm}$$

**26** (*c*) Given, wavelength of blue light = 4360Å
and wavelength of green light = 5460 Å
Position of $n$th maximum from central maxima is
$$x_n = \frac{n\lambda D}{d} \Rightarrow x_n \propto \lambda$$
As, $\lambda_{blue} < \lambda_{green}$ (given)
So, $x$(blue) < $x$ (green)

**27** (*a*) Given, wavelength, $\lambda = 589.3 \text{ nm} = 589.3\times 10^{-9}$ m
Distance of screen, $D = 100 \text{ cm} = 1$ m
Position of 10th bright fringe, $x = 12 \text{ mm} = 12\times 10^{-3}$ m

Position of 10th bright fringe $= \dfrac{10\lambda D}{d}$
$$\Rightarrow \frac{10\lambda D}{d} = 12 \text{ mm} \Rightarrow d = \frac{10\lambda D}{12\times 10^{-3}}$$
$\therefore$ The separation between the slits,
$$d = \frac{10\times 589.3\times 10^{-9}\times 1}{12\times 10^{-3}}$$
$$= 4.9\times 10^{-4} \text{ m} = 0.49 \text{ mm}$$

**28** (*a*) Let $n_1$th fringe formed due to first wavelength and $n_2$th fringe formed due to second wavelength coincide i.e. their distance from common central maxima will be same
i.e. $y_{n_1} = y_{n_2}$
$$\Rightarrow \frac{n_1\lambda_1 D}{d} = \frac{n_2\lambda_2 D}{d} \Rightarrow \frac{n_1}{n_2} = \frac{\lambda_1}{\lambda_2} = \frac{520}{650} = \frac{4}{5}$$
Hence, distance of the point of coincidence from the central maxima is
$$y = \frac{n_1\lambda_1 D}{d} = \frac{n_2\lambda_2 D}{d} = \frac{4\times 650\times 10^{-9}\times 1.5}{0.5\times 10^{-3}} = 7.8 \text{ mm}$$

**29** (*b*) Given, wavelengths,
$\lambda_1 = 12000$ Å and $\lambda_2 = 10000$ Å,
Distance of screen, $D = 2$ m
and distance between slits, $d = 2 \text{ mm} = 2\times 10^{-3}$ m.
$$\text{We have, } \frac{\lambda_1}{\lambda_2} = \frac{12000}{10000} = \frac{6}{5} = \frac{n_2}{n_1} \quad [\because n_1\lambda_1 = n_2\lambda_2]$$
$$\text{As, } x = \frac{n_1\lambda_1 D}{d} = \frac{5\times 12000\times 10^{-10}\times 2}{2\times 10^{-3}} \quad [\because n_1 = 5]$$
$$= 5\times 1.2\times 10^4\times 10^{-10}\times 10^3 = 6 \text{ mm}$$

**30** (*d*) Given, $\lambda_1 = 6000$ Å, $n_1 = 16$ fringes
and $n_2 = 24$ fringes
Position of $n$th fringe, $x_n = \dfrac{nD\lambda}{d}$

Now for two fringes, $\dfrac{\lambda_1}{\lambda_2} = \dfrac{n_2}{n_1} \Rightarrow \dfrac{6000}{\lambda_2} = \dfrac{24}{16}$
$$\Rightarrow \lambda_2 = \frac{6000\times 16}{24} = \frac{96000}{24} = 4000 \text{ Å}$$

**31** (*b*) Given, wavelength, $\lambda = 500 \text{ nm} = 500\times 10^{-9}$ m
$d = 1 \text{ mm} = 1\times 10^{-3}$ m and $D = 1$ m
Distance of 3rd minima, i.e. $x_n = \left(n + \dfrac{1}{2}\right)\dfrac{\lambda D}{d}$
$$x_3 = \left(2 + \frac{1}{2}\right)\frac{\lambda D}{d} \quad [\because n = 2]$$
$$\Rightarrow x_3 = \frac{5\lambda D}{2d} = \frac{5\times 500\times 10^{-9}\times 1}{2\times 1\times 10^{-3}}$$
$$= 12.5\times 10^{-4} \text{ m} = 1.25 \text{ mm}$$

**32** (*c*) As we know, fringe width, $\beta_1 = \dfrac{\lambda D}{d}$
According to question, $\beta_2 = \lambda\left(\dfrac{2D}{2d}\right) = \dfrac{\lambda D}{d}$
[$\because$ slit width and distance of screen both are doubled]
$$\Rightarrow \beta_1 = \beta_2$$

**33** (*b*) Given, fringe width, $\beta_1 = 3.2\times 10^{-4}$ m
Wavelengths, $\lambda_1 = 5600$Å and $\lambda_2 = 4200$Å
$$\text{Now, } \frac{\beta_2}{\beta_1} = \frac{\lambda_2}{\lambda_1} = \frac{4200}{5600} \quad [\because \beta \propto \lambda]$$
$$\text{or } \beta_2 = \frac{6}{8}\times \beta_1 = \frac{6}{8}\times 3.2\times 10^{-4} = 2.4\times 10^{-4} \text{ m}$$
Decrease in fringe width, $\Delta\beta = \beta_1 - \beta_2$
$$= (3.2 - 2.4)\times 10^{-4} = 0.8\times 10^{-4} \text{ m}$$

**34** (*d*) Given, wavelengths, $\lambda_1 = 6.4\times 10^{-7}$ m
$\lambda_2 = 4.0\times 10^{-7}$ m
Now, in YDSE fringe width, $\beta = \dfrac{\lambda D}{d}$
$$\therefore \frac{\beta_2}{\beta_1} = \frac{\lambda_2 D_2 d_1}{\lambda_1 D_1 d_2}$$
$$\Rightarrow \beta_2 = \frac{\beta_1\times\lambda_2\times 2D_1\times d_1}{\lambda_1\times D_1\times d_1/2} \quad \left[\begin{array}{l}\therefore \text{Given } D_2 = 2D_1 \\ \text{and } d_2 = d_1/2\end{array}\right]$$
$$\Rightarrow \beta_2 = \beta_1\times\frac{\lambda_2}{\lambda_1}\times 4 = 1\times 10^{-4}\times\frac{4.0\times 10^{-7}}{6.4\times 10^{-7}}\times 4$$
[$\because \beta_1 = 1\times 10^{-4}$ m]
$$\Rightarrow \beta_2 = 2.5\times 10^{-4} \text{ m}$$

**35** (*a*) Given, distance between source and observer,
$D = 2$ m
Distance of 5th dark fringe from centre $= 4\times 10^{-3}$ m
Wavelength, $\lambda = 600 \text{ nm} = 6\times 10^{-7}$ m
$\because$ Distance of 5th dark fringe from centre
$$= \beta + \beta + \beta + \beta + \frac{\beta}{2} = \frac{9\beta}{2}$$
where, $\beta$ = fringe width.

According to question, $\frac{9\beta}{2} = 4 \times 10^{-3}$

$$\frac{9}{2} \times \frac{D\lambda}{d} = 4 \times 10^{-3}$$

$$\frac{9}{2} \times \frac{2 \times 6 \times 10^{-7}}{d} = 4 \times 10^{-3}$$

$$\Rightarrow \quad d = 1.35 \times 10^{-3} \text{ m} = 1.35 \text{ mm}$$

**36** (*a*) The situation is as shown below

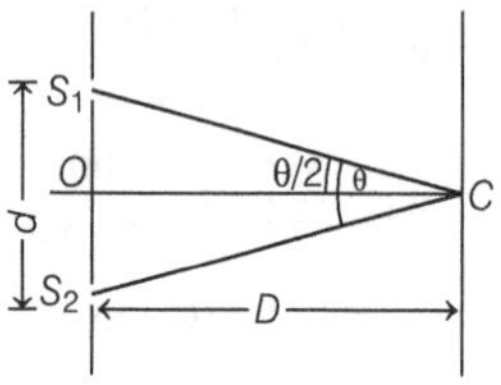


So, distance between two slits, i.e. $S_1$ and $S_2$ is

From $\Delta S_1OC, \frac{d/2}{D} = \tan\frac{\theta}{2} \Rightarrow d = (2\tan\theta/2)D$

For small angles $\theta$, $\tan\theta \approx \theta$

$$\Rightarrow \quad d = 2 \times \frac{\theta}{2} \times D = D\theta$$

or $$\frac{D}{d} = \frac{1}{\theta}$$

Fringe width, $\beta = \frac{\lambda D}{d} = \frac{\lambda}{\theta}$

**37** (*b*) In YDSE, angular width of a fringe is given as

$$\theta = \frac{\lambda}{d}$$

where, $\lambda$ is the wavelength of the light source and $d$ is the distance between the two slits.

$$\Rightarrow \quad \theta \propto \frac{1}{d} \text{ or } \frac{\theta_1}{\theta_2} = \frac{d_2}{d_1} \quad \ldots\text{(i)}$$

Given, $\theta_1 = 0.20°$, $\theta_2 = 0.21°$

and $d_1 = 2$ mm

Substituting the given values in Eq. (i), we get

$$\frac{0.20°}{0.21°} = \frac{d_2}{2 \text{ mm}}$$

$$\Rightarrow \quad d_2 = 2 \times \frac{0.20}{0.21} = \frac{0.40}{0.21} = 1.9 \text{ mm}$$

**38** (*a*) The fringe width, i.e. $\beta_1 = \frac{\lambda D_1}{d}$ and $\beta_2 = \frac{\lambda D_2}{d}$

So, $\beta_1 - \beta_2 = \frac{\lambda}{d}(D_1 - D_2)$

Here, $\beta_1 - \beta_2 = 3 \times 10^{-5}$ m, $d = 10^{-3}$ m

and $D_1 - D_2 = 5 \times 10^{-2}$ m

$$\Rightarrow \quad 3 \times 10^{-5} = \frac{\lambda}{10^{-3}}(5 \times 10^{-2})$$

$$\Rightarrow \quad \lambda = \frac{3}{5} \times \frac{10^{-8}}{10^{-2}} = 0.6 \times 10^{-6} \text{ m} = 6000 \text{ Å}$$

**39** (*a*) Given, wavelength of light, $\lambda = 500 \text{ nm} = 5 \times 10^{-7}$ m

Distance between slit and screen, $D = 1.8$ m

Distance between two slits, $d = 0.4 \text{ mm} = 4 \times 10^{-4}$ m

Velocity of screen $= \frac{dD}{dt} = 4$ m/s

Speed of first maxima $= \frac{d\beta}{dt} = ?$

In Young's double slit experiment, fringe width, $\beta = \frac{D\lambda}{d}$

Differentiating with respect to $t$, we get

$$\frac{d\beta}{dt} = \frac{\lambda}{d} \cdot \frac{dD}{dt} = \frac{5 \times 10^{-7}}{4 \times 10^{-4}} \times 4$$

$$= 5 \times 10^{-3} \text{ m/s} = 5 \text{ mm/s}$$

**40** (*b*) Since, when the source $S$ is on the perpendicular bisector, then the central bright fringe occurs at $O$, which is also on the perpendicular bisector.

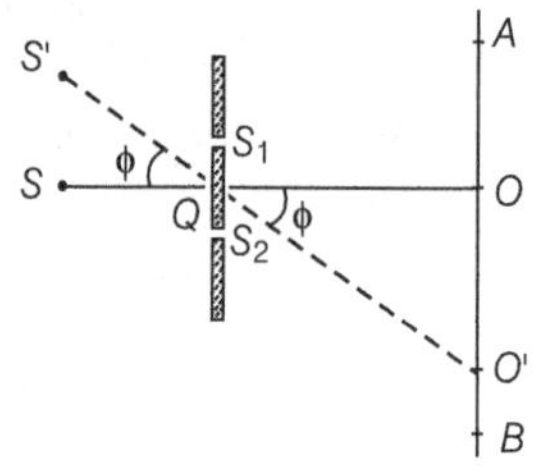


If $S$ is shifted by an angle $\phi$ to point $S'$, then the central fringe appears at a point $O'$ at an angle $-\phi$ (shown in figure below), which means that it is shifted by the same angle on the other side of the bisector. i.e. towards end $B$ of screen.

**41** (*b*) The resultant intensity, $I = I_0 \cos^2\frac{\phi}{2}$

Here, $I_0$ is the maximum intensity and $\phi = \frac{\pi}{2}$

$$I = I_0 \cos^2\left(\frac{\pi}{2 \times 2}\right) = I_0 \cos^2\frac{\pi}{4}$$

$$I = \frac{I_0}{2}$$

**42** (*c*) The contrast interference will occur when there is absolute darkness at the dark band due to destructive interference, i.e. $I_R = I_{min} = 0$ and there is complete (max) brightness at the bright band due to constructive interference, i.e. $I_R = I_{max} = 4I_0$, which is possible only when individual intensities are same.

So, $$I_1 = I_2 = I_0$$

**43** (*d*) The fringes are straight lines although $S_1$ and $S_2$ are point sources. If slits are used instead of point sources, each pair of points would have produced straight line fringes with increased intensities as shown below.

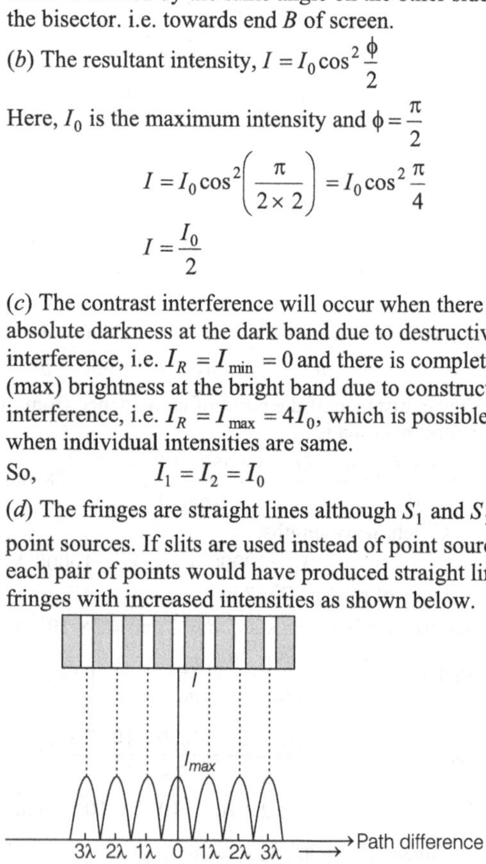


**45** (*a*) Here, wavelength of light used $(\lambda) = 5303$ Å.

Distance between two slit $(d) = 19.44$ μ-m

Width of single slit (a)= 4.05 μm

Here, angular width between first and second diffraction minima

$$\theta \simeq \frac{\lambda}{a}$$

and angular width of a fringe due to double slit is

$$\theta' = \frac{\lambda}{d}$$

∴Number of fringes between first and second

diffraction minima, $n = \frac{\theta}{\theta'} = \frac{\frac{\lambda}{a}}{\frac{\lambda}{d}} = \frac{d}{a} = \frac{19.44}{4.05} = 4.81$

or $\qquad n \simeq 5$

∴5 interfering bright fringes lie between first and second diffraction minima.

**46** (*d*) Given, $\lambda = 6000\ \text{Å} = 6000 \times 10^{-10}$ m, $d = 0.3$ mm

$$= 0.3 \times 10^{-3}\ \text{m}$$

For minima, $\quad d \sin\theta = m\lambda$

First minima means $(m = 1) \Rightarrow \sin\theta = \frac{\lambda}{d}$

Angular position of 1st minima,

$$\sin\theta = \theta = \frac{\lambda}{d} = \frac{6000 \times 10^{-10}}{0.3 \times 10^{-3}} = 2 \times 10^{-3}\ \text{rad}$$

So, angular position of first minima is $2 \times 10^{-3}$ rad.

**47** (*c*) For minima, $\quad a \sin\theta = n\lambda$

$$\Rightarrow \quad a \sin 30° = (1)\lambda \qquad (n = 1)$$

$$\Rightarrow \quad a = 2\lambda \left(\because \sin 30° = \frac{1}{2}\right) \quad \text{...(i)}$$

For 1st secondary maxima

$$\Rightarrow \quad a \sin\theta_1 = \frac{3\lambda}{2} \Rightarrow \sin\theta_1 = \frac{3\lambda}{2a} \quad \text{...(ii)}$$

Substitute value of *a* from Eq. (i) to Eq. (ii), we get

$$\sin\theta_1 = \frac{3\lambda}{4\lambda} \Rightarrow \sin\theta_1 = \frac{3}{4}$$

$$\theta_1 = \sin^{-1}\frac{3}{4}$$

**48** (*c*) For 2nd secondary maxima using red light,

$$d \sin\theta_1 = \frac{(2m+1)\lambda_1}{2}, \text{ where } m = 2$$

$$d \sin\theta_1 = \frac{5\lambda_1}{2} \Rightarrow \sin\theta_1 = \frac{5\lambda_1}{2d} \quad \text{...(i)}$$

When white light is used, for position of 3rd secondary maxima $(m = 3)$

$$d \sin\theta_2 = \frac{7\lambda_2}{2} \Rightarrow \sin\theta_2 = \frac{7\lambda_2}{2d} \quad \text{...(ii)}$$

Since, these positions coincide with each other for white and red light

$$\sin\theta_1 = \sin\theta_2 \Rightarrow \frac{5\lambda_1}{2d} = \frac{7\lambda_2}{2d}$$

$$\Rightarrow \quad \lambda_2 = \frac{5}{7}\lambda_1 = \frac{5}{7} \times 6500\ \text{Å}$$

Wavelength of white light, $\lambda_2 = 4642.85$ Å

**49** (*a*) Angular width of central maxima $(2\theta) = \frac{2\lambda}{d} = \frac{2\lambda}{e}$

$$\Rightarrow \quad 2\theta \propto \frac{\lambda}{e}$$

$$\Rightarrow \quad \theta \propto \frac{1}{e} \qquad (\text{for } \lambda = \text{constant})$$

Thus, on decreasing slit width *e*, θ will increases.

**50** (*b*) In YDSE, intensity at a point is given by

$$I_{net} = 2I_0(1 + \cos\phi) \quad \text{.... (i)}$$

where, $\phi$ = phase difference between interfering waves.

Here, $\quad \phi = \frac{1}{8}\lambda = \frac{1}{8} \times 2\pi$

(a path difference of one wavelength is equivalent to a phase difference of 2π radians)

$$\Rightarrow \quad \phi = \frac{\pi}{4} \Rightarrow \cos\phi = \frac{1}{\sqrt{2}} \approx 0.7$$

So, $I_{net}$ at given point *P* is

$$I_{net} = I_1 = 2I_0(1 + 0.7) = 2 \times 1.7 I_0$$

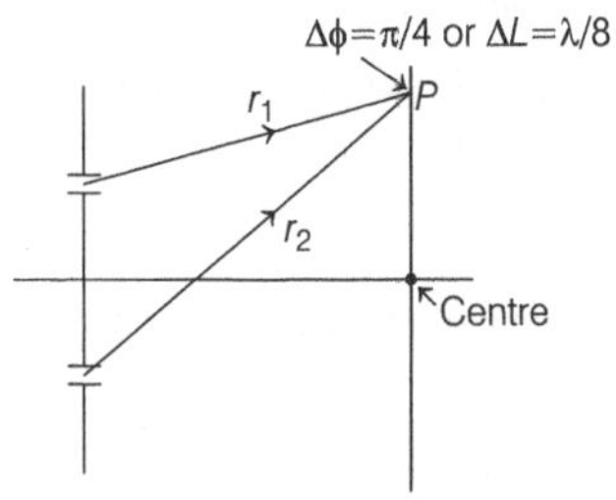


Also, intensity maxima at centre of a bright fringe is obtained by substituting $\phi = 0, 2\pi, \dots n\pi$ in Eq. (i).

So, $I_{net} = I_2 = 2I_0(1 + \cos 0°) = 4I_0$

Hence, required ratio is

$$\frac{I_1}{I_2} = \frac{2 \times 1.7 I_0}{4 I_0} = \frac{1.7}{2} = 0.853$$

**51** (*b*) If a converging lens of focal length *f* is placed after the slit and the fringe are obtained on screen, then separation of the central maximum from the first null of the diffraction pattern is $\lambda / a$.

If *f* is the focal length of the lens, then the size of screen will be $f\lambda / a$.

**52** (*a*) According to geometrical optics, the parallel beam from stars will get focussed to a point. But due to diffraction, they gets focused to a spot of finite area.

The pattern obtained, consist of a central bright region surrounded with concentric dark and bright rings as shown below.

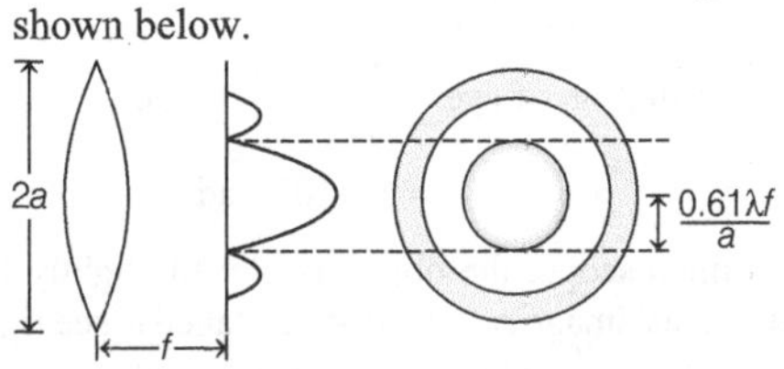


For two stars to be just resolved, the radius of the central bright region is given by

$$r_0 \approx \frac{1.22\lambda f}{2a} = \frac{0.61\lambda f}{a}$$

**53** (b) Given, $2a = 100$ inch

$\Rightarrow \quad a = 50 \text{ inch} = 50 \times 2.54 \text{ cm} = 127 \text{ cm}$

Wavelength, $\lambda = 6000 \text{Å} = 6 \times 10^{-7} \text{m} = 6 \times 10^{-5} \text{cm}$

Limit of resolution, $\Delta\theta = \frac{0.61\lambda}{a} = \frac{0.61 \times 6 \times 10^{-5}}{127}$

$= 2.9 \times 10^{-7}$ rad

**54** (a) Given, diameter, $D = 6 \text{cm} = 6 \times 10^{-2}$ m,

and wavelength, $\lambda = 540 \text{ nm} = 540 \times 10^{-9}$ m

Resolving power of telescope (RP) $= \frac{1}{\Delta\theta} = \frac{D}{1.22\lambda}$

where, $D$ = diameter of objective and $\lambda$ = wavelength of light.

$\Rightarrow \quad \text{RP} = \frac{6 \times 10^{-2}}{1.22 \times 540 \times 10^{-9}} \text{rad}^{-1}$

$= \frac{6000 \times 10^4}{540 \times 1.22} \text{rad}^{-1} = 9.1 \times 10^4 \text{ rad}^{-1}$

**55** (c) From Rayleigh's criteria, two objects will be just resolved when their angular separation,

$$\theta_R = 1.22 \frac{\lambda}{a}$$

where, $\lambda$ = wavelength and $a$ = aperture size.

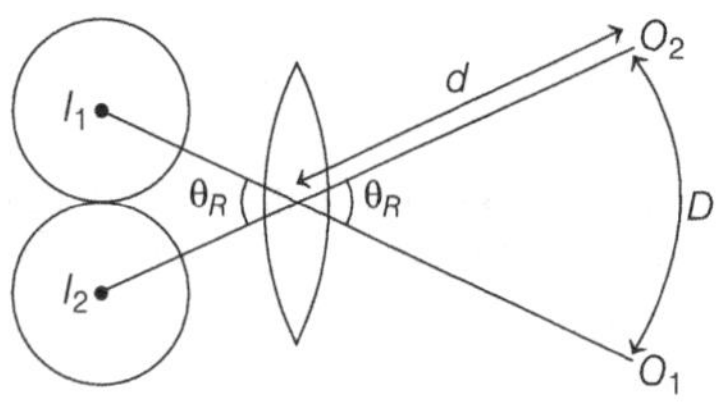


Two objects are just resolved when their central maximas does not overlap.

From diagram, distance between objects in object plane,

$$D = \theta_R \cdot d = 1.22 \frac{\lambda d}{a}$$

Here, $a = 5$ m, $d = 4 \times 10^5 \text{ km} = 4 \times 10^8$ m,

$\lambda = 5500 \text{ Å} = 5500 \times 10^{-10}$ m

So, $\quad D = \frac{(1.22 \times 5500 \times 10^{-10} \times 4 \times 10^8)}{5} \approx 54$ m

From options, minimum separation distance is 60 m.

**56** (a) Given, width of separation between two dark bands,

$$d = 0.01 \text{ cm}$$

Distance of screen from eye, $D = 2$ m

Limit of resolution of eye $= \frac{d}{D} = \frac{0.01}{2} = 0.005$

$= 5 \times 10^{-3}$ rad

**57** (c) In a microscope, the object is placed slightly beyond $f$, so that a real image is formed at a distance $v$ [see figure]. The magnification ratio of image size to object size is given by $m \approx v/f$. It can be seen from figure that

$$D/f \approx 2 \tan\beta \quad \text{...(i)}$$

where, $\beta$ is the angle subtended by the diameter of the objective lens at the focus of the microscope.

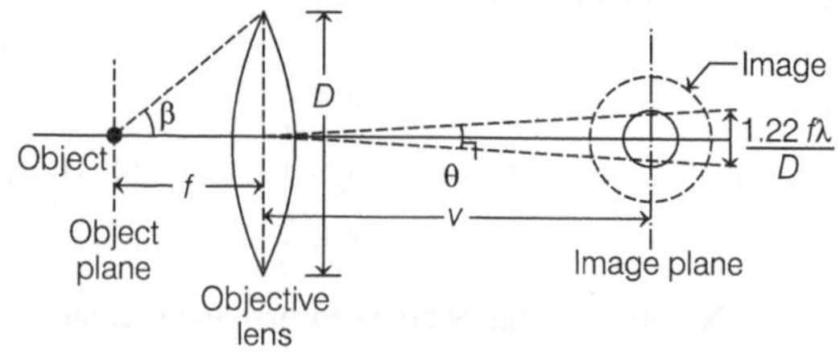


**58** (c) Given, wavelengths, $\lambda_1 = 4000 \text{Å}$ and $\lambda_2 = 6000 \text{Å}$

As, resolving power of a microscope, $\text{RP} = \frac{1}{\Delta\theta} = \frac{D}{1.22\lambda}$

$\Rightarrow \quad (\text{RP}) \propto \frac{1}{\lambda(\text{wavelength})}$

$\therefore \quad \frac{(\text{RP}_1)}{(\text{RP}_2)} = \frac{\lambda_2}{\lambda_1} = \frac{6000}{4000} = \frac{3}{2}$

**59** (d) For compound microscope, resolving power is

i.e. $\quad \text{RP} = \frac{2n \sin\beta}{1.22\lambda}$

(i) $\because$ RP $\propto n$

when the refractive index $n$ of the medium between the object and the objective lens increased keeping the wavelength constant, the resolving power increased.

(ii) $\because$ RP $\propto \frac{1}{\lambda}$

For the same medium on decreasing the wavelength of light used, the resolving power of microscope increases.

**60** (a) Given, aperture, $a = 3 \text{mm} = 3 \times 10^{-3}$ m

and wavelength, $\lambda = 500 \text{nm} = 5 \times 10^{-7}$ m

According to Fresnel distance, $Z_F$,

$$Z_F = \frac{a^2}{\lambda} = \frac{(3 \times 10^{-3})^2}{5 \times 10^{-7}} = \frac{9 \times 10^{-6}}{5 \times 10^{-7}} = 18 \text{ m}$$

**62** (a) In unpolarised beam, vibrations are probable in all directions in a plane perpendicular to the direction of propagation. Therefore, $\theta$ can have any value from 0 to $2\pi$.

$$[\cos^2\theta]_{av} = \frac{1}{2\pi}\int_0^{2\pi} \cos^2\theta \, d\theta$$

$$= \frac{1}{2\pi}\int_0^{2\pi} \left(\frac{1+\cos 2\theta}{2}\right) d\theta$$

$$= \frac{1}{4\pi}\left([\theta]_0^{2\pi} + \left[\frac{\sin 2\theta}{2}\right]_0^{2\pi}\right) = \frac{1}{4\pi} \times 2\pi = \frac{1}{2}$$

So, using law of Malus, $I = I_0 \cos^2\theta \Rightarrow I = I_0 \times \frac{1}{2} = \frac{I_0}{2}$

**63** (c) The situation is as shown below

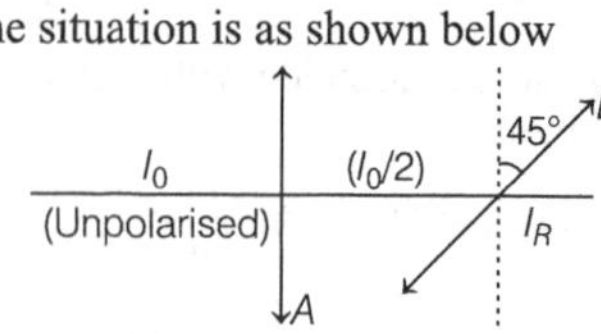

According to Malus' law,

$$I_R = \left(\frac{I_0}{2}\right)\cos^2(45°) = \frac{I_0}{2} \times \frac{1}{2} = \frac{I_0}{4}$$

**64** (*a*) Suppose $I_0$ be the intensity of polarised light after passing through the first polariser $P_1$. Then, the intensity of light after passing through second polariser will be

$$I = I_0 \cos^2 \theta \qquad \text{[Malus's law]}$$

where, $\theta$ is the angle between pass axes of $P_1$ and $P_2$. Since, $P_1$ and $P_3$ are crossed the angle between the pass axes of $P_2$ and $P_3$ will be $\left(\frac{\pi}{2} - \theta\right)$.

Hence, the intensity of light emerging from $P_3$ will be

$$I = I_0 \cos^2 \theta \cos^2\left(\frac{\pi}{2} - \theta\right)$$

$$= I_0 \cos^2 \theta \cdot \sin^2 \theta = \left(\frac{I_0}{4}\right)\sin^2 2\theta$$

Intensity is maximum, when $\sin 2\theta = 1$

$\Rightarrow 2\theta = \pi/2$ or $\theta = \pi/4$

Therefore, the transmitted intensity will be maximum when $\theta = \pi/4$.

**65** (*d*) Here, angle between $P_1$ and $P_2 = 60°$.

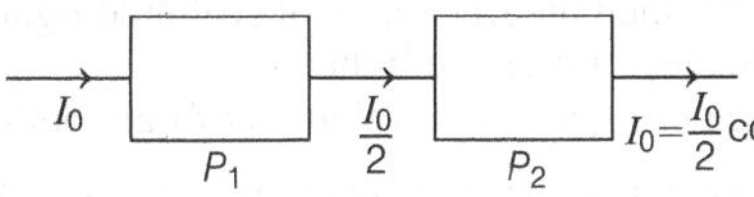


Intensity of light emerging from $P_2$ is $I = \frac{I_0}{2}\cos^2\theta$

where, $\quad \theta$ = angle between $P_1$ and $P_2$.

So, $\quad I = \frac{I_0}{2}\cos^2 60° = \frac{I_0}{2}\left(\frac{1}{2}\right)^2 = \frac{I_0}{8}$

**66** (*c*) According to the question, the diagram is

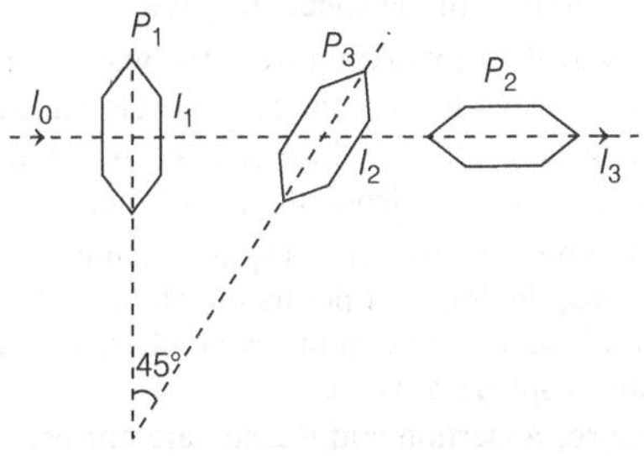


From the above diagram, the intensity of transmitted light through $P_1$, $I_1 = I_{0/2}$

The intensity of transmitted light through $P_3$

$$I_2 = I_1 \cos^2 45° = \frac{I_0}{2}\cos^2 45°$$

$$\Rightarrow \quad I_2 = \frac{I_0}{2} \times \left(\frac{1}{\sqrt{2}}\right)^2 \text{Þ} I_2 = \frac{I_0}{4}$$

Angle between polaroids $P_3$ and $P_2 = 90 - 45 = 45°$

$\therefore$ Intensity transmitted through $P_2$,

$$I_3 = I_2 \cos^2 45° = \frac{I_0}{4}\cos^2 45°$$

$$\Rightarrow \quad I_3 = \frac{I_0}{4} \times \left(\frac{1}{\sqrt{2}}\right)^2$$

$$\Rightarrow \quad I_3 = \frac{I_0}{4} \times \frac{1}{2} \Rightarrow I_3 = \frac{I_0}{8}$$

**67** (*c*)

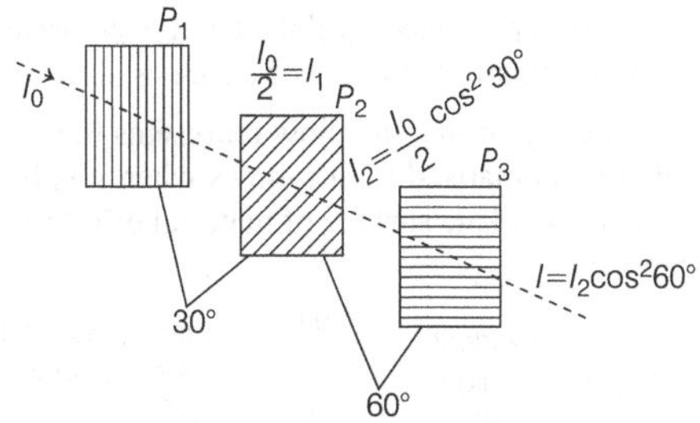


When unpolarised light pass through polaroid $P_1$, intensity obtained is

$$I_1 = \frac{I_0}{2}$$

where, $I_0$ = intensity of incident light.

Now, this transmitted light is polarised and it pass through polariser $P_2$. So, intensity $I_2$ transmitted is obtained by Malus law.

$$\Rightarrow \quad I_2 = I_1 \cos^2 \theta$$

As angle of pass axis of $P_1$ and $P_3$ is 90° and angle of pass axis of $P_2$ and $P_3$ is 60°, so angle between pass axis of $P_1$ and $P_2$ is $(90° - 60°) = 30°$. So,

$$I_2 = \frac{I_0}{2}\cos^2 30° = \frac{I_0}{2} \times \left(\frac{\sqrt{3}}{2}\right)^2 = \frac{3}{8}I_0$$

When this light pass through third polariser $P_3$, intensity $I$ transmitted is again obtained by Malus law.

So, $\quad I = I_2 \cos^2 60° = \left(\frac{3}{8}I_0\right)\cos^2 60°$

$$= \frac{3}{8}I_0 \times \left(\frac{1}{2}\right)^2 = \frac{3}{32}I_0$$

So, ratio $\frac{I_0}{I} = \frac{32}{3} = 10.67$

**68** (*b*) Polaroids can be used to control the intensity in sunglasses, window-panes, etc. The intensity can be further controlled from 50% to zero of the incident intensity by adjusting the angle between the pass-axes of two polaroids.

**69** (*a*) Under the influence of the electric field of the incident wave, the electrons in the molecules of the atmosphere acquire components of motion in both directions.

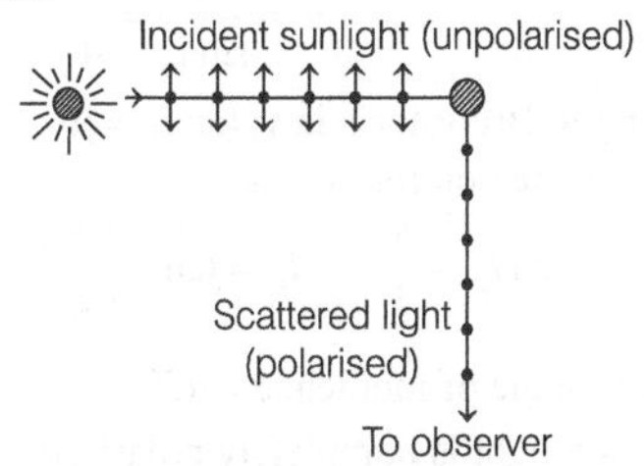

When the observer is looking at 90° to the direction of the sun, then the charges accelerating parallel to the double arrows do not radiate energy towards the observer, since their acceleration has no transverse component. The radiation scattered by the molecule is therefore represented by dots, i.e. it is polarised perpendicular to the plane of figure as shown.

**70** (*b*) The figure shown below represents the course of path an unpolarised light follows when it is incident from air on plane surface of material of refractive index $\mu$.

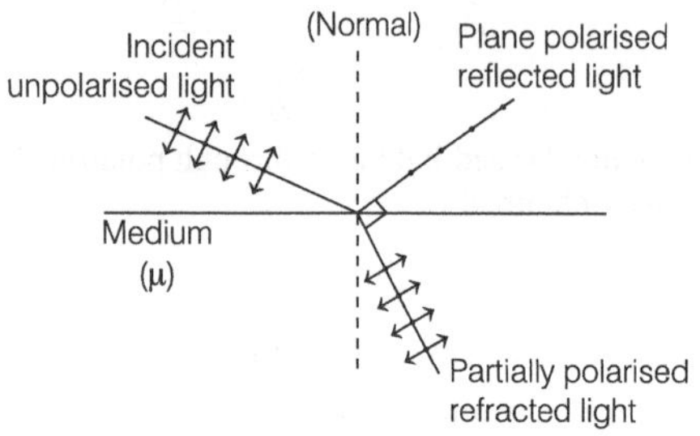


When the beam of unpolarised light is reflected from a medium (refractive index $= \mu$) and if reflected and refracted light are perpendicular to each other. Then, the reflected light is completely plane polarised at a certain angle of incidence. This means, the reflected light has electric vector perpendicular to incidence plane.

**71** (*a*) If unpolarised light is incident at polarising angle, then reflected light is completely or 100% polarised and it is perpendicular to the plane of incidence.

**72** (*d*) Given, $i + r = \pi / 2$

According to Brewster's law, we get

$$\tan i_B = \mu = 1.5$$

So, $$i_B = \tan^{-1}(1.5) \Rightarrow i_B = 57°$$

i.e. this is the Brewster's angle for air to glass interface.

**73** (*b*) Given, refractive index of medium, $\mu = 1$

Using Brewster's law, $\tan i_p = \mu$

$$\tan i_p = 1$$
$$i_p = \tan^{-1}(1) = 45°$$

As $i_p + r = 90°$

$$r = 90° - i_p = 90° - 45° \Rightarrow r = 45°$$

**74** (*a*) Given, critical angle, $i_c = \sin^{-1}\left(\frac{4}{5}\right)$

$\therefore$ $$\sin i_c = \frac{4}{5}$$

As $$\mu = \frac{1}{\sin i_c} = \frac{5}{4}$$

According to Brewster's law, $\tan i_p = \mu$ where, $i_p$ is the polarising angle.

$\therefore$ $$\tan i_p = \frac{5}{4} \Rightarrow i_p = \tan^{-1}\left(\frac{5}{4}\right)$$

**75** (*b*) Given, angle of incidence = 45°

As reflected light is completely polarised, therefore

$$i_p = 45°$$

From Brewster's law, $\mu = \tan i_p = \tan 45° = 1$

As, $$\mu = \frac{c}{v} \Rightarrow v = \frac{c}{\mu} = \frac{3 \times 10^8}{1} \Rightarrow v = 3 \times 10^8 \text{ ms}^{-1}$$

**76** (*c*) In the special situation, one of the two perpendicular components of the electric field is zero. At other angles, both components are present but one is stronger than the other.

When such light is viewed through a rotating analyser, one sees a maximum and a minimum of intensity but not complete darkness. This kind of light is called partially polarised.

**77** (*a*) The branch of optics in which one completely neglects the finiteness of the wavelength is called geometrical optics. The wavelength of light is very small as compared to the dimensions of objects (such as mirror, lenses, etc.) and hence it can be neglected and assumed to travel in a straight line.

Therefore, Assertion and Reason are correct and Reason is the correct explanation of Assertion.

**78** (*b*) Reflection and refraction arise through interaction of incident light with constituents of matter. When monochromatic light is incident on a surface separating two media, then the frequency of scattered light equals the frequency of incident light.

The speed of light and wavelength of light both changes in refraction, then the frequency of light $v = \frac{c}{\lambda}$ remains constant.

Therefore, Assertion and Reason are correct but Reason is not the correct explanation of Assertion.

**79** (*a*) According to Huygens' principle, each point of the wavefront is the source of a secondary disturbance and the wavelets emanating from these point spread out in all directions with the speed of wave.

These wavelets emanating from the wavefront are usually referred to as secondary wavelets and if we draw a common tangent to all these spheres, we obtain the new position of the wavefront at a later time.

So, for a point source emitting waves uniformly in all directions, the locus of points which have the same amplitude and vibrate in the same phase are spheres and we have a spherical wave.

Therefore, Assertion and Reason are correct and Reason is the correct explanation of Assertion.

**80** (*c*) Since, the speed of light waves is less in glass, the lower portion of the incoming wavefront (which travels through the greatest thickness of glass) will get delayed resulting a tilt in the emerging wavefront as shown.

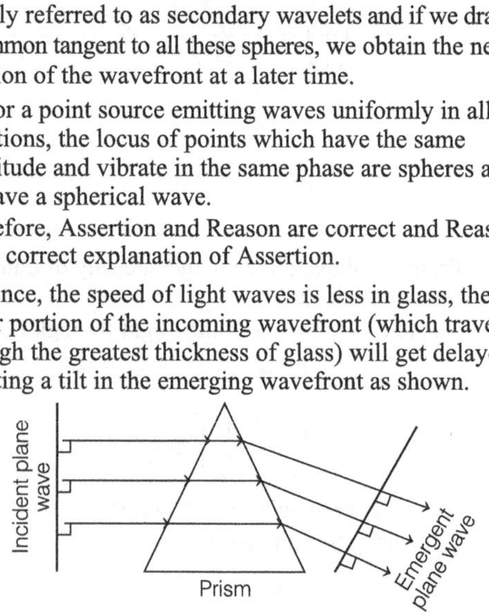


Therefore, Assertion is correct but Reason is incorrect.

**81** (*c*) Speed of light is independent of its colour only in vacuum.

As wavelength of red colour ($\lambda_1$) is higher than wavelength of blue colour ($\lambda_2$).

For light travelling from medium 1 to medium 2, $\frac{v_1}{v_2} = \frac{\lambda_1}{\lambda_2}$

where, $v_1$ and $v_2$ are velocities.

So, red colour light travels faster than violet colour light in glass.

Therefore, Assertion is correct but Reason is incorrect.

**82** (*a*) Increase in wavelength of light, when the source move away from the observer due to doppler's effect is called red shift.

The visible regions shifts towards red end of electromagnetic spectrum and hence called red shift.

Therefore, Assertion and Reason are correct and Rcason is the correct explanation of Assertion.

**83** (*a*) As we know, fringe width, $\beta = \frac{\lambda D}{d}$

So, smaller the distance between the slits $d$, the larger will be fringe width $\beta$.

Thus, single fringe will cover whole screen and pattern will not be visible, i.e. no interference pattern is detected.

Therefore, Assertion and Reason are correct and Reason is the correct explanation of Assertion.

**84** (*c*) If $I_1$ and $I_2$ are intensity of waves coming from two sources, then the maximum intensity,

$$I_{max} = (\sqrt{I_1} + \sqrt{I_2})^2$$
$$= (2\sqrt{I_0})^2 = 4I_0$$
$$[\because I_1 = I_2 = I_0 \text{ (given)}]$$

The minimum intensity observed at dark band is given by $\quad I_{min} = (\sqrt{I_1} - \sqrt{I_2})^2$

If $\quad I_1 = I_2 = I_0, I_{min} = 0$

If $\quad I_1 \neq I_2;\ I_{min} \neq 0$

Therefore, intensity of dark band is not always zero.

Therefore, Assertion is correct but Reason is incorrect.

**85** (*c*) In Young's double slit experiment, fringe width for dark and white fringes are same, while in the same experiment, when a white light (as source) is used, the central fringe is white around which few coloured fringes are observed on either side.

Therefore, Assertion is correct but Reason is incorrect.

**86** (*c*) Diffraction determines the limitations of the concept of light rays.

A beam of width $a$ travels a distance $\frac{a^2}{\lambda}$, called the Fresnel distance, before it starts to spread out due to diffraction.

Therefore, Assertion is correct but Reason is incorrect.

**87** (*b*) The statements I and II are correct but rest is incorrect and it can be corrected as,

At a large distance from the source, the small portion of the sphere can be considered as plane wave as shown in figure

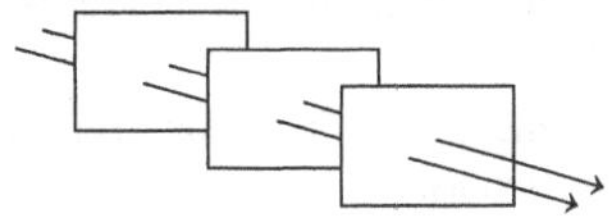

**88** (*c*) Figure shows that $AB$ is incident wavefront, so $A$ and $B$ are in same phase.

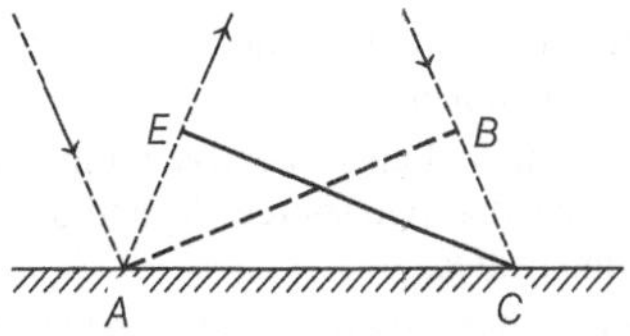


By the time $B$ reaches $C$, secondary wavelet from $A$ reaches $E$.

So, points $C$ and $E$ are same time intervals apart as they are in same phase.

So, statements III and IV are correct, rest are incorrect.

**91** (*c*) The statements I and II are correct but III is incorrect and it can be corrected as,

Diffraction is a wave phenomenon. It is observed in electromagnetic and longitudinal waves as well.

**92** (*c*) The statements I and III are correct but II is incorrect and it can be corrected as,

As the figure shows, the reflected light is linearly polarised perpendicular to the plane of the figure (represented by dots). This can be checked by looking at the reflected light through an analyser.

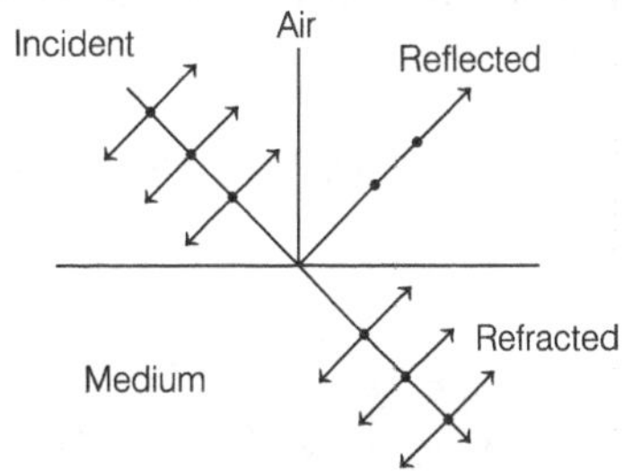


**93** (*d*) Given, $S_1P = 7\lambda$ and $S_2P = 9\lambda$

For constructive interference at any point, the path difference between waves should be $2\lambda$.

Now, we have, $\quad S_2P - S_1P = 9\lambda - 7\lambda$

$\Rightarrow \quad S_2P - S_1P = 2\lambda$

Therefore, the sources $S_1$ and $S_2$ are said to interfere constructively.

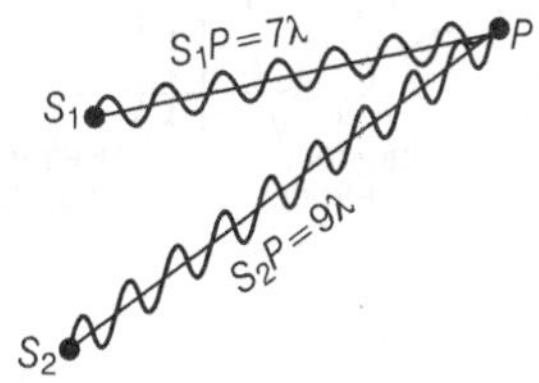

The waves emanating from $S_1$ will arrive exactly two cycles earlier than the waves from $S_2$ as $S_2P - S_1P = 2\lambda$ and will again be in phase.

Thus, the statement given in option (d) is incorrect, rest are correct.

**94** (*d*) The statement given in option (d) is incorrect and it can be corrected as,

The motion of interferring waves can be in same or in opposite direction.

Rest statements are correct.

**95** (*c*) Consider the pattern of the intensity shown in the figure given in question.

(i) As intensities of all successive minima is zero, hence we can say that two sources $S_1$ and $S_2$ are having same intensities.

(ii) Regular pattern shows constant phase difference between sources $S_1$and $S_2$ but these source are not in same phase.

(iii) We are using monochromatic light in YDSE to avoid overlapping and to have very clear pattern on the screen. So, wavelength of $S_1$ and $S_2$ will be same.

Thus, the statement given option (c) is incorrect, rest are correct.

**96** (*b*) Given, width of pinhole $= 10^3$ Å $= 1000$ Å

We know that, wavelength of sunlight ranges from 4000Å to 8000 Å.

As the width of slit is comparable to that of wavelength, hence light is diffracted from the hole.

Due to diffraction of the sunlight, the image formed on the screen will be different from the geometrical image and overlapping of colour occur.

Thus, the statement given in option (b) is correct, rest are incorrect.

**97** (*c*)

A. A light rays diverging from a point source.

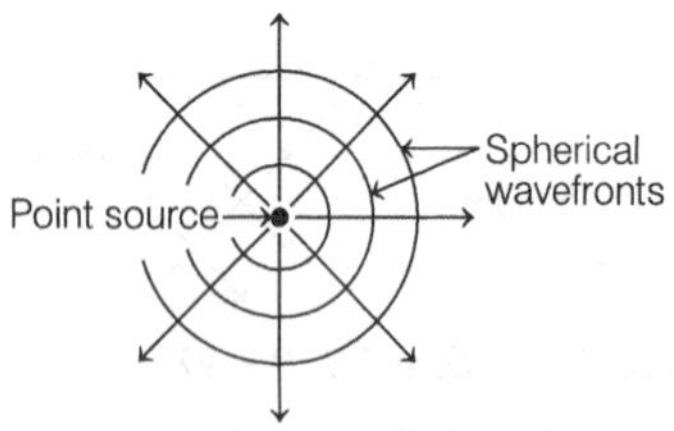


B. A light ray emerging out of convex lens when a point source is placed at its focus.

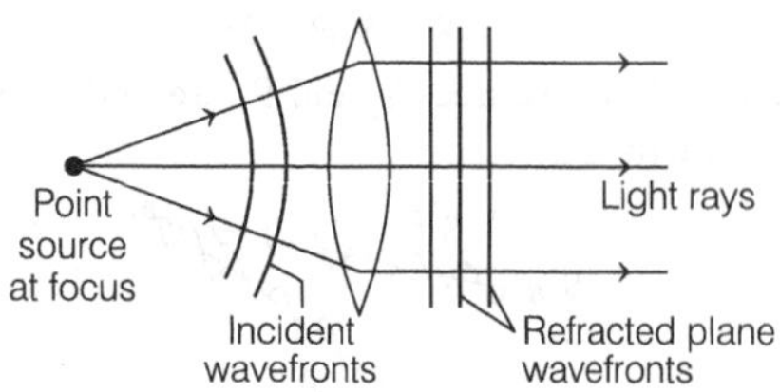


C. A plane wave incident on a thin prism.

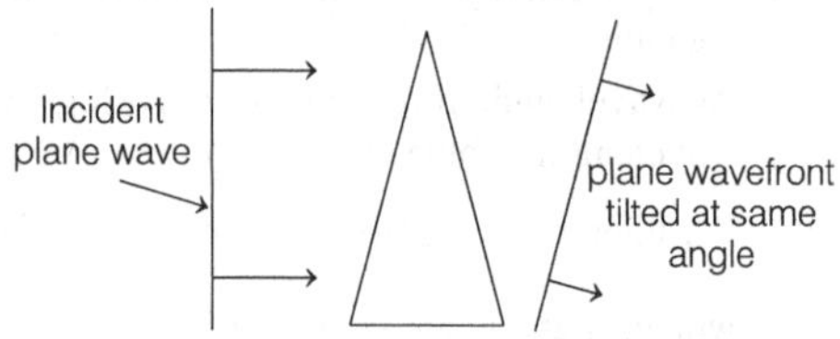


Hence, A $\to$ 2, B $\to$ 1 and C $\to$ 3.

**98** (*c*) A constructive interference is produced when waves overlaps such that a crest meets a crest and trough meet a trough, i.e. waves are in phase.

In destructive interference, waves are out of phase.

For maxima (or constructive interference), path difference $= n\lambda (n = 0, 1, 2, 3, \ldots)$

For minima (or destructive interference), path difference $= (2n+1)\dfrac{\lambda}{2}$ $(n = 0, 1, 2, 3, \cdots)$

Hence, A $\to$ 3, B $\to$ 4, C $\to$ 1 and D $\to$ 2.

**99** (*c*)

A. If screen is moved away from the plane of slits, i.e. on increasing the value of $D$, the angular separation of the fringes $(= \lambda / d)$ remains constant. But the actual separation of the fringes $\left(\dfrac{\lambda D}{d}\right)$ increases.

B. The separation of the fringes (and also angular separation) decreases, when source is replaced by another source of shorter wavelength.

C. When medium is water, then

$$\lambda' = \frac{\lambda_{air}}{4/3} = \frac{3}{4}\lambda_{air}$$

$$\therefore \text{Fringe width, } \beta' = \frac{\lambda' D}{d} = \frac{3}{4}\left(\frac{\lambda D}{d}\right) = \frac{3}{4}\beta$$

Therefore, fringe width becomes $\left(\dfrac{3}{4}\right)$th.

D. Fringe width, $\beta = \dfrac{D\lambda}{d}$

As, $\beta \propto \dfrac{1}{d}$, so when distance between slits is reduced, linear width of fringes $\beta$ increases.

Hence, A $\to$ 2, B $\to$ 3, C $\to$ 4 and D $\to$ 1.

**100** (*a*) Consider the figure given below

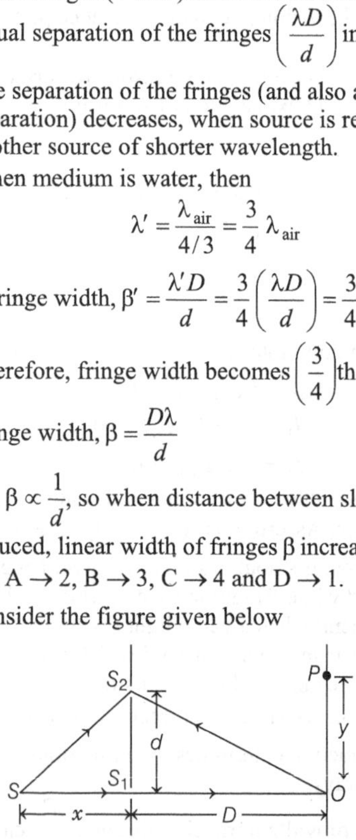


Path difference, $P = (S\,S_2 + S_2O) - (S\,S_1 + S_1O)$ ...(i)

Now from figure, $S\,S_2 = \sqrt{x^2 + d^2}$

$$= x\left(1+\frac{d^2}{x^2}\right)^{1/2} = x\left(1+\frac{d^2}{2x^2}\right) \quad (\because d <<< x)$$

Similarly, $S_2O = \sqrt{(D^2+d^2)} = D\left(1+\frac{d^2}{2D^2}\right) \quad (\because d << D)$

Also, $S\,S_1 + S_1O = x + D$

From Eq. (i),

∴ Path difference,

$$P = x\left(1+\frac{d^2}{2x^2}\right) + D\left(1+\frac{d^2}{2D^2}\right) - (x+D)$$

$$= x + \frac{d^2}{2x} + D + \frac{d^2}{2D} - x - D$$

$$\Rightarrow \quad P = \frac{d^2}{2}\left(\frac{1}{x}+\frac{1}{D}\right) \quad \text{...(ii)}$$

For dark fringe, path difference, $P = \frac{\lambda}{2}$

$$\left[\because P = \frac{(2n-1)\lambda}{2};\ n = 1\right]$$

From Eq.(ii), $\frac{\lambda}{2} = \frac{d^2}{2}\left(\frac{1}{x}+\frac{1}{D}\right)$ or $d = \sqrt{\frac{\lambda x D}{(x+D)}}$

When $x = D$, $d = \sqrt{\frac{\lambda D}{2}}$ and when $x = D/2$, $d = \sqrt{\frac{\lambda D}{3}}$

At $x = D$, fringe width, $\beta = \frac{\lambda D}{d} = \frac{\lambda D}{\sqrt{\lambda D/2}} = 2d$

$$\left(\because \text{at } x = D, d = \sqrt{\frac{\lambda D}{2}}\right)$$

Distance of consecutive bright and dark band

$$= \frac{\text{Fringe width}}{2} = \frac{2d}{2} = d$$

Hence, A → 4, B → 3, C → 2 and D → 1.

**101** (*a*) Fringe width, $\beta = \frac{D\lambda}{d}$

where, $D$ = distance between slits and screen,

$d$ = distance between slits

and $\lambda$ = wavelength of light.

A. If sodium light ($\lambda = 589$nm) is replaced by red light ($\lambda = 700$nm) of same intensity, then $\lambda$ is increases, so $\beta$ also increases.

B. White light produces coloured fringes with centred bright fringe.

C. If $D$ is doubled and $d$ is halved, then $\beta$ $\left(\frac{2\lambda D}{d/2} = \frac{4\lambda D}{d}\right)$ becomes four times.

D. If intensity of either slit is reduced by covering any slit with cellophone paper, the bright fringes became less bright.

Hence, A → 4, B → 1, C → 2 and D → 3.

**102** (*a*) Given, wavelength of light,

$\lambda = 589 \text{ nm} = 589 \times 10^{-9}$ m

Refractive index of water, $\mu_w = 1.33$

(i) **For reflected light**

(a) Wavelength of reflected light, $\lambda = 589 \times 10^{-9}$ m

(b) Frequency of reflected of light,

$$\nu = \frac{c}{\lambda} = \frac{3\times 10^8}{589\times 10^{-9}} \quad (\because c = 3\times 10^8 \text{ ms}^{-1})$$

$\nu = 5.09 \times 10^{14}$ Hz

(c) As the reflection takes place in the same medium so

Speed of reflected light, $c = 3\times 10^8$ ms$^{-1}$

(ii) **For refracted light** (In this process wavelength and speed changes but frequency remains the same)

Wavelength of refracted light,

$$\lambda' = \frac{\lambda}{\mu} = \frac{589\times 10^{-9}}{1.33} = 4.42\times 10^{-7} \text{ m}$$

∴ Velocity of refracted light,

$$v = \frac{c}{\mu} = \frac{3\times 10^8}{1.33} = 2.25\times 10^8 \text{ms}^{-1}$$

**103** (*b*) Given, separation between slits,

$d = 0.28 \text{ mm} = 0.28\times 10^{-3}$ m

Distance between screen and slit, $D = 1.4$ m

Distance between central bright and fourth fringe,

$x = 1.2 \text{ cm} = 1.2\times 10^{-2}$m

Number of fringes, $n = 4$

For constructive interference, $x = n\frac{D\lambda}{d}$

$$1.2\times 10^{-2} = \frac{4\times 1.4\times \lambda}{0.28\times 10^{-3}}$$

$$\text{Wavelength, } \lambda = \frac{1.2\times 10^{-2}\times 0.28\times 10^{-3}}{4\times 1.4}$$

$$\Rightarrow \quad \lambda = 6\times 10^{-7} \text{ m}$$

**104** (*a*) Given, wavelength, $\lambda_1 = 650 \text{ nm} = 650\times 10^{-9}$m

& $\lambda_2 = 520 \text{ nm} = 520\times 10^{-9}$m

and slit width, $d = 2 \text{mm} = 2\times 10^{-3}$m

(i) For third bright fringe, $n = 3, D = 1.2$ m

The distance of third bright fringe from central maximum,

$$x = \frac{n\lambda_1 D}{d} = \frac{3\times 650\times 10^{-9}\times 1.2}{2\times 10^{-3}}$$

$$= 1.17\times 10^{-3} \text{ m}$$

(ii) Let $n$th bright fringe due to wavelength, $\lambda_2 = 520$ nm, coincide with $(n+1)$th bright fringe due to wavelength, $\lambda_1 = 650$ nm.

i.e. $n\lambda_2\frac{D}{d} = (n-1)\lambda_1\frac{D}{d}$

$n\times 520\times 10^{-9} = (n-1)\,650\times 10^{-9}$

or $4n = 5n - 5$ or $n = 5$

Thus, the least distance,

$$x = n\lambda_2 \frac{D}{d} = 5\times 520\times 10^{-9}\times \frac{D}{d}$$

$$= 2600\times \frac{1.2\times 10^{-9}}{2\times 10^{-3}} \text{ m}$$

$$= 1.56\times 10^{-3} \text{ m}$$

**105** (*b*) Given, wavelength of light,

$\lambda = 5000\text{Å} = 5000\times 10^{-10}$ m

On the reflection, there is no change in wavelength and frequency, so wavelength of reflected light will be 5000Å.

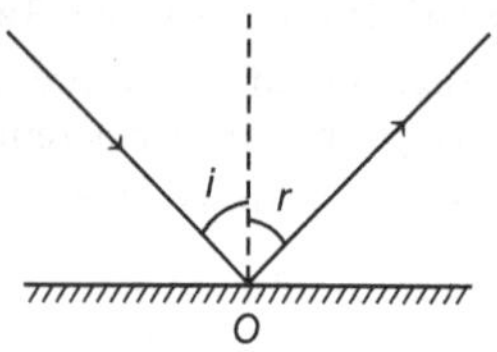


Frequency of the incident light,

$$\nu = \frac{c}{\lambda} = \frac{3\times 10^8}{5\times 10^{-7}} = 6\times 10^{14} \text{ Hz}$$

When reflected ray is normal to the incident ray,

i.e. $BO \perp AO$

$\therefore$ $\angle i + \angle r = 90°$

For reflection, $\angle i = \angle r$

$\therefore$ $2\angle i = 90°$

$\angle i = 45°$

Thus, the angle of incidence is 45°.

**106** (*b*) Given, wavelength of $H_\alpha$,

$\lambda = 6563\text{Å} = 6563\times 10^{-10}$ m

Red shift, $\Delta\lambda$=15Å

Since, the star is found to be red-shifed, hence star is receding away from earth and Doppler's shift is negative.

$$\Delta\lambda = -\frac{v\lambda}{c}$$

$$\Rightarrow \quad v = -\frac{\Delta\lambda\cdot c}{\lambda} = -\frac{15\times 10^{-10}\times 3\times 10^8}{6563\times 10^{-10}}$$

$$v = -6.86\times 10^5 \text{ ms}^{-1}$$

Negative sign shows that the star is receding away from earth.

**107** (*c*) Given, wavelength of light, $\lambda = 600\text{nm} = 600\times 10^{-9}$ m

Angular width of fringe, $\theta = 0.1° = \frac{0.1\pi}{180}$ rad

Using the formula, $\theta = \frac{\lambda}{d}$

$\therefore$ Spacing between the slits,

$$d = \frac{\lambda}{\theta} = \frac{600\times 10^{-9}\times 180}{0.1\times \pi}$$

$$d = 3.44\times 10^{-4} \text{ m}$$

Thus, the spacing between the two slits is $3.44\times 10^{-4}$ m.

**108** (*a*) There is no obstruction by the hill to the spreading of the radio beams, provided the radial spread of the beam over the hill 20 km away must not exceed 50 m.

i.e. $Z_F$ (Fresnel's distance) $= 20 \text{ km} = 20\times 10^{-3}$ m

and $a = 50$ m

$$\therefore \quad Z_F = \frac{a^2}{\lambda}$$

$$\Rightarrow \quad \lambda = \frac{a^2}{Z_F} = \frac{50\times 50}{20\times 10^3} = 1250\times 10^{-4} \text{ m}$$

Thus, the longest wavelength of radiowaves is 0.125 m.

**109** (*c*) Given, wavelength of light,

$\lambda = 500 \text{ nm} = 500\times 10^{-9}$ m

$D = 1$ m, $n = 1$,

and $x = 2.5 \text{ mm} = 2.5\times 10^{-3}$ m

Distance of *n*th minima from the centre, $x = \frac{nD\lambda}{d}$

$$d = \frac{nD\lambda}{x} = \frac{1\times 1\times 500\times 10^{-9}}{2.5\times 10^{-3}} = 2\times 10^{-4} \text{ m}$$

$\Rightarrow$ $d = 0.2$ mm

Thus, the width of slit of 0.2 mm.

**110** (*a*) Given, width of the slit $= 10^4$ Å $= 10000$Å

$= 10^4 \times 10^{-10}$ m $= 10^{-6}$ m $= 1\mu$m

Wavelength of (visible) sunlight varies from 4000 Å to 8000 Å.

As the width of slit is comparable to that of wavelength, hence, diffraction occurs with maxima at centre. So, at the centre all colours appear, i.e. mixing of colours form white patch at the centre.

Therefore, the image seen through the slit shall be a fine sharp slit white in colour at the centre.

**111** (*c*) In a Young's double slit experiment, when one of the holes is covered by a red filter and another by a blue filter. In this case due to filteration, only red and blue lights are present. In Young's double slit experiment, monochromatic light is used for the formation of fringes on the screen. Hence, in given case, there shall be no interference fringes.

**112** (*d*) As given, there is a hole at point $P_2$ (minima). The hole will act as a source of fresh light source for the slits $S_3$ and $S_4$. Therefore, there will be a regular two slit pattern on the second screen.

**113** (*a*) Given, angular resolution of human eye,

$\phi = 5.8\times 10^{-4}$ rad and printer prints 300 dots per inch.

The linear distance between two dots is

$$l = \frac{2.54}{300} = 0.84\times 10^{-2} \text{ cm}$$

At a distance of *z* cm, this subtends an angle,

$$\phi = \frac{l}{z}$$

$$\therefore \quad z = \frac{l}{\phi} = \frac{0.84\times 10^{-2}}{5.8\times 10^{-4}} = 14.5 \text{ cm}$$

CHAPTER > 11

# Dual Nature of Radiation and Matter

## KEY NOTES

## Electron Emission

- It is the phenomenon of emission of electrons from the surface of the metal.
- For this phenomenon to take place a certain minimum amount of energy is required for an electron to pull it out from the surface of the metal. This minimum energy required by an electron to escape from the metal surface is called the **work function** of the metal.

Following are few important points related to work function

(i) It is generally denoted by $\phi_0$ and measured in eV (electron-volt). One electron volt is the energy gained by an electron when it has been accelerated by a potential difference of 1 V, so that $1\ \text{eV} = 1.602 \times 10^{-19}\text{J}$.

(ii) $\phi_0$ depends on the properties of the metal and the nature of its surface.

- The minimum energy required for the electron emission from the metal surface can be supplied to the free electrons by any of the following physical processes

(i) **Thermionic Emission** By suitable heating, sufficient thermal energy can be imparted to the free electrons to enable them to come out of the metal.

(ii) **Field Emission** By applying a very strong electric field (of the order of $10^8\ \text{Vm}^{-1}$) to a metal, electrons can be pulled out of the metal, as in a spark plug.

(iii) **Photoelectric Emission** When light of suitable frequency illuminates a metal surface, electrons are emitted from the metal surface. These photo (light) generated electrons are called **photoelectrons**.

## Photoelectric Effect

It is the phenomenon of emission of photoelectrons from the surface of metal when radiation of suitable frequency fall on it. It is based upon the following observations

- **Hertz's Observations** He observed that high voltage sparks across the detector loop were enhanced when the emitter plate was illuminated by ultraviolet light from an arc lamp.
- **Hallwach's and Lenard's Observations**

(i) Lenard observed that when ultraviolet radiations were allowed to fall on emitter plate of an evacuated glass tube enclosing two electrodes (metal plates), current flows in the circuit. As soon as the ultraviolet radiations were stopped, the current flow also stopped.

(ii) Lenard and Hallwach also observed that when ultraviolet light fall on the emitter plate, no electrons were emitted at all when the frequency of the incident light was smaller than a certain minimum value called **threshold frequency**. This minimum frequency depends on the nature of the material of the emitter plate.

### Experimental Study of Photoelectric Effect

- On the basis of the experimental arrangement used for studying photoelectric effect, the variations of photo current with intensity of radiation, frequency of radiation and the potential difference between the plates $A$ and $C$ of photoelectric with intensity are as follows

- **Effect of Intensity of Light on Photocurrent** It is found that the photocurrent increases linearly with intensity of incident light as shown in the graph below

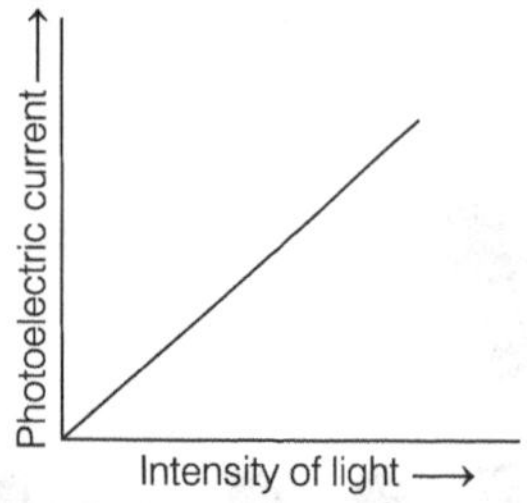


- **Effect of Potential on Photoelectric Current** Variation of photocurrent with collector plate potential for different intensity of incident radiation is as shown below

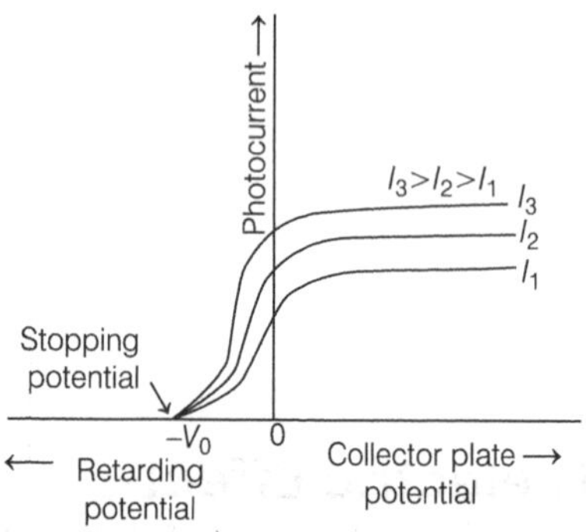


From the above graph, we can observe that,

(i) For a fixed frequency and intensity of incident radiation, the photocurrent increases with increase in the potential applied to the collector. At some stage, the photoelectric current becomes maximum or saturates.

(ii) Maximum value of photoelectric current is called **saturation current**. It corresponds to the case when all the photoelectrons emitted by the emitter plate reaches the collector plate.

(iii) For a particular frequency of incident radiation, the minimum negative (retarding) potential $V_0$ given to the plate $A$ for which the photocurrent stops or becomes zero is called the **cut-off** or **stopping potential**.

(iv) Photoelectric current is zero when the stopping potential is sufficient to repel even the most energetic photoelectrons with the maximum kinetic energy $K_{max}$, so that $K_{max} = eV_0$.

(v) Thus, for a given frequency of the incident radiation, the stopping potential is independent of its intensity. In other words, the maximum kinetic energy of photoelectrons depends on the light source and the emitter plate material, but it is independent of intensity of incident radiation.

- **Effect of Frequency of Incident Radiation on Stopping Potential** Variation of photoelectric current with collector plate potential for different frequencies of incident radiation is as shown below

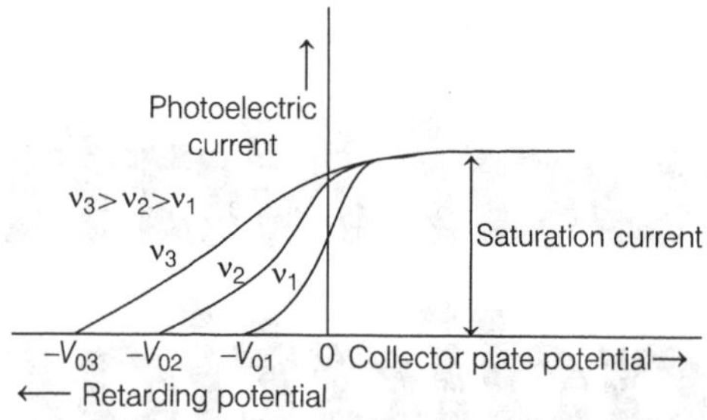


From the above graph, we observe that,

(i) Stopping potentials are in the order of $V_{03} > V_{02} > V_{01}$, if the frequencies are in the order of $\nu_3 > \nu_2 > \nu_1$.

(ii) This implies that greater the frequency of incident light, greater is the maximum kinetic energy of photoelectrons.

(iii) Consequently, greater retarding potential is required to stop them completely.

- The graph between the frequency of incident radiation and the corresponding stopping potential for different metals $A$ and $B$ is straight line as shown below

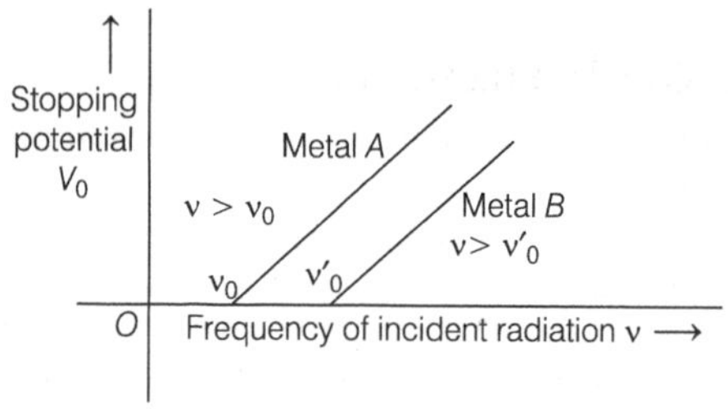


This graph shows that,

(i) The stopping potential $V_0$ varies linearly with the frequency of incident radiation for a given photosensitive material.

(ii) There exist a certain minimum cut-off frequency $\nu_0$ for which the stopping potential is zero.

**Note** The stopping potential is more negative for higher frequencies of incident radiations.

- From the above observations, two implications are as follows

(i) The maximum kinetic energy of the photoelectrons varies linearly with the frequency of incident radiation, but it is independent of its intensity.

(ii) For a frequency $\nu$ of incident radiation, lower than the cut-off frequency $\nu_0$, no photoelectric emission is possible even, if the intensity is large.

This minimum, cut-off frequency $\nu_0$ is called the **threshold frequency**. It is different for different metals.

KEY NOTES

**Note Threshold wavelength** $\lambda_0$ is the maximum wavelength of incident radiation above which no photoelectric emission takes place.

- If frequency of incident radiation exceeds the threshold frequency, the photoelectric emission starts instantaneously without any apparent time lag ($\sim 10^9$ s or less) even, if the incident radiation is very dim.

## Photoelectric Effect and Wave Theory of Light

- According to wave picture of light, photoelectric effect can be explained as follows
  - (i) Free electrons at the surface of the metal absorb the radiant energy continuously. Thus, maximum kinetic energy of the photoelectrons on the surface is then expected to increase with the increase in intensity.
  - (ii) No matter what the frequency of incident radiation is, a sufficiently intense beam of radiation (over sufficient time) should be able to impart enough energy to the electrons, so that they exceed the minimum energy needed to escape from the metal surface.
  - (iii) Since, the absorption of energy by electron takes place continuously, so it can take hours or more for a single electron to pick up sufficient energy to overcome the work function and come out of the metal.

  Thus, wave picture is unable to explain the basic features of photoelectric emission.

### Energy Quantum of Radiation : Einstein's Photoelectric Equation

- Radiation energy is build up of discrete units the so called **quanta of energy** of **radiation**. Each quantum of radiant energy has energy $h\nu$, where $h$ is Planck's constant and $\nu$ is frequency of light.
- Einstein's photoelectric equation is given as

$$K_{max} = \frac{1}{2}mv_{max}^2 = h(\nu - \nu_0) = h\nu - \phi_0$$

  where, $\nu$ is the frequency of incident radiation and $K_{max}$ is the maximum kinetic energy of the emitted electron.
- In terms of threshold wavelength $\lambda_0$, above equation can be written as $eV_0 = hc\left(\frac{1}{\lambda} - \frac{1}{\lambda_0}\right)$

  where, $\lambda$ is the wavelength of incident photon.

### Particle Nature of Light : The Photon

- Photoelectric effect gave evidence that light in interaction with matter behaved as if it was made of quanta or packets of energy, each of energy $h\nu$, this packet of energy is associated with a particle called photon.
- **Properties of photon** are as follows
  - (i) Each photon has energy $E(= h\nu)$ and momentum $p\left(= \frac{h\nu}{c}\right)$ and speed of light $c$.
  - (ii) All photons of light at a particular frequency $\nu$ or wavelength $\lambda$, have the same energy $E(= h\nu = hc/\lambda)$ and momentum $p(= h\nu/c = h/\lambda)$, whatever the intensity of radiation may be.
  - (iii) Photons are electrically neutral and are not deflected by electric and magnetic fields.
  - (iv) In a photon-particle collision (such as photon-electron collision), the total energy and total momentum are conserved. However, the number of photons may not be conserved in a collision. The photon may be absorbed or a new photon may be created.

## Wave Nature of Matter

- With the experimental verification, it was proved that light has dual nature, in some phenomena, it behaves like wave and in some phenomena, it behaves like particle depending upon the dimensions of object with which the particle interacts.

### Matter Waves : de-Broglie Waves

- Louis Victor de-Broglie put forward a hypothesis that moving particles of matter should display wave-like properties under suitable conditions.
- de-Broglie thus proposed that the wavelength $\lambda$ associated with a moving particle is given as

$$\lambda = \frac{h}{p} = \frac{h}{mv}$$

  where, $m$ is the mass of the particle and $v$ is speed.
- For charged particles accelerated through a potential $V$,

  wavelength, $\lambda = \frac{h}{\sqrt{2mqV}}$

  Specifically, for electrons $\lambda = \frac{1.227}{\sqrt{V}}$ nm.

### Heisenberg's Uncertainty Principle

- According to this principle, it is impossible to measure position and momentum of the microscopic particle simultaneously. Mathematically, it is represented as

$$\Delta x \cdot \Delta p \geq \frac{\hbar}{2}, \text{ where } \hbar = \frac{h}{2\pi}$$

  i.e. If $\Delta x = 0$, then $\Delta p = \infty$ and if $\Delta p = 0$, then $\Delta x = \infty$.

### Davisson and Germer Experiment

- The wave nature of electron was experimentally verified by Davisson-Germer experiment.
- It was observed in the experiment that, intensity of scattered beam of electrons was found to be maximum when angle of scattering is 50° and accelerating potential is 54 V.
- The wavelength of the matter wave found experimentally and from the de-Brogile relation is approximately equivalent, i.e. 0.167 nm. Thus, there is an excellent agreement between theoretical and the experimental obtained value of de-Broglie wavelength.

KEY NOTES

# Mastering NCERT

## MULTIPLE CHOICE QUESTIONS

## TOPIC 1 ~ *Electron Emission*

**1** At sufficiently low pressure of about 0.001mm of Hg column in a discharge tube, a fluorescent glow appears which was attributed due to

(a) positive rays
(b) cathode rays
(c) $\gamma$-rays
(d) discharge tube gases only

**2** Value of $e/m$ (specific charge) of the cathode ray particles

(a) depends on potential difference of cathode and anode
(b) depends on nature of gas in the discharge tube
(c) is independent of material of cathode or gas in tube
(d) depends on nature of metal used for cathode

**3** The specific charge of a proton is $9.6\times10^{7}\text{C kg}^{-1}$. The specific charge of an $\alpha$-particle will be

(a) $9.6\times10^{7}\text{ C kg}^{-1}$
(b) $19.2\times10^{7}\text{ C kg}^{-1}$
(c) $4.8\times10^{7}\text{ C kg}^{-1}$
(d) $2.4\times10^{7}\text{ C kg}^{-1}$

**4** R.A Millikan during his oil-drop experiment on the precise measurement of the charge on electron found that

(a) elementary charge, $e$ on an oil droplet is $1.76\times10^{-19}$ C
(b) charge on an oil droplet was always an even multiple of $e$
(c) electric charge is quantised
(d) charge on an oil droplet was always an odd multiple of $e$

**5** Free electrons cannot normally escape out of metal surface because

(a) free electrons are bounded to one or other ionic core
(b) free electrons lie deep inside atom
(c) free electrons are held in metal lattice by attraction force of protons
(d) if an electron attempts to come out of metal, the metal surface acquires a positive charge and pulls the electron back to metal

**6** Work-function is

(a) maximum possible energy acquired by an electron
(b) energy of electrons in valence shell
(c) minimum energy required by an electron to move out of metal surface
(d) maximum energy which is given to electron to move it out of metal surface

**7** Amongst the given metals in the following option, work function is least for which metal?

(a) Cesium
(b) Aluminium
(c) Silver
(d) Platinum

**8** Work-function $\phi_0$ for a metal will change, if it is

(a) heated
(b) cooled
(c) coated with some other metal
(d) All of the above

**9** Energy required by an electron for electron emission can be supplied to it by

I. hammering the metal surface.
II. heating the metal surface.
III. applying electric field.
IV. applying magnetic field.

(a) Only I
(b) I, II and IV
(c) II, III and IV
(d) Both II and III

## TOPIC 2 ~ Photoelectric Effect and Wave Theory

**10** Lenard observed that when UV-radiations were allowed to fall on the emitter plate $C$ of an evacuated glass tube enclosing two electrodes $C$ and $A$ as shown in figure, then current flows in the circuit as

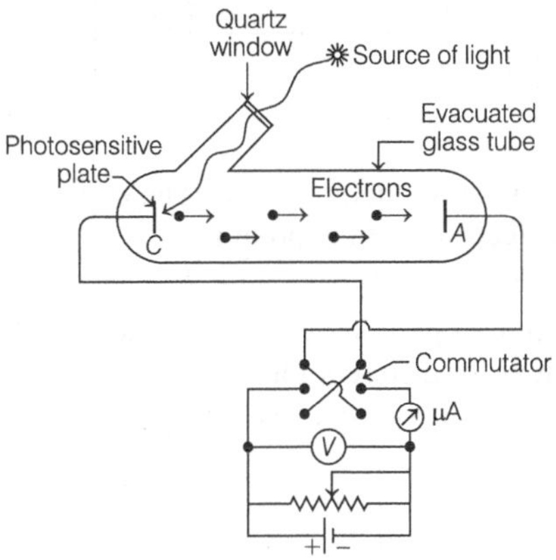


(a) UV-light ionises the traces of gas left after evacuation
(b) UV-light produces ionisation of cathode and anode
(c) UV-light causes ejection of electrons from the emitter plate $C$
(d) Both (a) and (b)

**11** Hallwach connected an uncharged zinc plate to an electroscope as shown below.

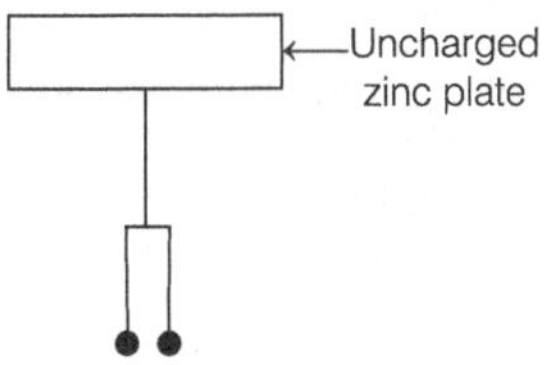


This plate is then irradiated by UV-light. The result of this is correctly shown in

(a) 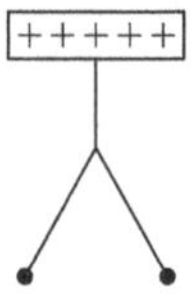


(b) 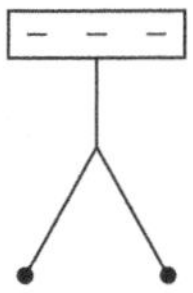


(c) 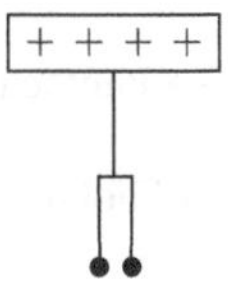


(d) 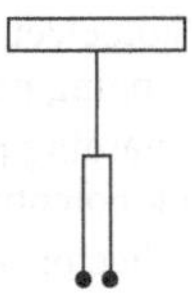

**12** Lenard observed that no free electrons are emitted when frequency of light is less than a certain minimum frequency. This minimum frequency depends on
(a) potential difference of emitter and collector plates
(b) distance between collector and the emitter plate
(c) size (area) of the emitter plate
(d) material of the emitter plate

**13** When yellow light is incident over a metal surface, no electrons are emitted while green light can emit electrons. If red light is incident over the surface, then
(a) no electrons are emitted
(b) more electrons are emitted
(c) electrons of higher energy are emitted
(c) electrons of lower energy are emitted

**14** Photoemission with visible light is possible in
(a) rubidium (b) cesium
(c) zinc (d) Both (a) and (b)

**15** Photoelectric effect involves conversion of
(a) nuclear energy into electrical energy
(b) atomic energy into electrical energy
(c) electronic energy into electrical energy
(d) light energy into electrical energy

**16** Variation of photoelectric current with intensity of light is correctly depicted in

(a) 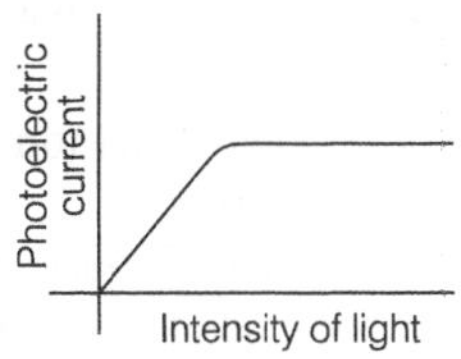


(b) 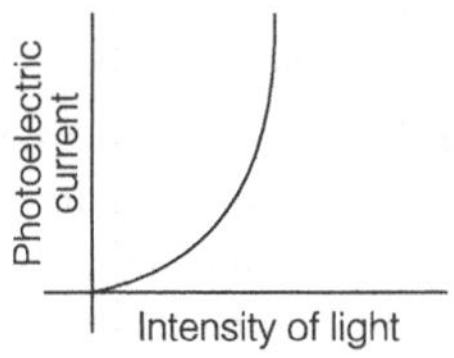


(c) 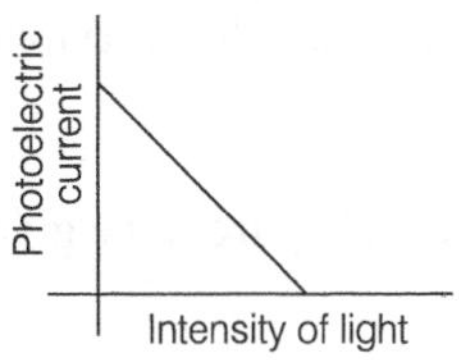


(d) 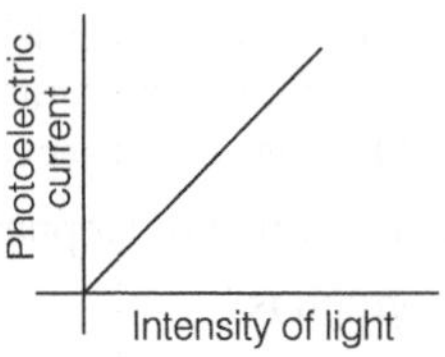


**17** Number of photoelectrons emitted per second from the surface of the metal during photoemission keeping the frequency constant, increases with
(a) increase in intensity of incident radiation
(b) increase in angle of incidence of incident radiation
(c) decrease in potential difference between collector and emitter plates
(d) Both (a) and (b)

**18** A point source of light is used in an experiment on photoelectric effect. Which of the following curves best represents the variation of photoelectric current $i$ with distance $d$ of the source from the emitter?

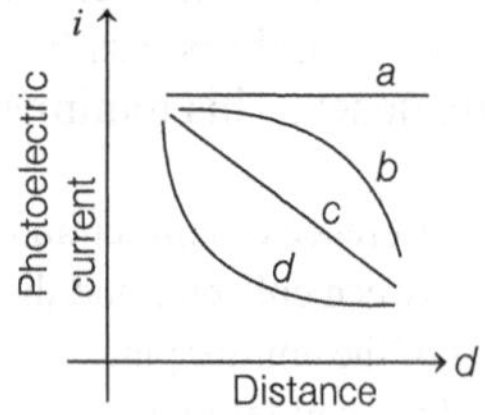


(a) $a$ (b) $b$ (c) $c$ (d) $d$

**19** In an experiment on photoelectric effect with the increase in potential difference of emitter and collector plate, keeping the frequency and intensity fixed of the incident light the photoelectric current

(a) increases
(b) decreases
(c) remains constant
(d) increases initially and then becomes constant

**20** At a certain positive potential of collector plate, photocurrent becomes maximum or saturates. Saturation current corresponds to the case when

(a) half of photoelectrons emitted by emitter plate reaches collector
(b) more than half of photoelectrons emitted by emitter plate reaches collector
(c) less than half of photoelectrons emitted by emitter plate reaches collector
(d) All of the photoelectrons emitted by emitter plate reaches the collector plate

**21** For a particular frequency of incident radiation, cut-off or stopping potential refers to

(a) maximum possible negative potential of collector
(b) minimum possible negative potential of collector for which photoelectric current is zero
(c) collector potential for which photoelectric current is saturated at constant intensity
(d) None of the above

**22** The value of stopping potential $V_0$ from the given graph is

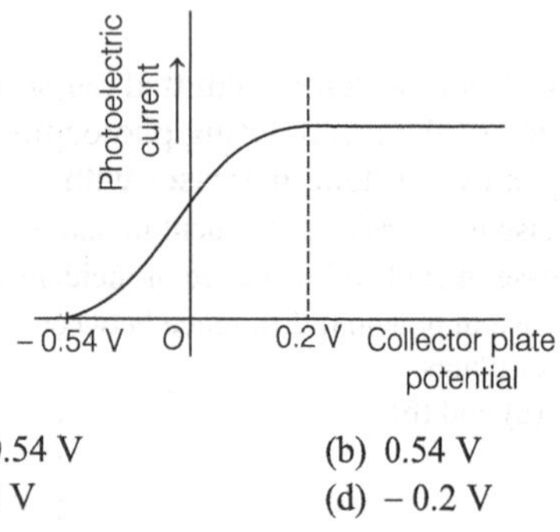


(a) – 0.54 V (b) 0.54 V
(c) 0.2 V (d) – 0.2 V

**23** In photoelectric effect experiment, collector plate is made negative with respect to emitter plate as shown in figure below till it reach a certain potential $V_0$, when photocurrent is zero.

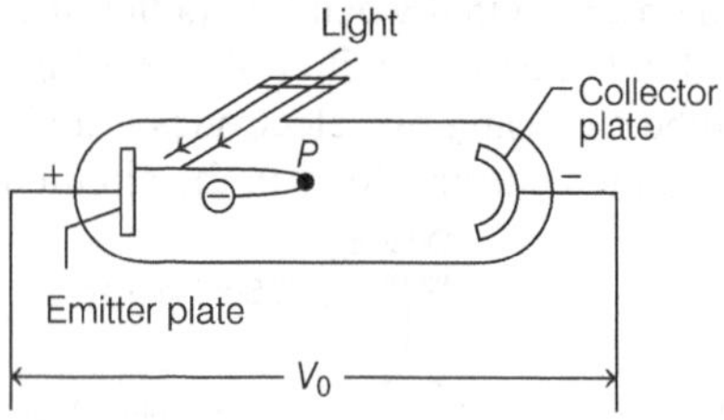


If $K$ indicates kinetic energy of an emitted photoelectron, then at point $P$

(a) $K > eV_0$ (b) $K < eV_0$ (c) $K = eV_0$ (d) $0 \le K \le eV_0$

**24** Variation of photoelectric current with collector plate potential for different intensities $I_1$ and $I_2$ (such that $I_1 > I_2$) at a fixed frequency is

(a) 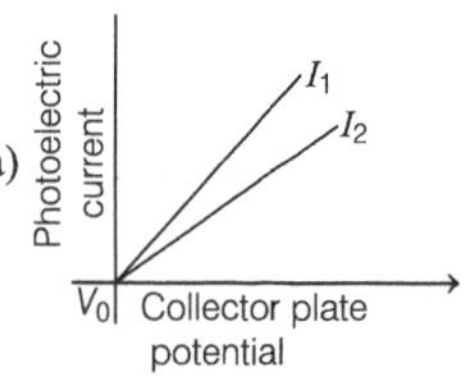


(b) 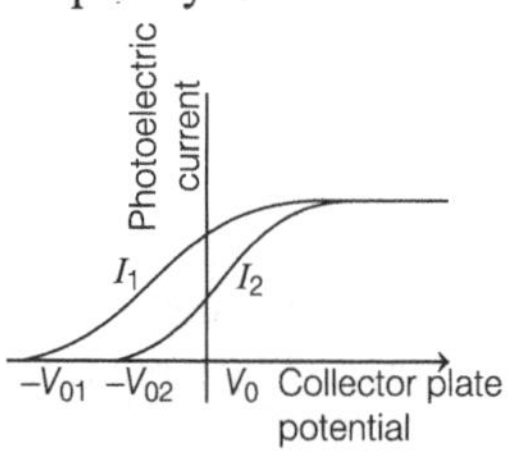


(c) 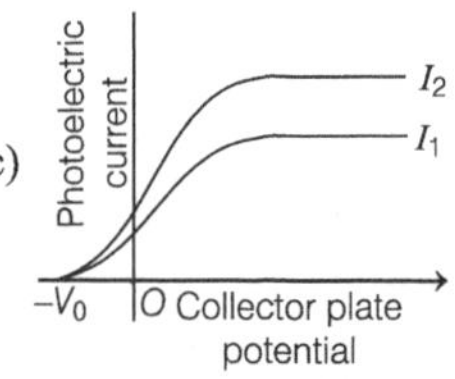


(d) 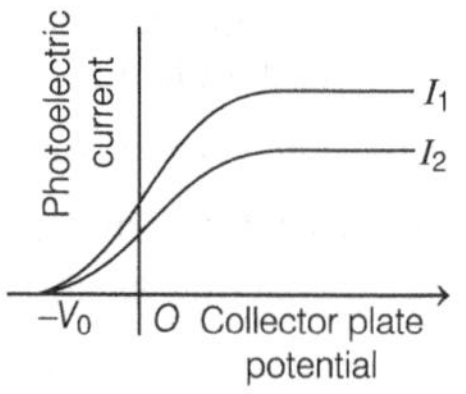


**25** A student plot a graph while performing an experiment on photoelectric effect using an evacuated glass tube, for light radiation of same intensity at various frequencies as shown below

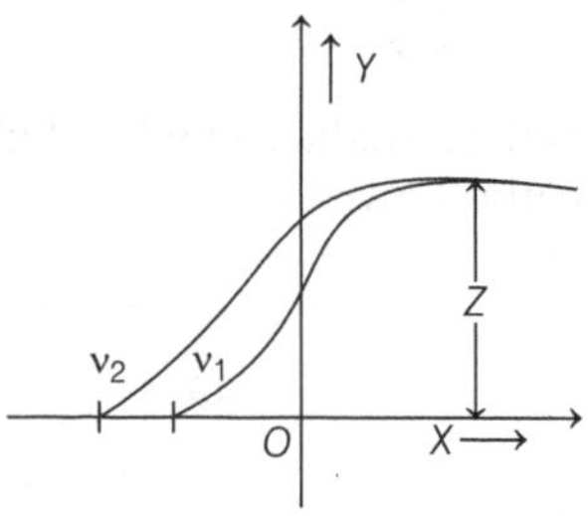


Here,

(a) $X \to$ Photoelectric current; $Y \to$ Retarding potential; $Z \to$ Stopping potential
(b) $X \to$ Retarding plate potential; $Y \to$ Photocurrent; $Z \to$ Stopping potential
(c) $X \to$ Collector plate potential; $Y \to$ Photocurrent; $Z \to$ Saturation current
(d) $X \to$ Retarding plate potential; $Y \to$ Photocurrent; $Z \to$ Saturation current

**26** Maximum kinetic energy of emitted photoelectrons depends on
(a) intensity of incident radiation only
(b) frequency of incident radiation only
(c) both on intensity and frequency of incident radiation
(d) does not depends on frequency or intensity of incident radiation

**27** Photoelectric emission of electrons
(a) is instantaneous process with a time lag of less than nano second
(b) is slow with a large time lag
(c) shows a time lag of around 1 s
(d) show time lag of less than 1 s

**28** A beam of wavelength $\lambda$ and intensity $I$ falls over a clean surface of sodium metal. If $N$ photoelectrons are emitted with kinetic energy $E$, then
(a) $N \propto I$ and $E \propto I$ (b) $N \propto I$ and $E \propto \frac{1}{\lambda}$
(c) $N \propto \lambda$ and $E \propto I$ (d) $N \propto \frac{1}{\lambda}$ and $E \propto \frac{1}{I}$

**29** The cathode of a photoelectric cell is changed such that the work function changes from $W_1$ to $W_2$ $(W_2 > W_1)$. If the current before and after are $i_1$ and $i_2$, respectively and all other conditions remaining unchanged, then
(a) $i_1 = i_2$ (b) $i_1 < i_2$ (c) $i_1 > i_2$ (d) $\frac{W_1}{W_2} = \frac{i_1}{i_2}$

**30** Experimental study of photoelectric effect shows that
I. photocurrent $\propto$ intensity of light.
II. saturation current $\propto$ intensity of light.
III. photoemission occurs only at frequency greater than threshold frequency.
IV. photoemission is an instantaneous process.
Which of the following observation(s) is/are correct?
(a) Both I and II
(b) I, II and III
(c) I, III and IV
(d) I, II, III and IV

**31** Phenomena of interference, diffraction and polarisation are explained by considering light as
(a) wave (b) particle
(c) corpuscles (d) quarks

**32** Experimental observation of photoelectric effect, if explained in terms of wave picture of light, then
(a) maximum kinetic energy first increases then decreases with frequency
(b) a sufficiently intense beam of radiation is enough for photoemission
(c) time required for photoemission is around $10^{-9}$ s
(d) maximum kinetic energy of photo electrons decreases with intensity

## TOPIC 3 ~ Energy Quantum of Radiation : Einstein's Photoelectric Equation

**33** According to Einstein's photoelectric equation, $K_{max} \propto \nu$ because an electron is emitted
(a) after absorption of few photons when absorbed energies exceeds certain value
(b) after number of photons falling over surface exceeds of critical volume
(c) only, if it is present on the surface
(d) after absorption of a single photon of sufficient energy

**34** The work function of platinum is 6.35 eV. The threshold frequency of platinum is
(a) $15.32 \times 10^{14}$ Hz (b) $15.32 \times 10^{16}$ Hz
(c) $15.32 \times 10^{19}$ Hz (d) $15.32 \times 10^{18}$ Hz

**35** The work function of a photosensitive material is 4.0 eV. This longest wavelength of light that can cause photon emission from the substance is (approximately) ***NEET (Odisha) 2019***
(a) 3100 nm (b) 966 nm (c) 31 nm (d) 310 nm

**36** The photoelectric threshold frequency of a metal is $\nu$. When light of frequency $6\nu$ is incident on the metal, the maximum kinetic energy of the emitted photoelectron is
(a) $4h\nu$ (b) $5h\nu$ (c) $3h\nu$ (d) $(3/2)\,h\nu$

**37** Light of wavelengths $\lambda_A$ and $\lambda_B$ falls on two identical metal plates $A$ and $B$, respectively. The maximum kinetic energy of photoelectrons is $K_A$ and $K_B$ respectively, then which one of the following relations is true? (Take, $\lambda_A = 2\lambda_B$)
(a) $K_A < \frac{K_B}{2}$ (b) $2K_A = K_B$
(c) $K_A = 2K_B$ (d) $K_A > 2K_B$

**38** A metallic surface is irradiated by a monochromatic light of frequency $\nu_1$ and stopping potential is found to be $V_1$. If the light of frequency $\nu_2$ irradiates the surface. The stopping potential will be
(a) $V_1 + \frac{h}{e}(\nu_1 + \nu_2)$ (b) $V_1 - \frac{h}{e}(\nu_1 - \nu_2)$
(c) $V_1 + \frac{e}{h}(\nu_2 - \nu_1)$ (d) $V_1 - \frac{h}{e}(\nu_1 + \nu_2)$

**39** Photoelectric emission is observed from a metallic surface for frequencies $\nu_1$ and $\nu_2$ of the incident light $(\nu_1 > \nu_2)$. If the maximum values of kinetic energy of the photoelectrons emitted in the two cases are in the ratio $1:n$, then the threshold frequency of the metallic surface is

(a) $\frac{(\nu_1 - \nu_2)}{(n-1)}$ (b) $\frac{(n\nu_1 - \nu_2)}{(n-1)}$

(c) $\frac{(n\nu_2 - \nu_1)}{(n-1)}$ (d) $\frac{(\nu_1 - \nu_2)}{n}$

**40** For a certain metal, incident frequency $\nu$ is five times the threshold frequency $\nu_0$ and the maximum velocity of emitted photoelectrons is $8 \times 10^6$ ms$^{-1}$. If $\nu = 2\nu_0$, the maximum velocity of photoelectrons will be

(a) $4 \times 10^6$ ms$^{-1}$ (b) $6 \times 10^6$ ms$^{-1}$

(c) $8 \times 10^6$ ms$^{-1}$ (d) $1 \times 10^6$ ms$^{-1}$

**41** The photoelectric threshold wavelength of silver is $3250 \times 10^{-10}$ m. The velocity of the electron ejected from a silver surface by ultraviolet light of wavelength $2536 \times 10^{-10}$ m is (given, $h = 4.14 \times 10^{-15}$ eV-s and $c = 3 \times 10^8$ ms$^{-1}$) **NEET 2017**

(a) $\approx 6 \times 10^5$ ms$^{-1}$ (b) $\approx 0.6 \times 10^6$ ms$^{-1}$

(c) $\approx 61 \times 10^3$ ms$^{-1}$ (d) $\approx 0.3 \times 10^6$ ms$^{-1}$

**42** In a photoelectric experiment, the wavelength of the light incident on a metal is changed from 300 n-m to 400 n-m. The decrease in the stopping potential is close to $\left(\frac{hc}{e} = 1240 \text{ n-mV}\right)$ **JEE Main 2019**

(a) 0.5 V (b) 2.0 V (c) 1.5 V (d) 1.0 V

**43** A photoelectric surface is illuminated successively by monochromatic light of wavelengths $\lambda$ and $\lambda/2$. If the maximum kinetic energy of the emitted photoelectrons in the second case is 3 times than in the first case, the work function of the surface of the material is ($h$ = Planck's constant and $c$ = speed of light) **CBSE AIPMT 2015**

(a) $\frac{hc}{2\lambda}$ (b) $\frac{hc}{\lambda}$ (c) $\frac{2hc}{\lambda}$ (d) $\frac{hc}{3\lambda}$

**44** In a photoelectric effect experiment, the threshold wavelength of light is 380 nm. If the wavelength of incident light is
260 nm, the maximum kinetic energy of emitted electrons will be

Given, $E \text{ (in eV)} = \frac{1237}{\lambda\text{(in nm)}}$ **JEE Main 2019**

(a) 15.1 eV (b) 3.0 eV (c) 1.5 eV (d) 4.5 eV

**45** In an experiment with rubidium photocell, the following lines from a mercury source were used $\lambda_1 = 3650$ Å, $\lambda_2 = 4047$ Å, $\lambda_3 = 4358$ Å, $\lambda_4 = 5461$ Å, $\lambda_5 = 6907$ Å and stopping potentials respectively are $V_{01} = 1.28$ V, $V_{02} = 0.95$ V, $V_{03} = 0.74$ V, $V_{04} = 0.16$ V, $V_{05} = 0$ V.
Threshold frequency and work-function of metal respectively are

(a) $4 \times 10^{14}$ Hz, 1.5 eV

(b) $4.3 \times 10^{14}$ Hz, 1.8 eV

(c) $4 \times 10^{14}$ Hz, 3 eV

(d) $1.5 \times 10^{14}$ Hz, 5 eV

**46** When the light of frequency $2\nu_0$ (where, $\nu_0$ is threshold frequency) is incident on a metal plate, the maximum velocity of electrons emitted is $v_1$. When the frequency of the incident radiation is increased to $5\nu_0$, the maximum velocity of electrons emitted from the same plate is $v_2$. The ratio of $v_1$ to $v_2$ is

(a) 4 : 1 (b) 1 : 4 **NEET 2018**

(c) 1 : 2 (d) 2 : 1

**47** When a metallic surface is illuminated with radiation of wavelength $\lambda$, the stopping potential is $V$. If the same surface is illuminated with radiation of wavelength $2\lambda$, the stopping potential is $\frac{V}{4}$. The threshold wavelength for the metallic surface is **NEET 2016**

(a) $5\lambda$ (b) $\frac{5}{2}\lambda$

(c) $3\lambda$ (d) $4\lambda$

**48** Light of wavelength 600 nm from a sodium lamp falls on a photocell and causes the emission of photoelectrons for which the stopping potential is 0.5 V and with the light of wavelength 400 nm from a sodium lamp, the stopping potential is 1.5 V. With this data, the value of $h/e$ is

(a) $4 \times 10^{-59}$ V-s (b) $0.25 \times 10^{15}$ V-s

(c) $4 \times 10^{-15}$ V-s (d) $4 \times 10^{-8}$ V-s

**49** Radiation of wavelength $\lambda$ is incident on a photocell. The fastest emitted electron has speed $v$. If the wavelength is changed to $\frac{3\lambda}{4}$, the speed of the fastest emitted electron will be

(a) $> v\left(\frac{4}{3}\right)^{1/2}$ (b) $< v\left(\frac{4}{3}\right)^{1/2}$

(c) $= v\left(\frac{4}{3}\right)^{1/2}$ (d) $= v\left(\frac{3}{4}\right)^{1/2}$

**50** Monochromatic radiation of wavelength 640.2 nm (1 nm $=10^{-9}$m) from a neon lamp irradiates photosensitive material made of caesium on tungsten. The stopping voltage is measured to be 0.54 V.

The source is replaced by an iron source and its 427.2 nm line irradiates the same photocell. Predict the new stopping voltage.

(a) 1.51 V (b) 3.5 V
(c) 0.05 V (d) 2.03 V

**51** In an experiment on photoelectric effect, the frequency ν of incident light is plotted against the stopping potential $V_0$. The work-function of the photoelectric surface is given by

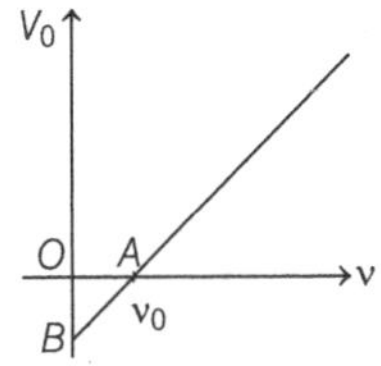


(a) $OB \times e$ (in eV) (b) $OB$ (in volts)
(c) $OA$ (in eV) (d) the slope of line $AB$

**52** In an experiment on photoelectric effect, the slope of the cut-off voltage *versus* frequency of incident light is found to be $4.12 \times 10^{-15}$ V-s. Calculate the value of Planck's constant.

(a) $6.0 \times 10^{-34}$ Js (b) $0.63 \times 10^{-34}$ Js
(c) $6.59 \times 10^{-34}$ Js (d) 0

**53** In a photoelectric experiment, the relation between applied potential difference $V$ between cathode and anode and the photoelectric current $I$ to be as shown in the graph below. If Planck's constant $h = 6.6 \times 10^{-34}$ J-s, the frequency of incident radiation would be nearly (in $s^{-1}$)(use, $\phi = 1$eV)

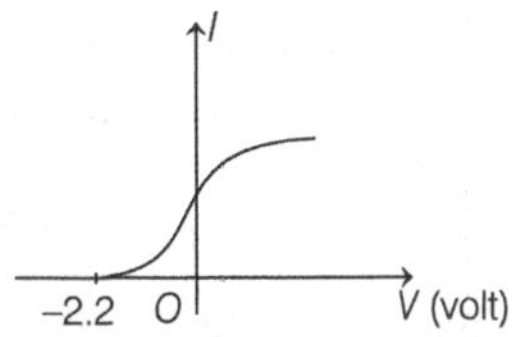


(a) $0.436 \times 10^{18}$ (b) $0.436 \times 10^{17}$
(c) $0.775 \times 10^{15}$ (d) $0.775 \times 10^{16}$

**54** For photoelectric effect with incident photon wavelength λ, the stopping potential is $V$. Identify the correct variations of $V_0$ with $\frac{1}{\lambda}$.

**JEE Main 2013**

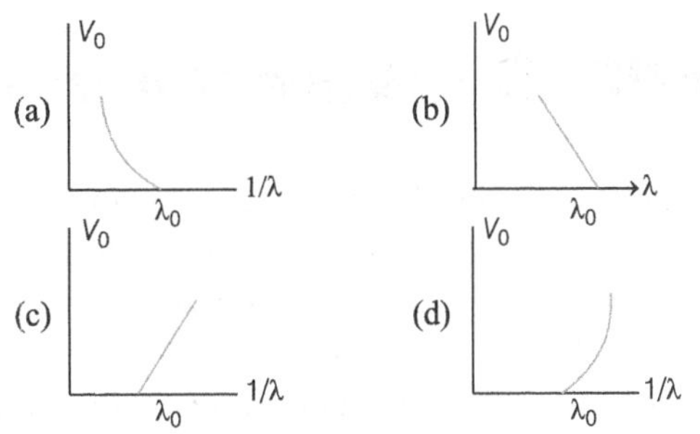


# TOPIC 4 ~ Particle Nature of Light : The Photon

**55** Photoelectric effect gave evidence that light in interaction with matter,

(a) is converted into particles of same size
(b) is converted into particles of same energy
(c) is converted into mass following $E = mc^2$
(d) behaves as if it was made of packets of energy, each of energy $h\nu$

**56** All photons present in a light beam of single frequency have

(a) same frequency but different momentum
(b) same momentum but different frequency
(c) different frequency and momentum
(d) same frequency and momentum

**57** When intensity of a light beam is increased,

(a) energy of photons present increases
(b) momentum of photons present increases
(c) wavelength of photons present increases
(d) number of photons crossing a unit area per second increases

**58** The linear momentum of a 6 MeV photon is

(a) 0.01 eV $sm^{-1}$ (b) 0.02 eV $sm^{-1}$
(c) 0.03 eV $sm^{-1}$ (d) 0.04 eV $sm^{-1}$

**59** If $h$ is Planck's constant, the momentum of a photon of wavelength 1 Å is

(a) $10^{10}\,h$ (b) $h$
(c) $10^2\,h$ (d) $10^{12}\,h$

**60** Monochromatic light of frequency $6.0 \times 10^{14}$ Hz is produced by a laser. The energy of a photon in the light beam is

(a) $5 \times 10^{-15}$ J (b) $3.98 \times 10^{-19}$ J
(c) $2.54 \times 10^{-14}$ J (d) $5.16 \times 10^{-14}$ J

**61** An X-ray tube produces a continuous spectrum of radiation with its short wavelength end at 0.55Å. The maximum energy of a photon in the radiation is

(a) 32.6 eV (b) 22.6 keV (c) 15.2 keV (d) 12.8 keV

**62** In an accelerator experiment on high-energy collisions of electrons with positrons, a certain event is interpreted as annihilation of an electron-positron pair of total energy 10.2 eV into two $\gamma$-rays of equal energy. What is the wavelength associated with each $\gamma$-ray?

(a) $2.436\times 10^{-7}$ m (b) $1.436\times 10^{-7}$ m
(c) $2.436\times 10^{+7}$ m (d) $1.436\times 10^{+7}$ m

**63** A laser beam of frequency $6.0\times 10^{14}$ Hz is emitted from a source of $2\times 10^{-3}$ W. Number of photons emitted per second is

(a) $3.\times 10^{10}$ (b) $6\times 10^{14}$ (c) $9\times 10^{19}$ (d) $5\times 10^{15}$

**64** The energy flux of sunlight reaching the surface of the earth is $1.52\times 10^{3}$ Wm$^{-2}$. The photons in the sunlight have an average wavelength of 520 nm. How many photons per square metre are incident on the earth per second?

(a) $4\times 10^{21}$ (b) $4\times 10^{34}$ (c) $4\times 10^{31}$ (d) $4\times 10^{28}$

**65** There are two sources of light, each emitting with a power of 200 W. One emits X-rays of wavelength 2 nm and the other visible light of 400 nm. The ratio of number of photons of X-rays to the photons of visible light of the given wavelength is

(a) 1:100 (b) 1:200
(c) 1:500 (d) 1:300

**66** A beam of 10.6 eV photons of intensity 2.0 Wm$^{-2}$ falls on a surface of platinum of surface area $1.0\times 10^{-4}$ m$^{2}$ and the work-function of the material is 5.6 eV. Given that, 0.53% of the incident photons eject photoelectrons. The number of photoelectrons emitted per second and their maximum kinetic energy is

(a) $7\times 10^{11}$, 0.0 eV
(b) $6.25\times 10^{11}$, 5.0 eV
(c) $9\times 10^{10}$, 6.0 eV
(d) $11\times 10^{11}$, 2.5 eV

# TOPIC 5 ~ *Wave Nature of Matter*

**67** The de-Broglie wavelength associated with an electron moving with a speed of $5.4\times 10^{6}$ ms$^{-1}$ is

(a) 0.135 nm (b) 0.125 nm
(c) 0.150 nm (d) 0.145 nm

**68** Which of the following figures represent the variation of particle's momentum $p$ and the associated de-Broglie wavelength $\lambda$?

(a) 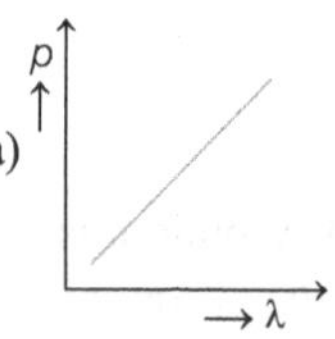


(b) 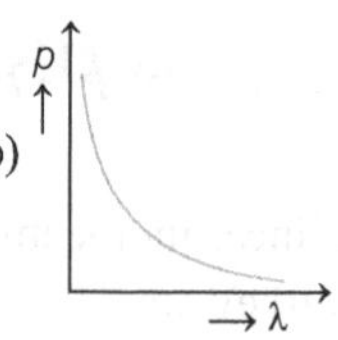


(c) 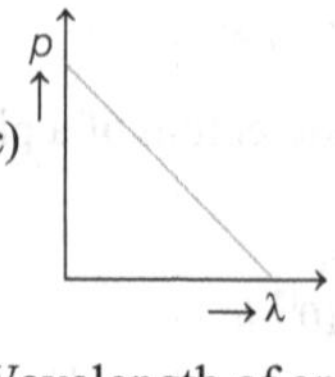


(d) 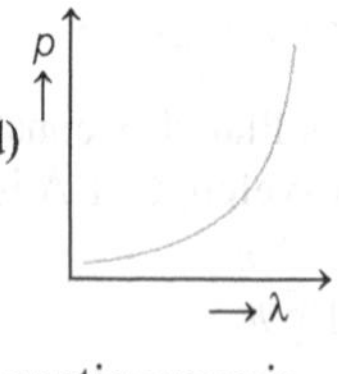


**69** Wavelength of an electromagnetic wave is

(a) more than de-Broglie wavelength of its photon
(b) less than de-Broglie wavelength of its photon
(c) there is no relation between wavelength of a radiation and de-Broglie wavelength of its photon
(d) is equal to de-Broglie wavelength of its photon

**70** If $\alpha$-particle, proton and electron move with the same momentum, then their respective de-Broglie wavelengths $\lambda_\alpha, \lambda_p, \lambda_e$ are related as

(a) $\lambda_\alpha = \lambda_p = \lambda_e$ (b) $\lambda_\alpha < \lambda_p < \lambda_e$
(c) $\lambda_\alpha > \lambda_p > \lambda_e$ (d) $\lambda_p > \lambda_e > \lambda_\alpha$

**71** A particle is moving three times as fast that of an electron. The ratio of de-Broglie wavelength of particle to that of the electron is $1.83\times 10^{-4}$, then the particle is

(a) a meuon
(b) a proton
(c) a neutron
(d) Either a neutron or proton

**72** If de-Broglie wavelength of

I. a bullet of mass 0.40 kg travelling with speed of 1.0 kms$^{-1}$ is $\lambda_1$.
II. a ball of mass 0.60 kg moving at a speed of 1.0 ms$^{-1}$ is $\lambda_2$.
III. a dust particle of mass $1.0\times 10^{-9}$ kg drifting with a speed of 2.2 ms$^{-1}$ is $\lambda_3$.

Then,

(a) $\lambda_1 > \lambda_2 > \lambda_3$ (b) $\lambda_2 > \lambda_1 > \lambda_3$
(c) $\lambda_1 < \lambda_2 < \lambda_3$ (d) $\lambda_3 > \lambda_1 > \lambda_2$

**73** Electrons with de-Broglie wavelength $\lambda$ fall on the target in an X-ray tube. The cut-off wavelength of the emitted X-ray is

(a) $\lambda_0 = \frac{2mc\lambda^2}{h}$ (b) $\lambda_0 = \frac{2h}{mc}$

(c) $\lambda_0 = \frac{2m^2c^2\lambda^3}{h^2}$ (d) $\lambda_0 = \lambda$

**74** The graph between the energy ($\log E$) of an electron and its de-Broglie wavelength ($\log \lambda$) will be **AIIMS 2018, JEE Main 2017**

(a) 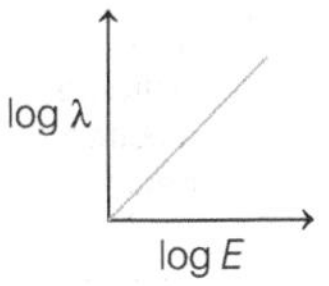


(b) 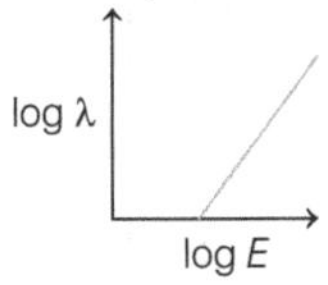


(c) 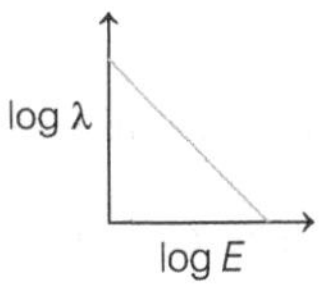


(d) 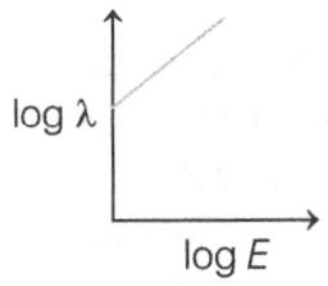


**75** A particle moving with kinetic energy $E$ has de Broglie wavelength $\lambda$. If energy $\Delta E$ is added to its energy, the wavelength become $\lambda / 2$. Value of $\Delta E$ is **JEE Main 2020**

(a) $2E$ (b) $4E$
(c) $3E$ (d) $E$

**76** If the kinetic energy of the particle is increased to 16 times its previous value, the percentage change in the de-Broglie wavelength of the particle is **NEET 2013**

(a) 25 (b) 75
(c) 60 (d) 50

**77** A proton and an $\alpha$-particle are accelerated from rest to the same energy. The de-Broglie wavelengths $\lambda_p$ and $\lambda_\alpha$ are in the ratio **NEET (Odisha) 2019**

(a) $2:1$ (b) $1:1$
(c) $\sqrt{2}:1$ (d) $4:1$

**78** Two particles $A_1$ and $A_2$ of masses $m_1$ and $m_2$ ($m_1 > m_2$) have the same de-Broglie wavelength, then

(a) their momenta are the same
(b) their energies are the same
(c) momenta of $A_1$ is less than the momenta of $A_2$
(d) energy of $A_1$ is more than energy of $A_2$

**79** The de-Broglie wavelength of a photon is twice the de-Broglie wavelength of an electron. The velocity of the electron is $v_e = \frac{c}{100}$, then

(a) $\frac{E_e}{E_p} = 10^{-4}$ (b) $\frac{E_e}{E_p} = 10^{-2}$

(c) $\frac{p_e}{m_e c} = 10^{-3}$ (d) $\frac{p_e}{m_e c} = 10^{-4}$

**80** The wavelength of light from the spectral emission line of sodium is 589 nm. Find the kinetic energy at which

(i) an electron
(ii) a neutron would have the same de-Broglie wavelength.

(a) $KE_e = 5 \times 10^{-25}$ J, $KE_n = 6.5 \times 10^{-28}$ J
(b) $KE_e = 6.96 \times 10^{-25}$ J, $KE_n = 3.82 \times 10^{-28}$ J
(c) $KE_e = 1.25 \times 10^{-25}$ J $KE_n = 6.23 \times 10^{-28}$ J
(d) $KE_e = 3.26 \times 10^{+25}$ J, $KE_n = 4.06 \times 10^{+28}$ J

**81** An electron of mass $m$ and a photon have same energy $E$. The ratio of de-Broglie wavelengths associated with them is ($c$ being velocity of light) **NEET 2016, JEE Main 2020**

(a) $\left(\frac{E}{2m}\right)^{1/2}$ (b) $c(2mE)^{1/2}$

(c) $\frac{1}{c}\left(\frac{2m}{E}\right)^{1/2}$ (d) $\frac{1}{c}\left(\frac{E}{2m}\right)^{1/2}$

**82** An electron of mass $m$ and charge $e$ initially at rest gets accelerated by a constant electric field $E$. The rate of change of de-Broglie wavelength of the electron at time $t$ ignoring relativistic effect is **JEE Main 2020**

(a) $\frac{-h}{e\,Et^2}$ (b) $\frac{-eEt}{h}$

(c) $\frac{-mh}{e\,Et^2}$ (d) $\frac{-h}{eE}$

**83** An electron is accelerated through a potential difference of 10000 V, its de-Broglie wavelength is ($m_e = 9 \times 10^{-31}$ kg) **NEET (National) 2019**

(a) $12.2 \times 10^{-12}$m (b) $12.2 \times 10^{-14}$m
(c) 12.2 nm (d) $12.2 \times 10^{-13}$m

**84** de-Broglie wavelength associated with an electron, accelerating through a potential difference of 100 V lies in the region of

(a) gamma rays (b) X-rays
(c) ultraviolet (d) visible region

**85** A proton and an $\alpha$-particle are accelerated through the same potential difference. The ratio of de-Broglie wavelength $\lambda_p$ to that of $\lambda_\alpha$ is

(a) $\sqrt{2}:1$ (b) $\sqrt{4}:1$
(c) $\sqrt{6}:1$ (d) $\sqrt{8}:1$

**86** Electrons used in an electron microscope are accelerated by a voltage of 25 kV. If the voltage is increased to 100 kV, then the de-Broglie wavelength associated with the electrons would

(a) decrease by 2 times (b) decrease by 4 times
(c) increase by 4 times (d) increase by 2 times

**87** When photon of energy 4.25 eV strike the surface of a metal $A$, the ejected photoelectrons have maximum kinetic energy $T_A$ eV and de-Broglie wavelength $\lambda_A$. The maximum kinetic energy of photoelectrons liberated from another metal $B$ by photon of energy 4.70 eV is $T_B = (T_A - 1.5)$ eV. If the de-Broglie wavelength of these photoelectron is $\lambda_B = 2\lambda_A$, then **JEE Main 2020**

(a) the work function of $A$ is 3.25 eV
(b) the work function of $B$ is 2.20 eV
(c) $T_A = 3.00$ eV
(d) $T_B = 0.5$ eV

**88** The de-Broglie wavelength of a neutron in thermal equilibrium with heavy water at a temperature $T$ (Kelvin) and mass $m$, is **NEET 2017**

(a) $\frac{h}{\sqrt{mkT}}$ (b) $\frac{h}{\sqrt{3mkT}}$
(c) $\frac{2h}{\sqrt{3mkT}}$ (d) $\frac{2h}{\sqrt{mkT}}$

**89** If $\Delta x$ is uncertainty in the specification of position and $\Delta p$ is the uncertainty in specification of momentum of an electron, then the product $\Delta x$ and $\Delta p$ is of the order of $\hbar$ such that $\hbar$

(a) $h/2\pi$ (b) $2h/\pi$
(c) $10^{-19}$ (d) $10^{-31}$

**90** An electron is confined to a 1 nm wide region. Find the uncertainty in momentum using Heisenberg uncertainty principle. (Take, $h = 6.63 \times 10^{-34}$ J-s)

(a) $1.05 \times 10^{-25}$ kg-ms$^{-1}$
(b) $2.03 \times 10^{-31}$ kg-ms$^{-1}$
(c) $3.05 \times 10^{-34}$ kg-ms$^{-1}$
(d) $2.49 \times 10^{-32}$ kg-ms$^{-1}$

**91** Davisson-Germer's electron diffraction arrangement is as shown

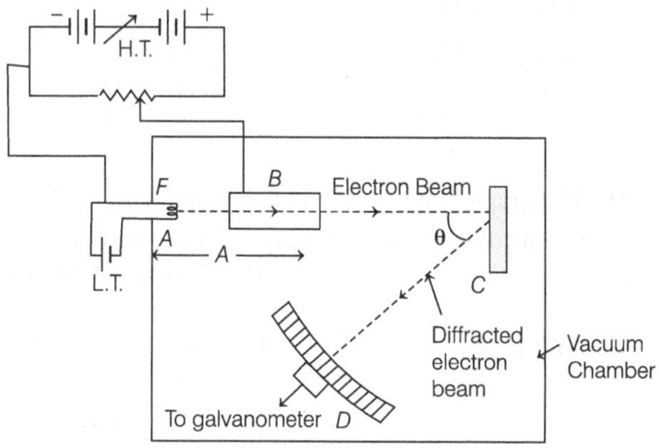


Here, correct labelling is

(a) $A$-electron source, $B$-metal crystal, $C$-reflector and $D$-detector
(b) $A$-electron source, $B$-hollow tube, $C$-wall and $D$-reflector
(c) $A$-electron gun, $B$-electron accelerator, $C$-detector and $D$- counter
(d) $A$-electron gun, $B$-collimating & accelerating tube, $C$-metal target and $D$ -movable collector

**92** In the Davisson and Germer experiment, the velocity of electrons emitted from the electron gun can be increased by

(a) increasing the potential difference between the anode and the filament
(b) increasing the filament current
(c) decreasing the filament current
(d) decreasing the potential difference between the anode and filament

# SPECIAL TYPES QUESTIONS

## I. Assertion and Reason

**Direction** (Q. Nos. 93-101) *In the following questions, a statement of Assertion is followed by a corresponding statement of Reason. Of the following statements, choose the correct one.*

(a) Both Assertion and Reason are correct and Reason is the correct explanation of Assertion
(b) Both Assertion and Reason are correct but Reason is not the correct explanation of Assertion
(c) Assertion is correct but Reason is incorrect
(d) Assertion is incorrect but Reason is correct

**93 Assertion** Cathode rays produce fluorescence in glass of a discharge tube and colour of glow depends on nature of glass.

**Reason** Cathode rays excite electrons of glass and they on de-excitation emits radiation in visible region.

**94 Assertion** Heinrich Hertz observed that high voltage spark across detector loop were enhanced when the emitter plate was illuminated by UV-light.

**Reason** Light shining on the metal surface facilitate the escape of free electrons.

**95** **Assertion** For $\nu > \nu_0$ (threshold frequency), photoelectric current is proportional to intensity.

**Reason** Greater the number of energy quanta available, greater is the number of electrons absorbing the energy quanta and greater is the number of electrons coming out of the metal.

**96** **Assertion** Photosensitivity of a metal is high, if its work-function is small.

**Reason** Work-function $= h\nu_0$, where $\nu_0$ is the threshold frequency.

**97** **Assertion** Light quantum of energy is associated with a photon.

**Reason** Light quantum of energy is also associated with momentum.

**98** **Assertion** Macroscopic objects in our daily life show wave like properties prominent.

**Reason** Wavelength of macroscopic objects is very small.

**99** **Assertion** The photocells, inserted in the door light electric circuit are used as automatic door opener.

**Reason** Abrupt change in photocurrent, helps to open the door.

**100** **Assertion** According to Born's approximation, the intensity of the matter wave at a point determines probability density of the particle at that point.

**Reason** If the intensity of the matter wave is large in a certain region, there is a greater probability of particle being found there than where intensity is small.

**101** **Assertion** Electron has the smallest wavelength as compare to proton and $\alpha$-particle (if they have same kinetic energy).

**Reason** de-Broglie wavelength is inversely proportional to $(\text{mass})^{\frac{1}{2}}$.

## II. Statement Based Questions

**102** When certain metals were irradiated by ultraviolet light or when certain metals are heated to high temperature, then emitted charged particle

I. travels with speed ranging from about 0.1 to 0.2 times the speed of light.

II. $e/m$ is approximately equals to $1.76 \times 10^{11}$ C/kg.

Which of the following statement(s) is/are correct?

(a) Only I (b) Only II
(c) Both I and II (d) Neither I nor II

**103** I. In the experimental set-up used for studying the photoelectric effect, a monochromatic light from the source of sufficiently short wavelength is used.

II. The intensity of the light incident on the emitter plate is independent of the distance of the light source from the emitter.

Which of the following statement(s) is/are incorrect?

(a) Only I (b) Only II
(c) Both I and II (d) Neither I nor II

**104** In the experimental set-up used for studying photoelectric effect, if the emitter plate is illuminated by a light of fixed frequency and fixed intensity. Then on applying negative (retarding) potential to the collector ($A$) plate w.r.t. emitter ($C$) plate and gradually increase it.

I. Only slow moving electrons reach collector plate.

II. Photocurrent decreases rapidly until it drops to zero at a certain sharply defined, critical value of potential $V_0$ on the collector plate.

Which of the following statement(s) is/are correct?

(a) Only I (b) Only II
(c) Both I and II (d) Neither I nor II

**105** For the graph as shown below

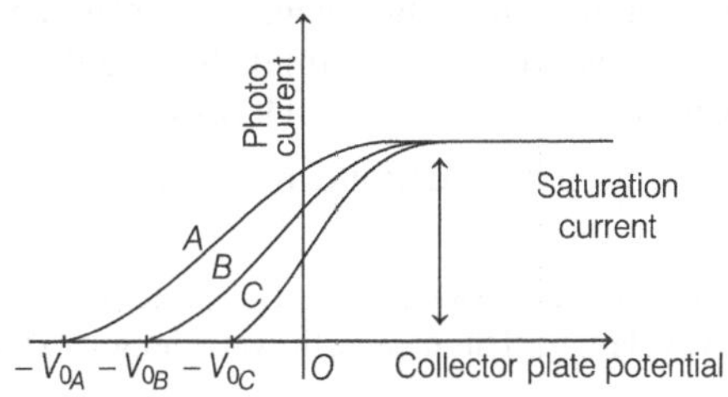


Which of the following statement(s) is/are correct?

I. Stopping potential are in the order $V_{OA} > V_{OB} > V_{OC}$, if the frequencies are in the order $\nu_A > \nu_B > \nu_C$.

II. Greater the frequency of incident radiation, lesser would be the maximum kinetic energy of photoelectrons.

(a) Only I (b) Only II
(c) Both I and II (d) Neither I nor II

**106** Variation of stopping potential $V_0$ with frequency $\nu$ of incident radiation for photosensitive materials $A$ and $B$ are as shown below

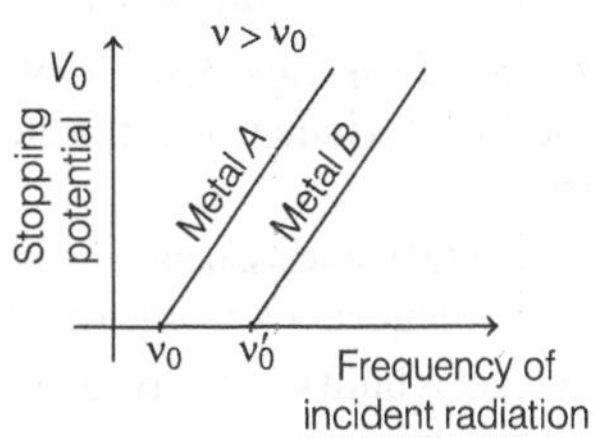

Which of the following statement(s) is/are correct?

I. Stopping potential varies linearly with frequency of incident radiation.

II. There exists a maximum frequency, above which only the photoemission could takes place.

(a) Only I (b) Only II
(c) Both I and II (d) Neither I nor II

**107** I. An electron is emitted from a metal surface when $h\nu < \phi_0$.

II. Energy $E$ of emitted photoelectrons ranges from $0 < E < K_{max}$.

III. Since, $K_{max}$ must be non-negative, so implies that photoelectric emission is only possible, if $h\nu = \phi_0$.

Which of the following statement(s) is/are correct?

(a) Both I and II (b) Both II and III
(c) Both I and III (d) Only II

**108** I. Low intensity means delay in electron emission.

II. Millikan in his experiment measured the slope of straight line in stopping potential *versus* frequency of incident radiation graph obtained for sodium by using which he calculated the value $h = 6.626 \times 10^{-34}$ J-s.

III. Stopping potential $V_0$ *versus* frequency $\nu$ curve's slope is independent of the nature of the material.

Which of the following statement(s) is/are correct?

(a) Both I and III (b) Both II and III
(c) Only II (d) I, II and III

**109** If an electron has a definite momentum $p$ (i.e. $\Delta p = 0$), then

I. by de-Broglie wavelength, it has indefinite wavelength. This means, the wavelength extends all over space.

II. By Born's probability interpretation with indefinite wavelength, the electron will not be localised in any finite region. Thus, it is inconsistent with uncertainity principle.

III. The wave would be as follows

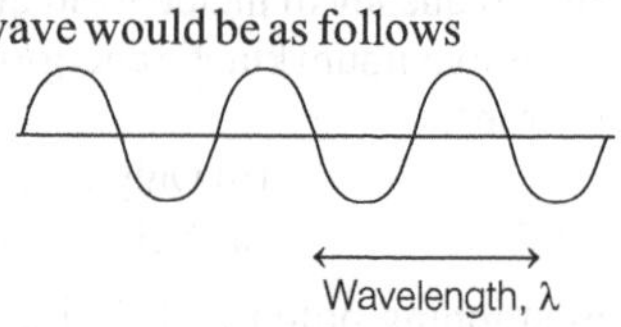


Which of the following statement(s) is/are correct?

(a) Both I and II (b) Both II and III
(c) Only III (d) Only II

**110** I. Whether a particle or wave description is best suited for an experiment is independent of the nature of experiment.

II. In general, matter wave associated with an electron is a wave packet extended over some finite region of space.

III. A wave packet is built up of wavelengths spread around some central wavelength.

Which of the following statement(s) is/are correct?

(a) Only I (b) Only II
(c) Only III (d) Both II and III

**111** In Davisson and Germer experiment,

I. a strong beam appeared in the intensity $I$ of the scattered electron for an acceleration voltage of 54V at a scattering angle $\theta = 50°$.

II. The appearance of the peak in the particular direction is due to resonance caused by scattered electrons.

Which of the following statement(s) is/are correct?

(a) Only I (b) Only II
(c) Both I and II (d) Neither I nor II

**112** I. The wavelength of electron obtained using de-Broglie is more as compare to wavelength obtained in electron diffraction measurement.

II. Photocells are used in the reproduction of sound in motion pictures and in the television camera for scanning and telecasting scenes.

III. Electron microscope utilised the wave properties of electron and has better resolution than optical microscope.

Which of the following statement(s) is/are correct?

(a) Both I and II (b) Both II and III
(c) Both I and III (d) Only I

**113** From graph of photoelectric current *versus* collector plate potential for a given frequency of incident radiation as shown below

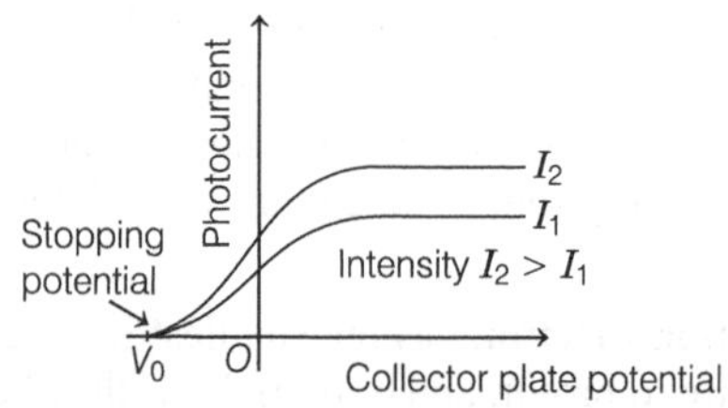


Choose the incorrect statement (s) regarding the maximum kinetic energy of emitted photoelectrons.

(a) It is dependent on intensity of incident radiation.
(b) It depends on the frequency.
(c) It is independent of the light source and the emitter plate material.
(d) All of the above

**114** Which of the following statement is correct for a photosensitive material like, selemium, zinc, copper, etc?

(a) All these materials shows photosensitivity to same extent.
(b) Selenium is less sensitive than zinc or copper.
(c) UV-light gives rise to photoelectric effect in copper while green light does not.
(d) All of these materials are sensitive to green light and show photoelectric effect.

**115** The graph between $1/\lambda$ and stopping potential $V$ of three metals having work functions $\phi_1, \phi_2$ and $\phi_3$ in an experiment of photoelectric effect is plotted as shown in the graph below. Which of the following statement(s) is/are correct? (Here, $\lambda$ is the wavelength of the incident ray)

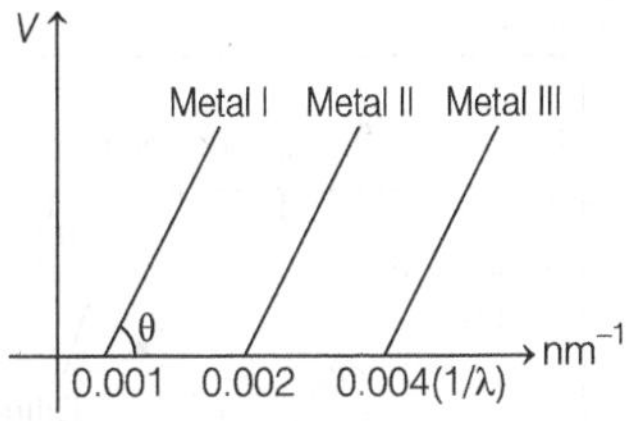


(a) Ratio of work function $\phi_1 : \phi_2 : \phi_3 = 1:2:4$
(b) Ratio of work function $\phi_1 : \phi_2 : \phi_3 = 4:2:1$
(c) $\tan\theta$ is inversely proportional to $hc/e$, where $h$ is Planck's constant and $c$ is the speed of light
(d) The violet colour light can eject photoelectrons from metals II and III.

**116** Choose the incorrect statement from the following.
(a) Particle like behaviour of light was confirmed by the experiment on the scattering of X-rays from electrons.
(b) Photons are deflected by electric fields.
(c) In photon-electron collision, total energy and momentum are conserved.
(d) Both (a) and (b)

**117** Choose the correct statement regarding photocell.
(a) A photocell of lead sulphide sensitive to infrared radiation is used in electronic ignition circuits.
(b) It converts current into light intensity.
(c) Light metres in cameras uses photocell to measure change in potential difference.
(d) Both (a) and (b)

## III. Matching Type

**118** Match the Column I (quantities) with Column II (expression of the values) and select the correct answer from the codes given below.

| | Column I | | Column II |
|---|---|---|---|
| A. | Maximum number of photons absorbed by an electron | 1. | $\frac{\phi_0}{\nu_0}$ |
| B. | Number of photons emitted by a source per second | 2. | $\frac{P}{E}$ |
| C. | Energy of a photon $E$ | 3. | $h\nu$ |
| D. | Planck's constant $h$ | 4. | 1 |

| | A | B | C | D |
|---|---|---|---|---|
| (a) | 2 | 4 | 1 | 3 |
| (b) | 4 | 2 | 3 | 1 |
| (c) | 2 | 4 | 3 | 1 |
| (d) | 3 | 1 | 2 | 4 |

**119** Match the Column I (fundamental experiment) with Column II (conclusion) and select the correct answer from the codes given below. ***JEE Main 2015***

| | Column I | | Column II |
|---|---|---|---|
| A. | Frank-Hertz experiment | 1. | Particle nature of light |
| B. | Photoelectric experiment | 2. | Discrete energy levels of atom |
| C. | Davisson-Germer experiment | 3. | Wave nature of electron |
| | | 4. | Structure of atom |

| | A | B | C | | A | B | C |
|---|---|---|---|---|---|---|---|
| (a) | 2 | 1 | 3 | (b) | 4 | 2 | 3 |
| (c) | 2 | 4 | 3 | (d) | 3 | 1 | 2 |

# NCERT & NCERT Exemplar

## MULTIPLE CHOICE QUESTIONS

## NCERT

**120** The minimum wavelength of X-rays produced by 30 kV electrons will be
(a) 0.0414 nm (b) 0.414 nm
(c) 4.14 nm (d) 41.4 nm

**121** The photoelectric cut-off voltage in a certain experiment is 1.5 V. What is the maximum kinetic energy of photoelectrons emitted?
(a) $4.4 \times 10^{-19}$ J (b) $5.4 \times 10^{-19}$ J
(c) $2.4 \times 10^{-19}$ J (d) $3.4 \times 10^{-19}$ J

**122** Monochromatic light of wavelength 632.8 nm is produced by a helium-neon laser. The power emitted is 9.42 mW. The energy and momentum of each photon in the light beam will be
(a) $3.14\times 10^{-19}$ J, $1.05\times 10^{-27}$ kg-ms$^{-1}$
(b) $2.4\times 10^{-19}$ J, $2.05\times 10^{-27}$ kg-ms$^{-1}$
(c) $4.2\times 10^{-19}$ J, $3\times 10^{16}$ kg-ms$^{-1}$
(d) $4.2\times 10^{-18}$ J, $1.05\times 10^{-26}$ kg-ms$^{-1}$

**123** The energy flux of sunlight reaching the surface of the earth is $1388 \times 10^3$ $Wm^{-2}$. How many photons (nearly) per square metre are incident on the earth per second? Assume that the photons in the sunlight have an average wavelength of 550 nm.

(a) $10^{21}$ (b) $10^{20}$ (c) $10^{10}$ (d) $10^{40}$

**124** A 100 W sodium lamp radiates energy uniformly in all directions. The lamp is located at the centre of a large sphere that absorbs all the sodium light which is incident on it. The wavelength of the sodium light is 589 nm. At what rate are the photons delivered to the sphere?

(a) $3 \times 10^{21}$ photon/s (b) $3 \times 10^{20}$ photon/s
(c) $3 \times 10^{22}$ photon/s (d) $3 \times 10^{24}$ photon/s

**125** The threshold frequency for a certain metal is $3.3 \times 10^{14}$ Hz. If light of frequency $8.2 \times 10^{14}$ Hz is incident on the metal, predict the cut-off voltage for the photoelectric emission.

(a) 2.03 V (b) 4.04 V (c) 5.04 V (d) 6.04 V

**126** Light of frequency $7.21 \times 10^{14}$ Hz is incident on a metal surface. Electrons with a maximum speed of $6.0 \times 10^5$ $ms^{-1}$ are ejected from the surface. What is the threshold frequency for photoemission of electrons?

(a) $4.74 \times 10^{14}$ Hz (b) $5.47 \times 10^{14}$ Hz
(c) $6.47 \times 10^{14}$ Hz (d) $6.47 \times 10^{14}$ Hz

**127** Light of wavelength 488 nm is produced by an argon laser, which is used in the photoelectric effect. When light from this spectral line is incident on the emitter, the stopping (cut-off) potential of photoelectrons is 0.38 V. Find the work function of the material from which the emitter is made.

(a) 3.17 eV (b) 2.17 eV (c) 4.17 eV (d) 5.17 eV

## NCERT Exemplar

**128** A particle is dropped from a height $H$. The de-Broglie wavelength associated with particle is proportional to

(a) $H$ (b) $H^{1/2}$ (c) $H^0$ (d) $H^{-1/2}$

**129** Wavelengths of a photon needed to remove a proton from a nucleus which is bound to the nucleus with 1 MeV energy is nearly

(a) 1.2 nm (b) $1.2 \times 10^{-3}$ nm
(c) $1.2 \times 10^{-6}$ nm (d) $1.2 \times 10^{1}$ nm

**130** Consider a beam of electrons (each with energy $E_0$) incident on a metal surface kept in an evacuated chamber. Then,

(a) no electrons will be emitted as only photons can emit electrons
(b) electrons can be emitted but all with energy $E_0$
(c) electrons can be emitted with any energy, with a maximum of $E_0 - \phi$ ($\phi$ is the work-function)
(d) electrons can be emitted with any energy, with a maximum of $E_0$

**131** Consider figure given below. Suppose the voltage applied to $A$ is increased. The diffracted beam will have the maximum at a value of $\theta$ that

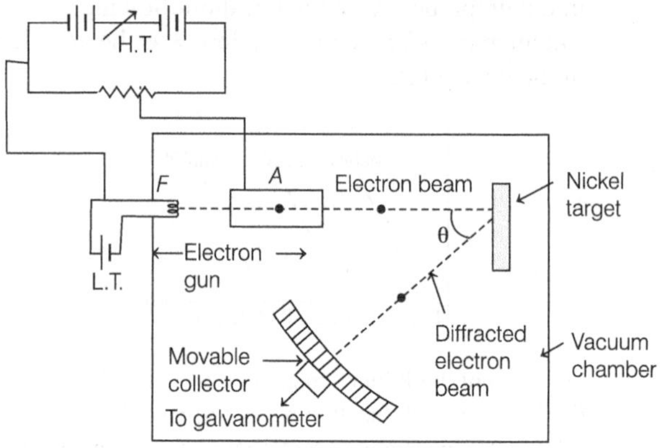


(a) will be larger than the earlier value
(b) will be the same as the earlier value
(c) will be less than the earlier value
(d) will depend on the target

**132** A proton, a neutron, an electron and an $\alpha$-particle have same energy, then their de-Broglie wavelengths compare as

(a) $\lambda_p = \lambda_n > \lambda_e > \lambda_\alpha$ (b) $\lambda_\alpha < \lambda_p = \lambda_n < \lambda_e$
(c) $\lambda_e < \lambda_p = \lambda_n > \lambda_\alpha$ (d) $\lambda_e = \lambda_p = \lambda_n = \lambda_\alpha$

**133** An electron moving with an initial velocity $\mathbf{v} = v_0 \hat{\mathbf{i}}$ and is in a magnetic field $\mathbf{B} = B_0 \hat{\mathbf{j}}$. Then, its de-Broglie wavelength

(a) remains constant
(b) increases with time
(c) decreases with time
(d) first increases decreases

**134** An electron (mass $m$) with an initial velocity $\mathbf{v} = v_0 \hat{\mathbf{i}}$ $(v_0 > 0)$ is in an electric field $\mathbf{E} = -E_0 \hat{\mathbf{i}}$ ($E_0$ = constant > 0) field. Its de-Broglie wavelength at time $t$ is given by **NEET 2018**

(a) $\dfrac{\lambda_0}{\left(1 + \dfrac{eE_0}{m}\dfrac{t}{v_0}\right)}$ (b) $\lambda_0 \left(1 + \dfrac{eE_0 t}{mv_0}\right)$

(c) $\lambda_0$ (d) $\lambda_0 t$

**135** An electron (mass $m$) with an initial velocity $\mathbf{v} = v_0 \hat{\mathbf{i}}$ is in an electric field $\mathbf{E} = E_0 \hat{\mathbf{j}}$. If $\lambda_0 = h/mv_0$, its de-Broglie wavelength at time $t$ is given by

(a) $\lambda_0$ (b) $\lambda_0 \sqrt{1 + \dfrac{e^2 E_0^2 t^2}{m^2 v_0^2}}$

(c) $\dfrac{\lambda_0}{\sqrt{1 + \dfrac{e^2 E_0^2 t^2}{m^2 v_0^2}}}$ (d) $\dfrac{\lambda_0}{\left(1 + \dfrac{e^2 E_0^2 t^2}{m^2 v_0^2}\right)}$

# Answers

## > Mastering NCERT with MCQs

| | | | | | | | | | |
|---|---|---|---|---|---|---|---|---|---|
| 1 (b) | 2 (c) | 3 (c) | 4 (c) | 5 (d) | 6 (c) | 7 (a) | 8 (d) | 9 (d) | 10 (c) |
| 11 (a) | 12 (d) | 13 (a) | 14 (d) | 15 (d) | 16 (d) | 17 (a) | 18 (d) | 19 (d) | 20 (d) |
| 21 (b) | 22 (a) | 23 (c) | 24 (d) | 25 (c) | 26 (b) | 27 (a) | 28 (b) | 29 (a) | 30 (d) |
| 31 (a) | 32 (b) | 33 (d) | 34 (a) | 35 (d) | 36 (b) | 37 (a) | 38 (b) | 39 (b) | 40 (a) |
| 41 (a,b) | 42 (d) | 43 (a) | 44 (c) | 45 (b) | 46 (c) | 47 (c) | 48 (c) | 49 (a) | 50 (a) |
| 51 (a) | 52 (c) | 53 (c) | 54 (c) | 55 (d) | 56 (d) | 57 (d) | 58 (b) | 59 (a) | 60 (b) |
| 61 (b) | 62 (a) | 63 (d) | 64 (a) | 65 (b) | 66 (b) | 67 (a) | 68 (b) | 69 (d) | 70 (a) |
| 71 (d) | 72 (c) | 73 (a) | 74 (c) | 75 (c) | 76 (b) | 77 (a) | 78 (a) | 79 (b) | 80 (b) |
| 81 (d) | 82 (a) | 83 (a) | 84 (b) | 85 (d) | 86 (a) | 87 (d) | 88 (b) | 89 (a) | 90 (a) |
| 91 (d) | 92 (a) | | | | | | | | |

## > Special Types Questions

| | | | | | | | | | |
|---|---|---|---|---|---|---|---|---|---|
| 93 (a) | 94 (a) | 95 (a) | 96 (b) | 97 (a) | 98 (d) | 99 (a) | 100 (b) | 101 (d) | 102 (c) |
| 103 (b) | 104 (b) | 105 (a) | 106 (a) | 107 (d) | 108 (b) | 109 (c) | 110 (d) | 111 (a) | 112 (b) |
| 113 (a) | 114 (c) | 115 (a) | 116 (b) | 117 (a) | 118 (b) | 119 (a) | | | |

## > NCERT & NCERT Exemplar MCQs

| | | | | | | | | | |
|---|---|---|---|---|---|---|---|---|---|
| 120 (a) | 121 (c) | 122 (a) | 123 (a) | 124 (b) | 125 (a) | 126 (a) | 127 (b) | 128 (d) | 129 (b) |
| 130 (d) | 131 (c) | 132 (b) | 133 (a) | 134 (a) | 135 (c) | | | | |

# Hints & Explanations

**1** (*b*) At sufficiently low pressure of about 0.001 mm of Hg column in a discharge tube, a discharge took place between the two electrodes on applying electric field to the gas in the tube.

A fluorescent glow appeared on the glass opposite to cathode. The cause of this fluorescence was attributed to the radiations which appeared to be coming from cathode. These radiations were known as to cathode rays.

**3** (*c*) Given, for proton,

Specific charge $= \dfrac{e}{m} = 9.6\times 10^{7}\,\text{C kg}^{-1}$ ... (i)

As, α-particle $({}_2^4\text{He})$ have 2 electrons and its mass number is 4.

$$\therefore \text{Specific charge of } \alpha\text{- particle} = \frac{2e}{4m} = \frac{1}{2}\frac{e}{m}$$

$$= \frac{1}{2}\times 9.6\times 10^{7} \quad \text{[using Eq. (i)]}$$

$$= 4.8\times 10^{7}\,\text{C kg}^{-1}$$

**4** (*c*) In 1913, RA Millikan during his oil-drop experiment for the precise measurement of charge on an electron found that electric charge is quantised. Charge is always an integral multiple of an elementary charge

$$= 1.602\times 10^{-19}\ \text{C}$$

**5** (*d*) If an electron attempts to come out of the metal, the metal surface acquires a positive charge and pulls the electron back to the metal. The free electron is thus held inside the metal surface by the attractive forces of the ions.

**7** (*a*) Work function for cesium is 2.14 eV, aluminium is 4.28 eV, silver is 4.70 eV and platinum is 5.65 eV.

∴ The work function for platinum is highest while it is lowest for cesium.

**8** (*d*) The work function $\phi_0$ depends on the properties of the metal and the nature of its surface.

So, its value will vary with the presence of surface impurities and with increase/decrease in temperature of the surface.

Thus, the value of $\phi_0$ for a metal will change, if it is cooled or heated or coated with some other metal.

**9** (*d*) The minimum energy required for the electron emission from the metal surface can be supplied to the free electrons by any of the following physical processes

(i) **Thermionic emission** By suitably heating, sufficient thermal energy can be imparted to the free electrons to enable them to come out of the metal.

(ii) **Field emission** By applying a very strong electric field (of the order of $10^{8}\ \text{Vm}^{-1}$) to a metal, electrons can be pulled out of the metal, as in a spark plug.

Thus, we can say heating the metal surface or applying electric field could leads to electron emission.

However, hammering the metal surface or applying magnetic field is not going to supply sufficient amount of energy for electron emission.

**10** (c) When ultraviolet (UV)-radiations fall on the emitter plate $C$ electrons are ejected from it which are attracted towards the positive collector plate $A$ by the electric field. Thus, the electrons flow through the evacuated glass tube, results in the current flow.

**11** (a) Hallwach observed that when the uncharged zinc plate is irradiated by (UV)-rays, then it becomes positively charged. Positive charge on this plate is further enhanced when it was continuously illuminated by UV-light.This will make the leaves to move apart from each other, as same type of charge flows through the leaves of an electroscope, making it repel each other.

Thus, it can be concluded that, negatively charged particles are emitted from zinc plate under the action of UV-light, making the leaves to move apart from each other.

**12** (d) Hallwachs and Lenard observed that when ultraviolet light fall on the emitter plate, no electrons were emitted at all when the frequency of the incident light was smaller than a certain minimum value, called the threshold frequency. This minimum frequency depends on the nature of the material of the emitter plate.

**13** (a) For photoemission to takes place, incident light should have frequency greater than a minimum frequency called threshold frequency. As it is given that, green light can emit electrons but yellow light cannot, so no electrons will be emitted by red light also because the frequency of red light is least amongst the three colours.

**14** (d) It has been observed that certain metals like zinc, cadmium, magnesium, etc. responded only to ultraviolet light, having short wavelength to cause electron emission from the surface.

However, some alkali metals such as lithium, sodium, potassium, cesium and rubidium were even sensitive to visible light.

**15** (d) Photoelectric effect involves emission of electrons known as photoelectrons from the metal surface when light of suitable frequency illuminates this metal's surface. These emitted photoelectrons, when ejected from an emitter plate inside a glass tube by UV-radiations which are then attracted towards the collector plate by the electric field.

Thus, this electron flow through the glass tube results in flow of current.

This means, this effect involves conversion of light energy into electrical energy.

**17** (a) Photocurrent increases with increase in intensity. This photocurrent is directly proportional to the number of photoelectrons emitted per second. This implies that the number of photoelectrons emitted per second is directly proportional to the intensity of incident radiation.

Thus, we can conclude that number of photoelectrons emitted per second can be increased by increasing the intensity of incident radiation.

**18** (d) As the distance of source from the surface increases, intensity of radiation decreases.

$\because$ Intensity $I \propto \frac{1}{(\text{distance } d)^2}$ and

Since, photoelectric current ($i$) $\propto$ intensity ($I$)

Thus, the variation of $i$ *versus* $d$ is correctly depicted by the curve $d$ in the given figure.

**19** (d) Keeping the frequency ν of the light fixed and its intensity $I$ fixed, firstly with the increase in the positive potential of collector plate, the photoelectric current increases. At some stage, for a certain positive potential of collector plate, the photoelectric current becomes maximum or saturates. If we further increase the potential of collector plate, the photocurrent does not increase.

**22** (a) For a particular frequency of incident radiation, the minimum negative (retarding) potential $V_0$ given to the collector plate for which the photoelectric current stops or becomes zero is called the cut-off or stopping potential.

From given the graph,

The value of stopping potential, $V_0 = -0.54$V.

**23** (c) In the given condition, photocurrent is zero when the stopping potential is sufficient to repel even the most energetic photoelectrons, with the maximum kinetic energy $K$, so that $K = eV_0$.

Thus, at point $P$ (as shown in the given figure), it can be seen clearly that the electron must have been repelled, so maximum kinetic energy, $K = eV_0$.

**24** (d) Variation of photoelectric current with collector plate potential for different intensities is shown correctly below as

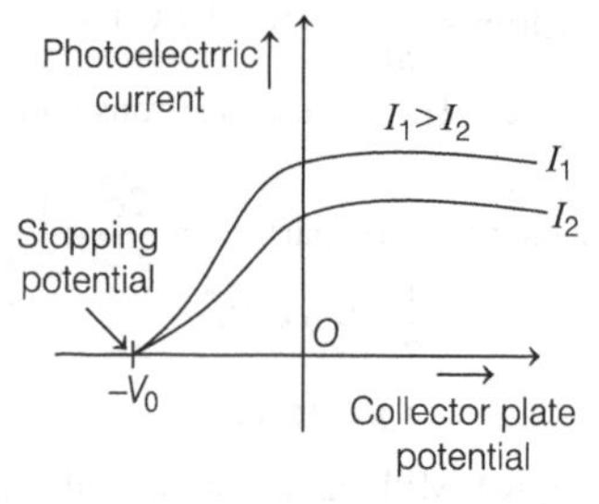


Here for greater intensity $I_1$, more photoelectrons are emitted and hence saturation current is more. Thus, graph corresponding to $I_1$ will be above than that of $I_2$. Since, the stopping potential is independent of intensity, hence the graphs converge at same value of stopping potential $V_0$ for both the intensities $I_1$ and $I_2$.

**25** (*c*) For incident radiation at same intensity and at varying frequencies, the given graph represents the variation of photocurrent with collector plate potential. This can be as shown in graph below

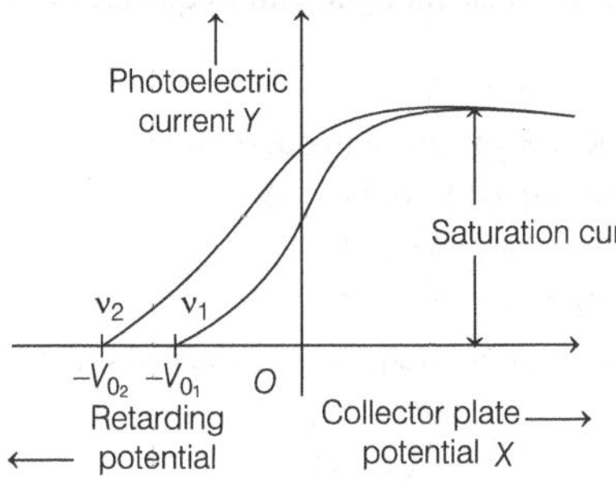


Thus, $X \rightarrow$ collector plate potential,

$Y \rightarrow$ photocurrent and $Z \rightarrow$ saturation current.

**26** (*b*) According to the observation of the experiment on photoelectric effect, it has been seen that, maximum kinetic energy of the photoelectrons varies linearly with the frequency of incident radiation, but is independent of its intensity.

**27** (*a*) Experimentally, it has been found that, if frequency of the incident radiation exceeds the threshold frequency, the photoelectric emission starts instantaneously without any apparent time lag, even if the incident radiation is very dim. It is now known that emission starts in a time of the order of $10^{-9}$ s or less.

**28** (*b*) We know that, for a given photosensitive material, $N$ number of photoelectrons emitted per second is directly proportional to the intensity $I$ of incident radiation.

$$\Rightarrow \quad N \propto I \quad \dots \text{(i)}$$

As, $\nu \cdot \lambda =$ speed of light $c$

(Here, $\lambda$ is wavelength and $\nu$ is frequency.)

$$\therefore \quad \nu \propto \frac{1}{\lambda}$$

Also, $E = h\nu \Rightarrow E \propto \nu \Rightarrow E \propto \dfrac{1}{\lambda}$ ...(ii)

So, from Eqs. (i) and (ii), we get

$$N \propto I \text{ and } E \propto \frac{1}{\lambda}$$

**29** (*a*) We know that, for a given photosensitive material and frequency of incident radiation (above the threshold frequency), the photoelectric current is directly proportional to the intensity of incident light only. Since, in the given case, only the work function has been changed from $W_1$ to $W_2$ keeping the intensity of the incident light same. This means, the value of photocurrent in both the cases will also remain same, i.e. $i_1 = i_2$.

**30** (*d*) Experimental study of photoelectric effect can be summarised as

(II) Saturation current is found to be proportional to the intensity of incident radiation.

(III) A certain minimum cut-off frequency of the incident radiation, called the threshold frequency, below which no emission of photoelectrons takes place, no matter how intense the incident light is.

For I and IV refer to sol. 29 and 27.

Thus, all the given observation are correct.

**31** (*a*) The wave nature of light can be explained by the phenomena of interference, diffraction and polarisation. According to this picture, light is an electromagnetic wave consisting of electric and magnetic fields with the continuous distribution of energy over the region of space over which the wave is extended.

**32** (*b*) According to the wave picture of light,

(i) the free electrons at the surface of the metal (over which the beam of radiation falls) absorb the radiant energy continuously. The greater the intensity of radiation, the greater are the amplitude of electric and magnetic fields. Consequently, the greater the intensity, the greater should be the energy absorbed by each electron. The maximum kinetic energy of the photoelectrons on the surface is then expected to increase with increase in intensity.

(ii) The absorption of energy by electron takes place continuously over entire wavefront of the radiation. Since, a large number of electrons absorb energy, the energy absorbed per electron per unit time turns out to be small. Thus, it can take hours or more for a single electron to pick up sufficient energy to overcome the work function and come out of the metal.

**33** (*d*) According to Einstein's photoelectric equation, basic elementary process involved in photoelectric effect is the absorption of a single light quantum (photon) of sufficient energy by a single electron.

Thus, $K_{max}$ depends linearly on the frequency $\nu$ of incident radiation (photon).

$$K_{max} \propto \nu$$

**34** (*a*) Given, work-function of platinum,

$$\phi_0 = h\nu_0 = 6.35 \text{ eV} = 6.35 \times 1.6 \times 10^{-19} \text{ J}$$

where, $\nu_0$ is threshold frequency of platinum and $h$ is Planck's constant.

$$\Rightarrow \quad \nu_0 = \frac{\phi_0}{h} = \frac{6.35 \times 1.6 \times 10^{-19}}{6.63 \times 10^{-34}} = 15.32 \times 10^{14} \text{ Hz}$$

**35** (*d*) The work function of material is given by

$$\phi = h\nu$$

$$\phi = \frac{hc}{\lambda} \text{ or } \lambda = \frac{hc}{\phi} \quad \dots \text{(i)} \left[\because \nu = \frac{c}{\lambda}\right]$$

where, $h =$ Planck's constant $= 6.63 \times 10^{-34}$ J-s,

$c =$ speed of light $= 3 \times 10^8 \text{ ms}^{-1}$

and $\lambda =$ threshold wavelength of light.

Given, $\phi = 4 \text{ eV} = 4 \times 1.6 \times 10^{-19}$ J.

Substituting the given values in Eq. (i), we get

$$\lambda = \frac{6.63 \times 10^{-34} \times 3 \times 10^{8}}{4 \times 1.6 \times 10^{-19}}$$

$$= 3.108 \times 10^{-7} \text{ m} \simeq 310 \text{ nm}$$

**36** (*b*) The maximum kinetic energy of the emitted photoelectron is given by

$$K_{max} = h\nu - \phi_0 = h\nu - h\nu_0 \quad \dots \text{(i)}$$

Here, frequency of incident light, $\nu = 6\nu$

Threshold frequency, $\nu_0 = \nu$

Substituting given values in Eq. (i), we get

$$K_{max} = h(6\nu) - h(\nu) = 5\, h\nu$$

**37** (*a*) Let the work function for both the metal plates $A$ and $B$, being identical is $\phi$.

According to Einstein photoelectric equation, for metal plate $A$,

$$K_A = \frac{hc}{\lambda_A} - \phi$$

$$\Rightarrow \quad 2K_A = \frac{2hc}{\lambda_A} - 2\phi$$

or

$$2K_A + \phi = \frac{2hc}{\lambda_A} - \phi \quad \dots \text{(i)}$$

It is given that, wavelength of metal $B$ is half of metal $A$, i.e. $\lambda_B = \frac{\lambda_A}{2}$, therefore

$$K_B = \frac{hc}{\lambda_B} - \phi = \frac{2hc}{\lambda_A} - \phi$$

$$\Rightarrow \quad 2K_A + \phi = K_B \quad [\because \text{using Eq. (i)}]$$

Thus, the relation between $K_A$ and $K_B$ can be expressed as

$$\Rightarrow \quad K_A < \frac{K_B}{2}$$

**38** (*b*) Maximum kinetic energy of a photoelectron emitted from a metallic surface,

$$K_{max} = \frac{1}{2}mv^2 = eV_0 \quad \dots \text{(i)}$$

where, $V_0$ is the stopping potential.

Given, frequency of incident light $= \nu_1$ and stopping potential $= V_1$, then according to Einstein's photoelectric equation

$$K_{max} = h\nu_1 + \phi_0$$

$$\Rightarrow \quad h\nu_1 = \phi_0 + eV_1 \quad [\text{using Eq. (i)}] \dots \text{(ii)}$$

Similarly, if light of frequency $\nu_2$ irradiates the surface, then

$$h\nu_2 = \phi_0 + eV_2 \quad \dots \text{(iii)}$$

From Eqs. (ii) and (iii), we get

$$\therefore \quad h(\nu_1 - \nu_2) = e(V_1 - V_2)$$

$$\frac{h}{e}(\nu_1 - \nu_2) = V_1 - V_2$$

$\therefore$ Stopping potential will be $V_2 = V_1 - \frac{h}{e}(\nu_1 - \nu_2)$

**39** (*b*) Einstein's photoelectric equation is given as

$$h\nu - h\nu_0 = K_{max}$$

$$\therefore \quad h(\nu_1 - \nu_0) = K_1 \quad \dots \text{(i)}$$

Similarly, for incident light with frequency $\nu_2$ equation will be

$$h(\nu_2 - \nu_0) = K_2 \quad \dots \text{(ii)}$$

Ratio of $K_1 / K_2$ is given in question $1/n$.

$\therefore$ Dividing Eq. (i) from Eq. (ii), we get

$$\frac{\nu_1 - \nu_0}{\nu_2 - \nu_0} = \frac{K_1}{K_2} = \frac{1}{n}$$

Thus, the threshold frequency of the metallic surface is

$$\nu_0 = \frac{(n\nu_1 - \nu_2)}{(n-1)}$$

**40** (*a*) According to Einstein's photoelectric equation,

$$h\nu = h\nu_0 + \frac{1}{2}mv_{max}^2$$

or

$$\frac{1}{2}mv_{max}^2 = h\nu - h\nu_0 \quad \dots \text{(i)}$$

According to the given problem,

**Case I** Maximum velocity of emitted photoelectrons

$$v_{max} = 8 \times 10^6 \text{ ms}^{-1}$$

Incident frequency, $\nu = 5\nu_0$

$$\Rightarrow \quad \frac{1}{2}m(8 \times 10^6)^2 = h(5\nu_0 - \nu_0) \quad \dots \text{(ii)}$$

**Case II** If $\nu = 2\nu_0$, then maximum velocity of photoelectrons

$$\frac{1}{2}mv_{max}^2 = h(2\nu_0 - \nu_0) \quad \dots \text{(iii)}$$

On dividing Eq. (ii) by Eq. (iii), we get

$$\frac{(8 \times 10^6)^2}{v_{max}^2} = \frac{4\nu_0}{\nu_0}$$

$$v_{max}^2 = \frac{(8 \times 10^6)^2}{4}$$

$$v_{max} = \frac{8 \times 10^6}{2} = 4 \times 10^6 \text{ ms}^{-1}$$

Thus, maximum velocity $v_{max}$ of photoelectrons will be $4 \times 10^6 \text{ ms}^{-1}$.

**41** (*a,b*) Given, threshold wavelength,

$$\lambda_0 = 3250 \times 10^{-10} \text{ m}$$

Wavelength of incident ultraviolet light,

$$\lambda = 2536 \times 10^{-10} \text{ m}$$

Let velocity of ejected electron be $v$.

Now, applying Einstein's photoelectric equation, we have

$$E = K + \phi_0$$

$$\Rightarrow \quad h\nu = \frac{1}{2}m_e v^2 + h\nu_0$$

$$\Rightarrow \quad \frac{1}{2}m_e v^2 = h\nu - h\nu_0 = hc\left(\frac{1}{\lambda} - \frac{1}{\lambda_0}\right)$$

$\Rightarrow$ Velocity of electron, $v = \sqrt{\dfrac{2hc}{m_e}\left(\dfrac{1}{\lambda} - \dfrac{1}{\lambda_0}\right)}$

$= \sqrt{\dfrac{2 \times 4.14 \times 10^{-15} \times 1.6 \times 10^{-19} \times 3 \times 10^{8}}{9.1 \times 10^{-31} \times 10^{-10}}\left(\dfrac{3250 - 2536}{3250 \times 2536}\right)}$

$= 0.6 \times 10^{6}\ \text{ms}^{-1} = 6 \times 10^{5}\ \text{ms}^{-1}$

**42** (*d*) Given, $\lambda_1 = 300$ n-m

$\lambda_2 = 400$ n-m

$\dfrac{hc}{e} = 1240$ n-m

Using Einstein equation for photoelectric effect,

$$E = h\nu = \phi + eV_0 \qquad \ldots(i)$$

(here, $\phi$ is work function of the metal and $V_0$ is stopping potential)

For $\lambda_1$ wavelength's wave,

$$E_1 = h\nu_1 = \phi + eV_{01}$$

or $$\frac{hc}{\lambda_1} = \phi + eV_{01} \qquad \ldots(ii)$$

Similarly, $$\frac{hc}{\lambda_2} = \phi + eV_{02} \qquad \ldots(iii)$$

From Eqs. (ii) and (iii), we get

$$hc\left[\frac{1}{\lambda_1} - \frac{1}{\lambda_2}\right] = e(V_{01} - V_{02})$$

or $$\frac{hc}{e}\left[\frac{1}{\lambda_1} - \frac{1}{\lambda_2}\right] = \Delta V$$

By using given values,

$$\Delta V = 1240\left[\frac{1}{300} - \frac{1}{400}\right]\frac{\text{nm-V}}{\text{nm}}$$

$$= 1240 \times \frac{1}{1200}\ \text{V}$$

$\Rightarrow$ $$\Delta V = 1.03\ \text{V} \approx 1\text{V}$$

**43** (*a*) According to Einstein photoelectric equation,

$$E = K_{\text{max}} + \phi$$

where, $K_{\text{max}}$ is maximum kinetic energy of emitted electron and $\phi$ is work function of an electron.

$\Rightarrow$ $$K_{\text{max}} = E - \phi = h\nu - \phi$$

So, in the given first case, if the photoelectric surface is illuminated by light of wavelength $\lambda$, so

$$K_{\text{max}} = \frac{hc}{\lambda} - \phi \qquad \ldots(i)$$

Similarly, in second case when the photoelectric surface is illuminated by light of wavelength $\lambda/2$ and maximum kinetic energy of emitted electron is 3 times that in first case so, we get

$$3K_{\text{max}} = \frac{hc}{\lambda/2} - \phi \qquad \ldots(ii)$$

Solving Eqs. (i) and (ii), we get work function of an emitted electron from a metal surface

$$\phi = \frac{hc}{2\lambda}$$

**44** (*c*) Given, threshold wavelength, $\lambda_0 = 380$ nm

Wavelength of incident light, $\lambda = 260$ nm

Using Einstein's relation of photoelectric effect,

$$(\text{KE})_{\text{max}} = eV_0 = h\nu - h\nu_0 \qquad \ldots(i)$$

But $h\nu = E = \dfrac{1237}{\lambda(\text{nm})}$ (Given)

$\therefore$ $$E_0 = \frac{1237}{\lambda_0(\text{nm})}\ \text{eV} \qquad \ldots(ii)$$

From Eqs. (i) and (ii), we get

$$(\text{KE})_{\text{max}} = E - E_0\left(\frac{1237}{\lambda} - \frac{1237}{\lambda_0}\right)\text{eV}$$

$$= 1237\left[\frac{1}{\lambda} - \frac{1}{\lambda_0}\right]\text{eV}\ (\lambda \text{ in nm}) \qquad \ldots(iii)$$

By putting values of $\lambda$ and $\lambda_0$ in Eq. (iii), we get

$$(\text{KE})_{\text{max}} = 1237\left(\frac{1}{260} - \frac{1}{380}\right)\text{eV}$$

$$= 1237 \times \left[\frac{380 - 260}{380 \times 260}\right]\text{eV}$$

$\Rightarrow$ $$(\text{KE})_{\text{max}} = 1.5\ \text{eV}$$

**45** (*b*) Frequency of lines can be found by using $\nu = \dfrac{c}{\lambda}$ and they are listed in tabular form as

| $\nu \times 10^{14}$ Hz | 8.219 | 7.412 | 6.884 | 5.493 | 4.343 |
|---|---|---|---|---|---|
| $V_0$ | 1.28 | 0.95 | 0.74 | 0.16 | 0 |

As, at $V_0 = 0$, $\nu = \nu_0$ (threshold frequency)

$\therefore$ From the table, frequency, $\nu_0 = 4.343 \times 10^{14}$ Hz

Work function, $\phi_0 = h\nu_0$

$$= 6.63 \times 10^{-34} \times 4.343 \times 10^{14}$$

$$= 2.88 \times 10^{-19}\ \text{J}$$

$$= \frac{2.88 \times 10^{-19}}{1.6 \times 10^{-19}}\ \text{eV} \approx 1.8\ \text{eV}$$

Thus, the threshold frequency and work function of the metal are $4.3 \times 10^{14}$ Hz and 1.8 eV, respectively.

**46** (*c*) According to the Einstein's photoelectric equation,

$$K_{\text{max}} = \frac{1}{2}mv_{\text{max}}^2 = h\nu - \phi_0 = h\nu - h\nu_0 \qquad \ldots(i)$$

where, $K_{\text{max}}$ is the maximum kinetic energy of photoelectrons having maximum velocity $v_{\text{max}}$.

When incident frequency of light, $\nu = 2\nu_0$ and $v_{\text{max}} = v_1$, then

Substituting the value of $\nu$ and $v_{\text{max}}$ in Eq. (i), we get

$$\frac{1}{2}mv_1^2 = h(2\nu_0) - h\nu_0 = 2h\nu_0 - h\nu_0 = h\nu_0 \qquad \ldots(ii)$$

If incident frequency of radiation, $\nu = 5\nu_0$ and $v_{\text{max}} = v_2$, then

Substituting the value of $\nu$ and $v_{\text{max}}$ in Eq. (i), we get

$$\frac{1}{2}mv_2^2 = h(5\nu_0) - h\nu_0 = 5h\nu_0 - h\nu_0 = 4h\nu_0 \qquad \ldots(iii)$$

On dividing Eq. (ii) by Eq (iii), we get

$$\frac{\frac{1}{2}mv_1^2}{\frac{1}{2}mv_2^2}=\frac{h\nu_0}{4h\nu_0}$$

$$\Rightarrow \quad \frac{v_1^2}{v_2^2}=\frac{1}{4} \text{ or } \frac{v_1}{v_2}=\frac{1}{2}$$

$$\therefore \quad v_1 : v_2 = 1:2$$

The ratio of $v_1$ to $v_2$ is 1 : 2.

**47** (*c*) In Ist case, when a metallic surface is illuminated with radiation of wavelength $\lambda$, the stopping potential is $V$. So, photoelectric equation can be written as

$$eV=\frac{hc}{\lambda}-\frac{hc}{\lambda_0} \quad \text{...(i)}$$

where, $\lambda_0$ is the threshold wavelength.

In IInd case, when the same surface is illuminated with radiation of wavelength $2\lambda$, the stopping potential is $\frac{V}{4}$. So, photoelectric equation can be written as

$$\frac{eV}{4}=\frac{hc}{2\lambda}-\frac{hc}{\lambda_0}$$

$$\Rightarrow \quad eV=\frac{4hc}{2\lambda}-\frac{4hc}{\lambda_0} \quad \text{...(ii)}$$

From Eqs. (i) and (ii), we get

$$\Rightarrow \quad \frac{hc}{\lambda}-\frac{hc}{\lambda_0}=\frac{4hc}{2\lambda}-\frac{4hc}{\lambda_0}$$

$$\Rightarrow \quad \frac{1}{\lambda}-\frac{1}{\lambda_0}=\frac{2}{\lambda}-\frac{4}{\lambda_0} \Rightarrow \lambda_0=3\lambda$$

The threshold wavelength for the metallic surface is $3\lambda$.

**48** (*c*) According to Einstein's photoelectric equation,

$$K_{max}=eV=h\nu-\phi_0=\frac{hc}{\lambda}-\phi_0$$

Here, when $\lambda = 600\,\text{nm} = 6\times10^{-7}$ m and $V_0 = 0.5$V, then

$$0.5e=\frac{hc}{6\times10^{-7}}-\phi_0$$

$$\Rightarrow \quad 0.5=\frac{h}{e}\left(\frac{c}{6\times10^{-7}}\right)-\frac{\phi_0}{e} \quad \text{...(i)}$$

Similarly, when $\lambda = 400\,\text{nm} = 4\times10^{-7}$ m and $V_0 = 1.5$V, then

$$1.5=\frac{h}{e}\left(\frac{c}{4\times10^{-7}}\right)-\frac{\phi_0}{e} \quad \text{...(ii)}$$

Subtracting Eq. (i) from Eq. (ii), we get

$$1=\frac{h}{e}\frac{c}{10^{-7}}\left[\frac{1}{4}-\frac{1}{6}\right]$$

$$\Rightarrow \quad \frac{h}{e}=\frac{12\times10^{-7}}{3\times10^{8}}$$

$$\frac{h}{e}=4\times10^{-15}\text{ V-s}$$

**49** (*a*) If the wavelength of radiation $\lambda$ is incident on a photocell, then according to Einstein's photoelectric emission of light,

$$E=(\text{KE})_{max}+\phi \text{ or } \frac{hc}{\lambda}=(\text{KE})_{max}+\phi \quad \text{... (i)}$$

Similarly, if the wavelength of radiation is changed to $\frac{3\lambda}{4}$, then

$$\frac{hc}{\left(\frac{3\lambda}{4}\right)}=(\text{KE})'_{max}+\phi$$

$$\Rightarrow \quad \frac{4}{3}\left(\frac{hc}{\lambda}\right)=(\text{KE})'_{max}+\phi \quad \text{...(ii)}$$

Substituting the value of $\left(\frac{hc}{\lambda}\right)$ from Eq. (i) in the above Eq. (ii), we get

$$\frac{4}{3}[(\text{KE})_{max}+\phi]=(\text{KE})'_{max}+\phi$$

$$\frac{4}{3}(\text{KE})_{max}+\frac{4}{3}\phi-\phi=(\text{KE})'_{max}$$

$$\frac{4}{3}(\text{KE})_{max}+\frac{\phi}{3}=(\text{KE})'_{max} \quad \text{...(iii)}$$

If $v$ and $v'$ are the maximum speed of emitted electron when the wavelength of light is $\lambda$ and $\frac{3\lambda}{4}$ respectively, so Eq. (iii) can be written as

$$\frac{1}{2}mv'^2=\frac{4}{3}\left(\frac{1}{2}mv^2\right)+\frac{\phi}{3}$$

i.e. $$v'>v\left(\frac{4}{3}\right)^{1/2}$$

**50** (*a*) Given, for neon lamp, wavelength of monochromatic radiation, $\lambda = 640.2\,\text{nm} = 640.2\times10^{-9}$ m

Stopping voltage, $V_0 = 0.54$ V

Let $\phi_0$ be the work-function.

$$\therefore \quad eV_0=\frac{hc}{\lambda}-\phi_0$$

Work-function of photosensitive material,

$$\phi_0=\frac{hc}{\lambda}-eV_0$$

$$=\frac{6.63\times10^{-34}\times3\times10^{8}}{640.2\times10^{-9}}-1.6\times10^{-19}\times0.54$$

$$=3.1\times10^{-19}-0.864\times10^{-19}$$

$$=2.236\times10^{-19}$$

$$=\frac{2.236\times10^{-19}}{1.6\times10^{-19}}=1.3975\simeq1.4\,\text{eV}$$

Now, if the material is irradiates by a iron source, then wavelength, $\lambda = 427.2\,\text{nm} = 427.2\times10^{-9}$ m

Let $V_0'$ be the new stopping potential.

$$\Rightarrow eV_0'=\frac{hc}{\lambda}-\phi_0=\frac{6.63\times10^{-34}\times3\times10^{8}}{427.2\times10^{-9}\times1.6\times10^{-19}}-1.4$$

$$=2.91-1.4=1.51\,\text{eV}$$

Required new stopping potential, $V_0'=1.51$V

**51** (*a*) By Einstein's equation,

we have $eV_0 = h\nu - \phi_0$

or $V_0 = \frac{h}{e}\nu - \frac{\phi_0}{e}$

Above is a equation of straight line ($y = mx - c$) with slope $\frac{h}{e}$ and $y$-intercept $= \frac{\phi_0}{e}$

$\therefore$ Work function, $\phi_0 = e \times y$-intercept (in eV)

As in the given graph $y$-intercept is $= OB$

$= e \times OB$ (in eV)

**52** (*c*) The graph of cut-off voltage *versus* frequency of incident light is as shown below

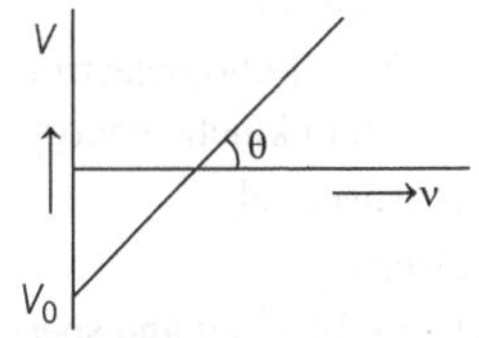


where, slope of graph, $\tan\theta = 4.12 \times 10^{-15}$ V-s

and charge on electron, $e = 1.6 \times 10^{-19}$ C

From graph, we can write $\tan\theta = \frac{\Delta V}{\Delta\nu}$

As we know, for photoelectric effect, $eV = h\nu - \phi$

Differentiating both sides of the above equation, we get

$$\frac{\Delta V}{\Delta\nu} = \frac{h}{e}$$

$$\therefore \quad \frac{h}{e} = 4.12 \times 10^{-15}$$

$$h = 1.6 \times 10^{-19} \times 4.12 \times 10^{-15}$$

$$= 6.592 \times 10^{-34}$$

$\therefore$ Value of Planck's constant is $6.592 \times 10^{-34}$ J-s.

**53** (*c*) From the graph, magnitude of stopping potential,

$$V_0 = 2.2\text{ V}$$

Since, for photoelectric effect

$$K_{max} = eV_0 = 2.2\text{ eV}$$

From Einstein's photoelectric equation

$$KE_{max} = h\nu - \phi$$

where, work function, $\phi = 1$ eV

$$\Rightarrow \quad h\nu = (2.2 + 1)\text{ eV} = 3.2\text{ eV}$$

$\therefore$ The frequency of incident radiation,

$$\nu = \frac{3.2 \times 1.6 \times 10^{-19}}{h}$$

$$= \frac{1.6 \times 10^{-19} \times 3.2}{6.6 \times 10^{-34}}$$

$$= 0.775 \times 10^{15}\text{ Hz}$$

**54** (*c*) As we know, from photoelectric equation,

$$K_{max} = \frac{hc}{\lambda} - \phi_0 \text{ and } K_{max} = eV_0$$

$$\Rightarrow \quad V_0 = \left(\frac{hc}{e}\right)\left(\frac{1}{\lambda}\right) - \frac{\phi_0}{e}$$

Thus, from the above equation, we can say that $V_0$ *versus* $\frac{1}{\lambda}$ graph is in the form $y = mx - c$.

So, we can conclude that, the graph between $V_0$ and $\frac{1}{\lambda}$ is as shown below

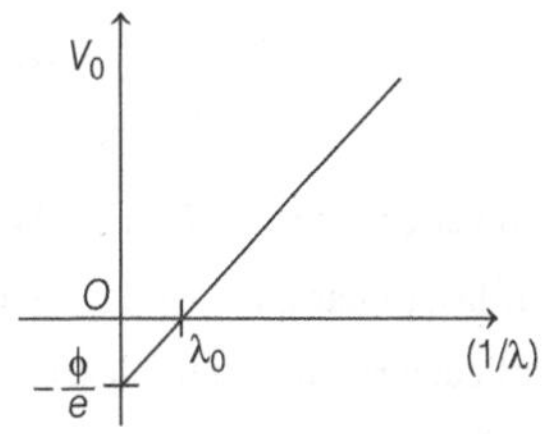


$\therefore$ Option (c) is correct.

**57** (*d*) All photons of light of a particular frequency $\nu$ or wavelength $\lambda$, have the same energy $E\left(= h\nu = \frac{hc}{\lambda}\right)$ and momentum $\left(= \frac{h\nu}{c} = \frac{h}{\lambda}\right)$, whatever the intensity of radiation may be. So, by increasing the intensity of light of given wavelength, there is only an increase in the number of photons per second crossing a given area, with each photon having the same energy and momentum.

**58** (*b*) Given, energy of a photon, $E = 6\text{ MeV} = 6 \times 10^6$ eV

Linear momentum of the photon, $p = \frac{E}{c}$

$$\Rightarrow \quad p = \frac{6 \times 10^6\text{ eV}}{3 \times 10^8\text{ ms}^{-1}} = 2 \times 10^{-2}\text{ eVs m}^{-1}$$

$$= 0.02\text{ eVs m}^{-1}$$

**59** (*a*) Momentum of photon, $p = \frac{E}{c} = \frac{h\nu}{c} = \frac{h}{\lambda}$

Given, wavelength, $\lambda = 1\text{ Å} = 1 \times 10^{-10}$ m

$$\Rightarrow \quad p = \frac{h}{\lambda} = \frac{h}{1 \times 10^{-10}} = 10^{10}\, h$$

**60** (*b*) Given, $\nu = 6 \times 10^{14}$ Hz

Energy of photons $= h\nu$

$= (6.626 \times 10^{-34}\text{ Js}) \times (6 \times 10^{14}\text{ s}^{-1}) = 3.98 \times 10^{-19}$ J

**61** (*b*) Given, $\lambda_{min} = 0.55\text{ Å} = 0.55 \times 10^{-10}$ m,

and $c = 3 \times 10^8\text{ ms}^{-1}$

Maximum energy of X-ray photon is

$$E_{max} = h\nu_{max} = \frac{hc}{\lambda_{min}} = \frac{6.63 \times 10^{-34} \times 3 \times 10^8}{0.55 \times 10^{-10} \times 1.6 \times 10^{-19}}\text{ eV}$$

$$= 22.6 \times 10^3\text{ eV} = 22.6\text{ keV}$$

**62** (*a*) Given, energy of 2 $\gamma$-rays $= 10.2$ eV

Energy of one $\gamma$-ray $= \frac{10.2}{2} = 5.1$ eV

$$= 5.1 \times 1.6 \times 10^{-19} \text{ J}$$
$$= 8.16 \times 10^{-19} \text{ J}$$

Let $\lambda$ be the wavelength.

Energy of each ray, $E = \frac{hc}{\lambda}$

or $\lambda = \frac{hc}{E} = \frac{6.63 \times 10^{-34} \times 3 \times 10^8}{8.16 \times 10^{-19}}$
$$= 2.436 \times 10^{-7} \text{ m}$$

**63** (*d*) Given, power of source $= 2 \times 10^{-3}$ W

Frequency of laser beam, $\nu = 6 \times 10^{14}$ Hz

$\Rightarrow$ Energy of each photon in the laser beam
$$= h\nu$$
$$= 6.63 \times 10^{-34} \times 6 \times 10^{14}$$

As, number of photons emitted per second
$$= \text{Power of the source/ Energy of one photon}$$
$$= \frac{2 \times 10^{-3}}{6.626 \times 10^{-34} \times 6 \times 10^{14}}$$
$$= 5 \times 10^{15} \text{ photons per second}$$

**64** (*a*) Given, $I = 1.52 \times 10^3 \text{ Wm}^{-2}$.

$\lambda = 520\text{nm} = 520 \times 10^{-9}$ m and $h = 6.63 \times 10^{-34}$ J-s

Number of photons incident on earth's surface per second per square metre is

$$n = \frac{I}{E} = \frac{I}{hc}\lambda \qquad \left[\because E = \frac{hc}{\lambda}\right]$$

$$\Rightarrow \quad n = \frac{1.52 \times 10^3 \times 520 \times 10^{-9}}{6.63 \times 10^{-34} \times 3 \times 10^8}$$
$$= 4 \times 10^{21} \text{ photons per metre}^2 \text{ per second}$$

**65** (*b*) Given, $P_1 = P_2 = P = 200$ W, $\lambda_1 = 2$ nm and $\lambda_2 = 400$ nm

Let $n_1$ and $n_2$ be the number of photons of X-rays and visible light emitted from the two sources.

As we know,

Power of the source
$$= \text{Number of photons emitted per second} \times \text{Energy of one proton}$$

$$\therefore \quad n_1 \frac{hc}{\lambda_1} = n_2 \frac{hc}{\lambda_2} \text{ or } \frac{n_1}{\lambda_1} = \frac{n_2}{\lambda_2}$$

$$\Rightarrow \quad \frac{n_1}{n_2} = \frac{\lambda_1}{\lambda_2} = \frac{2}{400} = \frac{1}{200} = 1:200$$

**66** (*b*) Given, energy of the incident photons, $E_i = 10.6$ eV
$$= 10.6 \times 1.6 \times 10^{-19} \text{ J}$$
$$\Rightarrow \quad E_i = 16.96 \times 10^{-19} \text{ J}$$

Energy incident per unit area per unit time (intensity) $I = 2.0 \text{Wm}^{-2} = 2$ J

$\therefore$ Number of photons incident on unit area in unit time
$$= \frac{I}{E_i} = \frac{2}{16.96 \times 10^{-19}} = 1.18 \times 10^{18}$$

Therefore, number of photons incident on given area $(1.0 \times 10^{-4} \text{ m}^2)$
$$= (1.18 \times 10^{18})(1.0 \times 10^{-4})$$
$$= 1.18 \times 10^{14}$$

But only 0.53% of incident photons emit photoelectrons.

$\therefore$ Number of photoelectrons emitted per second ($n$),
$$n = \left(\frac{0.53}{100}\right)(1.18 \times 10^{14}) = 6.25 \times 10^{11}$$

As, from Einstein's photoelectric equation
$$K_{\max} = E_i - \text{work-function} = (10.6 - 5.6) \text{ eV}$$
$$= 5.0 \text{ eV}$$

Thus, the number of photoelectrons emitted per second and their maximum kinetic energy will be $6.25 \times 10^{11}$ and 5.0 eV, respectively.

**67** (*a*) For an electron,

Mass, $m = 9.11 \times 10^{-31}$ kg and speed, $v = 5.4 \times 10^{-6} \text{ms}^{-1}$

Momentum, $p = mv = 9.11 \times 10^{-31} \times 5.4 \times 10^6$
$$= 4.92 \times 10^{-24} \text{kg ms}^{-1}$$

de-Broglie wavelength, $\lambda = \frac{h}{p} = \frac{6.63 \times 10^{-34} \text{ J-s}}{4.92 \times 10^{-24} \text{ kg ms}^{-1}}$
$$= 1.35 \times 10^{-10} \text{ m} = 0.135 \text{ nm}$$

**68** (*b*) The de-Broglie wavelength of a particle is given by
$$\lambda = h/p \Rightarrow p\lambda = h$$

This equation is in the form of $yx = c$, which is the equation of a rectangular hyperbola. Hence, the graph given in option (b) is the correct one.

**69** (*d*) Momentum of a photon of an electromagnetic radiation of energy $h\nu$,
$$p = \frac{h\nu}{c}$$

Therefore, $\frac{h}{p} = \frac{\nu}{c} = \lambda$ ...(i)

Also, by de-Broglie wavelength associated with a radiation formula, $\lambda = \frac{h}{p}$ ...(ii)

Thus, from Eqs. (i) and (ii), it was concluded that de-Broglie wavelength of a photon equals the wavelength of an electromagnetic radiation of which the photon is a quantum of energy and momentum.

**70** (*a*) de-Broglie wavelength, $\lambda = h/p$ ... (i)

where, $h$ is Planck's constant and $p$ is momentum.

Given, momentum is same for $\alpha$, proton and electron particles, so
$$p_\alpha = p_p = p_e$$
$$\therefore \quad \lambda_\alpha = \lambda_p = \lambda_e \qquad \text{[from Eq. (i)]}$$

**71** (*d*) Given, $\frac{\lambda_p}{\lambda_e} = 1.83 \times 10^{-4}$

where, $\lambda_p$ and $\lambda_e$ are de-Broglie wavelengths associated with particle and electron, respectively.

and $\frac{v_p}{v_e} = 3$ (where, $v_p$ and $v_e$ are velocities of the particle and electron, respectively.

Since, $\lambda = h/p = \frac{h}{mv}$

or $m = \frac{h}{\lambda v}$ ...(i)

Using Eq. (i), we get

$$\frac{m_e}{m_p} = \frac{\lambda_p v_p}{\lambda_e v_e} = (1.83\times 10^{-4})\times 3 = 5.49\times 10^{-4}$$

or $\frac{m_p}{m_e} = \frac{1}{5.49\times 10^{-4}}$

$\Rightarrow$ $m_p = \frac{m_e}{5.49\times 10^{-4}} = \frac{9.1\times 10^{-31}}{5.49\times 10^{-4}}$

$\approx 1.66\times 10^{-27}$ kg

So, the particle may be proton or a neutron.

**72** (c) As, de-Broglie wavelength is given as $\lambda = \frac{h}{p}$ and momentum $p = mv$

$\Rightarrow$ $\lambda \propto \frac{1}{p}$ ...(i)

According to the question,

$p_1 = m_1 v_1 = (0.4 \text{ kg} \times 1 \text{ kms}^{-1}) = 400 \text{ kg-ms}^{-1}$

$p_2 = m_2 v_2 = (0.60 \text{ kg} \times 1 \text{ ms}^{-1}) = 0.6 \text{ kg-ms}^{-1}$

$p_3 = m_3 v_3 = (1.0\times 10^{-9} \times 2.2 \text{ ms}^{-1})$

$= 2.2\times 10^{-9} \text{ kg-ms}^{-1}$

$\Rightarrow$ $p_1 > p_2 > p_3$

$\therefore$ $\lambda_1 < \lambda_2 < \lambda_3$ [using Eq. (i)]

**73** (a) Momentum of striking electrons, $p = \frac{h}{\lambda}$ ... (i)

$\therefore$ Kinetic energy of striking electrons,

$$K = \frac{p^2}{2m} = \frac{h^2}{2m\lambda^2}$$ [using Eq. (i)]

This is also, maximum energy of X-ray photons.

Therefore, $\frac{hc}{\lambda_0} = \frac{h^2}{2m\lambda^2}$

where, $\lambda_0$ is the cut-off wavelength $= \frac{2m\lambda^2 c}{h}$.

**74** (c) As we know that, wavelength of an electron,

$$\lambda = \frac{h}{\sqrt{2mE}} = \frac{h}{\sqrt{2m}}\cdot\frac{1}{\sqrt{E}}$$

Taking log on both sides of the above equation, we get

$\Rightarrow$ $\log\lambda = \log\left(\frac{h}{\sqrt{2m}}\cdot\frac{1}{\sqrt{E}}\right)$

$\Rightarrow$ $\log\lambda = \log\frac{h}{\sqrt{2m}} + \log\frac{1}{E^{1/2}}$

$\Rightarrow$ $\log\lambda = \log\frac{h}{\sqrt{2m}} - \frac{1}{2}\log E$

$\Rightarrow$ $\log\lambda = -\frac{1}{2}\log E + \log\frac{h}{\sqrt{2m}}$

So, the above equation is in the form of straight line equation $y = mx + c$.

Such that the slope of the graph between $\log\lambda$ and $\log E$ is $-\frac{1}{2}$. Since, the slope is negative, this implies, with the increase in $\log E$ the value of $\log\lambda$ decreases and *vice-versa*.

Thus, the graph between the energy $\log E$ of an electron and its de-Broglie wavelength $\log\lambda$ is correctly shown in option (c).

**75** (c) de-Broglie wavelength of a moving particle is given by

$$\lambda = \frac{h}{p} = \frac{h}{\sqrt{2mK}}$$

where, $K$ = kinetic energy of the particle.

Now, in given condition,

$$\lambda = \frac{h}{\sqrt{2mE}} \quad ...(i)$$

and $\frac{\lambda}{2} = \frac{h}{\sqrt{2m(E+\Delta E)}}$ ...(ii)

On dividing Eq. (i) by Eq. (ii), we get

$$2 = \frac{\sqrt{E+\Delta E}}{\sqrt{E}}$$

$\Rightarrow$ $\Delta E = 3E$

**76** (b) Initially let the particle has kinetic energy $K$.

So, initially de-Broglie wavelength,

$$\lambda_1 = \frac{h}{p} = \frac{h}{\sqrt{2mK}} \quad ...(i)$$

Now, if the KE is increased to 16 times its previous value, so

$$\lambda_2 = \frac{h}{\sqrt{2m\,(16K)}} = \frac{h}{4\sqrt{2mK}} = \frac{\lambda_1}{4}$$ [$\because$ using Eq. (i)]

$\lambda_2 = 25\%$ of $\lambda_1$

Thus, there is 75% change in the de-Broglie wavelength of the particle.

**77** (a) The de-Broglie wavelength associated with a charged particle is given by

$$\lambda = \frac{h}{p}$$

where, $h$ = Planck's constant and $p$ = momentum $= \sqrt{2m\text{KE}}$ (here, KE is the kinetic energy of the charged particle)

$\Rightarrow$ $\lambda = \frac{h}{\sqrt{2m\text{KE}}}$

For proton and α-particle, the wavelengths are respectively given as

$$\lambda_p = \frac{h}{\sqrt{2m_p\text{KE}_p}} \text{ and } \lambda_\alpha = \frac{h}{\sqrt{2m_\alpha\text{KE}_\alpha}}$$

$$\therefore \quad \frac{\lambda_p}{\lambda_\alpha} = \frac{\sqrt{2m_\alpha \text{KE}_\alpha}}{\sqrt{2m_p \text{KE}_p}} \quad \text{...(i)}$$

Here, $\text{KE}_\alpha = \text{KE}_p$ and $m_\alpha = 4m_p$

Substituting these above mentioned relations in Eq. (i), we get

$$\Rightarrow \quad \frac{\lambda_p}{\lambda_\alpha} = \sqrt{\frac{4m_p}{m_p}} = \frac{2}{1}$$

or $\quad \lambda_p : \lambda_\alpha = 2:1$

**78** (*a*) de-Broglie wavelength, $\lambda = \frac{h}{p} = \frac{h}{mv}$

$$\Rightarrow \quad p = \frac{h}{\lambda}$$

where, $h$ is a constant.

So, $\quad p \propto \frac{1}{\lambda}$

So, for two given particles $A_1$ and $A_2$, we can write

$$\frac{p_1}{p_2} = \frac{\lambda_2}{\lambda_1}$$

But $\quad \lambda_1 = \lambda_2 = \lambda$

Then, $\quad \frac{p_1}{p_2} = \frac{\lambda}{\lambda} = 1 \Rightarrow p_1 = p_2 \quad$ ... (i)

Thus, their momenta is same.

Also, $\quad E = \frac{1}{2}mv^2 = \frac{1}{2}\frac{mv^2 \times m}{m}$

$$= \frac{1}{2}\frac{m^2v^2}{m} = \frac{1}{2}\frac{p^2}{m}$$

Here, $p$ is constant $\quad$ [$\because$ from Eq. (i)]

$$\Rightarrow \quad E \propto \frac{1}{m}$$

$$\therefore \quad \frac{E_1}{E_2} = \frac{m_2}{m_1} < 1 \quad [\because m_1 > m_2]$$

$$\Rightarrow \quad E_1 < E_2$$

Thus, energy of particle $A_1$ is less than energy of particle $A_2$.

**79** (*b*) Suppose, mass of electron $= m_e$,

Mass of photon $= m_p$,

Velocity of electron $= v_e$

and velocity of photon $= v_p$

Given, velocity of electron, $v_e = \frac{c}{100}$

Thus, for electron, de-Broglie wavelength,

$$\lambda_e = \frac{h}{m_e v_e} = \frac{h}{m_e(c/100)} = \frac{100\,h}{m_e c} \quad \text{...(i)}$$

Kinetic energy, $E_e = \frac{1}{2}m_e v_e^2$

$$\Rightarrow \quad m_e v_e = \sqrt{2E_e m_e}$$

As, de-Broglie wavelength, $\lambda_e = \frac{h}{m_e v_e} = \frac{h}{\sqrt{2m_e E_e}}$

$$\Rightarrow \quad E_e = \frac{h^2}{2\lambda_e^2 m_e} \quad \text{...(ii)}$$

Now, for photon, with wavelength $\lambda_p$ energy,

$$E_p = \frac{hc}{\lambda_p} = \frac{hc}{2\lambda_e} \quad \text{...(iii)}$$

[$\because \lambda_p = 2\lambda_e$ (given)]

Dividing Eq. (iii) from Eq. (ii), we get

$$\therefore \quad \frac{E_p}{E_e} = \frac{hc}{2\lambda_e} \times \frac{2\lambda_e^2 m_e}{h^2}$$

$$= \frac{\lambda_e m_e c}{h} = \frac{100h}{m_e c} \times \frac{m_e c}{h} = 100$$

[using Eq. (i)]

So, $\quad \frac{E_e}{E_p} = \frac{1}{100} = 10^{-2}$

For electron, $\quad p_e = m_e v_e = m_e \times (c/100)$

So, $\quad \frac{p_e}{m_e c} = \frac{1}{100} = 10^{-2}$

**80** (*b*) Given, wavelength of light, $\lambda = 589\,\text{nm}$

$$= 589 \times 10^{-9}\,\text{m}$$

Mass of electron, $m_e = 9.1 \times 10^{-31}$ kg

Mass of neutron, $m_n = 1.67 \times 10^{-27}$ kg

Planck's constant, $h = 6.62 \times 10^{-34}$ J-s

Using formula, $\lambda = \frac{h}{\sqrt{2\text{KE}m_e}}$

(i) Kinetic energy of electron,

$$\text{KE}_e = \frac{h^2}{2\lambda^2 m_e} = \frac{(6.63 \times 10^{-34})^2}{2 \times (589 \times 10^{-9})^2 \times 9.1 \times 10^{-31}}$$

$$= 6.96 \times 10^{-25}\,\text{J}$$

(ii) Kinetic energy of neutron,

$$\text{KE}_n = \frac{h^2}{2\lambda^2 m_n} = \frac{(6.63 \times 10^{-34})^2}{2 \times (589 \times 10^{-9})^2 \times 1.67 \times 10^{-27}}$$

$$= 3.82 \times 10^{-28}\,\text{J}$$

**81** (*d*) de-Broglie wavelength for an electron will be given as

$$\lambda_e = \frac{h}{p} \quad \text{...(i)}$$

where, $h =$ Planck's constant

and $\quad p =$ linear momentum of electron.

As, kinetic energy of electron, $E = \frac{p^2}{2m}$

$$\Rightarrow \quad p = \sqrt{2mE} \quad \text{...(ii)}$$

From Eqs. (i) and (ii), we get

$$\lambda_e = \frac{h}{\sqrt{2mE}} \quad \text{...(iii)}$$

Energy of a photon can be given as

$$E = h\nu$$

$$\Rightarrow \quad E = \frac{hc}{\lambda_p}$$

where, $\lambda_p =$ de-Broglie's wavelength of photon.

$\Rightarrow \quad \lambda_p = \frac{hc}{E}$ ...(iv)

Now, dividing Eq. (iii) by Eq. (iv), we get

$$\frac{\lambda_e}{\lambda_p} = \frac{h}{\sqrt{2mE}} \cdot \frac{E}{hc}$$

$$\Rightarrow \quad \frac{\lambda_e}{\lambda_p} = \frac{1}{c} \cdot \sqrt{\frac{E}{2m}} = \frac{1}{c}\left(\frac{E}{2m}\right)^{\frac{1}{2}}$$

**82** (*a*) Here, initial velocity, $u = 0$

Acceleration of the electron in constant electric field $E$.

$a = \frac{eE}{m}$, final velocity, $v = ?$, time, $t = t$

Using kinematic equation of motion,

$$\therefore \quad v = u + at = 0 + \frac{eE}{m}t$$

$$\Rightarrow \quad v = \frac{eEt}{m}$$

de-Broglie wavelength, $\lambda = \frac{h}{mv} = \frac{h}{m(eEt/m)} = \frac{h}{eEt}$

Rate of change of de-Broglie wavelength,

$$\frac{d\lambda}{dt} = \frac{h}{eE}\left(-\frac{1}{t^2}\right) = \frac{-h}{eEt^2}$$

**83** (*a*) Given, potential difference, $V = 10000\text{ V}$

As we know, if an electron is accelerated through a potential of $V$ volt, then the wavelength associated with it is given by

$$\lambda = \frac{h}{\sqrt{2eVm_e}} \quad \text{... (i)}$$

where, $h$ = Planck's constant $= 6.63 \times 10^{-34}$ J-s,

$e$ = electronic charge $= 1.6 \times 10^{-19}$C

and $m_e$ = mass of electron $= 9 \times 10^{-31}$ kg.

Substituting these values in Eq. (i), we get

$$\lambda = \frac{12.27}{\sqrt{V}}\text{Å} = \frac{12.27}{\sqrt{10000}} \times 10^{-10}\text{ m}$$

$$= \frac{12.27 \times 10^{-10}}{100}\text{m} = 12.27 \times 10^{-12}\text{ m}$$

**84** (*b*) de-Broglie wavelength associated with an electron is

$$\lambda = \frac{1.227}{\sqrt{V}}\text{nm}$$

Substituting given value of potential difference,

$V = 100\text{ V}$

We get, $= \frac{1.227}{\sqrt{100}} = 0.123\text{ nm}$

This range of wavelength lies in X-ray region.

**85** (*d*) As de-Broglie wavelength, $\lambda = \frac{h}{\sqrt{2\,mqV}}$

Since, $V$ is same, for both the given particles, so

$$\therefore \quad \lambda \propto \frac{1}{\sqrt{mq}}$$

$$\therefore \quad \frac{\lambda_p}{\lambda_\alpha} = \frac{\sqrt{m_\alpha q_\alpha}}{\sqrt{m_p q_p}}$$

As we know, $m_\alpha = 4m_p$, $q_\alpha = 2q_p$

$\Rightarrow$ The ratio of $\frac{\lambda_p}{\lambda_\alpha} = \frac{\sqrt{4m_p \times 2e}}{\sqrt{m_p \times e}} = \sqrt{8} : 1$

**86** (*a*) de-Broglie wavelength associated with an electron accelerated through a potential of $V$ volts is given as

$$\lambda = \frac{12.27}{\sqrt{V}}\text{Å}$$

$$\therefore \quad \frac{\lambda_1}{\lambda_2} = \sqrt{\frac{V_2}{V_1}}$$

$$\text{or} \quad \lambda_2 = \lambda_1\sqrt{\frac{V_1}{V_2}}$$

Given, $V_1 = 25\text{ kV}$ and $V_2 = 100\text{ kV}$

$$\Rightarrow \quad \lambda_2 = \lambda_1\sqrt{\frac{25}{100}}$$

$$\Rightarrow \quad \lambda_2 = \lambda_1\sqrt{\frac{1}{4}} = \frac{\lambda_1}{2}$$

$\therefore$ de-Broglie wavelength associated with the electrons decreases by 2 times.

**87** (*d*) As, maximum kinetic energy, $K_{\max} = E - \phi_0$ ... (i)

Given, energy of photon that strikes metal $A$,

$E_A = 4.25\text{ eV}$

$\therefore$ Maximum kinetic energy for metal $A$,

$$T_A = 4.25 - (\phi_0)_A \quad \text{...(ii)}$$

[using Eq. (i)]

Energy of photon, $E_B = 4.70\text{ eV}$

$\therefore$ Maximum kinetic energy for metal $B$,

$$T_B = 4.70 - (\phi_0)_B \quad \text{...(iii)}$$

[using Eq. (ii)]

It is given that,

$$T_B = T_A - 1.5$$

So, Eq. (iii) can be given as

$$T_A - 1.5 = 4.70 - (\phi_0)_B$$

Substituting the value of $T_A$ from Eq. (ii) in the above equation, we get

$$4.25 - (\phi_0)_A - 1.5 = 4.70 - (\phi_0)_B$$

$$(\phi_0)_B - (\phi_0)_A = (4.70 - 2.75)$$

$$(\phi_0)_B - (\phi_0)_A = 1.95\text{ eV} \quad \text{... (iv)}$$

As, de-Broglie wavelength,

$$\lambda = \frac{h}{\sqrt{2meV}} \text{ or } \lambda = \frac{h}{\sqrt{2mK}}$$

$$\Rightarrow \quad \lambda \propto \frac{1}{\sqrt{K}}$$

$$\text{or} \quad \frac{\lambda_B}{\lambda_A} = \sqrt{\frac{K_A}{K_B}} = \sqrt{\frac{T_A}{T_B}} = \sqrt{\frac{T_A}{T_A - 1.5}}$$

Given, $\lambda_B = 2\lambda_A$

$\Rightarrow$ $$2 = \sqrt{\frac{T_A}{T_A - 1.5}} \Rightarrow T_A = 2\,\text{eV}$$

and $T_B = T_A - 1.5 = (2 - 1.5)\,\text{eV} = 0.5\,\text{eV}$

$T_B = 0.5\,\text{eV}$

Also, from Eqs. (i), we get

$$2 = 4.25 - (\phi_0)_A$$

$\Rightarrow$ $\phi_A = 2.25\,\text{eV}$

Now, from eq. (iv), we get

$$(\phi_0)_B - 2.25 = 1.95 \Rightarrow (\phi_0)_B = 4.50\,\text{eV}$$

**88** (*b*) de-Broglie wavelength associated with a moving particle can be given as

$$\lambda = \frac{h}{p} = \frac{h}{\sqrt{2m((\text{KE})}}$$

At thermal equilibrium, temperature of neutron and heavy water will be same.

This common temperature is given as $T$.

As we know, kinetic energy of the neutron KE $= \frac{3}{2}kT$

$\therefore$ de-Broglie wavelength of the neutron,

$$\lambda = \frac{h}{p} = \frac{h}{\sqrt{2m(\text{KE})}} = \frac{h}{\sqrt{2m \times \frac{3}{2}kT}} = \frac{h}{\sqrt{3mkT}}$$

**89** (*a*) There is always some uncertainty $\Delta x$ in the specification of position and some uncertainty $\Delta p$ in the specification of momentum. The product of $\Delta x$ and $\Delta p$ is of the order of $\hbar$ (with $\hbar = h/2\pi$), i.e. $\Delta x \cdot \Delta p \approx \hbar$

**90** (*a*) Given, $\Delta x = 1\,\text{nm} = 10^{-9}$ m

By Heisenberg uncertainty principle, $\Delta x \cdot \Delta p = \hbar$

$\therefore$Uncertainty in momentum,

$$\Delta p = \frac{\hbar}{\Delta x} = \frac{h}{2\pi\,\Delta x} \qquad \left(\because \hbar = \frac{h}{2\pi}\right)$$

$$= \frac{6.63 \times 10^{-34}}{2 \times \pi \times 10^{-9}}$$

$$\approx 1.05 \times 10^{-25}\ \text{kg-ms}^{-1}$$

**92** (*a*) The experimental arrangement used by Davisson and Germer consists of an electron gun which comprises of a tungsten filament $F$, coated with barium oxide and heated by a low voltage power supply. Electrons emitted by the filament are accelerated to a desired velocity by applying suitable potential/voltage from a high voltage power supply. Thus, the velocity of the electron gun can be increased by increasing the potential difference between the anode and the filament.

**93** (*a*) Cathode ray particles when strike the electrons of glass atom, the electrons of glass atom are excited and move to higher energy levels. On de-excitation, they fall to their ground state and release energy. This released energy appears in the form of visible energy, that we usually regard as fluorescence.

As energy levels are characteristics of a glass, so the colour of the glow depends on the type of glass used in discharge tube, e.g. it is yellowish-green for soda glass.

Therefore, Assertion and Reason are correct and Reason is the correct explanation of Assertion.

**94** (*a*) In a detector loop, when the light falls on the metal surface, i.e. the emitter plate, some electrons near the surface absorb enough energy from incident radiations, i.e. from UV light, to overcome the attraction of the positive ions in the material of the surface.

After gaining sufficient energy from UV-light, the electrons escape from the emitter plate into the surrounding space, thus enhancing the high voltage spark across detector loop.

Therefore, Assertion and Reason are correct and Reason is the correct explanation of the Assertion.

**95** (*a*) According to the Einstein's picture of photoelectric effect, intensity of the radiation is proportional to the number of energy quanta per unit area per unit time. Thus, greater the number of the energy quanta available, greater is the number of electrons the absorbing the energy quanta and therefore greater is the number of electrons coming out of the metal.

Hence, for $\nu > \nu_0$ photoelectric current is proportional to intensity.

Therefore, Assertion and Reason are correct and Reason is the correct explanation of Assertion.

**96** (*b*) Work function is the minimum energy required to eject the photoelectron from photosensitive metal.

Hence for metal to have high photosensitivity, the work-function should be small.

Mathematically, work function is given as $h\nu_0$, where $\nu_0$ is the threshold frequency.

Therefore, Assertion and Reason are correct but Reason is not the correct explanation of Assertion.

**97** (*a*) Light when interacted with matter behaves as, if it is made up of quanta or packet of energy, each packet of energy $h\nu$. Einstein later arrived at a important result that, the light quantum can also be associated with momentum $\left(\frac{h\nu}{c}\right)$.

Thus, a definite value of energy as well as momentum is a strong sign that the light quantum can be associated with a particle. This particle was later named photon.

Therefore, Assertion and Reason are correct and Reason is the correct explanation of Assertion.

**98** (*d*) Wavelength associated with a particle is given as

$$\lambda = \frac{h}{p} = \frac{h}{mv} \Rightarrow \lambda \propto \frac{1}{m} \text{ or } \lambda \propto \frac{1}{v}$$

Thus, we can say that wavelength $\lambda$ is smaller for heavier particles. As, macroscopic objects are heavier in nature so their wavelength is so small, that it is beyond any measurement.

This is the reason why macroscopic objects in our daily life do not show wave like properties.

Therefore, Assertion is incorrect but Reason is correct.

**99** (*a*) A person approaching a doorway may intrupt a light beam which is incident on photocell.

This interuption will leads to abrupt change in the amount of photocurrent. Thus, this change in photocurrent helps to start a motor which is fitted in the doorway which opens the door.

Therefore, Assertion and Reason are correct and Reason is the correct explanation of Assertion.

**100** (*b*) Max Born suggested a probability interpretation to the matter wave amplitude. According to this, the intensity (square of amplitude) of the matter wave at a point determines the probability density of the particle at that point. Probability means probability per unit volume.

Thus, if $A$ is the amplitude of the wave at that point, $IA^2 / \Delta V$ is the probability of the particle being found in a small volume $\Delta V$ around that point.

Thus, if intensity of matter wave is large in a certain region, there is a greater probability of the particle being found there than where the intensity is small.

Therefore, Assertion and Reason are correct but Reason is not the correct explanation of Assertion.

**101** (*d*) de-Broglie wavelength, $\lambda = \frac{h}{\sqrt{2mK}}$, where $K$ is kinetic energy.

For same kinetic energy, $\lambda \propto \frac{1}{\sqrt{m}}$ ... (i)

As $\alpha$-particle is heaviest amongst the given particles. From relation (i), $\lambda$ is smallest for $\alpha$-particle.

Therefore, Assertion is incorrect but Reason is correct.

**102** (*c*) In 1887, it was found that certain metals, when irradiated by ultraviolet light, emitted negatively charged particles having small speeds . Also, certain metals when heated at a high temperature were found to emit negatively charged particles. The value of $e/m$ of these particles was found to be same as that of cathode ray particles.

Thus, these observations established that all these particles, although produced under different conditions, were identical in nature. Since, cathode ray particles, travels with the speed ranging from about 0.1 to 0.2 times the speed of light ($3 \times 10^8$ m/s). So, these particles also travel with this speed.

Also, the $e/m$ value of this particle is $1.76 \times 10^{11}$ C/kg.

So, all statements are correct.

**103** (*b*) The statement II is incorrect and it can be corrected as,

In the experimental set-up used for studying the photoelectric effect, the intensity of light can be varied by changing the distance of the light source from the emitter. As,

$$\text{Intensity} \propto \frac{1}{(\text{Distance})^2}$$

Rest statement is correct.

**104** (*b*) The statement II is correct but rest is incorrect and it can be corrected as,

Under the given condition of the experimental set-up, it has been observed that, if accelerating potential is applied at collector $A$, then photoelectric current firstly increases, then eventually saturates or becomes constant. However, when a retarding potential is applied at $A$ and gradually increased, then on reversing the polarity, the electrons are repelled and only the most energetic electrons are able to reach the collector $A$.

Thus, the photocurrent is found to decrease rapidly until it drops to zero at a certain sharply defined, critical value of the negative potential $V_0$ on plate $A$.

**105** (*a*) The statement I is correct but rest is incorrect and it can be corrected as,

From the given graph, we can say that we have obtained different values of stopping potential but the same value of the saturation current for incident radiation of different frequencies. Since, the energy of emitted electron depends on the frequency of the incident radiations. This means, stopping potential is more negative for higher frequencies of incident radiation. Thus, we can conclude that,

I. stopping potential are in the order $V_{OA} > V_{OB} > V_{OC}$, if the frequencies are in the order $\nu_A > \nu_B > \nu_C$.

II. This further implies that greater the frequency of incident radiation, greater is the maximum kinetic energy of the photoelectrons. Consequently we need greater retarding potential to stop them completely.

**106** (*a*) The statement I is correct but rest is incorrect and it can be corrected as,

We know that, there exists a certain minimum cut-off frequency $\nu_0$ for which the stopping potential is zero.

This means, for a frequency $\nu$ of incident radiation lower than the cut-off frequency $\nu_0$, no photoelectric emission is possible even, if the intensity is large.

**107** (*d*) The statement II is correct but rest are incorrect and these can be corrected as,

Einstein's photoelectric equation is $K_{max} = h\nu - \phi_0$

where, $K_{max}$ = maximum KE of emitted electrons,

$h$ = Planck's constant, $\nu$ = frequency of incident radiation and $\phi_0$ = work function of metal.

Since, $K_{max}$ must be non-negative.

So, photoelectric emission is possible only, if

$$h\nu > \phi_0 \quad \text{or} \quad \nu > \nu_0,$$

where, $\nu_0 = \frac{\phi_0}{h}$

In other words, in photoelectric effect, an electron absorbs a quantum of energy of radiation. If this quantum of energy absorbed exceeds the minimum energy needed for the electron to escape from the metal surface, the electron emitted with maximum kinetic energy, $K_{max}$.

So, more tightly bound electrons will emerge with kinetic energies less than the maximum value.

∴Kinetic energy of emitted photoelectrons lies between 0 to $K$.

**108** (*b*) The statements II and III are correct but I is incorrect and it can be corrected as,

In Einstein's picture, the basic elementary process involved in photoelectric effect is the absorption of a light quantum of an electron. This process is instantaneous. Thus, whatever may be the intensity, i.e. the number of quanta of radiation per unit area per unit time, photoelectric emission is instantaneous.

Hence, low intensity does not mean delay in electron emission.

From photoelectric equation, $eV_0 = h\nu - \phi_0$

or $$V_0 = \left(\frac{h\nu}{e}\right) - \frac{\phi_0}{e}$$

So, it predicts that $V_0$ *versus* ν curve is a straight line with slope $\left(=\frac{h}{e}\right)$ and is independent of the nature of the material.

**109** (*c*) The statement III is correct but rest are incorrect and these can be corrected as,

If an electron has a definite momentum $p$ (i.e. $\Delta p = 0$), then by the de-Broglie relation, it has a definite wavelength λ. As, a wave of definite (single) wavelength extends all over space.

So, by Born's probability interpretation this means that the electron is not localised in any finite region of space, i.e. its position uncertainity is infinite ($\Delta x \to \infty$), which is consistent with the uncertainity principle.

As per uncertainity principle, $\Delta x\, \Delta p \propto \hbar$ which means, if $\Delta p$ is zero, then $\Delta x$ must be infinite.

Also, the wave pattern will be as shown below

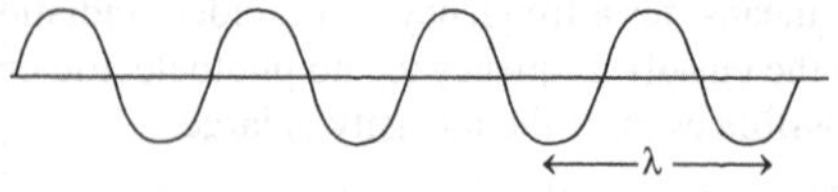


**110** (*d*) The statements II and III are correct but I is incorrect and it can be corrected as,

Whether a particle or wave description is best suited for understanding an experiment depends on the nature of the experiment. e.g. In the familiar phenomenon of seeing an object by our eye, both the following given descriptions are important.

Firstly, the gathering and focussing mechanism of light by the eye lens is well described in the wave picture.

Secondly, its absorption by rods and cones (of the retina) requires the photon picture of light.

**111** (*a*) The statement I is correct but II is incorrect and it can be corrected as,

The appearance of the peak in the particular direction in Davisson and Germer experiment is due to the constructive interference of electrons scattered from different layers of the regularly spaced atoms of the crystals.

**112** (*b*) From the electron diffraction measurement, the wavelength of matter was found to be 0.165 nm.

In Davisson-Germer experiment, a strong peak is appeared for the scattered electron for an accelerating voltage of 54V.

So, the de-Broglie wavelength λ associated with electron for $V = 54$ V is given by

$$\lambda = \frac{h}{p} = \frac{1.227}{\sqrt{V}}\text{ nm}$$

$$\lambda = \frac{1.227}{\sqrt{54}} = 0.167\text{ nm}$$

Thus, there is an excellent agreement between the theoretical value and the experimentally obtained value of de-Broglie wavelength.

So, statements II and III are correct but I is incorrect.

**113** (*a*) The statement given in option (a) is incorrect and it can be corrected as,

From the given graph, it can be concluded that for a given frequency of incident radiation, the stopping potential is independent of its intensity. But, we know that, $K_{max} = eV_0$, where $K_{max}$ = maximum kinetic energy. So, in other words, maximum kinetic energy of photoelectrons, depends on the light source and the emitter plate material but is independent of intensity of incident radiation.

Rest statements are correct.

**114** (*c*) The statement given in option (c) is correct but rest are incorrect and these can be corrected as,

Different photosensitive materials respond differently to light. e.g. Selenium is more sensitive than zinc or copper.

Also the same photosensitive substance gives different response to light of different wavelengths. e.g. Ultraviolet light gives rise to photoelectric effect in copper while green or red light does not.

**115** (*a*) According to Einstein's equation,

$$\frac{hc}{\lambda} - \phi = eV$$

or $$V = \frac{hc}{e\lambda} - \frac{\phi}{e}$$

From the given graph,

This equation is in the form of $y = mx - c$, such that slope of the graph,

$$\tan\theta = \frac{hc}{e}$$

| Metal I | Metal II | Metal III |
|---|---|---|
| $\frac{\phi_1}{hc} = 0.001$ | $\frac{\phi_2}{hc} = 0.002$ | $\frac{\phi_3}{hc} = 0.004$ |

$\Rightarrow \quad \phi_1 : \phi_2 : \phi_3 = 1:2:4$

For metal I, threshold wavelength

$$\lambda = \frac{hc}{\phi_1} = \frac{hc}{0.001hc} = \frac{1000}{1} = 1000\,\text{nm}$$

For metal II, threshold wavelength

$$\lambda = \frac{hc}{\phi_2} = \frac{hc}{0.002hc} = \frac{1000}{2} = 500\,\text{nm}$$

For metal III, threshold wavelength

$$\lambda = \frac{hc}{\phi_3} = \frac{hc}{0.004hc} = \frac{1000}{4} = 250\,\text{nm}$$

Since, the wavelength of violet colour light is approximately 430 nm, so it will eject photoelectrons from metal II not from metal III.

$\therefore$ The statement given in option (a) is correct.

**116** (*b*) The statement given in option (b) is incorrect and it can be corrected as,

Photons are electrically neutral and are not deflected by electric and magnetic fields.

However, particle like behaviour of light was confirmed by the experiment of AH Compton on scattering of X-ray from electrons.

Rest statements are correct.

**117** (*a*) The statement given in option (a) is correct but rest are incorrect and these can be corrected as,

Photocell is a device whose electrical properties are affected by light.

It converts a change in intensity of illumination into change in photocurrent.

This current can be used to operate control systems and in light measuring devices. So, light meters in photographic cameras make use of photocell to measure the intensity of light.

**118** (*b*) An electron absorbs a single photon of light and if the energy carried by photon is more than work function of the metal, the electron is emitted.

From source,

Total emitted power = $P$ (total energy per second)

Energy carried by each photon = $E$

So, number of photons emitted by the source = $P / E$

Energy of a photons $E = h\nu$.

For the cut-off or threshold frequency, the energy $h\nu_0$ of the incident radiation must be equal to work function $\phi_0$, so that $h\nu_0 = \phi_0 \Rightarrow h = \frac{\phi_0}{\nu_0}$

Hence, A $\rightarrow$ 4, B $\rightarrow$ 2, C $\rightarrow$ 3 and D $\rightarrow$ 1.

**120** (*a*) Given, voltage, $V = 30\text{ kV} = 30 \times 10^3\text{ V}$

and $\quad e = 1.6 \times 10^{-19}\text{C}$

Using formula for energy, $E = eV = h\nu$

$$\nu = \frac{eV}{h} = \frac{1.6 \times 10^{-19} \times 30 \times 10^3}{6.63 \times 10^{-34}}$$
$$= 7.24 \times 10^{18}\text{ Hz}$$

Minimum wavelength of X-ray,

$$\lambda = \frac{c}{\nu} = \frac{3 \times 10^8}{7.24 \times 10^{18}}$$

$$\Rightarrow \quad \lambda = 0.414 \times 10^{-10}$$
$$= 0.0414 \times 10^{-9}\text{ m}$$
$$= 0.0414\text{ nm}$$

**121** (*c*) Given, cut-off voltage, $V_0 = 1.5\text{ V}$

Use the formula for maximum kinetic energy,

$$\text{KE}_{\max} = eV_0 = 1.5\,\text{eV}$$
$$= 1.5 \times 1.6 \times 10^{-19}$$
$$= 2.4 \times 10^{-19}\text{ J}$$

**122** (*a*) Given, wavelength of monochromatic light,

$$\lambda = 632.8\,\text{nm}$$
$$= 632.8 \times 10^{-9}\text{ m}$$

Power, $P = 9.42\,\text{mW} = 9.42 \times 10^{-3}\text{ W}$

Energy of each photon,

$$E = \frac{hc}{\lambda} = \frac{6.63 \times 10^{-34} \times 3 \times 10^8}{632.8 \times 10^{-9}}$$
$$= 3.14 \times 10^{-19}\text{ J}$$

We know that, momentum of each photon, $p = \frac{h}{\lambda}$

$$p = \frac{6.63 \times 10^{-34}}{632.8 \times 10^{-9}}$$
$$= 1.05 \times 10^{-27}\text{ kg-ms}^{-1}$$

**123** (*a*) Given, energy per unit area per second,

$$P = 1.388 \times 10^3\text{ Wm}^{-2}$$

Let $n$ be the number of photons incident on the earth per square metre.

Wavelength of each photon, $\lambda = 550\,\text{nm} = 550 \times 10^{-9}\text{ m}$

Energy of each photon, $E = hc/\lambda$, where $h$ is the Planck's constant.

$$= \frac{6.63 \times 10^{-34} \times 3 \times 10^8}{550 \times 10^{-9}}$$
$$= 3.616 \times 10^{-19}\text{ J}$$

Number of photons incident on the earth's surface,

$$n = \frac{P}{E} = \frac{1.388 \times 10^3}{3.616 \times 10^{-19}} = 3.838 \times 10^{21}$$
$$= 3.838 \times 10^{21}\text{ photon/m}^2\text{-s}$$
$$\simeq 10^{21}\text{ photons}$$

**124** (*b*) Given, power of lamp, $P = 100\text{ W}$

Wavelength of the sodium light, $\lambda = 589\,\text{nm}$

$$= 589 \times 10^{-9}\text{ m}$$

Planck constant, $h = 6.63 \times 10^{-34}\text{ J-s}$

Energy of each photon,

$$E = \frac{hc}{\lambda} = \frac{6.63\times 10^{-34} \times 3\times 10^{8}}{589\times 10^{-9}} \quad (\because c = 3\times 10^{8}\text{ m/s})$$

$$= 3.38\times 10^{-19}\text{ J}$$

$$= \frac{3.38\times 10^{-19}}{1.6\times 10^{-19}} = 2.11\text{ eV}$$

Let $n$ photons are delivered per second, so

$$n = \frac{\text{Power}}{\text{Energy of each photon}} \quad (\text{from } P = En)$$

$$= \frac{100}{3.38\times 10^{-19}} = 3\times 10^{20}\text{ photon/s}$$

$= 3\times 10^{20}$ photon/s are delivered

**125** (*a*) Given, threshold frequency for a metal,

$$\nu_0 = 3.3\times 10^{14}\text{ Hz}$$

Frequency of light, $\nu = 8.2\times 10^{14}$ Hz

Let $V_0$ be the cut-off voltage.

Using the formula for kinetic energy,

$$(\text{KE})_{\max} = eV_0 = h\nu - h\nu_0$$

$$V_0 = \frac{h(\nu - \nu_0)}{e}$$

$$= \frac{6.63\times 10^{-34}(8.2\times 10^{14} - 3.3\times 10^{14})}{1.6\times 10^{-19}}$$

$$= \frac{6.63\times 10^{-34}\times 10^{14}\times 4.9}{1.6\times 10^{-19}} = 2.03\text{ V}$$

**126** (*a*) Given, frequency of light,

$$\nu = 7.21\times 10^{14}\text{ Hz}$$

Mass of electron, $m = 9.1\times 10^{-31}$ kg

Maximum speed of electrons, $v_{\max} = 6.0\times 10^{5}\text{ ms}^{-1}$

Let $\nu_0$ be the threshold frequency.

Using the formula for kinetic energy,

$$(\text{KE})_{\max} = \frac{1}{2}mv_{\max}^2 = h\nu - h\nu_0$$

i.e. $\frac{1}{2}\times 9.1\times 10^{-31}\times 6.0\times 10^{5}\times 6.0\times 10^{5}$

$$= 6.63\times 10^{-34}(\nu - \nu_0)$$

or $\nu - \nu_0 = \frac{36\times 9.1\times 10^{-21}}{2\times 6.63\times 10^{-34}} = 2.47\times 10^{14}$

$$\nu_0 = 7.21\times 10^{14} - 2.47\times 10^{14}$$

$$= 4.74\times 10^{14}\text{ Hz}$$

**127** (*b*) Given, wavelength of light,

$$\lambda = 488\text{ nm} = 488\times 10^{-9}\text{ m}$$

Cut-off potential, $V_0 = 0.38$ V, $e = 1.6\times 10^{-19}$ C

Planck constant, $h = 6.62\times 10^{-34}$ J-s

and velocity of light, $c = 3\times 10^{8}\text{ ms}^{-1}$

Let $\phi_0$ be the work function.

Using the formula for kinetic energy,

$$(\text{KE})_{\max} = eV_0 = \frac{hc}{\lambda} - \phi_0$$

$$\Rightarrow 1.6\times 10^{-19}\times 0.38 = \frac{6.63\times 10^{-34}\times 3\times 10^{8}}{488\times 10^{-9}} - \phi_0$$

or $6.08\times 10^{-20} = 40.75\times 10^{-20} - \phi_0$

or $\phi = (40.75 - 6.08)\times 10^{-20}$

$$= 34.67\times 10^{-20}\text{ J}$$

or $= \frac{34.67\times 10^{-20}}{1.6\times 10^{-19}}\text{ eV} = 2.17\text{ eV}$

**128** (*d*) A particle is dropped from height $H$, so its initial velocity $u$ will be zero and its final velocity is $v$.

From motion of equation, $v^2 - u^2 = 2gH$,

As, $u = 0$

and $g$ = acceleration due to gravity,

$\Rightarrow$ Velocity gained by particle, $v = \sqrt{2gH}$

$\therefore$ $$\lambda = \frac{h}{p} = \frac{h}{mv} = \frac{h}{m\sqrt{2gH}}$$

$\Rightarrow$ $$\lambda \propto H^{-1/2}$$

**129** (*b*) Given, $E = 1$ MeV

Energy of a photon, $E = \frac{hc}{\lambda}$

$$\Rightarrow \lambda = \frac{6.63\times 10^{-34}\times 3\times 10^{8}\times 10^{9}\text{eV-nm}}{1.6\times 10^{-19}\times 1\text{ MeV}}$$

$$\simeq \frac{1240\text{ eV-nm}}{1\text{ MeV}} = 1.2\times 10^{-3}\text{ nm}$$

**130** (*d*) When an electron with energy $E_0$ strikes another electron, it is one of the possibility that it replaces one of $e^-$ of metal by transferring complete energy. As there is no change in total number of electrons, no work has to be done as no attraction is produced by metal atoms.

$\therefore$ Electrons will be emitted with maximum energy $= E_0$.

**131** (*c*) In Davisson-Germer experiment, the de-Broglie wavelength associated with electron is

$$\lambda = \frac{12.27}{\sqrt{V}}\text{ Å} \quad \text{...(i)}$$

where, $V$ is the applied voltage.

If there is a maxima of the diffracted electrons at an angle $\theta$, then $2d\sin\theta = \lambda$ ...(ii)

So, $V$ will increase with the decrease in the $\lambda$. {using Eq. (i)}

So, with the decrease in $\lambda$, $\theta$ will also decrease. {from Eq. (ii)}

Thus, when the voltage applied to $A$ is increased. The diffracted beam will have the maximum at a value of $\theta$ that will be less than the earlier value.

**132** (*b*) We know that, the relation between $\lambda$ and $K$ is given by

$$\lambda = \frac{h}{\sqrt{2mK}}$$

Here, for the given value of energy $K$, $\frac{h}{\sqrt{2K}}$ is a constant.

Thus, $\lambda \propto \frac{1}{\sqrt{m}}$

$\therefore$ $\lambda_p : \lambda_n : \lambda_e : \lambda_\alpha$

$\Rightarrow$ $= \frac{1}{\sqrt{m_p}} : \frac{1}{\sqrt{m_n}} : \frac{1}{\sqrt{m_e}} : \frac{1}{\sqrt{m_\alpha}}$

Since, $m_p = m_n$, hence $\lambda_p = \lambda_n$

As, $m_\alpha > m_p$, therefore $\lambda_\alpha < \lambda_p$

As, $m_e < m_n$, therefore $\lambda_e > \lambda_n$

Hence, $\lambda_\alpha < \lambda_p = \lambda_n < \lambda_e$

**133** (*a*) Given, $\mathbf{v} = v_0\hat{\mathbf{i}}$ and $\mathbf{B} = B_0\hat{\mathbf{j}}$

Force on charged particle moving in external magnetic field $= q(\mathbf{v} \times \mathbf{B})$

or $\mathbf{F} = q(v_0\hat{\mathbf{i}} \times B_0\hat{\mathbf{j}})$

or $\mathbf{F} = qB_0 v_0\hat{\mathbf{k}}$

$\because$ $q$ is negative, force is in direction of $-\hat{\mathbf{k}}$ (into the plane of paper). For force perpendicular to velocity, the particle executes uniform circular motion. Since, magnitude of velocity $v_0$ is constant.

de-Broglie wavelength,

$$\lambda = \frac{h}{mv} = \frac{h}{mv_0} = \text{constant}$$

**134** (*a*) Initial de-Broglie wavelength of electron,

$$\lambda_0 = \frac{h}{mv_0} \quad \ldots(\text{i})$$

Given, $E = -E_0\hat{\mathbf{i}}$

Force on electron in electric field,

$$\mathbf{F} = -e\mathbf{E} = -e[-E_0\hat{\mathbf{i}}] = eE_0\,\hat{\mathbf{i}}$$

Acceleration of electron, $\mathbf{a} = \frac{\mathbf{F}}{m} = \frac{eE_0\hat{\mathbf{i}}}{m}$

Given, initial velocity, $u = v_0\hat{\mathbf{i}}$

Velocity of electron after time $t$,

$$v = v_0\,\hat{\mathbf{i}} + \left(\frac{eE_0\,\hat{\mathbf{i}}}{m}\right)t \quad (\text{using, } v = u + at)$$

$$= \left(v_0 + \frac{eE_0}{m}t\right)\hat{\mathbf{i}} = v_0\left(1 + \frac{eE_0}{mv_0}t\right)\hat{\mathbf{i}}$$

de-Broglie wavelength associated with electron at time $t$,

$$\lambda = \frac{h}{mv} = \frac{h}{m\left[v_0\left(1 + \frac{eE_0}{mv_0}t\right)\right]}$$

$$= \frac{\lambda_0}{\left[1 + \frac{eE_0}{mv_0}t\right]} \quad \left(\because \frac{h}{mv_0} = \lambda_0\right)$$

**135** (*c*) Initial de-Broglie wavelength of electron,

$$\lambda_0 = \frac{h}{mv_0}$$

Force on electron in electric field,

$$\mathbf{F} = -eE = -eE_0\,\hat{\mathbf{j}}$$

Acceleration of electron,

$$\mathbf{a} = \frac{\mathbf{F}}{m} = -\frac{eE_0}{m}\hat{\mathbf{j}}$$

It is acting along negative $Y$-axis.

The initial velocity of electron along $X$-axis, $\mathbf{v}_{x0} = v_0\,\hat{\mathbf{i}}$,

Initial velocity of electron along $Y$-axis, $\mathbf{v}_{y0} = 0$.

and velocity of electron after time $t$ along $X$-axis,

$$\mathbf{v}_x = v_0\hat{\mathbf{i}}$$

(since, there is no acceleration of electron along $X$-axis)

Velocity of electron after time $t$ along $Y$-axis,

$$\mathbf{v}_y = 0 + \left(-\frac{eE_0}{m}\hat{\mathbf{j}}\right)t = -\frac{eE_0}{m}t\,\hat{\mathbf{j}}$$

Magnitude of velocity of electron after time $t$,

$$|\mathbf{v}| = \sqrt{v_x^2 + v_y^2}$$

$$= \sqrt{v_0^2 + \left(\frac{-eE_0}{m}t\right)^2}$$

$$= v_0\sqrt{1 + \frac{e^2E_0^2t^2}{m^2v_0^2}}$$

de-Broglie wavelength,

$$\lambda = \frac{h}{mv} = \frac{h}{mv_0\sqrt{\frac{1 + e^2E_0^2t^2}{(m^2v_0^2)}}}$$

$$\left(\text{here, } \lambda_0 = \frac{h}{mv_0}\right)$$

$$= \frac{\lambda_0}{\left(1 + \frac{e^2E_0^2t^2}{m^2v_0^2}\right)^{1/2}}$$

# CHAPTER > 12

# Atoms

## KEY NOTES

- The first model of atom was proposed by JJ Thomson is 1898.
- According to this model, the positive charge of the atom is uniformly distributed throughout the volume of the atom and the negatively charged electrons are embedded in it like seeds in a watermelon.
- This model was called **plum pudding model** of the atom.

## α-Particle Scattering and Rutherford's Nuclear Model of Atom

- The schematic arrangement of Rutherford's α-particle scattering experiment (also called Geiger-Marsden experiment) is shown below.

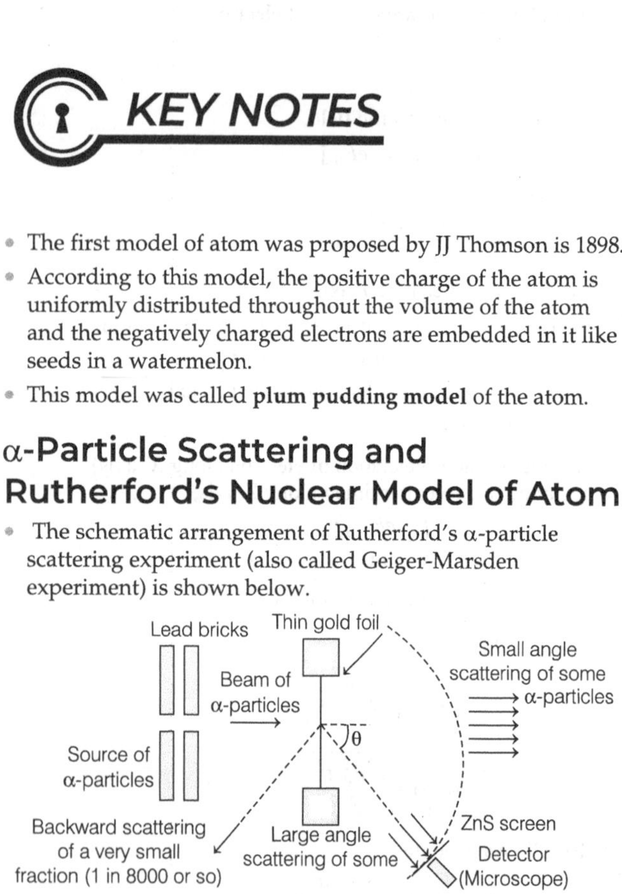


- Observations of the this experiment are as follows
  - (i) Many of the α-particles pass through the foil without deflection.
  - (ii) Only about 0.14% of the incident α-particles scattered by more than 1°.
  - (iii) Only about one α-particle in every 8000 α-particles deflected by more than 90°.
- Based on these observation, Rutherford proposed a nuclear model of atom, are called planetary model, in which entire positive charge and most of the mass of atom is concentrated in a small volume called the **nucleus** with electron revolving around the nucleus as planets revolve around the sun.
- Rutherford's experiment suggested that the size of nucleus to be about $10^{-15}$ m to $10^{-14}$ m. From kinetic theory, the size of an atom is known to be $10^{-10}$ m. Thus, most of an atom is empty space.
- The atomic electrons being so light, does not affect the α-particles.

## α-Particle Trajectory

The trajectory of the α-particle can be computed employing Newton's second law of motion and the Coulomb's law for electrostatic force of repulsion between the α-particle and the positively charged nucleus. The magnitude of this force is

$$F = \frac{1}{4\pi\varepsilon_0} \frac{(2e)(Ze)}{r^2}$$

where, $r$ is the distance between the α-particle and the nucleus.

- The magnitude and direction of the force on an α-particle continuously changes as it approaches nucleus and recedes away from it.
- Trajectory of the α-particles in the coulomb field of a target nucleus is as shown in the figure below

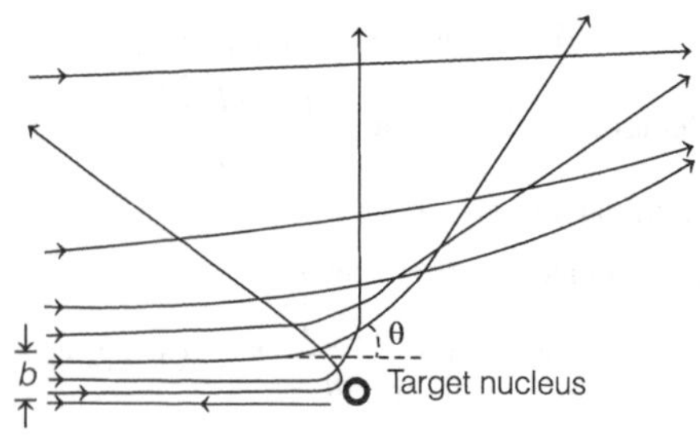


- **Impact parameter ($b$)** is the perpendicular distance of the initial velocity vector of the α-particle from the centre of the nucleus.
- In case of head-on-collision, the impact parameter $b$ is minimum and the α-particle rebounds back ($\theta \cong \pi$). For a large $b$, the α-particle goes nearly undeviated and has a small deflection ($\theta \cong 0$).

- The minimum distance from the nucleus upto which the α-particle approaches is called the **distance of closest approach** ($r_0$).
  At this distance, whole of kinetic energy of α-particle converts into electrostatic potential energy.
  It is given as $r_0 = \frac{4RZe^2}{mv^2}$

## Electron Orbits

- For a dynamically stable orbit in a hydrogen atom, electrostatic force of attraction $F_e$ between the revolving electrons provides the requisite centripetal force $F_c$.
- The relation between the orbit radius and the electron velocity is $r = \frac{e^2}{4\pi\varepsilon_0 mv^2}$.
- The kinetic energy $K$ and electrostatic potential energy $U$ of the electron in hydrogen atom are
  $$K = \frac{e^2}{8\pi\varepsilon_0 r} \text{ and } U = \frac{-e^2}{4\pi\varepsilon_0 r}$$
- The total energy $E$ of the electron in a hydrogen atom is
  $$E = \frac{-e^2}{8\pi\varepsilon_0 r}$$
  Negative sign implies the fact that electron is bound to the nucleus.
- **Drawbacks of Rutherford's atomic model** are given below
  (i) According to classical electromagnetic theory, an accelerating charged particle emits radiation in the form of EMW.
  The energy of an accelerating electron should therefore, continuously decrease. The electron would spiral inward and eventually fall into the nucleus. Thus, such an atom cannot be stable.
  (ii) Rutherford model was unable to explain line spectrum.

## Atomic Spectra

- Each element has a characteristic spectrum of radiation which it emits.
- When an atomic gas or vapour is excited at low pressure, usually by passing an electric current through it, the emitted radiation has a spectrum containing specific wavelengths only.
  This type of spectrum consists of bright lines on a dark background and is termed as **emission line spectrum.**
- When white light passes through a gas and the transmitted light consists of some dark lines in the spectrum. These dark lines correspond precisely to those wavelengths which were found in the emission spectrum of gas. This is called the **absorption spectrum** of the gas.

## Bohr Model of the Hydrogen Atom

- Bohr combined classical and early quantum concepts and gave his theory in form of three postulates.
  These are as given below
  (i) An electron in an atom could revolve in certain stable orbits (known as stationary orbits) without the emission of radiant energy. Thus, each atom has definite states and each possible state has definite total energy.
  (ii) The electron revolves around the nucleus only in those orbits for which the angular momentum is some integral multiple of $\frac{h}{2\pi}$, where $h$ is called the Planck's constant ($6.6 \times 10^{-34}$ J-s).
  Thus, the angular momentum $L$ of the orbiting electron is quantised, i.e.
  $$L = \frac{nh}{2\pi}$$
  (iii) An electron might make a transition from one of its specified non-radiating orbits to another of lower energy. When it does so, a photon is emitted having energy equal to the energy difference between the initial and final states. The frequency of the emitted photon is given by
  $$\nu = \frac{E_i - E_f}{h}$$
  where, $E_i$ and $E_f$ are the energies of the initial and final states and $E_i > E_f$.
- The **orbital velocity** of an electron in $n$th orbit is
  $$v_n = \frac{1}{n}\frac{e^2}{4\pi\varepsilon_0}\frac{1}{\left(\frac{h}{2\pi}\right)} \Rightarrow v_n \propto \frac{1}{n}$$
  Velocity of electron in first orbit ($n = 1$) of H-atom is
  $$v_1 = \frac{e^2}{2h\varepsilon_0} = \frac{c}{137}$$
- The **orbital radius** of an electron in $n$th orbit is
  $$r_n = \left(\frac{n^2}{m}\right)\left(\frac{h}{2\pi}\right)^2 \frac{4\pi\varepsilon_0}{e^2} \Rightarrow r_n \propto n^2$$
  Bohr's radius ($n = 1$) is given as
  $$r_1 = a_0 = \frac{h^2\varepsilon_0}{\pi me^2} = 5.29 \times 10^{-11} \text{m}$$
- **Total energy of electron** in the stationary states of hydrogen atom is given as
  $$E_n = \frac{-me^4}{8n^2\varepsilon_0^2 h^2}$$
  $$= \frac{-2.18 \times 10^{-18}}{n^2} \text{J} = \frac{-13.6}{n^2} \text{eV}$$

KEY NOTES

## Energy Level

- The energy of an atom is the least (largest negative value) when its electron is revolving in an orbit closest to the nucleus.
- The lowest state of the atom is called **ground state**.
- The minimum energy required to free the electron from the ground state of the hydrogen atom is 13.6 eV and is called the **ionisation energy of the hydrogen atom.**

**Note** At room temperature, most of the hydrogen atoms are in ground state.

- The highest energy state corresponds to $n = \infty$ and has an energy of 0 eV. This is the energy of the atom when the electron is completely removed from the nucleus and is at rest.
- When an atom makes a transition from the higher energy state with quantum number $n_i$ to the lower energy state with quantum $n_f$ $(n_f < n_i)$ in hydrogen like atom, the difference of energy is carried away by a photon of frequency $\nu_{if}$ such that

$$\nu_{if} = \frac{me^4}{8\varepsilon_0^2 h^2}\left(\frac{1}{n_f^2} - \frac{1}{n_i^2}\right) \text{ or } \nu_{if} = R\left(\frac{1}{n_f^2} - \frac{1}{n_i^2}\right)$$

where, $R = 1.03 \times 10^7 \text{ m}^{-1}$.

## Hydrogen Spectrum or Line Spectra of Hydrogen Atom

- In hydrogen spectrum, there is no resemblance of order or regularity in spectral lines but the spacing between lines within certain sets of spectrum decreases in regular way. Each of these sets is called a **spectral series**.
- Emission spectrum of hydrogen is as shown below

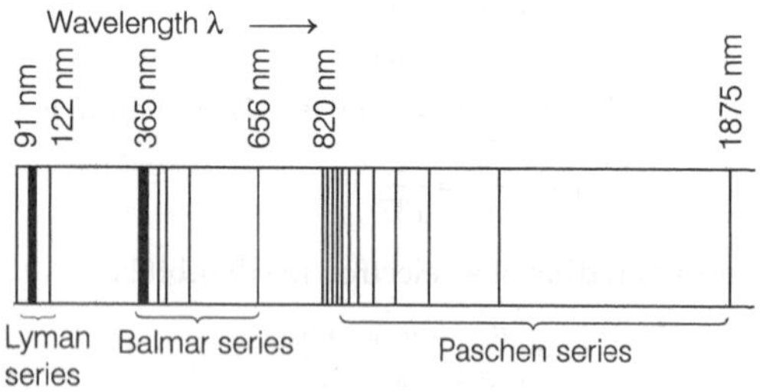


- Various series of spectra for hydrogen are represented as follows

(i) **Lyman series** $\frac{1}{\lambda} = R\left(\frac{1}{1^2} - \frac{1}{n^2}\right)$, where $n = 2, 3, 4, \ldots$.

It lies in ultraviolet region.
where, $R$ is called Rydberg's constant $= 1.097 \times 10^7 \text{ m}^{-1}$.

(ii) **Balmer series** $\frac{1}{\lambda} = R\left(\frac{1}{2^2} - \frac{1}{n^2}\right)$, where $n = 3, 4, \ldots$.

It lies in visible region of H-spectrum.

**Note** In Balmer series, the line with largest wavelength 656.3 nm in red is called $H_\alpha$; the next line with wavelength 486.1 nm in blue is called $H_\beta$ and so on.

(iii) **Paschen series** $\frac{1}{\lambda} = R\left(\frac{1}{3^2} - \frac{1}{n^2}\right)$, where $n = 4, 5, 6, \ldots$.

It lies in infrared region.

(iv) **Brackett series** $\frac{1}{\lambda} = R\left(\frac{1}{4^2} - \frac{1}{n^2}\right)$, where $n = 5, 6, 7, \ldots$.

It lies in infrared region.

(v) **Pfund series** $\frac{1}{\lambda} = R\left(\frac{1}{5^2} - \frac{1}{n^2}\right)$, where $n = 6, 7, 8, \ldots$.

It lies in infrared region.

**Note** The reciprocal of wavelength is called wave number, i.e. $\frac{1}{\lambda} = \bar{v}$.

- In terms of frequency, Balmer formula can be written as

$$\nu = Rc\left(\frac{1}{2^2} - \frac{1}{n^2}\right)$$

- The smallest wavelength of a given series of hydrogen spectrum is called **series limit**.
- The various lines in the atomic spectra are produced when electrons jump from higher energy state to a lower energy state and photons are emitted. These spectral lines are called **emission lines**.

**Note** Number of transition possible for an electron in $n$th orbit is $n(n-1)/2$.

- When an atom absorbs a photon that has same energy needed by the electron in a lower energy state to make transitions to a higher energy state, the process is called **absorption**.

## de-Broglie's Explanation of Bohr's Second Postulate of Quantisation

- According to de-Broglie, a stationary orbit is that which contains an integral number of de-Broglie standing waves associated with the revolving atom.

So, for permissible orbit, $2\pi r_n = n\lambda = \frac{nh}{mv_n}$ or $mv_n r_n = \frac{nh}{2\pi}$

i.e. angular momentum of electron revolving in $n$th orbit must be an integral multiple of $\frac{h}{2\pi}$, which is quantum condition proposed by Bohr in second postulate.

- **Limitations of Bohr's model** as given below
  (i) Bohr model is applicable to hydrogenic atoms only. It cannot be extended even to two electrons atom such as helium.
  (ii) Bohr's model is unable to explain the relative intensities of the frequencies in the spectrum.

# Mastering NCERT

## MULTIPLE CHOICE QUESTIONS

## TOPIC 1 ~ Rutherford's Model of Atom and Atomic Spectra

**1** In Geiger-Marsden experiment, detection of scattered $\alpha$-particles at a particular angle is done by

(a) counting flashes produced by $\alpha$-particles on a ZnS coated screen
(b) counting spots produced on a photographic film
(c) using a galvanometer detector
(d) using a Geiger-counter

**2** A beam of fast moving $\alpha$-particles was directed towards a thin gold foil. The parts $A'$, $B'$ and $C'$ of the transmitted and reflected beams corresponding to incident parts $A$, $B$ and $C$ of the beam are as shown in the figure below.

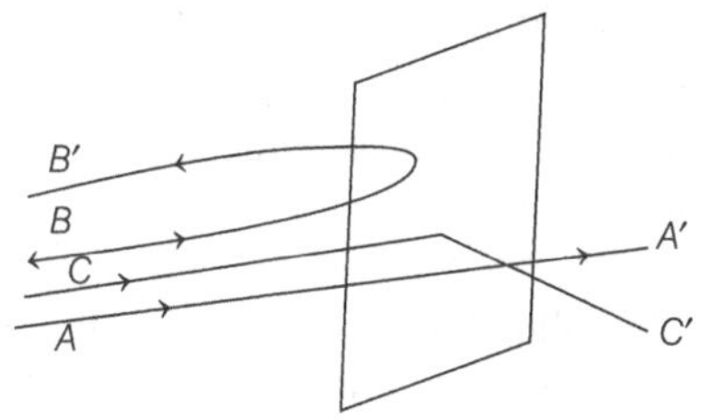


Then, the number of $\alpha$-particles in part

(a) $B'$ will be minimum and $C'$ is maximum
(b) $A'$ will be minimum and $B'$ is maximum
(c) $C'$ will be minimum and $A'$ is maximum
(d) $A'$ will be maximum and $B'$ will be minimum

**3** Atoms consists of a positively charged nucleus is a consequence from the following observation of Geiger-Marsden experiment that

(a) most of $\alpha$-particles do not pass straight through the gold foil
(b) very few $\alpha$-particles scattered through an angle more than 90°
(c) very large number of $\alpha$-particles are deflected by large angles
(d) None of the above

**4** Graph of total number of $\alpha$-particles scattered at different angles is **JEE Main 2020**

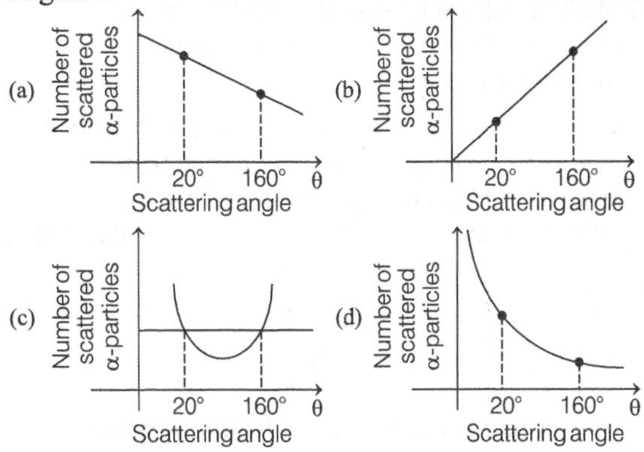


**5** Trajectory of an $\alpha$-particle is shown in the figure below, then impact parameter is

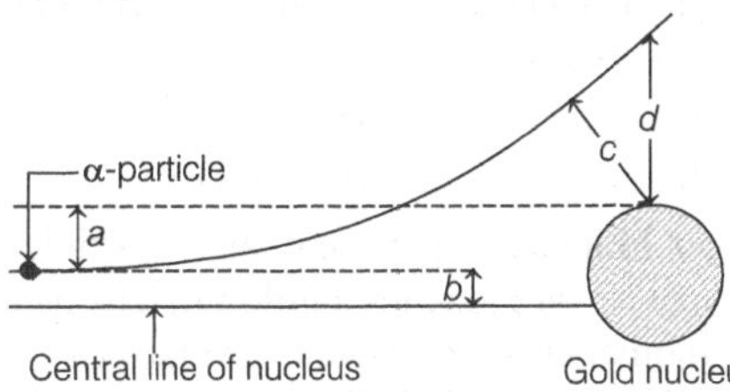


(a) $a$ (b) $b$ (c) $c$ (d) $d$

**6** A given beam of $\alpha$-particles has a distribution of impact parameter $b$, so that the beam is scattered in various direction as shown below.

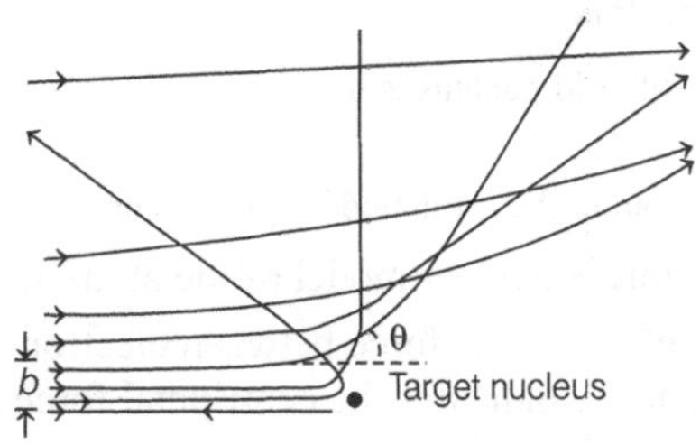


Then,

(a) $b$ = minimum and $\theta \cong 0$ (b) $b$ = maximum and $\theta \cong \pi$
(c) $b$ = minimum and $\theta \cong \pi$ (d) $b$ = maximum and $\theta$ at all values

**7** The fact that only a small fraction of the number of incident particles rebound back in Rutherford scattering indicates that

(a) number of $\alpha$-particles undergoing head-on collision is small
(b) mass of the atom is concentrated in a small volume
(c) mass of the atom is concentrated in a large volume
(d) Both (a) and (b)

**8** If the dimensions of the solar system had the same proportion as those of the atoms, then it implies that (the radius of earth's orbit is about $1.5 \times 10^{11}$ m and radius of the sun is $7 \times 10^{8}$ m)

(a) empty space in atom and solar system is comparable
(b) empty space in solar system > that in atom
(c) empty space in atom > that in solar system
(d) None of the above

**9** When an $\alpha$ -particle of mass $m$ moving with velocity $v$ bombards on a heavy nucleus of charge $Ze$, its distance of closest approach from the nucleus depends on $m$ as **NEET 2016**

(a) $\frac{1}{\sqrt{m}}$ (b) $\frac{1}{m^2}$
(c) $m$ (d) $\frac{1}{m}$

**10** In $\alpha$-scattering experiment, in a head-on-collision between an $\alpha$-particle and a gold nucleus, the minimum distance of separation or closest approach is $4 \times 10^{-14}$ m. The energy of $\alpha$-particle is (take, atomic number of gold = 79)

(a) 5.68 MeV (b) 8 MeV
(c) 4.47 MeV (d) 7.24 MeV

**11** In a Geiger-Marsden experiment, an $\alpha$-particle of 7.7 MeV energy approaches a gold nucleus head-on, it comes momentarily to rest and reverses its direction. We can conclude that (for gold, $Z = 79$)

(a) radius of gold nucleus is $3.0 \times 10^{-14}$ m
(b) radius of gold nucleus is more than $3.0 \times 10^{-14}$ m
(c) radius of gold nucleus is less than $3.0 \times 10^{-14}$ m
(d) radius cannot be estimated by given data

**12** In Rutherford's nuclear model of the atom, if $F_e$ indicates electrostatic force between electron and nucleus and $F_c$ indicates the centripetal force on revolving electron, then

(a) $F_e = F_c$
(b) $F_e > F_c$
(c) $F_e < F_c$
(d) $F_e = \infty$ and $F_c = 0$

**13** If the orbital radius of the electron in a hydrogen atom is $4.7 \times 10^{-11}$ m. Compute the kinetic energy of the electron in hydrogen atom.

(a) 15.3 eV (b) – 15.3 eV
(c) 13.6 eV (d) –13.6 eV

**14** The orbital radius of an electron in H-atom is increased by factor of 2, then the total energy of the electron will be

(a) decreased by 4 (b) increased by 2
(c) decreased by 2 (d) remain same

**15** It is found experimentally that 13.6 eV energy is required to separate a hydrogen atom into a proton and an electron. Then, the velocity of the electron in a hydrogen atom is

(a) $10^{15}$ m/s (b) $2.2 \times 10^{6}$ m/s
(c) $4.2 \times 10^{8}$ m/s (d) $3 \times 10^{8}$ m/s

**16** An emission line spectrum is observed when

(a) a gas is excited by heating or by passing a electric discharge through it
(b) a gas is excited by applying pressure
(c) a gas under low pressure condition is excited by heating or by passing a electric discharge through it
(d) a gas under high pressure condition is excited by heating or by passing a electric discharge through it

**17** When white light is passed through an unexcited gas, then transmitted light consists of

(a) few bright lines in dark background
(b) few dark lines in bright background
(c) alternate dark and bright lines
(d) alternate dark and bright bands

**18** Which of the following correctly shows the Balmer series in the emission spectrum of hydrogen?

(a)
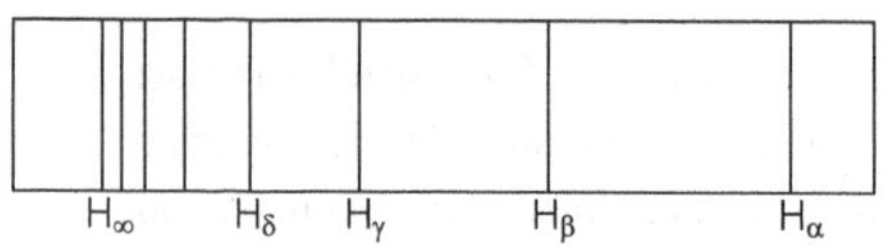


(b)
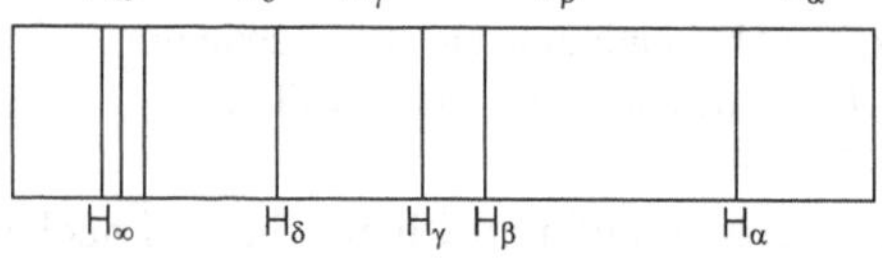


(c)
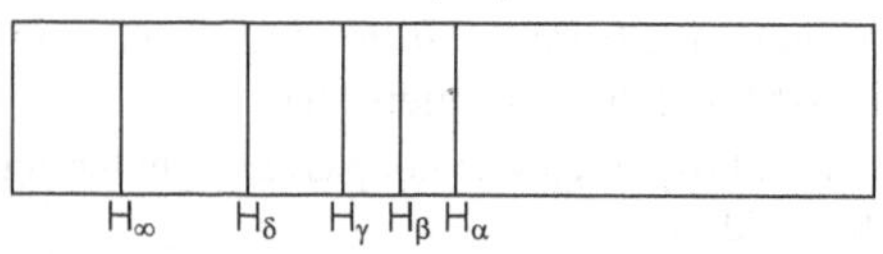


(d)
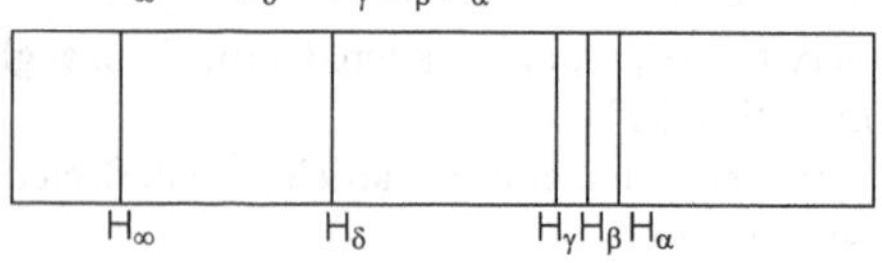


**19** Shortest wavelength or series limit in Balmer series is

(a) 128 nm (b) 486 nm (c) 656 nm (d) 364 nm

**20** If the series limit frequency of the Lyman series is $v_L$, then the series limit frequency of the Pfund series is **JEE Main 2018**

(a) $25 v_L$ (b) $16 v_L$ (c) $\frac{v_L}{16}$ (d) $\frac{v_L}{25}$

**21** If wavelength of the first line of the Balmer series of hydrogen is 6561Å, the wavelength of the second line of the series should be

(a) 13122Å (b) 3280Å (c) 4860Å (d) 2187Å

**22** The ratio of wavelengths of the last line of Balmer series and the last line of Lyman series is **NEET 2017**

(a) 2 (b) 1 (c) 4 (d) 0.5

**23** In Pfund series, ratio of maximum to minimum wavelength of emitted spectral lines is

(a) $\frac{\lambda_{max}}{\lambda_{min}} = \frac{4}{3}$ (b) $\frac{\lambda_{max}}{\lambda_{min}} = \frac{9}{5}$

(c) $\frac{\lambda_{max}}{\lambda_{min}} = \frac{16}{7}$ (d) $\frac{\lambda_{max}}{\lambda_{min}} = \frac{36}{11}$

**24** In the spectrum of hydrogen, the ratio of the largest wavelength in the Lyman series to the largest wavelength in the Balmer series is **CBSE AIPMT 2015, NEET 2013**

(a) $\frac{4}{9}$ (b) $\frac{9}{4}$ (c) $\frac{27}{5}$ (d) $\frac{5}{27}$

## TOPIC 2 ~ *Bohr Model of Hydrogen Atom*

**25** Rutherford's atomic model could account for

I. concept of stationary orbits.
II. the positively charged central core of an atom.
III. origin of spectra.
IV. stability of atoms.

Choose the correct option from the codes given below.

(a) Only I (b) Only II
(c) I, III and IV (d) I, II, III and IV

**26** According to classical electromagnetic theory, the electron revolving around a proton in hydrogen atom in an orbit of radius $5.3 \times 10^{-11}$ m with velocity $2.2 \times 10^6 \text{ ms}^{-1}$ is

(a) does not emit any radiation
(b) emits radiation of varying frequencies
(c) emits radiation of initial frequency $6.6 \times 10^{15}$ Hz
(d) emits radiation of initial frequency $2.2 \times 10^{15}$ Hz

**27** When an electron in hydrogen atom is excited from its 4 th to 5 th stationary orbit, the change in angular momentum of electron is

(a) $4.6 \times 10^{-34}$ J-s (b) $3.32 \times 10^{-34}$ J-s
(c) $1.05 \times 10^{-34}$ J-s (d) $2.08 \times 10^{-34}$ J-s

**28** The graph of radius $r_n$ of electron in $n$th orbit *versus* atomic number $Z$ is

(a) 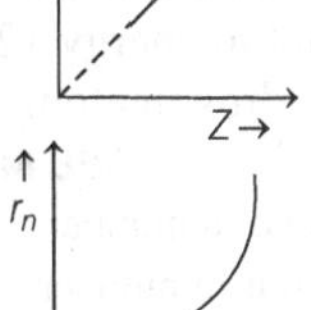


(b) 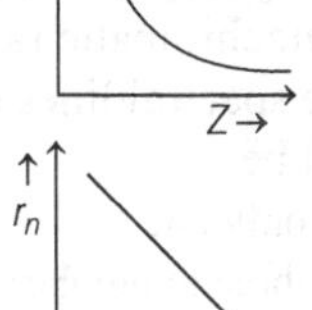


(c) 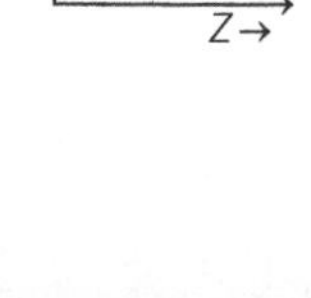


(d) 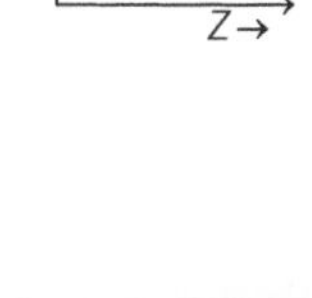


**29** If radius of first Bohr orbit is $r$, then the radius of second Bohr's orbit will be

(a) $2r$ (b) $8r$
(c) $4r$ (d) $2\sqrt{2}r$

**30** The ratio of speed of an electron in ground state in Bohr's first orbit of hydrogen atom to the velocity of light in air is

(a) $\frac{e^2}{2\pi hc}$ (b) $\frac{2e\pi}{hc}$

(c) $\frac{e^3}{2\pi hck}$ (d) $\frac{2\pi e^2 k}{hc}$

**31** The speed of an electron in ground state energy level is $2.6 \times 10^6 \text{ ms}^{-1}$, then its speed in third excited state will be

(a) $5.6 \times 10^6 \text{ ms}^{-1}$ (b) $6.5 \times 10^5 \text{ ms}^{-1}$
(c) $5.5 \times 10^5 \text{ ms}^{-1}$ (d) $4.3 \times 10^6 \text{ ms}^{-1}$

**32** Time of revolution of an electron around a nucleus of charge $Ze$ in $n$ th Bohr orbit is proportional to

(a) $n$ (b) $\frac{n^3}{Z^2}$

(c) $\frac{n^2}{Z}$ (d) $\frac{Z}{n}$

**33** In Bohr's atomic model, speed and time period of revolution of an electron in $n = 3$ level are respectively

(a) $v = 1.09 \times 10^6 \text{ ms}^{-1}, T = 1.52 \times 10^{-16}\text{s}$
(b) $v = 2.19 \times 10^6 \text{ ms}^{-1}, T = 1.52 \times 10^{-16}\text{s}$
(c) $v = 7.31 \times 10^6 \text{ ms}^{-1}, T = 1.22 \times 10^{-16}\text{s}$
(d) $v = 7.31 \times 10^5 \text{ ms}^{-1}, T = 4.08 \times 10^{-15}\text{s}$

**34** The orbital frequency of an electron in the hydrogen atom

(a) increases as $n$ increases
(b) decreases as $n$ increases
(c) remains same throughout
(d) first increases then decreases with increase in $n$.

**35** The electron in a hydrogen atom makes a transition $n_1 \rightarrow n_2$, where $n_1$ and $n_2$ are principal quantum numbers of these two states. Assuming Bohr's model to be valid, if the time period of an electron in the initial state is eight times of that in the final state, the possible values of $n_1$ and $n_2$ are

(a) $n_1 = 4, n_2 = 2$ (b) $n_1 = 8, n_2 = 2$
(c) $n_1 = 8, n_2 = 1$ (d) $n_1 = 6, n_2 = 5$

**36** The ratio of kinetic energy to the total energy of an electron in a Bohr orbit of the hydrogen atom, is **NEET 2018**

(a) 2 : – 1 (b) 1 : – 1
(c) 1 : 1 (d) 1 : – 2

**37** The total energy of an electron in an atom in an orbit is –3.4 eV. Its kinetic and potential energies are, respectively **NEET (National) 2019**

(a) –3.4 eV, –6.8 eV (b) 3.4 eV, –6.8 eV
(c) 3.4 eV, 3.4 eV (d) – 3.4 eV, –3.4 eV

**38** As the electron in Bohr orbit of hydrogen atom passes from state $n = 2$ to $n = 1$, the kinetic energy $K$ and potential energy $U$ changes as

(a) $K$ becomes one-fourth and $U$ becomes four times
(b) $K$ becomes four times, $U$ becomes twice
(c) Both $K$ and $U$ become four times
(d) Both $K$ and $U$ become twice

**39** Energy required to remove an electron in a hydrogen atom from $n = 10$ state is

(a) 13.6 eV (b) 1.36 eV
(c) 0.136 eV (d) 0.0136 eV

**40** If energy of a hydrogen atom in $n$th orbit is $E_n$, then energy of a singly ionised helium ($He^+$) atom in $n$th orbit will be

(a) $4E_n$ (b) $E_n/4$ (c) $2E_n$ (d) $E_n/2$

**41** If the binding energy of the electron in a hydrogen atom is 13.6 eV, then energy required to remove electron from first excited state of $Li^{2+}$ is

(a) 122.4 eV (b) 30.6 eV
(c) 13.6 eV (d) 3.4 eV

**42** Energy of an electron in an excited state of hydrogen atom is – 3.4 eV. Its angular momentum will be

(a) $1.11 \times 10^{34}$ J-s (b) $1.51 \times 10^{-31}$ J-s
(c) $2.10 \times 10^{-34}$ J-s (d) $3.72 \times 10^{-34}$ J-s

**43** An hydrogen atom and $Li^+$ ion are both in the second excited state. If $L_H$ and $L_{Li}$ are their respective electronic angular momentums and $E_H$ and $E_{Li}$ their respective energies, then

(a) $L_H > L_{Li}$ and $|E_H| > |E_{Li}|$ (b) $L_H = L_{Li}$ and $|E_H| < |E_{Li}|$
(c) $L_H = L_{Li}$ and $|E_H| > |E_{Li}|$ (d) $L_H > L_{Li}$ and $|E_H| < |E_{Li}|$

**44** A $He^+$ ion is in its first excited state. Its ionisation energy is **JEE Main 2019**

(a) 54.40 eV (b) 13.60 eV
(c) 48.36 eV (d) 6.04 eV

**45** The minimum energy to ionise an atom is the energy required to

(a) add one electron to the round state of atom
(b) excite the atom from its ground state to its first excited state
(c) remove one outermost electron from the round state of atom
(d) remove one innermost electron from the round state of atom

**46** As excitation of hydrogen atom increases (i.e. $n$ increases), the value of minimum energy required to free the electron from the excited state

(a) increases (b) decreases
(c) remains same
(d) first increases then decreases

**47** An electron with kinetic energy 5eV is incident on a H-atom in its ground state. The collision

(a) must be elastic
(b) must be partially elastic
(c) must be completely inelastic
(d) may be completely inelastic

**48** In a mixture of H—$He^+$ gas ($He^+$ is singly ionized He atom), H-atoms and $He^+$ ions are excited to their respective first excited states. Subsequently, H atoms transfer their total excitation energy to $He^+$ ions (by collisions). The quantum number $n$ of the state finally populated in $He^+$ ions is

(a) 2 (b) 3 (c) 4 (d) 5

**49** An energy of 15 eV is given to an electron in 4th orbit of H-atom. Then, find its final energy when it comes out of H-atom. **AIIMS 2019**

(a) 14.15 eV (b) 13.6 eV (c) 12.08 eV (d) 15.85 eV

**50** Ionisation potential of hydrogen atom is 13.6 eV. If a hydrogen atom in ground state is excited by monochromatic radiation of photon energy 12.1 eV. The spectral lines emitted by hydrogen atom will be **JEE Main 2013**

(a) only one (b) two in number
(c) three in number (d) four in number

# TOPIC 3 ~ Hydrogen Spectrum and de-Broglie's Explanation of Bohr's Second Postulate

**51** An atom makes a transition from higher energy state $n_i$ to the lower energy state, $n_f$, then the difference of energy of a photon is (where, $n$ stands for principal quantum number)

(a) $h = -E_{n_i} - E_{n_f}$ (b) $h\nu_{if} = E_{n_i} + E_{n_f}$
(c) $h\nu_{if} = E_{n_f} - E_{n_i}$ (d) $h\nu_{if} = E_{n_i} - E_{n_f}$

**52** Frequency of radiation emitted when a hydrogen atom is de-excites from level $n$ to level $(n-1)$ is **JEE Main 2013**

(a) $\dfrac{me^4 (2n-1)}{(4\pi)^3 \varepsilon_0^2 (h/2\pi)^3 n^2 (n-1)^2}$ (b) $\dfrac{me^4 (1-2n)}{4\pi^2 \varepsilon_0^2 (h/2\pi)^3 n^2}$

(c) $\dfrac{me^4 (1-2n)}{4\pi^2 \varepsilon_0 (h/2\pi)^2 (n-1)^2}$ (d) None of these

**53** An electron jumps from the 4th orbit to 2nd orbit of the hydrogen atom. Given, Rydberg's constant $R = 10^7 \text{ cm}^{-1}$, the frequency in hertz of emitted radiation will be

(a) $\frac{3}{16} \times 10^5$ (b) $\frac{3}{16} \times 10^{15}$ (c) $\frac{9}{16} \times 10^{15}$ (d) $\frac{3}{4} \times 10^{15}$

**54** Transition from state $n = 4$ to $n = 3$ in a hydrogen like atom results in a UV radiation. IR radiation will be obtained in the transition, when

(a) $2 \to 1$ (b) $3 \to 2$ (c) $1 \to 2$ (d) $5 \to 4$

**55** Some energy levels of a molecule are shown in the figure. The ratio of the wavelengths $r = \lambda_1 / \lambda_2$ is given by **JEE Main 2017**

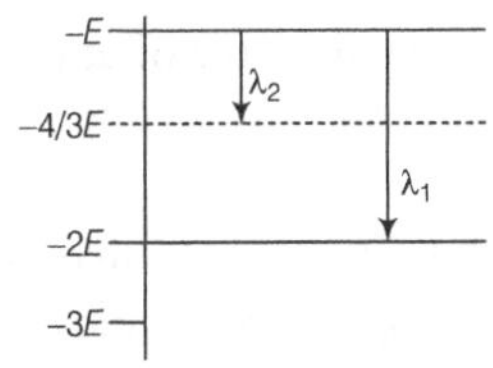


(a) $r = \frac{2}{3}$ (b) $r = \frac{3}{4}$ (c) $r = \frac{1}{3}$ (d) $r = \frac{4}{3}$

**56** Energy levels $A$, $B$ and $C$ of a certain atom corresponding to increasing values of energy, i.e. $E_A < E_B < E_C$. If $\lambda_1, \lambda_2$ and $\lambda_3$ are wavelengths of photon corresponding to transitions shown below.

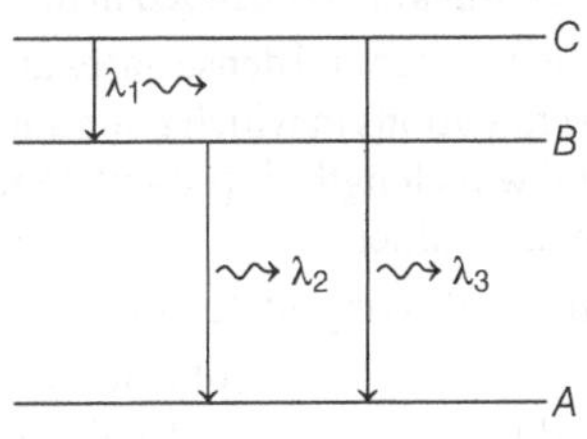


Then,

(a) $\lambda_3 = \lambda_1 + \lambda_2$ (b) $\lambda_3 = \lambda_1\lambda_2 / \lambda_1 + \lambda_2$
(c) $\lambda_1 + \lambda_2 + \lambda_3 = 0$ (d) $\lambda_3^2 = \lambda_1^2 + \lambda_2^2$

**57** Radiation coming from transitions $n = 2$ to $n = 1$ of hydrogen atoms fall on $He^+$ ions in $n = 1$ and $n = 2$ states. The possible transition of helium ions as they absorb energy from the radiation is **JEE Main 2019**

(a) $n = 2 \to n = 3$ (b) $n = 1 \to n = 4$
(c) $n = 2 \to n = 5$ (d) $n = 2 \to n = 4$

**58** Ratio of wavelengths emitted for transition $2 \to 1$ in $Li^{2+}$, $He^{2+}$ and H is

(a) 1 : 2 : 3 (b) 1 : 4 : 9 (c) 4 : 9 : 36 (d) 3 : 2 : 1

**59** Electron in hydrogen atom first jumps from third excited state to second excited state and then from second excited to the first excited state. The ratio of the wavelengths $\lambda_1 : \lambda_2$ emitted in the two cases is **CBSE AIPMT 2012**

(a) $\frac{7}{5}$ (b) $\frac{27}{20}$ (c) $\frac{27}{5}$ (d) $\frac{20}{7}$

**60** Hydrogen atom from excited state comes to the ground state by emitting a photon of wavelength $\lambda$. If $R$ is the Rydberg constant, then the principal quantum number $n$ of the excited state is

(a) $\sqrt{\frac{\lambda R}{\lambda R - 1}}$ (b) $\sqrt{\frac{\lambda}{\lambda R - 1}}$ (c) $\sqrt{\frac{\lambda R^2}{\lambda R - 1}}$ (d) $\sqrt{\frac{\lambda R}{\lambda - 1}}$

**61** The radius of the orbit of an electron in a Hydrogen like atom is 4.5 $a_0$, where $a_0$ is the Bohr radius. Its orbital angular momentum is $\frac{3h}{2\pi}$. It is given that $h$ is Planck constant and $R$ is Rydberg constant. Which amongst the following is not a possible wavelength(s), when the atom de-excites?

(a) $\frac{9}{32R}$ (b) $\frac{9}{16R}$ (c) $\frac{9}{5R}$ (d) $\frac{1}{3R}$

**62** Hydrogen $({}_1H^1)$, deuterium $({}_1H^2)$, singly ionised helium $({}_2He^4)^+$ and doubly ionised lithium $({}_3Li^8)^{2+}$ all have one electron around the nucleus. Consider an electron transition from $n = 2$ to $n = 1$. If the wavelengths of emitted radiation are $\lambda_1, \lambda_2, \lambda_3$ and $\lambda_4$, respectively for four elements, then approximately which one of the following is correct?

(a) $4\lambda_1 = 2\lambda_2 = 2\lambda_3 = \lambda_4$ (b) $\lambda_1 = 2\lambda_2 = 2\lambda_3 = \lambda_4$
(c) $\lambda_1 = \lambda_2 = 4\lambda_3 = 9\lambda_4$ (d) $\lambda_1 = 2\lambda_2 = 3\lambda_3 = 4\lambda_4$

**63** The wavelength of yellow line of sodium is 5896 Å, its wave number is

(a) 50883 per cm (b) 16961 per cm
(c) 17581 per cm (d) 50833 per cm

**64** The energy levels of the hydrogen spectrum is shown in figure.

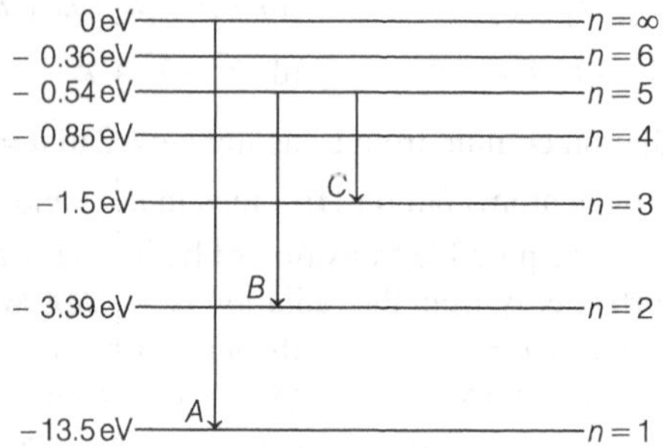


Transitions $A$, $B$ and $C$ represent

(a) first member of Lyman series, third spectral line of Balmer series and second line of Paschen series
(b) ionisation potential of hydrogen, second spectral line of Balmer series and third spectral line of Paschen series
(c) series limit of Lyman series, third spectral line of Balmer series and second spectral line of Paschen series
(d) series limit of Lyman series, second spectral line of Balmer series and third spectral line of Paschen series

**65** If $\nu_1$ is the frequency of the series limit of Lyman series, $\nu_2$ is the frequency of the first line of Lyman series and $\nu_3$ is the frequency of the series limit of the Balmer series. Then,

(a) $\nu_1 - \nu_2 = \nu_3$ (b) $\nu_1 = \nu_2 - \nu_3$
(c) $\frac{1}{\nu_2} = \frac{1}{\nu_1} + \frac{1}{\nu_3}$ (d) $\frac{1}{\nu_1} = \frac{1}{\nu_2} + \frac{1}{\nu_3}$

**66** A 12.75 eV electron beam is used to bombard gaseous hydrogen at room temperature, then wavelengths emitted are in

(a) either Paschen, Lyman or Balmer series
(b) either Brackett, Pfund or Lyman series
(c) either Paschen, Brackett or Balmer series
(d) either Lyman, Balmer or Brackett series

**67** An electron of a hydrogen like atom is in excited state. If total energy of the electron is – 4.6 eV, then de-Broglie wavelength of the electron is

(a) 6.27 nm (b) 2.48 mm (c) 0.57 nm (d) 8 nm

**68** The Bohr model for the spectra of a H-atom

(a) will not be applicable to hydrogen in the molecular form
(b) is applicable for multi-electron atom
(c) explain continuous as well as discrete spectral lines
(d) Both (a) and (b)

# SPECIAL TYPES QUESTIONS

## I. Assertion and Reason

**Direction** (Q. Nos. 69-72) *In the following questions, statement of Assertion is followed by a corresponding statement of Reason. Of the following statements, choose the correct one.*

(a) Both Assertion and Reason are correct and Reason is the correct explanation of Assertion
(b) Both Assertion and Reason are correct but Reason is not the correct explanation of Assertion
(c) Assertion is correct but Reason is incorrect
(d) Assertion is incorrect but Reason is correct

**69** **Assertion** Light emitted from rarefied gases heated in a flame or excited electrically in a glow tube like mercury vapour light has only certain discrete wavelengths.

**Reason** In such gases, the average spacing between atom is large.

**70** **Assertion** Most of an atom is an empty space.

**Reason** From Rutherford's experiment, size of the nucleus is $10^{-10}$ m and from kinetic theory, size of the atom is $10^{-15}$ m.

**71** **Assertion** If the electrons in an atom were stationary, then they would fall into the nucleus.

**Reason** Electrostatic force of attraction acts between negatively charged electrons and positive nucleus.

**AIIMS 2018**

**72** **Assertion** Highest energy state corresponds with the electron revolving in an orbit closest to the nucleus.

**Reason** Energies of the excited states come closer and closer together as $n$ increases.

## II. Statement Based Questions

**73** I. According to J.J. Thomson's atomic model, negative charge of the atom is uniformly distributed and positive charges are embedded in it.

II. Condensed matter and dense gases at all temperatures emit EM radiations in which a continuous distribution of several wavelengths is present, though with different intensities.

Which of the following statement(s) is/are correct?

(a) Only I (b) Only II
(c) Both I and II (d) Neither I nor II

**74** In Geiger-Marsden experiment,

I. $\alpha$-particles emitted from a $^{214}_{83}U$ radioactive source at a thin metal foil made of gold.

II. $\alpha$-particle were collimated into a narrow beam by their passage through lead bricks.

Which of the following statement(s) is/are correct?

(a) Only I (b) Only II
(c) Both I and II (d) Neither I nor II

**75** I. If total energy of an electron in a H-atom is positive, then the electron will be bound to the nucleus.

II. In Geiger-Marsden experiment, gold foil is very thin, so alpha-particle will not suffer not more than one scattering during their passage through it.

Which of the following statement(s) is/are correct?

(a) Only I (b) Only II
(c) Both I and II (d) Neither I nor II

**76** Study the following statements.

I. Bohr model is valid for both one electron atom and multi-electron atom.

II. Orbital is the one electron wave function that gives the probability of finding an electron in a certain region of space around the nucleus.

III. Orbital is similar to the orbit defined for an electron in the Bohr model.

Which of the following statement(s) is/are correct?

(a) Only I (b) Both II and III
(c) Only II (d) I, II and III

**77** Let $E_n = \dfrac{-me^4}{8\varepsilon_0^2 n^2 h^2}$ be the energy of $n$th level of H-atom. If all the H-atoms are in ground state and radiation of frequency $(E_2 - E_1)/h$ falls on it, then

I. it will not be absorbed at all.

II. some of electron moves to first excited state.

III. all electrons will be excited to $n = 2$ state.

IV. no electron makes a transition to the $n = 3$ state.

Which of the following statement(s) is/are correct?

(a) Only I (b) Both II and III
(c) Both II and IV (d) Both III and IV

**78** I. At room temperature, most of the hydrogen atoms are in the ground state.

II. Franck-Hertz's experiment concluded that the basic idea of discrete energy levels in atoms and the process of photoemission is not correct.

III. Magnetic moment of an $e^-$ revolving around the nucleus is directly proportional to $n$.

Which of the following statement(s) is/are correct?

(a) Both I and II (b) Only II
(c) Both I and III (d) Only III

**79** I. The acronym LASER stands for light amplification by stimulated emission of radiation.

II. If there are $N$ atoms in a source of laser light and each atom is emitting light with intensity $I$, then total intensity produced by it is $N^2 I$.

Which of the following statement(s) is/are correct?

(a) Only I (b) Only II
(c) Both I and II (d) Neither I nor II

**80** Choose the incorrect statement regarding the observations of Geiger-Marsden experiment.

(a) Many of $\alpha$-particles pass straight through the gold foil.
(b) Only about 0.2% of $\alpha$-particles scatter by more than 1°.
(c) About 1 in 8000 of $\alpha$-particles is deflected more than 90°.
(d) Very few $\alpha$- particles are reflected back.

**81** Choose the correct statement from the following regarding Geiger-Marsden experiment.

(a) Atomic electrons being heavier in nature can appreciably affect the $\alpha$-particles.
(b) Trajectory of an $\alpha$-particle is explained using Coulomb's and Gauss's law.
(c) Nucleus of gold is about 50 times heavier than an $\alpha$-particle, that remains stationary throughout scattering process.
(d) Both (b) and (c)

**82** Which amongst the following statements are incorrect regarding Bohr's atomic model?

(a) Electrons in a quantised orbit will not radiate the energy.
(b) Mass of electrons varies during revolution.
(c) Emission or absorption of energy results to transition of electron from one orbit to another.
(d) Angular momentum of orbital electron is quantised.

**83** As an electron makes a transition from an excited state to the ground state of a hydrogen like atom/ion. Then, choose the correct statement.

(a) Its kinetic energy increases but potential energy and total energy decrease.
(b) Kinetic energy, potential energy and total energy decrease.
(c) Kinetic energy decreases, potential energy increases but total energy remains same.
(d) Kinetic energy and total energy decrease but potential energy increases.

**84** Choose the incorrect statement from the following.

(a) Line spectra of a material can serve as a type of finger print for identification of the gas.
(b) In Balmer series, $H_\alpha$-line is of violet colour.
(c) Balmer formula can be used to reposing spectral of singly ionised helium and double ionised lithium.
(d) Beyond series limit no distinct lines appear, instead only a faint continuous spectrum is seen.

**85** Choose the incorrect statement from the following option regarding hydrogen atom.
(a) Linear momentum is directly proportional to the number of orbit $n$.
(b) Bohr's model assumed that electronic orbits are circular.
(c) When a hydrogen atom receives energy by processes such as electron emission, it is said to be in excited state.
(d) de-Broglie wavelength of an electron in first Bohr's orbit is equal to circumference of orbit.

## III. Matching Type

**86** For Balmer spectral series of hydrogen, match the Column I (Balmer spectral series) with Column II (wavelength) and select the correct answer from the codes given below .

| Column I | Column II |
|---|---|
| A. $H_\alpha$ | 1. 410.2 nm |
| B. $H_\beta$ | 2. 434.1 nm |
| C. $H_\gamma$ | 3. 486.1 nm |
| D. $H_\delta$ | 4. 656.3 nm |

| | A | B | C | D | | A | B | C | D |
|---|---|---|---|---|---|---|---|---|---|
| (a) | 4 | 1 | 2 | 3 | (b) | 4 | 3 | 2 | 1 |
| (c) | 2 | 2 | 4 | 4 | (d) | 2 | 3 | 4 | 1 |

**87** For spectral series of hydrogen atom, match the Column I (spectral series) with Column II (region) and select the correct answer from the codes given below.

| Column I | Column II |
|---|---|
| A. Lyman series | 1. Ultraviolet region |
| B. Balmer series | 2. Infrared region |
| C. Paschen series | 3. Visible region |
| D. Brackett series | |

| | A | B | C | D | | A | B | C | D |
|---|---|---|---|---|---|---|---|---|---|
| (a) | 1 | 3 | 2 | 2 | (b) | 2 | 1 | 2 | 2 |
| (c) | 2 | 1 | 3 | 3 | (d) | 1 | 2 | 2 | 3 |

**88** Match the Column I (spectral series) with Column II (maximum wavelength of the series) and select the correct answer from the codes given below.
($R$ = Rydberg's constant)

| Column I | Column II |
|---|---|
| A. Lyman series | 1. $400/9R$ |
| B. Balmer series | 2. $144/7R$ |
| C. Paschen series | 3. $36/5R$ |
| D. Brackett series | 4. $4/3R$ |

| | A | B | C | D | | A | B | C | D |
|---|---|---|---|---|---|---|---|---|---|
| (a) | 2 | 3 | 4 | 1 | (b) | 3 | 4 | 1 | 2 |
| (c) | 4 | 3 | 2 | 1 | (d) | 4 | 3 | 1 | 2 |

**89** For atomic model of hydrogen atom given by Niels Bohr, match the Column I (quantities) with Column II (proportionalities) and select the correct answer from the codes given below. ***JEE Main 2019***

| Column I | Column II |
|---|---|
| A. Angular momentum | 1. $1/n$ |
| B. Velocity of electron | 2. $n^2$ |
| C. Radius of electron | 3. $1/n^2$ |
| D. Energy of electron | 4. $n$ |

| | A | B | C | D |
|---|---|---|---|---|
| (a) | 1 | 2 | 3 | 4 |
| (b) | 4 | 3 | 2 | 1 |
| (c) | 4 | 1 | 3 | 2 |
| (d) | 4 | 1 | 2 | 3 |

# NCERT & NCERT Exemplar

## MULTIPLE CHOICE QUESTIONS

### NCERT

**90** What is the shortest wavelength present in the Paschen series of spectral lines?
(a) 411.85 nm (b) 873.8 nm
(c) 11.85 nm (d) 22.65 nm

**91** A difference of 2.3 eV separates two energy levels in an atom. What is the frequency of radiation emitted when the atom makes a transition from the upper level to the lower level?
(a) $3.4 \times 10^2$ Hz
(b) $1.25 \times 10^{10}$ Hz
(c) $5.6 \times 10^{14}$ Hz
(d) None of the above

**92** A hydrogen atom initially in the ground level absorbs a photon which excites it to the $n = 4$ level. Determine the wavelength of photon.
(a) $9.74 \times 10^{-8}$ m (b) $5.25 \times 10^{-5}$ m
(c) $1.05 \times 10^{-3}$ m (d) $7.15 \times 10^{-2}$ m

**93** The radius of the innermost electron orbit of a hydrogen atom is $5.3 \times 10^{-11}$ m. What are the radii of $n = 2$ and $n = 3$ orbits?

(a) $2.12 \times 10^{-19}$ m, $5.5 \times 10^{-10}$ m
(b) $2.12 \times 10^{-10}$ m, $4.77 \times 10^{-10}$ m
(c) $1.05 \times 10^{-10}$ m, $1.25 \times 10^{-19}$ m
(d) None of the above

**94** A 12.5 eV electron beam is used to bombard gaseous hydrogen at room temperature. What series of wavelength will be emitted?

(a) Lyman series with 225 Å
(b) Brackett series with 845 Å
(c) Lyman series with 993 Å
(d) Pfund series with 105 Å

**95** In accordance with the Bohr's model, find the quantum number that characterises the earth's revolution around the sun in an orbit of radius $1.5 \times 10^{11}$ m with orbital speed $3 \times 10^4$ ms$^{-1}$.
(Mass of earth $= 6.0 \times 10^{24}$ kg) ***JEE Main 2016***

(a) $2.6 \times 10^{74}$ (b) $8.5 \times 10^{80}$
(c) $4.34 \times 10^{100}$ (d) $3.2 \times 10^{8}$

**96** The total energy of an electron in the first excited state of the hydrogen atom is about −3.4 eV. What is the kinetic energy of the electron in this state?

(a) −3.4 eV (b) + 3.4 eV
(c) 6.8 eV (d) − 6.8 eV

**97** A 10 kg satellite revolves around the earth in circular orbit in every 2 h in an orbit of radius 8000 km. Then, according to Bohr's theory, quantum number of orbit of satellite is

(a) 100 (b) 204
(c) $5.3 \times 10^{45}$ (d) $5.3 \times 10^{12}$

## NCERT Exemplar

**98** Taking Bohr radius, $a_0 = 53$ pm, the radius of $Li^{2+}$ ion in its ground state on the basis of Bohr model will be

(a) 53 pm (b) 27 pm
(c) 18 pm (d) 13 pm

**99** The simple Bohr model cannot be directly applied to calculate the energy levels of an atom with many electrons as

(a) electrons are not subjected to a central force
(b) electrons are colliding with each other
(c) electrons are subjected to screening effects
(d) force between nucleus and electrons is in comparison to electron-electron electronic force.

**100** For ground state, the electron H-atom has an angular momentum is = *h*, according to the simple Bohr model. As angular momentum is a vector and hence there will be infinitely many possibilities of orbits with vector pointing in all possible directions. In actuality, this is not true

(a) because Bohr model gives incorrect values of angular momentum
(b) because only one of these would have a minimum energy
(c) angular momentum must be in the direction of spin of electrons
(d) because electrons go around in horizontal orbits only

**101** Two H-atoms in the ground state collide inelastically. The maximum amount by which their combined kinetic energy is reduced is

(a) 10.20 eV (b) 20.40 eV (c) 13.6 eV (d) 27.2 eV

**102** A set of atoms in an excited state decays

(a) in general to any of the states with lower energy
(b) into a lower state only when excited by an external electric field
(c) all together simultaneously into a lower state
(d) to emit photons only when they collide

**103** The binding energy of a H-atom, considering an electron moving around a fixed nuclei (proton) is

$$B = -\frac{me^4}{8n^2\varepsilon_0^2h^2} \quad (m = \text{electron mass})$$

If one decides to work in a frame of reference, where the electron is at rest, the proton would be moving around it. Similarly, binding energy would be

$$B = -\frac{Me^4}{8n^2\varepsilon_0^2h^2} \quad (M = \text{proton mass})$$

The last expression is incorrect, because

(a) *n* would not be integral
(b) Bohr-quantisation applies only two electrons
(c) The frame in which the electron is at rest is not inertial
(d) The motion of the proton would not be in circular orbits, even approximately

# Answers

## Mastering NCERT with MCQs

| | | | | | | | | | |
|---|---|---|---|---|---|---|---|---|---|
| 1 (a) | 2 (d) | 3 (b) | 4 (d) | 5 (b) | 6 (c) | 7 (d) | 8 (c) | 9 (d) | 10 (a) |
| 11 (c) | 12 (a) | 13 (a) | 14 (c) | 15 (b) | 16 (c) | 17 (b) | 18 (a) | 19 (d) | 20 (d) |
| 21 (c) | 22 (c) | 23 (d) | 24 (d) | 25 (b) | 26 (c) | 27 (c) | 28 (d) | 29 (c) | 30 (d) |
| 31 (b) | 32 (b) | 33 (d) | 34 (b) | 35 (a) | 36 (b) | 37 (b) | 38 (c) | 39 (c) | 40 (a) |
| 41 (b) | 42 (c) | 43 (b) | 44 (b) | 45 (c) | 46 (b) | 47 (a) | 48 (c) | 49 (a) | 50 (c) |
| 51 (d) | 52 (a) | 53 (c) | 54 (d) | 55 (c) | 56 (b) | 57 (d) | 58 (c) | 59 (d) | 60 (a) |
| 61 (b) | 62 (c) | 63 (b) | 64 (c) | 65 (a) | 66 (a) | 67 (c) | 68 (a) | | |

## Special Types Questions

| | | | | | | | | | |
|---|---|---|---|---|---|---|---|---|---|
| 69 (a) | 70 (c) | 71 (a) | 72 (d) | 73 (b) | 74 (b) | 75 (b) | 76 (c) | 77 (c) | 78 (c) |
| 79 (c) | 80 (b) | 81 (c) | 82 (b) | 83 (a) | 84 (b) | 85 (a) | 86 (b) | 87 (a) | 88 (c) |
| 89 (d) | | | | | | | | | |

## NCERT & NCERT Exemplar MCQs

| | | | | | | | | | |
|---|---|---|---|---|---|---|---|---|---|
| 90 (b) | 91 (c) | 92 (a) | 93 (b) | 94 (c) | 95 (a) | 96 (b) | 97 (c) | 98 (c) | 99 (d) |
| 100 (a) | 101 (a) | 102 (a) | 103 (c) | | | | | | |

# Hints & Explanations

**1** (*a*) In Geiger-Marsden experiment, the scattered $\alpha$-particles were observed through a rotatable detector consisting of zinc sulphide (ZnS) coated screen and a microscope. These $\alpha$-particles on striking the screen produce brief light flashes or scintillations. Thus, deflection of scattered $\alpha$-particles at a particular angle is done by counting these flashes using microscope.

**2** (*d*) In $\alpha$-scattering experiment,

(i) most of the $\alpha$-particles pass through the foil, which can corresponds to $AA'$ part in the figure.

(ii) only 0.14% are scatter by more than 1°, which can corresponds to $CC'$ part in the figure.

(iii) 1 out of 8000 is deflect by more than 90°, which can corresponds to $BB'$ part in the figure.

Thus, in the given figure, the number of $\alpha$-particles in $A'$ will be maximum and $B'$ will be minimum.

**3** (*b*) In Geiger-Marsden experiment, most of the $\alpha$-particles pass straight through the gold foil, a few scattered through small angles and a very few about (1 in 8000) were scattered through an angle more than 90° (or large angles).

These observations gave the conclusion that the atom consists of a positively charged nucleus at the centre, which provide repulsive force to $\alpha$-particles.

**5** (*b*) The impact parameter is the perpendicular distance of the initial velocity vector of the $\alpha$-particle from the central line of the nucleus of atom, when the $\alpha$-particle is far away from the nucleus.

Thus, in given figure impact parameter is represented by distance $b$.

**6** (*c*) It has been seen that, an $\alpha$-particle close to the nucleus suffers large scattering angle.

In case of head-on-collision, the impact parameter is minimum and the $\alpha$-particle rebounds back ($\theta \cong \pi$). For large impact parameter, the $\alpha$-particle goes nearly undeviated and has a small deflection ($\theta \cong 0$).

**7** (*d*) In case of head-on-collision, the impact parameter is minimum and the $\alpha$-particle rebounds back. So, the fact that only a small fraction of the number of incident particles rebound back indicates that the number of $\alpha$-particles undergoing head-on is small.

This in turn implies that the mass of the atom is concentrated in a small volume.

Hence, options (a) and (b) are correct.

**8** (*c*) The ratio of the radius of the electron's orbit to the radius of the nucleus is $\dfrac{10^{-10}\text{ m}}{10^{-15}\text{ m}} = 10^5$.

i.e. the radius of the electron's orbit is $10^5$ times larger than the radius of nucleus.

If the dimensions of the solar system had the same proportion as those of the atoms, then the radius of the earth's orbit around the sun must be $10^5$ times larger than the radius of the sun, so the radius of the earth's orbit would be $10^5 \times 7 \times 10^8\text{ m} = 7 \times 10^{13}\text{m}$.

This is 100 times greater than the actual orbital radius of the earth ($1.5 \times 10^{11}$ m). Thus, the earth would be much farther away from the sun.

It implies that the empty space in atom is greater than that in solar system.

**9** (*d*) At the distance of closest approach ($r_0$), whole of the KE of $\alpha$-particle converts into electrostatic potential energy, i.e. $\dfrac{1}{2}mv^2 = \dfrac{1}{4\pi\varepsilon_0}\dfrac{(Ze)(2e)}{r_0}$

[$\because$ charge on $\alpha$-particle is $+2e$ and charge on nucleus is $Ze$, where $Z$ is atomic number]

$$\Rightarrow \quad r_0 = \frac{1}{4\pi\varepsilon_0} \frac{2Ze^2}{\left[\frac{1}{2}mv^2\right]} = \frac{4k\,Ze^2}{mv^2} \quad \left[\because k = \frac{1}{4\pi\varepsilon_0}\right]$$

$$\Rightarrow \quad r_0 \propto \frac{1}{m}$$

**10** (*a*) At distance of minimum separation,

Kinetic Energy [KE] = Potential Energy [PE].

$$\Rightarrow \quad \text{KE} = \frac{k\,Ze \cdot 2e}{r}$$

where, $k = 9 \times 10^9$ N-m$^2$C$^{-2}$

$Z = 79$, $e = 1.6 \times 10^{-19}$ C and $r = 4 \times 10^{-14}$ m

$$\text{or} \quad \text{KE} = \frac{9 \times 10^9 \times 79 \times 2 \times (1.6 \times 10^{-19})^2}{4 \times 10^{-14}} \text{ J}$$

$$= \frac{9 \times 10^9 \times 79 \times 2\,(1.6 \times 10^{-19})^2}{4 \times 10^{-14} \times 1.6 \times 10^{-19}} \text{ eV}$$

$$= 5.68 \text{ MeV}$$

**11** (*c*) Let $d$ be the centre-to-centre distance between the $\alpha$-particle and the gold nucleus when the $\alpha$-particle is at its stopping point. Then, we can write the conservation of energy $E_i = E_f$ as $K = \frac{1}{4\pi\varepsilon_0}\frac{(2e)(Ze)}{d} = \frac{2Ze^2}{4\pi\varepsilon_0 d}$

Thus, the distance of closest approach $d$ is given by

$$d = \frac{2Ze^2}{4\pi\varepsilon_0 K} = \frac{2kZe^2}{K}$$

where, $k = 9 \times 10^9$ N-m$^2$C$^{-2}$, $e = 1.6 \times 10^{-19}$C

and $K = 7.7$ MeV $= 7.7 \times 1.6 \times 10^{-19}$ J $= 1.2 \times 10^{-12}$ J

$$\Rightarrow \quad d = \frac{(2)\,(9.0 \times 10^9 \text{ Nm}^2/\text{C}^2)\,(1.6 \times 10^{-19}\text{C})^2\,Z}{1.2 \times 10^{-12} \text{ J}}$$

$$= 3.84 \times 10^{-16}\,Z \text{ m}$$

The atomic number of foil material of gold is $Z = 79$, so that $d\,(\text{Au}) = 3.84 \times 10^{-16} \times 79 = 3.0 \times 10^{-14}$ m $= 30$ fm.

The radius of gold nucleus is therefore less than $3.0 \times 10^{-14}$ m.

**12** (*a*) The electrostatic force of attraction, $F_e$ between the revolving electrons and the nucleus provides the requisite centripetal force $F_c$ to keep them in their orbits.

Thus, for a dynamically stable orbit in a hydrogen atom $F_e = F_c$.

**13** (*a*) Kinetic energy of the electron in hydrogen atom,

$$\text{KE} = \frac{e^2}{8\pi\varepsilon_0 r}$$

where, $e = 1.6 \times 10^{-19}$ C, $\frac{1}{4\pi\varepsilon_0} = 9 \times 10^9$ N-m$^2$C$^{-2}$

and $r = 4.7 \times 10^{-11}$ m.

$$= \frac{(9 \times 10^9 \text{ Nm}^2\text{ C}^{-2})\,(1.6 \times 10^{-19}\text{C})^2}{(2)\,(4.7 \times 10^{-11}\text{m})}$$

$$= 2.45 \times 10^{-18} \text{ J} = \frac{2.45 \times 10^{-18}}{1.6 \times 10^{-19}} \text{ eV}$$

$$= 15.3 \text{ eV}$$

**14** (*c*) Total energy of the electron in H-atom is given as

$$E = \frac{-e^2}{8\pi\varepsilon_0 r} \Rightarrow E \propto \frac{1}{r}$$

$\therefore$ If $r$ is increased by 2, then $E$ is decreased by 2.

**15** (*b*) Total energy of the electron in hydrogen atom is $-13.6$ eV.

$$\Rightarrow \quad E = -13.6 \times 1.6 \times 10^{-19}\text{J} = -2.2 \times 10^{-18}\text{J}$$

As we know, total energy $E$ of the electron in a hydrogen atom is

$$E = \frac{-e^2}{8\pi\varepsilon_o r} \Rightarrow r = \frac{-e^2}{8\pi\varepsilon_0 E}$$

where, $e = 1.6 \times 10^{-19}$ C, $\frac{1}{4\pi\varepsilon_0} = 9 \times 10^9$ N-m$^2$C$^{-2}$

$$\Rightarrow \quad r = -\frac{(9 \times 10^9) \times (1.6 \times 10^{-19})^2}{2 \times (-2.2 \times 10^{-18})}$$

$$= 5.3 \times 10^{-11} \text{ m}$$

As, electron revolves around the nucleus, by centripetal force, so the electrostatic force required to separate electron from atom should be equal to this centripetal force.

i.e. $\quad F_c = F_e$

$$\Rightarrow \quad \frac{mv^2}{r} = \frac{1}{4\pi\varepsilon_0}\frac{e^2}{r^2} \Rightarrow v = \frac{e}{\sqrt{4\pi\varepsilon_0 mr}}$$

where, $\varepsilon_0 = 8.85 \times 10^{-12}$ C$^2$/N-m$^2$, $m = 9.1 \times 10^{-31}$ kg

and $r = 5.3 \times 10^{-11}$m.

$$\Rightarrow v = \frac{1.6 \times 10^{-19}}{\sqrt{4 \times 3.14 \times 8.85 \times 10^{-12} \times 9.1 \times 10^{-31} \times 5.3 \times 10^{-11}}}$$

$$= 2.2 \times 10^6 \text{ m/s}$$

**16** (*c*) When an atomic gas or vapour under low pressure is excited usually by heating or passing an electric current through it, the emitted radiation has a spectrum which contains certain specific wavelengths only. A spectrum of this kind is termed as emission line spectrum.

**17** (*b*) When white light is passed through an unexcited state gas and the transmitted light is analysed using a spectrometer, then it is found that transmitted light consists of few dark lines in bright background. These dark lines correspond precisely to those wavelengths which were found in the emission line spectrum of the gas. This is called the absorption spectrum of the material of the gas.

**18** (*a*) In Balmer series, the line with longest wavelength. is called $H_\alpha$-line, the next line is called $H_\beta$; the third line is called $H_\gamma$ and so on.

As the wavelength decreases, the lines appear closer together and weaker in intensity. Thus, the emission spectrum of hydrogen is correctly depicted in option (a).

**19** (*d*) Balmer's empirical formula is given as

$$\frac{1}{\lambda}=R\left(\frac{1}{2^2}-\frac{1}{n_2^2}\right)$$

For shortest wavelength, $n_1=2$ and $n_2=\infty$

$$\Rightarrow \quad \frac{1}{\lambda}=R\left(\frac{1}{2^2}-\frac{1}{\infty^2}\right)=1.097\times10^7\times\left(\frac{1}{4}\right)$$

$$\Rightarrow \quad \lambda=364\text{ nm}$$

**20** (*d*) Series limit occurs in the transition $n_2=\infty$ to $n_1=1$ in Lyman series and $n_2=\infty$ to $n_1=5$ in pfund series.

For Lyman series,

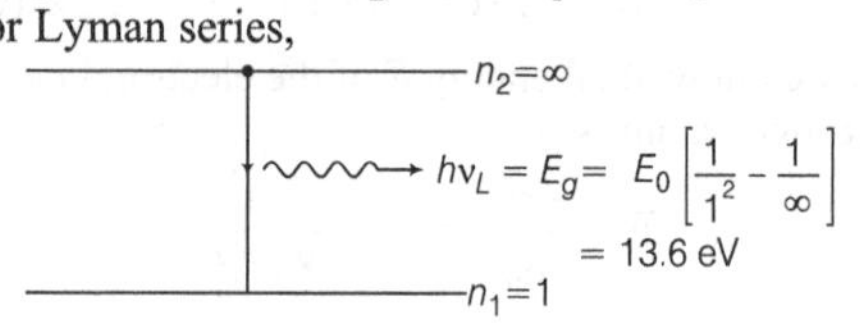


$$h\nu_L=13.6 \quad \text{...(i)}$$

In *pfund* series

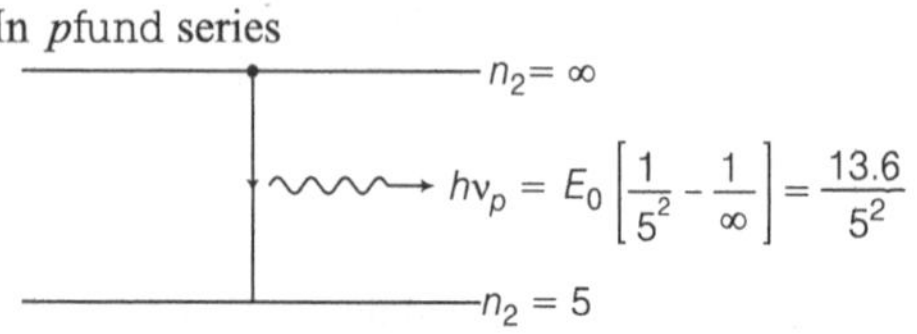


$$h\nu_p=\frac{13.6}{5^2} \quad \text{...(ii)}$$

From Eqs. (i) and (ii), we get

$$25h\nu_p=h\nu_L$$

$$\therefore \quad \nu_p=\frac{\nu_L}{25}$$

**21** (*c*) Wavelength of spectral line in Balmer series is given by

$$\frac{1}{\lambda}=R\left(\frac{1}{2^2}-\frac{1}{n^2}\right)$$

For first line of Balmer series, $n=3$

$$\Rightarrow \quad \frac{1}{\lambda_1}=R\left(\frac{1}{2^2}-\frac{1}{3^2}\right)=\frac{5R}{36} \quad \text{...(i)}$$

For second line, $n=4$

$$\Rightarrow \quad \frac{1}{\lambda_2}=R\left(\frac{1}{2^2}-\frac{1}{4^2}\right)=\frac{3R}{16} \quad \text{...(ii)}$$

Dividing Eq. (i) by Eq. (ii), we get

$$\therefore \quad \frac{\lambda_2}{\lambda_1}=\frac{20}{27}\Rightarrow \lambda_2=\frac{20}{27}\times\lambda_1$$

$$\Rightarrow \quad \lambda_2=\frac{20}{27}\times 6561=4860\text{ Å}$$

**22** (*c*) Wavelength of spectral lines are given by

$$\frac{1}{\lambda}=R\left(\frac{1}{n_1^2}-\frac{1}{n_2^2}\right)$$

For last line of Balmer series, $n_1=2$ and $n_2=\infty$

$$\Rightarrow \quad \frac{1}{\lambda_B}=R\left(\frac{1}{2^2}-\frac{1}{\infty^2}\right)=\frac{R}{4} \quad \text{... (i)}$$

Similarly, for last line of Lyman series,

$$n_1=1 \text{ and } n_2=\infty$$

$$\Rightarrow \quad \frac{1}{\lambda_L}=R\left(\frac{1}{1^2}-\frac{1}{\infty^2}\right)=R \quad \text{... (ii)}$$

From Eqs. (i) and (ii), we get

$$\frac{\frac{1}{\lambda_B}}{\frac{1}{\lambda_L}}=\frac{R/4}{R}$$

$$\Rightarrow \quad \frac{\lambda_L}{\lambda_B}=\frac{1}{4}\Rightarrow\frac{\lambda_B}{\lambda_L}=4$$

**23** (*d*) In Pfund series, $\frac{1}{\lambda}=R\left(\frac{1}{5^2}-\frac{1}{n^2}\right)$; where $n=6,7,\ldots$

Maximum wavelength is given by

$$\frac{1}{\lambda_{\text{max}}}=R\left(\frac{1}{5^2}-\frac{1}{6^2}\right)=R\left(\frac{11}{900}\right) \quad \text{...(i)}$$

Minimum wavelength is given by

$$\frac{1}{\lambda_{\text{min}}}=R\left(\frac{1}{5^2}-\frac{1}{\infty}\right)=\frac{R}{25} \quad \text{(ii)}$$

From Eqs. (i) and (ii), we get

$$\frac{\lambda_{\text{max}}}{\lambda_{\text{min}}}=\frac{R/25}{11R/900}=\frac{36}{11}$$

The ratio maximum to minimum wavelength of emitted spectral lines is 36 : 11 .

**24** (*d*) In hydrogen atom, wavelength of characteristic spectrum, $\frac{1}{\lambda}=R\left[\frac{1}{n_1^2}-\frac{1}{n_2^2}\right]$

For largest wavelength in Lyman series, $n_1=1$ and $n_2=2$

$$\frac{1}{\lambda_1}=R\left[\frac{1}{1^2}-\frac{1}{2^2}\right] \quad \text{...(i)}$$

For largest wavelength in Balmer series, $n_1=2$ and $n_2=3$

$$\frac{1}{\lambda_2}=R\left[\frac{1}{2^2}-\frac{1}{3^2}\right] \quad \text{...(ii)}$$

Dividing Eq. (ii) by Eq. (i), we get

$$\frac{\lambda_1}{\lambda_2}=\frac{R\left[\frac{1}{4}-\frac{1}{9}\right]}{R\left[1-\frac{1}{4}\right]}=\frac{\frac{5}{36}}{\frac{3}{4}}$$

$$\frac{\lambda_1}{\lambda_2}=\frac{5}{36}\times\frac{4}{3}=\frac{5}{27}$$

**25** (*b*) The model of atom proposed by Rutherford, assumes that the atom consisting of a central nucleus, i.e. positively charged central core and revolving electron is stable much like sun planet system.

But it was not able to explain the concept of stationary orbit, origin of spectra or stability of atoms.

So, option (b) is correct.

**26** (c) The frequency of the electron moving around the proton is

$$\nu = \frac{v}{2\pi r}$$

Given, $v = 2.2\times 10^6\ \text{ms}^{-1}$

and $r = 5.3\times 10^{-11}$ m

$$= \frac{2.2\times 10^6\ \text{ms}^{-1}}{2\pi\,(5.3\times 10^{-11}\ \text{m})} \approx 6.6\times 10^{15}\ \text{Hz}$$

According to the classical electromagnetic theory, the frequency of the electromagnetic waves emitted by the revolving electrons is equal to the frequency of its revolution around the nucleus.

Thus, the initial frequency of the light emitted is $6.6\times 10^{15}$ Hz.

**27** (c) Change in angular momentum,

$$\Delta L = L_2 - L_1 = \frac{n_2 h}{2\pi} - \frac{n_1 h}{2\pi} = \frac{h}{2\pi}(n_2 - n_1)$$

Given, $n_1 = 4$, $n_2 = 5$ and $h = 6.6\times 10^{-34}$ J-s

$$\Delta L = \frac{6.6\times 10^{-34}}{2\times 3.14}\times(5-4) = 1.05\times 10^{-34}\ \text{J-s}$$

**28** (d) Radius of electron in $n$th orbit,

$$r_n = \frac{n^2 h^2 \varepsilon_0}{\pi m Z e^2} \Rightarrow r_n \propto \frac{1}{Z}$$

So, the relation between $r_n$ and $Z$ is correctly depicted in option (d), as the value of $Z$ increases, $r_n$ decreases.

**29** (c) According to Bohr atomic radius, $r_n \propto n^2$

where, $n$ is principal quantum number.

$$\therefore \quad \frac{r_n}{r_1} = \frac{n^2}{1^2} \Rightarrow r_n = n^2 r \qquad [\because r_1 = r]$$

So, for $n = 2$, $r_2 = 2^2 r = 4r$

**30** (d) Speed of an electron in $n$th orbit, $v_n = \dfrac{2\pi Z e^2 k}{nh}$

For hydrogen atom in ground state, ($n = 1$, $Z = 1$)

$$v = \frac{2\pi e^2 k}{h} \Rightarrow \frac{v}{c} = \frac{2\pi e^2 k}{hc}$$

**31** (b) According to Bohr's model, velocity of electron,

$$v = \frac{2\pi k e^2 Z}{nh} \quad \text{or} \quad v \propto \frac{1}{n}$$

$$\therefore \quad \frac{v_A}{v_B} = \frac{n_B}{n_A}$$

$$\therefore \quad v_B = v_A \times \frac{n_A}{n_B}$$

Given, $n_A = 1$, $n_B = 4$ and $v_A = 2.6\times 10^6\ \text{ms}^{-1}$

$$v_B = 2.6\times 10^6 \times \frac{1}{4} = 6.5\times 10^5\ \text{ms}^{-1}$$

**32** (b) $T_n$ = Period of revolution $= \dfrac{2\pi r_n}{v_n}$

Here, $r_n = \dfrac{n^2 h^2 \varepsilon_0}{\pi m Z e^2}$ and $v_n = \dfrac{Z e^2}{2\varepsilon_0 nh}$

So, $T_n = 2\pi \times \dfrac{n^2 h^2 \varepsilon_0}{\pi m Z e^2} \times \dfrac{2\varepsilon_0 nh}{Z e^2} = \dfrac{4\varepsilon_0^2 n^3 h^3}{m Z^2 e^4}$

$$\Rightarrow \quad T_n \propto \frac{n^3}{Z^2}$$

**33** (d) Velocity of electron, $v_n = \dfrac{2\pi k Z e^2}{nh}$

Here, $n = 3$, $Z = 1$ (for hydrogen atom)

$k = 9\times 10^9\ \text{N-m}^2/\text{C}^2$, $e = 1.6\times 10^{-19}$ C

and $h = 6.6\times 10^{-34}$ J-s.

$$v_3 = \frac{2\times \pi \times 9\times 10^9 \times 1\times (1.6\times 10^{-19})^2}{3\times 6.6\times 10^{-34}}$$

$$= 7.31\times 10^5\ \text{ms}^{-1}$$

Also, time period, $T_n = \dfrac{2\pi r_n}{v_n}$

where, $r_n = \dfrac{n^2 h^2 \varepsilon_0}{\pi m Z e^2}$

Here, $n = 3$, $m = 9.1\times 10^{-31}$ kg and $\varepsilon_0 = 8.85\times 10^{-12}$ J-s

$$\Rightarrow r_3 = \frac{(3)^2 \times (6.6\times 10^{-34})^2 \times 8.85\times 10^{-12}}{\pi\times 9.1\times 10^{-31}\times 1\times (1.6\times 10^{-19})^2}$$

$$= 4.75\times 10^{-10}\ \text{m}$$

$$\Rightarrow \therefore T_3 = \frac{2\pi r_3}{v_3} = \frac{2\pi \times 4.75\times 10^{-10}}{7.31\times 10^5}$$

$$= 4.08\times 10^{-15}\ \text{s}$$

**34** (b) As, time period, $T \propto n^3$

$$\Rightarrow \quad \frac{1}{\text{Frequency}\ (f)} \propto n^3 \qquad \left[\because T = \frac{1}{f}\right]$$

$$\Rightarrow \quad f \propto n^{-3}$$

Thus, $f$ decreases with increase in $n$ or *vice-versa.*

**35** (a) For hydrogen atom, $T \propto n^3 \Rightarrow \dfrac{T_1}{T_2} = \left(\dfrac{n_1}{n_2}\right)^3$

Given, $T_1 = 8T_2$

$$\Rightarrow \quad \frac{8}{1} = \left(\frac{n_1}{n_2}\right)^3 \Rightarrow n_1 = 2n_2$$

From the given option, the above condition is only satisfied, when $n_1 = 4$ and $n_2 = 2$.

**36** (b) The velocity of an electron in $n$th orbit of Bohr's atom is

$$v_n = \frac{e^2 \times 2\pi}{4\pi\varepsilon_0 nh} = \frac{e^2}{2\varepsilon_0 nh}$$

$$\therefore \text{Kinetic energy, KE}_n = \frac{1}{2} m v_n^2 = \frac{1}{2} m \left(\frac{e^2}{2\varepsilon_0 nh}\right)^2$$

$$= \frac{m e^4}{8\varepsilon_0^2 n^2 h^2} \qquad \ldots\text{(i)}$$

Total energy of electron in hydrogen atom is

$$TE_n = -\frac{e^2}{8\pi\varepsilon_0 r_n}$$

where, $r_n = \frac{n^2h^2 4\pi\varepsilon_0}{m4\pi^2e^2} = \frac{n^2h^2\varepsilon_0}{\pi me^2}$

$$\Rightarrow \quad TE_n = -\frac{e^2}{8\pi\varepsilon_0} \times \frac{\pi me^2}{n^2h^2\varepsilon_0} = \frac{-me^4}{8n^2\varepsilon_0^2h^2} \quad \text{...(ii)}$$

From Eqs. (i) and (ii), we get

$$KE_n : TE_n = \frac{me^4}{8\varepsilon_0^2n^2h^2} : \frac{-8\varepsilon_0^2n^2h^2}{me^4} = 1 : -1$$

The ratio of kinetic energy tothe total energy in a Bohr orbit of the hydrogen atom is $1 : -1$.

**37** (*b*) According to Bohr's model, the kinetic energy of electron is given by

$$KE = \frac{me^4}{8\varepsilon_0^2n^2h^2} \quad \text{... (i)}$$

where, $h$ = Planck's constant

and $n$ = principal quantum number.

Similarly, potential energy is given by

$$PE = -\frac{me^4}{4\varepsilon_0^2n^2h^2} \quad \text{... (ii)}$$

$$\therefore \text{Total energy, } E = PE + KE = -\frac{me^4}{8\varepsilon_0^2n^2h^2}$$

[from Eqs. (i) and (ii)]

$\Rightarrow$ $KE = -E$ and $PE = 2E$

Given, $E = -3.4$ eV

$\therefore$ $KE = -(-3.4) = 3.4$ eV

and $PE = 2(-3.4) = -6.8$ eV

**38** (*c*) As we know, potential energy $U = 2E$, kinetic energy $K = -E$

and total energy of electron, $E = -\frac{13.6}{n^2}$

For state $n = 1$, $E_1 = \frac{-13.6}{(1)^2} = -13.6$

For state $n = 2$, $E_2 = \frac{-13.6}{(2)^2} = \frac{E_1}{4} \Rightarrow E_1 = 4E_2$

But $K = -E$, so $K_1 = 4K_2$

Similarly, for state $n = 1$, potential energy,

$$U_1 = 2E_1 = 2 \times \frac{-13.6}{(1)^2} = -2 \times 13.6$$

For state $n = 2$, potential energy,

$$U_2 = 2E_2 = 2 \times \frac{-13.6}{(2)^2} = \frac{-2 \times 13.6}{4}$$

$$U_2 = \frac{U_1}{4} \Rightarrow U_1 = 4U_2$$

Thus, both kinetic energy $K$ and potential energy $U$ becomes four times.

**39** (*c*) Energy of electron in a hydrogen atom in $n = 10$ state

$$= -\frac{13.6}{10^2} \text{ eV} = -0.136 \text{ eV}$$

$\therefore$ Energy required to remove it

$$= -E_n = 0.136 \text{ eV}$$

**40** (*a*) Energy in $n$th orbit, $E_n = -\frac{13.6}{n^2} Z^2 \cdot \text{eV}$

So, for a singly ionised helium atom, $Z = 2$;

$$\Rightarrow \quad E_{He_n} = E_{H_n} 2^2 = 4E_n$$

**41** (*b*) Energy of electron in first excited state of hydrogen

$$E_n = -\frac{13.6}{n^2} = -\frac{13.6}{2^2}$$

$= -3.40$ eV ($n = 2$ in first excitation)

So, for doubly ionised lithium ($Z = 3$), energy will be

$$E_n = (E_H)Z^2 = -3.40 \times 9$$

$$= -30.6 \text{ eV}$$

$\therefore$ Energy required to remove electron, $-E_n = 30.6$ eV

**42** (*c*) Using $E_n = -\frac{13.6}{n^2} = -3.4$ eV, we get

$$n = \sqrt{\frac{13.6}{3.4}} = 2$$

$$\therefore \text{Angular momentum, } L = \frac{nh}{2\pi} = \frac{2 \times 6.6 \times 10^{-34}}{2 \times 3.14}$$

$$= 2.10 \times 10^{-34} \text{ J-s}$$

**43** (*b*) Angular momentum of an electron, $L = \frac{nh}{2\pi}$

As, in 2nd excited state, $n = 3$, for both H and $Li^+$.

$$\Rightarrow \quad L_H = L_{Li}$$

Energy, $E_n = \frac{-Z^2(13.6)}{n^2} \Rightarrow E_n \propto Z^2$

As, $Z_H = 1$ and $Z_{Li^+} = 3$

$\therefore$ $|E_{Li}| > |E_H|$

**44** (*b*) Energy of a hydrogen atom like ion by Bohr's model is $E_n = -13.6\frac{Z^2}{n^2}$

where, $Z$ = atomic number

and $n$ = principal quantum number.

For a $He^+$ ion in first excited state,

$$n = 2, Z = 2$$

$$\therefore \quad E_2 = -13.6 \times \frac{4}{4} = -13.6 \text{ eV}$$

So, it's ionisation energy $= -(E_2) = 13.6$ eV.

**45** (*c*) The minimum energy to ionise an atom is the energy required to free an electron in the atom from its ground state corresponding to $n = 1$.

When this occurs, one outermost electron from the atom is just removed from the ground state of atom.

**46** (*b*) Energy in *n*th state of orbit, $E_n = -\frac{13.6\,Z^2}{n^2}$ eV

For hydrogen atom, $E_n = \frac{-13.6}{n^2}$eV $\quad [\because Z = 1]$

At ground state, $n = 1 \Rightarrow E_1 = -13.6$ eV

This is the minimum energy required to free the electron from ground state.

For $n = 2$, $E_2 = -\frac{13.6}{2^2} = -3.4$ eV

Similarly, $E_3 = \frac{-13.6}{3^2} = -1.51$ eV

Thus, as the excitation of hydrogen atom increases, i.e. *n* inorease, the value of minimum energy required to free the electron from the excited state decreases.

**47** (*a*) In hydrogen atom, excitation energy for first excited state

$$E_2 - E_1 = -13.6\left(\frac{1}{2^2} - \frac{1}{1^2}\right) = -13.6 \times \left(\frac{-3}{4}\right) = 10.2 \text{ eV}$$

Since, 5 eV < 10.2 eV

The electron cannot excite the hydrogen atom.

∴The collision must be elastic.

**48** (*c*) Energy of electron, $E_n = \frac{-13.6}{n^2} Z^2$

In first excited state, $E_{H_2} = \frac{-13.6(1)^2}{(2)^2} = -3.4$ eV

and $E_{He} = \frac{-13.6(2)^2}{(2)^2} = -13.6$ eV

$H_2$ atom gives excitation energy $(13.6 - 3.4 = 10.2$ eV) to helium atom.

Now energy of He ion $= -13.6 + 10.2 = -3.4$ eV

Again, $\quad E = \frac{-13.6}{n^2} \times Z^2 \Rightarrow n = \sqrt{\frac{13.6 \times 4}{3.4}} = 4$

**49** (*a*) Given, total energy given to electron in 4th orbit,

$$E_4 = 15 \text{ eV}$$

Energy of *n*th orbit of H-atom, $E' = -\frac{13.6}{n^2}$

So, energy of 4th orbit of H-atom,

$$E'_4 = \frac{-13.6}{4^2} = -0.85 \text{ eV}$$

∴Ionisation energy of 4th orbit $= -E'_4$

$$= -(-0.85 \text{ eV}) = 0.85 \text{ eV}$$

∴Final energy of electron, when it comes out of H-atom, $E$ = Total energy given to electron − Ionisation energy of 4th orbit

$\Rightarrow \quad E = 15 - 0.85 = 14.15$ eV

**50** (*c*) When electron absorbs photon of energy 12.1 eV, energy of electron will be = initial + absorbed energy

$$E = -13.6 \text{ eV} + 12.1 \text{ eV}$$

$\Rightarrow \quad E = -1.5$ eV

which corresponds to $n = \sqrt{\frac{13.6}{1.5}} = 3$ excited state.

So, number of possible transitions are

$$N = \frac{n(n-1)}{2} = 3$$

i.e. $\quad n = 3 \to n = 2$

$\Rightarrow \quad n = 2 \to n = 1$

$\Rightarrow \quad n = 3 \to n = 1$

∴The spectral lines emitted by hydrogen atom are three in number.

**52** (*a*) Emitted frequency ν is given by $h\nu = \Delta E$

$\Rightarrow \quad \nu = \frac{\Delta E}{h} = \frac{E_n - E_{n-1}}{h}$

As, $\quad E_n = \frac{-me^4}{(4\pi)^3 \varepsilon_0^2 \frac{h^2}{(2\pi)^3}} \left(\frac{1}{n^2}\right)$

$\Rightarrow \quad \nu = -\frac{me^4}{(4\pi)^3 \varepsilon_0^2 \left(\frac{h}{2\pi}\right)^3} \left\{\frac{1}{n^2} - \frac{1}{(n-1)^2}\right\}$

$$= \frac{me^4 (2n-1)}{(4\pi)^3 \varepsilon_0^2 \left(\frac{h}{2\pi}\right)^3 n^2 (n-1)^2}$$

**53** (*c*) We know that, frequency, $\nu = \frac{c}{\lambda} = c \cdot R\left(\frac{1}{n_1^2} - \frac{1}{n_2^2}\right)$

Given, $c = 3 \times 10^8$ ms$^{-1}$, $R = 10^7$ cm$^{-1}$

$n_1 = 2$ and $n_2 = 4$

$$= 3 \times 10^8 \times 10^7 \left(\frac{1}{2^2} - \frac{1}{4^2}\right) = \frac{9}{16} \times 10^{15} \text{Hz}$$

**54** (*d*) As, $\lambda_{IR} > \lambda_{UV}$.

Also, wavelength of emitted radiation, $\lambda \propto \frac{1}{\Delta E}$

$\Rightarrow \quad \lambda \propto n^2 \quad [\because E \propto 1/n^2]$

As, for UV-radiation, the transition is from $n = 4$ to $n = 3$, so for IR (infrared) radiation, the transition will be from

$$5 \to 4$$

**55** (*c*) We have, $\lambda = \frac{hc}{\Delta E}$

∴ So, ratio of wave lengths

$$\frac{\lambda_1}{\lambda_2} = \frac{hc/\Delta E_1}{hc/\Delta E_2} = \frac{\Delta E_2}{\Delta E_1} = \frac{\left(\frac{4}{3}E - E\right)}{2E - E} = \frac{1}{3}$$

**56** (*b*) Energy at *A*, *B* and *C* state is $E_A$, $E_B$ and $E_C$, then

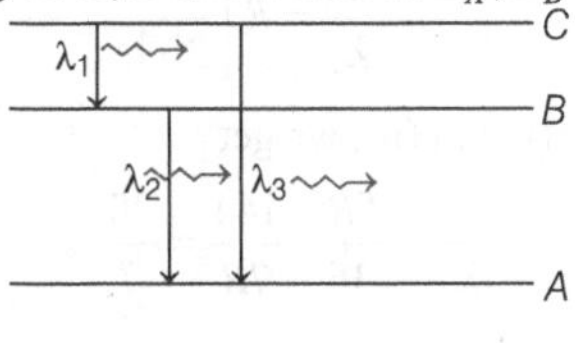

$$(E_C - E_B) + (E_B - E_A) = E_C - E_A$$

or $$\frac{hc}{\lambda_1} + \frac{hc}{\lambda_2} = \frac{hc}{\lambda_3} \Rightarrow \frac{1}{\lambda_1} + \frac{1}{\lambda_2} = \frac{1}{\lambda_3}$$

$$\Rightarrow \quad \lambda_3 = \lambda_1\lambda_2/\lambda_1 + \lambda_2$$

**57** (*d*) De-excitation energy of hydrogen electron in transition $n = 2$ to $n = 1$ is

$$E = 13.6 \times \left(\frac{1}{n_f^2} - \frac{1}{n_i^2}\right) \text{eV} = 13.6\left(\frac{1}{1^2} - \frac{1}{2^2}\right) = 10.2 \text{ eV}$$

Now, energy levels of helium ion's ($He^+$) electron are (For helium, $Z = 2$)

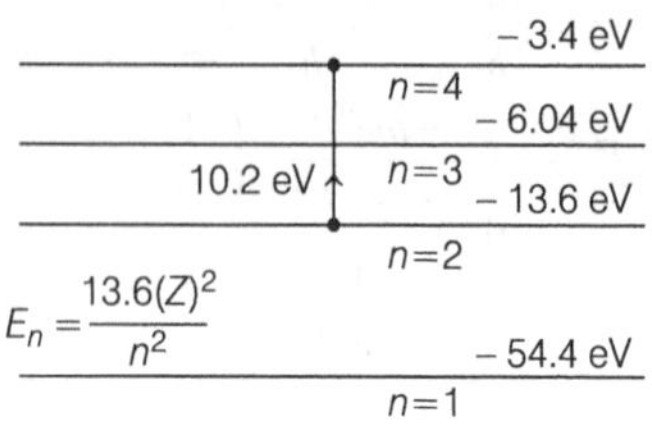


So, a photon of energy 10.2 eV can cause a transition $n = 2$ to $n = 4$ in a $He^+$ ion.

**58** (*c*) According to Rydberg's formula,

$$\frac{1}{\lambda} = RZ^2\left(\frac{1}{n_1^2} - \frac{1}{n_2^2}\right)$$

where, $R =$ Rydberg's constant and $Z =$ atomic number.

$$\Rightarrow \quad \lambda \propto \frac{1}{Z^2}$$

So, for transition, $2 \to 1$, in $Li^{2+}$, $He^{2+}$ and H, the ratio of wavelengths, $\frac{1}{3^2} : \frac{1}{2^2} : \frac{1}{1^2} = \frac{1}{9} : \frac{1}{4} : \frac{1}{1} = 4:9:36$

Hence, $\lambda_{Li^{2+}} : \lambda_{He^{2+}} : \lambda_H = 4:9:36.$

**59** (*d*) Here, for wavelength $\lambda_1, n_1 = 4$ and $n_2 = 3$ and for $\lambda_2, n_1 = 3$ and $n_2 = 2$

We have, $$\frac{1}{\lambda} = -R\left[\frac{1}{n_2^2} - \frac{1}{n_1^2}\right]$$

So, for $\lambda_1$, $n_2 = 4$ and $n_1 = 3$

$$\Rightarrow \quad \frac{1}{\lambda_1} = -R\left[\frac{1}{(4)^2} - \frac{1}{(3)^2}\right]$$

$$\frac{1}{\lambda_1} = R\left[\frac{7}{144}\right] \quad \text{...(i)}$$

Similarly, for $\lambda_2$, $n_2 = 3$ and $n_1 = 2$

$$\Rightarrow \quad \frac{1}{\lambda_2} = -R\left[\frac{1}{(3)^2} - \frac{1}{(2)^2}\right]$$

$$\frac{1}{\lambda_2} = R\left[\frac{5}{36}\right] \quad \text{...(ii)}$$

From Eqs. (i) and (ii), we get

$$\frac{\lambda_1}{\lambda_2} = \frac{5R}{36} \times \frac{144}{7R} = \frac{20}{7}$$

**60** (*a*) According to Rydberg's formula, $\frac{1}{\lambda} = R\left(\frac{1}{n_f^2} - \frac{1}{n_i^2}\right)$

Here, $n_f = 1$, $n_i = n$

$$\frac{1}{\lambda} = R\left(\frac{1}{1^2} - \frac{1}{n^2}\right)$$

$$\frac{1}{\lambda} = R\left(1 - \frac{1}{n^2}\right) \quad \text{...(i)}$$

$$\Rightarrow \quad \frac{R}{n^2} = R - \frac{1}{\lambda} \Rightarrow n = \sqrt{\frac{\lambda R}{\lambda R - 1}}$$

**61** (*b*) As, angular momentum, $L$

$$\frac{nh}{2\pi} = \frac{3h}{2\pi} \Rightarrow n = 3$$

Also, $$a_0 \times \frac{n^2}{Z} = 4.5a_0 \Rightarrow \frac{n^2}{Z} = 4.5 \Rightarrow Z = 2$$

According to Rydberg's formula, $\frac{1}{\lambda} = RZ^2\left(\frac{1}{n_1^2} - \frac{1}{n_2^2}\right)$

$$\Rightarrow \quad \frac{1}{\lambda_1} = R \times 2^2\left[\frac{1}{1^2} - \frac{1}{3^2}\right] \Rightarrow \frac{1}{\lambda_1} = 4R \times \frac{8}{9}$$

$$\frac{1}{\lambda_2} = 4R\left[\frac{1}{2^2} - \frac{1}{3^2}\right] \Rightarrow \frac{1}{\lambda_2} = 4R \times \frac{5}{36}$$

$$\frac{1}{\lambda_3} = 4R\left[\frac{1}{1^2} - \frac{1}{2^2}\right] \Rightarrow \frac{1}{\lambda_3} = 4R \times \frac{3}{4}$$

$$\Rightarrow \quad \lambda_1 = \frac{9}{32R}, \lambda_2 = \frac{9}{5R}, \lambda_3 = \frac{1}{3R}$$

Hence, $\lambda = \frac{9}{16R}$ is not possible.

**62** (*c*) For transition from $n = 2$ to $n = 1$, using Rydberg's formula, we get

$$\frac{1}{\lambda} = R(Z)^2\left(\frac{1}{1^2} - \frac{1}{2^2}\right)$$

For ${}_1H^1$, $$\frac{1}{\lambda_1} = R(1)^2\left(\frac{3}{4}\right)$$

For ${}_1H^2$, $$\frac{1}{\lambda_2} = R(1)^2\left(\frac{3}{4}\right)$$

For ${}_2He^4$, $$\frac{1}{\lambda_3} = R(2)^2\left(\frac{3}{4}\right)$$

For, ${}_3Li^8$, $$\frac{1}{\lambda_4} = R(3)^2\left(\frac{3}{4}\right)$$

$$\Rightarrow \quad \frac{1}{\lambda_1} = \frac{1}{\lambda_2} = \frac{1}{4\lambda_3} = \frac{1}{9\lambda_4}$$

$$\therefore \quad \lambda_1 = \lambda_2 = 4\lambda_3 = 9\lambda_4$$

**63** (*b*) Wave number = Number of wavelength in unit length

$$\frac{1}{\lambda} = \frac{1}{5896 \times 10^{-10}} = \frac{1.6961 \times 10^{-4}}{10^{-10}}$$

$$= 1.6961 \times 10^6 \text{m}^{-1} = 16961 \text{ per cm}$$

**64** (*c*) **Transition $A$,** $n = \infty$ to $n = 1$, series limit of Lyman series.

**Transition $B$,** $n = 5$ to $n = 2$, third spectral line of Balmer series.

**Transition $C$,** $n = 5$ to $n = 3$, second spectral line of Paschen series.

**65** (*a*) We know that, frequency, $\nu = Rc\left(\frac{1}{n_1^2} - \frac{1}{n_2^2}\right)$

where, $R =$ Rydberg's constant
and $c =$ speed of light.

For series limit of Lyman series, $n_1 = 1, n_2 = \infty$.

$$\nu_1 = Rc\left(\frac{1}{1^2} - \frac{1}{\infty}\right) = Rc$$

For first line, $n_1 = 1, n_2 = 2$.

$$\nu_2 = Rc\left(\frac{1}{1^2} - \frac{1}{4}\right) = \frac{3}{4}Rc$$

For series limit of Balmer series, $n_1 = 2, n_2 = \infty$.

$$\nu_3 = Rc\left(\frac{1}{4} - \frac{1}{\infty}\right) = \frac{Rc}{4}$$

$$\Rightarrow \quad \nu_1 - \nu_2 = Rc - \frac{3}{4}Rc = \frac{Rc}{4} = \nu_3$$

**66** (*a*) When electron beam of 12.75 eV is incident over H-gas,

Energy of H-atoms will be $= -13.6 + 12.75 = -0.85$ eV

This corresponds to $n = \sqrt{\frac{13.6}{0.85}} = 4$ excited state.

Various transitions, for $n = 4$ are shown below

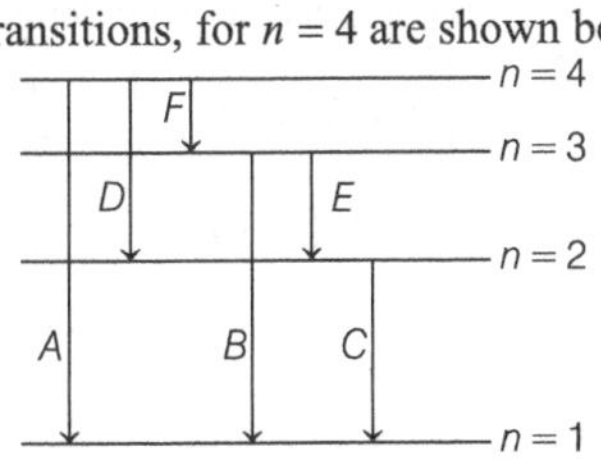


Clearly, emitted radiation lies in Lyman series ($A$, $B$ and $C$).

Few emissions are in Balmer series ($D$ and $E$).

One of them ($n = 4$ to $n = 3$) lies in Paschen series ($F$).

**67** (*c*) Given, $E = -4.6\,\text{eV}$

$$K = -E = -(-4.6) = 4.6\,\text{eV}$$

de-Broglie wavelength,

$$\lambda = \frac{h}{\sqrt{2mK}} = \frac{6.6 \times 10^{-34}}{\sqrt{2 \times 9.1 \times 10^{-31} \times 4.6 \times 1.6 \times 10^{-19}}}$$

$$= 0.57 \times 10^{-9}\,\text{m} = 0.57\,\text{nm}$$

**68** (*a*) The Bohr model for the spectra of a H-atom will not be applicable to hydrogen in the molecular form.

Also, it will not be applicable for multi-electron atom.

As, it is only able to explain discrete spectral lines for one-electron atom.

**69** (*a*) Light emitted from rarefied gases heated in a flame or excited electrically in a glow tube such as the familiar neon or mercury vapour light has only certain discrete wavelengths. The spectrum appears as a series of bright lines.

As, in such gases, the average spacing between atoms is large. Hence, the radiation emitted can be considered due to individual atoms rather than because of interaction between atoms or molecules.

Therefore, Assertion and Reason are correct and Reason is the correct explanation of Assertion.

**70** (*c*) Rutherford's experiment suggested the size of the nucleus to be about $10^{-15}$m to $10^{-14}$ m. From kinetic theory, the size of an atom was known to be $10^{-10}$m, about 10,000 to 100,000 times larger than the size of the nucleus.

Thus, the electrons would seem to be at a distance from the nucleus about 10,000 to 1,00,000 times the size of the nucleus itself.

Hence, most of an atom is an empty space.

Therefore, Assertion is correct but Reason is incorrect.

**71** (*a*) In an atom, electron revolves around nucleus, for this required centripetal force is provided by electrostatic force of attraction between negatively charged electron and positive nucleus.

If the electrons were stationary, then the electrostatic force will remain unbalanced, which leads to the electron to fall into the nucleus.

Therefore, Assertion and Reason are correct and Reason is the correct explanation of Assertion.

**72** (*d*) The energy of an atom is the least when its electron is revolving in an orbit closest to the nucleus, i.e. the one for which $n = 1$. However, the highest energy state corresponds to $n = \infty$ and has an energy of 0 eV.

Since, $\quad E_n \propto \frac{1}{n^2}$

So, as the value of $n$ increases, energies of excited states comes closer and closer.

Therefore, Assertion is incorrect but Reason is correct.

**73** (*b*) The statement II is correct but I is incorrect and it can be corrected as,

According to J.J. Thomson's atomic model, the positive charge of the atom is uniformly distributed throughout the volume of the atom and the negatively charged electrons are embedded in it like seeds in a watermelon.

**74** (*b*) The statement II is correct but I is incorrect and it can be corrected as,

In Geiger-Marsden experiment, H Geiger and E Marsden directed a beam of 5.5. MeV $\alpha$-particles emitted from a $^{214}_{83}$Bi radioactive source at a thin metal foil made of gold.

**75** (*b*) The statement II is correct but I is incorrect and it can be corrected as,

Total energy $E$ of an electron in a hydrogen atom is

$$E = \frac{-e^2}{8\pi\varepsilon_0 r}$$

So, $E$ is negative.

This implies the fact that the electron is bound to the nucleus. If $E$ is positive, an electron will not follow a closed orbit around the nucleus.

**76** (*c*) The statement II is correct but I and III are incorrect and these can be corrected as,

Bohr model is valid for only one electron atom.

The solution of the schrodinger wave equation, obtained for hydrogen like atoms called the wave function or orbital that gives the information about the probability of finding an electron in various regions around the nucleus.

This orbital has no resemblance whatsoever with the orbit defined for an electron in Bohr model.

**77** (*c*) As incident energy $(E = h\nu)$ is $E_2 - E_1$, so it is sufficient for an electron to make a transition from $n = 1$ to $n = 2$ state. So, depending on the number of incident photons. Some of the electrons moves to first excited state, i.e. $n = 2$ state and no electron makes a transition to $n = 3$ state.

So, statements II and IV are correct but I and III are incorrect.

**78** (*c*) The statements I and III are correct but II is incorrect and it can be corrected as,

Franck and Hertz gave the experimental verification of Bohr's basic idea of discrete energy levels in atoms and the process of photo emission.

Magnetic moment of an electron, i.e. $\mu = IA = \frac{e}{T}\cdot \pi r^2$

Here, $T = \frac{2\pi r}{v} \Rightarrow \mu = \frac{ev}{2\pi r}\cdot \pi r^2 = \frac{e}{2}\cdot vr = \frac{e}{2m}\cdot mvr$

$\mu = \frac{e}{2m}\cdot\frac{nh}{2\pi}$ $\quad [\because L = mvr = \frac{nh}{2\pi}]$

$\therefore \quad \mu \propto n$

Therefore, magnetic moment of $e^-$ revolving around the nucleus is directly proportional to $n$.

**80** (*b*) According to Geiger-Marsden experiment, only about 0.14% of the incident $\alpha$-particles scatter by more than 1°.

The statement given in option (b) is incorrect, rest are correct.

**81** (*c*) The atomic electrons, being so light, do not appreciably affect the $\alpha$-particles. Since, $\alpha$-particles are nuclei of helium atom and therefore carries two units of positive charge. But the charge on gold nucleus is 79*e*. So, nucleus of gold is about 50 times heavier than an $\alpha$-particle and it remains stationary throughout the scattering process.

Under these assumptions, the trajectory of an alpha particle can be explained by employing Newton's second law of motion and the Coulomb's law for electrostatic force of repulsion between the alpha particle and the positively charged particle.

Thus, the statement given in option (c) is correct, rest are incorrect.

**82** (*b*) The statement given in option (b) is incorrect and it can be corrected as,

According to Bohr's atomic model, electron revolves in stationary stable orbits without emission of radiant energy. Thus, its mass remains constant.

Rest statements are correct.

**83** (*a*) As we know that, kinetic energy of an electron is

$$\text{KE} \propto \left(\frac{Z}{n}\right)^2$$

When the electron makes transition from an excited state to the ground state, then $n$ decreases and KE increases.

We know that PE is lowest for ground state. Also, TE = – KE, so TE also decreases as $n$ increases.

Thus, the statement given in option (a) is correct, rest are incorrect.

**84** (*b*) Balmer's empirical formula is given as

$$\frac{1}{\lambda} = R\left(\frac{1}{2^2} - \frac{1}{n^2}\right)$$

where, $\lambda$ is wavelength and $R$ is a constant called the Rydberg's constant ($1.097\times 10^7$ m$^{-1}$).

Taking, $n = 3$, one obtains the wavelength of $H_\alpha$-line,

$$\frac{1}{\lambda} = 1.097\times 10^7\left(\frac{1}{2^2} - \frac{1}{3^2}\right)$$

$$= 1.522\times 10^6 \text{m}^{-1}$$

or $\quad \lambda = \frac{1}{1.522\times 10^6} = 656.3 \text{ nm}$

This means, $H_\alpha$ is red in colour.

Thus, the statement given in option (b) is incorrect, rest are correct.

**85** (*a*) As, linear momentum, $p = mv \Rightarrow p \propto v$

As $\quad v \propto \frac{1}{n} \Rightarrow p \propto \frac{1}{n}$

i.e. linear momentum is inversely proportional to the number of orbit $n$.

Thus, the statement given in option (a) is incorrect,rest are correct.

**86** (*b*) For Balmer series, $\frac{1}{\lambda} = R\left(\frac{1}{2^2} - \frac{1}{n^2}\right)$

where, $R = 1.097\times 10^7 \text{m}^{-1}$

A. For $n = 3$,

$(H_{\alpha\text{-series}})\frac{1}{\lambda_\alpha} = 1.097\times 10^7\left(\frac{1}{2^2} - \frac{1}{3^2}\right)$

$$= 1.097 \times 10^7 \times \frac{5}{36} \text{ m}^{-1}$$

$$\lambda_\alpha = \frac{36}{5 \times 1.097 \times 10^7} = 656.3 \times 10^{-9} \text{ m} = 656.3 \text{ nm}$$

B. For $n = 4$,

$$(H_{\beta\text{-series}}) \frac{1}{\lambda_\beta} = 1.097 \times 10^7 \left(\frac{1}{2^2} - \frac{1}{4^2}\right) = 1.097 \times 10^7 \times \frac{3}{16} \text{ m}^{-1}$$

$$\Rightarrow \quad \lambda_\beta = \frac{16}{3 \times 1.097 \times 10^7}$$

$$= 486.1 \times 10^{-9} \text{ m} = 486.1 \text{ nm}$$

C. For $n = 5$,

$$(H_{\gamma\text{-series}}) \frac{1}{\lambda_\gamma} = 1.097 \times 10^7 \left(\frac{1}{2^2} - \frac{1}{5^2}\right)$$

$$= 1.097 \times 10^7 \times \frac{21}{100} \text{ m}^{-1}$$

$$\Rightarrow \quad \lambda_\gamma = \frac{100}{21 \times 1.097 \times 10^7}$$

$$= 434.1 \times 10^{-9} \text{ m} = 434.1 \text{ nm}$$

D. For $n = 6$,

$$(H_{\delta\text{-series}}) \frac{1}{\lambda_\delta} = 1.097 \times 10^7 \left(\frac{1}{2^2} - \frac{1}{6^2}\right)$$

$$= 1.097 \times 10^7 \times \frac{8}{36} \text{ m}^{-1}$$

$$\Rightarrow \quad \lambda_\delta = \frac{36}{8 \times 1.097 \times 10^7}$$

$$= 410.2 \times 10^{-9} \text{ m}$$

$$= 410.2 \text{ nm.}$$

Hence, A $\rightarrow$ 4, B $\rightarrow$ 3, C $\rightarrow$ 2 and D $\rightarrow$ 1.

**88** (*c*) Lyman series

$$\frac{1}{\lambda} = R\left(\frac{1}{1^2} - \frac{1}{n^2}\right), \; n = 2, 3, 4, ..$$

Balmer series

$$\frac{1}{\lambda} = R\left(\frac{1}{2^2} - \frac{1}{n^2}\right), \; n = 3, 4, 5, ...$$

Paschen series

$$\frac{1}{\lambda} = R\left(\frac{1}{3^2} - \frac{1}{n^2}\right), \; n = 4, 5, 6, ..$$

Brackett series $\quad \frac{1}{\lambda} = R\left(\frac{1}{4^2} - \frac{1}{n^2}\right), \; n = 5, 6, 7, ..$

To obtain maximum wavelengths we substitute $n =$ number of next possible orbit.

$\therefore$ $n = 2$ in Lyman
$n = 3$ in Balmer
$n = 4$ in Paschen
$n = 5$ in Brackett

$$\therefore \text{For Lyman; } \lambda_{max} = \frac{1}{R\left(1 - \frac{1}{4}\right)} = \frac{4}{3R}$$

$$\text{For Balmer; } \lambda_{max} = \frac{1}{R\left(\frac{1}{4} - \frac{1}{9}\right)} = \frac{36}{5R}$$

$$\text{For Paschen; } \lambda_{max} = \frac{1}{R\left(\frac{1}{9} - \frac{1}{16}\right)} = \frac{144}{7R}$$

$$\text{For Brackett; } \lambda_{max} = \frac{1}{R\left(\frac{1}{16} - \frac{1}{25}\right)} = \frac{400}{9R}$$

Hence, A $\rightarrow$ 4, B $\rightarrow$ 3, C $\rightarrow$ 2 and D $\rightarrow$ 1.

**89** (*d*) As we know,

A. Angular momentum, $L = \frac{nh}{2\pi} \Rightarrow L \propto n$

B. Velocity of electrons, $v_n = \frac{1}{n} \frac{e^2}{4\pi\varepsilon_0} \cdot \frac{1}{(h/2\pi)}$

$$\Rightarrow \quad v_n \propto 1/n$$

C. Radius of an electron, $r_n = \frac{n^2}{m}\left(\frac{h}{2\pi}\right)^2 \frac{4\pi\varepsilon_0}{e^2}$

$$\Rightarrow \quad r_n \propto n^2$$

D. Potential energy, $\quad U_n = -\frac{me^4}{4n^2\varepsilon_0^2 h^2}$

$$\therefore \quad U_n \propto 1/n^2$$

Hence, A $\rightarrow$ 4, B $\rightarrow$ 1, C $\rightarrow$ 2 and D $\rightarrow$ 3.

**90** (*b*) For Paschen series, $n_1 = 3$ and $n_2 = \infty$ for shortest wavelength, using the formula, $\frac{1}{\lambda} = R\left[\frac{1}{n_1^2} - \frac{1}{n_2^2}\right]$

[where, $R$ is the Rydberg constant]

$$\frac{1}{\lambda} = 1.03 \times 10^7 \left[\frac{1}{3^2} - \frac{1}{\infty^2}\right]$$

or $\quad \frac{1}{\lambda} = \frac{1.03 \times 10^7}{9}$ or $\lambda = \frac{9}{1.03 \times 10^7}$

$$= 8.738 \times 10^{-7} \text{ m} = 873.8 \text{ nm}$$

**91** (*c*) Given, difference in energy level,

$$E = 2.3 \text{ eV}$$
$$= 2.3 \times 1.6 \times 10^{-19} \text{ J}$$

Planck's constant, $h = 6.63 \times 10^{-34}$ J-s

Let $\nu$ be the frequency, then $E = h\nu$

or $\quad \nu = \frac{E}{h} = \frac{2.3 \times 1.6 \times 10^{-19}}{6.63 \times 10^{-34}} = 5.6 \times 10^{14}$ Hz

**92** (*a*) For ground state, $n_1 = 1$ to $n_2 = 4$

Energy absorbed by photons $E$

$$= + 13.6\left(\frac{1}{n_1^2} - \frac{1}{n_2^2}\right) \times 1.6 \times 10^{-19} \text{ J}$$

$$= 13.6\left(\frac{1}{1^2} - \frac{1}{4^2}\right) \times 1.6 \times 10^{-19}$$

$$= 13.6\times 1.6\times 10^{-19}\left(\frac{15}{16}\right) = 20.4\times 10^{-19}$$

or $E = h\nu = 20.4\times 10^{-19}$

Frequency, $\nu = \frac{20.4\times 10^{-19}}{h} = \frac{20.4\times 10^{-19}}{6.63\times 10^{-34}}$

$\therefore \quad = 3.076\times 10^{15}$ Hz

Wavelength of photon, $\lambda = \frac{c}{\nu} = \frac{3\times 10^{8}}{3.076\times 10^{15}}$

$= 9.74\times 10^{-8}$ m

**93** (*b*) Given, the radius of the innermost electron orbit of a hydrogen, $r_1 = 5.3\times 10^{-11}$ m

As we know that, $r_n = n^2 r_1$

For $n = 2$,

radius, $r_2 = 2^2 r_1 = 4\times 5.3\times 10^{-11} = 2.12\times 10^{-10}$ m

For $n = 3$,

radius, $r_3 = 3^2 r_1 = 9\times 5.3\times 10^{-11} = 4.77\times 10^{-10}$ m

**94** (*c*) Energy of electron beam,

$$E = 12.5\text{ eV} = 12.5\times 1.6\times 10^{-19}\text{ J}$$

Planck's constant, $h = 6.63\times 10^{-34}$ J-s

Velocity of light, $c = 3\times 10^{8}$ ms$^{-1}$

Using the relation, $E = \frac{hc}{\lambda}$

$$\lambda = \frac{hc}{E} = \frac{6.63\times 10^{-34}\times 3\times 10^{8}}{12.5\times 1.6\times 10^{-19}}$$

$$= 0.993\times 10^{-7}\text{ m}$$

$$= 93\times 10^{-10}\text{ m} = 993\text{ Å}$$

This wavelength falls in the range of Lyman series (912 Å to 1216 Å), thus we conclude that Lyman series of wavelength 993 Å is emitted.

**95** (*a*) Given, radius of orbit, $r = 1.5\times 10^{11}$ m,

Orbital speed, $v = 3\times 10^{4}$ ms$^{-1}$ and mass of earth,

$m = 6.0\times 10^{24}$ kg

Angular momentum, $L = mvr = \frac{nh}{2\pi}$

or $\quad n = \frac{2\pi vrm}{h}$

where, $n$ is the quantum number of the orbit.

$$\Rightarrow n = \frac{2\times 3.14\times 3\times 10^{4}\times 1.5\times 10^{11}\times 6.0\times 10^{24}}{6.63\times 10^{-34}}$$

$$= 2.57\times 10^{74} \text{ or } n = 2.6\times 10^{74}$$

**96** (*b*) The kinetic energy of electron (KE) $= \frac{kZe^2}{2r}$

Potential energy of electron (PE) $= \frac{-kZe^2}{r}$

$\Rightarrow \quad$ PE = $-2$ KE

Total energy = PE + KE = $-$ 2KE + KE = $-$ KE

In the first excited state, total energy = $-$ 3.4 eV

KE = $-$ ($-$3.4) = 3.4 eV

**97** (*c*) Given, $m = 10$ kg and $r_n = 8\times 10^{6}$ m.

We have, the time period $T$ of the circuling satellite as 2 h, i.e.

$$T = 2\times 3600 = 7200\text{ s}$$

Thus, the velocity, $v_n = 2\pi r_n / T$

Also, $\quad mv_n r = \frac{nh}{2\pi}$

The quantum number of the orbit of satellite,

$$n = \frac{2\pi m v_n r_n}{h} = (2\pi\, r_n)^2 \times m/(T\times h)$$

Substituting the values, we get

$$n = (2\pi\times 8\times 10^{6}\text{ m})^2\times 10\,/\,(7200\text{ s}\times 6.64\times 10^{-34}\text{ J-s})$$

$$= 5.3\times 10^{45}$$

**98** (*c*) For a hydrogen like atom

$$r_n = 0.53\frac{n^2}{Z}\text{ Å} = 53\frac{n^2}{Z}\text{ pm}$$

For ground state $Li^{2+}$, $n = 1$ and $Z = 3$

$$\therefore \quad r_1 = \frac{53}{3} \approx 18\text{ pm}$$

**99** (*d*) Bohr model is applicable to hydrogenic atoms. It cannot be applied to many electrons atoms such as helium. As force between nucleus and electron is in comparison to electron-electron electronic force.

However, the formulation of Bohr model involves electrical force only between positively charged nucleus and electron.

**100** (*a*) In the simple Bohr model, only the magnitude of angular momentum is kept equal to some integral multiple of $\frac{h}{2\pi}$, where $h$ is Planck's constant and thus, the Bohr model gives incorrect values of angular momentum.

**101** (*a*) The total energy associated with the two H-atoms in the ground state collide in elastically

$$E = 2\times 13.6\text{ eV} = 27.2\text{ eV}$$

The maximum amount by which their combined kinetic energy is reduced when anyone of them goes into first excited state after the inelastic collision.

The total energy associated with the two H-atoms after the collision

$$E' = E_2 + E_1 = \left(\frac{13.6}{2^2}\right) + \frac{13.6}{2} = 17.0\text{ eV}$$

Therefore, maximum loss of their combined kinetic energy

$$\Delta E = E - E'$$

$$= 27.2 - 17.0 = 10.2\text{ eV}$$

**103** (*c*) When one decides to work in a frame of reference, where the electron is at rest, the given expression is incorrect as this frame for binding energy is not inertial.

# CHAPTER > 13

# Nuclei

## KEY NOTES

- In every atom, the positive charge and mass are densely concentrated at the centre of the atom forming its **nucleus**.
- Radius of nucleus is smaller than the radius of an atom by a factor of about $10^4$. Thus, its volume is $10^{-12}$ times the volume of the atom. It also contains most of the mass of an atom.

## Atomic Masses and Composition of Nucleus

- Mass of an atom is very small compared to a kilogram, so **atomic masses** are expressed using a different unit. This unit is **atomic mass unit** (u), defined as $\frac{1}{12}$th of the mass of the carbon ($^{12}$C) atom, i.e. $1\text{ u} = 1.660539 \times 10^{-27}$ kg.
- All the electrons of an atom are outside the nucleus.
- **Discovery of Neutron** Nucleus of an atom in addition with protons consists of neutral particles called **neutrons**.

  Mass of neutron is given as

  $$m_n = 1.00866\text{ u} = 1.6749 \times 10^{-27}\text{ kg}$$

- A free neutron, unlike a free proton, is unstable. It decays into a proton, an electron and a anti-neutrino. It is, however, stable outside the nucleus.
- The composition of the nucleus can be described using the following terms and symbols.

  $Z \rightarrow$ atomic number = number of protons

  = number of electrons

  $N \rightarrow$ number of neutrons

  $A \rightarrow$ mass number $= Z + N$

  = total number of protons and neutrons.

**Note** The term nucleons refers to a proton or a neutron.

- Different types of atoms of the same element, which exhibits same chemical properties but differ in mass is called **isotopes,** e.g. hydrogen ($^1_1$H), deuterium ($^2_1$H) and tritium ($^3_1$H).

  They occur at the same place in the periodic table of elements.

**Note** Tritium nuclei, being unstable do not occur naturally.

- All nuclides with same mass number $A$ are called **isobars**. e.g. $^3_1$H and $^3_2$He.
- Nuclides with same neutron number $N$ but different atomic number $Z$ are called **isotones**. e.g. $^{198}_{80}$Hg and $^{197}_{79}$Au.

## Size of Nucleus

- The radius of a nucleus depends only on its mass number $A$, according to the relation $R = R_0\, A^{1/3}$

  where, $R_0$ is a constant having a value $1.2 \times 10^{-15}$ m or 1.2 fm.

- Density of nucleus is a constant, independent of $A$, for all nuclei, according to the relation, $\rho = \dfrac{3m}{4\pi R_0^3}$.

  Its value is approximately $2.3 \times 10^{17}$ kg m$^{-3}$.

## Mass Energy

- Einstein showed that mass is another form of energy and can convert mass-energy into other forms of energy, says kinetic energy and *vice-versa*. The **mass-energy equivalence relation** is

  $$E = mc^2$$

- The conservation law of energy states that the initial and the final energies are equal, provided the energy associated with mass is also included.

- Difference between the sum of the masses of the nucleons constituting a nucleus and the rest mass of nucleus is called **mass defect** ($\Delta m$). It is expressed as

$$\Delta m = Zm_{\text{proton}} + (A - Z)m_{\text{neutrons}} - m_{\text{nucleus}}$$

## Nuclear Binding Energy

- If a certain number of neutrons and protons are brought together to form a nucleus of a certain charge and mass, an energy $E_b$ will be released. This energy $E_b$ is called the **binding energy of the nucleus.**
- Thus, the binding energy of a nucleus may be defined as the energy equivalent to the mass defect of the nucleus.

  If $\Delta m$ is the mass defect, then according to Einstein's mass-energy relation,

$$E_b = \Delta mc^2$$

- **Binding energy per nucleon** $E_{bn}$ is the ratio of the binding energy $E_b$ of a nucleus to the number of the nucleons, $A$ in that nucleus.

$$E_{bn} = \frac{E_b}{A}$$

**Note** $E_{bn}$ can also be defined as the average energy per nucleon needed to separate a nucleus into its individual nucleons.

### Binding Energy Curve

- A graph of the binding energy per nucleon $E_{bn}$ *versus* the mass number $A$ for a large number of nuclei is as shown below

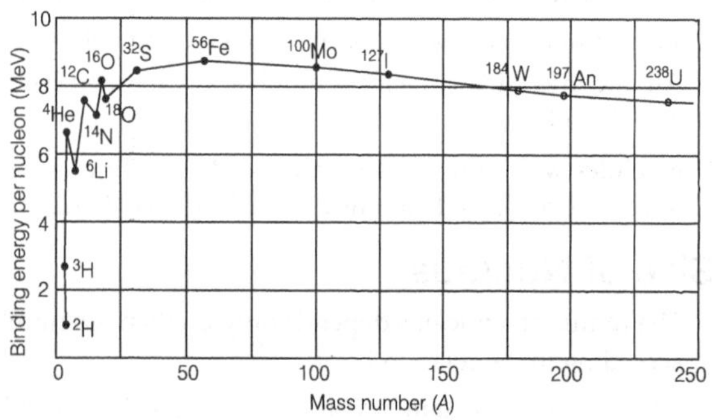


Main features of the graph are as follows

(i) The binding energy per nucleon $E_{bn}$ is practically constant, i.e. independent of $A$ for nuclei of middle mass number ($30<A<170$).

The curve has maximum 8.75 MeV for $A = 56$.

(ii) $E_{bn}$ is lower for both light nuclei ($A < 30$) and heavy nuclei ($A > 170$).

- Conclusions from the observations of $E_{bn}$ *versus* $A$ graph are as follows

(i) Force is attractive and sufficiently strong to produce $E_{bn}$ of a few MeV per nucleon.

(ii) The constancy of $E_{bn}$ in the range $30 < A < 170$ is the consequence of the fact that the nuclear force is short-ranged.

(iii) A very heavy nucleus (say $A = 240$) can be split into two lighter nuclei (say $A = 120$) near the flat maximum of the curve, such that $E_{bn}$ will increase. Hence, energy will be released in this process of fission.

(iv) Two very light nuclei ($A \leq 10$) joined together to form a heavier nucleus. $E_{bn}$ of the fused nuclei is more than that of lighter nuclei. Again, energy would be released in such as process of fusion.

## Nuclear Force

- The forces that holds the nucleons together inside the nucleus of an atom are called nuclear forces.
- Following are few important properties of the nuclear forces

(i) It is much stronger than the Coulomb force acting between the charges or the gravitational force between masses.

(ii) Nuclear force between two nucleons falls rapidly to zero as their distance is more than few femtometres (fm). This leads to saturation of forces or a large sized nucleus, which is the reason for constancy of binding energy per nucleon $E_{bn}$.

(iii) Nuclear force between neutron-neutron, proton-neutron and proton-proton is approximately the same. It does not depend on the electric charge.

## Radioactivity

- It is a nuclear phenomenon in which an unstable nucleus undergoes a decay with the emission of some particles ($\alpha$, $\beta$) and electromagnetic radiation ($\gamma$-rays). This is also referred as **radioactive decay.**
- Three types of radioactive decay occur in nature which are as follows

### $\alpha$-Decay

- It is the phenomenon of emission of nucleus of helium ($_2He^4$) from a radioactive nucleus.
- In this decay, the mass number of the product nucleus is four less than that of decaying nucleus, while the atomic number decreases by two, i.e.

$$^{A}_{Z}X \longrightarrow ^{A-4}_{Z-2}Y + {}_2He^4$$

- The disintegration energy $Q$ of a nuclear reaction is the difference between the initial mass energy and the total mass energy of the products.

$$Q = (m_X - m_Y - m_{He})c^2$$

where, $Q > 0$ for exothermic processes such as $\alpha$-decay.

### $\beta$-Decay

- It is a spontaneous emission of an electron ($\beta^-$-decay) or positron ($\beta^+$-decay).

KEY NOTES

- The emission of electron ($\beta^-$-decay) is accompained by the emission of anti-neutrino $\bar{\nu}$ and emission of positron ($\beta^+$-decay) is accompained by neutrino $\nu$.
- In both, the mass number $A$ remains unchanged.
- In $\beta^-$ decay $Z$ of the nucleus goes up by 1 while in $\beta^+$-decay $Z$ goes down by 1.

  Basic nuclear reactions for

  (i) $\beta^-$ decay, $n \longrightarrow p + e^- + \bar{\nu}$

  where, $\bar{\nu}$ is anti-neutrino.

  (ii) $\beta^+$ decay, $p \rightarrow n + e^+ + \nu$

  where, $\nu$ is neutrino.
- Neutrinos are neutral particles with mass comparable to $e^-$. They have only weak interaction with matter and difficult to detect.

**Note** Decay of proton to neutron is possible only inside the nucleus, since proton has smaller mass than neutron.

### $\gamma$-Decay

- When a nucleus in an excited state spontaneously decay to its ground state, a photon is emitted with energy equal to the difference in the two energy levels of the nucleus. This is called $\gamma$-decay.

  The energy (MeV) corresponds to radiation of extremely short wavelength shorter than the hard $X$-ray region.

**Note** A $\gamma$-ray is emitted when a $\alpha$ or $\beta$ decay results in a daughter nucleus in an excited state. This then return to the ground state by a single photon transition or successive transitions involving more than one photon.

## Law of Radioactive Decay

- According to this law, the rate of decay of radioactive nuclei at any instant is proportional to the number of nuclei present at that instant.

$$\frac{dN}{dt} \propto N \text{ or } \frac{dN}{dt} = -\lambda N$$

  where, $\lambda$ = decay constant and $N$ = number of active nucleus present in the sample at any instant $t$.

  Also, $N = N_0 e^{-\lambda t}$, where $N_0$ = number of nuclei present in the sample initially.
- The **total decay rate or activity** $R$ of a sample is the number of nuclei disintegrating per unit time. It is given as

$$R = \frac{-dN}{dt} = R_0 e^{-\lambda t}, \; R = \lambda N$$

- The SI unit of activity is **Becquerel (Bq).**

$$1 \text{ Becquerel (Bq)} = 1 \text{ disintegration/second}$$

$$1 \text{ Curie (Ci)} = 3.7 \times 10^{10} \text{ decays/second}$$

- **Half-life period** of a radionuclide is the time it takes for a sample that has initially, say $N_0$ radionuclei to reduce to $N_0/2$. It is given as, $T_{1/2} = \frac{\ln 2}{\lambda} = \frac{0.693}{\lambda}$
- After $n$ half-lives, the quantity of a radionuclei left intact (undecayed) is given by

$$N = N_0 \left(\frac{1}{2}\right)^n = N_0 \left(\frac{1}{2}\right)^{\frac{t}{T_{1/2}}}$$

- **Average life** or **mean life** is the average of all the lives of all the nuclei in a radioactive substance. It is given as

$$\tau = \frac{1}{\lambda}$$

- The **relation between $T_{1/2}$, $\lambda$ and $\tau$** is given as

$$T_{1/2} = \frac{\ln 2}{\lambda} = \tau \ln 2$$

## Nuclear Energy

- It is the energy released during the transformation of nuclei with less total binding energy to nuclei with greater binding energy.
- Two distinct ways of obtaining energy from nucleus are as follows

  (i) Nuclear fission (ii) Nuclear fusion

### Nuclear Fission

- It is the process of splitting of a heavy nucleus ( $_{92}U^{235}$ or $_{94}U^{239}$) into two lighter nuclei of comparable masses alongwith the release of a large amount of energy after bombarded by slow neutrons.
- When a uranium isotope $^{235}_{92}U$ bombarded with a neutron breakes into two intermediate mass nuclear fragments.

$$^1_0n + {}^{235}_{92}U \longrightarrow {}^{236}_{92}U \longrightarrow {}^{144}_{51}Ba + {}^{89}_{36}Kr + 3^1_0n + Q \text{ (energy released)}$$

- The disintegration energy in fission events first appears as the kinetic energy of the fragments and neutrons. Eventually it is transferred to the surrounding matter appearing as heat.
- The source of energy in nuclear reactor, which produces electricity, is **nuclear fission**.
- The enormous energy released in an atom bomb comes from uncontrolled nuclear fission.
- **Nuclear Reactor** It is a device which uses nuclear fission in sustained and controlled manner to generate energy.

  Following are the important terms related to reactor

  (i) **Core** is the site of nuclear fission. It is usually consists of $^{235}_{92}U$.

  (ii) Core contains a **moderator** which is a light nuclei used to slow down fast moving neutrons, e.g. water, heavy water ($D_2O$) and graphite.

  (iii) **Reflector** is used to reduce leakage.

  (iv) The reactor can be shut down by means of **control rods** (made of cadmium) that have high absorption rate of neutrons.

KEY NOTES

(v) The **coolant** transfers heat to a working fluid which in turn may produce steam. The steam drives turbines and generates electricity.

- **Multiplication factor** $k$ is the ratio of the number of fission produced by a given generation of neutrons to the number of fission of the proceeding generation.

  Following are the important points related to $k$
  
  (i) It is basically the measure of the rate of neutrons in the reactor.
  
  (ii) For $k = 1$, the reactor is said to be **critical**, which is the condition for steady power operation.
  
  (iii) For $k > 1$, the reaction rate and power increases exponentially. The reactor becomes **supercritical** and can even explode.
  
  (iv) $k$ can be brought down with the help of moderator. Also, it can be reduced rapidly to less than 1 with the use control and safety rods.

### Nuclear Fusion : Energy Generation in Stars

- It is a phenomenon in which two light nuclei fuse to form a larger nucleus and energy is released. e.g.

$$ {}_1^1\text{H} + {}_1^1\text{H} \longrightarrow {}_1^2\text{H} + e^+ + \nu + 0.42\,\text{MeV} $$

- In order to fuse two nuclei, they must come close enough, so that attractive short range nuclear force is able to affect them. For this, they must overcome coulomb repulsion of positive charges which could only be achieved at high temperature.
- When fusion is achieved by raising the temperature of the system, so that particles have enough kinetic energy to overcome the coulomb repulsive behaviour due to which it is also called **thermonuclear fusion**.

  It is the source of energy output in the interior of stars.
- In **controlled fusion reactors**, steady power is generated by heating the nuclei fuel to a temperature in the range of $10^8$ K. At these temperature, the fuel is a mixture of positive ions and electrons (plasma).
- Uncontrolled release of large nuclear energy is called an **atomic expression**. High temperature conditions for fusion reactions can be created by exploding a **fission bomb**.

# Mastering NCERT

## MULTIPLE CHOICE QUESTIONS

## TOPIC 1 ~ *Composition of Nucleus*

**1** Isotopes of an atom

(a) have different electronic structure
(b) does not occupy same place in periodic table
(c) contains same number of protons
(d) have same number of neutrons

**2** Chlorine has two isotopes having masses 34.98 u and 36.98 u with relative abundance of 75.4% and 24.6 %, respectively. The average atomic mass of chlorine is

(a) 34.98 u (b) 36.98 u (c) 35.47 u (d) 35 u

**3** Masses of the nuclei of hydrogen, deuterium and tritium are in ratio

(a) 1 : 2 : 3 (b) 1 : 1 : 1 (c) 1 : 1 : 2 (d) 1 : 2 : 4

**4** The ratio of the nuclear radii of the gold isotope ${}_{79}^{197}\text{Au}$ and silver isotope ${}_{47}^{107}\text{Ag}$, is

(a) 1.23 (b) 0.216 (c) 2.13 (d) 3.46

**5** If the nuclear radius of ${}^{27}\text{Al}$ is 3.6 fm, the approximate nuclear radius of ${}^{64}\text{Cu}$ in fm is **CBSE AIPMT 2012**

(a) 2.4 (b) 1.2
(c) 4.8 (d) 3.6

**6** Surface area of a nucleus (assuming it to be a perfect sphere), is (where, $A$ = mass number)

(a) $(1.8 \times 10^{-29})\, A^{1/3}$ (b) $(1.8 \times 10^{-29})\, A^2$
(c) $(1.8 \times 10^{-29})\, A^{2/3}$ (d) $(1.8 \times 10^{-29})\, A^3$

**7** Density of a nucleus is

(a) more for lighter elements and less for heavier elements
(b) more for heavier elements and less for lighter elements
(c) very less compared to ordinary matter
(d) a constant

**8** Mass of iron nucleus is 55.85 u and $A = 56$. Its nuclear density will be (take, $R_0 = 1.2 \times 10^{-15}$ m)

(a) $4.28 \times 10^{17}$ kg m$^{-3}$ (b) $2 \times 10^{10}$ kg m$^{-3}$
(c) $2.29 \times 10^{17}$ kg m$^{-3}$ (d) $9 \times 10^{10}$ kg m$^{-3}$

**9** The ratio of mass densities of nuclei of ${}^{40}\text{Ca}$ and ${}^{16}\text{O}$ is close to **JEE Main 2019**

(a) 5 (b) 2
(c) 0.1 (d) 1

## TOPIC 2 ~ Mass-Energy and Nuclear Binding Energy

**10** Energy equivalent of 2 g of a substance is
(a) $18\times10^{13}$ mJ (b) $18\times10^{13}$ J
(c) $9\times10^{13}$ mJ (d) $9\times10^{13}$ J

**11** How much mass has to be converted into energy to produce electric power of 200 MW for one hour?
(a) $2\times10^{-6}$ kg (b) $8\times10^{-6}$ kg
(c) $1\times10^{-6}$ kg (d) $3\times10^{-6}$ kg

**12** Mass defect of helium ($_2He^4$) is [use, mass of proton, $m_p=1.0072676$ u, mass of neutrons, $m_n=1.008665$ u and mass of $_2He^4=4.001506$ u]
(a) 0.016767 u
(b) 1.00726 u
(c) 2.00686 u
(d) 0.0303592 u

**13** Given, mass of a neutron = 1.00866 u, mass of a proton = 1.00727 u and mass of $^{16}_{8}O$ = 15.99053 u. Then, the energy required to separate $^{16}_{8}O$ into its constituents is
(a) 12.7 MeV (b) 931.5 MeV
(c) $1.49\times10^{-10}$ J (d) 127.5 MeV

**14** Given, $m(^{56}_{26}Fe)=55.934939$ u, $m_{proton}=1.007825$ u and $m_{neutron}=1.008665$ u. Then, BE per nucleon of Fe is,
(a) 8.790 MeV
(b) 7.75 MeV
(c) 7.5 MeV
(d) 9.28 MeV

**15** If the binding energy per nucleon of deuteron is 2.115 MeV, then its mass defect in atomic mass unit is
(a) 0.0024 u (b) 25.62 u
(c) 96.264 u (d) 0.0045 u

**16** The curve of binding energy per nucleon as function of atomic mass number has a sharp peak for lithium nucleus. This implies that lithium
(a) can be used as fissionable material
(b) is highly radioactive
(c) can be easily broken up
(d) is very stable

**17** Consider the binding energy of $_{17}Cl^{35}$ and $_{15}P^{31}$ are 287.67 MeV and 262.48 MeV, respectively. Then,
(a) $_{17}Cl^{35}$ is more stable than $_{15}P^{31}$
(b) $_{15}P^{31}$ is more stable than $_{17}Cl^{35}$
(c) Stability of both the elements is equivalent
(d) cannot be estimated from the given data

**18** If a nucleon can have a maximum of $p$-neighbours within the range of nuclear force, then on increasing $A$ by adding nucleons, the binding energy per nucleon
(a) increases abruptly
(b) decreases gradually
(c) remains constant
(d) first increases sharply and then decreases gradually

**19** Which amongst the following is a correct graph of potential energy $U$ of a pair of nucleons as a function of their separation $r$?

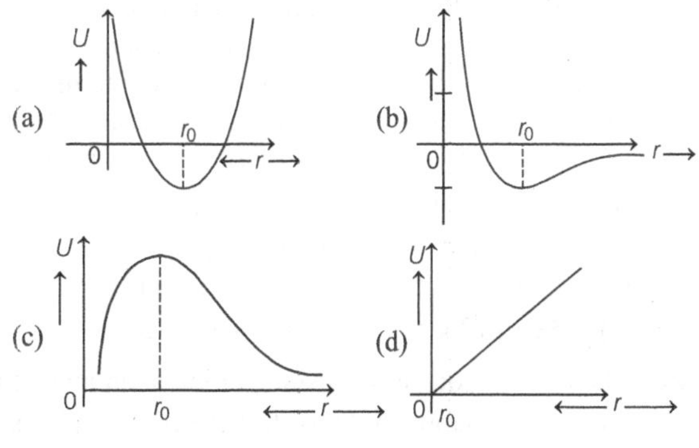


## TOPIC 3 ~ Radioactivity

**20** A nuclear phenomenon in which an unstable nucleus undergoes a decay is called
(a) Rutherford's scattering
(b) radioactive decay
(c) Geiger-Marsden scattering
(d) Nuclear energy

**21** Radioactive material $A$ has decay constant $8\lambda$ and material $B$ has decay constant $\lambda$. Initially, they have same number of nuclei. After what time, the ratio of number of nuclei of material $B$ to that $A$ will be $\frac{1}{e}$? **NEET 2017, JEE Main 2019**
(a) $\frac{1}{\lambda}$ (b) $\frac{1}{7\lambda}$ (c) $\frac{1}{8\lambda}$ (d) $\frac{1}{9\lambda}$

**22** Samples of two radioactive nuclides $A$ and $B$ are taken. $\lambda_A$ and $\lambda_B$ are the disintegration constants of sample $A$ and $B$, respectively. In which of the following cases, the two samples can simultaneously have the same decay rate at any time?
(a) Initial rate of decay of $A$ is twice the initial rate of decay of $B$ and $\lambda_A = \lambda_B$
(b) Initial rate of decay of $A$ is twice the initial rate of decay of $B$ and $\lambda_A > \lambda_B$
(c) Initial rate of decay of $B$ is twice the initial rate of decay of $A$ and $\lambda_A > \lambda_B$
(d) $\frac{\lambda_A}{\lambda_B} = \frac{N_B}{N_A}$

**23** The plot of the number $N$ of decayed nuclei *versus* activity $R$ of a radioactive sample is correctly depicted in graph

(a)
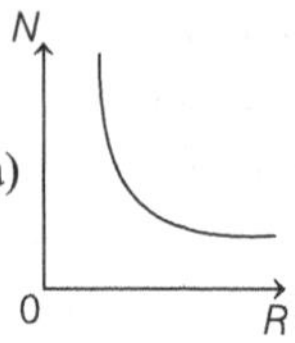


(b)
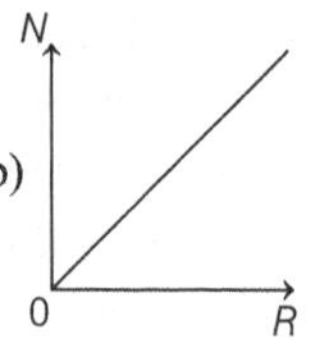


(c)
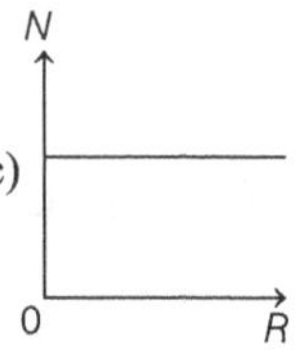


(d)
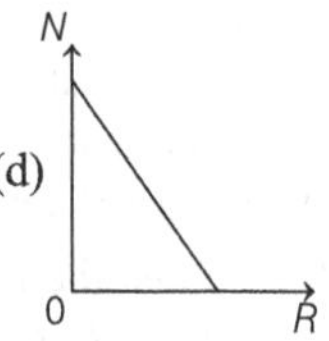


**24** A mixture consists of two radioactive materials $A_1$ and $A_2$ with half-lives of 20 s and 10 s, respectively. Initially, the mixture has 40 g of $A_1$ and 160 g of $A_2$. The amount of the two in the mixture will become equal after **CBSE AIPMT 2012**
(a) 60 s (b) 80 s (c) 20 s (d) 40 s

**25** A sample of $^{67}$Ga with a half-life of 78 h has a mass of 3.4 g. Initial decay rate of sample is
(a) $8.00 \times 10^{16}\ s^{-1}$ (b) $6.27 \times 10^{16}\ s^{-1}$
(c) $7.53 \times 10^{16}\ s^{-1}$ (d) $8.53 \times 10^{15}\ s^{-1}$

**26** The rate of radioactive disintegration at an instant for a radioactive sample of half-life $2.2 \times 10^9$ s is $10^{10}\ s^{-1}$. The number of radioactive atoms in that sample at that instant is **NEET (Odisha) 2019**
(a) $3.17 \times 10^{20}$ (b) $3.17 \times 10^{17}$ (c) $3.17 \times 10^{18}$ (d) $3.17 \times 10^{19}$

**27** A sample of radioactive material $A$, that has an activity of 10 mCi (1 Ci $= 3.7 \times 10^{10}$ decays/s) has twice the number of nuclei as another sample of a different radioactive material $B$ which has an activity of 20 mCi. The correct choices for half-lives of $A$ and $B$ would, then be respectively **JEE Main 2019**
(a) 20 days and 10 days (b) 5 days and 10 days
(c) 10 days and 40 days (d) 20 days and 5 days

**28** If half-life of an element is 69.3 h, then how much of its per cent will decay in 10th to 11th hours? Initial activity $= 50\ \mu$Ci. **AIIMS 2019**
(a) 1% (b) 2% (c) 3% (d) 4%

**29** A radio isotope $X$ with a half-life $1.4 \times 10^9$ yr decays to $Y$ which is stable. A sample of the rock from a cave was found to contain $X$ and $Y$ in the ratio 1 : 7. The age of the rock is **CBSE AIPMT 2014**
(a) $1.96 \times 10^9$ yr (b) $3.92 \times 10^9$ yr
(c) $4.20 \times 10^9$ yr (d) $8.40 \times 10^9$ yr

**30** At a given instant, there is 25% undecayed radioactive nuclei in a sample. After 10s, the number of undecayed nuclei reduces to 12.5%. The mean life of the nuclei is
(a) 9.2836 s (b) 14.428 s
(c) 12 s (d) 2.61 s

**31** Tritium having a half-life of 12.5 yr, undergo $\beta$-decay. Fraction of sample remaining undecayed after 25 yr will be
(a) $\frac{1}{8}$ (b) $\frac{1}{2}$ (c) $\frac{1}{4}$ (d) $\frac{1}{16}$

**32** The activity of a radioactive sample falls from $700 s^{-1}$ to $500 s^{-1}$ in 30 min. Its half-life is close to **JEE Main 2020**
(a) 62 min (b) 66 min (c) 72 min (d) 52 min

**33** The counting rate observed from a radioactive source at $t = 0$ s, was 1600 count/s and at $t = 8$ s, it was 100 count/s. The counting rate observed at $t = 6$ s was **JEE Main 2019**
(a) 400 (b) 300 (c) 200 (d) 150

**34** For a radioactive material, half-life is 10 min. If initially, there are 600 number of nuclei, the time taken (in min) for the disintegration of 450 nuclei is **NEET 2018**
(a) 30 (b) 10 (c) 20 (d) 15

**35** A radioactive nucleus $A$ with a half-life $T$, decays into a nucleus $B$. At $t = 0$, there is no nucleus $B$. After sometime $t$, the ratio of the number of $B$ to that of $A$ is 0.3. Then, $t$ is given by **JEE Main 2017**
(a) $t = T \frac{\log 1.3}{\log_e 2}$ (b) $t = T \log 1.3$
(c) $t = \frac{T}{\log 1.3}$ (d) $t = \frac{T \log_e 2}{2 \log 1.3}$

**36** Ba-122 has half-life of 2 min. Experiment is done using Ba-122 and it takes 10 min to set-up the experiment. If initially 80 g of Ba-122 was taken, then how much Ba was left, when experiment was started? **JIPMER 2019**
(a) 2.5 g (b) 5 g (c) 10 g (d) 20 g

**37** Out of the following radioactive substances, which are not found naturally on earth?

(i) Tritium (ii) Deuterium
(iii) Uranium (iv) Plutonium

(a) (i) and (ii) (b) (ii) and (iii)
(c) (i) and (iv) (d) (i) and (iii)

**38** $\alpha$-particle consists of ***NEET (National) 2019***

(a) 2 electrons, 2 protons and 2 neutrons
(b) 2 electrons and 4 protons
(c) 2 protons only
(d) 2 protons and 2 neutrons

**39** In the $\alpha$-decay of ${}^{238}_{92}\text{U} \longrightarrow X + {}^{4}_{2}\text{He}$

The nucleus $X$ is

(a) ${}^{234}_{90}\text{Th}$ (b) ${}^{235}_{90}\text{U}$
(c) ${}^{237}_{91}\text{Pa}$ (d) ${}^{140}_{54}\text{Xe}$

**40** Given, atomic masses are ${}^{238}_{92}\text{U} = 238.05079\text{u}$, ${}^{4}_{2}\text{He} = 4.00260\text{u}$ and ${}^{234}_{90}\text{Th} = 234.04363\text{u}$. The energy released during the $\alpha$-decay of ${}^{238}_{92}\text{U}$ is ***JEE Main 2014***

(a) 4.25 MeV (b) 4.5 MeV
(c) 6 MeV (d) 5 MeV

**41** The binding energy per nucleon of ${}^{7}_{3}\text{Li}$ and ${}^{4}_{2}\text{He}$ nuclei are 5.60 MeV and 7.06 MeV, respectively. In the nuclear reaction ${}^{7}_{3}\text{Li} + {}^{1}_{1}\text{H} \rightarrow {}^{4}_{2}\text{He} + {}^{4}_{2}\text{He} + Q$, the value of energy $Q$ released is ***CBSE AIPMT 2014***

(a) 19.6 MeV (b) −2.4 MeV
(c) 8.4 MeV (d) 17.3 MeV

**42** Ratio of charge on positron to mass of positron is approximately ***JIPMER 2019***

(a) $+2\times10^{11}$ C/kg (b) $+5\times10^{12}$ C/kg
(c) $-2\times10^{11}$ C/kg (d) $-5\times10^{11}$ C/kg

**43** For the radioactive nuclei that undergo either $\alpha$ or $\beta$ decay, which one of the following cannot occur?

(a) Isobar of original nucleus
(b) Isotopes of original nucleus
(c) Nuclei with higher atomic number than that of the original nucleus.
(d) Nuclei with lower atomic number than that of original nucleus.

**44** When a nucleus is in an excited state,

(a) it can stay in excited state
(b) it gives excess energy to surrounding electrons and comes to a lower energy state
(c) it can make a transition to a lower energy state by emission of a photon
(d) it can emit a proton or neutron with excess kinetic energy and so achieves a lower energy state

**45** In which radiation, atomic number and mass number are not affected? ***JIPMER 2019***

(a) $\alpha$-radiation (b) $\gamma$-radiation
(c) $\beta$- radiation (d) None of these

**46** Consider the following reaction

$${}_{Z}X^{A} \rightarrow {}_{Z+1}Y^{A} \rightarrow {}_{Z-1}K^{A-4} \rightarrow {}_{Z-1}K^{A-4}$$

Radioactive particles emitted is/are

(a) $\alpha$ and $\beta$ only (b) $\alpha$ and $\gamma$ only
(c) $\alpha, \beta$ and $\gamma$ (d) Only $\alpha$ and $\beta$

# TOPIC 4 ~ *Nuclear Energy*

**47** If the energy released in a nuclear reaction is $E_N$ and energy released from a chemical source is $E_C$, then

(a) $E_N \simeq E_C$ (b) $E_N > E_C$
(c) $E_N < E_C$ (d) $E_N = E_C$

**48** If a nucleus with mass number $A = 240$ with $E_{\text{bn}} = 7.6$ MeV breaks into two fragments of $A = 120$ and $E_{\text{bn}} = 8.5$ MeV, then released energy is around

(a) 216 MeV
(b) 200 MeV
(c) 100 MeV
(d) Cannot be estimated from given data

**49** Pick out the unmatched pair for a nuclear reactor.

(a) Moderator $\rightarrow$ Heavy water
(b) Nuclear fuel $\rightarrow {}_{92}\text{U}^{235}$
(c) Control rods $\rightarrow$ Carbon
(d) At critical condition, $k \rightarrow 1$

**50** Schematic diagram of a nuclear reactor based on thermal neutron fission is as shown below

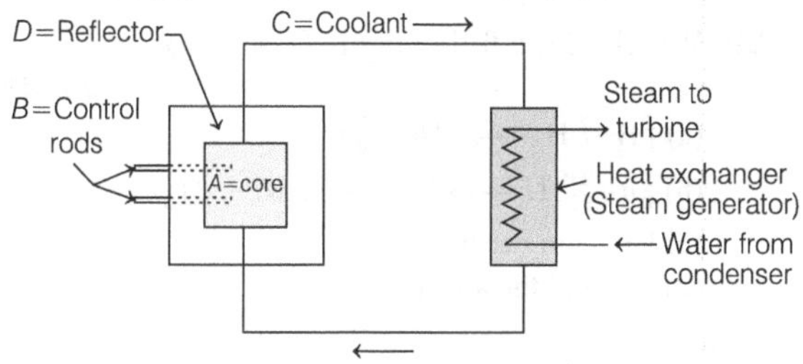


Here, the correct purpose of the parts specified by $A$, $B$, $C$ and $D$ is given in

(a) $A \rightarrow$ reduces leakage.
(b) $B \rightarrow$ can shut down the reactor
(c) $C \rightarrow$ site of nuclear fission and contains ${}^{235}_{92}\text{U}$
(d) $D \rightarrow$ transfers heat to a working fluid which in turn may produce steam

**51** On bombarding $U^{235}$ by slow neutron, 200 MeV energy is released. If the power output of atomic reactor is 1.6 MW, then the rate of fission will be

(a) $5\times 10^{22}\ s^{-1}$ (b) $5\times 10^{16}\ s^{-1}$
(c) $8\times 10^{16}\ s^{-1}$ (d) $20\times 10^{16}\ s^{-1}$

**52** In a nuclear reactor, the fuel is consumed at the rate of $1\ mgs^{-1}$. The power generated (in kW) is

(a) $9\times 10^{14}$ (b) $9\times 10^{7}$
(c) $9\times 10^{8}$ (d) $9\times 10^{12}$

**53** An atomic power nuclear reactor can deliver 300 MW. The energy released due to fission of each nucleus of uranium atoms $U^{238}$ is 170 MeV. The number of uranium atoms fissioned per hour will be **JIPMER 2018**

(a) $30\times 10^{25}$ (b) $4\times 10^{22}$
(c) $10\times 10^{20}$ (d) $5\times 10^{15}$

**54** Suppose that a reactor using U-235 has an output of 700 MW and is 20% efficient. An atom of $U^{235}$ undergoes fission in a reactor liberating 200 MeV energy.

Then, energy generated from the reactor per fission and mass of uranium it consume in 24 h are

(a) $3.2\times 10^{-11}$ J, 3.7 kg
(b) $6.4\times 10^{-11}$ J, 2.5 kg
(c) $3.2\times 10^{-12}$ J, 2.5 kg
(d) $6.4\times 10^{-12}$ J, 3.7 kg

**55** Which of the following is/are fission reaction(s)?

I. ${}_0^1n+{}_{92}^{235}U\rightarrow{}_{92}^{236}U\rightarrow{}_{56}^{144}Ba+{}_{36}^{89}Kr+3{}_0^1n$
II. ${}_0^1n+{}_{92}^{235}U\rightarrow{}_{92}^{236}U\rightarrow{}_{51}^{133}Sb+{}_{41}^{99}Nb+4{}_0^1n$
III. ${}_0^1n+{}_{92}^{235}U\rightarrow{}_{54}^{1.40}Xe+{}_{38}^{94}Sr+2{}_0^1n$
IV. ${}_1^2H+{}_1^2H\rightarrow{}_2^3He+{}_0^1n$

(a) I, II and IV (b) Both III and IV
(c) II, III and IV (d) I, II and III

**56** Which of the following is fusion process? **JIPMER 2018**

(a) ${}_1^2H+{}_1^2H\longrightarrow{}_2^4He+{}_0^1n$
(b) ${}_0^1n+{}_{92}^{235}U\longrightarrow{}_{56}^{141}Ba+{}_{36}^{92}Kr+3({}_0^1n)$
(c) Uranium decay
(d) None of the above

**57** Fusion reaction takes place at high temperature as

(a) molecules break at higher temperature
(b) nuclei break at high temperature
(c) atoms get ionised at high temperature
(d) kinetic energy is enough to overcome the coulomb repulsive force between nuclei

**58** A nucleus of uranium decays at rest into nuclei of thorium and helium, then **CBSE AIPMT 2015**

(a) the helium nucleus has more kinetic energy than the thorium nucleus
(b) the helium nucleus has less momentum than the thorium nucleus
(c) the helium nucleus has more momentum than the thorium nucleus
(d) the helium nucleus has less kinetic energy than the thorium nucleus

**59** Fusion reaction in the sun involves the following reactions.

$${}_1^1H+{}_1^1H\longrightarrow{}_1^2H+e^{+}+\nu+0.42\text{ MeV} \quad (A)$$
$$e^{+}+e^{-}\longrightarrow\gamma+\gamma+1.02\text{ MeV} \quad (B)$$
$${}_1^2H+{}_1^1H\longrightarrow{}_2^3He+\gamma+5.49\text{ MeV} \quad (C)$$
$${}_2^3He+{}_2^3He\longrightarrow{}_2^4He+{}_1^1H+{}_1^1H+12.86\text{ MeV} \quad (D)$$

For the fourth reaction (D) to occur, the combination of the above three reactions will be

(a) A + B + C (b) 2A + 2B + C
(c) A + 2B + C (d) 2A + 2B + 2C

**60** Nuclear fusion is common to the pair

I. uranium based reactor
II. hydrogen bomb
III. energy production in sun
IV. atom bomb

(a) Both I and II (b) Both II and III
(c) Both III and IV (d) Both II and IV

**61** In the fusion reaction,

$${}_1^2H+{}_1^2H\longrightarrow{}_2^3He+{}_0^1n$$

the masses of deuteron, helium and neutron expressed in amu are 2.015, 3.017 and 1.009, respectively. If 1 kg of deuterium undergoes complete fusion, find the amount of total energy released. (1 amu = 931.5 MeV.) **JIPMER 2018**

(a) $9.0\times 10^{13}$ J (b) $20\times 10^{5}$ J
(c) $5\times 10^{16}$ J (d) $8\times 10^{5}$ J

# SPECIAL TYPES QUESTIONS

## I. Assertion and Reason

**Direction** (Q. Nos. 62-73) *In the following questions, a statement of Assertion is followed by a corresponding statement of Reason. Of the following statements, choose the correct one.*

(a) Both Assertion and Reason are correct and Reason is the correct explanation of Assertion.
(b) Both Assertion and Reason are correct but Reason is not the correct explanation of Assertion.
(c) Assertion is correct but Reason is incorrect.
(d) Assertion is incorrect but Reason is correct.

**62** **Assertion** Kilogram is preferred over u in order to measure mass of an atom.

**Reason** Atomic mass unit is equal to $\frac{1}{12}$th of the mass of the carbon ($^{12}C$) atom.

**63** **Assertion** Ratio of mass of nucleus of H-atom and proton is approximately 1.

**Reason** Nucleus of the lightest atom of hydrogen is usually known as proton.

**64** **Assertion** Initially, it was believed that mass and energy are conserved separately in a nuclear reaction.

**Reason** According to Einstein, one can convert mass-energy into other forms of energy.

**65** **Assertion** Nuclear force between neutron-neutron, proton-neutron and proton-proton is approximately the same.

**Reason** The nuclear force does not depend on the electric charge.

**66** **Assertion** Binding energy per nucleon is practically constant for middle mass numbers ($30 < A < 170$).

**Reason** Nuclear force is short ranged in nature.

**67** **Assertion** For an element generally $N \geq Z$ ($N$ = number of neutrons, $Z$ = atomic number).

**Reason** Neutrons do not always experience attractive nuclear force. ***AIIMS 2019***

**68** **Assertion** In both, radioactivity and photoelectric effects, electrons may be ejected.

**Reason** Both photoelectric effect and radioactivity emission occurs only because of unstable elements. ***AIIMS 2019***

**69** **Assertion** Radioactive nuclei emits $\beta^-$-particles.

**Reason** Electrons exist inside the nucleus. ***AIIMS 2018***

**70** **Assertion** The detection of neutrino particles is extremely difficult.

**Reason** Neutrinos interact very weakly with matter.

**71** **Assertion** Heavy water is used to slow down fast moving neutrons in nuclear reactor.

**Reason** It does not react with slow neutron. ***AIIMS 2019***

**72** **Assertion** Thermal neutrons are much likely to cause fission of $^{235}_{92}U$ than fast neutrons.

**Reason** Fast neutrons can escape the nuclear reactor.

**73** **Assertion** Thermonuclear fusion reaction is not possible on earth naturally.

**Reason** For thermonuclear fusion reaction to take place, extreme condition of temperature and pressure are required.

## II. Statement Based Questions

**74** Study the following statements.

I. Atomic masses of various elements expressed in atomic mass unit are close to being integral multiple of the mass of H-atom.
II. Mass spectrometer is used to measure accurate atomic masses.
III Tritium nuclei is highly stable and hence occur naturally.

Choose the correct statement.

(a) Both I and III (b) Both II and III
(c) Both I and II (d) Only I

**75** Study the following statements regarding neutrons.

I. It is stable like a proton.
II. It decays into a proton and electron.
III. It has a mean life of 1000s.

Choose the correct statement.

(a) Only I
(b) Both II and III
(c) Both I and II
(d) Only III

**76** Study the following statements.

I. In α-scattering experiment, if the α-particles of higher energies than 5.5 MeV is used, then scattering will solely be due to coulombic repulsive force.

II. Nucleon in an atom refers to its mass number.

III. Neutron is stable inside the nucleus.

Choose the correct statement.

(a) Only I
(b) Both II and III
(c) Only III
(d) Both I and II

**77** For binding energy per nucleon *versus* mass number curve, which of the following statements are correct?

I. Binding energy per nucleon $E_{bn}$ is independent of mass number in range $30 < A < 170$.

II. Binding energy is lower for both light nuclei ($A < 30$) and heavy nuclei ($A > 170$).

III. Binding energy is maximum of about 8.75 MeV for $A = 56$.

IV. In region $0 < A < 80$, binding energy increases with mass number.

(a) I, II, III and IV
(b) I, II and IV
(c) II, III and IV
(d) I, II and III

**78** In which of the following situations, the binding energy of the formed nucleus is more than the initial nucleus and are tightly bound?

I. When a very heavy nucleus, $A = 240$ is broken into two $A = 120$ nuclei.

II. When two very light nuclei ($A \leq 10$) joined to form a heavier nucleus.

(a) Only I
(b) Only II
(c) Both I and II
(d) Neither I nor II

**79** I. $\Delta N$ is always negative and $dN$ is positive for any radioactive sample.

II. Spontaneous decay like α-decay are only possible when the total mass of the decay products is less than the mass of initial nucleus.

III. If the initial nucleus is at rest, then $Q$-value is the potential energy of the products.

Choose the incorrect statement from the following.

(a) Both I and II
(b) Only II
(c) Both II and III
(d) Both I and III

**80** Study the following statements regarding nuclear reactor.

I. Average $2\frac{1}{2}$ neutrons released per fission of $^{235}_{92}U$, leads to the possibility of chain reaction.

II. Control rods creates a situation similar to the one in which marble hits head on an identical marble at rest.

III. Apsara reactor at BAARC uses heavy water ($D_2O$) as moderator.

Choose the correct statement.

(a) Both I and II
(b) Only III
(c) Both II and III
(d) Only I

**81** A nuclear reactor in India involves the following stages for power generation given in random order.

I. Pu-239 uses fast moving neutrons for sustaining chain reactions.

II. $^{238}_{92}U$ captures neutrons and produces highly radioactive nuclei.

III. use of fast breeder reactor to produce fissile U-233 from Th-232.

The correct order of the following stages is

(a) I, II and III
(b) II, III and I
(c) II, I and III
(d) I, III and II

**82** In the interior of star, as the hydrogen in the core gets depleted and becomes helium, the core starts to cool. Then, star begins to collapse under its gravity which leads to

I. decrease in temperature of the core and star may fall

II. increase in temperature of the core and if $T > 10^8$ K, then fusion takes place of helium into carbon

Choose the correct statement.

(a) Only I
(b) Only II
(c) Both I and II
(d) Neither I nor II

**83** Choose the incorrect statement from the following.

(a) Radius of nucleus is smaller than radius of an atom by factor of about $10^4$.
(b) Ratio of mass of nucleus and the mass of an atom is approximately 1.
(c) Volume of nucleus is about $10^{-20}$ times volume of the atom.
(d) Both (a) and (c)

**84** Which of the following statement is incorrect about the nucleus?

(a) Positive charge in the nucleus is due to the protons.
(b) All the electrons of an atom are outside the nucleus.
(c) Charge of nucleus is $+Ze$.
(d) Charge of electrons may be greater than $-Ze$ in an atom.

**85** Choose the correct statement.

(a) Calculated mass of a nucleus is always less than the actual nuclear mass.
(b) In $_{88}Ra^{226}$ nucleus, there are 138 neutrons and 226 protons.
(c) Binding energy is the energy released when neutrons and protons are brought together to form nucleus.
(d) Both (a) and (c)

**86** Choose the correct statement from the following regarding nuclear force.
(a) It is either attractive or repulsive and sufficient to produce $E_{bn}$ of few eV per nucleon.
(b) Saturation property of the forces is the reason for constancy of $E_{bn}$.
(c) The order of the strength of force is $F_{gravitation} > F_{nuclear} > F_{electrostatics}$
(d) Nuclear force can be simply represented mathematically.

**87** If half-life of a radioactive substance is 1 month, then which of the following statement(s) is/are correct?
(a) (7/8)th part of the substance disintegrate in 3 months.
(b) (1/8)th part of the substance disintegrate in 4 months.
(c) The substance disintegrates completely in 4 months.
(d) The substance disintegrates completely in 2 months.

**88** Choose the correct statement from the following.
(a) Atomic energy level spacings are of the order of MeV while nuclear energy levels are of the order of eV.
(b) Decay of proton to neutron is possible only inside the nucleus.
(c) $Q > 0$ for endothermic process.
(d) Radioactive nuclides and light bulbs both follows exponential decay law.

**89** The variation of decay rate of two radioactive samples *A* and *B* with time is shown in figure.

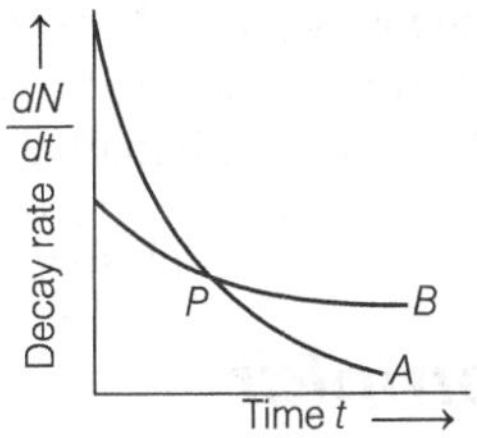


Which of the following statement(s) is/are correct?
(a) Decay rate of *A* is always greater than that of *B*, hence *A* always decays faster than *B*.
(b) Decay rate of *B* is greater than that of *A*.
(c) Decay rate of *B* is smaller than that of *A* but still its decay rate becomes equal to that of *A* at an instant.
(d) Both (a) and (c)

**90** Select the correct statement.
(a) $\beta^-$-decay is the process in which an positron is emitted from an unstable atom whose atomic number *Z* remains unchanged.
(b) $\gamma$-decay is the process in which daughter nucleus has atomic number one unit more than the parent nucleus.
(c) $\alpha$-decay is the process in which an unstable atom emits helium atom.
(d) $\beta^+$-decay is the process in which an electron emitted.

**91** Choose the correct statement from the following.
(a) Neutron-induced nuclear reaction is fusion.
(b) The disintegration energy in fission first appears as kinetic energy of fragments and neutrons.
(c) Greater the binding energy, greater is the mass of bound system.
(d) None of the above

**92** Choose the incorrect statement.
(a) In controlled nuclear reactor, steady power is generated by heating fuel to temperature range of $10^8$ K.
(b) Hydrogen bomb involves fusion of isotopes of hydrogen, deuterium and tritium.
(c) After the estimated life of sun, it's outer envelope that will convert into red giant.
(d) Research nuclear reactors can be the source of generation of electricity only.

## III. Matching Type

**93** Match the Column I (nuclei) with Column II (characteristic) and select the correct answer from the codes given below.

| Column I | | Column II |
|---|---|---|
| A. Isotopes | 1. | ${}^7_3Li, {}^7_4Be$ |
| B. Isobars | 2. | ${}^{18}_8O, {}^{19}_9F$ |
| C. Isotones | 3. | ${}^{17}_8O, {}^{17}_9F$ |
| D. Mirror nuclei | 4. | ${}^{235}_{92}U, {}^{238}_{92}U$ |

| | A | B | C | D |
|---|---|---|---|---|
| (a) | 1 | 3 | 4 | 2 |
| (b) | 4 | 1 | 3 | 2 |
| (c) | 4 | 3 | 2 | 1 |
| (d) | 3 | 4 | 2 | 1 |

**94** Match the Column I (reaction) with Column II (type of decay) and select the correct answer from the codes given below.

| Column I | Column II |
|---|---|
| A. ${}^{242}_{94}Pu \rightarrow {}^{238}_{92}U + {}^4_2He$ | 1. $\beta^+$-decay |
| B. ${}^{210}_{83}Bi \rightarrow {}^{210}_{84}Po + e^- + \overline{\nu}$ | 2. $\beta^-$-decay |
| C. ${}^{97}_{43}Te \rightarrow {}^{97}_{42}Mo + e^+ + \nu$ | 3. $\alpha$-decay |

| | A | B | C |
|---|---|---|---|
| (a) | 3 | 2 | 1 |
| (b) | 1 | 3 | 2 |
| (c) | 2 | 1 | 3 |
| (d) | 3 | 1 | 2 |

# NCERT & NCERT Exemplar

## MULTIPLE CHOICE QUESTIONS

## NCERT

**95** The three stable isotopes of neon, $^{20}_{10}Ne$, $^{21}_{10}Ne$ and $^{22}_{10}Ne$ have respective abundances of 90.51%, 0.27% and 9.22%. The atomic masses of the three isotopes are 19.99 u, 20.99 u and 21.99 u, respectively. Obtain the average atomic mass of neon.

(a) 20.18 u (b) 12.5 u
(c) 8.55 u (d) 1.257 u

**96** Obtain the binding energy (in MeV) of a nitrogen nucleus ($^{14}_{7}N$), given $m(^{14}_{7}N) = 14.00307$ u. (Use $m_p = 1.007834$ u and $m_n = 1.00867$ u)

(a) 210 MeV (b) 104.67 MeV
(c) 83.5 MeV (d) 75.25 MeV

**97** A given coin has a mass of 3.0 g. Calculate the nuclear energy that would be required to separate all the neutrons and protons from each other. For simplicity, assume that the coin is entirely made of $^{63}_{29}Cu$ atoms (of mass 62.92960 u).

(a) $2.5 \times 10^{25}$ MeV
(b) $0.5 \times 10^{12}$ MeV
(c) $1.58 \times 10^{25}$ MeV
(d) $7.5 \times 10^{12}$ MeV

**98** Amount of $^{60}_{27}Co$ necessary to provide a radioactive source of 8.0 mCi strength, if half-life of $^{60}_{27}Co$ is 5.3 yr, is

(a) $7.12 \times 10^{-6}$ g (b) $1.2 \times 10^{-5}$ g
(c) $4.5 \times 10^{-6}$ g (d) $3.5 \times 10^{-5}$ g

**99** The half-life of $^{90}_{38}Sr$ is 28 yr. What is the disintegration rate of 15 mg of this isotope?

(a) $5.7 \times 10^{10}$ Bq
(b) $7.877 \times 10^{10}$ Bq
(c) $4.3 \times 10^{10}$ Bq
(d) $2.34 \times 10^{10}$ Bq

**100** The fission properties of $^{239}_{94}Pu$ are very similar to those of $^{235}_{92}U$. The average energy released per fission is 180 MeV. How much energy, in MeV is released, if all the atoms in 1 kg of pure $^{239}_{94}Pu$ undergo fission?

(a) $4.53 \times 10^{26}$ MeV (b) $1.25 \times 10^{26}$ MeV
(c) $8.06 \times 10^{26}$ MeV (d) $3.75 \times 10^{26}$ MeV

**101** Boron has two stable isotopes $^{10}_{5}B$ and $^{11}_{5}B$. Their respective masses are 10.01294 u & 11.00931 u and the atomic mass of boron is 10.811 u. Then, the abundances of $^{10}_{5}B$ and $^{11}_{5}B$ are

(a) 19.9%, 80.1% (b) 80.1%, 19.9%
(c) 92.5%, 7.5% (d) 7.5%, 92.5%

**102** The binding energy per nucleon of the nuclei $^{56}_{26}Fe$ in units of MeV is $[m(^{56}_{26}Fe) = 55.934939$ u$]$

(a) 7.20 MeV (b) 8.79 MeV
(c) 10.2 MeV (d) 13.6 MeV

**103** A 1000 MW fission reactor consumes half of its fuel in 5 yr. The reactor operates 80% of the time that all the energy generated arises from the fission of $^{235}_{92}U$ and this nuclide is consumed only by the fission process.The power of reactor $P = 1000$ MW. How much $^{235}_{92}U$ did it contain initially?

(a) 2050 kg (b) 3070 kg
(c) 4000 kg (d) 5000 kg

## NCERT Exemplar

**104** Suppose we have a large number of containers, each containing initially 10000 atoms of a radioactive material with a half-life of 1 yr. After 1 yr,

(a) all containers will have 5000 atoms of the material
(b) all containers will contains same number of atoms of the material but that number will only be approximately 5000
(c) the containers will general have different number of the atoms but their average is around 5000
(d) None of the containers have more than 5000 atoms

**105** When nucleus in an atom undergoes a radioactive decay, the electronic energy level of the atom

(a) do not change for any type of radioactivity
(b) change for $\alpha$ and $\beta$-radioactivity but not for $\gamma$-radioactivity
(c) change for $\alpha$-radioactivity but not for others
(d) change for $\beta$-radioactivity but not for others

**106** $M_X$ and $M_Y$ denote the atomic masses of the parent and the daughter nuclei, respectively in a radioactive decay. The $Q$-value for a $\beta^-$-decay is $Q_1$ and that for a $\beta^+$ decay is $Q_2$. If $M_e$ denotes the mass of an electron, then which of the following statements is correct?

(a) $Q_1 = (M_X - M_Y)\ c^2$ and $Q_2 = (M_X - M_Y - 2M_e)\ c^2$
(b) $Q_1 = (M_X - M_Y)\ c^2$ and $Q_2 = (M_X - M_Y)\ c^2$
(c) $Q_1 = (M_X - M_Y - 2M_e)\ c^2$ and $Q_2 = (M_X - M_Y + 2M_e)\ c^2$
(d) $Q_1 = (M_X - M_Y + 2M_e)\ c^2$ and $Q_2 = (M_X - M_Y + 2M_e)\ c^2$

**107** Heavy stable nuclei have more neutrons than protons. This is because of the fact that

(a) neutrons are heavier than protons
(b) electrostatic force between protons is repulsive
(c) neutrons decay into protons by beta decay
(d) nuclear force between neutrons is weaker than protons

**108** In a nuclear reactor, moderator slow down the neutrons which come out in a fission process. The moderator have light nuclei. Heavy nucleus will not serve the purpose because

(a) they will break up
(b) elastic collision of neutrons with heavy nuclei will not slow them down
(c) weight of reactor is appreciably high
(d) substances with heavy nuclei are not liquids or gases at room temperature

**109** The gravitational force between a H-atom and another particle of mass $m$ will be given by Newton's law

$$F = G\,\frac{M \cdot m}{r^2}, \text{ where } r \text{ is in km and}$$

(a) $M = m_{\text{proton}} + m_{\text{electron}}$
(b) $M = m_{\text{proton}} + m_{\text{electron}} - \dfrac{B}{c^2} (B = 13.6 \text{ eV})$.
(c) $M$ is not related to the mass of the hydrogen atom.
(d) $M = m_{\text{proton}} + m_{\text{electron}} - \dfrac{|V|}{c^2}$ ($|V|$ = magnitude of the potential energy of electron in the H-atom.

**110** Tritium is an isotope of hydrogen whose nucleus Triton contains 2 neutrons and 1 proton. Free neutrons decay into $p + \bar{e} + \bar{\nu}$. If one of the neutrons in Triton decays, it would transform into ${}_2\text{He}^3$ nucleus. This does not happen. This is because,

(a) Triton energy is less than that of a $2\text{He}^3$ nucleus
(b) the electron created in the beta decay process cannot remain in the nucleus
(c) Both the neutrons in Triton have to decay simultaneously resulting in a nucleus with 3 protons which is not a $2\text{He}^3$ nucleus
(d) free neutrons decay due to external perturbations which is absent in a Triton nucleus

# Answers

## > Mastering NCERT with MCQs

| | | | | | | | | | |
|---|---|---|---|---|---|---|---|---|---|
| 1 (c) | 2 (c) | 3 (a) | 4 (a) | 5 (c) | 6 (c) | 7 (d) | 8 (c) | 9 (d) | 10 (b) |
| 11 (b) | 12 (d) | 13 (d) | 14 (a) | 15 (d) | 16 (d) | 17 (b) | 18 (c) | 19 (b) | 20 (b) |
| 21 (b) | 22 (d) | 23 (d) | 24 (d) | 25 (c) | 26 (d) | 27 (d) | 28 (a) | 29 (c) | 30 (b) |
| 31 (c) | 32 (a) | 33 (c) | 34 (c) | 35 (a) | 36 (a) | 37 (c) | 38 (d) | 39 (a) | 40 (a) |
| 41 (d) | 42 (a) | 43 (b) | 44 (c) | 45 (b) | 46 (c) | 47 (b) | 48 (a) | 49 (c) | 50 (b) |
| 51 (b) | 52 (b) | 53 (b) | 54 (d) | 55 (d) | 56 (a) | 57 (d) | 58 (a) | 59 (d) | 60 (b) |
| 61 (a) | | | | | | | | | |

## > Special Types Questions

| | | | | | | | | | |
|---|---|---|---|---|---|---|---|---|---|
| 62 (d) | 63 (a) | 64 (b) | 65 (a) | 66 (a) | 67 (d) | 68 (c) | 69 (c) | 70 (a) | 71 (a) |
| 72 (a) | 73 (a) | 74 (c) | 75 (d) | 76 (c) | 77 (d) | 78 (c) | 79 (d) | 80 (d) | 81 (c) |
| 82 (b) | 83 (c) | 84 (d) | 85 (c) | 86 (b) | 87 (a) | 88 (b) | 89 (c) | 90 (c) | 91 (b) |
| 92 (d) | 93 (c) | 94 (a) | | | | | | | |

## > NCERT & NCERT Exemplar MCQs

| | | | | | | | | | |
|---|---|---|---|---|---|---|---|---|---|
| 95 (a) | 96 (b) | 97 (c) | 98 (a) | 99 (b) | 100 (a) | 101 (a) | 102 (b) | 103 (b) | 104 (c) |
| 105 (b) | 106 (a) | 107 (b) | 108 (b) | 109 (b) | 110 (a) | | | | |

# Hints & Explanations

**2** (c) The average atomic mass of a chlorine atom is obtained by the weighted average of the masses of the two isotopes, which is

Average atomic mass

$$= \frac{75.4 \times 34.98 + 24.6 \times 36.98}{100}$$

$$= 35.47\text{u}$$

**3** (a) Since, the nuclei of deuterium ($_1H^2$) and tritium $_1H^3$ are isotopes of hydrogen $_1H^1$, they must contain only one proton each.

But the masses of the nuclei of hydrogen, deuterium and tritium are in the ratio of 1 : 2 : 3, because of presence of neutral matter in deuterium and tritium nuclei.

**4** (a) Given, atomic mass of gold isotopes, $A_1 = 197$ and atomic mass of silver isotope, $A_2 = 107$

As, radius of nucleus, $R = R_0 A^{1/3}$

$$\therefore \quad \frac{R_1}{R_2} = \left(\frac{A_1}{A_2}\right)^{1/3} = \left(\frac{197}{107}\right)^{1/3} = 1.225 \simeq 1.23$$

The ratio of nuclear radii of the gold isotope $^{197}_{97}\text{Au}$ and silver isotope istope $^{107}_{47}\text{Ag}$ is 1.23.

**5** (c) Nuclear radius, $R \propto A^{1/3}$, where $A$ is mass number.

For $^{27}\text{Al}$,

$$R = R_0 A^{1/3} = R_0 (27)^{1/3} = 3R_0$$

$$R_0 = \frac{3.6}{3} = 1.2 \text{ fm} \qquad [\because R = 3.6 \text{ fm, for Al}]$$

For $^{64}\text{Cu}$, $\quad R = R_0 A^{1/3} = 1.2 (64)^{1/3} = 4.8 \text{ fm}$

**6** (c) Surface area of nucleus,

$$S_A = 4\pi R^2 = 4\pi (R_0 A^{1/3})^2$$

$$= 4\pi R_0^2 \cdot A^{2/3}$$

Substituting constant ($R_0$) value $= 1.2 \times 10^{-15}$ m, we get

$$S_A = 4 (3.14)(1.2 \times 10^{-15})^2 A^{2/3}$$

$$= (1.8 \times 10^{-29}) A^{2/3}$$

**7** (d) Density of a nucleus, $\rho = \frac{\text{Mass}}{\text{Volume}}$

$$= \frac{mA}{\frac{4}{3}\pi R_0^3 A} = \frac{3m}{4\pi R_0^3} \qquad \ldots(\text{i})$$

As, mass of nucleon, $m = m_p = m_n = 1.67 \times 10^{-27}$ kg and $R_0 = 1.2 \times 10^{-15}$ m.

Substituting these values in Eq. (i), we get

$$\rho = \frac{3 \times 1.67 \times 10^{-27}}{4 \times 3.14 \times (1.2 \times 10^{-15})^3} \approx 2.3 \times 10^{17} \text{ kg m}^{-3}$$

which is a constant and independent of $A$, for all nuclei. Also, density of nuclear matter is very large as compared to ordinary matter.

**8** (c) Given, $m_{\text{Fe}} = 55.85$,

$$u = 55.85 \times 1.66 \times 10^{-27} \text{ kg}$$

$$= 9.27 \times 10^{-26} \text{ kg},$$

$$A = 56 \text{ and } R_0 = 1.2 \times 10^{-15} \text{ m}$$

$$\text{Nuclear density} = \frac{\text{Mass}}{\text{Volume}} = \frac{m_{\text{Fe}}}{(4/3)\pi R_0^3 A}$$

$$= \frac{9.27 \times 10^{-26}}{(4\pi/3)(1.2 \times 10^{-15})^3} \times \frac{1}{56}$$

$$= 2.29 \times 10^{17} \text{ kg m}^{-3}$$

**9** (d) Mass density of nuclear matter is a constant quantity for all elements. It does not depends on element's mass number or atomic radius.

$\therefore$ The ratio of mass densities of $^{40}\text{Ca}$ and $^{16}\text{O}$ is 1 : 1.

**10** (b) According to mass-energy equivalence relation,

Energy, $E = mc^2$

By substituting,

$m = 2\text{g} = 2 \times 10^{-3}$ kg, $c = 3 \times 10^8 \text{ ms}^{-1}$

We have, $\quad E = 2 \times 10^{-3} \times (3 \times 10^8)^2$

$$\Rightarrow \quad E = 2 \times 10^{-3} \times 9 \times 10^{16} = 18 \times 10^{13} \text{ J}$$

**11** (b) We have,

$P = 200 \text{ MW} = 200 \times 10^6 \text{ W} = 2 \times 10^8 \text{ W}$,

$$t = 1\text{h} = 3600 \text{ s}$$

As, energy, $\quad E = P \times t = 2 \times 10^8 \times 3600$

Also, $\quad E = mc^2$

$$\Rightarrow \quad m = \frac{E}{c^2} = \frac{2 \times 10^8 \times 3600}{(3 \times 10^8)^2} = 8 \times 10^{-6} \text{ kg}$$

**12** (d) Mass defect, $\Delta m$

$$= \text{Mass of nucleons} - \text{Mass of nucleus}$$

$$= [Zm_p + (A - Z) m_n] - M_n$$

Here, mass number, $A = 4$

Atomic number, $Z = 2$

Number of protons $= 2$

Number of neutrons $= A - Z = 4 - 2 = 2$

$m_p = 1.0072676$ u

$m_n = 1.008665$u and $M_n = 4.001506$ u

Mass of nucleon in $_2\text{He}^4$ =

Mass of 2 protons + Mass of 2 neutrons

So, mass of nucleons $= 2 \times 1.0072676 + 2 \times 1.008665$

$$\therefore \quad \Delta m = 2 \times 1.0072676 + 2 \times 1.008665 - 4.001506$$

$$= 0.0303592 \text{ u}$$

**13** (*d*) Given, mass of a neutron = 1.00866 u,
Mass of a proton = 1.00727 u, mass of $^{16}_{8}O = 15.99053$ u
Mass number, $A = 16$,
Atomic number, $Z = 8$
Number of protons = 8
Number of neutrons = $(A - Z) = 8$
Therefore, mass of 8 neutrons = $8 \times 1.00866$ u
Mass of 8 protons = $8 \times 1.00727$ u
So, the expected mass of $^{16}_{8}O$ nucleus,

$$= 8 \times (1.00866 + 1.00727)$$
$$= 8 \times 2.01593 \text{ u} = 16.12744 \text{ u}$$

Thus, mass defect, $\Delta m = 16.12744 \text{ u} - 15.99053 \text{ u} = 0.13691 \text{ u}$

Since, 1u = 931.5 MeV.
For $^{16}_{8}O$, $\Delta m = 0.13691$ u

$$= 0.13691 \times 931.5$$
$$= 127.5 \text{ MeV}$$

Thus, the energy required to separate $^{16}_{8}O$ into its constituents is 127.5 MeV.

**14** (*a*) Given, $m(^{56}_{26}Fe) = 55.934939$ u,

$$m_p = 1.007825 \text{ u}$$
and $$m_n = 1.008665 \text{ u}$$

Also, $A = 56$ and $Z = 26$
$^{56}_{26}$ Fe nucleus has 26 protons and $(56 - 26) = 30$ neutrons.

$$\therefore \text{Mass defect} = [Zm_p + (A - Z)\, m_n] - M$$
$$= (26m_p + 30\, m_n) - m(^{56}_{26}Fe)$$
$$= [26(1.007825) + 30(1.008665)] - 55.934939$$
$$= 56.46340 - 55.934939$$
$$= 0.528461 \text{ u}$$

$$\text{Total BE} = \Delta m \times 931 \text{ MeV}$$
$$= 0.528461 \times 931.5 \text{ MeV}$$
$$= 492.26 \text{ MeV}$$

$\therefore$ Binding energy per nucleon

$$= \frac{\text{Total BE}}{A} = \frac{492.26}{56} = 8.790 \text{ MeV}$$

**15** (*d*) Binding energy per nucleon is $E_{bn} = \frac{E_b}{A}$ ... (i)

For deuteron, atomic mass, $A = 2$ and $E_{bn} = 2.115$ MeV
Substituting these values in Eq. (i), we get

$$2.115 = \frac{E_b}{2}$$
$$E_b = 2 \times 2.115 \text{ MeV}$$

As, $$E_b = \Delta mc^2$$

$$\Rightarrow \Delta m = \frac{E_b}{c^2}$$

$$\therefore \text{Mass defect, } \Delta m = \frac{2 \times 2.115\, u}{931.5}$$
$$= 0.0045 \text{ u}$$

**16** (*d*) The elements higher on the BE *versus* mass number plot, i.e. representing the high peaks on the plot are very tightly bound and hence are stable. However, the elements on lower end of this plot, are less tightly bound and hence are unstable.

As, lithium nucleus shows a peak on this plot, so it implies that it is very stable.

**17** (*b*) Given, $E_{bn}$ of chlorine = 287.67 MeV
and $E_{bn}$ of phosphorus = 262.48 MeV
Binding energy per nucleon of $_{17}Cl^{35}$

$$= \frac{\text{Binding energy}}{\text{Mass number}}$$
$$= \frac{287.67}{35} = 8.22 \text{ MeV}$$

Similarly,

$$\text{Binding energy per nucleon of } _{15}P^{31} = \frac{262.48}{31}$$
$$= 8.47 \text{ MeV}$$

Since, BE of $_{15}P^{31}$ is more than BE of $_{17}Cl^{35}$, hence $_{15}P^{31}$ is more stable.

**18** (*c*) If a nucleon can have a maximum of *p*-neighbours within the range of nuclear force, then its binding energy would be proportional to *p*.
If we increase, *A* by adding nucleons, they will not change the binding energy of a nucleon, inside.
This is because most of the nucleons in a large nucleus reside inside it and not on the surface, so the change in binding energy per nucleons would be small.
$\therefore$The binding per nucleons remains constant in the given case.

**19** (*b*) Potential energy of a pair of nucleons as a function of their separation is correctly depicted in option (b). For a separation greater than $r_0$, the force is attractive, so potential energy decreases and for a separation less than $r_0$, the force is strongly repulsive, so potential energy increases for $r < r_0$.

**21** (*b*) Let initial number of nuclei in *A* and *B* is $N_0$.
Number of nuclei of *A* after time *t* is

$$N_A = N_0 e^{-8\lambda t} \quad ...(i)$$

Similarly, number of nuclei of *B* after time *t* is

$$N_B = N_0 e^{-\lambda t} \quad ...(ii)$$

It is given that

$$\frac{N_B}{N_A} = \frac{1}{e} \quad ...(iii)$$

Now, from Eqs. (i), (ii) and (iii), we get

$$\frac{N_A}{N_B} = \frac{e^{-8\lambda t}}{e^{-\lambda t}} = e$$

Rearranging this, we get

$$\Rightarrow \quad e^1 = e^{-7\lambda t}$$

$$\Rightarrow \quad 7\lambda t = -1$$

As, time cannot be negative, so

$$\text{Time, } t = \frac{1}{7\lambda}$$

**22** (*d*) Total decay rate of a sample is

$$R = -\frac{dN}{dt}$$

As, rate of change of $N$ is $\frac{dN}{dt} = -\lambda N$

$$\Rightarrow \quad R = \lambda N$$

So, in order to have the same decay rate for both the samples $A$ and $B$, we have

$$\lambda_A N_A = \lambda_B N_B$$

$$\frac{\lambda_A}{\lambda_B} = \frac{N_B}{N_A}$$

**23** (*d*) We know that,

Number of undecayed nuclei, $N = N_0 e^{-\lambda t}$

where, $N_0$ = number of nuclei at $t = 0$

and decay rate, $R = R_0 e^{-\lambda t}$

where, $R_0 = \lambda N_0$ is the decay rate at $t = 0$.

$$\Rightarrow \quad R = \lambda N_0 e^{-\lambda t}$$

$\therefore$ Number of decayed nuclei,

$$N_{\text{decayed}} = N_0 - N = N_0 - N_0 e^{-\lambda t}$$

$$= N_0 - \frac{R}{\lambda}$$

This represents equation of a straight line with negative slope. So, graph between $N$ and $R$ is correctly depicted in option (d).

**24** (*d*) For 40 g of $A_1$,

$$40\,\text{g} \xrightarrow[\text{half-life}]{20\text{s}} 20\,\text{g} \xrightarrow{20\text{s}} 10\,\text{g}$$

For 160 g of $A_2$,

$$160\,\text{g} \xrightarrow[\text{half-life}]{10\,\text{s}} 80\,\text{g} \xrightarrow{10\,\text{s}} 40\,\text{g} \xrightarrow{10\,\text{s}} 20\,\text{g} \xrightarrow{10\,\text{s}} 10\,\text{g}$$

So, after 40s, the amount of the two radioactive materials $A_1$ and $A_2$ in the mixture will become equal.

**25** (*c*) Given, $T_{1/2} = 78\,\text{h}$

Mass = 3.4 g and molar mass = 67

Initial decay rate, $R_0 = \lambda N_0$

Here, decay constant, $\lambda = \frac{\log 2}{T_{1/2}} = \frac{\log 2}{78\,\text{h}} = \frac{0.693}{78\,\text{h}}$

$$= 8.89 \times 10^{-3}\,\text{h}^{-1}$$

$$= \frac{8.89 \times 10^{-3}}{3600}\,\text{h}^{-1}$$

$$= 2.47 \times 10^{-6}\,\text{s}^{-1}$$

$N_0$ contains in 3.4 g of sample of $^{67}\text{Ga}$

$$= \frac{\text{mass}}{\text{molar mass}} \times N_A = \frac{3.4}{67} \times 6 \times 10^{23}$$

$$\Rightarrow \quad N_0 = 3.05 \times 10^{22}$$

$$\therefore \quad R_0 = (2.47 \times 10^{-6})(3.05 \times 10^{22})$$

$$= 7.53 \times 10^{16}\,\text{s}^{-1}$$

**26** (*d*) Given, half-life, $T_{1/2} = 2.2 \times 10^9$ s

Rate of disintegration, $R = 10^{10}\ \text{s}^{-1}$

If $N$ be the number of nuclei present, then the rate of disintegration is,

$$\frac{dN}{dt} = \lambda N \qquad (\lambda = \text{decay constant})$$

$$\Rightarrow \quad R = \frac{dN}{dt} = \lambda N \quad \text{or} \quad N = \frac{R}{\lambda} \qquad \text{...(i)}$$

Also, the half-life of an element is given by

$$T_{1/2} = \frac{0.693}{\lambda}$$

$$\Rightarrow \quad \lambda = \frac{0.693}{T_{1/2}} \qquad \text{...(ii)}$$

From Eqs. (i) and (ii), we get

$$N = \frac{R}{0.693} \times T_{1/2}$$

$$= \frac{10^{10} \times 2.2 \times 10^9}{0.693} = 3.17 \times 10^{19}$$

**27** (*d*) Activity of a radioactive material is given as

$$R = \lambda N$$

where, $\lambda$ is the decay constant and $N$ is the number of nuclei in the radioactive material.

For substance $A$,

$$R_A = \lambda_A N_A = 10\,\text{mCi}$$

For substance $B$,

$$R_B = \lambda_B N_B = 20\,\text{mCi} \qquad \text{...(i)}$$

As given in the question,

$$N_A = 2N_B \Rightarrow R_A = \lambda_A (2N_B) = 10\,\text{mCi} \qquad \text{...(ii)}$$

$\therefore$ Dividing Eq. (ii) and Eq.(i), we get

$$\frac{R_A}{R_B} = \frac{\lambda_A (2N_B)}{\lambda_B (N_B)} = \frac{10}{20}$$

$$\text{or} \quad \frac{\lambda_A}{\lambda_B} = \frac{1}{4} \qquad \text{...(iii)}$$

As, half-life of a radioactive material is given as

$$T_{1/2} = \frac{0.693}{\lambda}$$

$\therefore$ For material $A$ and $B$, we can write

$$\frac{(T_{1/2})_A}{(T_{1/2})_B} = \frac{\frac{0.693}{\lambda_A}}{\frac{0.693}{\lambda_B}} = \frac{\lambda_B}{\lambda_A}$$

Using Eq. (iii), we get

$$\frac{(T_{1/2})_A}{(T_{1/2})_B} = \frac{4}{1}$$

Hence, from the given options, only option (d) satisfies this ratio.

Therefore, $(T_{1/2})_A = 20$ days

and $(T_{1/2})_B = 5$ days

**28** (*a*) Given, half-life of an element, $T_{1/2} = 69.3$ h

Let initial amount (at $t = 0$) of radioactive element be $N_0$, then

Active nuclei at $t = 10$ h,

$$N_1 = N_0\, e^{-10\lambda}$$

Active nuclei at $t = 11$ h, $N_2 = N_0 e^{-11\lambda}$

$$\%\text{ decay in 10th to 11th hours} = \frac{N_1 - N_2}{N_1} \times 100$$

$$= \frac{N_0 e^{-10\lambda} - N_0 e^{-11\lambda}}{N_0\, e^{-10\lambda}} \times 100 = (1 - e^{-\lambda}) \times 100$$

$$= \left[1 - e^{\frac{-0.693}{69.3}}\right] \times 100 \qquad \left[\because \lambda = \frac{0.693}{T_{1/2}}\right]$$

$$= [1 - e^{-0.01}] \times 100$$

$$= [1 - 0.99] \times 100 = 0.01 \times 100 = 1\%$$

**29** (*c*) Given, ratio of $X : Y = 1 : 7$

$$\Rightarrow \quad \frac{m_X}{m_Y} = \frac{1}{7} \text{ or } m_X = \frac{m_Y}{7}$$

Let the initial total mass is $m$.

$$\Rightarrow \quad m_X + m_Y = m$$

$$\Rightarrow \quad \frac{m_Y}{7} + m_Y = m$$

$$\Rightarrow \quad \frac{8m_Y}{7} = m$$

$$\Rightarrow \quad m_Y = \frac{7}{8} m$$

$\therefore$ Only $\left(\frac{1}{8}\right)$th part remains of $X$.

$$\Rightarrow \quad 1 \xrightarrow{T_{1/2}} \frac{1}{2} \xrightarrow{T_{1/2}} \frac{1}{4} \xrightarrow{T_{1/2}} \frac{1}{8}$$

So, time taken to become $\left(\frac{1}{8}\right)$th unstable part

$$= 3 \times T_{1/2} = 3 \times 1.4 \times 10^9 \quad [\because T_{1/2} = 1.4 \times 10^9 \text{ yr}]$$

$$= 4.20 \times 10^9 \text{ yr}$$

**30** (*b*) As, 25% of undecayed nuclei reduces to 12.5% in 10s. So, half-life of the sample is 10s, i.e. $T_{1/2} = 10$ s

But $\quad T_{1/2} = \frac{0.6931}{\lambda}$

where $\lambda$ is decay constant.

Thus, $\quad \lambda = \frac{0.6931}{T_{1/2}} = \frac{0.6931}{10}$

The mean life of the nuclei is

$$\tau = \frac{1}{\lambda} = \frac{10}{0.6931} = 14.428 \text{ s}$$

**31** (*c*) Fraction of radioactive substance remains undecayed $= \frac{N}{N_0} = \left(\frac{1}{2}\right)^n$

where, $n = t / T_{1/2}$

$$\Rightarrow \quad N = N_0 \left(\frac{1}{2}\right)^{\frac{t}{T_{1/2}}}$$

Given, $T_{1/2} = 12.5$ yr, $t = 25$ yr

$$\Rightarrow \quad N = N_0 \left(\frac{1}{2}\right)^{\left(\frac{25}{12.5}\right)}$$

$$N = N_0 \left(\frac{1}{2}\right)^2$$

$$\frac{N}{N_0} = \frac{1}{4}$$

Thus, fraction of sample $(N / N_0)$ remaining undecayed after 25 yr will be $\frac{1}{4}$.

**32** (*a*) Activity of a sample after time $t$ is related to initial activity as

$$A = A_0 \left(\frac{1}{2}\right)^n \qquad \ldots \text{(i)}$$

where, $A$ = present activity

and $A_0$ = initial activity.

Also, $n = t / T_{1/2}$ $\qquad \ldots$ (ii)

where, $t$ = time interval

and $T_{1/2}$ = half-life time.

Here, $A = 500\ \text{s}^{-1}$, $A_0 = 700\ \text{s}^{-1}$,

$t = 30$ min

So, from Eq. (i), we get

$$500 = 700 \left(\frac{1}{2}\right)^n$$

$$\Rightarrow \quad 2^n = 1.4 \approx \sqrt{2}$$

So, $\quad n = \frac{1}{2}$

Now, from Eq. (ii), we get

$$\frac{t}{T_{1/2}} = \frac{1}{2} \Rightarrow T_{1/2} = 2 \times t = 60 \text{ min}$$

which is close to 62 min.

**33** (c) Given, $N_0 = 1600$ count/s

$N = 100$ count/s, at $t = 8$s

Number of nuclei left undecayed

after $n$ half-lives, $N = N_0\left(\frac{1}{2}\right)^n$

$$\Rightarrow \quad 100 = 1600\left(\frac{1}{2}\right)^n$$

$$\Rightarrow \quad \left(\frac{1}{2}\right)^n = \frac{1}{16}$$

$$\Rightarrow \quad \left(\frac{1}{2}\right)^n = \left(\frac{1}{2}\right)^4$$

$$n = 4 \text{ half-lives}$$

Four half-lives equivalent to 8s, so 6 s is equals to three half-lives.

Counting rate observed after 6 s will be,

$$= N_0\left(\frac{1}{2}\right)^3 = \frac{1600}{8} = 200$$

**34** (c) After $n$ half-life, the number of nuclei left undecayed is given as $\quad N = N_0\left(\frac{1}{2}\right)^n$

where, $n = \frac{t}{T_{1/2}}$

Given, initially number of nuclei, $N_0 = 600$

Number of disintegrated or decayed nuclei, $N' = 450$

$\therefore$ Number of nuclei left undecayed,

$$N = N_0 - N'$$
$$= 600 - 450 = 150$$

Half-life, $T_{1/2} = 10$ min

As, $\quad \frac{N}{N_0} = \left(\frac{1}{2}\right)^{t/T_{1/2}}$

Substituting the given values, we get

$$\frac{150}{600} = \left(\frac{1}{2}\right)^{t/10} \text{ or } \frac{1}{4} = \left(\frac{1}{2}\right)^{t/10}$$

or $\quad \left(\frac{1}{2}\right)^2 = \left(\frac{1}{2}\right)^{t/10}$ or $\frac{t}{10} = 2$

$$\Rightarrow \quad t = 20 \text{ min}$$

**35** (a) Decay scheme is ,

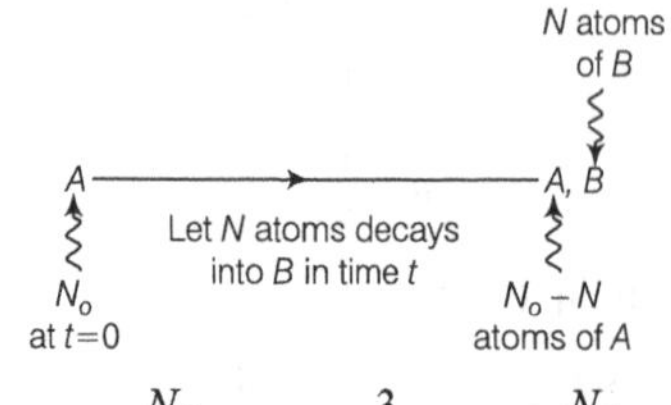


Given, $\quad \frac{N_B}{N_A} = 0.3 = \frac{3}{10} \Rightarrow \frac{N_B}{N_A} = \frac{30}{100}$

So, $\quad N_0 = 100 + 30 = 130$ atoms

By using $\quad N = N_0 e^{-\lambda t}$

We have, $\quad 100 = 130 e^{-\lambda t}$

$$\Rightarrow \quad \frac{1}{1.3} = e^{-\lambda t}$$

$$\Rightarrow \quad \log 1.3 = \lambda t$$

If $T$ is half-life, then $\lambda = \frac{\log_e 2}{T}$

$$\Rightarrow \quad \log 1.3 = \frac{\log_e 2}{T} \cdot t$$

$$\therefore \quad t = \frac{T \cdot \log (1.3)}{\log_e 2}$$

**36** (a) As, $T_{1/2} = 2$ min $\Rightarrow$ 10 min = 5 half-lives

We know that, fraction of sample remains undecayed,

$$\frac{N}{N_0} = \left(\frac{1}{2}\right)^n$$

where, $N$ = remaining number of nuclei

$N_0$ is the initial number of nuclei = 80

and $n$ is the number of half-lives = 5

$$\Rightarrow \quad N = \frac{N_0}{2^n} = \frac{80}{2^5} = \frac{80}{32} = 2.5 \text{ g}$$

**37** (c) Those radioactive elements, whose half-life is short compared to the age of the universe (13.7 billion years), are not found in observable quantities in nature today. However, they have been seen in the laboratories in nuclear reactions. Tritium and plutonium belongs to this category, which are not found naturally on earth.

**38** (d) α-particles are doubly ionised helium nucleus ($He^{2+}$) which are emitted in any radioactive process. So, they have two protons, 2 neutrons in its nucleus and no electron.

**40** (a) The α-decay of $^{238}_{92}U$ is given by equation

$$^{238}_{92}U \rightarrow ^{234}_{90}Th + ^4_2He + Q$$

The energy released in this process is given by

$$Q = \Delta mc^2 = (M_U - M_{Th} - M_{He})c^2$$

Here, $M_U = 238.05079$ u, $M_{He} = 4.00260$ u

and $\quad M_{Th} = 234.04363$ u

Substituting the given values of atomic masses, we get

$$Q = (238.05079 - 234.04363 - 4.00260)\text{u} \times c^2$$
$$= (0.00456 \text{ u})c^2$$
$$= (0.00456 \text{ u})(931.5 \text{ MeV/u}) = 4.25 \text{ MeV}$$

**41** (d) Given, nuclear reaction is

$$^7_3Li + ^1_1H \longrightarrow ^4_2He + ^4_2He + Q$$

The binding energy for $_1H^1$ is around zero and it is not given in the question, so we can ignore it.

$$\Rightarrow \quad Q = 2 \times 4\,(E_{bn} \text{ of He}) - 7(E_{bn} \text{ of Li})$$

where, $E_{bn}$ is the binding energy per nucleon.
Given, $E_{bn}$ of He = 7.06 MeV
and $E_{bn}$ of Li = 5.60 MeV
Substituting given values, we get
Energy released, $Q = 2(4 \times 7.06) - 7 \times 5.60$
$= (8 \times 7.06) - (7 \times 5.60)$
$= (56.48 - 39.2)$ MeV
$= 17.28 \text{ MeV} \simeq 17.3$ MeV

**42** (*a*) Charge on positron, $q = e = 1.6 \times 10^{-19}$ C

Mass of positron, $m$ = Mass of electron $= 9.1 \times 10^{-31}$ kg

Ratio of, charge on positron to mass of positron is,

$$\frac{q}{m} = \frac{1.6 \times 10^{-19}}{9.1 \times 10^{-31}} = 1.75 \times 10^{11}$$
$$\simeq 2 \times 10^{11} \text{ C/kg}$$

**43** (*b*) Isotopes are the atoms of same element having same number of protons but different number of neutrons, i.e. $Z$ remains same.

However, in $\alpha$ or $\beta$ decay, the nuclei hence formed always have $Z$ either decreased by 2 or increased by 1. Thus, isotopes of the original nuclei cannot be occured in these decay.

**45** (*b*) In $\gamma$-radiation, no change occurs in mass number and atomic number, because rest mass of $\gamma$-particle is zero and it is neutral for electric and magnetic field.

In $\alpha$-radiation, atomic number is decreased by 2 units and atomic mass is decreased by 4 units.

In $\beta$-radiation, no change occurs in atomic mass but atomic number increases by 1 unit.

**46** (*c*) The given reaction can be written as

$$_Z X^A \xrightarrow{\beta^-} {}_{Z+1}Y^A \xrightarrow{\alpha} {}_{Z-1}K^{A-4} \xrightarrow{\gamma} {}_{Z-1}K^{A-4}$$

So, all the three particles, i.e. $\alpha, \beta$ and $\gamma$ are emitted.

**47** (*b*) In chemical reaction, the energy released is in the range of eV. However, in nuclear reaction the energy release is of the order of MeV. Thus, for the same quantity of matter, nuclear source produce a million times more energy than a chemical source.

$$\therefore \qquad E_N > E_C$$

**48** (*a*) Given,

$E_{bn}$ for $A = 240$ nucleus is about 7.6 MeV.

$E_{bn}$ for the two $A = 120$ fragment nuclei is about 8.5 MeV.

So, gain in binding energy of nucleon is about (8.5–7.6 =) 0.9 MeV.

Hence, the total gain in binding energy is $240 \times 0.9 = 216$ MeV.

**49** (*c*) The options (a), (b) and (d) are correctly matched pair, while option (c) is unmatched pair. As control rods are present in the reactor to absorb neutrons and are usually made up of cadmium or boron.

**50** (*b*) The schematic diagram of a nuclear reactor is as shown below

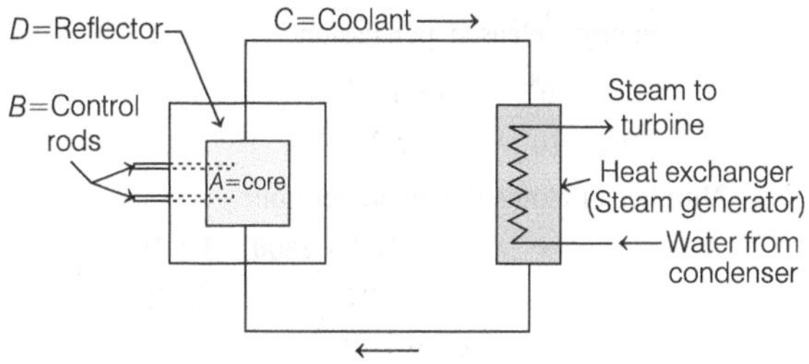


Here,

$A \rightarrow$ is the core of the reactor, which is the site of nuclear fission. It contains elements in suitably fabricated form. The fuel may be enriched uranium say $^{235}_{92}$U.

$D \rightarrow$ is a reflector surrounding the core which is used to reduce the leakage.

$C \rightarrow$ is coolant which helps in removing the energy (heat) released in fission. It transfers heat to working fluid which in turn may produce steam.

$B \rightarrow$ is control rods that shut down the reactor as they have high absorption of neutrons.

Hence, only option (b) shows correct purpose of given part.

**51** (*b*) Energy released on bombarding $U^{235}$ by neutron
$= 200$ MeV
$= 200 \times 10^6 \times 1.6 \times 10^{-19}$ J

Power output of atomic reactor = 1.6 MW
$= 1.6 \times 10^6$ W

$$\therefore \text{ Rate of fission } = \frac{1.6 \times 10^6}{200 \times 10^6 \times 1.6 \times 10^{-19}} = 5 \times 10^{16} \text{ s}^{-1}$$

**52** (*b*) From Einstein's mass-energy relation, the energy released is

$$\Delta E = \Delta mc^2$$

where, $\Delta m$ is mass and $c$ is speed of light.

Given, $\Delta m = 1 \text{ mg} = 1 \times 10^{-6}$ kg, $c = 3 \times 10^8$ ms$^{-1}$

$$\therefore \qquad \Delta E = 1 \times 10^{-6} \times (3 \times 10^8)^2 = 9 \times 10^{10} \text{ J}$$

The rate at which energy is dissipated is known as power,

$$\Rightarrow \qquad P = \frac{\Delta E}{t} = \frac{9 \times 10^{10}}{1} = 9 \times 10^{10} \text{ W}$$
$$\therefore \qquad P = 9 \times 10^7 \text{ kW}$$

**53** (*b*) Power $= \frac{\text{Energy}}{\text{Time}} = 300$ MW
$= 300 \times 10^6 \text{ W} = 3 \times 10^8$ Js$^{-1}$

Energy released due to fission,
$= 170 \text{ MeV} = 170 \times 10^6 \times 1.6 \times 10^{-19}$
$= 27.2 \times 10^{-12}$ J

Number of atoms fissioned per second

$$= \frac{\text{Energy released per second}}{\text{Energy released per fission}}$$

$$= \frac{3\times 10^8}{27.2\times 10^{-12}} = \frac{3\times 10^{20}}{27.2}$$

Number of atoms fissioned per hour

$$= \frac{3\times 10^{20}\times 3600}{27.2} = \frac{3\times 36}{27.2}\times 10^{22}$$

$$\approx 4\times 10^{22}$$

**54** (*d*) Energy released in each fission of $U^{235} = 200$ MeV

$$= 200\times 10^6\times 1.6\times 10^{-19}\text{J} = 3.2\times 10^{-11}\text{ J}$$

Since, only 20% of this is utilized efficiently, therefore energy generated from the reactor per fission

$$= 3.2\times 10^{-11}\times 0.20 = 6.4\times 10^{-12}\text{ J}$$

As, the output of the reactor is 700 MW, the number of fission required per second

$$= \frac{700\times 10^6\text{ Js}^{-1}}{6.4\times 10^{-12}\text{J}} = 1.1\times 10^{20}\text{s}^{-1}$$

The number of U-235 atoms consumed in 24 h

$$= 24\times 60\times 60\times 1.1\times 10^{20} = 9.5\times 10^{24}$$

There are $6.02\times 10^{26}$ atoms in 235 kg of U-235. Therefore, the mass of U-235 consumed during 24 h

$$= \frac{9.5\times 10^{24}}{6.02\times 10^{26}}\times 235\text{ kg} = 3.7\text{ kg}$$

Thus, energy generated from the reactor per fission and mass of uranium are $6.4\times 10^{-12}$ J and 3.7 kg.

**55** (*d*) Reactions I, II and III represent fission of uranium isotope $^{235}_{92}$U, when bombarded with neutrons that breaks it into two intermediate mass nuclear fragments. However, reaction IV represents two deuterons fuses together to form the light isotope of helium.

**56** (*a*) When two lighter nuclei combine to form a heavier nucleus, the process is called nuclear fusion. e.g.

$$^2_1\text{H} + ^2_1\text{H} \longrightarrow ^4_2\text{He} + ^1_0n$$

**57** (*d*) For fusion reaction to take place, the two nuclei must come close enough, so that the attractive short range nuclear force is able to affect them. Also, they must have suitable energy, enough to overcome the coulomb repulsive force barrier.

This could only be achieved, if the reaction is taking place at high temperature, which could make the particles of the system to have enough kinetic energy to overcome coulomb repulsive force between nuclei.

**58** (*a*) According to question,

$$_{92}\text{U}^{238} \longrightarrow {}_{90}\text{Th}^{234} + {}_2\text{He}^4$$

According to law of conservation of linear momentum, we have,

$$|p_{\text{Th}}| = |p_{\text{He}}| = p$$

As, kinetic energy of an element, $\text{KE} = \frac{p^2}{2M}$

where, $M$ is mass of an element.

Thus, $\text{KE} \propto \frac{1}{M}$

As, $M_{\text{He}} < M_{\text{Th}}$

$\Rightarrow$ $\text{KE}_{\text{He}} > \text{KE}_{\text{Th}}$

i.e. The helium nucleus has more kinetic energy than that the thorium.

**59** (*d*) For the fourth reaction (D) to occur, the first three reaction must occur twice in which two light helium nuclei ($^3_2$He) unite to form ordinary helium nucleus ($^4_2$He).

So, the correct combination will be 2A + 2B + 2C.

**60** (*b*) Energy produced in sun and hydrogen bomb is due to nuclear fusion.

However, energy released in uranium based nuclear reactor and atom bomb is due to nuclear fission.

**61** (*a*) Given, fusion reaction is,

$$^2_1\text{H} + ^2_1\text{H} \longrightarrow ^3_2\text{He} + ^1_0n$$

Mass of deuteron ($^2_1$H) = 2.015 amu

Mass of helium ($^3_2$He) = 3.017 amu

and mass of neutron ($^1_0n$) = 1.009 amu

Mass defect, $\Delta m$ = Mass of 2($^2_1$H) − Mass of ($^3_2$He + $^1_0n$)

$$\Delta m = 2(2.015) - (3.017 + 1.009)$$
$$= 0.004\text{ amu}$$

$\therefore$ Energy released = (0.004 × 931.5) MeV

$$= 3.726\text{ MeV}$$

Energy released per deuteron $= \frac{3.726}{2} = 1.863$ MeV

Number of deuterons in 1 kg $= \frac{6.02\times 10^{26}}{2} = 3.01\times 10^{26}$

$\therefore$ Energy released per kg of deuterium fusion

$$= (3.01\times 10^{26}\times 1.863)$$
$$= 5.6\times 10^{26}\text{ MeV}$$
$$= 5.6\times 10^{26}\times 10^6\times 1.6\times 10^{-19}\text{ J}$$
$$\approx 9.0\times 10^{13}\text{ J}$$

**62** (*d*) Mass of an atom is very small, compared to a kilogram. So, it not a very convenient unit to measure such small quantities. Therefore, a different mass unit is used for expressing atomic mass.

This unit is the atomic mass unit (u or amu), defined as (1/12) th of the mass of the carbon ($^{12}$C) atom.

1 u or amu is equal to $1.67\times 10^{-27}$ kg.

Therefore, Assertion is incorrect but Reason is correct.

**63** (*a*) The nucleus of the lightest atom of hydrogen, which has a relative abundances of 99.985% is called the proton.

Thus, we can conclude that ratio of the mass of nucleus of H-atom and the proton is approximately 1.

Therefore, Assertion and Reason are correct and Reason is the correct explanation of Assertion.

**64** (*b*) Before the advent of special theory of relativity, it was presented that mass and energy were conserved separately in a reaction.

However, Einstein showed that mass is a another form of energy and one can convert mass-energy into other forms of energy, say kinetic energy and *vice-versa*.

Therefore, Assertion and Reason are correct but Reason is not the correct explanation of Assertion.

**65** (*a*) Since, nuclear force does not depend on the electric charge.

So, nuclear force between neutron-neutron, proton-neutron and proton-proton is approximately the same.

Therefore, Assertion and Reason are correct and Reason is the correct explanation of Assertion.

**66** (*a*) The constancy of the binding energy per nucleon in the range $30 < A < 170$ is a consequence of the fact that the nuclear force is short ranged in nature.

Therefore, Assertion and Reason are correct and Reason is the correct explanation of Assertion.

**67** (*d*) For an element, it is not always true that atomic number $Z$ are less than or equal to number of neutrons ($N$). e.g. In hydrogen atom, there is no neutron but it has one proton.

Therefore, atomic number of hydrogen atom is greater than number of neutrons.

i.e. $Z > N$, for hydrogen atom.

The force between neutrons in the nucleus is attractive only upto the range of 1 fm. But this attractive force become highly repulsive when neutrons came nearer to about 0.6 to 0.7 fm. Therefore, neutrons do not always experience attractive nuclear force.

Therefore, Assertion is incorrect but Reason is correct.

**68** (*c*) In radioactivity, electrons are emitted in the form of $\beta({}_{-1}e^0)$ particles. It is also called negative $\beta$-decay, where an unstable nucleus emit energetic electron, anti-neutrino and neutron.

In photoelectric effect, when light rays of frequency greater than or equal to threshold frequency incident on metal surface, then electrons are emitted from metal surface.

For photoelectric emission, it is not necessary that element is unstable but for radioactivity, it is necessary that element is unstable.

Therefore, Assertion is correct but Reason is incorrect.

**69** (*c*) For a $\beta^-$-decay, ${}_0^1n \longrightarrow {}_1^1p + {}_{-1}^{0}e$

Thus, radioactive nuclei can emits $\beta^-$ particles, i.e. electrons.

However, electrons do not exist inside the nucleus.

Therefore, Assertion is correct but Reason is incorrect.

**70** (*a*) Neutrinos are neutral particles with very small mass compared to electrons. They interact very weakly with matter.

Therefore, they are very difficult to detect, since they can even penetrate the large quantity of matter (earth) without any interaction.

Therefore, Assertion and Reason are correct and Reason is the correct explanation of Assertion.

**71** (*a*) As, heavy water ($D_2O$) does not react with slow moving neutrons. So, it is used as a moderator to slow down the fast moving neutron in nuclear reactor.

Therefore, Assertion and Reason are correct and Reason is the correct explanation of Assertion.

**72** (*a*) It is known experimentally that neutrons/thermal neutrons are much likely to cause fission in ${}_{92}^{235}U$ than fast neutrons.

Because, fast neutrons liberated in fission can escape nuclear reactor instead of causing fission reaction.

Therefore, Assertion and Reason are correct and Reason is the correct explanation of Assertion.

**73** (*a*) Thermonuclear reaction are those fusion reactions which is achieved by raising the temperature and pressure of the system so that particles have enough kinetic energy to overcome the coulomb repulsive barrier.

Since, at earth such conditions are very difficult to attain naturally, so this reaction is not possible.

Therefore, Assertion and Reason are correct and Reason is the correct explanation of Assertion.

**74** (*c*) The statements I and II are correct but III is incorrect and it can be corrected as,

Tritium is one of the isotope of hydrogen. It being highly unstable, do not occur naturally and are produced artificially in laboratories.

**75** (*d*) The statement III is correct but I and II are incorrect and these can be corrected as,

A free neutron, unlike a free proton, is unstable.

It decays into a proton, an electron and a anti-neutrino (another elementary particle) and has a mean life of about 1000s.

**76** (*c*) The statement III is correct but I and II are incorrect and these can be corrected as,

In $\alpha$-scattering experiment, it has been observed that, if a beam of 5.5 MeV is used. Then, the scattering of the particles could be understood by assuming that the Coulomb force was solely responsible for scattering.

However, if we use α-particles of higher energies than 5.5 MeV then at some point, the scattering will begin to be affected by short range nuclear forces.

Also, term nucleon is used for a proton or a neutron that is stable inside the nucleus. Thus, the number of nucleons in an atom is its mass number $A$.

**77** (*d*) The statements I, II and III are correct but IV is incorrect and it can be corrected as,

In the region $0 < A < 30$, binding energy increases with mass number.

**78** (*c*) A very heavy nucleus, say $A = 240$ has lower binding energy per nucleon compared to that of nucleus with $A = 120$.

Thus, if a nucleus, $A = 240$ breaks into two $A = 120$ nuclei, nucleons get more lightly bound.

Similarly, consider two very light nuclei $[A \leq 10]$ joining to form a heavier nucleus. The binding energy per nucleon of the fused heavier nuclei is more than the BE per nucleon of the lighter nuclei. Thus, in both the cases BE of formed nucleus is more than that of initial nucleus.

**79** (*d*) The statements I and III are incorrect and these can be corrected as,

$\Delta N$ is the number of nuclei that decay and hence is always positive. $dN$ is the change in $N$, which may have either positive or negative sign. Here, for a radioactive sample, it is negative because out of original $N$ nuclei, $\Delta N$ have decayed leaving $(N - \Delta N)$ nuclei.

The disintegration value or $Q$-value of a nuclear reaction is the difference between the initial mass-energy and the total mass-energy of the decay products. It is also the net kinetic energy gained in the process. If initial nucleus $X$ is at rest, then $Q$-value is the kinetic energy of the products.

Rest statement is correct.

**80** (*d*) The statement I is correct but II and III are incorrect and these can be corrected as,

In a nuclear reactor, to sustain a chain reaction, fast moving neutrons are slowed by elastic collision with light nuclei.

It could be done by elastic collision with hydrogen, so that the neutron almost comes to rest and proton carries the energy. This is the same situation as when a marble hits head-on an identical marble at rest. Therefore, in reactor light nuclei called moderators are provided with fissionable nuclei for slowing down fast neutrons.

However, control rods are used to absorb neutrons to control reaction rate.

Also, Apsara reactor in BAARC uses water as moderator. Although other reactors in India uses heavy water $(D_2O)$.

**81** (*c*) The three-stage strategy of nuclear power generation are as follows

In the first stage, $^{238}_{92}U$ captures neutrons and produces highly radioactive nuclei.

The plutonium-239 obtained from reprocessing of the discharged fuel from the reactors, then serves as a fuel for the second stage in the fast breeder reactors. They are so called because they use fast neutrons for sustaining the chain reaction (hence no moderator is needed) and besides generating power, also breed more fissile species (plutonium) than they consume.

The third stage, most significant in the long term, involves using fast breeder reactors to produce fissile U-233 from Th-232 and to build power reactors based on them.

So, the correct order of the stages is II, I and III.

**82** (*b*) The statement II is correct but I is incorrect and it can be corrected as,

As hydrogen in the core gets depleted and becomes helium, the core starts to cool. The star begins to collapse under its own gravity which increases the temperature of the core.

If this temperature increases to about $10^8$ K, fusion takes place again, this time fusion take place of helium nuclei into carbon. This kind of process through fusion can generate higher mass elements.

**83** (*c*) The statement given in option (c) is incorrect and it can be corrected as,

Experiments of scattering of α-particle demonstrated that the radius of a nucleus was smaller the radius of an atom by a factor of about $10^4$.

This means the volume of a nucleus is about $10^{-12}$ times the volume of the atom.

Also, the nuclues contains more than 99.9% of the mass of an atom.

$$\therefore \quad \frac{\text{Mass of nucleus}}{\text{Mass of an atom}} = \frac{99.9}{100} \simeq 1$$

Rest statements are correct.

**84** (*d*) The statement given in option (d) is incorrect and it can be corrected as,

We know that, the number of electrons outside the nucleus of the atom is Z, the atomic number.

The total charge of the atomic electrons is thus $(-Ze)$ and since the atom is neutral, the charge of the nucleus is $(+Ze)$.

Rest statements are correct.

**85** (*c*) The statement given in option (c) is correct but rest are incorrect and these can be corrected as,

If a certain number of neutrons and protons are brought together to form a nucleus of a certain charge and mass, an energy $E_b$ will be released in the process. The energy $E_b$ is called the binding energy of the nucleus.

However, it has been seen that calculated mass of the nucleus is always greater than the actual nuclear mass.

Also, $\quad {}_{Z}X^{A} = {}_{88}\text{Ra}^{226}$

Number of protons = $Z = 88$

Number of neutrons = $A - Z = 226 - 88 = 138.$

**86** (*b*) The statement given in option (b) is correct but rest are incorrect and these can be corrected as,

The nuclear force between two nucleons falls rapidly to zero as their distance is more than a few femtometers. This leads to saturation of forces in a medium or a large sized nucleus, which is the reason for the constancy of the binding energy per nucleon.

However, nuclear force is attractive and sufficiently strong to produce binding energy of a few MeV per nucleon.

The order of the strength of the force is

$$F_{\text{nuclear}} > F_{\text{Coulomb}} > F_{\text{gravitation}}$$

Unlike Coulomb's law and Newton's law of gravitation, there is no simple mathematical form of the nuclear force.

**87** (*a*) As, $\frac{N}{N_0} = \left(\frac{1}{2}\right)^n = \left(\frac{1}{2}\right)^{\frac{t}{T_{1/2}}}$

Given, $T_{1/2} = 1$ month

For $t = 3$ months, $\left(\frac{N}{N_0}\right) = \left(\frac{1}{2}\right)^3 = \frac{1}{8}$

$\therefore$ Disintegrated part in 3 months $= 1 - \frac{1}{8} = \frac{7}{8}$th part

Thus, the statement given in option (a) is correct, rest are incorrect.

**88** (*b*) The statement given in option (b) is correct but rest are incorrect and these can be corrected as,

Atomic energy level spacings are of the order of eV, while the difference in nuclear energy levels is of the order of MeV.

As $Q$ is the difference between the initial mass-energy and the total mass-energy of the decay products.

So, $Q < 0$ for endothermic process
and $Q > 0$ for exothermic process.

If we test 100 bulbs for their life (time span before they burn out or fuse), it will be expected that they will decay (i.e. burn out) at more or less the same time. However, decay of radioactive nuclides are different. They follow law of radioactive decay.

**89** (*c*) Initially, the decay rate $\left(\frac{dN}{dt}\right)$ of $B$ is smaller than that of $A$ but still its decay rate becomes equal to that of $A$ at an instant (intersection point $P$). After point $P$, the decay rate of $B$ becomes greater than that of $A$.

$\therefore$ The statement given in option (c) is correct.

**90** (*c*) The statement given in option (c) is correct but rest are incorrect and these can be corrected as,

In $\alpha$-decay, the mass number of the product nucleus is four less than that of the decaying nucleus while the atomic number decreases by two. Thus, in this decay an unstable atom emits helium atom $({}_2H^4)$.

In both $\beta^+$ and $\beta^-$ decay, the mass number $A$ remains unchanged. In $\beta^-$ decay, the atomic number $Z$ of the nucleus goes up by 1 and an electron is emitted, while in $\beta^+$ decay $Z$ goes down by 1 and an positron is emitted.

In $\gamma$-decay, a nucleus in an excited state spontaneously decays to its ground state without change in its atomic number and a photon is emitted with the energy equal to the difference in the two energy levels of the nucleus.

**91** (*b*) The statement given in option (b) is correct but rest are incorrect and these can be corrected as,

The disintegration energy in fission first appears as kinetic energy of the fragments and neutrons. Eventually, it will be transferred to surrounding matter, which appears as heat.

Neutron-induced nuclear reaction is fission.

Also, greater the binding energy, the less is the total mass of system, like nuclei.

**92** (*d*) The statement given in option (d) is incorrect and it can be corrected as,

Research nuclear reactors not only generate power but also it is an excellent source for production of a variety of radioactive isotopes that find application in diverse fields, industry, medicine and agriculture.

Rest statements are correct.

**93** (*c*) For ${}^{235}_{92}U$ and ${}^{238}_{92}U$, $Z = 92$, i.e. number of protons are equal. Thus, they are isotopes.

In ${}^{17}_{8}O$ and ${}^{17}_{9}F$, $A = 17$, i.e. mass number are same. Thus, they are isobars.

In ${}^{18}_{8}O$ and ${}^{19}_{9}F$, number of neutrons, $N = A - Z = 10$, for both nuclei.

So, these are isotones.

In ${}^{7}_{3}Li$, $A = 7$ and $Z = 3 \Rightarrow N = A - Z = 4$

In ${}^{7}_{4}Be$, $A = 7$ and $Z = 4 \Rightarrow N = A - Z = 3.$

As in both elements $N$ and $Z$ values are interchanged (i.e., lateral inversion of image of a plane mirror).

So, they are mirror nuclei.

Hence, A $\rightarrow$ 4, B $\rightarrow$ 3, C $\rightarrow$ 2 and D $\rightarrow$ 1.

**94** (*a*) $\alpha$-decay is the process in which a helium nucleus ${}^{4}_{2}He$ is emitted.

$\beta^+$-decay is the process in which positron is emitted, while $\beta^-$-decay is the process in which electron is emitted. So, these processes are as follow,

$${}^{242}_{94}Pu \longrightarrow {}^{238}_{92}U + {}^{4}_{2}He \quad (\alpha\text{-decay})$$

$${}^{210}_{83}Bi \longrightarrow {}^{210}_{84}Po + e^- + \bar{\nu} \quad (\beta^-\text{-decay})$$

$${}^{97}_{43}Te \longrightarrow {}^{97}_{42}Mo + e^+ + \nu \quad (\beta^+\text{-decay})$$

Hence, A $\rightarrow$ 3, B $\rightarrow$ 2 and C $\rightarrow$ 1.

**95** (*a*) Average atomic mass $(m)$ = Weighted average of all isotopes

$$\Rightarrow m = \frac{90.51 \times 19.99 + 0.27 \times 20.99 + 9.22 \times 21.99}{90.51 + 0.27 + 9.22}$$

$$= \frac{1809.29 + 5.67 + 202.75}{100} = \frac{2017.7}{100} = 20.18 \text{ u}$$

Thus, the average atomic mass of neon is 20.18 u.

**96** (*b*) Given, mass of proton, $m_p = 1.007834$ u

Mass of neutron, $m_n = 1.00867$ u

${}^{14}_{7}N$ nucleus contains 7 protons and 7 neutrons.

Mass defect $(\Delta m)$ = mass of nucleons – mass of nucleus

$$= 7m_p + 7m_n - m_N$$
$$= 7 \times 1.00783 + 7 \times 1.00867 - 14.00307$$
$$= 7.05481 + 7.06069 - 14.00307 = 0.11243 \text{ u}$$

Binding energy of nitrogen nucleus

$$= \Delta m \times 931 \text{MeV}$$
$$= 0.11243 \times 931 \text{ MeV} = 104.67 \text{ MeV}$$

**97** (*c*) Given, mass of coin = 3.0 g and $m_{Cu} = 62.92960$ u

$$\text{Number of atoms in 1 g of Cu} = \frac{6.023 \times 10^{23}}{63}$$

$$\text{Number of atoms in 3 g of Cu} = \frac{6.023 \times 10^{23}}{63} \times 3$$
$$= 2.868 \times 10^{22}$$

Number of protons in Cu atom = 29

Number of neutrons in Cu atom = 63 – 29 = 34

Mass defect in each atom,

$$\Delta m = 29 \times m_p + 34 \times m_n - m_{Cu}$$

Here, $m_p = 1.00783$ u

and $m_n = 1.00867$ u

$$\Rightarrow \Delta m = 29 \times 1.00783 + 34 \times 1.00867 - 62.92960$$
$$= 0.59225 \text{ u}$$

$\therefore$ Total mass defect in all atoms

$$= 0.59225 \times 2.868 \times 10^{22} = 1.6985 \times 10^{22} \text{ u}$$

Binding energy = Mass defect × 931 MeV

$$= 1.6985 \times 10^{22} \times 931$$
$$= 1.58 \times 10^{25} \text{ MeV}$$

Thus, the energy required to separate all the neutrons and protons is $1.58 \times 10^{25}$ MeV, i.e. equal to binding energy.

**98** (*a*) Activity, $\frac{dN}{dt} = 8 \text{ mCi} = 8 \times 10^{-3} \times 3.7 \times 10^{10}$

$$= 8 \times 3.7 \times 10^{7} \text{ disintegration/s}$$

$(\because 1 \text{ Ci} = 3.7 \times 10^{10}$ disintegration/s)

Half-life of ${}^{60}_{27}Co$, $T_{1/2} = 5.3$ yr

$$= 5.3 \times 365 \times 24 \times 60 \times 60$$
$$= 1.67 \times 10^{8} \text{ s}$$

We know that,

$$\lambda = \frac{0.693}{T_{1/2}} = \frac{0.693}{1.67 \times 10^{8}} = 4.14 \times 10^{-9}/\text{s}$$

$$\text{Activity,} \quad \frac{dN}{dt} = \lambda N$$

$$\text{or} \quad N = \frac{dN/dt}{\lambda}$$

$$= \frac{8 \times 3.7 \times 10^{7}}{4.14 \times 10^{-9}} = 7.149 \times 10^{16}$$

By using the concept of Avogadro number,

Mass of $6.023 \times 10^{23}$ atoms of ${}^{60}_{27}Co$ = 60 g

$$\text{Mass of } 7.133 \times 10^{16} \text{ atoms of } {}^{60}_{27}Co = \frac{60 \times 7.133 \times 10^{16}}{6.023 \times 10^{23}}$$
$$= 7.12 \times 10^{-6} \text{ g}$$

Thus, the required mass of ${}^{60}_{27}Co$ is $7.12 \times 10^{-6}$ g.

**99** (*b*) Given, half-life of ${}^{90}_{38}Sr$, $T_{1/2} = 28$ yr

$$= 28 \times 365 \times 24 \times 60 \times 60 \text{ s}$$

According to Avogadro number concept,

90 g of Sr contains = $6.023 \times 10^{23}$ atom

$$\text{15 mg of Sr contains} = \frac{6.023 \times 10^{23} \times 15 \times 10^{-3}}{90}$$

Number of atoms, $N = 1.0038 \times 10^{20}$

$$\text{Activity,} \quad \frac{dN}{dt} = \lambda N$$

$$\text{or} \quad \frac{dN}{dt} = \left(\frac{0.693}{T_{1/2}}\right) N \quad \left(\because \lambda = \frac{0.693}{T_{1/2}}\right)$$

$$= \frac{0.6931 \times 1.0038 \times 10^{20}}{28 \times 365 \times 24 \times 60 \times 60}$$

$$\Rightarrow \frac{dN}{dt} = 7.877 \times 10^{10} \text{ disintegration/s}$$

$$= 7.877 \times 10^{10} \text{ Bq}$$

Thus, the disintegration rate of 15 mg of this isotope is $7.877 \times 10^{10}$ Bq.

**100** (*a*) According to the concept of Avogadro number,

The number of atoms in 239 g of ${}^{239}_{94}Pu = 6.023 \times 10^{23}$

Number of atoms in 1 kg of ${}^{239}_{94}Pu$

$$= \frac{6.023 \times 10^{23} \times 1000}{239} = 2.52 \times 10^{24}$$

The average energy released in one fission = 180 MeV

So, total energy released in fission of 1 kg of ${}^{239}_{94}Pu$

$$= 180 \times 2.52 \times 10^{24} = 4.53 \times 10^{26} \text{ MeV}$$

**101** (*a*) Given, mass of ${}^{10}_{5}B = 10.01294$ u

Mass of ${}^{11}_{5}B = 11.00931$ u

Atomic mass of boron = 10.811 u

Let the abundance of ${}^{10}_{5}B$ be $x$%.

So, the abundance of $^{11}_{5}B$ be $(100 - x)\%$.

Atomic mass = Weighted average of the isotopes

$$10.811 = \frac{x \times 10.01294 + (100 - x) \times 11.00931}{(x + 100 - x)}$$

$$\Rightarrow \quad x = \frac{19.831}{0.99637} = 19.9$$

$\therefore$ Abundance of $^{10}_{5}B$, $x = 19.9\%$

and Abundance of $^{11}_{5}B$, $(100 - x) = 100 - 19.9 = 80.1\%$

Thus, the abundances of $^{10}_{5}B$ and $^{11}_{5}B$ are 19.9% and 80.1%, respectively.

**102** (*b*) Given, mass of proton $m_p = 1.00783$ u

Mass of neutron, $m_n = 1.00867$ u

$^{56}_{26}Fe$ contains 26 protons and $(56 - 26) = 30$ neutrons.

Mass defect $(\Delta m)$ = Mass of nucleons – Mass of nucleus of $^{56}_{26}Fe$

$$\text{Mass defect, } \Delta m = 26 \times m_p + 30 \times m_n - m_N$$
$$= 26 \times 1.00783 + 30 \times 1.00867 - 55.934939$$
$$= 26.20345 + 30.25995 - 55.934939$$
$$= 0.528461 \text{ u}$$

$$\text{Total binding energy} = \Delta m \times 931 \text{MeV}$$
$$= 0.528461 \times 931.5$$
$$= 492.26 \text{ MeV}$$

Average binding energy per nucleon of $^{56}_{26}Fe$

$$= \frac{\text{Binding energy}}{\text{Total number of nucleons}} = \frac{492.26}{56} = 8.79 \text{ MeV}$$

**103** (*b*) Given, $P = 1000 \text{ MW} = 1000 \times 10^6$ W

Let $x$ kg of $^{235}U$ is used.

According to Avogadro number concept,

235 g of $^{235}U$ contains $= 6.023 \times 10^{23}$ atoms

$$\therefore \quad x \text{ kg of } ^{235}U \text{ contains} = \frac{6.023 \times 10^{23}}{235 \times 10^{-3}} \times x \text{ atoms}$$

As half fuel is used in 5 yr and each atoms gives energy of 200 MeV, so energy given by fuel is

$$= \frac{1}{2} \times \left( \frac{6.023 \times 10^{23}}{235 \times 10^{-3}} \times x \text{ atoms} \right) \times 200 \text{ MeV}$$

$$= \frac{6.023 \times 10^{23} \times x \times 200 \times 1.6 \times 10^{-13}}{235 \times 2 \times 10^{-3}} \text{ J} \quad \ldots\text{(i)}$$

Energy produced in reactor in 5 yr as 80%

$$= 1000 \times 10^6 \times 5 \times 365 \times 24 \times 60 \times 60 \times \frac{80}{100}$$

(from formula, $E = Pt$) ...(ii)

By equating Eqs. (i) and (ii), we get

$$\frac{6.023 \times 10^{23} \times 200 \times 1.6 \times 10^{-13} x}{235 \times 2 \times 10^{-3}} = \frac{10^9 \times 5 \times 365 \times 24 \times 3600 \times 80}{100}$$

$$\Rightarrow \quad x = \frac{5 \times 365 \times 24 \times 36 \times 80 \times 235 \times 2 \times 10^{-3} \times 10^9}{6.023 \times 10^{10} \times 200 \times 1.6}$$

$$= 3071.5 \text{ kg}$$

The initial amount of $^{235}_{92}U$ is 3071.5 kg or approximately 3070 kg.

**104** (*c*) Disintegration laws are based on an statistical model and give average values. Since, $T_{1/2}$ is the time radioactive nuclide takes for a sample that has initially say $N_0$ radioactive nuclide to reduce to $N_0 / 2$. Thus, the containers will in general have different number of atoms, but on an average will have around 5000 atoms.

**105** (*b*) As an α-particle carrier 2 units of positive charge and a β-particle carrier one unit of negative charge, but γ-particle carrier no charge.

Therefore, electronic energy levels of the atom change for α and β-radioactivity, but not for γ-radioactivity.

**106** (*a*) For a $\beta^-$-decay,

$$^{A}_{Z}X \longrightarrow {}^{A}_{Z+1}Y_1 + {}^{0}_{-1}e + \nu^-$$

$\therefore$ $Q_1$ value of decay

$$= [(M_X - Z M_e) - \{M_Y - (Z + 1) M_e\} - M_e]c^2$$
$$= (M_X - M_Y)c^2$$

For $\beta^+$-decay, $^{A}_{Z}X \longrightarrow {}^{A}_{Z-1}Y_2 + {}^{0}_{1}e + \nu$

$\therefore$ $Q_2$ value of decay

$$= [(M_X - ZM_e) - \{M_Y - (Z - 1) M_e\} - M_e]c^2$$
$$= (M_X - M_Y - 2M_e)c^2$$

**107** (*b*) Force between neutrons is only attractive and is short range force. However, force between protons is replusive.

$\therefore$ For stable nuclides, either $\frac{N}{Z} = 1$ or $\frac{N}{Z} > 1$

So, heavy stable nuclei have more neutrons than protons.

**108** (*b*) For efficient energy transfer between two bodies, both must be of comparable masses.

If a heavy nucleus is taken, then neutron rebounds with nearly same energy. So, the purpose of the moderator will not be served as the elastic collision neutrons with heavy nuclei will not slow them down.

**110** (*a*) Tritium $({}_1H^3)$ contains 1 proton and 2 neutrons.

A neutron decays as

$$n \longrightarrow p + e^- + \bar{\nu}$$

So, the nucleus may have two protons and 1 neutron. This means that tritium will transform into ${}_2He^3$ with two protons and 1 neutron. However, such a case does not happen as the Triton energy is less than that of ${}_2He^3$ nucleus.

CHAPTER > 14

# Semiconductor Electronics; Materials, Devices and Simple Circuits

## KEY NOTES

### Classification of Metals, Insulators and Semiconductors

- **On the basis of the relative values of electrical conductivity** ($\sigma$) or resistivity ($\rho = 1/\sigma$), solids are broadly classified as

  (i) **Metals** They possess very low resistivity (or high conductivity).

  $$\rho \sim 10^{-2}\text{-}10^{-8}\ \Omega\text{-m}$$
  $$\sigma \sim 10^{2}\text{-}10^{8}\ \text{Sm}^{-1}$$

  (ii) **Semiconductors** They have resistivity or conductivity intermediate to metals and insulators.

  $$\rho \sim 10^{-5}\text{-}10^{6}\ \Omega\text{-m}$$
  $$\sigma \sim 10^{5}\text{-}10^{-6}\ \text{Sm}^{-1}$$

  (iii) **Insulators** They have high resistivity (or low conductivity).

  $$\rho \sim 10^{11}\text{-}10^{19}\Omega\text{-m}$$
  $$\sigma \sim 10^{-11}\text{-}10^{-19}\ \text{Sm}^{-1}$$

- Inside the crystal of a solid each electron has a unique position and no two electrons see exactly the same pattern of surrounding charges. Because of this each electron will have a different energy level. These different energy levels with continuous energy variation are called **energy bands.**

  These bands are of two types

  (i) **Valence band** The energy band which includes the energy levels of the valence electrons.

  (ii) **Conduction band** The energy band above the valence band.

**Note** With no external energy, all the valence electrons will reside in the valence band.

- The gap between the top of the valence band and the bottom of the conduction band is called the **energy band gap**. It may be large, small or zero depending upon the material.
- **On the basis of the energy band gap,** solids can be classified as

  (i) **Metals** In these, either conduction band is partially filled and valence band is empty or conduction and valence bands overlap. When there is overlap electrons from the valence band can easily move into the conduction band. Therefore, resistance of such materials is low or the conductivity is high. It is as shown in the figure below

(i)

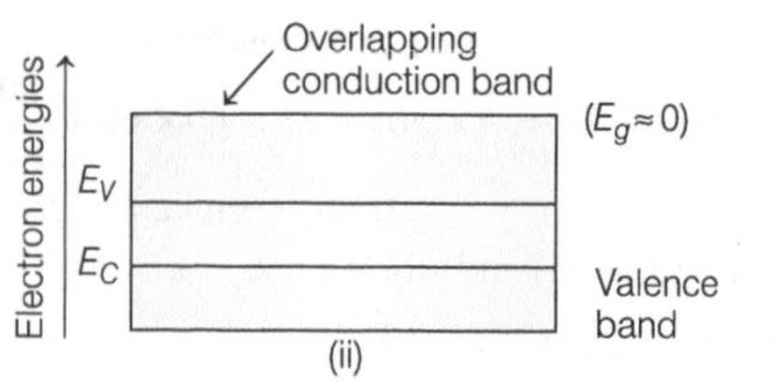


(ii)

(ii) **Insulators** In these, a large band gap $E_g$ exists ($E_g > 3$ eV). There are no electrons in the conduction band and therefore no electrical conduction is possible. It is as shown in the figure below

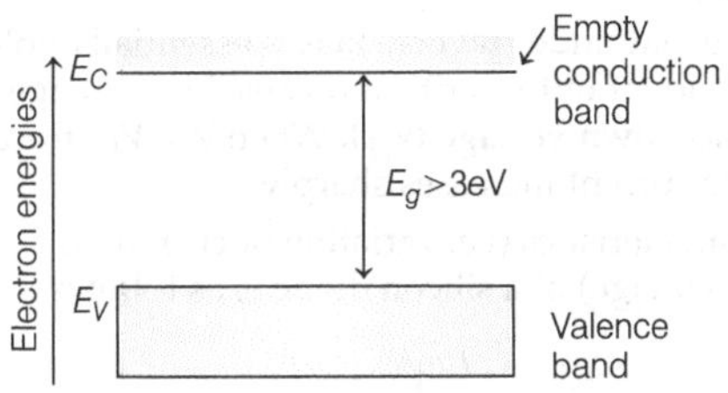


(iii) **Semiconductors** In these, finite but small energy gap exists ($E_g < 3$ eV). Because of this at room temperature, some electrons from the valence band can move to conduction band.

Thus, the resistance of these materials is not as high as that of the insulators. It is as shown in the figure below

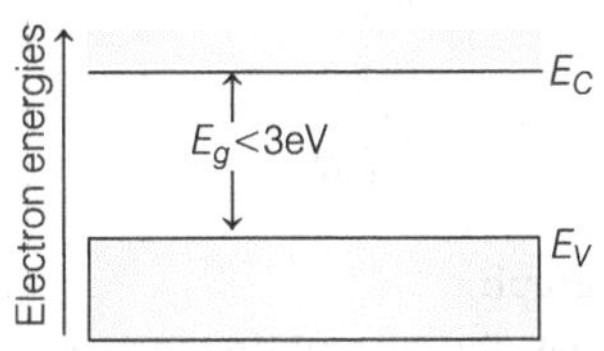


## Intrinsic Semiconductor

- These type of semiconductors are pure semiconductors without any impurity, e.g. silicon, germanium, etc. Following are few important points related to this
  (i) In intrinsic semiconductors, the number of free electrons$n_e$ is equal to the number of holes $n_h$.
  i.e. $n_e = n_h = n_i$
  where, $n_i$ is called **intrinsic carriers concentration**.

**Note** Semiconductor possess the unique property in which, apart from electrons, the holes also move.

  (ii) The free electron moves completely independently as conduction electron and give rise to an electron current $I_e$, under an applied field. Also, under the field affect, holes move towards negative potential giving rise to hole current $I_h$, so
  Total current, $I = I_e + I_h$
  (iii) At equilibrium, the rate of generation is equal to the rate of recombination of charge carriers. The recombination occurs due to the electron colliding with a hole.
  (iv) At $T = 0$ K, it behaves like an insulator. However, at $T > 0$ K, some electrons excite from valence to conduction band, leaving equal number of holes there.

## Extrinsic Semiconductor

- When a small amount (a few parts per million ppm) of a suitable impurity is added to the pure semiconductor, its conductivity increases manifold.
  Such materials are known as extrinsic semiconductors or impurity semiconductors.
- The deliberate addition of a desirable impurity is called **doping** and the impurity atoms are called **dopants**.
  So, extrinsic semiconductors are also called **doped semiconductors**.
- For doping, size of the dopant and the semiconductor atoms should be nearly same.
- There are two types of dopants used in doping the tetravalent Si or Ge
  (i) **Pentavalent** (valency 5) ; like arsenic (As), antimony (Sb), phosphorous (P), etc.
  (ii) **Trivalent** (valency 3); like indium (In), boron (B), aluminium (Al), etc.
- Based on impurity (or dopant) extrinsic semiconductor is of two types: *n*-type and *p*-type semiconductors
  (i) ***n*-type semiconductors** This type of semiconductor is obtained when Si or Ge is doped with pentavalent element. In this, electrons are majority carriers and holes are minority carriers, so the rate of recombination of holes would increase. Thus, we have, $n_e >> n_h$.
  (ii) ***p*-type semiconductors** This type of semiconductor is obtained when Si or Ge is doped with a trivalent impurity like Al, B, In, etc. In this, holes are the majority carriers and electrons are minority carriers, so the recombination process will reduce $n_i$ to $n_e$. Thus, we have, $n_h >> n_e$.
- Doped crystal maintains an overall charge neutrality as the charge of additional charge carriers is just equal and opposite to that of the ionised cores in the lattice.
- Semiconductors energy band structure is also affected by doping. In the case of
  (i) ***n*-type semiconductor,** the donor energy level $E_D$ exists slightly below the bottom $E_C$ of conduction band and electrons from this level move into the conduction band with very small supply of energy.
  (ii) ***p*-type semiconductor,** an acceptor energy level $E_A$ exists slightly above the top $E_V$ of the valence band with very small supply of energy an $e^-$ from valence band can jump to level $E_A$ and ionise the acceptor negatively.
- The electron and hole concentration in an extrinsic semiconductor in thermal equilibrium is given by
  $$n_e n_h = n_i^2$$

## *p-n* Junction

- It is the basic building block of many semiconductor devices like diodes, transistors, etc.
- **Current** flowing across a *p-n* junction are of two types as follows
  (i) **Diffusion Current** ($I_{df}$) Current flowing from *p*-side to *n*-side due to diffusion of electrons and holes because of concentration difference.

KEY NOTES

(ii) **Drift Current** ($I_{dr}$) Current flowing from $n$-side to $p$-side due to electrons-hole pairs (because of thermal collision) created in depletion region.

- It is formed when diffusion current is equal to drift current. Thus, under equilibrium, there is no net current.
- The loss of $e^-$ from the $n$-region and the gain of $e^-$ by the $p$-region causes a difference of potential across the junction of two regions. This potential tends to prevent the movement of electrons from $n$-region to $p$-region and is often called a **barrier potential.**

## Semiconductor Diode

- It is a basically $p$-$n$ junction with metallic contacts provided at the ends for application of an external voltage (bias). It is a two terminal device. Diode and its symbol is as shown below

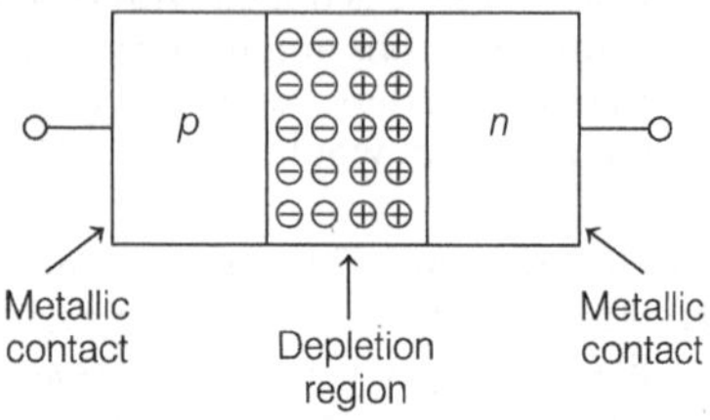


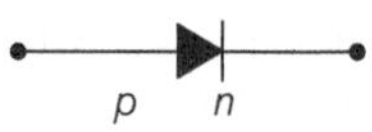


- When an external voltage $V$ is applied across a semiconductor diode such that $p$-side is connected to the positive terminal of the battery and $n$-side to the negative terminal of the battery, then it is said to be **forward biased**.

  Following are few important points related to it
  (i) The effective barrier height under forward bias is $(V_0 - V)$.
  (ii) Due to applied voltage, electrons from $n$-side cross the depletion region and $p$-side. Similarly, holes from $p$-side cross the junction and reach the $n$-side. This process under forward bias is called **minority carrier injection.**
  (iii) The total diode forward current is the sum of hole diffusion current and conventional current of electron diffusion. The magnitude of current is in mA.

- When an external voltage ($V$) is supplied across the diode such that $n$-side is connected to positive terminal of the battery and $p$-side is connected to negative terminal of the battery, then it is said to be **reverse biased.**

  Following are few important points related to it
  (i) The direction of applied voltage is same as the direction of barrier potential. So, effective barrier height will be $(V_0 + V)$.
  (ii) The current under reverse bias is essentially voltage independent upto a critical reverse bias voltage, known as **breakdown voltage ($V_{br}$).** When $V = V_{br}$, the diode reverse current increases sharply.

- The $V$-$I$ characteristics (i.e. variation of current as a function of applied voltage) of a silicon diode is as follows

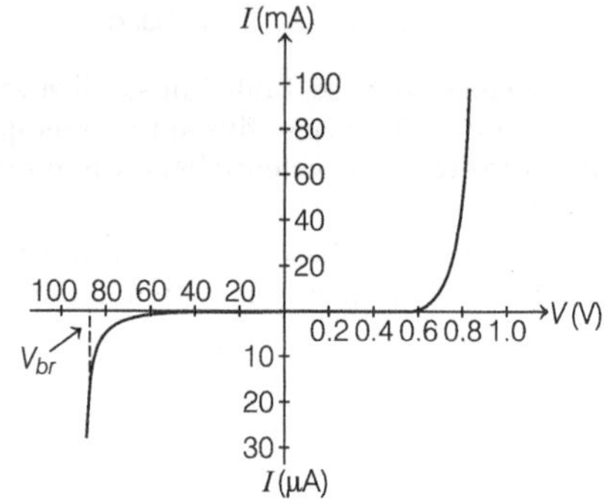


From the above graph,
(i) In forward bias, the current first increases very slowly, almost negligible till the voltage across the diode crosses a certain value. After the characteristic voltage, the diode current increases significantly. This voltage is called the **threshold voltage or cut-in voltage** (~ 0.2 V for Ge and ~ 0.7 V for Si).
(ii) In reverse bias, the current is very small (~ μA) and almost remains constant with change in bias. It is called **reverse saturation current**. However, at breakdown voltage the current suddenly increases.
(iii) **Dynamic resistance is** defined as the ratio of small change in voltage $\Delta V$ to a small change in current $\Delta I$,

i.e. $$r_d = \frac{\Delta V}{\Delta I}$$

**Note** $p$-$n$ junction diode primarily allows the flow of current only in one direction (forward bias).

## Junction Diode as a Rectifier

- If an alternating voltage is applied across a diode, the current flows only in that part of the cycle when the diode is forward biased. This property is used to rectify alternating voltages and the circuit used for this purpose is called a **rectifier.**

  It is of two types : half-wave rectifier and full-wave rectifier.

- **Half-wave rectifier** It is a circuit in which, an alternating voltage is applied across a diode in series with a load. A pulsating voltage will appear across the load only when the half cycles of the AC input when the diode is forward biased. It is as shown in figure below

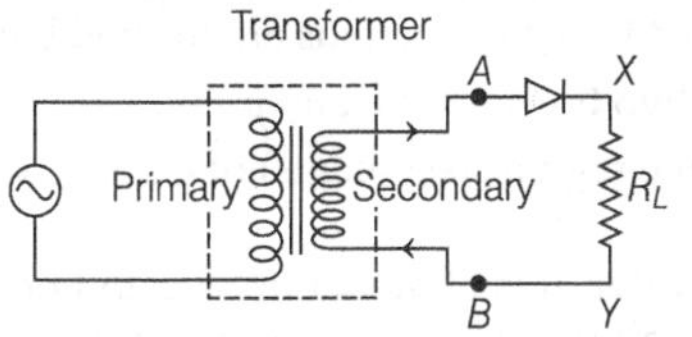


- Input AC voltage and output voltage waveforms from the half-wave rectifier are shown below

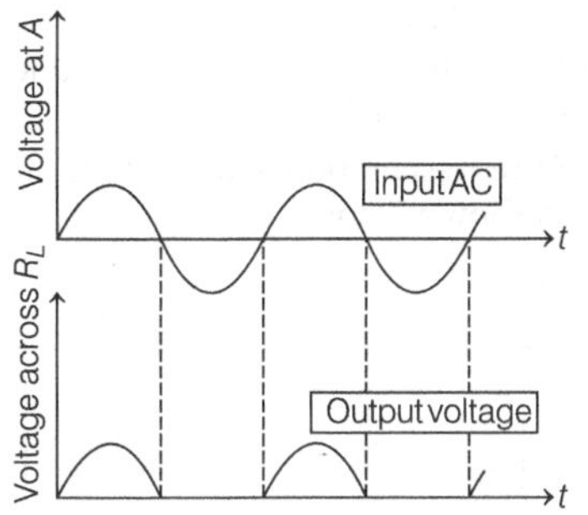


- **Full-wave rectifier** It is a circuit which gives output rectified voltage corresponding to both the positive as well as negative half of the AC cycle. It is as shown below

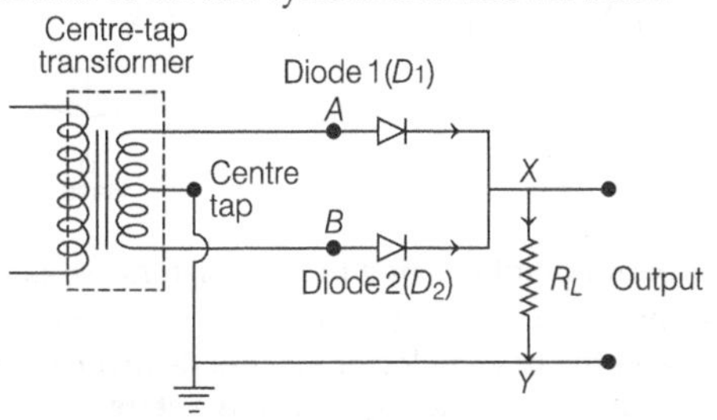


- Input waveforms given to $D_1$ at $A$, $D_2$ at $B$ and output across $R_L$ connected in a full-wave rectifier is as follows

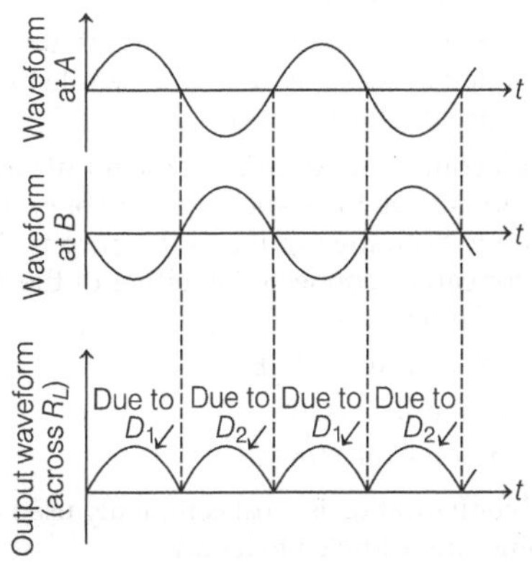


- Rectified voltage in the rectifier are in the form of pulses. To get steady DC output from pulsating voltage, a capacitor in parallel or an inductor in series is connected with $R_L$.

Since, the additional circuit filter out the AC ripple and gives pure DC voltage, so they are called **filters**.

- The rate of fall of voltage across the capacitor in filters depends inversely upon the product of capacitance $C$ and the effective resistance $R_L$ used in the circuit and is called **time constant**.

## Special Purpose *p-n* Junction Diodes

There are many types of *p-n* junction diode based on their purpose some of them are discussed here.

### Zener Diode

- It is a special purpose semiconductor diode designed to operate under reverse bias in the breakdown region.
- Its symbol is as shown below

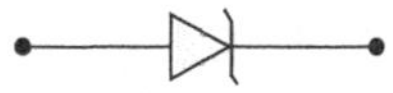

- *I-V* characteristics is as follows

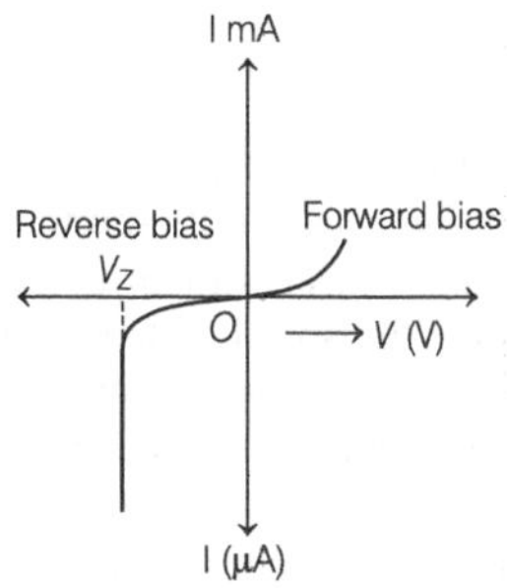


- Zener diode can be used as a **voltage regulator**, in which the constant output voltage is taken across a load resistance connected in parallel with it.

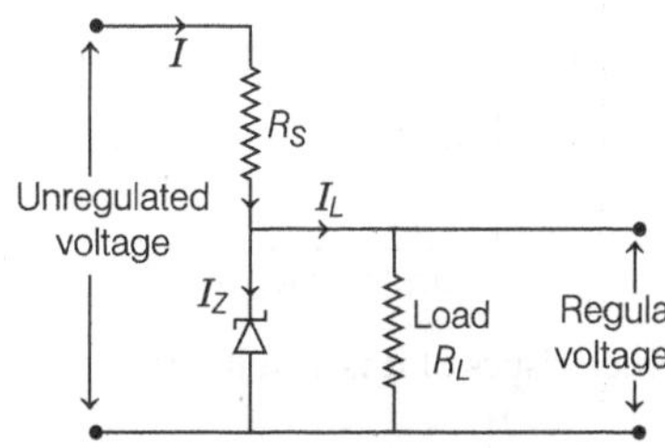


## Optoelectronic Junction Devices

Semiconductor diodes in which carriers are generated by photons (photo excitation) are called optoelectronic devices. e.g. Photodiodes, LED, photovoltaic cell (solar cell), etc.

- **Photodiode** is a junction diode fabricated with a transparent window to allow light to fall on the diode. It is operated under reverse bias.

(i) An illuminated photodiode under reverse bias is as shown below

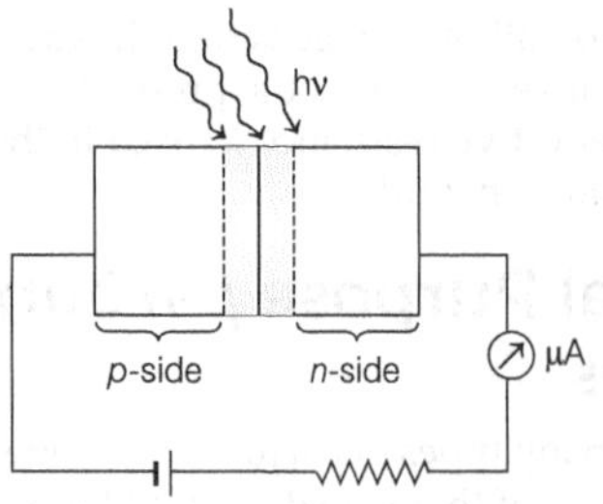


(ii) It can be used as photodetectors to detect optical signals.

(iii) *I-V* characteristics of it is as follows

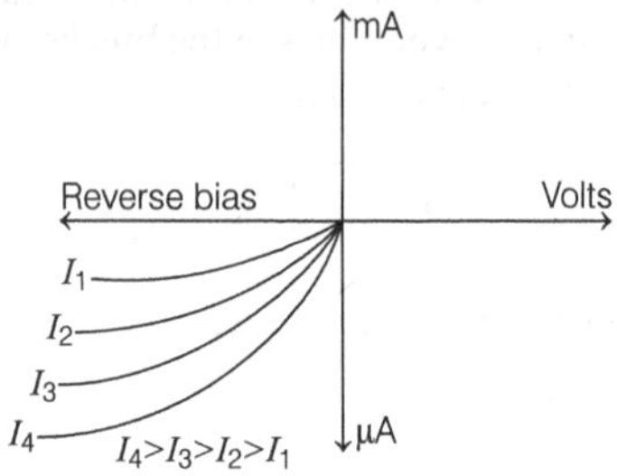


- **Light emitting diode** (LED) is a heavily doped *p-n* junction which under forward bias emits spontaneous radiation.
  - (i) Its *I-V* characteristics is similar to that of a Si-junction diode. But the threshold voltages are much higher and slightly different for each colour.
  - (ii) These are used in remote controls, burglar alarm systems, etc.
- **Solar cell** is basically a *p-n* junction which generates emf when solar radiation falls on it.
  - (i) An illuminated *p-n* junction solar cell is shown below as

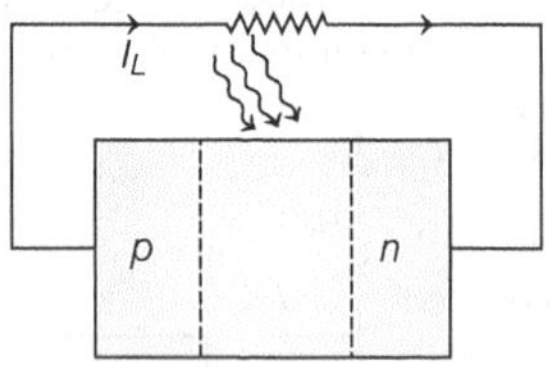


(ii) *I-V* characteristics of it is as follows

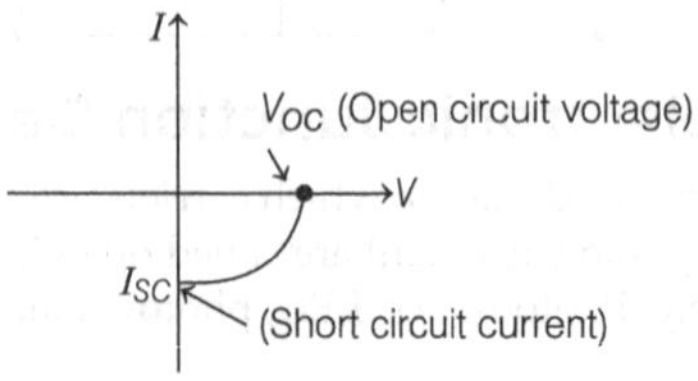


**Note** Semiconductors with band-gap close to 1.5 eV are ideal materials for solar cell fabrication.

KEY NOTES

## Junction Transistor

- It is a three terminal device having three doped regions such that two *p-n* junctions are formed between them.
  It is of two types : *n-p-n* and *p-n-p*.
- The three regions (segments) of a transistor are as follows
  - (i) **Emitter** It is of moderate size and heavily doped. It supplies a large number of majority carriers for current flow through the transistor.
  - (ii) **Base** It is the central segment and is very thin and lightly doped.
  - (iii) **Collector** It is moderately doped and larger in size as compared to the emitter.
- Schematic representation and symbols of *n-p-n* and *p-n-p* is as shown below

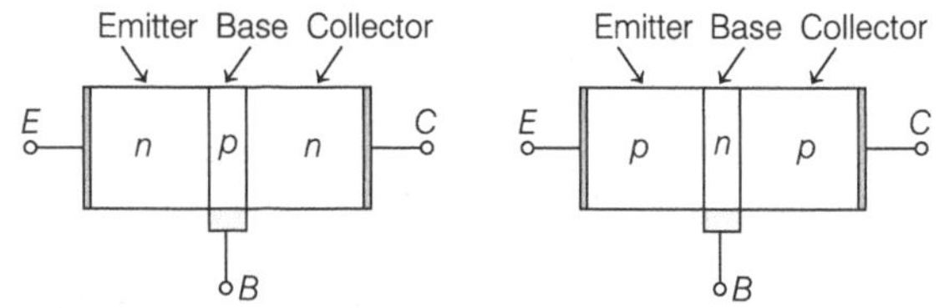


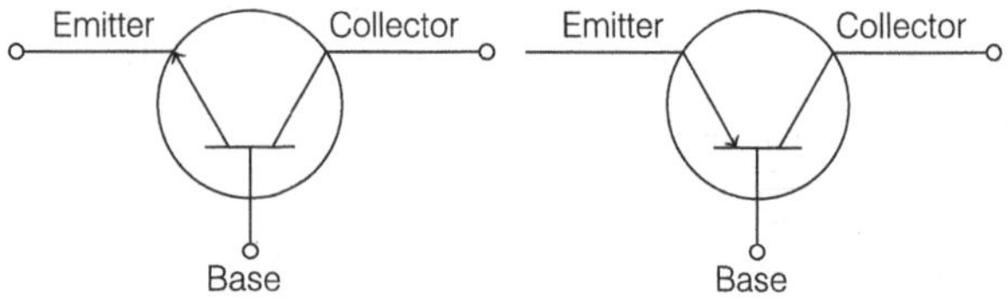


- Important points related to a transistors are as follows
  - (i) When its emitter-base junction is forward biased and the base-collector junction is reverse biased, the transistor is said to be in **active state**.
  - (ii) Generally, $I_E = I_C + I_B$
    (we also see that, $I_C \approx I_E$)
  - (iii) In active state of the transistor, emitter-base junction acts as low resistance while base-collector acts as a high resistance.
- In circuit containing transistor, the input/output connections have to be such that one of *E*, *B* or *C* is common to both the input and the output. Accordingly, transistor can be connected in either of the following three configurations
  - (i) Common Emitter (CE)
  - (ii) Common Base (CB)
  - (iii) Common Collector (CC)

**Note** CE configuration is most commonly used and further points are restricted to it only.

- When a transistor is used in CE configuration, the input is between the base & the emitter and the output is between the collector & the emitter.

Following are few important points related to it

(i) The circuit of *n-p-n* transistor in CE configuration is as follows

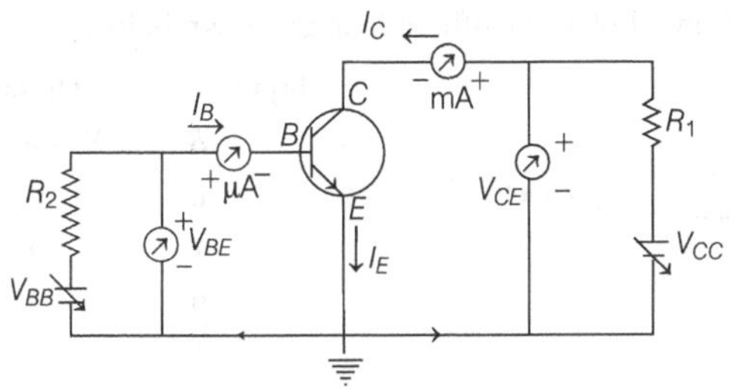


(ii) **Input characteristics** represents the variation of base current $I_B$ with the base - emitter voltage $V_{BE}$. It is as shown below

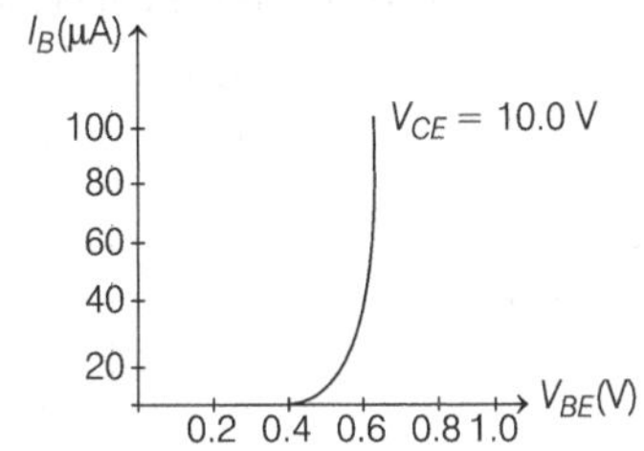


**Note** Increase in $V_{CE}$ appear as increase in $V_{CB}$, its effect on $I_B$ is negligible. As a consequence, input characteristics for various values of $V_{CE}$ will give identical curves.
Hence, it is enough to determine only one input characteristics.

(iii) **Output characteristics** represents variation of the collector current $I_C$ with the collector emitter voltage $V_{CE}$. It is as shown below

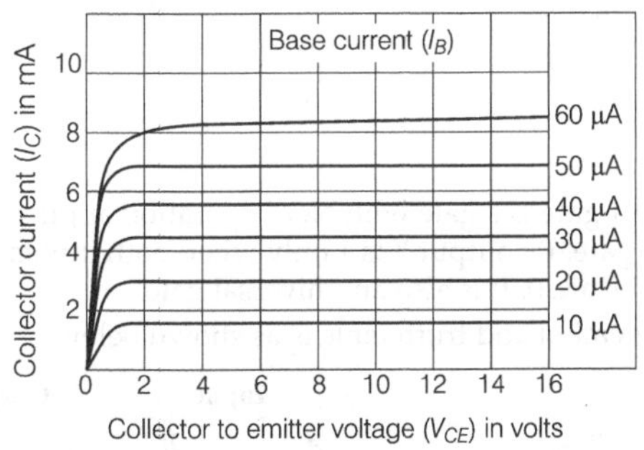


(iv) Linear segments of both input and output characteristics can be used to calculate following AC parameters

(a) **Input resistance** is defined as the ratio of change in base-emitter voltage ($\Delta V_{BE}$) to the resulting change in base current ($\Delta I_B$) at constant collector-emitter voltage ($V_{CE}$). It is given as

$$r_i = \left(\frac{\Delta V_{BE}}{\Delta I_B}\right)_{V_{CE}}$$

(b) **Output resistance** is defined as the ratio of change in collector emitter voltage ($\Delta V_{CE}$) to the change in collector current ($\Delta I_C$) at a constant base current $I_B$.

Mathematically, it is given as $r_o = \left(\frac{\Delta V_{CE}}{\Delta I_C}\right)_{I_B}$

**Note** Reciprocal of the slope of linear part of the output characteristic gives the value of $r_o$.

(c) **Current amplification** factor is defined as the ratio of the change in collector current to the change in base current at a constant collector-emitter voltage ($V_{CE}$) when transistor is in active state. It is given as

$$\beta_{AC} = \left(\frac{\Delta I_C}{\Delta I_B}\right)_{V_{CE}}$$

This is also known as **small signal current gain** and its value is very large.

Also, $\beta_{DC} = \frac{I_C}{I_B}$

## Transistor as a Device

- The transistor can be used as a device application depending on the configuration used, the biasing of *EB* and *BC* junction and the operation region namely cut-off, active region and saturation.
- Transfer characteristic ($V_o$ *vs* $V_i$ curve) of the base biased transistor depicting the operation regions is as follows

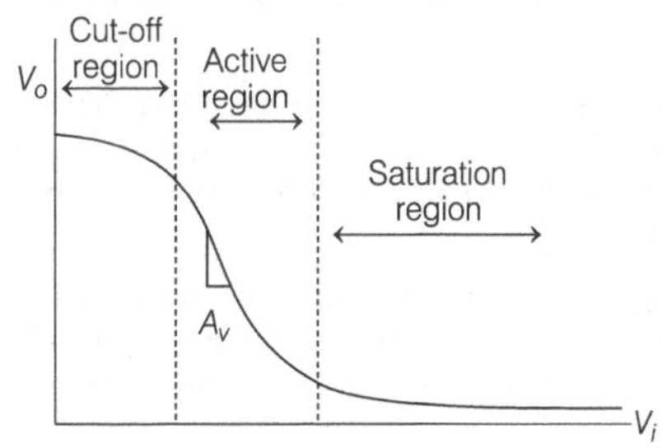


- When the transistor is used in the cut-off region or saturation state, it acts as a **switch**.
- When transistor is used in active state, it can be used as an **amplifier**.
- Some important terms related to it are as given below

(i) $\beta_{AC} = \frac{\Delta I_C}{\Delta I_B} = \frac{i_c}{i_b}$, is also known as AC **current gain** $A_i$ of amplifier.

(ii) **Voltage gain** of the amplifier, $A_V = \frac{V_o}{V_i} = -\frac{\beta_{AC} R_L}{r}$

Negative sign represents that output voltage is opposite with phase with input voltage.

(iii) Mathematically, **power gain** $A_P = \beta_{AC} \times A_V$

KEY NOTES

- Transistor can be used as an **oscillator** which produces electrical oscillation of constant frequency and amplitude without any external input signal.
- External input is necessary to sustain AC signal in the output of an amplifier. In case of oscillator, an AC output is received without any external input signal.

**Note** The feedback in the oscillators can be achieved by inductive coupling on *LC* or *RC* networks.

## Digital Electronics and Logic Gates

- Signals in the form of continuous, time-varying voltage or current is known as **analog signals**, as shown in the figure below

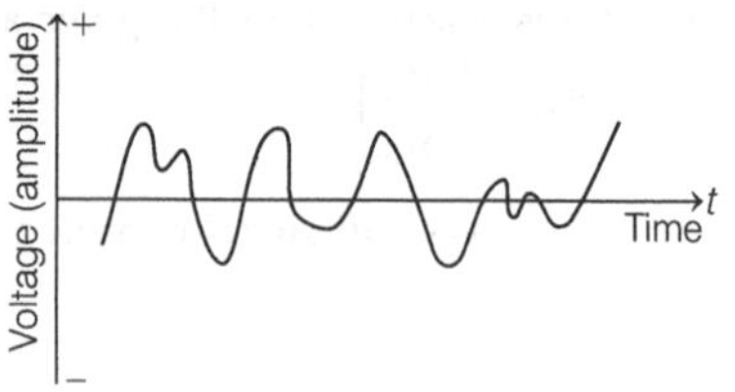


- **Digital signals** are the signals, whose waveform has only discrete values of voltages/current. In digital electronics, only two voltages are used as shown in the figure below

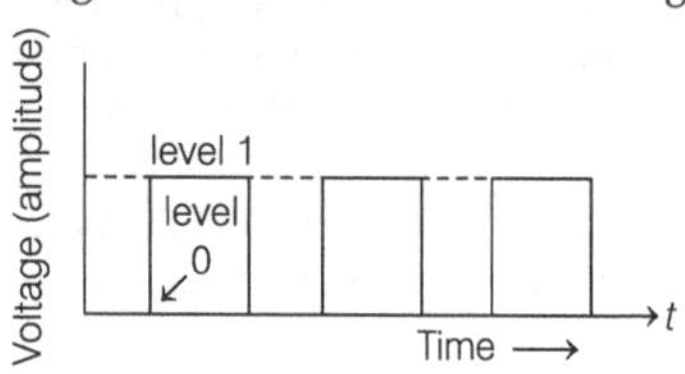


- **Logic gates** are the basic building blocks of digital electronics, which process the digital signals in a specific manner. It follows a logical relationship between the input and output voltages.

**Note** Each logic gate is indicated by a symbol and its function is defined by a truth table that rows are the possible input logic level combinations with their respective output logic levels.

- Various types of logic gates are as follows
  - (i) **NOT gate** is a basic logic gate with one input *A* and one output *Y*. It produces an inverted version of input at its output and is known as **inverter**.

    Its symbol and truth table is as shown below

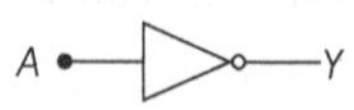


| Input | Output |
|---|---|
| $A$ | $Y = \overline{A}$ |
| 0 | 1 |
| 1 | 0 |

  - (ii) **OR gate** has two or more inputs with one output. The output is 1 when either input *A* or *B* both are 1s, i.e. if any of the input is high output is high.

    Its symbol and truth table is as shown below

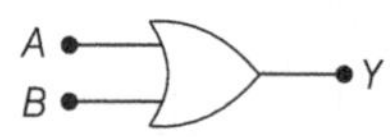


| Input | | Output |
|---|---|---|
| $A$ | $B$ | $Y = A + B$ |
| 0 | 0 | 0 |
| 0 | 1 | 1 |
| 1 | 0 | 1 |
| 1 | 1 | 1 |

  - (iii) **AND gate** has two or more inputs and one output. The output *Y* is 1 only when both inputs *A* and *B* are 1.

    Its symbol and truth table is as shown below

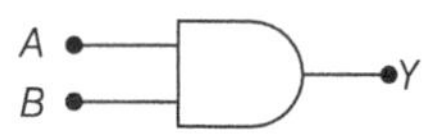


| Input | | Output |
|---|---|---|
| $A$ | $B$ | $Y = A \cdot B$ |
| 0 | 0 | 0 |
| 0 | 1 | 0 |
| 1 | 0 | 0 |
| 1 | 1 | 1 |

  - (iv) **NAND gate** is an AND gate followed by NOT gate. If both inputs *A* and *B* are 1, the output *Y* will not be 1. It is also called **universal gate,** since by using it, other basic gates OR, AND and, NOT can be realised.

    Its symbol and truth table is as shown below

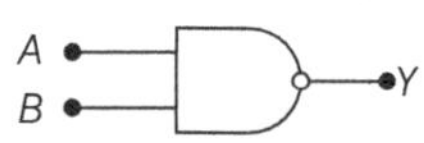


| Input | | Output |
|---|---|---|
| $A$ | $B$ | $Y = \overline{A \cdot B}$ |
| 0 | 0 | 1 |
| 0 | 1 | 1 |
| 1 | 0 | 1 |
| 1 | 1 | 0 |

  - (v) **NOR gate** is a gate with NOT operation applied after OR gate. Its output *Y* is 1 only when both inputs *A* and *B* are 0. It is also an universal gate.

    Its symbol and truth table is as shown below

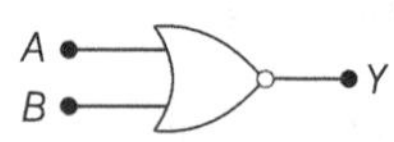


| Input | | Output |
|---|---|---|
| $A$ | $B$ | $Y = \overline{A + B}$ |
| 0 | 0 | 1 |
| 0 | 1 | 0 |
| 1 | 0 | 0 |
| 1 | 1 | 0 |

# Mastering NCERT

## MULTIPLE CHOICE QUESTIONS

## TOPIC 1 ~ Classification of Metals, Conductors and Semiconductors

**1** For the flow of electrons in a vacuum tube like vacuum diode, vacuum is required in the inter-electrode space otherwise,

(a) electrons will not be ejected from cathode
(b) extraction of electrons from remaining gas molecules or atoms will increase abruptly
(c) work function of cathode will be reduced
(d) electrons may lose their energy on collision with air molecules in their path

**2** In solid state semiconductors, the number and the direction of the flow of charge carriers can be altered by

(a) illuminating light
(b) supplying heat
(c) varying applied voltage
(d) All of the above

**3** Correct order of relative values of electrical conductivity $\sigma$ for different types of solid is

(a) $\sigma_{\text{semiconductor}} > \sigma_{\text{insulator}} > \sigma_{\text{metal}}$
(b) $\sigma_{\text{metal}} > \sigma_{\text{semiconductor}} > \sigma_{\text{insulator}}$
(c) $\sigma_{\text{semiconductor}} > \sigma_{\text{metal}} > \sigma_{\text{insulator}}$
(d) $\sigma_{\text{insulator}} > \sigma_{\text{semiconductor}} > \sigma_{\text{metal}}$

**4** Energy band gap of a certain material is neither very large nor zero. However, with the help of some external energy conduction takes place due to (i.e. at room temperature)

(a) vacancies in conduction band
(b) electrons in valence band
(c) electrons in conduction band
(d) all electrons are bound to valence band

**5** Which amongst the following diagram shows energy band positions in a semiconductor (silicon) at 0 K correctly?

(a)
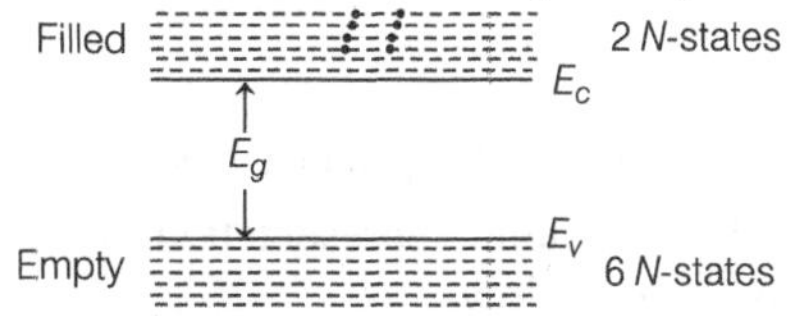


(b)
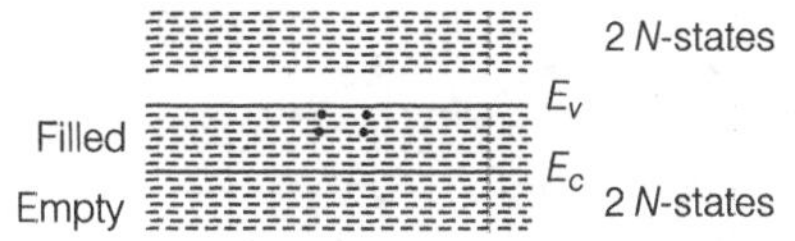


(c)
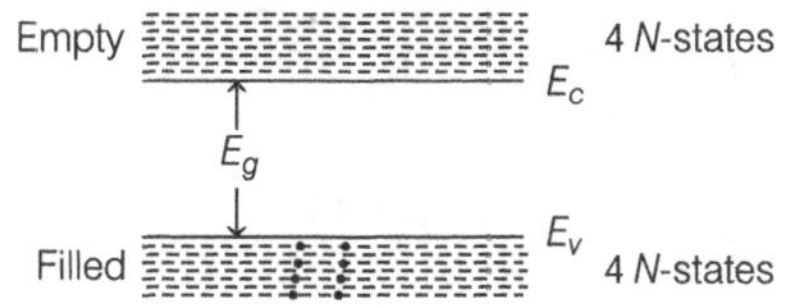


(d)
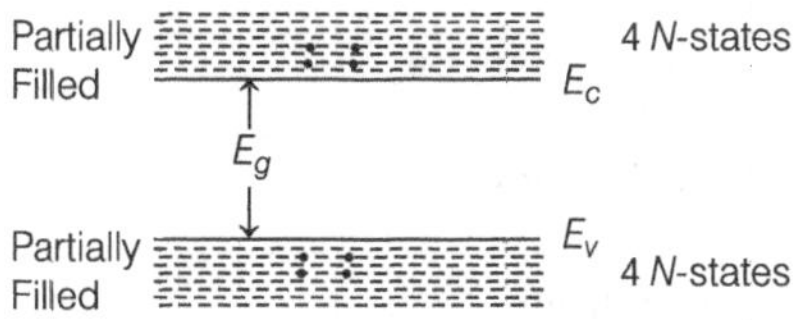


**6** In the energy band diagram of material $A$, conduction band is partially filled and valence band is partially empty and for material $B$, both conduction and valence band overlaps each other. Then, material $A$ and $B$ respectively are

(a) insulator and semiconductor
(b) semiconductor and conductor
(c) Both are semiconductors
(d) Both are conductors

**7** If the energy of a photon of sodium light ($\lambda = 580\,\text{nm}$) equals the band gap of semiconductor, the minimum energy required to create hole-electron pair is

(a) 1.5 eV (b) 3.2 eV (c) 2.1 eV (d) 4.1 eV

# TOPIC 2 ~ Intrinsic and Extrinsic Semiconductors

**8** A piece of semiconductor (say Si) is connected in series with a battery. On increasing the temperature of Si piece, the current in the circuit
(a) decreases
(b) increases
(c) remains unchanged
(d) first increases and then decreases

**9** At an elevated temperature ($T > 0$ K), few of covalent bonds of Si or Ge are broken and a vacancy in the bond is created. Effective charge of the vacancy or hole is
(a) positive
(b) negative
(c) neutral
(d) either positive or negative depending upon the material

**10** Schematic two-dimensional representation of Si structure is as shown below

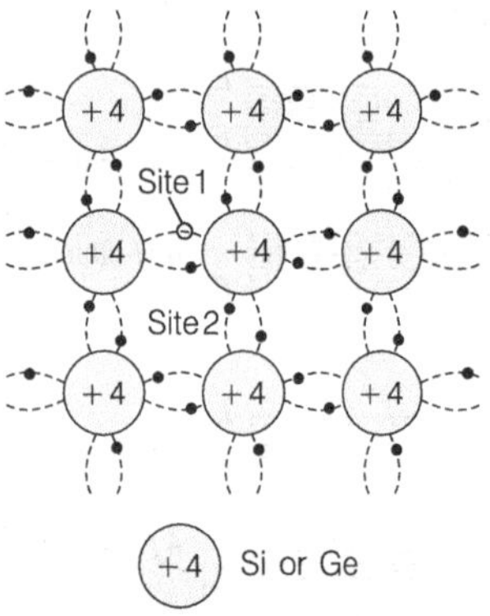


Here, ∘ → hole, • → electrons
Then, at $T = 0$K
(a) it behaves as an insulator
(b) a free electron might have been thermally exicted from site 1
(c) an electron from site 2 can jump to site 1
(d) Both (b) and (c)

**11** At equilibrium condition, the rate of generation of electron-hole pairs in Ge crystal
(a) is more than rate of recombination of electron and hole pairs
(b) is less than rate of recombination of electron and hole pairs
(c) equals to rate of recombination of electron and hole pairs
(d) is always zero

**12** Energy band gap $E_g$ diagram for an intrinsic semiconductor at temperature $T > 0$ K is
(Here, $E_C$ is energy for conduction band and $E_V$ is energy for valence band.)

(a) 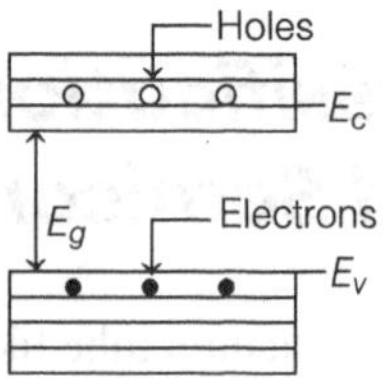


(b) 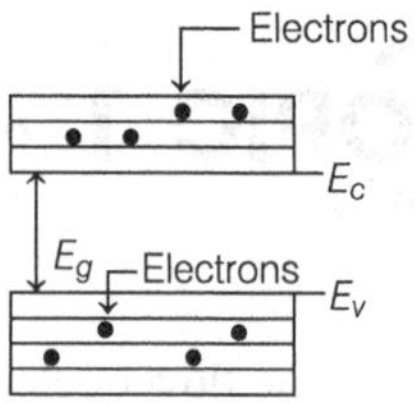


(c) 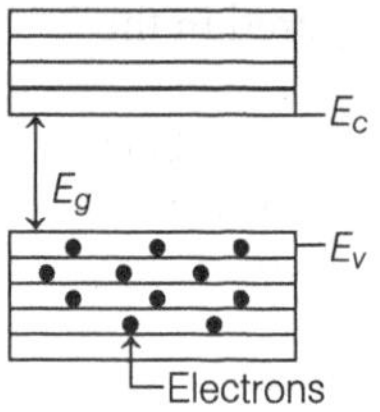


(d) 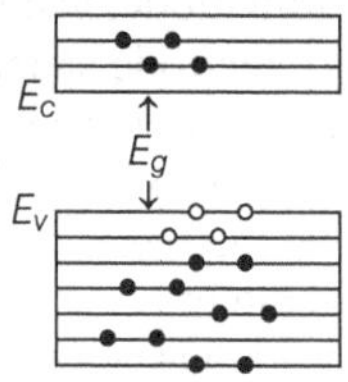


**13** The lattice structure of carbon (C), silicon (Si) and germanium (Ge) is same, then
(a) Si is an insulator
(b) C and Si are intrinsic semiconductors
(c) Si and Ge are intrinsic semiconductors
(d) only C is intrinsic semiconductor

**14** In extrinsic *p*- and *n*-type semiconductor materials, the ratio of the impurity atoms to the pure semiconductor atoms is about
(a) 1 (b) $10^{-1}$ (c) $10^{-4}$ (d) $10^{-7}$

**15** Number of electrons available for conduction due to doping atom in an extrinsic semiconductor
(a) increases with increase of temperature
(b) decreases with increase of temperature
(c) is independent of change in ambient temperature
(d) reduces to zero at temperature above room temperature

**16** If $n_e$ and $n_h$ are the number of electrons and holes in a semiconductor heavily doped with phosphorous, then
(a) $n_e >> n_h$ (b) $n_e << n_h$ (c) $n_e \le n_h$ (d) $n_e = n_h$

**17** $n_e$ and $v_d$ be the number of free electrons and drift velocity in a semiconductor. When the temperature is increased, then
(a) $n_e$ increases and $v_d$ decreases
(b) $n_e$ decreases and $v_d$ increases
(c) both $n_e$ and $v_d$ increase
(d) both $n_e$ and $v_d$ decrease

**18** *p*-type semiconductor is obtained when the doping impurity and the pure semiconductor, respectively are

I germanium and silicon

II arsenic and germanium

III indium and silicon

IV antimony and germanium

(a) Both I and III (b) Both II and III
(c) Only III (d) Only IV

**19** For an extrinsic semiconductor,

(a) the conduction band and valence band overlaps
(b) the gap between conduction band and valence band is more than 16 eV
(c) the gap between conduction band and valence band is near about 1 eV
(d) the gap between conduction band and valence band will be 100 eV and more

**20** The energy band diagrams for three semiconductor samples of silicon are as shown. We can then assert that,

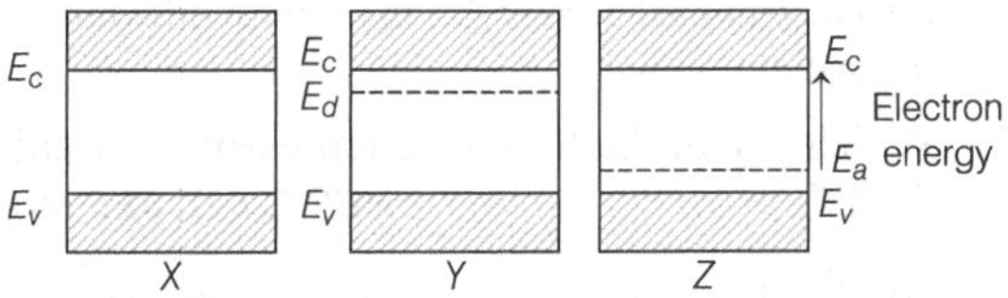


(a) sample *X* is undoped while samples *Y* and *Z* have been doped with a third group and a fifth group impurity, respectively
(b) sample *X* is undoped while both samples *Y* and *Z* have been doped with a fifth group impurity
(c) sample *X* has been doped with equal amounts of third and fifth group impurities while samples *Y* and *Z* are undoped
(d) sample *X* is undoped while samples *Y* and *Z* have been doped with a fifth and a third group impurity, respectively

**21** A Ge specimen is doped with Al. The concentration of acceptor atom is $\sim 10^{21}$ atoms per $m^3$. Given that the intrinsic concentration of electron-hole pairs is $\sim 10^{19}$ per $m^3$, the concentration of electrons in the specimen is

(a) $10^{17}$ per $m^3$ (b) $10^{15}$ per $m^3$
(c) $10^4$ per $m^3$ (d) $10^2$ per $m^3$

**22** Let a pure Si crystal has $5 \times 10^{28}$ atoms $m^{-3}$. It is doped by parts per million concentration of pentavalent arsenic. If number of intrinsic pairs is $1.5 \times 10^{16}\, m^{-3}$, then number of holes in doped crystal is

(a) $4.5 \times 10^9\ m^{-3}$ (b) $\sim 10^{16}\ m^{-3}$
(c) $2.25 \times 10^{32}\ m^{-3}$ (d) $5 \times 10^{22}\ m^{-3}$

**23** The number of silicon atoms per $m^3$ is $5 \times 10^{28}$. This is doped simultaneously with $5 \times 10^{22}$ atoms per $m^3$ of arsenic and $5 \times 10^{20}$ per $m^3$ atoms of indium. Given that, $n_i = 1.5 \times 10^{16} m^{-3}$.

Number of electrons and holes (in per metre cube of sample) are respectively,

(a) $4.95 \times 10^{22}, 4.54 \times 10^9$
(b) $4.54 \times 10^9, 4.54 \times 10^9$
(c) $4.54 \times 10^9, 4.95 \times 10^{22}$
(d) $4.95 \times 10^{22}, 4.95 \times 10^{22}$

**24** Three semiconductors are arranged in the decreasing order of their energy gap as follows. The correct arrangement is

(a) tin, germanium, silicon
(b) tin, silicon, germanium
(c) silicon, germanium, tin
(d) silicon, tin, germanium

# TOPIC 3 ~ *p-n* Junction and Semiconductor Diode

**25** The electrical resistance to the diffusion current in a *p-n* junction because of depletion layer, is large due to the presence of

(a) strong electric field
(b) large number of charge carriers
(c) electrons as charge carriers
(d) holes as charge carriers

**26** Which of these graphs shows potential difference between *p*-side and *n*-side of a *p-n* junction in equilibrium?

(a)
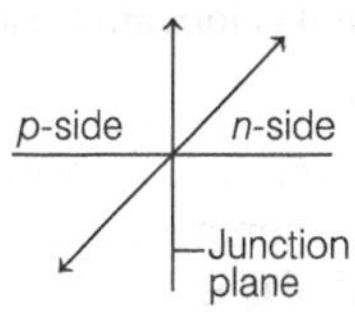


(b)
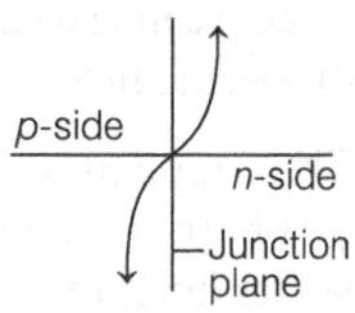


(c)
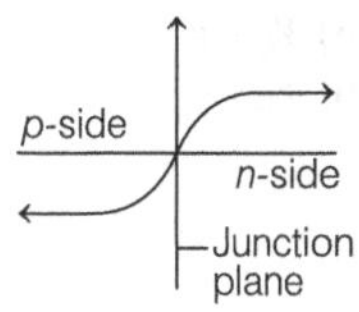


(d)
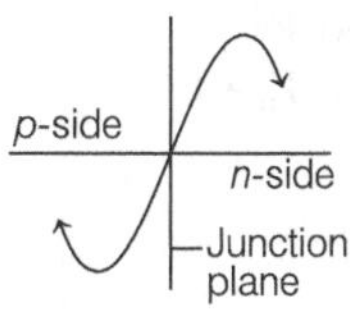

**27** The barrier potential of a $p$-$n$ junction depends on
(i) type of semiconductor material
(ii) amount of doping
(iii) temperature

Which one of the following is correct?

***CBSE AIPMT 2014***

(a) (i) and (ii)  (b) (ii)
(c) (ii) and (iii)  (d) (i), (ii) and (iii)

**28** A Si based $p$-$n$ junction has a depletion layer of thickness $1\,\mu m$ and barrier potential difference of $n$-side and $p$-side is 0.6 V.
The electric field in the depletion region is
(a) $0.6\ Vm^{-1}$  (b) $6\times 10^{-4}\ Vm^{-1}$
(c) $6\times 10^{5}\ Vm^{-1}$  (d) $6\times 10^{4}\ Vm^{-1}$

**29** If no external voltage is applied across $p$-$n$ junction, there would be
(a) no electric field across the junction
(b) an electric field pointing from $n$-type to $p$-type side across the junction
(c) an electric field pointing from $p$-type to $n$-type side across the junction
(d) a temporary electric field during formation of $p$-$n$ junction that would subsequently disappear

**30** A diode is a
(a) three-terminal device
(b) piece of a semiconductor crystal with metallic contacts provided at two ends
(c) $p$-$n$ junction with metallic contacts provided at two ends
(d) piece of a metal which is sprayed over by a semiconductor

**31** The arrow in the symbol of a $p$-$n$ junction diode as shown below indicates the direction of

(a) drift current
(b) flow of holes
(c) conventional current (when it is forward biased)
(d) electric field

**32** If $V$ is applied potential difference in forward bias and $V_0$ is barrier potential of a $p$-$n$ junction, then effective barrier height under forward bias is
(a) $V$  (b) $V_0 - V$
(c) $V_0 + V$  (d) $V_0$

**33** In the case of forward biasing of $p$-$n$ junction, which one of the following figures correctly depicts the process of minority charge injection.
Here, $e^-$ refers to electron and ∘ refers to hole

(a)
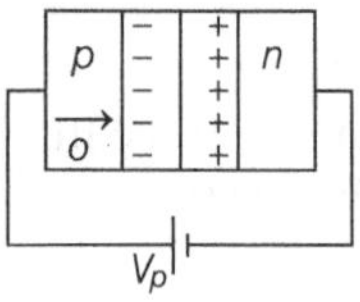


(b)
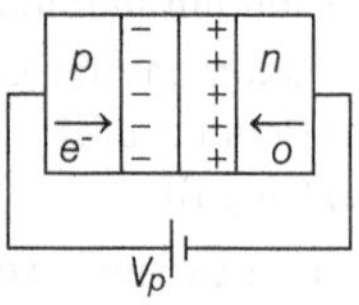


(c)
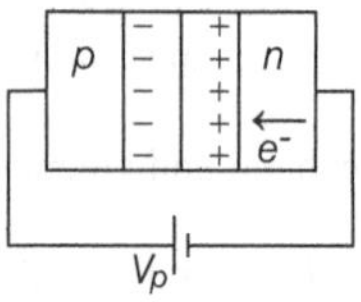


(d) Both (a) and (c)

**34** In forward bias, total forward current obtained from the $p$-$n$ junction diode is due to
(a) injection of electrons in $p$-side
(b) injection of holes in $n$ -side
(c) conventional current due to electron diffusion
(d) All of the above

**35** Which one of the following represents forward bias diode?

***NEET 2017, JEE Main 2014***

(a)
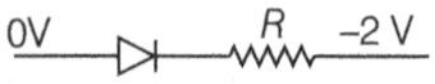


(b)
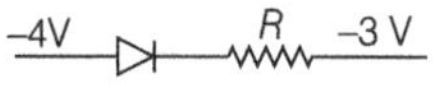


(c)
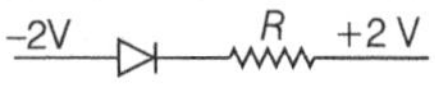


(d)
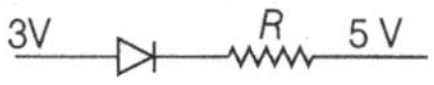


**36** If $V_0$ is the barrier potential of a unbiased $p$-$n$ junction diode, then on reverse bias the barrier potential is correctly depicted in

(a)
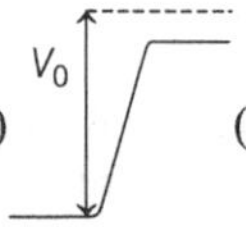


(b)
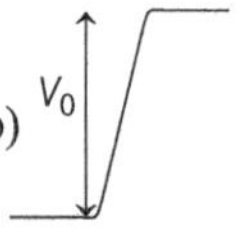


(c)
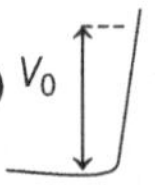


(d)
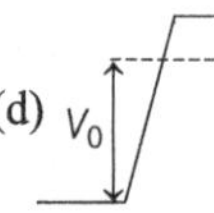


**37** In a reverse biased $p$-$n$ junction diode,
(a) current is not much dependent on applied voltage
(b) current is directly proportional to applied voltage
(c) current initially depends on applied voltage, then it becomes independent
(d) no current flows in reversed bias

**38** If reverse biasing potential is increased beyond a certain critical (breakdown) value, then
(a) diode gets destroyed due to overheating
(b) no current flows through the diode
(c) a heavy current flows from $p$ to $n$-side
(d) potential barrier becomes zero

**39** The breakdown in a reverse biased *p-n* junction diode is more likely to occurs due to

(a) small velocity of minority charge carriers, if the doping is small

(b) large velocity of minority charge carriers, if the doping is large

(c) strong electric field in depletion region, if the doping concentration is small

(d) strong electric field in depletion region, if the doping concentration is large

**40** The *V-I* characteristic curve of a silicon diode is

(a)
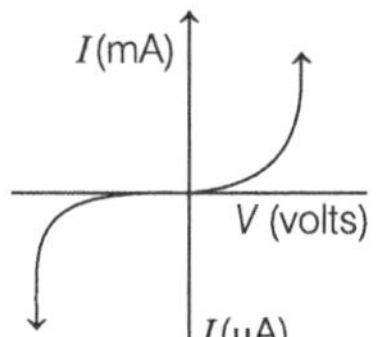


(b)
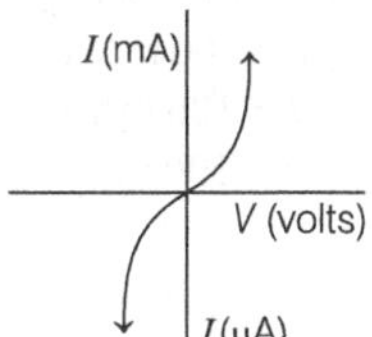


(c)
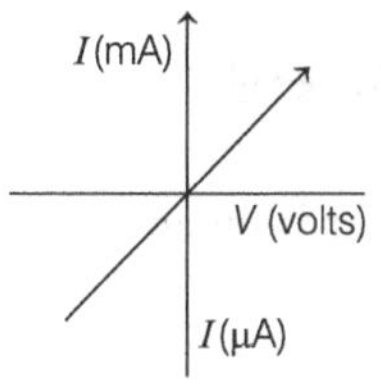


(d)
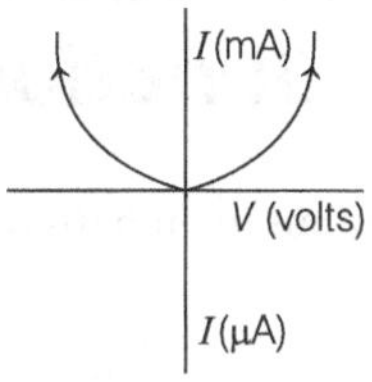


**41** In a *p-n* junction diode, change in temperature due to heating **NEET 2018**

(a) does not affect resistance of *p-n* junction

(b) affects only forward resistance

(c) affects only reverse resistance

(d) affects the overall *V-I* characteristics of *p-n* junction

**42** When the voltage drop across a *p-n* junction diode is increased from 0.65 V to 0.70 V, the change in the diode current is 5 mA. The dynamic resistance of diode is

(a) 5 Ω (b) 10 Ω (c) 20 Ω (d) 25 Ω

**43** *V-I* characteristics of a silicon diode is shown.

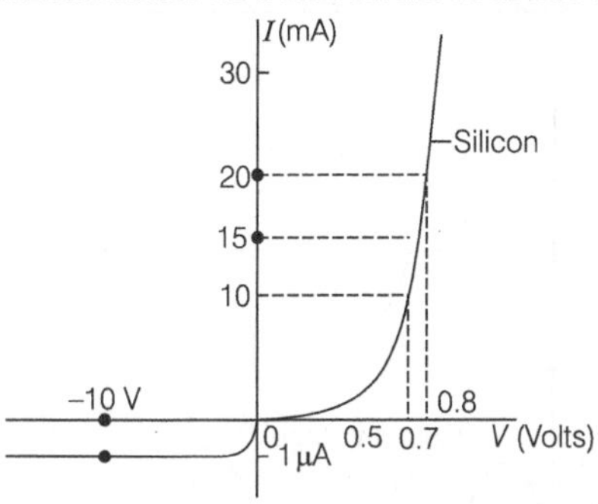


The ratio of resistance of diode at $I_D = 15$ mA and $V_D = -10$ V, is

(a) $10^{-3}$ (b) $10^{-4}$ (c) $10^{-5}$ (d) $10^{-6}$

**44** Consider the junction diode as ideal. The value of current flowing through *AB* is **NEET 2016**

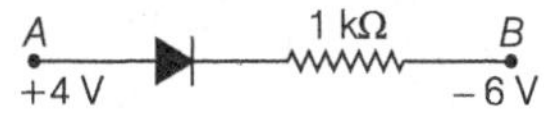


(a) $10^{-2}$ A (b) $10^{-1}$ A (c) $10^{-3}$ A (d) 0 A

**45** The diode *S* shown in the circuit is a silicon diode. The potential difference between the points *A* and *B* will be **JEE Main 2018**

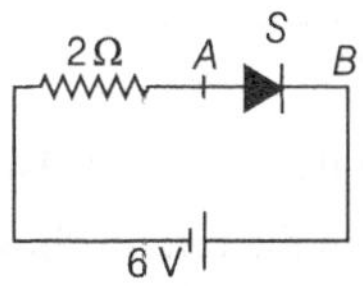


(a) 6 V (b) 0.6 V (c) 0.7 V (d) 0 V

**46** In the figure, potential difference between *A* and *B* is **JEE Main 2020**

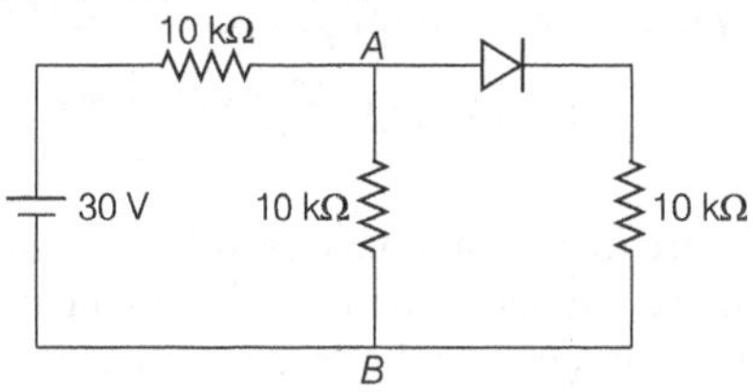


(a) zero (b) 10 V (c) 5 V (d) 15 V

**47** At 0.3V and 0.7 V, the diodes Ge and Si become conductor respectively. In given figure, if ends of diode Ge overturned, the change in potential $V_0$ will be **JEE Main 2019**

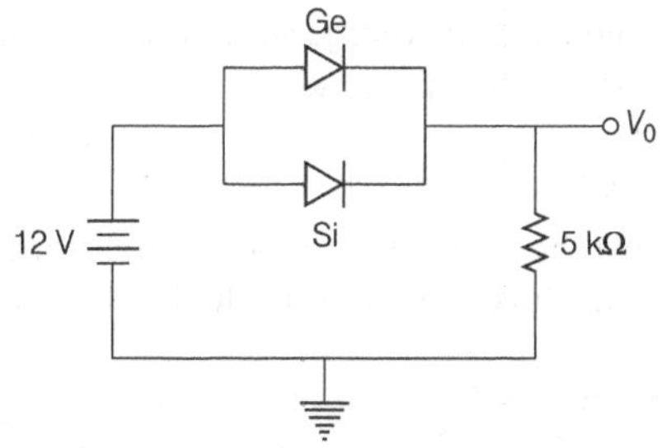


(a) 0.2 V (b) 0.6V (c) 0.4 V (d) 0.8V

**48** In the given figure, a diode *D* is connected to an external resistance $R = 100\,\Omega$ and an emf of 3.5 V. If the barrier potential developed across the diode is 0.5 V, the current in the circuit will be **CBSE AIPMT 2015**

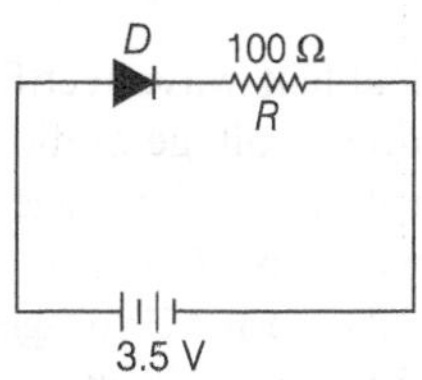


(a) 30 mA (b) 40 mA (c) 20 mA (d) 35 mA

**49** The current $i$ in the network is ***JEE Main 2020***

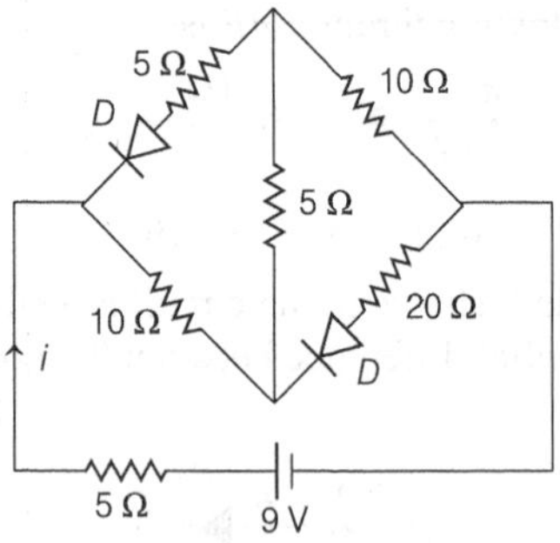


(a) 0.2A (b) 0 A (c) 0.6 A (d) 0.3 A

**50** The diode used in the circuit shown in the figure has a constant voltage drop of 0.5 V at all currents and a maximum power rating of 100 milliwatt. The value of maximum current in the circuit is $I$ when voltage across resistance $R$ is $V_R$, then ***AIIMS 2018, JEE Main 2018***

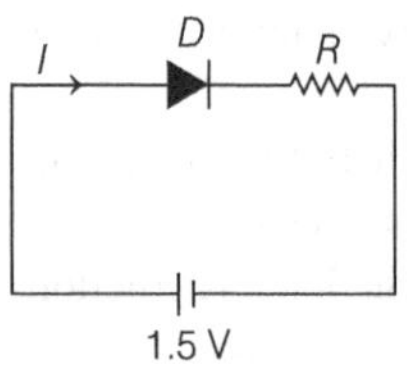


(a) $I = 200\text{ mA}, V_R = 1\text{ V}$
(b) $I = 200\text{ mA}, R = 10\,\Omega$
(c) $I = 100\text{ mA}, V_R = 2\text{ V}$
(d) $I = 100\text{ mA}, R = 10\,\Omega$

# TOPIC 4 ~ *Application of Junction Diode as a Rectifier and Special Purpose p-n Junction Diodes*

**51** Diodes are most commonly used for rectification of alternating voltages. Then, which amongst the following property of diode is responsible for this?
(a) Current flow in one direction
(b) Forward bias resistance is low compared to reverse bias
(c) Reverse current is independent of voltage applied
(d) Both (a) and (b)

**52** If an alternating voltage is applied across a diode in series with a load, then
(a) a continuous DC voltage appears across load
(b) an AC voltage appears across load
(c) a pulsating AC voltage appears across load
(d) no voltage appears across load

**53** Amongst the following circuits, which correctly represents the circuit used for half-wave rectification?

(a)
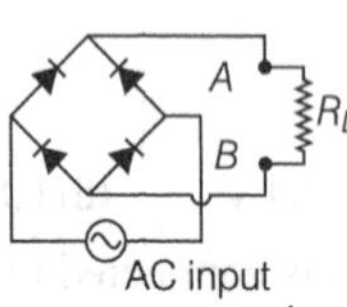


(b)
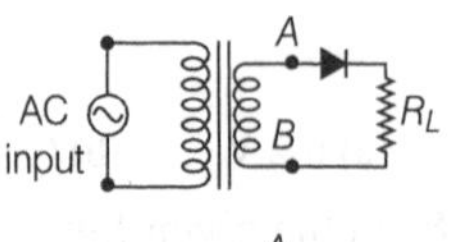


(c)
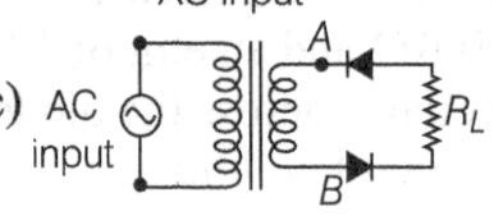


(d)
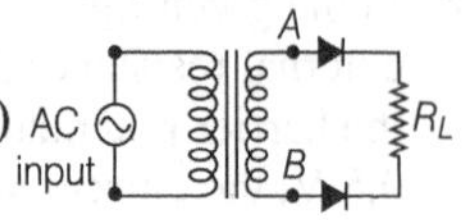


**54** For any practical half-wave rectifier circuit, the reverse breakdown voltage of diode
(a) greater than the peak AC voltage
(b) greater than the rms AC voltage
(c) greater than the mean AC voltage
(d) smaller than the rms AC voltage

**55** Input to an half-wave rectifier is given as follows

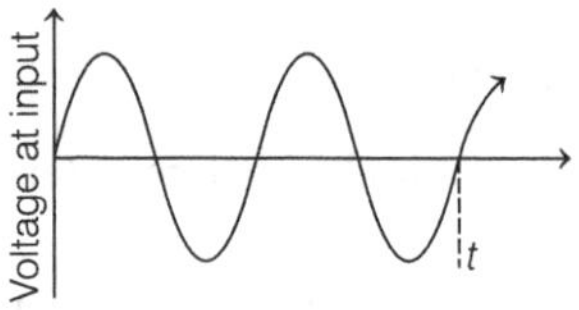


Its output (voltage across load resistance, $R_L$) will be

(a)
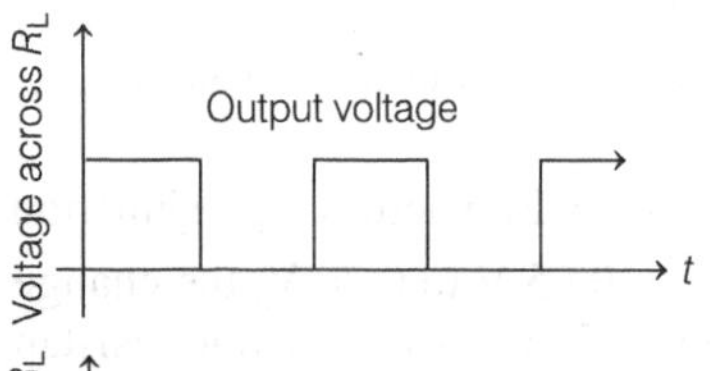


(b)
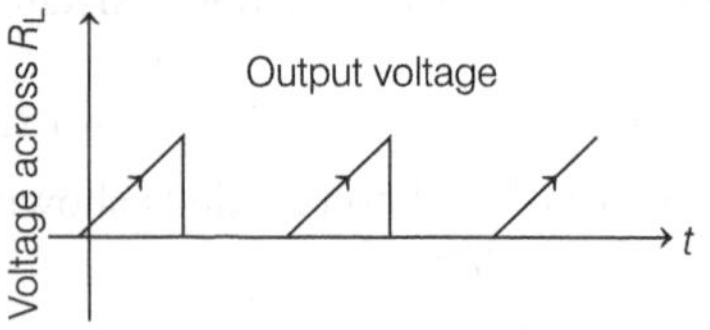


(c)
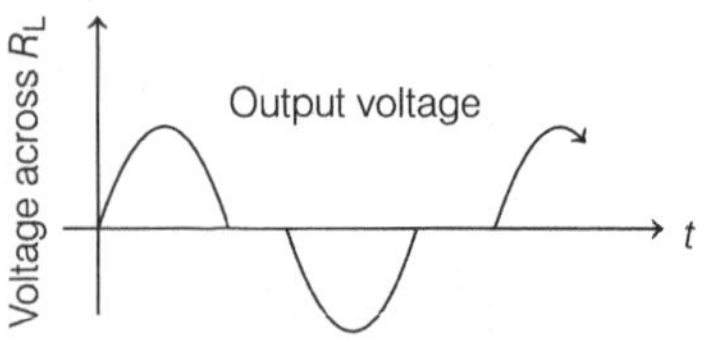


(d)
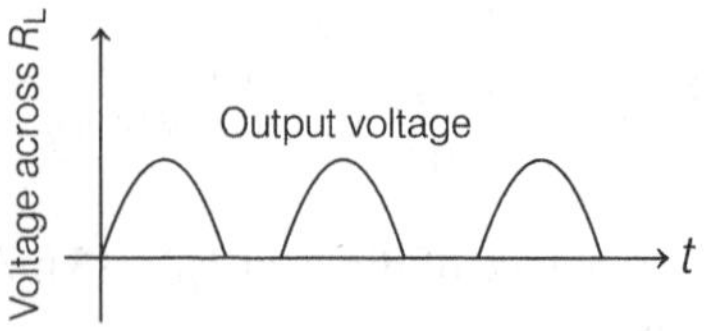

**56** If in a $p$-$n$ junction, a square input signal of 10 V is applied as shown

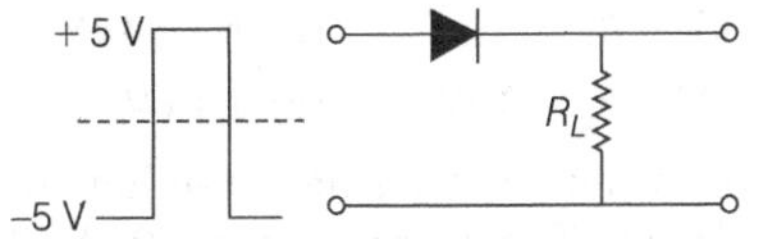


Then, the output across $R_L$ will be

(a) 

(b) 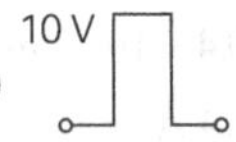


(c) 

(d) 

**57** If a centre tap transformer is used with 2 diodes for full wave rectification, then output voltage of rectifier w.r.t. each diode is

(a) $2\times$ secondary voltage of transformer
(b) $(2/3)\times$ secondary voltage of transformer
(c) $(1/2)\times$ secondary voltage of transformer
(d) $(3/2)\times$ secondary voltage of transformer

**58** In the given circuit,

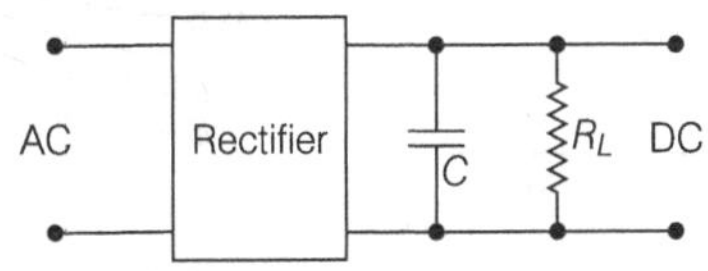


Capacitor $C$ is used

(a) for storing potential energy
(b) as a bypass to DC component to get AC in $R_L$
(c) to remove sparking
(d) as a bypass to AC component to get DC in $R_L$

**59** If in a full wave rectifier circuit, a capacitor $C$ is connected in parallel to the load $R_L$ (across output terminal), then rate of fall of voltage across $C$ is proportional to

(a) $R_L C$ (b) $\frac{C}{R_L}$
(c) $\frac{1}{R_L C}$ (d) $\frac{R_L}{C}$

**60** The ratio of output frequencies of full wave rectifier and a half-wave rectifier, when an input of frequency 50 Hz is fed at input is

(a) 1:2 (b) 2:1
(c) 4:1 (d) 1:4

**61** During the internal field emission or field ionisation in zener diode,

(a) valence electrons accelerated from $n \to p$
(b) holes are accelerated from $p \to n$
(c) resistance of the diodes becomes infinite
(d) valence electrons accelerates from $p \to n$

**62** Correct circuit diagram depicting the use of zener diode as a voltage regulator is

(a) 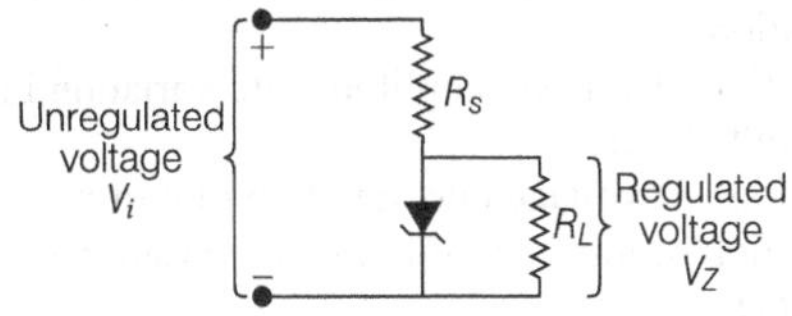


(b) 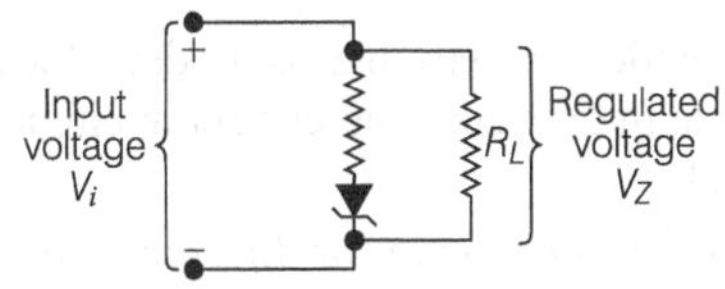


(c) 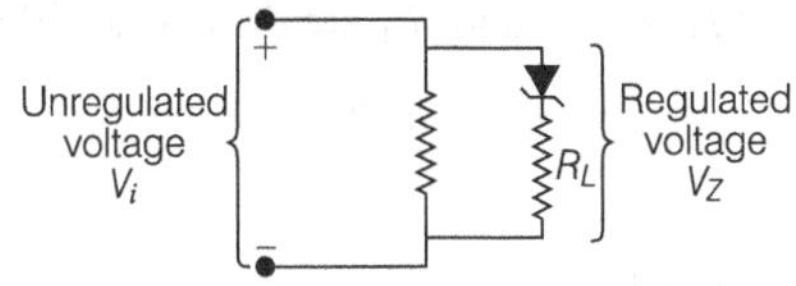


(d) 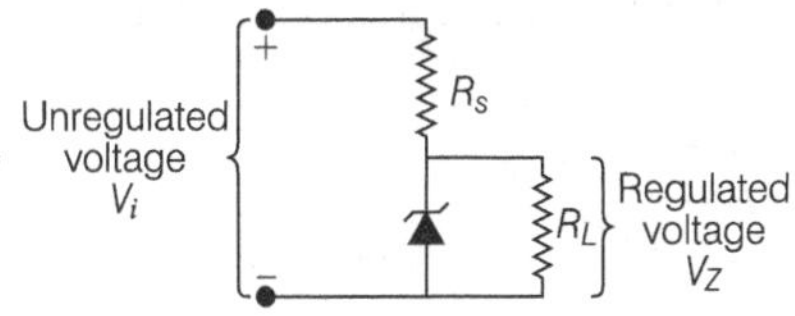


**63** $I$-$V$ characteristic curve of a zener diode is as shown below

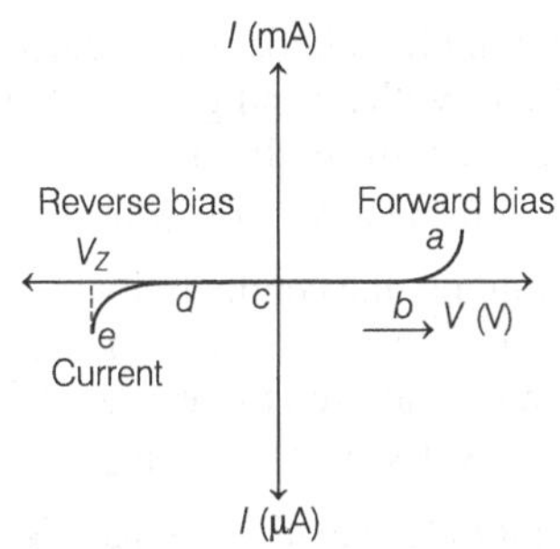


Which amongst the parts labelled above is suitable for the operation of voltage regulation?

(a) *ab* (b) *bc* (c) *cd* (d) *de*

**64** Figure shows a DC voltage regulator circuit, with a Zener diode of breakdown voltage = 6 V. If the unregulated input voltage varies between 10 V to 16 V, then what is the maximum Zener current? ***JEE Main 2019***

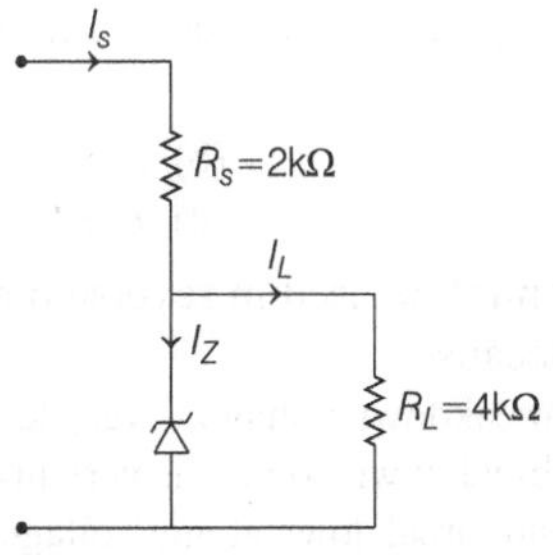


(a) 2.5 mA (b) 1.5 mA (c) 7.5 mA (d) 3.5 mA

**65** A photodiode converts
(a) variation in intensity of light into current amplitude variation
(b) variation of current amplitude into variation in intensity of emitted light
(c) variation of voltage into variation of current
(d) variation of intensity of light into variation of volume

**66** A photodiode in reverse bias is irradiated with light of suitable frequencies and current in circuit is measured.

Characteristics of diode for different illumination intensities $I_1, I_2, I_3$ and $I_4$ are drawn as follows.

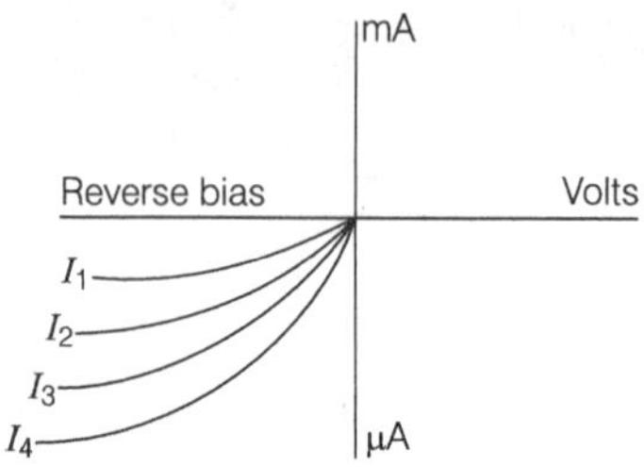


Then, the maximum intensity amongst these is
(a) $I_1$ (b) $I_2$ (c) $I_3$ (d) $I_4$

**67** A *p-n* junction photodiode is fabricated from a semiconductor with a band gap of 2.8 eV. It can detect a wavelength nearly to
(a) 5200 Å (b) 4400 Å (c) 6200 Å (d) 7500 Å

**68** For a photodiode, the conductivity increases when a wavelength less than 620 nm is incident on it. The band gap of crystal used to fabricate the diode is
(a) 1.12 eV (b) 1.8 eV (c) 2.0 eV (d) 1.62 eV

**69** A *p-n* photodiode is made of a material with a band gap of 2 eV. The minimum frequency of the radiation that can be absorbed by the material is nearly (take, $hc = 1240$ eV-nm)
(a) $1\times10^{14}$ Hz (b) $20\times10^{14}$ Hz
(c) $10\times10^{14}$ Hz (d) $5\times10^{14}$ Hz

**70** When LED is forward biased and $E_g$ is the energy gap of the semiconductor used for its fabrication. Then, released energy $E$ due to electron-hole recombination in it will be
(a) $E = E_g$ (b) $E > E_g$
(c) $E \le E_g$ (d) $E \ge E_g$

**71** An LED cannot be used in reverse bias as a voltage regulator because
(a) reverse breakdown voltage is very low for them
(b) reverse breakdown voltage is very high for them
(c) they do not breakdown at any voltage
(d) None of the above

**72** Semiconductors used to fabricate LED to produce visible light must have energy gap $E_g$ such that
(a) $1.1\text{ eV} < E_g$ (b) $E_g > 3\text{ eV}$
(c) $1.8\text{ eV} < E_g < 3\text{ eV}$ (d) $1.1\text{ eV} < E_g < 2.8\text{ eV}$

**73** An LED is constructed from a *p-n* junction diode using GaAsP. The energy gap is 1.9 eV. The wavelength of the light emitted will be equal to ***NEET (Odisha) 2019***
(a) $10.4\times10^{-26}$ m (b) 654 nm
(c) 654 Å (d) $654\times10^{-11}$ m

**74** The *I-V* characteristics of an LED is ***JEE Main 2013***

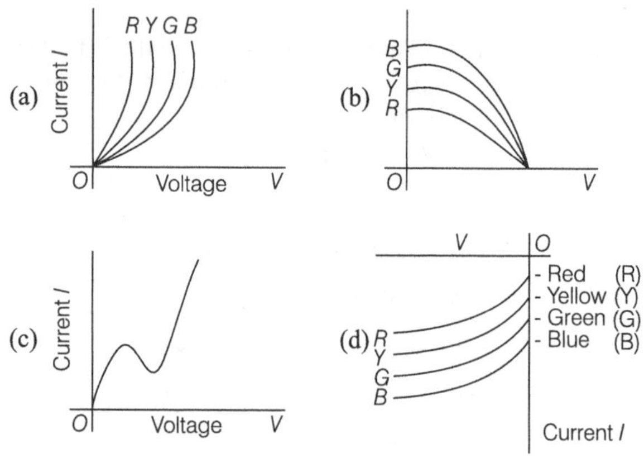


**75** LED's are not used for room lighting (although they are used for automobile bulbs and in industrial lighting) because
(a) our eyes are not comfortable with very intense light
(b) our eyes are not comfortable with monochromatic light
(c) LED's are much costlier than bulbs tubelights and CFL's
(d) LED manufacture in mass production will be a very polluting process

**76** A typical *p-n* junction solar cell is as shown in the figure below

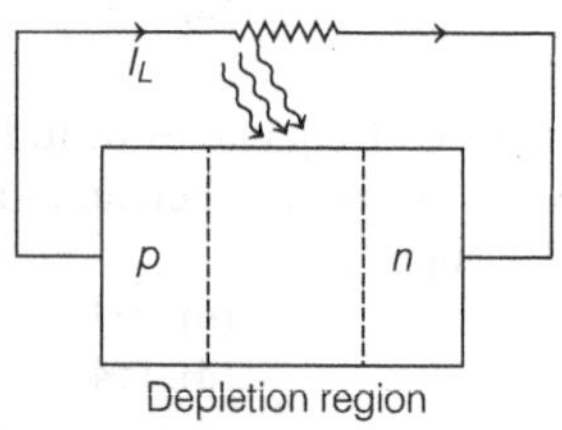


The generation of emf by it, is due to
(a) generation of electrons and holes due to light
(b) separation of electrons and holes due $E$ electric field of depletion region
(c) collection of electrons $e^-$ at *n*-side and holes at *p*-side
(d) All of the above

**77** Important criteria for the selection of a material for solar cell fabrication is

I band gap (~ 1.0 to 1.8 eV)

II sunlight

III high optical absorption ($\sim 10^4\,cm^{-1}$)

IV. electrical conductivity

(a) I, II and III
(b) III, II and IV
(c) I, III and IV
(d) All of the above

**78** Identify the semiconductor devices whose characteristics are as given below in the order (i), (ii) and (iii). ***JEE Main 2016***

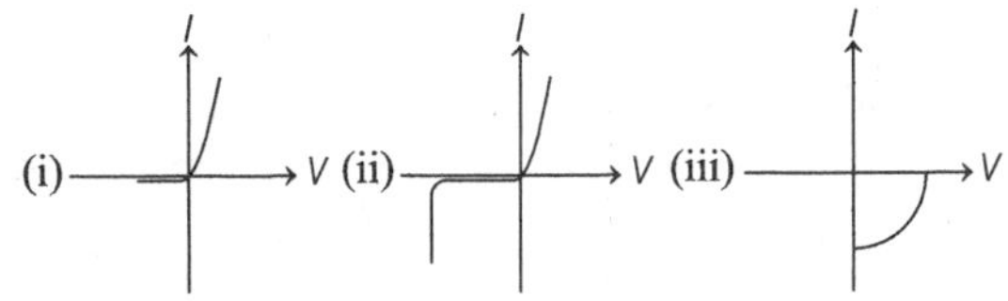


(a) Zener diode, simple diode, solar cell
(b) Solar cell, zener diode, simple diode
(c) Zener diode, solar cell, simple diode
(d) Simple diode, zener diode, solar cell

# TOPIC 5 ~ *Junction Transistor*

**79** For an *n-p-n* transistor shown below

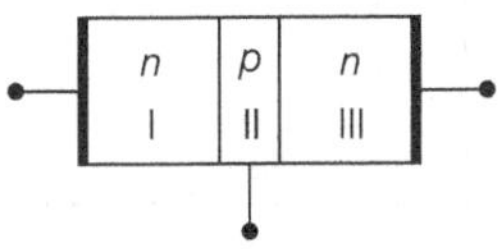


Regions marked I, II and III are respectively,

(a) emitter,collector, base (b) base, collector, emitter
(c) emitter, base, collector (d) collector, emitter, base

**80** A transistor is biased as follows

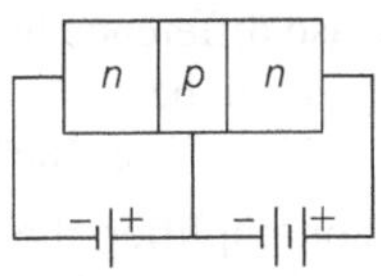


Then, it is said to be in a

(a) solid state (b) active state
(c) saturation state (d) cut-off state

**81** Let '•' shows an electron and '∘' shows a hole, then which of the following shows correct direction of motion of charge carriers?

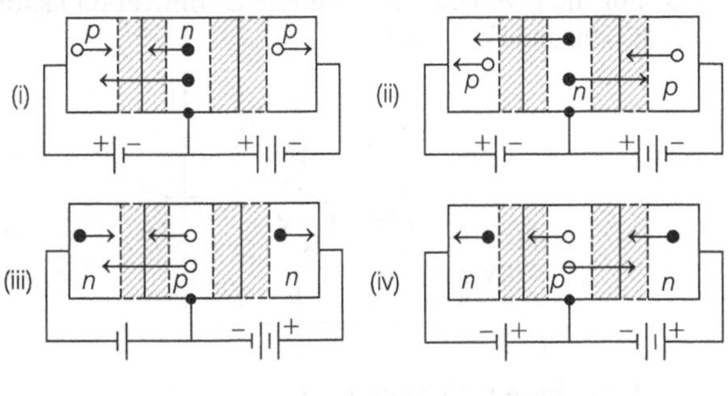


(a) (i) and (ii) (b) (ii) and (iii)
(c) (i) and (iii) (d) (ii) and (iv)

**82** Correct circuit to study input-output characteristics of an *n-p-n* transistor in CE configuration is

(a)
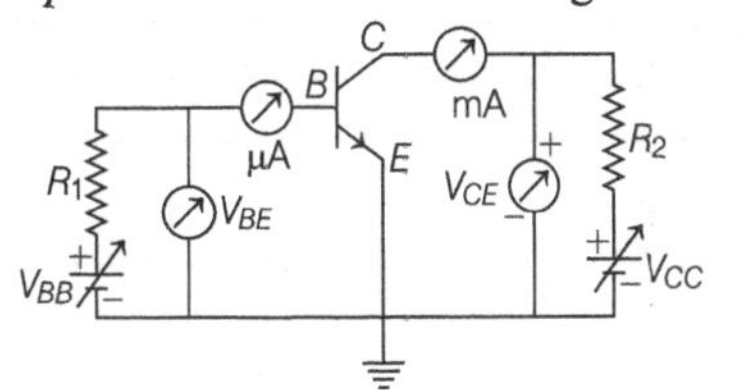


(b)
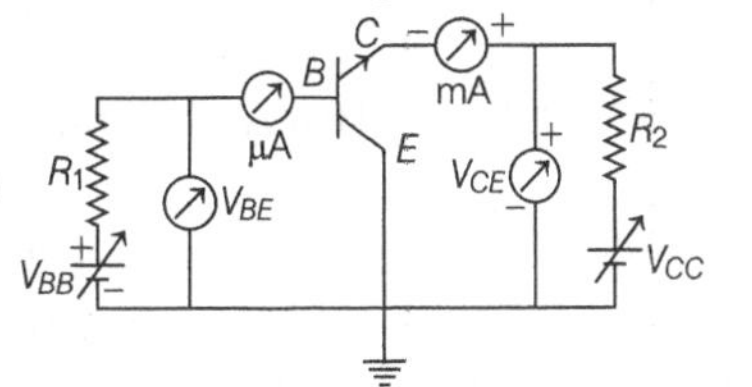


(c)
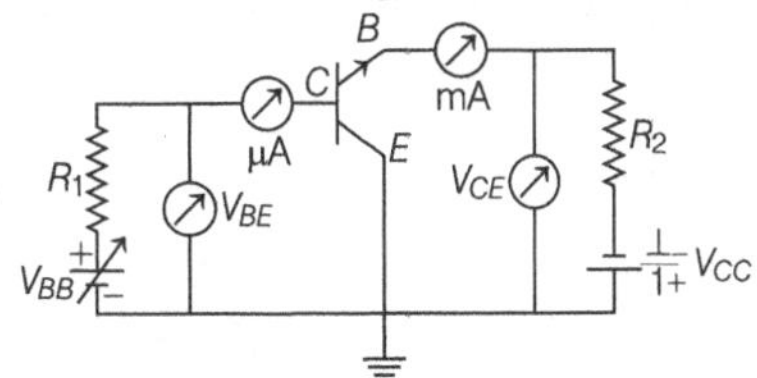


(d)
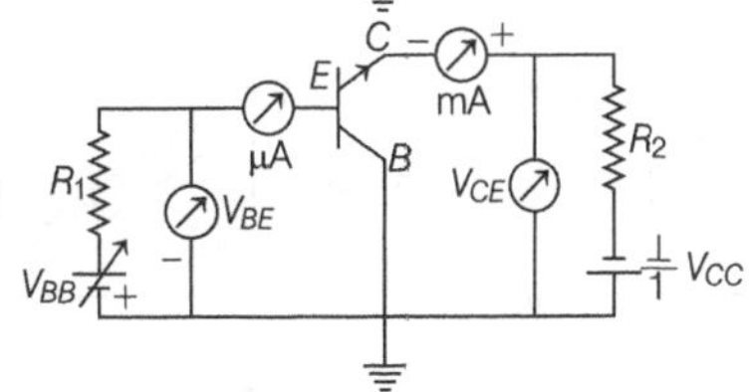


**83** In an *n-p-n* transistor, the collector current is 24 mA. The possible emitter current (in mA) is

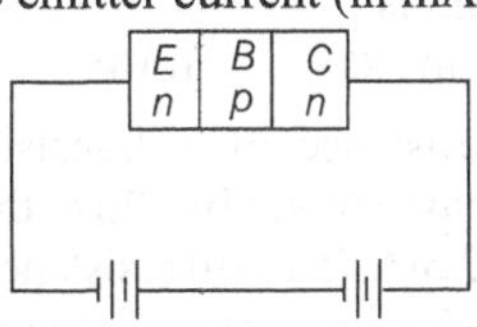


(where, $E$ = emitter, $B$ = base and $C$ = collector.)

(a) 36 (b) 20 (c) 16 (d) 6

**84** For an *n-p-n* transistor in CE configuration, correct graph showing input characteristics, i.e. variation $I_B$ *versus* base-emitter voltage $V_{BE}$ is

(a) 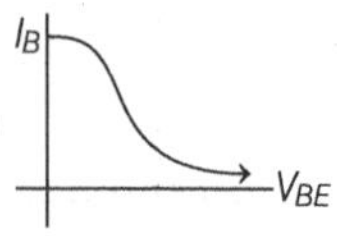


(b) 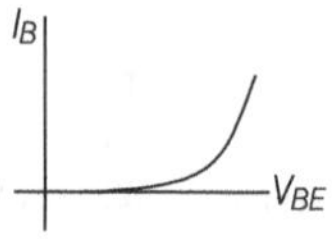


(c) 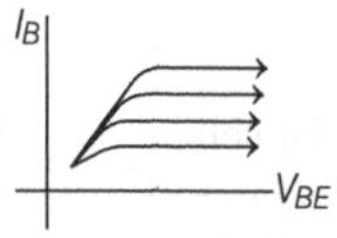


(d) 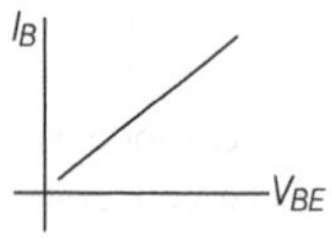


**85** The output characteristics of a typical *n-p-n* transistor in CE configuration are shown.

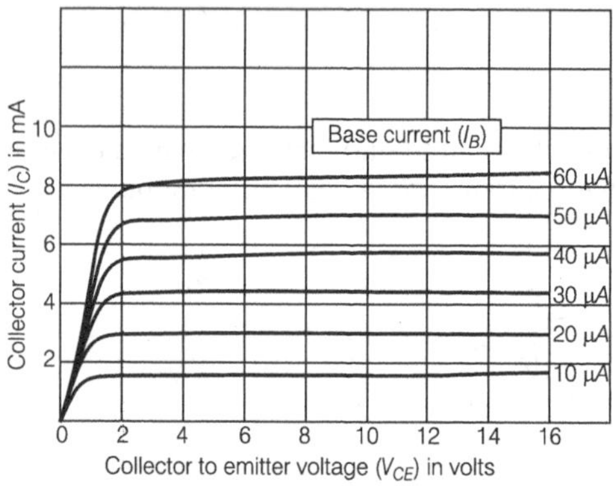


When $V_{CE} = 10$ V and $I_C = 4.0$ mA, then ratio of $\beta_{AC}$ and $\beta_{DC}$ is

(a) 1 (b) 2 (c) 3 (d) 4

**86** For a common-emitter *n-p-n* transistor which amongst the following represents the true relationship between $I_B$ and $I_C$, in active region.

(a) 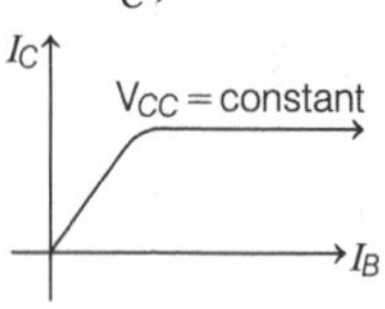


(b) 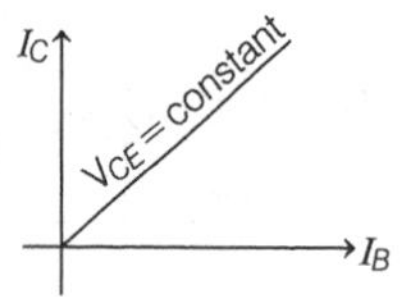


(c) 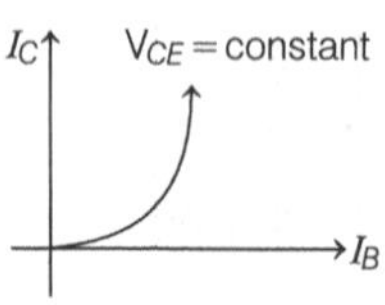


(d) 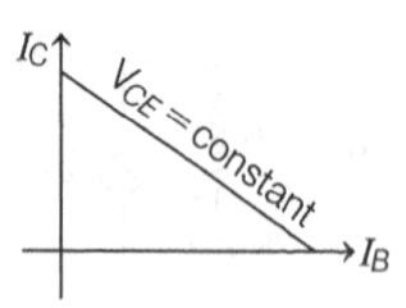


**87** A transistor has a current gain of 30. If the collector resistance is 6 kΩ and input resistance is 1 kΩ, then its voltage gain is

(a) 90 (b) 180 (c) 45 (d) 360

**88** The input resistance of a transistor is 1000 Ω on charging its base current by 10 μA, the collector current increases by 2 mA. If a load resistance of 5 kΩ is used in the circuit, the voltage gain of the amplifier is

(a) 100 (b) 500 (c) 1000 (d) 1500

**89** For the given circuit an *n-p-n* transistor is being used as a switch in CE configuration.

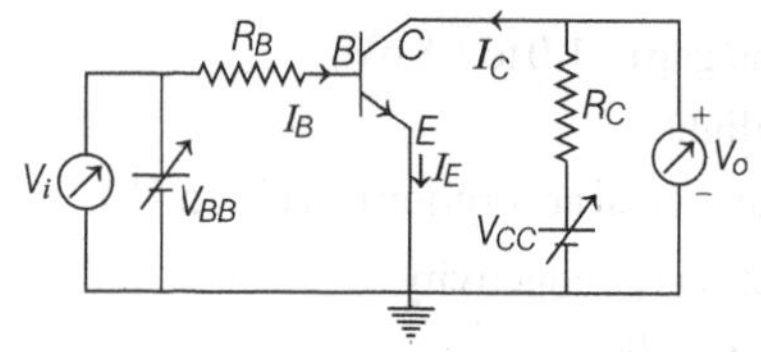


Which of the following are correct?

I. $V_{BB} = I_B R_B + V_{BE}$ II. $V_{CE} = V_{CC} - I_C R_C$

III. $V_i = I_B R_B + V_{BE}$ IV. $V_o = V_{CC} - I_C R_C$

(a) I, II and IV (b) II, III and IV
(c) I, II and III (d) I, II, III and IV

**90** Transfer characteristics of an *n-p-n* transistor in CE configuration is as shown below

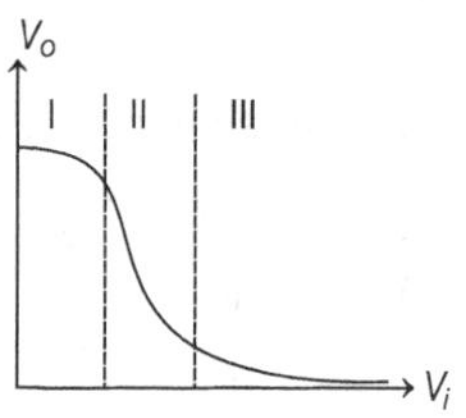


Cut-off region, active region and saturation region respectively, are

(a) II, III and I (b) III, I and II
(c) III, II and I (d) I, II and III

**91** In a common emitter amplifier circuit using an *n-p-n* transistor, the phase difference between the input and the output voltages will be **JEE Main 2017**

(a) 90° (b) 135° (c) 180° (d) 45°

**92** For a CE transistor amplifier, the audio signal voltage across collector resistance of 2.0 kΩ is 2.0 V. Suppose the current amplification factor of the transistor is 100. What should be the value of $R_B$ in series with $V_{BB}$ supply of 2.0 V, if DC base current has to be 10 times the signal current? (Take, $V_{BE} = 0.6$ V)

(a) 14 kΩ (b) 24 kΩ
(c) 34 kΩ (d) 44 kΩ

**93** For the given circuit, if current amplification factor $\beta = 90$ and $V_{BE} = 0.7$ V

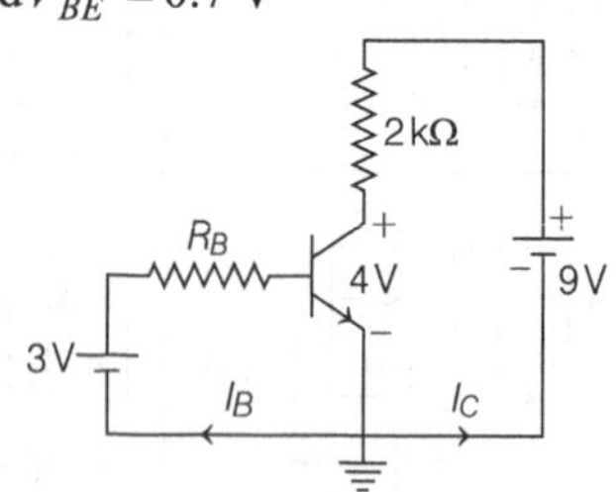


Then, base resistance $R_B$ is

(a) 180 kΩ (b) 185 kΩ (c) 82 kΩ (d) 190 kΩ

**94** The given transistor operates in saturation region, then what should be the value of $V_{BB}$? **NEET 2019**

$[R_{out} = 200\Omega, R_{in} = 100\text{ k}\Omega, V_{CC} = 3\text{V}, V_{BE} = 0.7\text{ V}, V_{CE} = 0 \text{ and } \beta = 200]$

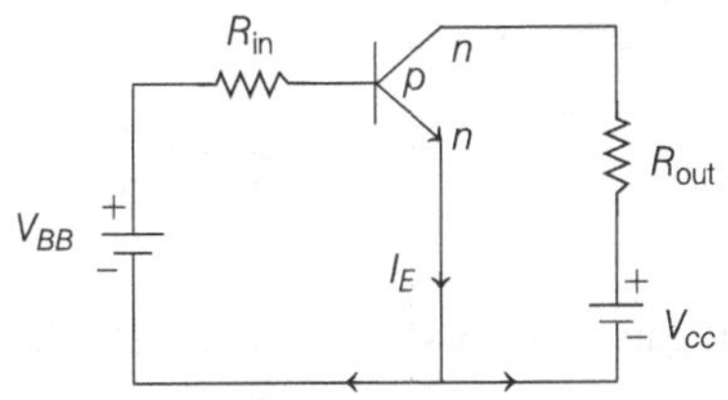


(a) 4.1 V (b) 7.5 V (c) 8.2 V (d) 6.8 V

**95** In the circuit shown in the figure, the input voltage $V_i$ is 20 V, $V_{BE} = 0$ V and $V_{CE} = 0$ V. The values of $I_B$, $I_C$ and $\beta$ are given by

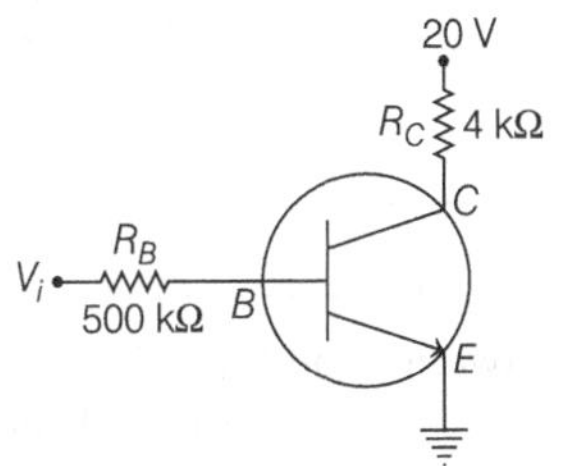


**NEET 2018**

(a) $I_B = 20\,\mu\text{A}, I_C = 5\text{ mA}, \beta = 250$
(b) $I_B = 25\,\mu\text{A}, I_C = 5\text{ mA}, \beta = 200$
(c) $I_B = 40\,\mu\text{A}, I_C = 10\text{ mA}, \beta = 250$
(d) $I_B = 40\,\mu\text{A}, I_C = 5\text{ mA}, \beta = 125$

**96** In a common emitter transistor amplifier, the audio signal voltage across the collector is 3 V. The resistance of collector is 3 kΩ. If current gain is 100 and the base resistance is 2 kΩ, the voltage and power gain of the amplifier is **NEET 2017, JEE Main 2019**

(a) 200 and 1000 (b) 15 and 200
(c) 150 and 15000 (d) 20 and 2000

**97** An *n-p-n* transistor is connected in common emitter configuration in a given amplifier. A load resistance of 800 Ω is connected in the collector circuit and the voltage drop across it is 0.8V. If the current amplification factor is 0.96 and the input resistance of the circuits is 192 Ω, the voltage gain and the power gain of the amplifier will be **NEET 2016**

(a) 3.69, 3.84
(b) 4, 4
(c) 4, 3.69
(d) 4, 3.84

**98** The input signal given to a CE amplifier having a voltage gain of 150 is $V_i = 2\cos\left(15t + \frac{\pi}{3}\right)$. The corresponding output signal will be **CBSE AIPMT 2015**

(a) $300\cos\left(15t + \frac{\pi}{3}\right)$

(b) $75\cos\left(15t + \frac{2\pi}{3}\right)$

(c) $2\cos\left(15t + \frac{5\pi}{3}\right)$

(d) $300\cos\left(15t + \frac{4\pi}{3}\right)$

**99** For tuned collector oscillator, using an *n-p-n* transistor its circuit diagram and rise & fall (or built up) of $I_C, I_E$ current graphs, respectively are as shown below. Initially when switch $S_1$ is ON, it can be concluded

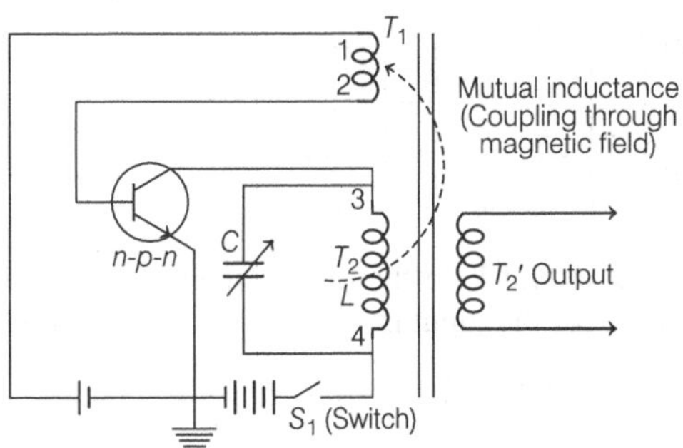


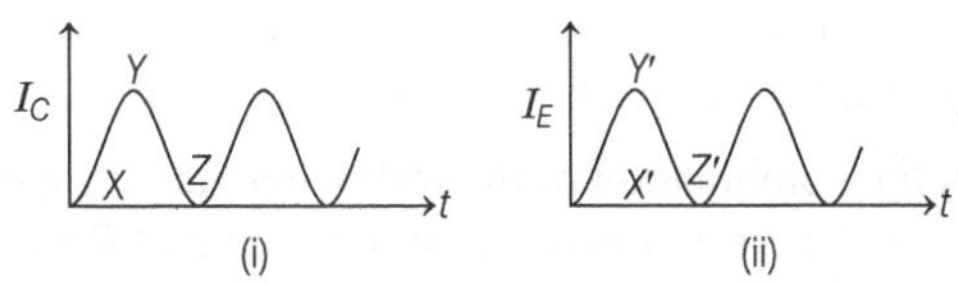


(a) Both $I_C, I_E$ increase initially
(b) Both $I_C, I_E$ decrease
(c) initially $I_C$ increases but $I_E$ decreases
(d) initially $I_C$ decreases but $I_E$ increases

# TOPIC 6 ~ Digital Electronics and Logic Gates

**100** Amongst the following, which correctly represents voltage amplitude ($V$) *versus* time ($t$) waveform for an analog signal.

(a) 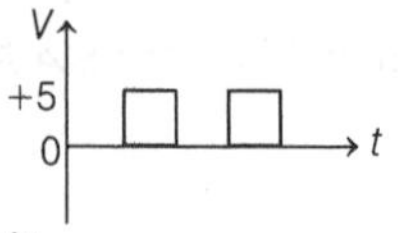
(b) 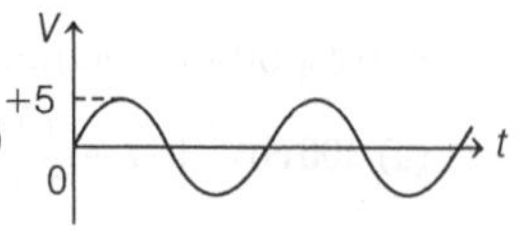

(c) 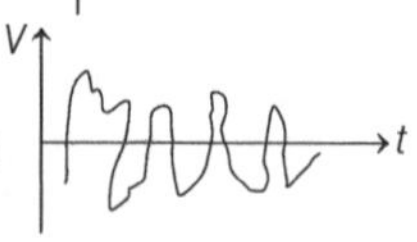
(d) Both (b) and (c)

**101** Inputs to a NAND gate $A$ and $B$ are as below.

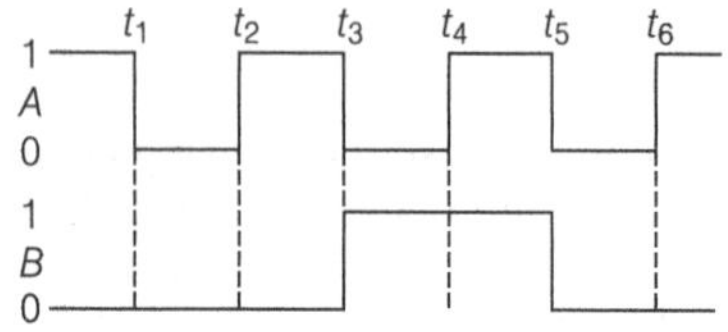

Output of the NAND gate is

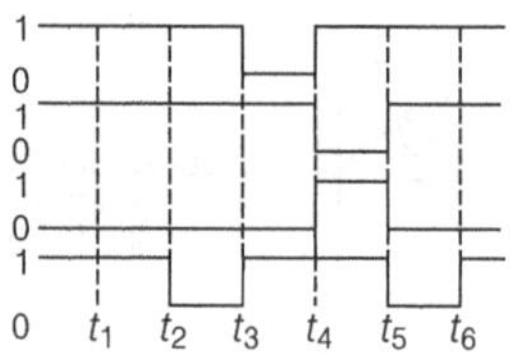

**102** To get output 1 for the following circuit, the correct choice for the input is **NEET 2016**

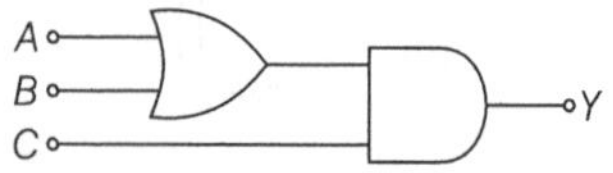

(a) $A=1, B=0, C=0$ (b) $A=1, B=1, C=0$
(c) $A=1, B=0, C=1$ (d) $A=0, B=1, C=0$

**103** In the combination of the following gates, the output $Y$ can be written in terms of inputs $A$ and $B$ as

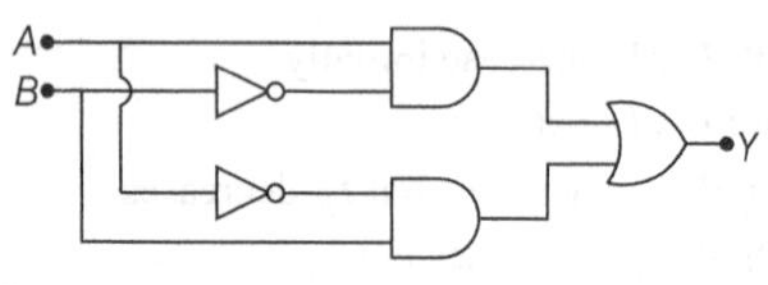

**NEET 2018**

(a) $\overline{A\cdot B}+A\cdot B$ (b) $A\cdot\overline{B}+\overline{A}\cdot B$
(c) $\overline{A\cdot B}$ (d) $\overline{A+B}$

**104** The output $X$ of the logic circuit shown in figure will be **NEET 2013, JIPMER 2019**

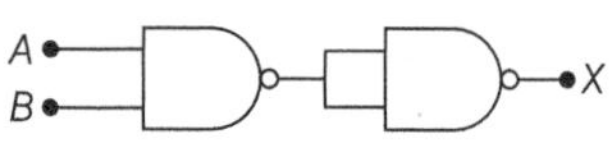

(a) $X=\overline{\overline{A}}\cdot\overline{\overline{B}}$ (b) $X=\overline{A\cdot B}$ (c) $X=A\cdot B$ (d) $X=\overline{A+B}$

**105** Following circuit will act as **JIPMER 2019**

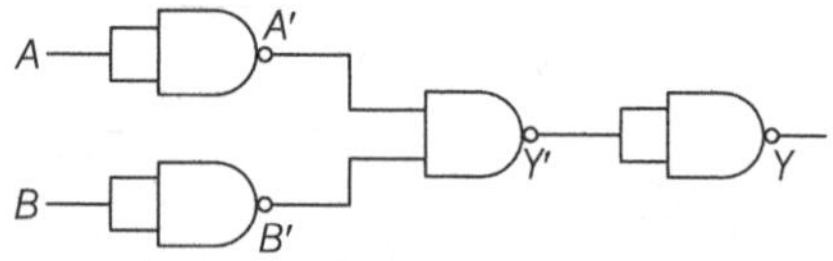

(a) NOR gate (b) NAND gate (c) AND gate (d) OR gate

**106** In the given circuit, value of $Y$ is **JEE Main 2020**

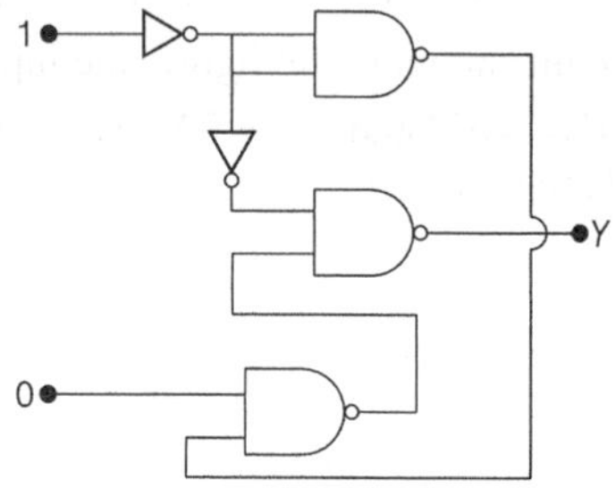

(a) toggles between 0 and 1 (b) 1
(c) 0 (d) will not execute

**107** The given electrical network is equivalent to **NEET 2017**

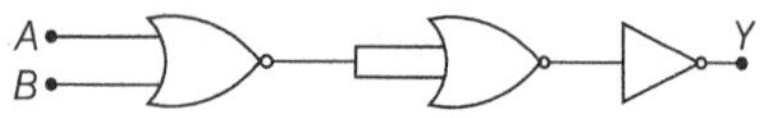

(a) AND gate (b) OR gate
(c) NOR gate (d) NOT gate

**108** The circuit diagram shown here corresponds to the logic gate, **NEET (Odisha) 2019**

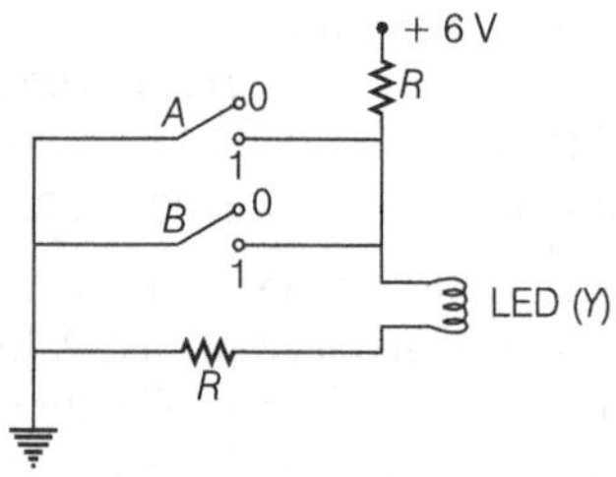

(a) NOR (b) AND (c) OR (d) NAND

**109** The correct Boolean operation represented by the circuit diagram drawn is

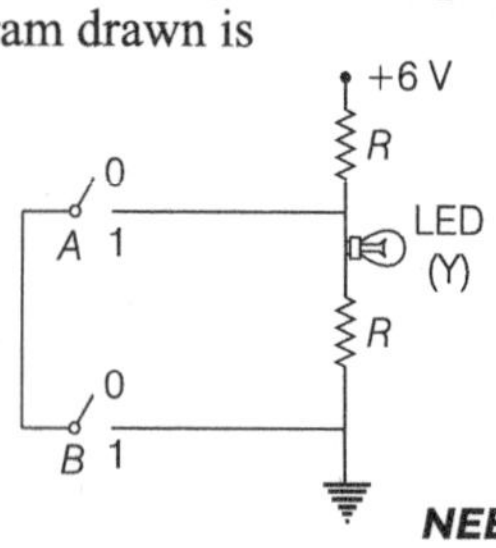

**NEET (National) 2019**

(a) OR (b) NAND (c) NOR (d) AND

# SPECIAL TYPES QUESTIONS

## I. Assertion and Reason

**Direction** (Q. Nos. 110-124) *In the following questions, a statement of Assertion is followed by a corresponding statement of Reason. Of the following statements, choose the correct one.*

(a) Both Assertion and Reason are correct and Reason is the correct explanation of Assertion.
(b) Both Assertion and Reason are correct but Reason is not the correct explanation of Assertion.
(c) Assertion is correct but Reason is incorrect.
(d) Assertion is incorrect but Reason is correct.

**110** **Assertion** Large evacuated tubes and external heating is required in semiconductor devices in order to flow charge carriers as compare to vacuum diode.

**Reason** Flow of charge carriers in the semiconductors are within the solid itself.

**111** **Assertion** Nature of electrons motion in a solid is different from that in an isolated atom.

**Reason** Outer orbits of electrons from neighbouring interact with each other or could even overlap in a solid.

**112** **Assertion** The total current $I$ in a semiconductor is the sum of electron current and hole current.

**Reason** In a semiconductor, $I_h$ arises due to the motion of holes towards positive potential and free electrons under an applied electric field.

**113** **Assertion** The conductivity of an intrinsic semiconductor depends on its temperature.

**Reason** The conductivity of an intrinsic semiconductor is slightly higher than that of a lightly doped *p*-type semiconductor.

**114** **Assertion** In a *n*-type semiconductor, number of holes gets reduced.

**Reason** Rate of recombination of holes is increased.

**115** **Assertion** An electron on *p*-side of a $p-n$ junction moves to *n*-side just an instant after drifting of charge carriers occurs across junction plane.

**Reason** Drifting of charge carriers reduces the concentration gradient across junction plane.

**116** **Assertion** In a *p-n* junction, under equilibrium, no net current flows across the junction.

**Reason** A stable *p-n* junction is only formed when drift and diffusion currents are equal.

**117** **Assertion** The applied voltage (in forward bias of a *p-n* junction) mostly drops across the depletion region and the voltage drop across the *p*-side and *n*-side of the junction is negligible.

**Reason** Resistance of depletion region is large compared to resistance of *n* or *p*-side.

**118** **Assertion** A zener diode is used to obtain voltage regulation.

**Reason** When zener diode is operated in reverse bias after a certain voltage (breakdown voltage), the current suddenly increases, but potential difference across diode remains constant.

**119** **Assertion** Photodiodes are operated under reverse bias.

**Reason** Current in forward bias is known to more (~ mA) than the current in reverse bias (~ μA).

**120** **Assertion** PbS or CdS are not preferred materials as compare to Si and GaAs for solar cell fabrication.

**Reason** For photo-excitation semiconductor with $E_g \leq 1.5$ eV gives solar conversion efficiency.

**121** **Assertion** Output in a CE configuration transistor when used as an amplifier is out of phase with input.

**Reason** Slope of active region of transfer characteristic is negative.

**122** **Assertion** Resistance at the initial part of the output characteristic when transistor (under CE configuration) is in saturation is very low.

**Reason** High magnitude of output resistance is due to the reverse biased state.

**123** **Assertion** In an oscillator, positive feedback implies the feedback is in the same phase.

**Reason** If the feedback voltage in an oscillator is in opposite phase, then it can never work as amplifier.

**124** **Assertion** A logic gate is a digital circuit.

**Reason** They are called gates, because they do not allow current to pass through them.

## II. Statement Based Questions

**125** Solid-state semiconductors has an edge ahead over vacuum tubes because they

I. are small in size and highly reliable.
II. consume high power and operate low voltage.
III. can only control the number of charge carriers.

Which of the following statement(s) is/are correct?

(a) Both I and II
(b) Both II and III
(c) Only I
(d) Both I and III

**126** I. In metals, when conduction band is partially empty, electrons from its lower level can move to higher level making conduction possible.

II. In insulators, energy gap is, so large that electrons cannot be excited from the valence band to the conduction band by thermal excitation.

Which of the following statement(s) is correct?

(a) Only I (b) Only II
(c) Both I and II (d) Neither I nor II

**127** Commonly used schematic representation of $p$-type material is as shown below

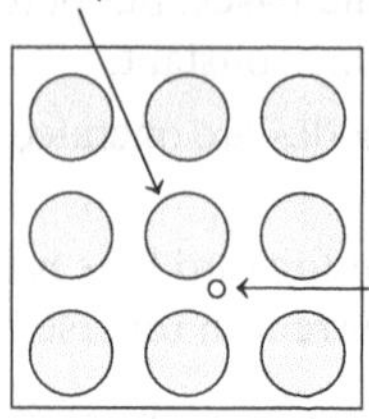


I. The dopant can be treated as core of one positive charge.

II. With very small supply of energy an electron can jump from valence band to the acceptor energy level.

Which of the following statement(s) is/are correct?

(a) Only I (b) Only II
(c) Both I and II (d) Neither I nor II

**128** I. Number of electrons made available for conduction by dopant atoms is independent of any increase in dopant level

II. When the electrical conductivity of semiconductor is due to breaking of its covalent bond, then the semiconductor is extrinsic in nature.

III. At room temperature, the density of holes in the valence band is predominately due to impurity in the extrinsic semiconductor.

Which of the following statement(s) is/are incorrect?

(a) Only I (b) Both I and II
(c) Only III (d) I, II and III

**129** Due to diffusion of electrons from $n$ to $p$-side,

I. a layer of negative charge is developed on $n$-side.

II. an ionised acceptor is left in the $p$-region.

III. an ionised donor is left in the $n$-region.

Which of the following statement(s) is/are correct?

(a) Both I and II (b) Both II and III
(c) Only III (d) Both II and III

**130** Study the following statements regarding reverse biased $p$-$n$ junction.

I. The direction of the applied voltage is opposite to the barrier potential, thus its width decrease.

II. Diffusion current and drift current is large as compare to that in forward bias.

Which of the following statement(s) is/are correct?

(a) Only I (b) Only II
(c) Both (a) and (b) (d) Neither (a) nor (b)

**131** A full wave rectifier circuit (centre tap circuit) with its output waveform across $R_L$ is as shown below

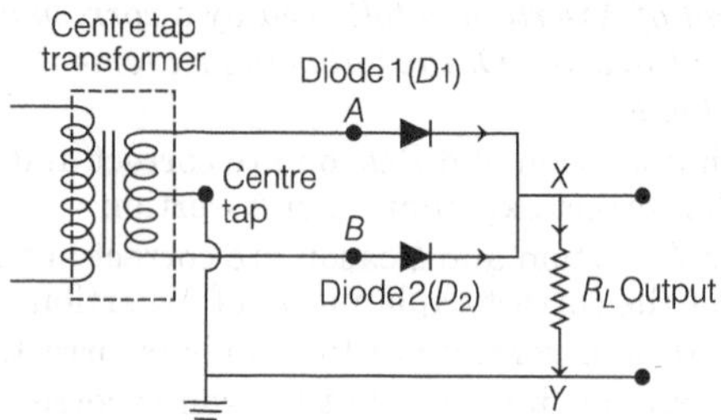


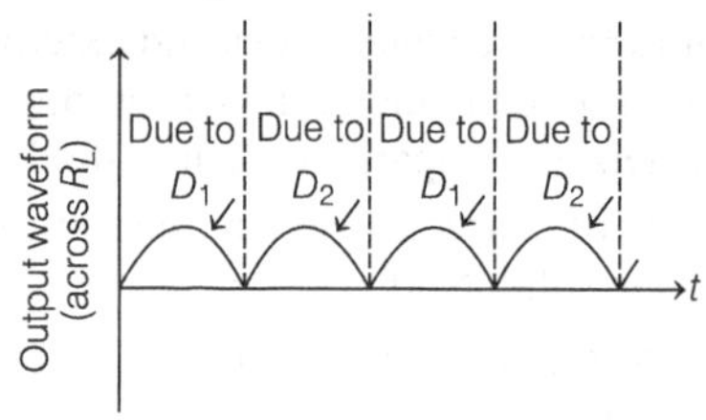


I Input waveform given to the diode $D_2$ at $B$ is

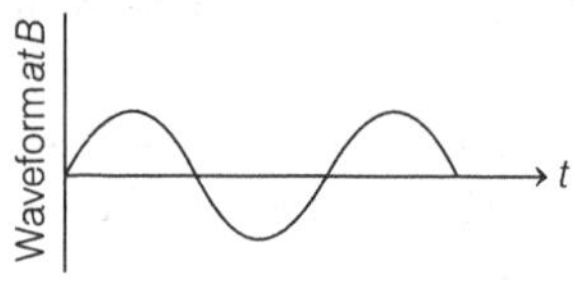


II This circuit is more efficient for getting rectified voltage as compared to half-wave rectifier circuit.

Which of the following statement(s) is/are correct?

(a) Only I (b) Only II
(c) Both I and II (d) Neither I nor II

**132** What happens during regulation action of a zener diode?

I. The current through series resistance $R_s$ changes.

II. The zener resistance is constant.

III. The resistance offered by zener changes.

Which of the following statement(s) is/are correct?

(a) Both II and III (b) Only II
(c) Both I and III (d) Only III

**133** I Zener diode is fabricated by heavily doped $p$ and $n$-sides both.

II The electric field required for field ionisation in zener diode is of the order of $10^6$ V / m.

III Optoelectronic devices are CFL's, discharge tubes, etc.

Which of the following statement(s) is/are correct?

(a) Both I and III (b) Both I and II
(c) Both II and III (d) I, II and III

**134** I. When zener diode is used as DC voltage regulator, then with the increase in input voltage $V_i$, zener voltage $V_Z$ remains unchanged. But when $V_i$ decreases, then $V_Z$ also decreases.

II LED's requires low operational voltage and less power as compare to incandescent lamp.

III. Ga As$_{0.6}$ P$_{0.4}$ is used for red LED.

Which of the following statement(s) is/are incorrect?

(a) Only I (b) Only II
(c) Only III (d) Both II and III

**135** Consider an *p-n-p* transistor, such that its emitter-base junction is forward biased and base-collector junction is reverse biased.

I Minority charge carries from emitter region enters base region.

II Holes in base cross the junction to enter into collector.

III Base current is small fraction of emitter current.

Which of the following statement(s) is/are correct?

(a) Only I (b) Only II
(c) Both I and III (d) Both II and III

**136** Base biased CE transistor has the transfer characteristics as shown

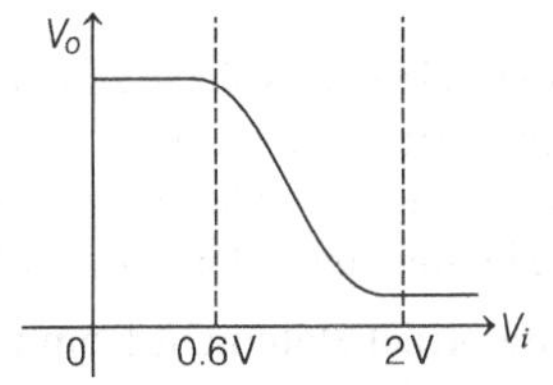


I. At $V_i = 0.4$ V, transistor is in active state.

II. At $V_i = 1$ V, transistor can be used as an amplifier.

III. At $V_i = 0.5$V, it can be used as a switch turned OFF.

IV. At $V_i = 2.5$V, it can be used as a switch turned ON.

Which of the following statement(s) is/are correct?

(a) I, II and III (b) II, III and IV
(c) I, II and IV (d) I, III and IV

**137** I When transistor is operated as a switch, then a low input switch the transistor ON and a high input switch the transistor OFF.

II In linear IC's the output is more or less directly proportional to input.

Which of the following statement(s) is/are incorrect?

(a) Only I (b) Only II
(c) Both I and II (d) Neither I nor II

**138** In an oscillator,

I. we get AC output without any external input signal.

II. output is not self sustained.

III. feedback can be achieved by inductive coupling (through mutual inductance) or *LC* or *RC* networks.

Incorrect statement is

(a) Only I (b) Only II
(c) Only III (d) None of these

**139** Choose the incorrect statement(s) from the following.

(a) Vacuum tetrode has four electrodes.
(b) Liquid crystal display monitors has been replaced by cathode rays tube.
(c) Vacuum diode is usually referred as valves.
(d) Both (a) and (b)

**140** Figure given below shows various regions attained by atoms in a Ge crystal when they are close enough (~ 2 to 3 Å), so that electrons not only interact with each other but also with neighbouring atomic cores.

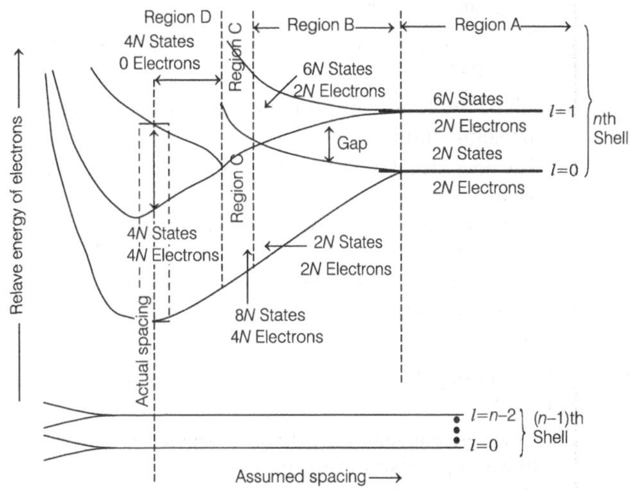


Now, choose the correct statement.

(a) In region A, 8*N* energy states has been re-apportioned between two bands.
(b) Region C represents energy levels in an isolated atom.
(c) Region B represents that 6*N* and 2*N*-states spread out and seperated by an energy gap.
(d) Region D represents no gap exists when the lower and upper states gets mixed.

**141** Choose the incorrect statement from the following.

(a) In intrinsic semiconductor, number of free electrons is equal to number of holes.
(b) If the lowest level in the conduction band is lower than the highest level of valence band, then conduction takes place.
(c) In semiconductors apart from electrons, holes also move.
(d) Size of the dopant can be greater than the semiconductor atoms.

**142** For a $p$-type semiconductor, which of the following statements is true? ***NEET (National) 2019***
(a) Holes are the majority carriers and trivalent atoms are the dopants.
(b) Holes are the majority carriers and pentavalent atoms are the dopants.
(c) Electrons are the majority carriers and pentavalent atoms are the dopants.
(d) Electrons are the majority carriers and trivalent atoms are the dopants.

**143** In an $n$-type semiconductor, which of the following statement is true? ***NEET 2013***
(a) Electrons are majority charge carriers and trivalent atoms are the dopants.
(b) Electrons are minority charge carriers and pentavalent atoms are the dopants.
(c) Holes are minority charge carriers and pentavalent atoms are the dopants.
(d) Holes are majority charge carriers and trivalent atoms are the dopants.

**144** Which of the following statement is correct for an $n$-type semiconductor?
(a) The donor level lies below the bottom of the conduction band.
(b) The donor level lies closely above the top of the valence band.
(c) The donor level lies at the halfway mark of the forbidden energy gap.
(d) None of the above

**145** Which amongst the following statement(s) is/are correct?
(a) $n$-type germanium is negatively charged and $p$-type silicon is positively charged.
(b) Both $n$-type and $p$-type germanium are neutral.
(c) $n$-type germanium is positively charged and $p$-type germanium is negatively charged.
(d) Both $n$-type silicon and $p$-type germanium are negatively charged.

**146** Which of the following statement concerning the depletion zone of an unbiased $p$-$n$ junction is (are) true?
(a) The width of the depletion region is independent on the densities of the dopants (impurities).
(b) The width of the depletion region is dependent on the densities of the dopants.
(c) The electric field in the zone is produced by the electrons in the conduction band and the holes in the valence band.
(d) Both (b) and (c)

**147** Choose the correct statement from the following.
(a) The thickness of the depletion region is of the order of $10^{-1}$ m.
(b) Diffusion and drift current flows in the same direction.
(c) $p$-$n$ junction cannot be formed by joining physically one slab of $p$-type and other of $n$-type semiconductor.
(d) Both (a) and (c)

**148** A semiconductor $X$ is made by doping a germanium crystal with arsenic ($Z = 33$). A second semiconductor $Y$ is made by doping germanium with indium ($Z = 49$). The two are joined end to end and connected to a battery as shown. Which of the following statement is correct?

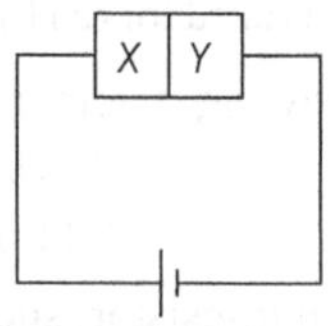


(a) $X$ is $p$-type, $Y$ is $n$-type and the junction is forward biased.
(b) $X$ is $n$-type, $Y$ is $p$-type and the junction is forward biased.
(c) $X$ is $p$-type, $Y$ is $n$-type and the junction is reverse biased.
(d) $X$ is $n$-type, $Y$ is $p$-type and the junction is reverse biased.

**149** For the given circuit of $p$-$n$ junction diode, which of the following statement is correct?

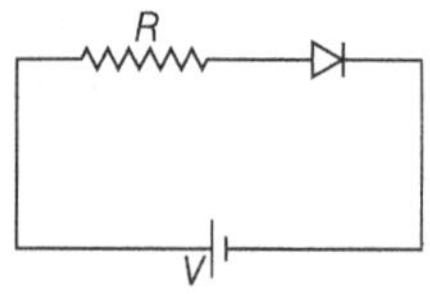


(a) In forward biasing, the voltage across $R$ is $V$.
(b) In forward biasing, the voltage across $R$ is $2V$.
(c) In reverse biasing, the voltage across $R$ is $V$.
(d) In reverse biasing, the voltage across $R$ is $2V$.

**150** Choose the incorrect statement from the following.
(a) A full wave rectifier circuit requires four diode to replace centre tapping.
(b) Output of a full wave rectifier is pulses of the half input AC voltage.
(c) Solar cell and photodiode when externally biased to generates electric current.
(d) Both (b) and (c)

**151** The given graph represents $V$-$I$ characteristic for a semiconductor device. ***CBSE AIPMT 2014***

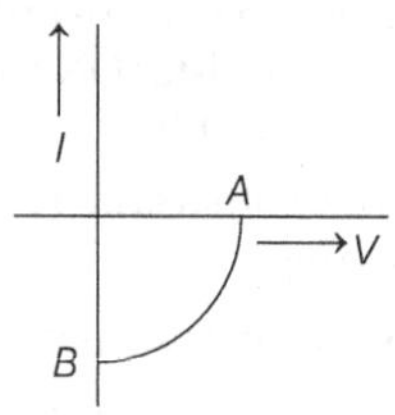


Which of the following statement is correct?
(a) It is $V$-$I$ characteristic for solar cell, where point $A$ represents open circuit voltage and point $B$ short circuit current.
(b) It is for a solar cell and points $A$ & $B$ represent open circuit voltage & current, respectively.

(c) It is for a photodiode and points $A$ & $B$ represent open circuit voltage & current, respectively.
(d) It is for a LED and points $A$ & $B$ represents open circuit voltage & short circuit current, respectively.

**152** Choose the correct statement from the following.
(a) Base, emitter and collector regions should have similar size and doping concentration.
(b) The base region must be very thin and lightly doped.
(c) For a transistor amplifier, voltage gain is high at high frequencies.
(d) Both (b) and (c)

**153** Consider an $n$-$p$-$n$ transistor with its base-emitter junction forward biased and collector-base junction reverse biased, which of these are correct?
(a) Electrons cross over from emitter to collector.
(b) Electrons move from base to emitter
(c) Electrons from emitter move out of base without going to collector.
(d) None of the above

**154** Choose the incorrect statement from the following.
(a) $\beta_{DC}$ and $\beta_{AC}$ are not equal.
(b) Different output characteristics are obtained for different values of $I_B$.
(c) Resonance frequency of formed collector oscillator determines the frequency at which the oscillator will oscillate.
(d) External input is necessary to sustain AC signal in the output of an amplifier.

## III. Matching Type

**155** Match the Column I (approximate energy band diagram) with Column II (type of solid) and select the correct answer from the codes given below.

| | Column I | | Column II |
|---|---|---|---|
| A. | 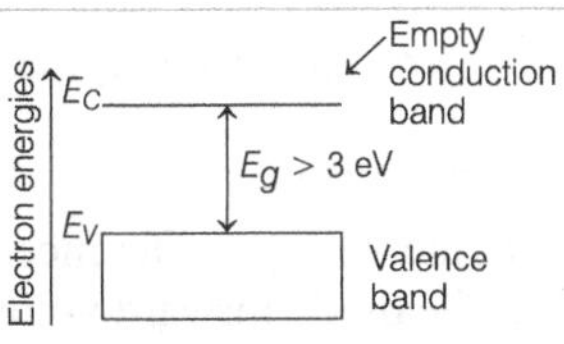 | 1. | Germanium |
| B. | 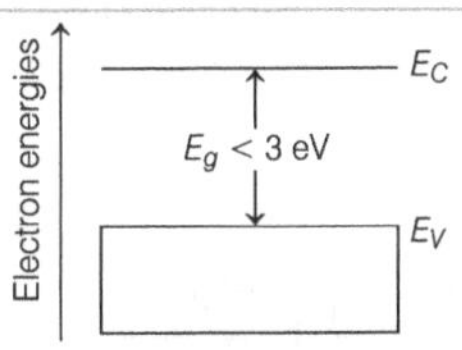 | 2. | Copper |
| C. | 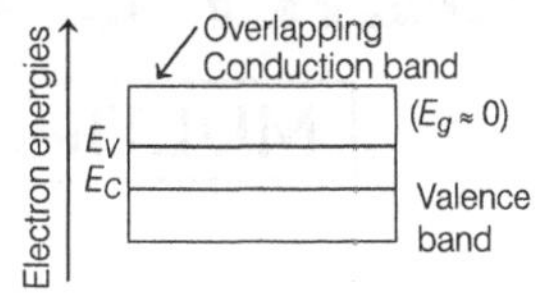 | 3. | Plastic |

| | A | B | C |
|---|---|---|---|
| (a) | 1 | 2 | 3 |
| (b) | 3 | 2 | 1 |
| (c) | 2 | 1 | 3 |
| (d) | 3 | 1 | 2 |

**156** Match the Column I (characteristic) with Column II (regions of transistor) and select the correct answer from the codes given below.

| Column I | Column II |
|---|---|
| A. Moderate size and heavily doped | 1. Base |
| B. Very thin and lightly doped | 2. Collector |
| C. Moderately doped and of large size | 3. Emitter |

| | A | B | C |
|---|---|---|---|
| (a) | 1 | 2 | 3 |
| (b) | 1 | 3 | 2 |
| (c) | 3 | 1 | 2 |
| (d) | 3 | 2 | 1 |

**157** Match the Column I (symbol) with Column II (their respective logic gate) and select the correct answer from the codes given below.

| Column I | | Column II |
|---|---|---|
| A. | 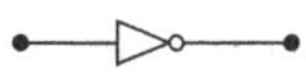 | 1. OR |
| B. | 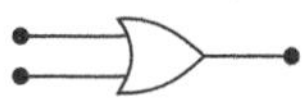 | 2. AND |
| C. | 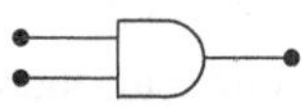 | 3. NAND |
| D. | 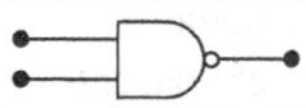 | 4. NOR |
| E. | 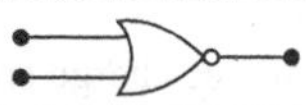 | 5. NOT |

| | A | B | C | D | E |
|---|---|---|---|---|---|
| (a) | 1 | 2 | 3 | 4 | 5 |
| (b) | 3 | 1 | 2 | 4 | 5 |
| (c) | 5 | 1 | 2 | 5 | 4 |
| (d) | 5 | 1 | 2 | 3 | 4 |

# NCERT & NCERT Exemplar

## MULTIPLE CHOICE QUESTIONS

### NCERT

**158** When a forward bias is applied to a *p-n* junction. It
(a) raises the potential barrier
(b) reduces the majority carrier current to zero
(c) lowers the potential barrier
(d) None of the above

**159** In an unbiased *p-n* junction, holes diffuse from the *p*-region to *n*-region because
(a) free electrons in the *n*-region attract them
(b) they move across the junction by the potential difference
(c) hole concentration in *p*-region is more as compared to *n*-region
(d) All of the above

**160** Two amplifiers are connected one after the other in series (cascaded). The first amplifier has a voltage gain of 10 and the second has a voltage gain of 20. If the input signal is 0.01 V, calculate the output AC signal.
(a) 2 V (b) 3 V
(c) 4 V (d) 5 V

### NCERT Exemplar

**161** A hole is
(a) an anti-particle of electron
(b) a vacancy created when an electron leaves a covalent bond
(c) absence of free electrons
(d) an artificially created (particle)

**162** Conductivity of a semiconductor increases with increase in temperature, because
(a) number density of free charge carriers increases
(b) relaxation time increases
(c) both number density of free charge carriers and relaxation time increases
(d) number density of free charge carriers increases and relaxation time decreases but effect of decrease in relaxation time is much less than increase in number density

**163** Assuming diodes to be ideal,

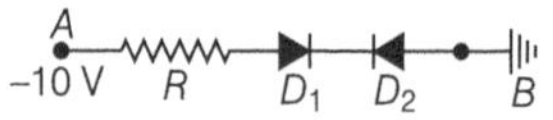


(a) $D_1$ is forward biased and $D_2$ is reverse biased, so current flows from $A$ to $B$
(b) $D_2$ is in forward bias and $D_1$ is in reverse bias and hence no current flows from $B$ to $A$ and *vice-versa*
(c) $D_1$ and $D_2$ both are in forward bias, so current flows from $A$ to $B$
(d) $D_1$ and $D_2$ both are in reverse bias, so no current flows from $A$ to $B$

**164** A 220 V AC supply is connected between points $A$ and $B$.

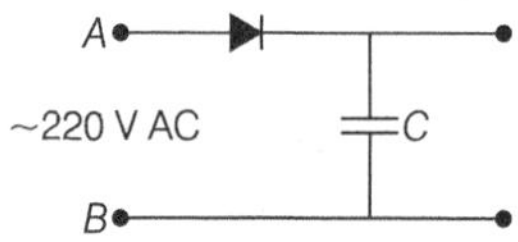


What will be potential difference across capacitor $C$?
(a) 220 V (b) 110 V
(c) 0 V (d) $220\sqrt{2}$ V

**165** Output of the circuit shown below will be

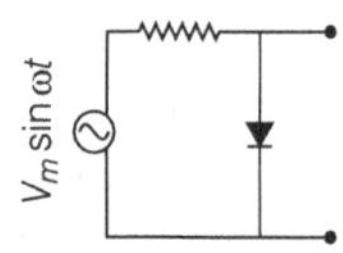


(a) zero all the times
(b) like half wave rectifier with positive cycles in output
(c) like half wave rectifier with negative cycles in output
(d) like that of a full wave rectifier

**166** In the circuit shown, the voltage difference between $A$ and $B$, if the diode forward voltage drop is 0.3V is

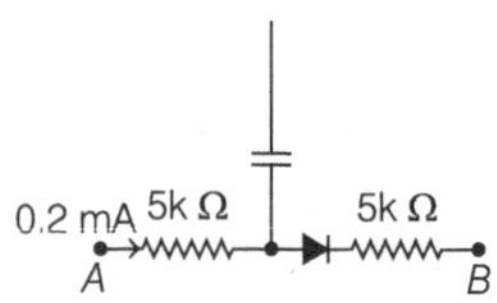


(a) 1.3 V (b) 2.3 V
(c) 0 (d) 0.5 V

**167** For given circuit, truth table is

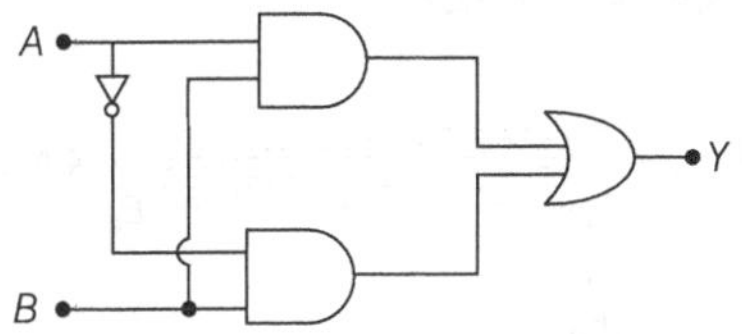


(a)

| $A$ | $B$ | $Y$ |
|---|---|---|
| 0 | 0 | 1 |
| 0 | 1 | 0 |
| 1 | 0 | 1 |
| 1 | 1 | 0 |

(b)

| $A$ | $B$ | $Y$ |
|---|---|---|
| 0 | 0 | 1 |
| 0 | 1 | 0 |
| 1 | 0 | 0 |
| 1 | 1 | 1 |

(c)

| $A$ | $B$ | $Y$ |
|---|---|---|
| 0 | 0 | 0 |
| 0 | 1 | 1 |
| 1 | 0 | 0 |
| 1 | 1 | 1 |

(d)

| $A$ | $B$ | $Y$ |
|---|---|---|
| 0 | 0 | 0 |
| 0 | 1 | 1 |
| 1 | 0 | 1 |
| 1 | 1 | 0 |

**168** In figure given below $V_o$ is the potential barrier across a *p-n* junction. When no battery is connected across the junction, then

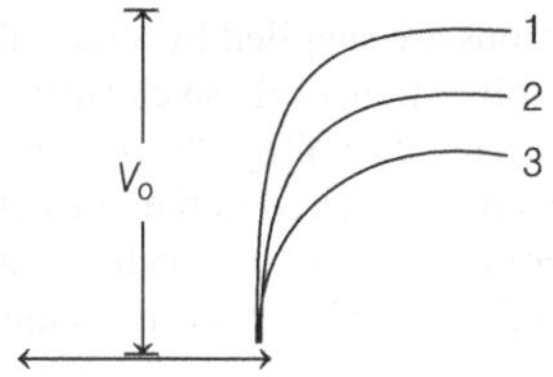


(a) 1 and 3 both correspond to forward bias of junction
(b) 3 corresponds to forward bias of junction and 1 corresponds to reverse bias of junction
(c) 1 corresponds to forward bias and 3 corresponds to reverse bias of junction
(d) 3 and 1 both correspond to reverse bias of junction

# Answers

## > Mastering NCERT with MCQs

| | | | | | | | | | |
|---|---|---|---|---|---|---|---|---|---|
| 1 (d) | 2 (d) | 3 (b) | 4 (c) | 5 (c) | 6 (d) | 7 (c) | 8 (b) | 9 (a) | 10 (a) |
| 11 (c) | 12 (d) | 13 (c) | 14 (d) | 15 (c) | 16 (a) | 17 (a) | 18 (c) | 19 (c) | 20 (d) |
| 21 (a) | 22 (a) | 23 (a) | 24 (c) | 25 (a) | 26 (c) | 27 (d) | 28 (c) | 29 (b) | 30 (c) |
| 31 (c) | 32 (b) | 33 (d) | 34 (d) | 35 (a) | 36 (d) | 37 (a) | 38 (a) | 39 (d) | 40 (a) |
| 41 (d) | 42 (b) | 43 (d) | 44 (a) | 45 (a) | 46 (b) | 47 (c) | 48 (a) | 49 (d) | 50 (a) |
| 51 (d) | 52 (c) | 53 (b) | 54 (a) | 55 (d) | 56 (d) | 57 (c) | 58 (d) | 59 (c) | 60 (b) |
| 61 (d) | 62 (d) | 63 (d) | 64 (d) | 65 (a) | 66 (d) | 67 (b) | 68 (c) | 69 (d) | 70 (c) |
| 71 (a) | 72 (c) | 73 (b) | 74 (a) | 75 (b) | 76 (d) | 77 (c) | 78 (d) | 79 (c) | 80 (b) |
| 81 (c) | 82 (a) | 83 (a) | 84 (b) | 85 (a) | 86 (b) | 87 (b) | 88 (c) | 89 (d) | 90 (d) |
| 91 (c) | 92 (a) | 93 (c) | 94 (c) | 95 (d) | 96 (c) | 97 (d) | 98 (d) | 99 (a) | 100 (d) |
| 101 (b) | 102 (c) | 103 (b) | 104 (c) | 105 (a) | 106 (c) | 107 (c) | 108 (a) | 109 (b) | |

## > Special Types Questions

| | | | | | | | | | |
|---|---|---|---|---|---|---|---|---|---|
| 110 (d) | 111 (a) | 112 (c) | 113 (c) | 114 (a) | 115 (c) | 116 (b) | 117 (a) | 118 (a) | 119 (b) |
| 120 (b) | 121 (a) | 122 (a) | 123 (c) | 124 (c) | 125 (c) | 126 (b) | 127 (b) | 128 (b) | 129 (c) |
| 130 (d) | 131 (b) | 132 (c) | 133 (b) | 134 (a) | 135 (d) | 136 (b) | 137 (a) | 138 (b) | 139 (b) |
| 140 (c) | 141 (d) | 142 (a) | 143 (c) | 144 (a) | 145 (b) | 146 (b) | 147 (c) | 148 (d) | 149 (a) |
| 150 (c) | 151 (a) | 152 (b) | 153 (a) | 154 (a) | 155 (d) | 156 (c) | 157 (d) | | |

## > NCERT and NCERT Exemplar MCQs

| | | | | | | | | | |
|---|---|---|---|---|---|---|---|---|---|
| 158 (c) | 159 (c) | 160 (a) | 161 (b) | 162 (d) | 163 (b) | 164 (d) | 165 (b) | 166 (b) | 167 (c) |
| 168 (b) | | | | | | | | | |

# Hints & Explanations

**1** (*d*) In a vacuum tube like the vacuum diode, the electrons are supplied by a heated cathode and the controlled flow of these electrons in vacuum is obtained by varying the voltage between its different electrodes. Vacuum is required in the inter-electrode space; otherwise the moving electrons may lose their energy on collision with the air molecules in their path.

**3** (*b*) The values of conductivity and resistivity for different types of solids are as follows

(i) **Metal**

$$\rho \sim 10^{-2}\text{-}10^{-8}\ \Omega\text{-m}$$
$$\sigma \sim 10^{2}\text{-}10^{8}\ \text{Sm}^{-1}$$

(ii) **Semiconductor**

$$\rho \sim 10^{-5}\text{-}10^{6}\ \Omega\text{-m}$$
$$\sigma \sim 10^{5}\text{-}10^{-6}\ \text{Sm}^{-1}$$

(iii) **Insulator**

$$\rho \sim 10^{11}\text{-}10^{19}\ \Omega\text{-m}$$
$$\sigma \sim 10^{-11}\text{-}10^{-19}\ \text{Sm}^{-1}$$

As $10^8 > 10^{-6} > 10^{-19}$, so $\sigma_{\text{metal}} > \sigma_{\text{semiconductor}} > \sigma_{\text{insulator}}$.

**4** (*c*) If there is some gap between the conduction band and the valence band which is neither very large nor zero. Then, some of the electrons from the valence band may gain external energy to cross the gap between the conduction band and valence band. Thus, these electrons will move into the conduction band. At the same time, they will create vacant energy levels in the valence band where other valence electrons can move.

Thus, the process creates the possibility of conduction due to electrons in conduction band as well as due to vacancies in the valence band.

**5** (*c*) For Si, the outermost orbit is, in which the number of electrons is 4. If $N$ is the total number of atoms in Si crystal, then the total number of outer electrons in the crystal is $4N$.

The maximum possible number of electrons in the outer orbit is 8. So, for the $4N$ valence electrons there are $8N$ available energy states.

These $8N$ discrete energy levels can either form a continuous band or they may be grouped in different bands depending upon the distance between the atoms in the crystal. As per the distance between the atoms in the crystal lattice of Si. The energy band of these $8N$ states is split apart into two which are separated by an energy gap $E_g$.

The lower band will be completely occupied by the $4N$ valence electrons at temperature of absolute zero, which is the valence band. The other band consisting $4N$ energy states which is the conduction band, is completely empty at absolute zero. Thus, the energy band position in Si at 0K is as follows

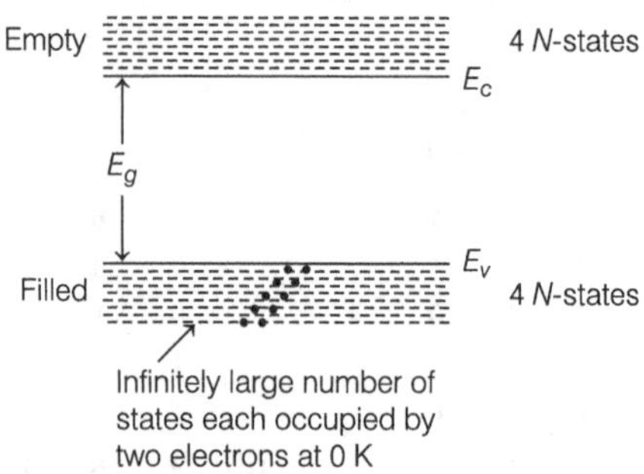


**6** (*d*) In the energy band diagram for a conductor, either the conduction band is partially filled and valence band is partially empty or conduction band and valence band overlap.

However, in the case of insulators, there are no electrons in the conduction band.

In the case of semiconductors, very few electrons are in the conduction band, after gaining sufficient energy.

Therefore, both the materials $A$ and $B$ are conductors.

**7** (*c*) If the energy of a photon of sodium light equals the band gap of semiconductor, so $E = E_g = \dfrac{hc}{\lambda}$

Here, Planck's constant $= 6.6\times 10^{-34}$ J-s,

Speed of light, $c = 3\times 10^{8}\,\text{ms}^{-1}$

and wavelength of sodium light,

$$\lambda = 580\ \text{nm} = 580\times 10^{-9}\ \text{m}$$

$$\Rightarrow\quad E_g = \frac{6.6\times 10^{-34}\times 3\times 10^{8}}{580\times 10^{-9}}\,\text{J}$$

$$= \frac{6.6\times 10^{-34}\times 3\times 10^{8}}{580\times 10^{-9}\times 1.6\times 10^{-19}}\,\text{eV} = 2.1\,\text{eV}$$

Thus, the minimum energy required to create hole-electron is 2.1 eV.

**8** (*b*) At low temperatures, semiconductors like Si and Ge behaves like insulators. However, as the temperature increases more thermal energy becomes available to the electrons, which may break away their covalent band. Thereby increase in the free electrons gradually contributes to increase in conduction. Hence, in the given case, with the increase in the temperature, the current in the circuit increases.

**9** (*a*) In semiconductor as the temperature increases, more thermal energy becomes available to electrons and some of these electrons may break away their covalent bond (becoming free electrons contributing to conduction).

Thus, the thermal energy effectively ionises only a few atoms in the crystalline lattice and creates a vacancy in the bond . These holes behaves as an apparent free particle with effective positive charge.

**10** (*a*) An intrinsic semiconductor, i.e. Si crystal will behave as an insulator at $T = 0$K.

However, at $T > 0$K, due to thermal energy electrons can be thermally excited. This would create a vacancy/hole at that position.

So, in the given structure this might have happened at site 1, thus an electron from the covalent bond at site 2 may jump to the vacant site 1.

Thus, after jump, the hole will be at site 2 and the site 1 will now have an electron.

**12** (*d*) In an intrinsic semiconductor, at $T > 0$K, due to thermal energy some electrons from the valence band excites to conduction band and partially occupy it. Thus, in the energy band diagram shown below, some electrons are in the conduction band. These have come from valence band leaving equal number of holes there.

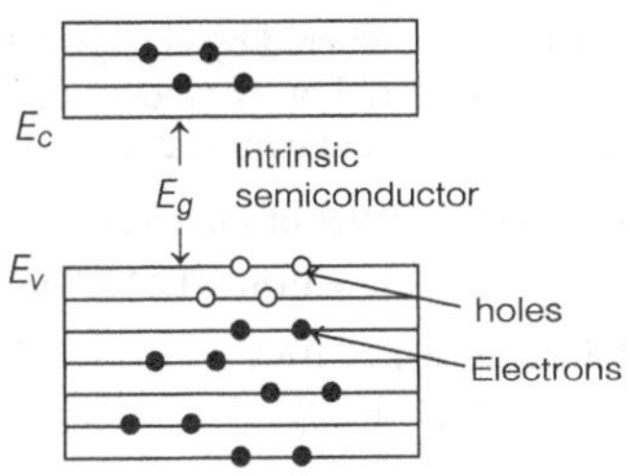


At $T>0$ K

$\therefore$ Option (d) is correct.

**13** (*c*) The 4 bonding electrons of C, Si or Ge are, respectively, in the second, third and fourth orbits. Hence, energy required to take out an electron from these atoms (i.e. ionisation energy $E_g$) will be least for Ge, followed by Si and highest for C. Hence, number of free electrons for conduction in Ge and Si are significant but negligibly small for C.

Thus, C is an insulator while Si and Ge are intrinsic semiconductors.

**14** (*d*) Dopant concentration in extrinsic semiconductor is usually 1 to 10 ppm (parts per million).

So, the ratio of impurity atoms to the pure semiconductor atom is

$$\frac{1}{10^7} \approx 10^{-7}$$

**15** (*c*) The number of electrons made available for conduction by dopant atoms depends strongly upon the doping level and is independent of any increase in ambient temperature. However, number of free electrons generated in semiconductor atom increases weakly with temperature.

**16** (*a*) Phosphorous is a pentavalent impurity. When a semiconductor is doped with it, then electrons becomes the majority charge carriers and holes the minority carriers,

$\therefore$ $$n_h << n_e$$

**17** (*a*) Increase in temperature causes more electrons to leave valence band and reach conduction band, so $n_e$ increases.

But increase in temperature causes more lattice vibrations, so $v_d$ decreases as number of collisions increases.

**18** (*c*) When a pure semiconductor, e.g. silicon and germanium are doped with trivalent impurity, i.e. elements of third group, e.g. indium, boron, aluminium, etc. then *p*-type semiconductors are obtained.

However, if they are doped with pentavalent impurity, i.e. elements of group fifth, e.g. arsenic, antimony, phosphorous, etc. then *n*-type semiconductors are obtained.

Thus, *p*-type semiconductor could be obtained in III only.

**19** (*c*) In an extrinsic semiconductors, due to doping, energy states due to donor impurity $E_d$ and acceptor impurity $E_a$ also exist which facilitates the conduction at room temperature. However, the energy gap between the conduction and valence band remains the same (about 0.72 eV for Ge and 1.1 eV for Si).

Thus, in extrinsic semiconductor, the gap between the conduction band and valence band is near about 1 eV.

**20** (*d*) An intrinsic semiconductor has an energy band diagram as given $X$, so sample $X$ is undoped.

However, when an intrinsic semiconductor is doped with a pentavalent impurity, e.g. As, Sb, P, etc. i.e. fifth group impurity an additional donor energy level $E_d$ exist, which is slightly below the bottom $E_c$ of conduction band.

Similarly, when an intrinsic semiconductor is doped with trivalent impurity, e.g. In, B, Al, etc. i.e. third group impurity an additional acceptor energy level $E_a$ exist, which is slightly above the top of the valence band.

So, sample $Y$ have been doped with a fifth group impurity and sample $Z$ have been doped with third group impurity.

**21** (*a*) For extrinsic semiconductor,

$$n_i^2 = n_e n_h$$

Here, number of intrinsic electron hole pairs,

$$n_i = 10^{19} \text{ per m}^3$$

Number acceptor atom of Al

$\approx$ Number of holes, $n_h = 10^{21}$ per m$^3$

$$\Rightarrow \quad (10^{19})^2 = n_e (10)^{21}$$

$$\Rightarrow \quad n_e = \frac{10^{38}}{10^{21}}$$

$$= 10^{17} \text{ per m}^3$$

**22** (*a*) Thermally generated electrons ($n_i \sim 10^{16}$ m$^{-3}$) are negligibly small as compared to those produced by doping.

Therefore, $n_e \approx N_D$

Given, total number of atoms $= 5\times 10^{28}\ \text{m}^{-3}$

$$n_i = 1.5\times 10^{16}\ \text{m}^{-3}$$

For every $10^6$ atoms, 1 As is doped.

1 As contributely $1e^-$

$$n_e = 5\times 10^{28}/10^6 = 5\times 10^{22}$$

Since, $\quad n_e n_h = n_i^2$

The number of holes

$$n_h = \frac{n_i^2}{n_e} = \frac{(1.5\times 10^{16})^2}{5\times 10^{22}}$$

$$= \frac{2.25\times 10^{32}}{5\times 10^{22}}$$

$$= 4.5\times 10^9\ \text{m}^{-3}$$

**23** (*a*) Given, $\quad N = 5\times 10^{28}\ \text{atoms/m}^3$,

Number of acceptor atom of indium,

$$N_A = 5\times 10^{20}\ \text{per m}^3 = 0.05\times 10^{22}\ \text{atoms/m}^3$$

Number of donor atoms of arsenic,

$$N_D = 5\times 10^{22}\ \text{atoms/m}^3$$

$$n_i = 1.5\times 10^{16}\ \text{m}^{-3}$$

$\Rightarrow\ N_D - N_A = n_e - n_h$

and $\quad n_e n_h = n_i^2$

$\therefore\ N_D - N_A = n_e - \dfrac{n_i^2}{n_e}$

$$\Rightarrow \quad n_e^2 - (N_D - N_A)n_e - n_i^2 = 0$$

$$\Rightarrow \quad n_e = \frac{(N_D - N_A) + \sqrt{(N_D - N_A)^2 - 4n_i^2}}{2}$$

On substituting values, the given values in the above equation, we get, $n_e = 4.95\times 10^{22}/\text{m}^3$

$$\therefore \quad n_h = \frac{n_i^2}{n_e} = \frac{(1.5\times 10^{16})^2}{4.95\times 10^{22}}$$

$$= 4.54\times 10^9\ \text{m}^{-3}$$

$\therefore$ Number of electrons $n_e$ and holes $n_h$ are $4.95\times 10^{22}/\text{m}^3$ and $4.54\times 10^9\ \text{m}^{-3}$, respectively.

**Note** Observe that doping level $\dfrac{10^{22}}{10^{28}} = 10^{-6}$ ppm nearly.

**24** (*c*) Energy gap $E_g$ for silicon, germanium and tin respectively is as follows $(E_g)_{\text{Si}} = 1.1$ eV

$$(E_g)_{\text{Ge}} = 0.74\ \text{eV and } (E_g)_{\text{Sn}} \approx 0\ \text{eV}$$

Thus, $\quad (E_g)_{\text{Si}} > (E_g)_{\text{Ge}} > (E_g)_{\text{Sn}}$

**25** (*a*) Depletion region is the space-charge region on the either side of the junction together. One to the positive space charge region on *n*-side and negative space charge region on *p*-side, due to which an electric field directed from the positive charge towards negative charge develops.

This field give rise to drift current which is opposite to diffusion current.

Due to this only, the resistance to the diffusion current by this region is large because of the presence of strong electric field.

**26** (*c*) In a *p-n* junction, the loss of electrons from the *n*-region and the gain of electrons by the *p*-region causes a difference of potential across the junction of the two regions. The polarity of this potential is such as to oppose further flow of carriers, so that a condition of equilibrium exists. Since, the *n*-side has lost electrons and *p*-side has acquired electrons. So, the potential of *n*-side is thus positive to that *p*-side.

So, the potential across the junction is correctly depicted in option (c).

**27** (*d*) Barrier potential of a *p-n* junction is the potential that tends to prevent the movement of electron from the *n*-region into the *p*-region.

Thus, it can easily be altered by changing the semiconducting material or by amount of doping or by increasing/decreasing the temperature.

**28** (*c*) Given, depletion layer of thickness,

$$d = 1\ \mu\text{m} = 1\times 10^{-6}\ \text{m}$$

Potential difference, $V = 0.6$ V

Using $\quad \left|\dfrac{dV}{dr}\right| = E$

Thus, electric field in the depletion region,

$$E = \frac{V}{d} = \frac{0.6}{1\times 10^{-6}}$$

$$= 6\times 10^5\ \text{Vm}^{-1}$$

**30** (*c*) A semiconductor diode is basically a *p-n* junction with metallic contacts provided at the ends for the application of an external voltage. It is a two terminal device.

**32** (*b*) When *p-n* junction is forward biased, the direction of the applied voltage is opposite to the built in potential $V_0$ (i.e., barrier potential). As a result, the depletion layer width decreases and the barrier height is reduced.

$\therefore$ The effective barrier height under forward bias is $V_0 - V$.

**33** (*d*) In the minority injection process, in the case of forward biasing, due to the applied voltage, electrons from *n*-side cross the depletion region and reach *p*-side (where they are minority). Similarly, holes from *p*-side cross the junction and reach the *n*-side (where they are minority carriers).

$\therefore$ This has been correctly depicted in both options (a) and (c).

**34** (*d*) Due to the minority carrier injection process, at the junction boundary, at each side, the minority carriers concentration increases significantly compared to the locations far from the junctions. Due to concentration gradient, the injected electrons on *p*-side diffuse from the junction edge of *p*-side to the other end of *p*-side.

Likewise, the injected holes on $n$-side diffuse from the junction edge of $n$-side to the other end of $n$-side.

This motion of charged carriers on either side gives rise to current. Thus, forward current is sum of hole diffusion current and the conventional current due to electron diffusion.

**35** (*a*) In the forward biasing of $p$-$n$ junction, $p$-side of junction diode is connected to higher potential and $n$-side of junction diode is connected to lower potential. Since, amongst the given options, only in option (a) this condition has been satisfied.
Hence, the option (a) is correct answer.

**36** (*d*) In reverse bias, the direction of applied voltage is same as the direction of barrier potential. As a result, the barrier height increases and the depletion region widens due to the change in the electric field. Thus, the effective barrier height under reverse bias is $V_0 + V$.
This has been correctly depicted in option (d).

**37** (*a*) In a reverse biased $p$-$n$ junction diode, current is not very much dependent on the applied voltage. In it even a small voltage is sufficient to sweep the minority charge carriers from one side of the junction to the other side of the junction.

$\therefore$ The current is not limited by the magnitude of the applied voltage but is limited due to the concentration of the minority charge carrier on either side of the junction.

**38** (*a*) The current under reverse bias is essentially voltage independent upto a critical reverse bias voltage, known as breakdown voltage ($V_{br}$). When $V = V_{br}$, the diode reverse current increases sharply from $n$ to $p$-side.

Even a slight increase in the bias voltage causes large change in the current. If the reverse current is not limited by an external circuit, the $p$-$n$ junction will get destroyed due to overheating.

**39** (*d*) In reverse biased $p$-$n$ junction diode, the minority charge carriers will be accelerated due to reverse biasing which on colliding with atoms cause ionisation that results in secondary electrons. Thus, this increases the number of charge carriers.
Thus, when doping concentration is large, there will be large number of ions in the depletion region, which will give rise to a strong electric field.

**41** (*d*) Due to increase in temperature because of heating, thermal collision between the electron and holes increases. Thus, net electron-hole pairs increase.
This leads to increase in the current in diode and overall resistance of the diode changes.
This in turn changes both the forward biasing and the reverse biasing.
Thus, the overall $V$-$I$ characteristics of $p$-$n$ junction diode gets affected.

**42** (*b*) Dynamic resistance, $r_d = \dfrac{\Delta V}{\Delta I}$

Here, $\Delta V = 0.70\text{V} - 0.65\text{V} = 0.05$ V

$$\Delta I = 5\text{mA} = 5 \times 10^{-3}\text{ A}$$

$$\therefore \quad r_d = \frac{0.05}{5 \times 10^{-3}} = 10\,\Omega$$

**43** (*d*) From the graph, given in question at $I = 20$ mA, $V$ is 0.8 V and at $I = 10$ mA, $V = 0.7$ V

$$r_{FB} = \frac{\Delta V}{\Delta I} = \frac{0.8-0.7}{20-10} = \frac{0.1\text{ V}}{10\text{ mA}} = \frac{0.1\text{ V}}{10 \times 10^{-3}\text{ A}} = 10\,\Omega$$

Also, at $V = -10$ V, $I = -1\,\mu$A

$$r_{RB} = \frac{10\text{ V}}{1\,\mu\text{A}} = \frac{10\text{ V}}{1 \times 10^{-6}\text{A}} = 1 \times 10^7\,\Omega$$

$$\therefore \quad \text{Ratio} = \frac{r_{FB}}{r_{RB}} = \frac{10}{10^7} = 10^{-6}$$

**44** (*a*) Let us assume that current through the diode is $I$.
From the given condition,

$$I = \frac{V_A - V_B}{R} = \frac{4-(-6)}{1\,\text{k}\Omega} = \frac{10}{1 \times 10^3} = 10^{-2}\text{ A}$$

Thus, value of current $I$ flowing through diode is $10^{-2}$ A.

**45** (*a*) Diode in the given circuit is reverse biased, because $p$-side of the diode is connected to negative terminal of the battery and $n$-side of the diode connected to positive terminal of the battery.
So, current across it is equal to 0, thus $V_{AB} = 6$ V.

**46** (*b*) In ideal condition, a forward biased $p$-$n$ junction acts like a conductive path of zero resistance. So, given circuit is equivalent to circuit shown below.

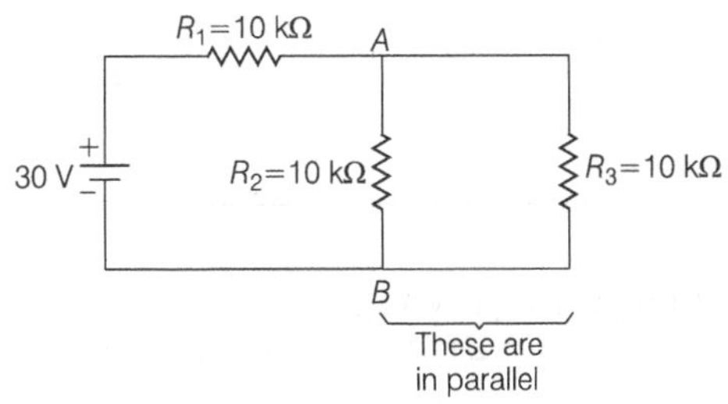


Total resistance of circuit,
$R = (R_3$ in parallel with $R_2)$ and these are in series with $R_1$

$$= \frac{10 \times 10}{10 + 10} + 10 = 15\text{ k}\Omega$$

So, total current in the circuit,

$$I = \frac{V}{R} = \frac{30}{15\text{ k}\Omega} = 2 \times 10^{-3}\text{ A}$$

Now, potential drop across $R_1$, $(10\text{k}\Omega)$,

$V_{R_1} = IR_1 = 2 \times 10^{-3} \times 10 \times 10^3 = 20\text{ V}$

Hence, remaining potential appears across parallel combination of $R_2$ and $R_3$,

$V_{AB} = 30 - 20 = 10\text{ V}$

This can be seen in the figure below.

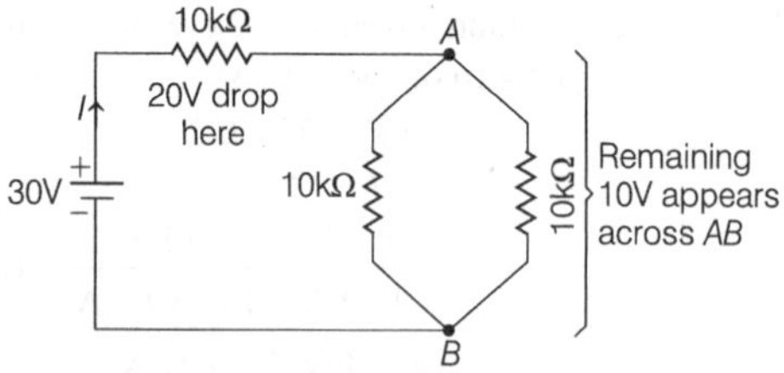


**47** (c) Initially Ge and Si are both forward biased. So, current will effectively pass through Ge diode with a voltage drop of 0.3 V.

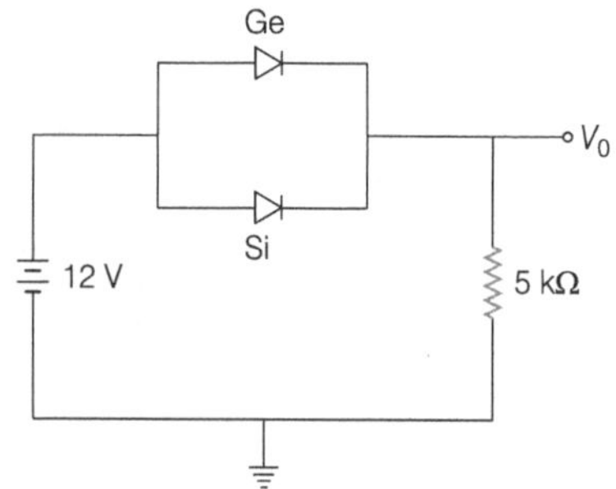


∴ Initial output voltage,

$$V_0 = 12 - 0.3 = 11.7\text{ V}$$

If Ge is reversed biased, then only Si diode will work. In this condition, output voltage

$$V_0 = 12 - 0.7 = 11.3\text{ V}$$

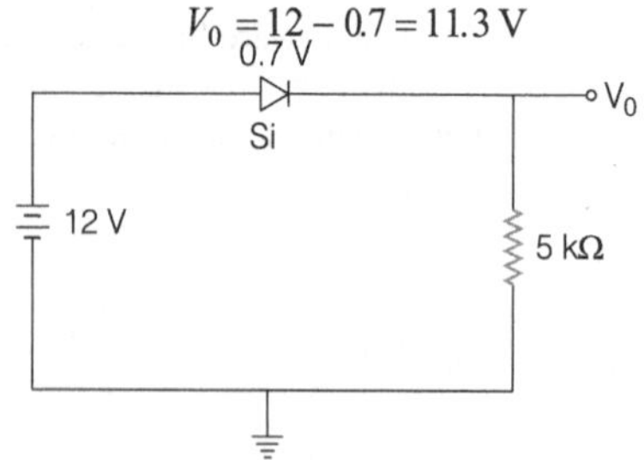


∴ Change in output voltage

$$= 11.7 - 11.3 = 0.4\text{ V}$$

**48** (a) Given, $R = 100\,\Omega$

and net potential drop across the resistor

$$= 3.5 - 0.5 = 3.0\text{V}$$

Current in the circuit, $I = \frac{V}{R} = \frac{3.0}{100} = 0.03\text{ A} = 30\text{ mA}$

**49** (d) As both diodes are in reverse bias, so given circuit is equivalent to the circuit as shown in the figure.

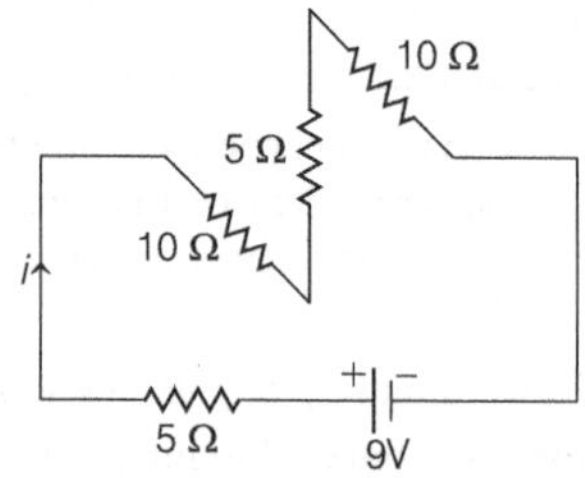


Equivalent resistance of circuit,

$$R = 5 + 10 + 5 + 10 = 30\Omega$$

∴Circuit current,

$$I = \frac{V}{R} = \frac{9}{30} = 0.3\text{A}$$

**50** (a) Current in circuit, $I = \frac{P}{V_d} = \frac{100 \times 10^{-3}}{0.5}$

(where, $V_d$ = voltage drop across diode)

$$= 200 \times 10^{-3}\text{ A}$$
$$= 200\text{ mA}$$

Voltage across resistance $R$,

$$V_R = 1.5 - 0.5 = 1.0\text{ V}$$

Thus, resistance, $R = \frac{V_R}{I}$

$$= \frac{1.0}{200 \times 10^{-3}} = 5\,\Omega$$

**51** (d) p-n junction diode primarily allows the flow of current only in one direction, i.e. only in forward bias condition.

Also, the forward bias resistance is low as compared to reverse bias resistance.

So, if an alternating voltage is applied across a diode, the current flows only in that part of the cycle when diode is forward biased. This property is thus used to rectify alternating voltages.

**54** (a) As we know, the reverse saturation current of a diode is negligible and can be considered equal to zero for practical purposes. Thus, the reverse breakdown voltage of the diode in a rectifier circuit must be sufficiently greater than the peak AC voltage at the secondary of the transformer to protect the diode from reverse breakdown.

**55** (d) The secondary of transformer in the half-wave rectifier circuit (refer to the circuit diagram given in synopsis) supplies the desired voltage to the diode. If the p-terminal of diode is positive, it is forward biased and it conducts. However, if that terminal is negative, the diode is reverse biased and it does not conduct. Therefore, in the positive half AC cycle there is current through $R_L$ but in negative half cycle there is no current.

Therefore, for the given input waveform in the question the respective output waveform for a half-wave rectifier is as shown below

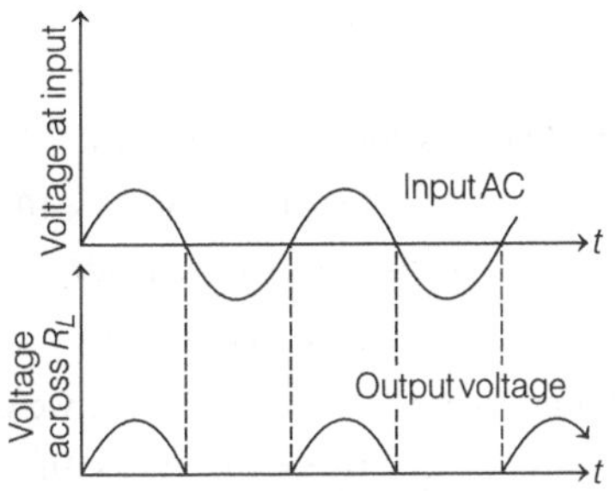


**56** (*d*) As, the given diode is forward biased, it means, it can only conduct the positive half of the signal.

So, the correct output across $R_L$ is shown in option (d).

**57** (*c*) From the circuit diagram of transformer of full wave rectifier (refer to circuit diagram given in synopsis,) each diode rectifier only half the cycle, but the two do, so for alternate cycle. Thus, the voltage rectified by each diode is only half the total secondary voltage of transformer.

**58** (*d*) In the given circuit, for the first half cycle of rectified output when the voltage across $C$ will rise, it gets charged. Then, in the presence of $R_L$, it gets discharged through it and the voltage begins to fall.

In the next-half cycle of rectified output, it again gets charged to peak value and will similarly gets discharged through $R_L$. Thus, a steady DC output from pulsating voltage is obtained. In other words, capacitor helps to filter out the AC ripple and give a pure DC voltage or bypass AC component to get DC.

This could be as shown in the waveform below

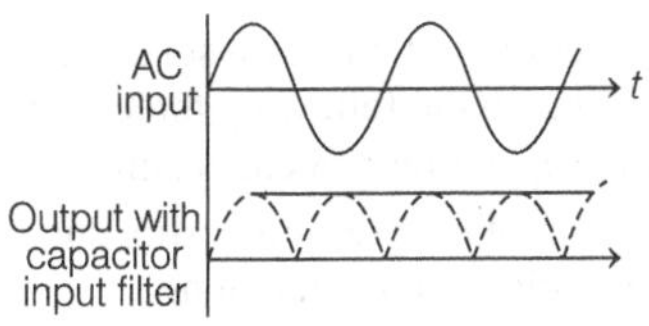


**59** (*c*) For a full wave rectifier circuit with a capacitor $C$ filter, the rate of fall of the voltage across the capacitor depends inversely upon the product of capacitance $C$ and the effective resistance $R_L$ used in the circuit.

**60** (*b*) For full wave rectifier,

output frequency = 2 × input frequency

and for half-wave rectifier,

output frequency = input frequency.

Given, input frequency = 50 Hz

∴ Ratio of output frequency of full wave rectifier to

$$\text{half-wave rectifier} = \frac{2 \times 50}{50} = \frac{2}{1}$$

**61** (*d*) Zener diodes works under reverse bias. So, as soon as the reverse bias voltage is increased, the electric field at the junction becomes significant. When the reverse bias voltage $V = V_Z$, then the electric field strength is high enough to pull valence electrons from the host atoms on the $p$-side which are accelerated to $n$-side. These electrons account for high current observed at the breakdown. Thus, the emission of electrons from the host atoms due to the high electric field is known as internal field emission or field ionisation.

**63** (*d*) For voltage regulation operation, increase/decrease in applied voltage should not affect the voltage drop across zener diode.

For this feature to be satisfied, from the graph given for zener diode the part $de$ is best suitable for its operation.

**64** (*d*) In given voltage regulator circuit breakdown of Zener occurs at 6 V.

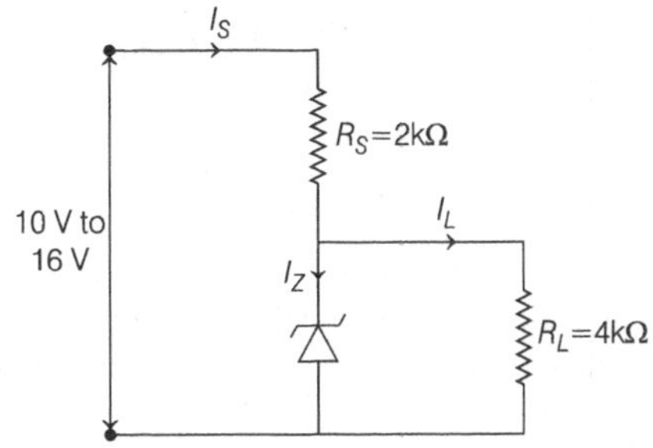


After breakdown, voltage across load resistance ($R_L = 4\text{k}\Omega$) is,

$$V = V_Z = 6\text{V}$$

∴ Load current after breakdown,

$$I_L = \frac{V_Z}{R_L} = \frac{6}{4000} = 1.5 \times 10^{-3}\text{ A}$$

When unregulated supply is of 16 V, potential drop occurring across series resistance ($R_S = 2\text{ k}\Omega$) is;

$$V_S = V - V_Z = 16 - 6 = 10\text{ V}$$

So, current across series resistance is

$$I_S = \frac{V_S}{R_S} = \frac{10}{2 \times 10^3} = 5 \times 10^{-3}\text{ A}$$

So, current across Zener diode is

$$I_Z = I_S - I_L = 5 \times 10^{-3} - 1.5 \times 10^{-3}$$

$$= 3.5 \times 10^{-3}\text{ A}$$

**65** (*a*) In a photodiode, the variation of intensity of incident light results in change in number of incident photons (per second). Hence a corresponding change in generation rate of electron and holes occurs. This causes a change in current amplitude.

**66** (*d*) The magnitude of the photocurrent depends on the intensity of the incident light as the photocurrent is proportional to incident light intensity. So, when intensity is increased, reverse saturation current also increases.

For the given, $I$-$V$ characteristics of a photodiode for different illumination intensity, the order will be

$$I_4 > I_3 > I_2 > I_1$$

Thus, maximum illumination of intensity is $I_4$.

**67** (*b*) Given, $E_g = 2.8$ eV

As, we know that wavelength

$$\lambda = \frac{hc}{E_g} = \frac{1240 \text{ eV-nm}}{2.8 \text{ eV}} = 440 \text{ nm} \approx 4400 \text{ Å}$$

**68** (*c*) Given, $\lambda = 620$ nm

Band gap of crystal used to fabricate the diode,

$$E_g = \frac{hc}{\lambda} = \frac{1240 \text{ eV - nm}}{620 \text{ nm}} = 2.0 \text{ eV}$$

**69** (*d*) Given, $E_g = 2$ eV

Wavelength of radiation corresponding to this energy,

$$\lambda = \frac{hc}{E_g} = \frac{1240 \text{ eV-nm}}{2\text{eV}} = 620 \text{ nm}$$

Frequency, $\nu = \frac{c}{\lambda} = \frac{3 \times 10^8 \text{ ms}^{-1}}{620 \times 10^{-9} \text{m}} = 5 \times 10^{14}$ Hz

**70** (*c*) When the LED is forward biased, then on recombination of electron-hole the energy is released in the form of photons. These photons are emitted with energy equal to or slightly less than the band gap, i.e. $E \le E_g$.

**71** (*a*) LED's operate under forward bias. They are biased such that the light emitting efficiency is maximum. Since, the reverse breakdown voltages of LEDs are very low, typically around 5 V. So, care should be taken that high reverse voltages do not appear across them.

**72** (*c*) As, emitted energy of radiation = Band gap value, when an electron moves from conduction band to valance band.
If $\lambda$ is wavelength of emitted radiation, then

$$E_g = \frac{hc}{\lambda}$$

$$\Rightarrow \quad \lambda = \frac{hc}{E_g} \quad \text{...(i)}$$

Taking $hc = 1240$ eV-nm and $450 \text{ nm} \le \lambda \le 750$ nm for visible region, we get

$$450 \le \lambda \le 750$$

$$\Rightarrow \quad 450 \le \frac{hc}{E_g} \le 750 \quad \text{[using Eq. (i)]}$$

$$\Rightarrow \quad E_g \le \frac{hc}{450} = \frac{1240 \text{ eV - nm}}{450 \text{ nm}} \approx 2.8 \text{ eV}$$

and $$E_g \ge \frac{hc}{750} = \frac{1240 \text{ eV nm}}{750 \text{ nm}} \approx 1.7 \text{ eV}$$

So, we have

$$1.8 \le E_g \le 3 \text{ eV}$$

**73** (*b*) The energy of emitted light of wavelength $\lambda$ is given by

$$E_g = h\nu = \frac{hc}{\lambda}$$

$$\Rightarrow \quad \lambda = \frac{hc}{E} \quad \text{...(i)}$$

Here, $h$ = Planck's constant $= 6.63 \times 10^{-34}$ J-s,

$c$ = speed of light $= 3 \times 10^8$ m/s

and $E_g$ = energy gap = 1.9 eV $= 1.9 \times 1.6 \times 10^{-19}$ J

Substituting the given values in Eq. (i), we get

$$\Rightarrow \quad \lambda = \frac{6.63 \times 10^{-34} \times 3 \times 10^8}{1.9 \times 1.6 \times 10^{-19}}$$

$$= 6.54 \times 10^{-7} \text{ m} \approx 654 \text{ nm} = 6540 \text{ Å}$$

Thus, the wavelength of light emitted from LED will be 654 nm.

**74** (*a*) *I-V* characteristics of an LED is similar to that of a Si junction diode. But the threshold voltages are much higher and different for each colour. So, for same value of current, higher value of voltage is required for higher frequency.
Since, we know that

$$\nu_{\text{red}} < \nu_{\text{yellow}} < \nu_{\text{green}} < \nu_{\text{blue}}$$

So, the corresponding voltages for these colours for the same value of currect will be

$$V_{\text{red}} < V_{\text{yellow}} < V_{\text{green}} < V_{\text{blue}}$$

Thus, this has been correctly depicted in option (a).

**75** (*b*) Light from an LED lies in the bandwidth 100Å to 500Å or in other words, it is nearly monochromatic light. Since, our eyes are not very comfortable with these types of light. Thus, they are not used for room lighting.

**76** (*d*) The generation of emf by a solar cell, when light falls on, is due to the following three basic processes : generation, separation and collection.

(i) generation of electron-hole pairs due to light (with $h\nu > E_g$) close to the junction.

(ii) separation of electrons and holes due to electric field $E$ of the depletion region. Electrons are swept to *n*-side and holes to *p*-side.

(iii) the electrons reaching the *n*-side are collected by the front contact and holes reaching *p*-side are collected by the back contact.

Thus *p*-side becomes positive and *n*-side becomes negative giving rise to photovoltage.
Hence, all the given options are correct.

**77** (*c*) The important criteria for the selection of a material for solar cell fabrication are

(i) band gap (~ 1.0 to 1.8 eV),

(ii) high optical absorption (~ $10^4$ cm$^{-1}$),

(iii) electrical conductivity,

(iv) availability of the raw material and

(v) cost.

However, sunlight is not always required for a solar cell. As any light with photon energies greater than the band gap required will do the working properly.

Hence, the criteria given in I, III and IV are correct.

**78** (*d*) For a simple diode, in forward biased, it carries current. However, in reverse biased, it carries negligible current. So, its correct *I-V* characteristics is depicted in (i).

Zener diode is designed to operate under reverse bias in the breakdown region. So, its *I-V* characteristics is correctly depicted in (ii).

Solar cell generates emf when solar radiation falls on the *p-n* junction. Its *I-V* characteristics lies in the fourth quadrant of the coordinate axes. Thus, it has been correctly depicted in (iii).

**79** (*c*) For an *n-p-n* transistor, two segments of *n*-type semiconductor (emitter *E* and collector *C*) are separated by a segment of *p*-type semiconductor (base *B*) as shown below.

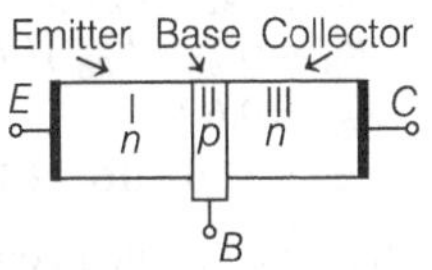


So, region marked I represents emitter and region II represents base while region III represents collector.

**80** (*b*) The transistor is said to be in active state, if its emitter-base junction is forward biased and base-collector junction is reversed biased.

Since, in the given transistor, emitter-base junction is forward biased and base-collector junction is reversed biased. So, it said to be in active state.

**81** (*c*) In a transistor, the heavily doped emitter has a high concentration of majority carriers, which will be holes in an *p-n-p* transistor and electrons in an *n-p-n* transistor. These majority carriers enter the base region in large numbers. The base is thin and lightly doped. So, the majority carriers there would be few.

In a *p-n-p* transistor the majority carriers in the base are electrons since base is of *n* -type semiconductor. The large number of holes entering the base from the emitter swamps the small number of electrons there. As the base collector-junction is reverse-biased, these holes which appear as minority carriers at the junction, can easily cross the junction and enter the collector. Similarly in *n-p-n* transistor, in the base collector junction, electrons can easily cross it and enter the collector.

So, the direction of motion of charge carriers is correctly depicted in (i) and (iii).

**82** (*a*) For *n-p-n* transistor in CE-configuration emitter (*E*) is common to both the input and the output, i.e. the input is between the base and the emitter and the output is between the collector and the emitter.

$\therefore$ This has been correctly depicted in the circuit diagram given in option (a).

**83** (*a*) As we know, $i_E = i_B + i_C \Rightarrow i_E > i_C$

Since, collector current, $i_C = 24$ mA

So, according to the question, the possible emitter current $i_E$ is 36 mA.

**84** (*b*) For the variation of base current $I_B$ with base-emitter voltage $V_{BE}$, collector-emitter voltage $V_{CE}$ is usually kept constant. Since, to obtain input characteristics, the transistor has to be in active region. So, the $V_{CE}$ is kept large enough to make the base collector junction reverse biased.

So, for a given collector voltage, $I_B$ increases rapidly with increase in $V_{BE}$.

Since, the increase in $V_{CE}$ appears as increase in $V_{CB}$, its effect on $I_B$ is negligible. As a consequence, input characteristics for various values of $V_{CE}$ will give almost identical curves. Hence, it is enough to determine only one input characteristics. Thus, the input characteristics of a transistor is correctly shown in option (b).

**85** (*a*) For determining $\beta_{AC}$ and $\beta_{DC}$ at the stated values of $V_{CE}$ and $I_C$ one can proceed as follows.

Consider any two characteristics for two values of $I_B$ which lie above and below the given value of $I_C$.

Here $I_C = 4.0$ mA.

(Choose characteristics for $I_B = 30$ and $20\mu$A).

At $V_{CE} = 10$ V we read the two values of $I_C$ from the graph given in question. Then, $\Delta I_B = (30-20)\,\mu$A $= 10\,\mu$A

$$\Delta I_C = (4.5 - 3.0) \text{ mA} = 1.5 \text{ mA}$$

$$\text{Therefore, } \beta_{AC} = \left(\frac{\Delta I_C}{\Delta I_B}\right)_{V_{CE}} = 1.5 \text{ mA} / 10\,\mu\text{A} = 150$$

For determining $\beta_{DC}$, either estimate the value of $I_B$ corresponding to $I_C = 4.0$ mA at $V_{CE} = 10$ V or calculate the two values of $\beta_{DC}$ for the two characteristics chosen and find their mean.

Therefore, for $I_C = 4.5$ mA and $I_B = 30\,\mu$A.

$$\beta_{DC} = \frac{I_C}{I_B} = \frac{4.5 \text{ mA}}{30\,\mu\text{A}} = 150$$

and for $I_C = 3.0$ mA and $I_B = 20\,\mu$A

$$\beta_{DC} = 3.0 \text{ mA}/20\,\mu\text{A} = 150$$

Hence, $\beta_{DC} = (150 + 150)/2 = 150$

Thus, ratio of $\beta_{AC}$ to $\beta_{DC}$ is 1.

**86** (*b*) Transfer characteristic is the graph between collector current and the base current $I_B$ at constant values of $V_{CE}$. In active region, it is almost a straight line. Thus, graph given in option (b) is correct.

**87** (*b*) Voltage gain $=$ Current gain $\times$ Resistance gain

$$= \text{Current gain} \times \frac{R_C}{R_i}$$

Given, collector resistance, $R_C = 6\text{ k}\Omega$

Input resistance, $R_i = 1\text{ k}\Omega$

Current gain = 30

$\Rightarrow$ Voltage gain $= 30 \times \frac{6}{1} = 180$

**88** (c) Current gain, $\beta = \left(\frac{\Delta I_C}{\Delta I_B}\right)_{V_{CE}}$

and voltage gain, $A_V = \frac{\beta \times R_{out}}{R_{in}}$

Given, $R_{in} = 1000\,\Omega, \Delta I_B = 10\,\mu A = 10^{-5}$ A

$R_{out} = 5\,k\Omega = 5 \times 10^3\,\Omega$

$\Delta I_C = 2$ mA $= 2 \times 10^{-3}$ A

$\beta = \frac{2 \times 10^{-3}}{10^{-5}} = 200$

Hence, $A_V = \frac{200 \times 5 \times 10^3}{1000} = 1000$

Thus, voltage gain of the amplifier is 1000.

**89** (d) Applying Kirchhoff's voltage rule to the input and output sides of the circuit given in the question, we get

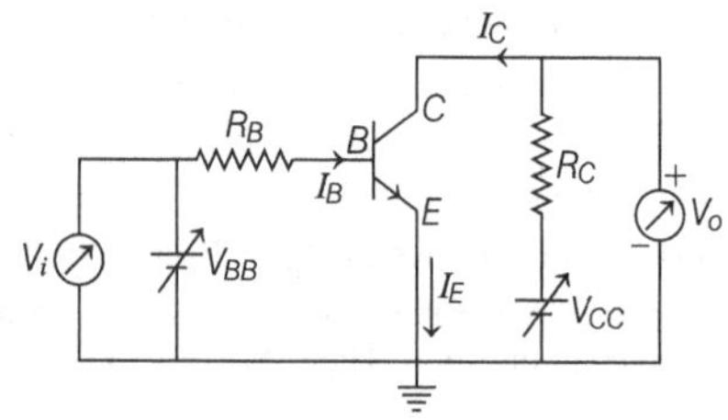


$V_{BB} = I_B R_B + V_{BE}$

and $V_{CE} = V_{CC} - I_C R_C$

Let us treat $V_{BB}$ as the DC input voltage $V_i$ and $V_{CE}$ as the DC output voltage $V_o$. So, we have

$V_i = I_B R_B + V_{BE}$

and $V_o = V_{CC} - I_C R_C$.

Thus, all the four relations are correct.

**90** (d) We know that, for $n$-$p$-$n$ transistor in CE-configuration

$V_i = I_B R_B + V_{BE}$ and $V_o = V_{CC} - I_C R_C$

In the case of Si transistor, as long as input $V_i$ is less than 0.6 V, the transistor will be in cut-off state and current $I_C$ will be zero.

Hence, $V_o = V_{CC}$

When $V_i$ becomes greater than 0.6 V, the transistor is in active state with some current $I_C$ in the output path and the output $V_o$ decrease as the term $I_C R_C$ increases.

With increases of $V_i$, $I_C$ increases almost linearly and so $V_0$ decreases linearly till its value becomes less than about 1.0 V.

Beyond this, the change becomes non-linear and transistor goes into saturation state. With further increase in $V_i$, the output voltage is found to decrease further towards zero though it may never becomes zero. Thus, from $V_0$ *versus* $V_i$ curve plot.

I. $\rightarrow$ Cut-off region II. $\rightarrow$ Active region
III. $\rightarrow$ Saturation region

**91** (c)

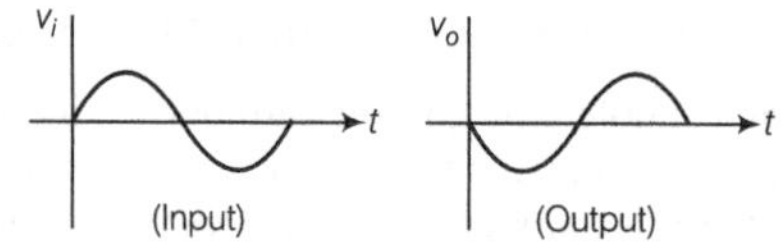


In a CE $n$-$p$-$n$ transistor amplifier output is 180° out of phase with input.

**92** (a) Given, the output AC voltage is 2.0 V.

Collector resistance = 2.0 k$\Omega$ = 2000 $\Omega$

So, the AC collector current, $i_C = 2.0/2000 = 1.0$ mA.

The signal current through the base is, therefore given by

$$i_B = \frac{i_C}{\beta} = \frac{1.0 \text{ mA}}{100} = 0.010 \text{ mA}.$$

The DC base current has to be $10 \times 0.010 = 0.10$ mA.

We know that, $V_{BB} = V_{BE} + I_B R_B$

$\Rightarrow R_B = (V_{BB} - V_{BE})/I_B$

As $V_{BE} = 0.6$ V

$\Rightarrow R_B = (2.0 - 0.6)/0.10 = 14\,k\Omega$

**93** (c) From Kirchhoff's loop rule in output loop, of the given circuit in question, $9 - 4 = I_C R_C$

As, $R_C = 2k\Omega$,

$\Rightarrow I_C = \frac{5}{2 \times 10^3} = 2.5$ mA $= 2.5 \times 10^{-3}$ A

$\Rightarrow I_B = \frac{I_C}{\beta} = \frac{2.5 \times 10^{-3}}{90} = 0.0278 \times 10^{-3}$ A

$= 27.8\,\mu A$

Given, $V_{BE} = 0.7$ V

From Kirchhoff, loop rule in input loop, of the given circuit

$$I_B R_B = 3 - 0.7 \Rightarrow R_B = \frac{2.3}{27.8 \times 10^{-6}} = 82\,k\Omega$$

Thus, base resistance $R_B$ is 82 k$\Omega$.

**94** (c) Given, $R_{out} = 200\Omega$, $R_{in} = 100k\Omega = 10^5\Omega$

$V_{CC} = 3V, V_{BE} = 0.7V, V_{CE} = 0$ V, $\beta = 200$

Applying KVL at output side, in given circuit diagram in question, we get

$V_{CE} = V_{CC} - I_C R_{out}$

$\Rightarrow I_C = \frac{3-0}{200} = 0.015$ A $= 15$ mA

$\therefore I_B = \frac{I_C}{\beta} = \frac{15 \times 10^{-3}}{200} = 75 \times 10^{-6}$A

$\therefore I_B = 75\,\mu A$

Again, applying KVL at input side, in given circuit diagram in question

$V_{BE} = V_{BB} - I_B R_{in}$

$\Rightarrow V_{BB} = V_{BE} + I_B R_{in} = 0.7 + 75 \times 10^{-6} \times 10^5$

$= 0.7 + 7.5 = 8.2$ V

**95** (*d*) Given, $V_{BE} = 0$ V, $V_{CE} = 0$ V and $V_i = 20$ V

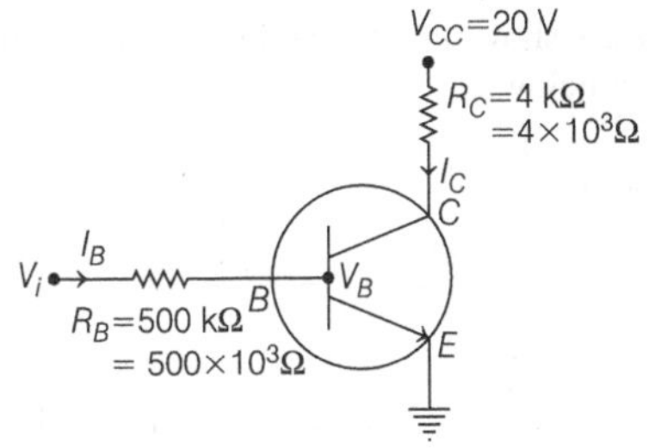


Applying Kirchhoff's law to the base-emitter loop, we get

$$V_i = I_B R_B + V_{BE}$$

Substituting the given values, we get

$$20 = I_B \times (500 \times 10^3) + 0$$

$$\Rightarrow \quad I_B = \frac{20}{500 \times 10^3} = 0.04 \times 10^{-3}$$

$$= 40 \times 10^{-6} = 40\,\mu\text{A} \quad \ldots\text{(i)}$$

Similarly, $V_{CC} = I_C R_C + V_{CE}$

Substituting the given values, we get

$$20 = I_C \times (4 \times 10^3) + 0$$

$$\Rightarrow \quad I_C = \frac{20}{4 \times 10^3} = 5 \times 10^{-3} = 5\text{mA} \quad \ldots\text{(ii)}$$

Current gain is given as, $\beta = \frac{I_C}{I_B}$

Substituting the value of $I_B$ and $I_C$ from Eqs. (i) and (ii) in the above, we get

$$\beta = \frac{5 \times 10^{-3}}{40 \times 10^{-6}}$$

$$= 0.125 \times 10^3 = 125$$

**96** (*c*) Given,

Voltage across collector, $V = 3$ V

Resistance of collector, $R_C = 3\,\text{k}\Omega = 3 \times 1000\,\Omega$

Collector current, $i_C = \frac{V}{R_C} = \frac{3}{3 \times 10^3} = 10^{-3}$ A

Current gain, $\beta = 100$

Base resistance, $R_B = 2\,\text{k}\Omega = 2 \times 10^3\,\Omega$

Now, base current, $i_B = \frac{i_C}{\beta} = \frac{10^{-3}}{100} = 10^{-5}$ A

As, input voltage, $V_{\text{in}} = i_B R_B$

$$\therefore \quad V_{\text{in}} = 10^{-5} \times 2 \times 10^3 = 2 \times 10^{-2}\text{ V}$$

So, voltage gain, $A_V = \frac{V_{\text{out}}}{V_{\text{in}}} = \frac{3}{2 \times 10^{-2}} = 150$

Power gain $= A_V \times \beta = 150 \times 100 = 15000$

Thus, the voltage gain and power gain of the amplifier is 150 and 15000, respectively.

**97** (*d*) Given, resistance across load, $R_L = 800\,\Omega$

Voltage drop across load, $V_L = 0.8$ V

Input resistance of circuit, $R_i = 192\,\Omega$

Collector current is given by

$$I_C = \frac{V_L}{R_L} = \frac{0.8}{800} = \frac{8}{8000} = 1 \times 10^{-3}\text{ A} = 1\text{ mA}$$

$$\because \text{Current amplification} = \frac{\text{Output current}}{\text{Input current}} = \frac{I_C}{I_B} = 0.96$$

$$\Rightarrow \quad I_B = \frac{1\text{ mA}}{0.96}$$

$$\because \quad \text{Voltage gain, } A_V = \frac{V_L}{V_{\text{in}}} = \frac{V_L}{I_B R_i} = \frac{0.8 \times 0.96}{10^{-3} \times 192} = 4$$

$$\Rightarrow \quad A_V = 4$$

and power gain,

$$A_P = \frac{I_C^2 R_L}{I_B^2 R_i} = \left(\frac{I_C}{I_B}\right)^2 \cdot \frac{R_L}{R_i} = (0.96)^2 \times \frac{800}{192}$$

$$A_P = 3.84$$

Thus, the voltage gain and power gain of the amplifier is 4 and 3.84, respectively.

**98** (*d*) Input signal of a CE amplifier, $V_i = 2\cos\left(15t + \frac{\pi}{3}\right)$

Voltage gain, $A_V = 150$

As CE amplifier gives phase difference of $\pi$ between input and output signals.

So, $\quad A_V = \frac{V_o}{V_i}$

$$\Rightarrow \quad V_o = A_V V_i$$

$$V_o = 150 \times 2\cos\left(15t + \frac{\pi}{3} + \pi\right)$$

Output signal, $V_o = 300\cos\left(15t + \frac{4\pi}{3}\right)$

**99** (*a*) When $S_1$ is put ON to apply proper bias for the first time in the given circuit. Obviously, a surge of collector current flows in the transistor. This current flows through the coil $T_2$ where terminals are numbered 3 and 4 in the circuit. This current does not reach full amplitude instantaneously but increases from $X$ to $Y$, as shown in Fig (i). The inductive coupling between coil $T_2$ and coil $T_1$ now causes a current to flow in the emitter circuit (**Note** that this actually is the 'feedback' from input to output). As a result of this positive feedback, this current (in $T_1$; emitter current) also increases from $X'$ to $Y'$ as shown in Fig (ii). Thus, initially, both $I_C$ and $I_E$ increase.

**100** (*d*) The signal (current or voltage) in the form of continuous, time-varying voltage or current are called continuous or analog signals.

However, a pulse waveform in which only discrete values of voltages are possible are called digital signals.

So, option (a) represents digital signal, whereas option (b) and (c) represents analog signal.

**101** (*b*) For an NAND gate, if inputs $A$ and $B$ are both 1 the output will not be 1. So, the truth table for the given input is

| | Input | | Output |
|---|---|---|---|
| | $A$ | $B$ | $Y$ |
| $t_1$ | 0 | 0 | 1 |
| $t_2$ | 1 | 0 | 1 |
| $t_3$ | 0 | 1 | 1 |
| $t_4$ | 1 | 1 | 0 |
| $t_5$ | 0 | 0 | 1 |
| $t_6$ | 1 | 0 | 1 |

$\therefore$ The correct output is represented in option (b).

**102** (*c*) The resultant Boolean expression for the given logic circuit will be

$$Y = (A + B) \cdot C$$

So, we can see that amongst the given options, only option (c) is the correct choice for input, i.e.

Output $Y = 1$ only, when inputs

$$A = 1, B = 0 \text{ and } C = 1.$$

**103** (*b*) According to the question, the figure of combination of gates in terms of inputs and outputs can be given as

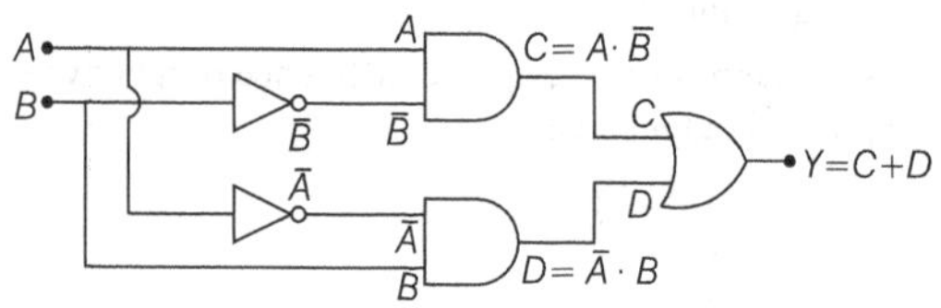


Thus, $Y = C + D = A \cdot \overline{B} + \overline{A} \cdot B$

**104** (*c*) The output $X$ of the logic circuit will be

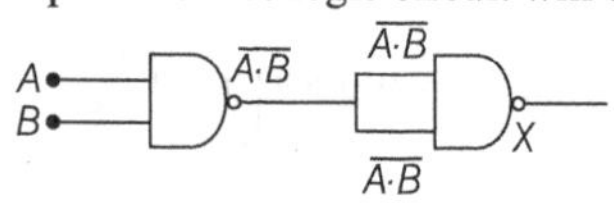


$$X = \overline{\overline{\overline{A \cdot B} \cdot \overline{A \cdot B}}} = \overline{\overline{A \cdot B}} + \overline{\overline{A \cdot B}}$$

(using de-Morgan's theorem, $\overline{x \cdot y} = \overline{x} + \overline{y}$)

$$= (A \cdot B) + (A \cdot B) \quad [\because x + x = x]$$
$$= A \cdot B$$

This represents the Boolean expression of AND gate.

**105** (*a*) The given circuit can be shown as

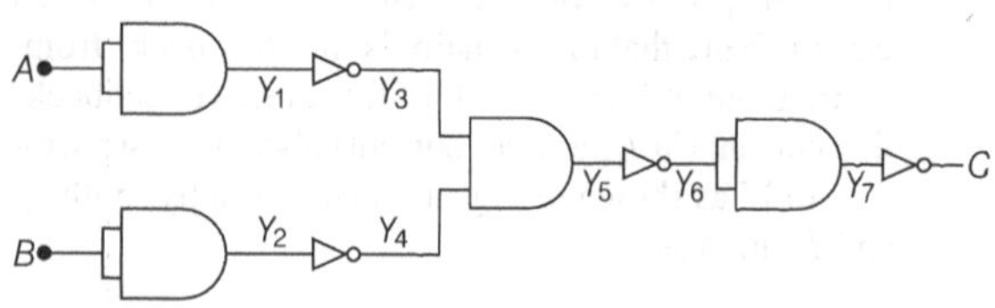


Truth table for the given circuit is

| $A$ | $B$ | $Y_1$ | $Y_2$ | $Y_3$ | $Y_4$ | $Y_5$ | $Y_6$ | $Y_7$ | $C$ |
|---|---|---|---|---|---|---|---|---|---|
| 0 | 0 | 0 | 0 | 1 | 1 | 1 | 0 | 0 | 1 |
| 0 | 1 | 0 | 1 | 1 | 0 | 0 | 1 | 1 | 0 |
| 1 | 0 | 1 | 0 | 0 | 1 | 0 | 1 | 1 | 0 |
| 1 | 1 | 1 | 1 | 0 | 0 | 0 | 1 | 1 | 0 |

Output of the given circuit matches with the output of NOR gate. Thus, this is NOR gate.

**106** (*c*) We are showing intermediate outputs on the circuit as given below.

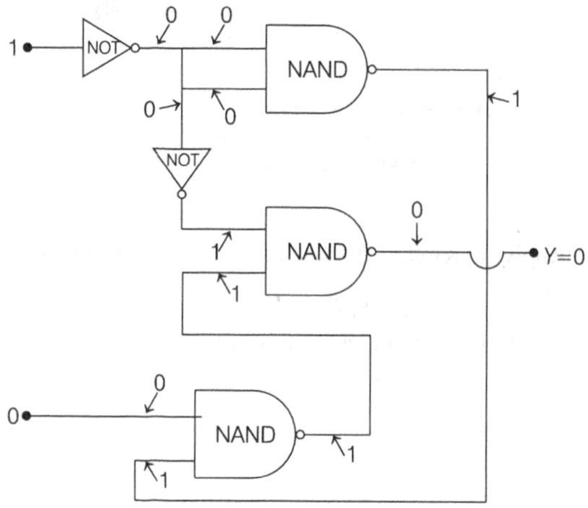


Output of above circuit at given inputs is zero.

To calculate output, we uses truth tables of NAND and NOT gates.

NAND gate

| $A$ | $B$ | $Y$ |
|---|---|---|
| 0 | 0 | 1 |
| 1 | 0 | 1 |
| 0 | 1 | 1 |
| 1 | 1 | 0 |

NOT gate

| $Y_{in}$ | $Y_{out}$ |
|---|---|
| 0 | 1 |
| 1 | 0 |

**107** (*c*) Truth table for given network is

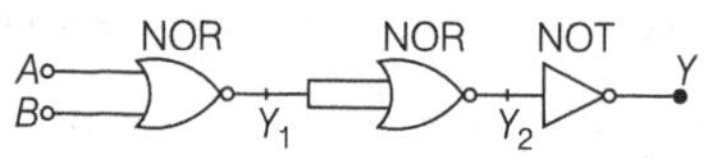


| $A$ | $B$ | $Y_1$ | $Y_2$ | $Y$ |
|---|---|---|---|---|
| 0 | 0 | 1 | 0 | 1 |
| 1 | 0 | 0 | 1 | 0 |
| 0 | 1 | 0 | 1 | 0 |
| 1 | 1 | 0 | 1 | 0 |

Output $Y$ of network matches with that of NOR gate. So, the given electrical network is equivalent to NOR gate.

**108** (*a*) From the circuit diagram given in the question, it can be seen that the current will flow to ground if any of the switch is closed. Also, the LED will only glow when current flows through it.

Thus, the truth table for the circuit diagram can be formed as given below

| $A$ | $B$ | $Y$ |
|---|---|---|
| 0 | 0 | 1 |
| 0 | 1 | 0 |
| 1 | 0 | 0 |
| 1 | 1 | 0 |

∴ The output $Y$ is equivalent to that of NOR gate.

**109** (*b*) The LED in the given circuit in the question will glow when the current flows through it, i.e. when the voltage across it is high.

∴ The truth table can be formed from this circuit diagram is

| $A$ | $B$ | $Y$ |
|---|---|---|
| 0 | 0 | 1 |
| 0 | 1 | 1 |
| 1 | 0 | 1 |
| 1 | 1 | 0 |

The output $Y$ is same as that come for NAND gate.

Thus, the correct Boolean operation represented by the given circuit diagram is drawn by NAND gate.

**110** (*d*) Flow of charge carriers in the semiconductor devices are within the solid itself,
while in vacuum diode the mobile electrons are obtained from a heated cathode and they are made to flow in an evacuated space or vacuum. Thus, no external heating or large evacuated tube is required by the semiconductor devices.

Therefore, Assertion is incorrect but Reason is correct.

**111** (*a*) According to Bohr atomic model, in an isolated atom the energy of any of its electrons is decided by the orbit in which it revolves. But when the atoms come together to form a solid they are close to each other. So the outer orbits of electrons from neighbouring atoms would come very close or could even overlap.

This would make the nature of electrons motion in a solid very different from that in an isolated atom.

Therefore, Assertion and Reason are correct and Reason is the correct explanation of Assertion.

**112** (*c*) In a semiconductor at $T > 0K$, electrons originally set free are not involved in the process of hole motion. These free electrons moves completely as conduction electron and give rise to an electron current $I_e$, under an applied electric field. However, motion of hole is only a convenient way of describing the actual motion of bound electrons wherever there is an empty bond anywhere in the crystal. Thus, under the action of an electric field these holes move towards negative potential giving the hole current $I_h$.

The total current $I$ is thus the sum of the electron current $I_e$ and hole current $I_h$,

$$I = I_e + I_h$$

Therefore, Assertion is correct but Reason is incorrect.

**113** (*c*) The conductivity of an intrinsic semiconductor is less than that of a lightly doped *p*-type semiconductor. Also, with the increase in the temperature, conductivity of intrinsic semiconductor increase.

Therefore, Assertion is correct but Reason is incorrect.

**114** (*a*) In *n*-type semiconductor, the pentavalent dopant is donating one extra electron for conduction.

Thus, the number of electrons will now be due to the electrons contributed by donors and those generated intrinsically. However, holes will only be due to the intrinsic source.

So, the rate of recombination of holes would increase due to the increase in number of electrons. As a result, the number of holes would get reduced further.

Therefore, Assertion and Reason are correct and Reason is the correct explanation of Assertion.

**115** (*c*) In a *p-n* junction, due to diffusion of electrons, a positive space-charge region on *n*-side of the junction and a negative space charge region on *p*-side of the junction is formed due to this, an electric field directed from positive charge towards negative charge develops. (Electric field is from *n*-side to *p*-side). Due to this field, an electron on *p*-side of the junction moves to *n*-side and a hole on *n*-side of the junction moves to *p*-side. This motion of charge carriers due to the electric field is called drift. Thus, a drift current, which is opposite in direction to the diffusion current starts. However,concentration gradient is due to doping of sides. It is not affected by drift of charge carriers.

Therefore, Assertion is correct but Reason is incorrect.

**116** (*b*) Initially, in a *p-n* junction diffusion current is large and drift current is small.

As the diffusion process continues through the space, charge regions on the either side of the junction extend. Thus, this increases the electric field strength and hence the drift current.

This process continues until the diffusion current equals the drift current. Thus, a *p-n* junction is formed.

∴ In a *p-n* junction under equilibrium, there is no net current as drift current is opposite to the diffusion current.

Therefore, Assertion and Reason are correct but Reason is not the correct explanation of Assertion.

**117** (*a*) As resistance of depletion region is large, potential drop occurs mainly in depletion region.

So, the applied voltage mostly drop across the depletion region and the voltage drop across the *p*-side and *n*-side of the junction is negligible.

Therefore, Assertion and Reason are correct and Reason is the correct explanation of Assertion.

**118** (*a*) It has been observed that when the applied reverse bias voltage $V$ reaches the breakdown voltage $V_Z$ of the zener diode, there is large change in current.

Also, after the $V_Z$, a large change in current can be produced by almost insignificant change in reverse bias voltage. In other words, zener voltage remains constant, even though current through zener diode varies over a wide range. This property of the zener diode is used for regulating supply voltages, so that they are constant.

Therefore, Assertion and Reason are correct and Reason is the correct explanation of Assertion.

**119** (*b*) Consider the case of an *n*-type semiconductor. Obviously, the majority carrier density $n$ is considerably larger than the minority hole density $p$ (i.e. $n >> p$). On illumination, let the excess electrons and holes generated be $\Delta n$ and $\Delta p$, respectively, $n' = n + \Delta n$, $p' = p + \Delta p$

Here, $n'$ and $p'$ are the electron and hole concentrations at any particular illumination and $n$ and $p$ are carriers concentration when there is no illumination. Remember $\Delta n = \Delta p$ and $n >> p$. Hence, the fractional change in the majority carriers (i.e., $\Delta n / n$) would be much less than that in the minority carriers (i.e., $\Delta p / p$). In general, we can state that the fractional change due to the photo-effects on the minority carrier dominated reverse bias current, which is more easily measurable than the fractional change in the forward bias current.

Hence, photodiodes are preferably used in the reverse bias condition for measuring light intensity.

Therefore, Assertion and Reason are correct but Reason is not the correct explanation of Assertion.

**120** (*b*) The solar radiation spectrum received by us is shown in figure below

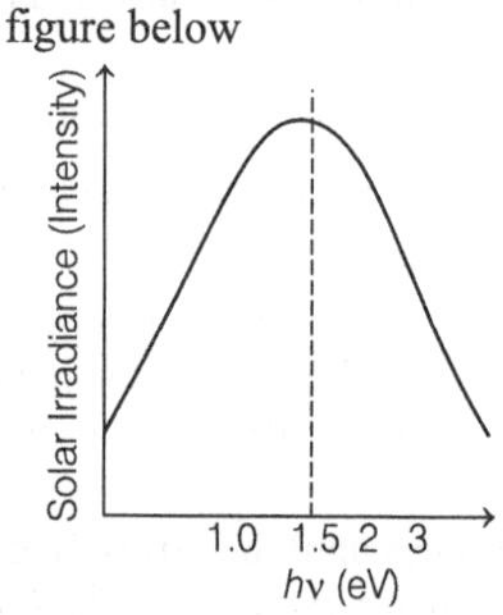


The maxima is near 1.5e V. For photo-excitation, $h\nu > E_g$. Hence, semiconductor with band gap ~ 1.5 eV or lower is likely to give better solar conversion efficiency.

Silicon has $E_g$ ~ 1.1 eV, while for GaAs it is ~ 1.53 eV. In fact, GaAs is better (in spite of its higher band gap) than Si because of its relatively higher absorption coefficient. If we choose materials like CdS or CdSe ($E_g$ ~ 2.4 eV), we can use only the high energy component of the solar energy for photo-conversion and a significant part of energy will be of no use.

In the case of PbS ($E_g$ ~ 0.4 eV) most of the solar radiation will be absorbed on the top-layer of solar cell and will not reach in or near the depletion region. For effective electron-hole separation, due to the junction field, we want the photo-generation to occur in the junction region only.

Therefore, Assertion and Reason are correct but Reason is not the correct explanation of Assertion.

**121** (*a*) For using the transistor as an amplifier, active region of the $V_o$ *versus* $V_i$ curve is used. The slope of the linear part of the curve represents the rate of change of the output with the input. It is negative because the output is $V_{CC} - I_C R_C$ and not $I_C R_C$.

That is why as input voltage of the CE amplifier increases, its output voltage decreases and the output is said to be out of phase with the input.

Therefore, Assertion and Reason are correct and Reason is the correct explanation of Assertion.

**122** (*a*) The output characteristics of a transistor in CE-configuration shows that initially for very small values of $V_{CE}$, $I_C$ increases almost linearly. This happens because the base-collector junction is not reverse biased and the transistor is not in active state. In fact, the transistor is in the saturation state and the current is controlled by the supply voltage $V_{CC} (= V_{CE})$ in this part of the characteristic. When $V_{CE}$ is more than that required to reverse bias the base-collector junction, $I_C$ increases very little with $V_{CE}$. The reciprocal of the slope of the linear part of the output characteristic gives the values of output resistance. The output resistance of the transistor is mainly controlled by the bias of the base-collector junction.

So, the high magnitude of the output resistance (of the order of $100\text{k}\Omega$) is thus due to the reverse-biased state of this diode.

This also explains that the resistance at the initial part of the characteristic, when the transistor is in saturation state, is very low.

Therefore, Assertion and Reason are correct and Reason is the correct explanation of Assertion.

**123** (*c*) In an oscillator, the feedback is in the same phase, i.e., positive feedback.

If the feedback voltage is in opposite phase, i.e. negative feedback, the gain is less than one and it can never work as oscillator. It will be an amplifier with reduced gain.

Therefore, Assertion is correct but Reason is incorrect.

**124** (*c*) A gate is a digital circuit that follows certain logical relationship between the input and output voltages. Therefore, they are generally known as logic gates because they control the flow of information. The five common logic gates used are NOT, AND, OR, NAND and NOR.

Therefore, Assertion is correct but Reason is incorrect.

**125** (*c*) The statement I is correct but rest are incorrect and these can be corrected as,

Some solid state semiconductors and their junctions have the possibility of controlling the number and the direction flow of charge carriers through them.

However, in vacuum tube, only the charge carriers, i.e. electrons are controlled by varying the voltage between the electrons and has only one direction.

Also, vacuum tubes are bulky, consume high power and operates generally at high voltages (~ 100 V).

**126** (*b*) The statement II is correct but I is incorrect and it can be corrected as,

In metals/conductors, although conduction band is partially filled but as valence band is partially empty, electrons from its lower level can move to higher level making conduction possible.

**127** (*b*) In a *p*-type semiconductor, the dopant has one valence electron less than pure semiconductor (Si or Ge). Therefore, this atom can form covalent bonds with neighbouring three Si atoms but does not have any electron to offer fourth Si atom. So the bond between the fourth neighbour and trivalent atom has a vacancy or hole. So, the trivalent foreign atom (the doping impurity) becomes effectively negatively charged when it shares fourth electron with neighbouring Si atom (pure semiconductor). Therefore, the dopant atom of *p*-type material can be treated as core of one negative charge along with its associated hole, as shown

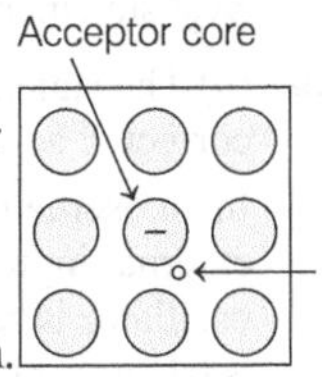


Also, for *p*-type semiconductor, in the energy band diagram $E_A$ is slightly above the top of $E_V$ of the valence band. So, with very small supply of energy an electron from the valence band can jump to the level $E_A$ and ionise the acceptor negatively.

So, statement II is correct but I is incorrect.

**128** (*b*) The statements I and II are incorrect and these can be corrected as,

Number of electrons made available for conduction by dopant atoms depends strongly upon the doping level and is independent of any increase in ambient temperature. With proper doping level, the number of conduction electrons can be made much larger than the number of holes.

In intrinsic semiconductor, conductivity occurs due to excitation of electrons when they absorb energy and break covalent bonds.

However, the conductivity in the extrinsic semiconductor is majorly due the doping impurity.

Rest statement is correct.

**129** (*c*) The statement III is correct but rest are incorrect and these can be corrected as,

When an electron diffuses from $n \rightarrow p$, it leaves behind an ionised donor (species which has become ion by donating electron) on *n*-side.

This ionised donor (positive charge) is immobile as it is bonded to the surrounding atoms. As the electrons continue to diffuse from $n \rightarrow p$, a layer of positive charge (or positive space-charge region) on *n*-side of the junction is developed. However, when hole diffuses from $p \rightarrow n$, it leaves behind an ionised acceptor.

**130** (*d*) When a *p-n* junction is reverse biased, the direction of applied voltage is same as the direction of barrier potential. As a result, the barrier height increases and the depletion region widens due to the change in the electric field.

This suppresses the flow of electrons from $n \rightarrow p$ and holes from $p \rightarrow n$. Thus, diffusion current decreases enormously compared to the diode under forward bias.

The electric field direction of the junction is such that if electrons on *p*-side or holes on *n*-side in their random motion come close to the junction, they will be swept to its majority zone. This drift of carriers gives rise to current. The drift current is of the order of a few $\mu$A. This is quite low because it is due to the motion of carriers from their minority side to their majority side across the junction. The drift current is also there under forward bias but it is negligible ($\mu$A) when compared with current due to injected carriers which is usually in mA.

So, all statements are incorrect.

**131** (*b*) From the output waveform, it is clear that initially $D_1$ has to be forward biased for the first half-cycle and subsequently for rest half-cycle. This is because, if the input voltage $A$ w.r.t. centre tap at any instant is + ve, then at that instant voltage at $B$ being out of phase will be negative.

Only, then $D_1$ is forward biased and conduct and $D_2$ is reverse biased. Hence, during this positive half cycle we get an output current. In the course of the AC cycle when the voltage at $A$ becomes negative with respect to centre tap, voltage of $B$ would be positive. In this part of the cycle $D_1$ would not conduct but $D_2$ would give output waveform. Thus, we get output as shown in the question.

So, the input waveform at $B$ and $A$ will be

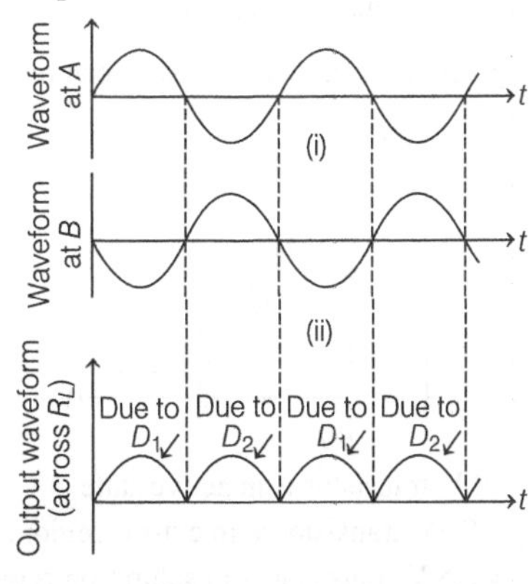


Since, we get output voltage during both the positive as well as the negative half cycle of the voltage. Thus, this is more efficient circuit for getting rectified voltage or current than the half-wave rectifier.

So, statement II is correct but I is incorrect.

**132** (*c*) The statements I and III are correct but II is incorrect and it can be corrected as,

During regulation action of zener diode :

Resistance of zener diode changes at breakdown voltage and current through $R_s$ increases after breakdown due to increased movement of minority carrier.

**133** (*b*) The statements I and II are correct but III is incorrect and it can be corrected as,

Semiconductor diodes in which charge carriers are generated by photons (photo-excitation) are called optoelectronic devices. Optoelectronic devices are photodiodes, Light Emitting Diodes (LED) and photovoltaic devices.

**134** (*a*) The statement I is incorrect and it can be corrected as,

When zener diode is used DC voltage regulator, then it is connected through a series resistance $R_s$ such that the zener diode is reverse biased. If the input voltage increases, the current through $R_s$ and zener diode also increases. This increases the voltage drop across $R_s$ without any change in the voltage across the zener diode. This is because in the breakdown region, zener voltage remains constant even though the current through the zener diode changes. Similarly, if the input voltage decreases, the current through $R_s$ and zener diode also decreases. The voltage drop across $R_s$ decreases without any change in the voltage across the zener diode.

Thus, any increase/decrease in the input voltage results in, increase/decrease of the voltage drop across $R_s$ without any change in voltage across the zener diode.

Rest statements are correct.

**135** (*d*) The statements II and III are correct but I is incorrect and it can be corrected as,

The heavily doped emitter has a high concentration of majority carriers, which will be holes in a *p-n-p* transistor. These majority carriers enter the base region in large numbers. The base is thin and lightly doped. So, the majority carriers there would be few.

**136** (*b*) The given transfer characteristics can be as shown below

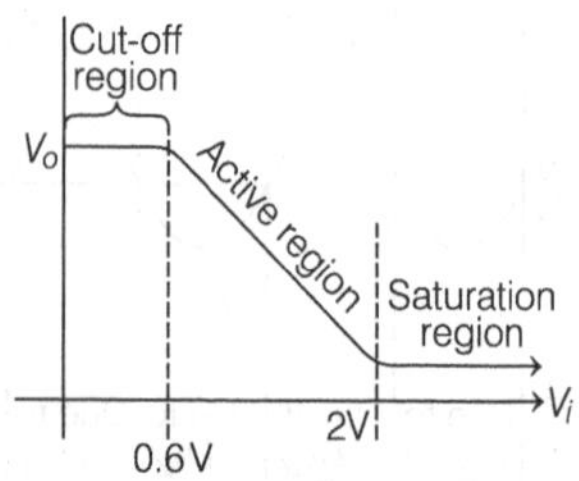


$\therefore$ At $V_i = 1$ V, transistor is in active state.

At $V_i = 0.5$V, transistor is in cut-off region.

At $V_i = 2.5$ V, transistor is in saturation region.

Since, the transistor circuit in active state can be used as an amplifier and in saturation or cut-off region, it can be used as a switch.

So, statements II, III and IV are correct but I is incorrect.

**137** (*a*) The statement I is incorrect and it can be corrected as,

When the transistor is used in cut-off or saturation state, it acts as a switch. Considering the CE-configuration as long as $V_i$ is low and unable to forward bias, the transistor, $V_o$ is high (at $V_{CC}$). If $V_i$ is high enough to drive the transistor into saturation, then $V_o$ is low, very near to zero. When the transistor is not conducting it is said to be switched OFF and when it is driven into saturation it is said to be switched ON.

This shows that if we define low and high states as below and above certain voltage levels corresponding to cut off and saturation of the transistor, then we can say that a low input switches the transistor OFF and a high input switches it ON.

Rest statement is correct.

**138** (*b*) The statement II is incorrect and it can be corrected as,

In an oscillator, we get AC output without any external input signal. In other words, the output in an oscillator is self-sustained. To attain this, an amplifier is taken. A portion of the output power is returned back (feedback) to the input in phase with the starting power (this process is termed as positive feedback) as shown in figure. The feedback can be achieved by inductive coupling (through mutual inductance) or *LC* or *RC* networks.

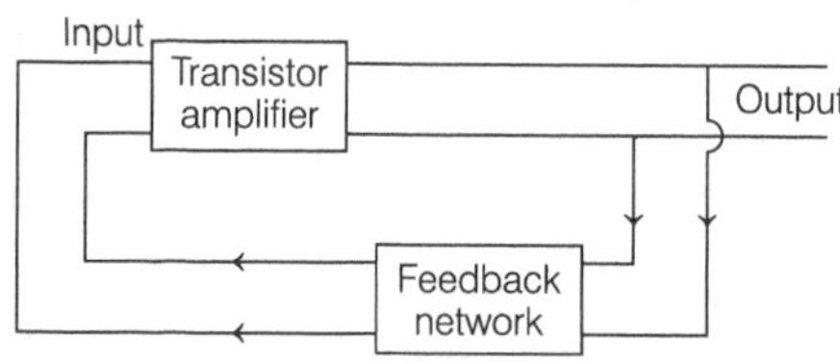


Rest statements are correct.

**139** (*b*) The statement given in option (b) is incorrect and it can be corrected as,

Cathode ray tubes (CRT) used in television and computer monitors work on the principle of vacuum tubes are being replaced by liquid crystal display (LCD) monitors with supporting solid state electronics.

Rest statements are correct.

**140** (*c*) For Ge outermost orbit is fourth ($n = 4$). The number of electrons in the outermost orbit is 4 (2*s* and 2*p* electrons). Hence, the total number of outer electrons in the crystal is $4N$. The maximum possible number of outer electrons in the orbit is 8 ($2s + 6p$ electrons).

So out of the $4N$ electrons, $2N$ electrons are in the $2N$ *s*-states (orbital quantum number $l = 0$) and $2N$ electrons are in the available $6N$ *p*-states. Obviously, some *p*-electron states are empty as shown in the extreme right of figure. This is the case of well separated or isolated atoms [region A of figure].

Suppose these atoms start coming nearer to each other to form a solid. The energies of these electrons in the outermost orbit may change (both increase and decrease) due to the interaction between the electrons of different atoms.

The $6N$ states for $l = 1$, which originally had identical energies in the isolated atoms, spread for and form an energy band [region B in figure]. Similarly, the $2N$ states for $l = 0$, having identical energies in the isolated atoms, split into a second band (carefully see the region B of figure) separated from the first one by an energy gap.

At still smaller spacing, however, there comes a region in which the bands merge with each other. The lowest energy state that is a split from the upper atomic level appears to drop below the upper state that has come from the lower atomic level. In this region (region C in figure), no energy gap exists where the upper and lower energy states get mixed.

Finally, if the distance between the atoms further decreases, the energy bands again split apart and are separated by an energy gap $E_g$ (region D in figure). The total number of available energy states $8N$ has been re-apportioned between the two bands ($4N$ states each in the lower and upper energy bands).

The statement given in option (c) is correct but rest are incorrect and these can be corrected as,

**141** (*d*) The statement given in option (d) is incorrect and it can be corrected as,

For doping a semiconductor, dopant has to be such that it does not distort the original pure semiconductor lattice. It occupies only a very few of the original semiconductor atoms sites in the crystal. A necessary condition to attain is that the sizes of the dopant and the semiconductor atoms should be nearly the same.

Rest statements are correct.

**142** (*a*) These statement given in option (a) is correct but rest are incorrect and these can be corrected as,

*p*-type semiconductors are obtained when a trivalent impurity (e.g. boron, aluminium, gallium or indium) is added to a intrinsic semiconductor. (e.g. germanium or silicon).

In other words, the dopants in *p*-type semiconductor is trivalent atom.

Thus, this addition creates deficiencies of valence electron which are most commonly known as holes. These are the majority charge carriers in this type of semiconductor.

However, in *n*-type semiconductors, the dopants are pentavalent impurities. Also, the majority charge carriers are electrons.

**143** (*c*) The statement given in option (c) is correct but rest are incorrect and these can be corrected as,

*n*-type semiconductor can be produced by doping an impurity atom of valency 5, i.e. a pentavalent atoms. This means, elements from group fifth e.g. arsenic, antimony, phosphorous, etc. In these, electrons are majority charge carriers and holes are minority carriers.

**144** (*a*) The statement given in option (a) is correct but rest are incorrect and these can be corrected as,

In energy band diagram of *n*-type semiconductor, the donor level $E_D$ is slightly below the bottom of $E_C$ of the conduction band and electrons from this level move into the conduction band with very small supply of energy. This is as shown in the figure below

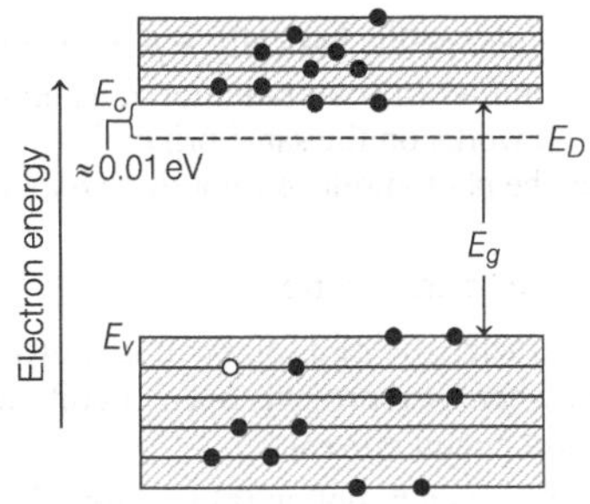


**145** (*b*) The statement given in option (b) is correct but rest are incorrect and these can be corrected as,

Extrinsic semiconductor crystals either *n*-type or *p*-type maintains an overall charge neutrality as the charge of additional charge carriers is just equal and opposite to that of the ionised core in the lattice.

**146** (*b*) The statement given in option (b) is correct but rest are incorrect and these can be corrected as,

The space-charge region on the either side of the junction developed due to the diffusion of electrons from $n \rightarrow p$ and holes from $p \rightarrow n$ is the depletion region. Since, these holes and electrons can be increased/decreased by varying the densities of the impurities.

$\therefore$ Its width can be varied by increasing/decreasing the densities of the dopant.

Also, due to this positive space charge region on *n*-side of the junction and negative space charge region on *p*-side of the junctions, an electric field directed from positive charge towards negative charge develops.

**147** (*c*) The statement given in option (c) is correct but rest are incorrect and these can be corrected as,

The thickness of the depletion region is of the order of one tenth of a micrometre, i.e. around $10^{-7}$ m.

Also, diffusion current and drift current flows in opposite direction to each other.

**148** (*d*) The statement given in option (d) is correct but rest are incorrect and these can be corrected as,

Arsenic is pentavalent, so $X$ is *n*-type and indium is trivalent, so $Y$ is *p*-type.

Thus, *p*-side to the diode is connected to negative terminal and *n*-side of the diode is connected to positive terminal. So, the junction is in reverse bias.

**149** (*a*) The statement given in option (a) is correct but rest are incorrect and these can be corrected as,

In forward biasing a negligible potential drop occurs in diode, so potential drop across resistance $R$ is $V$.

**150** (*c*) The statement given in option (c) is incorrect and it can be corrected as,

Photodiode is a special purpose *p-n* junction diode, which produces *e-h* pairs when illuminated by light. This when connected to external load, flow of current takes place.

Similarly, solar cell is basically a *p-n* junction which generates emf when solar radiation falls on the *p-n* junction. It works on the same principle (photovoltaic effect) as the photodiode, except that no external bias is applied.

Rest statements are incorrect.

**151** *(a)* *V-I* characteristic of a solar cell is shown in the question, where *A* represents open circuit voltage and *B* shows short circuit current.

∴ Statement given in option (a) is correct but rest are incorrect.

**152** *(b)* The statement given in option (b) is correct but rest are incorrect and these can be corrected as,

In a transistor,

**Emitter** is of moderate size and heavily doped.

**Base** is very thin and lightly doped.

**Collector** is moderately doped and larger in size as compared to the emitter.

For a transistor amplifier, the voltage gain is low at high and low frequencies and constant at mid frequencies.

**153** *(a)* The statement given in option (a) is correct but rest are incorrect and these can be corrected as,

In an *n-p-n* transistor, electrons moves from emitter to base and initially crossing the base, these electrons reaches the collector.

**154** *(a)* The statement given in option (a) is incorrect and it can be corrected as,

As, current gain, $\beta_{AC} = \left(\frac{\Delta I_C}{\Delta I_B}\right)_{V_{CE}}$ and

Current gain, $\beta_{DC} = \frac{I_C}{I_B}$

Since, $I_C$ increases with $I_B$ almost linearly and $I_C = 0$, when $I_B = 0$, the values of both $\beta_{DC}$ and $\beta_{AC}$ are nearly equal.

Both $\beta_{AC}$ and $\beta_{DC}$ vary with $V_{CE}$ and $I_B$ (or $I_C$) slightly.

Rest statements are correct.

**158** *(c)* When a forward bias is applied across the *p-n* junction, the applied voltage opposes the barrier voltage. Due to this, the potential barrier across the junction is lowered.

**159** (c) In an unbiased *p-n* junction, the diffusion of charge carriers across the junction takes place from higher concentration to lower concentration. Therefore, holes diffuses from the *p*-region to *n*-region because its concentration in *p*-region is more as compared to *n*-region.

**160** *(a)* Given, voltage gain of first amplifier, $A_{V_1} = 10$

Voltage gain of second amplifier, $A_{V_2} = 20$

Input voltage, $V_i = 0.01$ V

Total voltage gain, $A_V = \frac{V_o}{V_i} = A_{V_1} \times A_{V_2}$

$$\therefore \quad \frac{V_o}{0.01} = 10 \times 20$$

$$V_o = 2\text{ V}$$

**163** *(b)* In the given circuit, *p*-side of *p-n* junction, diode $D_1$ is connected to lower voltage and *n*-side of $D_1$ to higher voltage. Thus, $D_1$ is reverse biased. Similarly, the *p*-side of *p-n* junction diode $D_2$ is at higher potential and *n*-side of $D_2$ is at lower potential. Thus, $D_2$ is forward biased.

Hence, no current flows through the junction *B* to *A*.

**164** *(d)* As *p-n* junction conducts during positive half cycle only, the diode connected here will work in positive half cycle. Thus, the potential difference across $C$ = peak voltage of the given AC voltage

$V_0 = V_{rms}\sqrt{2} = 220\sqrt{2}$ V.

**165** *(b)* For each positive half cycle of the input voltage, the diode will be forward biased but for each negative half cycle, it will be reverse biased. So, the output of the given circuit will only contains positive halves of voltage like a half-wave rectifier.

**166** *(b)* Since, diode is forward biased, so no current goes through side branch after capacitor is charged.

$$\therefore \quad V_{AB} = I_{AB} \times R_{AB}$$
$$= 0.2 \times 10^{-3} \times 10 \times 10^3$$
$$= 2\text{ V}$$

Option (b) is most appropriate, extra 0.3 V occurs due to voltage drop of diode.

**167** *(c)* The given circuit is a combination of AND, OR and NOT gates as shown below.

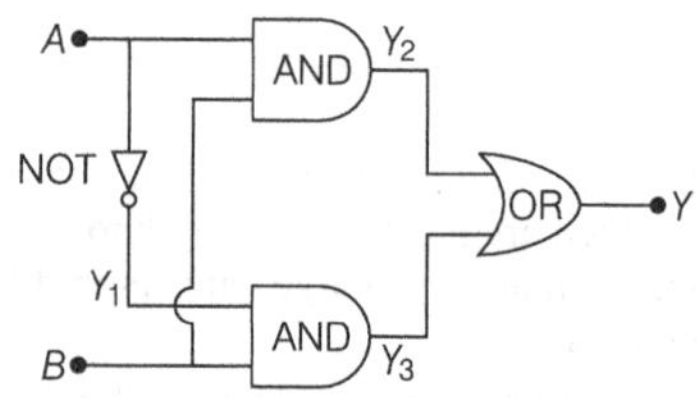


Thus, the truth table is

| *A* | *B* | $Y_1$ | $Y_2$ | $Y_3$ | *Y* |
|---|---|---|---|---|---|
| 0 | 0 | 1 | 0 | 0 | 0 |
| 0 | 1 | 1 | 0 | 1 | 1 |
| 1 | 0 | 0 | 0 | 0 | 0 |
| 1 | 1 | 0 | 1 | 0 | 1 |

**168** *(b)* When *p-n* junction is forward biased, it opposes the potential barrier junction resulting in decrease in barrier. When *p-n* junction is reverse biased, it supports the potential barrier junction, resulting increase in potential barrier across the junction.

So, the curve 3 in given graph (question) represents the forward bias of junction and curve 1 represents reverse bias of junction.